PHOTOMORPHOGENESIS IN PLANTS

2nd edition

The cover illustrates the tall cucumber long-hypocotyl mutant which is deficient in the photoreceptor phytochrome B and its short isogenic wild type after 14 days incubation in a 16-h light/8-h dark cycle.

PHOTOMORPHOGENESIS IN PLANTS

2ND EDITION

edited by

R. E. KENDRICK

Department of Plant Physiology,
Wageningen Agricultural University,
Wageningen, The Netherlands
and
Laboratory for Photoperception and Signal Transduction,
Frontier Research Program,
Institute of Physical and Chemical Research (RIKEN),
Wako City, Saitama, Japan

and

G. H. M. KRONENBERG

Department of Plant Physiology,
Wageningen Agricultural University,
Wageningen, The Netherlands

KLUWER ACADEMIC PUBLISHERS

DORDRECHT / BOSTON / LONDON

Library of Congress Cataloging-in-Publication Data

Photomorphogenesis in plants / edited by R.E. Kendrick and G.H.M. Kronenberg.
p. cm.
Includes bibliographical references and index.
ISBN 0-7923-2550-8 (alk. paper)
1. Plants--Photomorphogenesis. I. Kendrick, Richard E. II. Kronenberg, G. H. M.
QK757.P45 1993
581.19'153--dc20 93-32993

ISBN 0-7923-2550-8 (HB)
ISBN 0-7923-2551-6 (PB)

Published by Kluwer Academic Publishers,
P.O. Box 17, 3300 AA Dordrecht, The Netherlands.

Kluwer Academic Publishers incorporates
the publishing programmes of
D. Reidel, Martinus Nijhoff, Dr W. Junk and MTP Press.

Sold and distributed in the U.S.A. and Canada
by Kluwer Academic Publishers,
101 Philip Drive, Norwell, MA 02061, U.S.A.

In all other countries, sold and distributed
by Kluwer Academic Publishers Group,
P.O. Box 322, 3300 AH Dordrecht, The Netherlands.

Printed on acid-free paper

Printed in the Netherlands

This book is dedicated to the memory of

Harold William (Bill) Siegelman,

a pioneer in phytochrome research, who died on December 3rd, 1992.

Preface to the first edition

It is perhaps not surprising that plants have evolved with a mechanism to sense the light environment around them and modify growth for optimal use of the available 'life-giving' light. Green plants and ultimately all forms of life depend on the energy of sunlight, fixed in the process of *photosynthesis.* By appreciating the *quality, quantity, direction* and *duration* of light, plants are able to optimize growth and control such complex processes as germination and flowering. To perceive the environment a number of receptors have evolved, including the red/far-red light absorbing *phytochrome,* the blue/UV-A light absorbing *cryptochrome* and a UV-B *light-absorbing pigment.* The isolation and characterization of phytochrome is a classical example of how use of photobiological techniques can predict the nature of an unknown photoreceptor. The current knowledge of phytochrome is found in Part 2 and that of cryptochrome and other blue/UV absorbing receptors in Part 3. Part 4 concerns the light environment and its perception. Part 5 consists of selected physiological responses: photomodulation of growth, phototropism, photobiology of stomatal movements, photomovement, photocontrol of seed germination and photocontrol of flavonoid biosynthesis. Further topics in Part 6 are the photobiology of fungi, a genetic approach to photomorphogenesis and coaction between pigment systems.

Our plan was to produce an advanced textbook which took a broad interdisciplinary approach to this field of *photomorphogenesis.* In particular to bring together in one volume work on phytochrome, cryptochrome and other blue/UV responses. In addition it was conceived as a book available in paperback at a price the individual could afford, as well as in a hardback. While we may not have succeeded on all fronts, we hope that we have come some way to our goals and that the end result is a useful and stimulating volume for the young research worker entering the field of photomorphogenesis.

The book is equipped with a complete index, selected further reading and references and provides a useful reference text for student and teacher alike. The 23 chapters are written by leading experts from Europe, Israel, Japan and the USA, who between them have a wealth of research and teaching experience.

During the production of the book we have attempted to standardize terminology. This is always difficult in a multi-author text, but we hope we have approached consistency. We have not ignored the problems where they exist and have attempted to clarify confusion, *e.g.* terminology of light measurement. In photomorphogenesis as in all dynamic areas of research there are divergent opinions and interpretations of data. In the final analysis the views expressed are those of the individual authors. We hope that the reader will find the chapters stimulating and an encouragement for future research.

In producing the book we set ourselves, what we originally described as a 'realistic' schedule. In this was room for a few authors to be late in submitting material at the various production deadlines. A word of thanks to those authors who sent manuscripts ahead of schedule, thus enabling us to cope with the inevitable latecomers.

We thank the members of the Laboratory of Plant Physiological Research, Wageningen, for support and encouragement during production of this volume. Assistance of Ad. C. Plaizier on behalf of the publishers, Martinus Nijhoff/Dr. W. Junk is gratefully acknowledged. Last but not least a special word of thanks to Chris Kendrick for assistance in editing.

R.E. Kendrick and G.H.M. Kronenberg
Wageningen, 1986

Laboratory of Plant Physiological Research
Generaal Foulkesweg 72
6703 BW Wageningen
The Netherlands

Preface to the second edition

Seven years has elapsed since the first edition of this book was published. During the planning stages the many suggestions and comments from friends and colleagues in the field of photomorphogenesis were taken into account. While some of the basic structure has been retained several new chapters and authors have been recruited to make it more complete and to accommodate the recent developments in the field. Since photomorphogenesis is an active research area it means that most chapters have completely changed. We have set out to make the book as complete as possible with a comprehensive index and have taken care to standardize terminology. Reflecting the development of plant science research much more emphasis is placed on molecular approaches, but the physiological aspects of the subject are still covered in depth. We hope that this balance will assist the molecular biologist and physiologist alike.

This second edition is dedicated to the memory of Harold William (Bill) Siegelman who died on the December 3rd, 1992. Bill was one of the pioneers in phytochrome research and in the late 1950's and early 1960's was the first to partially purify phytochrome. One of us (R.E.K.), while a Postdoctoral Research Fellow at Brookhaven National Laboratory, USA, has fond memories of discussions with Bill when he was Chairman of the Biology Department.

The production of this book was not without difficulty. However, the majority of authors were sympathetic to our editorial tasks. We would like to thank authors who finished there work ahead of schedule enabling us to cope with the late (and very late!) arrivals. Our production schedule has been kept on target and the book will appear on time. This edition has been printed from camera ready copy which has been produced single-handed by Chris Kendrick. We admire her skill and patience in dealing with the material which arrived in variable condition, from those that were nearly perfect to those which at best could be described as poor. The possibility for us to have our hands on the copy to a late stage has enabled very recent developments to be included.

We would like to thank Kluwer Academic Publishers, and in particular Ad. C. Plaizier with whom we have worked over the years and Gilles Jonker who took over the production of this edition. We appreciate your help and support during this project.

It is hoped that this edition will meet our objective, to stimulate research in the fascinating field of plant photomorphogenesis.

R.E. Kendrick[1,2] and G.H.M. Kronenberg[1]
Wageningen and Wako City, 1993

[1]Department of Plant Physiology
Wageningen Agricultural University
Arboretumlaan 4
NL-6703 BD Wageningen
The Netherlands

[2]Laboratory for Photoperception and Signal Transduction
Frontier Research Program
Institute of Physical and Chemical Research (RIKEN)
Hirosawa 2-1
Wako City
Saitama 351-01
Japan

Contents

Preface to the first edition VII

Preface to the second edition VIII

Abbreviations XXXI

Part 1 Introduction

1. Introduction
by Lars Olof Björn

1.1 A developing research field 3
1.2 Plant vision 3
1.3 The discovery of phytochrome 5
1.4 The surprises of phytochrome 9
1.5 Perception of B and UV light 10
1.6 Other photoreceptors and coaction between photoreceptors 11
1.7 Molecular biology and genetics 13
1.8 Conclusion and outlook 13
1.9 Further reading 13
1.10 References 13

Part 2 Quantification of light

2. Quantification of light
by Lars Olof Björn and Thomas C. Vogelmann

2.1 Basic concepts 17
2.2 The wavelength problem 18
2.3 The problem of direction and shape 19
2.4 Biological weighting functions and units 22
2.5 Further reading 24
2.6 References 24

Part 3 Instrumentation in photomorphogenesis research

3. Instrumentation in photomorphogenesis research
by Masaki Furuya and Yasunori Inoue

3.1 Introduction 29
3.2 Spectrographs: analyses of wavelength effects 29
3.2.1 The Beltsville spectrograph 30
3.2.2 The Okazaki large spectrograph 30
3.2.3 Safelight 34
3.3 Microbeam irradiators: spatial analyses 34
3.3.1 Pioneering work 35
3.3.2 Microbeam apparatus having multiple light sources 35
3.4 Sequential observation of elementary processes: temporal analysis 36
3.4.1 Time-lapse recorders 37
3.4.2 Computer analysis of temporal processes with micro-images 39
3.4.3 Flash photolysis 40
3.4.4 Low-temperature spectroscopy 41
3.5 Spectrophotometers 42
3.5.1 *In-vivo* spectrophotometry 42
3.5.2 Micro-spectrophotometry 43
3.6 Concluding remarks 45
3.7 Further reading 45
3.8 References 46

Part 4 Phytochrome

4.1 The phytochrome chromophore
by Wolfhart Rüdiger and Fritz Thümmler

4.1.1 Introduction 51
4.1.2 The structure of the phytochrome chromophore 52
4.1.3 Differences between the Pr and the Pfr chromophore 57
4.1.4 Intermediates 60
4.1.5 Chromophore biosynthesis 64
4.1.6 Problems to be solved 65
4.1.7 Further reading 67
4.1.8 References 67

4.2 Phytochrome genes and their expression
by Peter H. Quail

4.2.1 Introduction 71
4.2.2 Structure and evolution of *phy* genes 71
4.2.2.1 Multiple *phy* genes 72
4.2.2.2 Nomenclature 73
4.2.2.3 Structure of *phy* genes 76
4.2.3 Biological functions of multiple *phy* genes 81
4.2.4 Regulation of *phy* gene expression 84
4.2.4.1 Contrasting regulation of *phyA*, *phyB*, and *phyC* genes 85
4.2.4.2 Diversity in patterns of *phyA* regulation between plant species 87
4.2.4.3 Autoregulation of monocot *phyA* genes as a model system 89
4.2.4.4 The control of *phyA* expression by *cis*-acting elements and *trans*-acting factors 93
4.2.5 Conclusions 100
4.2.6 Further reading 101
4.2.7 References 101
4.2.8 Appendix 104

4.3 Assembly and properties of holophytochrome
by Masaki Furuya and Pill-Soon Song

4.3.1 Introduction 105
4.3.2 Biogenesis of phytochrome apoprotein and chromophore 106
4.3.2.1 Biosynthesis of phytochrome apoproteins 106
4.3.2.2 Biosynthetic pathway of phytochromobilin 107
4.3.3 Assembly of PHY with chromophores 108
4.3.3.1 Assembly in wild-type and transgenic plants 108
4.3.3.2 Assembly *in vitro* with recombinant apophytochrome 109
4.3.3.3 Failure of chromophore assembly in mutants 112
4.3.4 Physical properties of holophytochromes 113
4.3.4.1 The molecular structure and properties 113
4.3.4.2 Spectroscopic characteristics 116
4.3.5 Photochemistry and photophysics 119
4.3.5.1 Mechanism of the Pr to Pfr phototransformation 119
4.3.5.2 Quantum yields 120
4.3.5.3 Fluorescence properties and primary photoprocesses 120

4.3.5.4 Intermediates and kinetic models 123
4.3.5.5 Non-photochemical transformation 124
4.3.6 Photo-induced conformational changes 126
4.3.6.1 Chromophore topography 126
4.3.6.2 Proton transfer 127
4.3.6.3 Photoreversible conformational changes 127
4.3.7 Phytochrome structure and function relationship 130
4.3.7.1 Functional domain or active site 130
4.3.7.2 Dichroic effects and putative receptors 131
4.3.8 Concluding remarks 133
4.3.9 Further reading 134
4.3.10 References 134

4.4 Phytochrome degradation
by Richard D. Vierstra

4.4.1 Introduction 141
4.4.2 Properties of phytochrome degradation 142
4.4.2.1 Etiolated plants 142
4.4.2.2 Light-grown plants 145
4.4.2.3 Relationship of degradation to phytochrome sequestering 146
4.4.3 Mechanisms of phytochrome degradation 147
4.4.3.1 Possible involvement of ubiquitin-dependent proteolysis 147
4.4.3.2 Potential mechanisms for Pfr recognition. 154
4.4.4 Use of transgenic plants to study phytochrome degradation. 156
4.4.5 Physiological function(s) of phytochrome degradation .. 158
4.4.6 Concluding remarks 159
4.4.7 Further Readings 160
4.4.8 References 160

4.5 Distribution and localization of phytochrome within the plant
by Lee H. Pratt

4.5.1 Introduction 163
4.5.2 Phytochrome assays 164
4.5.2.1 Biological assay 164
4.5.2.2 Spectrophotometric assay 165
4.5.2.3 Immunochemical assay 166
4.5.3 Intercellular distribution 168
4.5.3.1 Biological assay 168
4.5.3.2 Spectrophotometry 169
4.5.3.3 *In vivo* spectrofluorometry 171

4.5.3.4 Immunochemistry 172
4.5.3.5 Summary 176
4.5.4 Intracellular localization 177
4.5.4.1 Microbeam irradiation 177
4.5.4.2 *In vitro* responses 178
4.5.4.3 Microspectrophotometry 178
4.5.4.4 Subcellular fractionation 179
4.5.4.5 Immunolocalization 180
4.5.4.6 Summary 183
4.5.5 Concluding remarks 184
4.5.6 Further reading 184
4.5.7 References 184

4.6 Signal transduction in phytochrome responses
by Stanley J. Roux

4.6.1 Introduction 187
4.6.2 Intrinsic protein kinase activity 188
4.6.3 G-protein involvement 189
4.6.4 Changes in metabolism of inositol phosphoslipids 192
4.6.5 Role of protein phosphorylation 193
4.6.6 Induction of rapid ion fluxes 197
4.6.6.1 Calcium signalling 198
4.6.6.1.1 Specific targets of calcium action 202
4.6.7 Interactions with growth regulators 203
4.6.8 Genetic approaches 204
4.6.9 Conclusions 206
4.6.10 Further reading 208
4.6.11 References 208

4.7 The physiology of phytochrome action
by Alberto L. Mancinelli

4.7.1 Introduction 211
4.7.2 Light and phytochrome 213
4.7.2.1 Photochemical properties of purified phytochrome 213
4.7.2.1.1 Absorption and action spectra 213
4.7.2.1.2 Kinetics of phytochrome photoconversion 216
4.7.2.1.3 Pfr/P, φ, *k*, and *H*. 217
4.7.2.1.4 Phytochrome photoconversion under dichromatic irradiation. 218
4.7.2.1.5 Intermediates in phytochrome photoconversion. 223

4.7.2.1.6 Phytochrome dimers. 223
4.7.2.2 Photochemical properties of phytochrome *in vivo* 223
4.7.2.3 Light, Pfr/P, φ, *k*, *H*, and expression of phytochrome-mediated responses 225
4.7.3 The state of phytochrome *in vivo* 228
4.7.3.1 Intercellular and intracellular distribution of phytochrome 229
4.7.3.2 Synthesis of phytochrome 229
4.7.3.3 Phytochrome destruction 230
4.7.3.4 Dark reversion 233
4.7.3.5 Photoconversion, dark reversion, destruction and state of phytochrome 234
4.7.4 Phytochrome-mediated responses 235
4.7.4.1 Red-far red reversible low fluence responses .. 236
4.7.4.1.1 Pfr/P and LFR 239
4.7.4.1.2 Kinetics of phytochrome action and LFR 241
4.7.4.2 The very low fluence responses 242
4.7.4.3 The high irradiance responses (prolonged irradiation responses) 245
4.7.4.3.1 The spectral sensitivity of the HIR ... 246
4.7.4.3.2 Continuous and cyclic irradiations and reciprocity failure of the HIR 249
4.7.4.3.3 The irradiance dependence of the HIR 251
4.7.4.3.4 HIR and LFR 253
4.7.4.3.5 Photosynthesis and the HIR 253
4.7.4.3.6 Models for phytochrome action in the HIR 254
4.7.5 Interaction between phytochrome and other photomorphogenic photoreceptors 255
4.7.6 Analysis of phytochrome action 258
4.7.6.1 Multiplicity of display of phytochrome action .. 258
4.7.6.2 Phytochrome, membranes and light signal-transduction pathways 259
4.7.6.3 Labile and stable phytochrome and time course of phytochrome action 260
4.7.6.4 The HIRs: R-HIR and FR-HIR 260
4.7.6.5 Phytochrome synthesis and destruction 261
4.7.7 The future 262
4.7.8 Further reading 263
4.7.9 References 263
4.7.10 Appendix 266

4.8 The use of transgenic plants to examine phytochrome structure/function

by Joel R. Cherry and Richard D. Vierstra

4.8.1 Introduction 271
4.8.2 Creating transgenic plants expressing functional phytochrome 272
4.8.2.1 Choice of coding sequence 272
4.8.2.2 Choice of promoter 274
4.8.2.3 Choice of host plant 275
4.8.3 Analysis of plants expressing heterologous phytochromes 276
4.8.3.1 Expression in tobacco 276
4.8.3.2 Expression in tomato 278
4.8.3.3 Expression in *Arabidopsis* 279
4.8.3.4 Expression in lower plants 279
4.8.3.5 Comments on phytochrome overexpression 280
4.8.4 Application of transgenics to phytochrome structural analysis 281
4.8.4.1 Chromophore attachment 281
4.8.4.2 Photoreversibility and spectral stability 282
4.8.4.3 Dimerization 283
4.8.4.4 Pfr degradation 284
4.8.5 Use of transgenic expression as an assay for biological activity 285
4.8.5.1 Considerations in transgenic assay development 285
4.8.5.2 Relationship of phytochrome dose to phenotypic response 286
4.8.5.3 Possible mutant classes based on dose/response . 288
4.8.5.4 Biological activity of phytochrome mutants 289
4.8.6 Concluding remarks 294
4.8.7 Further Reading 295
4.8.8 References 295

Part 5 Blue-light and UV-receptors

5.1 Diversity of photoreceptors

by Horst Senger and Werner Schmidt

5.1.1 Introduction 301
5.1.2 Historical aspects 302
5.1.3 Blue light UV responses 303

5.1.3.1 Phototropism of *Phycomyces* 303
5.1.3.2 Light-induced absorbance changes 303
5.1.3.3 Hair whorl formation in *Acetabularia* 306
5.1.3.4 Reactivation of nitrate reductase 306
5.1.3.5 Germination of spores in *Pteris* 306
5.1.3.6 Perithecial formation in *Gelasinospora* 307
5.1.3.7 Synthesis of 5-aminolevulinic acid 307
5.1.3.8 Phototropism in oats 307
5.1.3.9 Respiration enhancement in *Scenedesmus* 307
5.1.3.10 Inhibition of indole acetic acid 308
5.1.3.11 Chloroplast rearrangement in *Funaria* 308
5.1.3.12 Cortical fibre reticulation in *Vaucheria* 308
5.1.3.13 Photoreactivation 308
5.1.3.14 Loss of carbohydrates in *Chlorella* 309
5.1.3.15 Carotenoid synthesis in *Neurospora* 309
5.1.4 Concerted action of photoreceptors 309
5.1.4.1 Chlorophyll synthesis in *Scenedesmus* 309
5.1.4.2 Conidiation in *Alternaria* 310
5.1.4.3 Morphogenic index in the fern *Dryopteris* 310
5.1.4.4 Geotropism in maize roots 310
5.1.4.5 High irradiance response of phytochrome 311
5.1.4.6 Red and blue interaction in maize coleoptiles .. 311
5.1.5 Energy requirements 311
5.1.6 The nature of B/UV photoreceptors 314
5.1.7 Methodological problems 318
5.1.8 Terminology 321
5.1.9 Ecological aspects and outlook 322
5.1.10 References 322

5.2 Properties and transduction chains of the UV and blue light photoreceptors
by Benjamin A. Horwitz

5.2.1 Introduction 327
5.2.2 Excited state chemistry of the chromophores 329
5.2.2.1 Flavins 329
5.2.2.2 Carotenoids 331
5.2.2.3 Pterins 332
5.2.3 Kinetic properties of the blue light photoreceptors 332
5.2.4 Rapid effects of blue light and their relevance to transduction 334
5.2.4.1 Light-induced absorbance changes 335
5.2.4.2 Electrical consequences of blue light reception .. 336
5.2.5 The biochemistry of transduction: intracellular signalling 340

5.2.5.1 Transmembrane signalling and transduction by G-proteins 340
5.2.5.2 Effectors and second messengers 341
5.2.5.3 Can blue light biochemically stress? 346
5.2.6 Concluding remarks 347
5.2.7 Further reading 348
5.2.8 References 348

Part 6 Coaction between pigment systems

6. Coaction between pigment systems
by Hans Mohr

6.1 Sensor pigments in higher plants 353
6.2 A unifying model of coaction 354
6.3 Photomorphogenesis of the milo seedling (*Sorghum vulgare* Pers., cv. Weider-hybrid) 356
6.3.1 Accumulation of plastid GPD (glyceraldehyde-3-phosphate dehydrogenase, EC 1.2.1.13) in the shoot (mainly primary leaf) 356
6.3.2 Synthesis of anthocyanin 357
6.4 Photomorphogenesis of the Scots pine seedling (*Pinus sylvestris* L.) 360
6.4.1 Axis (hypocotyl) straight growth 360
6.4.2 Synthesis of plastid Fd-GOGAT (ferredoxin-dependent glutamate synthase, EC 1.4.7.1) in the cotyledonary whorl 363
6.5 Photosensors involved in light control of stem elongation in seedlings of angiosperm plants 364
6.5.1 Hypocotyl elongation in the mustard (*Sinapis alba* L.) seedling 364
6.5.2 Hypocotyl elongation in the cucumber (*Cucumis sativus* L.) seedling 366
6.6 Gene expression in the tomato phytochrome-deficient *aurea* mutant 368
6.7 Coaction between photoreceptors in phototropism: sesame seedling, *Sesamum indicum* L. 369
6.7.1 Hypocotyl straight growth 369
6.7.2 Phytochrome and phototropism 370
6.8 Conclusion 371
6.9 Further reading 372
6.10 References 372

Part 7 The light environment

7.1 Sensing the light environment: the functions of the phytochrome family

by Harry Smith

7.1.1 Introduction 377
7.1.2 The function of informational photoreceptors 378
7.1.3 Information in the light environment 379
7.1.4 Light-quantity perception: theoretical aspects 380
7.1.5 Light-quality perception: theoretical aspects 381
7.1.6 The complexity of spectral information 382
7.1.6.1 R:FR, φ_e and φ_c; phytochrome related parameters 383
7.1.7 The natural radiation environment 386
7.1.7.1 The daylight spectrum 386
7.1.7.2 Diurnal fluctuations in daylight quality 387
7.1.7.3 Light quality within vegetation canopies 387
7.1.7.4 Light quality underwater 388
7.1.7.5 The light environment under the soil 389
7.1.8 The phytochromes as sensors of environmental R:FR . 390
7.1.8.1 Sensitivity considerations 390
7.1.8.2 Plant strategies in response to shade 390
7.1.8.3 R:FR perception and the induction of shade-avoidance reactions 393
7.1.8.4 Proximity perception, or the detection of neighbours 396
7.1.8.5 End-of-day effects 398
7.1.8.6 Seed germination and seedling establishment in nature 399
7.1.9 Eco-physiological functions of the members of the phytochrome family 401
7.1.9.1 The physiological response modes 401
7.1.9.2 The members of the phytochrome family 402
7.1.9.3 Approaches for identifying the physiological functions of the phytochromes 403
7.1.9.4 The physiological function of phytochrome A . 404
7.1.9.5 The physiological function of phytochrome B . . 407
7.1.9.6 Which phytochrome mediates the shade-avoidance syndrome? 410
7.1.10 Concluding remarks 412
7.1.11 Further Reading 413
7.1.12 References 414

7.2 Light direction and polarization

by Manfred Kraml

7.2.1 Introduction 417
7.2.2 Physical aspects of light direction and polarization 418
7.2.2.1 The rectilinear propagation of light 418
7.2.2.2 Polarization 420
7.2.3 Mechanisms for the perception of unilateral light 422
7.2.3.1 Perception of light direction by attenuation 422
7.2.3.2 Perception of light direction by refraction (= lens effect) 422
7.2.3.3 Spatial and temporal sensing of an internal light gradient 423
7.2.4 Biological examples for perception of light direction by attenuation and lens effect 423
7.2.4.1 Induction of polarity by unilateral light 424
7.2.4.2 Phototropism of *Phycomyces* 425
7.2.5 Action dichroism and polarized light 427
7.2.5.1 Characterization of dichroic pigment orientation by polarized light 427
7.2.5.2 Perception of 'light direction' by dichroic orientated photoreceptors 429
7.2.5.3 The formation of tetrapolar gradients by pigment dichroism 429
7.2.5.3.1 Spherical cells. 430
7.2.5.3.2 Cylindrical cells. 431
7.2.6 Biological examples for action dichroism 431
7.2.6.1 Dichroism and induction of polarity 432
7.2.6.2 Effects of polarized light in *Phycomyces* 432
7.2.6.3 Flip-flop dichroism of phytochrome 432
7.2.6.4 Action dichroism of phytochrome in *Mougeotia* . 433
7.2.6.5 Phytochrome dichroism in fern and moss protonemata 437
7.2.6.6 Wavelength-dependent action dichroism of flavin-mediated photo-responses 439
7.2.6.7 Wavelength-dependent action dichroism for B/UV-A in low-irradiance movement of *Mougeotia* 439
7.2.6.8 Orientation of the B/UV-A photoreceptor in *Mougeotia* as analyzed by microbeam irradiations 440
7.2.7 Concluding remarks 441
7.2.8 Further reading 443
7.2.9 References 443

7.3 The duration of light and photoperiodic responses
by Daphne Vince-Prue

7.3.1 Introduction ... 447
7.3.2 Circadian rhythms ... 447
7.3.3 Seasonal responses ... 451
7.3.4 General aspects of photoperiodism ... 452
7.3.5 Photoperiodic timekeeping ... 457
7.3.5.1 External coincidence models ... 460
7.3.6 Photoperception ... 465
7.3.6.1 The night-break reaction ... 466
7.3.6.2 Phase-setting ... 468
7.3.6.3 The dusk signal ... 469
7.3.6.4 The Pfr-requiring reaction ... 471
7.3.6.5 Do different phytochromes control flowering in SDP? ... 473
7.3.6.5.1 Phytochrome destruction ... 473
7.3.6.5.2 Phytochrome reversion ... 474
7.3.7 Photoperiodic induction under long photoperiods ... 477
7.3.7.1 Responses to light quantity ... 477
7.3.7.2 Responses to light quality ... 478
7.3.7.3 Timekeeping ... 479
7.3.7.4 Possible mechanisms ... 481
7.3.8 The action of phytochrome in photoperiodism ... 483
7.3.8.1 Changes in gene expression ... 485
7.3.8.2 Changes at the membrane level ... 486
7.3.8.3 The nature of the floral stimulus ... 488
7.3.9 Further Reading ... 489
7.3.10 References ... 489

7.4 Light within the plant
by Thomas C. Vogelmann

7.4.1 Introduction ... 491
7.4.2 Physical aspects of light propagation in plants ... 491
7.4.2.1 Light as a particle versus wave; limitations when considering plant optics ... 491
7.4.2.2 Cells as lenses ... 492
7.4.2.3 Absorption and the sieve effect ... 499
7.4.2.4 Fluorescence effects ... 501
7.4.2.5 Light scattering ... 502
7.4.2.6 Plants as optical waveguides ... 507
7.4.3 Plants as light traps ... 511
7.4.3.1 Internal fluence rates within plants ... 511
7.4.3.2 Calculation of light gradients within tissues ... 514

7.4.3.3 Experimental measurement of light gradients with a fibre-optic probe 519
7.4.4 Light gradients and photomorphogenesis 521
7.4.4.1 Light gradients and phytochrome 522
7.4.4.2 Light gradients and phototropism 524
7.4.4.3 Light gradients in leaves and photosynthesis 528
7.4.4.4 How plants control the penetration of light 530
7.4.5 Summary 532
7.4.6 Further reading 533
7.4.7 References 533

7.5 Modelling the light environment
Lars Olof Björn

7.5.1 Introduction to natural light 537
7.5.2 Modification of sunlight by the earth's atmosphere ... 537
7.5.3 The SPCTRAL2 model of Bird and Riordan 539
7.5.4 Computation of photosynthetically active radiation ... 541
7.5.5 Underwater daylight 542
7.5.5.1 Transmission through the water surface 543
7.5.5.2 Water types 544
7.5.5.3 Modelling water properties 545
7.5.5.4 Underwater light direction 546
7.5.6 Effects of ground and vegetation 548
7.5.7 Effects of clouds 550
7.5.8 Further reading 550
7.5.9 References 550
7.5.10 Appendix 552

Part 8 A molecular and genetic approach to photomorphogenesis

8.1 The molecular biology of photoregulated genes
by Alfred Batschauer, Philip M. Gilmartin, Ferenc Nagy and Eberhard Schäfer

8.1.1 Introduction 559
8.1.2 Properties of phytochrome action 560
8.1.2.1 Kinetics of the phytochrome system 560
8.1.2.2 Kinetics of signal transduction 561
8.1.3 Regulation of chalcone synthase expression in mustard and parsley 563
8.1.3.1 Summary 570

8.1.4 Regulation of ribulose-1,5-bisphosphate carboxylase/ oxygenase ... 571
8.1.4.1 Ribulose-1,5-bisphosphate carboxylase/ oxygenase (Rubisco) gene (*RbcS*) organization and expression ... 571
8.1.4.2 Light-responsive element localization and dissection ... 576
8.1.4.3 Identification and characterization of *trans*-acting factors ... 578
8.1.4.3.1 GT-binding proteins ... 581
8.1.4.3.2 G-box binding proteins ... 582
8.1.4.3.3 GATA-binding proteins ... 583
8.1.4.3.4 AT-rich binding proteins ... 584
8.1.4.4 Summary ... 585
8.1.5 Circadian-clock, tissue-specific and light-regulated expression of *Cab* genes in higher plants ... 586
8.1.5.1 Function, structure and organization ... 586
8.1.5.1.1 Function ... 586
8.1.5.1.2 Structure and organization ... 587
8.1.5.2 Light-regulated expression of *Cab* genes ... 587
8.1.5.2.1 Phytochrome-mediated *Cab* gene expression ... 587
8.1.5.2.2 Blue-light induced *Cab* gene expression ... 588
8.1.5.2.3 Circadian clock-regulated *Cab* gene expression ... 589
8.1.5.3 The regulated expression of *Cab* genes by *cis* and *trans*-acting elements ... 590
8.1.5.4 Signal-transduction chains for *Cab* gene expression ... 592
8.1.5.4.1 Photoreceptors ... 592
8.1.5.4.2 Second messengers ... 592
8.1.5.4.3 Transcription factors ... 592
8.1.5.5 Summary ... 593
8.1.6 Further reading ... 593
8.1.7 References ... 593

8.2 Photomorphogenic mutants of higher plants
by Maarten Koornneef and Richard E. Kendrick

8.2.1 Introduction ... 601
8.2.2 General aspects of the genetic and molecular analysis of mutants ... 602
8.2.2.1 The cloning and transfer of genes ... 602
8.2.2.2 General aspects of mutant isolation ... 603

8.2.3 Photomorphogenic mutants 605
8.2.3.1 Phytochrome-deficient mutants 606
8.2.3.2 Phytochrome overexpressors 609
8.2.3.3 Light-response mutants 610
8.2.3.3.1 Mutants with a constitutive light phenotype 610
8.2.3.3.2 Light hyper-responsive mutants 611
8.2.3.4 Blue-light mutants 612
8.2.3.5 Putative transduction-chain mutants with reduced light responsiveness 614
8.2.4 The use of mutants in understanding photomorphogenesis 614
8.2.4.1 Seed germination 615
8.2.4.2 The inhibition of hypocotyl and internode elongation 616
8.2.4.2.1 The role of the light-labile and light-stable phytochrome 616
8.2.4.2.2 The effect of B and UV-A photoreceptors 618
8.2.4.3 Phototropism 619
8.2.4.4 Chlorophyll synthesis and chloroplast development 620
8.2.4.5 The induction of flowering 620
8.2.4.6 Photomorphogenic mutants and plant hormones . 622
8.2.5 Conclusions 623
8.2.6 Further Reading 624
8.2.7 References 624

Part 9 Selected topics

9.1 Photomodulation of growth

by Daniel J. Cosgrove

9.1.1 Introduction 631
9.1.1.1 Light gives the plant cues about the environment 631
9.1.1.2 Light effects on growth depend on the type of organ 633
9.1.1.3 Light-growth responses are adaptive and controlled developmentally 634
9.1.2 Photobiology of plant growth 635
9.1.2.1 Action spectra reveal the diversity of continuous light-growth responses 636

9.1.2.1.1 Phytochrome 638
9.1.2.1.2 Specific blue light responses 641
9.1.2.2 Plants also exhibit growth responses to brief light pulses 643
9.1.2.3 Other photoreceptors 644
9.1.3 Mechanisms of action 644
9.1.3.1 Light can affect cell number and cell size 645
9.1.3.2 Light may affect growth quickly (in seconds) or more slowly (hours) 646
9.1.3.3 Electrical and biochemical changes in membranes often precede or accompany light-growth responses 648
9.1.3.4 Light affects cell wall yielding properties 649
9.1.3.5 Hormones may be involved in some light-growth responses 652
9.1.3.5.1 Auxin 653
9.1.3.5.2 Gibberellins 654
9.1.3.6 Other possible mediators of light-growth responses 655
9.1.4 Summary 656
9.1.5 Further reading 656
9.1.6 References 657

9.2 Phototropism
by Richard D. Firn

9.2.1 What is phototropism? 659
9.2.2 Scope of this chapter 662
9.2.3 An historical summary of some key concepts of phototropism 662
9.2.4 Ways of inducing a phototropic response 664
9.2.5 Measuring the phototropic response 665
9.2.5.1 The inadequacy of angle of curvature measurements 666
9.2.5.1.1 The same degree of curvature can be produced by many different types of response 666
9.2.5.1.2 The location of organ curvature 668
9.2.5.1.3 Temporal complexity 668
9.2.5.1.4 The autotropic straightening response 668
9.2.5.2 The phototropic responses of individual cells within the organ 668
9.2.6 The basic elements of the phototropic response 669
9.2.6.1 The perception phase 670

9.2.6.2 The latent period 670
9.2.6.3 Patterns of differential growth 671
9.2.6.4 The autotropic phase 672
9.2.7 Fluence-response curves for phototropism 672
9.2.8 Models of phototropism 674
9.2.8.1 The Cholodny-Went models 674
9.2.8.1.1 Auxin as a longitudinal messenger: model 1 674
9.2.8.1.2 Auxin as a lateral co-ordinator: model 2 674
9.2.8.1.3 Differential rates of longitudinal auxin movement giving rise to auxin gradients: model 3 676
9.2.8.1.4 Differential rates of auxin destruction giving rise to auxin gradients: model 4 676
9.2.8.1.5 Auxin as a local messenger: model 5 676
9.2.8.1.6 Auxin sensitivity arguments: model 6 677
9.2.8.1.7 Multi-regulator control: model 7 ... 677
9.2.8.2 The Blaauw model of phototropism 677
9.2.9 Phototropism and gravitropism 679
9.2.10 Conclusions 679
9.2.11 Further reading 680
9.2.12 References 680

9.3 The photobiology of stomatal movements
by Eduardo Zeiger

9.3.1 Introduction 683
9.3.2 Light as an environmental signal for stomatal movements 684
9.3.3 Direct response of stomata to light 685
9.3.4 The photobiological components of the light response of stomata 687
9.3.5 Action spectroscopy 689
9.3.6 Properties of the guard-cell chloroplast 691
9.3.7 Properties of the stomatal response to blue light 693
9.3.8 Localization of the blue light photoreceptor 699
9.3.9 Sensory transduction of the stomatal response to light .. 701
9.3.10 Regulatory aspects of the light response of stomata in the intact leaf 701
9.3.11 Ecophysiological and agricultural implications of the light response of stomata 702

9.3.12 Further reading 705
9.3.13 References 705

9.4 Photomovement
by Wolfgang Haupt and Donat-P. Häder

9.4.1 Introduction 707
9.4.2 Photomovement in motile micro-organisms 707
9.4.2.1 Motile bacteria 708
9.4.2.2 Photosynthetic flagellates 711
9.4.2.3 Slime moulds 715
9.4.2.4 Ecological consequences of photomovement 717
9.4.3 Photoregulation of intracellular movement 718
9.4.3.1 Photodinesis 719
9.4.3.2 Light-regulated chloroplast redistribution 721
9.4.3.3 *Mougeotia*, a special case of chloroplast movement 724
9.4.4 Synopsis 727
9.4.5 Comparative conclusions 730
9.4.6 Further reading 730
9.4.7 References 731

9.5 Photocontrol of flavonoid biosynthesis
by Christopher J. Beggs and Eckard Wellmann

9.5.1 Introduction 733
9.5.2 Flavonoid biosynthesis 734
9.5.3 The photoreceptors and effective wavebands 737
9.5.4 Coactions between the photoreceptors 742
9.5.5 Mode of action of light-induced flavonoid synthesis 744
9.5.6 The problem of correlation between enzyme activities and flavonoid accumulation 745
9.5.7 Significance of light induction of flavonoids and anthocyanins 747
9.5.8 Further reading 750
9.5.9 References 750

9.6 Photomorphogenesis in fungi
by Gérard Manachère

9.6.1 Introduction 753
9.6.2 Photo-induction of fruiting 754
9.6.2.1 General data 754
9.6.2.2 Photo-induction in basidiomycetes 756

9.6.2.3 Variability of light requirements: interaction of factors 757
9.6.3 Photomorphogenesis of sporophores and photo-sporogenesis 758
9.6.3.1 General data 758
9.6.3.2 Photomorphogenesis and photosporogenesis in basidiomycetes 759
9.6.3.2.1 Coprinus congregatus: *a model species* 759
9.6.3.2.2 Control of fruit-body growth by hymenial cells: a possible relationship to phototropic curvature 763
9.6.3.2.3 Interactions of light and temperature: consequences for photoperiodic responses 768
9.6.4 Photocontrol of fruiting rhythms 769
9.6.4.1 Circadian rhythms 769
9.6.4.2 Low frequency rhythms 773
9.6.5 Action spectra and photoreceptors 775
9.6.5.1 Survey 775
9.6.5.2 Photoreception in basidiomycetes 776
9.6.5.3 Variability in light-quality requirements 778
9.6.6 Concluding remarks 779
9.6.7 Further reading 780
9.6.8 References 780

9.7 Photobiology of ferns
by Masamitsu Wada and Michizo Sugai

9.7.1 Introduction 783
9.7.2 Spore germination 784
9.7.2.1 Photoregulation 784
9.7.2.2 Signal transduction 786
9.7.3 Protonemal growth 787
9.7.3.1 Apical growth 787
9.7.3.2 Apical swelling 788
9.7.3.3 Cytoskeleton and microfibrils 789
9.7.4 Phototropism and polarotropism 790
9.7.4.1 Tropic responses 790
9.7.4.2 Localization and orientation of photoreceptors 792
9.7.4.3 Recognition of light direction 793
9.7.4.4 Mechanism of tropic response 794
9.7.5 Cell division and its orientation 795
9.7.5.1 Cell cycle 795
9.7.5.2 Localization of photoreceptors 795

9.7.5.3 Orientation of cell division ... 797
9.7.6 Chloroplast photo-orientation ... 798
9.7.6.1 Survey ... 798
9.7.6.2 Low fluence response ... 799
9.7.6.3 High fluence response ... 799
9.7.6.4 Mechanism of photomovement ... 799
9.7.7 Concluding remarks ... 800
9.7.8 Further reading ... 801
9.7.9 References ... 801

Index ... 803

Abbreviations

We have attempted to use SI units throughout and all non-SI abbreviations are defined on first usage in each chapter. This list consists of those abbreviations used on pages other than those on which they are defined.

Abbreviation	Meaning
A	absorbance
Δ*A*	difference in absorbance
ΔΔ*A*	difference in absorbance difference
ABO	*Arabidopsis* phytochrome B overproducer, *Arabidopsis* transformed with *Arabidopsis phyB*
ALA	5-aminolevulinic acid
ANS	8-anilinonaphthalene 1-sulphonate
bis-ANS	*bis*-anilinonaphthalene 8-sulphonate
ATP	adenosine triphosphate
au	*aurea* mutant of tomato
B	blue light
BHF	blue light high fluence rate
B-HIR	blue light-high irradiance response
BLF	blue light low fluence rate
blu	blue light-insensitive mutant of *Arabidopsis*
bp	base pair
bR	bacteriorhodopsin
Cab	gene coding for chlorophyll *a*/*b*-binding protein
Cab-PSII	gene coding for chlorophyll *a*/*b*-binding protein of PSII
CAB or Cab	chlorophyll *a*/*b*-binding protein
CCCP	carbonyl cyanide 3-chlorophenyl hydrazone
CD	circular dichroism
cDNA	copy deoxyribonucleic acid
Chl	chlorophyll
CHS	chalcone synthase
Chs	gene coding for chalcone synthase
CNL	critical night length
cop	constitutively-photomorphogenic mutant of *Arabidopsis*
CT	circadian time
C-W	Cholodny-Went (model)

DAYLIGHT PAR	computer program to calculate photon irradiance integrated over the interval 400-700 nm (PAR)
DAYLIGHT VIS-IR	computer program to calculate daylight spectral irradiance
DCMU	3-(3,4-dichlorophenyl)-1,1-dimethylurea
det	de-etiolated mutant of *Arabidopsis*
dim	dim-sighted mutant of *Trichoderma*
DN	daylength neutral
DNA	deoxyribonucleic acid
E	photon energy
E-vector	electrical vector of light
ε	extinction coefficient
ELISA	enzyme-linked immunosorbent assay
EM	electron microscopy
EOD	end-of-day (responses)
EODFR	end-of day far-red light
FAD	flavin adenine dinucleotide
Fd-GOGAT	ferredoxin-dependent glutamate synthase
fhy	far-red light-elongated hypocotyl mutant of *Arabidopsis*
FR	far-red light
fre	far-red light elongated mutant of *Arabidopsis*
FR-HIR	far-red light-high irradiance response
G-protein	guanosine triphosphate binding protein
GA	gibberellic acid
GPD	glyceraldehyde-3-phosphate dehydrogenase
GTP	guanosine triphosphate
GUS	ß-glucuronidase
H	cycling rate of phytochrome
h	Planck's constant
HIR	high irradiance response
hp	high-pigment mutant of tomato
HPLC	high performance liquid chromatography
hR	halorhodopsin
hy	long-hypocotyl mutant of *Arabidopsis*
I	fluence rate
I/I_0	relative internal fluence rate
I_{700}	phytochrome intermediate with absorption peak around 700 nm
IAA	indole-3-acetic acid (auxin)
IP_3	inositol 1,4,5-trisphosphate
IR	infra-red (> 700 nm)
KM	Kubelka-Munk theory
λ	wavelength
LD	long day

LDP	long day plant(s)
LF(R)	low fluence (response)
lh	long hypocotyl mutant of cucumber
LHCI	light-harvesting Chl associated with PSI
LHCII	light-harvesting Chl associated with PSII
LIAC	light-induced absorbance change
LRE	light regulatory element
lv	long-stemmed mutant of pea
lw	dwarf mutant of pea
lx	lux (lumen m^{-2}), unit for illuminance
mad	phototropic mutant of *Phycomyces*
mRNA	messenger ribonucleic acid
N	photon flux
n	refractive index
ν	frequency
NAD	nicotinamide adenine dinucleotide
NADP	nicotinamide adenine dinucleotide phosphate
NADPH	reduced form of NADP
NB_{max}	time of maximum sensitivity to a night break
NMR	nuclear magnetic resonance
P	total phytochrome (Pr + Pfr)
PAL	phenylalanine ammonia lyase
PAR	photosynthetically active radiation (400-700 nm)
PCB	phycocyanobilin
PFB	paraflagellar body
PFD	photon flux density
Pfr	far-red light-absorbing form of phytochrome
φ	proportion of P as the Pfr form at photoequilibrium
Φ	quantum efficiency
PHY	phytochrome apoprotein, immunochemically detectable product of a *phy* gene
phy	gene coding for phytochrome
phyA	gene coding for phytochrome A
PHYA	phytochrome apoprotein of phytochrome A
PI	type I phytochrome (light-labile) or phosphatidylinositol
PII	type II phytochrome (light-stable)
PIP_2	phosphatidylinositol 1,4-bisphosphate
PPFD	photosynthetic photon flux density
Pr	red light-absorbing form of phytochrome
PSI	photosynthesis photosystem I
PSII	photosynthesis photosystem II
$\Delta\psi$	water potential difference
Quin-2	fluorescence probe for Ca^{2+}
R	reflectance

R	red light
RbcS	gene coding for ribulose-1,5-bisphosphate carboxylase/oxygenase small subunit
RBO	rice phytochrome B overproducer, *Arabidopsis* transformed with rice *phyB*
R:FR	red light:far-red light photon ratio (ζ)
R-HIR	red light-high irradiance response
Rubisco	ribulose-1,5-bisphosphate carboxylase/oxygenase
σ_1 and σ_2	photoconversion cross sections of Pr and Pfr, respectively
SD	short day or standard deviation
SDP	short day plant(s)
SDS	sodium dodecyl sulphate
SDS-PAGE	sodium dodecyl sulphate polyacrylamide gel electrophoresis
T	transmission of light or turgor pressure
T-DNA	transfer deoxynucleic acid
Δt	difference in time
$t_{1/2}$	half-life of a reaction
θ	angle between the electronic transition moment of the molecule and the E-vector of the incident light
Ub-P	ubiquitin-phytochrome conjugate
UV	ultra violet light
UV-A	320-400 nm UV
UV-B	280-320 nm UV
UV-C	$<$ 280 nm UV
UV-HIR	UV high irradiance response
VLF(R)	very low fluence (response)
W	white light
wc	white-collar mutant of *Neurospora*
WT	wild type
X	reaction partner for Pfr
Y	yield threshold
yg	yellow-green mutant of tomato
ζ	red light:far-red light photon ratio (R:FR)

Part 1 Introduction

1. Introduction

Lars Olof Björn

Section of Plant Physiology, Lund University,
Box 7007, S-220 07 Lund, Sweden

1.1 A developing research field

I ended the introduction to the first edition of this book by saying that plant photomorphogenesis is an active research field, and we need not fear to reach the end in our lifetime. The 7 years that have elapsed since then have strengthened my conviction on this point. A comparison of the contents of the first edition with that of the present book reveals that there is more new than old material. There are 9 major sections instead of 6, and 28 chapters instead of 22. Eleven new authors have been recruited.

1.2 Plant vision

When an animal is hungry or thirsty, it goes searching for something to eat or drink. If the sun feels hot, the animal moves to the shade, and if the wind or the precipitation is too fierce, the animal seeks shelter. Most plants cannot do any of these things; they have to stay where they have germinated. A plant must collect its energy and building materials on the spot, and cope with the hardships that may be encountered there. It cannot flee its enemies but has to defend itself. The plant must use physiology where the animal can resort to behaviour, and this is one reason why plant physiology is such a fascinating subject.

Instead of moving about, a plant can adapt its growth habit, and thus its body shape, to the environment. To be able to do this, it has to sense the environment. Because plants do not move rapidly, and are not designed to communicate with us, at first glance they may appear passive, or even non-living (my 3-year-old daughter, pers. comm.). With the new possibilities of monitoring signal transmitters such as calcium ions in intact cells, we are getting impressive evidence of their ability to register immediately what is going on around them (Fig. 1).

For plants, just as for us, vision is one of the most important senses. Since the plant's 'vision' is so different from ours, we usually talk about 'light sensing',

R.E. Kendrick & G.H.M. Kronenberg (eds.), Photomorphogenesis in Plants - 2nd Edition

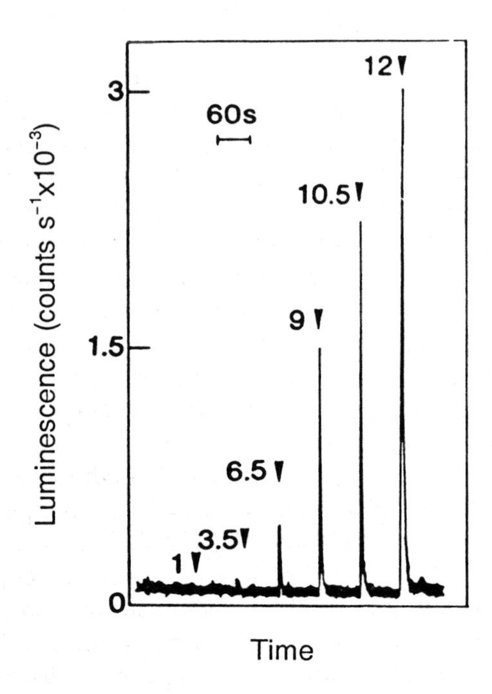

Figure 1. Pulses of calcium ions set free in the cytosol by wind gusts of different strength (the values on the curve are wind forces in Newtons). The plant was *Nicotiana plumbaginifolia* engineered to produce the protein aequorin, which emits light when it comes in contact with calcium ions. After Knight *et al.* (1992).

or 'light perception', and use many words beginning with 'photo': photoperiodism (Chapter 7.3), photomorphogenic (Chapter 8.2), photomodulation (Chapter 9.1), phototropism (Chapter 9.2), photomovement (Chapter 9.4), photocontrol (Chapter 9.5) and photobiology (Chapters 9.6 and 9.7).

In general, light in human or animal vision acts only as a medium for transferring information about position and movement, shape and colour of material objects. The human and animal interest is centred on food, enemies, seeking other members of the same species for reproduction, *etc.*

Light for the plant is not only an information medium, but also food, and therefore of interest in itself, and it is important for the plant to obtain information about its amount, main direction and spectral composition. Plants, however, also use light sensing for other purposes, such as detecting competitors and keeping track of time.

No wonder then, that as a result of evolution the plant's 'vision' developed in quite another direction from ours; in a way so different that it took humans a long time to discover it. It is still very difficult even for specialists in 'plant vision' to understand how it works, and what the ecological significance is of the strange things we detect. Chapter 7.1 focuses on some aspects of the latter question.

Figure 2 shows weeds growing at the edge of a field of oats, and also within the crop. We can see that the weed plants outside the crop are short, and their flower stalks are also short. The flowers are far below the level of the panicles. To send the flowers higher into the air would be an unnecessary waste of resources, since they are readily visible to the insects in any case. With longer stalks the plants would also be more vulnerable to damage by wind or large animals. The weeds within the crop, on the other hand, send their flowers up above the panicles, this being necessary if insects are to see and visit them. No doubt their response is beneficial to the weeds, but how do they know where they are: at the edge of the field or among the oat plants?

Perhaps it is even incorrect to talk about 'the' sensing of light, for the plant has more than one system for light perception, and this is one area where great progress has been made in the period since the first edition of this book. While for a long time plant physiologists were working in a paradigm with two photoreceptors, phytochrome and 'the' blue (B)/UV-A light receptor (also called cryptochrome, although this name is no longer recommended, *cf.* Chapter 5.1), we now realize the existence of a whole 'phytochrome family' (Chapter 4.2) with members coded for by different genes (Chapter 4.2) and having different functions, which are now being sorted out using genetic dissection (Reed *et al.* 1993). 'The' B receptor has turned out to be even more diverse (Chapters 5.1 and 5.2), and plants are able to monitor the level of UV-B radiation separately. Pigments closely related to our own visual pigments have been established as photoreceptors in algae (Chapter 9.4).

We should not be misled by experts specializing on different light sensing systems into believing that the different photoreceptors work independently of one another. Just as the signals from the different kinds of light sensors in our eyes are combined to give us a composite colour picture of our surroundings, so the plant combines the signals from the various photoreceptors and other information channels to a unity. The plant has no nerves and no brain, but this does not mean that it lacks an information-processing system. The information processing is quite sophisticated, but so different from ours that we have to look hard to see it (Chapter 6).

1.3 The discovery of phytochrome

Although effects of daylight on plant development had already been noted in the last century (Henfrey 1852; Kjellman 1885), the experiments by Garner and Allard (1920) initiated a rapid development of knowledge in this field. Some plants (short-day plants) require sufficiently long nights for induction of flowering. It was found that flowering, that would otherwise be induced by long

Figure 2. The common weed, *Matricaria indora* (known under several English names: horse daisy, corn mayweed, scentless mayweed) growing at the edge (*top*) and within (*bottom*) an oat field. The weed plants growing within the crop are taller.

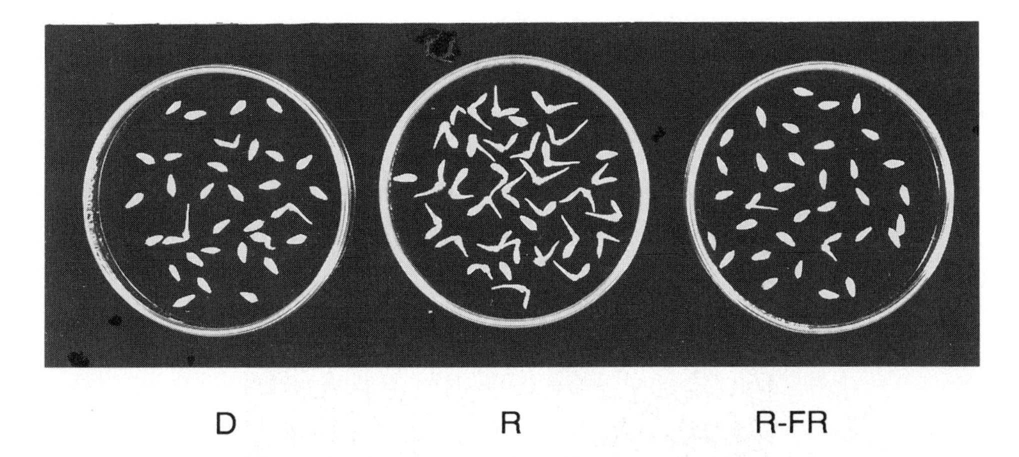

Figure 3. The germination of lettuce (*Lactuca sativa*) achenes in the dark (D), after 3 min red light (R) and 3 min R followed by 3 min far-red light (R-FR). The irradiations were given after 2-h imbibition and the photograph taken after a further 36 h incubation in D at 25°.

nights, could be prevented if the dark period was interrupted by a short pulse of light (*night-break*). Red light (R) was the most efficient kind of light for this (Parker and Hendricks 1946) and, in some cases, the effect of R could be cancelled by a subsequent pulse of far-red light (FR). The reader is referred to Chapter 7.3 for more details about *photoperiodism.*

Seeds of some plants can lie dormant, yet viable, in the soil for a long time (Ødum 1965), but can be reactivated by light. In this case R was also found to be the most efficient wavelength (Fig. 3; Flint and McAlister 1937; Borthwick *et al.* 1952).

A third line of investigation concerned the effect of short light pulses on the growth of plants cultivated in darkness. As the reader may now have surmised, R showed the highest effectiveness, and its effect could be cancelled by subsequent FR. The idea arose that all these effects are mediated by a common type of regulator, which is activated by R, but can be inactivated again by FR. It was also shown that both light processes are reversible; the response after an alternating sequence of alternating R and FR is only determined by the last irradiation (Fig. 3). The explanation, advanced by S.B. Hendricks, is that the regulator is synthesized in darkness in an inactive R-absorbing form (now called Pr) which is able to absorb R and thereby be transformed into the active form (now called Pfr). The Pfr form absorbs FR, and is thereby reconverted to the inactive Pr form.

In a classical paper and an outstanding example from which generations of photobiologists can learn, Withrow *et al.* (1957) determined the action spectra (*i.e.* quantified the efficiency of different wavebands of light) for both the R effect and for its reversal using the straightening of the plumular hook of

Figure 4. The photomorphogenic responses of the bean plant, which were used for determination of phytochrome action spectra. All the plants are genetically similar and of the same age. All plants except the one to the left were grown in darkness except for brief exposures as described on 3 consecutive days. W = white light; D = total darkness; R = red light; FR = far-red light and R-FR = 10 min R followed by 10 min FR.

etiolated bean plants as indicator reaction (Fig. 4). Others determined similar action spectra for seed germination (Borthwick *et al.* 1952; Shropshire Jr. *et al.* 1961) and photoperiodic induction of flowering (Borthwick *et al.* 1948). It was postulated that the two forms of the regulator (the inactive Pr and the active Pfr form) would have absorption spectra resembling these action spectra, just as the absorption spectrum of rhodopsin, the light sensitive pigment in the rod cells of the eye, resembles the action spectrum for triggering of signals from them.

The thrilling story of how this 'photomorphogenic pigment', now called phytochrome (meaning 'plant colour') was finally tracked down, and how the lability of Pfr (*cf.* Chapter 4.4 and Section 4.7.3.3) was discovered, has been vividly told by W.L. Butler (1980), a member of the Beltsville team which succeeded in doing it. Other contributors to that success were S.B. Hendricks, H.A. Borthwick, H.A. Siegelman and K. Norris. In a few of the most productive hours of the history of plant science, phytochrome was detected by *in vivo* spectroscopy, extracted and found to be a protein. From the shape of the action spectra Borthwick *et al.* (1952) had much earlier suggested that the *chromophore* enabling phytochrome to absorb visible light was an open chain tetrapyrrole, and this has now been confirmed. However, the details of the differences between the Pr and Pfr forms of phytochrome, as regards both chromophore (Section 4.1.3) and protein (Section 4.3.6), have been much more difficult to elucidate.

Karl Norris made an important contribution to the discovery of phytochrome by constructing a new type of spectrophotometer with high sensitivity. Subsequently, research in photomorphogenesis has also relied heavily on specialized instrumentation (Chapter 3).

1.4 The surprises of phytochrome

The decades which were to follow the spectrophotometric detection of phytochrome brought forth many new facts about the chromoprotein, some of which will be found later in this book. The following are only a few of the surprises.

(i) Biologists have searched for the mechanism which links phytochrome photoconversion to physiological or biochemical responses, but it appears now that there is no single mechanism. Some responses, such as modulation of growth rate (Chapter 9.1) take place so rapidly that activation or inactivation of genes cannot be involved. On the other hand, in most phenomena, gene regulation is no doubt involved (Chapter 8.1); in many cases photomodulation of mRNA synthesis has been measured directly.

(ii) Gene-mediated phytochrome effects fall into three categories: *very low fluence responses* (VLFRs), *low fluence responses* (LFRs), and *high irradiance responses* (HIRs) (Section 4.7.4). The classical R-induced, FR-reversible responses are LFRs. For an LFR to occur, a high amount of the active phytochrome form, Pfr, is required, but only for a relatively short time. To take place, LFRs require about 1-1000 μmol m^{-2} of R. On the other hand, VLFRs are induced by only 10^{-4}-10^{-2} μmol m^{-2} of R. It is peculiar that such a low fluence does not cause measurable conversion of Pr to Pfr (it can be estimated that by 10^{-4} μmol m^{-2} of R only about 0.01% is converted). The VLFRs are not reversible by FR; on the contrary they are induced by FR alone (*i.e.* no R is necessary). The explanation of this is that Pr absorbs FR (although not as strongly as Pfr does), so that a few percent of the phytochrome may be converted to Pfr by FR, as compared to a maximum conversion of about 80% by R.

Typical HIRs in etiolated plants require long irradiation times (many hours with FR or B); in etiolated plants they are not effectively induced with continuous R. To understand why R is ineffective in this case, we should consider the basic scheme describing phytochrome *synthesis, conversion* and breakdown (*destruction*):

$$\rightarrow \text{Pr} \underset{k_2}{\overset{k_1}{\rightleftarrows}} \text{Pfr} \rightarrow \text{destruction}$$

In R the rate constant k_1 is four times the rate constant k_2, while in FR k_1 is only about 4% of k_2 [the photochemical rate constants are the products of *photon fluence rate* (Chapter 2), *quantum yield*, and the wavelength dependent *absorption coefficient* of the reactant pigment form]. Except in darkness or very weak light, the photochemical rates are much higher than the rates of synthesis or destruction, and therefore a *photo-'stationary' equilibrium* of the phytochrome system is established. This means that the ratio of the concentrations of Pfr and Pr is equal to the ratio of the rate constants, k_l/k_2.

For HIRs to occur, Pfr must be present over an extended time, but only at a relatively low concentration (typically a few percent of the concentration of Pr present in a non-irradiated plant). This situation cannot be produced in etiolated plants by continuous R (except possibly by very weak light), since Pfr is unstable; it is destroyed by a non-photochemical reaction. When R is applied, it converts most of the phytochrome to the Pfr form which is destroyed, whereupon the remaining phytochrome is converted and destroyed and cannot induce an HIR.

(iii) It seems that HIRs are not determined by Pfr concentration alone. They increase with fluence rate above that required to establish photoequilibrium (Section 4.7.4.3.3). It is possible that *phytochrome cycling rate* or the concentrations of intermediates in the Pr $\rightleftarrows$ Pfr photoconversion are also important, but other explanations have also been advanced (Section 4.7.4.3.6).

(iv) It seems that in phytochrome-mediated modulation of extension growth, the ratio (φ, phi) of Pfr to total phytochrome (P) is more important than the concentration of Pfr *per se* (Chapter 7. 1).

(v) Most of the above applies primarily to the type of phytochrome first discovered, *i.e.* that present in dark-grown plants ('etiolated-type' phytochrome). Other phytochrome types, coded for by different genes (Chapter 4.2) have now been discovered, and most probably they have different functions in the plant (Sections 7.1.9 and 7.3.6.5).

(vi) Phytochrome occurs not only in various types of plant organs and cell types, but also in different parts of the cells (Chapter 4.5). It seems to be membrane bound, at least in the active form. One reason for believing this, is that in some cases phytochrome conversion results in a rapid change in membrane potential and ion fluxes (Chapter 4.6 and Section 9.1.3). Another reason is the effect of *plane-polarized light* (Chapter 7.2).

1.5 Perception of B and UV light

Phytochrome is the best known of the pigment systems by which plants gather information about the light environment. Before the existence of the R/FR sensitive phytochrome system was suspected, other light reactions of plants had been studied; in particular *phototropism*, the plant's orientation with respect to light direction. The mechanism of phototropism is still subject to research (Chapter 9.2). It was found that light of short wavelength (< 500 nm) is required to elicit phototropic bending both in plants and fungi (Chapters 5.1, 5.2 and 9.7), while, e.g, mosses also react to R. Soon it was also found that a number of other responses are also caused specifically by B, or, to be more exact, by B, violet and long wavelength UV (UV-A) light. Figure 5 shows a 'classic' example for fern morphogenesis. The term B receptor (or B/UV-A receptor) was coined for the unknown pigment system(s) mediating this type of effect.

Detailed action spectra indicated that the chromophoric group of B receptors could be either *flavins or carotenoids.* It is now recognized that a choice between the two groups of substances cannot be made based only on comparisons of spectra. More recently a third group of compounds, the *pterins* (Section 5.2.2.3) has also made its entrance into the discussion.

From an evolutionary point of view the B receptors are older than phytochrome. In higher plants they seem to have been specialized for sensing light direction and for regulating certain processes related to photosynthesis: chloroplast development, stomatal movement (Chapter 9.3), chloroplast position (Section 9.4.3.2), the chemical fate of carbon in photosynthesis.

1.6 Other photoreceptors and coaction between photoreceptors

Phytochrome and B photoreceptors do not work independently; in many cases there is a coaction (Chapters 6 and 9.3). Furthermore, they are not the only photomorphogenically active pigments in higher plants. In many plants a specific UV-B photoreceptor operates (Chapter 9.5), and many of the effects of UV-B radiation (a 'hot' topic due to concern for the stratospheric ozone) can be more properly described as photomorphogenesis rather than damage. Fungi are diverse and treated in a separate chapter (Chapter 9.6), while photomorphogenic effects described so far in ferns (Chapter 9.7) seem to be mediated by phytochrome and B receptors. Some cyanobacteria have a pigment system which senses the balance between R and green light, and allows these organisms to adjust their pigment antennae accordingly for maximum efficiency of photosynthesis. In some respects this pigment system resembles the phytochrome system. Another pigment system mediates light-dependent germination of akinetes (cyanobacterial resting cells) (Braune 1979).

In eukaryotic algae still other photoreceptor systems operate (as well as phytochrome and blue-light receptors in some of them). The pigment enabling some free-swimming flagellates to orientate in relationship to light direction (*phototaxis,* Section 9.4.2.2) is a close relative to the rhodopsin by which we ourselves see (Foster *et al.* 1984).

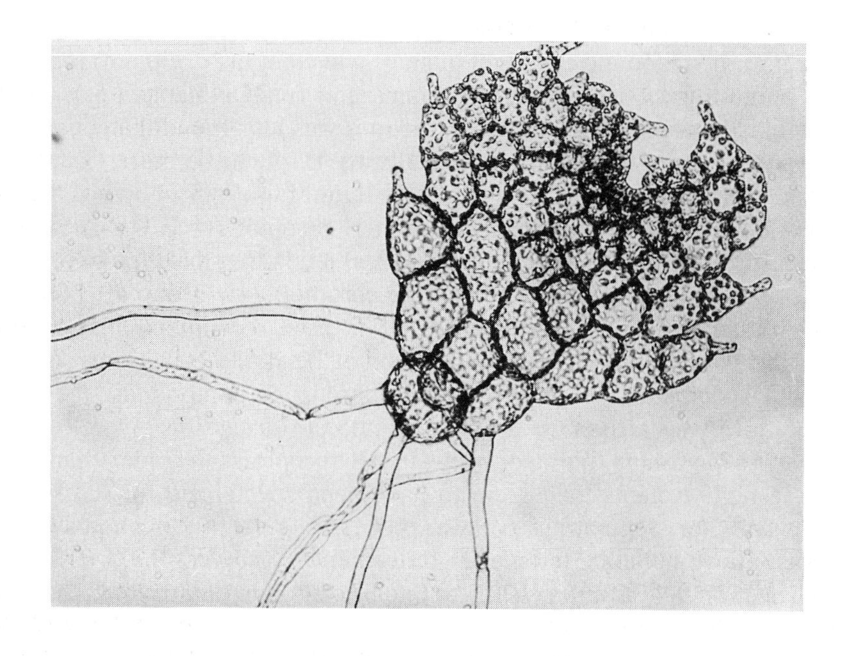

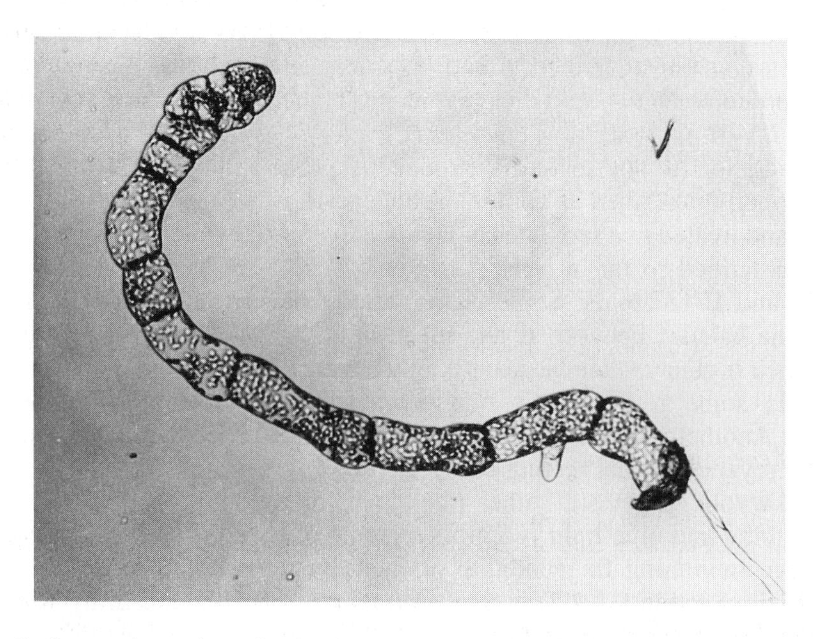

Figure 5. Gametophytes (sporelings) of the fern, *Dryopteris filix-mas,* grown for 7 days in continuous blue light (B, *top*) or continuous red light (R, *bottom*). Note B is necessary for two-dimensional growth.

1.7 Molecular biology and genetics

As in many other areas of biology, molecular biology and genetics have provided important techniques for dissecting the processes of photomorphogenesis (Chapters 4.2, 4.8, 8.1, 8.2 and Section 4.6.8). By acquiring mutants deficient in various types of photoreceptors it should be possible to define their roles more clearly. This type of investigation started long ago with the B receptors of fungi, proceeded with the successful identification of the rhodopsin-like receptor in *Chlamydomonas,* and is now in full action for higher plants (Chapters 4.8 and 8.2). Molecular biology is also offering new opportunities for elucidating the transduction chains (Chapters 4.6 and 5.2).

1.8 Conclusion and outlook

The molecular mechanisms of photomorphogenesis are fascinating, as are the methods used for elucidating them, but what matters for the plant, and also for the people involved in plant cultivation, is how the plant reacts in its light environment, and it is also important to understand what this light environment is (Part 7). The processes regulated by light are so diverse that we have not been able to describe everything in this book. We have, for instance, largely refrained from treating photomorphogenesis at the organelle level. On the other hand, we have included chapters on photoperiodism and rhythms (Chapters 7.3 and 8.1), because we consider these important aspects closely related to photomorophogenesis in the strict sense, and because they are areas of very 'hot' research.

Plant photomorphogenesis has clearer than many other fields of plant science demonstrated how complex, advanced, and well adapted to their environment our green co-inhabitants of this planet are.

1.9 Further reading

Sage L.C. (1992) *Pigment of the Imagination - A History of Phytochrome.* 562 pp., Academic Press, New York.

1.10 References

Borthwick H.A., Hendricks S.B. and Parker M.W. (1948) Action spectrum for the photoperiodic control of floral initiation of a long day plant, winter barley *(Hordeum vulgars). Bot. Gaz.* 110: 103-118.

Borthwick H.A., Hendricks S.B., Parker M.W., Toole E.M. and Toole V.K. (1952) A reversible photoreaction controlling seed germination. *Proc. Natl. Acad. Sci. USA* 38: 662-666.

Braune W (1979) C-phycocyanin, the main photoreceptor in the light dependent germination process of *Anabaena* akinetes. *Arch. Microbiol.* 122: 289-295.

Butler W.L. (1980) Remembrances of phytochrome twenty years ago. In: *Photoreceptors and plant development.* Proceedings of the Annual European Photomorphogenesis Symposium, pp. 3-7, De Greef J. (ed.) Antwerp University Press.

Flint L.H. and McAlister E.D. (1935) Wavelengths of radiation in the visible spectrum inhibiting the germination of light-sensitive lettuce seed. *Smithsonian Misc. Collect.* 94: 1-11.

Flint L.H. and McAlister E.D. (1937) Wavelengths of radiation in the visible spectrum promoting the germination of light-sensitive lettuce seed. *Smithsonian Misc. Collect.* 96: 1-8.

Foster K.W., Saranak J., Patel N., Zarilli G., Okabe M., Kline T. and Nakanishi K. (1984) A rhodopsin is the functional photoreceptor for phototaxis in the unicellular eukaryote *Chlamydomonas. Nature* 311: 756-759.

Garner W.W. and Allard H.A. (1920) Effect of the relative length of day and night and other factors of the environment on growth and reproduction in plants. *J. Agric. Res.* 18: 553-606.

Henfrey A. (1852) *The Vegetation of Europe, its Conditions and Causes,* pp.37-39, J. van Voorot, London.

Kjellman F.R. (1885) Aus dem Leben der Polarpflanzen. In: *Studien und Forschungen. Veranlasst von meinen Reisen im hohen Norden. V.,* pp. 443-521, Nordenskiöld A.E. (ed.) Leipzig.

Knight M.R., Smith S.M. and Trewavas A.J. (1992) Wind-induced plant motion immediately increases cytosolic calcium. *Proc. Natl. Acad. Sci. USA* 89: 4967-4971.

Ødum S. (1965) Germination of ancient seeds. Floristical observations end experiments with archaeologically dated soil samples. *Dansk Bot. Arkiv.* 24: 2.

Parker M.W., Hendricks S.B., Borthwick H.A. and Scully N.J. (1946) Action spectrum for the photoperiodic control of floral initiation of a short day plant. *Bot. Gaz.* 108: 1-26.

Reed J.W., Nagpai P., Poole D.S., Furuya M. and Chory J. (1993) Mutations in the gene for the red/far-red light receptor phytochrome B alter cell elongation and physiological responses throughout *Arabidopsis* development. *Plant Cell* 5: 147-157.

Shropshire W. Jr., Klein W.H. and Elstad W.B. 1961. Action spectra of photomorphogenetic induction and photoinactivation of germination in *Arabidopsis thaliana. Plant Cell Physiol.* 2: 63-69.

Withrow R.B., Klein W.H. and Elstad V.B. (1957) Action spectra of photomorphogenesis and its inactivation. *Plant Physiol.* 32: 453-462.

Part 2 Quantification of light

2. Quantification of light

Lars Olof Björn[1] and Thomas C. Vogelmann[2]

[1]*Section of Plant Physiology, Lund University, Box 7007, S-220 07 Lund, Sweden*
[2]*Department of Botany, University of Wyoming, Laramie, WY 82071 USA*

2.1 Basic concepts

The topic of light quantification seems confusing, not only to the layman and the student, but also to the expert. Some reasons for this confusion are as follows:

(i) The amount or intensity of light is often regarded as something that can be completely described by a number. Such a view disregards the following facts. (a) Light consists of components with different wavelength. A full description of the light would thus give information about the 'amount' of light of each wavelength. (b) Light has direction. The simplest case is that all the light being considered has the same direction, *i.e.* the light is collimated (the rays are all parallel). Another case is that light is isotropic, *i.e.* all directions are equally represented. Between these extremes there is an infinite number of possible distributions of directional components. (c) Light may be polarized, either circularly or plane polarized (Chapter 7.2). In the rest of this chapter this complication will be disregarded, but one should always be aware of the fact that a device such as a photocell may be differentially sensitive to components of different polarization, and polarization may be introduced by part of an experimental setup, such as a monochromator or a reflecting surface. (d) Light may be more or less coherent, *i.e.* the photons more or less 'in step'. This complication, which is important in some experiments, *e.g.* with laser light, will also be neglected. (e) Finally, people often disregard, neglect or confuse the time factor. It should be decided whether to express an *instantaneous* or a *time-integrated* quantity, *e.g. fluence rate* or *fluence*, *power* or *energy*. Power means energy per unit time.

(ii) Light is of interest to people investigating or working in widely different fields. Experts in different fields have used different concepts and different nomenclature, partially depending on which properties of light have been of

R.E. Kendrick & G.H.M. Kronenberg (eds.), Photomorphogenesis in Plants - 2nd Edition

interest to them, and partly on the whims of historical development. Only recently have there been serious attempts to achieve a uniform nomenclature, and the process is not yet complete.

2.2 The wavelength problem

Sometimes it is not practical to give the complete spectral distribution of light, and we have to quantify it in some simpler way. From the purely physical viewpoint, there are two basic ways. Either a quantity is used that is related to the number of photons, or one related to the energy content of light. For light of a single wavelength the energy of a photon is inversely proportional to the wavelength as given by:

$$E = \mathrm{h} \cdot c / \lambda$$

where h = Planck's constant and c = the velocity of light. From this it follows that equivalent numbers of photons at 350 nm have twice the energy of those at 700 nm. It depends upon the application as to whether it is more appropriate to quantify light in terms of energy or number of photons.

The ultimate way to calibrate for the energy emitted by light sources makes use of a hollow heat radiator of known temperature, which radiates in a way predictable by basic laws of physics. However, most laboratories do not have a hollow heat radiator of known temperature, but may have a standard lamp which has been compared to such a radiator using a photothermal device. Photothermal devices, such as a thermopile or a bolometer, measure the heat given off by absorbed light, and their sensitivity is independent of wavelength. A standard lamp can consist of a 1 kW tungsten lamp that is run at a precisely controlled electric current. For the ultraviolet region calibrated deuterium lamps are also available.

The unit for energy is the joule (J). Energy per time is power and a joule per second is a watt (W). Both can be expressed per unit area (*e.g.* J m^{-2}, W m^{-2}). In rare circumstances it may be desirable to express energy or power as function of volume (*cf.* Table 1).

Although measurement of the energy content of light is valid for some applications, in others it is more desirable to express the amount of light as number of photons. There are several ways to construct photon counters. One of the simplest ways is by actinometry in which light is passed through a solution of molecules that react to light:

$$A + \text{photon} \rightarrow B$$

absorption of a photon converts molecule '*A*' to molecule '*B*'. Measuring the number of '*B*' molecules gives an estimate of the number of photons absorbed by the solution. Actinometry is in some cases very accurate. Unfortunately, the ability of photons to cause a photochemical response varies with wavelength. This variable quantum yield means that actinometers are only useful under special circumstances where the spectral quality of the light matches the absorption characteristics of the photochemical reactant and the quantum yield is well known. Thus, actinometers, as well as other types of photon counters must be calibrated against standards previously calibrated against photothermal devices.

Photomultiplier tubes can also be used as photon counters, but one should be aware that they do not, strictly speaking, count photons. Instead they count electric pulses caused by photons. The principle of operation of a photomultiplier is relatively simple. It contains a series of charged metal plates inside an evacuated glass envelope. A photon that enters the tube strikes a photocathode and knocks out an electron. The electron is accelerated by an electric field until it strikes the next plate. The energy added by the field enables the first electron to knock out several other electrons from the second plate, and the process is repeated at subsequent plates. One photon eventually gives rise to an avalanche of electrons, which can be measured as an electric pulse. Photomultipliers are efficient and sensitive. Unfortunately some of the electric pulses are not caused by incident photons, but rather by electrons ejected from the photocathode by thermal vibrations. This can be minimized by cooling the photocathode. Even more limiting is the fact that photomultiplier tubes have different counting efficiency at different wavelengths. Thus, they must be calibrated against standards that have been previously calibrated against photothermal devices.

The units for expressing light as photons are: (i) photons (number of photons); (ii) mole of photons (the symbol is mol; there are $6.02217 \cdot 10^{23}$ photons in a mole). The unit micromole (μmol) is commonly used in plant physiology and is equivalent to $6.02217 \cdot 10^{17}$ photons. Numbers of photons and moles of photons can be expressed per unit time, area, or volume.

Even though light may be expressed in these units, sometimes this is not enough. For example, a value followed by (W m^{-2}) gives no information about the direction of light. Are the rays parallel, or is the light diffuse? This brings us to the topic of the next section.

2.3 The problem of direction and shape

Most light measuring systems are calibrated using light of approximately a single direction, *i.e.* collimated light. However, light in nature and in most experimental situations is not collimated. When the sky is cloudless and unobstructed the rays coming directly from the sun are quite well collimated, but in addition there is skylight and the light reflected from the ground and other objects. A plant physiologist who wants to understand how plants use and react to light has to take this into account.

Traditionally, most measuring devices have a flat sensitive surface, and when we calibrate the instrument we generally position this surface perpendicular to a collimated calibration beam. A plant leaf is also flat, so in the first approximation we can measure light in single-leaf experiments with a flat device with the same direction as the leaf. But a whole plant is far from flat (except in very special cases). Different surfaces on the plant have different orientations. Ideally we should know the detailed directional (and spectral) distribution of the light impinging on the plant, but this is not possible in practice. Since a plant is a three-dimensional object, it would be better to quantify the light using a device having a spherical shape and being equally sensitive to light from any direction. This brings us to the distinction between: (i) irradiance, *i.e.* radiation power incident on a flat surface of unit area; (ii) energy fluence rate (or fluence rate for short): radiation power incident on a sphere of unit cross section. The term fluence rate was introduced by Rupert (1974).

Both these concepts have their counterparts in photon terms. For case (i) the nomenclature is not settled, but it would be logical to use the term photon irradiance. Many people, especially in the photosynthesis field, use the term photon flux density, and the abbreviation PFD (PPFD for photosynthetic photon flux density, see below). However, Holmes *et al.* (1985) have given very good reasons for not using this term. For case (ii) the term photon fluence rate is generally accepted among plant physiologists, but hardly at all among scientists in general.

Energy fluence is the energy fluence rate integrated over time. Fluence is taken to mean the same as energy fluence.

We shall now compare irradiance and energy fluence for different directional distributions of light (*cf.* Fig. 1):

(i) Collimated light falling perpendicularly to the irradiance reference surface. In this case the flat surface of unit area and the sphere of unit cross sectional area will intercept the light equally, and irradiance will be the same as fluence rate.

(ii) Collimated light falling at an angle x to the normal of the irradiance reference plane. In this case the light intercepted by the flat surface of unit area will be less than that intercepted by the sphere of unit cross sectional area. The irradiance will be $\cos(x)$ times the fluence rate. Since $\cos(x)$ is less than unity, the irradiance in this case will be lower than the fluence rate.

(iii) Completely diffuse light falling from one side only. The ratio of irradiance to fluence in this case will be an average of $\cos(x)$ for all angles x from 0 to $+\pi/2$ weighted by $|\sin(x)|$, *i.e.* the absolute value of the sine of x. The mathematical expression for this weighted value is $\int|\sin(x)|\cdot\cos(x)\cdot dx/\int|\sin(x)|\cdot dx$

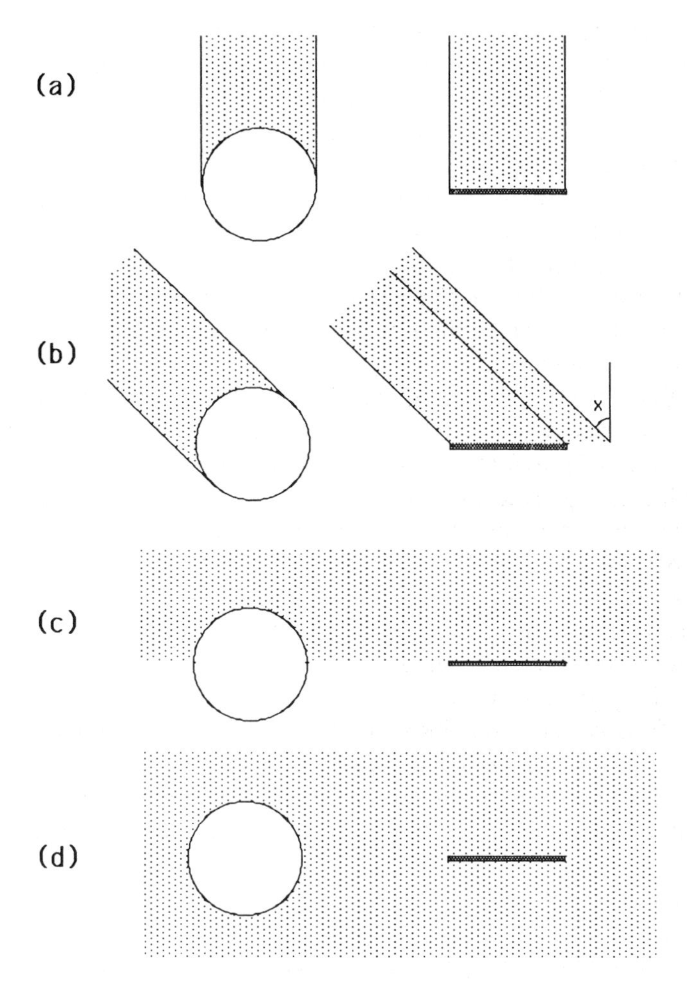

Figure 1. The concepts of fluence rate and irradiance. Comparison is made between a spherical receiver (recording fluence rate, to the *left*), and a flat receiver of the same cross sectional area (recording irradiance, to the *right*). In (a) collimated light strikes the flat receiver perpendicularly. In this case fluence rate and irradiance are the same. In (b) collimated light strikes the flat receiver at an angle x with the normal. The beam is drawn to the same size as in (a), to make it easy to understand that the flat receiver will miss some light intercepted by the spherical receiver. In this case the irradiance is $\cos(x)$ times the fluence rate. In (c) diffuse (isotropic) light is coming from above, and the irradiance is half of the fluence rate. In (d) isotropic light comes from all directions. The irradiance is one quarter of the fluence rate.

with the integration from 0 to $+\pi/2$, and this is equal to 1/2. Thus the irradiance in this case is half the fluence rate. The reason for having to weight $\cos(x)$ by $\sin(x)$ is that all values of x are not equally common. In other words they do not have the same probability. The various directions may be thought of as corresponding to points on a big sphere, the centre of which is the point of measurement. The sphere can be thought of as divided into a pile of rings, and each ring, corresponding to a value of x, has a radius, and hence a circumference proportional to $\sin(x)$.

(iv) Completely diffuse light from both sides. The sphere is hit by isotropic light over its whole surface, but for the flat receiver only one surface (irradiance is defined in this way). Therefore the irradiance is one quarter of the fluence rate in this case. We can easily remember this if we consider that the area of a circle is one quarter of the area of a sphere with the same radius.

Table 1 shows the various quantities associated with light measurements.

Table 1. Concepts and units for the quantification of light

Sensor geometry	Instantaneous values		Time integrated values	
	Energy system	Photon system	Energy system	Photon system
Flat	(energy) irradiance unit: W m^{-2}	photon irradiance (photon flux density; term not recommended) unit: mol m^{-2} s^{-1}	time integrated (energy) irradiance unit: J m^{-2}	time integrated photon irradiance unit: mol m^{-2}
Spherical	(energy) fluence rate unit: W m^{-2}	photon fluence rate unit: mol m^{-2} s^{-1}	(energy) fluence unit: J m^{-2}	photon fluence unit: mol m^{-2}

In all the cases we add the word spectral before the various terms if we wish to describe the spectral variation of the quantity. We may thus write spectral photon fluence rate on the vertical axis of a spectrum of light received by a spherical sensor. A suitable unit for spectral photon fluence rate would be mol m^{-2} s^{-1} nm^{-1}. It would be correct but less easily understood to express it in mol m^{-3} s^{-1}.

The term space irradiance, introduced by Grum and Becherer (1979) is synonymous with fluence rate. Some authors use the term vector irradiance instead of irradiance, and scalar irradiance instead of fluence rate. Spherical irradiance (*e.g.* Jerlov 1968) is one quarter of the fluence rate. The term dose should, as is standard for ionizing radiation, be used only for absorbed radiation.

2.4 Biological weighting functions and units

Sections 2.2 and 2.3 considered the physical quantification of light. However, there has been a need for additional concepts in connection with organisms and biological problems. Traditionally there has been a special system related to human perception of light. Here we can limit ourselves to illuminance, which is expressed in lux (lx). Neglecting the historical development, we can say that lux is the integrated spectral irradiance weighted by a special weighting function. This weighting function is precisely defined (Fig. 2), but can be thought of as the average (photopic) eye spectral sensitivity for a large number of people.

Similarly we may, for purposes other than vision, such as light for reading or working, weight the spectral irradiance by other functions. These functions approximate various photobiological action spectra. One special function pertains to photosynthesis and is zero below 400 nm and above 700 nm, and unity from 400 to 700 nm. By definition this describes the photosynthetically active radiation (PAR). Usually one uses the spectral photon irradiance to weight by this function, and this is the meaning of the often used term PPFD, photosynthetic photon flux density. The assumption of zero photosynthetic action outside the range 400-700 nm, and the same action for all components within the range is, of course an approximation. However, people have agreed on this approximation.

Another weighting function is used for 'sunburn meters' to yield sunburn units, but in this field we have to watch out for various 'units' used by different

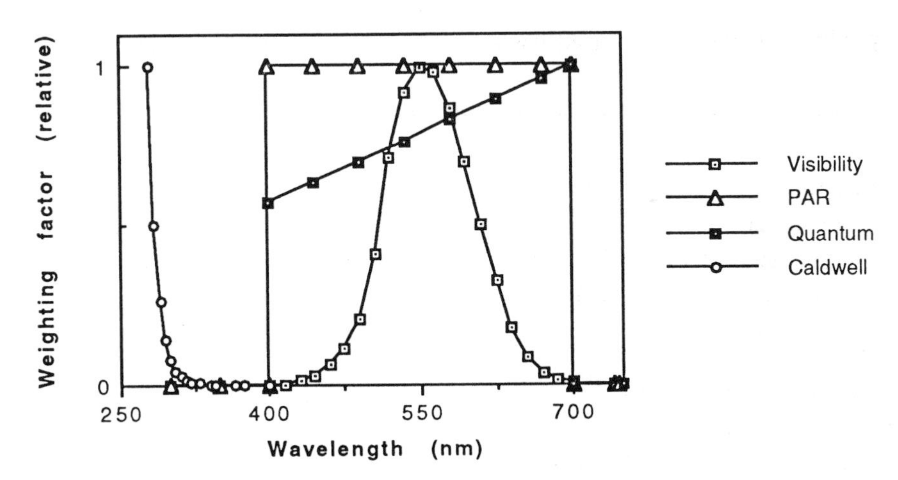

Figure 2. Examples of weighting functions used for converting energy units to other quantifications of light, all with relative ordinates. 'Visibility' is the function used for converting energy units to photometric units (such as lumen and lux). In absolute units the maximum of the curve corresponds to 683 lumens W^{-1} or 683 lux $(W\ m^{-2})^{-1}$. 'PAR' is photosynthetically active radiation in W m^{-2}. This function is unity from 400 to 700 nm and zero outside these limits. 'Quantum PAR' is photosynthetically active radiation in μmol m^{-2} s^{-1}. This function is proportional to wavelength from 400 to 700 nm and zero at lower and higher wavelengths. 'Caldwell' is a function devised by Rundel (1983) based on data by Caldwell *et al.* (1986) for plant active UV-B. It is an example of a group of functions used for computing various types of biologically active UV-B. Usually they are all normalized to unity at 300 nm, but for clarity the curve is drawn to another scale here.

people. One kind of sunburn meter used much in the past is the Robertson-Berger meter, but recently a new agreement has been reached for using a weighting spectrum that more closely resembles the true sunburn action spectrum of caucasian skin.

There are also numerous other weighting spectra in use for estimating radiation with other biological actions. One of particular interest for plant scientists is the 'generalized plant action spectrum' defined by Caldwell (1971), to estimate the damaging and inhibiting effects on plants by UV and an action spectrum for inhibition of photosynthetic carbon dioxide uptake in intact plants (Rundel 1983; Caldwell *et al.* 1986). In many cases, such as for photosynthetic radiation and the damaging effect of UV, one regards the weighting function as dimensionless, so that the weighted units will be the same as the unweighted, *e.g.* W m^{-2} or mol m^{-2} s^{-1}. This calls for caution to avoid misunderstanding. It must also be recognized that the same action spectrum, such as the generalized plant action spectrum can be given different height by different researchers. For example, they may be normalized to unity at 300 nm by one person and in another way by another scientist, yielding, in effect, different unit systems.

Finally it should be noted that some meters are constructed so that they measure lux, sunburn units, or PAR directly. They are designed so that their spectral response approximates the particular weighting function of interest. When the sensor of one of these devices is placed under a light source, the meter gives the desired units without further complications. However, these meters are often used inappropriately. For example, some PAR meters give accurate results only when placed under a light that has a spectral composition that is similar to sunlight. These meters can give highly erroneous results when used to measure monochromatic light, or any light whose spectral composition deviates significantly from that of the sun. When working on special problems such as a photobiological response that has an unknown action spectrum, or one that varies among organisms it is best to measure the amount of light for each wavelength. This can be done with a spectroradiometer. Although once a tedious task, computerization of such instrumentation has made measurement of light and associated computational procedures such as application of weighting functions relatively easy (Chapter 7.1).

2.5 Further reading

Diffey B.L. (ed.) (1989) *Radiation Measurement in Photobiology.* Academic Press, London.

Krizek D.T. (1982) Guidelines for measuring and reporting environmental conditions in controlled-environment studies. *Physiol. Plant.* 56: 231-235.

2.6 References

Caldwell M.M. (1971) Solar UV irradiation and the growth and development of higher plants. In: *Photophysiology*, Vol. VI, pp.131-177, Giese A.C. (ed.). Academic Press, New York.

Caldwell M.M., Camp L.B., Warner C.W. and Flint S.D. (1986) Action spectra and their key role in assessing the biological consequences of solar UV-B radiation change. In: *Stratospheric Ozone Reduction, Solar Ultraviolet, Radiation and Plant Life*, pp. 87-111, Worrest R.C. and Caldwell M.M. (eds.) NATO ASI Series G: Ecological Sciences, Vol. 8.

Grum F. and Becherer R.J. (1979) *Optical Radiation Measurements* 1, pp. 14-15, Academic Press, New York.

Holmes M.G., Klein W.H. and Sager J.C. (1985) Photons, flux, and some light on philology. *Hort. Sci.* 20: 29-31.

Jerlov N.G. (1968) *Optical Oceanography*. Elsevier Publishing Company, Amsterdam. Library of Congress Catalog Card Number 68-12475.

Rundel R.D. (1983) Action spectra and estimation of biologically effective UV radiation. *Physiol. Plant.* 58: 360-366.

Rupert C.S. (1974) Dosimetric concepts in photobiology. *Photochem. Photobiol.* 20: 203-212.

Part 3 Instrumentation in photomorphogenesis research

3. Instrumentation in photomorphogenesis research

Masaki Furuya[1] and Yasunori Inoue[2]

[1]Advanced Research Laboratory, Hitachi Ltd., Hatoyama, Saitama, 350-03, Japan
[2]Department of Applied Biological Science, Faculty of Science and Technology, Science University of Tokyo, Noda 278, Japan

3.1 Introduction

New instruments, particularly in photobiology, have always led to advancement in our knowledge. We measure the molecular properties of photoreceptor pigments and their photo-induced labile intermediates by spectrophotometry. The photoperceptive site within individual cells, tissues or organs is determined by microbeam-irradiation techniques and orientation of the photoreceptor pigments by using polarized light as a probe. Plants utilize light not only as an energy source for photosynthesis but also as a source of information to 'search' for seasonal, diurnal or sudden changes in the light environment. During their evolution plants have acquired the ability to use different properties of light such as irradiance, wavelength, timing and duration of irradiation, direction of exposure, and plane of polarization to sense the environment about them. Researchers into photomorphogenesis have investigated the capacity of plants to measure the light stimuli and regulate molecular and subcellular processes in cells, utilizing not only commonly available instruments but ones specially designed and developed. In this chapter the major instruments will be introduced which have contributed to photomorphogenesis research, including the background for their development.

3.2 Spectrographs: analyses of wavelength effects

When we find an unknown light-dependent phenomenon and wish to know more about the photoreceptor involved, we determine the effectiveness of

R.E. Kendrick & G.H.M. Kronenberg (eds.), Photomorphogenesis in Plants - 2nd Edition

different wavelengths on the response action spectrum. To do this plants need to be exposed to monochromatic light of different wavelengths and fluence rates. The many complex photoreactions in plants can be separated into their elementary reactions by irradiation with monochromatic light of different wavelengths. Considering that biological, living material is fairly large in size, we have to provide sufficient spectral dispersion of monochromatic wavelengths to account for the sample size. A high power output is required over areas of the sample and sufficient purity of radiation needed to permit detection of a weak action at each wavelength. Thus, large-scale spectrographs were designed and built at the Agricultural Experimental Station, USDA (Parker *et al.* 1945) and Argonne National Laboratory (Monk and Ehret 1956) in the USA; Le Phytotron, CNRS, in France (Jacques *et al.* 1964) and Okazaki National Laboratory in Japan (Watanabe *et al.* 1982). The one at Okazaki is still actively being used for international collaboration by workers who have applied for its use on a time-share basis.

3.2.1 The Beltsville spectrograph

In 1945 S.B. Hendricks and his colleagues designed and constructed the first spectrograph (Fig. 1) in the workshops at the Plant Industry Station, USA in Beltsville (Borthwick 1972). The instrument consisted of two prisms, and a 10-kW DC carbon-arc lamp, powered by a 20-horsepower motor-generator unit. The prisms were of flint glass, *ca.* 20 cm high, the spectrum at the focal plane was *ca.* 10 cm wide and 1.5-2 m from the blue (B) to far-red (FR) extremes. There was a distance of 14 m between the last light-scattering optical surface in the instrument and the focal plane (Parker *et al.* 1945). It was installed in a dark room, 'the rainbow room' which was about 18 m long (Fig. 1b), and was connected to two plant-growth rooms. The whole setup, including the power source, was assembled during World War II from materials available in the USDA or on loan and at no cost other than for services of the USDA workshops (McGee 1987). During its lifetime, the instrument was widely used to investigate a variety of photobiological problems in plants and animals.

The most important contribution of this spectrograph were the action spectra which led to the observation of the 'red light (R)-FR photoreversible reaction' in lettuce seed germination (Borthwick *et al.* 1952) and which eventually resulted in the discovery of phytochrome (Butler *et al.* 1959).

3.2.2 The Okazaki large spectrograph

The classic spectrographs built in the 1950's and 60's were extremely useful, but it was a very time-consuming and tedious job to set up samples at the focal

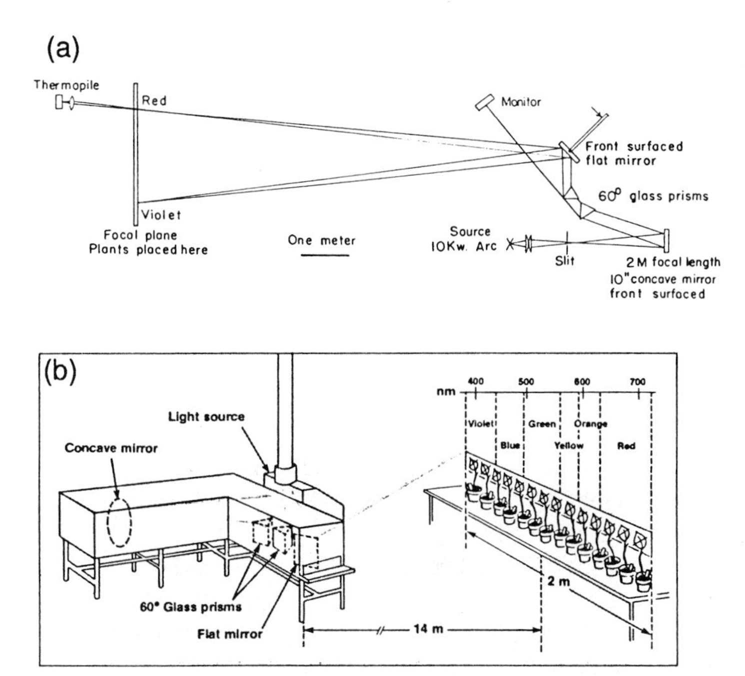

Figure 1. A schematic drawing of (a) the Beltsville spectrograph (After Parker *et al.* 1946) and (b) Experimental layout for determining the action spectrum for the inhibition of flowering in soybeans (McGee 1987).

plane (Fig. 1b) using a safelight and determine the fluence rates of different wavelengths of monochromatic light in each experiment. In addition, although the samples should not be exposed to any light except the experimental monochromatic light, they were lined up at the focal plane without any protection from scattered light, so that this inevitably influenced the results (Fig. 1b). It was also difficult to irradiate simultaneously with monochromatic light of different wavelengths and fluence rates for determining action spectra. Plant materials, may also respond differently to the same light treatment if irradiated at different times of day, because of circadian and other rhythms.

To solve these problems, a computer-operated, large-scale spectrograph (Fig. 2) was designed and constructed in the early 1980's, at the Okazaki National Institute, in Japan (Watanabe *et al.* 1982). Samples can be set in microcomputer-controlled threshold boxes (Fig. 2b, B2), that are transferred to the stage at the focal curve (Fig. 2a, B1) by a computer-operated carrier

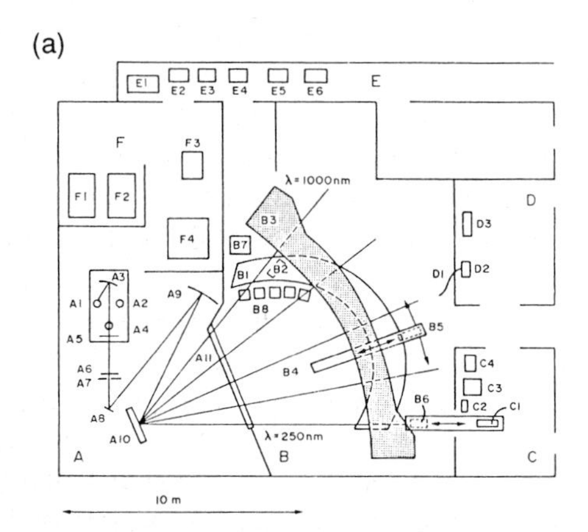

Figure 2. Spatial arrangement (a) and a view of the irradiation room (b) of the Okazaki large spectrograph (after Watanabe *et al.* 1982). A, monochromator room; A1, 30-kW Xe short-arc lamp; A2, 6-kW Xe short-arc lamp; A3, rotatable condensing mirror; A4, medium pressure Hg lamp; A5, shutter; A6, heat-absorbing filter; A7, entrance slit; A8, plane mirror; A9, condensing mirror; A10, double-blazed plane grating, which is a mosaic of 36 replicas with aluminum surfaces backed with BK-7 glass. B, the irradiation room; B1, focal curve stage; B2, sample box; B3, a horizontally arc-shaped, fixed x axis frame; B4, (y-axis) frame, which moves toward the centre of the grating at a speed of 5 m min^{-1}; B5, arm, hanging under and moving along the y-axis frame; B6, origin of the automatic carrier system; B7, interface for entrance slit control and mirror cover drive; B8, interface for connector drive. C, Sample box preparation room; C1, trolley; C2, control panel; C3, CRT terminal; C4, printer. D, Optical-fibre room; D1, optical-fibre bundle; D2, optical-fibre outlet unit; D3, panel for monitoring the fluence rated in the sample boxes and for wavelength indication of the optical fibre system. E, Microcomputer room; E1, host microcomputer (DSC 23, Hitachi Ltd.); E2, data typewriter; E3, CRT terminal; E4, printer; E5, numerical control device; E6, NC interface. F, Power supply room; F1, air cooling unit for lamps; F2, power supply for lamps; F3, control panel for lamps; F4, water-cooling unit for lamps.

(Fig. 2b, B3-5) and exposed to monochromatic light, of the appropriate wavelength, photon fluence rate, photon fluence and timing of irradiation, automatically-controlled according to the pre-programmed schedule of the researchers. Light emitted from a 30-kW Xe short-arc lamp is diffracted by a grating which is double-blazed at 250 and 500 nm (Fig. 2a, A10) and focused on the horseshoe-shaped plane (Fig. 2a, B3). Monochromatic light ranged from 250-1000 nm is available with a wavelength dispersion of 0.8 nm cm^{-1}. The fluence rate of monochromatic light reaches 1016 photons cm^{-2} s^{-1} nm^{-1} at 500 nm and this fluence rate is about 20 times higher than that in sun light. The level of stray light is lower than 1.6×10^{-4} of the main light throughout the whole wavelength range. In particular, the threshold box (Fig. 3) was carefully designed so that the fluence rate of a chosen monochromatic light at a cross section in the compartment was highly uniform (see Fig. 6 in Watanabe *et al.* 1982).

This spectrograph has contributed uniquely in providing UV of high fluence rates for determining action spectra of UV dependent processes, which previously were not possible to measure for technical reasons (Senger 1980). For example, the peaks of action spectra in the UV region with perithecial formation in *Gelasinospora* (Inoue and Watanabe 1984) and primordial initiation and development in *Coprinus* (Durand and Furuya 1985) were discovered, for the first time, using this spectrograph (Fig. 11 in Chapter 9.6).

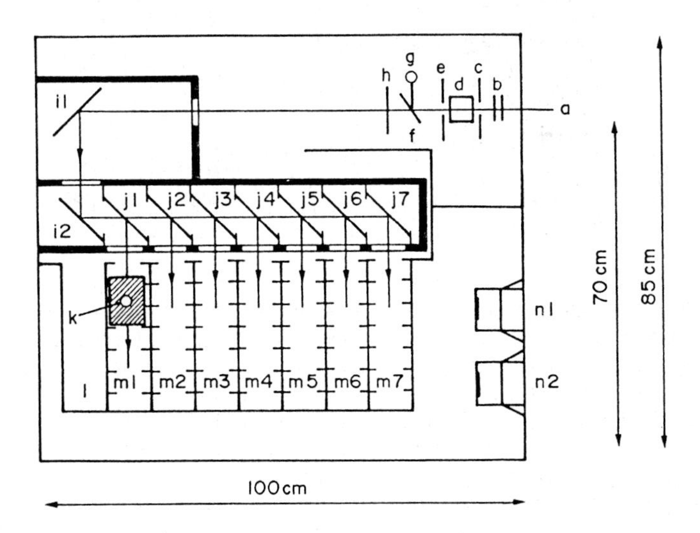

Figure 3. Design of the threshold box (after Watanabe *et al.* 1982). (a) optical axis; (b) sharp cut-off filters for 360-580 and 581-1000 nm, respectively; (c) box slit; (d) optical mixer, which is a bundle of quartz rods; (e) attenuator; (f) quartz plate; (g) 1st photodiode;; (h) shutter; (i) 1 and 2, first-surface mirrors; (j) 1-7, 1st-7th partially transmitting mirrors; (k) 2nd photodiode; (l) dark compartment; (m) 1-7, 1st-7th compartment for placing sample (26 cm wide, 9 cm deep and 30 cm high); (n) 1 and 2, connector receptacles.

3.2.3 Safelight

Plants respond so sensitively to environmental light that 'room light' always causes noise which effects the experimental results. Therefore, experiments on photomorphogenesis are usually performed in a dark room, where we need 'safelight' to manipulate samples during experimentation. A broader range of wavelengths is perceptible to plants than to human beings, so that a monochromatic light has to be chosen as a safelight that has the least influence on plant materials but which can still be observed by our eyes. Since the early 1960's (Hillman 1965), a dim green light has been used as safelight in dark rooms, because the action spectra of photomorphogenesis in plants and fungi generally show the minimum effectiveness in the green spectral region (see Chapters 9.6 and 9.7), whereas those of human vision show maximum sensitivity in the green region.

However, in some cases such as the very low fluence responses (Mandoli and Briggs 1981), dim green light is no longer safe enough. In such cases, infra-red light of 820 nm or longer wavelength is used as a safelight and we observe samples by an image converter from infra-red light to visible light (Kadota and Furuya 1977) or an infra-red sensitive TV camera (Fig. 6). Currently infra-red emitting diodes are available as a safe cool-light source, to reduce the damage of samples by heat in infra-red light obtained from incandescent lamps.

What is most important in experiments on photomorphogenesis is to reduce the period of observation of the sample to as short a time as possible so minimizing any possible effects of the safelight.

3.3 Microbeam irradiators: spatial analyses

It is crucial in studies of photoregulation to find out which intracellular or cellular part of a plant or plant organ captures photons for the induction of a particular response. Biochemical and immunological detection of a photoreceptor in a certain region of the plant does not mean that the detected photoreceptor really results in the response of interest. However, 'spot' illumination tells us the precise localization of the active photoreceptor(s) which induce a response. For instance, immunocytochemical studies (Coleman and Pratt 1974) indicated the possible association of phytochrome with the plasma membrane, endoplasmic reticulum, nuclear envelope and mitochondria, but this does not mean that all this detected phytochrome induces any relevant response. In contrast, local irradiation techniques using a microbeam irradiator provide us with crucial evidence of specific photoreceptive site(s).

When we want to expose a sample to 'spot' illumination, we have to observe the sample under a microscope without any significant influence of the observing light. Plant cells, however, often respond to visible light as sensitively as photographic films, so that, if cells are once observed under a traditional microscope using white light, the cell can no longer be used to study photomorphogenesis, just like films already exposed to light. This technical difficulty has meant that not many investigations using microbeam irradiation have been published.

3.3.1 Pioneering work

In the 1970's, pioneering work using the microbeam irradiation method was extensively carried out by W. Haupt (see his review 1980) to study phytochrome mediated movement of the chloroplast in the green filamentous alga *Mougeotia*. Microbeam light was obtained using the condenser lens of a microscope to produce a reduced image of the small diaphragm put at the position of the field stop. The whole image of a sample was observed by green safelight projected through the objective like in an epi-fluorescence microscope.

Using polarized R and FR microbeam irradiation, Haupt (1970) discovered that the position of chloroplasts was controlled by the gradient of the FR-absorbing form of phytochrome (Pfr) in the outer cytoplasm or plasma membrane and that phytochrome molecules in the R-absorbing form of phytochrome (Pr) are arranged parallel to the cell surface and those in the Pfr form are oriented normal to the cell surface (see Chapter 7.2).

3.3.2 Microbeam apparatus having multiple light sources

A convenient microbeam-irradiation apparatus, having two light sources (Fig. 4) was custom-designed and built by Kadota and Furuya (1977). Wavelengths of both the light sources were adjustable depending upon the experimental approach. For example, the whole image of a sample cell can be observed by a safe infra-red light of the light source II, while the position of the microbeam of actinic light is adjusted using a safelight of light source I, prior to the microbeam irradiation with the chosen wavelength. Then, the monochromatic filter of source I is changed to that for actinic irradiation with either R, FR or B, and this microbeam is irradiated at the pre-chosen position of the sample cell. The sample cell and the position of microbeam are photographed or recorded on video tape using a safelight. Using this microbeam irradiator, the photoreceptive site of phytochrome and the B-absorbing photoreceptor, regulating the timing of cell division (Wada and Furuya 1978) and controlling apical growth and apical swelling in protonemata of *Adiantum* (see Chapter 9.7), and that of the B/UV effect on perithecial induction (Inoue and Furuya 1978) have been demonstrated.

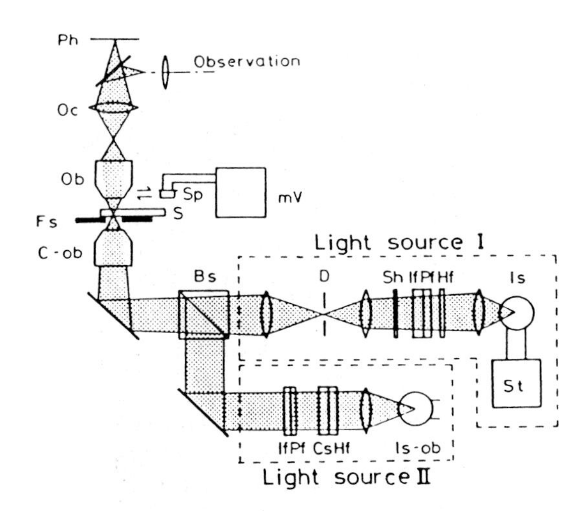

Figure 4. Diagram of the microbeam irradiator (after Wada and Furuya 1978). Ph, photographic camera; Oc, ocular lens; Ob, objective lens; S, specimen; Fs, focusing stage; C-ob, condenser objective lens; Sp, silicon photocell; mV, millivolt meter; Bs, beam splitter; D, diaphragm; Sh, shutter; If, interference filter; Pf, plastic filter; Hf, heat filter; Is, irradiation source; St, stabilizer; Cs, $CuSO_4$ solution; Is-ob, irradiation source for observation.

Recently, a Nomarski time-lapse microscope for observing at 900 nm was designed and constructed (Fig. 5), so that cells can be continuously maintained in any optical section of the micro-images during photomorphogenic responses. The microscope was connected with a computer-operated image analyzer, microspectrophotometer, and a microbeam irradiator. The details of this instrument (Fig. 6) will be described below. Using this new microbeam irradiator, Nick *et al.* (1993) discovered that, in phytochrome-induced formation of anthocyanin in *Sinapis* cotyledons, individual cells exhibit an all-or-none type response which is subsequently integrated by inhibitory signals that transmit through tissues to result in a stochastic pattern.

3.4 Sequential observation of elementary processes: temporal analysis

When we want to observe sequentially with a single sample, the least influence of observing light is a prerequisite for temporal analysis of a light-induced response. If the observing light significantly affects the final response during time course studies, we have to prepare a large number of the same type of sample and discard them once they have been observed. In such studies, the relevant light-induced processes must synchronously progress in every sample of the population, but this is not so easy to establish in experimental plant

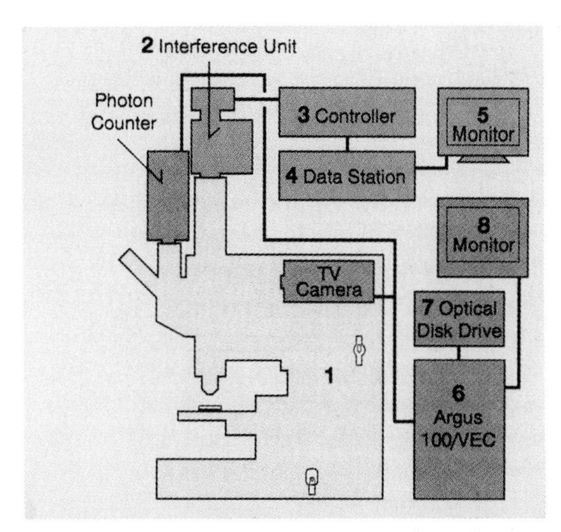

Figure 5. Diagram of the visible and infrared light Nomarski microscope with micro-image analyzer, micro-spectrophotometer and microbeam irradiator (after Furuya *et al.* 1991).Image acquisition and processing, including all standard subroutines as well a time-lapse series, were based on an infra-red-sensitive television camera with image intensifier, the two-dimensional photoncounter, and the subsequent image processing computer system ARGUS 200 (Hamamatsu Photonics), including an extra-fine-pitch colour television-screen.

systems. Hence, several techniques to observe without secondary effects have been developed for temporal analysis with a single sample.

3.4.1 Time-lapse recorders

Years ago the temporal processes after light treatments were recorded with a photomicrograph using infra-red light that was provided by a tungsten lamp with an infra-red filter and a heat-cut filter. The developed film was magnified with a photographic enlarger and the images of cells were measured with a ruler (Fig. 6 in Kadota and Furuya 1977). This method, however, was enormously labour intensive but even so it was still not sufficient to record continuously.

The next improvement was the use of a video camera equipped with an infra-red sensitive tube, connected to a video tape recorder and a video monitor, so that it became possible to monitor, precisely and continuously, changes in the cellular and subcellular images of the same sample (Fig. 1 in Furuya *et al.* 1980). The recorder and the monitor are controlled by a time-lapse controller and a timer, showing the real time on a screen. Preparation of the sample and operation of the equipment should be carried out under a safelight. Using this type of instrument, several photomorphogenic phenomena were studied, such as

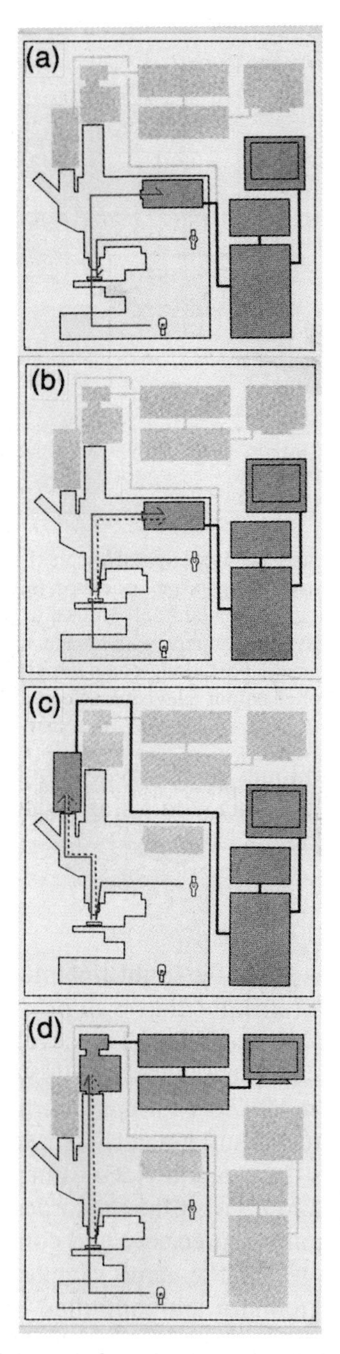

Figure 6. Diagrams of (a) Light path for microbeam irradiation; (b) micro-image analysis for temporal change; (c) photon counting of microscopic sample; (d) micro-spectrophotometry (after Furuya *et al.* 1991).

phototactic movement in flagellate algae (Uematsu-Kaneda and Furuya 1982); apical growth of protonemal cells (Kadota and Furuya 1977) and cell cycle progression (see Chapter 9.7; Wada and Furuya 1978); and pre-mitotic positioning of the nucleus (Mineyuki and Furuya 1980).

This technique is still useful for preliminary work and demonstrating to students, because of the low cost of the instruments, especially since cheap CCD TV cameras that have excellent sensitivity in the infra-red region are available. Infra-red emitting diodes can also be used to provide 'cool' safelight for the observation as described above in the section on safelights.

3.4.2 Computer analysis of temporal processes with micro-images

In the past two decades, the digital-image processing technique has rapidly progressed and has been applied widely in biological and medical electronics. This is because the technique has great advantages in the analysis of static and also dynamic images. The technique was developed for the simultaneous detection of changes in organelle movements in different regions of a cell (Mineyuki *et al.* 1983). The intracellular movements were monitored and recorded at chosen intervals of time under a safelight with a video-tape recorder, as described in the previous section. The micro-images can be recorded on a video tape, converted to digital images by a digitizer and stored in a multichannel-image memory (Takagi and Onoe 1981). The changes in organelle movements are observed as sequential dynamic digital images in terms of brightness change at each pixel (see Fig. 1 in Mineyuki *et al.* 1983). A certain number of the sequential dynamic images are selected and processed for a final image and the brightness change at each pixel is displayed on a colour monitor, in terms of the I_r value of standard error:

$$I_r = \sqrt{\sum_{\lambda=1}^{n} \frac{(I_i - \langle I_i \rangle)^2}{15 \times 14} \times k^2}$$

where I_i is the value of brightness at a pixel in the i-th image, $\langle I_i \rangle$ is the mean value of I_i and k is a constant.

Using this method the movement of organelles in the nuclear region of a single-celled protonema in *Adiantum* was examined, finding temporal and spatial change in organelle movements throughout the progression of its cell cycle, in particular the pre-mitotic positioning of the nucleus (Mineyuki *et al.* 1984). A similar method was applied to analysis of phototaxis in *Euglena* (Häder and Häder 1988).

In contrast to the above method, an unlimited number of digital micro-images observed by a FR microscope with a time-lapse video camera or photon counter (Fig. 5) can be stored on an erasable optical disk unit, combined to an image processor and a hard disk unit (Fig. 6b), and temporal changes in subcellular movement analysed in terms of the I_r value of standard errors of each pixel described above for any micro-images, that are selected from data on the optical disk. The processing of chosen images is now easily done with commercially available software such as Argus 200 (Hamamatsu Photonics). Using this equipment, we are able to analyse rapid milli-second responses and combined with the microbeam irradiation technique (Fig. 6a), it is possible to analyse rapid responses to light. Studies of such processes as autonomy of irradiated cells, as well as intra- and inter-cellular signal transduction are promising applications of this technique.

3.4.3 Flash photolysis

The chromophore of a photoreceptor is believed to capture photons within femto-seconds, remain excited in a time range of pico-seconds and then relax (see Chapters 4.1 and 4.3). The early events of the chromophore eventually result in changes in structure and function of the entire chromoprotein. Current laser technology makes it possible to analyse the behaviour of pigments in the order of femto- and pico-seconds by laser flash photolysis (Song *et al.* 1989) and the elementary processes of phototransformation pathways between Pr and Pfr have been studied in detail (see Chapters 4.1 and 4.3).

However, such a fast flash photolysis instrument is not always suitable for the analysis of phenomena which take place in the order of micro- and milli-seconds, which are important steps for transmitting a signal from the pigment to the interacting compound (receptor) in cells. Hence a photometer for recording flash-light-induced transient absorbance spectra with micro-second time resolution in the wavelength range from 350 to 800 nm with 0.5 nm resolution was custom-designed (Fig. 7) and built (Furuya *et al.*. 1984). This instrument is a two laser flash apparatus, of which wavelengths are tunable and the duration of both the flashes is controllable. The data obtained using this equipment demonstrates that the dimer molecule of phytochrome A is transformed from Pr to Pfr through four intermediates within 2 s at 24°C (Inoue *et al.*. 1990) and these intermediates can be photoconverted back to the Pr form by a second laser flash irradiation (Pratt *et al.*. 1984).

This double laser flash irradiator can measure not only transparent samples but turbid ones with a 100 μs time resolution (Inoue and Furuya 1985). Using this equipment,, we can determine at which step during phototransformation from Pr to Pfr a signal is transduced from the phytochrome molecule to the receptor in the cell. For example, the change of action dichroism between Pr

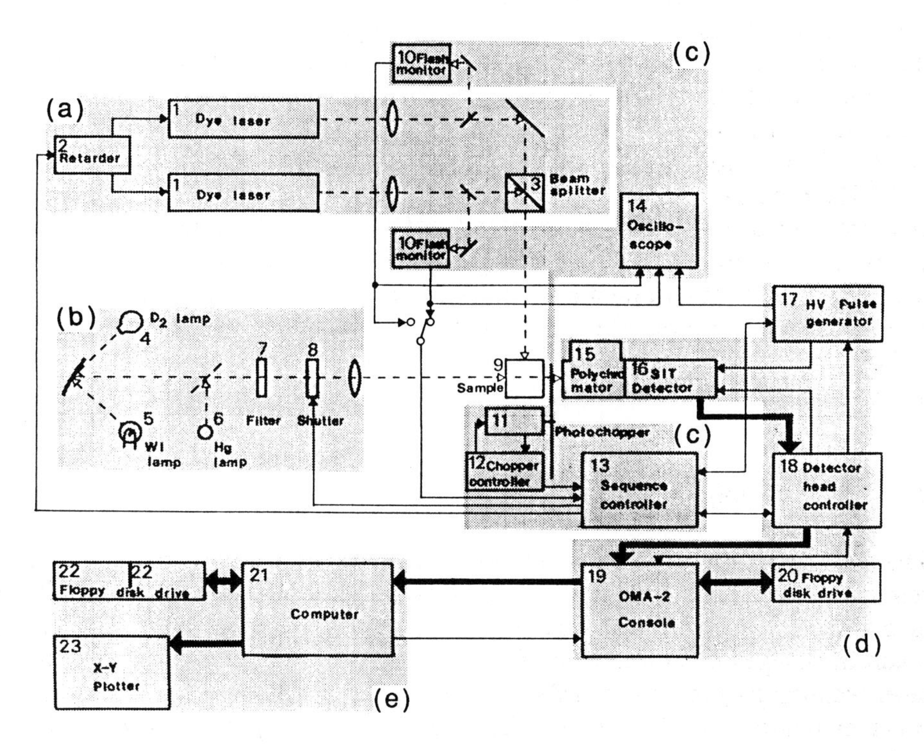

Figure 7. A schematic diagram of the multi-channel transient spectrum analyzer (after Furuya *et al.* 1989). (a) Flash excitation apparatus; (b) measuring light assembly; (c) fluence controller; (d) detector assembly; (e) computer setup.

and Pfr is well known in fern protonema (see Chapter 9.7) and the turning point of action dichroism in the phototransformation pathway from Pr to Pfr was determined in this double-flash equipment using polarized lights (Fig. 8). The change of transition moment of phytochrome chromophore occurs between 2 ms and 30 s after a R flash, suggesting that a change in the orientation of the chromophore takes place at the final phototransformation step between the intermediate Ibl and Pfr (Kadota *et al.* 1986).

3.4.4 Low-temperature spectroscopy

Similar or the same labile intermediates of phytochrome which occur in the phototransformation pathways in both the directions as discussed above were detectable by spectrophotometry at low temperature (Spruit and Kendrick 1973; Spruit *et al.* 1975). Cooling of phytochrome as either Pr or Pfr to −196°C with a specially constructed glass cryostat resulted in a marked increase in extinction

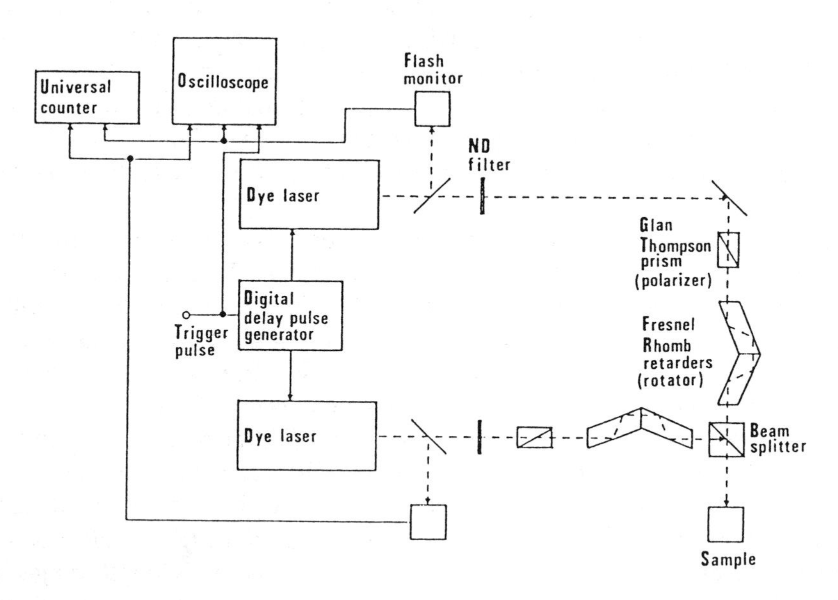

Figure 8. Diagram of the polarization plane-rotatable laser flash irradiator (after Kadota *et al.* 1986).

coefficients, but no such increase in the oscillator strengths (Sasaki *et al.* 1986). On warming, at least two intermediates were detected in both directions of phytochrome phototransformation. Comparison of the data for labile intermediates obtained by the different methods of flash photolysis and low temperature spectrophotometry, provided convincing evidence for the same intermediates being detected (Furuya 1983).

3.5 Spectrophotometers

Spectrophotometers have been a major tool in determining the amount, state and distribution of photoreceptor pigments both *in vivo* and *in vitro,* as photoreceptor pigments intrinsically show characteristic absorption, fluorescence or circular dichroism spectra. Hence, unique spectrophotometers have been designed and used for studies of photomorphogenic photoreceptors.

3.5.1 In-vivo *spectrophotometry*

The predicted spectral properties of phytochrome (R-FR reversibility) led, in the 1950's, to the development and construction of a newly designed spectrophotometer at Beltsville. A recording, single-beam spectrophotometer was custom-built for detecting small samples of turbid material. By placing the cuvette directly in front of the end-window multiple-type phototube, spectral measurements could be made on light-scattering samples of apparent absorbances between 0 and 6 (Butler *et al.* 1959). This custom-made spectrophotometer was used for the discovery of phytochrome, detecting the R-FR photoreversible absorption changes for its interconversion for the first time (Butler *et al.* 1959).

A dual-wavelength spectrophotometer was built for measuring directly the difference in absorbance, ΔA, between two fixed wavelengths for both *in vivo* and *in vitro* measurements (Butler *et al.* 1959). Using this instrument the amount of phytochrome was determined in terms of $\Delta(\Delta A)$ measured after subsequently given, active irradiations of R and FR. In 1963 this spectrophotometer was commercially marketed as the Ratio-spect R-2 and promoted the early research on phytochrome (Hillman 1964; Furuya and Hillman 1964; Briggs and Siegelman 1965). The phytochrome signal determined simultaneously by this method, however, may possibly show a distortion caused by the phototransformation of protochlorophillide to chlorophillide. Currently, difference spectra between Pr and Pfr in the wavelength range from 500 to 800 nm are calculated with the aid of a computer and used to adjust the signal distortion. At present, commercially available modern spectrophotometers can perform multiple wavelength and multi-component analysis utilizing a built-in microcomputer.

3.5.2 Micro-spectrophotometry

A spectrophotometer provides an average concentration in a sample of plant cells and tissues. However, the substance to be measured is not necessarily distributed homogeneously in a cell or a tissue, and may for example exist in particular localizations of a cell. Therefore, it would be useful to determine the content of the relevant pigment utilizing a micro-spectrophotometer. However, it has been technically very difficult using conventional double beam spectrophotometry, because no proper control is available.

In the early days, phytochrome distribution in an individual *Cucurbita* seed was measured and compared with the photoperceptive site for light-induced seed germination (Boisard and Malcoste 1970). Phytochrome concentration was found to be the highest in the hypocotyl-radicle complex, which was also the photosensitive site. However, further more detailed investigation is difficult with such instruments.

A visible- and infra-red-light Nomarski microscope with an image analyzer and micro-spectrophotometer was custom-built to obtain micro-absorption spectra; micro-fluorescence spectra and micro-photographs for micro-picture

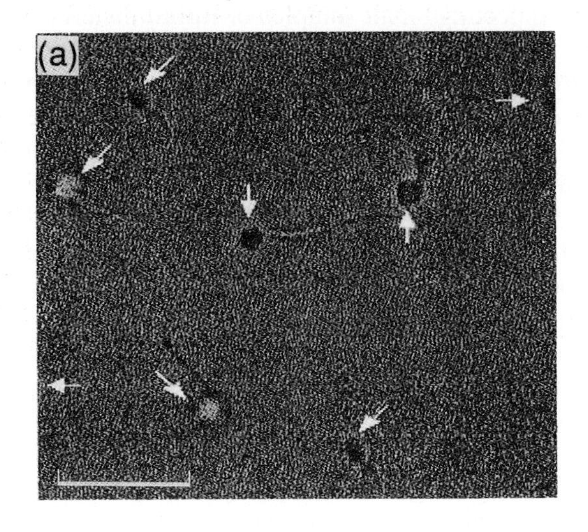

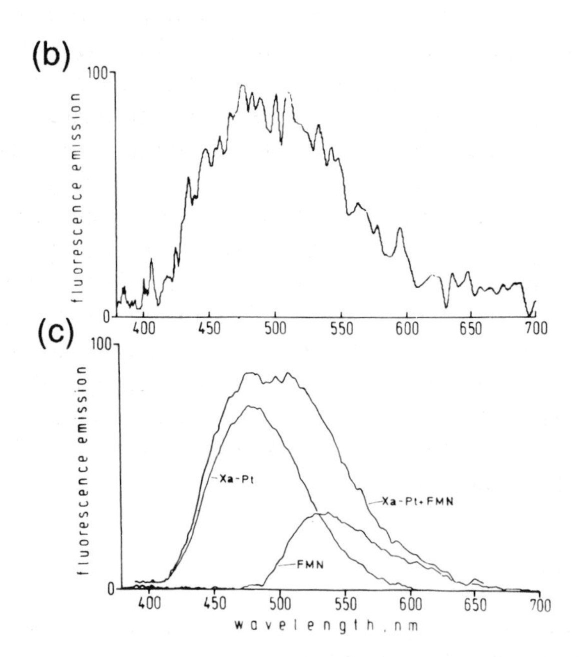

Figure 9. Micro-spectrofluorescence spectrophotometry of *Euglena gracilis* (after Schmidt 1990). (a) Isolated flagella with attached paraflagellar body. Bar = 10 μm, (b) Fluorescence-emission spectrum of the paraflagellar body of a wild-type *Euglena* cell (excitation 365 nm) and (c) fluorescence-emission spectra of flavin adenine mononucleotide (10^{-5} *M*), xanthopterin (10^{-5} *M*), and a mixture of flavin adenine mononucleotide and xanthopterin (excitation at 365 nm). The concentration ratio and normalization factor were chosen to minimize the difference spectrum shown in B.

precessing (Fig. 6d). The highest possible resolution was obtained with an objective of 100 times magnification and a numerical aperture of 1.30 based on oil immersion, with transmission in the near-UV for fluorescence excitation. A novel interferometer monochromator (based on a Wollaston-prism) with image intensifier (Hitachi U-6000) set on top of the microscope generates the proper interference fringes on a diode array. A wavelength range from 380 to 900 nm is covered by the monochromator. The signal emerging from the diode array is converted to the corresponding spectrum *via* a Fast Fourier transform routine and can be further mathematically processed, plotted and stored. Using this instrument, fluorescence-emission spectra (excitation at 365 nm) of single paraflagellar bodies of isolated flagella of *Euglena gracilis* were measured and the maxima near 470 and 520 nm were observed, indicating the presence of pterins and flavins (Fig. 9; Schmidt *et al.* 1990).

3.6 Concluding remarks

Phytochrome was once termed 'a pigment of the imagination' (McGee 1987), but this imagination gradually became a reality when experimental evidence was provided by newly developed instruments such as the spectrograph and the dual-wavelength difference spectrophotometer for turbid samples. The history of phytochrome studies has demonstrated that the construction and/or improvement of instruments is a powerful driving force for the progress of scientific knowledge. This principle is just as valid today and will remain so in the future.

Recent rapid progress of studies on photoregulation in plants has clearly resulted from molecular biological and genetic approaches. However, Northern blot analysis with gene transcripts and Western blot and immunochemical detection techniques of pigments, for example, are not sufficient to understand the biological activity of a photoreceptor, since the optical activity of the molecule containing its chromophore is crucial for its function. Hence, there is still an urgent need for the development of spectrophotometric and other physical instruments in photobiology.

3.7 Further reading

Borthwick H. (1972) History of phytochrome. In: *Phytochrome*, pp. 3-23, Mitrakos K. and Shropshire W. Jr. (eds.) Academic Press, London.

Butler W.L. (1980) Remembrances of phytochrome twenty years ago. In: *Photoreceptors and Plant Development*, pp. 3-7, De Greef J. (ed.) Antwerp University Press.

Sage L.C. (1992) *Pigment of the Imagination. A history of Phytochrome Research.* Academic Press, Inc, New York.

3.8 References

Boisard J. and Malcoste R. (1970) Analyse spectrophotométrique du phytochrome dans l'embryon de Courge (*Cucurbita pepo*) et de Potiron (*Cucurbita maxima*). *Planta* 91: 54-67.

Borthwick H.A., Hendricks S.B., Parker M.W., Toole E.H. and Toole V.K. (1952) A reversible photoreaction controlling seed germination. *Proc. Nat. Acad. Sci. USA* 38: 662-666.

Briggs W.R. and Siegelman H.W. (1965) Distribution of phytochrome in etiolated seedlings. *Plant Physiol.* 40: 934-941.

Butler W.L., Norris K.H., Siegelman H.W. and Hendricks S.B. (1959) Detection, assay, and preliminary purification of the pigment controlling photoresponsive development of plants. *Proc. Nat. Acad. USA* 45: 1703-1708.

Coleman R.A. and Pratt L.H. (1974). Electron microscopic localization of phytochrome in plants using an indirect antibody-labelling method. *Histochem. Cytochem.* 11: 1039-1047.

Durand R. and Furuya M. (1985) Action spectra for stimulatory and inhibitory effects of UV and blue light on fruit-body formation in *Coprinus congregatus. Plant Cell Physiol.* 26: 1175-1183.

Furuya M. (1983) Molecular properties of phytochrome. *Phil. Trans. R. Soc. Lond. B* 303: 361-375.

Furuya M. and Hillman W.S. (1964) Observations on spectrophotometrically assayable phytochrome *in vivo* in etiolated *Pisum* seedlings. *Planta* 63: 32-42.

Furuya M., Inoue Y. and Maeda Y. (1984) A multichannel transient spectrum analyser for absorption changes measurement with one microsecond resolution. *Photochem. Photobiol.* 40: 771-774.

Furuya M., Wada M. and Kadota A. (1980) Regulation of cell growth and cell cycle by blue light in *Adiantum* gametophytes. In: *The Blue Light Syndrome*, pp. 119-132, Senger H. (ed.) Springer Verlag, Berlin.

Furuya M., Nagatani A., Dosaka S., Yamagishi S, Kamiya K., Uchiyama S., Toyama K. and Matsui S. (1991) Microspectrophotometry and microimage analysis. *Plant Science Tomorrow* 3: 14-15.

Häder D.-P. and Häder M. (1988) Ultraviolet-B inhibition of motility in green and dark bleached *Euglena gracilis. Current Microbiol.* 17: 215-220.

Haupt W. (1970) Über den Dichroismus von Phytochrom 660 und Phytochrom 730 bei *Mougeotia. Z. Pflanzenphysiol.* 62: 287-298.

Haupt W. (1980) Microbeam irradiation in *Mougeotia*. In: *Handbooks of Physiological Methods*, pp. 195-204, Gantt E. (ed.) Cambridge University Press.

Hillman W.S. (1964) Endogenous circadian rhythms and the response of *Lemna perpusilla* to skeleton photoperiods. *Amer. Naturalist* 98: 324-328.

Hillman W.S. (1965) Phytochrome conversion by brief illumination and the subsequent elongation of etiolated *Pisum* stem segments. *Physiol. Plant.* 18: 346-358.

Inoue Y. (1984) Re-examination of action spectroscopy in blue/near-UV light effects, In: *Blue Light Effects in Biological Systems*, pp. 110-117, Senger H. (ed.) Springer-Verlag, Berlin.

Inoue Y. and Furuya M. (1978) Perithecial formation in *Gelasinospora reticulispora.* VI. Inductive effect of microbeam irradiation with blue light. *Planta* 143: 255-259.

Inoue Y. and Furuya M. (1985) Phototransformation of the red-light-absorbing form to the far-red-light-absorbing form of phytochrome in pea epicotyl tissue measured by a multichannel transient spectrum analyser. *Plant Cell Physiol.* 26: 813-819.

Inoue Y., Rüdiger W., Grimm R. and Furuya M. (1990) The phototransformation pathway of dimeric oat phytochrome from the red-light-absorbing form to the far-red-light-absorbing form at physiological temperature is composed of four intermediates. *Photochem. Photobiol.* 52: 1077-1083.

Inoue Y. and Watanabe M. (1984) Perithecial formation in *Gelasinospora reticulispora* VII: Action spectra in UV region for the photoinduction and the photoinhibition of photoinductive effect brought by blue light. *Plant Cell Physiol.* 25: 107-113.

Jacques R. Chabbal R., Chouard P. and Jacquinot P. (1964) Mise au point d'un illuminateur spectral à usage biologique. *C. R. Ac. Sci. Paris* 259: 1581-1584.

Kadota A. and Furuya M. (1977) Apical growth of protonemata in *Adiantum capillus-veneris* I. Red far-red reversible effect on growth cessation in the dark. *Develop. Growth Differ.* 19: 357-365.

Kadota A., Inoue Y. and Furuya M. (1986) Dichroic orientation of phytochrome intermediates in the pathway from Pr to Pfr as analyzed by double laser flash irradiations in polarotropism of *Adiantum protonemata. Plant Cell Physiol.* 27: 867-873.

Mandoli D.F. and Briggs W.R. (1981) Phytochrome control of two low-irradiance responses in etiolated oat seedling. *Plant Physiol.* 67: 733-739.

McGee H. (1987) *A pigment of the Imagination*, USDA, Albany, California.

Mineyuki Y. and Furuya M. (1980) Effect of centrifugation on the development and timing of premitotic positioning of the nucleus in *Adiantum* protonemata. *Develop. Growth Differ.* 22: 867-874.

Mineyuki Y., Takagi M. and Furuya M. (1984) Changes in organelle movement in the nuclear region during the cell cycle of *Adiantum* protonema. *Plant Cell Physiol.* 25: 297-308.

Mineyuki Y., Yamada M., Takagi M., Wada M. and Furuya M. (1983) A digital image processing technique for the analysis of particle movements: Its application to organelle movements during mitosis in *Adiantum* protonemata. *Plant Cell Physiol.* 24: 225-234.

Monk G.S. and Ehret C.F. (1956) Design and performance of a biological spectrograph. *Radiation Research* 5: 88-106.

Nick P., Ehmann B., Furuya M. and Schäfer E. (1993) Stochastic responses of individual cells and cell communication determine the pattern of phytochrome-induced biosynthesis of anthocyanin in mustard cotyledons. *Plant Cell* 5: in press.

Parker M.W., Hendricks S.B., Borthwick H.A. and Scully N.J. (1945) Action spectrum for the photoperiodic control of floral initiation in *Biloxi* soybean. *Science* 102:152-155.

Pratt L.H., Inoue Y. and Furuya M. (1984) Photoactivity of transient intermediates in the pathway from the red-absorbing to the far-red-absorbing form of *Avena* phytochrome as observed by a double-flash transient-spectrum analyzer. *Photochem. Photobiol.* 39: 241-246.

Sasaki N, Oji Y., Yoshizawa T., Yamamoto K.T. and Furuya M. (1986) Temperature dependence of absorption spectra of 114 kDa pea phytochrome and relative quantum yield of its phototransformation. *Photobiochem. Photobiophys.* 12: 243-251.

Schmidt W., Galland P., Senger H. and Furuya M. (1990) Microspectrophotometry of *Euglena gracilis. Planta* 182: 375-381.

Senger H. ed., (1980) *The Blue Light Syndrome*, Springer-Verlag, Berlin

Song P.-S., Singh B.R., Tamai N., Yamazaki T., Yamazaki I., Tokutomi S. and Furuya M. (1989) Primary photoprocesses of phytochrome. Picosecond fluorescence kinetics of oat and pea phytochromes. *Biochemistry* 28: 3265-3271.

Spruit C.J.P. and Kendrick R.E. (1973) Phytochrome intermediates *in vivo* - II Characterization of intermediates by difference spectrophotometry. *Photochem. Photobiol.* 18: 145-152.

Spruit C.J.P., Kendrick R.E.and Cooke R.J. (1975) Phytochrome intermediates in freeze-dried tissue. *Planta* 127: 121-132.

Takagi M. and Onoe M. (1981) Colour display for image processing with multiple functions. In: *Real-Time/Parallel Computing Image Analysis*, pp. 361-370, Onoe M. Preston K. Jr. and Rosenfeld S. (eds.) Plenum Press, New York.

Uematsu-Kaneda H. and Furuya M. (1982) Effects of viscosity on phototactic movement and period of cell rotation in *Cryptomonas* sp. *Physiol. Plant.* 56: 194-198.

Wada M. and Furuya M. (1978) Effects of narrow-beam irradiations with blue and far-red light on the timing of cell division in *Adiantum* gametophytes. *Planta* 138: 85-90.

Watanabe M., Furuya M., Miyoshi Y., Inoue Y., Iwahashi I. and Matsumoto K. (1982) Design and performance of the Okazaki Large Spectrograph for photobiological research. *Photochem. Photobiol.* 36: 491-498.

Part 4 Phytochrome

4.1 The phytochrome chromophore

Wolfhart Rüdiger and Fritz Thümmler

Botanisches Institut der Universität München,
Menzinger Str. 67, D-80638 München 19, Germany

4.1.1 Introduction

Absorption of light quanta is the precondition for any photochemical reaction and hence for photobiological phenomena. Starting with white light, one can consider a 'black body' the ideal absorber in a physical sense because all wavelengths (*i.e.* all energy levels) of the light are equally absorbed. However, this situation is not realized in plants. The actual absorbers preferentially absorb certain wavelengths of visible light but are more or less transparent for other wavelengths. After such an absorption, the remaining light looks 'coloured' to the human eye. The absorbers are therefore called *chromophores* (= carriers of colours). Molecules which have the properties of such absorbers are called *pigments.* These terms are used here in such a sense that the entire molecule is called pigment whereas the particular part of the molecule which absorbs light is called its chromophore.

The chromophore of the pigment (photoreceptor) *phytochrome* is a bilin or bile pigment, *i.e.* an open-chain tetrapyrrole (Section 4.1.2) named *phytochromobilin.* Phytochrome is a photochromic pigment, which exists in two forms, Pr [red light (R)-absorbing] and Pfr [far-red light (FR)-absorbing]. The Pr and Pfr forms are mutually interconvertible by appropriate irradiation; Pfr is considered to be the physiologically active form of phytochrome (Chapters 4.6 and 4.7). The names of the phytochrome forms indicate their maxima of absorption. The absorption bands of the two froms are broad (Fig. 1) and because of their overlap a photoequilibrium is obtained by irradiation with monochromatic light. A simplified scheme for this photoconversion is:

$$\mathrm{Pr} \underset{\mathrm{FR}}{\overset{\mathrm{R}}{\rightleftarrows}} \mathrm{Pfr}$$

R.E. Kendrick & G.H.M. Kronenberg (eds.), Photomorphogenesis in Plants - 2nd Edition

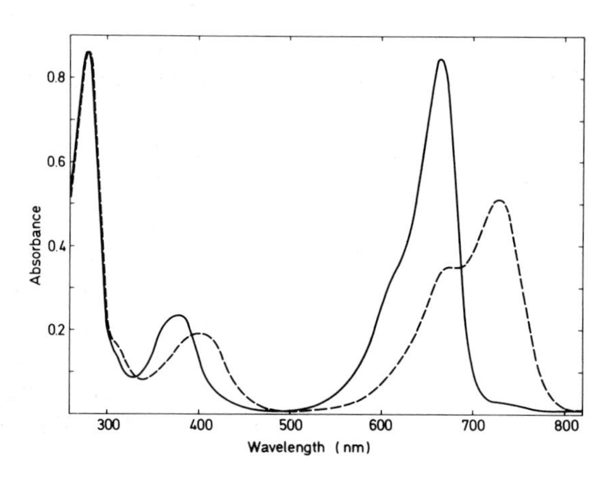

Figure 1. Absorption spectra of 124-kDa Avena phytochrome. Absorption spectra were measured after red light (- - -), *ca.* 85% Pfr and 15% Pr and far-red light (——), *ca.* 100% Pr. After Grimm R. and Rüdiger W. (1986) *Z. Naturforsch.* 41c: 988-992.

more details of the photoconversion including intermediate steps will be dealt with in Section 4.1.4. Absorption differences at the absorption maxima of Pr and Pfr after irradiation with R or FR are used for detection and quantitation of phytochrome (Chapter 4.5).

The spectral properties of Pr are distinct from those of Pfr (Fig. 1). It is therefore obvious to assume that spectral differences between Pr and Pfr are somehow reflected in structural differences of the chromophore. The basic chemical structure of phytochromobilin will be treated first (Section 4.1.2). Section 4.1.3 then deals with differences between the Pr and the Pfr chromophore. Several aspects of the phytochrome chromophore have been reviewed (Rüdiger 1992; Rüdiger and Scheer 1983; Rüdiger *et al.* 1985; Song 1985; Schaffner *et al.* 1990; Rüdiger and Thümmler 1991).

4.1.2 The structure of the phytochrome chromophore

The similarity of the absorption spectra of the Pr form of phytochrome and phycocyanin, an accessory pigment of photosynthesis occurring in red and blue-green algae (= cyanobacteria) was taken, as early as 1950, as evidence that their chromophores were closely related. This correlation did not contribute much to the knowledge of the Pr chromophore since at that time it was only known that the phycocyanin chromophore was a bile pigment. The structure of phycocyanobilin was elucidated 17 years later. The chromophore was cleaved off, albeit with poor yield, from phycocyanin with boiling methanol and investigated spectroscopically and by oxidative degradation. Chromic acid or dichromate oxidation is a simple, powerful micro-method for structural investigations in the field of bile pigments and biliproteins (Klein and Rüdiger 1978). Its application to free phycocyanobilin (Fig. 2) yielded four typical degradation products from all four pyrrole rings A, B, C and D.

When the degradation method was applied to the chromoprotein phycocyanin only the products derived from rings B, C and D (Fig. 2) were obtained. A problem arose as a result of the covalent linkage between the chromophore (at ring A) and the protein (by a thiol group of cysteine, Fig. 3). This linkage is not cleaved by the chromic acid oxidation. The thioether is oxidized to the sulphone as detected in detailed model studies with synthetic compounds. The sulphone can then be cleaved by mild alkali or ammonia (Fig. 3). This method ('CrO_3-NH_3-degradation') applied to the investigation of phycocyanin (Klein and Rüdiger 1978) proved the structure of ring A including the thioether linkage (Fig. 3).

The first rigorous proof of the bile pigment nature of the phytochrome chromophore was obtained by oxidative degradation (Rüdiger and Correll 1969). The 'CrO_3-NH_3-degradation' yielded the same results as with phycocyanin indicating the same thioether linkage in phytochrome (Klein *et al.* 1977). The only difference was a substituent at ring D which is an ethyl group in phycocyanobilin and a vinyl group in phytochromobilin (Fig. 2). Whereas elimination of the chromophore by boiling methanol proved to be unsatisfactory for phytochrome, the cleavage with cold HBr gas was successful if the starting material was a chromopeptide, obtained from phytochrome by proteolytic digestion, rather than intact phytochrome (Rüdiger *et al.* 1980). The reaction product was identical with a product of total synthesis which was treated in the same way. These experiments were the final proof for the structure of phytochromobilin as given in Fig. 2.

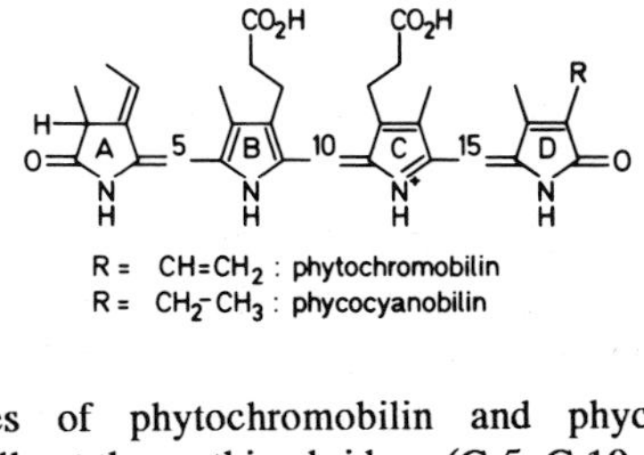

CO_2H CO_2H R H O A 5 B 10 C 15 D O N H N H N H N H

R = $CH{=}CH_2$: phytochromobilin
R = CH_2-CH_3 : phycocyanobilin

Figure 2. Chemical structures of phytochromobilin and phycocyanobilin. Chromic acid degradation occurs preferentially at the methine bridges (C-5, C-10 and C-15). The substituents at the ß-pyrrolic positions are still present in the oxidation products. Phycocyanobilin has an ethyl substituent at ring D instead of a vinyl group but is otherwise identical with phytochromobilin.

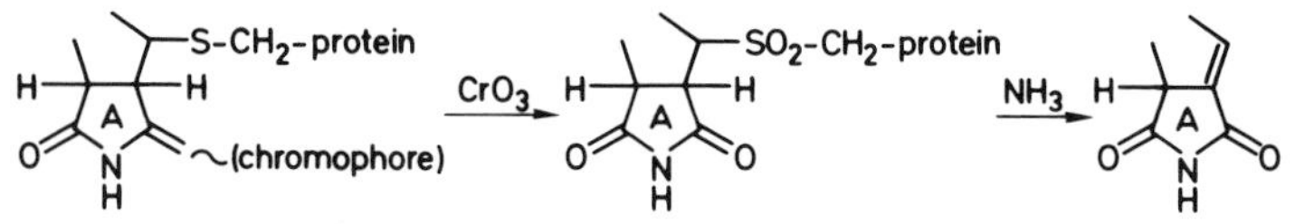

Figure 3. Chromic acid-ammonia degradation of biliproteins (Klein and Rüdiger 1978). Only ring A is shown.

The same chromophore structure including the thioether linkage at ring A (Fig. 3) was independently derived from high-resolution ^1H NMR studies of a Pr chromopeptide (Lagarias and Rapoport 1980). Assignment of the resonance signals was mainly based on a previous investigation of a chromopeptide obtained from phycocyanin, free phycocyanobilin and a synthetic peptide which had the same sequence as the chromopeptide, but lacked the chromophore.

These data have been obtained with small chromopeptides, therefore a direct conclusion can only be drawn on their chromophore structure. It can be argued that the chemical structure of the chromophore is presumably unchanged by proteolysis, nevertheless, a comparison of the chromophore in native phytochrome and in the chromopeptide(s) is necessary.

The absorption spectra of native Pr and a chromopeptide obtained from it by pepsin digestion are shown in Fig. 4. Both preparations have two absorption bands at about 370-380 and 660-670 nm. The absorption spectrum of the chromopeptide is identical with that of Pr after denaturation with urea or other reagents which cause unfolding of the native peptide chain. It should be noted that the chromopeptide and denaturated Pr are investigated here at $pH < 3$ where the chromophore occurs as a cation (structure in Fig. 2). The free base of the chromophore observed at pH 6-8 has an absorption band at about 600 nm. The coincidence of the peak position of the chromopeptide cation and native Pr has been taken as an indication that the chromophore is protonated in native Pr (Lagarias and Rapoport 1980). The presence of a protonated chromophore in native phytochrome was later supported by resonance-Raman and Fourier transform infra-red spectra (Fodor *et al.* 1990; Siebert *et al.* 1990).

A remarkable difference between native Pr and the protonated 'free' chromophore (in the chromopeptide) concerns the shape of the absorption curves: the long-wavelength band of native Pr is sharper and has about a 3-fold higher extinction coefficient (ε) than the corresponding absorption band of the chromopeptide. An explanation can be derived from phycocyanin and model tetrapyrrole chromophores. The absorbance of the long-wavelength band of phycocyanobilin is also high in native phycocyanin and comparatively low after proteolytic digestion or protein denaturation (Fig. 5 *top*). The general idea of

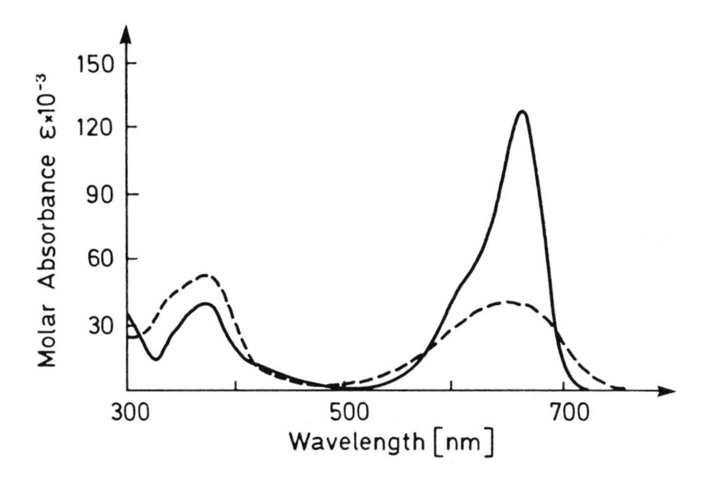

Figure 4. Absorption spectra of native Pr (——) in 10 m*M* potassium phosphate buffer, pH 7.8, and of Pr chromopeptide (- - -) in aqueous formic acid. The molar extinction coefficient (ε_{665}[Pr] = 132 000) was determined by Lagarias J.C., Kelly J.M., Cyr K.L. and Smith W.O. (1987) *Photochem. Photobiol.* 46: 5-13.

these experiments was that denaturation (*i.e.* unfolding of the native protein) removes the specific perturbation of the chromophore by the protein. Under carefully controlled conditions, denaturation of phycocyanin can be reversed. The 'renatured' phycocyanin has the same spectral properties as the original native one including the high extinction coefficient. Perturbation of the chromophore by the protein is also removed by proteolysis albeit irreversibly. Such investigations had already been applied to a number of biliproteins: the general conclusion was that bile pigment chromophores in a denatured protein or a chromopeptide behave like free bile pigments. Comparison of absorption spectra of denatured biliproteins and free bile pigments therefore allows certain conclusions to be made about the chemical structure of the bile pigment chromophore. The reversible spectral changes of phycocyanin obtained by denaturation and renaturation can best be explained by conformational changes of the tetrapyrrole chromophore (Rüdiger and Scheer 1983). Free bile pigments preferentially adopt cyclic-helical conformations as determined by X-ray studies of the crystalline compounds (Falk 1989). The spectral properties calculated for cyclic conformations agree with those observed for the small chromopeptides and for denatured phycocyanin whereas the spectral properties of native phycocyanin agree with those calculated for extended conformations (reviewed by Rüdiger and Scheer 1983).

Recently, the extended conformation of phycocyanobilin in the native state has been directly demonstrated by X-ray analysis of crystalline phycocyanins (Fig. 5 *middle*; Schirmer *et al.* 1987; Duerring *et al.* 1991). The extended conformation of phycocyanobilin results from specific interactions between the

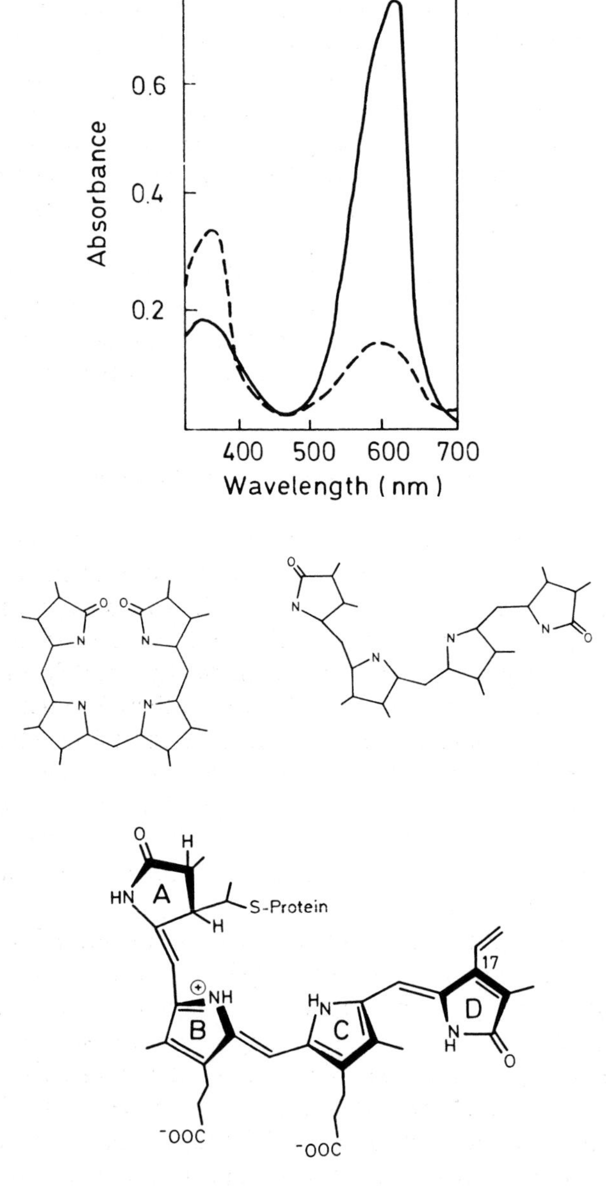

Figure 5. Absorption spectra and conformations of linear tetrapyrroles. *Top:* absorption spectra of native phycocyanin (——) and phycocyanin denaturated in 8*M* urea/pH 7.5 (- - -). Redrawn from W. Kufer, Dissertation Univ. München (1980). The absorption spectrum of native phycocyanin agrees with that of linear tetrapyrroles in extended conformation; that of denaturated phycocyanin with that of linear tetrapyrroles in cyclic conformation. *Middle:* cyclic and extended conformation of linear tetrapyrroles. Redrawn from X-ray structures. (See Falk 1989; Duerring *et al.* 1991). *Bottom:* proposed structure of the chromophore of native phytochrome in the Pr-form. The extended conformation corresponds to that in native phycocyanin.

chromophore and amino acid residues. Such interactions have been precisely defined in crystalline phycocyanins but not yet in phytochrome. Nevertheless, 'native' phytochromobilin might have a similar conformation to 'native' phycocyanobilin: the arginine residue five amino acid residues upstream of the covalently bound cysteine is conserved, not only in several biliproteins, but also in all known phytochrome sequences (summarized by Rüdiger 1992). The positively charged arginine interacts with the negatively charged propionic side chain of ring B in phycocyanins; it may well function in a similar manner in phytochrome to 'stretch' the chromophore.

In conclusion, the chromophore in native Pr can be assumed to have an extended conformation similar to that of native phycocyanin (Fig. 5 *bottom*). According to the cited spectral data it is very likely that the Pr chromophore is kept protonated in the native protein, independent of the pH of the solvent.

4.1.3 Differences between the Pr and the Pfr chromophore

Investigations of the phytochrome chromophore have focused on the difference between the Pfr and the Pr chromophore. It is obvious from the absorption spectra (Fig. 1) that the chromophore in native Pfr is somehow different in its electronic state from the chromophore in native Pr. It was shown in the preceeding section that the native protein can dramatically modify the electronic properties of tetrapyrrole chromophores. One may therefore ask whether the chromophore has a different chemical structure in Pr and Pfr or whether it is the same chromophore, only in a different protein environment. It has to be appreciated that the light which is active in phytochrome photoconversion is absorbed by the chromophore and not by the protein. Therefore it is probable that the chromophore, starting from its excited state, undergoes a photochemical reaction prior to any change in the protein.

Early speculations on the differences between the Pr and Pfr chromophore have been reviewed by Rüdiger and Scheer (1983). An essential step towards further knowledge of the Pfr chromophore was achieved by preparation of small chromopeptides which differed if they were prepared from Pfr (Pfr chromopeptides) or Pr (Pr chromopeptides) (Thümmler and Rüdiger 1983). The absorption spectra of Pr and Pfr chromopeptides are shown in Fig. 6. Irradiation transforms the Pfr chromopeptide into the Pr one; the latter is stable against irradiation. The chromophore of the Pfr chromopeptide was demonstrated, by a variety of methods, including high-resolution ^{1}H NMR spectroscopy (Rüdiger *et al.* 1983; Thümmler *et al.* 1983), to be the 15E isomer, whereas the chromophore of the Pr chromopeptide is the 15Z isomer (Fig. 7). Otherwise the Pfr chromophore (in the small peptide) has the same chemical structure as the Pr chromophore, including the covalent linkage to the peptide.

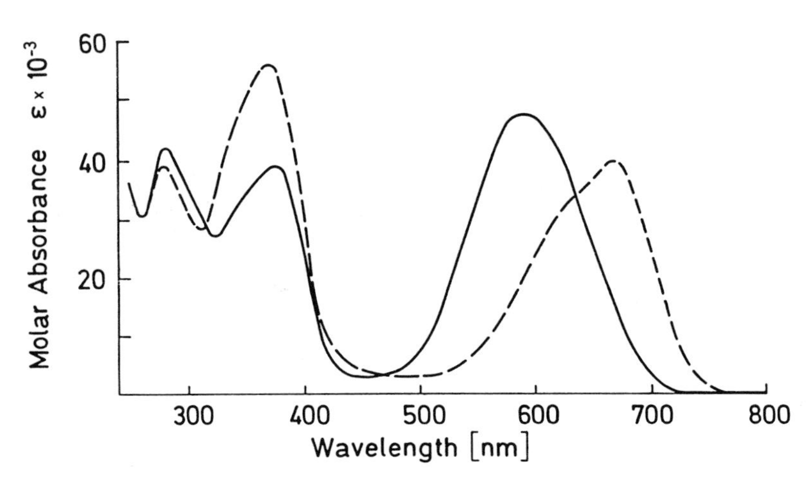

Figure 6. Absorption spectra of Pfr chromopeptide (——) and Pr chromopeptide (- - -) in aqueous formic acid.

To understand this finding in more detail, some general aspects of Z,E isomerism (commonly also called *cis-trans* isomerism) should be considered. Rotation around a single bond (*e.g.* C–C) has a low activation energy and is therefore easily achieved at ambient temperature ('free rotation'). Rotation around a double bond (*e.g.* C=C) has a high activation barrier and cannot therefore normally be achieved as a thermal reaction at ambient temperature. However, in many cases it can be achieved as a photochemical reaction if free rotation around this bond is possible in the excited state. The photoproducts are then stable isomers (designated Z and E in Fig. 7) because the activation energy for the thermal back reaction is high. In bile pigments, only the 4,5 and 15,16 double bonds have such properties, *i.e.* typical double bond character. Rotation around the neighbouring single bonds (5,6 and 14,15) is easily achieved without irradiation; this certainly contributes a lot to the conformational flexibility of bile pigments. A different situation exists at the middle methine bridge (C-9, C-10, C-11). This part of the chromophore behaves as a conjugated system where almost no difference between single and double bond character exists. In addition to *tautomerism* (Fig. 7), Z,E isomerization at C-9/C-10 or C-10/C-11 can be achieved photochemically but the reaction products undergo thermal reversion (Schaffner *et al.* 1990). These (and possibly even more) ways of internal conversion seem to compete efficiently with the photochemistry at the double bonds 4,5 or 15,16 in free bilins such as phytochromobilin. However, such photochemistry at these double bonds could be achieved after chemical modificaton at C-10 (reviewed by Falk 1989) which blocks the ways of internal conversion at this site.

Therefore, it can be concluded that either the native protein has to prevent processes of internal conversion in the chromophore or no photoproduct is

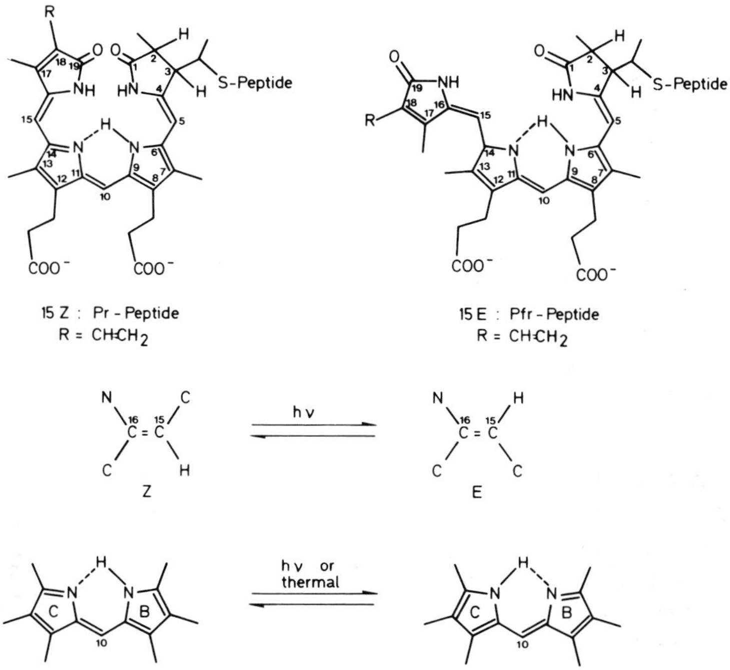

Figure 7. Upper part: structures of the chromophores in small chromopeptides of phytochrome. 15Z = Pr chromophore, 15E = Pfr chromophore. Because of steric hindrance, the Pfr chromophore is drawn in a semi-extended conformation. *Lower* part: photochemical Z,E isomerization at a double bond (*e.g.* at C-15, C-16) and photochemical or thermal tautomerization (*e.g.* at rings B,C).

obtained. It is not yet known how this is achieved. The absolute quantum yield for photoconversion of Pr to Pfr was determined in several laboratories to be between 15 and 20% and the fluorescence yield of Pr at ambient temperature to be < 1% (Schaffner *et al.* 1990). This indicates that there is still a considerable percentage (> 80%) of ill-defined ways of radiationless de-activation ('heat-loss') which apparently cannot be avoided.

When considering the relationship between the 15E chromophore in the small chromopeptide and the chromophore in native Pfr it can be concluded that the configuration is 15E not only in the chromopeptide, but also in native Pfr. The activation barrier for Z,E isomerization is not achieved during preparation of the chromopeptide (by proteolysis of Pfr in the dark). Even if it were achieved, one would not obtain the 15E configuration in the chromopeptide if the chromophore had the 15Z configuration in the parent Pfr, because the direction of the thermal reaction is always E → Z. This view is supported by direct spectroscopy of Pfr: resonance Raman spectra (Fodor *et al.* 1990; Mizutani *et al.* 1991) and Fourier transform infra-red spectra (Siebert *et al.* 1990) point to a 15E chromophore in native Pfr. A Z,E isomerization is the best candidate for the primary photoreaction; previous assumptions of a proton transfer as the primary reaction had to be rejected due to lack of a deuterium isotope effect within the primary steps of photoconversion (Eilfeld *et al.* 1986; Brock *et al.* 1987; Song *et al.* 1989).

As outlined above, rotation around single bonds leading to conformational changes of the chromophore can easily be achieved by thermal processes. According to the extinction coefficient (ε) of the long-wavelength band which is higher in native Pfr than in the Pfr chromopeptide (*cf.* Figs. 1 and 6) the chromophore conformation must be more extended in native Pfr than in the chromopeptide. However, this difference in the conformation does not explain the *bathochromic shift* of this band (from λ_{max} about 600 nm in the chromopeptide to λ_{max} about 730 nm in native Pfr). The bathochromic shift must be induced in the chromophore by the native protein. As Falk (1989) has pointed out, many possibilities for induction of such a red shift in bile pigments exist.

It is now well established, that the chromophore moves within the protein during photoconversion from Pr to Pfr. The first indication of spatial re-orientation came from the phytochrome-mediated chloroplast movement in the alga *Mougeotia* (Haupt 1972). Whereas this could have been re-orientation of entire phytochrome molecules within the algal cell, a re-orientation of the chromophore in isolated, immobilized phytochrome could be demonstrated by measurement of linear dichroism (Ekelund *et al.* 1985).

It was shown in a number of investigations (reviewed by Song 1985) that the chromophore is more exposed in the Pfr than in the Pr form. This was mainly demonstrated by means of oxidizing or reducing reagents. The difference in the reaction rate was larger than expected considering the difference in chemical reactivity of the Pfr and Pr chromophore, and the reactivity was much larger in partially degraded than in intact Pfr. These data indicate that the native protein partially shields the chromophore.

4.1.4 Intermediates

A characteristic feature of phytochrome is the formation of *intermediates* for both photoconversions, Pr → Pfr and Pfr → Pr. The initial photoreaction of the chromophore is followed by a series of relaxation processes of the protein which are independent of light. Conventionally, only products with differing spectral properties have been recognized as intermediates. Such changes of the absorption spectra reflect changes in the electronic state of the chromophore. Therefore, the relaxation processes of the classical intermediates are those which also involve the chromophore in some way. In addition, there might be other relaxation steps of the protein which have not yet been detected. There have been several experimental approaches to study these intermediates: (i) determination of rapid kinetics after flash irradiation; (ii) low temperature studies; (iii) dehydration of phytochrome; (iv) limited proteolysis of phytochrome; (v) spectral changes after continuous irradiation. Intermediates of the Pr → Pfr pathway are different from those of the pathway from Pfr → Pr. The different methods, *e.g.* rapid absorption measurements at room temperature and steady-state measurements at low temperature, appear to reveal the same intermediates.

No uniform terminology is used for phytochrome intermediates in the literature. Many authors only characterize intermediates by the approximate position of their long-wavelength absorption peak as derived from absorbance difference spectra. For example, I_{700} means an intermediate with absorption peak around 700 nm. This is unsatisfactory since the entire absorption spectra of intermediates between 300 and 800 nm are now known (Eilfeld and Rüdiger 1985). We use a terminology here (Fig. 8 and Table 1) which has been adopted from that used for other sensory pigments (Kendrick and Spruit 1977).

At low temperature, the photoconversion of phytochrome does not proceed to completion, but stops at intermediates which are stable under these conditions. Which intermediate is stabilized depends on the particular temperature. In Fig. 8 the known intermediates and temperature ranges of their stabilization are given. Starting with Pr, one can produce and stabilize the first intermediate, lumi-R, by irradiation at any temperature below −100°C. The next intermediate, meta-Ra, can either be formed by warming a lumi-R containing sample to any temperature between −100°C and −65°C or by irradiating Pr within this temperature range. With native (124-kD) phytochrome, the next intermediate is meta-Rc which is stable between −65°C and −25°C. The bleached intermediate meta-Rb (also called Pbl) predominates with partially degraded phytochrome (see below) in this temperature range. The intermediate meta-Rb is formed from native Pr only during irradiation, especially at high fluence rates, but immediately relaxes to meta-Rc in the dark. The final product, Pfr, is formed from

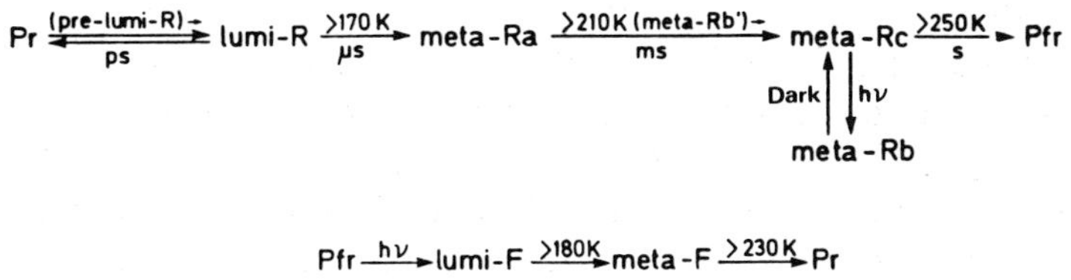

$$\text{Pr} \underset{\text{ps}}{\overset{\text{(pre-lumi-R)}\rightarrow}{\rightleftharpoons}} \text{lumi-R} \xrightarrow[\mu\text{s}]{>170\,\text{K}} \text{meta-Ra} \xrightarrow[\text{ms}]{>210\,\text{K (meta-Rb}')\rightarrow} \text{meta-Rc} \xrightarrow[\text{s}]{>250\,\text{K}} \text{Pfr}$$

$$\text{meta-Rc} \underset{\text{Dark}}{\overset{h\nu}{\rightleftharpoons}} \text{meta-Rb}$$

$$\text{Pfr} \xrightarrow{h\nu} \text{lumi-F} \xrightarrow{>180\text{K}} \text{meta-F} \xrightarrow{>230\,\text{K}} \text{Pr}$$

Figure 8. Scheme of phytochrome intermediates as derived from time-resolved and low-temperature studies (Rüdiger and Thümmler 1991).

Table 1. Spectral data of different species of phytochrome, determined at –140°C in glycerol buffer solution. Wavelength maxima λ^1_{max} and λ^2_{max} and their extinction coefficient ε^1 and ε^2, respectively. After Eilfeld and Rüdiger (1985).

Species	λ^1_{max} nm	ε^1 $dm^3\ mol^{-1}\ cm^{-1}$	λ^2_{max} nm	ε^2 $dm^3\ mol^{-1}\ cm^{-1}$
Pathway Pr → Pfr				
Pr	667	140 000	380	50 000
lumi-R	693	190 000	384	47 000
meta-Ra	663	86 000	386	52 000
meta-Rb	665	40 000	380	57 000
meta-Rc	725	80 000	387	50 000
Pfr	741	119 000	403	41 000
Pathway Pfr → Pr				
Pfr	741	119 000	403	41 000
lumi-F	673	105 000	388	44 000
meta-F	660	133 000	381	55 000
Pr	667	140 000	380	50 000

meta-Rc or eventually from meta-Rb above –25°C. Only two intermediates, lumi-F and meta-F, have been fully characterized for the Pfr → Pr transformation. As judged from absorption difference spectra at low temperatures, properties of phytochrome intermediates in intact plant tissue are the same as those given in Fig. 8 for isolated, native phytochrome.

Absorption bands of compounds are narrowed by decreasing the temperature. The molar extinction coefficients increase correspondingly. The peak position can be bathochromically shifted by decreasing temperature. This is especially true for Pfr, but not for Pr. Spectral data of phytochrome intermediates, normalized to –140°C, are summarized in Table 1.

Flash photolysis of Pr has been carried out in the temperature range 0-25°C. First order kinetics have been found for decay or formation of the single intermediates and Pfr. Each step is well separated from the subsequent steps on the time scale. Nevertheless, kinetic analysis is complicated, because two or three parallel pathways exist for the single steps. The kinetics depend on temperature and the solvent, the reactions are slowed down by decreasing temperature or increasing viscosity (*e.g.* addition of glycerol or ethylene glycol to the aqueous buffer). The data given here refer to aqueous buffer at 0-4°C. According to fluorescence measurements, the main excited-state component of Pr has a lifetime of 48 ps. The intermediate lumi-R appears within less than 1 ns. Two components with lifetimes of about 20 and 200 ns have been determined for its decay (Schaffner *et al.* 1990). Most authors do not distinguish between meta-Ra, meta-Rb, or meta-Rc but combine these as Ibl. The formation of Ibl is synonymous with decay of lumi-R, and the decay of Ibl, at the millisecond or longer time scale, synonymous with formation of Pfr. An additional intermediate meta-Rb' was described in the pathway between meta-Ra and meta-Rc (Inoue *et al.* 1990). Picosecond absorption spectroscopy of flash irradiated Pr revealed precursors of lumi-R; it was suggested that they are derived from excited Pr by rotations at the single bonds in the chromophoric methine bridges (Lippitsch *et al.* 1993).

Dehydration experiments have been performed with either freeze-dried plant tissues containing phytochrome or with isolated phytochrome dissolved in glycerol buffer. Although not all intermediate steps have been investigated in detail, it can be concluded that the last step of photoconversion of Pr to Pfr, namely the formation of Pfr, is prevented. It has been suggested that dehydration restricts the mobility of the protein and by this means prevents Pfr formation (Kendrick and Spruit 1977). Rehydration restores full photoreversibility. Formation of intermediates and their dark relaxations may be of physiological consequence in dry or partly imbibed seeds. Phytochrome reactions in such seeds have been reviewed by Kendrick and Spruit (1977).

Limited proteolysis also affects the last steps of the photoconversion of Pr to Pfr. Until 1982, only products of limited proteolysis of native phytochrome, namely 'small' (60-kD) and 'large' (114/118-kD) phytochrome were known. All available evidence suggests that the chromophore site in Pr and also the formation of the first intermediates is identical in all of these species (Schaffner *et al.* 1990). In the 60- and 114/188-kD fragments, however, the final product of irradiation (λ_{max} = 720 nm) is more similar to meta-Rc (Table 1) than to native Pfr (λ_{max} = 730 nm at room temperature). Further proteolysis of Pfr leads to a photoreversible 39-kD fragment which yields meta-Rb as the final product of Pr photoconversion. Low temperature spectroscopy shows that the intermediates lumi-R and meta-Ra are formed in the same way as in native phytochrome. Proteolysis apparently removes that part of the protein which is responsible for the typical Pfr absorption at 730 nm.

Any irradiation which lasts longer than the formation of the first intermediate will establish a *photoequilibrium* between Pr and intermediates or Pfr. Continuous irradiation will then lead to a steady photoconversion of Pr to Pfr and *vice versa* ('pigment cycling'); photoequilibria are formed between Pr, Pfr and all phototransformable intermediates assuming the light is absorbed by these species. White light causes accumulation of considerable amounts of meta-Rb because: (i) the subsequent relaxation step is the slowest step of the whole cycle; (ii) meta-Rb is the most photostable intermediate (Kendrick and Spruit 1977). Relaxation to Pfr and Pr *via* intermediates can be observed immediately after transfer of the irradiated phytochrome sample into the dark.

4.1.5 Chromophore biosynthesis

Tetrapyrrole biosynthesis is found in all organisms. The first specific precursor, 5-aminolevulinate (ALA), is formed from succinyl-CoA and glycine ('Shemin -pathway') in animals and some photosynthetic bacteria but from glutamate ('C-5 pathway') in plants (Fig. 9). Characteristic for biosynthesis of chlorophylls is insertion of magnesium into the 'ring-closed' tetrapyrrole protoporphyrin. It is unlikely that phytochromobilin synthesis occurs by the 'magnesium pathway': the structure of phytochromobilin is closely related to linear tetrapyrroles of animals which are known to be products of the 'iron pathway': they are derived from haem by oxidative opening of the porphyrin ring (haem oxygenase reaction). As outlined below, biliverdin IXα, a normal product of haem breakdown in animals, is a precursor of phytochromobilin in plants.

The first insight into early steps of phytochromobilin biosynthesis came from inhibitor experiments with gabaculine (Gardner and Corton 1985). Gabaculine (5-amino-1,3-cyclohexadienylcarboxylic acid) inhibits the transaminase step of the C-5 pathway of tetrapyrrole biosynthesis (Fig. 9). Gardner and Corton (1985) were the first to demonstrate that phytochromobilin biosynthesis is inhibited by gabaculine indicating that at least the bulk of the chromophore is formed by the C-5 pathway (and not by the Shemin pathway). Interestingly, phytochrome apoprotein synthesis is virtually uninhibited under these conditions (Jones *et al.* 1986) so that etiolated seedlings treated with gabaculine or related transaminase inhibitors are a good source for the phytochrome apoprotein. Such apoprotein preparations have been used for reconstitution of the holoprotein (see Chapter 4.3) with phytochromobilin, related tetrapyrroles and precursors *in vivo* (Elich and Lagarias 1987) and *in vitro* (Terry and Lagarias 1991) in order to study later steps of phytochromobilin biosynthesis (see below).

The later steps of biosynthesis are better known for the structurally related phycobilins which can be considered as models for phytochromobilin (Fig. 10). Haem IX and biliverdin IXα were identified as obligatory intermediates. The enzymes for the single steps have been investigated in extracts from the unicellular rhodophyte *Cyanidium caldarium* (Beale and Cornejo 1991a, b). The haem oxygenase step and subsequent hydrogenation steps are all ferredoxin-linked but make use of NADPH *via* ferredoxin-$NADP^+$ reductase. Biliverdin is at first reduced at a methine bridge to yield 15,16-dihydrobiliverdin before ring A is reduced. The product bearing a vinyl group at the reduced ring is unstable; it isomerizes spontaneously to the Z-ethylidene compound. Isomerization to the E-ethylidene compound requires reduced glutathione as a cofactor whereas isomerization of phycoerythrobilin does not require low molecular weight cofactors besides the respective enzyme.

Biliverdin IXα is also an intermediate for phytochromobilin synthesis. Here it is assumed that ring A of biliverdin is reduced first. The reduction occurs in the

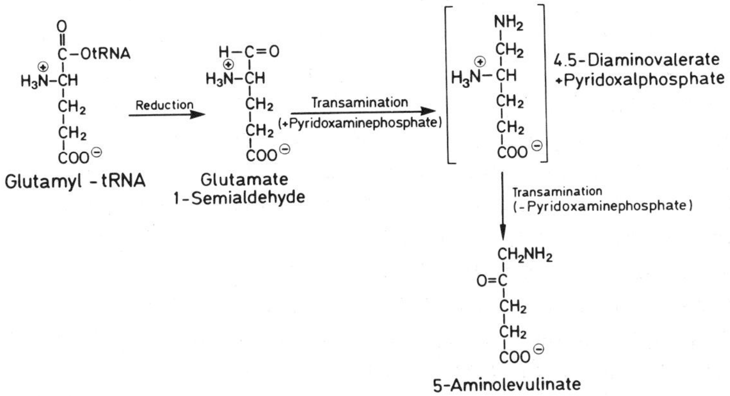

Figure 9. Biosynthesis of 5-aminolevulinate (ALA) from glutamyl-tRNA *via* glutamate 1-semialdehyde (GSA) and diaminovalerate bound to the transaminase. After Smith M.A., Kannangara C.G., Grimm B. and v. Wettstein D. (1991) *Eur. J. Biochem.* 202: 749-797.

plastids (etioplasts or etiochloroplasts): it is stimulated by NADPH or a system which regenerates NADPH within the plastids (Terry and Lagarias 1991). Phytochromobilin formation was detected by incorporation into the apoprotein with formation of photoreversible phytochrome. This incorporation occurs outside the plastids. Added biliverdin can apparently penetrate the plastid envelope and the phytochromobilin which is formed inside the plastids is then excreted. Terry and Lagarias (1991) assume that excretion of phytochromobilin from plastids into the cytoplasm also occurs *in vivo.*

4.1.6 Problems to be solved

There are several sets of data which indicate that the long-wavelength absorption of the Pfr chromophore is due to a specific chromophore-protein interaction (reviewed by Rüdiger *et al.* 1985). This interaction can be abolished by chemical modification of phytochrome, especially by intercalation of chemicals, or by partial proteolysis which removes that part of the protein which is the carrier of this interaction. In both cases, bleached phytochrome is obtained instead of Pfr. Apparently, only the specific interaction of the Pfr chromophore with the protein is concerned because the preceding intermediates are still observable. As outlined in Section 4.1.3 these data do not yet show the nature of this specific interaction. The FR absorption band of Pfr still awaits explanation at the molecular level.

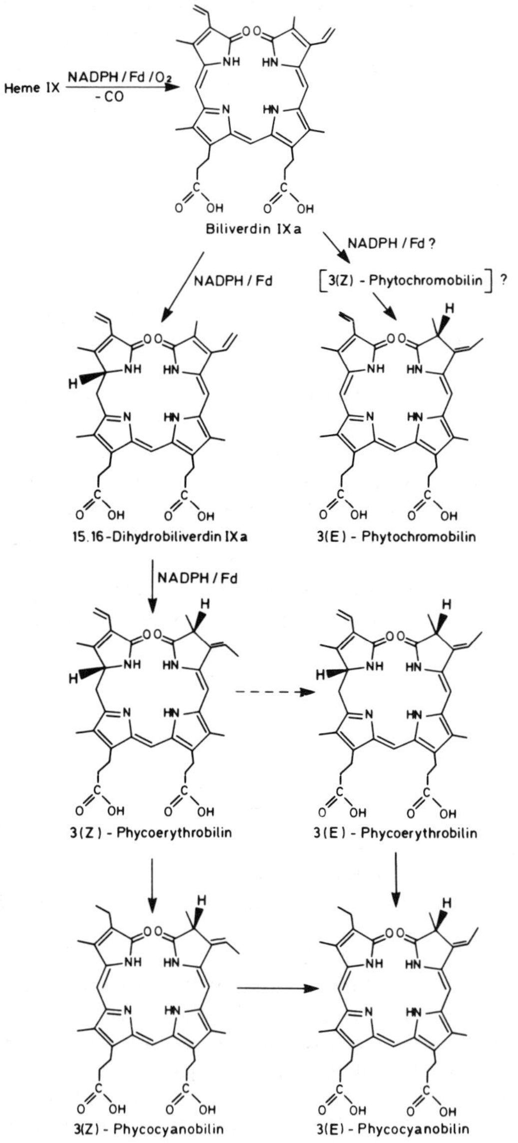

Figure 10. Biosynthesis of phycocyanobilin. Synthesis of phytochromobilin was assumed to occur in analogous steps. After Beale and Cornejo (1991a).

Enzymes and cofactors for chromophore biosynthesis have yet to be isolated and characterized. It is not yet clear from which source the phytochrome chromophore is derived after photodestruction of plastids (*e.g.* in carotenoid-free mutants or after inhibition of carotenoid biosynthesis by Norflurazon) and in mutants which lack the C-5 pathway in plastids. In all these cases the chlorophylls are almost completely lost, but spectrally active phytochrome is present in virtually normal amounts. Are all these mutants leaky, is this inhibition always incomplete or does a second tetrapyrrole pathway also exist outside the plastids? These questions have to be addressed in further research.

4.1.7 Further reading

Rüdiger W. (1992) Events in the phytochrome molecule after irradiation. *Photochem. Photobiol.* 56: 803-809.

Rüdiger W. and Scheer H. (1983). Chromophores in photomorphogenesis. In: *Encyclopedia of Plant Physiology,* New Series, Vol. 16A. *Photomorphogenesis,* pp. 119-151, Shropshire Jr. W. and Mohr H. (eds.) Springer Verlag, Berlin.

Rüdiger W. and Thümmler F. (1991) Phytochrome, the visual pigment of plants. *Angew. Chemie Int. Ed. Engl.* 30: 1216-1228.

Rüdiger W., Eilfeld P. and Thümmler F. (1985) Phytochrome, the visual pigment of plants: chromophore structure and chemistry of photoconversion. In: *Optical Properties and Structure of Tetrapyrroles,* pp. 349-366, Blauer G. and Sund H. (eds.) W. de Gruyter, Berlin.

Schaffner K., Braslavsky S.E. and Holzwarth A.R. (1990) Photophysics and photochemistry of phytochrome. *Advances in Photochemistry* 15: 229-277.

Song P.S. (1985) The molecular model of phytochrome deduced from optical probes. In: *Optical Properties and Structure of Tetrapyrroles,* pp. 331-348, Blauer G. and Sund H. (eds.) W. de Gruyter, Berlin.

4.1.8 References

Beale S.I. and Cornejo J. (1991a) Biosynthesis of phycobilins. 15,16-dihydrobiliverdin IXα is a partially reduced intermediate in the formation of phycobilins from biliverdin IXα. *J. Biol. Chem.* 266: 22341-22345.

Beale S.I. and Cornejo J. (1991b) Biosynthesis of phycobilins. 3(Z)-phycoerythrobilin and 3(Z)-phycocyanobilin are intermediates in the formation of 3(E)-phycocyanobilin from biliverdin IXα. *J. Biol. Chem.* 266: 22333-22340.

Brock H., Ruzsicska B.P., Arai T., Schlamann W., Holzwarth A.R., Braslavsky S.E. and Schaffner K. (1987) Fluorescence lifetimes and relative quantum yields of 124-kilodalton oat phytochrome in H_2O and D_2O solutions. *Biochemistry* 26: 1412-1417.

Duerring M., Schmidt G.B. and Huber R. (1991) Isolation, crystallization, crystal structure analysis and refinement of constitutive C-phycocyanin from the chromatically adapting cyanobacterium Fremyella diplosiphon at 1.66 Å resolution. *J. Mol. Biol.* 217: 577-592.

Eilfeld P. and Rüdiger W. (1985) Absorption spectra of phytochrome intermediates. *Z. Naturforsch.* 40c: 109-114.

Eilfeld P., Eilfeld P. and Rüdiger W. (1986) On the primary photoprocess of 124-kilodalton phytochrome. *Photochem. Photobiol.* 44: 761-769.

Ekelund N.G.A., Sundqvist Ch., Quail P.H. and Vierstra R.D. (1985) Chromophore rotation in 124-k Dalton *Avena sativa* phytochrome as measured by light-induced changes in linear dichroism. *Photochem. Photobiol.* 41: 221-223.

Elich T.D. and Lagarias J.C. (1987) Phytochrome chromophore biosynthesis. Both 5-aminolevulinic acid and biliverdin overcome inhibition by gabaculine in etiolated *Avena sativa* L. seedlings. *Plant Physiol.* 84: 304-310.

Falk H. (1989) *The Chemistry of Linear Oligopyrroles and Bile Pigments.* Springer-Verlag, Wien.

Fodor S.P.A., Lagarias J.C. and Mathies R.A. (1990) Resonance Raman analysis of the Pr and Pfr forms of phytochrome. *Biochemistry* 29: 11141-11146.

Gardner G. and Corton H.L. (1985) Inhibition of phytochrome synthesis by gabaculine. *Plant Physiol.* 77: 540-543.

Haupt W., (1972) Short-term phenomena controlled by phytochrome. In: *Phytochrome,* pp.349-368, Mitrakos K. and Shropshire Jr. W. (eds.) Academic Press, London.

Inoue Y., Rüdiger W., Grimm R. and Furuya M. (1990) The phototransformation pathway of dimeric oat phytochrome from the red-light-absorbing form to the far-red-light-absorbing form at physiological temperature is composed of four intermediates. *Photochem. Photobiol.* 52: 1077-1083.

Jones A.M., Allen C.D., Cardner G. and Quail P.H. (1986). Synthesis of phytochrome apoprotein and chromophore are not coupled obligatorily. *Plant Physiol.* 81: 1014-1016.

Kendrick R.E. and Spruit C.J.P. (1977) Phototransformations of phytochrome. *Photochem. Photobiol.* 26: 201-204.

Klein G. and Rüdiger W. (1978) Über die Bindungen zwischen Chromophor und Protein in Biliproteiden. V. Stereochemie von Modell-Imiden. *Liebigs Ann. Chem.* 267-279.

Klein G., Grombein S. and Rüdiger W. (1977) On the linkage between chromophore and protein in biliproteins. VI. Structure and protein linkage of the phytochrome chromophore, *Hoppe-Seyler's Z. Physiol. Chem.* 358: 1077-1079.

Lagarias J.C. and Rapoport H. (1980) Chromopeptides from phytochrome. The structure and linkage of the Pr form of the phytochrome chromophore. *J. Amer. Chem. Soc.* 102: 4821-4828.

Lippitsch M.E., Hermann G., Brunner H., Müller E. and Aussenegg F.R. (1993) Picosecond events in the phototransformation of phytochrome - a time resolved absorption study. *J. Photochem. Photobiol.* in press.

Mizutani Y., Tokutomi S., Aoyagi K., Horitsu K. and Kitagawa T. (1991) Resonance Raman study on intact phytochrome and its model compounds: Evidence for proton migration during the phototransformation. *Biochemistry* 30: 10693-10700.

Rüdiger W. and Correll D.L. (1969) Über die Struktur des Phytochrom-Chromophors und seine Protein-Bindung. *Liebigs Ann. Chem.* 723: 208.

Rüdiger W., Brandlmeier T., Blos I., Gossauer A. and Weller J.P. (1980) Isolation of the phytochrome chromophore. The cleavage reaction with hydrogen bromide. *Z. Naturforsch.* 35c: 763-769.

Rüdiger W., Thümmler F., Cmiel E. and Schneider S. (1983) Chromophore structure of the physiologically active form (Pfr) of phytochrome. *Proc. Nat. Acad. Sci. USA* 80: 6244-6248.

Schirmer T., Bode W. and Huber R. (1987) Refined three-dimensional structures of two cyanobacterial C-phycocyanins at 2.1 and 2.5 Å resolution. A common principle of phycobilin-protein interaction. *J. Mol. Biol.* 196: 677-695.

Siebert F., Grimm R., Rüdiger W., Schmidt G. and Scheer H. (1990) Infrared spectroscopy of phytochrome and model pigments. *Eur. J. Biochem.* 194: 921-928.

Song P.-S., Singh B.R., Tamai N., Yamazaki T., Yamazaki I., Tokutomi S. and Furuya M. (1989) Primary photoprocesses of phytochrome. Picosecond fluorescence kinetics of oat and pea phytochromes. *Biochemistry* 28: 3265-3271.

Terry M.J. and Lagarias J.C. (1991) Holophytochrome assembly. Coupled assay for phytochromobilin synthetase in *Organello. J. Biol. Chem.* 266: 22215-22221.

Thümmler F. and Rüdiger W. (1983) Models for the photoreversibility of phytochrome. Z,E-isomerization of chromopeptides from phycocyanin and phytochrome. *Tetrahedron* 39: 1943-1951.

Thümmler F., Rüdiger W., Cmiel E. and Schneider S. (1983) Chromopeptides from phytochrome and phycocyanin. NMR studies of the Pfr and Pr chromophores of phytochrome and E,Z-isomeric chromophores of phycocyanin. *Z. Naturforsch.* 38c: 359-368.

4.2 Phytochrome genes and their expression

Peter H. Quail

U.C. Berkeley/USDA Plant Gene Expression Center,
800 Buchanan Street, Albany, CA 94710, USA

4.2.1 Introduction

The molecular cloning of phytochrome (*phy*) genes has unveiled a rich source of information on the structure, evolution and biological functions of the photoreceptor. In addition, it has provided a powerful experimental system for exploring the central unresolved question in phytochrome research, namely, the molecular mechanism by which the photoreceptor regulates gene expression and thereby plant growth and development.

Of pivotal importance to the conceptual framework that is evolving from these studies was the discovery that phytochrome, rather than representing a single, homogeneous population of molecules, is instead a family of photoreceptors encoded by multiple, divergent and differentially regulated genes. This realization has brought into sharp focus the twin questions of whether individual phytochrome family members have discrete physiological or photosensory functions, and whether each has a distinct primary mechanism of action. The potential for functional specialization among multiple phytochromes has provided a possible mechanistic explanation for the large diversity of phytochrome-regulated responses that have been difficult to reconcile with the action of a single molecular species of the photoreceptor. In addition, the availability of recombinant clones representing the multiple *phy* sequences has provided exquisite new tools and experimental strategies with which to approach these fundamental questions. This chapter summarizes the current status of a rapidly expanding body of data on the structure, function and expression of the *phy* gene family (see also Quail 1991).

4.2.2 Structure and evolution of *phy* genes

Like other eukaryotic genes, *phy* genes are composed of two major regions: the promoter region responsible for regulation of expression, and the coding region

R.E. Kendrick & G.H.M. Kronenberg (eds.), Photomorphogenesis in Plants - 2nd Edition

which specifies the amino-acid sequence of the phytochrome polypeptide. Analysis of the promoter regions of *phy* genes has begun to identify regulatory DNA sequences and protein factors involved in the expression of these genes, whereas examination of the coding regions has provided insight into the structure and evolution of the phytochrome family.

4.2.2.1 Multiple phy *genes*

Direct molecular evidence that angiosperms contain several species of phytochrome encoded by a small, multigene family was provided initially from studies with *Arabidopsis* (Sharrock and Quail 1989). Five phytochrome-related sequences were detected by genomic Southern blot analysis, and full-length cDNAs representing three of these genes, designated *phyA*, *phyB* and *phyC* were isolated and sequenced. These genes have been mapped to chromosomes 1, 2 and 5, respectively (Chang *et al.* 1988; E. Meyerowitz pers. comm.). Subsequently, the two remaining *Arabidopsis* genes, *phyD* and *phyE* have been isolated and partially characterized (R.A. Sharrock pers. comm.), thus verifying the presence of five single-copy *phy* genes in this smallest of known plant genomes. Sequence comparisons among these genes shows that the encoded phytochrome polypeptides are substantially different from each other. The *phyA*, *phyB* and *phyC* polypeptides exhibit only 50% amino-acid sequence identity in each pairwise comparison (Sharrock and Quail 1989), and initial analysis indicates that the *phyD* and *phyE* encoded proteins are also significantly divergent from each other and the other three family members (R.A. Sharrock, pers. comm.). The majority of published phytochrome sequences from other angiosperms have been categorized as *phyA* family members because they are more closely related to the *Arabidopsis phyA*-encoded polypeptide (65 - 80% amino-acid sequence identity) than to the *phyB* and *phyC* polypeptides (50% identity). Conversely, a rice phytochrome sequence that has 73% amino-acid sequence identity with the *Arabidopsis phyB* polypeptide and only 50% identity with the *phyA* and *phyC* encoded products has been classified as representing rice *phyB* (Dehesh *et al.* 1991). A potato *phyB* sequence has also recently been reported (Heyer and Gatz 1992).

The phylogenetic relationship between these published angiosperm phytochrome sequences is summarized in Fig. 1. The data imply that the divergence of the *phyA*, *phyB* and *phyC* subfamilies was an ancient evolutionary event, preceding the divergence of the monocots and dicots over 100 million years ago. It seems likely, therefore, that all angiosperms have the five phytochrome family members described for *Arabidopsis*. Only limited information is yet available on *phy* sequences from other plant groups. Preliminary data for the gymnosperm *Pinus* indicates the presence of at least two sequences most closely related to *phyA* and *phyC*, respectively (J. Silverthorne pers. comm.). A

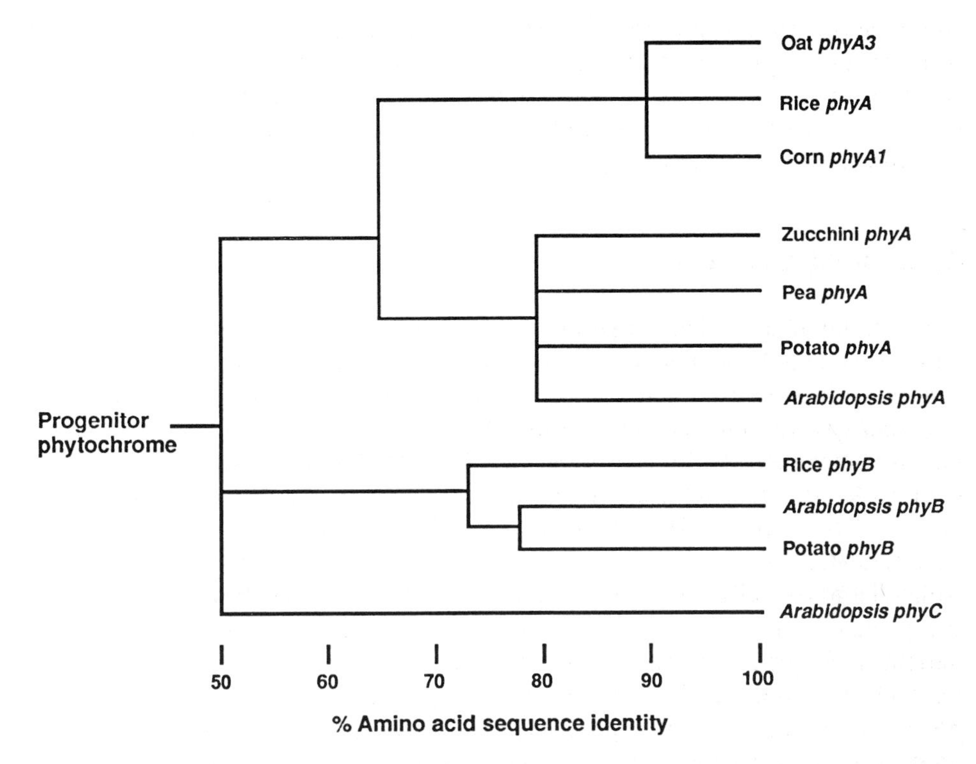

Figure 1. Deduced phylogeny of angiosperm phytochrome polypeptides. Percent amino-acid sequence identity between each pairwise comparison of the various phytochromes was used to group related sequences as a measure of phylogenetic distances in constructing the tree. Modified from Dehesh *et al.* (1991) by addition of potato *phyA* (Heyer A. and Gatz C. (1992) *Plant Mol. Biol.* 18: 535-544) and potato *phyB* (Heyer and Gatz 1992) sequences.

full-length sequence for the fern *Selaginella* has been reported, and the encoded polypeptide is about equally related (56 to 62% amino-acid sequence identity) to each of the higher plant subfamily members (Hanelt *et al.* 1992). A partial phytochrome sequence for the alga *Mougeotia* (Winands *et al.* 1992) and a novel hybrid sequence for the moss *Ceratodon* (Rüdiger and Thümmler 1991; Thümmler *et al.* 1992) have also been presented. The phylogenetic relationships between these various sequences and those of all five angiosperm phytochromes when available will be of great interest to the question of phytochrome evolution.

4.2.2.2 Nomenclature

The rapid proliferation of studies on multiple phytochromes in diverse plant species has led to a certain degree of nonuniformity in the noenclature used to designate the various genes and their encoded products. Table 1 gives an outline of the terminology used here [but the reader should be aware of alternative terminology that is gaining favour (4.2.8 Appendix)]. Where needed, a simple prefix can specify the plant species of origin for the molecular entity under discussion (*e.g. Arabidopsis* phytochrome A). This is flexible enough to accommodate future sequences, including those which do not readily fall into a predefined subfamily on the basis of sequence similarity. The term 'form' here is reserved for the different photochemical species of phytochrome [*i.e.* the PrA form (for the red light (R)-absorbing form of phytochrome A); the PfrB form (for the far-red light (FR)-absorbing form of phytochrome B), *etc.*], and the term 'type' is used for the different molecular species of phytochrome (*e.g.* type A phytochromes; type I and type II phytochromes, *etc.* see below).

Before the identification of multiple *phy* genes, a variety of studies over a number of years had provided evidence for the existence of more than one type

Table 1. Phytochrome nomenclature.

Molecular entity	Designation	Description
Wild-type genes	*phyA*	Sub-family A
	phyA1	Member number 1 of sub-family A where there are multiple closely related wild-type sequences of this sub-family at multiple loci in a single genome (*e.g.*, oats where there are at least four identified *phyA* sequences that are 98% identical).
	phyB	Sub-family B, *etc.*
Mutant genes (alleles)	*phyA-1*	Mutant allele number 1 at the A locus
	phyB1-1	Mutant allele number 1 at the B1 locus, *etc.*
Transcripts	*phyA* mRNA	Mature transcript encoded by the *phyA* gene, *etc.*
Apoprotein	*phyA* apoprotein or apophytochrome A*	Polypeptide encoded by the *phyA* gene, *etc.*
Holoprotein	phytochrome A*	Fully assembled chromoprotein (holoprotein) with chromophore covalently attached to *phyA* apoprotein, *etc.*
Photochemical forms	PrA	Red light-absorbing form of phytochrome A
	PfrA	Far-red light-absorbing form of phytochrome A
	PrB	Red light-absorbing form of phytochrome B, *etc.*

*The symbol PHYA etc. has been suggested as an abbreviation for the apoprotein, and the symbol phyA *etc.* for the holoprotein. However, because of the potential for ambiguity in verbal communication, we prefer to encourage the routine use of the more complete designations above for written text as well as for verbal communication. For brevity in the body of tables and figures, the gene prefix *phyA etc.* can be used in isolation as an adjective where the remainder (-encoded apoprotein) or (-encoded mRNA) is understood and is clear from the context.

Table 2. Correspondence between phytochrome molecular species predicted from gene sequences and those detected by physicochemical and physiological means.

Gene	Chromoprotein		
	Predicted*	Spectrophotometric† Immunochemical Biochemical	Physiological‡
phyA	phytochrome A	type I‖	light-labile (Pfr)
phyB	phytochrome B	type II¶	light-stable (Pfr)
phyC	phytochrome C	type II¶	light-stable (Pfr)
phyD	phytochrome D	?	?
phyE	phytochrome E	?	?

*From nucleotide sequence. †Direct detection of phytochrome molecular species by physicochemical methods. ‡Behaviour of phytochrome species involved in a particular response deduced from physiological studies. ‖Also termed 'etiolated-tissue phytochrome'. ¶Also termed 'green-tissue phytochrome'

of phytochrome (Smith and Whitelam 1990). Physiological studies indicated that certain responses were under the control of a 'light-labile' type of phytochrome (*i.e.* a species that is unstable and rapidly inactivated in the cell once converted to the Pfr form), whereas other responses appeared to be under the control of a 'light-stable' type of phytochrome (*i.e.* a species that remains stable and active for long periods as Pfr in the cell). In parallel, spectroscopic, biochemical, and immunochemical studies had led to the concept of two operationally defined molecular species of the photoreceptor (Furuya 1989): type I or 'etiolated-tissue' phytochrome; and type II or 'green-tissue' phytochrome. Type I phytochrome is the historically familiar species that is abundant in etiolated tissue, is rapidly proteolytically degraded as Pfr *in vivo* and is the molecule that has been purified and extensively characterized *in vitro* over a number of years. Type II phytochrome is detected at low abundance in etiolated tissue (1-2% of type I), but is stable as Pfr and therefore becomes the predominant species, albeit at relatively low levels, in green tissue as a consequence of the selective depletion of the light-labile type I molecule. Strong correlative evidence supports the widely-held belief that the physicochemically detectable type I phytochrome corresponds to the physiologically active 'light-labile' photoreceptor. Conversely, the apparent Pfr stability of the spectrally and immunochemically detectable type II phytochrome is consistent with its possible correspondence to the physiologically defined 'light-stable' photoreceptor.

Our current understanding of the relationship of these physiologically and physicochemically defined phytochromes to those encoded by the various molecularly defined *phy* genes is summarized in Table 2. Protein microsequencing of purified type I phytochrome has established unequivocally that

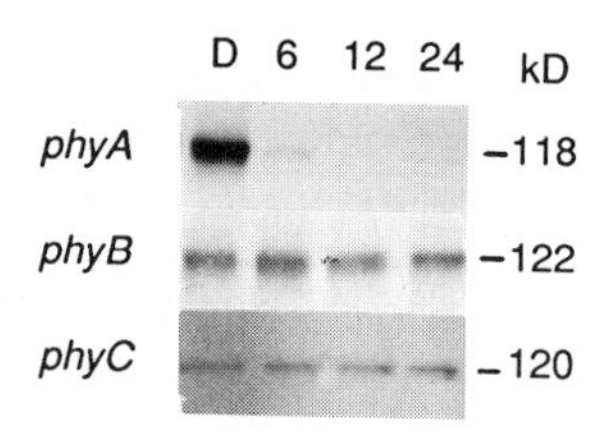

Figure 2. Immunoblot analysis of phytochromes A, B, and C in *Arabidopsis*. Phytochrome-enriched extracts from 7-day-old dark-grown seedlings treated before harvest with 0 (D), 6, 12, or 24 h of continuous red light. Immunoblot lanes were loaded with ammonium sulphate-precipitated fractions from 300 μg (rows *phyA* and *phyB*) and 1.2 mg (row *phyC*) of crude extract protein, and the blots were probed with the appropriate type-selective monoclonal antibody. From Somers *et al.* (1991).

Table 3. Relative levels of phytochromes A, B, and C in dark-grown and red light-irradiated *Arabidopsis* seedlings detected by immunoblot analysis. Modified from Somers *et al.* (1991).

Photoreceptor species	Dark	Red light*
Phytochrome A	500	5
Phytochrome B	10	10
Phytochrome C	1	1

*Seven-day-old, dark-grown seedlings were exposed to 24 h of continuous red light before extraction.

this photoreceptor species is encoded by the *phyA* gene, and is, therefore, phytochrome A. Because in *Arabidopsis phyA* is a single-copy gene and phytochrome A is the predominant, if not the only, light-labile phytochrome (Fig. 2 and Table 3), it is highly likely that phytochrome A is also the physiologically functional light-labile photoreceptor species. Phytochromes B and C, on the other hand, are of low abundance and light-stable (Fig. 2 and Table 3). Thus, these latter two molecular species have the properties of type II phytochrome and the physiologically defined light-stable photoreceptor. However, definitive linkage of a specific phytochrome subfamily member to a particular physiological response requires the isolation of genetic mutants defective in the function of individual phytochromes (see below).

4.2.2.3 Structure of phy *genes*

Figure 3 depicts the prototypical *phyA3* gene from oats and its relationship to the encoded phytochrome polypeptide. The gene is 5.94 kbp long and contains six exons and five introns, including one intron in each of the two untranslated

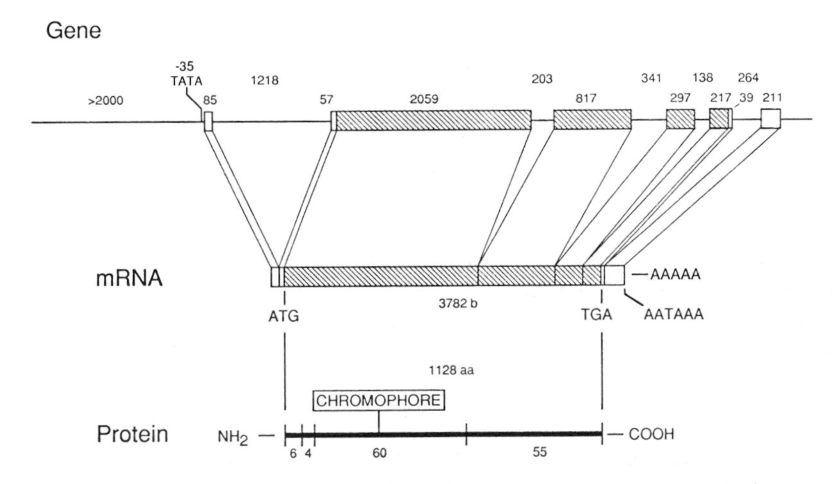

Figure 3. Schematic representation of the oat *phyA3* gene (*top*), its mature transcript (*middle*), and polypeptide product (*bottom*) (Hershey H.P., Barker R.F., Idler K.B., Murray M.G. and Quail P.H. (1987) *Gene* 61: 339-348). Gene: exons are indicated as boxes (open = untranslated; shaded = translated); introns and 5' flanking DNA are indicated as lines. The length of each segment is indicated above in base pairs (bp). The position of the TATA box (-35) is indicated relative to the transcription start site. mRNA: the overall length of the mature transcript is indicated below in bases (b), with the locations of the original introns marked by vertical lines. Initiation (ATG) and stop (TGA) codons are indicated together with the putative poly(A)-addition signal (AATAAA) and poly$(A)^+$ tail. Protein: the mature phytochrome chromoprotein (1128 amino acids (aa) long = 125 kD) with chromophore covalently linked at Cys-321 is aligned with the mRNA coding region. Short, vertical lines indicate the locations of three proteolytically vulnerable sites in the polypeptide chain leading to peptides of the size (in kD) indicated below. Modified from Quail P.H., Gatz C., Hershey H.P., Jones A.M., Lissemore J.L., Parks B.M., Sharrock R.E., Barker R.F., Idler K., Murray M.G., Koornneef M. and Kendrick R.E. (1987) In: *Phytochrome and Photoregulation in Plants*, pp. 23-38, Furuya M. (ed.) Academic Press, NY, USA.

regions at the 5' and 3' ends of the sequence. The transcription start site (designated as +1) is located 35 bp downstream of tandem TATA boxes, and a number of other short sequence motifs of potential importance to *phy* gene expression are present at various positions further upstream in the 5' flanking DNA (see below). The mature mRNA produced by the splicing out of the intron sequences from the primary transcript is 3.78 kb long before poly$(A)^+$ tail addition at the 3' end, and consists of 142 b, 3387 b, and 252 b of 5'-untranslated, protein-coding and 3'-untranslated sequence, respectively. The mature transcript contains an ATG translation-initiation codon, tandem TGA stop codons, and an AATAAA poly(A)-addition signal 35 b upstream of the poly(A)-addition site. The mature, encoded phytochrome polypeptide is 1128 amino acids in length (125 kD) with a single chromophore attachment site at Cys-321. The polypeptide folds into two major domains: an ≈ 70 kD, chromophore-bearing NH_2-terminal domain and an ≈ 55 kD COOH-terminal domain.

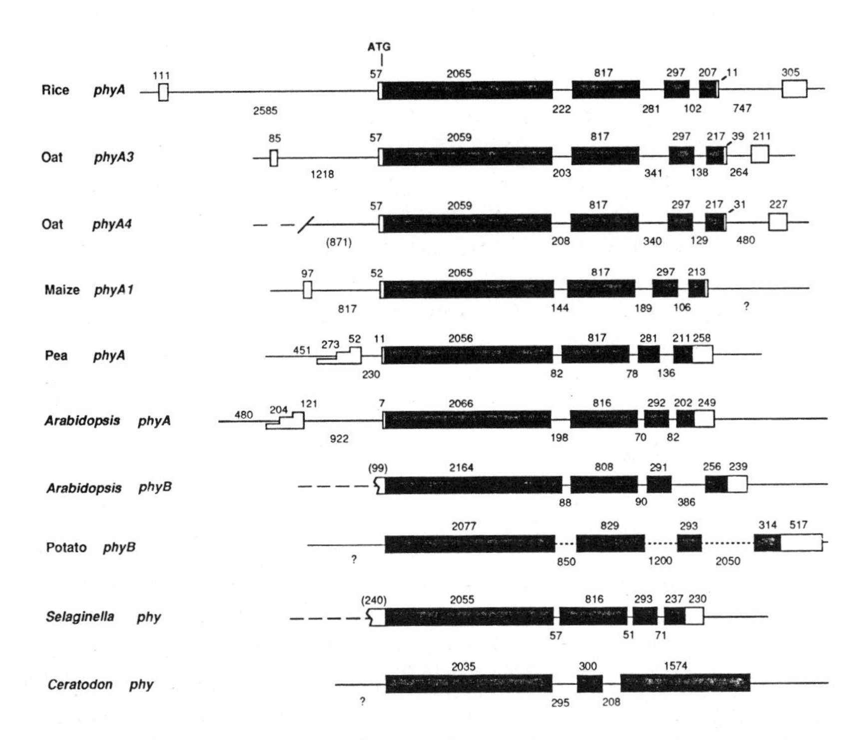

Figure 4. Schematic representation of *phy* genes from various plant species. Solid lines = introns and flanking DNA; boxes = exons (open = untranslated; solid = translated); dashed or dotted lines = unsequenced. Sizes in base pairs of exons are indicated above and of introns below each gene. The sizes of the introns for the potato *phyB* gene have only been estimated by restriction mapping and are therefore represented as dotted lines and not drawn to scale. References for the gene sequences: rice *phyA* (Kay S.A., Keith B., Shinozaki K., Chye M.-L. and Chua N.-H. (1989) *Plant Cell* 1: 351-360); oat *phyA3* (Hershey H.P., Barker R.F., Idler K.B., Murray M.G. and Quail P.H. (1987) *Gene* 61: 339-348); oat *phyA4* (K. B. Idler, H. P. Hershey and P. H. Quail, unpublished data); maize *phyA1* (Christensen A.H. and Quail P.H. (1989) *Gene* 85: 381-90); pea *phyA* (Sato 1988); *Arabidopsis phyA* (R.A. Sharrock, K.Dehesh, D.Somers and P.H. Quail, unpublished data); *Arabidopsis phyB* (R.A. Sharrock and P.H. Quail, unpublished data); potato *phyB* (Heyer and Gatz 1992); *Ceratadon phy* (Thümmler *et al.* 1992); *Selaginella phy* (Hanelt *et al.* 1992).

The NH_2-terminal domain corresponds approximately to the major 5' exon of the gene (Fig. 3), consistent with the pattern frequently observed for multi-domain proteins. Based on this pattern, it has been proposed that multidomain proteins may have arisen during evolution through the assembly of pre-existing sequences (exons) into a contiguous stretch of DNA encoding a single transcript. The COOH-terminal domain of the phytochrome polypeptide is encoded by three separate exons, but it remains unclear whether these represent distinct structural or functional sub-domains.

Figure 4 compares the structures of all *phy* genes for which complete, or almost complete, sequence is currently available. From these data, it is clear that with the apparent, striking exception of *Ceratodon* the general structural features of these genes are strongly conserved. The *Ceratodon* sequence aside, the number of protein coding exons is constant and their sizes are, in general, highly similar in the remaining genes. Likewise, the number and sites of intron insertions are conserved. One variation on this theme is the presence of one additional intron in the 3' untranslated region of the rice and oat (and possibly maize) genes not apparent in the other genes. The sizes of the introns, particularly those in the untranslated regions, vary considerably, but the significance of this observation, if any, is unknown. The monocot *phyA* genes that have been investigated have only a single, detectable transcription start site, whereas their counterparts in the dicots, pea and *Arabidopsis*, appear to have three transcription start sites. This finding has interesting implications for the regulated expression of these genes as each gene has a different TATA box and TATA-proximal promoter region. Yet more intriguing is the strongly divergent structure reported for the *phy*-related sequence from *Ceratodon*. The large 5' exon and the 100 amino acids present in the central, small exon of this sequence show sequence similarity to the corresponding regions of the other *phy* genes. However, the large 3' exon of 530 amino acids has no homology to any known phytochrome. Instead, it contains a 300 amino-acid region that bears striking similarity to the catalytic domain of eukaryotic protein kinases (Thümmler *et al.* 1992). This result raises the possibility of the presence of an, as yet undetected, homolog of the *Ceratodon* sequence in other plant species. The presence of two apparent introns in new locations on either side of the chromophore-attachment site in the preliminary partial sequence recently reported for *Mougeotia* (Winands *et al.* 1992) further suggests the possibility of novel *phy* gene sequences in lower plants.

The availability of multiple phytochrome sequences provides the opportunity to identify regions and individual amino-acid residues in the polypeptide that have been conserved during evolution. Invariant residues defined in this way are those most likely to be involved in critical structural or functional properties of the molecule. In this respect, comparison of the *phyA*, *phyB*, and *phyC* subfamilies is the most informative because these represent the greatest evolutionary distance between the phytochromes thus far described. Figure 5 depicts the distribution of invariant amino-acid residues along the polypeptide for all available angiosperm phytochromes. Addition of the *Selaginella* sequence does not greatly alter this pattern. In general, sequence identity is highest in the central region of the NH_2-terminal domain surrounding the chromophore-attachment site. This observation suggests that the segments of the polypeptide involved in protein-chromophore interactions may be subject to strong evolutionary pressure to conserve the structural environment required for correct perception of light signals by the chromophore. The lower level of sequence identity in the remainder of the protein, especially toward the NH_2- and COOH-terminal ends, could indicate either less rigid structural constraints

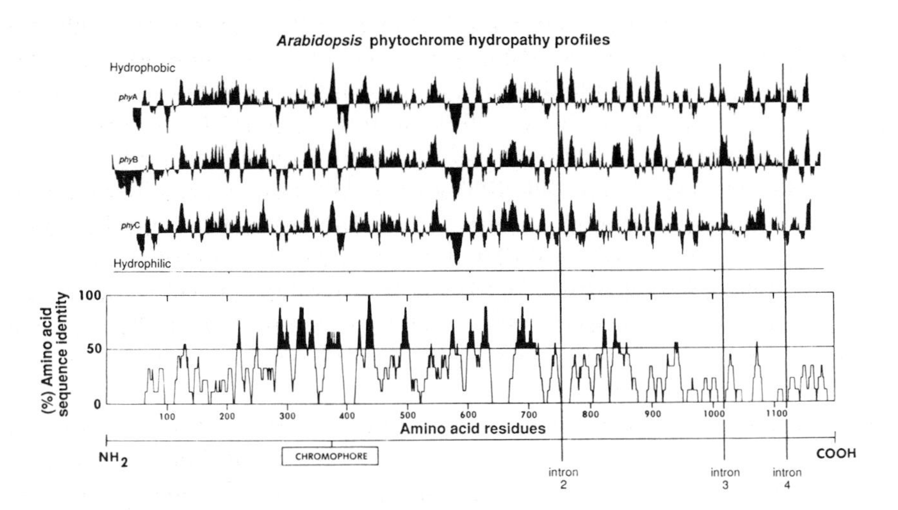

Figure 5. Top: hydropathy profiles of *Arabidopsis phyA*, *phyB*, and *phyC* polypeptides. Modified after Sharrock and Quail (1989). *Bottom*: distribution of invariant amino-acid residues along the aligned polypeptides of all angiosperm *phy* sequences currently available. The number of invariant residues within a moving window of nine amino acids is expressed as percent identity and plotted at the middle position of the window. Shaded portions indicate regions of > 50% identity. A schematic diagram of the longest polypeptide, *Arabidopsis phyB*, with chromophore attachment site is indicated below the plots. The approximate positions of the three introns that interrupt the coding region are indicated (see Fig. 4). Modified from Quail *et al.* (1991).

in these regions or functionally important differences related to potential specialized activities of the individual phytochromes. It is significant therefore that many of the amino-acid substitutions that have occurred are conservative in nature. This observation is reflected in the general similarity of the hydropathy profiles of phytochromes A, B, and C (Fig. 5) which indicates that the overall three-dimensional structure of the molecule has been conserved. Thus, the residues most likely to be involved in any potential differential activities between the different phytochrome subfamilies are those which are variant between the subfamily consensus sequences. Of significance in this regard are the NH_2- and COOH-terminal extensions of the rice and *Arabidopsis* phytochrome B polypeptides which overall are 40 to 60 amino acids longer than phytochromes A and C.

The availability of phytochrome sequences also provides the opportunity to search for sequence similarities to other proteins in databases. Such similarities to proteins of established function in other systems can provide strong indications as to potential mechanisms of action. Until recently, however, no convincing homology to any sequences in available databases had been reported. Earlier proposals that the angiosperm phytochromes might be related to eukaryotic protein kinases are not strongly supported by the available data because all these phytochrome sequences, and that of *Selaginella*, lack the canonical motifs strongly conserved in the literally hundreds of eukaryotic protein kinases sequenced to date. However, an interesting new computer analysis points to some intriguing sequence similarities between a C-terminal region of the phytochromes and a family of bacterial sensory proteins (Schneider-Poetsch *et al.* 1991). These proteins have sensory-stimulus regulated histidine protein kinase activity that catalyses phosphorylation of a partner regulator (frequently a transcription factor) whose activity is controlled by this phosphorylation. Since no sequence similarity is apparent with the large family of eukaryotic protein kinases, the bacterial proteins appear to represent a distinct class of protein kinases. The fascinating possibility that phytochrome is a photoregulated eukaryotic descendent of these prokaryotic sensory molecules is clearly worthy of investigation.

The unusual *Ceratodon* sequence (Fig. 4) provides an additional dimension to this protein kinase question. The polypeptide predicted from this sequence represents a chimera between a partial phytochrome polypeptide and a eukaryotic protein kinase. As indicated above, the first two exons are co-linear with the NH_2-terminal two-thirds of other phytochromes, whereas the large 3' exon encodes a sequence with the conserved features of a eukaryotic protein kinase (Thümmler *et al.* 1992). Should the protein predicted from this sequence be verified to exist in *Ceratodon* cells, the possibility emerges that the phytochrome photoreceptor family has evolved such that the chromophore-bearing photoresponsive module is coupled to different COOH-terminal modules with a variety of regulatory mechanisms of action.

4.2.3 Biological functions of multiple *phy* genes

There is emerging evidence that individual members of the phytochrome family may indeed have specialized regulatory roles in controlling the overall response of plants to the complexities of the light environment (Smith and Whitelam 1990). This evidence has come primarily from studies on photoreceptor mutants and transgenic plants overexpressing specific phytochromes (Kendrick and Nagatani 1991; Quail 1991). The key to this problem initially was the demonstration that the *hy3* mutant of *Arabidopsis* and the *lh* mutant of cucumber are selectively deficient in phytochrome B (Somers *et al.* 1991; López-Juez *et al.* 1992). In the case of the *hy3* mutant, both *phyB* mRNA and phytochrome B apoprotein levels are reduced without a change in the corresponding *phyA* or *phyC* gene products, thereby indicating either a reduction in *phyB* gene expression or production of an aberrant transcript (Fig. 6). Genomic mapping studies had shown earlier that *phyB* and *hy3* map close to each other on chromosome 2 (E. Meyerowitz pers. comm.; Quail 1991) and recent sequencing has verified that the *hy3* locus is the *phyB* structural gene (Reed *et al.* 1993). These mutants

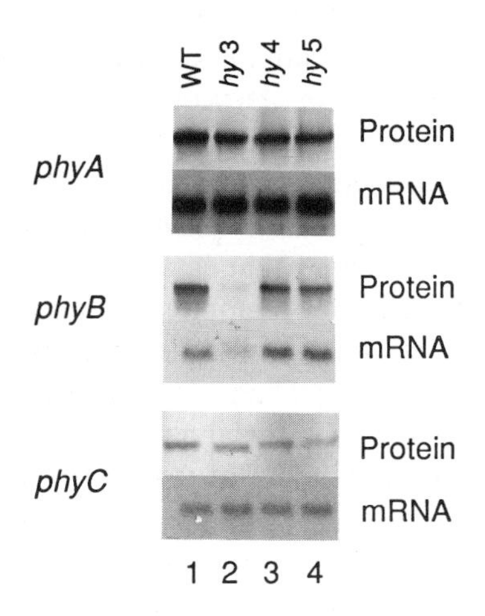

Figure 6. Levels of *phyA*, *phyB*, and *phyC* mRNAs and proteins in *hy3*, *hy4*, *hy5*, and wild-type *Arabidopsis*. Phytochrome-enriched protein extracts and total RNA preparations were made from wild-type (WT, lane 1), *hy3* (lane 2), *hy4* (lane 3), and *hy5* (lane 4) *Arabidopsis* seedlings grown for 7 days in the dark. Immunoblot lanes (= protein) were loaded with ammonium sulphate-precipitated fractions from 300 (*phyA* and *phyB*) and 600 (*phyC*) μg of crude extract protein and then probed with type-selective monoclonal antibodies against phytochromes A (*phyA*), B (*phyB*) and C (*phyC*). RNA blots (= mRNA) containing 5 μg of total RNA per lane were hybridized with the appropriate transcript-specific single-stranded DNA probe. From Somers *et al.* (1991).

exhibit a long hypocotyl relative to wild-type plants when grown in the light because of a severe reduction in the ability to perceive R. Because the mutant seedlings have normal levels of phytochromes A and C (Fig. 6) (and probably phytochromes D and E), it would appear that these other four phytochromes do not contribute greatly to the perception of red photons in the control of hypocotyl growth. This particular response to this specific region of the spectrum appears to be primarily under the control of phytochrome B. The principal consequence of the loss of sensitivity to R in the *hy3* and *lh* mutants is a loss of the capacity to monitor the ratio of R to FR in the environment, and thereby a loss of the capacity to detect shading and neighbour proximity. Thus, phytochrome B appears to assume the primary responsibility for this specific biological function of the photoreceptor family in light-grown plants.

Conversely, phytochrome A has now been shown to mediate responsiveness to continuous FR (the so-called 'FR high irradiance response', FR-HIR) observed in etiolated seedlings as inhibition of hypocotyl elongation and enhanced cotyledon expansion. The likelihood of this possibility emerged initially primarily on the basis of accumulated circumstantial evidence from physiological, mutant and transgenic-overexpressor studies. This evidence is as follows. The responsiveness of the *hy3* mutant to continuous FR is quantitatively indistinguishable from that of wild-type *Arabidopsis* (McCormac *et al.*

1993). Thus, a phytochrome other than B would appear to mediate this response. The *hy1*, *hy2*, and *hy6* mutants of *Arabidopsis* and the *au* mutant of tomato are defective in responsiveness to both R and FR (Kendrick and Nagatani 1991). These mutants have been shown to be severely deficient in photo-active phytochrome A, consistent with a potential role for this phytochrome species in the FR component of this response. However, because the accompanying impaired responsiveness of these mutants to R suggests reduced phytochrome B activity as well, deficiencies in the other phytochromes, C, D, and E, are also possible, and the correlation with phytochrome A is therefore not definitive. Indeed, the *hy1*, *hy2*, and *hy6 Arabidopsis* mutants have been shown to be blocked in chromophore synthesis and are therefore probably deficient in all phytochromes (Chory 1991; Parks and Quail 1991). On the other hand, transgenic plants overexpressing phytochrome A retain the FR-HIR when fully de-etiolated, whereas plants overexpressing phytochrome B lose this response upon de-etiolation in a manner identical to that of wild-type plants (McCormac *et al.* 1993; Whitelam *et al.* 1992). This result strongly suggests that the capacity to mediate the FR-HIR is an intrinsic property of the phytochrome A molecule not possessed by phytochrome B.

Definitive evidence for this conclusion has come recently from the isolation of a new class of *Arabidopsis* long-hypocotyl mutants, *hy8* (Parks and Quail 1993). These mutants are selectively deficient in functional phytochrome A (Fig. 7). Moreover, because two classes of *hy8* alleles have been identified, one with no detectable phytochrome A (*hy8*-1 and *hy8*-2) and one with wild-type levels of spectrally normal phytochrome A (*hy8*-3), it was initially proposed that the *hy8* locus is likely to correspond to the *phyA* structural gene itself (Parks and Quail 1993). This proposal has now been confirmed by detection of mutations in the *PhyA* genes of three independent *hy8* mutant lines through direct sequencing (K. Dehesh and P. Quail unpublished data). The *hy8* mutants are defective in the capacity to respond to continuous FR (Parks and Quail 1993). This result provides direct evidence that phytochrome A is responsible for the FR-HIR and that none of the other phytochromes B, C, D, or E can substitute in this capacity. Conversely, the *hy8* mutants exhibit wild-type responsiveness to continuous R or white light (W), indicating that phytochrome A does not play a major role in mediating responses to these irradiation conditions. Thus, as summarized in Table 4, the available evidence indicates that phytochromes A and B have complementary sensitivities to the R and FR regions of the spectrum with antagonistic effects on hypocotyl elongation in FR-enriched light environments such as encountered under vegetative canopies. It appears, therefore, that phytochromes A and B have distinct photosensory functions in regulating seedling development. Although no data are yet available on possible functions of phytochromes C, D, or E, the probability has now increased that these remaining members of the phytochrome family also have discrete photosensory activities. The expression of antisense sequences in

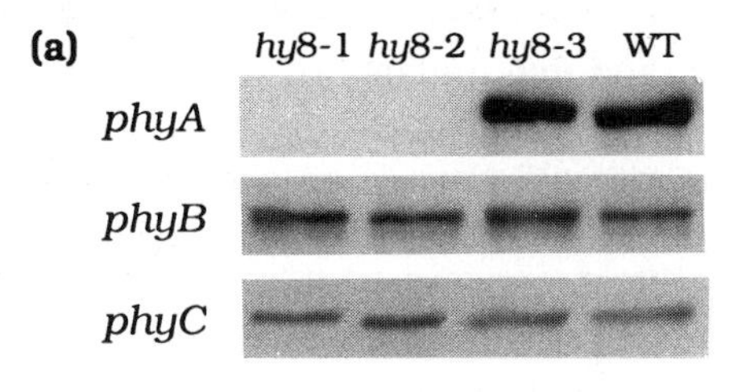

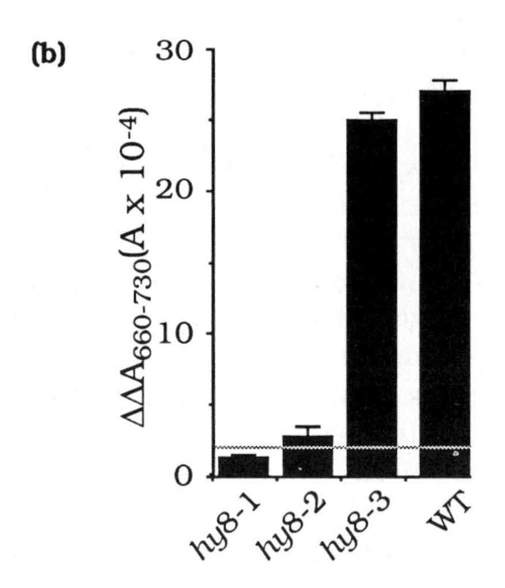

Figure 7. Phytochrome in phytochrome A mutants of *Arabidopsis*. (a) *phyA*-, *phyB*-, and *phyC*-encoded protein levels in wild type and three *hy*8 mutants of *Arabidopsis*. Phytochrome-enriched protein extracts were made from etiolated *Arabidopsis* seedlings carrying wild-type (WT) or one of three *hy*8 mutant alleles (*hy*8-1, *hy*8-2, and *hy*8-3). Immunoblot lanes were loaded with ammonium sulphate-precipitated protein (17.5 μg per lane for *phyA* and *phyB* immunoblots, and 35 μg per lane for the *phyC* immunoblot) and then probed with type-selective monoclonal antibodies against phytochromes A, B, or C. (b) Spectrally detectable phytochrome in crude extracts of etiolated *Arabidopsis* seedlings. Crude extracts were prepared from etiolated *Arabidopsis* seedlings and phytochrome was measured spectrophotometrically. The dashed horizontal line denotes the detection limit for the instrument. Modified from Parks and Quail (1993).

transgenic plants is one feasible targeted approach to this problem. Another is the isolation of mutants aberrant in phytochrome-regulated responses that are normal in the *hy8* and *hy3* mutants.

4.2.4 Regulation of *phy* gene expression

The availability of cloned phytochrome sequences has provided sensitive and specific tools with which to explore the spatial and temporal patterns of *phy*

Table 4. Summary of the responsiveness of mutant and transgenic overexpressor lines of *Arabidopsis* to continuous red (Rc) or far-red (FRc) irradiation. From Parks and Quail (1993).

Arabidopsis line*	Phytochrome levels†,‡	Responsiveness‡	
		Rc	FRc
WT	A^+B^+	+	+
*hy*8	A^-B^+	+	–
*hy*3	A^+B^-	–	+
*hy*1, *hy*2	A^-B^-	–	–
AOX	$A^{++}B^+$	++	++
BOX	A^+B^{++}	++	+

*WT, wild type; *hy*, long-hypocotyl mutants; AOX, phytochrome A overexpressor; BOX, phytochrome B overexpressor. †A, phytochrome A; B, phytochrome B. ‡ +, WT; ++, enhanced; –, deficient.

gene expression throughout the plant life cycle. In particular, the capacity to distinguish between individual members of the photoreceptor family affords the opportunity to monitor for potential differences in expression patterns that might have functional implications, and to begin to dissect the molecular basis of transcriptional regulation of these genes.

4.2.4.1 *Contrasting regulation of* phyA, phyB, *and* phyC *genes*

With gene-specific probes it has been shown that the *phyA*, *phyB*, and *phyC* genes of *Arabidopsis* are differentially expressed, both quantitatively and qualitatively (Fig. 8). The *phyA* transcript is the most abundant species in etiolated seedlings and is down-regulated in continuous W. By contrast, the *phyB* and *phyC* mRNAs are present at relatively low levels in the dark and are not significantly regulated by light. This pattern of expression matches that observed at the protein level (Fig. 2). The predominance of the phytochrome A protein in dark-grown tissue is directly correlated with a high level of *phyA* mRNA. The light-induced decrease in the amount of *phyA* mRNA, superimposed on the decrease in photoreceptor protein resulting from rapid degradation of the Pfr form (Fig. 2), indicates that phytochrome A abundance is regulated at both the protein and mRNA levels in *Arabidopsis*. It is not possible to determine from these data, however, whether mRNA abundance is photoregulated at the transcriptional or post-transcriptional levels. It is also unclear which photoreceptor is responsible for this response since a R pulse is relatively ineffective compared to continuous W in this plant. The absence of significant light effects on *phyB* and *phyC* mRNA levels (Fig. 8) is consistent with the apparent constitutive steady state levels of the corresponding photoreceptor proteins (Fig. 2). Similar results have been obtained for the *phyA* and *phyB*

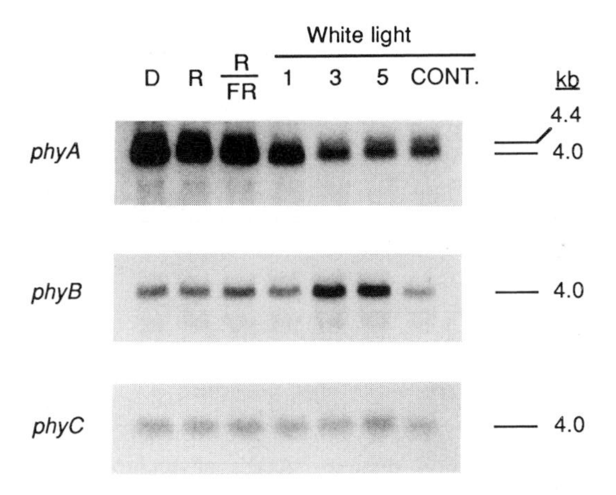

Figure 8. Blot hybridization analysis of the *phy* mRNAs present in total RNA from *Arabidopsis* seedlings. The RNA was isolated from seedlings that were grown for 5 days completely in the dark (D), given a pulse of red (R) or R followed by far-red light (R/FR) 3 h prior to harvest, transferred to white light for 1, 3, or 5 h prior to harvest, or grown for 5 days in continuous white light (CONT). Blots were probed with ssDNA transcript-specific probes for the *phyA*, *phyB*, and *phyC* mRNAs. From Sharrock and Quail (1989).

genes of the monocot rice indicating evolutionary conservation of this pattern of differential expression (Dehesh *et al.* 1991).

The availability of the promoter regions of *phyA*, *phyB*, and *phyC* genes enables the assembly of chimeric reporter constructs for exploring any differences in tissue-specific and developmentally-regulated patterns of expression, in addition to the light effects described above. Thus far, only one such study has been reported (Komeda *et al.* 1991). This study involved the creation of transgenic petunia plants expressing a chimeric gene consisting of a pea *phyA* promoter fused to the bacterial β-glucuronidase (GUS) gene. The levels of GUS expression in dark-grown seedlings were high in the apical region containing the hook, cotyledons and shoot apex, lower in a declining gradient down the hypocotyl, and barely detectable in the root (Fig. 9). This pattern of expression generally reflects that determined previously for the presumptive phytochrome A molecule in dicots using spectroscopic and immunochemical methods (Chapter 4.5). Growth of the transgenic seedlings in light suppressed GUS accumulation, indicating negative regulation of the pea *phyA* promoter in a manner qualitatively similar to that previously observed for the native pea *phyA* gene at the level of mRNA abundance. This result establishes that the pea *phyA* gene is regulated, at least in part, at the transcriptional level. Comparison with the behaviour of the analogous *phyB* and *phyC* chimeric constructs in the future will be of great interest.

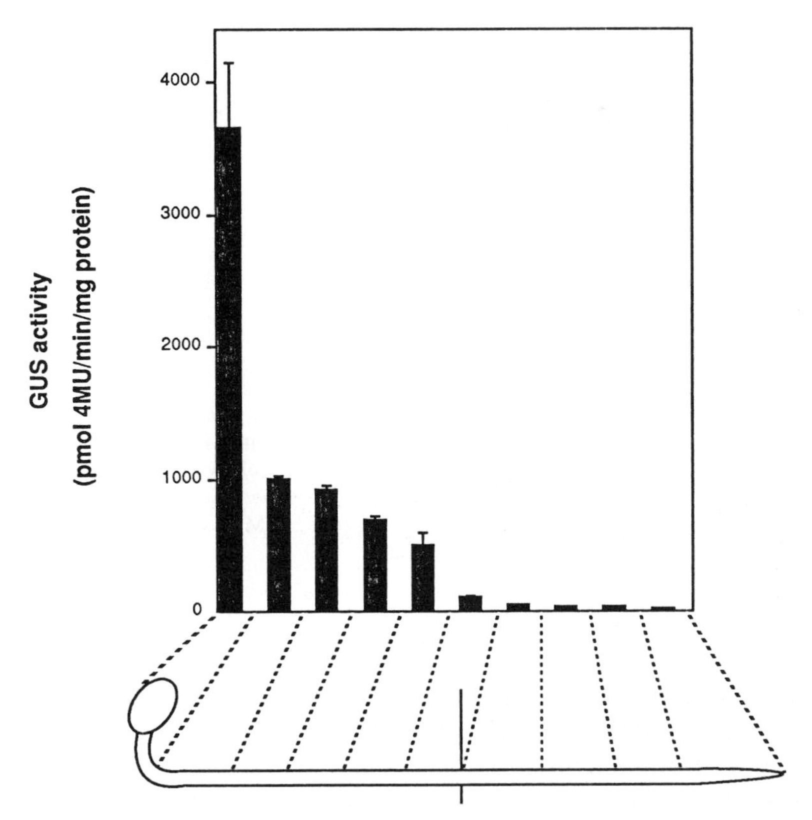

Figure 9. The distribution of β-glucuronidase (GUS) activity in transgenic petunia plants. The T2 plants of transgenic line 12-4 were grown in the dark for 10 days. Each seedling was cut into 0.5 cm segments from the top of the seedling (the apical hook region was expanded by hand) and segments were examined for GUS activity. The root started from the region 3.0 cm from the top. From Komeda *et al.* (1991).

4.2.4.2 Diversity in patterns of phyA *regulation between plant species*

Northern blot analysis of the effect of R pulses on phytochrome mRNA abundance in etiolated seedlings has revealed a considerable spectrum of quantitatively different responses among various plant species (Fig. 10). Although *phyA* coding-region probes were used for the analysis depicted here, it is likely that the data reflect primarily, if not exclusively, *phyA* mRNA levels, both because *phyA* transcripts predominate in etiolated tissue (Fig. 8) and probe cross-hybridization to other family members will be minimal under the conditions used. The data show that a R pulse induces a decline in *phyA* mRNA levels over the subsequent 3 h in darkness that ranges from very strong in barley and oats to relatively weak in *Arabidopsis* and undetectable in tomato. In general, the extent of this down regulation appears to be greatest in the monocots thus far tested and least in the dicots. This correlation is not absolute, however, as

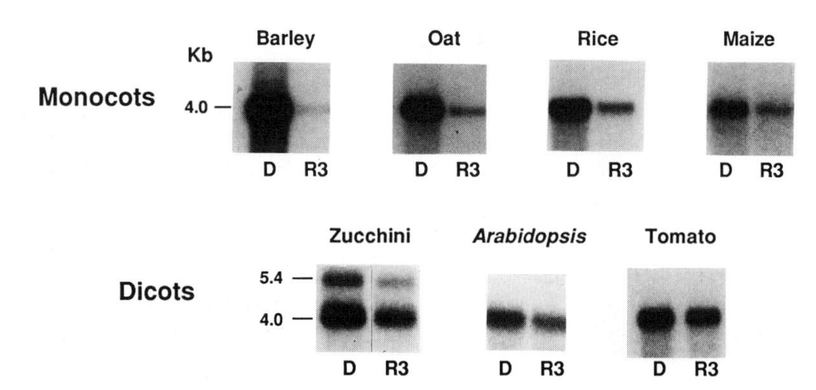

Figure 10. An RNA blot analysis of the effects of red light (R) on *phyA* mRNA abundance in various plant species. The RNA was isolated from dark-grown seedlings (D) or from dark-grown seedlings given saturating R pulses 3 h prior to extraction (R3). Following electrophoresis and transfer to nitrocellulose filters, the RNA was hybridized with either an oat *phyA* cDNA probe, pAP3.2 (barley, oat, rice, maize) or a zucchini cDNA probe, pFMD1 (zucchini, *Arabidopsis*, tomato). J.L. Lissemore and P.H. Quail, unpublished data.

decreases similar to those for maize have been reported, for example, for pea (Furuya 1989) and cucumber (Cotton *et al.* 1990). Moreover, the kinetics of the change in mRNA abundance and the photoreceptor responsible also appear to be variable. Usually, minimal *phyA* mRNA levels are approached within 3 h of irradiation indicating rapid turnover of the mRNA. However, whereas this minimal level is sustained for 16 h in pea (Furuya 1989) and in excess of 10 h for oats (Lissemore and Quail 1988), the decrease is sharply transient for cucumber where the minimum is at 2 h and re-accumulation is complete by 10 h (Cotton *et al.* 1990). In the species where R induced decreases occur, at least partial FR reversibility has been demonstrated, thus indicating phytochrome involvement. The need for continuous W to observe a decrease in *phyA* mRNA in *Arabidopsis* and tomato, on the other hand, raises the possibility of involvement of other photoreceptors, such as the blue/UV-A receptor.

Yet another level of complexity is observed for the *phyA* genes of pea and *Arabidopsis*, the only dicot genes to be characterized thus far. Each of these genes has three distinct transcription start sites (Fig. 4) thereby giving rise to three mRNAs of increasing size, designated mRNA1, mRNA2, and mRNA3, respectively. In pea, the best characterized case, mRNA1 is the most abundant transcript in etiolated seedlings, and its abundance decreases in response to light in a R/FR reversible fashion (Sato 1988; Tomizawa *et al.* 1989). Light has little or no effect on the lower abundance *phyA* mRNA1 and mRNA2 species. The appearance in response to W of a lower abundance 4.4 kb *Arabidopsis phyA* transcript concomitant with the decrease of the initially major 4.0 kb mRNA (Fig. 8) is consistent with there being a similar pattern of regulation in *Arabidopsis*. These results underscore the differences between monocot and dicot

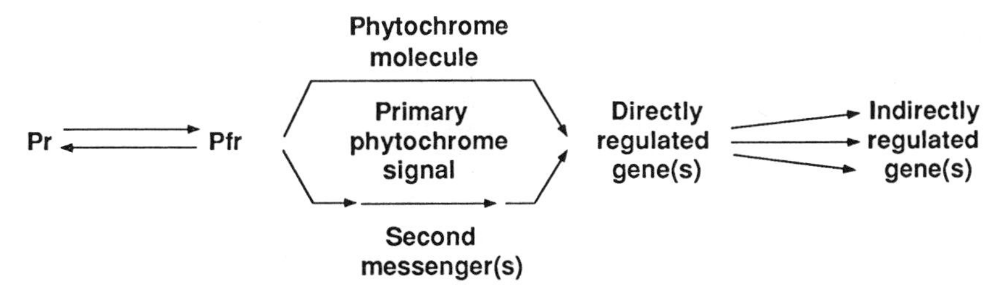

Figure 11. Scheme depicting the two formal classes of genes subject to phytochrome regulation; and two formal pathways of primary phytochrome signal transduction. From Lissemore and Quail (1988).

phyA genes and raise the possibility that the differences in promoter structure may be related to the apparent differences in photoregulation of expression.

4.2.4.3 Autoregulation of monocot phyA *genes as a model system*

There is now substantial, direct evidence for the fundamental notion that phytochrome regulates plant growth and development *via* differential gene expression (Gilmartin *et al.* 1990; Quail 1991). However, the molecular mechanism by which this regulation is implemented remains to be elucidated. As in all regulatory cascades, it is possible to define two formal classes of phytochrome-responsive genes: 'directly' regulated genes and 'indirectly' regulated genes (Fig. 11). 'Directly' here refers to those genes that respond to the primary phytochrome transduction signal, which in turn is defined as the signal transduced from photoreceptor to responsive gene without the intervention of altered expression of an intermediary regulatory gene as an integral step in the transduction process. Conversely, the term 'indirectly' here refers to those genes which do respond to phytochrome activation *via* altered expression of one or more intermediary regulatory genes in cascade fashion. There are also two formal classes of mechanisms by which the primary phytochrome signal might be transduced to the promoters of directly regulated genes (Fig. 11): the photoreceptor molecule itself could interact directly with the DNA or other components of the transcriptional machinery of these genes; or Pfr formation could trigger some form of second messenger cascade, the terminal member of which would interact with the promoters of responsive genes.

From the above considerations, it is clear that in order to understand the molecular mechanism of primary phytochrome signal transduction it is ultimately necessary to utilize directly regulated genes. The autoregulatory control that phytochrome exerts over the expression of its own *phyA* genes in monocots provides an excellent model system for this purpose. Figure 12 shows that a R pulse triggers a rapid decrease in oat *phyA* transcription that is detectable within

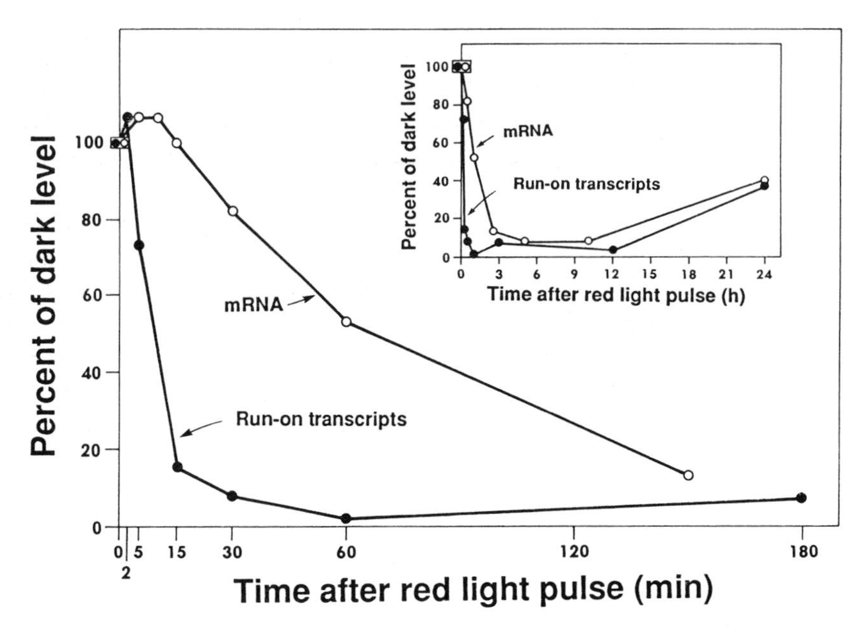

Figure 12. Time-course of the effect of a red-light (R) pulse on oat *phyA* gene transcription and *phyA* mRNA levels. Four-day-old dark-grown oat seedlings were given a 5 s saturating pulse of R and returned to darkness for the period indicated. Run-on transcription reactions were conducted with isolated nuclei and the labelled products were hybridized to an oat *phyA* DNA probe. From Lissemore and Quail (1988). RNA was isolated and subjected to Northern blot analysis using an oat *phyA* cDNA clone as a probe. From Colbert J.T., Hershey H.P. and Quail P.H. (1985) *Plant Mol. Biol.* 5: 91-102. Data are expressed as per cent of the level detected in dark control seedlings.

5 min of the irradiation. The level of *phyA* mRNA then declines with a half-life of ≈ 1 h. This decline is the result of the high intrinsic turnover rate of the mRNA as available evidence indicates that there is no direct photoregulation of the stability of these transcripts. The rate of transcription and mRNA abundance begin to increase again sometime after 10-12 h of darkness following the initial R pulse such that by 24 h about 40% of the original levels are reached (Fig. 12, inset). A pulse of FR alone also triggers a decrease in *phyA* transcription, with the initial kinetics and extent of the decline being indistinguishable from those following a R pulse (Fig. 13). However, the duration of transcriptional repression is much shorter following a FR pulse than a R pulse, whether or not the FR pulse is preceded by a R pulse. These results indicate that the initial repression of *phyA* transcription is saturated by a low level of Pfr (the ≤ 1% formed by saturating FR), but that the duration of repression is determined by the size of the initial Pfr pool established by the terminal light pulse. These data establish, therefore, that transcriptional repression of the oat *phyA* gene is imposed by a light-labile phytochrome, most likely phytochrome A. A schematic summary of these observations is presented in Fig. 14. Importantly, in

addition to being rapid, this repression of *phyA* transcription occurs unimpeded in the absence of significant new protein synthesis following Pfr formation (Lissemore and Quail 1988). This result indicates that all essential components of the signal transduction chain exist in the cell prior to light perception by the photoreceptor and that *phyA*, therefore, is a directly regulated gene by the criteria outlined in Fig. 11.

The identity of the molecular species that transduce the regulatory signal from phytochrome to *phyA* remains open, but all available data indicate that it is unlikely to be the photoreceptor molecule itself. Rigorously performed immunocytochemical and cell fractionation studies have established that phytochrome A is a soluble, cytoplasmically localized molecule that does not relocate to the nucleus upon Pfr formation (Chapter 4.5). In addition, no binding of the purified molecule to the promoter of the *phyA* gene is detected in either

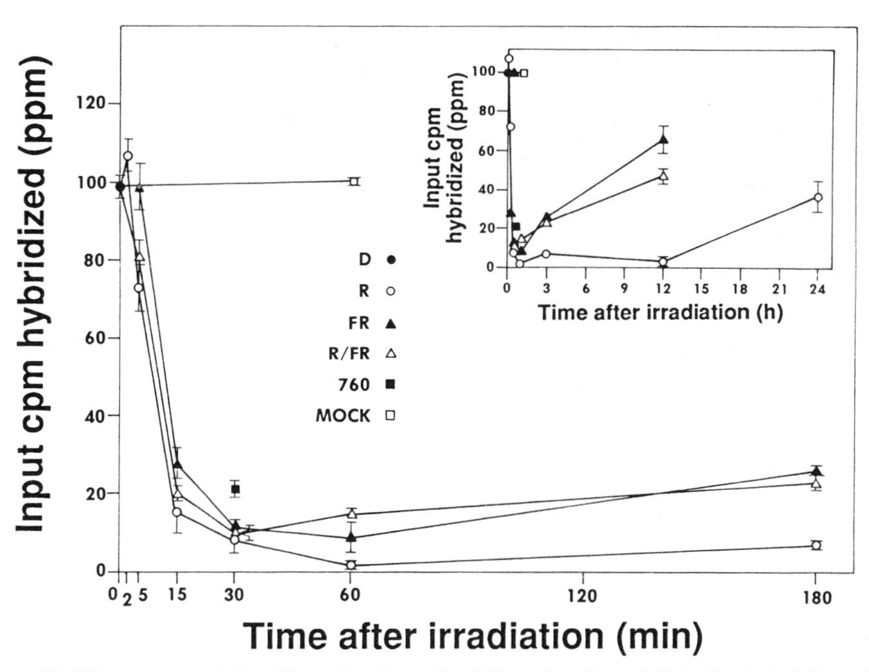

Figure 13. Time-course of the effect of pulses of red light (R), far-red light (FR), R followed by FR (R /FR), and 760 nm-light on transcription of *phyA* in dark-grown oat seedlings. Four-day-old dark-grown oat seedlings were given one of the following light treatments and returned to darkness: D, not irradiated; R, 5 s of R; FR, 5 s of FR; R/FR, 5 s of R followed immediately by 5 s of FR; 760, 20 s of 760 nm light; MOCK, mock irradiation. Mock-irradiated seedlings were briefly exposed to the overhead green safelight only, to simulate the conditions experienced by R-, FR-, and R/FR-irradiated plants, and were returned to darkness. Tissue was harvested at the indicated times after the light treatments, nuclei were isolated, and transcription reactions were conducted. ^{32}P-labelled run-on transcripts were hybridized to single-stranded M13mp19 containing sense and antisense inserts of oat *phyA* DNA, and the parts per million of input [^{32}P]RNA that hybridized specifically with the *phyA* antisense probe were determined. From Lissemore and Quail (1988).

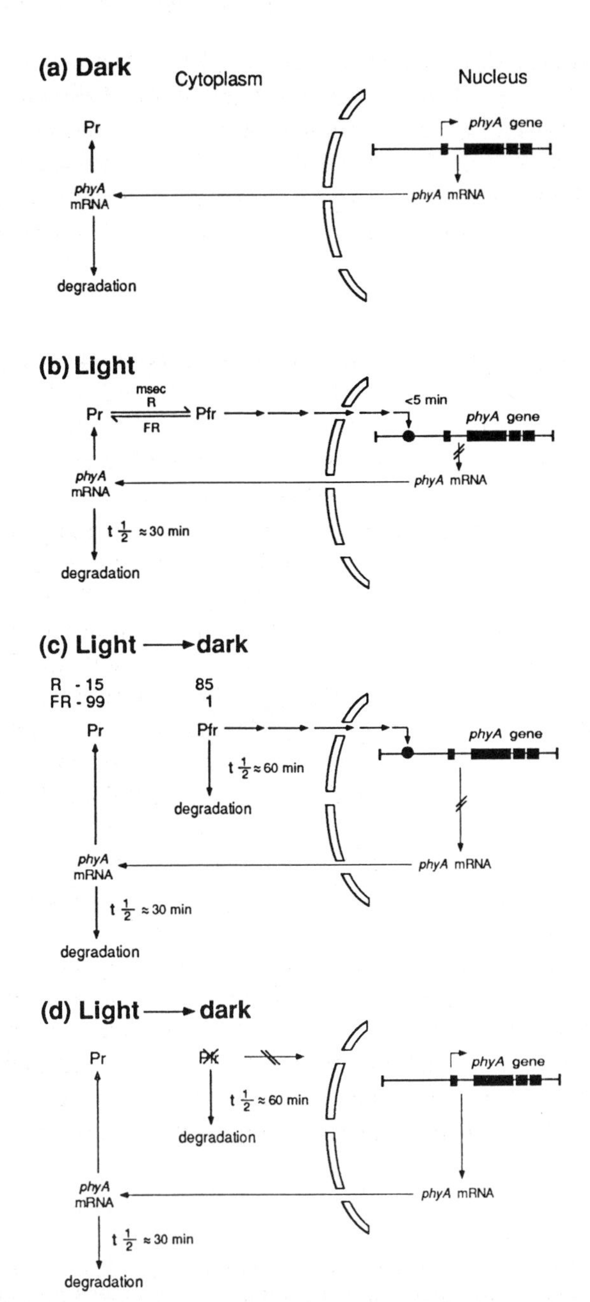

Figure 14. Schematic depiction of phytochrome regulated expression of its own *phyA* genes in monocots. (a) In dark-grown tissue *phyA* genes are actively transcribed, high steady-state mRNA levels result, and phytochrome A is synthesized in its inactive, relatively stable ($t_{½} \approx 100$ h) Pr form which accumulates in the cytoplasm. (b) Exposure to light triggers conversion in milliseconds to the active Pfr form which initiates transcriptional repression within 2-5 min by a mechanism that neither requires new protein synthesis nor detectable relocation of phytochrome

the Pr or Pfr forms when examined by *in vitro* DNA-binding assays (Quail 1991). Older data suggesting that addition of purified phytochrome A to isolated nuclei caused specific changes in run-on transcription *in vitro* have since been shown to be artifactual. Thus, a minimum of one cellular component in addition to the photoreceptor appears necessary to carry the signal from Pfr in the cytoplasm to the *phyA* promoter in the nucleus of the living cell (Figs. 11 and 14).

4.2.4.4 The control of phyA *expression by* cis-*acting elements and* trans-*acting factors*

One strategy aimed at defining the mechanism of signal transduction from phytochrome to nuclear genes is first to identify the phytochrome-responsive DNA sequence elements in target promoters and then the protein factors that bind specifically to these elements. This information can then be used in principle as a starting point to trace back sequentially through the interacting components of the signal-transduction chain to the photoreceptor. A combination of three principal procedures is generally used to define important sequence elements in a promoter. First, preliminary indications of potentially significant elements are frequently sought by performing computer searches for conserved sequences in other genes. Second, sequences which interact specifically with protein factors extracted from isolated nuclei are defined by a variety of *in vitro* DNA-binding assays. Third, promoter elements necessary and sufficient for the expression and regulation of the gene under study are identified by functional assay *in vivo* following transfer of chimeric promoter-reporter gene fusion constructs into living plant cells.

The results of such an analysis of the oat *phyA3* promoter are summarized schematically in Fig. 15 (Bruce *et al.* 1991). A set of sequences designated boxes I, II, and III and a pair of identical GT1-boxes were initially identified as being conserved in the *phyA* promoters of oat, rice, and maize. Subsequent DNase I protection ('footprint') analysis has defined a series of nine to ten sites within the first 400 bp upstream of the transcription start site that bind factors

---

into the nucleus. The *phyA* mRNA levels then decline with a $t_{1/2}$ of *ca.* 30 min. (c) When light treated tissue is returned to darkness, transcriptional repression is sustained until the residual pool of Pfr drops below some critical level as a result of the relatively rapid turnover of this form in the cell ($t_{1/2} \approx 60$ min). The duration of this repression is determined by the size of the residual pool of Pfr established by the final light treatment prior to darkness: red light (R) establishes a large pool (85 relative units of Pfr) that sustains repression for > 12 h, whereas far-red light (FR) establishes a small pool (1 relative unit of Pfr) leading to derepression in < 3 h. (d) In prolonged darkness Pfr drops below the critical level, high transcriptional activity resumes, and *phyA* mRNA and Pr re-accumulate. Modified from Quail (1991).

present in nuclear extracts. However, as no change in binding pattern is observed in response to prior R irradiation of dark-grown seedlings, there is no evidence from these data that any of these factors are involved in photo-responsiveness through changes in DNA-binding activity. A transient expression assay system involving direct microprojectile-mediated introduction of *phyA3* promoter constructs into the living cells of intact etiolated rice seedlings has permitted identification of the functionally active sequence elements in this promoter (Bruce *et al.* 1991). Initial 5'-deletion derivative analysis showed that 400 bp of promoter DNA contains all the sequence information necessary for maximal expression and Pfr-imposed repression. A combination of deletion and linker-substitution mutagenesis has further resolved the functionally active regions to three sequence elements designated PE1 (positive element-1), PE3 (positive element-3), and RE1 (repressor element-1) (Fig. 15). The data show that PE1 and PE3 act synergistically to support maximal expression under low Pfr (derepressed) conditions, as mutagenesis of either element decreases expression to basal levels. By contrast, mutagenesis of the RE1 element results in constitutively maximal expression under either high or low Pfr conditions. This establishes that RE1 and its presumptive associated repressor factor-1,

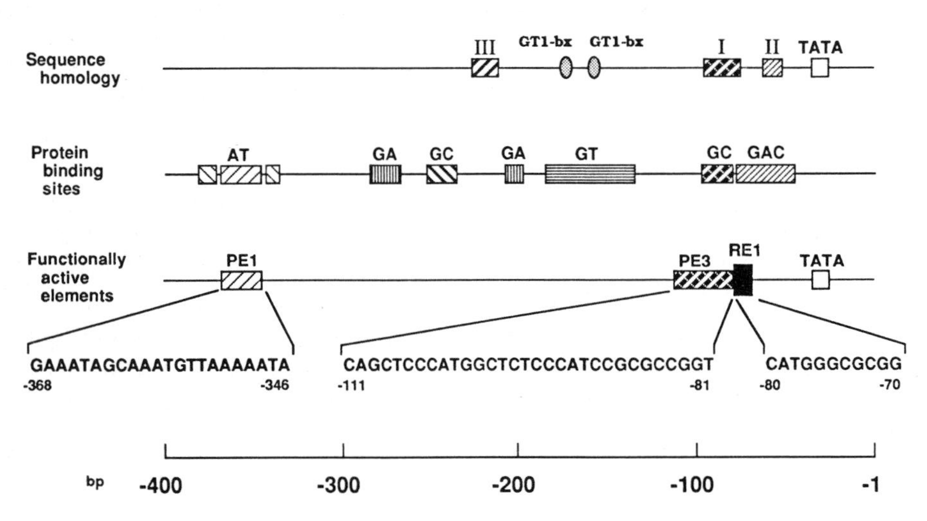

Figure 15. Schematic representation of sequence elements in the oat *phyA3* minimal promoter. *Top:* elements defined by sequence similarity between *phyA* promoters from oat, maize, and rice (Christensen A.H. and Quail P.H. (1989) *Gene* 85: 381-90; Kay S.A., Keith B., Shinozaki K., Chye M.-L. and Chua N.-H. (1989) *Plant Cell* 1: 351-360). *Middle:* sequences identified as protein binding sites *in vitro* by DNaseI protection assays using oat nuclear extracts. The designations above each element (AT, GA, GT, *etc.*) loosely denote the predominant nucleotides in the element. *Bottom:* elements identified by gene-transfer experiments in rice as being functionally active in the expression or light-regulation of the *phyA3* promoter. PE1: positive element-1; PE3: positive element-3; RE1: repressor element-1. From Bruce *et al.* (1991).

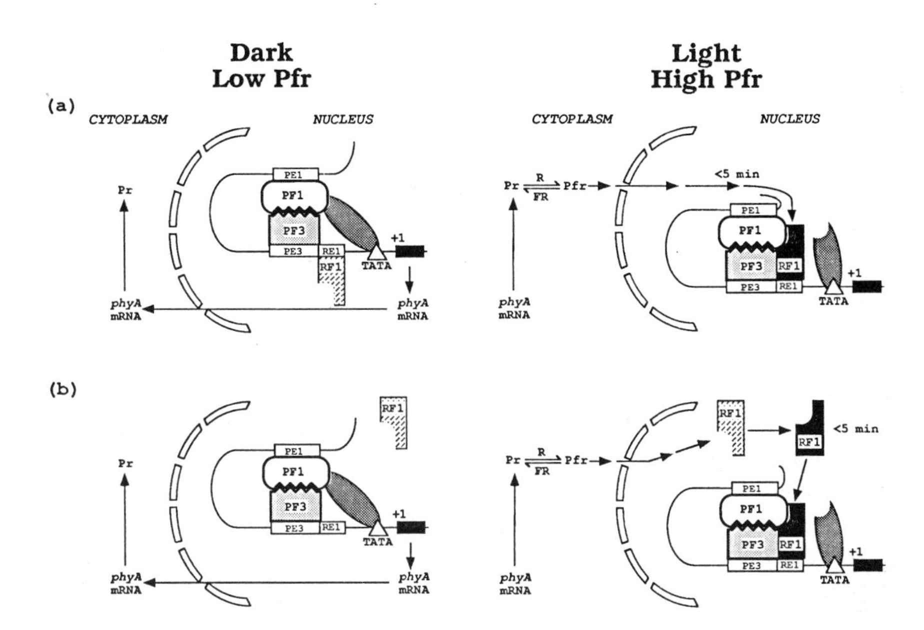

Figure 16. Schematic representation of two possible alternative mechanisms by which phytochrome might regulate expression of the oat *phyA3* gene. In both models, the positive factors PF1 and PF3 are constitutively bound to their respective cognate binding sites, PE1 and PE3, and cooperate to support high levels of transcription under low Pfr conditions (dark). (a) Repressor factor-1 (RF1) is bound to its target site, RE1, but under low (or zero) Pfr conditions does not negatively affect transcription. Red light (R)-induced generation of high Pfr levels causes a molecular change in RF1 that leads to suppression of the capacity of PF1 and PF3 to enhance transcription. (b) RF1 is not bound to RE1 under low Pfr conditions. Generation of high Pfr levels induces a change in RF1, enabling it to bind to its target site and suppress transcription. Other aspects of the two models are identical. The phytochrome molecule is known to be synthesized as Pr and appears to transmit its regulatory signal to target genes upon Pfr formation without relocating from cytoplasm to nucleus. Initial transcriptional repression occurs in less than 5 min from Pfr formation and is sustained as long as Pfr levels remain above a critical minimum. Derepression occurs in darkness when Pfr levels drop below this minimum as a result of rapid ($t_{1/2} \approx 60$ min), selective degradation of this active form of the photoreceptor. In the present models, the drop in Pfr levels in darkness would in turn cause reversal or loss of repressor activity by RF1. The transcription initiation complex is represented as the oval symbol bound to the TATA box. The transcription start site is indicated as +1. The direct contacts shown between the various factors are schematic and not intended to imply current evidence of such direct interactions. Modified from Bruce *et al.* (1991).

RF1, comprise the molecular switch at the end of the primary signal transduction pathway through which phytochrome operates to impose transcriptional repression on the *phyA3* gene. Conversely, the data also demonstrate that the PE1 and PE3 elements and their cognate positive factors, PF1 and PF3, are not direct targets of Pfr action.

Two alternative formal mechanisms by which the *phyA3* promoter might be regulated are depicted in the molecular models in Fig. 16. In both models, the positive factors PF1 and PF3 are constitutively bound to their respective cognate binding sites, PE1 and PE3, and co-operate to support high levels of transcription under low (or zero) Pfr conditions (dark). In model (a), repressor factor-1, RF1, is bound to its target site, RE1, but under low (or zero) Pfr conditions does not negatively affect transcription. The R-induced generation of high Pfr levels causes a molecular change in RF1 that leads to suppression of the capacity of PF1 and PF3 to enhance transcription. In model (b), RF1 is not bound to RE1 under low Pfr conditions and could be either cytoplasmically or nuclear localized. Generation of high Pfr levels induces a change in RF1, enabling it to bind to its target site and suppress transcription. Other aspects of the two models are identical and incorporate the overall features of *phyA3* regulation depicted in Fig. 14. The derepression that occurs in darkness when Pfr levels drop below the critical minimum necessary for repression (Figs. 12 and 13) would in the present models result from reversal or loss of repressor activity by RF1.

Sequence comparisons show that the PE3 and RE1 elements defined functionally in the oat *phyA3* promoter, have strongly conserved counterparts in the *phyA* promoters of rice and maize (Fig. 17). These sequences overlap the previously noted box I sequence (Fig. 15). This strong sequence similarity in the three monocot *phyA* promoters is highly suggestive of functional equivalence. By contrast, the rice and maize promoters contain no elements with high sequence similarity to the AT-rich, upstream oat PE1 element (Fig. 15). Instead, the rice *phyA* promoter contains a triplet of similar, but non-identical, GT-elements, designated GT1-box, GT2-box, and GT3-box (Figs. 18 and 19), that have been shown by gene transfer experiments to be functionally equivalent to PE1 (Dehesh *et al.* 1990). This result is highly significant because a pair of GT1-boxes in the oat *phyA3* promoter (Figs. 15 and 19) have been shown to be inactive in gene transfer experiments. Thus, the positively acting factor that interacts with components of the rice *phyA* GT-box complex to support high

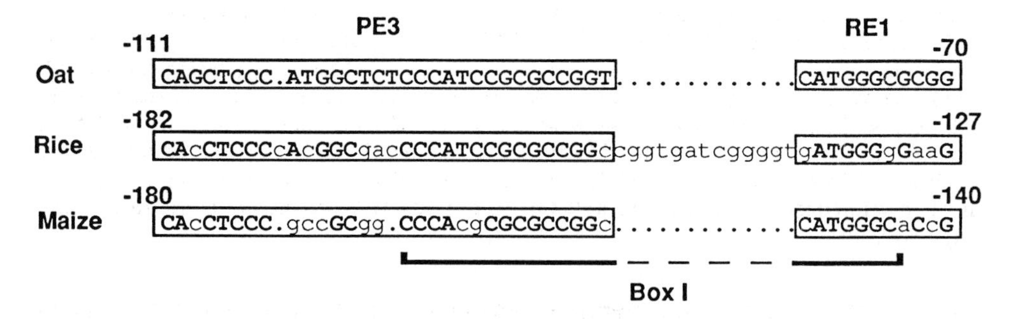

Figure 17. Sequence comparisons of PE3- and RE1-homologous regions in the promoters of the *phyA* genes from oat, rice, and maize. Box I refers to a previously identified, conserved sequence motif found in these three *phyA* promoters (see Fig. 15). Modified from Bruce *et al.* (1991).

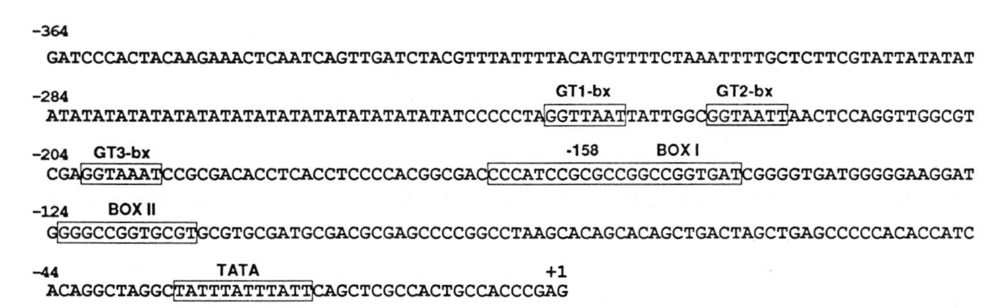

Figure 18. Nucleotide sequence of the rice *phyA* promoter region necessary for maximum expression (Dehesh *et al.* 1990). Sequence motifs of interest are boxed. The core motifs of the three related GT boxes are designated GT1-bx, GT2-bx, and GT3-bx. Box I and box II are conserved sequences present in the promoters of oat, maize, and rice *phyA* genes (Christensen A.H. and Quail P.H. (1989) *Gene* 85: 381-90; Kay S.A., Keith B., Shinozaki K., Chye M.-L. and Chua N.-H. (1989) *Plant Cell* 1: 351-360; Hershey H.P., Barker R.F., Idler K.B., Murray M.G. and Quail P.H. (1987) *Gene* 61: 339-348). TATA = TATA element. Transcription start site is designated as +1. From Dehesh *et al.* (1992).

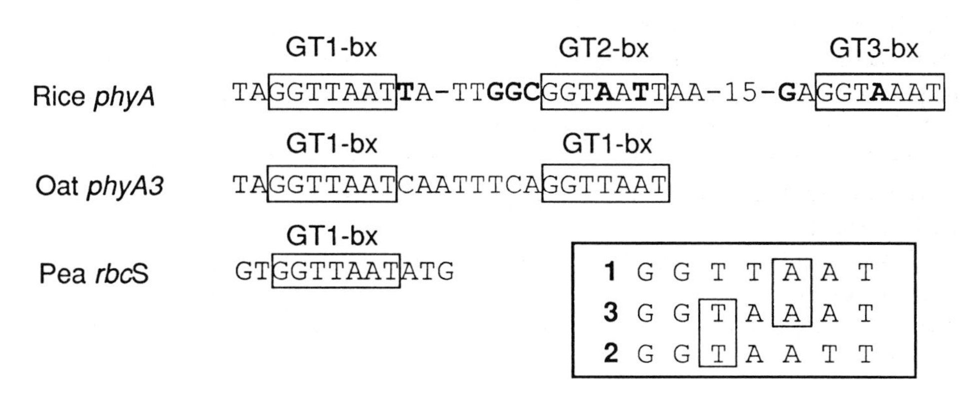

Figure 19. Sequence comparisons of GT-motifs found in the promoters of the rice *phyA* (Kay S.A., Keith B., Shinozaki K., Chye M.-L. and Chua N.-H. (1989) *Plant Cell* 1: 351-360), oat *phyA3* (Hershey H.P., Barker R.F., Idler K.B., Murray M.G. and Quail P.H. (1987) *Gene* 61: 339-348), and pea *rbc*S-3A genes (Gilmartin *et al.* 1990). The core motifs are boxed, and the three different motifs in the rice *phyA* promoter are designated GT1-bx, GT2-bx, and GT3-bx. Fifteen base pairs (15) have been omitted between GT2-bx and GT3-bx. The bold letters indicate nucleotide differences between the GT2-bx and GT3-bx of rice *phyA* and the oat *phyA3* sequence. Inset shows a direct comparison of the GT1-bx, GT2-bx, and GT3-bx core sequences from the rice *phyA* promoter.

rates of transcription is apparently incapable of recognizing the closely related oat *phyA3* GT1-box pair. Such a transcription factor has been detected in rice

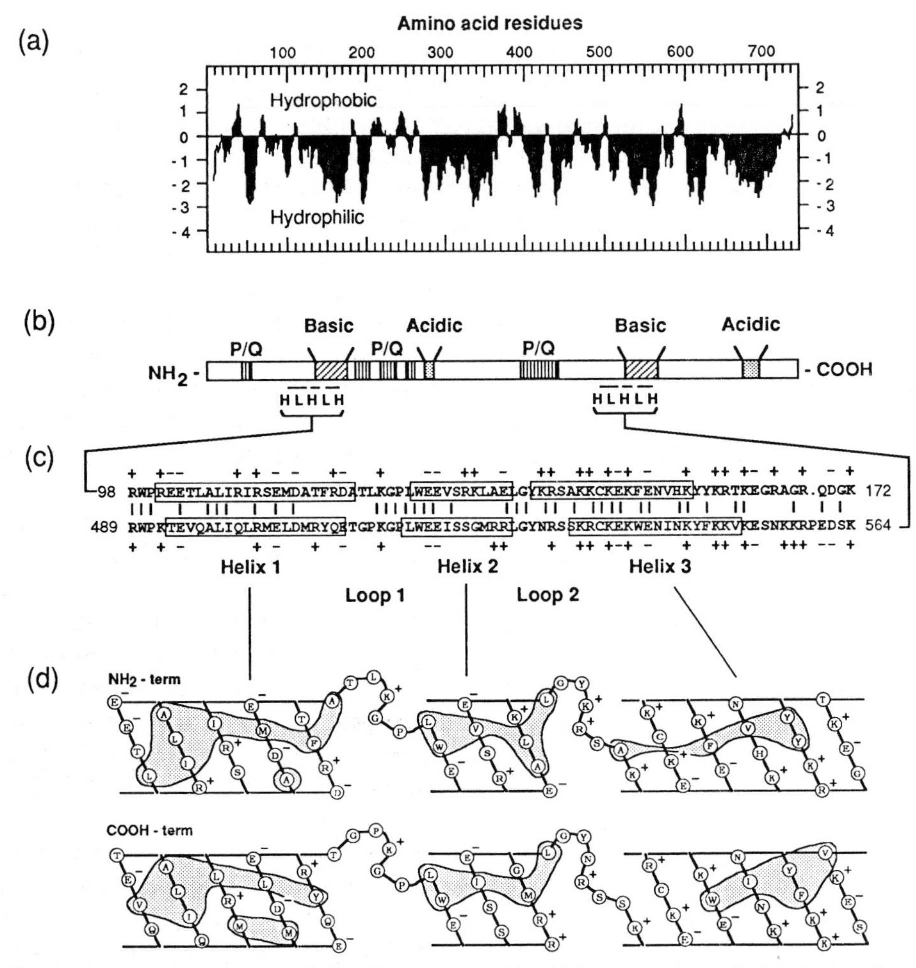

Figure 20. Structural features of rice GT-2 polypeptide. (a) Hydropathy analysis using a window of nine amino acids. (b) Schematic representation of major domains in the GT-2 polypeptide. Regions rich in basic, acidic, and proline (P) + glutamine (Q) residues are indicated. HLHLH (helix-loop-helix-loop-helix) = regions predicted to form amphipathic helices separated by a short segment of random coil (designated trihelix regions). (c) Amino-acid sequence comparison of the twin basic/trihelix regions present in NH_2-terminal (*top*) and COOH-terminal (*bottom*) halves of the polypeptide. Residue co-ordinates are at the ends of each segment. Identical residues are indicated by vertical bars. Boxed sequences indicate predicted helices 1, 2, and 3 separated by non-helical regions labelled loops 1 and 2. Positively (+) and negatively (–) charged residues are indicated. (d) Open helical net depiction of the trihelix region in the NH_2-terminal (*top*) and COOH-terminal (*bottom*) domains. Each helix is presented as the flattened outer surface of the helix cylinder generated by slitting the cylinder on one side parallel to the long axis and unrolling. For ease of comparison, the extent and locations of the helices in the NH_2-terminal trihelix region have been made identical to those in the COOH-terminal region, even though locations predicted by the algorithm, especially for helix 3, do not precisely coincide. Hydrophobic residues in the helices are enclosed in shaded zones. Charged residues are indicated. Single letter code for amino

nuclear extracts and has been molecularly cloned (Dehesh *et al.* 1992). This factor is called GT-2.

The principal structural features of the GT-2 protein are shown in Fig. 20. The polypeptide is 737 amino acids long (80 kD), is strongly hydrophillic (Fig. 20a), and exhibits a number of features commonly found in transcription factors. Among these are separate regions rich in acidic or proline + glutamine residues (Fig. 20b) shown in other proteins to function as transcriptional activation domains. The most striking feature, however, is a duplicated sequence of 75 amino acids present once in the NH_2-terminal half and once in the COOH-terminal half of the polypeptide (Fig. 20b,c). Each copy of this duplicated sequence is basic and is predicted to form three amphipathic α-helices separated from each other by two short loops (Fig. 20c,d). This structural unit, termed the trihelix motif, is highly likely to be involved in binding to the GT-box DNA target elements in the rice *phyA* promoter. This is because the two halves of the GT-2 polypeptide when expressed separately as individual domains have been shown to bind autonomously to these target elements (Dehesh *et al.* 1992). Moreover, whereas the NH_2-terminal domain has high affinity for the GT3-box, the COOH-terminal domain has high affinity for the GT2-box (Fig. 19). Neither domain has high affinity for the GT1-box. Together these data indicate that the GT-2 protein is a transcription factor with twin autonomous DNA-binding domains that can discriminate with high resolution between the three GT motifs. Thus, it appears that GT-2 functions as an efficient transcriptional activator of the rice *phyA* promoter because of its capacity to recognize the GT2- and GT3-boxes. Conversely, the inactivity of the pair of GT1-boxes in the oat *phyA3* promoter may be because GT-2 is unable to recognize this sequence element efficiently.

The observation that the downstream positive element PE3 is conserved between monocot *phyA* promoters (Fig. 17) whereas the upstream element is divergent (PE1 in oat; GT-boxes in rice; Figs. 15 and 18) implies that different upstream transcription factors (PF1 and GT-2) can synergize with the same or similar downstream factors (PF3) to activate transcription in these promoters. This suggestion is consistent with the apparent modular structure of many other eukaryotic promoters. Indeed, it has been proposed that almost any two *cis*-acting elements can synergize to activate transcription in the appropriate cellular context (Herbomel 1990). The key to understanding the autoregulation of *phyA* expression is to determine the mechanism by which the repressor factor RF1 abrogates the activity of the positive factors, and the molecular change in

---

acids: A, alanine; R, arginine; N asparagine; D, aspartic acid; B, asparagine + aspartic acid; C, cysteine; Q, glutamine; E, glutamic acid; Z, glutamine + glutamic acid; G, glycine; H, histidine; I, isoleucine; L, leucine; K, lysine; M, methionine; F, phenylalanine; P, proline; S, serine; T, threonine; W, tryptophan; Y, tyrosine; V, valine. From Dehesh *et al.* (1992).

RF1 induced by the Pfr-triggered signal transduction pathway that enables it to perform its negative regulatory function. The resolution of this question has implications beyond the *phy* gene system itself because transcriptional repression mechanisms are poorly understood in eukaryotic cells in general.

4.2.5 Conclusions

It has been recognized for some time that the complexity and diversity of responses attributed to phytochrome control are inconsistent with the action of a single molecular species of the photoreceptor. The realization that phytochrome is in fact a family of photoreceptors encoded by multiple, divergent genes has afforded the opportunity to explore whether the structural diversity, differential expression and/or potential differences in spatial location of the different family members can account for the diversity of functional roles attributed to the photoreceptor. The initial striking evidence that phytochromes A and B have discrete physiological roles in regulating seedling development provides strong support for this notion (Parks and Quail, 1993; Somers *et al.* 1991). The isolation of additional photomorphogenic mutants deficient in specific phytochromes and the expression of antisense constructs representing individual family members in transgenic plants is expected to provide greater insight into this issue in the near future. Comparisons among the currently available phytochrome sequences have identified evolutionarily conserved domains and residues which are likely critical to structure and function. However, neither these comparisons nor those of phytochrome with other proteins in the databases have yet led to elucidation of the photoreceptor's molecular mechanism of action, at least in angiosperms. On the other hand, provocative proposals arising from sequence similarities, such as that suggesting that phytochrome may function as a homolog of the bacterial sensory protein kinases (Schneider-Poetsch *et al.* 1991), are subject to functional testing using *in vitro* mutagenesis coupled to transgenic plant technology (Boylan and Quail 1991; Wagner *et al.* 1991). Similarly, the recent report that a novel *Ceratodon* phytochrome contains a COOH-terminal domain homologous to eukaryotic protein kinases (Thümmler *et al.* 1992) raises the possibility that the phytochrome family has evolved in modular fashion in a variety of ways in different plants such that the chromophore-bearing, photoresponsive domain is spliced to a diversity of COOH-terminal domains with a spectrum of regulatory mechanisms. The identification of a single *cis*-acting element, RE1, in the oat *phyA3* promoter that functions to repress transcriptional activity in less than 5 min in response to Pfr (Bruce *et al.* 1991) furnishes the opportunity to identify the *trans*-acting factor that binds to this element. This factor, RF1, is a component of the molecular switch that is the target at the terminus of the phytochrome signal transduction chain that controls *phyA* transcription. The cloning of RF1 will provide the opportunity for identification of other upstream primary transduction pathway components that link the photoreceptor with the transcriptional machinery.

Acknowledgements. I am grateful to the many colleagues who contributed to the work from this laboratory cited here, to Jim Tepperman for figure preparation and to Ron A. Wells for preparing and editing this manuscript. Our research is supported by National Institutes of Health grant no. GM47475; Department of Energy grant no. PR03-92ER13742; USDA National Research Initiative Competitive Grants Program grant no. 92-37301-7678; National Science Foundation grant no. MCB 9220161; and USDA/ARS CRIS grant no. 5335-21000-006-00D.

4.2.6 Further reading

Furuya M. ed. (1987) *Phytochrome and Photoregulation in Plants.* Academic Press, New York, Tokyo.

Kendrick R.E. and Nagatani A. (1991) Phytochrome mutants. *Plant J.* 1: 133-139.

Quail P.H. (1991) Phytochrome: A light-activated molecular switch that regulates plant gene expression. *Annu. Rev. Genet.* 25: 389-409.

Quail P.H., Hershey H.P., Idler K.B., Sharrock R.A., Christensen A.H., Parks B.M., Somers D., Tepperman J., Bruce W.B. and Dehesh K. (1991) *phy*-gene structure, evolution, and expression. In: *Phytochrome Properties and Biological Action*, pp. 13-38, Thomas B. and Johnson C. (eds.) Springer-Verlag, Berlin.

Rüdiger W. and Thümmler F. (1991) Phytochrome, the visual pigment of plants. *Angew. Chem. Int. Ed. Engl.* 30: 1216-1228.

Thomas B. and Johnson C. eds. (1991) *Phytochrome Properties and Biological Action*, Springer-Verlag, Berlin.

4.2.7 References

Boylan M.T. and P.H. Quail. (1991) Phytochrome A overexpression inhibits hypocotyl elongation in transgenic *Arabidopsis*. *Proc. Natl. Acad. Sci. USA* 88: 10806-10810.

Bruce W.B., Deng X.-W. and Quail P.H. (1991) A negatively acting DNA sequence element mediates phytochrome-directed repression of *phyA* gene transcription. *EMBO J.* 10: 3015-3024.

Chang C., Bowman J.L., DeJohn A.W. and Lander E.S. (1988) Restriction fragment length polymorphism linkage map for *Arabidopsis thaliana*. *Proc. Natl. Acad. Sci. USA* 85: 6856-6860.

Chory J. (1991) Light signals in leaf and chloroplast development: photoreceptors and downstream responses in search of a transduction pathway. *New Biologist* 3: 538-548.

Cotton J.L.S., Ross C.W., Byrne D.H. and Colbert J.T. (1990) Down-regulation of phytochrome mRNA abundance by red light and benzyladenine in etiolated cucumber cotyledons. *Plant Mol. Biol.* 14: 707-714.

Dehesh K., Bruce W.B. and Quail P.H. (1990) A trans-acting factor that binds to a GT-motif in a phytochrome gene promoter. *Science* 250: 1397-9.

Dehesh K., Tepperman J., Christensen A.H. and Quail P.H. (1991) *phyB* is evolutionarily conserved and constitutively expressed in rice-seedling shoots. *Mol. Gen. Genetics* 225: 305-313.

Dehesh K., Hung H., Tepperman J.M. and Quail P.H. (1992) GT-2: A transcription factor with twin autonomous DNA-binding domains of closely related but different target sequence specificity. *EMBO J.* 11: 4131-4144.

Furuya M. (1989) Molecular properties and biogenesis of phytochrome I and II. *Adv. Biophys.* 25: 133-67.

Gilmartin P.M., Sarokin L., Memelink J. and Chua N.-H. (1990) Molecular light switches for plant genes. *Plant Cell* 2: 369-78.

Hanelt S., Braun B., Marx S. and Schneider-Poetsch H. (1992) Phytochrome evolution: a phylogenetic tree with the first complete sequence of phytochrome A of a cryptogamic plant (*Selaginella martensii* Spring). *Photochem. Photobiol.* 56: 751-758.

Herbomel P. (1990) Synergistic activation of eukaryotic transcription: the multiacceptor target hypothesis. *New Biologist* 2: 1063-1070.

Heyer A. and Gatz C. (1992) Isolation and characterization of a cDNA-clone encoding for potato type B phytochrome. *Plant Mol. Biol.* 20: 589-600.

Komeda Y., Yamashita H., Sato N., Tsukaya H. and Naito S. (1991) Regulated expression of a gene-fusion product derived from the gene for phytochrome I from *Pisum sativum* and the *uidA* gene from *E. coli* in transgenic *Petunia hybrida*. *Plant Cell Physiol.* 32: 737-743.

López-Juez E., Nagatani A., Tomizawa K.-I., Deak M., Kern R., Kendrick R.E. and Furuya M. (1992) The cucumber long hypocotyl mutant lacks a light-stable phyB-like phytochrome. *Plant Cell* 4: 241-251.

Lissemore J.L. and Quail P.H. (1988) Rapid transcriptional regulation by phytochrome of the genes for phytochrome and chlorophyll a/b-binding protein in *Avena sativa*. *Mol. Cell. Biol.* 8: 4840-4850.

McCormac A.C., Wagner D., Boylan M.T., Quail P.H., Smith H. and Whitelam G.C. (1993) Photoresponses of transgenic *Arabidopsis* seedlings expressing introduced phytochrome B-encoding cDNAs: Evidence that phytochrome A and phytochrome B have distinct photoregulatory functions. *Plant J.*, in press.

Parks B.M. and Quail P.H. (1991) Phytochrome-deficient *hy*1 and *hy*2 long hypocotyl mutants of *Arabidopsis* are defective in phytochrome chromophore biosynthesis. *Plant Cell* 3: 1177-1186.

Parks B.M. and Quail P.H. (1993) *hy*8, a new class of *Arabidopsis* long hypocotyl mutants deficient in functional phytochrome A. *Plant Cell* 5: 39-48.

Reed J.W., Nagpal P., Poole D.S., Furuya M. and Chory J. (1993) Mutations in the gene for the red/far-red light receptor phytochrome B alter cell elongation and physiological responses throughout *Arabidopsis* development. *Plant Cell* 5: in press.

Sato N. (1988) Nucleotide sequence and expression of the phytochrome gene in *Pisum sativum*: Differential regulation by light of multiple transcripts. *Plant Mol. Biol.* 11: 697-710.

Schneider-Poetsch H.A.W., Sensen C. and Hanelt S. (1991) Are bacterial sensory systems models for phytochrome action? Hydrophobic cluster analysis of the phytochrome module related to bacterial transmitter modules. *Z. Naturforsch.* 46c: 750-758.

Sharrock R.A. and Quail P.H. (1989) Novel phytochrome sequences in *Arabidopsis thaliana*: Structure, evolution, and differential expression of a plant regulatory photoreceptor family. *Genes Develop.* 3: 1745-57.

Smith H. and Whitelam G.C. (1990) Phytochrome, a family of photoreceptors with multiple physiological roles. *Plant Cell Environ.* 13: 695-707.

Somers D.E., Sharrock R.A., Tepperman J.M. and Quail P.H. (1991) The *hy*3 long hypocotyl mutant of *Arabidopsis* is deficient in phytochrome B. *Plant Cell* 3: 1263-1274.

Thümmler F., Dufner M., Kreisl P. and Dittrich P. (1992) Molecular cloning of a novel phytochrome gene of the moss *Ceratodon purpureus* which encodes a putative light-regulated protein kinase. *Plant Mol. Biol.* 20: 1003-1017.

Tomizawa K.-I., Sato N., Furuya M. (1989) Phytochrome control of multiple transcripts of the phytochrome gene in *Pisum sativum*. *Plant Mol. Biol.* 12: 295-299.

Wagner D., Tepperman J.M. and Quail P.H. (1991) Overexpression of phytochrome B induces a short hypocotyl phenotype in transgenic Arabidopsis. *Plant Cell* 3: 1275-1288.

Whitelam G.C., McCormac A.C., Boylan M.T. and Quail P.H. (1992) Photoresponses of *Arabidopsis* seedlings expressing an introduced oat *phyA* cDNA: persistence of etiolated plant type responses in light-grown plants. *Photochem. Photobiol.* 56: 617-622.

Winands A., Wagner G., Marx S. and Schneider-Poetsch H.A.W. (1992) Partial nucleotide sequence of phytochrome from the zygnematophycean green alga *Mougeotia*. *Photochem. Photobiol.* 56: 765-770.

4.2.8 Appendix

Table 1. Alternative phytochrome nomenclature*.

Molecular entity	Designation	Description
Wild-type genes	*PHYA*	Sub-family A
	PHYA1	Member number 1 of sub-family A where there are multiple closely related wild-type sequences of this sub-family at multiple loci in a single genome (*e.g.*, oats where there are at least four identified *PHYA* sequences that are 98% identical).
	PHYB	Sub-family B, *etc.*
Mutant genes (alleles)	*phyA-1*	Mutant allele number 1 at the A locus
	phyB1-1	Mutant allele number 1 at the B1 locus, *etc.*
Transcripts	*PHYA* mRNA	Mature transcript encoded by the *PHYA* gene, *etc.*
Apoprotein	*PHYA* apoprotein or apophytochrome A or PHYA	Polypeptide encoded by the *PHYA* gene, *etc.*
Holoprotein	phytochrome A or phyA	Fully assembled chromoprotein (holoprotein) with chromophore covalently attached to *PHYA* apoprotein, *etc.*
Photochemical forms	PrA	Red light-absorbing form of phytochrome A
	PfrA	Far-red light-absorbing form of phytochrome A
	PrB	Red light-absorbing form of phytochrome B, *etc.*

*Based on the *Arabidopsis* gene nomenclature system (which in turn is derived from the yeast system).

4.3 Assembly and properties of holophytochrome

Masaki Furuya[1] and Pill-Soon Song[2]

[1]*Advanced Research Laboratory, Hitachi Ltd., Hatoyama, Saitama 350-03, Japan*
[2]*Department of Chemistry, University of Nebraska, Lincoln, NE 68588-0304, USA*

4.3.1 Introduction

As described in Chapter 4.2, phytochrome apoproteins (PHYs) are a family of 120-130 kD soluble proteins that result from the expression of diverse phytochrome genes (*phy*). A linear tetrapyrrole chromophore (see Chapter 4.1) is covalently attached to a cysteinyl residue of each PHY, forming a holophytochrome monomer. The native holophytochrome is always found as a dimer of the holophytochrome in extracts from plant tissues and shows spectrally photoreversible forms, the red light (R)-absorbing form (Pr) and the far-red light (FR)-absorbing form (Pfr). Neither PHY nor the chromophore alone show any photoreversible functions, and only spectrally active phytochrome (holophytochrome) is biologically active. Hence, it has been one of the major open questions in phytochrome studies as to what molecular mechanism results in photoreversible conformation changes between Pr and Pfr upon environmental light exposure. To answer this question, we will have to experimentally modify the chemical structure of either PHY or the chromophore. Until recently, it has been very difficult to carry out such experiments with phytochrome samples isolated and purified from plants, because of the covalent binding of the chromophore with PHY. However, Cornejo *et al.* (1992) have described a simple procedure for preparing phytochromobilin from photoerythrobilin.

As far as PHY preparation is concerned, the modern techniques of molecular biology and gene engineering provide us with new approaches to solve this problem. Namely, we can transform intact or mutated *phy* genes into bacteria, yeast or higher plants, and express them to produce either intact or recombinant PHYs, and can also assemble *in vitro* the expressed PHY to a native or chemically modified tetrapyrrole chromophore to produce R/FR reversible holophyto-

R.E. Kendrick & G.H.M. Kronenberg (eds.), Photomorphogenesis in Plants - 2nd Edition

chrome. This is a really useful tool for the structure/function analysis of phytochrome molecules.

Since the discovery of phytochrome (Butler *et al.* 1959), physical and chemical properties of phytochrome molecules have been well studied by various techniques (see Chapter 3). The major findings on molecular properties of holophytochrome are reviewed in this chapter.

4.3.2 Biogenesis of phytochrome apoprotein and chromophore

In the early days of phytochrome studies, phytochrome was believed to be synthesized and accumulated in the Pr form in plant tissues grown in darkness. When the dark-grown tissues are exposed to light, the resulting Pfr rapidly disappears through the processes of Pfr decay [destruction (Chapter 4.4)] and/or Pfr reversion.This story, however, is only true for type I phytochrome (PI) such as phytochrome A, because type II phytochrome (PII) such as phytochrome B and phytochrome C are now known to be constitutively produced, and are physiologically and spectrophotometrically quite stable in plant tissues irrespective of environmental light conditions (Furuya 1993). Hence we must always be very cautious when discussing phytochrome types (molecular species) in studies of phytochromes, although we do not yet know why there are different types of phytochrome biogenesis and stability in a single plant species (*vide infra*).

4.3.2.1 Biosynthesis of phytochrome apoproteins

The content of spectrophotometrically active phytochrome and its immunochemically detectable PHY can be determined with each molecular species in plant tissues (Konomi *et al.* 1987). For example, the content of phytochromes in embryonic axes of pea seeds increases during imbibition for 12 h in the dark, so that PHYA increases from 3.5 to 33 ng/mg fresh weight (fr wt), while PHYB remains more or less constant with a similar level of 8.3-10 ng/mg fr wt. When imbibed in the light, however, the amount of PHYA stays at the PHYB level. Probably, because the expression of *phyA* gene is down-regulated and holophytochrome A is labile in the light while that of *phyB* and *phyC* are constitutive and the products, holophytochrome B and C are photostable (see Chapter 4.2). This is generally true for holophytochromes *in vivo,* whereas apoproteins alone behave differently in chromophore-deficient mutants as discussed below.

Furthermore, it is also possible to overexpress cloned *phy* genes in transgenic higher plants such as tobacco (Kay *et al.* 1989; Keller *et al.* 1989), tomato (Boylan and Quail 1989), and *Arabidopsis* (Wagner *et al.* 1991), as well as its mutant forms (Stockhaus *et al.* 1992). In all these cases, overexpressed PHYs assemble to the chromophore in host plants and produce biologically and spectrophotometrically active phytochrome (see Furuya 1993 for review).

Considering that the amino-acid homology between PI and PII is usually 70-80% in tested plant species, some specific region of the amino-acid sequences (or, even a single amino acid site) may result in the difference found among the phytochrome types. To examine this problem, full-length or recombinant *phy* cDNAs are transformed and expressed in bacteria (Tomizawa *et al.* 1991) and yeast (Ito *et al.* 1991). This approach is particularly promising for apophytochrome studies, if the chromophore is covalently bound to such experimentally produced PHY either *in vitro* or *in vivo*. In this method, however, PHY in the absence of chromophore is very labile and tends to aggregate with itself, so that it is not so easy to isolate PHY from yeast or bacterial cells and purify it biochemically. Thus, several trials have been made to solubilize a large amount of PHY and then refold it (Furuya *et al.* 1991).

4.3.2.2 Biosynthetic pathway of phytochromobilin

The chemical structure of the phytochrome chromophore, phytochromobilin, was identified as a linear tetrapyrrole by Siegelman *et al.* (1966) and Rüdiger and Correll (1969). The biogenesis pathway of phytochromobilin has mainly been studied using metabolic inhibitors rather than tracer techniques except that ^{14}C-5-aminolevulinic acid (ALA) was incorporated into pea holophytochrome A (Bonner 1967). It should be noted that ALA is a mutual precursor for not only phytochromobilin but also haems and chlorophylls, and the amounts of the latter two are significantly larger than phytochromobilin.

Gabaculine (5-amino-1,3-cyclohexadienyl-carboxylic acid), a transaminase inhibitor, prevents the synthesis of the phytochrome chromophore, but not the biogenesis of phytochrome apoprotein in oat (Jones *et al.* 1986) and pea (Konomi and Furuya 1986). This is in contrast to the cases of haemoglobin, cytochrome c, and tryptophan pyrrolase, of which biosynthesis of each part is co-ordinately regulated. An inhibitor for the assembly of succinyl CoA to glycine, 4-amino-5-hexynoic acid, showed an inhibitory effect on phytochrome A synthesis (Elich and Lagarias 1989). However, in contrast, biliverdin IX, a metabolic product from haem, can overcome the inhibitory effects of these inhibitors, indicating that biliverdin IX may be one of the intermediates in the phytochromobilin synthesis.

Recently, both N-phenylimide S-23142, an inhibitor for protoporphyrin synthesis, and N-methyl mesoporphyrin IX, an inhibitor of Fe chelatase for haem, have been reported to decrease the amount of phytochrome in pea (Konomi *et al.* in press), suggesting that protoporphyrinogen IX, protoporphyrin IX and haem are the intermediates in phytochromobilin biogenesis. Experiments using metabolic inhibitors of the chromophore indicate the lack of co-ordinated regulation of chromophore and PHYA synthesis in dicots (Konomi and Furuya 1986) and monocots (Elich and Lagarias 1987).

4.3.3 Assembly of PHY with chromophores

A covalent linkage between the A ring of the chromophore and a cysteine residue of PHY was first shown with oat phytochrome A (Lagarias and Rapoport 1980). This was later confirmed by amino-acid sequencing of oat PHYA that was deduced from its cDNA sequence (Hershey *et al.* 1985), indicating that a single chromophore is attached at Cys_{321} of oat PHYA. Until recently, however, it was obscure whether: (i) the PHY of a plant can assemble only species-specifically with the chromophore of that plant, or PHY can assemble with a wide variety of chromophores from other sources, and (ii) PHY can assemble with a chromophore without any catalytic process(es) or requires one or many enzymes. The answers to these questions have been provided by recent research using transgenic techniques with recombinant *phy* genes and *in vitro* assembly experiments.

4.3.3.1 Assembly in wild-type and transgenic plants

In 1989, three research groups independently succeeded in expressing the full-length *phy* cDNA in transgenic plants. Namely, rice *phyA* cDNA was overexpressed in tobacco (Kay *et al.* 1989), and oat *phyA* cDNA in tobacco (Keller *et al.* 1989) and tomato (Boylan and Quail 1989). All of them found that the overexpressed monocot PHYs assembled *in vivo* with dicot phytochromobilin of the host plants. This technology was immediately extended using deleted or site-specific mutated *phy* cDNA to explore the mechanism of chromophore assembly *in vivo*.

If oat *phyA*, of which the Cys_{321} codon was point-mutated to a Ser codon, was overexpressed in homozygous transgenic *Arabidopsis*, this site-specific mutation at the chromophore attachment site resulted in a 124-kD PHYA that lacked photoreversibility (Boylan and Quail 1991), indicating that the Cys_{321} residue is crucial for chromophore assembly. On the other hand, when the first 10 Ser codons in N-terminus of rice *phyA* were changed to Ala codons and this mutant *phyA* was transferred in tobacco plants, the overexpressed rice PHYA assembled with the tobacco chromophore, as efficiently as wild-type (WT) rice PHYA transgenics (Stockhaus *et al.* 1992).

The question as to how long and which region of chromophoric domain of PHY is required for chromophore assembly can be answered by a similar transgenic method using deletion mutants of the *phy* gene. When a 118-kD oat *phyA* lacking codons 7-69 is overexpressed in tobacco, the resulting oat PHYA exists as a homodimer having covalently bound chromophore and shows R/FR reversible spectral changes, but does not induce any biological activity (Cherry *et al.* 1992). In contrast, when a rice *phyA* lacking codons 1-80 of the N-terminus is overexpressed in tobacco, the N-terminus deleted PHYA fails to assemble with the chromophore and shows no biological activity (J. Stockhaus and A. Nagatani unpublished data). Hence, a crucial region of PHYA for the assembly appears to lie somewhere between amino acid 69-80.

The studies on chromophore assembly using transgenic plants has recently started so that we must wait for some years before crucial conclusions can be drawn.

4.3.3.2 Assembly in vitro *with recombinant apophytochrome*

When oat seedlings were exogenously treated with non-natural bilitrienes such as biliverdin IXα and phycocyanobilin, PHY assembles with them resulting in a spectrally active, non-natural phytochrome (Elich *et al.* 1989) Therefore other bile pigments other than phytochromobilin can be the prosthetic group for phytochrome.

The first evidence for holophytochrome assembly *in vitro* was provided with PHY obtained from tetrapyrrole-deficient oat seedlings (Elich and Lagarias 1989) and then with PHY prepared by transcription and translation of oat *phyA* cDNA in a reticulocyte lysate system (Lagarias and Lagarias 1989). In both cases, the synthetic phytochromes were constructed with phycocyanobilin as an analog of the natural chromophore *in vitro* and the phycocyanobilin-adducted PHYAs exhibit a photoreversible absorption change. The results indicate that the PHYA itself has a structure that allows chromophore binding autocatalytically. This pioneering work, however, could not provide sufficient PHY for further analysis of the assembly process.

In contrast, the use of expression system in bacteria or yeast can provide us a large amount of both WT and recombinant PHYs. For example, an *E. Coli* transformant with a plasmid containing a full-length pea *phyA* cDNA produced 230 mg of PHYA in 1 g fresh weight of cells at 32°C, in which PHYA is immunochemically detected, not only in a typical inclusion body, but throughout the cytoplasmic region of the transformant (Tomizawa *et al.* 1991). Recombinant *phyA* cDNAs of oats (Wahleithner *et al.* 1991) and pea (Deforce *et al.* 1991) were also successfully expressed in yeast, in which there were no inclusion bodies but only 40 mg/g fresh weight of pea PHYA (Ito *et al.* 1991) and 28 mg/g fresh weight of oat PHYA (Wahleithner *et al.* 1991) were produced. All the PHYAs obtained were assembled *in vitro* with phycocyanobilin to produce a photoreversible phytochrome-like adduct, which shows repeated R/FR reversibility (Fig. 1).

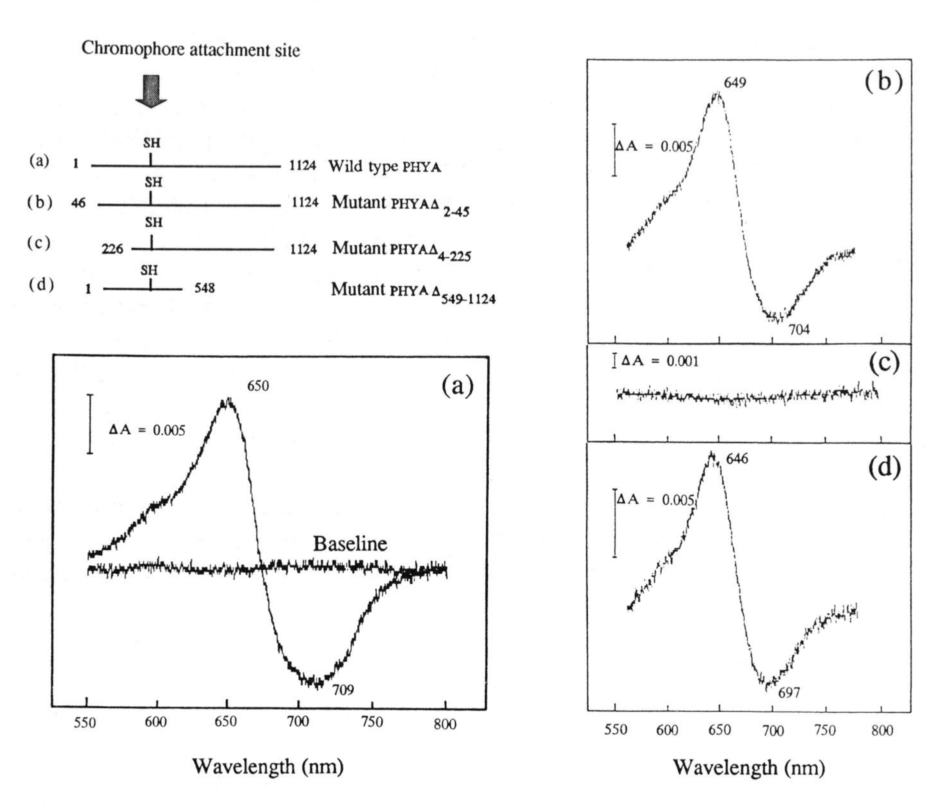

Figure 1. Difference in absorption spectrum (far-red *minus* red) of wild-type pea apophytochrome A (PHYA)-phycocyanobilin (PCB) adduct (modified from Deforce *et al.* 1991).

Truncation of the N-terminal tail of phytochrome A to residue 46 demonstrates that this region is not critical to bilin attachment, but when truncated to residue 80 and 225 fails to yield holophytochrome *in vitro* under the same conditions (Fig. 1). A mutant, comprising of a deletion of the C terminus to residue 548, showed bilin incorporation and R/FR photoreversibility, indicating that bilin-apophytochrome assembly still occurred even when the entire C-terminal domain was truncated (Fig. 1).

Although reconstitution experiments using bacteria- or yeast-produced PHY and phycocyanobilin have yielded a covalently bound adduct that undergoes photoreversible spectral changes, the difference spectrum differed from that of native phytochrome (Wahleithner *et al.* 1991; Deforce *et al.* 1991). Although it is difficult to isolate and purify enough phytochromobilin from higher plants for *in vitro* assembly experiments, methanolysis of the unicellular rhodophyte, *Porphyridium cruentum,* and the filamentous cyanobacterium, *Calothrix* sp., cells produce a ready supply of 2(R), 3(E)-phytochromobilin. When incubated in this recombinant oat PHYA yields a covalent adduct that shows spectral changes indistinguishable from native oat phytochrome A (Cornejo *et al.* 1992).

As these alga do not contain phytochromobilin, this appears to be derived from phycoerythrobilin-containing proteins during the methanolysis.

Using these *in vitro* assembly systems, we now can characterize the requirements for their assembly by modifying the structure of either the PHY or the bilin precursor. First, the linear tetrapyrrole chromophore analogues were compared for their ability to covalently bind to PHY and to show photoreversible absorbance changes (Li and Lagarias 1992). It became clear that: (i) ethylidene containing analogues such as 3(E)-phycocyanobilin readily form covalent adducts with PHYA, while chromophores lacking a C15 double bond between the rings C and D such as 3(E)-phycoerythrobilin are poor substrates for attachment. Furthermore, the phytochromobilin and phycocyanobilin-PHYA adducts show a photoreversible spectral change similar to the native phytochrome, while the phycoerythrobilin-PHYA adducts locked into a photochemically inactive, Pr-like conformation.

Very recently, T. Kunkel (pers. comm.) and colleagues have expressed a full-length tobacco *phyB* in yeast by the same technique as for *phyA*, and assembled it with phycocyanobilin, finding typical photoreversible spectral

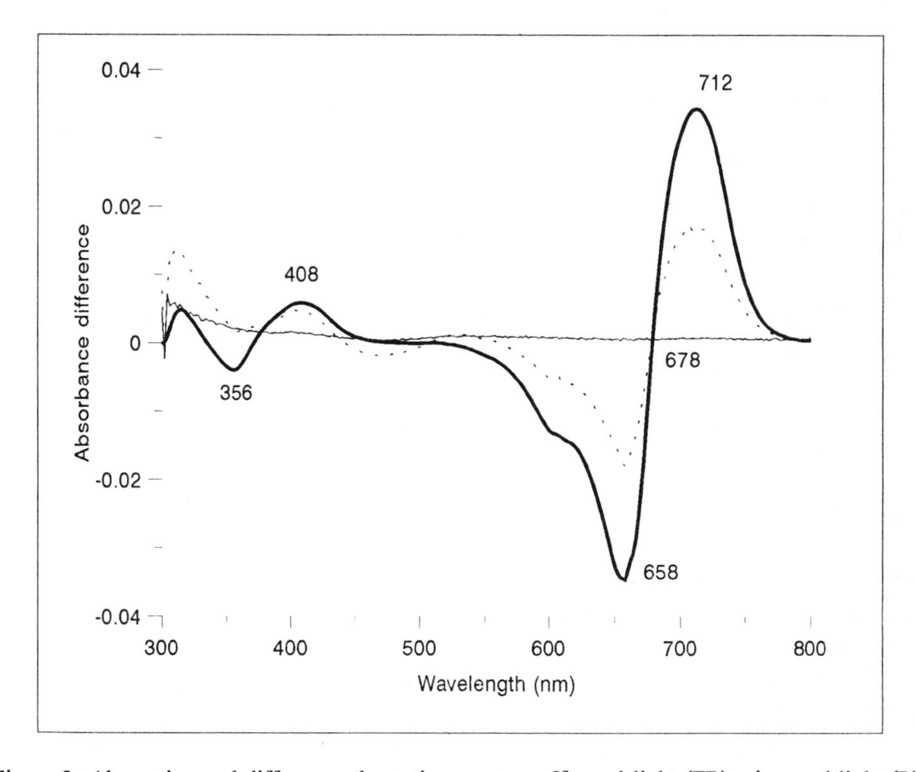

Figure 2. Absorption and difference absorption spectrum [far-red light (FR) *minus* red light (R)] of wild-type tobacco apophytochrome B (PHYB)-phycocyanobilin (PCB) adduct. Thick line, determined after irradiation with saturating R; thin line, taken after subsequent irradiation with saturating FR; dashed line, determined after second R irradiation (T. Kunkel unpublished data).

changes (Fig. 2). The PHYB-phycocyanin adduct has an absorption maxima of 658 and 356nm in the Pr form and those of 712 and 408 nm in the Pfr form. Under saturating concentration of phycocyanobilin the kinetic analysis of the *in vitro* assembly with PHYB revealed a pseudo first-order rate constant of $2.8 \times 10^2\ s^{-1}$, which is very similar to that for PHYA-phycocyanobilin assembly (Li and Lagarias 1992). Hence it is likely that both PHYA and PHYB assemble with the chromophore in a very similar manner.

4.3.3.3 Failure of chromophore assembly in mutants

Among phytochrome-related mutants, different types of chromophore assembly-related mutants have been characterized in higher plants. If some step(s) is defective in phytochromobilin biosynthesis, such a plant may cause a global functional deficiency in all molecular species of phytochrome. Examples of this are the *hy1* and *hy2* mutants of *Arabidopsis* (Parks and Quail 1991). When *hy1* and *hy2* seedlings are grown in white light on a medium containing biliverdin IXα or phycocyanobilin (Fig. 1), a phenotype is restored which is indistinguishable from the WT. Moreover, the phytochrome A which accumulates in the presence of biliverdin declines *in vivo* when the seedlings are transferred to white light. These results indicate that the biliverdin-dependent phytochrome A is photochemically functional and decays normally. However, it is still an open question whether PHYB and other apophytochromes lack their chromophore in *hy1* and *hy2* seedlings and whether exogenously applied chromophore can restore their biological activities.

In the tomato *aurea* mutant, PHYA is immunochemically detectable although reduced, but phytochrome A is not detectable spectrophotometrically, whereas PHYB and phytochrome B are more or less the same as in the WT (R.P. Sharma pers. comm.). Hence the mutant appears to be a chromophore-assembly mutant having its greatest effect on phytochrome A.

The data available in the literature which refers to the assembly of genetically engineered PHY to a chromophore *in vitro* and in transgenic plants (Fig. 3) are summarized as follows: (i) the non-chromophoric domain (C-terminal half of PHYA) is not required for chromophore assembly in both *in vitro* and in transgenic plants; (ii) recombinant monocot PHYA can assemble with phytochromobilin in all tested dicot transgenic plants and with phycocyanobilin prepared from algae; (iii) the N-terminal portion of up to 70 amino-acid residues of PHYA appears not to be crucial for the assembly, but the deletion of N-terminal 80 amino acids or more shows no assembly both *in vivo* and *in vitro;* (iv) biological activity of the transgenes has usually only been demonstrated when the full-length *phy* gene is overexpressed in transgenic plants (Chapter 4.8), and (v) PHYB can assemble with the chromophore in a similar manner to PHYA.

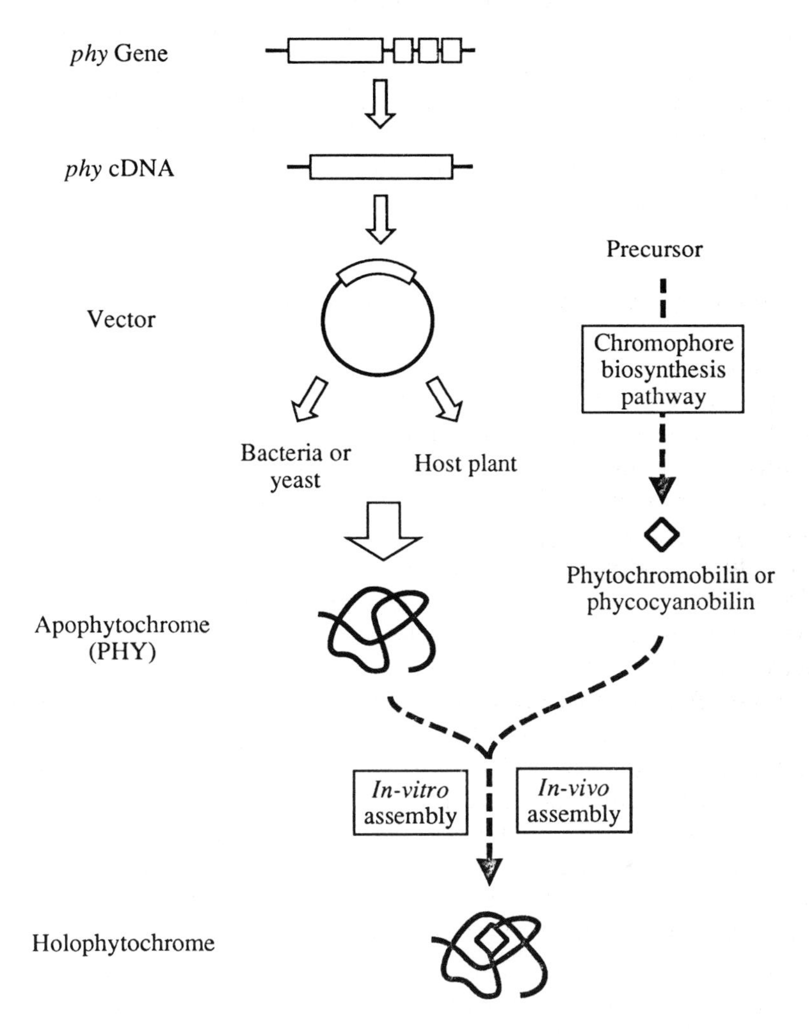

Figure 3. Schematic diagram of *in vivo* and *in vitro* assembly of apophytochrome (PHY) with its tetrapyrrole chromophore (A. Nagatani, original).

4.3.4 Physical properties of holophytochromes

4.3.4.1 The molecular structure and properties

The molecular mass of phytochromes from different plant species that have been characterized so far are in the range of 121-129 kD. Here, we describe primarily 124-kD *Avena* holophytochrome A (hereafter referred to as phytochrome A, unless specified otherwise), because this chromoprotein has been most extensively studied. The 121-kD *Pisum* phytochrome A will be referred to as pea phytochrome A.

The primary structures of several phytochrome species have been determined on the basis of the *phy* cDNA sequence (see Furuya 1989; Quail 1991; Romanowski and Song 1992 for reviews), the first one being that of oat phytochrome (Hershey *et al.* 1985). Grimm *et al.* (1988) also reported sequence determination of proteolytic fragments of 124-kD phytochrome, confirming the DNA sequence-derived primary structure of PHYA. The *Avena* PHYA consists of 46.8% polar amino-acid residues, with a mean hydropathy index of 1.06 (Hershey *et al.* 1985). The isoelectric point of phytochrome A is 5.9 (Vierstra and Quail 1982). So this protein is comparable to several well known soluble globular proteins in terms of percent polar amino-acid residues and mean hydropathy index.

The secondary structure of phytochrome A has been studied by several laboratories since the early days of phytochrome isolation [first phytochrome purified for circular dichroism (CD) study was a 60-kD proteolytic fragment; Anderson *et al.* (1970)] (Vierstra *et al.* 1987; Chai *et al.* 1987). Table 1 presents the secondary structure of phytochrome (Sommer and Song 1990). At the present time, CD is probably the only practical method available for determining the secondary structure of a large protein such as phytochrome in solution. We note two striking features of the secondary structure of phytochrome A (Table 1); extensive α-helical folding and apparent lack of ß-sheet conformation in the native holoproteins (both Pr and Pfr forms). In contrast, both the Chou-Fasman sequence-based prediction (Table 1) and the CD analysis of a

Table 1. Secondary structure (in percent) of etiolated *Avena* phytochrome in 20 m*M* Tris buffer, pH 7.8 (secondary structure, particularly α-helix, was found to be sensitive to the nature of buffer; Sommer and Song 1990) and apophytochrome A (PHYA) [Parker W. and Song P.-S. (1990) Location of helical regions in tetrapyrrole-containing proteins by a helical hydrophobic moment analysis. *J. Biol. Chem.* 265(29): 17568-17575].

Phytochrome form	α-Helix	ß-Sheet*	ß-Turn	Random coil
Pr	45.5	0.0	21.0	33.5
Pfr	50.5	0.0	17.5	32.0
PHYA	35.0	27.0	31.0	7.0

*Calculated from the CD spectra according to the method of Yang *et al.* [Yang J.T., Wu C.-S.C. and Martinez H.M. (1986) Calculation of protein conformation from circular dichroism. *Meth. Enzymol.* 130: 208-269]. Using the method of Manavalan and Johnson [Manavalan P. and Johnson W.C. Jr. (1987) Variable selection method improves the prediction of protein secondary structure from circular dichroism spectra. *Anal. Biochem.* 167: 76-85], the Pr CD spectrum yielded 47 ± 1% α-helix, 5 ± 2% anti-parallel ß-sheet, 1 ± 1% parallel ß-sheet, 20 ± 1% ß-turn, and 28 ± 2% random coil: The Pfr CD spectrum yielded 56 ± 2% α-helix, 2 ± 3% anti-parallel ß-sheet, 0% parallel ß-sheet, 22 ± 2% ß-turn and 20 ± 2% random coil. (D. Sommer unpublished data).

100-kD *Pisum* PHYA (Furuya *et al.* 1991) show significant amounts of ß-sheet conformation.

As the secondary structure of phytochrome A depends somewhat on the chromophore-PHY interactions, proteolytic patterns for the holo- and apoprotein of phytochrome A appear to show a noticeable effect of the chromophore on the protein conformation, in terms of pattern (for Pfr) and rate profile (for Pr) of proteolysis by subtilisin (Li and Lagarias 1992). This result is not surprising, because at least the secondary structure of the apoprotein is significantly different from that of native holophytochrome, according to CD and theoretical analyses (Table 1). The solubility of PHYA is also noticeably lower than that of the native protein, so it would seem reasonable to expect a significant difference in conformation/tertiary structure between PHYA and native phytochrome A.

Phytochrome exists primarily as a dimer (Lagarias and Mercurio 1985; Jones and Quail 1986). *bis*-Anilinonaphthalene 8-sulphonate (*bis*-ANS) monomerizes 124-kD phytochrome dimers and oligomers (Choi *et al.* 1990). The Stokes' radius of dimeric phytochrome ranges from 56 Å according to steric exclusion chromatography (Lagarias and Mercurio 1985; Jones and Quail 1986) to about 80 Å based on quasi-elastic light scattering (Sarkar *et al.* 1984). The reason for the discrepancy is unknown at this time. The relatively high concentration of phytochrome and ionic strength used may affect the molecular size determination of the protein, because phytochrome can form oligomers under such conditions (Choi *et al.* 1990). However, electron microscopy (EM) revealed a tripartite 'Y' shaped structure composed of three globular domains each having a radius of 35-40 Å for the *Avena* phytochrome (Jones and Erickson 1989). The distance between any two of the equilaterally spaced domains is estimated at 150 Å. Thus, the Stokes' radius of 80 Å is not unreasonable for a dimeric phytochrome species. The EM-derived molecular shape and dimensions of 114-kD pea phytochrome are similar to those of the oat phytochrome (Nakasako *et al.* 1990). Small-angle X-ray scattering yielded 53.8 Å for the radius of gyration for the 114-kD ('large') pea phytochrome.

Figure 4 shows the tertiary and quaternary/dimeric structure of the pea phytochrome A (Furuya 1989; Nakasako *et al.* 1990). The lower resolution model proposed for the *Avena* phytochrome A (Jones and Erickson 1989) is essentially in agreement with the small-angle X-ray-based model.

In the various models for dimeric phytochrome, it is generally accepted that dimerization contact occurs within the carboxyl terminal half of the monomer subunits. Based on partial proteolysis study, Jones and Quail (1986) identified residues 750 to the C-terminus as the subunit contact region. Yamamoto and Tokutomi (1989) also identified this region as containing dimerization sites (residues Phe_{744} and Gln_{1128}) for pea phytochrome. From a sequence-structure correlation study of several phytochromes, Romanowski and Song (1992) proposed a structural domain model for the dimerization region. The monomer

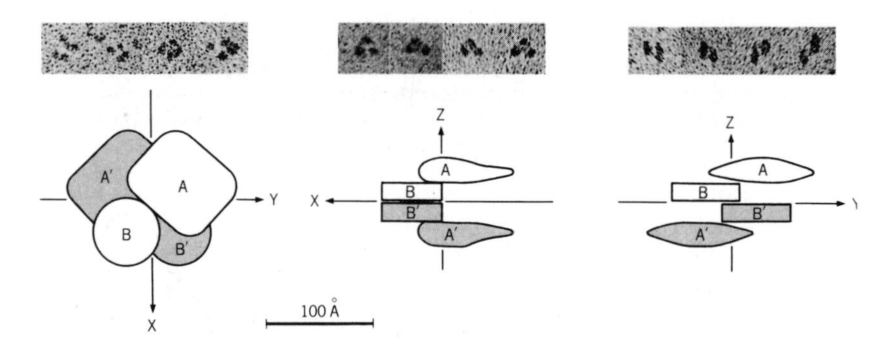

Figure 4. Three projection views of the 114-kD pea phytochrome dimer *below* and images by low-angle, rotary-shadowing electromicroscopy *above*. A: 59-kD chromophore domain. B: 55-kD C-terminal domain. (After Nakasako *et al.* 1990).

subunits interact by two antiparallel ß-strands, with charged residues providing the two-fold symmetry for the electrostatic interactions Val_{730} through Asn_{821}. However, the model can also be based on interactions between α-helical motifs.

ß-Sheet structures are often involved in dimerization (Chou 1989). If indeed the dimerization motif in phytochrome is ß-strands, its percentage must be relatively small because native phytochrome exhibits little ß-sheet conformation (Table 1). Figure 5 presents a domain structural model for phytochrome, including the dimerization motifs ('G'-domain; Parker *et al.* 1991). The G-domain interaction involves both hydrophobic and ionic interactions, with the latter providing the specific sequence match between the subunit G-domains.

It is possible that the monomerizing effect of *bis*-ANS (Choi *et al.* 1990) is attributable to its hydrophobic and electrostatic functional groups. This cartoon illustration is meant to highlight the dimerization domain, according to the data based on tryptic digestion and sequence autocorrelation analysis, but is not meant to rule out the possible involvement of other sequence segments (Edgerton and Jones 1992).

4.3.4.2 Spectroscopic characteristics

The absorption spectra of phytochrome Pr and phytochrome Pfr (Fig. 6), can be analyzed by assigning the R and FR absorbance bands to the S_0 to S_1 transition (A to Q_y in free-electron notations) at λ_{max} = 668 nm, ε = 132 mM^{-1} cm^{-1} (Lagarias *et al.* 1987) and 730 nm for Pr and Pfr, respectively. The second electronic transition, S_0 to S_2 or A to Q_x, is represented by a shoulder absorbance band at 608 nm and *ca.* 674 nm for Pr and Pfr, respectively (Fig. 5). The A to Q_y transition dipole lies along the long axis of the phytochrome chromophore molecule, with a near maximum degree of polarization (Song and Yamazaki

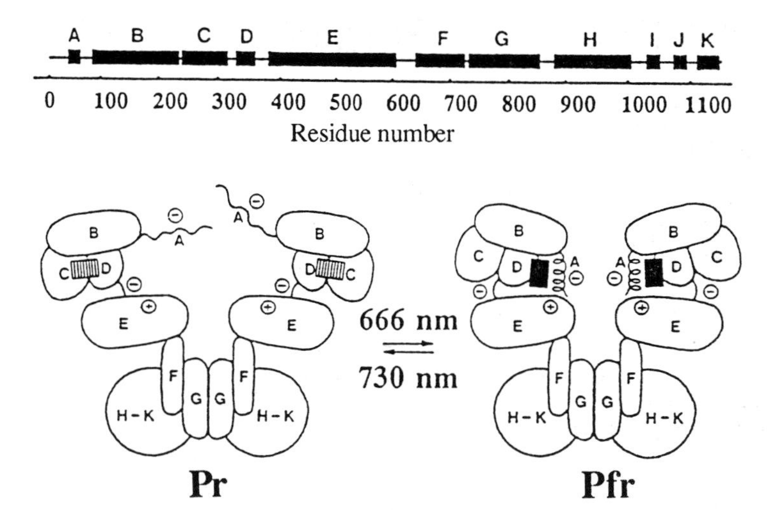

Figure 5. A structural domain model of 124-kD oat phytochrome. The chromophore domain contains regions A through E (box; chromophore), and the C-terminal domain includes G through K. The linker region is F. Approximate locations of net negative charges are located. The negatively charged PEST sequence occurs between the D and E domains. (After Parker *et al.* 1991; Romanowski and Song 1992).

1987), consistent with the linear dichroism of pea phytochrome (Tokutomi and Mimuro 1989).

The near-UV absorption ('Soret band') at 380 nm and 406 nm for Pr and Pfr, respectively, is made up of at least three separate electronic transitions, with A to B_x and A to B_y transitions contributing the major intensities. The net transition dipole of the Soret band in Pr is at about 30° relative to the Q_y transition dipole (Fig. 6), consistent with theoretical predictions (Song and Chae 1979; Song and Yamazaki 1987).

The oscillator strength ratio of the visible to the Soret bands, f_v/f_s, is a sensitive indicator of the conformation of a tetrapyrrole (Song *et al.* 1979; Song and Chae 1979). The Gaussian fitting of the phytochrome spectra (Fig. 6) followed by determination of the f_v/f_s for both forms of phytochrome demonstrated that the oscillator strength ratio lies between the fully linear and cyclic conformations. Molecular orbital calculations of almost 30 chromophores yielded an average ratio of 17.3 for extended conformations, 1.29 for semi-cyclic/-extended conformations and 0.21 for cyclic ones. Thus, one can conclude from these values that the native phytochrome chromophore maintains a semi-extended conformation for both Pr and Pfr forms. This is consistent with the recent molecular mechanics calculations and molecular modelling studies (Parker *et al.* 1993). However, precise determination of the chromophore conformation in the native phytochrome must await crystallographic analysis,

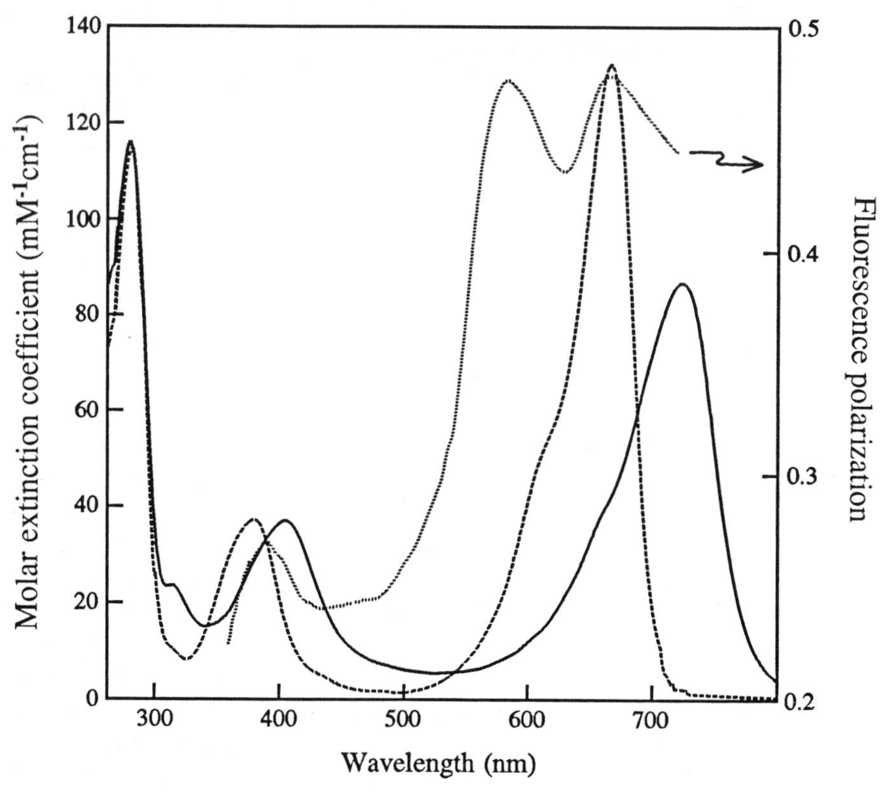

Figure 6. Absorption and fluorescence polarization spectra of 124-kD oat phytochrome. Broken line, Pr; solid line, Pfr.

especially with respect to non-planar conformations around the inter-pyrrole ring single bonds (Chapter 4.1; Fodor *et al.* 1990).

The absorbance band maxima of different molecular mass species of phytochrome are virtually identical for the Pr form, but the absorbance maximum for undegraded, 124-kD species is at slightly longer wavelength (730-732 nm) than that for degraded phytochromes (720-725 nm) (Vierstra and Quail 1982, 1983; Lagarias *et al.* 1987). In absolute energy units, the bathochromic shift of the Pfr absorption maximum in going from degraded species to undegraded protein corresponds to less than 200 cm^{-1} or 2.5 kJ mol^{-1}. This suggests that the difference in interaction forces between the chromophore and the apoprotein among different molecular mass species of phytochrome is quite small. The small energy difference reflects a subtle change in hydrogen bonding, hydrophobic force and/or electrostatic forces between the chromophore and its environment.

Phytochrome in the Pr and Pfr forms are spectrophotometrically detectable as stable forms at physiological temperatures, but numerous short-lived intermediates of phytochrome have been characterized in both directions of phototransformation *in vivo* and *in vitro* by flash kinetic spectroscopy and low-temperature spectroscopy (Furuya 1983).

Four intermediates in the Pr to Pfr pathway for 'small' (60-kD) proteolytically degraded phytochrome, were identified by kinetic analyses after flash photolysis (Linschitz *et al. 1966)* and a few more were subsequently added (Pratt *et al.* 1984; Inoue *et al.* 1990).

4.3.5 Photochemistry and photophysics

4.3.5.1 Mechanism of the Pr to Pfr phototransformation

A semi-extended conformation for the Pr and Pfr species of phytochrome meets both f_v/f_s ratio and polarization data shown in Fig. 6. An analysis of f_v/f_s ratios for phytochrome and phycobilins suggest that the chromophore in the former is less extended than the crystallographically known, fully extended chromophore conformation of the latter (Parker *et al.* 1993). Surface-enhanced resonance Raman studies of phytochrome in comparison with Z- and E-isomeric model phycocyanobilins (Farrens *et al.* 1989; Rospendowski *et al.* 1989), resonance Raman by lowest electronic excitation at 752 nm (Fodor *et al.* 1990), and by Fourier transform techniques (Hildebrandt *et al.* 1992) all support the original NMR-derived assignment, based on chromopeptide studies, that the Pr to Pfr phototransformation involves Z(15) to E(15) isomerization (Rüdiger *et al.* 1983). Figure 7 depicts the photo-isomerization of phytochrome, with a conservation of the semi-extended conformation to satisfy the spectral properties shown in Fig. 6. It should be noted that the chromophore of native

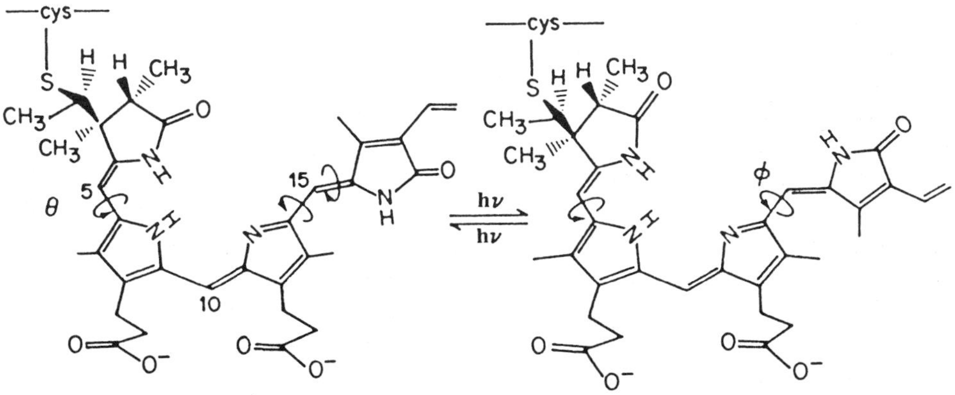

Figure 7. The presumed photo-isomerization mechanism with a conservation of the semi-extended conformation in the phototransformation of phytochrome. Dihedral angles 'zeta' (θ) and 'phi' (ϕ) around 5-6 and 14-15 single bonds are indicated.

phytochrome is not coplanar, as has been revealed by molecular mechanics calculations and molecular modelling (Parker *et al.* 1993) and by FT resonance Raman spectroscopy, especially with respect to the ring D which undergoes rotation about the C15=C16 bond during the Pr → Pfr phototransformation (Hildebrandt *et al.* 1992).

4.3.5.2 Quantum yields

The quantum yields for the Pr → Pfr phototransformation of phytochrome are 0.152-0.154 and 0.172-0.174 for oat and rye phytochrome As, respectively, and those for the Pfr → Pr photoreversion are 0.060-0.065 and 0.074-0.078, respectively, at 278 K and pH 7.8 (Lagarias *et al.* 1987). Although the forward reaction is fairly efficient, it is significantly lower than the quantum yield for the primary reaction step leading to the formation of intermediate lumi-R or I_{700} (Heihoff *et al.* 1987). This suggests that the Pr species is regenerated from I_{700} and some other subsequent intermediates *via* photochemical and thermal routes, thus partially accounting for the relatively low quantum yields for the Pr → Pfr phototransformation. The photoreversibility and photoequilibrium between Pr and I_{700} and other intermediates is possible under steady state actinic irradiation conditions because the absorption spectra of these species overlap significantly throughout the visible region.

The quantum yields for the Pfr → Pr photoconversion is even less efficient than the forward phototransformation. The low efficiency is probably due to (i) highly photo- and thermal-reversibility of the Pfr photo-intermediate(s) and/or (ii) radiationless non-photochemical relaxation of the Pfr-excited state, even though the primary photoprocess of Pfr may well be ultrafast as inferred from the fact that Pfr is non-fluorescent. Unfortunately, very little has been done to elucidate the mechanism of the phytochrome photoreversion.

4.3.5.3 Fluorescence properties and primary photoprocesses

In the 1960's and 1970's, fluorescence spectra of phytochrome *in vitro* were determined at ambient and low temperature. In these studies little attention was paid to the actinic effects of the excitation beam on the phototransformation of phytochrome. Hence, the nature of the fluorescence in the Pr form is not clear because Pr must partly be phototransformed to the intermediates in these early studies.

At liquid nitrogen temperature, fluorescence spectra of native rye phytochrome in the Pr form shows a major peak at 685 nm and a broad sub-peak around 515 nm. This spectrum was completely reversed by a subsequent irradiation at 700 nm (Inoue *et al.* 1985).

Excitation of Pr produces spontaneous emission (λ_{em} = 678 nm; Φ_f = 0.0029; radiative lifetime τ_{rad} = 14 ns). The fluorescence decay is non-exponential, with measured lifetimes of 44 ps (91%), 160 ps (8%) and 900 ps (1 %) in phosphate buffer at 275-278 K (Holzwarth *et al.* 1984; Schaffner *et al.* 1990 and references therein). In spite of the use of a sensitive near infra-red (IR) detector, no fluorescence was observed from Pfr (Song *et al.* 1989). The multi-exponential fluorescence lifetimes have also been observed with pea phytochrome (Pr) and their lifetimes and percent amplitudes are microviscosity-dependent. The multi-exponential fluorescence decay kinetics can be simplified to a two-component kinetic model involving an excited state reversible equilibrium (Song *et al.* 1986, 1989),

$$\mathrm{Pr} + h\nu \rightarrow \mathrm{Pr}^*; \quad \text{excitation}$$

$$\mathrm{Pr}^* \underset{k_{-1}}{\overset{k_1}{\rightleftarrows}} \mathrm{Pr}'; \quad \text{‘reversible primary reaction’}$$

where $k_1 = 2.7 \times 10^{10}$ and 2.1×10^{10} s^{-1} for oat 124-kD and pea 121-kD phytochromes, respectively, and $k_{-1} = 3.3 \times 10^8$ and 2.6×10^8 s^{-1}, respectively. The k_1 values in 67% glycerol are 2.0×10^{10} and 1.8×10^{10} s^{-1}, respectively and the k_{-1} values are 1.51×10^9 and 6.6×10^8 s^{-1}, respectively. It appears that glycerol slightly retards the forward reaction, whereas it significantly accelerates the reverse reaction. This is not easy to understand. One speculative explanation is that the forward reaction involves a partial conformational/configurational relaxation of the chromophore in the excited state, but the reverse reaction of Pr’, still an electronically excited state, is accelerated because the contained chromophore returns to Pr* more readily in the presence of glycerol. According to the above model, the observed fluorescence decay of phytochrome can be approximated reasonably well in terms of only two components, 38 ps and 1.11 ns (Song *et al.* 1989), but the latter is not due to a photochromically active component (Schaffner *et al.* 1990). Thus, the model is by no means quantitatively satisfactory.

The identity of the primary intermediate remains uncertain. It is either I_{700} or prelumi-R (696 nm) which may be formed at 77 K prior to lumi-R (Song *et al.* 1981) or by a picosecond pulse excitation at room temperature (Lippitsch *et al.* 1988). It was suggested that ‘an initial intermediate arising from the primary reaction is either lumi R or prelumi-R’ (Song *et al.* 1989). Schaffner *et al.* (1990) concluded that the formation of prelumi-R in equilibrium with the excited state Pr* is unlikely. A recent study showed that the fluorescence decays in oat phytochrome are resolved into four exponentials, not three, and that a new ultrafast decay component of a 4-16 ps lifetime predominates the fluorescence kinetics (Holzwarth *et al.* 1992). The four-component fluorescence kinetics are accommodated by assuming two emitting excited states in equilibrium. At first glance, this scheme is similar to the one (k_1, k_{-1}, above) used to simplify the three-component fluorescence decay kinetics.

$$Pr^* \underset{k_b}{\overset{k_a}{\rightleftarrows}} B^* \xrightarrow{k_c} C$$

Here, the state B* might be a configurational intermediate between the initial Z-configuration and the final E-configuration (Holzwarth *et al.* 1992).

The simplistic model for the complex fluorescence decay kinetics of phytochromes assumes that only the fastest, predominant lifetime component [37.9 ps and 89% amplitude (Song *et al.* 1989); 44 ps and 91 % amplitude (Holzwarth *et al.* 1984; Schaffner *et al.* 1990)] for oat phytochrome in phosphate buffer reflects directly both radiative and radiationless relaxations including the primary reaction of the Pr-excited state. If we disregard the longer lifetime components as unimportant because together they contribute only about 10% to the fluorescence emission, this simplistic model yields the measured fluorescence lifetime of 38-44 ps, which agrees with the calculated lifetime of $\tau_{calc} = \Phi_f \times \tau_{rad}$ = 41 ps. In this case, we obtain the rate constant of $\geq 1.3 \times 10^{10}$ s^{-1} for the primary reaction (lumi-R formation assumed with $\Phi_{reaction} \geq 0.5$, *vide infra*). This is not far off from the $k_1 = 2.7 \times 10^{10}$ s^{-1} cited earlier. Occam's razor suggests that perhaps the simplest model that can explain the data is most appropriate.

The question of the identify of a picosecond or even femtosecond earlier primary intermediate aside, the first intermediate readily observable upon flash photolysis of Pr at room temperature is I_{700} or lumi-R with absorbance maximum at *ca.* 700 nm and with a quantum yield of ≥ 0.5 (Heihoff *et al.* 1987). Presumably, configuration of the chromophore in I_{700} is E (Fig. 7), although there is no direct evidence for this. The sign of the circular dichroism in the visible spectral region reverses in going from Pr to lumi-R trapped at low temperatures, suggesting that configuration of the chromophore isomerized from a Z to an E configuration (Eilfeld and Eilfeld 1988). However, evidence is still equivocal because an excitonic coupling with aromatic residues of the apoprotein near the chromophore binding site predominates the induced optical activity of phytochrome in the visible wavelength region. In fact, a 500 ns 500 ms time-resolved CD study of phytochrome at higher temperature (283 K) showed no new CD band corresponding to the absorbance maximum of lumi-R (Björling *et al.* 1992). In this study, the first intermediate, lumi-R, had an absorbance peak at 696 nm, but its difference time-resolved CD spectrum shows a CD maximum at 660 nm, suggesting that the first species trapped in the low-temperature CD measurement is probably an unrelaxed lumi-R type intermediate.

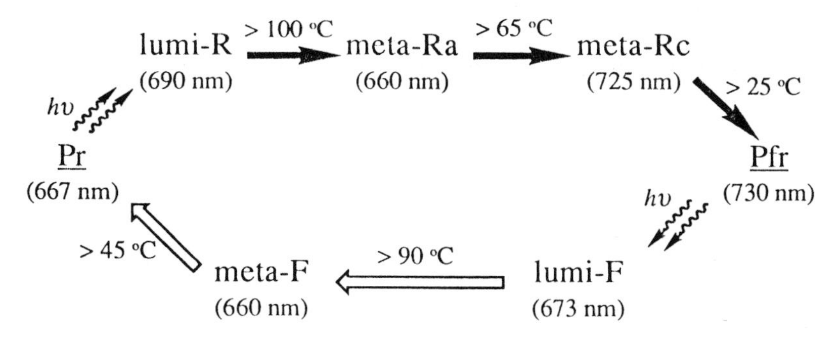

Figure 8. A photochemical cycle for phytochrome phototransformation deduced from low temperature spectrophotometry. The temperatures below which the specific intermediate step is blocked are indicated. (After Eilfeld P.H. and Rüdiger W. (1985) *Z. Naturforsch.* 40c: 109-114).

4.3.5.4 Intermediates and kinetic models

Numerous photolysis studies have been carried out with various phytochrome preparations. The reader is referred to selected key review articles on this subject by Kendrick and Spruit (1977), Furuya (1983), Braslavsky (1984), and Schaffner *et al.* (1990). In analogy to the freeze-irradiate-thaw spectrophotometry of rhodopsin, the phototransformation of phytochrome can be studied under low temperature steady-state conditions. Figure 8 shows the generally accepted scheme of phytochrome phototransformation derived from such studies.

However, a recent 100 ns-800 ms time-resolved spectroscopic study of the Pr → Pfr phototransformation kinetics using 7-ns laser pulse excitation and transient absorbance difference measurements over the UV-visible range revealed more complex kinetics at 283 K for the forward reaction pathway than represented in Fig. 8 (Zhang *et al.* 1992). Thus, a global analysis fitting of the time-resolved absorbance spectra of phytochrome entails at least five kinetic intermediates. The apparent lifetimes associated with these intermediates are 7.4 μs, 7.6 ms, 42.4 ms, and ≥ 266 ms. Figure 9 shows three plausible kinetic schemes consistent with the data. A crucial feature of these models involving either sequential (a) or parallel mechanism (b) and (c) is the occurrence of equilibria at certain stages of the phototransformation (Fig. 9). Figure 10 shows the absorption spectra of the five intermediates derived from the kinetic models shown in Fig. 9, with values of the rate constants given in the caption. It should be emphasized that at present there is no definitive mechanism for the Pr → Pfr pathway that can uniquely account for the observed kinetic and spectroscopic data. In contrast to the forward phototransformation of phytochrome, very little has been done to elucidate the Pfr → Pr photoreversion by employing laser flash photolysis (Inoue and Furuya 1985).

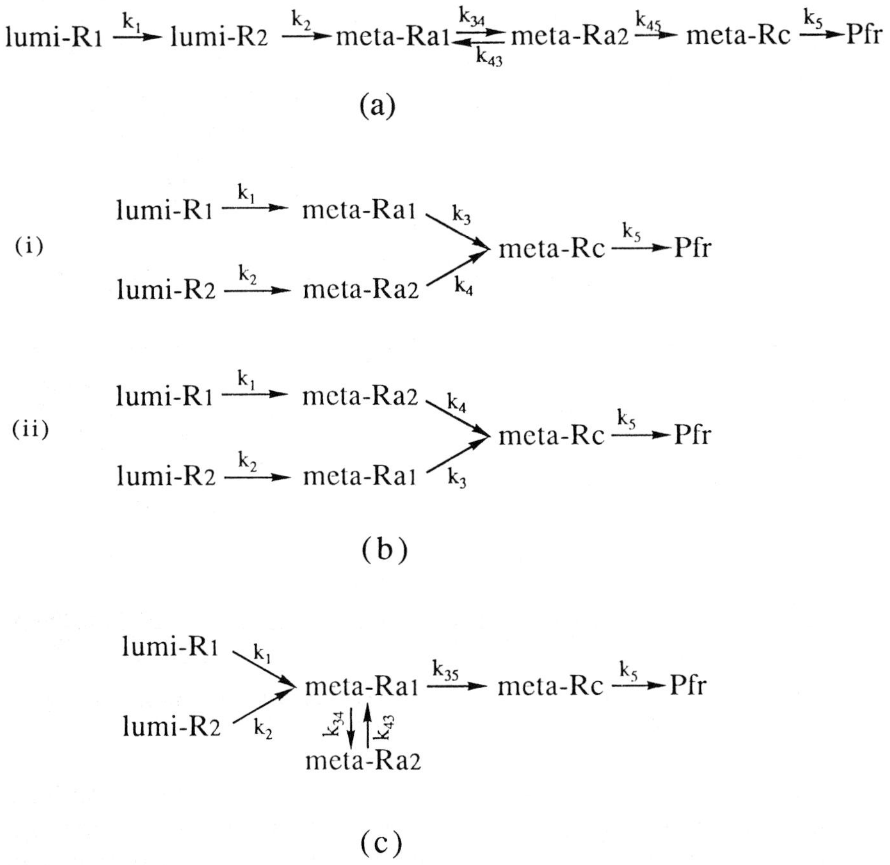

$$\text{lumi-R1} \xrightarrow{k_1} \text{lumi-R2} \xrightarrow{k_2} \text{meta-Ra1} \underset{k_{43}}{\overset{k_{34}}{\rightleftharpoons}} \text{meta-Ra2} \xrightarrow{k_{45}} \text{meta-Rc} \xrightarrow{k_5} \text{Pfr}$$

(a)

(i)

$$\text{lumi-R1} \xrightarrow{k_1} \text{meta-Ra1} \xrightarrow{k_3} \text{meta-Rc}$$
$$\text{lumi-R2} \xrightarrow{k_2} \text{meta-Ra2} \xrightarrow{k_4} \text{meta-Rc}$$
$$\text{meta-Rc} \xrightarrow{k_5} \text{Pfr}$$

(ii)

$$\text{lumi-R1} \xrightarrow{k_1} \text{meta-Ra2} \xrightarrow{k_4} \text{meta-Rc}$$
$$\text{lumi-R2} \xrightarrow{k_2} \text{meta-Ra1} \xrightarrow{k_3} \text{meta-Rc}$$
$$\text{meta-Rc} \xrightarrow{k_5} \text{Pfr}$$

(b)

$$\text{lumi-R1} \xrightarrow{k_1} \text{meta-Ra1}, \quad \text{lumi-R2} \xrightarrow{k_2} \text{meta-Ra1}$$
$$\text{meta-Ra1} \xrightarrow{k_{35}} \text{meta-Rc} \xrightarrow{k_5} \text{Pfr}$$
$$\text{meta-Ra1} \underset{k_{43}}{\overset{k_{34}}{\rightleftharpoons}} \text{meta-Ra2}$$

(c)

Figure 9. Kinetic schemes involving five intermediates. (a) Sequential reactions; (b) parallel reactions [(i) and (ii)], and (c) combined parallel-branched reactions. For details, see Zhang *et al.* (1992).

4.3.5.5 Non-photochemical transformation

The Pfr form of phytochrome reverts to the Pr form in darkness. The rate of dark reversion depends on the molecular integrity of the phytochrome molecule. The proteolytically truncated species (60-kD 'small phytochrome' and 114/118-kD 'large phytochrome') undergo significantly faster dark reversion than the full length, 124-kD species of phytochrome. This suggests that the Pfr form is thermodynamically unstable, compared to the Pr form. The activation energy for the dark reversion is 102 kJ mol^{-1} for 'small phytochrome' (Anderson *et al.* 1969).

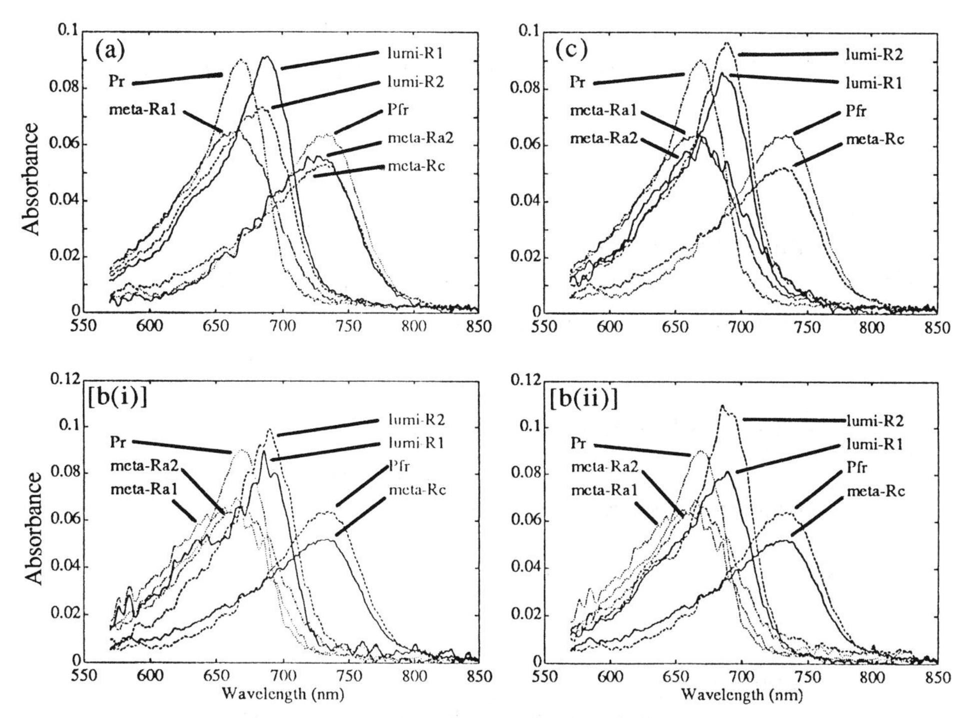

Figure 10. Absorption spectra of the intermediates derived from the schemes shown in Fig. 7. For details, see Zhang *et al.* (1992).

Interestingly, the dark reversion of phytochrome is accelerated by millimolar concentrations of reducing agents such as dithionite. Only the reducing agents with a negatively charged group appear to be 'catalytic'. The dark reversion is also pH-dependent and specific acid-catalyzed, as opposed to general acid-catalyzed (Hahn and Song 1981). The mechanism of the dark reversion and the catalytic effects of certain reducing agents are not known. Because the chromophore reduction by reducing agents is apparently not involved in the acceleration of the dark reversion, it is possible that a specific binding of the reducing agent at/near the chromophore binding site disrupts an electrostatic linkage between the chromophore (for example a propionate group) and the apoprotein, thus accelerating the dark reversion. Consistent with this interpretation, phytochrome 'Pfr-killer' (Shimazaki and Furuya 1975), 8-anilinonaphthalene 1-sulphonate (ANS), with its hydrophobic affinity and sulphonate group, interferes with the dithionite-catalyzed dark reversion of 118-kD phytochrome (Hahn and Song 1981; Song 1988).

4.3.6 Photo-induced conformational changes

4.3.6.1 Chromophore topography

According to the proposed topographic model for the phototransformation of phytochrome, the Pfr chromophore is more exposed than that of Pr (Song 1988). Tetranitromethane (Hahn *et al.* 1984b), sodium borohydride (Chai *et al.* 1987), and bilirubin oxidase (Singh *et al.* 1989) have been used to probe the differential topographies of the Pr and Pfr chromophores, all confirming the preferential exposure of the chromophore in the Pfr-state. However, the relative degree of exposure of the Pfr-chromophore depends on the size of the protein, indicating that the 6-kD peptide chain along the N-terminal sequences shields and interacts closely with the Pfr chromophore (Hahn *et al.* 1984b).

To estimate the extent of the chromophore movement during the Pr → Pfr phototransformation, Förster energy transfer from the N-terminal bound, fluorescently labelled Oat-25 Fab antibody fragment to the phytochrome chromophore has been measured. Results suggest that the chromophore moves by about 12 Å toward the N-terminus or the N-terminus moves toward the chromophore, or both, upon Pr → Pfr phototransformation (Farrens *et al.* 1992). Spectral evidence, both absorbance and CD, suggests that a synthetic 54-mer peptide having an amino-acid sequence identical to that of the N-terminal chain of oat phytochrome can mimic the specific interaction between the Pfr-chromophore and the native N-terminal chain of the phytochrome molecule. Thus, the 54-mer provides a model system for studying geometric relations between the tetrapyrrole chromophore and N-terminal region (Parker *et al.* 1992).

Both ANS and *bis*-ANS are well known for their fluorescence enhancement upon binding to a hydrophobic site on a protein. These hydrophobic fluorescence probes also bind to phytochrome. The binding causes bleaching of the chromophore absorbance bands, especially in the Pfr form of phytochrome (see Song 1988; Choi *et al.* 1990 for reviews). The spectral bleaching of the Pfr species apparently arises from a competitive binding of the probe at the chromophore binding site or its vicinity. The binding of the probe disrupts the interaction between the chromophore and its binding pocket and results in the formation of a cyclic conformation of the chromophore. The susceptibility of phytochrome to ANS bleaching is a sensitive function of the apoprotein integrity, particularly with respect to the 6/10-kD amino terminal peptide segment. This suggests that the amino terminus segment interacts and/or masks the Pfr chromophore of 124-kD phytochrome, thus making it less susceptible to cyclization in the presence of ANS than the Pfr form of degraded proteins.

The Pr → Pfr phototransformation-induced movement of the chromophore apparently exposes acidic amino-acid residues to the surface, possibly the nine acidic residues contained in the region Cys_{321}/chromophore-Glu_{354}, resulting in a higher negative net charge of the Pfr form of phytochrome (Schendel and Rüdiger 1989). It appears that the photochemistry of the chromophore brings about changes in both the chromophore topography and protein tertiary structure of the peptide chains near and away from the chromophore-binding pocket, *vide infra.*

4.3.6.2 Proton transfer

Ultra-fast intramolecular proton transfer indeed occurs prior to subsequent isomerization reactions in a phytochrome model compound, biliverdin. However, proton transfer does not appear to play an important role in the primary photoprocess (Holzwarth *et al.* 1984). It is not known at what stage of the phototransformation intramolecular or intermolecular proton transfer takes place along the thermal pathway leading to the Pfr form of phytochrome and what residue(s) are responsible for proton release/uptake in solution (Tokutomi *et al.* 1982, 1988). Unlike proteolytically truncated pea phytochromes (Tokutomi *et al.* 1982), proton release from the Pfr of pea phytochrome is suppressed in the full-length molecule, suggesting that the conjugate acid group near/at the chromophore binding pocket is shielded from the medium in the latter (Tokutomi *et al.* 1988).

At the molecular level, the observed proton release, preferentially from the Pfr form of phytochrome, is consistent with recent resonance Raman results. The protonation level of the intact pea phytochrome remains unchanged in going from Pr → Pfr (Mizutani *et al.* 1991). The ring-C nitrogen is protonated in the Pr form of oat and pea phytochromes (Fodor *et al.* 1990; Mizutani *et al.* 1991; Hildebrandt *et al.* 1992), but it is deprotonated in the Pfr form (Hildebrandt *et al.* 1992). The basic amino-acid residue that accepts this proton, thus producing a conjugate acid, is unknown. Overall, the Pfr form has slightly lower isoelectric points (5.80-5.85) than the Pr form (5.85-5.90), suggesting that a higher negative net charge of the former arises from the exposed acidic residues, *vide supra* (Schendel and Rüdiger 1989).

4.3.6.3 Photoreversible conformational changes

Several lines of evidence indicate that the N-terminus of phytochrome is preferentially exposed in the Pr form of phytochrome, compared to the Pfr form. It has also been shown that the oat phytochrome N-terminal chain (roughly 70 residues with amphiphilicity) assumes an α-helical conformation in the Pfr form. The secondary structure apparently results from a specific interaction between the N-terminal chain and the tetrapyrrole chromophore, preferentially when phytochrome is in the Pfr form.

To elucidate the interaction between the N-terminal region and the chromophore, particularly Pfrs, studies with a synthetic 54-mer peptide corresponding to the far N-terminal 54 residues of oat phytochrome were carried out to show that the synthetic peptide interacts specifically with N-terminal chain-truncated 118-kD phytochrome A Pfr, in a native fashion, but not with large phytochrome A Pr (Parker *et al.* 1992).

Additional evidence for the possible interaction between the chromophore and the N-terminus peptide chain has come from a CD study (Vierstra *et al.* 1987; Chai *et al.* 1987; Sommer and Song 1990). A photoreversible CD change occurs during the phototransformation of 124-kD phytochrome, corresponding to a few percent increase in α-helical folding in the Pfr form. However, no such photoreversible CD changes have been observed with proteolytically truncated phytochromes, primarily 60- and 114/118-kD species. Secondary structure analysis of the CD spectra is summarized in Table 1.

The photoreversible CD change, and the increased α-helicity in the Pfr form, are suppressed by monoclonal antibody binding along the N-terminal chain (Chai *et al.* 1987). The photoreversible CD spectral change can also be inhibited by sodium borohydride, which specifically bleaches the chromophore by reducing it, and by tetranitromethane, *vide supra*.

Amphiphilic helix formation along the N-terminal chain of phytochrome during Pr → Pfr phototransformation is supported by the presence of amphiphilic helical motifs in the N-terminal chain and CD-detectable α-helical folding of the chain in an amphiphilic environment [sodium dodecyl sulphate (SDS) micelles; Parker *et al.* 1992]. The amphiphilic α-helix formation is also consistent with the observation that the difference in helical content between Pr and Pfr was not observable by IR spectroscopy (Siebert *et al.* 1990). Under amphiphilic conditions such as the films used in IR measurements, helical formation should diminish the difference in α-helix between the Pr and Pfr forms of phytochrome, because the exposed N-terminal chain of the former would form a amphiphilic helix at the air/water interface.

From Table 1, a significant increase in α-helical folding upon Pr → Pfr photoconversion arises largely at the expense of either ß-turns or random coil. According to the method of Chou and Fasman (1978), the N-terminal sequences do tend to assume ß-turn and random coil conformations (Hershey *et al.* 1985; Quail *et al.* 1987). It is well known that prosthetic groups such as the tetrapyrrole haem of myoglobin (Beychok 1966) exhibit a marked propensity for facilitating α-helical conformations in peptide chains; perhaps, this type of interaction is responsible for the photoreversible α-helical folding in phytochrome. Although the mode of interactions between the Pfr-chromophore and the N-terminus sequence is not known, amino-acid residues near the N-terminus include serine and threonine clusters [Fig. 2; also see moss phytochrome (Thümmler *et al.* 1992)], and some of these hydroxyl amino-acid residues could form hydrogen bonds with the pyrrole nitrogens, carbonyl and carboxylate groups of the chromophore.

In a reversible conformational change, proteins cannot gain or lose significant amounts of volume. Thus, it is possible that the photoreversible phytochrome does not exhibit a gross conformational change, especially given the relatively small energy of 180 kJ mol^{-1} of R photons. Most hydrodynamic data tend to confirm this prediction. However, in addition to the photoreversible α-helix formation along the N-terminal chain of phytochrome, there are some recognizable changes in tertiary structure accompanying the phototransformation of phytochrome. For example, there is a significant difference in the number of exchangeable protons between the Pr and Pfr forms of phytochrome (Hahn *et al.* 1984a), which could be partially attributed to a change in secondary structure.

Peptide maps for 124-kD oat phytochrome deduced from partial proteolysis show the protein with two domains, one for the N-terminus/chromophore domain of *ca.* 74 kD and one for the C-terminus side domain of *ca.* 55 kD. It turns out that the N-terminus sequence on the chromophore domain is extremely susceptible to proteolysis in the Pr form but not in the Pfr form, whereas at least one site near the central hinge region between the two major domains becomes preferentially accessible to a proteolytic enzyme (Jones *et al.* 1985; Lagarias and Mercurio 1985). The Pfr form of phytochrome is preferentially degraded by endogenous proteases. The possible proteolytic target sites include PEST sequences at Cys_{321}/chromophore-Glu_{360} and residues Asn_{537}-Arg_{547} (Schendel and Rüdiger 1989).

There are ten Trp residues in 124-kD oat and 121-kD pea phytochromes. Fluorescence decay properties of the Trp residues did not reveal any large and extensive overall reorganization of the protein in its Pr *versus* Pfr form (Schaffner *et al.* 1991). However, quenching of the Trp fluorescence by cationic, anionic and neutral quenchers distinguished between the Pr and Pfr species of phytochrome (Singh *et al.* 1988; Singh and Song 1990). The Trp fluorescence quenching has also revealed a significant difference in Trp environment/exposure between the Pr and Pfr forms of pea phytochrome (M. Nakazawa, K. Manabe, T. Wells and P.-S. Song unpublished data). These results suggest that the phototransformation of phytochrome involves a subtle change in tertiary structure, which is apparently recognizable by protease mapping (Grimm *et al.* 1988) and in terms of topography of Trp residues. The Trp fluorescence quenching of phytochrome by Cs^+ and Tl^+ ions and a differential spectral analysis suggested a subtle conformation change in the peptide domain containing Trp_{569} (Singh and Song 1990) and Tyr_{572} (Singh *et al.* 1989). This region is also near the site of phosphorylation (Ser_{598}) in the Pfr form [Ser_{17} in the Pr form (McMichael and Lagarias 1990)]. Enzymatic ubiquitination also only occurs in the 558-688 segment in the Pfr form of phytochrome, but not in the Pr form (Shanklin *et al.* 1989), indicating that this region undergoes a conformational change during phytochrome phototransformation. Epitope-specific monoclonal antibodies also indicate significant differences in surface topography between the Pr and Pfr forms of phytochromes (Cordonnier 1989).

4.3.7 Phytochrome structure and function relationship

4.3.7.1 Functional domain or active site

Several structural domains of phytochrome can be considered as the possible location of 'active site(s)'. These include: (i) the N-terminal sequence which contains photoreversible α-helix motifs; (ii) the chromophore binding proper where a photoreversible chromophore exposure/re-orientation occurs, and exposure of the region around Glu_{354} (Schendel and Rüdiger 1989; Cordonnier 1989 for review); (iii) the 69/72-kD chromophore-52/55-kD non-chromophore linker region, around residue Leu_{630} (Lagarias and Mercurio 1985) where a subtle conformation/surface exposure accompanies Pr → Pfr phototransformation; (iv) subunit contact sequences for dimerization; (v) C-terminal domain with canonical sequence or structural motifs conserved for signal-transducer proteins. In this section, we will ascertain the possible functional roles of some of these structural motifs based on very limited amounts of information in the literature.

(i) *Amino-terminal sequence.* The only change in secondary structure detectable by CD seems to occur in the 6-kD amino terminal segment of the phytochrome molecule, *vide supra*, forming amphiphilic α-helix upon Pr → Pfr phototransformation (Parker *et al.* 1992). Substitution of the serine residues in the 6-kD amino-terminal segment with alanine forms a more helix favouring sequence, which produced a physiologically hyperactive phytochrome (Stockhaus *et al.* 1992). The amino terminus-truncated phytochrome does not exhibit full biological activity *in vivo* (Cherry *et al.* 1992). These results suggest that the amino terminal segment with its amphiphilic helix-forming sequences are potential active site(s) for phytochrome function. This suggestion is in contrast with the C-terminal homology of bacterial sensor proteins, which points toward the C-terminus as an active site or the receptor-binding domain (see below).

(ii) *Carboxyl-terminal sequence.* An analysis of phytochrome sequences from five plant species showed an amphiphilic C-terminal region (Partis and Grimm 1990). Further, they suggested that the C-terminal region might be responsible for interaction of phytochrome with membrane. However, neither the amino-terminal segment nor the C-terminal region of phytochrome meets the combined hydrophobic moment (*ca.* 0.15-0.4) and hydrophobicity (0.5-1.4) requirements for a transmembrane protein (Eisenberg *et al.* 1984). It is well known that phytochrome binds to membrane and liposomes *in vitro*, and it is possible that either or both of the amphiphilic terminal segments could be involved in membrane anchoring.

A C-terminal domain of phytochromes comprised of about 250 amino-acid residues has about 20% sequence homology with bacterial sensor proteins (Schneider-Poetsch 1992). The C-terminal side of moss (*Ceratodon purpureus*) phytochrome also contains amino-acid sequence homology with a protein kinase (Thümmler *et al.* 1992; Algarra *et al.* 1993). Wong *et al.* (1989) suggested the possibility that phytochrome itself might be a protein kinase. In spite of short sequence homologies found between phytochromes and protein kinases, the suggestion that phytochrome is a protein kinase remains only an interesting hypothesis, because all currently available phytochrome sequences lack the canonical motifs highly characteristic of eukaryotic protein kinases (Quail 1991) and an apparent phytochrome-associated kinase can be chromatographically and electrophoretically separated from phytochrome itself (Grimm *et al.* 1989; Kim *et al.* 1989).

(iii) *Heterodimers.* The C-terminal domain is involved in the quaternary structure in forming phytochrome dimers in solution. The Pr form of phytochrome exists as PrPr, whereas Pfr at its photostationary equilibrium (87.4% Pfr) established by R contains 76.4% PfrPfr, 22% PrPfr and 1.6% PrPr. Using LAS41 monoclonal antibody specific for the Pr species, Holdsworth and Whitelam (1987) determined the heterodimer at equilibrium to be 24.5%, in agreement with the expected dimer population distributions, assuming that each subunit chromophore of the dimers undergoes phototransformation independently, *i.e.* no co-operative effect for the phototransformation. It remains to be investigated if the phototransformation of each subunit is indeed non-cooperative. There is evidence that the phytochrome heterodimers play differential roles in mediating fluence-dependent physiological responses in plants (VanDerWoude 1987).

4.3.7.2 Dichroic effects and putative receptors

Since there appears to be a significant change in the topography of the Pr and Pfr chromophores accompanying phototransformation, the degree of chromophore movement/re-orientation relative to the fixed axis of the chromoprotein molecule can be measured by linear dichroic measurements. Ekelund *et al.* (1985) carried out a linear dichroism study of the immobilized phytochrome, yielding a re-orientation angle of 32° (or 148°). However, an analysis of linear dichroism of oriented pea phytochrome (114 kD) prepared by the gel squeezing method yielded a difference of only 7° between the Q_y transition dipoles in Pr and Pfr (Tokutomi and Mimuro 1989). However, the relatively low anisotropic orientation of the molecules in the gel makes this analysis equivocal. A functional implication of change in the chromophore dipole orientation during the phytochrome phototransformation has been discussed elsewhere (Haupt 1991).

The polarotropic response in fern protonema is regulated by phytochrome, since phytochrome in the Pr and Pfr forms is dichroically oriented parallel and normal to the cell surface (Chapter 9.7). This change during phototransformation between Pr and Pfr was analyzed by a polarization plane-rotatable, double flash irradiator (Chapter 3). The results (Kadota *et al.* 1986) indicated that the orientation of phytochrome intermediates is parallel to the cell surface as is the case for Pr until 2 ms after a R flash, whereas the orientation change of phytochrome molecule occurs between 2 ms and 30 s after the flash. The time scale of this change in the orientation of phytochrome is so slow that it cannot be ascribed to the conformation change of the chromophore. The change is therefore probably due to a change in the conformation of the protein.

We now know a great deal about the phytochrome molecule with respect to its structure and spectroscopic/photochemical properties, but we know little about its 'receptors' and its mode of action largely remains to be elucidated. For example, it is well known that the Pfr form of phytochrome tends to sequester without being associated with a particular organelle in plant cells (Pratt 1986), but its implications in the light signal transduction are still puzzling. Is it possible that the Pfr sequestering by itself drives the signal-transduction system either thermodynamically or kinetically, as proposed for transmembrane signalling in mast cells (Metzger 1992; VanDerWoude 1985). Is sequestering merely a means to store the light signal by protecting the sequestered Pfr from proteolytic degradation?

One can also ask where the 'functionally active' pool of phytochrome is localized within the cell. Is phytochrome a membrane-bound protein, at least in certain systems such as in the polarotropically sensitive *Mougeotia* (Haupt 1991) and in phototropism of the fern *Adiantum* protonemata (Chapters 7.2 and 9.7; Hayami *et al.* 1992). There is apparently no clear-cut, direct evidence for the membrane-bound phytochrome *in vivo*. However, there are several lines of indirect evidence that suggest the membrane as the site of phytochrome localization for certain rapid *in vitro* processes (see Roux 1983; Furuya 1987 for reviews). As mentioned earlier, an analysis of the amino-acid sequences of phytochromes in terms of hydrophobic moment plot, suggests that phytochromes could not be transmembrane proteins. However, this does not rule out the possibility that phytochromes attach to membranes as surface/peripheral proteins. In fact, this may explain the propensity of phytochrome for membrane and liposome associations (Lamparter *et al.* 1992).

Putative receptors for phytochrome are largely unknown. Phytochrome as a *trans*-acting transcriptional factor does not appear to involve DNA as its 'receptor' (Quail 1991). G-proteins are currently receiving attention as a possible signal transducer (Romero *et al.* 1991; Romero and Lam 1993; see Tretyn *et al.* 1991 for review).

4.3.8 Concluding remarks

We now know the primary structure of the phytochrome molecular species fairly accurately in terms of the *phy* gene sequences (see Chapter 4.2). Apart from phytochrome A, the molecular and spectrophotometric properties of the corresponding gene products, have remained obscure because it has been very difficult to prepare sufficient amounts of the purified samples of other phytochromes for major physical and chemical analysis. For example, 1 g of light-grown pea shoots contain only 0.15 μg of phytochrome B apoprotein, whereas for single measurements the amounts of phytochrome needed are 3 mg for X-ray scattering studies, 1 mg for Raman spectroscopy and 0.1 mg for infra-red spectroscopy (Furuya *et al.* 1991). In contrast 1 g of yeast cells transformed with *phyA* cDNA give 40 μg of phytochrome A apoprotein within a single day, and an *E. coli* transformant can produce much more than this. The reader will readily appreciate how much time, labour and money can be saved by preparing phytochrome samples by *in vitro* assembly techniques as described in Section 4.3.3. Alternatively if instruments such as the Raman or NMR spectrometers could be improved so that the sample required was several orders less than that needed at present, the molecular properties of the different phytochromes might at last be fully elucidated.

As far as the structural features of the phytochrome molecular species are concerned, they will only be elucidated by analysis of their crystal structure. However, it has proved difficult to crystallize phytochrome in either its Pr or Pfr form. It might be possible to crystallize transformant-produced recombinant phytochrome with a non-natural chromophore, that does not undergo a photoreversibly change in molecular structure. Work in this direction is certainly warranted at this time.

It is becoming evident that the modes of action of type I and type II phytochromes are significantly different, but the primary structure of their N-terminal chromophoric domains (Figs. 4 and 5) is quite similar. It is reasonable to assume that the N-terminal domain acts as a light-signal sensor motif for the chromophore antenna and its conformational change upon phototransformation from Pr to Pfr in essentially the same way in all molecular species of phytochrome, whereas the C-terminal domain is more heterogeneous in primary structure (Chapter 4.2), resulting in more strikingly different secondary and higher structures and thus eventually different molecular functions. At the present time we have little knowledge of partner compound(s), *phytochrome receptors,* which transmit the light-induced signal for phytochrome to the signal-transduction pathways. However, we suspect that diverse functions of phytochrome may result from different *receptors* for the corresponding molecular species of phytochromes, which eventually induce the wide variety of effects at the molecular, cellular and organ levels throughout the life cycle of a plant.

It is interesting that all the phytochromes that have been characterized are cytosolic proteins. However, considering the dichroic effect of phytochrome in photomorphogenesis (Chapters 7.2 and 9.7), some molecular species of phytochrome must be spatially fixed on or in some intra-cellular structure. If so, crystal structures of different phytochromes may reveal such potential receptor-binding motifs.

Acknowledgements: The authors' works described in this chapter were supported by grants from the Frontier Research Program/RIKEN (to M.F.) and US PHS NIH (No. GM-36956 to P.-S.S).

4.3.9 Further reading

Furuya M. (1989) Molecular properties and biogenesis of phytochrome I and II. *Adv. Biophys.* 25: 133-167.

Furuya M. (1993) Phytochromes: Their molecular species, gene families and functions. *Annu. Rev. Plant Physiol. Plant Mol. Biol.* 44: 617-645.

Sage L.C. (1992) *Pigment of the Imagination. A History of Phytochrome Research.* Academic Press, Inc. San Diego.

4.3.10 References

Algarra P., Linder S. and Thümmler F. (1993) Biochemical evidence that phytochrome of the moss *Ceratodon purpureus* is a light-regulated protein kinase. *FEBS Lett.* 315: 69-73.

Anderson G.R., Jenner E.L. and Mumford F.E. (1969) Temperature and pH studies on phytochrome *in vitro*. *Biochemistry* 8: 1182-1187.

Anderson G.R., Jenner E.L. and Mumford F.E. (1970) Optical rotary dispersion and circular dichroism spectra of phytochrome. *Biochim. Biophys. Acta* 221: 69-73.

Beychok S. (1966) Circular dichroism of biological macromolecules. *Science* 154:1288-1299.

Björling S.C., Zhang C.F., Farrens D.L., Song P.S. and Kliger D.S. (1992) Time-resolved circular dichroism of native oat phytochrome photointermediates. *J. Amer. Chem. Soc.* 114: 4581-4588.

Bonner B.A. (1967) Incorporation of delta aminolevulinic acid into the chromophore of phytochrome. *Plant Physiol.* 42(Suppl.), s-11.

Boylan M.T., Quail P.H. (1989) Oat phytochrome is biologically active in transgenic tomatoes. *Plant Cell* 1: 765-773.

Boylan M.T. and Quail P.H. (1991) Phytochrome A overexpression inhibits hypocotyl elongation in transgenic *Arabidopsis*. *Proc. Natl. Acad. Sci. USA* 88: 10806-10810.

Braslavsky S.E. (1984) The photophysics and photochemistry of the plant photosensor-pigment phytochrome. *Pure Appl. Chem.* 56: 1153-1165.

Butler W.L., Norris K.H., Siegelman H.W., Hendricks S.B. (1959) Detection, assay, and preliminary purification of the pigment controlling photoresponsive development of plants. *Proc. Natl. Acad. Sci. USA* 45: 1703-1708.

Chai Y.G., Song P.-S., Cordonnier M.-M. and Pratt L.H. (1987) A photoreversible circular dichroism spectral change in oat phytochrome is suppressed by a monoclonal antibody that binds near its N-terminus and by chromophore modification. *Biochemistry* 26: 4947-4952.

Cherry J.R., Hondred D., Walker J.M. and Vierstra R.D. (1992) Phytochrome requires the 6-kDa N-terminal domain for full biological activity. *Proc. Natl. Acad. Sci. USA* 89: 5039-5043.

Choi J.K., Kim I.S., Kwon T.I., Parker W. and Song P.-S. (1990) Spectral perturbations and oligomer/monomer formation in 124-kilodalton *Avena* phytochrome. *Biochemistry* 29: 883-6891.

Chou P.Y. (1989) Prediction of protein structural classes from amino acid compositions. In: *Prediction of Protein Structure and the Principles of Protein Conformation*, pp. 549-586, Fasman G.D. (ed.) Plenum, New York.

Chou P.Y. and Fasman G.D. (1978) Empirical predictions of protein conformation. *Annu. Rev. Biochem.* 47: 251-276.

Cordonnier M.-M. (1989) Monoclonal antibodies: Molecular probes for the study of phytochrome. *Photochem. Photobiol.* 49: 821-831.

Cornejo J., Beale S.I., Terry M.J., Lagarias J.C. (1992) Phytochrome assembly. The structure and biological activity of 2(R), 3(E)-phytochromobilin derived from phycobiliproteins. *J. Biol. Chem.* 267: 14790-14798.

Deforce L., Tomizawa K., Ito N., Farrens D., Song P.-S., Furuya M. (1991) *In vitro* assembly of apophytochrome and apophytochrome deletion mutants expressed in yeast with phycocyanobilin. *Proc. Natl. Acad. Sci. USA* 88: 10392-10396.

Edgerton M.D. and Jones A.M. (1992) Localization of protein-protein interactions between subunits of phytochrome. *Plant Cell* 4: 161-171.

Eilfeld P.H. and Eilfeld P.G. (1988) Circular dichroism of phytochrome intermediates. *Physiol. Plant.* 74: 169-175.

Eisenberg D., Schwarz E., Komaromy M. and Wall R. (1984) Analysis of membrane and surface protein sequences with the hydrophobic moment plot. *J. Mol. Biol.* 179: 125-142.

Ekelund N.G.A., Sundqvist C., Quail P.H. and Vierstra R.D. (1985) Chromophore rotation in 124-kilodalton *Avena* phytochrome as measured by light-induced changes in linear dichroism. *Photochem. Photobiol.* 41: 221-223.

Elich T.D., Lagarias J.C. (1987) Phytochrome chromophore biosynthesis. *Plant Physiol.* 84: 304-310.

Elich T.D. and Lagarias J.C. (1989) Formation of a photoreversible phycocyanobilin-apophytochrome adduct *in vitro*. *J. Biol. Chem.* 264: 12902-12908.

Elich T.D., Mcdonagh A.F., Palmas L.A. and Lagarias J.C. (1989) Phytochrome chromophore biogenesis. *J. Biol. Chem.* 264: 183-189.

Farrens D.L., Holt R.E., Rospendowski B.N., Song P.-S. and Cotton T.M. (1989) Surface-enhanced resonance Raman scattering spectroscopy applied to phytochrome and its model compounds. 2. Phytochrome and phycocyanin chromophores. *J. Amer. Chem. Soc.* 111: 9162-9169.

Farrens D.L., Cordonnier M.-M., Pratt L.H. and Song P.-S. (1992) The distance between the phytochrome chromophore and the N-terminal chain decreases during phototransformation. A novel fluorescence energy transfer method using labelled antibody fragments. *Photochem. Photobiol.* 56: 725-733.

Fodor S., Lagarias J.C. and Mathies R. (1990) Resonance Raman analysis of the Pr and Pfr forms of phytochrome. *Biochemistry* 29: 11141-11146.

Furuya M. (1983) Molecular properties of phytochrome. *Phil. Trans. R. Soc. Lond.* B 303: 361-375.

Furuya M. (ed.) (1987) *Phytochrome and Photoregulation in Plants*, Academic Press, New York.

Furuya M., Tomizawa K., Ito N., Sommer D., Deforce L., Konomi K., Farrens D. and Song P.-S. (1991) Biogenesis of phytochrome apoprotein in transgenic organisms and its assembly to the chromophore. In: *Phytochrome Properties and Biological Action*, Thomas B. and Johnson C.B. (eds.) NATO ASI Series, Vol. 50: 71-83.

Grimm R., Eckerskorn C., Lottspeich F., Zenger C. and Rüdiger W. (1988) Sequence analysis of proteolytic fragments of 124-kilodalton phytochrome from etiolated *Avena sativa* L.: Conclusions on the conformation of the native protein. *Planta* 174: 396-401.

Grimm R., Gast D. and Rüdiger W. (1989) Characterization of a protein-kinase activity associated with phytochrome from etiolated oat (*Avena sativa* L) seedlings. *Planta* 178: 199-206.

Hahn T.R. and Song P.-S. (1981) The hydrophobic properties of phytochrome as probed by 8-anilinonaphthalene 1-sulphonate fluorescence. *Biochemistry* 20: 2602.

Hahn T.R., Chae Q. and Song P.-S. (1984a) Molecular topography of intact phytochrome probed by hydrogen-tritium exchange measurements. *Biochemistry* 23: 1219-1224.

Hahn T.R., Song P.-S., Quail P.H. and Vierstra R.D. (1984b) Tetranitromethane oxidation of phytochrome chromophore as a function of spectral form and molecular weight. *Plant Physiol.* 74: 755.

Haupt W. (1991) Introduction to photosensory transduction chains. In: *Biophysics of Photoreceptors and Photomovements in Microorganisms,* pp. 7-19, Lenci F., Ghetti F, Colombetti G., Häder D.-P. and Song P.-S. (eds.) NATO ASI Series A211 Plenum, New York.

Hayami J., Kadota A. and Wada M. (1992) Intracellular dichroic orientation of the blue light-absorbing pigment and the blue-absorption band of the red-absorbing form of phytochrome responsible for phototropism of the fern *Adiantum* protonemata. *Photochem. Photobiol.* 56: 661-666.

Heihoff K., Braslavsky S.E. and Schaffner K. (1987) Study of 124-kilodalton oat phytochrome photoconversions *in vitro* with laser-induced optoacoustic spectroscopy. *Biochemistry* 26: 1422-1427.

Hershey H.P., Barker R.F., Idler K.B., Lissemore J.L. and Quail P.H. (1985) Analysis of cloned cDNA and genomic sequences for phytochrome: complete amino acid sequences for two gene products expressed in etiolated *Avena. Nucleic Acids Res.* 13: 8543-8559.

Hildebrandt P., Hoffman A., Lindemann P., Heibel G., Braslavsky S.E., Schaffner K. and Schrader B. (1992) Fourier transform resonance Raman spectroscopy of phytochrome. *Biochemistry* 31: 7957-7962.

Holdsworth M.L. and Whitelam, G.C. (1987) A monoclonal antibody specific for the red-absorbing form of phytochrome. *Planta* 172: 539-547.

Holzwarth A.R., Wendler J., Ruzsicska B.P., Braslavsky S.E. and Schaffner K. (1984) Picosecond time-resolved and stationary fluorescence of oat phytochrome highly enriched in the native 124 kDa protein. *Biochim. Biophys. Acta* 791: 265-273.

Holzwarth A.R., Venuti E., Braslavsky S.E. and Schaffner K. (1992) The phototransformation process in phytochrome. I. Ultrafast fluorescence component and kinetic models for the initial Pr → Pfr transformation steps in native phytochrome. *Biochim. Biophys. Acta* 1140: 59-68.

Inoue Y. and Furuya M. (1985) Phototransformation of the red-light-absorbing form to the far-red-light-absorbing form of phytochrome in pea epicotyl tissue measured by a multichannel transient spectrum analyser. *Plant Cell Physiol.* 26: 813-819.

Inoue Y., Hamaguchi H., Yamamoto K.T., Tasumi M. and Furuya M. (1985) Light induced fluorescence spectral changes in native phytochrome from *Secale cereale* L. at liquid nitrogen temperature. *Photochem. Photobiol.* 42: 423-427.

Inoue Y., Rüdiger W., Rudolf G. and Furuya M. (1990) The phototransformation pathway of dimeric oat phytochrome from the red-light-absorbing form to the far-red-light-absorbing form at physiological temperature is composed of four intermediates. *Photochem. Photobiol.* 52: 1077-1083.

Ito N., Tomizawa K., Furuya M. 1991. Production of full-length pea phytochrome A (type 1) apoprotein by yeast expression system. *Plant Cell Physiol.* 32: 891-895.

Jones A.M. and Quail P.H. (1986) Quaternary structure of 124-kilodalton phytochrome from *Avena sativa* L. *Biochemistry* 25: 2987.

Jones A.M. and Erickson H.P. (1989) Domain structure of phytochrome from *Avena sativa* visualized by electron microscopy. *Photochem. Photobiol.* 49: 479-483.

Jones A.M., Vierstra R.D., Daniels S.M. and Quail P.H. (1985) The role of separate molecular domains in the structure of phytochrome from etiolated *Avena sativa* L. *Planta* 164: 501-506.

Jones A.M., Allen C.D., Gardner G., and Quail P.H. (1986) Synthesis of phytochrome apoprotein and chromophore are not coupled obligatorily. *Plant.Physiol.* 81: 1014-1016.

Kadota A., Inoue Y. and Furuya M. (1986) Dichroic orientation of phytochrome intermediates in the pathway from PR to PFR as analyzed by double laser flash irradiations in polarotropism of *Adiantum* protonemata. *Plant Cell Physiol.* 27: 867-873.

Kay S.A., Nagatani A., Keith B., Deak M., Furuya M., Chua N.-H. (1989) Rice phytochrome is biologically active in transgenic tobacco. *Plant Cell* 1: 775-782.

Keller J.M., Shanklin J., Vierstra R.D., Hershey H.P. (1989) Expression of a functional monocotyledonous phytochrome in transgenic tobacco. *EMBO J.* 8: 1005-1012

Kendrick R.E. and Spruit C.J.P. (1977) Phototransformations of phytochrome. *Photochem. Photobiol.* 26: 201-214.

Kim I.S., Bai U. and Song P.-S. (1989) A purified 124 kDa oat phytochrome does not possess a protein kinase activity. *Photochem. Photobiol.* 49: 319-323.

Konomi K. and Furuya M. (1986) Effect of gabaculine on phytochrome synthesis during imbibition in embryonic axes of *Pisum sativum* L. *Plant Cell Physiol.* 27: 1507-1512.

Konomi K., Abe H., Furuya M. (1987) Changes in the content of phytochrome I and II apoproteins in embryonic axes of pea seeds during imbibition. *Plant Cell Physiol.* 28: 1443-1451.

Konomi K., Li H.-S, Kuno N., Furuya M. (1993) Effect of N-phenylimide S-23142 and N-methyl mesoporphyrin IX on the synthesis of phytochrome chromophore in pea embryonic axes. *Photochem. Photobiol.* in press.

Lagarias J.C., Rapoport H. (1980) Chromopeptides from phytochrome. The structure and linkage of the Pr form of the phytochrome chromophore. *J. Amer. Chem. Soc.* 104: 4821-4828.

Lagarias J. C. and Mercurio F. M. (1985) Structure function studies on phytochrome Identification of light-induced conformational changes in 124-kDa *Avena* phytochrome *in vitro. J. Biol. Chem.* 260: 2415-2413.

Lagarias J.C., Lagarias D.M. (1989) Self-assembly of synthetic phytochrome holoprotein *in vitro. Proc. Natl. Acad. Sci. USA* 86: 5778-5780.

Lagarias J.C., Kelly J.M., Cyr K.L. and Smith W.O. Jr. (1987) Comparative photochemical analysis of highly purified 124 kilodalton oat and rye phytochromes *in vitro. Photochem. Photobiol.* 46: 5-13.

Lamparter T., Lutterbuese P., Schneider-Poetsch H.A.W. and Hertel R. (1992) A study of membrane-associated phytochrome: Hydrophobicity test and native size determination. *Photochem. Photobiol.* 56: 697-707.

Li L. and Lagarias J.C. (1992) Phytochrome assembly: Defining chromophore structural requirements for covalent attachment and photoreversibility. *J. Biol. Chem.* 267: 19204-19210.

Lippitsch M.E.. Riegler H., Aussenegg F.R., Harmann G. and Müller E. (1988) Picosecond absorption and fluorescence studies on large phytochrome from rye. *Biochem. Physiol. Pflanz.* 183: 1-6.

McMichael R.W. Jr. and Lagarias J.C. (1990) Phosphopeptide mapping of *Avena* phytochrome phosphorylated by protein kinases *in vitro. Biochemistry* 29: 3872-3878.

Metzger H. (1992) Transmembrane signalling: The joy of aggregation. *J. Immunol.* 149: 1477-1487.

Mizutani Y., Tokutomi S., Aoyagi K., Horitsu K. and Kitagawa T. (1991) Resonance Raman study on intact pea phytochrome and its model compounds: Evidence for proton migration during the phototransformation. *Biochemistry* 30: 10693-10700.

Nakasako M., Wada M., Tokutomi S., Yamamoto K.T., Sakai J., Kataoka M., Tokunaga F. and Furuya M. (1990) Quaternary structure of pea phytochrome I dimer studied with small-angle X-ray scattering and rotary-shadowing electron microscopy. *Photochem. Photobiol.* 52: 3-12.

Parker W., Romanowski M. and Song P.-S. (1991) Conformation and its functional implications in phytochrome. In: *Phytochrome Properties and Biological Action,* Thomas B. and Johnson C.B. (eds.) NATO ASI Series Vol. 50: 85-112.

Parker W., Partis M. and Song, P.-S. (1992) N-Terminal domain of *Avena* phytochrome: Interactions with sodium dodecyl sulfate micelles and N-terminal chain truncated phytochrome. *Biochemistry* 31: 9413-9420.

Parker W., Goebel P., Song P.-S. and Stezowski J.J. (1993) Molecular modelling of phytochrome using constitutive C-phycocyanin from *Fremyella diplosiphon* as a structural template. *Bioconjugate Chem.* in press.

Parks B.M. and Quail P.H. (1991) Phytochrome-deficient *hy1* and *hy2* long hypocotyl mutants of *Arabidopsis* are defective in phytochrome chromophore biosynthesis. *Plant Cell* 3: 1177-1186.

Partis M.D. and Grimm R. (1990) Computer analysis of phytochrome sequences from five species: Implications for the mechanism of action. *Z. Naturforsch* 45c: 987-998.

Pratt L.H. (1986) Phytochrome: localization within the plant. In: *Photomorphogenesis in Plants*, pp. 61-81, Kendrick R.E. an Kronenberg G.H.M. (eds.) Martinus Nijhoff Publishers, Dordrecht.

Pratt L.H., Inoue,Y. and Furuya M. (1984) Photoactivity of transient intermediates in the pathway from the red-absorbing to the far-red-absorbing form of Avena phytochrome as observed by a double-flash transient-spectrum analyzer. *Photochem. Photobiol.* 39: 241-246.

Quail P.H. (1991) Phytochrome: A light-activated molecular switch that regulates plant gene expression. *Annu. Rev. Genet.* 25: 389-409.

Quail P.H., Barker R.F., Colbert J.T., Daniels S.M., Hershey H.P., Idler K.B., Jones A.M. and Lissemore J.L. (1987) Structural features of the phytochrome molecule and feedback regulation of the expression of its genes in *Avena*. In: *Molecular Biology of Plant Growth Control*, pp. 425-439, Fox J.F. and Jacobs M. (eds.) Alan R. Liss, New York.

Romero L.C. and Lam E. (1993) Guanine nucleotide binding protein involvement in early steps of phytochrome-regulated gene expression. *Proc. Natl. Acad. Sci. USA* 90: 1465-1469.

Romero L.C., Sommer D., Gotor C., and Song P.-S. (1991) G-protein(s) in etiolated *Avena* seedlings: Possible phytochrome regulation. *FEBS Lett.* 282: 341-346.

Romanowski M. and Song P.-S. (1992) Structural domains of phytochrome deduced from homologies in amino acid sequences. *J. Protein Chem.* 11: 139-155.

Rospendowski B.N., Farrens D.L., Cotton T.M. and Song P.-S. (1989) Surface enhanced resonance Raman scattering (SERRS) as a probe of the structural differences between the Pr and Pfr forms of phytochrome. *FEBS Lett.* 258: 1-4.

Roux S.J. (1983) A possible role of Ca^{2+} in mediating phytochrome responses. In: *The Biology of Photoreception*, pp. 561-508, Cosens D. and Vince-Prue D. (eds.) Cambridge University Press, Cambridge.

Rüdiger W. and Correll D.L. (1969) Über die Struktur des Phytochrom-Chromophors und seine Protein-Binding. *Liebigs. Ann. Chem.* 723: 208-212.

Rüdiger W., Thümmler F., Cmiel E. and Schneider S. (1983) Chromophore structure of the physiologically active form (Pfr) of phytochrome. *Proc. Natl. Acad. Sci. USA* 80: 6244-6248.

Sarkar H.K., Moon D.K., Song P.-S., Chang T. and Yu H. (1984) Tertiary structure of phytochrome probed by quasi-elastic light scattering and rotational relaxation time measurements. *Biochemistry* 23: 1882-1888.

Schaffner K., Braslavsky S.E. and Holzwarth A.R. (1990) Photophysics and photochemistry of phytochrome. *Adv. Photochem.* 15: 229-277.

Schaffner K., Braslavsky S.E. and Holzwarth A.R. (1991) Protein environment, photophysics and photochemistry of prosthetic biliprotein chromophores. In: *Frontiers in Supramolecular Organic Chemistry and Photochemistry*, pp. 421-452, Schneider H.J. and Dürr H. (eds.) VCH Verlagesellschaft.

Schendel R. and Rüdiger W. (1989) Electrophoresis and electrofocusing of phytochrome from etiolated *Avena sativa* L. *Z. Naturforsch* 44c: 12-18.

Schneider-Poetsch H.A.W. (1992) Signal transduction by phytochrome: phytochromes have a module related to the transmitter modules of bacterial sensor proteins. *Photochem. Photobiol.* 56: 839-846.

Shanklin J., Jabben M. and Vierstra R.D. (1989) Partial purification and peptide mapping of ubiquitin-phytochrome conjugates from oat. *Biochemistry* 28: 6028-6034.

Shimazaki Y. and Furuya M. (1975) Isolation of a naturally occurring inhibitor for dark Pfr reversion from etiolated *Pisum* epicotyls. *Plant Cell Physiol.* 16: 623-630.

Siebert F., Grimm R., Rüdiger W., Schmidt G. and Scheer H. (1990) Infrared spectroscopy of phytochrome and model pigments. *Eur. J. Biochem.* 194: 921-928.

Siegelman H.W., Turner B.C. and Hendricks S.B. (1966) The chromophore of phytochrome. *Plant Physiol.* 41: 1289-1292.

Singh B.R. and Song P.-S. (1990) A differential molecular topography of the Pr and Pfr forms of native oat phytochrome as probed by fluorescence quenching. *Planta* 181: 263-267.

Singh B.R., Chai Y.G., Song P.-S., Lee J. and Robinson G.W. (1988) A photoreversible conformational change in 124 kDa *Avena* phytochrome. *Biochim. Biophy. Acta* 936: 395-405.

Singh B.R., Choi J., Kwon T. and Song P.-S. (1989) Use of bilirubin oxidase for probing chromophore topography in tetrapyrrole proteins. *J. Biochem. Biophys. Meth.* 18: 135-148.

Sommer D. and Song P.-S. (1990) Chromophore topography and secondary structure of 124-kilodalton *Avena* phytochrome probed by Zn^{2+}-induced chromophore modification. *Biochemistry* 29: 1943-1948.

Song P.-S. (1988) The molecular topography of phytochrome: Chromophore and apoprotein. *J. Photochem. Photobiol.* Part B 2: 43-57.

Song P.-S. and Chae Q. (1979) The transformation of phytochrome to its physiologically active form. *Photochem. Photobiol.* 30: 117-123.

Song P.-S. and Yamazaki I. (1987) Structure-function relationship of the phytochrome chromophore. In: *Phytochrome and Photoregulation in Plants,* pp. 139-156, Furuya M. (ed.) Academic Press, New York.

Song P.-S., Chae Q. and Gardner J.G. (1979) Spectroscopic properties and chromophore conformations of the photomorphogenic receptor: Phytochrome *Biochim. Biophys. Acta* 576: 479-495.

Song P.-S., Sarkar H., Kim I.S. and Poff K.L. (1981) Primary photoprocesses of undegraded phytochrome excited with red and blue light at 77 K. *Biochim. Biophys. Acta* 635: 369-382.

Song P.-S., Tamai N. and Yamazaki I. (1986) Viscosity dependence of primary photoprocesses of 124 kdalton phytochrome. *Biophys. J.* 49: 645.

Song P.-S., Singh B.R., Tamai N., Yamazaki T., Yamazaki I., Tokutomi S. and Furuya M. (1989) Primary photoprocesses of phytochrome. Picosecond fluorescence kinetics of oat and pea phytochromes. *Biochemistry* 28: 3265-3271.

Stockhaus J., Nagatani A., Halfter U., Kay S., Furuya M. and Chua N.-H. (1992) Serine to alanine substitutions at the amino-terminal region of phytochrome A result in an increase in biological activity. *Genes Develop.* 6: 2364-2372.

Thümmler F., Dufner M., Kreisl P. and Dittrich P. (1992) Molecular cloning of a novel phytochrome gene of the moss *Ceratodon purpureus* which encodes a putative light-regulated protein kinase. *Plant Mol. Biol.* 20: 1003-1017.

Tokutomi, S. and Mimuro, M. (1989) Orientation of the chromophore transition moment in the 4-leaved shape model for pea phytochrome molecule in the red-light absorbing form and its rotation induced by the phototransformation to the far-red-light absorbing form. *FEBS Lett.*, 255: 350-353.

Tokutomi S., Yamamoto K.T., Miyoshi Y. and Furuya M. (1982) Photoreversible changes in pH of pea phytochrome solutions. *Photochem. Photobiol.* 35: 431-433.

Tokutomi S., Yamamoto K.T. and Furuya M. (1988) Photoreversible proton dissociation and association in pea phytochrome and its chromopeptides. *Photochem. Photobiol.* 47: 439-445.

Tomizawa K., Ito N,. Komeda Y,. Uyeda T.Q.P., Takio K., Furuya M. (1991) Characterization and intracellular distribution of pea phytochrome I polypeptides expressed in *E. coli. Plant Cell Physiol.* 32: 95-102

Tretyn A., Kendrick R.E. and Wagner G. (1991) The role(s) of calcium ions in phytochrome action. *Photochem. Photobiol.* 54: 1135-1155.

VanDerWoude W.J. (1985) A dimeric mechanism for the action of phytochrome: Evidence from photothermal interactions in lettuce seed germination. *Photochem. Photobiol.* 42: 655-661.

VanDerWoude W.J. (1987) Application of the dimeric model of phytochrome action to high irradiance responses. In: *Phytochrome and Photoregulation in Plants,* pp. 249-258, Furuya M. (ed.) Academic Press, New York.

Vierstra R.D. and Quail P.H. (1982) Proteolysis alters the spectral properties of 124 kdalton phytochrome from *Avena*. *Planta* 156: 158-165.

Vierstra R.D. and Quail P.H. (1983) Purification and initial characterization of 124-kilodalton phytochrome. *Biochemistry* 22: 2498-2505.

Vierstra R.D., Quail P.H., Hahn T.R. and Song P.-S. (1987) Comparison of the protein conformations between different forms (Pr vs. Pfr) of native (124 kDa) and degraded (118/114 kDa) phytochromes from *Avena*. *Photochem. Photobiol.* 45: 429-432.

Wagner D., Tepperman J.M. and Quail P.H. (1991) Overexpression of phytochrome B induces a short hypocotyl phenotype in transgenic *Arabidopsis*. *Plant Cell* 3: 1275-1288.

Wahleithner J.A., Li L., Lagarias J.C. (1991) Expression and assembly of spectrally active recombinant holophytochrome. *Proc. Natl. Acad. Sci. USA* 88: 10387-10391.

Wong Y.S., McMichael R.W. Jr. and Lagarias J.C. (1989) Properties of a polycation-stimulated protein kinase associated with purified *Avena* phytochrome. *Plant Physiol.* 91: 709-718.

Yamamoto K.T. and Tokutomi S. (1989) Formation of aggregates of tryptic fragments derived from the carboxyl-terminal half of pea phytochrome and localization of the site of contact between the fragments by amino-terminal amino acid sequence analysis. *Photochem. Photobiol.* 50: 113-120.

Zhang C.F., Farrens D.L., Björling S.C., Song P.-S. and Kliger D.S. (1992) Time-resolved absorption studies of native etiolated oat phytochrome. *J. Amer. Chem. Soc.* 114: 4569-4580.

4.4 Phytochrome degradation

Richard D. Vierstra

Department of Horticulture, University of Wisconsin-Madison, 1575 Linden Drive, Madison, WI 53706, USA

4.4.1 Introduction

A common feature of phytochrome in almost all plants examined is its rapid degradation upon photoconversion of the red light (R)-absorbing form (Pr) to the far-red light (FR)-absorbing form (Pfr) (Pratt 1979; Jabben and Holmes 1984). In etiolated *Cucurbita pepo* seedlings for example, this transformation increases the degradation rate of the photoreceptor over 100-fold, decreasing the half-life ($t_{1/2}$) from > 100 h for Pr to < 1 h for Pfr (Quail *et al.* 1973b). Breakdown involves the loss of both spectrally and immunologically detectable phytochrome, not merely masking of the chromoprotein. Paradoxically, R can be viewed not only as necessary for converting phytochrome to the physiologically active form, but also for initiating the rapid breakdown of that form. This intriguing phenomenon, also referred to in the literature as phytochrome destruction, has received much attention because of its potential relevance to phytochrome function(s) and because it is responsible for regulating, to a large extent, the amount of the photoreceptor *in vivo*. Furthermore, because this differential stability can be rapidly and synchronously manipulated *in vivo* with non-invasive light treatments, phytochrome degradation has provided a useful paradigm in understanding how intact cells selectively breakdown intracellular proteins.

This chapter reviews the current state of knowledge concerning phytochrome degradation and relates its importance to a proper function of the photoreceptor. Where possible, attempts will be made to associate previous data with the current realization that multiple forms of phytochrome exist, each with potentially different functions and stabilities (Chapter 4.2; Furuya 1989; Quail 1991). In fact, analysis of degradation kinetics provided one of the first indications that phytochrome is heterogeneous (Brockmann and Schäfer 1982). In keeping with the nomenclature of Furuya (1989), the more rapidly degraded form of phytochrome abundant in etiolated seedlings will be designated type I and is predicted to be the species encoded by the *phyA* gene(s). The more stable phyto-

R.E. Kendrick & G.H.M. Kronenberg (eds.), Photomorphogenesis in Plants - 2nd Edition

chrome that predominates in light-grown plants will be designated type II and includes those proteins encoded by the *phyB* and *C* genes and possibly also the *phyD* and *E* genes (Quail 1991). Because all phytochromes appear to some extent to be degraded once converted to Pfr, the use of the designations 'photo-labile' or 'photostable' will be avoided. Moreover, the discussion will be restricted to breakdown occurring rapidly after phototransformation to Pfr, despite the likelihood that plants also slowly turnover Pr prior to photo-conversion.

4.4.2 Properties of phytochrome degradation

4.4.2.1 Etiolated plants

Phytochrome degradation in etiolated (or dark-grown) plants was inadvertently discovered in 1959 during the first attempts to publicly demonstrate the existence of the photoreceptor. It provides one of the more amusing anecdotes in phytochrome folklore (Sage 1992). In an initial screen of plant materials for phytochrome, S.B. Hendricks, W.L. Butler, and co-workers from the USDA laboratory in Beltsville found that whereas green plants had generally low or undetectable levels, etiolated tissue of the same species had high levels of the photoreceptor. Thus, to insure a successful presentation at an international meeting in Montreal, they transported etiolated maize seedlings to the site from Beltsville in the trunk of their car, periodically checking the plants during the 2-day ride to insure their safety. Unfortunately, when finally assayed during the lecture, no phytochrome was detected in the seedlings and the audience left skeptical. In an attempt to understand the reasons behind the fiasco, they soon discovered that their brief examinations were sufficient to photoconvert all the Pr to Pfr, which in turn initiated destruction of the chromoprotein even before the plants were assayed. Once this instability was realized in maize, they soon detected it in etiolated seedlings from many other dicot and monocot species (Butler *et al.* 1963; Butler and Lane 1964).

Most of our current understanding on phytochrome degradation is derived from studies with etiolated seedlings. This is mainly because etiolated plants can have 10-100 times higher phytochrome levels than green plants and because they lack chlorophyll which interferes with the spectrophotometric assays of the chromoprotein. The high phytochrome levels result from active transcription of *phyA* genes in the dark and the extreme stability of Pr ($t_{1/2} >$ 100 h in *Cucurbita*) (Quail *et al.* 1973b; Furuya 1989; Quail 1991). In some etiolated tissues, *e.g.* the plumule hook in dicots and the coleoptile tip and mesocotyl node in monocots, phytochrome can account for up to 1% of the total protein if the tissues are kept in the dark. This pool represents mainly type I phytochrome with type II chromoproteins comprising only a minor fraction

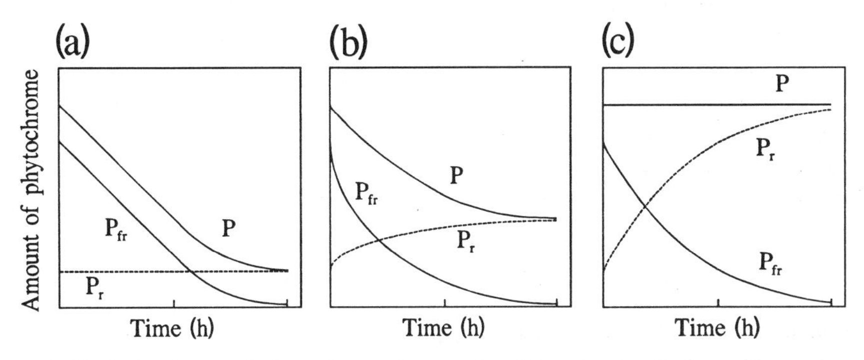

Figure 1. Typical patterns of phytochrome degradation following irradiation with a pulse of red light. (a) Etiolated monocot or *Amaranthus* seedlings where only Pfr degradation is observed. (b) Etiolated dicot seedlings where Pfr reversion to Pr occurs in addition to Pfr degradation. (c) Cauliflower inflorescences where Pfr degradation is not detectable and only Pfr reversion occurs. P = total phytochrome (Pr + Pfr).

(Tokuhisa and Quail 1987; Furuya 1989). Upon photoconversion to Pfr, phytochrome disappears rapidly with half of the pool degraded within 30 min to 2 h (Pratt *et al.* 1974; Schäfer *et al.* 1975; Brockmann and Schäfer 1982; Shanklin *et al.* 1987; Jabben *et al.* 1989b). This half-life varies depending on the species, tissue type, age, and degree of etiolation. Pratt *et al.* (1974) showed using immunochemical assays that destruction results in loss of the chromoprotein and is not a trivial consequence of compartmentation as was first proposed. Kinetic analyses of both Pr and Pfr pools indicate that only those molecules converted to Pfr are degraded (Fig. 1). If a brief pulse of R is used, the amount of phytochrome degraded reflects the level of Pfr created initially (up to 86-87% with saturating 660 nm R). If continuous R is used, phytochrome levels continue to drop until an equilibrium is reached between synthesis and degradation. Because Pfr also represses further transcription of *phyA* genes, especially in monocots (Quail 1991), this steady state level can represent as little as ≈ 1% of that prior to irradiation.

In an attempt to understand the basis for Pfr degradation, a multitude of inhibitor and kinetic studies have been reported (Pratt 1979). The temperature coefficient (or Q_{10}) of the reaction is ≈ 3, implying that destruction is an enzymatic process. Degradation is attenuated by respiratory inhibitors such as carbon monoxide, azide, cyanide, and anaerobic conditions, indicating that energy is also required. This finding is in agreement with parallel work showing that *in vivo* degradation of most proteins in animals, plants, and bacteria requires energy (Goldberg and St John 1976; Vierstra 1993). Originally, such an energy requirement for general proteolysis was surprising given that peptide bond hydrolysis is an exergonic reaction. However, as will be shown below, this energy assists in keeping proteolysis under control. Generally, Pfr degradation begins immediately implying that the degradation machinery pre-exists in plant cells. However, delays up to 40 min in the onset of destruction have been noted for very young seedlings and for those germinated in closed containers (Stone and Pratt 1978; Schäfer *et al.* 1975). These observations have been used to speculate that the potential to degrade phytochrome can be induced. For dicots, kinetic analyses following saturating R pulses indicate that degradation is first order with respect to phytochrome (Fig. 1b). This implies that the machinery used to degrade the dicot photoreceptor is in excess relative to Pfr. However, for monocots, zero order kinetics have been observed even at reduced initial Pfr levels (Fig. 1a). From these data, Schäfer *et al.* (1975) concluded that the degradation machinery in monocots is limiting but has a high catalytic efficiency. The independence of the degradation rate to Pfr concentrations in monocots allows for rapid phytochrome breakdown even at the low Pfr levels maintained under continuous FR (Jabben *et al.* 1989a)

As expected, rapid degradation of Pfr in etiolated tissues mainly involves the destruction of type I phytochrome (Furuya 1989). However, detailed analysis of Pfr disappearance in etiolated seedlings of several species indicate biphasic kinetics; in these cases, the well described fast destruction of the bulk of photoreceptors was followed by a slower loss of a substantially smaller pool. Half-lives for the slow degradation ranged from 2.5 h in *Brassica*, 5 h in *Amaranthus*, and 6-7 h in *Pharbitis* (Heim *et al.* 1981; Brockmann and Schäfer 1982). Such biphasic rates implied that a small pool of additional phytochromes are present in etiolated plants that are more stable than type I phytochrome. This proposal has since been supported by immunological data that identified type II phytochrome and showed it to be an apparently more stable species (Tokuhisa and Quail 1987; Furuya 1989). However, it must be noted that present estimates for the rate of type II phytochrome turnover actually represent a balance between synthesis and degradation. Given the recent observations that transcription of type II phytochromes genes are unaffected by Pfr (Sharrock and Quail 1989; Quail 1991), chromoprotein synthesis could potentially mask ongoing Pfr breakdown following irradiation.

Even though rapid degradation is generally assumed to be specific for Pfr, rapid degradation of Pr has also been observed under specific conditions in etiolated *Avena* (Pratt 1979). It requires that Pr be converted to Pfr and then back to Pr generating cycled Pr. Approximately one third of the cycled Pr pool can be degraded by this interconversion. Because cycled Pr degradation is affected by the same conditions and inhibitors as Pfr degradation, Stone and Pratt (1979) concluded that cycled Pr is broken down by the same mechanism. Cycled Pr destruction has subsequently been observed in several other plant species, including both monocots and dicots, suggesting that the phenomenon may be universal (Fukshansky and Schäfer 1984).

An additional non-photochemical reaction that competes with Pfr destruction in some species is dark reversion (Pratt 1979). This is the thermal reversion of Pfr back to Pr in the absence of light. In etiolated *Cucurbita* seedlings, up to

30% of Pfr will revert to Pr and avoid degradation (Schäfer and Schmidt 1974) (Fig. 1b). Dark reversion is prevalent in etiolated seedlings of dicots [with the exception of *Amaranthus* (Kendrick and Frankland 1968)], but is notably absent in etiolated grass seedlings. The basis for Pfr reversion is an enigma, especially considering the fact that *Cucurbita* phytochrome, purified intact from tissue that exhibits dark reversion *in vivo*, does not revert *in vitro* (Vierstra and Quail 1985). The physiological significance of dark reversion is also unknown, but it clearly protects a fraction of Pfr from degradation. This is best exemplified for light-grown cauliflower inflorescences, in which all Pfr appears to revert to Pr and no Pfr degradation is observed (Butler *et al.* 1963) (Fig. 1c).

4.4.2.2 Light-grown plants

While not analyzed to the same extent as etiolated seedlings, light grown plants also appear to degrade phytochrome after photoconversion to Pfr (Jabben and Holmes 1984). Most data measuring phytochrome loss *in vivo* have been obtained with Norflurazon treated seedlings to mitigate the interference of chlorophyll in phytochrome spectrophotometric assays (Jabben and Deitzer 1978; Jabben 1980). As a result, secondary effects of the herbicide on degradation cannot be eliminated. Immunological assays have since been developed to accurately measure phytochrome, even in the presence of chlorophyll (Hunt and Pratt 1979; Shimazaki *et al.* 1983). These assays have generated similar degradation kinetics even though only soluble phytochrome was detected. Despite these caveats, it appears that Pfr degradation does occur in most light-grown plants [with the exception of cauliflower inflouresences (Butler *et al.* 1963)], but is significantly slower than that observed in etiolated seedlings. For example, the half-life of Pfr in maize is 8 h in light-grown seedlings but only 1 h in etiolated seedlings (Jabben 1980). Similar slow destruction rates have been observed for *Avena* and several dicots (Heim *et al.* 1981; Jabben and Holmes 1984; Shimazaki *et al.* 1983).

The reasons for the slower Pfr destruction in light-grown plants is unclear. It may reflect the possibility that light-grown plants have inherently less capacity to degrade Pfr or that the type(s) of phytochrome present are less susceptible. In support of the latter, light grown plants are enriched in the apparently more stable type II phytochromes. This is a consequence of the constitutive expression of type II phytochromes genes (*i.e. phyB* and *C)* in conjunction with the light-repressed expression of type I phytochrome *(phyA)* and the rapid degradation of type I Pfr (Sharrock and Quail 1989; Furuya 1989; Quail 1991).

4.4.2.3 Relationship of degradation to phytochrome sequestering

One of the more unique responses of phytochrome after photoconversion to Pfr is phytochrome sequestering (Chapter 4.5; Mackenzie *et al.* 1975; Pratt 1986). It is observed in etiolated plants but may occur in light-grown plants as well. Sequestering involves the relocation of immunodetectable phytochrome from its disperse distribution within the cytoplasm into numerous discrete aggregates. These aggregates are approximately 1 μm in size and appear as amorphous granules by electron microscopy. While first proposed to represent phytochrome bound to organelles, it has been subsequently shown that the aggregates are neither associated with any recognizable subcellular structure nor delineated by a membrane (McCurdy and Pratt 1986a). Under special extraction conditions (*e.g.* inclusion of divalent cations), such sequestered areas can be retained *in vitro* as 'pelletable phytochrome' and precipitate with other particulate material (Quail *et al.* 1973a; McCurdy and Pratt 1986a). The aggregates contain type I phytochromes; whether type II phytochromes are also present remains to be demonstrated. The speed of the process is striking, occurring within seconds after Pfr formation (Quail and Briggs 1978; McCurdy and Pratt 1986b). It is also reversible, but takes considerably longer for cycled Pr to disaggregate [$t_{1/2} \approx 25$ min at 25°C for etiolated *Avena* (Pratt and Marmé 1976)]. Sequestering can be attenuated by low temperature and metabolic inhibitors implying that an energy-dependent, enzymatic reaction is required after Pfr formation (Pratt and Marmé 1976; Quail and Briggs 1978; McCurdy and Pratt 1986b).

The function(s) of sequestering is not known. Since it represents one of the fastest *in vivo* reactions attributed to the active Pfr form, much speculation has been generated as to a possible connection between phytochrome redistribution and action (Chapter 4.5; Pratt 1986). It has also been proposed that sequestering is involved in phytochrome degradation (Mackenzie *et al.* 1975). In support of this, sequestered phytochrome is the form predominantly lost during Pfr degradation (Boisard *et al.* 1974; MacKenzie *et al.* 1975). Moreover, phytochrome aggregates appear similar in structure to inclusion bodies resulting from the aggregation of abnormal proteins in both prokaryotes and eukaryotes (Goldberg and St John 1976). Inclusion bodies appear to be sites where such abnormal proteins are eventually catabolized. In contrast to phytochrome aggregation, formation of inclusion bodies is energy *independent* and may be a spontaneous reaction. Thus, it is tempting to speculate that the energy requirement for phytochrome sequestering reflects a mechanism that allows Pfr to aggregate even in the native state.

4.4.3 Mechanisms of phytochrome degradation

In an attempt to understand how phytochrome is rapidly degraded once converted to Pfr, a number of strategies have been explored. These include the analysis of phytochrome sequence and structure for instability domains (Rogers *et al.* 1986), the isolation of sequestered phytochrome (Hofmann *et al.* 1991), and the search for degradation intermediates (Shanklin *et al.* 1987; Jabben *et al.* 1989a, b). Major limitations have been the inability to develop an *in vitro* system capable of specifically degrading Pfr and the lack of mutants affecting the proteolytic pathway involved. Given the remarkable selectivity (especially for type I phytochrome), the proteolytic mechanism responsible must recognize structural differences between Pr and Pfr. Whether these differences are detected directly or indirectly as a consequence of phytchrome re-location following photoconversion is unknown. We can also not rule out the possibility that more than one pathway is responsible for phytochrome turnover.

4.4.3.1 Possible involvement of ubiquitin-dependent proteolysis

Shanklin *et al.* (1987) reported the first attempts to identify proteolytic cleavage products during Pfr degradation. Even extremely sensitive immunoblot analyses failed to detect any *in vivo* cleavage fragments in etiolated *Avena* seedlings. This indicated, based on the limits of detection, that such fragments represent less than 0.1% of the phytochrome pool and implied that breakdown is both fast and complete. However, such immunoblots did reveal a novel modification potentially involved in Pfr degradation, that of the small protein *ubiquitin* linked one or more times to the photoreceptor (Shanklin *et al.* 1987). These ubiquitin-phytochrome conjugates (Ub-P) were first identified as a ladder of immunodetectable phytochrome proteins with apparent molecular masses greater than that of the unmodified native phytochrome monomer during sodium dodecyl sulfate-polyacrylamide gel electrophoresis (SDS-PAGE). The increased mass was consistent with incremental additions of ubiquitin moieties. The presence of ubiquitin moieties was subsequently confirmed by immunorecognition of the ladder with anti-ubiquitin immunoglobulins. The known function of ubiquitin in protein degradation led to a possible mechanism for Pfr destruction.

Ubiquitin is a highly conserved, 76-amino acid protein that plays a prominent role in the degradation of many abnormal and short-lived eukaryotic proteins (Finley and Chau 1991; Hershko and Ciechanover 1992; Vierstra 1993). It was first discovered in 1980 by Hershko and co-workers during the search for components essential for protein breakdown in rabbit reticulocyte lysates and has since been shown to be important in plants as well (Vierstra 1993). The structure of ubiquitin consists of a compact globular core with a flexible,

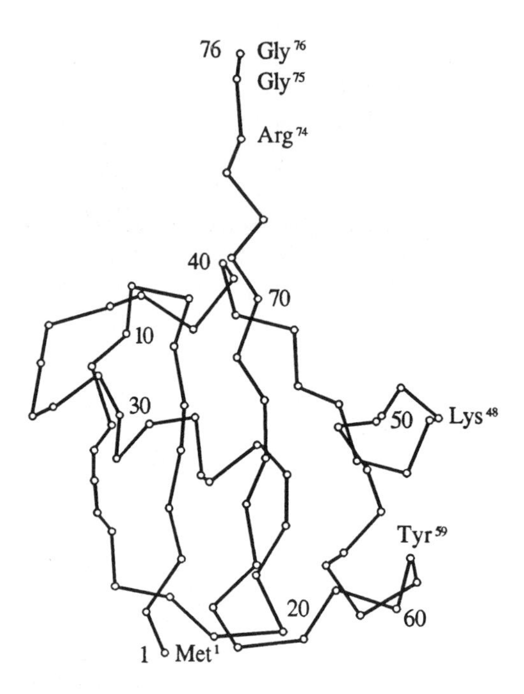

Figure 2. Three-dimensional structure of plant ubiquitin as determined by X-ray crystallography. From Vijay *et al.* (1987).

protruding carboxyl terminus (Fig. 2). Extensive hydrogen bonding within the molecule confers resistance to heat, proteases, and acidic or basic conditions (Vijay *et al.* 1987).

The primary function of ubiquitin is to become covalently bound to other cellular proteins through an isopeptide linkage between the carboxyl terminus of ubiquitin and free lysyl ε-amino groups on targeted proteins. Ubiquitin also becomes conjugated to itself generating multi-ubiquitin chains attached to the target (Finley and Chau 1991; Hershko and Ciechanover 1992). Here, attachment is through lysine^{48} within the ubiquitin moieties (Fig. 2). This added multi-ubiquitin chain then serves as a recognition signal for a specific proteolytic complex (termed the 26S proteasome) that degrades the target protein to amino acids with the release of ubiquitin in a free, functional form. As a result of the cycle depicted in Fig. 3, ubiquitin serves as a re-usable recognition signal for proteolysis. Conjugation requires the action of three enzymes E1, E2, and E3 with the latter two being responsible for substrate recognition. Multiple forms of E2 and E3 exist with potentially different functions and/or target protein specificities. The hydrolysis of ATP is required at two steps, activation of ubiquitin by E1 and degradation of conjugates by the 26S proteosome (Fig. 3). These energy-dependent steps potentially provide important control points for proteolysis. The pathway has been identified in a

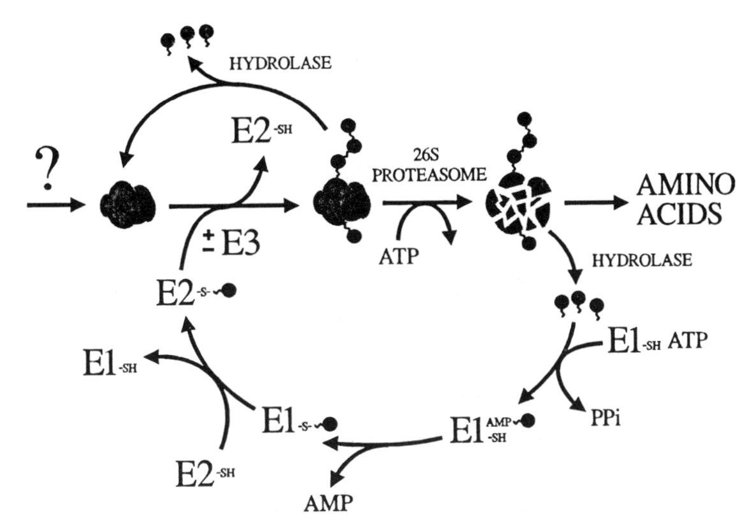

Figure 3. Pathway for ubiquitin-dependent proteolysis. Pathway begins with activation of ubiquitin with ATP by E1 followed by binding of ubiquitin to E1 with the release of AMP. Activated ubiquitin is then transferred to E2s which in turn ligates ubiquitin *via* an isopeptide bond to the target protein with or without the help of E3. Ubiquitin-protein conjugates are either degraded to amino acids by the ATP-dependent 26S proteosome with the release of ubiquitin undegraded, or disassembled by ubiquitin-protein hydrolases.

variety of eukaryotes including mammals, yeast, and higher plants and appears to contain many conserved elements (Finley and Chau 1991; Hershko and Ciechanover 1992; Vierstra 1993).

Given ubiquitin's involvement in proteolysis and the fact that Ub-P appear and disappear concomitant with Pfr degradation, Shanklin *et al.* (1987) proposed that Ub-P represent an intermediate step during type I phytochrome catabolism. The enzymatic pathways for ubiquitin attachment and conjugate degradation are consistent with the temperature and energy (ATP) dependence for Pfr degradation (Pratt 1979). Together, kinetic analyses, localization of Ub-P, and the analysis of Pfr degradation in a variety of etiolated seedlings, both monocots and dicots, further support this hypothesis (Shanklin *et al.* 1987; Jabben *et al.* 1989a, b; Cherry *et al.* 1991). Figure 4 illustrates the accumulation of Ub-P during Pfr degradation in etiolated tobacco seedlings; similar results have been obtained with *Avena*, *Cucurbita*, pea, corn, and rye (Jabben *et al.* 1989a, b). The Ub-P, which are not detectable in dark-grown plants, appear soon (< 5 min) after R irradiation and gradually disappear after Pfr degradation is complete. Phytochrome modified with as many as seven ubiquitins can be detected, in agreement with the requirement of multi-ubiquitin chains for protease recognition (Fig. 4). During Pfr degradation in *Avena*, Jabben *et al.* (1989a) estimated that Ub-P can comprise as much as 11% of the total phyto-

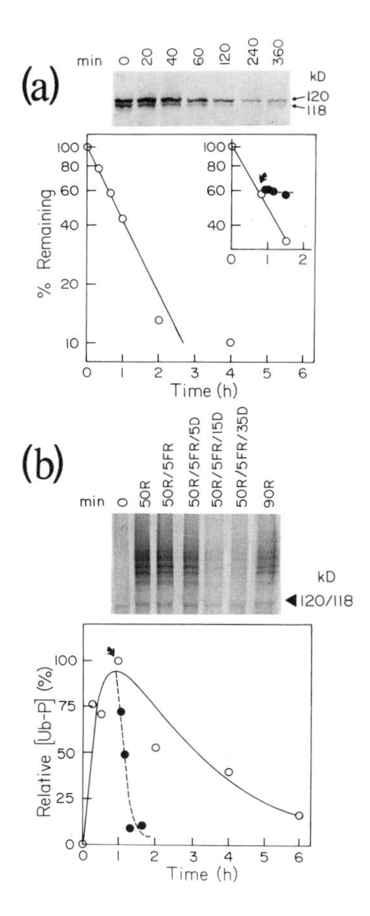

Figure 4. Phytochrome degradation and ubiquitin-phytochrome conjugate (Ub-P) accumulation in etiolated tobacco following irradiation. Seedlings were grown in the dark (D) for 5 days and at t = 0 either were irradiated continuously with red light (R, ○) or irradiated with R followed by a 5 min irradiation with far-red light (FR, ●) and a further incubation in D. At various times the seedlings were assayed for phytochrome content and Ub-P. (a) Kinetics of Pfr degradation: *upper* panel, loss of phytochrome measured by immunoblot analysis with anti-phytochrome immunoglobulins; *lower* panel, loss of phytochrome by R-minus-FR difference spectroscopy ($\Delta\Delta A$); *inset*, the effect of a 50 min R irradiation followed by a 5 min FR irradiation on phytochrome degradation. (b) Accumulation and loss of Ub-P during phytochrome degradation. The Ub-P were partially purified by immunoprecipitation with anti-phytochrome immunoglobulins and detected by immunoblot analysis with anti-ubiquitin immunoglobulins. *Upper* panel: immunoblot analysis of Ub-P. *Lower* panel: relative levels of Ub-P during phytochrome degradation determined from the immunoblots by reflective densitometry. Arrowheads indicate the position of unmodified 120/118-kD tobacco phytochrome. From Cherry *et al.* (1991).

chrome pool, implying that conjugation is not a trivial side reaction. Reducing Pfr levels by attenuating R flashes concomitantly lower both the amount of Pfr degraded and the accumulation of Ub-P. Moreover, Pfr degradation and Ub-P accumulation are indistinguishable using either R flashes to convert 86% of Pr to Pfr or continuous FR to maintain a constant level of Pfr at 4%. Taken together, the data indicate that the important feature for Ub-P accumulation is not the amount of Pfr, but the amount of phytochrome degraded. The Ub-P also appear during cycled Pr degradation in *Avena* if photocycling occurs at growth temperatures. However, if cycling occurs at 0-4°C, cycled Pr degradation does not occur and Ub-P do not accumulate. Consistent with Ub-P being intermediates of Pfr degradation, pulse-chase studies employing continuous R irradiations to make Pfr (and subsequently generate Ub-P) followed by a irradiation with a pulse of FR to regenerate Pr and halt degradation, show that Ub-P are continuously synthesized during phytochrome degradation and are rapidly turned over with a half-life 4-5 times shorter than that for Pfr (Fig. 4; Jabben *et al.* 1989a, b).

Lastly, in agreement with the hypothesis that sequestration is involved in phytochrome destruction (MacKenzie *et al.* 1975), Ub-P are found first and to the greatest extent in the pelletable (sequestered) phytochrome fraction during Pfr breakdown (Fig. 5). A similar co-localization of ubiquitin and phytochrome to sequestered areas was observed by immunocytochemistry (Speth *et al.* 1987). By analogy with inclusion bodies, sequestered areas could actually represent the locale for degrading most proteins by the ubiquitin pathway. Thus, other ubiquitin-protein conjugates in addition to Ub-P might be present in these aggregates. However, immunochemical analysis of partially purified sequestered phytochrome failed to support this possibility, indicating that phytochrome is the only abundant ubiquitinated species in such preparations (J. Shanklin and R.D. Vierstra unpublished data).

Following the detection of Ub-P, the potential involvement of the ubiquitin pathway in the breakdown of a number of other rapidly turned-over eukaryotic proteins has been studied. These included synthetic proteins and important cell regulators such as cyclins, oncogene products p53 and *mos*, yeast αMAT-2 repressor, and Sindbis virus RNA polymerase (Rechsteiner 1991; Hershko and Ciechanover 1992). In each case, the ubiquitin pathway has been implicated either by detection of ubiquitinated intermediates during degradation or by the ability of ubiquitin pathway mutants to stabilize the protein. Both the accumulation kinetics and size distribution of ubiquitinated intermediates were similar to that observed during phytochrome breakdown. These comparisons further support a role of Ub-P in phytochrome destruction and suggest that such kinetics are common to all ubiquitin pathway targets.

While most data are in agreement with the notion that type I phytochrome is degraded by the ubiquitin-system in etiolated plants, both the estimated pool size and the turnover rates of Ub-P are unable to account for the degradation of *all* Pfr (Jabben *et al.* 1989a). It is not yet clear whether this discrepancy suggests: (i) involvement of additional proteolytic systems, (ii) technical problems with Ub-P quantification, and or (iii) the possibility that two pools of Ub-P

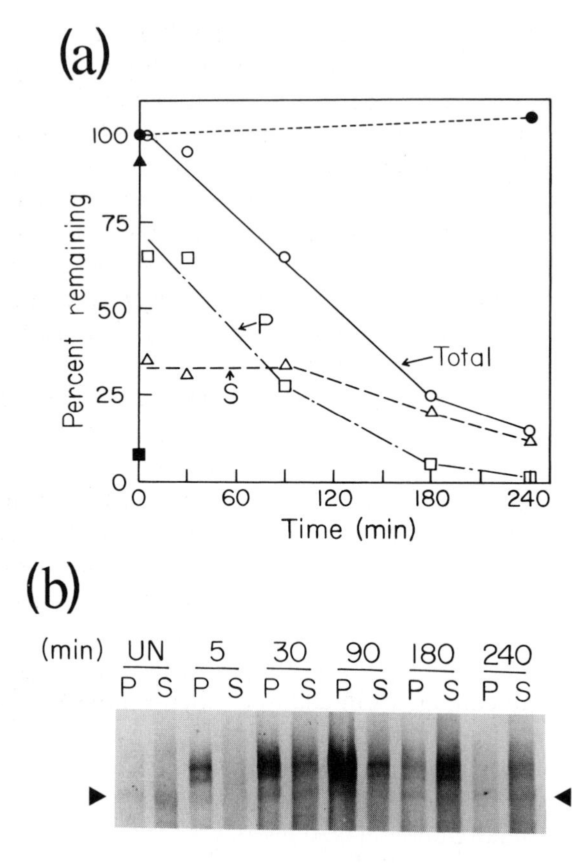

Figure 5. Association of ubiquitin-phytochrome conjugates (Ub-P) with pelletable and soluble phytochrome during Pfr degradation in etiolated *Avena*. Seedlings were either kept in the dark (closed symbols) or irradiated with 5 min of red light (open symbols) and at various times homogenized in a Mg^{2+}-containing buffer and the soluble and pelletable fractions separated by centrifugation. (a) Loss of total phytochrome (○, ●) and the soluble (△, ▲) and pelletable (□, ■) phytochrome fractions as measured by red-minus-far red difference spectroscopy (ΔΔA). (b) Accumulation and loss of Ub-P in the soluble (S) and pelletable (P) fractions. The Ub-P were isolated by immunoprecipitation with anti-phytochrome immunoglobulins and detected by immunoblot analysis with anti-ubiquitin immunoglobulins. UN, seedlings prior to irradiation. Arrowheads indicate the position of unmodified 124-kD *Avena* phytochrome. From Jabben *et al.* (1989a).

exist, one degraded more rapidly than the other. There are no data as yet for green plants concerning either the fate of type II Pfr or its potential conjugation with ubiquitin. The slower apparent turnover rates for type II phytochrome may reflect different degradative mechanism(s) in green plants or inherent structural differences between the type I and II chromoproteins that make the latter less susceptible to ubiquitination.

Assuming that the ubiquitin system is responsible for all type I Pfr breakdown, a minimal model for Pfr degradation can be proposed (Fig. 6). It includes a role for phytochrome sequestering and takes into account the estimated speeds of the relevant reactions determined from etiolated *Avena* seedlings. First, phytochrome is rapidly sequestered ($t_{1/2} \approx 2$ s at 25°C) after photoconversion to Pfr (Quail and Briggs 1978). This form subsequently becomes conjugated with ubiquitin in an ATP dependent process and is ultimately degraded by the 26S proteasome [$t_{1/2} \approx 15$ min (Jabben *et al.* 1989a)]. The energy required for sequestering could provide another contol point for Pfr degradation in addition to the two energy-dependent steps in the ubiquitin pathway (Fig. 3). In the model, no estimate is available for Ub-P formation. However, given that ubiquitination appears to be the rate limiting step for breakdown of artificial substrates both *in vitro* and *in vivo* (Hershko and Ciechanover 1992; Vierstra 1993), it is likely that Ub-P synthesis is slower than Ub-P degradation. If Pfr is cycled back to Pr, two competing reactions may occur; one of disaggregation and another of ubiquitin ligation to aggregated Pr. If disaggregation is fast, little cycled Pr degradation would be predicted. In *Avena* for example, the relative rates of each ($t_{1/2} \approx 25$ min for disaggregation *versus* ≈ 1 h for cycled Pr degradation) would

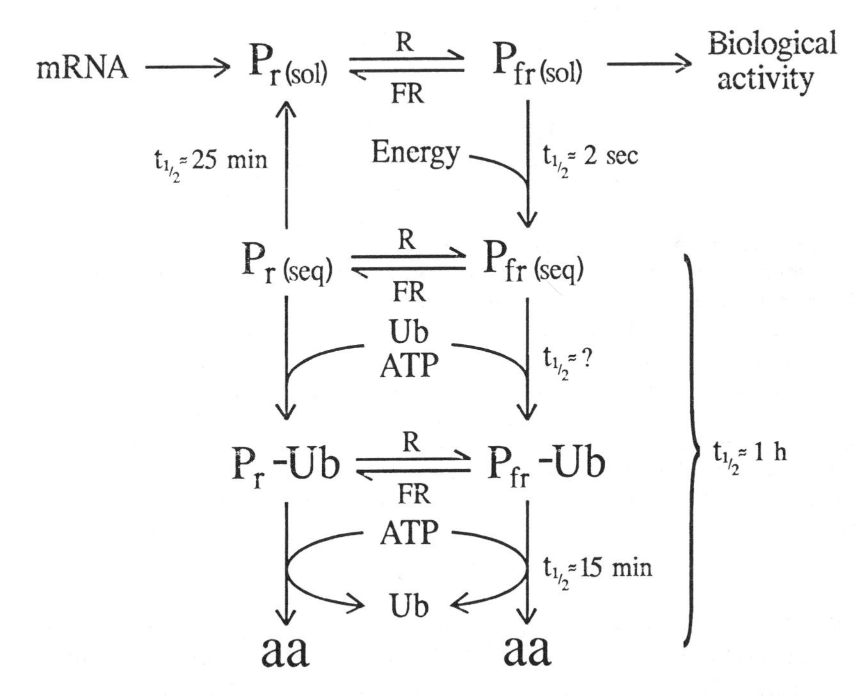

Figure 6. Proposed model for phytochrome degradation by the ubiquitin-dependent proteolytic pathway. Model assumes a role of phytochrome sequestering and where available takes into account the relative rates of the various reactions determined experimentally with etiolated *Avena* seedlings. Ub, ubiquitin; sol, soluble phytochrome; seq, sequestered phytochrome; aa, amino acids; R, red light; FR, far-red light.

predict that much of cycled Pr would escape degradation, as is observed experimentally (Pratt and Marmé 1976; Stone and Pratt 1979; Jabben *et al.* 1989a).

In its simplest form, the model predicts that Pfr degradation is ultimately determined by its localization and thus independent of form once sequestered. However, the prospect that destruction is both form and location dependent cannot be ruled out. Clearly, understanding the mechanism(s) responsible for Pfr aggregation is necessary. Conjugation of ubiquitin to sequestered phytochrome likely involves a specific E2 working in concert with an E3. This E2/E3 complex may be specific to phytochrome, given the physiological importance of this developmental regulator. A variety of ubiquitin conjugating enzymes (E1s, E2s, and E3s) have been purified from plants (Vierstra 1993). Unfortunately, even though these factors effectively work with model substrates, none of them recognize and conjugate purified phytochrome *in vitro*.

4.4.3.2 Potential mechanisms for Pfr recognition.

In an effort to understand why Pfr is selectively degraded (as well as why it becomes physiologically active), the structure of the photoreceptor has been analyzed in considerable detail (Chapters 4.3 and 4.8; Vierstra and Quail 1986). With respect to the potential involvement of the ubiquitin proteolytic pathway, Ub-P have been characterized biochemically, with special emphasis on mapping ubiquitin attachment sites (Shanklin *et al.* 1989). The hope is that the structure of such sites will reveal how Pfr is selectively modified. Partially purified Ub-P from etiolated *Avena* retained R/FR photoreversibility indicating that the gross structure of the phytochrome moiety has not been profoundly perturbed. However, the native molecular mass of Ub-P was substantially larger than the unmodified dimer (≈ 600 *versus* 350 kD) and that predicted from the combined mass of phytochrome and multiple ubiquitins (≈ 360-400 kD). This suggests that the ubiquitin moieties extend significantly from the chromoprotein surface.

Peptide mapping and antibody interference assays with *Avena* Ub-P suggested that the preferred ubiquitin attachment site is between amino-acid residues 742 and 790 (Fig. 7; Shanklin *et al.* 1989). Interestingly, this site becomes more exposed upon photoconversion of Pr to Pfr (Grimm *et al.* 1988; Vierstra and Quail 1986), suggesting that Pfr-specific ubiquitination may be regulated by conformation. Furthermore, this region displays substantial amino-acid sequence homology among all phytochrome A proteins analyzed [≈ 70% identity (Quail *et al.* 1991)]. It includes three invariant lysine residues at positions 744, 753, and 784 which may represent actual sites of ubiquitin attachment (Fig. 7). These three lysines are also present in phytochrome B of both rice and *Arabidopsis,* whereas one of the three is present in *Arabidopsis*

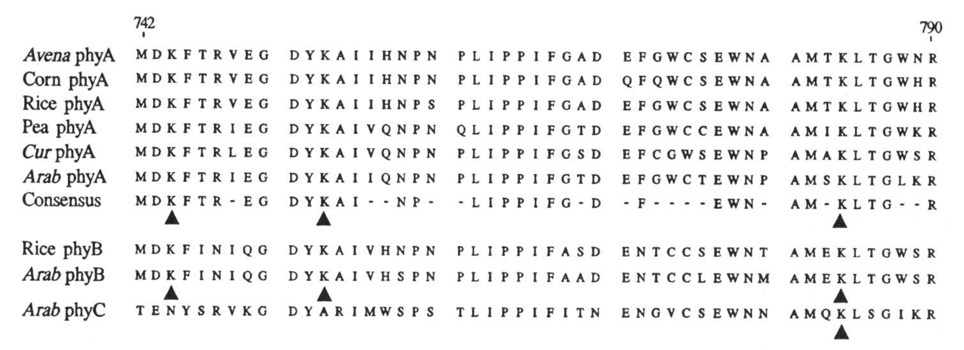

Figure 7. Amino-acid sequence comparison of phytochromes A, B, and C from various plant species of a domain potentially involved in ubiquitin conjugation. Arrowheads identify the conserved lysines within the region that may represent actual sites of ubiquitin attachment. Consensus represents amino-acid residues conserved among all phyA phytochromes sequenced to date. *Cur.*, *Cucurbita pepo*; *Arab.*, *Arabidopsis thaliana*. Abbreviations for the amino-acid residues are: A, Ala; C, Cys; D, Asp; E, Glu; F, Phe; G, Gly; H, His; I, Ile; K, Lys; L, Leu; M, Met; N, Asn; P, Pro; Q, Gln; R, Arg; S, Ser; T, Thr; V, Val; W, Trp; Y, Tyr.

phytochrome C (Fig. 7). Thus, their presence alone cannot account for the increased instability of type I Pfr relative to type II Pfr.

Another site that has been proposed to be important for Pfr degradation resides near the chromophore between amino-acid residues 323-360 (in *Avena* phytochrome A). This sequence was discovered by Rogers *et al.* (1986), using computer algorithms, as a motif common among most short-lived proteins but absent in long-lived proteins. The motif consists of a loosely defined hydrophilic sequence rich in amino acids Pro (P), Asp, Glu (E), Ser (S), and Thr (T); hence the region has been defined as the PEST domain. It is proposed that the PEST domain acts as a degradative signal for unspecified proteolytic pathway(s). Given the close proximity of the PEST domain to the phytochrome chromophore, one could envision that during the Pr to Pfr photoconversion, the natural re-orientation of the chromophore exposes this domain to the proteolytic machinery that degrades Pfr (Vierstra and Quail 1986). Consistent with this possibility, the PEST domain is more exposed in Pfr than Pr (Grimm *et al.* 1988). Furthermore, whereas all phytochrome A proteins contain PEST-like domains, similar locations in phytochrome B and C sequences are less well conserved, in potential agreement with their enhanced stabilities *in vivo* (Quail *et al.* 1991).

Unfortunately, the major limitation to accepting the PEST hypothesis is the lack of supportive biological data implicating the region directly in protein instability *in vivo* (Rechsteiner 1991). The PEST domain has been shown *not* to be essential for the degradation of several PEST-containing proteins. Thus, until such supportive data are presented, it is difficult to supersede the possible involvement of the ubiquitin system, which has experimental support, with this hypothesis. In addition, even though the PEST hypothesis has been argued as an alternative mechanism to ubiquitin modification for Pfr degradation (Quail 1991), there is no evidence that the two mechanism are mutually exclusive. For example, the PEST domain still could represent an important recognition site by the ubiquitin system.

4.4.4 Use of transgenic plants to study phytochrome degradation.

The cloning of phytochrome genes has recently been exploited in several laboratories for the analysis of phytochrome structure/function using *in vitro* mutagenesis and subsequent expression in transgenic plants (Chapter 4.8; Quail 1991; Cherry *et al.* 1992). Successful expression of *Avena*, rice, and *Arabidopsis* phytochrome genes have been reported in tobacco, tomato, and *Arabidopsis*. The most thoroughly characterized system involves the constitutive expression of an *Avena phyA* gene in tobacco (Keller *et al.* 1989; Cherry *et al.* 1991). By all criteria measured, the *Avena* chromoprotein synthesized is indistinguishable from that isolated from etiolated *Avena* seedlings indicating that proper assembly of the monocot photoreceptor occurs even in a dicot species. Interestingly, the transgenic plants have a radically altered phenotype (*e.g.* decreased stem elongation, increased chlorophyll content, reduced apical dominance, and delayed senescence), indicating that the introduced protein is biologically active. Using this light-exaggerated phenotype as an assay, a major effort is underway to map domains responsible for phytochrome activity (Cherry *et al.* 1992). For more detail of this approach, the reader is referred to Chapter 4.8.

With respect to phytochrome degradation, the transgenic approach also has the potential for locating domains necessary for Pfr recognition and for identifying the proteolytic pathways responsible. Similar to the endogenous photoreceptor, *Avena* phytochrome in etiolated tobacco seedlings is rapidly degraded upon photoconversion to Pfr (Cherry *et al.* 1991). Whereas the tobacco protein has a $t_{1/2} \approx 1$ h, the half-life of the transgenic *Avena* protein is ≈ 4 h (Fig. 8). Similar Pfr degradation is observed for *Avena* phytochrome A expressed in tomato and *Arabidopsis*; like phytochrome A in tobacco, the rate appears slower than the endogenous chromoprotein (Quail 1991). The reason behind the slightly increased stability of the transgenic chromoprotein is unknown, but may reflect in part the high rate of apoprotein synthesis driven by the constitutive promoter used. Breakdown of both tobacco and oat Pfr in transgenic tobacco is associated with the accumulation of Ub-P (Cherry *et al.* 1991). Consistent with its slower degradation rate, *Avena* Ub-P are turned over more slowly and persist over a longer time after R than tobacco Ub-P (Fig. 8). Taken together, these data provide additional support for the notion that Pfr is degraded, at least in part, by the ubiquitin-dependent proteolytic pathway. They

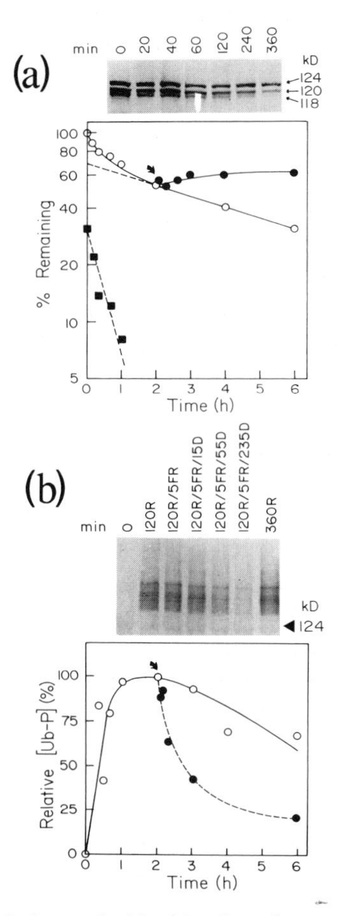

Figure 8. Phytochrome degradation and ubiquitin-phytochrome conjugate (Ub-P) accumulation in etiolated tobacco seedlings constitutively expressing *Avena* phyA phytochrome. Seedlings were grown in the dark (D) for 5 days and at t = 0 either were irradiated continuously with red light (R, ○) or irradiated with R followed by a 5 min irradiation with far-red light (FR, ●) and a further incubation in D. At various times the seedlings were assayed for phytochrome content and Ub-P. (a) Kinetics of Pfr degradation. *Upper* panel: loss of phytochrome measured by immunoblot analysis with anti-*Cucurbita* phytochrome immunoglobulins that detect both tobacco and *Avena* phytochromes. The migration position of 124-kD *Avena* and 118/120-kD tobacco phytochromes are indicated on the right. *Lower* panel: loss of phytochrome by R-minus-FR difference spectroscopy ($\Delta\Delta A$). Dashed lines denote the mathematical peeling of total phytochrome degradation to reveal the predicted degradation kinetic of tobacco phytochromes separately (■). (b) Accumulation and loss of *Avena* Ub-P during phytochrome degradation. *Avena* Ub-P were partially purified from tobacco seedlings by immunoprecipitation with a monoclonal antibody specific to *Avena* phytochrome and detected by immunoblot analysis with anti-ubiquitin immunoglobulins. *Upper* panel: immunoblot analysis of *Avena* Ub-P. *Lower* panel: relative levels of *Avena* Ub-P during phytochrome degradation determined from the immunoblots by reflective densitometry. Arrowheads indicate the position of unmodified 124-kD *Avena* phytochrome. From Cherry *et al.* (1991).

also demonstrate that the domains involved in selective proteolytic recognition of Pfr are conserved between the dicot and monocot chromoproteins.

Mapping domains relevant to Pfr degradation and ubiquitination is in progress. Regions of potential interest include the regions surrounding the proposed ubiquitin attachment site (Shanklin *et al.* 1989) and the PEST domain (Rogers *et al.* 1986). The N-terminus of the protein does not appear to be involved as deletion of amino acids 7-69 failed to alter rapid Pfr destruction (Cherry *et al.* 1992). However, it should be emphasized that any data generated by this approach should be interpreted with caution. Many alterations, by directly modifying the physicochemical properties of the chromoprotein, may profoundly affect phytochrome degradation only secondarily.

4.4.5 Physiological function(s) of phytochrome degradation

In accord with the importance of phytochrome to plant growth and development, the level of active photoreceptor is precisely controlled, allowing high levels to accumulate in etiolated plants while maintaining low levels in light grown plants. Phytochrome abundance is regulated by feed back repression of *phyA* gene transcription, rapid turnover of *phy* mRNA, and form dependent degradation of the chromoprotein (Furuya 1989; Quail 1991). In combination, these mechanisms result in the 10- to 100-fold drop in phytochrome during greening of etiolated seedlings (Pratt 1979; Jabben and Holmes 1984). Of these controls, the most important is Pfr degradation. This is especially true for dicots where light represses transcription of *phyA* genes only $\approx$ 2-fold even though photoreceptor levels fall precipitously during greening.

At first glance, this exquisite control appears wasteful as energy must be expended to maintain appropriate phytochrome levels. However, upon closer examination it may serve an important physiological function in enabling plants to adequately respond to the large ranges of irradiances found in the natural environment (Blaauw-Jensen 1983). In etiolated plants, increased levels of phytochrome presumably would confer a greater light sensitivity to seedlings; the greater the phytochrome level, the more likely that each photon will be absorbed by a Pr molecule, converting it to Pfr. Because only Pfr is biologically active, the seedling would be oblivious to the high Pr levels and only respond to the small amount of Pfr generated in such dim light. In fact, many photoresponses important for seedlings growing toward the soil surface, such as inhibition of hypocotyl or coleoptile growth and activation of geoperception, are sensitive to very low Pfr levels (Chapter 4.7; Kendrick and Kronenberg 1986).

In green plants where light is no longer limiting and Pfr would be continuously abundant, using small changes in Pfr would be impractical as a photodetector. Hence, green plants repress apoprotein synthesis and rapidly degrade

Pfr. Consistent with this, many photoresponses in green plants, such as pigmentation and stem elongation, become more sensitive to the ratio of Pfr to Pr and the photocycling between the two than to actual Pfr levels (Chapter 7.1; Kendrick and Kronenberg 1986). The importance of decreasing phytochrome levels during the transition from etiolated to green plants is demonstrated by the phenotypic sensitivity of transgenic plants to artificially elevated levels of the chromoprotein (Cherry *et al.* 1991; Quail 1991). In transgenic tobacco for example, an increase in phytochrome content of only 2- to 3-fold profoundly disrupts light sensitivity of many phytochrome controlled responses (McCormac *et al.* 1991; Cherry *et al.* 1992).

In addition to its role in depressing phytochrome levels over the short period of de-etiolation, continuous degradation of Pfr over the entire life of a plant may also be necessary for proper photoreceptor function. One can imagine that plants responding to ever changing light conditions would need to continuously purge a previous light signal (*i.e.* Pfr) in order to react to the next one. Consistent with the biogenesis of phytochrome, this would require that previous Pfr be degraded and new Pr be synthesized. Failure to degrade Pfr could leave the photomorphogenic systems controlled by phytochrome permanently activated even in the absence of light. Obviously, such blindness would be disadvantageous for organisms so dependent on light quantity and quality for morphogenic cues. It is tempting to speculate that whereas degradation likely represents an important mechanism for removing Pfr, it is possible that this reaction is not fast enough to allow plants to respond hourly or even daily to the environment. In etiolated seedlings, where over 100 times more Pfr than necessary to saturate a photoresponse may be present, many hours would be needed to degrade most of the signal. Thus, a more expedient mechanism like phytochrome sequestering may be required. This mechanism could allow the cell to rapidly inactivate excess Pfr by aggregation and put it in a form amenable to slower proteolytic removal. Thus, sequestering and destruction may work in tandem to effectively attentuate levels of physiologically active photoreceptor.

4.4.6 Concluding remarks

One of the more remarkable features of phytochrome (in addition to its ability to activate photomorphogenesis) is its rapid degradation once converted to Pfr. This degradation is likely to be essential for proper function of the chromoprotein by controlling the levels of the active form. Determining how Pfr is selectively degraded will represent a major advance in understanding how plant cells regulate phytochrome levels and will hopefully provide clues to understanding another mysterious process, that of phytochrome sequestering. To date, phytochrome represents a unique system for studying selective protein catabolism because of its extreme specificity and the facile way degradation can be initiated. While a large body of information exists for type I phytochrome destruction, little is known concerning the *in vivo* stability of type II molecules. The cloning of type II phytochrome genes, *phyB*, *C*, *D*, and *E* and the generation of isoform-specific antibodies will be useful in this direction (Furuya 1989; Quail 1991).

Recent studies on Pfr breakdown have implicated the ubiquitin proteolytic pathway, at least for type I phytochrome, and suggest that sequestration may represent an intermediate step. Whereas most of the data support such an involvement, a direct causal link between the formation of Ub-P and Pfr degradation is still needed. Hopefully through the use of transgenic plants and/or generation of mutants that fail to degrade the photoreceptor, the exact details of Pfr destruction for both type I and II chromoproteins will be forthcoming. This will include identifying the sites on phytochrome responsible for its form-specific degradation, the enzymes responsible for recognition, and the proteolytic machinery involved in catabolism.

4.4.7 Further Readings

Furuya M. (1989) Molecular properties and biogenesis of phytochrome I and II. *Arch Biophys.* 25: 133-137.

Pratt L.H. (1979) Phytochrome: Function and Properties. *Photochem. Photobiol. Rev.* 4: 59-123.

Quail P.H. (1991) Phytochrome: a light-activated molecular switch that regulates plant gene expression. *Annu. Rev. Genet.* 25: 389-409.

Sage L.C. (1992) *Pigment of the Imagination: History of Phytochrome Research.* Academic Press, New York.

Vierstra R.D. (1993) Protein degradation in plants. *Annu. Rev. Plant Physiol. Molec. Biol.* in press.

4.4.8 References

Blaauw-Jensen G. (1983) Thoughts on the possible role of phytochrome destruction in phytochrome-controlled responses. *Plant Cell Environ.* 6: 173-179.

Boisard J., Marmé D. and Briggs W.R. (1974) In vivo properties of membrane-bound phytochrome. *Plant Physiol.* 54: 272-276.

Brockmann J. and Schäfer E. (1982) Analysis of Pfr destruction in *Amaranthus caudatus* L. - evidence for two pools of phytochrome. *Photochem. Photobiol.* 35: 555-558.

Butler W.L., Lane H.C. and Siegelman H.W. (1963) Nonphotochemical transformations of phytochrome *in vivo. Plant Physiol.* 38: 514-519.

Butler W.L. and Lane H.C. (1964) Dark transformations of phytochrome *in vivo.* II. *Plant Physiol.* 40: 13-17.

Cherry J.R., Hershey H.P. and Vierstra R.D. (1991) Characterization of tobacco expressing oat phytochrome: domains responsible for the rapid degradation of Pfr are conserved between monocots and dicots. *Plant Physiol.* 96: 775-785.

Cherry J.R., Hondred D., Walker, J.M. and Vierstra R.D. (1992) Phytochrome requires the 6-kDa N-terminal domain for full biological activity. *Proc. Natl. Acad. Sci. USA* 89: 5039-5043.

Finley D. and Chau V. (1991) Ubiquitination. *Ann. Rev. Cell Biol.* 7: 25-69.

Fukshansky I. and Schäfer E. (1984) Models in photomorphogenesis. In: *Encyclopedia of Plant Physiology*, New Series, Vol 16A: *Photomorphogenesis*, pp. 69-95, Shropshire Jr. W. and Mohr H. (eds.) Springer, New York.

Goldberg A.L. and St John A.C. (1976) Intracellular protein degradation in mammalian and bacterial cells. *Annu. Rev. Biochem.* 45: 747-803.

Grimm R., Eckerskorn C., Lottspeich F., Zenger C. and Rudiger W (1988) Sequence analysis of proteolytic fragments of 124-kilodalton phytochrome from etiolated *Avena sativa* L.: conclusions on the conformation of the native protein. *Planta* 174: 396-401.

Heim B., Jabben M. and Schäfer E. (1981) Phytochrome destruction in dark- and light-grown *Amaranthus caudatus* seedlings. *Photochem. Photobiol.* 34: 89-93.

Hershko A. and Ciechanover A. (1992) The ubiquitin system for protein degradation. *Annu. Rev. Biochem.* 61: 761-807.

Hofmann E., Grimm R., Speth H.V. and Schafer E. (1991) Partial purification of sequestered particles of phytochrome from oat (*Avena sativa* L.) seedlings. *Planta* 183: 265-273

Hunt R.E. and Pratt L.H. (1979) Phytochrome radioimmunoassay. *Plant Physiol.* 64: 327-331.

Jabben M. (1980) The phytochrome system in light-grown *Zea mays* L. *Planta* 149: 91-96.

Jabben M. and Deitzer G. F. (1978) Spectrophotometric phytochrome measurements in light-grown *Avena sativa* L. *Planta* 143: 309-313.

Jabben M. and Holmes G. (1984) Phytochrome in light-grown plants. In: *Encyclopedia of Plant Physiology*, New Series, Vol 16B: *Photomorphogenesis*, pp. 704-722, Shropshire Jr. W. and Mohr H. (eds.) Springer, New York.

Jabben M., Shanklin J. and Vierstra R.D. (1989a) Ubiquitin-phytochrome conjugates: pool dynamics during *in vivo* phytochrome degradation. *J. Biol. Chem.* 264: 4998-5005.

Jabben M., Shanklin J. and Vierstra R.D. (1989b) Red-light accumulation of ubiquitin-phytochrome conjugates in both monocots and dicots. *Plant Physiol.* 90: 380-385.

Keller J.M., Shanklin J., Vierstra R.D. and Hershey H.P. (1989) Expression of a functional monocotyledonous phytochrome gene in transgenic tobacco plants. *EMBO J.* 8: 1005-1012.

Kendrick R.E. and Frankland B. (1968) Kinetics of phytochrome decay in *Amaranthus* seedlings. *Planta* 82: 317-320.

Kendrick R.E. and Kronenberg G.H.M. (1986) *Photomorphogenesis in Plants*. Martinus Nijhoff Publishers, Dordrecht.

MacKenzie J.M., Coleman R.A., Briggs W.R. and Pratt L.H. (1975) Reversible redistribution of phytochrome within the cell upon conversion to its physiologically active form. *Proc. Natl. Acad. Sci. USA* 72: 799-803.

McCormac A.C., Cherry J.R., Hershey H.P., Vierstra R.D. and Smith H. (1991) Photoresponses of transgenic tobacco plants expressing an oat phytochrome gene. *Planta* 185: 162-170.

McCurdy D.W. and Pratt L.H. (1986a) Immunological electron microscopy of phytochrome in *Avena*: identification of intracellular sites responsible for phytochrome sequestering and enhanced pelletability. *J. Cell Biol.* 103: 2541-2550.

McCurdy D.W. and Pratt L.H. (1986b) Kinetics of intracellular redistribution of phytochrome in *Avena* coleoptiles after its photoconversion to the active far-red-absorbing form. *Planta* 167: 330-336.

Pratt L.H. (1986) Phytochrome: localization within the plant. In: *Photomorphogenesis in Plants*, pp. 61-81, Kendrick R.E. and Kronenberg G.H.M. (eds.) Martinus Nijhoff Publishers, Dordrecht.

Pratt L.H. and Marmé D. (1976) Red-light enhanced phytochrome pelletability: re-examination and further characterization. *Plant Physiol.* 58: 686-692.

Pratt L.H., Kidd G.H. and Coleman R.A. (1974) An immunochemical characterization of the phytochrome destruction reaction. *Biochim. Biophys. Acta* 365: 93-107.

Quail P.H. and Briggs W.R. (1978) Irradiation-enhanced phytochrome pelletability: requirement for phosphorylative energy *in vivo*. *Plant Physiol.* 62: 773-778.

Quail P.H., Marme D. and Schäfer E. (1973a) Particle-bound phytochrome from maize and pumpkin. *Nature New Biol.* 245: 189-191.

Quail P.H., Schafer E. and Marmé D. (1973b) Turnover of phytochrome in pumpkin cotyledons. *Plant Physiol.* 52: 128-131.

Quail P.H., Hershey H.P., Idler K.B., Sharrock R.A., Christensen A.H., Parks B.M., Somers D., Tepperman J., Bruce W.A. and Dehesh K. (1991) Phy-gene structure, evolution, and expression. In: *Phytochrome Properties and Biological Action*, pp. 13-38. Thomas B. and Johnson C.B. (eds.) Springer-Verlag, Berlin.

Rechsteiner M. (1991) Natural substrates of the ubiquitin proteolytic pathway. *Cell* 66: 615-618.

Rogers S., Wells R. and Rechsteiner M. (1986) Amino acid sequences common to rapidly degraded proteins: the PEST hypothesis. *Science* 234: 364-368.

Schäfer E and Schmidt W. (1974) Temperature dependence of phytochrome dark reactions. *Planta* 116: 257-266.

Schäfer E., Lassig T.-U. and Schopfer P. (1975) Photocontrol of phytochrome destruction in grass seedlings: the influence of wavelength and irradiance. *Photochem. Photobiol.* 22: 193-202.

Shanklin J., Jabben M. and Vierstra R.D. (1987) Red light-induced formation of ubiquitin-phytochrome conjugates: identification of possible intermediates of phytochrome degradation. *Proc. Natl. Acad. Sci. USA* 84: 359-363.

Shanklin J., Jabben M. and Vierstra R.D. (1989) Partial purification and peptide mapping of ubiquitin-phytochrome conjugates in oat. *Biochemistry*. 28: 6028-6034.

Sharrock R.A. and Quail P.H. (1989) Novel phytochrome sequences in *Arabidopsis thaliana*: Structure, evolution, and differential expression of a plant regulatory photoreceptor family. *Genes Develop.* 3: 1745-1757.

Shimazaki Y., Cordonnier M.-M. and Pratt L.H. (1983) Phytochrome quantitation in crude extracts of *Avena* by enzyme-linked immunosorbent assay with monoclonal antibodies. *Planta* 159: 534-544.

Speth V., Otto V. and Schafer A. (1987) Intracellular localization of phytochrome and ubiquitin in red-light-irradiated oat coleoptiles by electron microscopy. *Planta* 171: 332-338.

Stone H.J. and Pratt L.H. (1978) Phytochrome destruction: Apparent inhibition by ethylene. *Plant Physiol.* 62: 922-923.

Stone H.J. and Pratt L.H. (1979) Characterization of the destruction of phytochrome in the red-absorbing form. *Plant Physiol.* 63: 680-682.

Tokuhisa J. and Quail P.H. (1987) The levels of two distinct species of phytochrome are regulated differentially during germination in *Avena sativa* L. *Planta* 172: 371-377.

Vierstra R.D. and Quail P.H. (1985) Spectral characterization and proteolytic mapping of native 120-kilodalton phytochrome from *Cucurbita pepo* L. *Plant Physiol.* 77: 990-998.

Vierstra R.D. and Quail P.H. (1986) Phytochrome: the protein. In: *Photomorphogenesis in Plants*, pp 35-60, Kendrick R.E. and Kronenberg G.H.M (eds.) Martinus Nijhoff Publishers, Dordrecht.

Vijay-Kumar S., Bugg C., Wilkinson K.D. Vierstra R.D., Hatfield P.M. and Cook W.J. (1987) Comparison of the three-dimensional structures of yeast and oat ubiquitin with human ubiquitin. *J. Biol. Chem.* 262: 6396-6399.

4.5 Distribution and localization of phytochrome within the plant

Lee H. Pratt

Botany Department, University of Georgia, Athens, Georgia 30602, USA

4.5.1 Introduction

Despite several decades of investigation, not only do the immediate molecular functions of phytochrome remain unidentified, but the transduction chains between these immediate molecular functions and ultimate biological responses remain inadequately described. While insufficient by itself, knowledge of the distribution and localization of phytochrome provides important information that will assist in both formulating and testing hypotheses concerning phytochrome function. The question of phytochrome distribution and localization within the plant can be divided into two aspects. On the one hand, there is the question of how phytochrome is partitioned among the tissues, organs and organ systems of a plant, that is of its *inter*cellular distribution. On the other hand, there is the question of where phytochrome is found within a single cell, that is of its *intra*cellular localization.

Information about the intercellular distribution of phytochrome is needed to understand fully the transduction chains of phytochrome-mediated responses (Part 9). For example, to understand the entire transduction chain connecting photoreception to an ultimate biological response, it is important to know whether the photoreceptor is in the same cell that exhibits the response. If not, then one must consider, in addition to other parameters, the mechanism for signal translocation. Moreover, it is becoming increasingly evident that the fluence rate and wavelength distribution of light incident upon a plant changes markedly as that light penetrates into different regions of the plant (Chapter 7.4). Consequently, the location of phytochrome determines to a significant extent the quantity and quality of light to which it responds.

Knowledge of the intracellular localization of phytochrome is important for elucidating and/or testing hypotheses concerning its primary mode of action. For example, the hypothesis that the far-red (FR)-absorbing form of phyto-

R.E. Kendrick & G.H.M. Kronenberg (eds.), Photomorphogenesis in Plants - 2nd Edition

chrome (Pfr) mediates directly the extent of gene transcription necessitates its presence within the nucleus, while the hypothesis that it modulates in a direct sense a membrane activity requires that it be membrane-associated. In fact, one of the driving forces behind many efforts to determine the intracellular localization of phytochrome was to test these two hypotheses directly.

The relatively recent discovery that within a single plant there are at least three phytochromes (Chapter 4.2) complicates greatly the determination of phytochrome distribution and localization. In some instances it is effectively impossible to know which phytochrome is being investigated. In other cases, however, the methodology used permits unambiguous identification of the gene product whose distribution is being determined. Inasmuch as is practical the precise nature of the phytochrome whose distribution and/or localization is being determined will be given.

Three general approaches are available for determining the distribution and localization of phytochrome. Because each approach can be used for investigating both its intercellular and intracellular distributions, these approaches will be presented first, together with a summary of the advantages and limitations of each. The distribution of phytochrome at an intercellular level and its localization at an intracellular level will then be discussed independently, in each case with a summary that attempts to accommodate as much of the available information as is possible. Throughout the emphasis will be on methodology.

4.5.2 Phytochrome assays

Three approaches have been widely used for investigating the distribution and localization of phytochrome within a plant: biological, spectrophotometric, and immunochemical. Each offers one or more unique advantages, as well as inherent limitations. Consequently, each approach provides important information concerning phytochrome distribution and localization.

4.5.2.1 Biological assay

Bio-assays have been used to infer the location of phytochrome at both inter- and intracellular levels. In the first instance, selective irradiation of isolated portions of a plant have been used to determine the organs or tissues in which phytochrome is found. In the second, information about the subcellular localization of phytochrome has been obtained by the imaginative, combined use of *microbeam* and *polarized* irradiations (Chapters 7.2 and 9.4), as well as by bio-assay of isolated subcellular fractions.

Bio-assays have the distinct advantage of being the only way to ascertain the location of biologically active phytochrome. They also have the advantage of being highly sensitive, especially in the case of the very low fluence responses (VLFRs)(Chapter 4.7). Bio-assays have, however, correspondingly significant limitations. It is difficult, if not often impossible, to determine which of the phytochromes is being detected. Moreover, by this approach it is impossible to determine the location of phytochromes that are not active in the biological response being investigated, either because they are of a different type or are present in cells or subcellular fractions that do not exhibit the response.

4.5.2.2 Spectrophotometric assay

Spectrophotometric assays of phytochrome derive from its unique *photoreversibility* between two forms with different *absorbance spectra* (Chapter 4.1). Absorbance spectra of intact, *etiolated* oat seedlings illustrate the principle behind spectrophotometric assays (Fig. 1). Following saturating irradiation with red light (R), absorbance is reduced in the R spectral region and enhanced in the FR. Conversely, saturating irradiation with FR enhances absorbance in the R and reduces it in the FR. These photoreversible absorbance changes can be visualized by subtracting from the absorbance spectrum obtained after R *actinic* irradiation that obtained after FR actinic irradiation (Fig. 1, *upper curve)*. The extent of these photoreversible absorbance changes can be expressed as:

$$\Delta\Delta A = |\Delta A_R| + |\Delta A_{FR}|$$

where $\Delta\Delta A$ is proportional to the amount of phytochrome present (Pratt 1983; Gross *et al.* 1984) and $|\Delta A_R|$ and $|\Delta A_{FR}|$ are the absolute values of the absorbance changes in the R and FR spectral regions respectively.

Spectrophotometers designed for phytochrome assay are not commercially available, although it is possible to adapt some to this application. Consequently, a few investigators have constructed custom-designed instruments, the most recent being fully automated, with a microcomputer to control all instrument functions and to accept and analyze data (Pratt *et al.* 1985). Such instruments not only increase the sensitivity of spectrophotometric assays, but also the efficiency with which they can be made. A photoreversibility measurement requires repeated irradiations with actinic R and FR, with intervening $\Delta\Delta A$ measurements, over a period of a few minutes. Since the microcomputer can perform all of these repetitive functions under software control, the experimenter is freed from this task, giving more time for other experimental manipulations and/or sample preparation.

Advantages of spectrophotometric assays include the possibility to use them with essentially intact tissue samples or with turbid subcellular fractions. These assays are also rapid, simple, and identify only photoreversible *holoproteins* (Chapter 4.3). On the negative side, there are several limitations that deserve

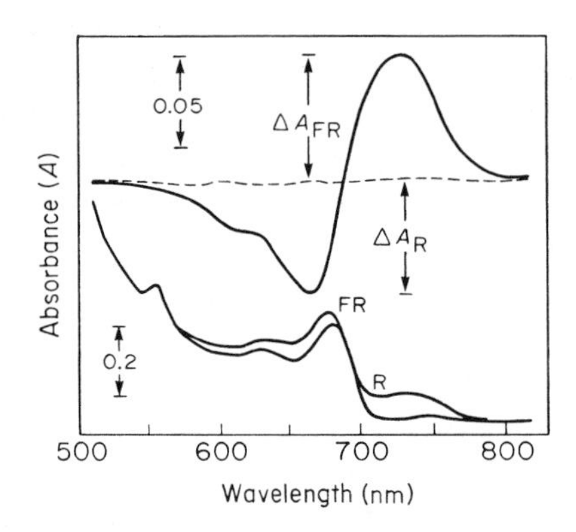

Figure 1. Absorbance spectra of 4-day-old, etiolated oat shoots. Shoots were harvested under dim green light, cut into small segments, packed into a 1-cm pathlength cuvette cooled with ice water, and given a saturating irradiation with red light (R) both to convert all protochlorophyll(ide) to chlorophyll(ide) and to saturate the photoconversion of Pr to Pfr. After measuring an absorbance spectrum (R), phytochrome was converted back to Pr by saturating irradiation with far-red light (FR), after which a second absorbance spectrum (FR) was recorded. A difference spectrum *(upper curve)* was calculated by subtracting spectrum FR from spectrum R. The difference spectrum is displayed with an expanded absorbance scale. Adapted from Pratt (1983). Original spectra obtained with the assistance of Y. Inoue and Y. Shimazaki, University of Tokyo.

emphasis here. (i) Photoreversibility assays cannot discriminate between biologically active and inactive pools of phytochrome, nor can they detect chromophore-free phytochrome *apoprotein*. (ii) Spectral assays can also not discriminate among the different phytochromes. (iii) *Chlorophyll* prevents phytochrome assay in green tissues (Pratt 1983). (iv) Spectral assays are relatively insensitive. In some cases, it has been shown that enough phytochrome to yield a biological response is present even though it is insufficient to detect by photoreversibility. (v) Precise quantitative comparisons among $\Delta\Delta A$ values obtained with samples prepared from different tissues can be difficult at best. The reason for this impasse is that the effective light path through a tissue sample is a function of the magnitude of light scattering within the sample itself, which is highly variable among different tissues and might even vary within a single tissue during its development (Butler 1964).

4.5.2.3 Immunochemical assay

Phytochrome is an excellent *antigen*, making it easy to develop *immunochemical assays* that overcome limitations inherent to spectral assays. Immunochemical assays take advantage of the specificity and high affinity of an *antibody* for its antigen in order to visualize phytochrome. The most useful assays include enzyme-linked immunosorbent assay *(ELISA)*, *immunoblotting* (Western blotting), and *immunocytochemistry*. Because the latter visualizes phytochrome *in situ*, it provides the most precise spatial information.

While ELISA, immunoblotting and immunocytochemistry might seem to be very different assay methods, they in fact share common methodology in that they all involve indirect labelling of phytochrome *via* antibodies (Pratt *et al.* 1986). One of many immunochemical labelling protocols is presented in Fig. 2. This protocol uses *monoclonal antibodies* to phytochrome (MAP) to identify the chromoprotein, either adsorbed to the well of a microtitre plate (ELISA), adsorbed to nitrocellulose (immunoblotting), or at its normal location in a tissue section (immunocytochemistry). The monoclonal antibodies, which in this case originate from a mouse, are then visualized by sequential application of rabbit antibodies to mouse immunoglobulins (RAM) and goat antibodies to rabbit immunoglobulins (GAR). The latter antibodies have a label (L) attached, which permits the indirect visualization of the original antigen, phytochrome. For

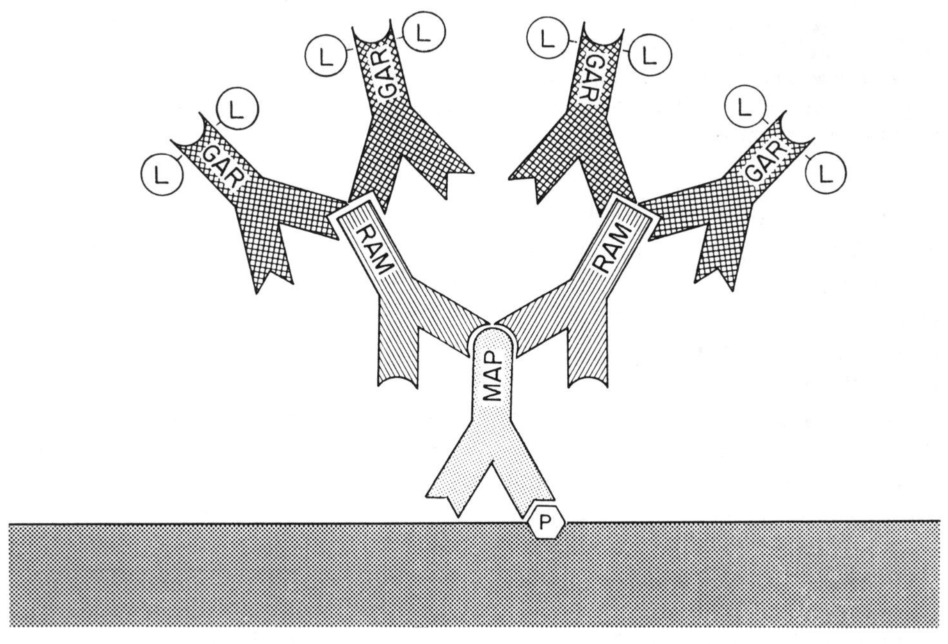

Figure 2. One of several protocols for labelling phytochrome. In this example, phytochrome (P) is initially detected by a monoclonal mouse antibody directed to phytochrome (MAP). The second antibody, which is rabbit antibody directed to mouse immunoglobulin (RAM), binds to MAP and increases the amount of label ultimately associated with each phytochrome molecule, thereby enhancing assay sensitivity. The third antibody, which is goat antibody to rabbit immunoglobulin (GAR), binds to RAM and also has covalently coupled to it the label (L). The specific nature of the label (enzyme, fluorophore, colloidal gold, *etc.*) depends upon whether the assay is intended to visualize phytochrome bound to the well of a microtiter plate (ELISA), bound to nitrocellulose (immunoblotting), or located at its normal site *in situ* (immunocytochemistry). Adapted from Pratt L.H. (1984) In: *Techniques in Photomorphogenesis*, pp. 201-226, Smith H. and Holmes M.G. (eds.) Academic Press, London.

ELISA or immunoblotting, the label is typically an enzyme such as *alkaline phosphatase* or *peroxidase*, either of which can provide a densely coloured reaction product. For immunocytochemistry, the label can also be an enzyme, but is now more typically a fluorophore such as *rhodamine* or *fluorescein*, permitting observation by light microscopy, or an electron dense material such as *colloidal gold*, permitting observation by electron microscopy.

Immunochemical assays offer many advantages. (i) They can easily be 1000-fold more sensitive than spectrophotometric assays. (ii) They are insensitive to the presence of chlorophyll, permitting their use with green-plant tissues. (iii) The specificity of antibodies permits discrimination among the different phytochromes, such that each phytochrome can be identified independently of the others. (iv) They provide the only practical method for identifying phytochrome apoprotein and protein fragments. (v) One of them, immunocytochemistry, provides spatial information not only on a cell-by-cell basis, but also within a single cell with the resolution of transmission electron microscopy. Immunochemical assays thereby constitute an important complement to biological and spectrophotometric assays. Nonetheless, they also have significant limitations, including their inability to discriminate between holoproteins and apoproteins, and between biologically active and inactive pools.

4.5.3 Intercellular distribution

Information about the distribution of phytochrome among organ systems, organs and tissues has been gained by several approaches, in particular by observing the biological consequences of selectively irradiating portions of a plant, and by direct spectrophotometric and immunochemical assay.

4.5.3.1 Biological assay

If a biological response is obtained by selective irradiation of an organ or tissue, then it can be inferred that the photoreceptor must reside in that organ or tissue. In experiments originally intended to elucidate the nature of the photoreceptor responsible for mediating the effects of irradiations given during the night in *photoperiod*ically sensitive plants (Chapter 7.3), the location of the responsible pigment was also identified. In order to determine *action spectra* for this night-break effect, Harry Borthwick, Sterling Hendricks, and their colleagues at the US Department of Agriculture laboratories in Beltsville, Maryland, irradiated a selected leaf on each plant. Thus, for this particular biological response, the photoreceptor, which is a phytochrome, must reside in the irradiated leaf. Because the ultimate response occurs in shoot apices, it can be deduced that there is a transmissible signal in the transduction chain for this response, providing communication between the leaf and the shoot apex.

Microbeam irradiations, which provide greater spatial resolution, have also been used to determine the location of phytochrome responsible for some biological responses. For example, Tepfer and Bonnett (1972) demonstrated that biologically active phytochrome is found in the root apex by showing that selective, microbeam irradiation of the apical 1 mm of a *Convolvulus arvensis* root induces it to become gravitropically sensitive. Similarly, the sensitivity of the maize coleoptile to phototropically active blue light is modified by R absorbed by phytochrome. This phytochrome must be located in the extreme apex of the coleoptile as this is the region most sensitive to R. Harry Smith and his colleagues have also demonstrated that the phytochrome which modulates the rate of internode extension is located in the growing internode itself, thus eliminating in this instance the need for a transmissible signal in the transduction chain (Chapter 7.1).

Interpretation of data obtained by selective irradiation, however, is not as simple as it might seem. Mandoli and Briggs (1982) have demonstrated that light can be propagated efficiently along plant axes, which means that an incident microbeam can be readily absorbed by a photoreceptor at an appreciable distance from the site of irradiation (Chapter 7.4). Thus, determination of the actual site of photoperception becomes significantly more complex. Nonetheless, by correcting for the efficiency of axial transmission as a function of position along an etiolated oat shoot, they were able to predict the location of phytochrome active in the elongation responses being studied (Fig. 3.). Mesocotyl growth was inhibited by light absorbed near the top of the mesocotyl, while coleoptile growth was stimulated by light absorbed in the same region in the mesocotyl and at a site just above the coleoptilar node. Interestingly, these predicted sites of photoperception do not necessarily coincide with the regions of highest phytochrome content as detected by spectrophotometric and immunochemical assays. Thus, one cannot assume that simply because phytochrome is abundant in a particular location that it is necessarily responsible for nearby photomorphogenic effects.

4.5.3.2 Spectrophotometry

The first detailed information about phytochrome distribution throughout entire dark-grown seedlings was provided by spectrophotometric assays. Because of their relative insensitivity, however, this method at best detects only the most abundant phytochrome, which is that encoded by the *phyA* gene (Chapter 4.2), thus effectively limiting its application to this gene product.

Because of the inability to detect phytochrome spectrophotometrically in tissues that contain chlorophyll (Pratt 1983) most measurements have been

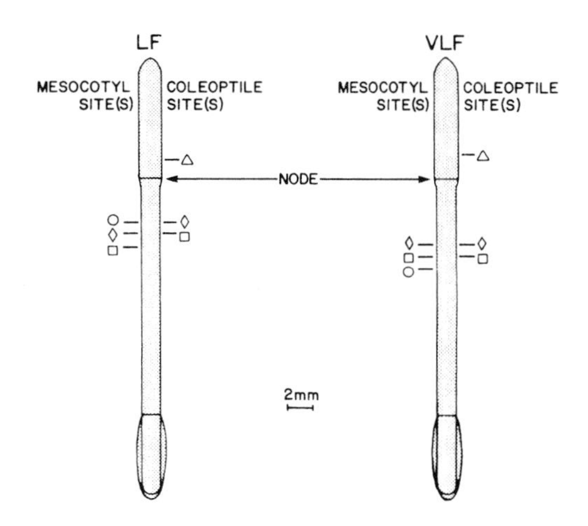

Figure 3. Locations of phytochrome active in modifying the elongation rate of the mesocotyl and the coleoptile in etiolated oat shoots (Δ, ○, ◊, □) as predicted from results of microbeam-irradiation experiments for both low fluence (LF) and very low fluence (VLF) responses, and from empirically determined optical properties of the tissue. The different symbols relate to slightly different ways by which the site of photoperception was predicted. Adapted from Mandoli and Briggs (1982).

made with etiolated seedlings, or with tissue that is normally achlorophyllous, such as underground stems and storage organs (Hillman 1964). An imaginative alternative that has proved useful under certain circumstances was introduced by Merten Jabben and Gerald Deitzer (1978) who utilized light-grown, normally chlorophyllous tissue, grown in the presence of the herbicide *Norflurazon*. This herbicide prevents chlorophyll accumulation, thereby permitting phytochrome assay. Unfortunately, however, the utility of this approach is limited because herbicide-bleached tissue is inevitably altered by the absence of photosynthetically produced nutrient, such that measurements need not accurately reflect phytochrome status in untreated plants.

Phytochrome has been found in almost all tissues examined spectrophotometrically, including leaves of both monocot and dicot plants, bulbs, roots, petioles, cotyledons, developing fruits, inflorescences, hypocotyls, and coleoptiles. Phytochrome has also been found in Norflurazon-bleached, mature leaves of the dicots *Sinapis alba* and *Impatiens parviflora*, and in secondary leaves of *Zea mays*. Systematic surveys of phytochrome distribution throughout single, light-grown plants have, however, not been made.

Similarly, detailed studies of phytochrome distribution throughout dark-grown seedlings have been rare. The first systematic surveys indicated that phytochrome was most abundant in meristematic and/or recently meristematic cells (Briggs and Siegelman 1965), as illustrated by representative data for both a dicot (Fig. 4) and a monocot (Fig. 5) seedling. Moreover, as Winslow Briggs

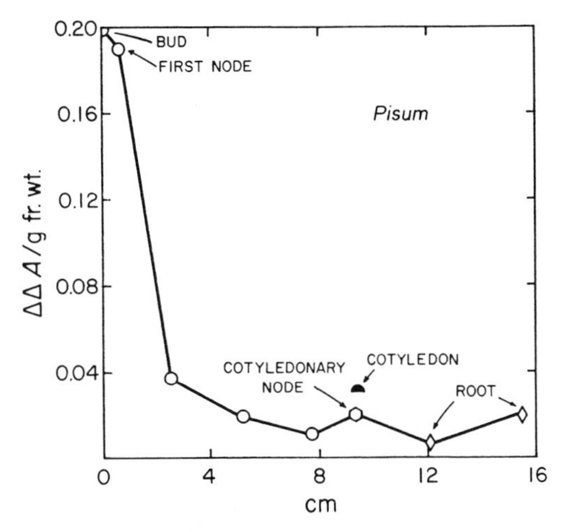

Figure 4. Phytochrome distribution throughout a 5-day-old etiolated pea seedling as determined by spectrophotometric assay ($\Delta\Delta A$/g fresh weight). Adapted from Briggs and Siegelman (1965).

and H.W. (Bill) Siegelman documented, phytochrome abundance exhibits its own specific pattern of expression, not reflecting that of protein in general. Subsequent measurements, while perhaps improving upon spatial resolution, have generally corroborated these initial observations.

4.5.3.3 In vivo *spectrofluorometry*

A potential alternative to spectrophotometric assays detects phytochrome by means of its unique fluorescence properties at cryogenic temperatures (Sinesh-chekov and Rüdiger 1992). This *spectrofluorometric assay* provides much greater sensitivity than does dual-wavelength spectrophotometric assay, permitting phytochrome detection and quantitation even in tissues with low levels of this chromoprotein. An additional advantage is the ability of this technique to resolve different phytochrome pools. As these authors themselves point out, however, additional work remains to be done to determine whether the different phytochrome pools identified by spectrofluorometric assay are equivalent to the different phytochromes that are encoded by different *phy* genes (Chapter 4.2).

Although spectrofluorometric assay has not been widely applied, it has been used to follow phytochrome content as a function of seedling age and of the transfer of etiolated seedlings to light, yielding data consistent with those obtained by other methods. By examining dissected plant parts, phytochrome has been quantitated independently in both shoot and root segments of cress

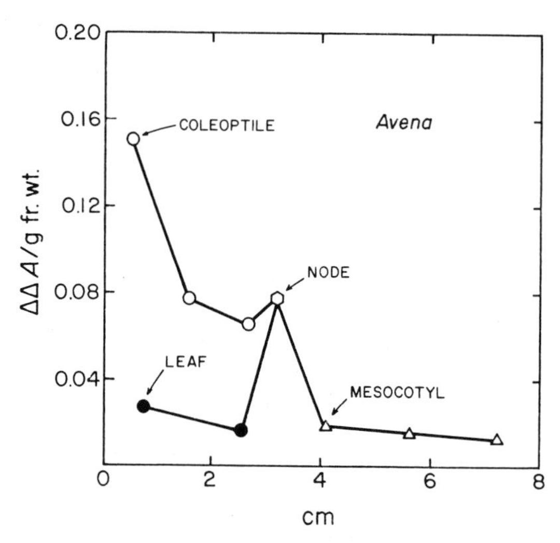

Figure 5. Phytochrome distribution throughout a 5-day-old etiolated oat seedling as determined by spectrophotometric assay ($\Delta\Delta A$/g fresh weight). Adapted from Briggs and Siegelman (1965).

seedlings, which contained 37 and 6.2 μg of phytochrome per gram fresh weight of tissue, respectively (Sineshchekov and Rüdiger 1992). Moreover, Sineshchekov and Rüdiger could deduce that under continuous illumination both shoots and roots should contain about 3.5 μg of phytochrome per gram fresh weight. As is the case for spectrophotometry, spectrofluorometric assay is also limited to application with achlorophyllous tissues. Consequently, it remains to be seen how extensively utilized this assay might become.

4.5.3.4 Immunochemistry

Antibodies are a powerful tool for determining the distribution of an antigen such as phytochrome. In addition to providing greater sensitivity than spectrophotometric assays, they also permit independent detection of the different phytochromes. Assays performed with extracts of pooled plant parts provide quantitative information at about the same level of resolution as spectrophotometry, while the technique of immunocytochemistry, which utilizes histologically prepared tissue sections, provides qualitative information on a cell-by-cell basis.

With monoclonal antibodies directed to phytochrome from etiolated oat shoots, Schwarz and Schneider (1987) determined phytochrome levels in different parts of etiolated maize *(Zea mays)* seedlings by ELISA. They found relatively high levels in the coleoptile tip, the root cap and the shoot apex (120, 80, and 70 μg/g fresh weight, respectively), with much lower levels in the

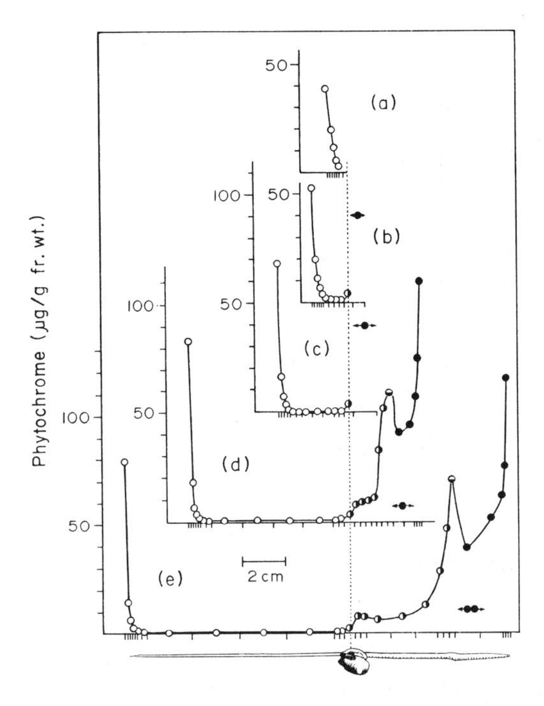

Figure 6. Phytochrome distribution in etiolated maize seedlings as a function of seedling age, as assayed by ELISA. Pooled segments were assayed for phytochrome content at 1.5 (a), 2 (b), 3 (c), 4 (d), and 5 (e) days of age, respectively. Tick marks on the abscissa indicate the size of tissue segments used for assay. Measurements were made for root (○), mesocotyl (◑), shoot apex (◒), coleoptile (●), and leaf (◂●▸) segments. When an entire organ was used for assay, its size is indicated by a horizontal bar. Adapted from Schwarz and Schneider (1987).

mesocotyl and leaves (< 10 µg/g fresh weight; Fig. 6). In addition, they also determined the distribution of phytochrome as a function of seedling age, finding no change in the general pattern over a 7-day period. Because they used antibodies directed to phytochrome from etiolated tissue, and worked with etiolated seedlings, it is almost certain that they detected only the *phyA* gene product (Chapter 4.2).

The discovery that a single plant contains three or more phytochromes (Chapter 4.2) has led to the development of monoclonal antibodies specific to each of the phytochromes so far identified. With these antibodies it has been possible to assess the distribution of each phytochrome independently, thus answering the question of whether the expression of each is regulated independently of the others. In the first use of such antibodies for examining the spatial distribution of three phytochromes in both dark- and light-grown seedlings, Wang *et al.* (1993) observed that each of the phytochromes exhibited the same general distribution among shoot, scutellum and root, and that with one exception each was more abundant when seedlings were grown in the dark than in the

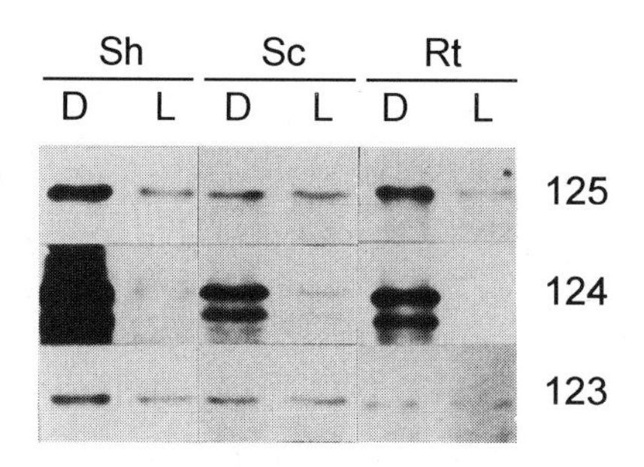

Figure 7. Phytochrome levels in 3-day-old dark- and light-grown oat seedlings as determined by immunoblot assay. Monoclonal antibodies specific to each of the three phytochromes detected in such seedlings were used to quantitate independently the 125-, 124- (*phyA* gene product), and 123-kD phytochromes. Phytochrome levels were determined for the shoot (Sh), scutellum (Sc) and root (Rt) grown either in darkness (D) or continuous white light (L). Only the phytochrome-containing portion of each blot is shown. Sample lanes were loaded with either 5% (shoot, scutellum) or 10% (root) of the protein obtained from a single organ or organ system. Quantitation was performed by video densitometry, with phytochrome levels ranging from less than 300 pg to more than 350 ng of 124-kD phytochrome in the scutellum of a light-grown seedling and the shoot of a dark-grown seedling, respectively. Adapted from Wang *et al.* (1993).

light (Fig. 7). The exception is that the two phytochromes predominating in light-grown tissue (125 kD and 123 kD in monomer size) were found at equivalent levels in the root, whether grown in darkness or light. No evidence could be found at this level of resolution for a substantial difference in spatial distribution among the three phytochromes. Of course, the possibility remains that these phytochromes are expressed in different cells within the same organ or organ system. Unfortunately, because of the exceedingly low levels of the 125- and 123-kD phytochromes, it will be difficult to determine their distributions on a cell-by-cell basis.

Immunocytochemical assay is powerful because it provides information at a cellular level. It is difficult to perform with the sensitivity of an ELISA or immunoblot assay and has been used so far only with antibodies directed to the phytochrome that is abundant in etiolated seedlings (presumably the *phyA* gene product, see Chapter 4.2). With rare exception it has been applied only to etiolated seedlings. Thus, information gained by this approach is so far presumably limited to detection of the phytochrome encoded by the *phyA* gene. As an example, immunocytochemical examination of phytochrome distribution throughout a young, etiolated oat seedling (Fig. 8) reveals that immunochemically detectable phytochrome is most abundant near, but not at, the tip of the coleoptile, in the region of the coleoptilar node, at the tip of adventitious roots that are still embedded within the shoot axis, and in the root cap, as opposed to either the root meristem *per se* or the zone of elongation on the side opposite from the cap. Moreover, it is evident that the cell type in which phytochrome is found is a function of position within the seedling; for example, phytochrome is

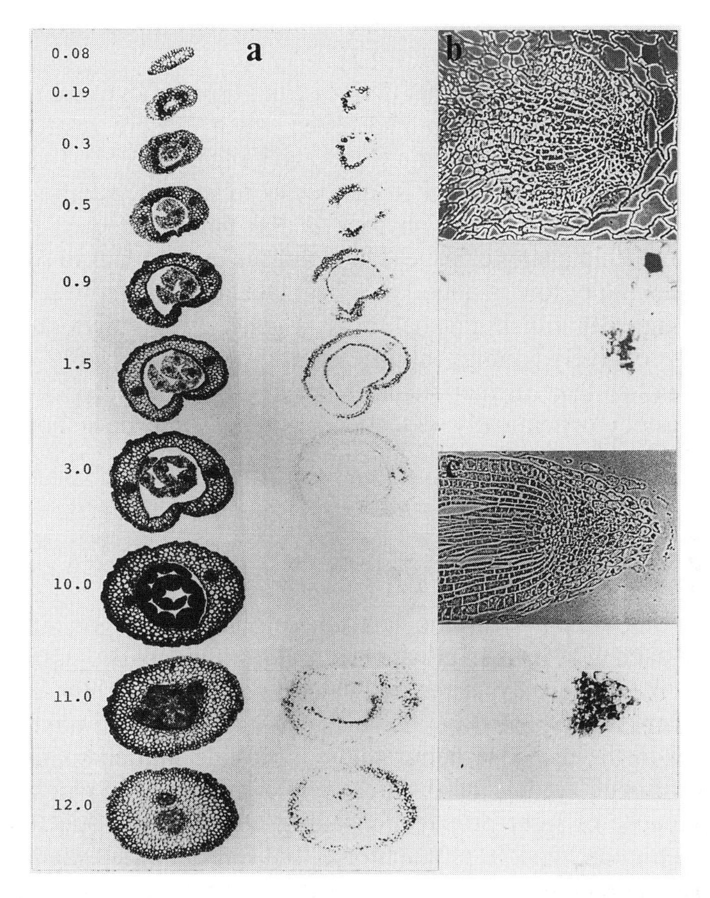

Figure 8. Distribution of phytochrome throughout a 3-day-old etiolated oat seedling as determined by immunocytochemical assay with peroxidase as the label. In this case, the presence of phytochrome is detected by a densely coloured reaction product produced by a peroxidase-catalyzed reaction. (a) Each section was photographed twice, once with dark-field optics to indicate structure *(left)* and once with bright-field optics to reveal immunocytochemical stain associated with the presence of phytochrome *(right)*. Distances in mm from the coleoptile apex are given. (b) Enlargement of part of a cross-section through the coleoptilar node (about 11 mm from the coleoptile tip). The section was photographed with phase-contrast optics to show structure *(upper)* and bright-field optics to reveal the location of phytochrome *(lower)*. (c) Longitudinal section through the apex of a primary root. The section was photographed with phase-contrast optics to show structure *(upper)* and bright-field optics to reveal phytochrome *(lower)*. Adapted from Pratt L.H. and Coleman R.A. (1971) *Proc. Natl. Acad. Sci. USA* 68: 2431-2435.

most abundant in non-epidermal cells within 0.5 mm of the coleoptile tip, while just below this region it is much more abundant in the epidermal layers (Fig. 8a). Similar, although less extensive, observations for the dicot pea indicate that in the epicotyl hook of an etiolated seedling phytochrome is most abundant in subepidermal, cortical cells, and is not detected in epidermal cells except for the guard cells, which contain considerable amounts of immunochemically detectable phytochrome (Saunders *et al.* 1983).

An important generalization that derives from immunocytochemical observations is that each species examined, even when closely related, exhibits a unique distribution of phytochrome (Pratt and Coleman 1974). Even though the pattern of distribution thus varies from species to species, additional generalizations are still possible. Most obvious is the heterogeneous distribution of phytochrome within a single tissue. Phytochrome is abundant in some cells, but apparently lacking or low in abundance in adjacent cells that, in some instances, are morphologically indistinguishable. As a general rule, phytochrome is most abundant in relatively young, rapidly expanding cells recently derived from meristems, rather than in meristems themselves. As with spectrophotometric assay, immunocytochemically detectable phytochrome does not parallel the distribution of total protein.

4.5.3.5 Summary

Knowledge about phytochrome distribution has been obtained by four approaches, each with its own advantages and limitations. Bio-assays, including microbeam irradiation experiments, detect only biologically active phytochrome and are highly sensitive; however, they fail to detect phytochrome that is not active in the response being studied, provide limited spatial resolution, fail to discriminate among the different phytochromes, and can be difficult to interpret because of light propagation along a plant axis. Spectrophotometry detects only photoreversible phytochrome, but offers limited spatial resolution, is relatively insensitive, detects phytochrome irrespective of its biological activity, and fails to discriminate among the different phytochromes. Spectrofluorometry by comparison offers enhanced sensitivity and the potential to discriminate among different phytochromes. Nonetheless, it too suffers somewhat from limited spatial resolution and detects phytochrome regardless of its biological activity. In contrast, immunochemical assays are not only sensitive, but also offer the greatest spatial resolution, and have the potential to detect specific phytochrome pools; the same assays, however, detect phytochrome regardless of its spectral or biological activity.

Interestingly, both spectral and immunochemical assays yield the same basic information. Namely, in etiolated seedlings phytochrome is most abundant in young, rapidly expanding cells, including those in root caps, epicotyl and hypocotyl hooks, and coleoptilar nodes. Somewhat unexpected, perhaps, is the observation that biologically active phytochrome need not be distributed in the same way as total phytochrome. It is evident that further attempts to correlate biological activity with bulk distribution are needed, as is further work with light-grown plants and with the question of the distribution of phytochromes other than the *phyA* gene product. However, because of the exceedingly low abundance of phytochrome in light-grown plants, and of phytochromes other than that encoded by the *phyA* gene, progress is likely to be slow, at least until methodologies have been significantly improved.

4.5.4 Intracellular localization

Numerous experimental approaches have provided information about the subcellular distribution of phytochrome. With rare exception they depend ultimately upon either spectrophotometric or immunochemical assay to verify the presence of phytochrome in a particular subcellular fraction or compartment.

4.5.4.1 Microbeam irradiation

To obtain information about the subcellular localization of phytochrome with microbeam irradiations effectively requires the study of relatively large single-celled organisms or organisms that are only one cell thick. Even so, however, this approach has provided unique information.

Low fluence rate R, *via* phytochrome, controls *chloroplast orientation* in the green alga *Mougeotia* (Chapters 7.2 and 9.4). Microbeam irradiation experiments indicate that the phytochrome controlling chloroplast orientation is associated with the periphery of the cytoplasm, rather than with the chloroplast itself. Moreover, the *absorbance dipole* of the active phytochrome is oriented with respect to the cell surface and undergoes a reversible change in that orientation upon phototransformation. Collectively, these observations indicate that the phytochrome responsible for photocontrol of chloroplast orientation in *Mougeotia* is probably located in or on its plasma membrane (Haupt 1970), a conclusion consistent with recent immunocytochemical observations. Whether this localization also occurs in higher plants, however, is not established. It is quite possible that *Mougeotia* and similar organisms represent a special case, in which the phytochrome system has evolved to provide directional information. Since the overwhelming majority of phytochrome-mediated responses in higher plants are unrelated to the direction from which light arrives, this apparent membrane association of phytochrome in *Mougeotia* might be an anomaly.

Similar experiments concerning *polarotropism* of fern protonemata, especially those of *Adiantum*, have led to similar conclusions (Chapter 9.7). Active phytochrome exhibits in this case as well a *dichroic* orientation, with the R-absorbing form of phytochrome (Pr) and Pfr exhibiting different orientations with respect to the plane of the cell wall. Microbeam irradiations have also shown that the phytochrome inducing polarotropism is confined to a region of from 5-15 μm behind the growing tip of an apical cell. As in the case of *Mouge-*

otia, however, one must ask whether conclusions derived from study of this somewhat special case can be generalized. In this regard, it is interesting to note that phytochrome also influences the duration of the cell cycle in the *Adiantum* protonemata, which is clearly not a directional response. In this case as well the active phytochrome exhibits a photoreversible, dichroic orientation, indicating that membrane-associated phytochrome might also be involved in non-directional responses. Nevertheless, attempts to establish a dichroic orientation of phytochrome in higher plants have never been fully convincing.

4.5.4.2 In vitro *responses*

The reversible photomodulation of a process or activity by R and FR in a purified, subcellular fraction is presumptive evidence that that fraction contains biologically active phytochrome. Phytochrome control of several activities in isolated mitochondria have been reported, including enhancement in the rate of NADP reduction, modulation of the extent of NADH dehydrogenase activity, alteration in the rate of calcium flux, and change in the rate of succinate uptake (Chapter 4.6). Phytochrome control of development, alteration in the ratio of acidic to nonacidic gibberellins, and changes in the rate of succinate uptake have similarly been reported for isolated etioplasts. More recently, it has also been reported that phytochrome can alter the rate of transcription in isolated nuclei. Other less well defined responses *in vitro* include changes in peroxidase and ATPase activities in crude, membrane-containing fractions. Taken at face value these reports indicate that biologically active phytochrome is associated with mitochondria, plastids and nuclei. Much of this work, however, has been at best difficult to repeat, and has often been difficult to interpret. Consequently, it is necessary to evaluate these reports with caution until one or more of them has been more thoroughly characterized.

4.5.4.3 Microspectrophotometry

In principle, *microspectrophotometry* (Part 3) would be a powerful tool for determining the subcellular localization of phytochrome within tissue sections. In practice, however, the few attempts to utilize this method have failed. In fact, theoretical considerations argue that such measurements are not feasible with currently available technology.

4.5.4.4 Subcellular fractionation

An obvious approach for determining the intracellular localization of phytochrome is to separate crude plant extracts into defined, subcellular fractions and follow one of two strategies. The more direct one is to ask whether a purified subcellular fraction contains phytochrome as assayed spectrophotometrically or immunochemically. The other is to ask whether added phytochrome will bind to a putative receptor that is an integral part of a subcellular fraction.

Phytochrome has been reported by spectrophotometric assay to be associated with amyloplasts, etioplasts, etioplast envelopes, chloroplast envelopes, mitochondria, nuclei, plasma membrane, and endoplasmic reticulum. Often, too few data are provided to permit reliable evaluation; in those instances where sufficient data are available, it is generally apparent that the specific activity of phytochrome in the purified fraction is about the same as, or lower than, that in the crude extract from which the fraction was prepared, indicating that the association might be spurious. At least three alternatives exist to explain the observed associations. (i) They are all artifacts. (ii) They are all relevant, indicating that phytochrome plays a role in a very large number of different subcellular compartments. (iii) While most are artifactual, a limited number are of biological significance, in which case one must decide how to determine which association is artifact and which is 'real'. At present, perhaps, it is best to remain sceptical about all such reported associations until firmer evidence becomes available.

Associations between phytochrome and subcellular fractions to which reference has already been made involve the small percentage of total phytochrome that is inherently associated with sedimentable material upon tissue homogenization (Rubinstein *et al.* 1969). If Pr is converted to Pfr prior to homogenization, however, as much as 60% or more of the total phytochrome can be associated with particulate material (Quail *et al.* 1973; Pratt and Marmé 1976). Unfortunately, this light-induced, particulate phytochrome requires relatively high levels of Mg^{2+} to remain associated with pelletable material. Because subcellular organelles and membranes co-aggregate spontaneously under these conditions, it has been difficult to determine to what phytochrome is bound, or even whether it is bound to any single organelle or membrane type. Nevertheless, it has been possible to achieve a substantial purification of these particles in recent years. Characterization of these particles, as well as their independent visualization by immunocytochemistry in crude preparations (Section 4.5.4.5), has, however, led to the conclusion that phytochrome does not associate with these particles because of any association with a membrane. Thus, the relatively common interpretation that has been offered many times over the years that this light-induced association of phytochrome with pelletable subcellular material involves binding of phytochrome to a membrane is not supported by currently available evidence.

The second strategy, which has been to isolate subcellular fractions and ask whether added phytochrome binds to them, is most useful only if the interaction meets certain criteria. (i) Binding affinity should be appropriately high, indicating the possibility that it can occur *in situ* where phytochrome concentrations are likely to be low. (ii) Binding should exhibit specificity not only for phytochrome, but also for its binding partner. (iii) Binding must be an integral step in the chain of events leading to biological activity. While it is comparatively easy to satisfy the first two criteria, the third has never been satisfied. Moreover, in at least one instance, an interaction that satisfies the first two criteria has been well documented to be an artifact. In this instance, which represents one of the initially described interactions between phytochrome and membranous subcellular fractions (Quail *et al.* 1973), the association was found to result from an artifactual association of phytochrome with degraded ribonucleoprotein particles (Quail 1975). Consequently, data that purport to document phytochrome-membrane or phytochrome-organelle interactions, in the absence of convincing evidence that the interaction is part of a transduction chain between Pfr and a biological response, must be interpreted with the greatest of caution (Quail 1982).

For reasons already given, virtually all, if not all, observations involve the subcellular localization of the *phyA* gene product. The recent availability of antibodies to three phytochromes in oat, however, has permitted a consideration of the hypothesis that one of the newly discovered, low-abundance phytochromes might be preferentially associated with membranes. In an initial test of this hypothesis, however, the results do not indicate any such association (Wang *et al.* 1993). Of course, it is virtually impossible to exclude the possibility of an association between a quantitatively minor fraction of one of the phytochromes and a membrane, or an association that is readily disrupted, leaving open the possibility that at least some biologically active phytochrome might be associated with membranes.

4.5.4.5 Immunolocalization

It is simple, at least in concept, to determine phytochrome localization by direct, visual observation after immunostaining with a suitable label. Numerous pitfalls exist, however, many of which are inherent to the methodology (Pratt 1976). For example, phytochrome must be immobilized by fixation prior to visualization. Thus, one must be satisfied with immunostaining of only the phytochrome still recognized by antibodies after fixation. In addition, one can never be certain that phytochrome did not move from its original location during the fixation process. Moreover, unanticipated interactions between antibodies and antigens other than phytochrome might be observed by immunocytochemistry even though they are not observed by other assays often used to

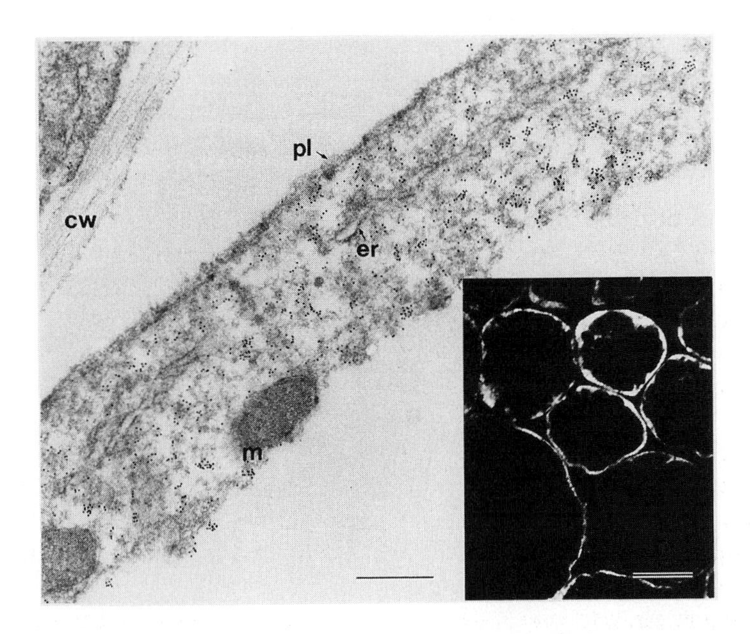

Figure 9. Phytochrome localization as Pr in dark-grown, non-irradiated oat coleoptiles, as revealed both by labelling with colloidal gold for observation by transmission electron microscopy and with the fluorophore, rhodamine, for observation by light microscopy *(inset)*. The Pr is diffusely distributed throughout the cytosol, showing no preferential association with the endoplasmic reticulum (er), mitochondria (m), or the plasma membrane (pl), which here has drawn away from the cell wall (cw) as a consequence of plasmolysis. Bar = 0.5 μm (inset bar = 20 μm). Adapted from McCurdy and Pratt (1986).

assess antibody specificity. Even with these and other limitations, however, immunocytochemistry remains a powerful tool.

When phytochrome is immunostained in non-irradiated, etiolated plant cells for observation by light or electron microscopy (Fig. 9), it appears to be distributed uniformly throughout the cytosol. While a few isolated observations have indicated that a small proportion of the immunostained phytochrome might be associated with nuclei, plasma membrane, endoplasmic reticulum, plastids and even mitochondria, these observations must be interpreted with a high degree of scepticism. Nonetheless, it is intriguing that this approach, like those discussed in the previous section (Section 4.5.4.4), indicate at least superficially a widespread distribution of phytochrome within the cell.

Perhaps the most intriguing result to derive from immunocytochemical observations is that phytochrome distribution within the cell is a function of its form (Mackenzie *et al.* 1975). While Pr is diffusely distributed throughout the cell, Pfr rapidly associates with amorphous structures of about 200 nm in size (Fig. 10). These particles are not apparent prior to the formation of Pfr, indicating that their formation is concomitant with the association of phytochrome

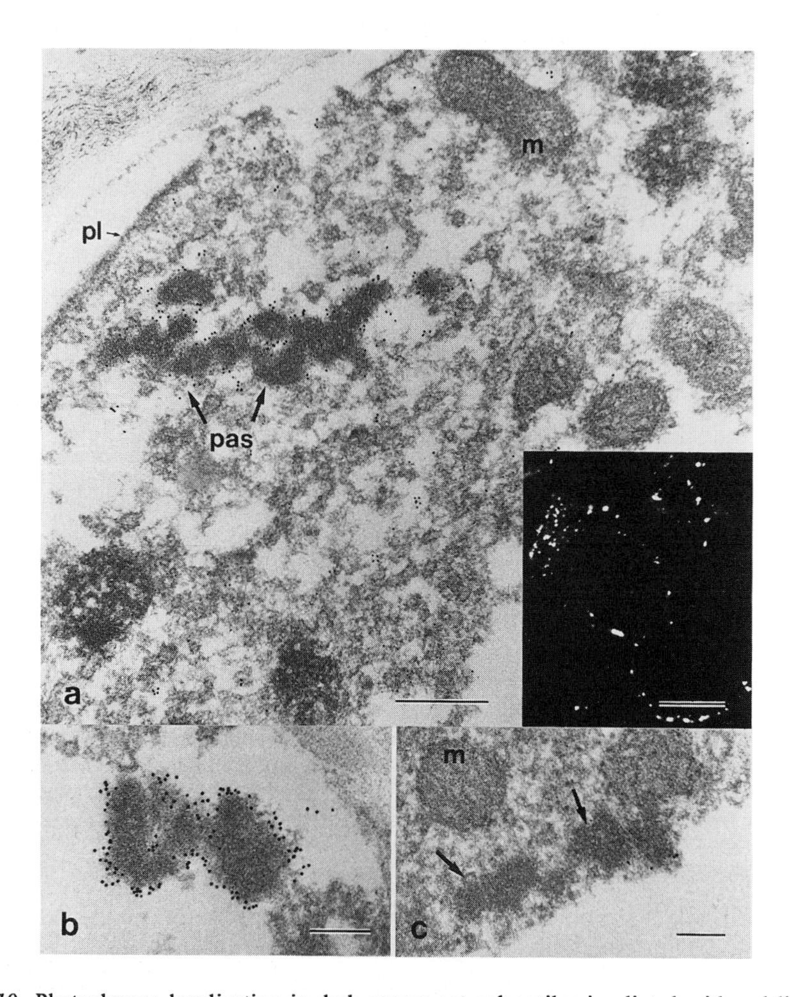

Figure 10. Phytochrome localization in dark-grown oat coleoptiles irradiated with red light (R) followed by far-red light (FR) and fixed immediately thereafter at 4°C. Phytochrome was revealed both by labelling with colloidal gold for observation by transmission electron microscopy and with the fluorophore, rhodamine, for observation by light microscopy *(inset).* (a) Phytochrome-specific label is associated with amorphous structures having a unit size of about 200 nm (pas). No immunolabel is found with mitochondria (m) or the plasma membrane (pl). The *inset* indicates how the distribution of this sequestered phytochrome appears at the light microscope level. Bar = 0.5 μm *(inset* bar = 20 μm). (b) View of two phytochrome-associated structures at higher magnification. Bar = 0.2 μm. (c) Control section incubated with non-immune mouse antibody in place of the monoclonal mouse antibodies directed to phytochrome (see Fig. 2). The phytochrome-associated structures *(arrows)* are not labelled with colloidal gold, indicating that the colloidal gold seen in (a) and (b) is associated with phytochrome as expected. Mitochondrion = m; bar = 0.2 μm. Adapted from McCurdy and Pratt (1986).

with them. Moreover, they do not possess any membrane, indicating further that this sequestration of phytochrome does not involve binding of Pfr to membrane.

If sequestered Pfr is converted back to Pr, phytochrome slowly resumes its initial diffuse distribution indicating that it is a reversible process. Presently available data indicate that this association is probably not part of a transduction chain between phytochrome and phytochrome-mediated responses. Instead, it appears that this sequestering of Pfr might be a step in the enhanced degradation of phytochrome after its photoconversion to Pfr (Chapter 4.4).

Immunocytochemically observed sequestering of phytochrome as Pfr correlates well with the enhanced pelletability of phytochrome that is seen following photoconversion of Pr to Pfr *in vivo* (Section 4.5.4.4). Both processes occur with comparable time courses, and both are equally reversible when Pfr is converted back to Pr. Moreover, immunocytochemical observation of the particles with which phytochrome is associated *in vitro* indicate that these structures are indistinguishable from those observed for sequestered phytochrome *in situ* (McCurdy and Pratt 1986). It thus appears safe to conclude that these two phenomena are probably different manifestations of the same intracellular event.

4.5.4.6 Summary

Perhaps the most important reason for determining the intracellular distribution of phytochrome, at least initially, was to test conflicting hypotheses concerning its mode of action. The two major hypotheses in question were that phytochrome functions: (i) by modulating one or more membrane properties, which implies a membrane localization, or (ii) by directly influencing gene transcription, which implies a nuclear association. Considerable data can be offered in support of both hypotheses. However, the net outcome of all efforts to establish the intracellular localization of phytochrome supports neither hypothesis well.

If one were to attempt to summarize observations concerning the intracellular distribution of phytochrome, the following points could be made. Within cells that have never been exposed to light, phytochrome as Pr is distributed uniformly throughout the cytosol as though it were a soluble protein. There is no completely convincing evidence to permit the conclusion that it exhibits a unique, biologically relevant association with any particular membrane or organelle, although it has been found associated with almost every subcellular component examined. Following photoconversion to Pfr, much or most of the phytochrome becomes associated with an as yet poorly characterized, amembranous structure. No biologically relevant binding of Pfr to a membrane or receptor has so far been rigorously documented. Moreover, there is as yet precious little information about the subcellular localization of phytochrome other than that which derives from the *phyA* gene. What information there is, however, fails to indicate any substantial association between a phytochrome and a particulate subcellular component.

4.5.5 Concluding remarks

The tone of this chapter has been consciously sceptical, at least with respect to the question of the subcellular localization of phytochrome. Too often, data have been interpreted to support a currently favourable hypothesis, even though they might better have been interpreted in an alternative, albeit possibly less interesting, fashion. Consequently, premature and potentially erroneous conclusions have often been reached, especially concerning its subcellular distribution. While more rigorous and more conservative interpretation of available data might have the disadvantage of appearing to slow progress, or at least of being less exciting, it does have the distinct advantage of preventing less fashionable, but perhaps equally viable, hypotheses concerning the mode of action of phytochrome from being prematurely discarded. For example, one of the earliest hypotheses is that Pfr possesses a unique enzyme activity lacking in Pr. Given that the simplest interpretation that satisfies the bulk of available data is that phytochrome is a soluble, cytosolic protein, this hypothesis remains quite reasonable, even if no longer as fashionable as when first proposed. Certainly, much more information is needed if we are to understand the distribution and subcellular localization of phytochrome, especially in plants grown in a natural environment, for which little information is currently available.

4.5.6 Further reading

Pratt L.H. (1983) Assay of photomorphogenic photoreceptors. In: *Encyclopedia of Plant Physiology*, New Series, 16A, *Photomorphogenesis*, pp. 152-177, Shropshire Jr. W. and Mohr H. (eds.) Springer-Verlag, Berlin.

Pratt L.H., McCurdy D., Shimazaki Y. and Cordonnier M.-M. (1986) Immunodetection of phytochrome: immunocytochemistry, immunoblotting, and immunoquantitation. In: *Modern Methods of Plant Analysis*, New Series, Vol. 4, pp. 50-74, Linskens H.-F. and Jackson J.F. (eds.) Springer-Verlag, Berlin.

Quail P.H. (1982) Intracellular location of phytochrome. In: *Trends in Photobiology*, pp. 485-500, Hélène C., Charlier M. and Montenay-Garestier T.L. (eds.) Plenum, New York.

4.5.7 References

Briggs W.R. and Siegelman H.W. (1965) Distribution of phytochrome in etiolated seedlings. *Plant Physiol.* 40: 934-941.

Butler W.L. (1964) Absorption spectroscopy in vivo: theory and application. *Annu. Rev. Plant Physiol.* 15: 451-470.

Gross J., Seyfried M., Fukshansky L. and Schäfer E. (1984.) *In vivo* spectrophotometry. In: *Techniques in Photomorphogenesis*, pp. 131-157, Smith H. and Holmes M.G. (eds.) Academic Press, London.

Haupt W. (1970) Localization of phytochrome in the cell. *Physiol. Vég.* 8: 551-563.

Hillman W.S. (1964) Phytochrome levels detectable by *in vivo* spectrophotometry in plant parts grown or stored in the light. *Amer. J. Bot.* 51: 1102-1107.

Jabben M. and Deitzer G.F. (1978) A method for measuring phytochrome in plants grown in white light. *Photochem. Photobiol.* 27: 799-802.

Mackenzie Jr. J.M., Coleman R.A., Briggs W.R. and Pratt L.H. (1975) Reversible redistribution of phytochrome within the cell upon conversion to its physiologically active form. *Proc. Natl. Acad. Sci. USA* 72: 799-803.

Mandoli D.F. and Briggs W.R. (1982) The photoperceptive sites and the function of tissue light-piping in photomorphogenesis of etiolated oat seedlings. *Plant Cell Environ.* 5: 137-145.

McCurdy D.W. and Pratt L.H. (1986) Immunogold electron microscopy of phytochrome in *Avena*: identification of intracellular sites responsible for phytochrome sequestering and enhanced pelletability. *J. Cell Biol.* 103: 2541-2550.

Pratt L.H. (1976) Immunological visualization of phytochrome. In: *Light and Plant Development*, pp. 75-94, Smith H. (ed.) Butterworths, London.

Pratt L.H. and Coleman R.A. (1974) Phytochrome distribution in etiolated grass seedlings as assayed by an indirect antibody-labelling method. *Amer. J. Bot.* 61: 195-202.

Pratt L.H. and Marmé D. (1976) Red light-enhanced phytochrome pelletability: a reexamination and further characterization. *Plant Physiol.* 58: 686-692.

Pratt L.H., Wampler J.E. and Rich E.S. (1985) An automated dual-wavelength spectrophotometer optimized for phytochrome assay. *Analytical Instrumentation* 13: 269-287.

Quail P.H. (1975) Particle-bound phytochrome: association with a ribonucleoprotein fraction from *Cucurbita pepo* L. *Planta* 123: 223-234.

Quail P.H., Marmé D. and Schäfer E. (1973) Particle-bound phytochrome from maize and pumpkin. *Nature New Biol.* 245: 189-190.

Rubinstein B., Drury K.S. and Park R.K. (1969) Evidence for bound phytochrome in oat seedlings. *Plant Physiol.* 44: 105-109.

Saunders M.J., Cordonnier M.-M., Palevitz B.A. and Pratt L.H. (1983) Immunofluorescence visualization of phytochrome in *Pisum sativum* L. epicotyls using monoclonal antibodies. *Planta* 159: 545-553.

Schwarz H. and Schneider H.A.W. (1987) Immunological assay of phytochrome in small sections of roots and other organs of maize *(Zea mays* L.) seedlings. *Planta* 170: 152-160.

Sineshchekov V.A. and Rüdiger W. (1992) Fluorescence spectroscopy of stable and labile phytochrome in stems and roots of etiolated cress seedlings. *Photochem. Photobiol.* 56: 735-742.

Tepfer D.A. and Bonnett H.T. (1972) The role of phytochrome in the geotropic behavior of roots of *Convolvulus arvensis*. *Planta* 106: 311-324.

Wang Y.-C., Cordonnier-Pratt M.-M. and Pratt L. H. (1993) Spatial distribution of three phytochromes in dark- and light-grown *Avena sativa* L. *Planta* 189: 384-390.

4.6 Signal transduction in phytochrome responses

Stanley J. Roux

Department of Botany, University of Texas at Austin,
Austin, TX 78713, USA

4.6.1 Introduction

The absorption of red light (R) by phytochrome initiates a cascade of causally related biochemical events in responsive cells that culminate in major changes in plant growth and development. This chain of biochemical events transduces the light signal into critical adaptive changes, and the study of these events is an active and exciting area of research in the phytochrome field today. The purpose of this chapter is to review the evidence on what may be some of the key transduction events in the cascade leading from the photoactivation of phytochrome to the many-varied displays of photomorphogenesis.

Starting with the light-induced generation of the far-red light (FR)-absorbing form of phytochrome (Pfr) from the R-absorbing form of phytochrome (Pr), which occurs on the millisecond time scale, the transduction chain for a phytochrome response can stretch out for many hours or even days (Mohr 1983), depending on the response. Different responses may also vary many-fold in the number of steps in the transduction chain. Given that there are several cellular phytochromes (Chapter 4.2; Quail 1991), and that any one phytochrome may exist in several different subcellular locales, it seems probable that there may be a variety of transduction chains emanating from a variety of Pfrs functioning in a variety of cellular micro-environments. Like rain drops hitting the surface of a pond, R may set off ripples of transduction chains throughout the cell, each with a different range and scope of influence. Furthermore, these light-initiated waves of biochemical changes can reinforce or be cancelled by the endogenous waves of change that occur rhythmically in the daily life of the cell; *i.e.* the effects of Pfr are influenced by what hour of the circadian clock they are initiated.

The overview presented above predicts that the transduction chains for phytochrome responses may be varied and complex. It may be that as our

R.E. Kendrick & G.H.M. Kronenberg (eds.), Photomorphogenesis in Plants - 2nd Edition

ignorance about this subject subsides in the coming years, this apparent complexity may turn out to be more illusory than real. However, at the present time the best that can be done is to identify probable biochemical events in the transduction chain for some specific phytochrome responses, and then speculate as to whether these events might be common to other phytochrome responses.

Every phytochrome transduction chain is unique in detail, but it should not be surprising if we find that many of them share common intermediary steps. Indeed, one of the major insights about signal transduction in plants and animals revealed by the discoveries of the past decade and a half is that there are common events that occur repeatedly in many different transduction chains, independent of the stimulatory signal or the eventual response. It would be surprising, then, if there were no role for such common transducing events as GTP-binding protein (G-protein) activation, inositol phospholipid turnover, change in intracellular free $[Ca^{2+}]_{cyt}$ or protein phosphorylation in any phytochrome responses. Still, any evidence purporting to implicate these common events in any specific phytochrome transduction chain should be viewed critically and on an equal footing with alternative explanations.

Space limitations preclude this chapter from having a comprehensive literature review on phytochrome signal transduction. The emphasis will be more on articles published since the last edition of this book (1986). Even with this focus the literature is still so voluminous that the reader will be referred frequently to excellent review articles that deal with selected relevant sub-topics in more detail.

Although it is premature to define a temporal sequence for any phytochrome transduction chain, the order of this chapter will be to discuss first those Pfr-triggered events that would be likely to occur on the time scale of seconds and then cover those that would be closer in time to the final physiological response.

4.6.2 Intrinsic protein kinase activity

The involvement of protein kinases in signal transduction in plants has been well documented (Trewavas and Gilroy 1991). Thus the finding that highly purified phytochrome had protein kinase activity closely associated with it (Wong *et al.* 1989) may have great significance for understanding early steps in phytochrome signal transduction. The proposal that phytochrome itself is a protein kinase has been challenged by reports that it is possible to remove protein kinase activity from purified phytochrome without spectrally denaturing it (Singh and Song 1990). On the other hand, authentic protein kinases can lose activity during their purification. Furthermore, phytochrome itself can bind ATP, which is a requisite for it being a kinase. Conclusive evidence on the question of whether phytochrome has intrinsic protein kinase activity awaits further studies.

If phytochrome proves to be a protein kinase and its kinase activity changes upon its photoconversion (Wong *et al.* 1989), then this consequence of R absorption would surely qualify as a prime candidate for being an early step in the transduction chain. There is precedence for receptor protein kinases in plants. However, these are typically intrinsic membrane proteins, and none of the phytochromes so far characterized have primary structures that predict intrinsic membrane associations. Whether phytochrome associates tightly with membranes or with membrane proteins as a consequence of its photoconversion remains a controversial issue (Roux 1986).

As discussed below, Pfr-induced protein kinase activity may also be a later event in signal transduction. Whether phytochrome activation directly or indirectly stimulates protein kinase activity, the significance of this event for signal transduction would still have to be proved. Thus, there are two critical questions to answer in order to evaluate the role of phosphorylation in phytochrome responses. (i) What is the phosphoprotein substrate? (ii) How does phosphorylation alter the function of this substrate?

4.6.3 G-protein involvement

The activation of G-proteins is often an early event required for stimulus-response coupling in animals and perhaps also in plants (Trewavas and Gilroy 1991). In their GTP-activated form, these proteins can bind to key metabolic enzymes and up- or down-regulate their activity, thus they can be important players in a signal-transduction cascade.

Two families of G-proteins have been characterized in both plants and animals: (i) small molecular weight proteins that appear to function as monomers; (ii) heterotrimeric proteins with α, β, and γ subunits, in which the α subunit binds GTP, then dissociates from the β and γ subunits and binds to target enzymes that it regulates. When the GTP is hydrolysed to GDP by the GTPase activity intrinsic to the α subunit, this subunit re-associates with the β and γ subunits and the G-protein is inactivated. A non-hydrolysable analogue of GTP can be made by substituting a S for a P in the terminal (γ) phosphate. Thus when GTP-γ-S binds to G-proteins it cannot be hydrolysed to GDP, and this constitutively activates G-proteins. G-proteins are also stabilized in an active form when their GTPase activity is blocked by ADP ribosylation, a reaction specifically induced by cholera toxin. In contrast, the binding of GDP-β-S [which substitutes a S for a P in the terminal (β) phosphate of GDP] to G-proteins blocks their activation because GDP-β-S cannot be phosphorylated to GTP.

The possible involvement of G-proteins in phytochrome signal transduction was first reported by Hasunuma *et al.* (1987) in their study of light-regulated flowering in *Lemna paucicostata.* Just as R and FR interruption of dark periods could block flowering in *Lemna,* both light treatments also slightly depressed the binding of [^{35}S]GTP-γ-S to G-proteins detected in crude extracts of *Lemna* 8 h after the irradiation. However, the light effects reported were small and difficult to relate with confidence to mechanisms of photoperiodic induction of flowering. Intriguingly, the authors noted that if the binding of [^{35}S]GTP-γ-S to G-proteins was assayed 4 h instead of 8 h after the light irradiation, then R stimulated rather than inhibited the level of binding relative to the dark control.

Bossen *et al.* (1990) implicated the involvement of a G-protein in the R-induced swelling of etiolated wheat protoplasts (Fig. 1). This study utilized a rather comprehensive array of agonists and antagonists to indicate that not only G-protein activation but also the activation of phospholipase C, calmodulin, protein kinase C (PKC)-like protein kinase and cyclic AMP-dependent enzymes could play important roles in the transduction chain leading from Pfr to the swelling response of wheat protoplasts. The authors were appropriately reserved in their interpretation of the mostly inhibitor-based results, nonetheless, it was clear that all their results could be rationalized in accord with G-protein initiated transduction schemes that are supported by abundant evidence in animal systems.

Romero *et al.* (1991a) presented evidence for G-protein involvement in Pfr-induced changes in the expression of chlorophyll *a*/*b*-binding (*Cab*) and phytochrome *(phy)* genes in etiolated oat seedlings. They found that cholera toxin, which stabilizes G-proteins in their GTP-activated form, could mimic the effects of Pfr by stimulating the expression of a *Cab* gene and inhibiting the expression of a *phyA* gene in the absence of any R signal. The major G-protein they detected in oat seedlings by Western blot analysis, using a GA/1 antibody that recognizes the GTP-binding site of several α-subunits, had a molecular mass of 24 kD, and an oat protein of this size was efficiently ADP ribosylated by cholera toxin in the presence of GTP. However, the authors had no evidence whether this was the G-protein that was involved in controlling the expression of *Cab* and *phy* genes.

Clark *et al.* (1993) have reported a R-FR reversible activation of monomeric G-protein activity in the envelope-matrix fraction of pea nuclei. Studies on the function of G-proteins in nuclei is a relatively new field in cell biology. Monomeric nuclear G-proteins have been proposed to play a role in the GTP-dependent fusion of membrane vesicles to the nuclear envelope during its reassembly after mitosis and in the regulation of transport across the nuclear pore. As discussed by Clark *et al.* (1993), there are reports of photoreversible control of both mitosis and nuclear transport, so it will be of interest to investigate whether the R regulation of these nuclear activities is related to its stimulation of nuclear G-protein activity.

The data reviewed above constitute initial evidence consistent with the hypothesis that G-protein activation may be an important early step in the transduction chain for at least some Pfr-induced responses. How can this

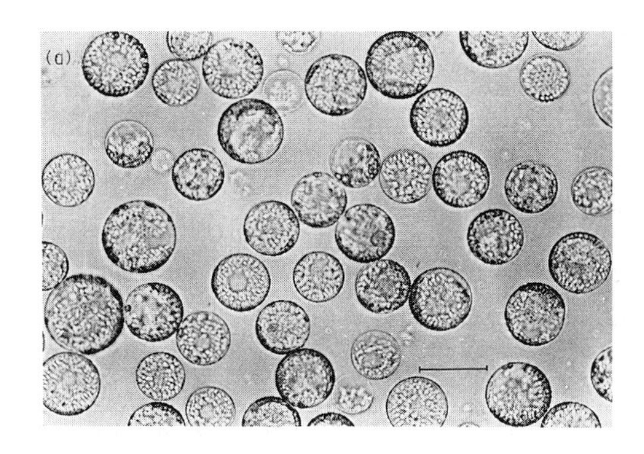

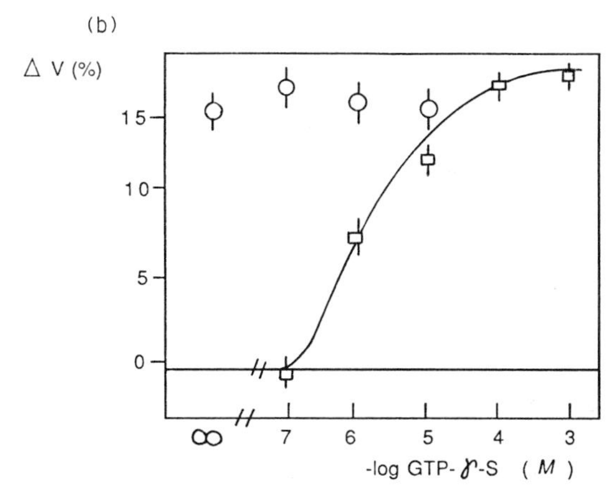

Figure 1. (a) Purified mesophyll protoplasts from dark-grown primary wheat leaves. These cells swell in response to red light (R), and are a useful model system for studying the early biochemical steps in the phytochrome transduction chain. Bar = 50 μm. (b) The promotion of the swelling of far-red light (FR)-(□) or R (○)-irradiated protoplasts by G-protein activator GTP-γ-S. Swelling is expressed as mean % volume change, ΔV% ± SE of three independent experiments, compared to the FR control. After Bossen *et al.* (1990).

hypothesis be more rigorously tested? One promising approach is that described by Neuhaus *et al.* (1993), who micro-injected agents that activate and inactivate heterotrimeric GTP-binding proteins into phytochrome-deficient cells of the *aurea* mutant of tomato (Chapter 8.2) to test their effect on three phytochrome responses (*Cab*-reporter gene expression, anthocyanin synthesis and chloroplast development) in those cells. They found that GTP-γ-S could substitute for R in inducing all three responses in a low but significant percentage of the trials. Micro-injected GDP-β-S when co-injected with oat phytochrome A blocked the ability of the phytochrome to induce the responses. The low percentage of successes in these experiments is troublesome, but the results clearly indicate that G-protein participation occurs early in the transduction chains for the R-induced responses tested.

There is evidence that at least some G-proteins regulate only a specific set of stimulus-response coupling pathways, so that the genetic elimination of one of these G-proteins would block only certain transduction chain(s) (Simon *et al.* 1991). If this is also true in plants there may be a G-protein specialized for transducing phytochrome responses, or even a different G-protein for different responses. Investigation of this question using genetic approaches is clearly warranted.

4.6.4 Changes in metabolism of inositol phosphoslipids

In animal cells, signal transduction often involves, as an early step, the phospholipase C-catalyzed conversion of phosphatidylinositol bisphosphate (PIP_2) to inositol 1,4,5-trisphosphate (IP_3) and diacylglycerol (DAG), both of which have well-described regulatory functions that can advance the transduction cascade further. As reviewed recently by Trewavas and Gilroy (1991), all the molecular components of this pathway appear to be present in plants, even though they may not be operative in all plant cells. While recognizing limitations in which cells actively maintain PIP_2-based transduction systems, and noting that there may be pathways of inositol-phosphate metabolism in plants that do not exist in animals, it can at least be said that for some plant cells PIP_2 turnover is an important signalling device.

Morse *et al.* (1987) were the first to measure light-induced changes in PIP_2 turnover in plants. However, they observed only small changes, and they did not characterize the photoreceptor involved. Basu *et al.* (1990) recorded R-induced changes in phosphatidylinositol turnover in *Brassica* hypocotyls, which were only slowly reversible by FR. Given the slow kinetics of the changes they observed, it was difficult to assess the physiological significance of these results. More recently, Guron *et al.* (1992) characterized rapid R-induced and FR reversible changes in PIP_2 turnover in leaves of *Zea mays.* Interestingly, their findings suggested that the most rapid membrane turnover change *(ca.* 15 s) was an increase in PIP_2 (Fig. 2). This effect could be expected to immediately impact at least two important cellular activities: it could activate a plasma membrane H^+-ATPase, and would provide substrate for increased production of DAG and IP_3. That IP_3 itself may be involved in phytochrome responses is suggested indirectly by the results of Shacklock *et al.* (1992). They showed that IP_3 released from a chemically caged form inside wheat protoplasts could substitute for R in inducing the Ca^{2+}-dependent swelling response in these cells. Confirmation of a photoreversible effect on PIP_2 turnover as a general phenom-

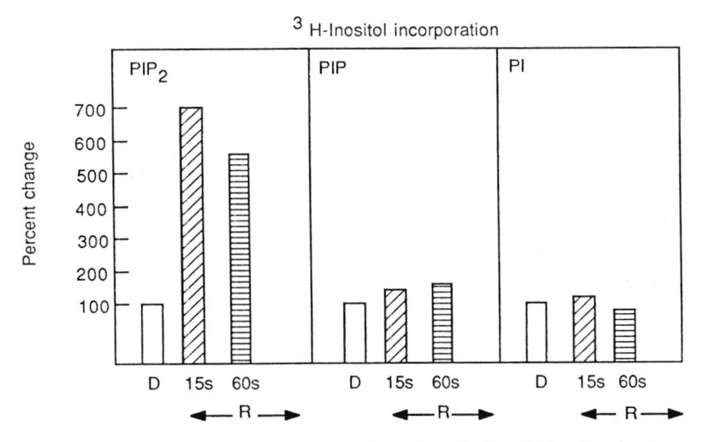

Figure 2. Red light (R)-induced rapid changes in phosphoinositide levels in maize leaves. Because R stimulates an increase in PIP_2, but not PIP or PI, phytochrome may be controlling a PIP kinase activity. After Guron *et al.* (1992).

enon in plant cell membranes would represent a major advance in our understanding of the phytochrome transduction chain.

4.6.5 Role of protein phosphorylation

Recent reviews have well documented the role of protein phosphorylation in plant growth and development (Trewavas and Gilroy 1991). Given the central role of this process in the activation of plant enzymes and given its participation in most well-documented transduction chains, it would be amazing if it were not also important in one or more of the steps leading from Pfr to photomorphogenesis. We discussed earlier the evidence for and implications of phytochrome itself being a protein kinase. In this section we review the evidence for phytochrome-induced protein phosphorylation without regard to whether the responsible kinase activity originates with phytochrome or some other protein.

There is a fairly extensive literature demonstrating the effects of light on protein phosphorylation. However, the first report of a light-induced phosphorylation of proteins that was R-FR reversible was by Datta *et al.* (1985). For these experiments, both the irradiations and the labelling reaction with [γ-^{32}P] ATP were carried out *in vitro* using highly purified nuclei from etiolated pea seedlings. The phosphorylation was dependent on Ca^{2+} and could be blocked by chlorpromazine, an antagonist of calmodulin action. Three bands, at 77, 64 and 47 kD, showed significantly enhanced incorporation of ^{32}P after R. Autoradiographic analysis of the nuclear preparation after R treatment indicated that virtually all of the ^{32}P was incorporated into the *nuclei* of the preparation. These results were consistent with the hypothesis that some aspects of phytochrome regulation of nuclear metabolism were modulated by phosphorylation.

More recently, Romero *et al.* (1991b) also reported R-induced phosphorylation of nuclear proteins. They used nuclei isolated from etiolated oat seedlings, and they found that both R and FR enhanced the phosphorylation of protein bands near 75 and 60 kD. Additionally, they showed that the phosphorylation pattern was changed by two modulators of G-proteins discussed above, cholera toxin and GDP-β-S. However, the same modulator had different effects on different bands, so the results were difficult to interpret in relation to any simple model.

Although the metabolism of isolated nuclei often closely reflects that which occurs in intact cells, one cannot assume this to be true. For this reason the results of Otto and Schäfer (1988) represented an important advance. They described phytochrome effects on protein phosphorylation *in vivo* (Fig. 3). They found that R given to etiolated oat coleoptile tips rapidly induced an increase in the phosphorylation of a protein band near 32 kD and a decrease in that of two bands near 29 and 33 kD, and that FR reversed this effect. The locale and identity of the bands that showed altered ^{32}P incorporation after R was unknown, but the results clearly indicated that phytochrome could regulate the phosphorylation state of several cellular proteins in living cells. The apparent rapidity of this response (half-time of the reaction for two of the bands was reported to be 3 s at 0°C) suggested that it would occur very near the primary responses induced by Pfr in cells. The half-time estimate was based on the unproven (and improbable) assumption that all phosphorylation/dephosphorylation activities were stopped immediately by the addition of sodium dodecyl sulphate (SDS) to the homogenate, and that these activities did not proceed further during the subsequent 45 s delay before the sample was transferred to a boiling water bath.

There have been other reports of R-induced protein phosphorylation in living cells in which FR reversibility was not demonstrated. For these reports either phytochrome or chlorophyll could be the photoreceptor for the response. For all the reports thus far on R-induced changes in phosphorylation, one has to withhold final judgment on the designation of the molecular masses of the phosphorylated proteins. Typically, protease inhibitor cocktails were not used during the homogenizations, and, as is evident from the history of phytochrome biochemistry, one should never underestimate the speed of proteolytic attack on plant proteins once cell breakage has been initiated.

To evaluate the significance of R-FR reversible changes in protein phosphorylation, it will be essential to know what kinase or phosphatase is being activated, what protein is being phosphorylated or dephosphorylated, and what is the effect of this change on that protein's function. An approach toward achieving this evaluation was described by Li *et al.* (1991), who have raised antibodies to one of the nuclear phosphoproteins whose phosphorylation was

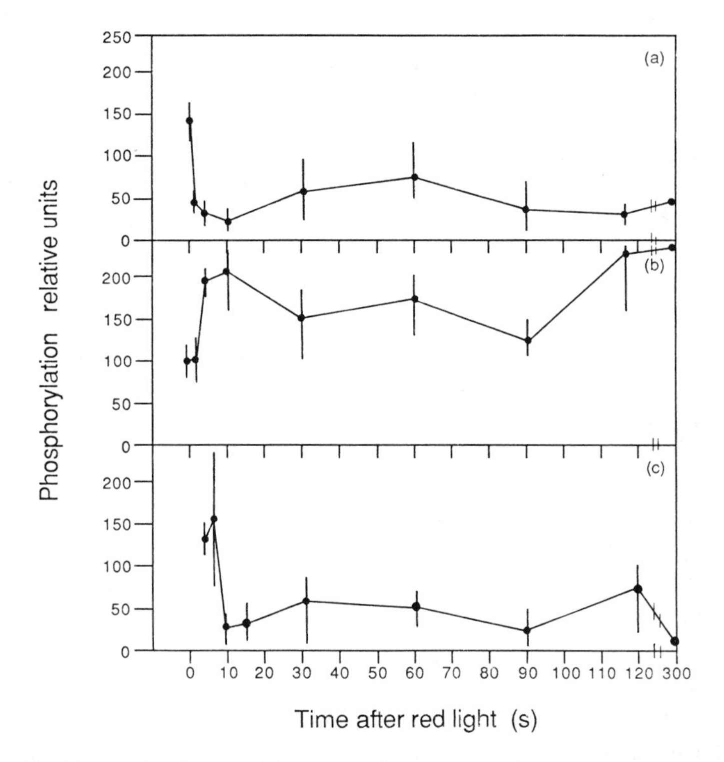

Figure 3. Rapid effects of red light (R) on protein phosphorylation in *Avena* coleoptiles. At various time points after R treatment and ^{32}P labelling, proteins were separated by SDS-PAGE and subjected to autoradiography. Densitometric readings (shown here) recorded changes in three bands with apparent molecular masses of (a) 33 kD (decreased labelling), (b) 32 kD (increased labelling), and (c) 29 kD (decreased labelling). After Otto and Schäfer (1988).

enhanced by calcium and by R in isolated nuclei. These antibodies have been used to immunoprecipitate a 43-kD phosphoprotein, from which direct amino-acid sequence data have been obtained, and to isolate a cDNA clone which may code for this protein (H. Li and S.J. Roux unpublished data). Apropos to the caveat on proteases given above, preliminary evidence suggests that the 43-kD phosphoprotein may be a breakdown product of the 77-kD phosphoprotein whose phosphorylation was also stimulated by light and by calcium. Determining the primary sequence of this protein will ultimately resolve its correct molecular mass and could also reveal important insights on its function.

More specific progress on identifying proteins whose phosphorylation may be regulated by light has come from the work of Datta and Cashmore (1989) and Klimczak *et al.* (1992). In both of these reports, evidence was shown that the binding of factors to promoters of phytochrome-regulated genes could be controlled by phosphorylation/dephosphorylation of the factor, and that the responsible protein kinase was likely to be a casein-kinase II type. In Datta and Cashmore (1989), phosphorylation of a crude preparation of the nuclear factor AT-1 resulted in inhibition of its specific DNA-binding activity; in Klimczak *et al.* (1992) phosphorylation of purified GBF1 enhanced its DNA-binding activity. In both papers the identity of the responsible kinase as a CK II-type was inferred by its ability to use GTP as well as ATP as phosphoryl donors and by its sensitivity to heparin inhibition. Klimczak *et al.* (1992) also partially purified a CK II-like kinase from broccoli that showed a high affinity for GBF1 as substrate and had a native molecular mass *(ca.* 128 kD) similar to that reported for animal CK II kinases.

The localization of the partially purified CK II kinase that phosphorylates GBF1 was not described, but the properties of this kinase are similar to those of a chromatin-associated CK II kinase purified from pea nuclei (Roux 1992). However, *cytoplasmic* CK II kinases have also been purified from plants and, in principle, factor-phosphorylating kinases, including the one that phosphorylated GBF1, could function effectively in either the cytoplasm or the nucleus.

One important mechanism by which phytochrome could regulate protein kinase activity would be to control the accumulation of mRNAs that code for protein kinases. The results shown in Fig. 4 are consistent with this possibility. They demonstrate that during light-induced greening in pea seedlings there are concurrent changes (accumulation and decline) in several transcripts whose translated sequences are characteristic of protein kinases. These data do not demonstrate R-FR reversibility of the light-induced transcript changes, but the well established role of phytochrome in de-etiolation makes it a likely candidate to be the main photoreceptor in this response.

Protein phosphatases are as essential as protein kinases for the control of protein phosphorylation. The purification and characterization of both a tyrosine protein phosphatase and a serine/threonine protein phosphatase in pea nuclei are currently underway (Y.-L. Guo and S.J. Roux unpublished data). Understanding the regulation of these phosphatases will contribute to a more

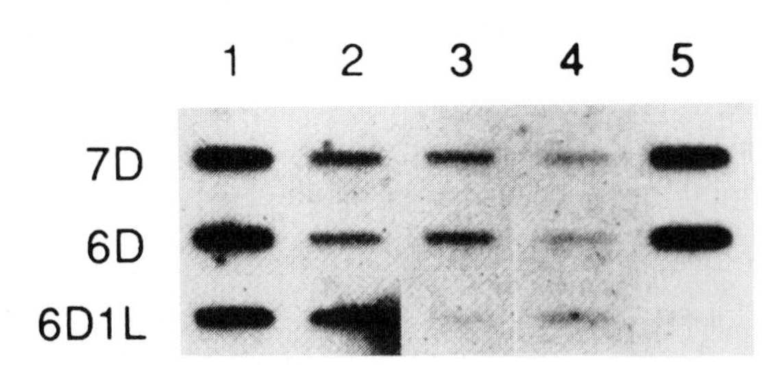

Figure 4. Slot blot analysis showing light-induced changes in the abundance of mRNAs encoding five different putative serine/threonine protein kinases (lanes 1-5) in pea seedlings. The mRNAs hybridized in lane 2 increased in abundance, those in lanes 3 and 5 decreased in abundance, and those in lanes 1 and 4 showed little change. After Lin X., Feng X.-H. and Watson J.C. (1991) *Proc. Natl. Acad. Sci. USA* 88: 6951-6955.

complete picture of nuclear protein phosphorylation and its control by phytochrome.

Of course, protein phosphorylation plays major regulatory roles throughout the cell, not just in nuclei. In particular, both in plants and animals, a major function of protein kinases and phosphatases is to regulate the activity of pumps and channels on the plasma membrane (PM). Given the evidence that phytochrome can stimulate changes in protein phosphorylation in intact cells, and the evidence (reviewed below) that it regulates both membrane potential and a variety of ion fluxes in plant cells, it will be of interest to test whether these two phenomena are linked: *i.e.* are there pumps and channels whose phosphorylation and subsequent function can be regulated by the action of Pfr?

To summarize the status of phosphorylation as a signalling step for phytochrome transduction, the evidence favouring this notion is mainly indirect. However, it seems almost certain that phosphorylation of transacting factors plays an important role in controlling the expression of phytochrome-regulated genes, so a strong expectation is that, directly or indirectly, Pfr will influence the phosphorylation status of these factors. Given the central role of phosphorylation as a signal transducing event in eukaryotes, one could also expect that it will be well represented in many if not most of the transduction chains initiated by Pfr. However, data have a way of changing expectations, and surely there is a crucial lack of information on what specific phosphorylation events are induced by phytochrome and are needed for the photoresponse. Experiments currently in progress should clarify these issues.

4.6.6 Induction of rapid ion fluxes

As reviewed by Roux (1986) there are numerous papers documenting the effects of Pfr on membrane potential and on the transport of various ions across membranes, including H^+, K^+, and Ca^{2+}. At least some of these light-induced transport changes could be initiated by mechanisms involving phosphorylation, for protein kinases play a central role in controlling the activity of membrane pumps and channels.

R.R. Lew, B.S. Serlin and colleagues have published a series of studies documenting R-FR reversible changes in the activity of a K^+ uptake channel in *Mougeotia,* as assayed by patch-clamping techniques (Lew *et al.* 1992). Their evidence indicates that this channel can be induced to open by Ca^{2+}, and that phytochrome is likely to be the photoreceptor. The channel takes several minutes to open so they assume there are a number of transduction steps leading to this event from Pfr. They point out that their data do not clarify whether K^+-channel activation is part of the transduction chain leading to the classic chloroplast rotation response. Channel activation may simply serve to voltage clamp the membrane to the Nerst potential for K^+.

Because of an increased interest in calcium as an agent for stimulus-response coupling, studies on this ion dominate much of the recent literature on phytochrome-induced ion transport. While acknowledging the central role of Ca^{2+} in signal transduction, one senses that other light-induced ion fluxes are not being adequately studied. Hopefully, as more is learned about the metabolic significance of proton and K^+ fluxes, their role in phytochrome physiology will begin to receive more of the attention it deserves.

4.6.6.1 Calcium signalling

Beginning in the late 1970's the role of calcium in phytochrome signal transduction has been an increasingly studied topic. Appropriately, this trend has both influenced and paralleled an increased interest in the role of calcium in plant growth and development generally (Trewavas and Gilroy 1991). Two recent reviews have been written on the phytochrome-calcium connection, (Tretyn *et al.* 1991; Roux 1992), and the reader is referred to these reviews for a comprehensive survey of the voluminous literature on this subject. The effort here will be to summarize some of the main findings, identify some of the principal unanswered questions, and propose some avenues for progress.

As emphasized by virtually all the reviews written thus far, calcium-mediated transduction pathways represent just one class of mechanisms for amplifying the Pfr signal. To decide whether calcium is helping to mediate a specific phytochrome response, the criteria would include demonstrating that: (i) R induces Ca^{2+} fluxes in the responding cells or organelles; (ii) Ca^{2+} chelators or antagonists of Ca^{2+}-binding proteins can block R effects on that phytochrome response; (iii) chemical agents that can induce Ca^{2+} uptake into cells can substitute for R in stimulating that phytochrome response. Here we will briefly review the experiments that have tested whether these three minimal criteria can be met.

As reviewed by Tretyn *et al.* (1991), investigators have used a variety of methods to measure R-FR reversible Ca^{2+} fluxes in cells that are induced to undergo growth and development changes by the same light treatments. Briefly, these methods have included such procedures as autoradiography, colorimetric assays with murexide, radiometric measurements of $^{45}Ca^{2+}$ uptake and release, atomic absorption spectroscopy measurements of changes in total cellular Ca^{2+}, cytochemical assays of intracellular Ca^{2+} localization, and fluorimetric assays of Ca^{2+}-Quin-2 complexes, among others. For Ca^{2+} to be a physiological effector, it needs to bind to one or more Ca^{2+}-binding proteins in cells, such as calmodulin or a Ca^{2+}-dependent protein kinase. Typically the Ca^{2+}-binding affinities of these proteins is such that they show half-maximal activation at or below 1.0 μM Ca^{2+}. Thus, a usual requirement to implicate Ca^{2+} as a signal transducer for a given stimulus is to show that the stimulus increases the

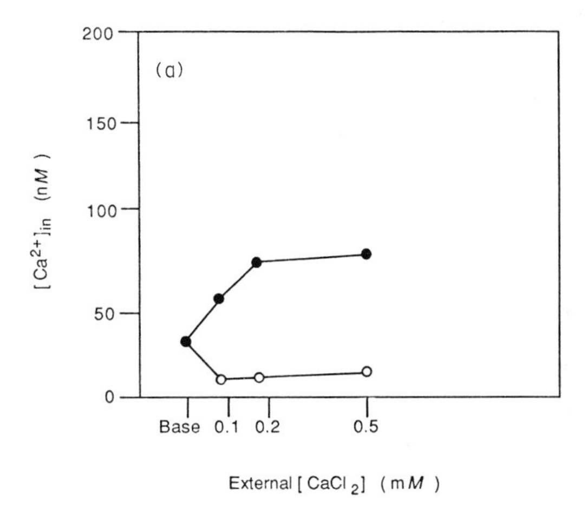

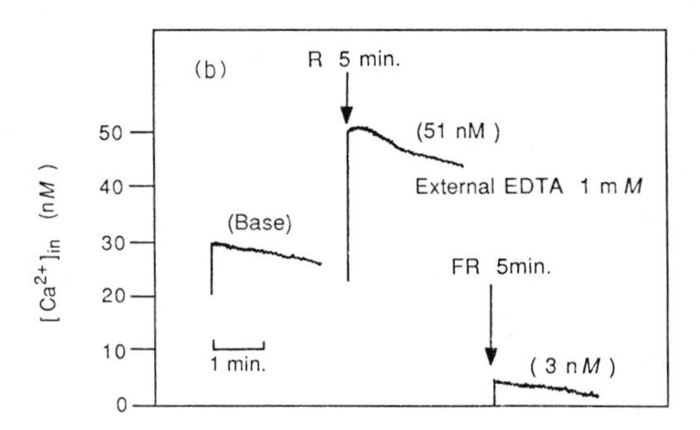

Figure 5. Red light (R), but not far-red light (FR), induces an increase in the $[Ca^{2+}]_{cyt}$ in oat protoplasts. (a) The increase is greater at greater $[Ca^{2+}]$ in the external medium. (b) R can induce an increase in $[Ca^{2+}]_{cyt}$ even in the absence of external Ca^{2+}. After Chae *et al.* (1990).

$[Ca^{2+}]_{cyt}$ of the responding cell to at least several hundred nanomolar. All of the initial reports of Pfr-induced changes in Ca^{2+} transport used methods that documented only a net change in Ca^{2+} transport without addressing the significant question of whether the $[Ca^{2+}]_{cyt}$ changed. The results of Chae *et al.* (1990) were the first to demonstrate that R can induce an increase in $[Ca^{2+}]_{cyt}$ under conditions in which FR has no effect (Fig. 5). Indicating the generality of such results, Shacklock *et al.* (1992) have shown that R stimulates a transient rise in $[Ca^{2+}]_{cyt}$ in wheat protoplasts in which R had induced swelling. The Pfr-induced $[Ca^{2+}]_{cyt}$ increase may prove to be a common signalling event.

The potential of calcium channel-blockers or chelators to block phytochrome responses, has also been repeatedly demonstrated (Tretyn *et al.* 1991). Additionally, antagonists of Ca^{2+}-binding proteins, such as phenothiazine drugs and the naphthalenesulphonamide derivatives, W-5 and W-7 also interfere with

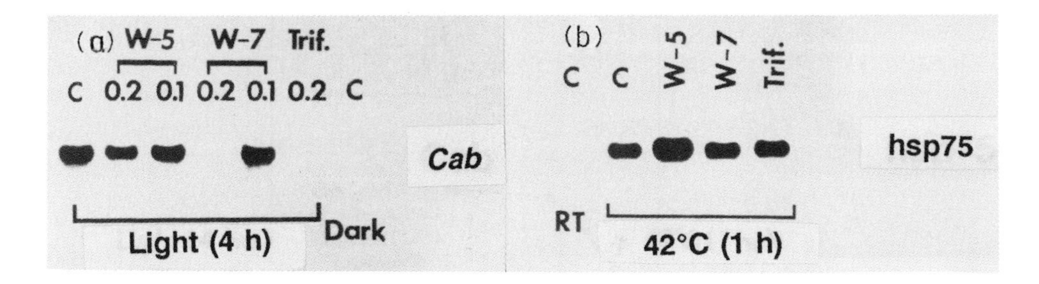

Figure 6. (a) White light-induced increase in *Cab* mRNA abundance, and the inhibition of this effect by antagonists of calcium-binding proteins. W-7 has a higher affinity for Ca^{2+}-binding proteins than its analog W-5, and is also a more potent antagonist of this response. Trif. = trifluoroperazine. (b) Control experiment showing that antagonists used in (a) have no effect on the expression of the gene for heat shock protein *hsp75*. RT = room temperature. A principal photoreceptor for white light promotion of *Cab* gene transcription is known to be phytochrome. After Lam E., Benedyk M. and Chua N.-H. (1989) *Mol. Cell. Biol.* 9: 4819-4823.

phytochrome transduction chains (Tretyn *et al.* 1991), including those leading to changes in gene expression (Fig. 6). The limited specificity of inhibitors is well renowned, and so studies with calcium and calmodulin antagonists have led to limited kinds of conclusions that are consistent with but do not prove a role for calcium in coupling Pfr to photomorphogenesis.

Inhibitor studies have also provided insights on the probable timing of calcium participation in phytochrome responses. Several authors have used Ca^{2+} chelators or inhibitors to show that Pfr is not directly coupled to Ca^{2+} uptake into cells (Tretyn *et al.* 1991). From this one would predict that there would be intervening steps between the production of Pfr in cells and the uptake of Ca^{2+} by these cells. Furthermore, although *Dryopteris* spores must take up Ca^{2+} from the medium in order to undergo R-induced germination, Pfr does not have to be present at the time Ca^{2+} is taken up for germination to proceed (Fig. 7). Interestingly, the period during which Ca^{2+} is required for R-induced germination of *Dryopteris* spores corresponds exactly with the period during which staurosporine, a potent inhibitor of Ca^{2+}-dependent protein kinases, most effectively blocks R-induced germination (Haas *et al.* 1991). This result suggests that one important target of Ca^{2+} action during the transduction chain leading to germination is a Ca^{2+}-dependent protein kinase.

As reviewed by Tretyn *et al.* (1991), calcium ionophores such as A23187 and ionomycin can substitute for Pfr in a number of well described phytochrome responses, including fern-spore germination, chloroplast rotation in *Mougeotia,* protoplast swelling, and transcription of the *Cab* message. Given that the ionophores induce Ca^{2+} uptake into the cytoplasm, these results indicate that a

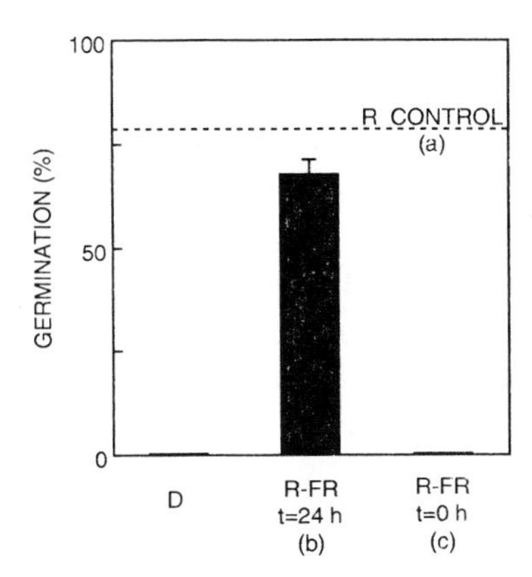

Figure 7. Medium Ca^{2+} can stimulate spore germination in the fern, *Dryopteris paleacea*, whether it is added to the medium immediately after red light (R) (a) or is added 24 h after R and immediately after far-red light (FR) (b). Note that in the situation of (b), the response has escaped from reversal control by FR, yet in the apparent absence of photoreversible Pfr at this time, exogenous Ca^{2+} can still stimulate germination. If the FR is given immediately after R, it reverses the promotive effects of R on germination (c). D = dark control. After Scheuerlein R., Wayne R. and Roux S.J. (1989) *Planta* 178: 25-30.

key step in the phytochrome transduction chain is an increase in $[Ca^{2+}]_{cyt}$, and that this step can be induced either chemically or by light with the same result.

One of the better studied model systems for testing Ca^{2+} involvement in a phytochrome response is the R-induced chloroplast rotation response in *Mougeotia.* As reviewed by Roux (1986), R promotes an uptake of $^{45}Ca^{2+}$ into this alga that is FR reversible, the polar application of A23187 can substitute for Pfr in inducing the rotation response, and calmodulin antagonists can block the induction of this response by R.

All of these results are consistent with the interpretation that a Ca^{2+}-dependent step is important for this phytochrome response. As reviewed by Tretyn *et al.* (1991), this interpretation is apparently countered by the observations that R-induced rotation can occur in the absence of Ca^{2+} in the external medium and in the presence of Ca^{2+}-channel blockers. The latter experiments do not address the possibility that Pfr could induce the release of Ca^{2+} from internal stores, an interpretation favoured by the results reviewed by Wagner and Klein (1981). Russ *et al.* (1991), using Quin-2 as a reporter dye, were unable to detect a significant change in $[Ca^{2+}]_{cyt}$ in *Mougeotia* after R, but they correlated the rate of chloroplast movement induced by blue light (B) with B-induced increases in $[Ca^{2+}]_{cyt}$.

Clearly, at the present time there are contradictory indications on whether there is a Ca^{2+}-dependent step in the transduction chain leading from Pfr to chloroplast rotation in *Mougeotia.* The interpretation favoured by Tretyn *et al.* (1991) is that Ca^{2+} serves to establish cytoskeletal competence for chloroplast movement rather than as a signal to induce the movement. If it can be established that R induces no change in $[Ca^{2+}]_{cyt}$ in *Mougeotia*, either by release of Ca^{2+} from internal stores or by opening plasma-membrane Ca^{2+} channels, then the interpretation that Ca^{2+} is a prerequisite for movement rather than an inducer of movement would be favoured. In the meantime, because R-induced changes in Ca^{2+} transport are known to regulate changes in cytoplasmic movements (Trewavas and Gilroy 1991), and B-induced increases in $[Ca^{2+}]_{cyt}$ correlate with increased rates of chloroplast rotation in *Mougeotia*, the hypothesis that R-induced chloroplast rotation also utilizes a Ca^{2+}-dependent regulatory mechanism should still be considered viable.

Given the diversity of phytochrome responses it would be surprising if all of them were mediated through calcium-dependent pathways. Evidence that at least one phytochrome response is achieved through a calcium-independent pathway is provided by the micro-injection work of Neuhaus *et al.* (1993) referred to above (Section 4.6.3). They showed that the micro-injection of neither calcium nor calcium-activated calmodulin could promote anthocyanin synthesis under conditions in which both of these agents could induce the expression of a *Cab*-reporter gene.

4.6.6.1.1 Specific targets of calcium action. Whether one is studying chloroplast rotation or any other R-induced response system, to have confidence that there is a Ca^{2+}-regulated step in the transduction chain will require identifying the specific target of Ca^{2+} action that has to be activated for the response to occur. As indicated above, because the resting level of $[Ca^{2+}]_{cyt}$ in cells is typically below 0.5 μM, most regulatory proteins that respond to Ca^{2+} show half maximal activation near or below 1.0 μM $[Ca^{2+}]$. At present the regulatory proteins in plants known to have a K_m for Ca^{2+} in this range include calmodulin, a variety of different Ca^{2+}-dependent protein kinases, and annexin-like proteins (Roux 1992).

All of these regulators modulate cell metabolism by stimulating enzymes or cellular processes. Some of the functions they stimulate are also controlled by phytochrome. Biochemical events that appear to be regulated both by phytochrome and by calcium in plant cells are summarized in Table 1. These events may serve as useful starting points for those interested in defining specific steps in a Pfr-initiated transduction chain. The cells in which these R/Ca^{2+}-regulated events occur should be tested to determine whether they also show a Pfr-induced increase in $[Ca^{2+}]_{cyt}$, and, if so, whether this increase precedes the Pfr-induced metabolic change. In this way one can begin to construct plausible transduction chains in which the coupling role of Ca^{2+} is defined.

One can reasonably speculate that the events listed in Table 1 are connected to other responses known to be induced by phytochrome. For example, NAD kinase activation could be the underlying basis for reports of R-induced

Table 1. Enzyme activities regulated by both Ca^{2+} and phytochrome.

NAD kinase (Roberts and Harmon 1992)*
Nuclear protein kinase (Roux 1992)
Nuclear NTPase (Roux 1992)
Ca^{2+}-ATPase (Roberts and Harmon 1992)

*References given are review chapters. Original documentation is cited in these reviews.

increases in NADP/NAD ratios. The Pfr-induced increase in the rate of Ca^{2+} export from cells into the cell wall could increase the $[Ca^{2+}]_{wall}$, which would be expected to decrease wall extensibility, and thus help account for R-induced decrease in mesocotyl growth (Roux 1986). The review by Roux (1992) discusses the possible connection between activation by Pfr of both Ca^{2+}-dependent protein kinases and Ca^{2+}-sensitive transcription events in the nuclei of legumes. It also notes the possible relationship between the ability of Pfr to stimulate a nucleoside triphosphatase (NTPase) localized on pea nuclear envelopes and its proposed ability to stimulate RNA export from pea nuclei.

In addition to the phenomena summarized in Table 1, it is of interest to note that both Pfr and Ca^{2+} affect xylem development in plants. One of the molecular targets of Ca^{2+} action in this process may be the newly discovered Ca^{2+}-binding protein annexin. A 35-kD annexin in peas is concentrated in cells undergoing xylogenesis, where it is thought to play a role in the secretion of wall materials (Clark *et al.* 1992). The expression of annexin in newly developing xylem cells is stimulated by hormonal signals that induce xylogenesis in cultured *Zinnia* cells (Fig. 8), a process known to require the uptake of extracellular Ca^{2+} (Clark *et al.* 1992). Phytochrome can stimulate the rate of xylem formation (Kleiber and Mohr 1967). Since the effects of light on xylogenesis, like hormone-induced xylogenesis, involve Ca^{2+} mediation, the proposal that annexin could participate in this process appears quite plausible. Although this proposed connection and all those in the preceding paragraph are speculative, they are also testable, and they illustrate how one can begin to evaluate the precise role of Ca^{2+} in transduction chains that extend from Pfr all the way to physiological changes.

4.6.7 Interactions with growth regulators

There is abundant literature on phytochrome interactions with growth regulators such as hormones (Roux and Serlin 1987), polyamines (Slocum *et al.* 1984) and other compounds (*e.g.* acetylcholine, Jaffe 1970). Clearly, to the extent that phytochrome can modulate the level of a growth regulator, this kind of induced chemical change can be a key intermediate step in coupling Pfr to altered growth and development. Increasing or decreasing the level of a growth

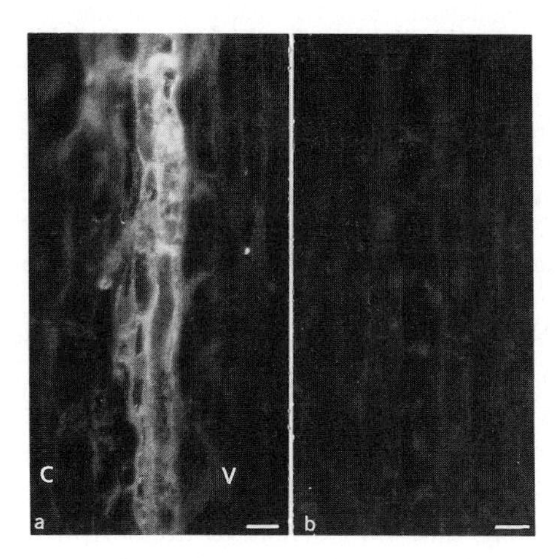

Figure 8. Immunocytochemical localization of annexin in the first millimetre of the pea root apex. This fluorescence micrograph of a longitudinal section shows preferential immunostaining of annexin in a strand of newly developing xylem cells that appear to be in the vascular cylinder (V). C = root cortex cells. Bar = 10 μm. After Clark *et al.* (1992).

regulator will itself initiate a distinct cascade of transduction events, one or more of which may be the same as or reinforce a coupling event in the phytochrome cascade. For example, both Pfr (see above) and auxin have been reported to stimulate an increased $[Ca^{2+}]_{cyt}$ and this could help explain how these two growth regulators are synergistic in some of their physiological effects, as illustrated in Fig. 9.

4.6.8 Genetic approaches

A most powerful tool for defining steps in the phytochrome transduction pathway is to obtain mutants defective in one or more crucial steps in that pathway. Progress in this area has been reviewed in detail recently by Quail (1991), so only a general overview will be given here.

An important breakthrough in the genetic approach to dissecting signal transduction was reported by Chory *et al.* (1989) who described a recessive de-etiolation mutant *(det)* in *Arabidopsis* that exhibits many of the phenotypic characters of a de-etiolated plant (short hypocotyls, expanded leaves, and accumulation of anthocyanin) even when grown in complete darkness (Fig. 10). It also resembles de-etiolated plants at the cellular, subcellular and gene expression levels (Chory *et al.* 1989). The simplest explanation of this mutation is that it is the result of a defect in a gene coding for a product (DET) that prevents de-etiolation in darkness, and that the level or function of DET can be suppressed by Pfr. A second promising mutant, phenotypically similar to *det,* is

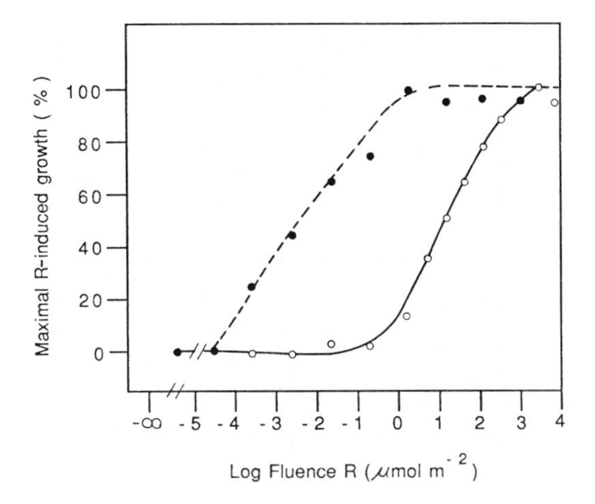

Figure 9. Fluence-response curve for red-light (R)-stimulated elongation of oat coleoptile sections. The two curves are with (dashed line) and without (solid line) addition of 6 *μM* indole-3-acetic acid (IAA) to incubation buffer. They indicate that IAA renders coleoptiles 10 000 times more sensitive to growth stimulation by R. After Shinkle J. and Briggs W.R. (1984) *Proc. Natl. Acad. Sci. USA* 81: 3742-3746.

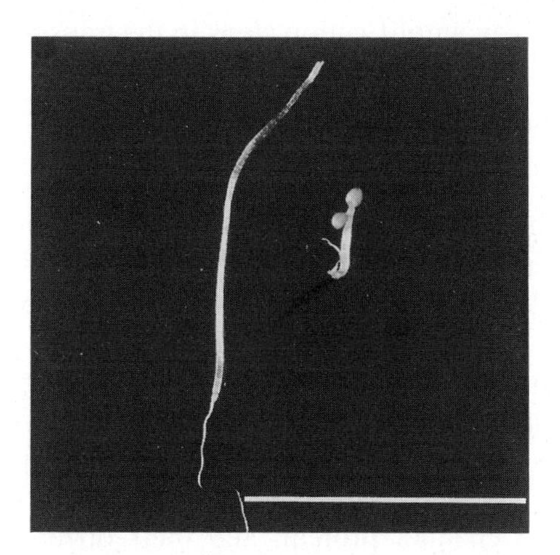

Figure 10. De-etiolated *(det)* mutant of *Arabidopsis thaliana,* that develops as a light-grown plant in the absence of light. Seven-day-old dark-grown (etiolated) wild type *(left)* and 7-day-old dark-grown *det2 (right).* Note that the mutant is short and has expanded leaves, as if it had been grown in the light. After Chory J., Nagpal P. and Peto C.A. (1990) *Plant Cell* 3: 445-459.

the constitutively photomorphogenic *(cop)* mutant (Quail 1991). It maps to a different locus from *det* but closely resembles it otherwise. Discovering the function of the proteins DET and COP will clarify key steps in the transduction chain leading from Pfr to de-etiolation. A major step toward this goal has been reported by Deng *et al.* (1992), who have cloned and sequenced the *cop1* gene. They found that it encodes a protein with both a zinc-finger motif, which suggests that it may bind to DNA, and a G-protein related domain. This raises the possibility that the COP1 protein can both function as a negative transcript ional regulator and can interact with proteins in a G-protein signalling pathway.

Mutants have also been found that partially resemble dark-grown plants even when they have been grown in the light. These include the long-hypocotyl mutant of *Arabidopsis (hy5)* and a pea mutant with long internodes *(lv)*. As reviewed by Quail (1991), these are not photoreceptor mutants. They are also not mutants with defects early in the transduction pathway common to all phytochromes, since they both show normal R-induced *Cab* and *ferredoxin (FedA)* gene expression. Nonetheless, they may be useful for revealing later steps in the transduction chain leading from Pfr to growth suppression.

Surely the mutants described above are only the first of many that will be isolated and characterized for analysis of transduction-chain components. Analyses of these many mutants will be tedious, and some of the results may end up revealing only some generic insight, such as that a kinase or a phosphatase is involved. Nonetheless, this genetic approach, in concert with the biochemical and physiological approaches described above, is certain to define what are some of the key amplification steps in the transduction of a R signal.

4.6.9 Conclusions

Because some photomorphogenetic events clearly require the presence of one kind of phytochrome, not all phytochromes are equivalent in what they regulate (Quail 1991). It is likely that there will be at least subtle differences in the transduction chains initiated by the different cellular phytochromes. Thus, insights gathered from experiments to date may not be equally applicable to all the phytochrome transduction pathways that can be expressed in a cell. However, it is also possible that all phytochrome transduction chains begin in a similar fashion, *e.g.* by activating a protein kinase or changing the transport state of a membrane channel protein, and their differences, due perhaps to differences in their subcellular locales, may only be expressed downstream from these initial events.

Given the above caveats, are there any generalizations that have emerged? At the least one can say that all attempts to disprove the hypothesis that phytochrome transduction chains contain components common to those described in many other plant and animal systems (activated G-proteins, inositol phospholipid turnover, enhanced calcium transport, and protein kinase activity) have failed. While this does not prove that these components are definitely operative, they provide a very strong rationale for testing the hypothesis more rigorously.

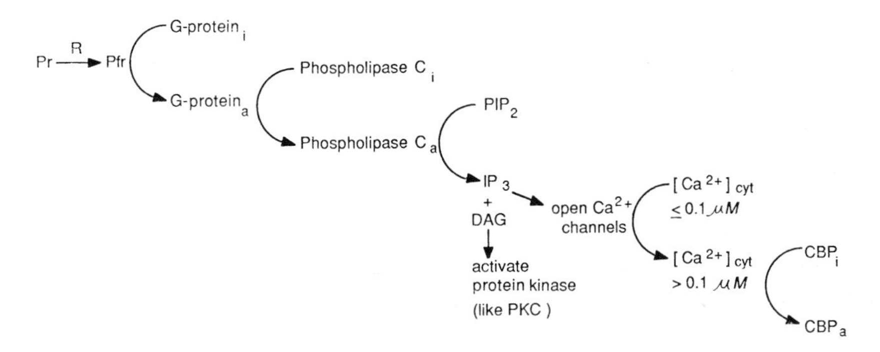

Figure 11. Speculative model of the early steps of a phytochrome transduction chain that accounts for and links together many of the rapid biochemical changes induced by Pfr in cells. Different phytochromes may induce different transduction chains, or several of them may induce steps such as those proposed here. The subscripts 'i' and 'a' denote the inactive and active forms, respectively, of a regulatory component. CBP = any one of a number of different Ca^{2+}-binding proteins, including Ca^{2+}-dependent protein kinases, as discussed in the text.

Some of these tests are now in progress. Based on abstracts and unpublished meeting reports, they include experiments that are addressing the following questions. Which kinase or phosphatase regulates the phosphorylation state of transacting factors that bind to promoters of light-regulated genes, and how is the activity of that kinase/phosphatase altered by R? Where are the G-proteins that are activated by Pfr and what are the enzymes these G-proteins activate? Can calcium-activated calmodulin rescue specific phytochrome transduction pathways (*e.g.* leading to gene expression changes) that are blocked genetically or physiologically? What are the specific proteins in nuclei or in the cytoplasm that show enhanced phosphorylation following R, and how does phosphorylation alter their function? In the process of answering these questions, the results obtained will also clarify (or disprove) the role of G-protein based pathways in phytochrome signal transduction.

Another valuable approach will be to discover the molecular basis of rhythmic changes in the sensitivity of a plant to Pfr action. Clearly some key component needed for phytochrome transduction chains becomes limiting once each day, and defining that component will represent a major advance. The elegant studies in the laboratories of S.A. Kay and of F. Nagy (reviewed in Chapter 8.1) are already producing valuable new insights in this area.

If experiments such as those above end up supporting the hypothesis that one or more of the phytochromes initiates a G-protein based transduction pathway, the kind of steps likely to be part of that pathway are illustrated in Fig. 11. This speculative model accounts for many of the results reviewed in this chapter. Tools for testing this model have already been developed to investigate stimulus-response coupling mechanisms in other systems, so the model is readily subject to disproof. Some alternative models that do not assume G-protein participation (phytochrome itself is a kinase; phytochrome alters proton pump activity; phytochrome alters the level of growth regulators) have been proposed here. Surely other plausible alternatives can be proposed. By systematically testing alternatives to those proposed in Fig. 11, we can expect new and more detailed models to be developed in the years ahead.

4.6.10 Further reading

Mohr H. (1983) Pattern specification and realization in photomorphogenesis. In: *Encyclopedia of Plant Physiology,* New Series, Vol. 16A, *Photomorphogenesis,* pp. 336-357, Shropshire Jr. W. and Mohr H. (eds.) Springer-Verlag, Berlin.

Quail P.H. (1991) Phytochrome: a light-activated molecular switch that regulates plant gene expression. *Annu. Rev. Genet.* 25: 389-409.

Roux S.J. (1992) Calcium-regulated nuclear enzymes: potential mediators of phytochrome-induced changes in nuclear metabolism? *Photochem. Photobiol.* 56: 811-814.

Tretyn A., Kendrick R.E. and Wagner G. (1991) The role(s) of calcium ions in phytochrome action. *Photochem. Photobiol.* 54: 1135-1155.

Trewavas A. and Gilroy S. (1991) Signal transduction in plant cells. *Trends in Genetics* 7: 356-361.

4.6.11 References

Basu A., Sethi U. and Guha-Mukherjee S. (1990) Phytochrome in control of differentiation and phosphatidylinositol turnover in *Brassica oleracea* cultures. *Phytochemistry* 29: 1539-1541.

Bossen M.E., Kendrick R.E. and Vredenberg W.J. (1990) The involvement of a G-protein in phytochrome-regulated, Ca^{2+}-dependent swelling of etiolated wheat protoplasts. *Physiol. Plant.* 80: 55-62.

Chae Q., Park H.J. and Hong S.D. (1990) Loading of quin2 into the oat protoplast and measurement of cytosolic calcium ion concentration changes by phytochrome action. *Biochim. Biophys. Acta* 1051: 115-122.

Chory J., Peto C., Feinbaum R., Pratt L. and Ausubel F. (1989) *Arabidopsis thaliana* mutant that develops as a light-grown plant in the absence of light. *Cell* 58: 991-999.

Clark G., Dauwalder M. and Roux S.J. (1992) Purification and immunolocalization of an annexin-like protein in pea seedlings. *Planta* 187: 1-9.

Clark G., Memon A.R., Tong C.-G., Thompson Jr. G.A. and Roux S.J. (1993) Phytochrome regulates GTP-binding protein activity in the envelope of pea nuclei. *Plant J.* in press.

Datta N. and Cashmore A. (1989) Binding of a pea nuclear protein to promoters of certain photoregulated genes is modulated by phosphorylation. *Plant Cell* 1: 1069-1077.

Datta N., Chen Y.-R. and Roux S.J. (1985) Phytochrome and calcium stimulation of protein phosphorylation in isolated pea nuclei. *Biochem. Biophys. Res. Commun.* 128: 1403-1408.

Deng X.W., Matsui M., Wei N., Wagner D., Chu A.M., Feldmann K.A. and Quail P.H. (1992) *COP1*, an *Arabidopsis* regulatory gene, encodes a protein with both a zinc-binding motif and a G_β homologous domain. *Cell* 71: 791-802.

Guron K., Chandok M.R. and Sopory S.K. (1992) Phytochrome-mediated rapid changes in the level of phosphoinositides in etiolated leaves of *Zea mays. Photochem. Photobiol.* 56: 691-696.

Haas C.J., Scheuerlein R. and Roux S.J. (1991) Phytochrome-mediated germination and early development in spores of *Dryopteris filix-mas* L: Phase-specific and non-phase specific inhibition by staurosporine. *J. Plant Physiol.* 138: 747-751.

Hasunuma H., Furukawa K., Funadera K., Kubota M. and Watanabe M. (1987) Partial characterization and light-induced regulation of GTP-binding proteins in *Leman paucicostata. Photochem. Photobiol.* 46: 531-535.

Jaffe M. (1970) Evidence for the regulation of phytochrome-mediated processes in bean roots by the neurohumor, acetylcholine. *Plant Physiol.* 46: 768-777.

Kleiber H. and Mohr H. (1967) Vom Einfluss des Phytochroms auf die Xylemdiferenzierung im Hypokotyl des Senfkeimlings (*Sinapis alba* L.). *Planta* 76: 85-92.

Klimczak L.J., Schindler U. and Cashmore A.J. (1992) DNA binding activity of the *Arabidopsis* G-box binding factor GBF1 is stimulated by phosphorylation by casein kinase II. *Plant Cell* 4: 87-98.

Lew R.R., Krasnoshtein F., Serlin B.S. and Schauf C.L. (1992) Phytochrome activation of K^+ channels and chloroplast rotation in *Mougeotia. Plant Physiol.* 98: 1511-1514.

Li H., Dauwalder M. and Roux S.J. (1991) Partial purification and characterization of a Ca^{2+}-dependent protein kinase from pea nuclei. *Plant Physiol.* 96: 720-727.

Morse M.J., Crain R.C. and Satter R.L. (1987) Light-stimulated inositol phospholipid turnover in *Samanea saman. Proc. Natl. Acad. Sci. USA* 84: 7075-7078.

Neuhaus G., Bowler C., Kern R. and Chua N.-H. (1993) Calcium/calmodulin-dependent and independent phytochrome signal transduction pathways. *Cell* 73: in press.

Otto V. and Schäfer E. (1988) Rapid phytochrome-controlled protein phosphorylation and dephosphorylation in *Avena sativa. Plant Cell Physiol.* 29: 1115-1121.

Roberts D.M. and Harmon A.C. (1992) Calcium-modulated proteins: targets of intracellular calcium signals in higher plants. *Annu. Rev. Plant Physiol. Plant Mol. Biol.* 43: 375-414.

Romero L.C., Biswal B. and Song P.-S. (1991a) Protein phosphorylation in isolated nuclei from etiolated *Avena* seedlings. Effects of red/far-red light and cholera toxin. *FEBS Lett.* 282: 347-350.

Romero L.C., Sommer D., Gotor C. and Song P.-S. (1991b) G-proteins in etiolated *Avena* seedlings. Possible phytochrome regulation. *FEBS Lett.* 282: 341-346.

Roux S.J. (1986) Phytochrome and membranes. In: *Photomorphogenesis in Plants,* pp. 115-134, Kendrick R.E. and Kronenberg G.H.M. (eds.) Martinus Nijhoff Publishers, Dordrecht.

Roux S.J. and Serlin B.S. (1987) Cellular mechanisms controlling light-stimulated gravitropism: role of calcium. *CRC Crit. Rev. Plant Sci.* 5: 205-236.

Russ U., Grolig F. and Wagner G. (1991) Changes of cytoplasmic free calcium in the green alga *Mougeotia scalaris*: monitored with indo-1 and effect on the velocity of the chloroplast movements. *Planta* 184: 105-112.

Shacklock P.S., Reed N.D. and Trewavas A.J. (1992) Cytosolic free calcium mediated red light-induced photomorphogenesis. *Nature* 358: 753-755.

Simon M.I., Strathmann M.P. and Gautam N. (1991) Diversity of G proteins in signal transduction. *Science* 252: 802-808.

Singh B.R. and Song P.-S. (1990) Phytochrome and protein phosphorylation. *Photochem. Photobiol.* 52: 249-254.

Slocum R.D., Kaur-Sawhney R. and Galston A.W. (1984) The physiology and biochemistry of polyamines in plants. *Arch. Biochem. Biophys.* 235: 283-303.

Wagner G. and Klein K. (1981) Mechanism of chloroplast movement in *Mougeotia. Protoplasma* 109: 169-185.

Wong Y.S., McMichael Jr. R.W. and Lagarias J.C. (1989) Properties of a polycation-stimulated protein kinase associated with purified *Avena* phytochrome. *Plant Physiol.* 91: 709-718.

4.7 The physiology of phytochrome action

Alberto L. Mancinelli

Department of Biological Sciences, Columbia University, New York, NY 10027, USA

4.7.1 Introduction

Plants possess the capability of adapting their patterns of growth and development to changes in the light conditions of the environment. The action of light on plant growth and development *(photomorphogenesis)* is mediated by specific photoreceptors. Light-induced changes in the state of the photoreceptors provide the signal for the induction/modulation of plant responses to light. A major plant photomorphogenic photoreceptor is phytochrome, a biliprotein pigment (Chapter 4.1) which is reversibly interconverted by light between an inactive form, Pr, and a physiologically active one, Pfr, with peaks of absorption in the red (R) and far-red (FR) regions of the spectrum, respectively. The reversible photoconversion can be schematized as follows:

$$\text{Pr} \underset{k_2}{\overset{\text{Light} \downarrow\downarrow\downarrow \; k_1}{\rightleftarrows\rightleftarrows\rightleftarrows\rightleftarrows}} \text{Pfr}$$

where k_1 and k_2 are the rate constants for the Pr → Pfr and Pfr → Pr photoconversions, respectively *(Note: see Appendix Table 1 for definition and symbols of phytochrome parameters)*. Red and FR light are the most effective for the Pr → Pfr and Pfr → Pr photoconversion, respectively. The reversible Pr ⇄ Pfr photoconversion is the basic process underlying phytochrome action in photomorphogenesis.

The existence of a photoreceptor capable of reversible photoconversion between two states was first postulated by Borthwick *et al.* (1952) after their discovery of the R-FR reversibility of plant responses to light. Under appropriate conditions, many plant responses can be effectively induced by a short light exposure, and R is the most effective spectral region. The inductive effect of R

R.E. Kendrick & G.H.M. Kronenberg (eds.), Photomorphogenesis in Plants - 2nd Edition

was lost or significantly reduced if a short exposure to FR was applied immediately after R, and this R-FR reversibility persisted over several cycles of alternating R-FR (Fig. 1). The R-FR reversibility is the consequence of photoconverting Pfr back to Pr before Pfr has had the time to complete its action on the responding system. The discovery of R-FR reversibility and the postulate derived from it were seminal events in plant photomorphogenesis research.

The existence and spectral properties of the postulated photoreceptor (*a pigment of the imagination*) were confirmed a few years later by the demonstration of R-FR reversible absorbance changes in etiolated seedlings (Butler *et al.* 1959), the isolation of phytochrome from etiolated oat seedlings (Siegelman and Firer 1964) and the determination of its photochemical properties *in vitro* (Butler *et al.* 1964). The alternation of short R and FR treatments, as used in the early experiments (Fig. 1), even though not comparable with the light conditions of the natural environment, is still a useful tool in studies of phytochrome action. The biological significance of phytochrome has been established beyond reasonable doubt. In higher plants, phytochrome-mediated effects of light have been reported for all phases of the life cycle, from seed germination to seed maturation, at all levels of organization, from the molecular to whole organisms, and for responses ranging from fully R-FR reversible ones to those in which R-FR reversibility is poor or nil. Recent research has shown that phytochrome consists of a family of photoreceptors, distinguishable in at least two groups, phytochrome type I and II (PI and PII); they share the property of

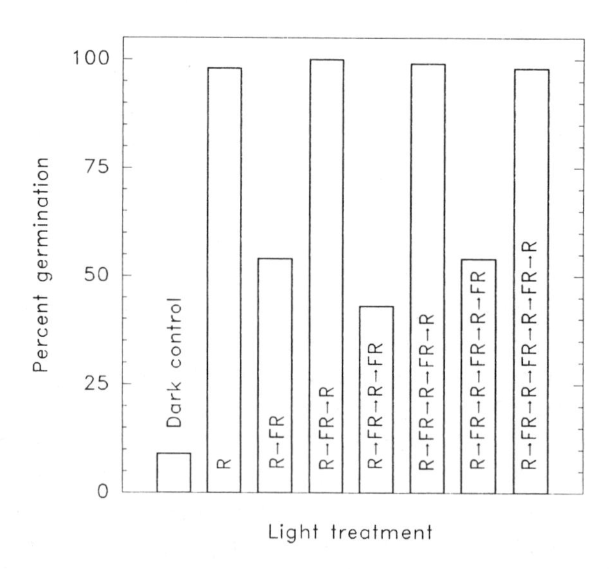

Figure 1. Photoreversibility of seed germination. Experimental protocol: sowing → 16 h dark → light treatment → 48 h dark; light treatments: R, 1 min red; FR, 4 min far-red. Redrawn from data of Borthwick *et al.* (1952). This particular experiment was the first to show the multiple R-FR reversibility of plant photomorphogenic responses.

reversible Pr ⇄ Pfr photoconversion, but differ in the stability of Pfr, immunochemical and spectral properties, proportions present in the plant at different stages (*e.g.* etiolated and de-etiolated seedlings), and contribution to the regulation of different responses. The mechanisms by which phytochrome interact with other molecules to regulate plant photoresponses are still unknown. The purpose of this chapter is to provide a general overview of what has been learned about the physiology of action of phytochrome since 1952.

4.7.2 Light and phytochrome

One of the basic steps in studies of the physiology of phytochrome action is the analysis of the quantitative relationships between the effects of light on the state of the photoreceptor and the expression of phytochrome-mediated responses. The analysis of these relationships cannot be carried out without a knowledge of the photochemical properties of phytochrome. Most of what is known about these properties has been derived from studies of phytochrome extracted and purified from dark-grown seedlings in which PI phytochrome is the predominant type. The photochemical properties of PII phytochrome, predominant in light-grown plants, are not known in as many details as those for PI.

4.7.2.1 Photochemical properties of purified phytochrome

4.7.2.1.1 Absorption and action spectra. In the spectral region between 300 and 800 nm (Fig. 2), Pr has an absorbance maximum (λ_{max}) at about 666 nm (PI; 655-657 nm in PII) and a secondary maximum at 380 nm; Pfr has a λ_{max} at about 730 nm and a secondary maximum at 400-405 nm. Both Pr and Pfr have a strong peak of absorption at 280 nm, consistent with the protein nature of phytochrome.

The rate of phytochrome photoconversion is a function of the extinction coefficient (ε) of the pigment and the quantum yield (Φ) of the process; the product, $2.3\varepsilon\Phi$ is called the photoconversion cross-section (σ) of the pigment *(see Appendix Table 1 for definitions and symbols of phytochrome parameters)*. Plots of values of the photoconversion cross-sections of Pr (σ_R) and Pfr (σ_{FR}) as a function of wavelength (Fig. 3) represent the action spectra for the Pr ⇄ Pfr photoconversions. Red (peak at about 666 nm) and FR (peak at about 730 nm) are the most effective for the Pr → Pfr and Pfr → Pr photoconversions, respectively (Fig. 3), but radiation in other spectral regions is not without effect, including 280-290 nm radiation (not shown in Fig. 3), as determined with direct measurements by Pratt and Butler (1970). The absorption (Fig. 2) and action (Fig. 3) spectra are similar, but, at any given wavelength, $\sigma_{FR\lambda}$, as % of σ_{R666}, is lower than $\varepsilon_{FR\lambda}$ as % of ε_{R666}: this is a reflection of the fact that Φ_{FR} is lower than Φ_R (Table 1).

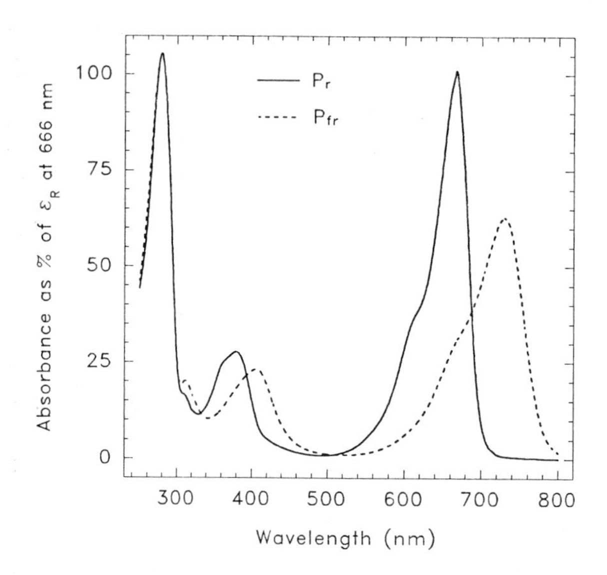

Figure 2. Relative absorption spectra of Pr and Pfr, based on average values of molar extinction coefficients of phytochrome purified from dark-grown oat and rye seedlings (Kelly and Lagarias 1985; Lagarias *et al.* 1987).

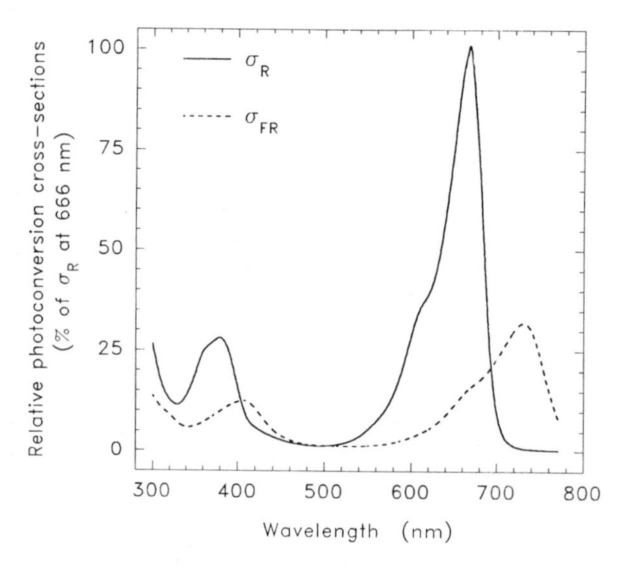

Figure 3. Wavelength dependence of the photoconversion cross-sections of Pr (σ_R) and Pfr (σ_{FR}). Relative values calculated from absolute average values of σ_R and σ_{FR} (Appendix Table 2) for phytochrome purified from dark-grown oat and rye seedlings (Kelly and Lagarias 1985; Lagarias *et al.* 1987).

Table 1. Photochemical parameters of phytochrome *in vitro* (extracted and purified from dark-grown oat and rye seedlings) and *in vivo* (cotyledons of dark-grown *Cucurbita* seedlings).

	Oat-1	Oat-2	Oat-3	Oat-4	Oat-5	Rye	*Cucurbita*
Photoconversion cross-sections (σ, $m^2\ mol^{-1}$)							
σ_{R450}	153	91	159	86	99	183	85
σ_{FR450}	264	110	196	127	187	225	115
σ_{R660}	6650	3979	4209	2677	3270	5014	5123
σ_{FR660}	1258	526	558	431	806	638	1531
σ_{R730}	46	27	36	14	16	33	7
σ_{FR730}	2890	1209	1219	1128	1589	1496	2192
$\sigma_{R660}/\sigma_{FR730}$	*2.30*	*3.29*	*3.45*	*2.37*	*2.06*	*3.35*	*2.34*
Quantum yields for photoconversion (Φ)							
Φ_R	0.254*	0.152†	0.154†	0.17*	-	0.173†	-
Φ_{FR}	0.165*	0.069†	0.063†	0.10*	-	0.076†	-
Φ_R/Φ_{FR}	1.54	2.20	2.43	1.70	1.5*	2.28	1.38‡
Pfr/P ratios at photoequilibrium (φ)							
φ_{450}	0.37	0.45	0.45	0.40	0.35	0.45	0.43
φ_{660}	0.84	0.88	0.88	0.86	0.80	0.89	0.77
φ_{730}	0.016	0.022	0.029	0.012	0.010	0.022	0.003

*Initial rate analysis (*IRA*); †approach-to-equilibrium analysis (*ATEA*); ‡based on φ value at the isosbestic point (686 nm; $\varphi_{iso} = 1/(1 + \Phi_{FR}/\Phi_R)$). Note that *IRA* tends to give $\Phi_R/\Phi_{FR} < 2.0$ and *ATEA* tends to give $\Phi_R/\Phi_{FR} > 2.0$. Summary of original data: oat-1 and -2, Kelly and Lagarias (1985); oat-3 and rye, Lagarias *et al.* (1987); oat-4, Vierstra and Quail (1983); oat-5, Butler *et al.* (1964); *Cucurbita*, Seyfried and Schäfer (1985b).
Note: some of the values of σ may be different than those reported in the listed papers. The symbol σ is used not only for the photoconversion cross-section ($\sigma = 2.3\varepsilon\Phi$), but also for the photoconversion coefficient ($\sigma = \varepsilon\Phi$). All values of σ reported in this table are photoconversion cross-sections.

4.7.2.1.2 Kinetics of phytochrome photoconversion. In a phytochrome solution exposed to monochromatic radiation of wavelength λ and photon flux N_λ, the rate of change of [Pfr] is described by the following equation:

$$d[\mathrm{Pfr}]/dt = N_\lambda \sigma_{R\lambda}[\mathrm{Pr}] - N_\lambda \sigma_{FR\lambda}[\mathrm{Pfr}] = k_{1\lambda}[\mathrm{Pr}] - k_{2\lambda}[\mathrm{Pfr}] \qquad (t = \mathrm{time}).$$

At any given time, the rate of the Pr → Pfr (Pfr → Pr) photoconversion is proportional to the amount of Pr (Pfr) remaining: both photoconversions are first order reactions. The reaction proceeds to a photoequilibrium state ($t = \infty$) at which

$$d[\mathrm{Pfr}]/dt = 0;\ N_\lambda \sigma_{R\lambda}[\mathrm{Pr}] = N_\lambda \sigma_{FR\lambda}[\mathrm{Pfr}]\ ;\ \text{and } [\mathrm{Pfr}]/[\mathrm{Pr}] = \sigma_{R\lambda}/\sigma_{FR\lambda}.$$

The fraction of total phytochrome (P = Pr + Pfr) in the Pfr form, Pfr/P, is a function of wavelength and photon flux before reaching photoequilibrium and a function of wavelength only at photoequilibrium: Pfr/P at photoequilibrium (φ) $= k_{1\lambda}/(k_{1\lambda}+k_{2\lambda}) = \sigma_{R\lambda}/(\sigma_{R\lambda} + \sigma_{FR\lambda})$. The rate of cycling ($H = (1-\varphi)k_1 = \varphi k_2 = (\varphi-\varphi^2)k$; $k = k_1 + k_2$) between Pr and Pfr at photoequilibrium is a function of wavelength and photon flux.

Under polychromatic irradiation, the rate constants for the Pr → Pfr and Pfr → Pr photoconversions are the integral across the spectral region of interest, thus:

$$k_1 = \sum_{i=\lambda_1}^{\lambda_2} N_i \sigma_{Ri} \qquad k_2 = \sum_{i=\lambda_1}^{\lambda_2} N_i \sigma_{FRi}$$

The kinetics of phytochrome photoconversion can be presented graphically in different ways; two examples are shown. The first (Fig. 4) is perhaps of more immediate impact: it shows the changes of Pfr/P as a function of exposure duration until photoequilibrium is established. Kinetics of photoconversion starting from low (0) and high (0.87) initial values of Pfr/P under light producing intermediate values of φ (410 and 690 nm) indicate that the rate with which photoequilibrium is approached is the same for both initial conditions. The second form of graphic presentation (Fig. 5) is generally preferred for the analysis of photoconversion kinetics: in this semilogarithmic plot the photoconversion curves are straight lines, as expected for first order reactions; the data for 410 and 690 nm show more immediately and clearly than in Fig. 4 that k is the same for photoconversion starting from low (0) or high (0.87) initial values of Pfr/P. The effectiveness of light (slope of the line) and the time required for 50% photoconversion (t½, time corresponding to a value of 0.5 on the vertical axis; $k = (\ln 2)/t_{½}$) are easier to determine than from the photoconversion curves of Fig. 4.

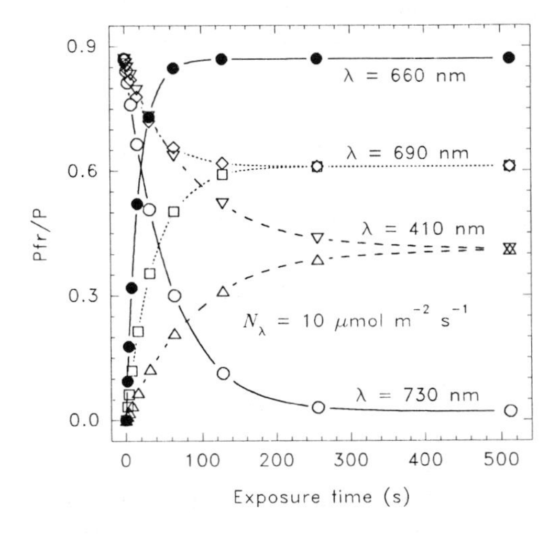

Figure 4. Phytochrome photoconversion kinetics at four wavelengths. Based on the equation for phytochrome photoconversion (Appendix Table 1) and values of σ in Appendix Table 2.

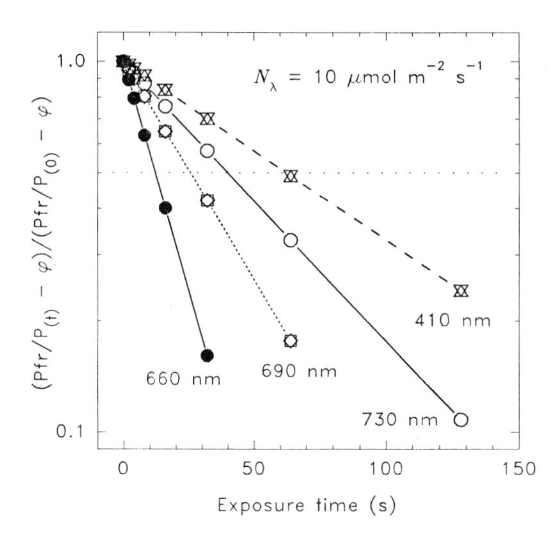

Figure 5. Phytochrome photoconversion kinetics. Same data as in Fig. 4, but presented on a semilogarithmic plot.

4.7.2.1.3 Pfr/P, φ, k, *and* H. These photochemical parameters are used in the analysis of the relationships between phytochrome state and expression of phytochrome-mediated responses (*e.g.*, percent germination *versus* Pfr/P, stem elongation *versus* φ and *H*). As shown above, Pfr/P is a function of light quality and photon flux before photoequilibrium is reached, and a function of light quality only at photoequilibrium. The maximum (≈ 0.85) and minimum (≈ 0.02) values of φ are in the R (600-670 nm) and FR (730-750 nm) regions, respectively (Fig. 6).

Both *k* and *H* are functions of wavelength and photon flux, but should not be confused with one another. The different effects of light on *k* and *H* illustrate the differences between these two parameters: for example, at a constant photon flux (Fig. 7), *H* is lower than *k*; the extent of the difference varies with wavelength and is a function of the wavelength dependence of φ and *k*. The relationships between *H* and φ (at a constant *k*) are shown in Fig. 8: *H* is maximum at φ = 0.5 ($H = 0.25k$) and decreases symmetrically at φ higher and lower than 0.5. Photon fluxes required to produce a constant *k* and a constant *H* at different wavelengths are shown in Fig. 9.

Some phytochrome-mediated responses (VLFR, Section 4.7.4) can be effectively induced at very low Pfr/P values (10^{-6} to 10^{-3}). The photon fluences required at four different wavelengths to produce Pfr/P values from 10^{-6} to 10^{-2} are shown in Fig. 10. Data for green (520 nm) light were included in the figure because green light is often used as a *safelight* in phytochrome research: *note that green light is not without effect on phytochrome photoconversion!*

4.7.2.1.4 Phytochrome photoconversion under dichromatic irradiation. The effects of dichromatic irradiations (simultaneous exposure to two wavelengths) on the state of phytochrome are considered in detail for several reasons. First, dichromatic R + FR treatments combining R and FR in appropriate proportions allow a high degree of flexibility in manipulating the state of phytochrome.

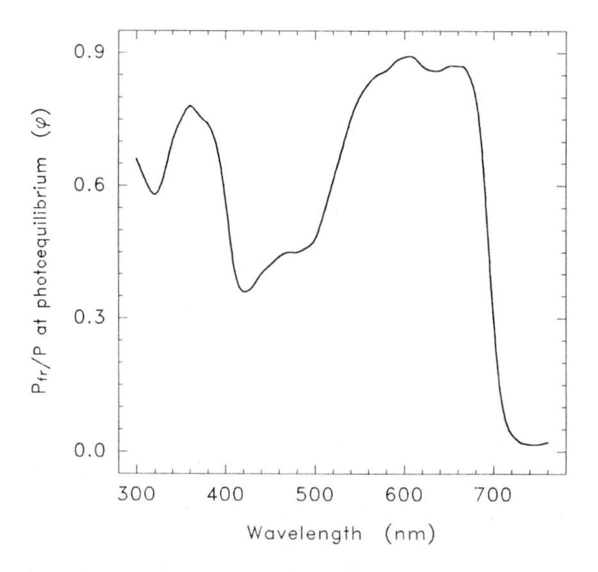

Figure 6. Wavelength dependence of Pfr/P ratio at photoequilibrium. Average values from data for purified oat and rye phytochrome (Appendix Table 2).

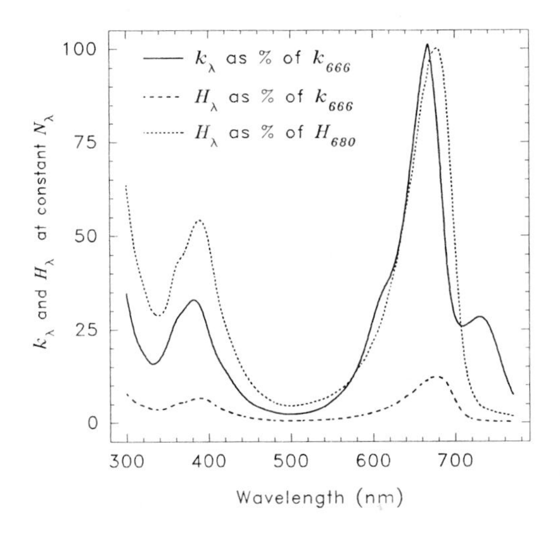

Figure 7. Wavelength dependence of k and H at a constant photon flux (N_λ). Values of k and H calculated from the data of Appendix Table 2. For N_λ = 10 μmol m^{-2} s^{-1}: k_{666} (= k_{max}) = 0.0613 s^{-1}, and H_{680} (= H_{max}) = 0.0075 s^{-1}.

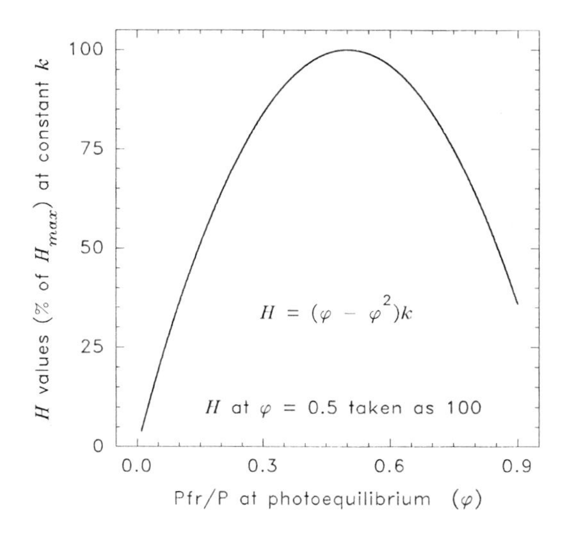

Figure 8. Relationships between H and φ at a constant value of k.

Second, dichromatic irradiations have been very useful in the study of several responses. For example, the effects of dichromatic R + FR treatments on hypocotyl elongation in dark-grown lettuce seedlings (Hartmann 1966) provided the first convincing evidence for the involvement of phytochrome in the photoregulation of non R-FR reversible responses to prolonged irradiations (Section 4.7.4.3.1). The use of dichromatic B + R and B + FR treatments have

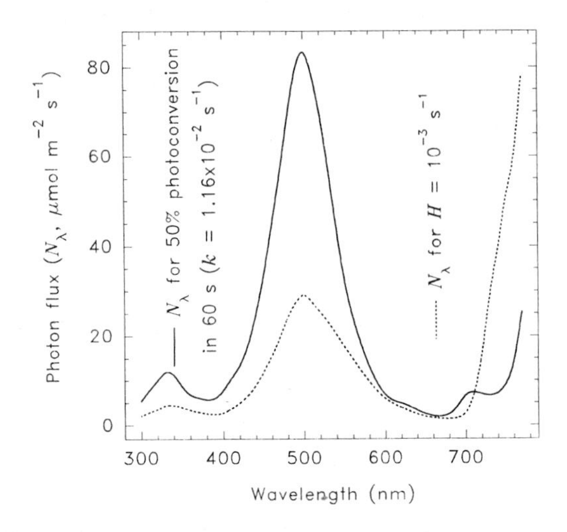

Figure 9. Wavelength dependence of photon fluxes required to obtain given values of k and H. Calculated values based on the parameters of Appendix Table 2.

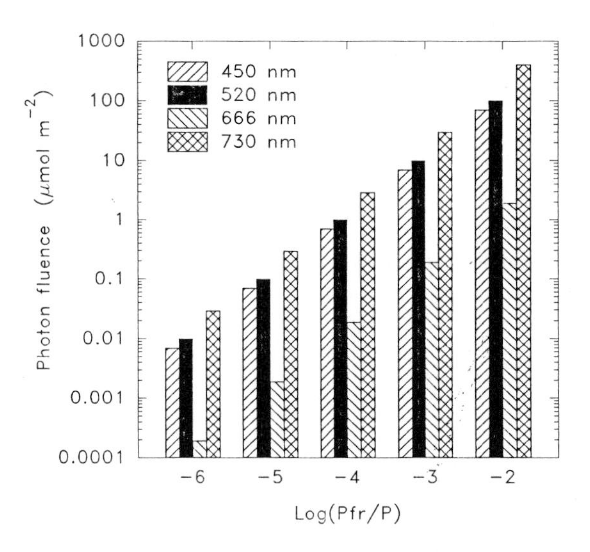

Figure 10. Photon fluences required at 450, 520, 666, and 730 nm to produce Pfr/P ratios from 10^{-6} to 10^{-2}. Calculated values based on the parameters of Appendix Table 2.

helped to obtain some evidence for the action and interaction between the blue (B)/UV-A photoreceptor and phytochrome (Section 4.7.5). Third, the effects of sunlight on the state of phytochrome are due mainly to the R and FR regions, and, in studies of phytochrome-mediated responses to light in the natural environment (Chapter 7.1), the R to FR photon flux ratio (R:FR or ζ) is often used as an index of the quality of natural light. On land, the R:FR of natural light varies from about 1.2 in full sunlight to about 0.1 in the shade of green vegetation canopies.

Under dichromatic irradiation (*e.g.* a mixture of 660 and 730 nm light), the values of k_1, k_2, and k are:

$$k_1 = N_{660} \times \sigma_{R660} + N_{730} \times \sigma_{R730}; \; k_2 = N_{660} \times \sigma_{FR660} + N_{730} \times \sigma_{FR730}$$

$$k = k_1 + k_2 = N_{660} \times \sigma_{660} + N_{730} \times \sigma_{730}$$

If one only wants to know the value of φ, it can be quickly calculated according to:

$$\varphi = \frac{0.87}{1 + \frac{0.295}{\zeta}}$$

where $\zeta = N_{660}{:}N_{730}$; this equation has been derived using the phytochrome photoconversion cross-sections given in *Appendix Table 2.*

Under dichromatic R + FR irradiation, φ is a function of R:FR only and decreases with decreasing R:FR (Fig. 11). In laboratory experiments, the method used to change R:FR has a significant effect on the patterns of change of k and H. The values of k (Fig. 12) and H (Fig. 13) may either decrease or

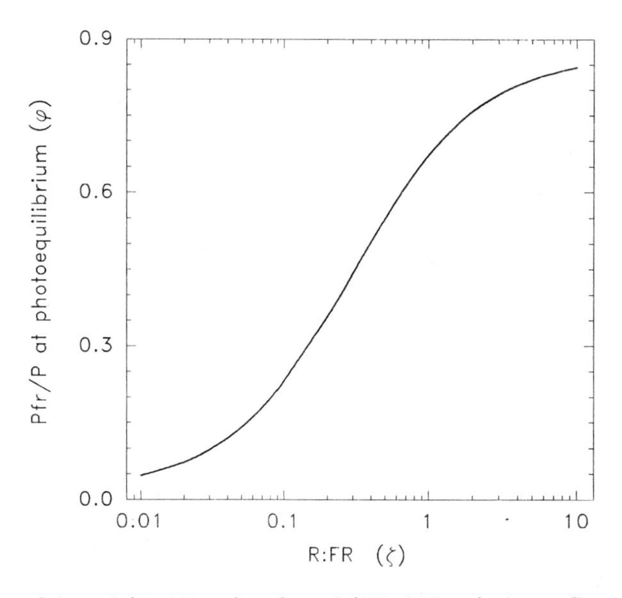

Figure 11. Effects of the red (R, 660 nm) to far-red (FR, 730 nm) photon flux ratio (R:FR or ζ) on Pfr/P at photoequilibrium (φ). Calculated values of φ based on σ_R and σ_{FR} values in Appendix Table 2. The decrease in φ is less pronounced for a decrease in R:FR from 10 to 1 (φ, 0.85 to 0.65; $\Delta\% \approx -23$) than for decreases in R:FR from 1 to 0.1 (φ, 0.65 to 0.20; $\Delta\% \approx -70$) and 0.1 to 0.01 (φ, 0.20 to 0.05; $\Delta\% \approx -75$).

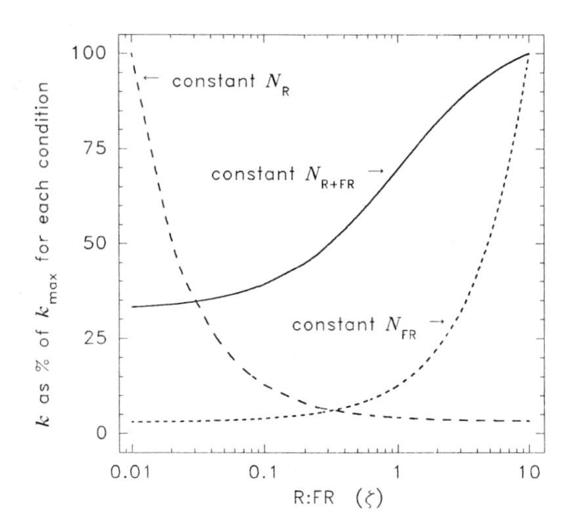

Figure 12. Pattern of change of k as a function of the red (R, 660 nm) to far-red (FR, 730 nm) photon flux ratio (R:FR or ζ) and the mode in which R and FR are mixed to obtain the required R:FR. For N_{R+FR} = c (c, constant): $N_R = c\zeta/(1+\zeta)$ and $N_{FR} = c/(1+\zeta)$; for N_R = c: $N_{FR} = c/\zeta$ and $N_{R+FR} = c + c/\zeta$; for N_{FR} = c: $N_R = c\zeta$ and $N_{R+FR} = c(1+\zeta)$.

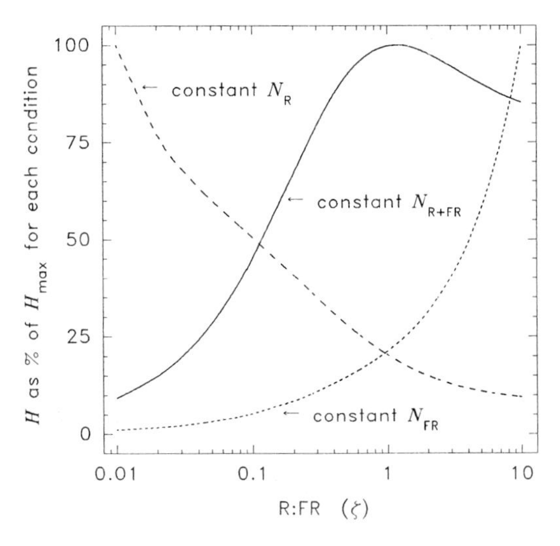

Figure 13. Pattern of change of H as a function of the red (R, 660 nm) to far-red (FR, 730 nm) photon flux ratio (R:FR or ζ) and the mode in which R and FR are mixed to obtain the required R:FR.

increase with decreasing R:FR, depending on whether R:FR is changed maintaining a constant N_{R+FR} total photon flux, or a constant N_{FR}, or a constant N_R.

4.7.2.1.5 Intermediates in phytochrome photoconversion. The Pr → Pfr and Pfr → Pr photoconversions are not single-step processes (Chapter 4.1). The role of photoconversion intermediates in the physiology of phytochrome action is not fully established. A relatively large fraction of photoconversion intermediates may accumulate under high irradiances of light that elicit intermediate values of φ and high rates of cycling between Pr and Pfr.

4.7.2.1.6 Phytochrome dimers. The photochemical properties of phytochrome summarized so far are those based on the monomeric form of the photoreceptor. There is evidence that phytochrome exists as a dimer in solution, and three dimeric forms are possible: PrPr, PrPfr, and PfrPfr. Certain aspects of phytochrome transformations *in vivo* (kinetics of dark reversion and dark destruction, different values of k for photoconversion starting from low and high initial values of Pfr/P) suggest that phytochrome may also behave as a dimer *in vivo* (Brockmann *et al.* 1987). One of the theoretical models of phytochrome action is based on the properties of dimeric phytochrome (VanDerWoude 1985, 1987). Phytochrome photoconversion for a dimeric system proceeds according to the following scheme:

$$\mathrm{PrPr} \underset{k_2}{\overset{2k_1}{\rightleftarrows}} \mathrm{PrPfr} \underset{2k_2}{\overset{k_1}{\rightleftarrows}} \mathrm{PfrPfr}$$

and the rates of change of the three dimers are:

$$d[\mathrm{PrPr}]/dt = -2k_1[\mathrm{PrPr}] + k_2[\mathrm{PrPfr}]; \quad d[\mathrm{PfrPfr}]/dt = k_1[\mathrm{PrPfr}] - 2k_2[\mathrm{PfrPfr}];$$
$$d[\mathrm{PrPfr}]/dt = 2k_1[\mathrm{PrPr}] + 2k_2[\mathrm{PfrPfr}] - (k_1+k_2)[\mathrm{PrPfr}]$$

At photoequilibrium, the mole fractions of the three dimers are a function of φ (Brockmann *et al.* 1987):

$$\mathrm{PrPr/P} = (1-\varphi)^2, \quad \mathrm{PrPfr/P} = 2(\varphi-\varphi^2), \quad \text{and } \mathrm{PfrPfr/P} = \varphi^2;$$

PrPr/P decreases with increasing φ, PfrPfr/P increases with increasing φ, and PrPfr/P is maximum at $\varphi = 0.5$ (Fig. 14).

4.7.2.2 Photochemical properties of phytochrome in vivo

Photoconversion kinetics of phytochrome *in vivo* have only been studied in detail in etiolated seedlings. The reasons for this are the limitations of the spectrophotometric assay of phytochrome *in vivo* (Pratt 1983). The assay can be

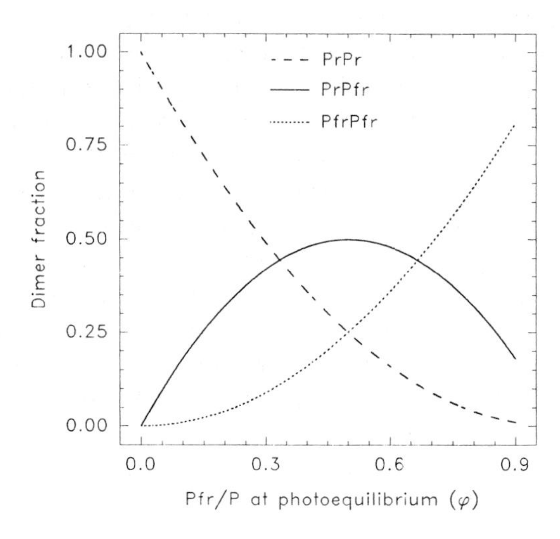

Figure 14. Relationships between Pfr/P at photoequilibrium and the relative concentrations of the dimers PrPr, PrPfr, and PfrPfr.

carried out with some confidence in dark-grown tissues with a relatively high phytochrome level, but interference by chlorophyll prevents its use with green tissues. A few spectrophotometric data for phytochrome in light-grown seedlings have been obtained using seedlings in which chlorophyll accumulation had been prevented by inhibitors (Jabben and Deitzer 1978), but, even with chlorophyll out of the way, the assay is still severely limited because of the low levels of phytochrome in light-grown seedlings. Because of its relatively low sensitivity and the possibility of measurement artifacts (Pratt 1983), the spectrophotometric assay of phytochrome *in vivo* is subject to limitations even under the best conditions of use (etiolated seedlings with relatively high levels of phytochrome), and a great deal of caution must be exercised in the interpretation of kinetic data from spectrophotometric assays *in vivo.*

It is not clear at present whether there are significant differences between the photoconversion cross-sections of phytochrome *in vivo* and *in vitro.* The comparison of $\sigma_{R\lambda}$ and $\sigma_{FR\lambda}$ values *in vitro* and *in vivo* (Table 1) is not very helpful because there are differences between the photochemical parameters of purified phytochrome (*see also Appendix Fig. 1*) and there is only one complete set of data for $\sigma_{R\lambda}$ and $\sigma_{FR\lambda}$ values *in vivo* (Seyfried and Schäfer 1985b).

Several differences from what was expected on the basis of the photochemical properties of purified phytochrome were observed in studies of phytochrome photoconversion *in vivo.* Deviations from first order photoconversion kinetics were reported by Boisard *et al.* (1971), Schäfer *et al.* (1971), and several others. In etiolated mustard seedlings, φ_{FR} (produced by the same FR source) decreases with increasing age (Schäfer *et al.* 1972); the rate constant for Pr → Pfr photoconversion under R increases and that for Pfr → Pr photoconversion under FR decreases with increasing age (Schäfer and Mohr 1980). Rhythmic oscillation in photoconversion rates were observed in cotyledons of dark-grown *Pharbitis* during the dark period following a saturating exposure to R (King *et al.* 1982). The initial value of Pfr/P apparently has a significant effect on the rate constant for phytochrome photoconversion *in vivo* (Mancinelli *et al.* 1992). These differences between phytochrome photoconversion kinetics *in vivo* and *in vitro* are not fully understood. They might represent real differences, but they could also result from measurement artifacts of the spectrophotometric assay *in vivo*. For example, according to Schäfer *et al.* (1971), deviations from first order photoconversion kinetics might be a consequence of the co-existence *in vivo* of kinetically distinguishable phytochrome populations, but, while this cannot be excluded in consideration of the recognized multiplicity of phytochrome types, they might also be brought about as a consequence of light attenuation in the samples used for the *in vivo* assay (Spruit and Kendrick 1972).

4.7.2.3 Light, Pfr/P, φ, k, H, *and expression of phytochrome-mediated responses*

In vivo measurements (when possible) of Pfr/P, φ, and k (H cannot be measured directly, either *in vivo* or *in vitro*) or projected values (when measurements *in vivo* are not possible) of these parameters for a given light treatment have a lower degree of confidence than that with which one can measure the effects of the same light treatment on the expression of photomorphogenic responses. Selective attenuation, scattering and trapping of light within plant tissues (Chapter 7.4) contribute to create differences in the state of phytochrome between cells at different depth within an organ (Kazarinova-Fukshansky *et al.* 1985; Seyfried and Schäfer 1985a). Thus, *in vivo* values of Pfr/P, φ and k, when measured in intact organs (*e.g.* cotyledon, hypocotyl) or entire small seedlings, as it is often done, are only average values that may not provide a good index of the state of phytochrome relevant for a given response at a given location. In those cases in which spectrophotometric assays *in vivo* cannot be made, researchers use projected values of Pfr/P, φ, k, and H, calculated on the basis of the photoconversion cross-sections of purified phytochrome and data for the spectral photon flux distribution of incident light. Projected values of Pfr/P, φ, k and H make it possible to compare different light situations using physiologically relevant parameters, but they may not correspond to values *in vivo* (Table 2). Another factor to be considered when using projected values is that conventional light measurements may not provide an accurate estimate of the light actually perceived by the plant (Holmes 1984). For example, consider the effects of low and high reflectance conditions in a light field on phytochrome photoconversion kinetics *in vivo* (Fig. 15): light measurements taken in a

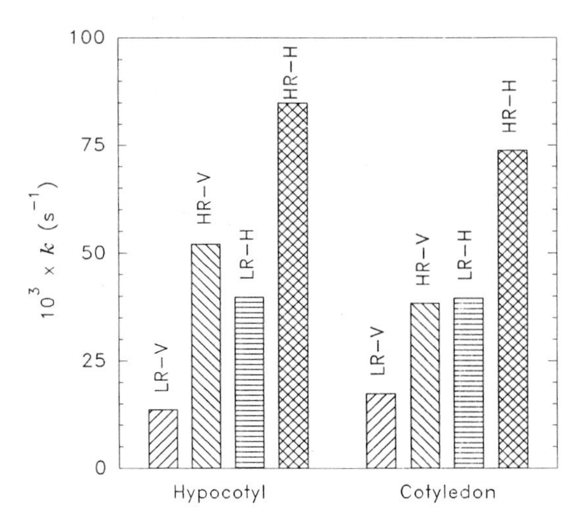

Figure 15. Effects of reflectance and sample orientation on phytochrome photoconversion in excised *Cucurbita* hypocotyl segments and cotyledons. LR, low reflectance light field (walls of irradiation box ultraflat black); HR, high reflectance light field (walls of irradiation box high gloss white); H, sample in horizontal position; V, sample in vertical position. White light (from a modified projector) entering the irradiation box from above. Data of A.L. Mancinelli and J. Lim (1989) *Photochem. Photobiol.* 50: 125-132.

conventional way [*cosine corrected (angle of acceptance ≈ 180°) spectroradiometric sensor aimed at the light source*] indicated a difference of only about 10% between the photon fluxes of the low and high reflectance light fields; but the differences in measured values of *k* for the same organ in the same position exposed to radiation in the two light fields were significantly higher than 10%; the effects of sample orientation are also clearly evident. The limitations of projected and measured *in vivo* values of φ, *k*, and *H* must be kept in mind in the analysis of the physiology of action of phytochrome in terms of its state in the plant.

The following example from research carried out almost 30 years ago might provide a demonstration of the importance of keeping in mind the limitations of the experimental methods. In the famous *Pisum* paradox (Hillman 1965, 1972), a FR reversibility of a R effect was observed several hours after the exposure to R (R → dark → FR) when there was no spectrophotometrically detectable Pfr. However, the physiological effects of exposures establishing different Pfr/P values suggested Pfr/P ≈ 0.2. The discrepancy between physiological and *in vivo* spectrophotometric data led Hillman (1965) to suggest the presence of two pools of phytochrome, *'bulk'* and *'active'*, the latter being only a small fraction of the total so that its changes could not be detected with the spectrophotometric assay in the presence of excess levels of bulk phytochrome. Hillman's suggestion was probably one of the determining factors in the development of the

Table 2. Comparison of projected and measured, *in vivo* values of φ and k for irradiation with broad-band blue (B), red (R), and far red (FR) light sources. Photon fluxes in $\mu mol\ m^{-2}\ s^{-1}$: B, 22.6; R, 2.9; FR, 19.5; temperature, 0-1°C. From A.L. Mancinelli, 1988, *Plant Physiol.* 86: 749-753.

	φ_B	φ_R	φ_{FR}	k_B	k_R $(10^3 \times s^{-1})$	k_{FR}
Projected values						
Oat-1	0.37	0.79	0.094	13.2	17.5	34.7
Oat-2	0.45	0.84	0.13	6.4	10.2	15.1
Measured values, *in vivo*						
Cabbage seedlings	0.45	0.84	0.081	5.8	35.9	49.9
Cucumber cotyledons	0.36	0.83	0.045	6.4	30.9	69.3
Measured to projected value ratio*						
Cabbage/Oat-1	1.22	1.06	0.86	0.44	2.05	1.44
Cabbage/Oat-2	1.00	1.00	0.62	0.91	3.52	3.30
Cucumber/Oat-1	0.97	1.05	0.48	0.48	1.77	2.00
Cucumber/Oat-2	0.80	0.99	0.35	1.00	3.03	4.59

*Note the significant variability in the measured to projected value ratios, depending on sample used for the *in vivo* spectrophotometric assay, spectral region, and set of photochemical parameters of purified phytochrome (oat-1 and -2; Table 1 and Appendix Fig. 1) used to calculate the projected values of φ and k.

research that resulted about 20 years later in the confirmation of the existence and identification of different phytochrome types and is a clear demonstration that even a paradox, when properly analyzed (taking into account the limitations of the methods) can lead to significant advancement.

4.7.3 The state of phytochrome *in vivo*

The state of phytochrome *in vivo* is a function of the interaction between photoconversion and non-photochemical reactions: biosynthesis, dark reversion of Pfr to Pr, dark destruction or degradation of phytochrome, and reaction(s) between Pfr and the still unknown reaction partner(s), *X*. Dark reversion and destruction have been observed *in vitro*, but most of the information on the two processes has been derived from studies *in vivo*. Destruction (Chapter 4.4) results in a loss of spectrophotometrically and immunochemically detectable phytochrome. The reaction between Pfr and *X* (Pfr action, referring specifically to the *first* molecular change caused by Pfr in the responding system) is the first step in the transduction of the signal; very little is known about this first step and successive ones in the signal transduction pathway.

Another factor which plays a significant role in determining the state and action of phytochrome *in vivo* is that phytochrome consists of a family of photoreceptors which share the property of reversible Pr ⇄ Pfr photoconversion, but differ in many other aspects, for example, stability of Pfr and contribution to photoregulation, depending on the response and the state of the responding system. The different phytochromes can be divided into at least two groups: phytochrome type I (PI), predominant in etiolated seedlings (*etiolated* phytochrome), and phytochrome type II (PII), predominant in light-grown seedlings (*green* phytochrome). Other terms are also used to distinguish different phytochromes: labile phytochrome (*ℓ*P, characterized by the rapid destruction of Pfr, *ℓ*Pfr) and bulk phytochrome (Hillman 1965) might be the same as PI; stable phytochrome (*s*P, with a long-lived Pfr, *s*Pfr) and the phytochrome present in seeds might be the same as PII. These attributions are consistent with some observations, but are not definitive and may be revised as we learn more about the nature of different phytochromes.

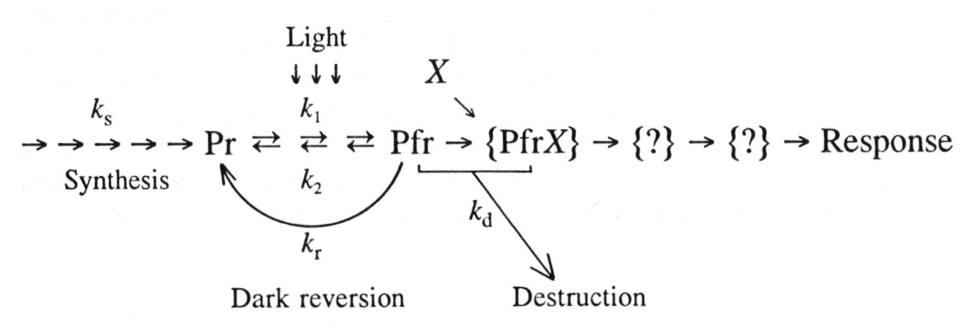

This scheme, whose many limitations will become evident as the reader progresses through this chapter, describes in first approximation the interaction between photochemical and non-photochemical processes.

4.7.3.1 Intercellular and intracellular distribution of phytochrome

Phytochrome is not uniformly distributed throughout the different organs and tissues of a plant (Chapter 4.5). The relationships between phytochrome content and expression of phytochrome-mediated responses in a given organ or tissue have not been fully defined. The phytochrome involved in the photoregulation of a given response is not necessarily only that present in the affected organs. For example, in the photoperiodic induction of flowering, the light signal responsible for the transformation of a vegetative apex into a flowering one is perceived in the leaves (Vince-Prue 1975); a significant fraction of the action of R on hypocotyl elongation is contributed by light absorbed by the cotyledons (Black and Shuttleworth 1974).

The intracellular distribution of phytochrome is affected by light (Chapter 4.5). In dark-grown seedlings, a change from a *free* (= homogeneously distributed through the cytosol, soluble) to a *bound* (= sequestered in small aggregates, pelletable) state is activated upon photoconversion of Pr to Pfr and is a very fast process ($t½ \approx 2$ s at 25°C); the release from the *bound* to the *free* state, after photoconversion of Pfr back to Pr, is much slower ($t½ \approx 25$ min at 25°C). The relevance of changes in intracellular distribution for phytochrome action remains to be determined. Correlations between pelletability and phytochrome destruction have been reported (Boisard *et al.* 1974).

4.7.3.2 Synthesis of phytochrome

Spectrophotometric assays *in vivo* (Fig. 16), recovery upon extraction (Quail *et al.* 1973), and immunochemical assays (Hilton and Thomas 1987; Tokuhisa and Quail 1987) show a several-fold increase of the phytochrome level (the result of *de novo* synthesis) during the first few days of growth in seedlings kept in darkness. Most of this increase is due to synthesis of PI; PII increases also, but its rate of synthesis is much lower than that of PI (Konomi *et al.* 1987). The synthesis of PI is negatively regulated by light; the effects of light are R-FR reversible; thus, PI biosynthesis is a phytochrome-mediated response; the synthesis of PII is apparently light-independent (Chapter 4.2). In dry seeds and during the initial phase of imbibition in darkness, PII may be more abundant than PI, but, if incubation in darkness continues, PI increases to levels at least 5-10 times higher than PII.

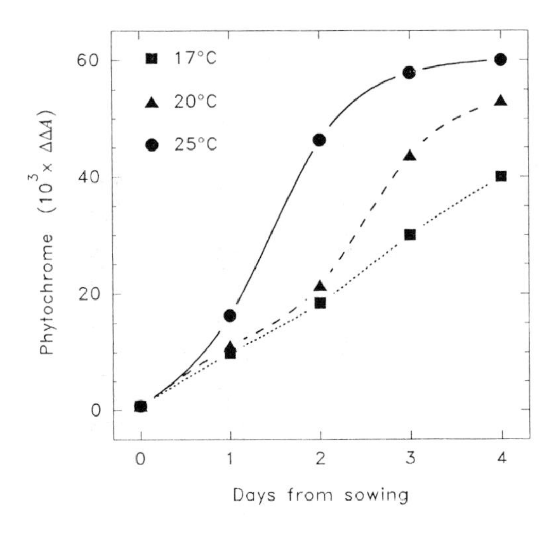

Figure 16. Spectrophotometrically detectable phytochrome in *Cucumis* cotyledons during incubation in darkness. Data of A.L. Mancinelli and A. Tolkowsky (1968) *Plant Physiol.* 43: 489-494.

Phytochrome levels are low in light-grown seedlings, and PII is the predominant type. However, in plants exposed to daily light-dark cycles, PII may not be the predominant type all the time and the [PI]/[PII] ratio might vary significantly, as suggested by the oscillations in the level of immunochemically detectable phytochrome in green seedlings exposed to light-dark cycles (Fig. 17); the phytochrome formed during the intervening dark periods is mostly ℓP which is rapidly destroyed during the successive light period.

4.7.3.3 Phytochrome destruction

Most of what is known about phytochrome destruction has been derived from studies in etiolated seedlings in which PI is the predominant type. The molecular aspects of the destruction process are analyzed in Chapter 4.4. Phytochrome destruction is activated by the photoconversion of Pr to Pfr.

In etiolated *Zea* and *Helianthus* seedlings (Fig. 18), the destruction of Pfr in darkness after a short R exposure goes to completion in about 2-3 h at 25°C. The rate of destruction is a function of temperature: in *Amaranthus* seedlings, the half-life of Pfr in darkness varies from about 10 min at 30°C to 85 min at 15°C (Kendrick and Frankland 1969). The loss of phytochrome is higher in systems in which there is no dark reversion (Fig. 18, *Zea*) than in those in which dark reversion is present (Fig. 18, *Helianthus*). Phytochrome destruction in darkness after a short light exposure is maximal ($\approx$ 80% loss of P in systems without dark reversion) after R, and minimal ($\leq$ 5%) after FR alone, but is not

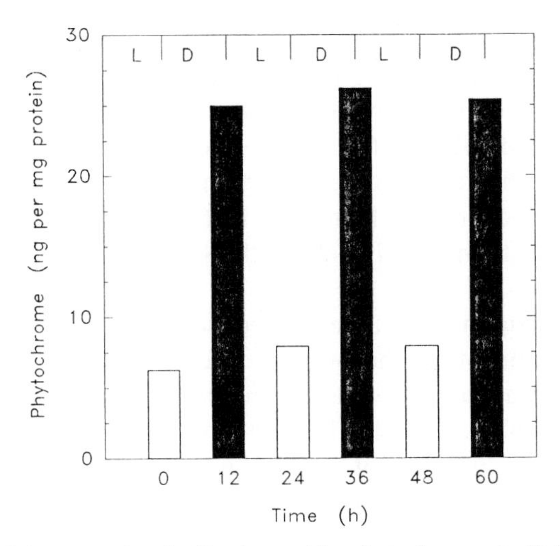

Figure 17. Levels of immunochemically detectable phytochrome in light-grown oat seedlings exposed to (12 h L → 12 h D) cycles. Measurements started 5 days after sowing (= 0 on abscissa). Redrawn from data of R.E. Hunt and L.H. Pratt (1980) *Plant Cell Environ.* 3: 91-95).

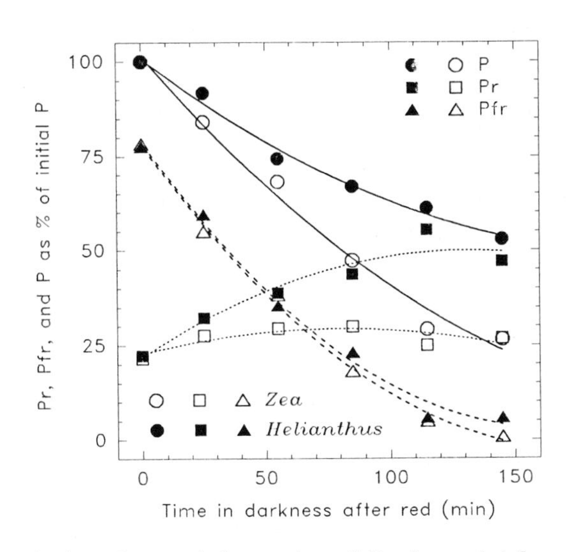

Figure 18. Kinetics of phytochrome dark reactions following a brief exposure to red light. Temperature, 24-25°C. Redrawn from data of Frankland (1972) In: *Phytochrome*, pp.195-225, K. Mitrakos and W. Shropshire, Jr. (eds.) Academic Press, London; and L.H. Pratt and W.R. Briggs (1966) *Plant Physiol.* 41: 467-474.

fully R-FR reversible within these limits. The loss of phytochrome after a short R-FR treatment is about 25-30% (Dooskin and Mancinelli 1968; Stone and

Pratt 1979). This loss is apparently a consequence of a destruction of Pr which starts only after phytochrome has been cycled through Pfr (Pfr-induced Pr destruction). The rate constants of Pfr destruction and Pfr-induced Pr destruction are about the same. The Pfr-induced Pr destruction has been observed in both monocots and dicots. Rates of phytochrome destruction in etiolated seedlings may vary with age: for example, the half-life for Pfr destruction in Cucurbita cotyledons decreases from 150 to 60 min between the 2nd and 5th day after sowing, and then remains about the same up to the 10th day (Schäfer 1978).

Destruction of phytochrome in etiolated seedlings exposed to continuous irradiation is a function of φ, photon flux, and temperature. Under continuous irradiation even FR may cause a considerable loss of total phytochrome, despite the low value of φ established (Fig. 19; 40% loss of [P] in about 5 h); these data are for dicots. In *Avena*, a monocot, under the same FR exposure, the loss of [P] is about 90%; destruction in monocots is saturated at a much lower φ and photon flux than in dicots. Phytochrome destruction under prolonged irradiation starts initially from Pfr, but, as the irradiation continues, the Pfr-induced Pr destruction might contribute to the process. One aspect of phytochrome destruction under continuous irradiation is that, in time, levels of Pfr might become higher for irradiations establishing the lower values of φ (Fig. 19). The rate of phytochrome destruction under continuous irradiation is a function of irradiance, but may decrease when the irradiance increases above certain levels, especially for light treatments establishing intermediate values of φ; under these

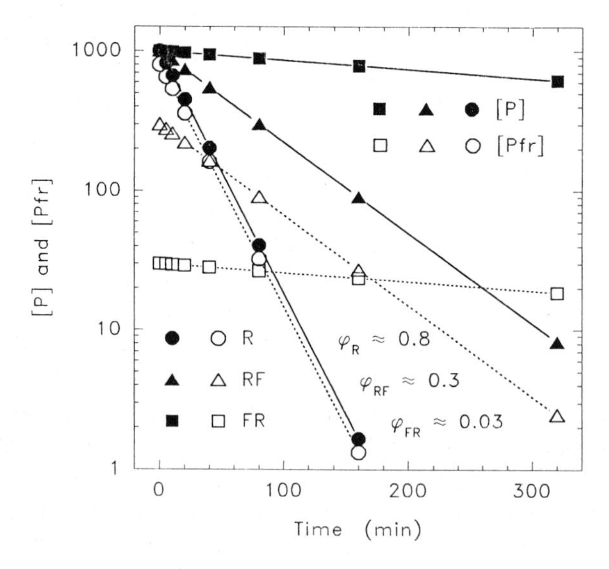

Figure 19. Kinetics of phytochrome destruction *in vivo* under continuous red (R), far-red (FR) and dichromatic R + FR (RF) irradiations. A theoretical example based on average values of measurements of phytochrome destruction in dicots.

conditions, as a consequence of the high rate of cycling between Pr and Pfr at photoequilibrium, there is a significant accumulation of photoconversion intermediates, apparently not subject to destruction, resulting in a *photoprotection* of phytochrome (Kendrick and Spruit 1972).

There is much less information on phytochrome destruction in light-grown seedlings. Spectrophotometric assays in light-grown seedlings in which chlorophyll synthesis had been prevented by inhibitors show a small decrease in phytochrome during the first 60 min after the light → dark transition, followed by a steady rate of increase for several hours (Jabben 1980; Jabben *et al.* 1980). The initial loss of phytochrome (10-30% of the level at the time of the light → dark transition) suggests that PI might constitute about 10-30% of the low level of phytochrome in light-grown seedlings. The decay of Pfr is fast during the first 30 min or so, and then continues at a much slower rate for several hours: spectrophotometrically detectable Pfr is still found 6-24 h after the light → dark transition. Perhaps, this slow decay might represent decay of stable Pfr; if so, the decay rate of *s*Pfr is at least 10 times slower than that of *ℓ*Pfr. Data from immunochemical assays in light-grown seedlings (Fig. 17; Johnson *et al.* 1991) suggest that the phytochrome formed during the dark period following exposure to light may be *ℓ*P because it is rapidly destroyed upon re-exposure to light.

4.7.3.4 Dark reversion

In some systems (Fig. 18, *Helianthus*), the kinetics of phytochrome transformation in darkness after a brief R exposure indicate a reappearance of Pr and suggest a dark reversion of Pfr to Pr. The situation is much clearer in cauliflower heads (Butler *et al.* 1963) and artichoke receptacles (Hillman 1964), which contain relatively high levels of phytochrome, probably an *s*P, even after prolonged storage in light. Phytochrome destruction is very slow or absent in these systems and the disappearance of Pfr in darkness following an exposure to R is fully or almost fully compensated by a reappearance of Pr; the half-life for the Pfr → Pr dark reversion is about 60 min in cauliflower heads. Dark reversion is apparently absent in monocots (Fig. 18, *Zea*) and members of the Centrospermae. In systems in which both dark reversion and destruction are present (Fig. 18, *Helianthus*), the fraction of Pfr reverting in darkness to Pr is smaller than the fraction that is destroyed. The fraction of Pfr that reverts in darkness to Pr is not affected by inhibitors that reduce or prevent destruction: this led to the suggestion that dark-reversible and destructible Pfr might be different pools. The fraction of dark-reversible Pfr is apparently a function of the value of Pfr/P established by the light treatment: in etiolated mustard seedlings, the relationship between Pfr/P and fraction of dark-reversible Pfr shows a pattern matching closely that between Pfr/P and the fraction of the PrPfr dimer (Brockmann *et al.* 1987); in this system, dark reversion is approximately 10 times faster than destruction and is virtually completed in about 10 min.

Prolonged exposures to continuous or cyclic FR are required to inhibit the germination of some dark-germinating seeds. The effects of cyclic FR decrease with increasing cycle duration and can be reversed by applying R immediately after FR in each cycle. It was suggested that prolonged exposures to continuous or cyclic FR were required to prevent the Pfr that was reappearing in the system from acting. Later on, a slow reappearance of Pfr during the dark period following a FR exposure was observed in dry and partially imbibed seeds using spectrophotometric assays (Spruit and Mancinelli 1969): this reappearance of Pfr was dubbed *'inverse dark reversion'*. Data from further studies led to the suggestion that, in dry and partially hydrated seeds, the photochemical conversion of Pfr may not go all the way to Pr, and the reappearance of Pfr starts from one of the photoconversion intermediates (Kendrick and Spruit 1974).

4.7.3.5 Photoconversion, dark reversion, destruction and state of phytochrome

The combined effects of photoconversion, dark reversion, and destruction on the rate of change of [Pfr] can be described as follows:

$$d[\mathrm{Pfr}]/dt = k_1[\mathrm{Pr}] - k_2[\mathrm{Pfr}] - k_r[\mathrm{Pfr}] - k_d[\mathrm{Pfr}].$$

At the photostationary state* ($t = \infty$), the situation will be:

$$d[\mathrm{Pfr}]/dt = 0;\ \ k_1[\mathrm{Pr}] = (k_2 + k_r + k_d)[\mathrm{Pfr}];\ \ \mathrm{Pfr/P} = \varphi_S = k_1/(k_1 + k_2 + k_r + k_d).$$

The above relationships indicate that the state of phytochrome determined by the interaction between light and dark reactions is different than that determined by photoconversion only: for example, φ is a function of light quality only; φ_S is a function of light quality, photon flux and temperature, the dependence on the latter being a consequence of the temperature dependence of k_d and k_r. A comparison of photoconversion kinetics *in vivo* at 0°C (negligible dark reversion and destruction) and 25°C (high rates of dark reversion and destruction) provided experimental confirmation of the difference between the two situations. The values of Pfr/P produced by a given light treatment were lower at 25°C than 0°C, and the differences were significantly larger for exposures to low irradiances of light of low efficiency for photoconversion than for exposures to high irradiances of light of high efficiency for photoconversion (Jabben

---

*The equilibrium state resulting from the interaction between light and dark reactions is called the *photostationary state* to distinguish it from the photoequilibrium established by light only; φ_S = Pfr/P at the photostationary state.

et al. 1982). The impact of the interaction between photochemical and non-photochemical reactions on the state of phytochrome *in vivo* is more pronounced in systems in which labile PI is the predominant phytochrome type.

4.7.4 Phytochrome-mediated responses

Phytochrome-mediated responses can be divided into six groups of photo-response modes, distinguishable on the basis of their most characteristic photobiological properties.

(i) Low fluence response (LFR). The LFRs are the classical phytochrome-mediated responses; they are effectively induced by a single short exposure to R (occasionally a few short exposures separated by long dark intervals) and show R-FR reversibility. The effectiveness of FR in reversing the inductive effects of R decreases and is eventually lost as progressively longer dark intervals are inserted between R and FR. The photon fluences required for saturation of the response by R vary from 1 to 1000 μmol m^{-2} s^{-1}. The LFR obeys the reciprocity law (extent of response induced by $N \times t$ = extent of response induced by $xN \times t/x$; N and t, irradiance and duration of the light treatments, respectively); deviations from reciprocity may occur for exposures longer than 20-30 min.

(ii) Very Low Fluence Response (VLFR). The VLFRs can be effectively induced by very low photon fluences (10^{-4} to 10^{-1} μmol m^{-2}) which elicit very low values of Pfr/P (10^{-6} to 10^{-3}; Fig. 10). In some cases, the VLFR is fully saturated at very low fluences; in other cases, a response may show a biphasic fluence-response curve, with significant induction in the very low fluence range, full induction in the low fluence range and a response plateau between the two ranges.

(iii) High irradiance response (HIR). The HIRs require prolonged exposures to light of relatively high photon flux for maximum expression. The extent of the HIR is a function of wavelength, irradiance, and duration of the light treatment. The spectral sensitivity and irradiance dependence of the HIR vary significantly, depending on the combination of response, species, and experimental conditions. The HIR does not show R-FR reversibility and does not obey the reciprocity law.

(iv) Photoperiodic responses. These are responses to the duration of the light and dark periods in a 24 h cycle. The role of phytochrome in these responses was first demonstrated through the R-FR reversibility of the effects of short middle-of-the-night light breaks on the induction of flowering. Photoperiodic responses are discussed in detail in Chapter 7.3.

(v) End-of-day (EOD) responses. These are responses of light-grown plants to the state of phytochrome established at the end of the daily light period. The sensitivity of EOD responses to the state of phytochrome at the end of the day is

significantly affected by daylength (Fig. 20). The EOD responses show R-FR photoreversibility (Fig. 21) and obey the reciprocity law.

(vi) The R:FR perception responses. These are responses to the R to FR photon flux ratio (R:FR; ζ) of incident radiation. The phytochrome-mediated perception of changes in R:FR translates into the capability of detecting nearby or shading vegetation, because chlorophyll-containing tissues reflect and transmit FR while absorbing most of the R. Shade-avoidance responses (*e.g.* increased rate of stem elongation in light-grown plants) to low daytime R:FR may increase the ability of a plant to compete efficiently for light. In some cases, these responses may show a high sensitivity to small variations in R:FR which produce small changes in φ (Fig. 11). The R:FR perception responses are discussed in detail in Chapter 7.1.

The above categories are convenient for discussion, but there is some overlap between the different groups. For example, in the photoperiodic induction of flowering, the effects of EOD and middle-of-night brief exposures to light are R-FR reversible LFRs; the effects of prolonged day-extensions show characteristics reminiscent of both R:FR perception and HIR responses; irradiance and wavelength dependence of the effects of prolonged middle-of-night breaks are within the range of those typical for the HIR; the same effect of a middle-of-night break can be produced with a brief exposure to R or a prolonged one to FR. Insofar as the R:FR perception response is concerned, the effects of a low daytime R:FR can be mimicked by a short FR applied at the end of high daytime R:FR light periods.

4.7.4.1 Red-far red reversible low fluence responses

The inductive, R-FR reversible LFRs include a large variety of responses varying from transient processes (*e.g.* chloroplast movements, nyctinastic leaf movements, ion fluxes; Chapters 4.6 and 9.4) to developmental changes (*e.g.* seed germination, de-etiolation, stem growth, leaf expansion, vegetative to floral transition; Chapters 7.3 and 9.1). The lag period (time between exposure to light and first detectable change) varies from seconds/minutes for transient processes to hours/days for developmental changes. Action spectra for induction and reversion show peaks of action between 650 and 670 nm and between 720 and 740 nm, respectively (Fig. 22). For irradiations shorter than those required to establish photoequilibrium values of Pfr/P, the extent of the response is a function of the fluence of incident light and may show a good correlation with Pfr/P and/or [Pfr] ([Pfr] = [P] × Pfr/P). Photon fluence requirements for response saturation by R vary over three orders of magnitude, from about 1 μmol m^{-2} (*e.g.* induction of plumular hook opening in bean seedlings) to 1000 μmol m^{-2} (*e.g.* effects of middle-of-night light breaks on flowering). The sensitivity to light of a given response in a given system can be affected by

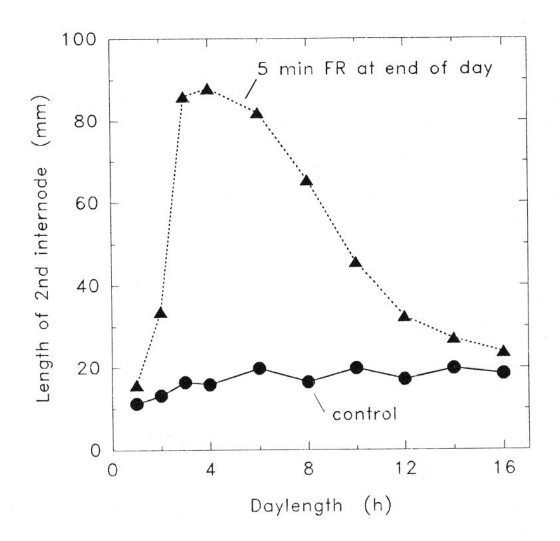

Figure 20. Effects of daylength and end-of-day far-red light (FR) on stem elongation in Pinto beans. Redrawn from data of R.J. Downs, S.B. Hendricks and H.A. Borthwick (1957) *Bot. Gaz.* 118: 199-208.

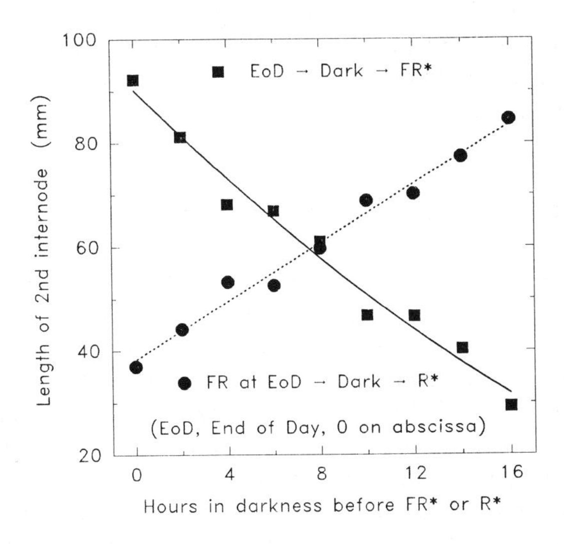

Figure 21. Length of second internode of Pinto bean in relation to the time of application of short red (R) or far-red (FR) exposures within a 16-h dark period. Redrawn from data of R.J. Downs, S.B. Hendricks and H.A. Borthwick (1957) *Bot. Gaz.* 118: 199-208.

several factors, for example, temperature. At temperatures below 20°C, the germination of dark-germinating tomato seeds can be effectively inhibited by a single exposure to FR applied at the appropriate time (Fig. 23); the effect of FR

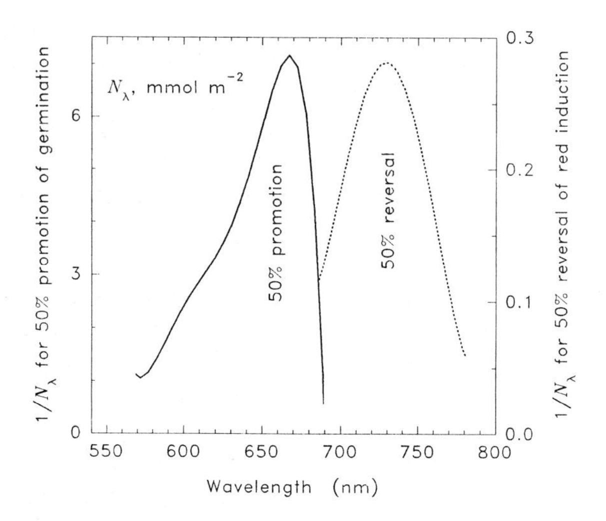

Figure 22. Action spectra for induction of germination and reversal of induction in light-requiring seeds. Redrawn from data of H.A. Borthwick, S.B. Hendricks, E.H. Toole and V.K. Toole (1954) *Bot. Gaz.* 115: 205-225.

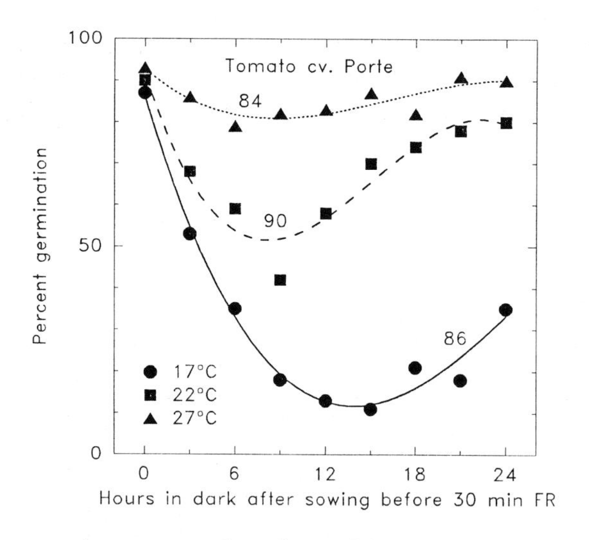

Figure 23. Time course and temperature dependence of the sensitivity to far-red light (FR) in the germination of tomato seeds. Numbers close to lines are % germination of dark controls; germination scored 4 days after sowing; effects of FR can be reversed by red light. Data of A.L. Mancinelli, Z. Yaniv and P. Smith (1967) *Plant Physiol.* 42: 333-337.

can be reversed by R; the response is a typical R-FR reversible LFR. At temperatures ≥ 20°C, prolonged exposures to continuous or cyclic FR are required to inhibit germination effectively. Inhibition by continuous FR is irradiance and duration dependent, and thus shows some characteristics in common with HIR responses; the effects of cyclic light treatment are R-FR reversible (Fig. 24), just as a typical LFR. The variability in light sensitivity may be considered a further complication, but, if properly exploited, can provide a great deal of information on the kinetics of phytochrome action.

4.7.4.1.1 Pfr/P and LFR. The LFRs have been used extensively to analyze the relationships between response extent and Pfr/P using either projected or measured *in vivo* (when possible) values of Pfr/P. Relatively good correlations between response expression and Pfr/P have been reported for some responses. For example, a good correlation was observed for the germination of partially light-requiring lettuce seeds (dark germination ≈ 25%). The fluence-response curves (Fig. 25) show that induction (inhibition) of germination requires lower R (FR) fluences when R (FR) is applied to seeds kept in darkness than after sFR (sR; s, saturating). This suggests that Pfr/P in seeds imbibing in darkness may be an intermediate value between those established by sFR (Pfr/P ≈ 0.02) and sR (Pfr/P ≈ 0.85) pretreatments. Germination percentages for the sFR → R and sR → FR treatments fall on the same line when plotted versus calculated Pfr/P values (Fig. 26) and indicate that 25% germination (same as in dark controls) is obtained at Pfr/P ≈ 0.3. Germination percentages for the Dark → R and Dark → FR treatments, when plotted *versus* Pfr/P (calculated using 0.3 as the initial

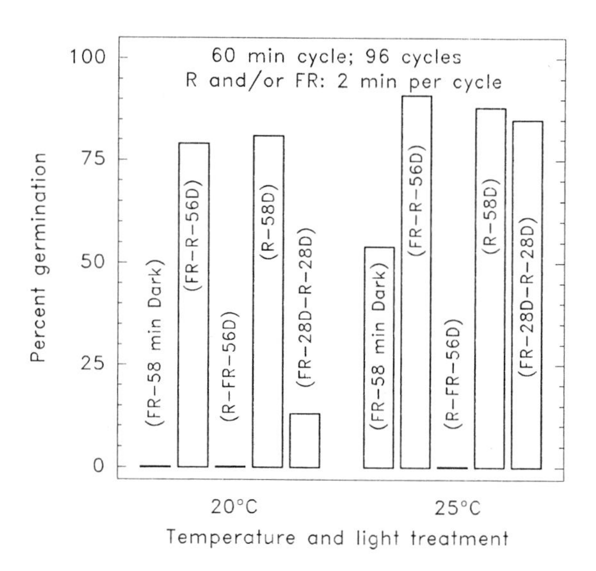

Figure 24. Effects of cyclic light treatments with red (R) and far-red (FR) light on the germination of tomato seeds. Germination of dark controls, 85-90%. Light treatments started at sowing. Data of Z. Yaniv, A.L. Mancinelli and P. Smith (1967) *Plant Physiol.* 42: 1479-1482.

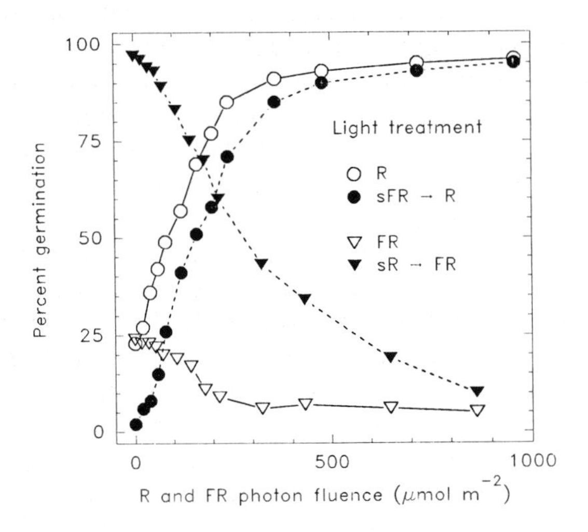

Figure 25. Fluence-response curves for induction and inhibition of germination in lettuce seeds. Experimental protocol: sowing → 6 h dark → light treatment → 66 h dark; temperature, 17°C; R and FR, red and far-red light, respectively; sR and sFR, saturating (10 min) R and FR, respectively. Data obtained by Columbia University students in a laboratory exercise for a course in Plant Physiology, 1966.

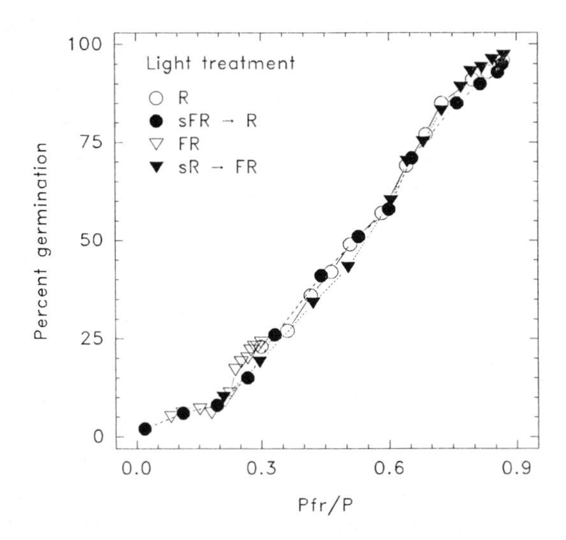

Figure 26. Germination percentages from Fig. 25 plotted versus calculated values of Pfr/P (R, FR = red and far-red light, respectively; sR, sFR = saturating R and FR, respectively).

value of Pfr/P) fall in line with the data for the sFR → R and sR → FR treatments. On the basis of these results, an estimate value of Pfr/P ≈ 0.3 during seed imbibition in darkness seems reasonable. Results of several studies of this type suggested the presence of endogenous Pfr in seeds imbibing in darkness, a suggestion confirmed later by spectrophotometric assays *in vivo.*

There are also cases in which the correlations are poor (*e.g. Pisum* paradox, Section 4.7.2.3), and an interpretation of data based solely on Pfr/P is not sufficient to provide a plausible explanation. For example, the rate of closure of the leaflets of *Mimosa* after transfer from light to dark at the end of the day is slowed down if a FR pulse is applied at the end of the light period; R applied immediately after FR restores a fast rate of closure, indicating that leaflets closing is enhanced at high Pfr/P; however, the leaflets are open during the day, when Pfr/P is high. This is one case in which phytochrome action may be complicated by the interaction with an endogenous rhythm, and the mode of action of Pfr is different during the light and dark phases of a 24 h cycle: leaflets' closing after the light → dark transition is an EOD/R-FR reversible LFR, while leaflets opening after a dark → light transition shows HIR characteristics. Another example of poor correlations is illustrated in Fig. 24. The effects of cyclic FR (cFR, 2 min FR-58 min dark per cycle) and cyclic R-FR (cR-FR, 2 min R-2 min FR-56 min dark per cycle) are not the same. At 25°C, cR-FR is more effective than cFR in preventing germination; at 20°C, the two treatments seem equally effective; however, the fluence of R required to promote germination after two to four days of exposure to cR-FR is higher than after exposure to cFR. The different effectiveness of the two treatments is difficult to explain in terms of differences in Pfr/P, which is the same (≈ 0.04) after FR and R-FR; however, it might be explained as a consequence of $[P]_{cR\text{-}FR}$ being lower than $[P]_{cFR}$, as determined with spectrophotometric assays *in vivo*, and, consequently, at equal Pfr/P, $[Pfr]_{cR\text{-}FR} < [Pfr]_{cFR}$. This is a case in which the extent of the response shows a good correlation with [Pfr], but not with Pfr/P.

4.7.4.1.2 Kinetics of phytochrome action and LFR. One aspect of the physiology of phytochrome action that has been extensively investigated using LFRs is the time course of Pfr action. The rate of escape of a response from R-FR reversibility (measured by determining the effects of varying periods of darkness between short exposures to R and FR) is the parameter used as an index of the time course of Pfr action. Rates of escape from photoreversibility vary depending on the combination of response, species and experimental conditions (*e.g.* temperature). As expected, Pfr action is fast (rapid escape from photoreversibility) in fast responses (*e.g.* leaflets closing at the end of the day in *Mimosa*). However, Pfr action varies from slow to fast in slow responses (*e.g.* seed germination, stem elongation, flowering induction), as indicated by the following examples of times required for 50% loss of photoreversibility: 9 h for the induction of germination in light-requiring lettuce seeds (Fig. 27; slow

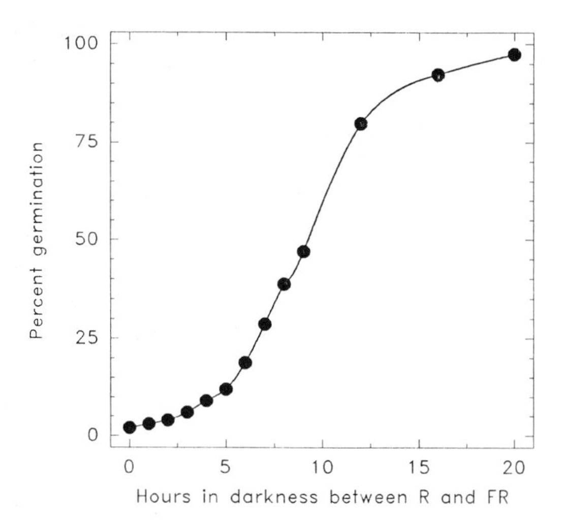

Figure 27. Effects of varying dark intervals between red (R) and far-red (FR) light on the photoreversibility of lettuce seed germination. Redrawn from data of H.A. Borthwick, S.B. Hendricks, E.H. Toole and V.K. Toole (1954) *Bot. Gaz.* 115: 205-225.

action); 8 h for the effects of EOD treatments on stem elongation (Fig. 21; slow action); 5 h for the effects of R on elongation of etiolated oat coleoptile segments (slow action); about 30 min for the effects of middle-of-night light breaks on flowering induction in *Xanthium* (relatively fast action); and 1.5 min (full loss in about 4 min, fast action) for the effects of middle-of-night light breaks on flowering induction in *Pharbitis* (Fredericq 1964). The developmental changes elicited by night breaks in *Pharbitis* do not become visible for several days, but Pfr action is completed in about 4 min, a time approximately equal to the lag period of some fast responses for which a detectable change can be measured 5 min after the exposure to R (*e.g.* leaflets closing at the end of day in *Mimosa*). Perhaps, some reports of photoreversibility failure in slow responses might be a consequence of a fast Pfr action: FR applied immediately after R cannot reverse the inductive effects of a R exposure if the latter is as long as the time required for Pfr action.

Short exposures and alternation of R and FR are only laboratory expedients; they have no counterpart in the light conditions of the natural environment. However, these laboratory expedients have added considerable informational background on the physiology of action of phytochrome.

4.7.4.2 The very low fluence responses

Studies carried out with seedlings grown in total darkness and seeds imbibed in total darkness and subject to appropriate treatments to reduce the level of endogenous Pfr have shown that a significant level of response can be elicited by R fluences between 10^{-4} and 10^{-1} μmol m^{-2} which establish Pfr/P between 10^{-6} and 10^{-3} (Fig. 10), significantly below those required for induction of the LFR. The VLFRs do not show R-FR reversibility (but see the '*Zea* paradox', below) and can be induced by very low fluences of any light, including FR and the '*green safelight*' used in photomorphogenesis research. Perhaps many VLFRs have been overlooked because of the use of green safelights. Fluence-response curves for the induction of seed germination in *Arabidopsis* (Fig. 28) and the effects of R pretreatments on greening in barley (Briggs *et al.* 1988) show a biphasic pattern, with a plateau between 0.1 to 1 μmol m^{-2}; about 35-40% of the total response is induced by 0.1 μmol m^{-2} of R (Pfr/P $\approx$ 0.001). Percent germinations elicited by R and FR fluences establishing Pfr/P between 10^{-6} and 10^{-2} fall on the same line when plotted *versus* Pfr/P (Fig. 29). Since FR can induce the VLFR, it is not surprising that the response induced by R cannot be reversed by a subsequent saturating FR which produces a higher Pfr/P than that established by a very low fluence R. One hypothesis suggested to explain the VLFR is based on the interaction between phytochrome dimers and a receptor *X*: calculated fluence requirements for the formation of PrPfr*X* and PfrPfr*X* match those required for the induction of the VLFR and LFR, respectively.

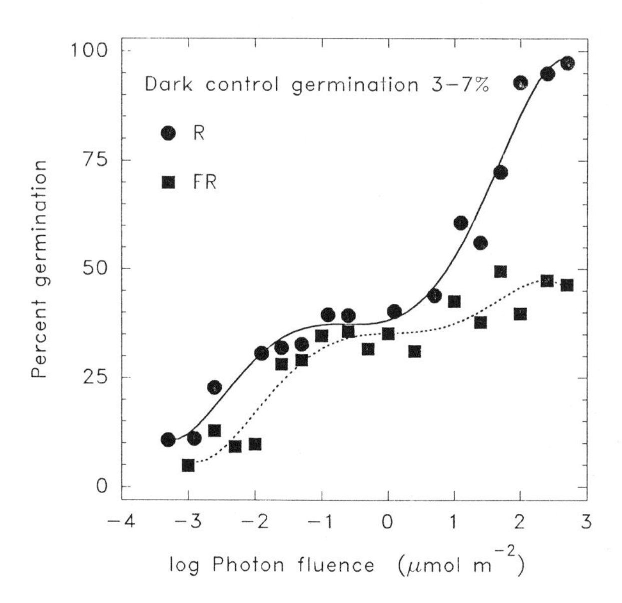

Figure 28. Fluence-response curves for induction of germination by red (R) and far-red (FR) light in *Arabidopsis thaliana* seeds. Experimental protocol: 8 days dark at 2°C → 1 day dark at 35°C → light treatment → 4 days dark at 20°C. The combination of low and high temperature before exposure to light was used to enhance the light sensitivity of germination and to reduce the level of endogenous Pfr. Redrawn from data of J.W. Cone, P.A.P.M. Jaspers and R.E. Kendrick (1985) *Plant Cell Environ.* 8: 605-612.

The '*Zea* paradox' has characteristics significantly different than those of the above VLFRs. The effect of a R pretreatment on the phototropic curvature induced by B is fully saturated at a fluence producing Pfr/P ≈ 0.001 and can be reversed by FR establishing Pfr/P about 20-fold higher than that established by R (Table 3). The hypothesis suggested to explain the '*Zea* paradox' was based

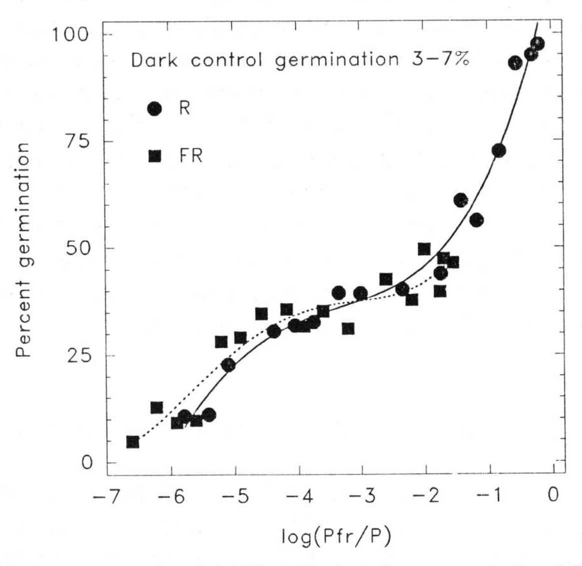

Figure 29. Germination percentages from Fig. 28 plotted versus calculated Pfr/P. (R, FR = red and far-red light, respectively) Note that about 30% germination is obtained at Pfr/P ≈ 0.0001 in this system and at Pfr/P ≈ 0.35 for the system of Fig. 26, a 3500-fold difference; however, the differences in Pfr/P for about 70% germination vary only by a factor of 2-2.5.

Table 3. Effects of red (R) and far-red (FR) light pretreatments on the phototropic curvature of *Zea.* Summary of data from H.P. Chon and W.R. Briggs (1966) *Plant Physiol.* 41: 1715-1724.

Pretreatment	Curvature (°)*	Pfr/P†
None (dark control)	3.4	0.0
R (0.18)‡	21.6	0.0009
R (0.18) → FR (≈ 1050)	3.6	0.0169
R (0.18) → FR (≈ 2800)	4.6	0.0199
FR (≈ 1750)	0.8	0.0191
FR (≈ 3500)	2.0	0.020
FR (≈ 21000)¶	13.3	0.020

*Induced by 436 nm light. †Calculated values at the end of the pretreatments, assuming 0 as the initial value of Pfr/P in darkness; calculation based on σ_{660} and σ_{730} values in Appendix Table 2 and indicated photon fluences. ‡Numbers in parenthesis are photon fluences, in μmol m^{-2}. ¶The fluence and exposure duration (60 min) required to elicit a significant effect with a FR-only pretreatment are close to the lower limit of fluence and exposure duration required to bring about the expression of some high irradiance responses.

on considerations of the interaction between Pfr, Pr and a receptor site *X*: first, the extent of the response is a function of [Pfr*X*]; second, [*X*] is small compared to [P] (*e.g.* [*X*]/[P] no larger than 0.001 in consideration of the fact that the effect of the R pretreatment is saturated at Pfr/P ≈ 0.001); third, phytochrome remains bound to *X* even if photoconverted back to Pr (*cf.* the slow rate of release of Pr from a bound state; Section 4.7.3.1). Very low fluences of R produce enough Pfr to saturate all the *X* sites. Exposure to saturating FR after R photoconverts Pfr*X* to physiologically inactive Pr*X* and, even though the level of Pfr is higher than that formed under very low fluence R, there is reversibility because there is no free *X* for binding with the free FR-produced Pfr. Exposure to FR alone, at fluences within the range of those required to establish φ_{FR} would have no effect on the response because Pfr*X* is being photoconverted to Pr*X* shortly after its formation. A unified hypothesis covering both the R-FR reversible and non-reversible VLFRs has not been developed so far; knowledge of the nature of *X* and its interaction with phytochrome is needed for further theoretical development and experimental verification.

4.7.4.3 The high irradiance responses (prolonged irradiation responses)

The HIRs [*called HER (High Energy Reaction) between 1960-70*] require prolonged exposures to light of relatively high irradiance for maximum expression (Mancinelli and Rabino 1978). The extent of the HIRs is a function of wavelength, irradiance and duration of the exposure.

Early action spectroscopy studies (1956-1959) of the effects of prolonged irradiations on seed germination, anthocyanin production, axis elongation, and flowering induction (night breaks) showed peaks of action in the B, R, and FR regions. The results of these studies were interpreted by Hendricks *et al.* (1959) as follows: *"The phenomena discussed respond both to the changes in form of the photomorphogenic pigment* (the name phytochrome started being used about two years later) *and to the continuous photoexcitation in those spectral regions 4000-5000 Å and 6000-8000 Å where both forms have appreciable absorptivities. It is difficult to disentangle completely the two effects, but the evidence for both actions is clear cut".* Since then, the experimental data base for the HIR has increased by a factor of at least 100 and several theoretical models (Section 4.7.4.3.6) have been proposed to explain phytochrome action in the HIR, but this class of responses is still poorly understood. Both dark- and light-grown seedlings respond to prolonged irradiations, but the photoresponse modes are different. The most common difference is the reduction or total loss of FR action in light-grown seedlings. Perhaps, as suggested by many, only the irradiance-dependent responses to prolonged exposures in etiolated seedlings should be considered true HIRs. In this chapter, the acronym HIR is used for irradiance-dependent responses to prolonged exposures in both etiolated and de-etiolated seedlings.

4.7.4.3.1 The spectral sensitivity of the HIR. Action maxima for responses to prolonged irradiations have been found in the UV-B, UV-A, B, R, and FR regions (Fig. 30; Chapter 9.5, Fig. 6). In recent times, it has become common to identify the spectral components of the HIR as FR-HIR, R-HIR, B-HIR and UV-HIR. The relative efficiencies of the different spectral regions vary significantly, depending on the combination of response, species, and experimental conditions. The HIR action spectra for inhibition of hypocotyl elongation in dark-grown lettuce seedlings [lettuce (a) and (b), Fig. 30] vary depending on age and exposure duration. The action spectra for dark-grown lettuce (a) and mustard (a) seedlings (Fig. 30) show significant action for both B and FR, but action in the R is high in mustard and almost nil in lettuce. The difference between mustard (a) and (b) (Fig. 30) provides a good example of the effects of de-etiolation on the spectral sensitivity of the HIR. Each of the several action

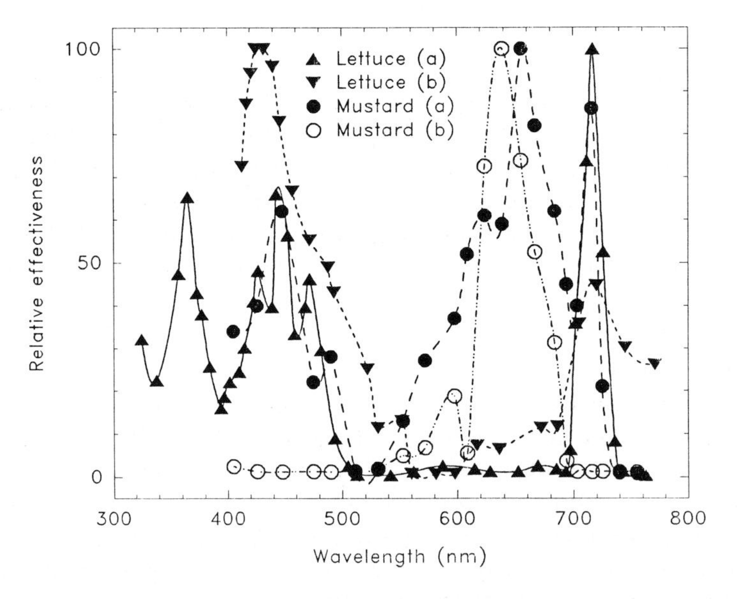

Figure 30. Wavelength dependence of the effects of prolonged irradiations on the inhibition of hypocotyl elongation. Experimental protocols: lettuce (a), sowing → 54 h dark (D) → 18 h monochromatic radiation (MCR); lettuce (b), sowing → 36 h D → 8 h MCR → 5 min red light → 16 h D (*Note:* a 50% reduction in the effectiveness of blue and an almost total loss of far-red action were observed when the same light treatment used for lettuce (b) was applied to 84-h-old dark-grown lettuce seedlings); mustard (a), sowing → 54 h dark → 24 h MCR; mustard (b), sowing → 54 h white light → 24 h MCR. Redrawn from original data: lettuce (a), K.M. Hartmann (1967) *Z. Naturforsch.* 22b: 1172-1175; lettuce (b), L.T. Evans, S.B. Hendricks and H.A. Borthwick (1965) *Planta* 64: 201-218; mustard (a) and (b), C.J. Beggs, M.G. Holmes, M. Jabben and E. Schäfer (1980) *Plant Physiol.* 66: 615-618.

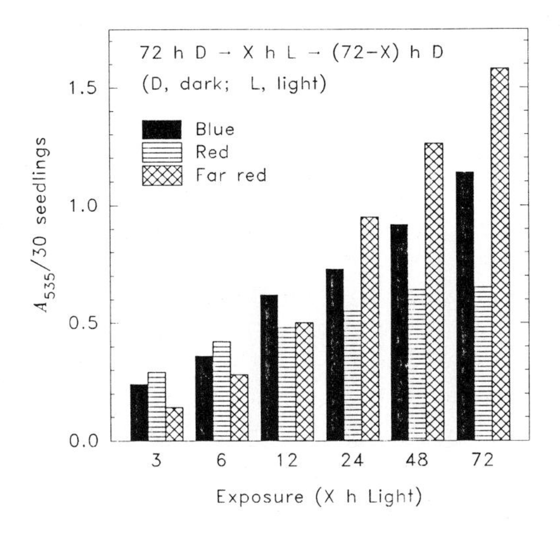

Figure 31. Effects of exposure duration on the extent and spectral sensitivity of HIR anthocyanin production in cabbage seedlings. (A_{535}, absorbance at 535 nm of extracts from 30 seedlings) Data of A.L. Mancinelli and L. Walsh (1979) *Plant Physiol.* 63: 841-846.

spectra for anthocyanin production (Chapter 9.5, Fig. 6) was measured under conditions different than those of the others; for example, the duration of the exposure varied from 4 h (followed by 5 min R and 24 h dark) for cabbage to 8 h (preceded by 2 h white light and followed by 5 min R and 24 h dark) for turnip to 24 h or longer for other species. A few examples of the effects of different experimental conditions on the spectral sensitivity of HIR anthocyanin production are shown in Figs. 31-33; these data suggest that the differences between the HIR action spectra of anthocyanin production in different species might be, in part, a reflection of differences in experimental conditions (Mancinelli 1985).

Action spectra are used to obtain information on the spectral characteristics of a photoreceptor as a first step toward its identification. The HIR action spectra were not very useful in this respect, partly because of their variability and partly for other reasons. For example, action in the B and R might have been a reflection of a requirement for photosynthesis (Section 4.7.4.3.5). Action in the B/UV region might be mediated by photomorphogenic photoreceptors specific for this region (Section 4.7.5; Chapters 5.1, 5.2 and 6), but phytochrome might also be involved (it absorbs and is photoconverted by B/UV light; Figs. 2-7). The effects of prolonged exposures in the 600-760 nm region might be mediated by phytochrome, but, in the early periods (1957-1965) of HIR research, many researchers were unsure of this: the HIR did not show R-FR reversibility and the action spectra in the 600-760 nm region did not match those of the R-FR reversible LFRs nor the absorption spectra of Pr and Pfr.

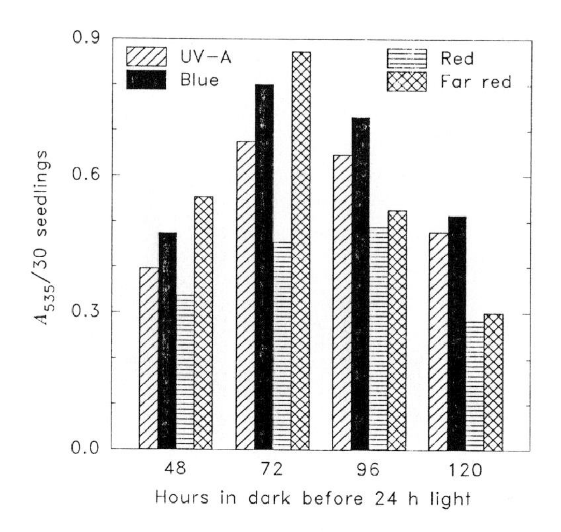

Figure 32. Effects of age on extent and spectral sensitivity of HIR anthocyanin production (A_{535}) in cabbage seedlings. Data of A.L. Mancinelli (1985) *Annali di Botanica (Roma)* 43: 21-36.

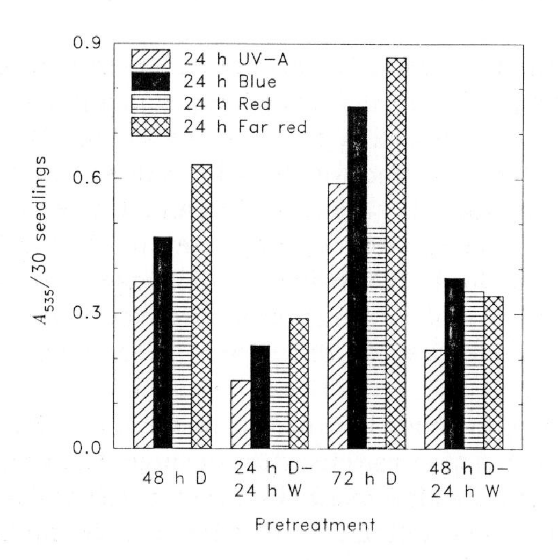

Figure 33. Effects of age and light pretreatments on extent and spectral sensitivity of HIR anthocyanin production (A_{535}) in cabbage seedlings. (D, dark; W, white light) Note that the effects of de-etiolation on anthocyanin production are about the same for the R-HIR ($\Delta\% \approx -50$) and FR-HIR ($\Delta\% \approx -53$) in 48-h-old seedlings, and significantly more pronounced for the FR-HIR ($\Delta\% \approx -60$) than the R-HIR ($\Delta\% \approx -27$) in 72-h-old seedlings. Data of A.L. Mancinelli (1984) *Plant Physiol.* 75: 447-453.

The results of experiments in which seedlings were exposed to dichromatic irradiation (Hartmann 1966) provided the first clear supporting evidence for the involvement of phytochrome in the mediation of the FR-HIR. The HIR action

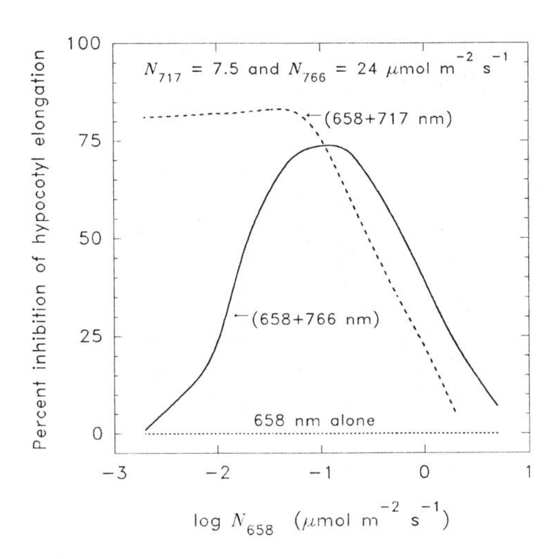

Figure 34. Effects of simultaneous irradiation with constant photon fluxes of far-red light (717 nm or 766 nm) and variable photon flux of red light (658 nm) on hypocotyl elongation in dark-grown lettuce seedlings. Redrawn from data of K.M. Hartmann (1966).

spectrum for the inhibition of hypocotyl elongation in lettuce (a) (Fig. 30) shows a peak of action at 717 nm and no action at 658 and 766 nm, two wavelengths effective in phytochrome photoconversion (Fig. 3). The effects of simultaneous exposures to 658 and 766 nm radiation on hypocotyl elongation varied depending on the N_{658}:N_{766} ratio (Fig. 34); an effect equal to that elicited by 717 nm light was obtained at an N_{658}:N_{766} ratio producing Pfr/P at photoequilibrium equal to that produced by 717 nm radiation. The effect of 717 nm radiation was nullified by the simultaneous application of 658 nm light of appropriate photon flux (Fig. 34). This was a clear demonstration of the involvement of phytochrome in the HIR using operational criteria based on the known photochemical properties of the photoreceptor. Hartmann (1966) suggested that 717 nm radiation and the appropriate mixture of 658 and 766 nm radiation provided the optimal Pfr/P ratio for action under prolonged irradiation (Section 4.7.6.4). Most doubts about the involvement of phytochrome in the mediation of the long-wavelength HIR disappeared shortly after the publication of these results.

4.7.4.3.2 Continuous and cyclic irradiations and reciprocity failure of the HIR. The full expression of HIR responses requires prolonged exposures, but not necessarily continuous ones. Downs and Siegelman (1963) reported that intermittent light treatments were highly effective in bringing about anthocyanin production, even though pigment formation was higher under continuous than cyclic treatments. These findings were confirmed and extended several years later when it was shown that intermittent and continuous irradiations could bring about the same level of anthocyanin production in young seedlings of cabbage and mustard when the experiments were carried out under precisely definite conditions: (i) continuous and cyclic irradiations applied over the same period of time; (ii) equal fluences applied with cyclic and continuous irradiations; (iii) duration of the dark interval between successive exposures in the cyclic treatments kept within certain limits, 2-5 min for FR, and no longer than 50 min for R. When all three conditions are observed, cyclic and continuous irradiations bring about the same level of anthocyanin production (Fig. 35). The decrease in anthocyanin production resulting from the increase in the duration of the dark interval between successive FR irradiations (Fig. 35) shows a correlation with the rate of Pfr decay in darkness. These observations and the conditions required to obtain equal effects with cyclic and continuous irradiations were later confirmed for another response in a different species (inhibition of hypocotyl elongation in mustard seedlings).

The reciprocity failure of the HIR was observed in early studies (Siegelman and Hendricks 1957) from a comparison of the effects of treatments of the following general type: (a) t h light at fluence rate N, and, (b) t/x h light at fluence rate xN followed by (t – t/x) h dark, for various values of x. Treatments (a) and (b) are equal in terms of light fluence applied and time allowed for the expression of the response; but the extent of the response is significantly larger for (a) than (b) (reciprocity failure). The two treatments are not equal in terms

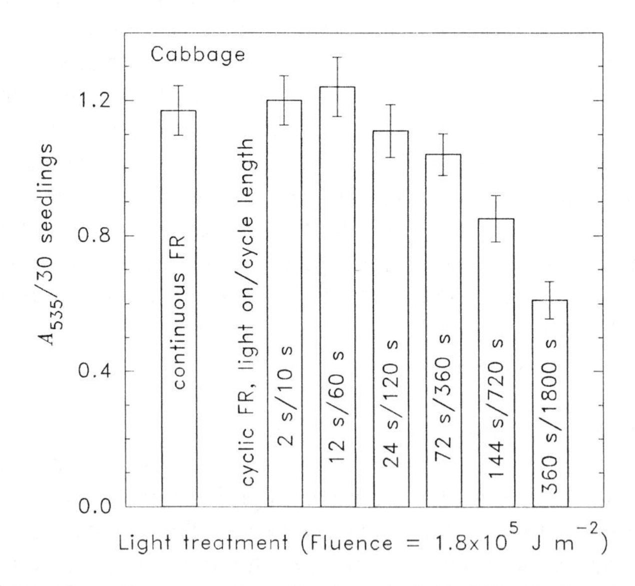

Figure 35. Effects of continuous and cyclic far-red light (FR) treatments on anthocyanin production (A_{535}) in cabbage seedlings. Experimental protocol: sowing → 96 h dark → 48 h continuous or cyclic FR; temperature, 20°C. Data of A.L. Mancinelli and I. Rabino (1975) *Plant Physiol.* 56: 351-355.

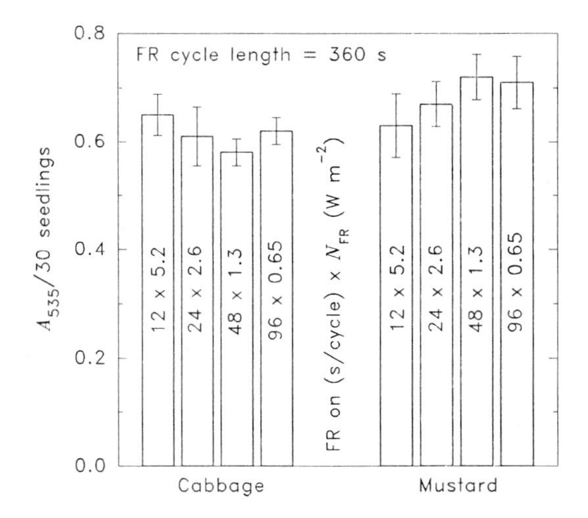

Figure 36. Effects of cyclic far-red (FR) light treatments of fixed period (360 s) and fixed fluence, but varying irradiances and exposure times on anthocyanin production (A_{535}) in cabbage and mustard seedlings. Experimental protocols: cabbage, sowing → 96 h dark → 48 h cyclic FR; mustard, sowing → 48 h dark → 48 h cyclic FR; temperature, 20°C. Data of A.L. Mancinelli and I. Rabino (1975) *Plant Physiol.* 56: 351-355.

of the state of phytochrome: there is some Pfr throughout the duration t of treatment (a) even though [Pfr] decreases in time as a consequence of Pfr destruction (Fig. 19); however, under treatment (b), due to Pfr destruction and dark reversion of Pfr to Pr, there is little or no Pfr left within a short time after the end of the exposure. Thus, the reciprocity failure can be plausibly explained as a consequence of differences in the state of phytochrome resulting from the contribution of the dark reactions. If this interpretation is correct, there should be no reciprocity failure of the HIR when the contribution of the dark reactions to the state of phytochrome is approximately the same, for example, under intermittent light treatments of fixed cycle length. The experimental results (Fig. 36) indicated that there was no reciprocity failure under intermittent light treatments, *QED*.

4.7.4.3.3 The irradiance dependence of the HIR. The irradiance dependence of the HIR is a common feature of plant responses to prolonged irradiations, but is more or less pronounced depending on the combination of response, species, light quality, and other experimental conditions. The effects of one of these factors, light quality, is shown in Fig. 37: the irradiance dependence varies depending on φ. Results of this type may be presented in a different way (Fig. 38) to provide a more direct visualization of the effects of light quality and irradiance in terms of parameters of the state of phytochrome (*e.g.* φ and *H*).

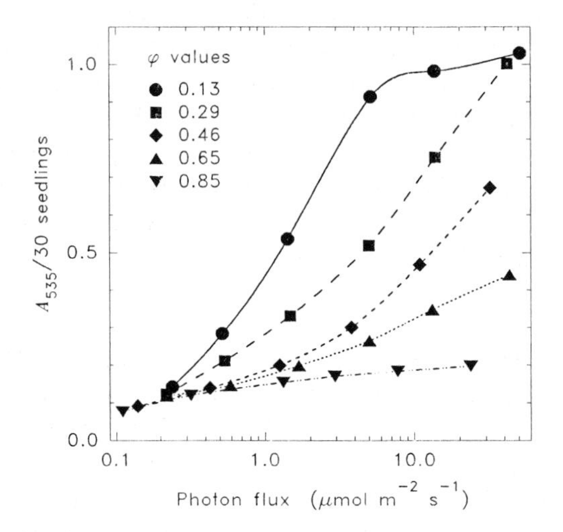

Figure 37. Effects of light quality and irradiance on anthocyanin production (A_{535}) in cabbage seedlings. Protocol: sowing → 96 h dark → 24 h light. Light quality was varied by mixing radiation from red and far-red light sources in the proportions required to obtain the indicated, measured values of φ *in vivo*. The cultivar of cabbage used in these experiments has a very low sensitivity to red for anthocyanin production. Data of A.L. Mancinelli (1990) *Plant Physiol.* 92: 1191-1195.

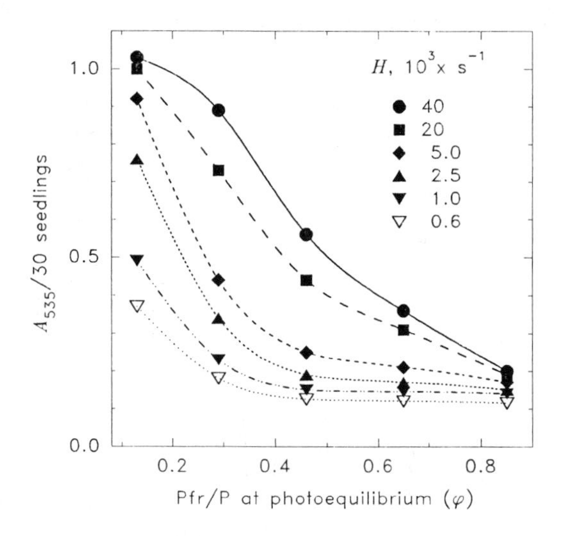

Figure 38. Effects of φ and *H* on anthocyanin production (A_{535}) in cabbage seedlings. Derived from the data in Fig. 36.

The effects of light quality and irradiance on anthocyanin production in etiolated cabbage seedlings are those typical for HIR responses in dark-grown seedlings. A comparison of these data with those for a response in de-etiolated seedlings provides another example of the differences between responses to prolonged irradiation in dark- and light-grown seedlings. The responses in etiolated (Fig. 38) and de-etiolated (Fig. 39) seedlings are almost the mirror image of each other when plotted against φ; for any given value of φ between 0.1 and 0.8, the extent of the response increases with increasing H (increasing irradiance) in both cases, but it decreases in etiolated seedlings and increases in de-etiolated seedlings with increasing φ.

4.7.4.3.4 HIR and LFR. Phytochrome-mediated HIR and LFR are not independent processes. An interaction between the two processes has been observed in several studies of the effects of prolonged irradiations on the extent of the R-FR reversible LFR. In several cases (*e.g.* anthocyanin production, cotyledon expansion, inhibition of hypocotyl elongation), the extent of the R-FR reversible LFR in etiolated seedlings is minimal and can be enhanced most effectively by prolonged light pretreatments. The irradiance and wavelength dependence of the effects of prolonged light pretreatments on the extent of the R-FR reversible LFR are within the range of those typical of HIR responses (Beggs *et al.* 1981).

4.7.4.3.5 Photosynthesis and the HIR. A contribution of photosynthesis was one of the arguments debated in early discussions of the HIR. Originally there

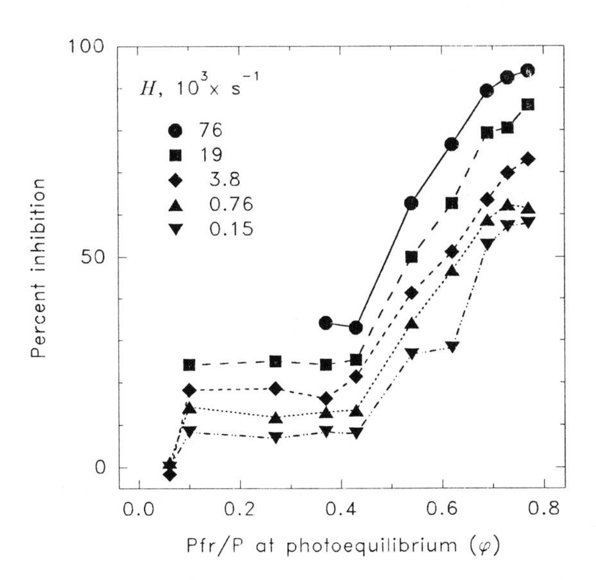

Figure 39. Effects of φ and H on hypocotyl elongation in de-etiolated cucumber seedlings. Protocol: sowing → 90 h dark → 30 h white light → 24 h experimental light at indicated values of φ and H. Redrawn from data of V. Gaba and M. Black (1985) *Plant Physiol.* 79: 1011-1014.

was evidence both for and against it, but the situation has since been clarified. For example, the products of photosynthesis are apparently required for HIR anthocyanin production in green systems (*e.g.* apple skin, cabbage and strawberry leaves) with a functional photosynthetic apparatus: DCMU, an inhibitor of photosynthesis inhibits anthocyanin production in green systems, and its effects can be counteracted by supplying sugars to the system. The situation is different in dark-grown seedlings which lack a functional photosynthetic apparatus at the time when the light treatments are started: DCMU has no apparent effect on HIR anthocyanin production in etiolated seedlings. Actually, a significant enhancement of HIR anthocyanin production in etiolated seedlings has often been observed in the presence of inhibitors that prevent the morphological and functional development of the photosynthetic apparatus. A review of the relationships between photosynthesis and HIR can be found in Mancinelli and Rabino (1978) and Mancinelli (1985).

4.7.4.3.6 Models for phytochrome action in the HIR. In the early years (1957-1965) of HIR research many doubted whether phytochrome was involved at all. Today, there is no doubt about the involvement of phytochrome in the HIR, but this group of photomorphogenic responses is still poorly understood. This situation is not the result of a lack of efforts, but is mainly a consequence of the complexity of this group of responses. Analysis of phytochrome action in the B/UV-HIR is complicated by the problem of the interaction between phytochrome and specific B/UV photoreceptors, the unknown nature of the B/UV photoreceptors and the unknown nature of the interaction (Section 4.7.5). But the analysis is difficult even when limited to the R/FR-HIR for which one can reasonably assume, at least for the present, that phytochrome is the only photomorphogenic photoreceptor involved. Another problem complicating the analysis is the difficulty of evaluating the significance of the variability in the irradiance and wavelength dependence of the HIRs; in many cases, one can only compare data for *AAA* (response *A* in system *A* under experimental conditions *A*) to *BBB*, *CCC*, *etc.*, without any of the data (*ABB, AAB, ABA, BAA, BBA, BAB, etc.*) necessary for a full comparison.

Three groups of theoretical *'models'* have been developed to provide an analysis of phytochrome action in the HIR. All the models are based on the properties of labile phytochrome and take into consideration the effects of the interaction between photoconversion, dark reversion, destruction, and reaction with an unknown *X* on the state of phytochrome. The basic hypothesis in the first group is that there are two forms of Pfr or Pfr*X*; one of them is a transient form decaying to a less transient one; the transient Pfr or Pfr*X* form, whose formation is irradiance- and wavelength-dependent, is the effector for the HIR. The basic hypothesis in the second group is that the irradiance and wavelength dependence of the formation of Pfr*X* is a function of the net rate of cycling between Pr and Pfr at photoequilibrium; the net rate of cycling is a function of φ, *k* and [P] (see Fukshansky and Schäfer 1983, for a review of these two model groups). The third model is based on the dimeric behaviour of phytochrome (VanDerVoude 1987). Each of the models is consistent with at least a few of the features of the HIR determined in different studies. For example, in a model of the first group, the calculated action spectrum for the amount of Pfr*X* formed under continuous irradiation matches closely the action spectra for inhibition of hypocotyl elongation in etiolated lettuce (a) seedlings (Fig. 30). In a model of the second group, the calculated action spectra for the cycling rate of phytochrome indicate a shift from the R to the FR with increasing duration of exposure (*cf.* changes in the effectiveness of R and FR for HIR anthocyanin production, Fig. 31). In the third model, the calculated action spectra for the formation of PfrPfr*X* (peak of action in the R) and PrPfr*X* (peak of action in the FR) match closely the peaks of action in the R and FR for inhibition of hypocotyl elongation in etiolated mustard (a) seedlings (Fig. 30). However, none of the models provide a satisfactory match for all the features of the HIR as determined for different combinations of response, species and experimental conditions. Some of the assumptions (*e.g.* interaction between phytochrome and *X*) in the models are still beyond the possibility of full experimental verification. Further development of the analysis of phytochrome action in the HIR must wait for additional experimental and theoretical work.

4.7.5 Interaction between phytochrome and other photomorphogenic photoreceptors

The properties of specific B and UV photoreceptors (UV-B photoreceptor; B/UV-A photoreceptor) involved in the mediation of photomorphogenic responses to 280-500 nm radiation and the interaction between them and phytochrome are described in Chapters 5.1, 5.2, and 6, and there is no need for repetition here. The only purposes of this section are, first, to remind the reader that studies of photoreceptor action and interaction in plant responses to B/UV are subject to significant limitations in both experimental techniques and interpretation of results (Gaba and Black 1987; Mancinelli 1989), and, second, to provide an example of methods that have been used to at least partially overcome the limitations. The limitations are due to the unknown nature of the B/UV photoreceptors and the fact that B/UV radiation excites not only the B/UV photoreceptors, but also phytochrome (Figs. 3-7). Thus, responses to B/UV may be mediated by phytochrome or B/UV photoreceptors or both, either interacting or independently of one another.

In the examples discussed below, the methods used to obtain evidence for an involvement of the B/UV-A photoreceptor (Figs. 40 and 41) and its interaction with phytochrome (Fig. 42) in the photoregulation of responses to B were based

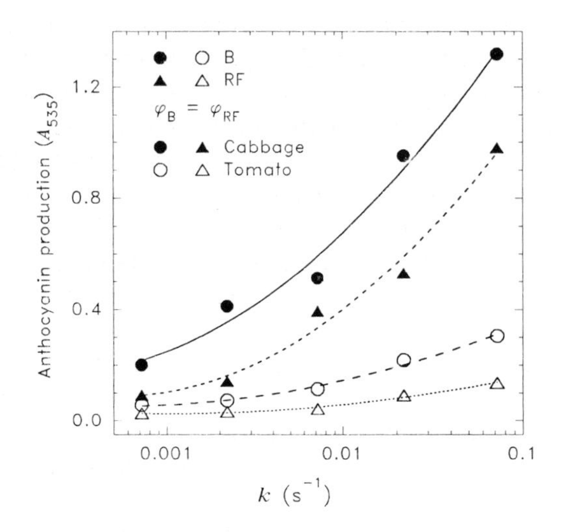

Figure 40. Effects of blue (B) and red + far-red (RF) light treatments eliciting equal state of phytochrome ($\varphi_B = \varphi_{RF}$ and $k_B = k_{RF}$) on anthocyanin production in cabbage and tomato seedlings. Protocols: cabbage, sowing → 72 h dark (D) → 2×(8 h B or RF → 16 h D); tomato, sowing → 96 h D → 2×(8 h B or RF → 16 h D). (A_{535} values for extracts from 30 seedlings for cabbage and 50 seedlings for tomato) Data of F. Sponga, G.F. Deitzer and A.L. Mancinelli (1986) *Plant Physiol.*, 82: 952-955.

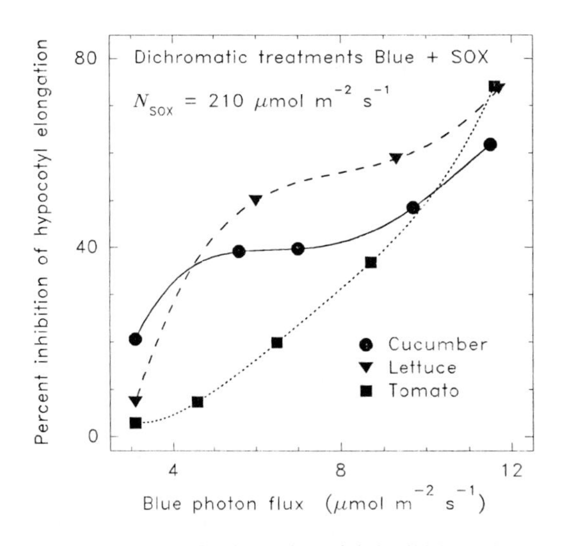

Figure 41. Fluence response curves for blue light (B) inhibition of hypocotyl elongation in de-etiolated seedlings exposed simultaneously to B and SOX radiation from low pressure sodium lamps. Under the experimental conditions used, the state of phytochrome elicited by the various combinations of background SOX and varying fluxes of B was not significantly different. Redrawn from data of B. Thomas and G.H. Dickinson (1979) *Planta* 146: 545-550.

on applications of the principle of equivalent light action. According to this principle, two light treatments, *x* and *y*, producing the same state of phytochrome (*e.g.* $\varphi_x = \varphi_y$ and $k_x = k_y$, and consequently, $H_x = H_y$) should be perceived by the responding system as being the same and elicit the same extent of the response if phytochrome is the only photoreceptor involved in the mediation of the response to light. The first example (Fig. 40) shows that anthocyanin production in cabbage and tomato is higher under B than RF (a mixture of R and FR) under conditions in which B and RF maintain the same state of phytochrome ($\varphi_B = \varphi_{RF}$ and $k_B = k_{RF}$). Since RF excites phytochrome, but not the B/UV-A photoreceptor (which has no significant absorption in the 600-760 nm region; Chapter 5.1) and B excites both photoreceptors, the differences in anthocyanin production between B and RF treatments that maintain the same state of phytochrome can be reasonably attributed to an involvement of the B/UV-A photoreceptor in the mediation of the response to B. The second example (Fig. 41; Thomas and Dickinson 1979) shows a marked B irradiance dependence of the inhibition of hypocotyl elongation when relatively low irradiances of B are applied simultaneously with high irradiance SOX light (95% of the SOX radiation is in a narrow band centred at 589 nm). The extent of the response should have been independent of N_B if phytochrome had been the only photoreceptor involved because, under the experimental conditions used, the addition of B to SOX had minimal effects on the state of phyto-

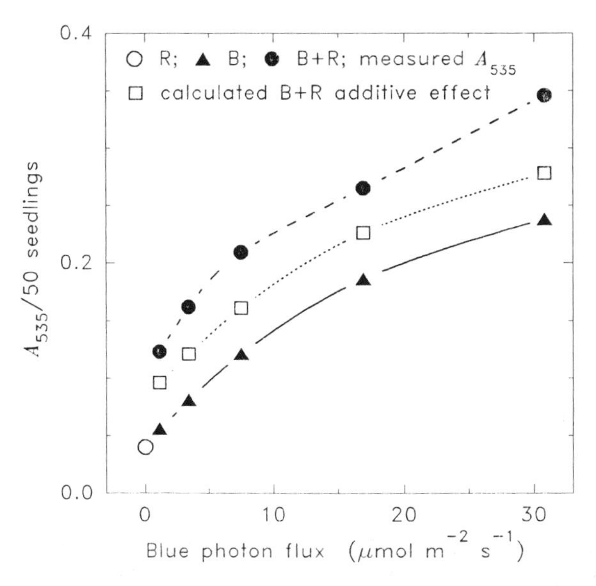

Figure 42. Effects of blue (B), red (R), and dichromatic B + R light exposures on anthocyanin production (A_{535}) in tomato seedlings. Protocol: sowing → 96 h dark → 24 h R or B or B + R; $\varphi_R \approx \varphi_{R+B} \approx 0.83$; $H_R \approx H_{R+B} \approx 0.048$ s^{-1}; $\varphi_B \approx 0.45$; H_B varying from ≈ 0.0001 to 0.0059 s^{-1} depending on photon flux. Data of A.L. Mancinelli, F. Rossi and A. Moroni (1991) *Plant Physiol.* 96: 1079-1085.

chrome. Thus, the B irradiance dependence of the response under dichromatic B + SOX treatments can be reasonably attributed to an involvement of the B/UV-A photoreceptor in the mediation of the response to B. The third example (Fig. 42) suggests both an involvement of the B/UV-A photoreceptor in the mediation of the response to B and an interaction between the B/UV-A photoreceptor and phytochrome: the extent of the response is a function of N_B under simultaneous exposure to B + R; *cf.* Figs. 41 and 42, and Table 4 for the effects of adding B to R on the state of phytochrome); the extent of the response to R and B + R should have been the same if phytochrome had been the only photoreceptor involved ($\varphi_B \approx \varphi_{B+R}$ and $H_B \approx H_{B+R}$); the extent of the response to B and B + R at equal N_B should have been the same if the B/UV-A photoreceptor had been the only photoreceptor involved; the synergistic effect (response to B + R larger than the calculated additive effects of B and R) suggests an interaction between the two photoreceptors.

Current hypotheses on photoreceptor interactions suggest that: (i) Pfr might be required for the expression of the B/UV-A photoreceptor-mediated effects of B; and, (ii) the B/UV-A photoreceptor might act by establishing/enhancing/maintaining the sensitivity of the responding system to Pfr. The two hypotheses are not mutually exclusive and both suggest that Pfr is the effector for the expression of the response. The nature of the interaction is still unknown. The interaction might occur at the level of the signal-transduction chain. A direct interaction between photoreceptors might also be considered: Sarkar and Song (1982) showed that, *in vitro*, the rate of phytochrome photoconversion under B is affected by the addition of flavin; flavins are considered likely candidates for the role of B-absorbing photoreceptors. Further development in studies of responses to B/UV and photoreceptor interaction should be significantly aided by the use of mutants and transgenic plants in photomorphogenesis research (Chapter 8.2).

4.7.6 Analysis of phytochrome action

4.7.6.1 Multiplicity of display of phytochrome action

Despite the multiplicity of display of the phytochrome-mediated action of light (at least six groups of photoresponse modes), the initial step is the same, the light-activated change in the state of the photoreceptor. The multiplicity of display of light-dependent developmental responses is probably a consequence of the process of integration in time and space (according to instructions encoded in the developmental program of the plant genome) of the information derived from changes in the state of phytochrome. The differential regulation of the expression of different phytochrome types (Chapter 4.2) and a difference in the role played by different phytochromes, depending on the response and the

Table 4. Effects of the addition of blue light (B, 450 nm) to red (R, 660 nm) or far-red (FR, 730 nm) light on the values of φ and k. Values of φ and k were calculated according to the phytochrome photoconversion cross-sections of Appendix Table 2.

Photon flux (μmol m^{-2} s^{-1})			φ	k
B	R	FR		($10^3 \times s^{-1}$)
10	0	0	0.42	3.45
0	10	0	0.87	57.1
1	10	0	0.87	57.4
10	10	0	0.84	60.5
0	0	10	0.020	17.4
0.1	0	10	0.021	17.4
1	0	10	0.028	17.7

The effects of adding B to R or FR on the state of phytochrome *in vivo* might be less pronounced than the above calculated ones because the B to R and B to FR effectiveness ratios might be lower for phytochrome photoconversion *in vivo* than *in vitro* (*cf.* Table 2).

state of the responding system, may contribute to the variability of the display. The multiplicity of display also invites consideration of alternative possibilities about the primary mechanism of Pfr action (first molecular change caused by Pfr in the responding system), either a single primary action leading to the formation of different secondary messages, or multiple primary actions, *e.g.* interaction of the same or different Pfr types with different receptor sites X. The debate on single primary action *versus* multiple primary actions is an old one and will probably not be resolved until after the identification of X(s).

4.7.6.2 Phytochrome, membranes and light signal-transduction pathways

The photoregulation of gene expression (Chapter 8.1) is the basic process underlying the phytochrome-mediated induction and/or modulation of developmental changes, but little is known about the pathways involved in the transduction of the light signal. The discovery of rapid, phytochrome-mediated responses (*e.g.* nyctinastic movements and changes in transmembrane potential and ion flux) led to the suggestion that membranes were the site of the primary action of phytochrome. The results of several studies suggest that a phytochrome-mediated modulation of Ca^{2+} flux may be a component of the signal transduction chain (Chapter 4.6). Photomorphogenic mutants and mutants in which the phytochrome-mediated perception of the light signal is uncoupled from the signal transduction pathway offer a great potential for the analysis of the mechanism of transduction of the light signal.

4.7.6.3 Labile and stable phytochrome and time course of phytochrome action

The time required for 50% loss of photoreversibility in two responses as diverse as the induction of seed germination (seeds kept in darkness before the brief exposure to R) and an EOD response in light-grown plants is very similar, 8-9 h (Figs. 21, 27). These slow rates of escape from photoreversibility suggest that the Pfr involved in the photoregulation of these responses must be a stable Pfr; it cannot be a labile Pfr that disappears almost completely in 2-3 h (Fig. 18). The involvement of an *s*P in EOD responses is supported by the observation that overexpression of biologically active *ℓ*P in transgenic seedlings has little effect on the extent of this response, even though it may significantly affect other responses. A rate of escape from photoreversibility significantly slower (*e.g.* 5 h for 50% loss of photoreversibility for light action on elongation of etiolated oat coleoptiles) than the rate of loss of *ℓ*Pfr suggests that *s*Pfr may play a significant role in the photoregulation of responses even in etiolated systems in which *ℓ*P might be the predominant phytochrome. This suggestion is consistent with recent reports on the effects of deficiency and overexpression of *s*P on the light-dependent inhibition of hypocotyl elongation in *Arabidopsis* (Chapters 4.8 and 8.2). A slow rate of escape from R-FR reversibility can be considered a reliable indicator of the involvement of *s*Pfr in the photoregulation of a response. However, a fast rate of escape from R-FR reversibility cannot be taken as an indicator of an involvement of *ℓ*Pfr or a lack of involvement of *s*Pfr in the photoregulation of a response because other factors may be responsible for it. For example, a fast rate of escape from R-FR reversibility may be the result of a situation under which neither [Pfr] nor the concentration of one or more *X* in the reaction chain(s) leading to the irreversible potentiation of the response are limiting factors for the rate of Pfr action.

4.7.6.4 The HIRs: R-HIR and FR-HIR

One of the puzzles faced by researchers during the early years of HIR research was the high effectiveness of FR despite the low values of Pfr/P established at photoequilibrium. Hartmann (1966) was the first to suggest that Pfr destruction played an essential role in the HIR: in etiolated seedlings exposed to continuous irradiation, FR maintained the best balance between the requirement for a Pfr level sufficiently high to maintain action over a prolonged period of time and the opposite requirement for a Pfr level sufficiently low to keep destruction at a minimum. This old hypothesis is still current today and is consistent with data from recent research. The extent of the FR-HIR is high in etiolated seedlings (Figs. 30, 33) in which [*ℓ*P] is high, and is reduced or absent in de-etiolated ones (Figs. 30, 33) in which [*ℓ*P] is low; is reduced or absent in etiolated seedlings of mutants in which [*ℓ*PI] is low, and is retained in light-grown transgenic seedlings overexpressing biologically active heterologous *ℓ*P (Chapter 8.2). At least for the present, it seem reasonable to assume that *ℓ*P may be the main phytochrome type involved in the photoregulation of the FR-HIR; the FR-HIR is extremely difficult to understand on any other basis than the involvement of a light-labile Pfr whose rate of destruction is a function of φ (Fig. 19). The effects of day-extension on flowering induction of LDPs (Johnson *et al.* 1991) show a peak of action in the FR (between 710 and 720 nm) and irradiance dependence within the range of the FR-HIR in etiolated seedlings, suggesting that the activity of *ℓ*P might not be restricted to etiolated seedlings. What is the situation for the R-HIR? Certain aspects of the HIR, for example, an effectiveness of R significantly lower than that of FR (lettuce (a) and (b), Fig. 30) and the decrease of the R to FR effectiveness ratio with increasing duration of exposure (Fig. 31) suggest that *ℓ*P may be involved in the R-HIR: [*ℓ*Pfr] would be significantly higher under FR than R during a considerable fraction of a prolonged exposure (Fig. 19). However, a contribution of an *s*P to the photoregulation of the R-HIR would seem highly likely in those cases in which the extent of the response to continuous irradiations longer than 12 h is about the same under R and FR [mustard (a), Fig. 30], and the R-HIR is reduced significantly less than the FR-HIR by de-etiolation [*cf.* mustard (a) and (b), Fig. 30]. At the same time, it should not be forgotten that, according to the model for phytochrome action in the HIR based on the dimeric behaviour of labile phytochrome (VanDerWoude 1987; Section 4.7.4.3.6), there would be no need for an involvement of stable phytochrome in the R-HIR in order to obtain the same extent of R-HIR and FR-HIR responses. It might not be unreasonable to consider the possibility that the contribution of *ℓ*P and *s*P to the HIR might vary depending on the species, duration of the exposure, seedling age and other factors, and this variability might be one the factors responsible for the observed variability in the irradiance and wavelength dependence of the HIR. A full comparison of the effects of prolonged R and FR on HIR responses in dark- and light-grown wild type, photomorphogenic mutants (deficient in *ℓ*P and/or *s*P) and transgenic (overexpressing biologically active *ℓ*P and/or *s*P) plants is necessary for the experimental verification of these hypotheses.

4.7.6.5 Phytochrome synthesis and destruction

The negative autoregulation of the synthesis of *ℓ*P (Chapter 4.2) in and its rapid destruction (Chapter 4.4) in light might represent mechanisms used by the plant to adapt the level of phytochrome to the prevailing light conditions of the environment. The relatively high rate of synthesis of *ℓ*P in dark-growing seedlings and during the dark period of daily light-dark cycles (Section 4.7.3.2) might represent dark-adaptation mechanisms, required to increase the sensitivity of perception of the first dark-to-light transition in etiolated seedlings and for photoperiodic monitoring in light-grown plants. The perception of the first dark-to-light transition is probably the most important aspect of phytochrome action in etiolated seedlings: a high level of phytochrome might be required to carry out this function optimally. But, after de-etiolation, a high level of phytochrome might work against normal plant development in light. In this respect, it is interesting to note that overexpression of ℓP seems to confer an *'etiolated-like'* character to light-grown transgenic seedlings (*e.g.* persistence of the FR-HIR). The destruction of ℓP and its decreased rate of synthesis in light might represent light-adaptation mechanisms required to reduce the interference of etiolated-like responses in light-grown plants. Again, a full comparison of the effects of light on several responses in etiolated and de-etiolated wild type, photomorphogenic mutants and transgenic plants is necessary for the experimental verification of these suggestions (Chapters 4.8, 7.1 and 8.2).

4.7.7 The future

From the discovery of R-FR reversibility to the determination and analysis of the basic properties of different groups of phytochrome-mediated responses, *'classical'* physiological studies of a black box process *(light → phytochrome excitation → ■ → response expression; only the first and last step measurable with a sufficient degree of precision)* established a firm foundation for further development. Research on a black box process is subject to considerable limitations and several ingenious photophysiological techniques were developed to at least partially overcome the limitations. In recent times, molecular biology and genetics have contributed enormously to the progress in photomorphogenesis research: identification of different phytochrome types and their genes, gene cloning and sequencing, determination of amino-acid composition and sequencing, identification of conserved regions in different phytochrome genes, negative autoregulation of phytochrome synthesis, phenotypic variations in photomorphogenic mutants and transgenic plants. The box is beginning to turn from black to grey, and molecular biology and genetics will play a predominant role in future photomorphogenesis research. The student reading this volume might wonder whether there is a future for physiological studies in photomorphogenesis research. The full realization of the potential offered by molecular biology and genetics cannot be realized without a full and comparative analysis of phytochrome-mediated responses in the systems (*e.g.* mutated and transgenic plants) made available by current developments. This analysis will require physiological tests designed to address specific questions, and the tests will require the proper application of appropriate photophysiological techniques. Thus, physiology will continue to play a fundamental role, and a close co-operation between physiologists, molecular biologists, and geneticists is essential for further progress in photomorphogenesis research. The student reading this volume should not forget that the progress made in Beltsville between 1952 (discovery of R-FR reversibility) and 1964 (extraction and purification of phytochrome and analysis of its photochemical properties) would have not been possible without the close co-operation between the physiologists, biochemists, biophysicists and engineers that came to be known collectively as the Beltsville group.

4.7.8 Further reading

Borthwick H.A. (1972) History of phytochrome - Biological significance of phytochrome. In: *Phytochrome*, pp. 3-44, Mitrakos K. and Shropshire W.Jr. (eds.) Academic Press, London.

Hendricks S.B. and VanDerWoude W.J. (1983) How phytochrome acts: perspectives on the continuing quest. In: *Encyclopedia of Plant Physiology,* New Series, Vol 16A, *Photomorphogenesis,* pp. 3-23, Shropshire W.Jr. and Mohr H. (eds.) Springer-Verlag, Berlin.

Hillman W.S. (1967) The physiology of phytochrome. *Annu. Rev. Plant Physiol.* 18: 301-324.

Mancinelli A.L. (1985) Light-dependent anthocyanin synthesis. *Bot. Rev.* 51: 107-157.

Toole E.H., Hendricks S.B., Borthwick H.A. and Toole V.K. (1956) Physiology of seed germination. *Annu. Rev. Plant Physiol.* 7: 299-324.

4.7.9 References

Beggs C.J., Geile W., Holmes M.G., Jabben M., Jose A.M. and Schäfer E. (1981) High irradiance response promotion of a subsequent light induction response in *Sinapis alba* L. *Planta* 151: 135-140.

Black M. and Shuttleworth J.E. (1974) The role of the cotyledons in the photocontrol of hypocotyl extension in *Cucumis sativus* L. *Planta* 117: 57-66.

Boisard J., Marmé D. and Schäfer E. (1971) The demonstration *in vivo* of more than one form of Pfr. *Planta* 99: 302-310.

Boisard J., Marmé D. and Briggs W.R. (1974) *In vitro* properties of membrane-bound phytochrome. *Plant Physiol.* 54: 272-276.

Borthwick H.A., Hendricks S.B., Parker M.W., Toole E.H. and Toole V.K (1952) A reversible photoreaction controlling seed germination. *Proc. Natl. Acad. Sci. USA* 38:662-666.

Briggs W.R., Mösinger E. and Schäfer E. (1988) Phytochrome regulation of greening in barley: effects on chlorophyll accumulation. *Plant Physiol.* 86: 435-440.

Brockmann J., Rieble S., Kazarinova-Fukshansky N., Seyfried M. and Schäfer E. (1987) Phytochrome behaves as a dimer *in vivo. Plant Cell Environ.* 10: 105-111.

Butler W.L., Lane H.C. and Siegelman H.W. (1963) Nonphotochemical transformations of phytochrome *in vivo. Plant Physiol.* 38: 514-519.

Butler W.L., Norris K.H., Siegelman H.W. and Hendricks S.B. (1959) Detection, assay, and preliminary purification of the pigment controlling photoresponsive development of plants. *Proc. Natl. Acad. Sci. USA* 45: 1703-1708.

Butler W.L., Hendricks S.B. and Siegelman H.W. (1964) Action spectra of phytochrome *in vitro. Photochem. Photobiol.* 3: 521-528.

Dooskin R.H. and Mancinelli A.L. (1968) Phytochrome decay and coleoptile elongation in *Avena* following various light treatments. *Bull. Torrey Bot. Club* 95: 474-487.

Downs R.J. and Siegelman H.W. (1963) Photocontrol of anthocyanin synthesis in Milo seedlings. *Plant Physiol.* 38: 25-30.

Fredericq H. (1964) Conditions determining effects of far-red and red irradiation on flowering response of *Pharbitis nil. Plant Physiol.* 39: 812-816.

Fukshansky L. and Schäfer E. (1983) Models in photomorphogenesis. In: *Encyclopedia of Plant Physiology,* New Series, Vol. 16A, *Photomorphogenesis,* pp. 69-95, Shropshire W.Jr. and Mohr H. (eds.) Springer-Verlag, Berlin.

Gaba V. and Black M. (1987) Photoreceptor interaction in plant photomorphogenesis: the limits of experimental techniques and their interpretation. *Photochem. Photobiol.* 45: 151-156.

Hartmann K.M. (1966) A general hypothesis to interpret 'high energy phenomena' of photomorphogenesis on the basis of phytochrome. *Photochem. Photobiol.* 5: 349-366.

Hendricks S.B., Toole E.H., Toole V.K. and Borthwick H.A. (1959) Photocontrol of plant development by the simultaneous excitations of two interconvertible pigments. III. Control of seed germination and axis elongation. *Bot. Gaz.* 121: 1-8.

Hillman W.S. (1964) Phytochrome levels detectable by *in vivo* spectrophotometry in plant parts grown or stored in the light. *Amer. J. Bot.* 51: 1102-1107.

Hillman W.S. (1965) Phytochrome photoconversion by brief illumination and the subsequent elongation of etiolated *Pisum* stem segments. *Physiol. Plant.* 18: 346-358.

Hillman W.S. (1972) On the physiological significance of *in vivo* phytochrome assay. In: *Phytochrome*, pp. 573-584, Mitrakos K. and Shropshire W.Jr. (eds.) Academic Press, London.

Holmes M.G. (1984) Radiation measurements. In: *Techniques in Photomorphogenesis*, pp. 81-107, Smith H. and Holmes M.G. (eds.) Academic Press, London.

Hilton J.R. and Thomas B. (1987) Photoregulation of phytochrome synthesis in germinating embryos of *Avena sativa* L. *J. Exp. Bot.* 38: 1704-1712.

Jabben M. (1980) The phytochrome system in light-grown *Zea Mays* L. *Planta* 149: 91-96.

Jabben M. and Deitzer G.F. (1978) A method for measuring phytochrome in plants grown in white light. *Photochem Photobiol.* 27: 799-802.

Jabben M., Heim B. and Schäfer E. (1980) The phytochrome system in light- and dark-grown dicotyledonous seedlings. In: *Photoreceptors and Plant Development*, pp. 145-158, DeGreef J. (ed.) Antwerp Univ. Press, Antwerp.

Jabben M., Beggs C. and Schäfer (1982) Dependence of Pfr/Ptot ratios on light quality and light quantity. *Photochem. Photobiol.* 35: 709-712.

Johnson C.B., Allsebrook S.M., Carr-Smith H. and Thomas B. (1991) A quantitative approach to the molecular biology of phytochrome action. In: *Phytochrome Properties and Biological action*, pp. 273-288, Thomas B. and Johnson C.B. (eds.) Springer-Verlag, Berlin.

Kazarinova-Fukshansky N., Seyfried M. and Schäfer E. (1985) Distortion of action spectra in photomorphogenesis by light gradient within the plant tissue. *Photochem. Photobiol.* 41: 689-702.

Kelly J.M. and Lagarias J.C. (1985) Photochemistry of 124-kilodalton *Avena* phytochrome under constant illumination *in vitro. Biochemistry* 24: 6003-6010.

Kendrick R.E. and Frankland B. (1969) The *in vivo* properties of *Amaranthus* phytochrome. *Planta* 86: 21-32.

Kendrick R.E. and Spruit C.J.P. (1972) Phytochrome decay in seedlings under continuous incandescent illumination. *Planta* 107: 341-350.

Kendrick R.E. and Spruit C.J.P. (1974) Inverse dark reversion of phytochrome: an explanation. *Planta* 120: 265-272.

King R.V., Schäfer E., Thomas B. and Vince-Prue D. (1982) Photoperiodism and rhythmic responses to light. *Plant Cell Environ.* 5: 395-404.

Konomi K., Abe H. and Furuya M. (1987) Changes in content of phytochrome I and II apoprotein in embryonic axes of pea seeds during imbibition. *Plant Cell Physiol.* 28: 1443-1451.

Lagarias J.C., Kelly J.M., Cyr K.L. and Smith W.O. (1987) Comparative photochemical analysis of highly purified 124-kilodalton oat and rye phytochromes *in vitro. Photochem. Photobiol.* 46: 5-13.

Mancinelli A.L. (1989) Interaction between cryptochrome and phytochrome in higher plant photomorphogenesis. *Amer. J. Bot.* 76: 143-154.

Mancinelli A.L. and Rabino I. (1978) The high irradiance responses of plant photomorphogenesis. *Bot. Rev.* 44: 129-180.

Mancinelli A.L., Rossi F. and Moroni A. (1992) Phytochrome photoconversion *in vivo*: effect of the initial Pfr/Ptot ratio. *Photochem. Photobiol.* 56: 593-598.

Pratt L.H. (1983) Assay of photomorphogenic photoreceptors. In: *Encyclopedia of Plant Physiology,* New Series, Vol. 16A, *Photomorphogenesis,* pp. 154-177, Shropshire W.Jr. and Mohr H. (eds.) Springer-Verlag, Berlin.

Pratt L.H. and Butler W.L. (1970) Phytochrome conversion by ultraviolet light. *Photochem. Photobiol.* 11: 503-509.

Quail P.H., Schäfer E. and Marmé D. (1973) *De novo* synthesis of phytochrome in pumpkin hooks. *Plant Physiol.* 52: 124-127.

Sarkar H.K. and Song P.-S. (1982) Blue light induced phototransformation of phytochrome in the presence of flavin. *Photochem. Photobiol.* 35: 243-246.

Schäfer E. (1978) Variation in the rates of synthesis and degradation of phytochrome in cotyledons of *Cucurbita pepo* L. during seedling development. *Photochem. Photobiol.* 27: 775-780.

Schäfer E. and Mohr H. (1980) Changes in the rate of photoconversion of phytochrome during etiolation in mustard seedlings. *Photochem. Photobiol.* 31: 495-500.

Schäfer E., Marchal B. and Marmé D. (1971) On the phytochrome phototransformation kinetics in mustard seedlings. *Planta* 101: 265-276.

Schäfer E., Marchal B. and Marmé D. (1972) *In vivo* measurements of the phytochrome photostationary state in far red light. *Photochem. Photobiol.* 15: 457-464.

Seyfried M. and Schäfer E. (1985a) Phytochrome macro-distribution, local photoconversion and internal photon fluence rate for *Cucurbita pepo* L. cotyledons. *Photochem. Photobiol.* 42: 309-318.

Seyfried M. and Schäfer E. (1985b) Action spectra of phytochrome *in vivo. Photochem. Photobiol.* 42: 319-326.

Siegelman H.W. and Firer E.M. (1964) Purification of phytochrome from oat seedlings. *Biochemistry* 3: 418-423.

Siegelman H.W. and Hendricks S.B. (1957) Photocontrol of anthocyanin formation in turnip and red cabbage seedlings. *Plant Physiol.* 32: 393-398.

Spruit C.J.P. and Kendrick R.E. (1972) On the kinetics of phytochrome photoconversion *in vivo. Planta* 103: 319-326.

Spruit C.J.P. and Mancinelli A.L. (1969) Phytochrome in cucumber seeds. *Planta* 83: 303-310.

Stone H.J. and Pratt L.H. (1979) Characterization of the destruction of phytochrome in the red-absorbing form. *Plant Physiol.* 63: 680-682.

Thomas B. and Dickinson H.G. (1979) Evidence for two photoreceptors controlling growth in de-etiolated seedlings. *Planta* 146: 545-550.

Tokuhisa J.G. and Quail P.H. (1987) The levels of two distinct species of phytochrome are regulated differently during germination in *Avena sativa* L. *Planta* 172: 371-377.

VanDerWoude W.J. (1985) A dimeric mechanism for the action of phytochrome: evidence from photothermal interactions in lettuce seed germination. *Photochem. Photobiol.* 42: 655-661.

VanDerWoude W.J. (1987) Application of the dimeric model of phytochrome action to high irradiance responses. In: *Phytochrome and Photoregulation in Plants*, pp.249-258, Furuya M. (ed.) Academic Press, Tokyo.

Vierstra R.D. and Quail P.H. (1983) Photochemistry of 124-kilodalton *Avena* phytochrome *in vitro. Plant Physiol.* 72: 264-267.

Vince-Prue D. (1975) *Photoperiodism in Plants,* McGraw-Hill, London.

4.7.10 Appendix

Appendix Table 1. Symbols and definitions of phytochrome parameters.

Symbol	Definition
Pr	Form of phytochrome with major absorbance peak in the red (R) region, between 650 and 670 nm; commonly called the R-absorbing form.
Pfr	Form of phytochrome with major absorbance peak in the far-red (FR) region, between 725 and 735 nm; commonly called the FR-absorbing form.
P	Total phytochrome, Pr + Pfr.
Pfr/P	Fraction of total phytochrome present as Pfr.
$\varepsilon_{R\lambda}$	Extinction coefficient* of Pr at wavelength λ.
$\varepsilon_{FR\lambda}$	Extinction coefficient* of Pfr at wavelength λ.
$\Delta\Delta A = (A_{\lambda x}-A_{\lambda y})^{FR} - (A_{\lambda x}-A_{\lambda y})^{R}$	Difference of the differences in phytochrome absorbance between λx and λy after saturating exposures to FR and R; λx = 730 nm and λy either 665 or 800 nm.
Φ_R	Quantum yield for Pr → Pfr photoconversion.
Φ_{FR}	Quantum yield for Pfr → Pr photoconversion.
$\sigma_{R\lambda} = 2.3\varepsilon_{R\lambda}\Phi_R$	Photoconversion cross-section† of Pr at wavelength λ.
$\sigma_{FR\lambda} = 2.3\varepsilon_{FR\lambda}\Phi_{FR}$	Photoconversion cross-section† of Pfr at wavelength λ.
$\sigma_\lambda = \sigma_{R\lambda} + \sigma_{FR\lambda}$	Photoconversion cross-section† of phytochrome at wavelength λ.
N_λ	Photon flux (= photon fluence rate, mol of photons $m^{-2}\,s^{-1}$) at wavelength λ. (Fluence = $N_\lambda \times t$; t = duration of irradiation).
$k_{1\lambda} = N_\lambda\sigma_{R\lambda}$	Rate constant of Pr → Pfr photoconversion at wavelength λ.
$k_{2\lambda} = N_\lambda\sigma_{FR\lambda}$	Rate constant of Pfr → Pr photoconversion at wavelength λ.
$k_\lambda = k_{1\lambda} + k_{2\lambda}$	Rate constant of phytochrome photoconversion at wavelength λ.
φ_λ	Value of the Pfr/P ratio at photoequilibrium at wavelength λ: $\varphi_\lambda = k_{1\lambda}/k_\lambda = \sigma_{R\lambda}/\sigma_\lambda = \varepsilon_{R\lambda}\Phi_R/(\varepsilon_{R\lambda}\Phi_R+\varepsilon_{FR\lambda}\Phi_{FR})$.
$t½ = (ln2)/k_\lambda$	Exposure time for 50% photoconversion; $Pfr/P_{(t½)} = 0.5(\varphi_\lambda + Pfr/P_{(0)})$; $Pfr/P_{(0)}$, initial value of Pfr/P.

(Appendix Table 1, continued)

$Pfr/P_{(t)} = \varphi + (Pfr/P_{(0)} - \varphi)e^{-kt}$	Equation‡ for phytochrome photoconversion; t, exposure time; wavelength subscript (λ) omitted.
$H = (1-\varphi)k_1 = \varphi k_2 = (\varphi-\varphi^2)k$	Light-dependent rate of cycling between Pr and Pfr at photoequilibrium; wavelength subscript (λ) omitted.
k_r	Rate constant of non-photochemical, temperature-dependent Pfr → Pr dark reversion.
k_d	Rate constant of temperature-dependent phytochrome destruction.

*The unit most commonly used for ε is liter (L = dm^3) mol^{-1} cm^{-1}. Values of ε for Pr and Pfr are often given in units of area mol^{-1} (*e.g.* cm^2 mol^{-1});

$$1.0\ L\ mol^{-1}\ cm^{-1} = (1.0\ L\ mol^{-1}\ cm^{-1} \times 1000\ cm^3\ L^{-1}) = 1000\ cm^2\ mol^{-1}.$$

The value of ε used to calculate the photoconversion cross-section (σ) is that expressed in units of area mol^{-1}.

†The photoconversion cross-section is defined as $\sigma = \varsigma\Phi$, where ς is the molar absorption cross-section (units: area mol^{-1}): $\varsigma = \varepsilon \times ln10$ ($ln10 \approx 2.3$). Note that, occasionally, the product $\varepsilon\Phi$ (photoconversion coefficient) is called photoconversion cross-section and the symbol σ is used for it; this may cause some confusion. The photoconversion cross-section is:

$$\sigma = \varsigma\Phi = 2.3\varepsilon\Phi \quad \text{(units: area mol}^{-1}\text{)}$$

‡The equation for phytochrome photoconversion is often shown in its logarithmic form, thus,

$$ln \frac{Pfr/P_{(t)} - \varphi}{Pfr/P_{(0)} - \varphi} = -kt$$

Appendix Table 2. Photoconversion cross-sections of Pr (σ_R), Pfr (σ_{FR}) and P ($\sigma = \sigma_R + \sigma_{FR}$) and Pfr/P ratios at photoequilibrium (φ) of type-I phytochrome. Average values calculated from data of Kelly and Lagarias (1985) and Lagarias *et al.* (1987) for phytochrome extracted and purified from dark-grown oat and rye seedlings.

Wavelength (nm)	σ_R (m^2 mol^{-1})	σ_{FR} (m^2 mol^{-1})	σ (m^2 mol^{-1})	φ
300	1404	728.3	2132	0.66
310	906.1	572.7	1469	0.61
320	677.7	481.0	1159	0.58
330	609.0	366.9	975.9	0.62
340	723.7	314.1	1038	0.70
350	981.2	328.0	1309	0.75
360	1301	376.9	1678	0.78
370	1423	450.9	1874	0.76
380	1492	526.0	2081	0.74
390	1310	599.6	1910	0.69
400	845.5	650.6	1496	0.57
410	453.9	658.3	1112	0.41
420	312.1	559.7	871.8	0.36
430	241.6	409.0	650.6	0.37
440	188.7	280.7	469.4	0.40
450	146.5	198.6	345.1	0.42
460	114.7	146.1	260.8	0.44
470	94.08	113.4	207.5	0.45
480	76.80	91.95	168.8	0.45
490	66,69	79.24	145.9	0.46
500	67.12	71.52	138.6	0.48
510	79.26	67.39	146.7	0.54
520	101.0	64.71	165.7	0.61
530	136.4	63.93	200.3	0.68
540	195.1	65.82	260.9	0.75
550	287.5	71.76	359.3	0.80
560	398.8	81.73	480.5	0.83
570	541.5	96.74	638.2	0.85
580	758.6	119.1	877.7	0.86
590	1084	149.9	1234	0.88
600	1508	190.1	1698	0.89
610	1870	237.7	2108	0.89
620	2069	296.5	2366	0.87
630	2403	376.2	2779	0.86
640	3062	482.0	3544	0.86
650	3975	610.9	4586	0.87
660	4963	743.9	5707	0.87
666	5313	817.6	6131	0.87
670	5230	854.5	6085	0.86
680	3770	945.7	4715	0.80
690	1647	1061	2708	0.61
700	515.8	1223	1739	0.30
710	160.2	1416	1576	0.102
720	65.45	1602	1667	0.039
730	35.53	1701	1737	0.020
740	26.63	1592	1619	0.016
750	20.39	1237	1257	0.016
760*	16.88	794.7	811.6	0.021
770*	13.40	443.5	456.9	0.029

*There is some uncertainty in the values of ε_R at wavelengths longer than 750 nm and, consequently, the values of σ_R and φ are also uncertain.

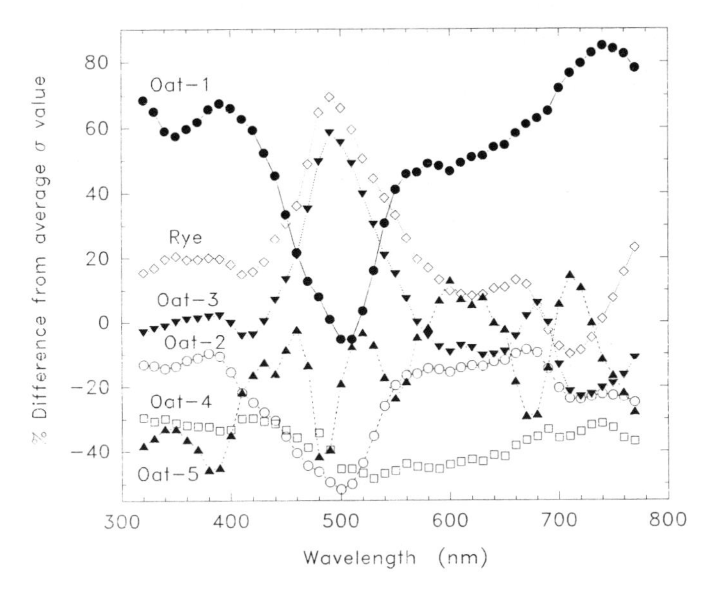

Appendix Figure 1. Comparison of phytochrome photoconversion cross-sections ($\sigma = \sigma_R + \sigma_{FR}$) of type-I phytochrome extracted and purified from dark-grown oat and rye seedlings. The average values of σ used for the comparison are those given in Appendix Table 2. Data for purified phytochrome from: oat-1 and -2, Kelly and Lagarias (1985); oat-3 and rye, Lagarias *et al.* (1987); oat-4, Vierstra and Quail (1983); oat-5, Butler *et al.* (1964).

4.8 The use of transgenic plants to examine phytochrome structure/function

Joel R. Cherry[1] and Richard D. Vierstra

Department of Horticulture, University of Wisconsin-Madison, 1575 Linden Drive, Madison, WI 53706, USA
[1]Present address: Novo Nordisk Biotech, Inc., 1445 Drew Avenue, Davis, CA 95616, USA

4.8.1 Introduction

The ability of phytochrome to act as a light-regulated molecular switch must initially result from conformational differences between the red light (R)-absorbing Pr and far-red light (FR)-absorbing Pfr forms of the chromoprotein. As a result, much effort has been directed towards characterizing the structure of purified phytochrome and locating domains that change upon photoconversion in attempts to understand how phytochrome functions. While many interesting structural domains have been identified to date (Chapter 4.3; Vierstra and Quail 1986; Quail 1991), elucidating the role(s) they play in phytochrome action has been hampered by the lack of an *in vitro* assay suitable for assessing the biological activity of the chromoprotein.

The cloning of phytochrome genes from various plant species has allowed the creation of transgenic plants that express heterologous phytochromes (Keller *et al.* 1989; Boylan and Quail 1989; Kay *et al.* 1989). When expressed to sufficient levels, the introduced proteins are biologically active, imparting a 'light exaggerated' phenotype in several plant species. By exploiting this system as an *in vivo* assay of phytochrome function, it is now possible to generate phytochrome variants *in vitro* using site-specific mutagenesis and to subsequently correlate perturbations in structure to biological activity (Boylan and Quail 1991; Cherry *et al.* 1992, 1993; Stockhaus *et al.* 1992). Such an approach allows us to assess the biological importance of many of phytochrome's physico-chemical properties including dimerization, chromophore/protein interactions, conformational changes between Pr and Pfr, Pfr stability, sequestering, and Pfr-enhanced degradation. It also allows us to examine the role of the various phytochrome isoforms [*i.e.* phytochrome A, phytochrome B, etc.

R.E. Kendrick & G.H.M. Kronenberg (eds.), Photomorphogenesis in Plants - 2nd Ediion

(Quail 1991)] present within plants by overexpressing each individually (Wagner *et al.* 1991). This chapter describes the development and use of transgenic plants in phytochrome structure/function research and summarizes the results obtained to date exploiting this approach. As will be seen, the use of transgenic plants offers the potential to unlock many of the secrets of phytochrome structure and function previously unavailable for analysis.

4.8.2 Creating transgenic plants expressing functional phytochrome

While the use of transgenic plants does have tremendous potential, it also has several disadvantages, the foremost being the long time required to transform plants and generate stable transgenic lines. Even with the rapid-cycling crucifer, *Arabidopsis thaliana*, many months are required (Boylan and Quail 1991; E. Jordan and R.D. Vierstra unpublished data). Success of the approach is also dependent on many complex processes, including faithful transcription of the gene in tissues that are responsive to the chromoprotein's action, proper excision of introns, efficient translation, and correct post-translational processing [*i.e.* chromophore attachment, phosphorylation, glycosylation, acetylation, etc. (see Chapter 4.3)]. Since alterations of phenotype may require the expression of functional phytochrome above a certain threshold, other processes such as protein and mRNA stability may also be of critical importance. As a result, many factors must be weighed before the transgenic approach is employed to insure success. From experience with a number of phytochrome deletions in transgenic tobacco, we also note that the amount of phytochrome that accumulates can vary greatly depending on the mutation and is often independent of mRNA levels (J. Colbert, J.R. Cherry and R.D. Vierstra unpublished data). In fact, some mutant proteins we have created never accumulated to levels sufficient for many biochemical and/or functional analyses [*e.g.* deletion CF (Table 1)]. This implies that certain apoprotein subdomains may improperly fold and/or lack sufficient stability to allow analysis by this approach.

4.8.2.1 Choice of coding sequence

Recent studies indicate that phytochrome is encoded by small divergent gene families in higher plants, designated *phyA-E* in *Arabidopsis* (Chapter 4.2; Sharrock and Quail 1989; Quail 1991). The *phyA* gene product (phytochrome A), previously referred to as etiolated or type I phytochrome, is expressed at high levels in etiolated plants, whereas the remaining light-grown or type II phytochrome gene products (phytochrome B and C, and posssibly phytochrome D and E) appear to be expressed at low constitutive levels in both etiolated and light-grown tissues. To date, a number of *phy* genes have been isolated, includ-

Table 1. Summary of mutant analysis.

Mutant	Chromophore attachment	Pr ⇄ Pfr	Dimer formation	Biological function	*phy* Source	*phy* Host	Reference
NA, Δ7-69	+	+	+	–	Oat *phyA*	Tobacco	Cherry *et al.* 1992
NB, Δ49-62	+	+	nd	+/–	"	"	J.R. Cherry & R.D. Vierstra unpublished data
NC, Δ6-47	+	+	nd	–	"	"	"
ND, Δ7-21	+	+	nd	+	"	"	"
NE, Δ2-5	+	+	nd	+	"	"	"
NF, Δ6-12	+	+	nd	+	"	"	"
CA, Δ1113-1129	+	+	+	–	"	"	Cherry *et al.* 1993
CB, Δ1094-1129	+	+	+	–	"	"	"
CC, Δ919-1129	+	+	–	–	"	"	"
CD, Δ786-1129	+	+	–	–	"	"	"
CE, Δ653-1129	+	+	–	–	"	"	"
CF, Δ472-1129	nd*	nd	–	nd	"	"	"
CG, Δ399-1129	+	+/–	–	–	"	"	"
Δ617-1129	+	+	nd	–	"	*Arabidopsis*	Boylan & Quail 1991
Cys_{322}→Ser	–	–	nd	–	"	"	"
Δ686-1129	+	+	nd	–	"	"	M.E. Boylan & P.H. Quail pers. comm.
Δ617-686	+	+	nd	–	"	"	"
Δ3-52	+	+	nd	+	"	"	"
Δ37-46	+	+	nd	+	"	Tomato	Quail 1991
S/A, Ser→Ala	nd	nd	nd	++	Rice *phyA*	Tobacco	Stockhaus *et al.* 1992
Δ2-45	+	+	nd	nd	Pea *phyA*	Yeast	Deforce *et al.* 1991
Δ4-225	–	–	nd	nd	"	"	"
Δ549-1124	+	+	nd	nd	"	"	"
Δ623-673	nd	nd	+	nd	Oat *phyA*	*E. coli*	Edgerton & Jones 1992
Δ1049-1129	nd	nd	+	nd	"	"	"

*nd = not determined

ing *phyA* sequences from oat, corn, rice, *Cucurbita*, potato, and pea; *phyB* sequences from rice, potato, pea, and *Arabidopsis*, and a *phyC* sequence from *Arabidopsis* (Quail *et al.* 1991; Quail 1991; Heyer and Gatz 1992). Recently, a complete *phy* sequence was reported for the cryptogam *Selaginella* (Hanelt *et al.* 1992) and a partial *phy* sequence was reported from the moss *Ceratodon purpureus* (Thümmler *et al.* 1990). Physiological experiments using mutants defective in specific *phy* genes or transgenic plant lines overexpressing *phyA* or *phyB* indicate that the different phytochrome isoforms may have distinct photosensory roles (Chory 1991; López-Juez *et al.* 1992; McCormac *et al.* 1992). Thus, plants overexpressing particular *phy* genes may display different phenotypes that will prove useful in determining the natural function of the various gene family members. All phytochrome genomic sequences examined to date contain a multitude of introns (Quail *et al.* 1991). Problems associated with incorrect intron splicing, especially when expressing monocot phytochromes in dicots, are avoided by using either *phy* cDNAs containing the entire coding region or chimeras of cDNA and/or genomic sequences in which all introns have been deleted.

4.8.2.2 Choice of promoter

The use of phytochrome overexpression as an assay of photoreceptor function requires that the introduced phytochrome gene be faithfully expressed in those plant tissues which can respond to Pfr. Moreover, those cells must synthesize the linear tetrapyrrole chromophore needed to assemble the spectrally active photoreceptor. In etiolated plants, phytochrome expression is concentrated in only a few tissue types. These include the mesocotyl node, coleoptile tip, and root cap of monocot seedlings and the plumule hook, cotyledons and root cap of dicot seedlings (Chapter 4.5; Pratt 1986). The tissue distribution of phytochrome in light-grown plants has not been confidently determined because its levels are near the limits of detection by immunocytochemistry. The chromoprotein is present in leaves as this tissue is responsive to R and FR and is frequently used as a source for purifying phytochrome from light-grown plants (Vince-Prue 1986; Wang *et al.* 1991).

At first glance, the most efficacious way to create transgenic plants that exhibit amplified phytochrome responses, yet preserve the normal tissue specificity of expression, would be to utilize phytochrome promoters. However, because *phy* mRNAs are expressed at relatively low levels (Quail 1991), it is not clear whether such promoters would be sufficiently active to substantially increase phytochrome content above wild-type levels. As yet, no attempts have been reported using such natural *phy* promoters to drive expression in a homologous plant. Initial attempts by Keller *et al.* (1989), using the oat *phyA* promoter to drive oat *phyA* expression in tobacco, failed to result in the accumulation of detectable oat *phyA* mRNA, presumably because the monocot *phy* promoter elements were not recognized by the dicot transcriptional apparatus. As an alternative approach, most authors have driven ectopic expression using the strong, constitutive 35S promoter from cauliflower mosaic virus (CaMV 35S) (Boylan and Quail 1989, 1991; Keller *et al.* 1989; Kay *et al.* 1989; Wagner *et al.* 1991). Although expression from this promoter likely differs from that of *phy* promoters in both developmental timing and tissue specificity, it does produce high levels of *phy* mRNA in tissues that are responsive to phytochrome action. It remains to be seen what effect the use of tissue- or cell-specific promoters will have on the overexpression phenotype. It is quite possible that a localized increase of phytochrome may activate only some phenotypic responses and not others. Such investigations would yield interesting information concerning the responsiveness of various cell types to altered phytochrome levels.

4.8.2.3 Choice of host plant

The best transgenic system for assaying the biological activity of phytochrome variants would be to introduce mutagenized *phy* genes back into the plant species from which they were isolated. To accentuate the phenotypic difference between expressing and non-expressing lines, the ideal recipients would be mutant lines deficient in the particular phytochrome isoform in use. The recent characterization of cucumber and *Arabidopsis* mutants deficient in *phyB* expression make this approach feasible for analysis of phytochrome B (Sommers *et al.* 1991; López-Juez *et al.* 1992). Only recently have other viable mutant plant lines deficient in the other phytochromes been isolated (Chapters 4.2 and 8.2). In their absence, a choice of host plant was made on the basis of other, more practical considerations such as transformation competence, fertility, seed yield, and ease of assaying phytochrome content. Because transformation in dicots such as tobacco, tomato, and *Arabidopsis* is relatively routine, these species have been employed as recipients in most studies to date. Tobacco offers advantages because the plants produce copious amounts of seeds and are relatively large, thus facilitating biochemical studies on the introduced protein (Cherry *et al.* 1991, 1992). *Arabidopsis,* in contrast, is rather small but can complete its life cycle in just 6 weeks (Boylan and Quail 1991). With the recent success in transforming rice, creating a monocot plant overexpressing phytochrome is now also possible (E. Jordan, P. Christou, and R.D. Vierstra unpublished data).

It is important to note that the nature and extent of phenotypic changes induced by phytochrome overexpression is likely to be strongly dependent on the species, and possibly even the cultivar, chosen as the recipient. This has already been demonstrated with two tobacco cultivars, Xanthi and SR1.

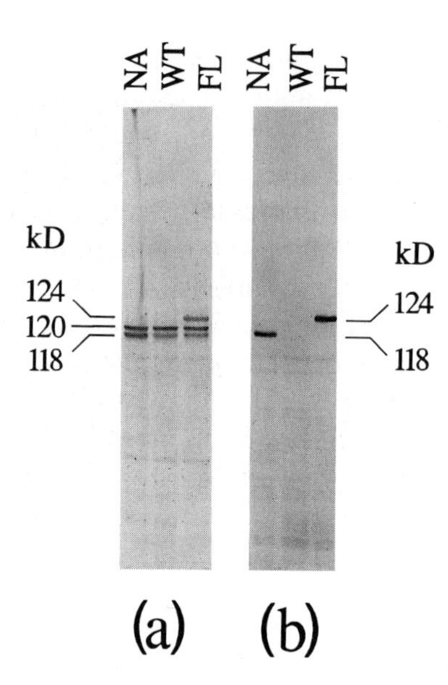

Figure 1. Synthesis of full length (FL) oat phytochrome A and an N-terminal truncation missing amino acids 7-69 (NA) in transgenic tobacco. Biochemical characteristics of FL and NA phytochromes are described in Table 1. Phytochromes were partially purified from etiolated seedlings and detected by immunoblot analysis following SDS-PAGE using either (a) anti-*Cucurbita* phytochrome immunoglobulins or (b) an oat phytochrome-specific monoclonal antibody (Oat-22). The migration positions of FL (124 kD), NA (118 kD) and tobacco phytochromes (118 and 120 kD) are indicated. WT = non-transformed tobacco. From Cherry *et al.* (1992).

Whereas a strong phenotype was observed when oat or rice *phyA* was expressed in Xanthi, little or no phenotype was detected when rice *phyA* was expressed in SR1 (Keller *et al.* 1989; Kay *et al.* 1989; Nagatani *et al.* 1991). Such variations not only complicate the choice of hosts but also question the general utility of phytochrome overexpression for agronomic benefit.

4.8.3 Analysis of plants expressing heterologous phytochromes

4.8.3.1 Expression in tobacco

Tobacco (*Nicotiana tabacum*, cv. Xanthi) was the first species reported to express a biologically active heterologous phytochrome (Keller *et al.* 1989). An intronless oat *phyA* gene was introduced which resulted in as much as 5- to 20-fold increases in total phytochrome content for dark- and light-grown seedlings, respectively. Biochemical analysis of the transgenically expressed oat phytochrome A indicated that it was indistinguishable from that purified from

Figure 2. Phenotypic changes associated with the expression of oat phytochrome A in (a) transgenic tobacco (*Nicotiana tabacum* cv. Xanthi) and (b) tomato (*Lycopersicon esculentum* cv. VF36). (a) WT = non-transformed tobacco; 9A4 = transgenic tobacco expressing high levels of oat phytochrome A; 9B2 = transgenic tobacco containing the oat *phyA* gene but not expressing detectable levels of the chromoprotein. From Cherry *et al.* (1991). (b) *Right* = non-transformed tomato; *left* = transgenic tomato expressing high levels of oat phytochrome A. From Boylan and Quail (1989).

etiolated oats (Cherry *et al.* 1991). A full-length 124-kD protein accumulated that was easily distinguished from the endogenous chromoproteins by both size and antibody recognition (Fig. 1). The molecule contained a bound chromophore and existed as a dimer with an apparent molecular mass of ≈ 300 kD. Heterodimers between the introduced oat and the endogenous tobacco phytochromes could be detected, showing that the dimerization domains of the two species are cross compatible (Cherry *et al.* 1991). Like oat phytochrome purified from oats, the transgenic chromoprotein had photoreversible difference spectra maxima at 665 and 730 nm and exhibited negligible dark reversion of Pfr to Pr. Following synthesis, the dynamics of the introduced photoreceptor was like that of all *phyA* phytochromes analyzed to date; it accumulated to substantial levels as Pr in the dark and was rapidly degraded upon photoconversion to Pfr (Chapter 4.5; Cherry *et al.* 1991). Degradation of the transgenic chromoprotein appears to also involve the ubiquitin proteolytic pathway.

Oat phytochrome A expression induced a number of phenotypic alterations that approximate what one might predict as a 'light-exaggerated' growth habit. These changes included a reduction in stem elongation for both young seedlings and mature plants, increased chlorophyll content, reduced apical dominance, and delayed leaf senescence (Fig. 2a; Keller *et al.* 1989; Cherry *et al.* 1991).

When grown under natural light, there was no difference in the time of flowering and seed yield per inflorescence. The extent of dwarfing was especially dramatic, resulting in plants ≈ 4-fold shorter than their wild-type (WT) counterparts at maturity. Later work by Nagatini *et al.* (1991) suggested that the short stature results from a reduction in cell elongation and not from a decrease in cell division. Dark-grown transgenic seedlings were morphologically identical to untransformed seedlings, despite accumulating high levels of oat Pr, providing additional evidence that Pr is biologically inactive.

Like the increase in chlorophyll content, transgenic tobacco expressing oat phytochrome A also had ≈ 2-fold more protein on a leaf area basis. Elevated levels of several photosynthetic carbon metabolism enzymes were detected, including fructose bisphosphatase, glyceraldehyde 3-phosphate dehydrogenase, and sucrose-phosphate synthase, but the level of ribulose bisphosphate carboxylase/oxygenase was unaltered (Sharkey *et al.* 1991). Despite these increases in photosynthetic enzymes, carbon fixation under ambient CO_2 concentrations was less in phytochrome-overexpressing plants compared to non-expressing plants (Sharkey *et al.* 1991). Accelerated rates of carbon fixation were observed for the transgenic plants only at elevated CO_2 levels. The reason behind the attenuated rate at ambient CO_2 is unclear. Electron microscopic examination revealed that chloroplasts in the transgenic plants were often cup-shaped, potentially slowing diffusion of CO_2 to the chloroplasts (Sharkey *et al.* 1991).

Kay *et al.* (1989) reported successful expression of functional rice phytochrome A in the tobacco cv. SR1. Its presence altered the normal phytochrome-controlled circadian rhythm of *Cab* expression and inhibited hypocotyl elongation in young light-grown seedlings (Kay *et al.* 1989; Nagatani *et al.* 1991). In more mature plants, rice phytochrome A failed to induce any of the light-exaggerated characteristics seen in the oat phytochrome A-expressing Xanthi cultivar [*e.g* dwarfism and increased chlorophyll content (Kay *et al.* 1989)]. This was surprising given that rice phytochrome A is functional in mature plants of the cv. Xanthi (Nagatani *et al.* 1991). It highlights the fact that the observed phenotype can be highly variable, even between cultivars of the same species.

4.8.3.2 Expression in tomato

Expression of oat *phyA* in tomato *(Lycopersicon esculentum,* cv. VF36) induced many of the same phenotypic characteristics found in transgenic tobacco (cv. Xanthi) described above (Boylan and Quail 1989). Transgenic plants had as much as 2- and 20-fold more spectrally detectable phytochrome in etiolated and light-grown plants, respectively, as compared with their non-transformed counterparts. Changes included dwarfing of both seedlings and mature plants, increased anthocyanin content in both leaves and fruits, and increased leaf chlorophyll (Fig. 2b). Transgenic tomatoes were distributed between two classes: those expressing high levels of the introduced oat protein and exhibiting the phenotype described above, and those expressing low but detectable levels of the oat phytochrome A apoprotein with unaltered phenotypes. This observation initially suggested that phytochrome levels must exceed a certain threshold level before plant growth habit is affected. This possibility was substantiated subsequently by phytochrome dose/phenotype response curves generated with transgenic tobacco (Section 4.8.5.2). Although not as well characterized as phytochrome A expressed in tobacco, the transgenic apoprotein in tomato does attach chromophore creating photoreceptors with absorbance characteristics indistinguishable from the purified oat chromoprotein (Boylan and Quail 1989).

4.8.3.3 Expression in Arabidopsis

Recently, *Arabidopsis thaliana* has been successfully transformed with several *phy* genes including oat *phyA*, and rice and *Arabidopsis phyB* (Boylan and Quail 1991; Wagner *et al.* 1991). In each case, a R/FR photoreversible chromoprotein of the expected size accumulated. In young seedlings, overexpression of either phytochrome A or phytochrome B induced substantial hypocotyl dwarfing under dim R. In contrast, mature *Arabidopsis* plants appear phenotypically normal despite having up to 16 times more photoreversible phytochrome (Boylan and Quail 1991). Whether this lack of phenotype reflected an insensitivity of mature *Arabidopsis* plants to elevated phytochrome levels in general, insensitivity to the type of phytochrome gene used in particular *(phyA* or *phyB)*, and/or inadequate expression of the introduced genes is unknown. It should be noted that the phenotypes of plants expressing phytochrome A or phytochrome B were identical, suggesting that these two members of the photoreceptor family are functionally equivalent in the regulation of hypocotyl elongation. More detailed investigations comparing the light responses of *phyA*- and *phyB*-overexpressing *Arabidopsis* indicate that whereas the corresponding phytochromes may induce similar phenotypes when overexpressed, they differ in their physiological roles (Chapter 8.2; McCormac *et al.* 1992; Whitelam *et al.* 1992).

4.8.3.4 Expression in lower plants

In the first report of an attempt to overexpress phytochrome in a lower plant, Thümmler *et al.* (1992) described the successful expression of oat *phyA* phytochrome in the moss *Ceratodon purpureus*. Transgenic moss protonema contained an immunoreactive oat phytochrome of the appropriate size

(124 kD), but failed to exhibit any phenotypic alterations during photomorphogenesis. The authors reported difficulty in obtaining sufficient transgenic tissue to determine whether a photoreversible oat chromoprotein was synthesized, thus preventing any conclusions regarding the biological activity of higher plant phytochromes in lower plants.

4.8.3.5 Comments on phytochrome overexpression

In addition to providing the first functional assay for phytochrome, studies of the various species expressing heterologous phytochromes also raise two interesting points. First, synthesis of functional phytochromes does not appear to be limited by chromophore availability. In fact, etiolated transgenic tobacco have been generated that accumulate as much as 20-fold more spectrally active phytochrome than untransformed plants (J.R. Cherry and R.D. Vierstra unpublished data). No authors have yet noted an accumulation of phytochrome apoprotein, suggesting that the chromophore is synthesized in the majority of plant tissues in which the CaMV 35S promoter is active. Second, it appears that the pathway responsible for the Pfr-specific degradation of phytochrome is not overwhelmed in plants expressing high levels of heterologous phytochrome as etiolated transgenic plants retain the ability to degrade the endogenous photoreceptor with unaltered kinetics (Boylan and Quail 1989; Cherry *et al.* 1991; Stockhaus *et al.* 1992). This is in spite of the fact that heterologous phytochrome appears to be degraded by the same pathway as the endogenous form (Cherry *et al.* 1991).

Even though the phenotype of phytochrome-overexpressing plants is consistent with that of plants with exaggerated light-sensing properties (Cherry *et al.* 1991; McCormac *et al.* 1991), it is still possible that the phenotype does not result from the natural activity of the photoreceptor. Since all transgenic experiments thus far have utilized the strong constitutive CaMV 35S promoters, phytochrome may be accumulating in tissues that contain little, if any, endogenous phytochrome. This ectopic distribution could be further accentuated by a greater stability of the transgenic chromoproteins and/or the expression in cells potentially ill equipped to degrade Pfr (Cherry *et al.* 1991). While the phenotype may not prove valid in assessing various aspects of phytochrome physiology, it is still useful in examining the structure and biological activity of phytochrome derivatives (Boylan *et al.* 1991; Cherry *et al.* 1991, 1992; Stockhaus *et al.* 1992). For a more detailed discussion of the physiology and light-responsiveness of transgenic plants expressing heterologous phytochromes, readers are referred to Chapter 8.2

4.8.4 Application of transgenics to phytochrome structural analysis

Until recently, phytochrome structural studies relied mainly on examining the biochemical properties of the full-length chromoprotein and various proteolytic degradation products. Efforts have been focused on localizing protein domains involved in conformational changes between Pr and Pfr and responsible for such physico-chemical properties as dimerization, photoreversibility, and chromophore accessibility/stability in an attempt to learn how phytochrome is assembled, photo-interconverts between Pr and Pfr, and functions (Vierstra and Quail 1986; Quail 1991). Using transgenic plant technology coupled with the techniques of molecular biology, it is now possible to more precisely identify specific protein sequences required for these functions, as well as to extend the examination to functions that require additional activities within the plant, such as chromophore attachment and Pfr-specific degradation. Recent data obtained with this approach have validated its potential. In many cases, the data derived from site-specific mutagenesis has confirmed and extended earlier studies with proteolytic fragments of the chromoprotein. A summary of the various mutagenic analyses to date are presented in Table 1.

4.8.4.1 Chromophore attachment

It has been known for many years that the linear tetrapyrrole chromophore is attached to phytochrome *via* a thiol-ether linkage to cysteine-322 (Vierstra and Quail 1986), but the mechanism of attachment has remained an enigma. Experiments showing that *in vitro*-translated oat phytochrome A apoprotein could attach a bilin chromophore without the addition of other plant proteins led Lagarias and Lagarias (1989) to conclude that covalent attachment of chromophore is an autocatalytic process (see Chapter 4.3). If so, this chromophore-lyase activity would represent the first enzymatic function identified for the phytochrome protein. When cysteine-322 was converted to serine and the resulting protein expressed in *Arabidopsis*, chromophore attachment was completely blocked (Boylan and Quail 1991). The resulting plants failed to show any increase in spectrally active phytochrome despite accumulating significant quantities of immunodetectable oat phytochrome A apoprotein. The results indicated that the thiol-ether linkage at residue 322 is essential for chromophore attachment and that no secondary sites can serve as replacements.

Attempts to define the minimal protein sequence required for chromophore-lyase activity have relied on the expression of deletion mutants in both plants and micro-organisms. Pea phytochrome A deletion mutants expressed in yeast, lacking residues 2-45 or 549-1124, assembled photoreversible holoprotein when mixed with phycocyanobilin *in vitro*, whereas a mutant lacking residues 4-225 failed to attach the chromophore (Table 1; Deforce *et al.* 1991). More recently, sequences necessary and sufficient for chromophore attachment *in vivo* have been further defined using oat phytochrome A deletion mutants expressed in tobacco (Cherry *et al.* 1992, 1993). Mutants lacking N-terminal residues 7-69 or C-terminal residues 399-1129 from oat phytochrome A covalently attached chromophore in transgenic tobacco (Fig. 3 and Table 1). Sequences necessary and sufficient for chromophore-lyase activity can therefore be predicted to lie in a relatively small N-terminal domain encompassing residues 69 through 399.

4.8.4.2 Photoreversibility and spectral stability

Previous work on proteolytically degraded forms of oat phytochrome indicated that the N-terminal 6-10 kD of the protein contain a domain potentially important to photoreceptor function. The region undergoes a dramatic conformational change during photoconversion between Pr and Pfr and is required for the

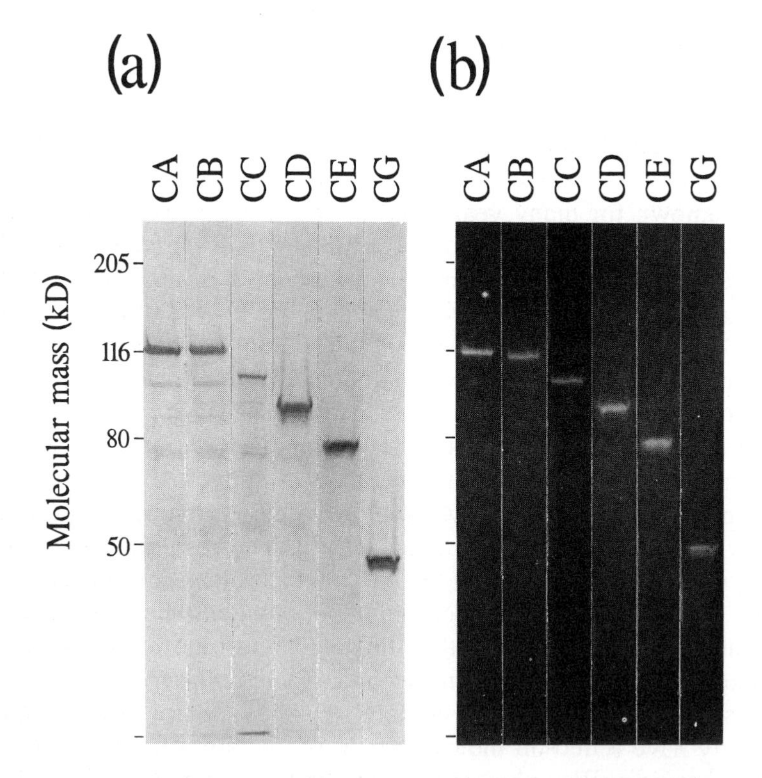

Figure 3. Synthesis of various C-terminal deletion mutants of oat phytochrome A expressed in transgenic tobacco. Deletion mutants are described in Table 1. Phytochromes were partially purified from etiolated seedlings and detected following SDS-PAGE either (a) by immunoblot analysis with anti-oat phytochrome A immunoglobulins or (b) by UV-induced chromophore fluorescence in the presence of Zn^{2+}. From Cherry *et al.* (1993).

stability and spectral integrity of Pfr (Vierstra and Quail 1986). Such effects suggested that this N-terminal region directly interacts with the chromophore. Transgenic tobacco expressing deletion mutants of oat phytochrome have recently been used to further define this domain to a region between residues 7 and 69. A phytochrome deletion mutant lacking this region (designated NA) was found to have spectral properties similar to proteolytically degraded phytochrome (Fig. 1 and Table 1; Cherry *et al.* 1992). The absorbance maxima of Pr and Pfr were shifted to shorter wavelengths and Pfr reverted non-photochemically to Pr at an increased rate relative to the full-length chromoprotein. Identification of specific N-terminal residues within the NA deletion that are responsible for the chromophore-protein interactions is currently underway.

Analysis of various C-terminal deletion mutants demonstrated that the spectral properties of phytochrome are not significantly affected by the loss of residues 652 to 1129 (Table 1; Cherry *et al.* 1993). All such mutants (CB, CC, CD, CE, and CF) had spectral properties indistinguishable from the full-length chromoprotein. This is in agreement with studies on proteolytic fragments of the purified chromoprotein, showing that much of the C-terminus can be removed without affecting spectral integrity (Jones *et al.* 1985; Vierstra and Quail 1986). However, further deletions to residue 398 (CG in Table 1) altered the absorbance spectra and reduced the efficiency of the Pr to Pfr phototransformation, although there was no apparent effect on chromophore attachment (Cherry *et al.* 1993). The Pfr absorbance spectrum for CG was substantially bleached and the absorbance maxima for both Pr and Pfr were shifted toward shorter wavelengths. Thus, while not directly required for chromophore attachment, the region between residues 399 and 651 is apparently involved in stabilizing the chromophore to maximize light absorption by Pfr.

4.8.4.3 Dimerization

Phytochrome exists *in vitro* and probably *in vivo* as a homodimer of ≈ 350 kD. The regions responsible for dimerization have been tentatively localized, based on proteolytic mapping studies, to the C-terminal half of the molecule (Vierstra and Quail 1985; Jones and Quail 1986). From structural modelling of phytochrome amino-acid sequences, Romanowski and Song (1992) have proposed that the dimerization contact sites are between residues 730 and 810 (oat phytochrome A). Dimerization sequences have been further refined empirically by expressing various C-terminal deletions of oat phytochrome A in tobacco (Cherry *et al.* 1993). Phytochrome lacking residues 1095 to 1129 (CB) existed in solution as dimers, whereas phytochrome lacking residues 920 to 1129 (CC) failed to dimerize (Table 1). Thus, a region between residues 920 and 1095 appears necessary for maintaining phytochrome-phytochrome interaction.

An alternative approach to identifying dimerization sites has recently been employed by Edgerton and Jones (1992), who fused various regions of the oat phytochrome C-terminus to a monomeric fragment of the lambda repressor. Since the lambda repressor acts optimally only after forming homodimers, phytochrome dimerization domains could be tentatively identified by assaying for lambda repressor activity in such fusions. Two regions, encompassing residues 623-673 and 1049-1129, were identified in oat phytochrome A that increased repressor activity and thus potentially were involved in dimerization. When taken together with the results obtained using deletion mutants expressed in transgenic plants, it suggests that a region between residues 1049 and 1094 is necessary, but not sufficient for phytochrome dimerization by itself and that a second site between residues 623 and 673 may also be required. Neither of these domains coincide with mathematical predictions of such dimer contact sites (Romanowski and Song 1992).

4.8.4.4 Pfr degradation

Phytochrome appears to be degraded by the ubiquitin proteolytic pathway after photoconversion from Pr to Pfr (Chapter 4.4; Shanklin *et al.* 1987; Jabben *et al.* 1989). How the pathway specifically recognizes Pfr is unknown but several regions are of potential interest, including sites for ubiquitin ligation [residues 742-790 (Shanklin *et al.* 1989; J.R. Cherry unpublished data)] and a PEST domain (Rogers *et al.* 1986) located adjacent to the chromophore attachment site between residues 323-360. Analysis of these and other sites has been hampered by the inability to develop a Pfr degradation system *in vitro*, but the transgenic approach appears to have the potential to overcome this barrier. In tobacco overexpressing oat phytochrome A, Cherry *et al.* (1991) found that the transgenic chromoprotein becomes modified with ubiquitin and rapidly degraded in etiolated seedlings after photoconversion to Pfr. This indicated that the mechanism that targets Pfr for rapid degradation by the ubiquitin system is conserved between monocots and dicots. The apparent breakdown rate was 4-fold slower than that for endogenous tobacco Pfr [4 h half-life for oat Pfr *versus* 1 h for tobacco Pfr (Cherry *et al.* 1991)]. Whether this slower rate reflects a reduced ability of tobacco to degrade oat Pfr or masking of ongoing breakdown by higher rates of apoprotein synthesis driven by the CaMV 35S promoter is unclear. Similar slow degradation rates may also exist for phytochrome A and phytochrome B phytochromes expressed in tomato and *Arabidopsis* under control of CaMV 35S (Boylan and Quail 1989, 1991; Wagner *et al.* 1991).

Although numerous phytochrome mutants have been constructed, few have been analysed to determine whether the protein alterations affect conjugation of ubiquitin to Pfr and/or subsequent degradation of the chromoprotein in transgenic plants. Both the oat deletion NA (Δ7-69) and the rice N-terminal mutation

S/A are degraded at an apparent rate similar to the full-length oat protein in tobacco, suggesting that the N-terminus is not involved in Pfr breakdown (Cherry *et al.* 1992; Stockhaus *et al.* 1992). Further examination of deletion and site-directed phytochrome mutants expressed in transgenic plants should allow the identification of domains involved in the selective turnover of Pfr in the near future.

4.8.5 Use of transgenic expression as an assay for biological activity

4.8.5.1 Considerations in transgenic assay development

One of the major goals in phytochrome research using transgenic plants has been to identify sequences critical to biological activity. It is hoped such sequences will help uncover the molecular mechanism of action. To this end, various site-directed and deletion mutants have been created and expressed, and the resulting plants examined for induction of altered phenotypes.

As mentioned in Section 4.8.2.3, the most direct way to assay phytochrome variants for biological activity would be to express them in plants defective in specific *phy* genes and look for complementation of a phytochrome-minus phenotype. In the absence of appropriate *phy* mutants (Chory 1991), most researchers have expressed mutants created *in vitro* in wild-type plants and scored for biological activity based on induction of the light-exaggerated phenotype. To be effective, the specific trait followed must be easily quantified and be strongly correlated with the level of transgene expression. There are numerous traits that are potentially suitable, including elevated anthocyanin content in tomato seedlings and leaf chlorophyll content in tobacco leaves, inhibition of hypocotyl elongation in any of the transgenic species, and inhibition of stem elongation in mature tobacco or tomato (Fig. 2; Boylan and Quail 1989, 1991; Cherry *et al.* 1991; Nagatani *et al.* 1991).

In practice, the most obvious and easily scored phenotypic change is the effect of phytochrome on plant height, either in seedlings or mature plants (where possible). In the case of tobacco, non-transformed plants typically grow to heights exceeding 100 cm, whereas plants expressing high levels of transgenic phytochrome reach heights of only 20-30 cm (Fig. 2; Cherry *et al.* 1991). Although the inhibition of hypocotyl elongation is attractive because it requires such a short time to manifest itself (Boylan and Quail 1991; Nagatani *et al.* 1991), it may not be technically practical. First of all, variation in seed germination makes hypocotyl length more difficult to measure as accurately and reproducibly as stem elongation in mature plants. Secondly, determining the relationship between the level of phytochrome expressed and the degree of hypocotyl growth inhibition requires either a way to measure phytochrome levels in individual seedlings or alternatively, the generation of a series of *stable* homozygous plant lines expressing various levels of the transgenic phytochrome (see below). Neither of these are easy to accomplish.

It should be noted that the degree of phenotypic alteration induced by phytochrome overexpression is sensitive to both light quantity and quality. Inhibition of hypocotyl elongation requires low fluence rates, generally in the order of 5-10 μmol m^{-2} s^{-1} (Nagatani *et al.* 1991; Boylan and Quail 1991). For mature tobacco, the dwarf phenotype is much less dramatic when plants are grown under a 12 h photoperiod in 300 μmol m^{-2} s^{-1} fluorescent light rather than under natural diurnal light in a greenhouse (J.R. Cherry and R.D. Vierstra unpublished data). In both cases, light eliciting an increased Pfr/P ratio, where P is the total phytochrome (Pr + Pfr), diminished the differences between non-expressing and expressing lines by selectively inhibiting growth of the non-expressing lines (McCormac *et al.* 1991)

4.8.5.2 Relationship of phytochrome dose to phenotypic response

To use the light-exaggerated phenotype as a quantitative assay of biological activity, it is critical to understand the relationship between phytochrome dose, *i.e.* the level of introduced protein, and the degree of phenotypic response. Using transgenic tobacco (cv. Xanthi) expressing oat phytochrome, Cherry *et al.* (1992) demonstrated that there is a strong non-linear correlation between phytochrome dose and plant height (Fig. 4). The shape of the phytochrome-dose/phenotype-response curve is intriguing in that the transition from tall plants to short plants occurs abruptly at relatively low phytochrome levels [measured either by spectrophotometrically or by immunoassays (Fig. 4)]. Plants containing as little as twice as much total phytochrome as untransformed plants showed a dramatic reduction in height, growing only a quarter the height of WT plants prior to flowering. Interestingly, higher levels of heterologous phytochrome have little or no additional effect indicating that the plants had become insensitive to further increases in Pfr beyond the dose threshold.

Similar response curves have been generated in the analysis of hormone/receptor binding (Goth 1981; Firn 1986). Thus, it is tempting to speculate that phytochrome may function in an enzymatic cascade analogous to the amplification systems that follow hormone-receptor interactions; In this case, phytochrome as Pfr would constitute the ligand. The extreme sensitivity of light-grown tobacco to phytochrome levels also provides an explanation for why etiolated plants reduce their phytochrome levels by 30-fold or more upon exposure to light (Chapter 4.5; Shanklin *et al.* 1987). Since the range of phytochrome content that is effective in modulating the growth habit of green plants is small, its level must be precisely controlled to respond to fluctuating light conditions.

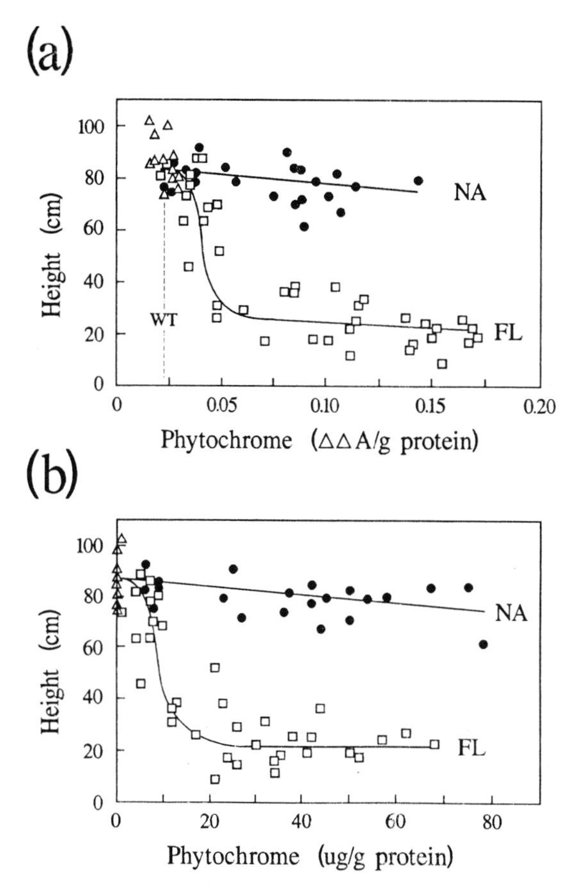

Figure 4. Phenotypic response of tobacco (cv. Xanthi) to various levels of full-length oat phytochrome A (FL) or an N-terminal truncation missing amino acids 7-69 (NA). Individual transgenic plants expressing various levels of FL and NA phytochromes were grown to maturity under natural light conditions. After detection of flower primordia, leaf tissue was collected from each plant and the amount of phytochrome measured either (a) by red-minus-far red difference spectroscopy (ΔΔA), detecting both endogenous tobacco phytochrome and the introduced oat chromoproteins, or (b) by sandwich enzyme-linked immunoadsorbent assay (ELISA) using the monoclonal antibody Oat-22 specific for the oat phytochrome. Amounts of phytochrome detected by the two assays were plotted against plant height at maturity. (Δ) Non-transformed plants (WT); (□) plants expressing FL phytochrome; (●) plants expressing NA phytochrome. From Cherry *et al.* (1992).

Although studies rigorously correlating phytochrome content to altered growth habits have yet to be carried out with other species, available data suggests that *Arabidopsis* and tomato have similar dose-response curves. Both species failed to exhibit an altered phenotype when expressing low but detectable levels of oat phytochrome but did so when higher levels accumulated (Boylan and Quail 1989, 1991). This phenotypic segregation could also be observed in progeny of heterozygous lines suggesting that levels of phytochrome in the progeny straddled both sides of the dose threshold.

4.8.5.3 Possible mutant classes based on dose/response

Based on the dose-response curves observed for transgenic tobacco, one can predict that mutant phytochromes will fall into one of five broad classes with respect to biological activity (Fig. 5; Firn 1986). The first class is composed of those chromoproteins that fail to alter plant phenotype at any level of expression and are here termed non-functional mutants. In the second class, some phytochrome variants may be fully-functional if the introduced mutations do not affect the biological activity of the protein. Fully-functional mutants will allow definition of the minimal phytochrome domain required for biological activity, and may serve to simplify future structure/function studies by allowing us to focus in on only the functionally essential features of the molecule. The third class, partially functional mutants, are those that either require a higher phytochrome dose to alter photomorphogenesis, or affect photomorphogenesis

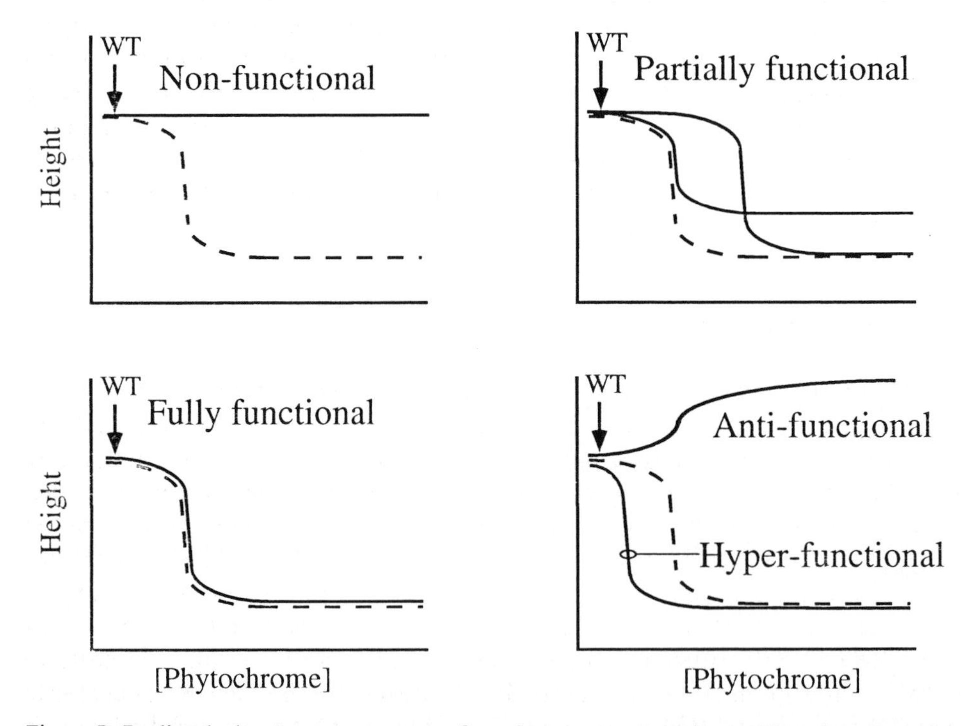

Figure 5. Predicted dose/response curves for plants overexpressing unaltered and mutant phytochromes. Values for non-transformed plants (WT) are indicated by arrows. Dashed lines, dose/response of plants to unaltered phytochrome; solid lines, predicted dose/response classes for various phytochrome mutants.

at the same dose but not to the same degree as the intact photoreceptor (*i.e.* intermediate dwarfism) (Fig. 5). According to models of hormone/receptor interactions (Firn 1986), the former type of partially functional mutants would interact with the next member of the signal transduction chain with less affinity; this change in affinity would increase the amount of complex required for a given level of response. The latter type would bind to its partner with the same affinity but create a complex that works less efficiently, thus altering the response capacity (Firn 1986). The last two classes can be termed anti-functional and hyper-functional (Fig. 5). Anti-functional mutants may act as dominant negative effectors by blocking the biological action of endogenous phytochromes. Plants expressing such anti-functional phytochromes may display phenotypes characteristic of etiolated plants with elongated hypocotyls/ stems and reduced pigmentation even when grown under sufficient light. Plants expressing hyper-functional mutants would exhibit the light-exaggerated phenotype at doses lower than that of the unmodified photoreceptor. The discovery of such mutants could potentially identify a phytochrome region which is normally involved in attenuating phytochrome function *in vivo*.

To distinguish between these classes (and any others that may exist), plants expressing a range of mutant phytochrome levels must be analysed to determine the relationship between protein levels and phenotypic response. A phytochrome-dose/phenotype response curve for unmodified phytochrome would serve as standard against which all mutants can be compared. Only by taking mutant analysis to this extreme can one accurately quantitate the relative biological activity of each mutant. Coupled with biochemical characterization of the mutant proteins, such analyses should help assign functional importance to various structural domains of the chromoprotein.

4.8.5.4 Biological activity of phytochrome mutants

A number of phytochrome mutant constructs have been expressed in transgenic tobacco and *Arabidopsis* and their potential biological activity assessed. As might be expected, most mutants fall into the non-functional class, with a few fully-functional, one partially functional, and one presumed hyper-functional (Table 1). One of the first non-functional mutants described involved mutation of cysteine-322 to serine, resulting in an apoprotein unable to bind the chromophore (Boylan and Quail 1991). When expressed in *Arabidopsis*, the phytochrome apoprotein that accumulated appeared to be biologically *in*active. This result agreed with previous studies using tetrapyrrole synthesis inhibitors to enhance apoprotein accumulation by blocking chromophore availability (Gardner and Gorton 1985; Jones *et al.* 1986) and the analysis of *Arabidopsis* mutants, *hy1* and *hy2*, defective in chromophore synthesis (Parks *et al.* 1989). In each case, plants had a phytochrome-deficient phenotype despite accumulating substantial amounts of apoprotein. For *hy1* and *hy2*, the normal phenotype could be rescued by adding the direct precursor of phytochromobilin (Parks and Quail 1991).

Several interesting mutants involve changes within the N-terminus. Because the N-terminus is required for the spectral integrity of phytochrome and undergoes a conformational change during the Pr to Pfr photoconversion, this region has been proposed to be important for phytochrome function (Vierstra and Quail 1986). In support, Cherry *et al.* (1992) showed that an N-terminal deletion (designated NA), lacking only residues 7 through 69, is biologically inactive (Cherry *et al.* 1992). Even for plants expressing up to 5 times more NA phytochrome than is required by unmodified phytochrome to induce a dwarf phenotype, no effect on plant height was observed (Figs. 1 and 4). As mentioned above, NA phytochrome assembled into a dimeric photoreversible chromoprotein but had spectral properties similar to large, 120-kD phytochrome in which the N-terminus has been proteolytically removed (Cherry *et al.* 1992). This demonstrated that the extreme N-terminus, which is essential for proper protein-chromophore interaction, is also required for biological activity.

Recent experiments to further define N-terminal residues required for function have relied on the construction of five smaller deletions falling within the first 69-amino acids (NB, NC, ND, NE, and NF in Table 1; J.R. Cherry and R.D. Vierstra unpublished data). They all bound chromophore and all but two had full biological activity in transgenic tobacco (Figs. 6 and 7). Of these, deletion NB, lacking residues 49 to 62, was partially active and resulted in an intermediate degree of dwarfing than is seen with fully active phytochromes. The NB deletion appeared to affect both the dose threshold and the degree of dwarfing (*i.e.* response capacity). Deletion NC (Δ6-47), on the other hand, had no detectable biological activity even when expressed at levels five times that required to saturate the response by fully-functional phytochromes.

In combination, these data locate an essential N-terminal domain between residues 21 and 47. This domain contains a short hydrophobic stretch followed by a region rich is hydrophilic amino acids but has no homology to any proteins of known function. Parker *et al.* (1991) have speculated that this domain could change from a random coil to an amphiphilic helix during Pr to Pfr conversion, but how it may regulate biological activity is not yet known. Experiments to determine how the NB and NC phytochrome deletions differ biochemically (*e.g.* spectral properties, stability) from the active N-terminal deletion mutants (ND, NE, NF) are in progress.

In contrast to the inactivity of oat phytochrome A deletions NA and NC in tobacco, an oat phytochrome A N-terminal deletion lacking residues 3-52 was active in inhibiting hypocotyl elongation in *Arabidopsis* (M. Boylan and P.H. Quail unpublished data). In addition, a mutant lacking residues 37 to 46 was able to induce the dwarf phenotype in mature tomato despite the loss of this conserved region (Quail 1991). Unfortunately, plants with a range of protein

Figure 6. Phenotypes of transgenic tobacco (cv. Xanthi) expressing high levels of various N-terminal deletions of oat phytochrome A. The biochemical properties of the various deletions are described in Table 1. Plants were grown to maturity under natural light conditions in a greenhouse. WT = non-transformed plant; FL = plants expressing full-length oat phytochrome A; NB, NC, ND, NE and NF = plants expressing N-terminal deletions of oat phytochrome A missing amino acids 49-62, 6-47, 7-21, 2-5, and 6-12, respectively. From J.R. Cherry and R.D. Vierstra unpublished data.

expression were not examined precluding assessment of the relative activity of the mutants as compared to the unmodified chromoprotein. Thus, it is not possible to determine how active these deletions were or whether the contradictory results using tobacco and *Arabidopsis*/tomato reflected differences in the sensitivity of different species to phytochrome overexpression. For example, the levels of expression insufficient to inhibit stem elongation in tobacco may inhibit hypocotyl elongation in *Arabidopsis*. Such discrepancies can only be resolved by expressing all mutants in the same plant species and examining the effects on the same trait.

Sequencing of several *phy* genes has identified a novel stretch within the N-terminus highly enriched in serine residues (Quail *et al.* 1991). The observations that phytochrome is a phosphoprotein and that several of its N-terminal serines are selectively phosphorylated *in vitro* has led to the h0pothesis that Pfr function can be affected by protein kinase activities (Wong *et al.* 1986; Wong and Lagarias 1989). In an effort to test this hypothesis, a mutant of rice phytochrome A was generated in which the first ten N-terminal serines (residues 2-4, 10-14, 19, and 20) were converted to alanines. In accord with functional analysis of N-terminal deletions of this region (see above), this mutant (desig-

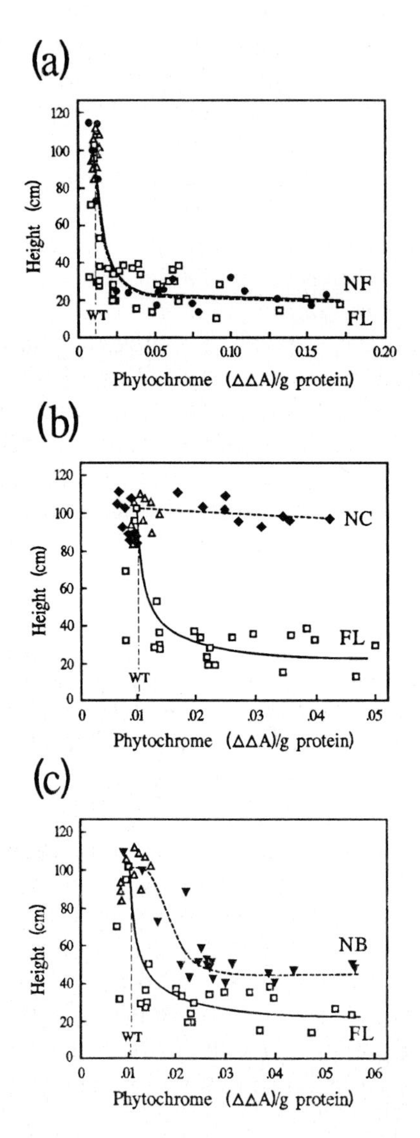

Figure 7. Phenotypic response of transgenic tobacco (cv. Xanthi) to various levels of full-length oat phytochrome A (FL) or the N-terminal truncations missing amino acids 6-12 (NF), 6-47 (NC), 49-62 (NB). Individual transgenic plants expressing various levels of FL, NF, NB, and NC phytochromes were grown to maturity under natural light conditions. After detection of flower primordia, leaf tissue was collected from each plant, protein extracted, and the amount of photoreversible phytochrome, including both the introduced oat chromoprotein and the endogenous tobacco phytochromes, was measured by red-minus-far red difference spectroscopy ($\Delta\Delta A$). Amount of phytochrome detected was then plotted against plant height at maturity. (Δ), Non-transformed plants (WT); (a) plants expressing NF phytochrome; (b) plant expressing NC phytochrome; (c) plants expressing NB phytochrome. From J.R. Cherry and R.D. Vierstra unpublished data.

nated S/A) was biologically active when assayed in transgenic tobacco (Stockhaus *et al.* 1992). In fact, the mutant appeared to be more active than unaltered rice phytochrome and possibly represented a hyper-functional mutant (Fig. 5). The results suggest that whereas the N-terminal serines are not essential for function, they may be involved in attenuating activity. The possibility that they interact with protein kinases leading to a repression of Pfr action in the light is appealing (Stockhaus *et al.* 1992).

While not required for the spectral properties of phytochrome (Jones *et al.* 1985), the C-terminus is required for biological activity (Cherry *et al.* 1993). Tobacco plants expressing various C-terminal deletion mutants of oat phytochrome A failed to exhibit any of the light-exaggerated traits seen when the unmodified chromoprotein was expressed (Table 1, Figs. 3 and 8). In the cases of deletions CA (Δ1113-1129), CC (Δ920-1129), and CF (Δ472-1129), this may be due to insufficient accumulation of the oat protein in either etiolated or green tobacco plants. The remaining mutants CB, CD, CE, and CG, although expressing at levels beyond the dose threshold, did not detectably alter the phenotype of mature plants (Cherry *et al.* 1993). In the case of CB, which is lacking only

Figure 8. Phenotypes of transgenic tobacco (cv. Xanthi) expressing high levels of various C-terminal deletions of oat phytochrome A. The biochemical properties of the various deletions are described in Table 1 and Fig. 4. Plants were grown under natural light conditions in a greenhouse to maturity. WT = non-transformed plant; FL = plants expressing full-length oat phytochrome A; CB, CD, CE and CG = plants expressing C-terminal deletions of oat phytochrome A missing amino acids 1094-1129, 919-1129, 786-1129, 653-1129, and 399-1129, respectively. From Cherry *et al.* (1993).

35 amino acids from the C-terminus and is biochemically identical to the full-length chromoprotein by all criteria tested, this is most surprising. In similar studies, Boylan and Quail (1991) reported that an oat phytochrome A mutant, missing residues 617-1129, failed to induce the short-hypocotyl phenotype in *Arabidopsis* despite accumulating substantial quantities of a spectrally active 68-kD protein. Oat phytochrome A deletions spanning residues 617-686, or 686-1129 also failed to affect the growth pattern of transgenic *Arabidopsis* or tomato. Whether this is because they are inherently non-functional or expressed to inadequate levels is unknown. From analysis of all the C-terminal mutants, it appears that a second functional domain resides near the C-terminus. This domain appears separate from that required for dimerization and may be essential for a heretofore unknown activity for the photoreceptor.

One of the most surprising observations is that none of the phytochrome mutants affect the activity of the endogenous photoreceptor, *i.e.* act as anti-functional mutants (Fig. 5). This is despite the fact that at least some of the mutants are nearly indistinguishable biochemically from functional phytochrome (Table 1). Assuming that Pfr interacts directly with a signal transduction component, it is reasonable to expect that some inactive forms could compete with the functional photoreceptor for interactions with its binding partner. While the creation of such mutants may in fact be lethal to plants, the possibility of anti-functional phytochrome mutants remains an intriguing possibility.

4.8.6 Concluding remarks

Transgenic plant technology has provided a new way to functionally dissect the phytochrome protein. In addition to facilitating the identification and definition of domains involved in the physico-chemical properties of this intriguing photoreceptor, it has the potential to relate these properties to biological activity. As described in this chapter, sequences necessary for chromophore attachment, light absorbance properties, dimerization, and biological activity have been identified or further defined using transgenic plants. This line of research holds much promise in future efforts to unravel the molecular mechanism of phytochrome action.

As more *phy* genes are cloned, it will be possible to better define conserved protein motifs that may be critical to biological function. Such motifs will serve as future candidates for studies employing site-directed mutagenesis and expression in transgenic plants. Moreover, domain swapping experiments in which a *phy* sequence from one *phy* gene family member is used to replace homologous sequences in another can now be performed and the resulting hybrid phytochrome assayed for biological activity. Finally, photomorphogenic mutants putatively lacking a specific phytochrome can be tested for their ability to be complemented by specific *phy* genes to restore a normal pattern of photomorphogenesis.

4.8.7 Further Reading

Boylan M.T. and Quail P.H. (1989) Oat phytochrome is biologically active in transgenic tomatoes. *Plant Cell* 1: 765-773.

Cherry J.R., Hondred D., Walker J.M. and Vierstra R.D. (1992) Phytochrome requires the 6-kDa N-terminal domain for full biological activity. *Proc. Natl. Acad. Sci. USA* 89: 5039-5043.

Keller J.M., Shanklin J., Vierstra R.D. and Hershey H.P. (1989) Expression of a functional monocotyledonous phytochrome in transgenic tobacco. *EMBO J.* 8: 1005-1012.

Nagatani A., Kay S.A., Deak M., Chua N.-H. and Furuya M. (1991) Rice type I phytochrome regulates hypocotyl elongation in transgenic tobacco seedlings. *Proc. Natl. Acad. Sci. USA* 88: 5207-5211.

Quail P.H. (1991) Phytochrome: A light-activated molecular switch that regulates plant gene expression. *Annu. Rev. Genet.* 25: 389-409.

4.8.8 References

Boylan M.T. and Quail P.H. (1991) Phytochrome A overexpression inhibits hypocotyl elongation in transgenic *Arabidopsis*. *Proc. Natl. Acad. Sci. USA* 88: 10806-10810.

Cherry J.R., Hershey H.P. and Vierstra R.D. (1991) Characterization of tobacco expressing functional oat phytochrome. *Plant Physiol.* 96: 775-785.

Cherry J.R., Hondred D., Walker J.M., Keller J., Hershey H.P. and Vierstra R.D. (1993) Carboxyl-terminal deletion analysis of oat phytochrome A reveals the presence of separate domains required for structure and biological avtivity. *Plant Cell* in press.

Chory, J. (1991) Light signals in leaf and chloroplast development: photoreceptors and downstream responses in search of a transduction pathway. *New Biologist.* 3: 538-548.

Deforce L., Tomizawa K-I., Ito N., Farrens D., Song P.-S. and Furuya M. (1991) *In vitro* assembly of apophytochrome and apophytochrome deletion mutants expressed in yeast with phycocyanobilin. *Proc. Natl. Acad. Sci. USA* 88: 10392-10396.

Edgerton M.D. and Jones A.M. (1992) Localization of protein-protein interactions between subunits of phytochrome. *Plant Cell* 4: 161-171.

Firn R.D. (1986) Growth substance sensitivity: the need for clearer ideas, precise terms, and purposeful experiments. *Physiol. Plant.* 67: 267-272

Gardner G. and Gorton H.L. (1985) Inhibition of phytochrome synthesis by gabaculine. *Plant Physiol.* 77: 540-543.

Goth A. (1981) *Medical Pharmacology*. 10th Edition, pp 7-14, C.V. Mosby Co, St Louis.

Hanelt S., Braun B., Marx S. and Scheider-Poetsch H.A.W. (1992) Phytochrome evolution: a phylogenetic tree with the first complete sequence of phytochrome from a cryptogamic plant (*Selaginella martensii* Spring). *Photochem. Photobiol.* 56: 751-758.

Heyer A. and Gatz C. (1992) Isolation and characterization of a cDNA-clone coding for potato type B phytochrome. *Plant Mol. Biol.* 20: 589-600.

Jabben M., Shanklin J. and Vierstra R.D. (1989) Ubiquitin-phytochrome conjugates: pool dynamics during *in vivo* phytochrome degradation. *J. Biol. Chem.* 264: 4998-5005.

Jones A.M. and Quail P.H. (1986) Quarternary structure of 124-kilodalton phytochrome from *Avena sativa* L. *Biochemistry* 25: 2987-2995.

Jones A.M., Vierstra R.D., Daniels S.M. and Quail P.H. (1985) The role of separate molecular domains in the structure of phytochrome from etiolated *Avena sativa* L. *Planta* 164: 501-506.

Jones A.M., Allen, C.D., Gardner G. and Quail P.H. (1986) Synthesis of phytochrome apoprotein and chromophore are not coupled obligatorily. *Plant Physiol.* 81: 1014-1016.

Kay S.A., Nagatani A., Keith B., Deak M., Furuya M. and Chua N.-H. (1989) Rice phytochrome is biologically active in transgenic tobacco. *Plant Cell* 1: 775-782.

Lagarias J.C. and Lagarias D.M. (1989) Self-assembly of synthetic phytochrome holoprotein *in vitro*. *Proc. Natl. Acad. Sci. USA* 86: 5778-5780.

López-Juez E., Nagatani A., Tomizawa K.-I., Deak M., Kern R., Kendrick R.E. and Furuya M. (1992) The cucumber long hypocotyl mutant lacks a light-stable PHYB-like phytochrome. *Plant Cell* 4: 241-251.

McCormac A.C., Cherry J.C., Hershey H.P., Vierstra R.D. and Smith H. (1991) Photoresponses of transgenic tobacco plants expressing an oat phytochrome gene. *Planta* 185: 162-170.

McCormac A.C., Whitelam G. and Smith H. (1992) Light-grown plants of transgenic tobacco expressing an introduced phytochrome A gene under the control of a constitutive viral promoter exhibit persistent growth-inhibition by far-red light. *Planta* 188: 173-181.

Parker W., Romanowski M. and Song P.-S. (1991) Conformation and its functional implications in phytochrome. In: *Phytochrome Properties and Biological Action*, pp. 85-112, Thomas B. and Johnson C.B. (eds.) Springer-Verlag, Berlin.

Parks B.M. and Quail P.H. (1991) Phytochrome-deficient *hy1* and *hy2* long hypocotyl mutants of *Arabidopsis* are defective in phytochrome chromophore biosynthesis. *Plant Cell* 3: 1177-1186.

Parks B.M., Shanklin J., Koornneef M., Kendrick R.E. and Quail P.H. (1989) Immunochemically detectable phytochrome is present at normal levels but is photochemically nonfunctional in the *hy1* and *hy2* long hypocotyl mutants of *Arabidopsis*. *Plant Mol. Biol.* 12: 425-437.

Pratt L.H. (1986) Phytochrome: localization within the plant. In: *Photomorphogenesis in Plants* pp. 61-81, Kendrick R.E. and Kronenberg G.H.M. (eds.) Martinus Nijhoff Publishers, Dordrecht.

Quail P.H., Hershey H.P., Idler K.B., Sharrock R.A., Christensen A.H., Parks B.M., Somers D., Tepperman J., Bruce W.A. and Dehesh K. (1991) Phy-gene structure, evolution, and expression. In: *Phytochrome Properties and Biological Action*, pp. 13-38, Thomas B. and Johnson C.B. (eds.) Springer-Verlag, Berlin.

Rogers S., Wells R. and Rechsteiner M. (1986) Amino acid sequences common to rapidly degraded proteins: the PEST hypothesis. *Science* 234: 364-368.

Romanowski M. and Song P.-S. (1992) Structural domains of phytochrome deduced from homologies in amino acid sequence. *J. Protein Chem.* 11: 139-155.

Shanklin J., Jabben M. and Vierstra R.D. (1987) Red light-induced formation of ubiquitin-phytochrome conjugates: identification of possible intermediates of phytochrome degradation. *Proc. Natl. Acad. Sci. USA* 84:359-363.

Shanklin J., Jabben M. and Vierstra R.D. (1989) Partial purification and peptide mapping of ubiquitin-phytochrome conjugates in oat. *Biochemistry*. 28: 6028-6034.

Sharkey T.D., Vassey T.L., Vanderveer P.J. and Vierstra R.D. (1991) Carbon metabolism enzymes and photosynthesis in transgenic tobacco (*Nicotiana tabacum* L.) having excess phytochrome. *Planta* 185: 287-296.

Sharrock R.A. and Quail P.H. (1989) Novel phytochrome sequences in *Arabidopsis thaliana*: Structure, evolution, and differential expression of a plant regulatory photoreceptor family. *Genes Develop*. 3: 1745-1757.

Sommers D.E., Sharrock R.A., Tepperman J.A. and Quail P.H. (1991) The *hy3* long hypocotyl mutant of *Arabidopsis* is deficient in phytochrome B. *Plant Cell* 3: 1263-1274.

Stockhaus J., Nagatani A., Halfter U., Kay S., Furuya M. and Chua N.-H. (1992) Serine-to-alanine substitutions at the amino-terminal region of phytochrome A result in an increase in biological activity. *Genes Develop*. 6: 2364-2372.

Thümmler F., Beetz A. and Rudiger W. (1990) Phytochrome in lower plants: Detection and partial sequence of a phytochrome gene in the moss *Ceratodon purpureus*. *FEBS Lett.* 275: 125-129.

Thümmler F., Schuster H. and Bonenberger J. (1992) Expression of the oat *phyA* gene in the moss *Ceratodon purpureus*. *Photochem. Photobiol.* 56: 771-776.

Vierstra R.D. and Quail P.H. (1985) Spectral characterization and proteolytic mapping of native 120-kilodalton phytochrome from *Cucurbita pepo* L. *Plant Physiol.* 77: 990-998.

Vierstra R.D. and Quail P.H. (1986) Phytochrome: the protein. In: *Photomorphogenesis in Plants*, pp 35-60, Kendrick R.E. and Kronenberg G.H.M (eds.) Martinus Nijhoff Publishers, Dordrecht.

Vince-Prue D. (1986) The duration of light and photoperiodic responses. In: *Photomorphogenesis in Plants*, pp 269-305, Kendrick R.E. and Kronenberg G.H.M (eds.) Martinus Nijhoff Publishers, Dordrecht.

Wagner D., Tepperman J.M. and Quail P.H. (1991) Overexpression of phytochrome B induces a short hypocotyl phenotype in transgenic *Arabidopsis*. *Plant Cell* 3: 1275-1288.

Wang Y.-C., Stewart S.J., Cordonnier M.-M. and Pratt L.H. (1991) *Avena sativa* L. contains three phytochromes, only one of which is abundant in etiolated tissue. *Planta* 184: 96-104.

Whitelam G., McCormac A.C., Boylan M.T. and Quail P.H. (1992) Photoresponses of *Arabidopsis* seedlings expressing an introduced oat *phyA* gene: persistence of etiolated plant type responses in light-grown plants. *Photochem. Photobiol.* 56: 617-622.

Wong Y.-S. and Lagarias J.C. (1989) Affinity labeling of *Avena* phytochrome with ATP analogs. *Proc. Natl. Acad. Sci. USA* 86: 3469-3473.

Wong Y.-S., Cheng H.-C., Walsh D.A. and Lagarias J.C. (1986) Phosphorylation of *Avena* phytochrome *in vitro* as a probe of light-induced conformational changes. *J. Biol. Chem.* 261: 12089-12097.

Part 5 Blue-light and UV receptors

5.1 Diversity of photoreceptors

Horst Senger and Werner Schmidt*

Fachbereich Biologie/Botanik, Philipps-Universität, 35032 Marburg, Lahnberge, Germany
*Present address: *Meisenweg 7, 78465 Konstanz l6, Germany*

5.1.1 Introduction

All life depends on light. Principally, the whole spectrum of sunlight is available to organisms. However, for molecular reasons photoperception over the whole visible spectrum cannot be accomplished by one single photoreceptor molecule. For energy conservation in the photosynthetic process chlorophylls, carotenoids and biliproteins serve as photoreceptors. Another prerequisite of life is the control of metabolic, morphological and directional responses by light. On a spectral basis most of these responses can be confined solely or in concert to the red, blue or near-UV region. The UV region of the spectrum is subdivided into UV-C (200-280 nm), UV-B (280-320 nm) and UV-A (320-400nm) (Fig. 2). The term 'near UV' commonly describes the UV above 300 nm. The present chapter outlines the diversity of blue light (B) and near-UV effects. Since photoperception of some physiological reactions of plants and fungi extends into the UV-region, recently the term 'B/UV photoreceptors' is used.

The so-called 'physiological B/UV effects' are phylogenetically among the oldest, but still amongst the least understood phenomena of photobiology. A tremendous number of diverse B/UV effects have been well documented and can be crudely classified as *metabolic*, *morphological* and *directional responses*. However, in most cases our knowledge does not exceed the bare observation of the phenomenon that B/UV is capable of triggering or influencing specific effects. The photoreceptor pigments have only been identified in a few cases and the sensory transduction chain, *i.e.* the mode of action, is not known in detail in any single case. Thus the definition of B/UV photoreceptors is still evolving. In addition, it is difficult to draw a precise borderline between B effects and those mediated by the UV and green part of the spectrum. As deduced from the variety of effects, the differences in action spectra and light saturation, we have to assume that we are dealing with several B/UV photo-

R.E. Kendrick & G.H.M. Kronenberg (eds.), Photomorphogenesis in Plants - 2nd Edition

receptors and mechanisms of action and in addition, most likely with a group of light-harvesting pigments supporting the B/UV photoreceptors. The diversity of problems combined with the ubiquitous appearance and the clear-cut physiological significance of B/UV effects throughout the living world makes B/UV research so fascinating not only to photobiologists, but also to an increasing number of scientists, from ecologists and physiologists to molecular biologists, biochemists and biophysicists.

5.1.2 Historical aspects

Scientific research on B/UV effects in plants was founded in 1864 by Julius Sachs, who demonstrated that bending of plants (phototropism) towards light is stimulated only by the B region of the spectrum (Sachs 1864). When searching for the mechanism mediating the phototropic curvature of oat (*Avena sativa*) coleoptiles, some eight decades later Galston and Baker (1949) discovered the B-induced photo-oxidation of the plant growth hormone *indole acetic acid* (IAA or 'auxin'). Particularly the discovery of specific effects on *carbon metabolism* raised new interest in the field of B physiology. The research on this topic, as holds true for photobiology in general, was substantially promoted by advances in technology including light sources of higher intensity, stability and improved spectral quality, *i.e.* narrower bandwidth of interference filters and monochromators, as well as by modern optical spectroscopy in general.

Subsequently a wide range of physiological B/UV effects has been studied: the influence of B/UV on various *metabolic effects* such as enzyme regulation, pigment biosynthesis, carbon metabolism, respiration enhancement, nucleic acid metabolism or protein biosynthesis; on *morphogenic effects* such as flowering, conidiation, growth inhibition or growth promotion, seed germination, cortical fibre reticulation before chloroplast aggregation; phase shifting of various circadian rhythms which cannot be categorized unequivocally under one of these topics: and, finally, on *directional responses* such as phototropism, phototaxis and intracellular chloroplast rearrangement and on *non-directional responses* such as leaflet closure (Chapters 9.2 and 9.4).

There are three main questions to be answered: (i) what is the chemical nature of the physiological B/UV receptor(s)? (ii) Where is (are) the photoreceptor(s) pigment(s) localized in the cell? (iii) What are the primary photophysical and photochemical reactions of the photoreceptor and the subsequent events of the signal chain (*sensory-transduction chain*) and what are the succeeding steps which finally lead to the particular response observed? For detailed discussion and further reading refer to the reviews (Senger and Briggs 1981; Schmidt 1984a; Senger 1987; Galland and Senger 1988a, b; Galland 1992), or to the proceedings of the first and second conferences on physiological B action, both held in Marburg, Germany (Senger 1980, 1984a). The present chapter only covers some representative topics of B/UV research, no comprehensive coverage of this subject was intended. For clarity, all spectra shown (Figs. 1 and 2) are reproduced as smooth curves without the actual data points.

5.1.3 Blue light UV responses

Figure 1 exemplifies the great variety of biological B/UV action on the basis of diverse *action spectra.* Many of these action spectra do not cover the whole UV-A region because corresponding interference filters or powerful light sources have not been available. True action spectra are determined by measuring fluence-response curves at various wavelengths. From such data, provided the Bunsen-Roscoe reciprocity law holds, one can obtain the number of quanta to produce a standard effect. A plot of the reciprocal of this number of quanta against wavelength gives the 'true' standard action spectrum. In contrast, the measurement of *preliminary* action spectra (measuring response for a constant number of incident quanta) are much less time consuming, but often provide the very information required: the extent of a specific response as a function of wavelength.

5.1.3.1 Phototropism of Phycomyces

Directional responses such as phototropism and *phototaxis* (the swimming of freely mobile organisms towards or away from the light source) endow physiological advantages to the organisms. These include more efficient photosynthetic activity, higher rates of cell reproduction or in the case of chloroplast re-arrangements, a more efficient utilization of light energy and protection against photodestruction.

Figure 1a shows the action spectrum of the phototropic balance for the wild-type *Phycomyces* sporangiophore. The sporangiophore is exposed to two opposed beams of light: a fixed 'reference' beam of broad B of 10^{-4} J m^{-2} s^{-1} and a 'test beam' of variable wavelength and irradiance. The ordinate represents the relative quantum efficiency determined from the reciprocal of the test beam irradiance that balanced the effectiveness of the reference beam (Lipson *et al.* 1984).

5.1.3.2 Light-induced absorbance changes

One possible assay for the B photoreceptor is seen in what is generally abbreviated as 'LIAC', the *Light-Induced Absorbance Change*. If the primary reaction of the light-response signal transduction chain is accompanied by a redox-

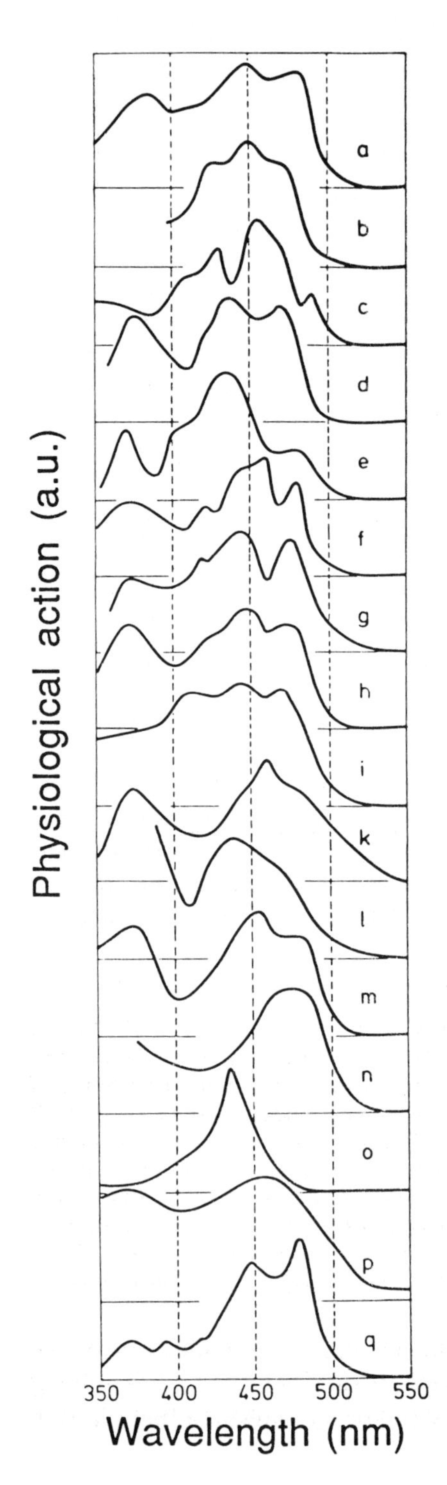

Figure 1. Blue (B)UV-A light action spectra (physiological action in arbitrary units, a.u.) demonstrating the diversity of B/UV-A responses. (a) Phototropism *Phycomyces* (Lipson *et al.* 1984). (b) Light-induced absorbance change (LIAC) (Widell *et al.* 1983). (c) Hair whorl formation *Acetabularia* (Schmid 1984). (d) Photoreactivation of nitrate reductase (Roldán and Butler 1980). (e) Germination of spores of *Pteris vittata* (Sugai *et al.* 1984). (f) Perithecial formation in *Gelasinospora reticulispora* (Inoue and Watanaba 1984). (g) Formation of 5-aminolevulinic acid (Oh-hama and Senger 1978). (h) Phototropism in *Avena*, 10 ° and (i) 0 ° (Shropshire Jr. and Withrow 1958). (k) Respiration enhancement in *Scenedesmus* (Brinkmann and Senger 1978a). (l) Inhibition of indole acetic acid (Galston and Baker 1949). (m) Chloroplast rearrangement in *Funaria* (Zurzycki 1967). (n) Cortical fibre reticulum in *Vaucheria* (Blatt and Briggs 1980). (o) DNA-photoreactivation (Saito and Werbin 1970). (p) Loss of carbohydrate in *Chlorella* (Kowallik and Schänzle 1980). (q) Carotenogenesis in *Neurospora* (DeFabo *et al.* 1976).

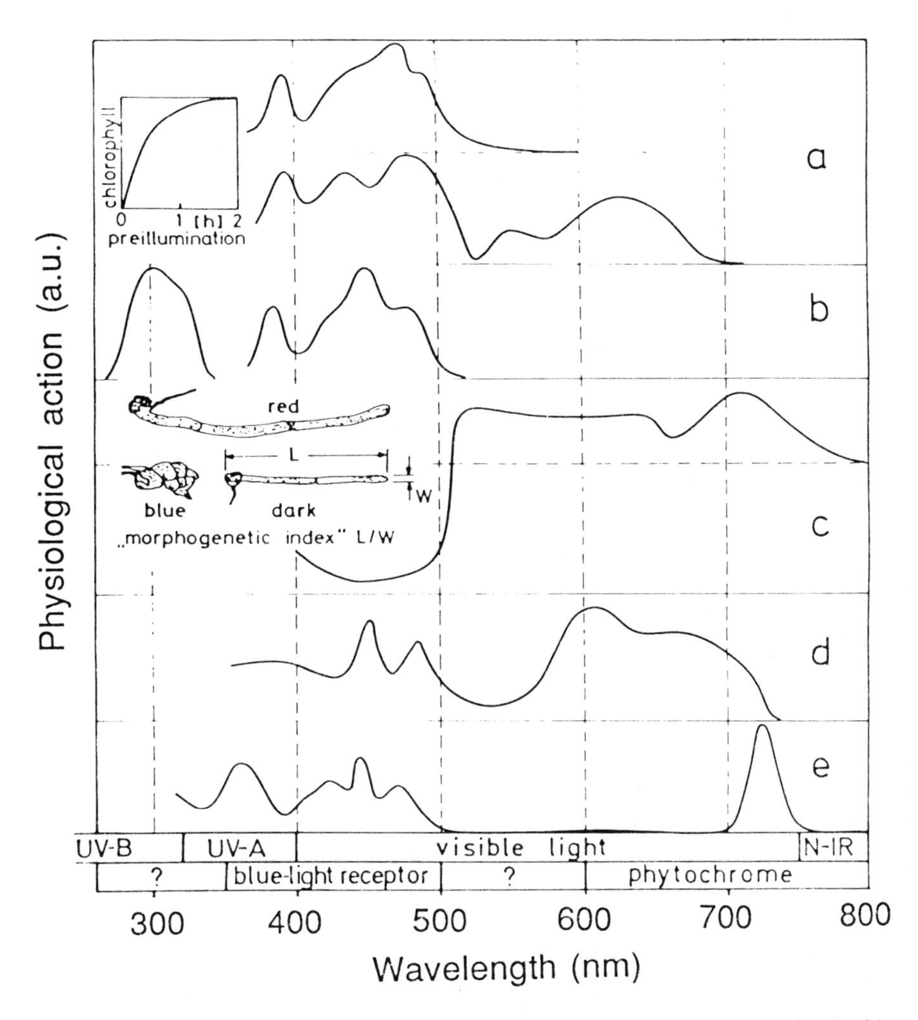

Figure 2. Action spectra (physiological action spectra in arbitrary units, a.u.) of: (a) *top,* chlorophyll synthesis in dark-grown *Scenedesmus* (Brinkmann and Senger 1978a); *bottom,* same after pre-irradiation with blue light (B) (Brinkmann and Senger 1978b); *inset,* time course of sensitizing effect of pre-irradiation with B. (b) Induction of conidiation in *Alternaria* by UV-B and its reversion by B (Kumagai 1983). (c) Morphogenetic index L/W in *Dryopteris filix-mas* as a function of wavelength. Under these conditions and within the experimental period, the sporelings remained filamentous throughout, even in B (Mohr 1956). (d) Light-induced sensitization to geotropic stimulus in maize roots (Klemmer and Schneider 1979). (e) Action spectrum for the high irradiance response (HIR) of light-inhibition of hypocotyl growth of *Lactuca sativa* is included, reflecting physiological activities in two spectrally distinct regions (Hartmann 1967).

process (*vide infra*), this should be verified by appropriate changes in the spectrum. Indeed, during the past 10 years various LIACs most often involving the dark reversible photoreduction of specific b-type cytochromes (cyt b), have been reported. The action spectrum for the cyt b-photoreduction in most cases reported resembles a typical B spectrum. Subsequently several researchers succeeded in isolating plasma membrane fractions from various B-sensitive organisms which still exhibit LIACs. Figure 1b exemplifies this by a LIAC observed in a plasma membrane fraction from cauliflower (Widell *et al.* 1983). For criticism of this procedure refer to Section 5.1.5.

5.1.3.3 Hair whorl formation in Acetabularia

Among the unicellular algae *Acetabularia* proves to be particularly suitable for photomorphogenic studies. The morphogenic events during the growth of the stalk of the cell are the production of lateral hair whorls in regular intervals and finally the formation of the cap marking the end of the vegetative growth phase. The formation of whorls and caps require light. A precise action spectrum for hair whorl formation has been worked out by Schmid (1984). The quantum fluence required for 50% whorl formation was adapted for this spectrum (Fig. 1c).

5.1.3.4 Reactivation of nitrate reductase

The first B action spectrum determined for a LIAC was measured in mycelia mats of the fungus *Neurospora crassa* (Munõz and Butler 1975). In this case the membrane-bound enzyme nitrate reductase, which contains both flavins (flavin adenine dinucleotide, FAD) and b-type cytochromes, might contribute to the LIAC. The inactive (reduced) form of the enzyme can be re-activated by light, again with a typical B action spectrum (Fig. 1d) (Roldán and Butler 1980). The first correlation of a LIAC and a physiological response was also described for *Neurospora*: the light-promoted conidiation, again with a B action spectrum (not shown). These findings suggest that nitrate reductase is the photoreceptor for nitrate assimilation in *Neurospora* and, therefore, of subsequent light-promoted conidiation.

5.1.3.5 Germination of spores in Pteris

Germination of spores in the fern *Pteris vittata* is induced by red light (R), the induction being abolished by subsequent B irradiation (see Chapter 9.7). Figure 1e exhibits the corresponding action spectrum, calculated from the reciprocal number of photons required for 50 % inhibition (Sugai *et al.* 1984).

5.1.3.6 Perithecial formation in Gelasinospora

Figure 1f shows the action spectrum for the photo-induction of perithecial formation in the sordariaceous fungus *Gelasinospora* (Inoue and Watanabe 1984). The induced perithecial formation is inhibited by UV-B (250-300nm); note the opposite spectral behaviour in light-induced conidiation in *Alternaria tomato* (Fig. 2b).

5.1.3.7 Synthesis of 5-aminolevulinic acid

In angiosperms, the late step in chlorophyll biosynthesis, the conversion of protochlorophyllide to chlorophyllide, requires light. In recent years, however, it has been found that the first step, the formation of 5-aminolevulinic acid (ALA), may become light dependent under certain circumstances. In addition, the formation of ALA is known to be the rate-limiting step in tetrapyrrole biosynthesis. The formation of ALA was found to be B dependent in several pigment mutants and wild-type cells of algae. In angiosperms the important photoreceptor phytochrome potentiates the formation of ALA by activation of enzyme synthesis (Kasemir and Mohr 1981). Thus the light regulation of ALA synthesis probably plays a key role in plant photomorphogenesis. The action spectrum of B-induced ALA-formation in the green alga *Chlorella protothecoides* is shown in Fig. 1g (Oh-hama and Senger 1978).

5.1.3.8 Phototropism in oats

The spectra (h) and (i) in Fig. 1 are normalized 'true' action spectra of phototropism of oat coleoptiles for different curvatures, (h) for 10° and (i) extrapolated for 0°. The dramatic difference in the UV-A region between these two spectra corresponding to different degrees of the observed response should warn us about over interpretation of even 'true' biological action spectra (Shropshire Jr. and Withrow 1958).

5.1.3.9 Respiration enhancement in Scenedesmus

Irradiation of the pigment mutant C-2A' of the unicellular green alga *Scenedesmus obliquus* induces oxygen uptake, respiration enhancement (Fig. 1k) and breakdown of starch (Brinkmann and Senger 1978a). This phenomenon has been observed in many green algae (Kowallik 1982). All action spectra are very similar (*cf.* Fig. 1k and 1p) and the reaction already saturates at very low irradiances (Table 1; Senger and Briggs 1981).

5.1.3.10 Inhibition of indole acetic acid

Phototropic curvature of plant organs have been originally explained by photolysis of the plant growth hormone IAA on the irradiated side of the organ, enhancing its relative growth on the shaded side, thereby inducing bending towards the light source. The theory of phototropism (the Blaauw theory) probably does not hold any longer (Chapter 9.2). Figure 1l shows the often cited action spectrum for decarboxylation of auxin: again a B action spectrum (Galston and Baker 1949).

5.1.3.11 Chloroplast rearrangement in Funaria

Another B effect is found throughout the plant kingdom and its physiological advantage is self explanatory. For example, the loci of photosynthesis, the disc-shaped chloroplasts of the moss *Funaria*, move within the cell in order to optimize their energy-absorbing system. To avoid exposure to very strong light and prevent photobleaching of chlorophyll, they turn their edge towards the light source. In weak light, however, their flat side is fully exposed. The corresponding action spectrum for this chloroplast re-arrangement is given in Fig. 1m (Zurzycki 1967).

5.1.3.12 Cortical fibre reticulation in Vaucheria

Interestingly, the chloroplast displacement is a passive process: prior to irradiation, longitudinal fibres along which the organelles appear to move (Chapter 9.4), can readily be seen through a light microscope with differential interference contrast optics. Upon irradiation, these fibres appear to become stabilized forming a 'cortical fibre reticulum'. This process always precedes chloroplast aggregation and studies of wavelength dependence revealed the B spectral region is most effective for the green alga *Vaucheria* (Fig. 1n) (Blatt and Briggs 1980).

5.1.3.13 Photoreactivation

Photoreactivation of UV-inactivated bacteria, viruses and eucaryotes is mediated by special light-activated enzymes 'DNA photolyases'. These enzymes bind to double- or single-stranded DNA containing pyrimidine dimers. Upon irradiation with B/UV or visible light they split the cyclobutane ring and restore the original state. The action spectrum for photorepair of the DNA of *Haemophilus* by a photolyase purified from the cyanobacterium *Anacystis nidulans* is given in Fig. 1o (Saito and Werbin 1970).

5.1.3.14 Loss of carbohydrates in Chlorella

One of the early observations of B action was the loss of carbohydrates (Kowallik 1969) and the enhancement of respiration (Kowallik 1982) in the green alga *Chlorella*. The loss of carbohydrates, *i.e.* the breakdown of storage starch follows a typical B action spectrum (Kowallik and Schänzle 1980).

5.1.3.15 Carotenoid synthesis in Neurospora

Much of our knowledge about molecular physiology of B responses has been deduced from light-induced carotenoid biosynthesis in lower organisms. Carotenoids are powerful anti oxidants that protect pigments such as chlorophyll in higher plants and also translucent micro-organisms from light damage: many non-photosynthetic organisms such as *Neurospora*, *Phycomyces, Fusarium* turn yellow as a result of carotenoid synthesis upon irradiation. The action spectrum of carotenogenesis in *Neurospora* is one of the most precise B action spectra known (Fig. 1q) (DeFabo *et al.* 1976). The missing UV-peak is often cited by the defenders of the carotenoid photoreceptor hypothesis (*cf.* Section 5.1.3.8).

5.1.4 Concerted action of photoreceptors

5.1.4.1 Chlorophyll synthesis in Scenedesmus

Chlorophyll synthesis in the mutant C-2A' of *Scenedesmus* requires light. If the culture is grown for 3 days in darkness, the wavelength-dependence clearly reveals a typical B action spectrum (Fig. 2a, *top*) (Brinkmann and Senger 1978a). However, pre-irradiation with white light (W) of 20 J m^{-2} s^{-1} for 2 h imposes a dramatic change: in addition to B, wavelengths between 525 and 700 nm have become effective (Fig. 2a, *bottom*) (Brinkmann and Senger 1978b). The inset of Fig. 2a shows the time-course of this sensitizing effect of light applied prior to the R-induced chlorophyll biosynthesis.

Both, the phototransformation of protochlorophyllide to chlorophyllide, as well as photosynthesis appear to be involved. Due to their Soret bands, the pigments mediating these two processes also strongly absorb in the B region of the spectrum. Therefore, the specific B effect primarily observed is only a *prerequisite* for the subsequent reactions. It regulates the protein formation of a controlling enzyme for the chlorophyll precursor ALA and the chlorophyll protein complex. In conclusion, here B only controls a rate-limiting step, but *not* the complete reaction as originally presumed (*cf.* Fig. 6).

5.1.4.2 Conidiation in Alternaria

Kumagai (1983) suggested a *mycochrome* system to be involved in the B and UV reversible photoreaction controlling both conidiophore maturation, and induction of conidiation in the fungus *Alternaria tomato*. Conidiation induction is essentially restricted to the UV-B, no action beyond 340 nm being observed. The effect is completely reversed by B (Fig. 2b). The final response depends on the quality of light received last. Two pigments P_B and $P_{UV\text{-}B}$ absorbing in the B and the UV-B regions respectively have been proposed to interact on a redox-basis. The $P_{UV\text{-}B}$ photoreceptor is a low molecular mass compound containing iron, whereas P_B is most likely a flavin. The reduced flavin is the active form, and its redox state is regulated by light (Kumagai 1983):

$$P_B\,(\text{oxidised}) \underset{\text{B}}{\overset{\text{UV-B } via \text{ } P_{UV\text{-}B}}{\rightleftarrows}} P_B\,(\text{reduced}) \longrightarrow \text{conidiation}$$

5.1.4.3 Morphogenic index in the fern Dryopteris

This example demonstrates the independent action of the B and 'longer wavelength photoreceptor(s)'. In R the sporeling remains filamentous but relatively longer than in darkness, whereas in B a more 'surface-like' structure develops (see inset Fig. 2c). The morphogenic index, length/width quantitatively describes the shape of chloronemata of the fern *Dryopteris filix-mas* as a function of wavelength and is plotted in Fig. 2c. The intersection with the control line ('darkness') is close to 500 nm: a kind of balance of opposite morphological R and B effects (Mohr 1956). Part of the longer wavelength response can probably be attributed to phytochrome.

5.1.4.4 Geotropism in maize roots

The roots of some varieties of maize are gravitropically sensitive to light. Without light-pretreatment the roots do not show gravitropic curvature. However, light is capable of 'sensitizing' the roots. Figure 2d shows the action spectrum which clearly is separated into a B and a 'longer wavelength' part.

Higher fluences and prolonged irradiation periods tend to level the troughs and peaks of the spectrum. The curvature proceeds concomitantly with a negative growth response (light partially inhibits growth). Based upon reversibility of the effect by far-red light (FR) and LIACs, phytochrome and a haemoprotein-like photoreceptor have been suggested (Klemmer and Schneider 1979).

5.1.4.5 High irradiance response of phytochrome

Hypocotyl lengthening in lettuce seedlings (*Lactuca sativa)* is inhibited by continuous irradiation. Figure 2e shows the true action spectrum throughout the whole visible spectrum as deduced from fluence rate-response curves, based upon 60-80% response (Hartmann 1967). The peak at 716 nm has been exclusively explained by phytochrome action, the portion in the B represents a typical B action spectrum and is probably not due to phytochrome. Its origin is still unknown.

5.1.4.6 Red and blue interaction in maize coleoptiles

Phototropism of grass coleoptiles, probably the 'classical' physiological B response is significantly modified by R pretreatment. With R, the sensitivity of the 'first positive curvature' in maize is decreased over 10-fold. The R action spectra show a marked peak near 660 nm, characteristic of phytochrome and the effect is fully reversed by FR (converting the active FR-absorbing form Pfr, into the inactive R-absorbing Pr form) (Chon and Briggs 1966).

5.1.5 Energy requirements

Physiolgoical B/UV responses are observed over an extraordinary range of fluences, extending from 10^{-9} J m^{-2} for phototropism in *Phycomyces* to more than 10^5 J m^{-2} required for the adaptation of the photosynthetic apparatus of *Scenedesmus*. Table 1 and Fig. 3 give a crude survey (as based upon randomly collected data) of fluences required to induce a great diversity of B/UV responses found in the literature (if required, fluence rates have been converted to fluences by multiplication with the time needed to induce the particular response; the validity of the law of reciprocity is implicitly assumed, even though this was not always demonstrated). The data are arranged in increasing order. For comparison, the range of peak values of B/UV fluence rates (same numerical value as fluence when applied for 1 s) of sunlight observed during July 1985 in Marburg, Germany, is indicated. More than 12 h per 24 h B fluence rate (measurements were performed with a broad-band B glass filter in

Table 1. Blue (B)/UV-A and UV-B light responses and the required light fluences as obtained from the literature. If the original paper only indicated the fluence rate, the fluence is calculated by multiplication with the irradiation time given in the last column. If possible the table indicates the type of responses *i.e.* saturation (sat), half-saturation (hs) or threshold (th) values.

Response	Fluence (J m^{-2})	Type	Irradiation time
Phototropism of *Phycomyces*	10^{-9}	th	–
Phototropism of *Pilobolus*	3.6×10^{-9}	th	–
Phototropism of corn coleoptile (1st positive curvature)	3.6×10^{-8}	sat	–
Oxygen uptake in *Chlorella*	10^{-7}	hs	–
Chloroplast development in *Euglena*	10^{-6}	–	–
Phaseshift of conidiation in *Neurospora*	1.5×10^{-5}	th	–
Retardation of flower opening in *Oenothera lamarckiana*	2.2×10^{-5}	–	1 h
Anthocyanin synthesis in *Sorghum*	3×10^{-4}	th	–
Sporangiophore initiation in *Phycomyces*	2.7×10^{-3}	–	1 h
Photoreactivation of nitrate reductase	2×10^{-1}	th	10 min
Inhibition of spore germination in *Pteris vittata*	1	th	–
Inhibition of B induction of perithecial formation in *Gelasinospora reticulispora*	10	th	–
Carotenogenesis in *Phycomyces* (low fluence)	20	th	–
Conidiophore formation in *Alternaria cichorii*	24	th	–
Cortical fibre reticulation in the alga *Vaucheria sessilis*	30	th	–
Induced lateral electrical potential in oat coleoptiles	10^{2}	th	10 min
Light-induced absorbance change (LIAC) in membrane fractions of corn and *Neurospora*	2×10^{2}	–	10 s
Current in *Vaucheria sessilis*	6×10^{2}	th	10 min
Carotenogenesis in *Phycomyces* (high fluence)	10^{3}	sat	–
Flavin-mediated transport of redox-equivalents across membranes	10^{3}	th	10 s
Inhibition of tobacco transketolase in the presence of 0.5 μM flavin-mononucleotide (FMN)	1.35×10^{3}	th	15 min
Inhibition of L-lactate dehydrogenase in the presence of 7.0 μM FMN	1.9×10^{3}	th	15 min
Induction of absorbance change in *Neurospora*	2×10^{3}	th	1 s
Induction of hatching of *Pectinophora* eggs	2.7×10^{3}	hs	–
Induction of stimulation of geo-responsiveness in corn roots	10^{5}	th	–
Adaptation of the photosynthetic apparatus of *Scenedesmus*	5.8×10^{5}	–	15 h
Phytochrome, high irradiance response	2×10^{5}	–	18 h
Phytochrome, low fluence response	10^{-1}-1.2×10^{3}	–	–

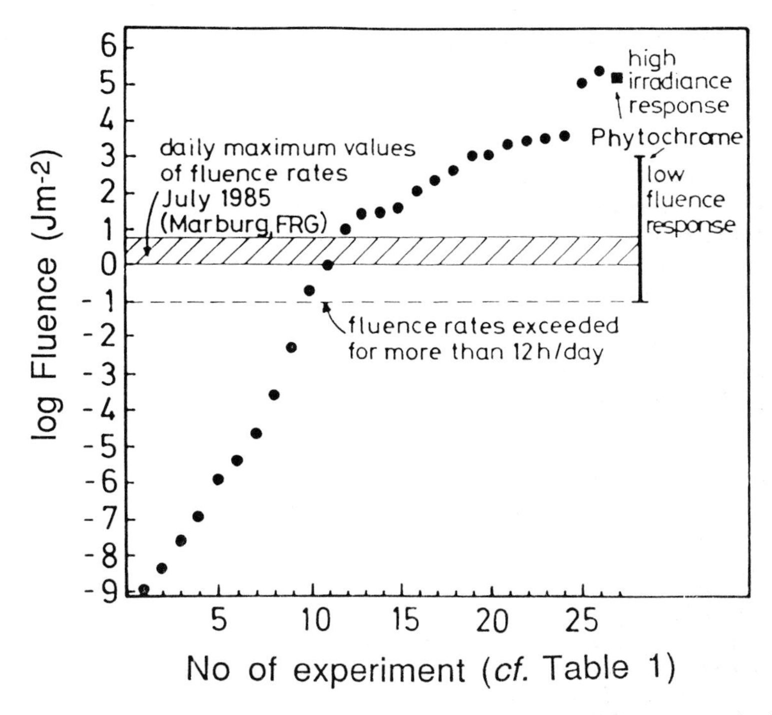

Figure 3. Illustration of Table 1, demonstrating the enormous range of fluences required to induce particular blue (B)/UV-A and UV-B light responses. The data have been randomly selected and were arranged in increasing order. For comparison, the variability in maximum and minimum values of B fluence rates observed during July 1985 in Marburg is indicated, as well as the minimum fluence rate observed for at least 12 h per day (NB the fluence indicated is therefore the fluence received in 1 s). The fluence required for a high irradiance response (HIR) of phytochrome and the range of the low fluence response of phytochrome to red light are also given for comparison.

front of a radiometer connected to a slow running chart recorder) exceeds 10^{-1} J m^{-2} s^{-1}. For further comparison, the range of common low fluence responses (LFR) and one high irradiance response (HIR) of phytochrome are included (Smith 1975), however, not the very low fluence response (VLFR) mediated by phytochrome (Chapter 4.7).

High fluences are typically administered over a long period of time (*e.g.* for adaptational processes), whereas low fluence responses often can be 'pulsed' (seconds or minutes being capable of inducing the response), as for photoinduced sporulation or suppression of dark germination.

Clearly, the fluence range exceeding 10 J m^{-2} is physiologically restricted to 4 orders of magnitude, the range below to as much as 7 orders. Correspondingly B/UV reactions can be crudely grouped into low and high fluence responses. Low fluence responses probably describe light reactions which only trigger pre-existing sensory-transduction chains, whereas in high fluence responses the utilization of the B/UV-energy must be considered. Low fluence responses are commonly correlated with high quantum efficiencies (close to unity). The LIACs are typically observed upon application of fluence rates of as much as 100 J m^{-2} s^{-1} administered for a few seconds. For example, LIACs have been described for human cells (HeLa) that most likely do not show any physiological B/UV reaction; LIACs in corn coleoptiles have never been observed *in vivo* but only in various kinds of preparations, probably as a result of mixing components which might normally present in different compartments. Therefore, most LIACs reported have to be interpreted with extreme caution.

5.1.6 The nature of B/UV photoreceptors

Abundant examples of UV/B action spectra restrict the possible photoreceptor molecules to only a few classes of pigments. For about half a century there has been a sporadically lively debate as to whether carotenoids or flavins are the most likely B photoreceptor. Detailed discussion appears elsewhere (Chapters 5.2 and 9.7) and needs not be amplified here. For several years (Galland and Senger 1988) pterins have been considered as possible candidates for B/UV photoperception in plants and experimental evidence has been provided that flavins and pterins occur together in the light absorbing organelle, the paraflagellar body, of *Euglena* (Schmidt *et al.* 1990, Galland *et al.* 1990) and in the light-sensitive portion of the sporangiophore *Phycomyces* (Hohl *et al.* 1992a). Rhodopsin was reported to be the B-photoreceptor for *Chlamydomonas* (Forster *et al.* 1988) and other, so far unknown, photoreceptors may yet emerge. Although we have to anticipate that a variety of B/UV photoreceptors are involved in B/UV action of the various organisms, the bulk of recent evidence favours flavins in cooperation with pterins as the most abundant of the B/UV photoreceptors.

The only B/UV photoreceptors which have been isolated and identified to be involved in the B/UV absorption are the flavin/pterin photoreceptors of the DNA photolyases. In this way UV-C induces the formation of pyrimidine dimers (mainly *cis-syn* thymine-thymine dimers) in DNA which prevents further DNA replication. Such dimers can be removed by DNA photolyase in procaryotes and eucaryotes. After absorption of light of wavelengths between 300 and 600 nm a DNA photolyase splits the pyrimidine dimers and restores the replication ability of the DNA. The DNA photolyase of *E. coli* contains FAD (Schuman Jorns *et al.* 1984) and a pterin, identified as 5,10-methenyltetrahydrofolyl-polyglutamate (Johnson *et al.* 1988). The same photoreceptors have been isolated from species such as *Saccharomyces* (Johnson *et al.* 1988) and *Neurospora*.

Evidence of the energy transfer between the pterin and the flavin suggest that pterin has an antenna function for the redox changes in the FADH/$FADH_2$ reaction (Chanderkar and Schuman Jorns 1991).

The B/UV mediated photorepair of DNA by DNA photolyase is a very specialized reaction. Recently a more general mode of action of B/UV photoreceptors has been reported, the B/UV photoreceptor mediated membrane protein phosphorylation (Short *et al.* 1992). From a pea membrane fraction a 120-KD protein could be isolated which is phosphorylated upon B/UV irradiation (Short and Briggs 1990). In addition membranes from dark-grown pea buds were found to contain a G protein which apparently mediates B/UV-enhanced binding of GTP (Warpeha *et al.* 1991a, b).

The phytoflagellate *Euglena* is the only organism in which the photoreceptor for its B/UV dependent phototaxis has been located. The photoreceptor is part of the paraflagellar body at the end of the flagella (Benedetti and Lenci 1977). In isolated paraflagellar bodies flavins and pterins can be identified by fluorescence (Fig. 4), (Schmidt *et al.* 1990). Pterins have been isolated and partially identified from fractions of flagellae with paraflagellar bodies (Geiss D., Galland P. and Senger H. unpublished data).

The fungus *Phycomyces* is a well known organism for studies of B/UV triggered phototropism. A number of flavins and pterins have been extracted from *Phycomyces* and it has been demonstrated that mutants having deviant photoresponses show a different pattern in flavin and pterin distribution (Fig. 5) (Hohl *et al.* 1992a, b).

A classical B/UV photoreceptor model has been derived from the fungus *Neurospora.* The B/UV dependent conidiation is mediated by nitrate reductase which contains FAD, a molybdo-pterin and cytochrome b (Klemm and Ninnemann 1978). Studying this system, it was discovered that the B/UV mediated cytochrome b reduction causes the observed LIAC (Munõz *et al.* 1975), which is still a relevant assay to identify B/UV photoreceptor systems (Widell and Larsson 1987).

For most B/UV reactions we can assume the following general scheme of action: the B/UV photoreceptors are membrane-bound and undergo B/UV-induced redox reactions. Further down the sensory transduction pathway towards the response side the signal chain diversifies. By analogy to better understood photobiological reactions, it is reasonable to assume that the photoreceptor relaxation (a term comprising all possible processes by which the electronically excited molecule dissipates its energy) is followed by a changed proton, electrical and/or a redox-gradient across a membrane, as induced by light-stimulated membrane permeability changes. However, some of the above mentioned results suggest a direct membrane phosphorylation without intermediate redox changes.

In addition to the redox reactions described above, B/UV-photoreceptors might represent the cofactor for a light-regulated enzyme. In a comprehensive

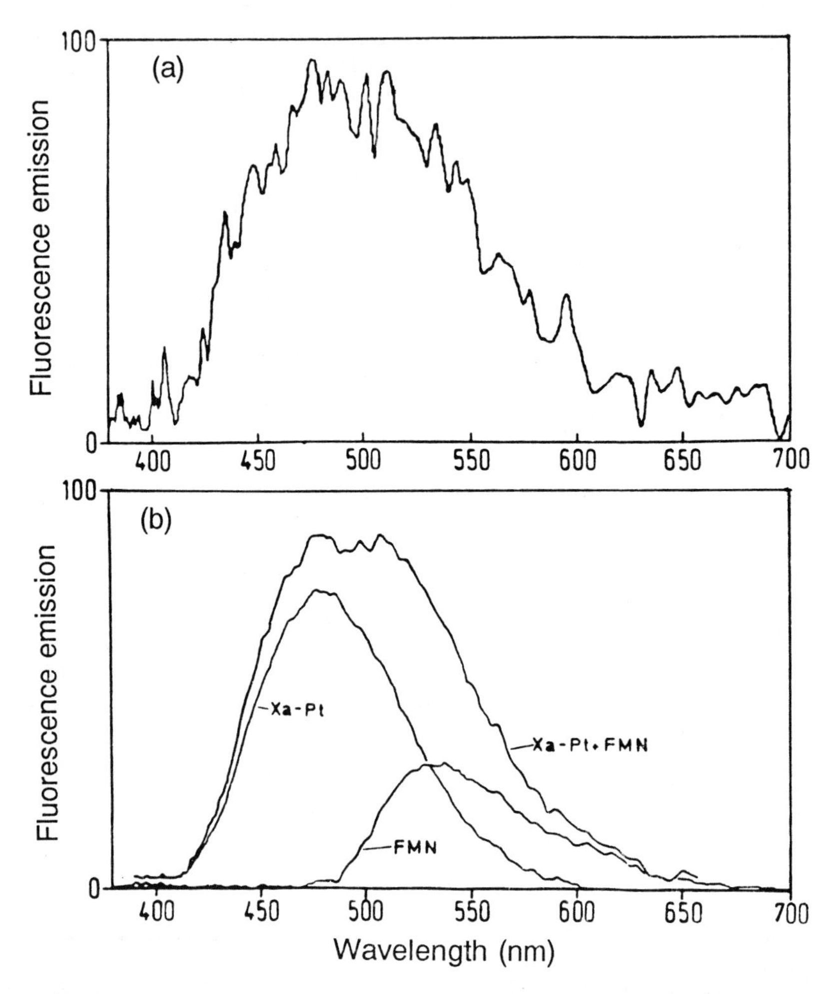

Figure 4. (a) Fluorescence emission spectrum of the isolated paraflagellar body of *Euglena gracilis.* Excitation wavelength was 365 nm. (b) Fluorescence emission spectrum composed of flavin mononucleotide (FMN, 10^{-5} *M*) and xanthopterin (Xa-Pt, 10^{-5} *M*) simulating the pigments of the paraflagellar body. (Modified from Schmidt *et al.* 1990).

review more than 30 enzymes influenced by B/UV are described (Ruyters 1984). However, only in a few cases have flavoproteins been shown to be directly influenced by B/UV. In such cases the relaxation process of the pigment to its ground state may be accompanied by conformational changes in the apoprotein (Schmid 1970) which, in turn, might trigger the subsequent sensory-transduction chain.

As a minimum molecular model, different amphiphilic flavins have been synthesized and anchored by means of long hydrocarbon chains within artificial membranes made from various phospholipids. It has been found that virtually

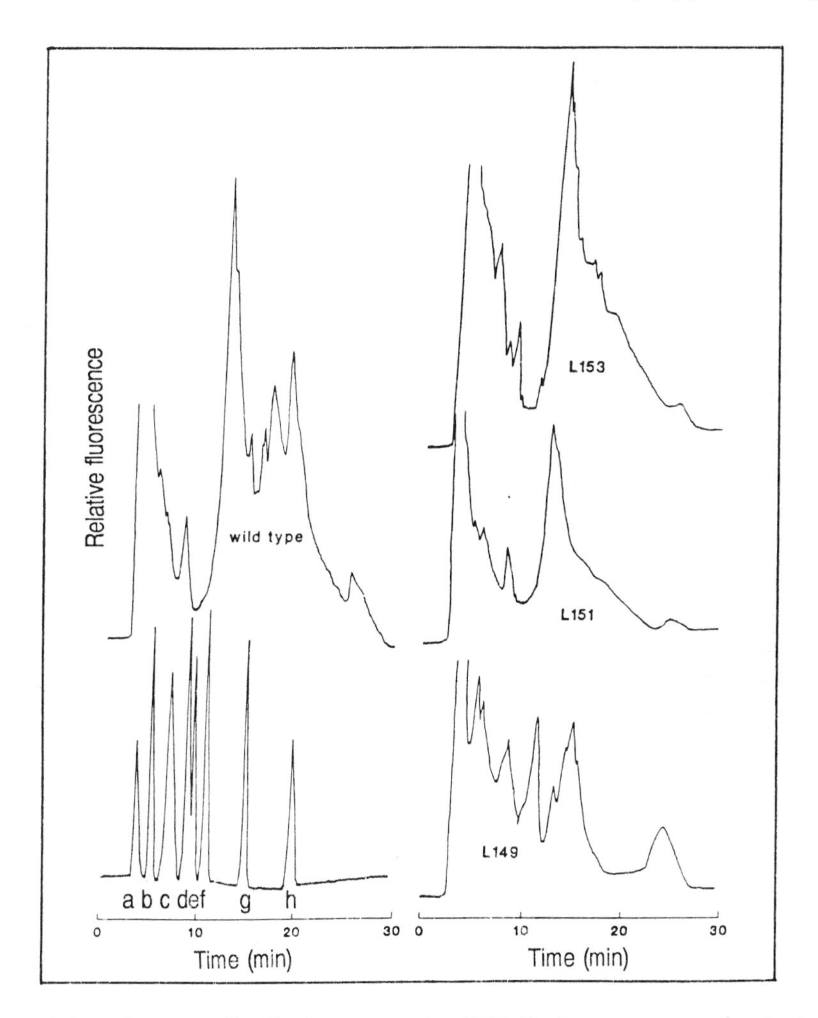

Figure 5. High performance liquid chromatography (HPLC) chromatograms of extracts from sporangiophores from the wild-type strain and three mutant strains of *Phycomyces* with deviating phototropic response. Fluorescence was excited at 360 nm and emission was measured at 440 nm. For comparison the following standard components were measured: a = 6-carboxypterin; b = neopterin; c = xanthopterin; d = biopterin; e = 6-hydroxymethylpterin; f = pterin; g = 6-methylpterin; h = 6,7-dimethylpterin (modified from Hohl *et al.* 1992).

all flavin properties known from flavin chemistry in solution are specifically modified under membrane-bound conditions by lipid type, lipid phase and specific orientation and localization of the flavin nucleus within the membrane (Schmidt 1984a). Membrane-bound flavins can mediate membrane transport of electrons and protons resulting in redox- and pH-gradients. The mechanisms involved are complex, comprising molecular species such as dihydroflavin, flavosemiquinone, superoxide and singlet oxygen (Schmidt 1984b). These results might serve to understand the principles underlying B/UV physiology.

5.1.7 Methodological problems

So far serious methodological problems have impeded clear-cut and straight forward scientific progress in B/UV physiology as known for other biological signal-transducing photoreceptors such as stentorin, phytochrome or rhodopsin. First of all, the concentration of the B/UV photoreceptor is probably low: it corresponds to an absorbance (*A*) of as little as 10^{-4} with respect to phototropism of the sporangiophore of *Phycomyces* (Bergmann 1972). Moreover, the absorption of the B/UV receptor coincides with the absorption of the abundant pigments such as flavins or carotenoids in the cell not being involved in UV/B responses. However, the action of those 'bulk pigments' as 'light-harvesting antennae', as is well established in photosynthesis, cannot be excluded; it remains one plausible explanation for the exceedingly high sensitivity of various physiological B/UV effects (Table 1).

Based upon kinetic analyses and various action spectra published over the years (Figs. 1 and 2) we have to assume the existence of more than one unique B/UV photoreceptor (*i.e.* different chromophores and/or different apoproteins) and/or their/its localization at different sites in the cell. This might even hold true in one and the same cell exhibiting various B/UV phenomena independent from each other. For example, the B/UV responses known in *Neurospora* are the inhibition of circadian rhythm of conidiation, induction of carotenogenesis, phase shifting of a circadian rhythm, induction of protoperithecia formation, phototropism of perithecial beaks, light-promoted conidiation of starved mycelia of the double mutant *albino band*, and LIACs.

More than one photoreceptor for B/UV and additional longer wavelength photoreceptors have to be anticipated for the pigment mutant C-2A' of *Scenedesmus.* Blue light of low irradiance enhances respiration and induces formation of soluble proteins, whereas B of high irradiance stimulates the formation of structural thylakoid proteins. Both R and B are necessary for protochlorophyllide photoreduction and a green light photoreceptor seems to regulate porphyrin biosynthesis (Fig. 6).

However, the assumption of several chemically distinct B/UV photoreceptors in one specific or in different organisms is not necessarily valid, since various factors might influence the spectral behaviour. Intermolecular interaction can induce changes in absorption of a pigment, *e.g.* the micro-environment of the lipid membrane has a great impact upon absorption of a solute (here the B/UV photoreceptor). In addition, binding of small molecules (co-enzymes) to (apo) proteins and/or membranes have a demobilizing effect, similar to that of low temperature, and will consequently change the absorption, *i.e.* specifically the potential action spectra. Sugai and co-workers (1984) demonstrated the modifications of absorption and action spectra by the cell wall. It was also shown that shorter wavelengths might damage the photoreceptor by photolysis thereby truncating the action spectrum (Inoue and Watanabe 1984). Finally we have to

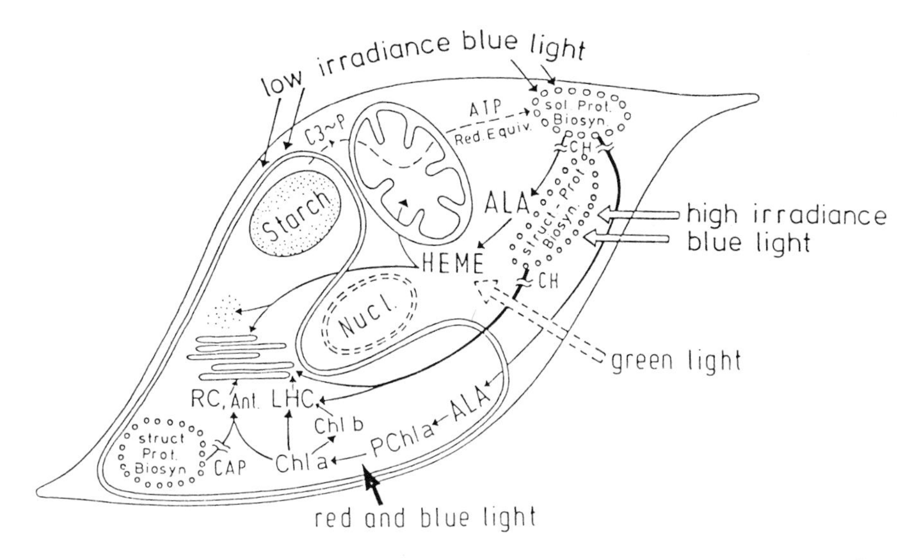

Figure 6. Schematic summary of the various light effects known in the green alga *Scenedesmus*. More than one photoreceptor for blue light and in addition pigments absorbing in other spectral regions have to be considered (see text for details).

take into account the possibility that the impinging light of a specific wavelength is converted into light of another, now physiologically active wavelength, by some physical processes, such as fluorescence.

Another experimental fact should be kept in mind when interpreting action spectra of directional responses. For example, in phototropism of grass coleoptiles the light direction is detected by monitoring the internal light distribution. This, in turn, is predetermined by bulk pigments not being primarily involved in phototropism which also absorb in the B region of the spectrum. As a result (in mathematical terms) the action spectrum represents the convolution of the absorption spectrum of the B photoreceptor with the absorption spectra of these bulk pigments. Finally, wavelength-dependent scattering effects have to be taken into account. Since none of these components is precisely known by itself, there is no unique way of 'reconstructing' action spectra; so far all attempts have failed (*cf.* Chapter 7.4).

Recent investigations (Bornman and Vogelmann 1991) have demonstrated that light distributions within a leaf undergo considerable changes. Hence, the level of light reaching the photoreceptor within the leaf might be significantly different from the irradiance of the incident light.

Several pigments absorb specifically in the B region of the spectrum (Fig. 7). Their absorption spectra might be superimposed by tails of the Soret bands of other pigments. For example, tails of the Soret bands of haemes or chlorophylls extend into the B region. Thus, neglecting such spectral interference might lead

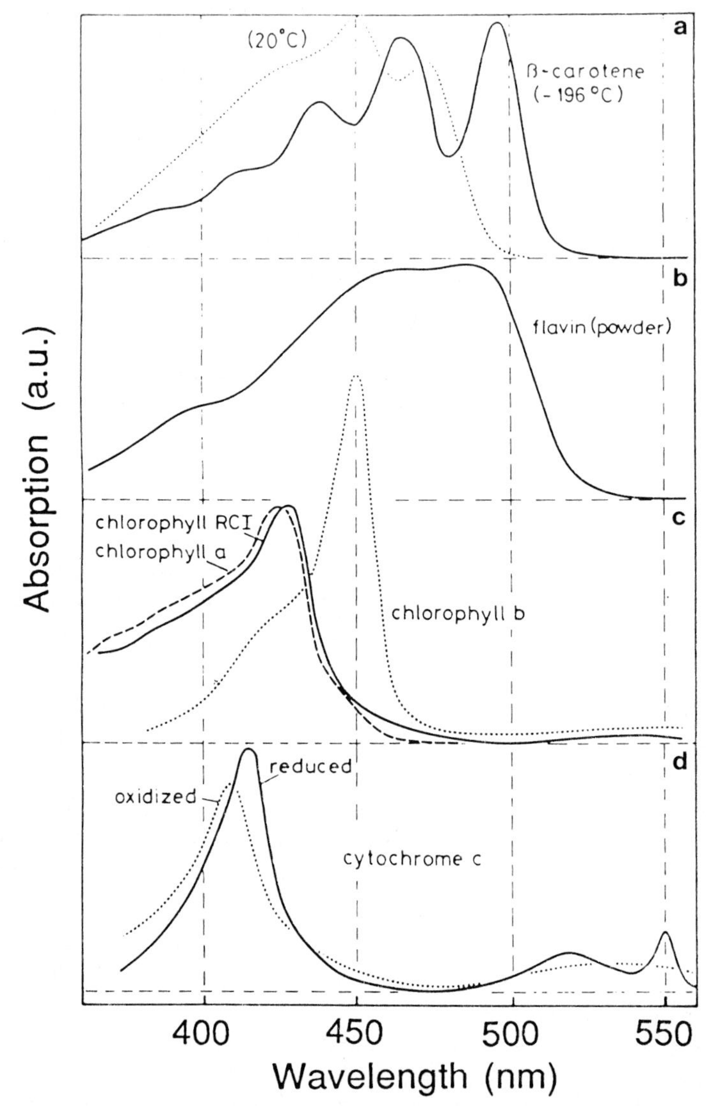

Figure 7. Absorption spectra (in arbitrary units, a.u.) of various pigments with respect to the UV-B, blue (B) and green parts of the spectrum possibly relevant in physiological B action. (a) Spectra of ß-carotene at room temperature and at liquid nitrogen temperature reveal dramatic differences due to the hindered rotational motility of the aliphatic chain at −196°C. In contrast, the corresponding absorption spectra of flavins differ much less. (b) Spectrum of a thin dry layer of crystalline flavin evaporated from ethanolic solution: the UV-A peak characteristic for and commonly taken to identify flavins is suppressed. (c) Chlorophyll-a, chlorophyll-b and chlorophyll RCI, which has absorption characteristics similar to chlorophyll-a (with the Soret-band at slightly shorter wavelength) and which is assumed to be intimately correlated, or even identical, with the reaction centre of photosystem I. (d) Absorption spectra of cytochrome-c in its oxidized and reduced forms (room temperature).

to erroneous interpretation of action spectra (*cf.* Section 5.1.4.1). Probably the best documented example for the co-operative involvement of several pigments in a physiological photoreaction is the interaction of phytochrome and the B photoreceptor in anthocyanin biosynthesis in seedlings of higher plants (Chapters 4.7, 6.0 and 9.5). In *Sorghum vulgare* B is indispensable for establishing the responsiveness to phytochrome, whereas in *Sinapis alba* it only amplifies the responsiveness to phytochrome (with respect to anthocyanin biosynthesis). In *Triticum aestivum* a UV-B photoreceptor replaces the B photoreceptor necessary for eliciting the phytochrome-controlled anthocyanin biosynthesis.

Summarizing, being aware of the uncertainties mentioned above a 'fine structure analysis' of action spectra, as frequently attempted, remains futile. To supplement action spectroscopy by additional, 'nonspectroscopic' data is an indispensable prerequisite when analyzing physiological B action.

5.1.8 Terminology

As long as the chemical nature of the B/UV photoreceptor(s) is (are) uncertain and the existence of one or more chromophores is still a matter of debate it is very difficult to introduce a proper terminology. Currently, photoreceptors absorbing in the blue region of the spectrum, but not beyond 520 nm are collectively called (physiologically) 'B photoreceptors'. Another critical point is the separation of 'pure' B/UV effects from physiological activities of light in the green and B/UV range.

There are specific UV-B photoreceptors with a single peak at 290 nm (Yatsuhashi *et al.* 1982) or at 300 nm as part of the mycochrome system. The mycochrome system, which controls the induction of fungal conidiation, consists of two photoconvertible systems, the UV-B and a B photoreceptor (Chapter 9.6). A final terminology for the different members of the family of B/UV photoreceptors has to be postponed until we know more about their mechanism and chemical nature.

The action spectra of some B/UV-A responses follow a characteristic pattern with peaks or shoulders at 420, 450 and 480 nm and sometimes an additional peak between 360 to 380 nm. Photoreceptors of this type have been proposed to be called 'cryptochrome' (*cf.* Senger 1984b). However, this term is obsolete and useless and should be avoided for several reasons: (i) it leads to the incorrect assumption that only one group or even one single photoreceptor is responsible for the B/UV absorption; (ii) most of the B-action spectra (Figs. 1 and 2) do not fit into the proposed absorption scheme; (iii) the term 'chryptochrome' has for a long time been used for a specific carotenoid (see Karrer and Jucker 1948).

5.1.9 Ecological aspects and outlook

Due to the sun's surface temperature (*ca.* 6000 K), the radiation reaching the earth covers the range between 290 and 1000 nm, including the whole visible range between 380 and 750 nm, and is further restricted by the ozone layer at the B-end and by water vapour and carbon dioxide at the R-end of the spectrum. In addition, plants growing under the canopy of other plants, or in deeper waters or oceans and lakes, experience a strongly modified sun spectrum in time and space (Chapter 7.1). In order to cope with the specific environmental conditions, a confusing complexity of photoreceptors absorbing B/UV has evolved.

The ecological relevance of B-reactions has clearly been demonstrated for the control of stomatal opening (Chapter 9.3.; Zeiger 1984), pigment adaptation of algae (Dring 1984; Jeffrey 1984), and adaptation of the photosynthetic apparatus to shade conditions (Humbeck *et al.* 1984). More findings of B responses involved in ecological adaptations have to be expected.

Whereas two decades ago 'B responses' were a minor topic in the general field of photobiology, nowadays they are subdivided into various disciplines: *e.g.* phototropism, carotenogenesis, light-induced respiration. However, this is by no means a cause for despair, but reflects the typical pattern of scientific progress: (i) recognition of a general phenomenon; (ii) analysis of a great number of seemingly unrelated aspects; (iii) identification of the causal connection of reactions leading from the triggering effect to the final event; (iv) the development of a 'general theory'. The field of B/UV research has now entered the third state of this development.

5.1.10 References

Benedetti P.A. and Lenci F. (1977) *In vivo* microspectrofluorometry of photoreceptor pigments in *Euglena gracilis*. *Photochem. Photobiol.* 26: 315-318.

Bergmann K. (1972) Blue light control of sporangiophore initiation in *Phycomyces*. *Planta* 107: 56-67.

Blatt M.R. and Briggs W.R. (1980) Blue light-induced cortical fibre reticulation concomitant with chloroplast aggregation in the alga *Vaucheria sessilis*. *Planta* 147: 355-362.

Bornman J.F. and Vogelmann T.C. (1991) The effect of UV-B radiation on leaf optical properties measured with fibre optics. *J. Exp. Bot.* 42: 547-554.

Brinkmann G. and Senger H. (1978a) The development of structure and function in chloroplasts of greening mutants of *Scenedesmus*. IV. Blue light-dependent carbohydrate and protein metabolism. *Plant Cell Physiol.* 19: 1427-1437.

Brinkmann G. and Senger H. (1978b) Light-dependent formation of thylakoid membranes during the development of the photosynthesis apparatus in pigment mutant C-2A' of *Scenedesmus obliquus*. In: *Chloroplast Development. Development in Plant Biology, 2,* pp. 201-206, Akoyunoglou G. and Argyroudi-Akoyunoglou J.H. (eds.) Elsevier/North Holland Biomedical Press, Amsterdam.

Chanderkar L.P. and Schuman Jorns M. (1991) Effect of structure and redox state on catalysis by and flavin-pterin energy transfer in *Escherichia coli* DNA photolyase. *Biochemistry* 30: 745-754.

Chon H.P. and Briggs W.R. (1966) The effect of red light on the phototropic sensitivity of corn coleoptiles. *Plant Physiol.* 41: 1715-1724.

DeFabo E.C., Harding R.W. and Shrophire Jr. W. (1976) Action spectrum between 260 and 800 nanometres in *Neurospora crassa. Plant Physiol.* 57: 440-445.

Dring M.J. (1984) Blue light effects in marine macroalgae. In: *Blue Light Effects in Biological Systems,* pp. 509-516, Senger H. (ed.) Springer-Verlag, Berlin.

Foster K.W., Saranak J., Derguini F. and Nakanishi K. (1988) In: *Molecular Physiology of Retinal Proteins,* pp. 61-66, Hara T. (ed.) Yamada Science Foundation, Osaka.

Galland P. (1992) 40 years of blue-light research and no anniversary. *Photochem. Photobiol.* 56: 847-853.

Galland P. and Senger H. (1988a) The role of flavins as photoreceptors. *J. Photochem. Photobiol. B* 1: 277-294.

Galland P. and Senger H. (1988b) The role of pterins in the photoreception and metabolism of plants. *Photochem. Photobiol.* 48: 811-820.

Galland P., Keiner P., Dörnemann D., Senger H., Brodhun B. and Häder D.-P. (1990) Pterin- and flavin-like fluorescence associated with isolated flagella of *Euglena gracilis. Photochem. Photobiol.* 51: 675-680.

Galston A.W. and Baker R.S. (1949) Studies on the physiology of light action. II. The photodynamic action of riboflavin. *Amer. J. of Bot.* 36: 773-780.

Hartmann K.M. (1967) Ein Wirkungsspektrum der Photomorphogenese unter Hochenergiebedingungen und seine Interpretation auf der Basis des Phytochroms (Hypokotylwachstumshemmung bei *Lactuca sativa L.*). *Z. Naturforsch.* 22b: 1172-1175.

Hohl N., Galland P. and Senger H. (1992a) Altered pterin patterns in photobehavioral mutants of *Phycomyces blakesleeanus. Photochem. Photobiol.* 55: 239-246.

Hohl N., Galland P. and Senger H. (1992b) Altered flavin patterns in photobehavioral mutants of *Phycomyces blakesleeanus. Photochem. Photobiol.* 55: 247-255.

Hohl N., Galland P., Senger H. and Eslava A.P. (1992) Altered pterin patterns in photoreceptor mutants of *Phycomyces blakesleeanus* with defective madl gene. *Bot. Acta* 6, 105: 441-448.

Humbeck K., Schumann R. and Senger H. (1984) The influence of blue light on the formation of chlorophyll-protein complexes in *Scenedesmus*. In: *Blue Light Effects in Biological Systems,* pp. 359-365, Senger H. (ed.) Springer-Verlag, Berlin.

Inoue Y. and Watanabe M. (1984) Perithecial formation in *Gelasinospora reticulispora*. VII. Action spectra in the UV region for the photoinduction and photoinhibition of photoinductive effect brought by blue light. *Plant Cell Physiol.* 25: 107-113.

Jeffrey S.W. (1984) Responses of unicellular marine plants to natural blue-green light environments. In: *Blue Light Effects in Biological Systems,* pp. 497-508, Senger H. (ed.) Springer Verlag, Berlin.

Johnson J.L., Hamm-Alvarez S. Payne G., Sancar G.B., Rajagopalan K.V. and Sancar A. (1988) Identification of the second chromophore of *Escherichia coli* and yeast DNA photolyases as 5,10-methenyltetrahydrofolate. *Proc. Natl. Acad. Sci. USA* 85: 2046-2050.

Karrer P. and Jucker E. (1948) *Carotinoide* Verlag Birkhäuser, Basel.

Kasemir H. and Mohr H. (1981) The involvement of phytochrome in controlling chlorophyll and 5-aminolevulinate formation in a gymnosperm seedling (*Pinus silvestris*). *Planta* 152: 369-373.

Klemm E. and Ninnemann H. (1978) Correlation between absorbance change and a physiological response induced by blue light in *Neurospora. Photochem. Photobiol.* 28: 227-230.

Klemmer R. and Schneider Hj. A.W. (1979) On a blue light effect and phytochrome in the stimulation of georesponsiveness of maize roots. *Z. Pflanzenphysiol.* 95: 189-197.

Kowallik W. (1969) Der Einfluβ von Licht auf die Atmung von *Chlorella* bei gehemmte Photosynthese. *Planta* 86: 50-62.

Kowallik W. (1982) Blue light effects on respiration. *Ann. Rev. Plant Physiol.* 33: 51-72.

Kowallik W. and Schänzle S. (1980) Enhancement of carbohydrate degradation by blue light. In: *The Blue Light Syndrome,* pp. 344-360, Senger H. (ed.) Springer-Verlag, Berlin.

Kumagai T. (1983) Action spectra for the blue and near ultraviolet reversible photoreaction in the induction of fungal conidiation. *Physiol. Plant.* 57: 468-471.

Lipson E.D., Galland P. and Pollock J.A. (1984) Blue Light receptors in *Phycomyces* investigated by action spectroscopy, fluorescence lifetime spectroscopy, and two-dimensional gel electrophoresis. In: *Blue Light Effects in Bioiogical Systems,* pp. 228-236, Senger H. (ed.) Springer-Verlag, Berlin.

Mohr H. (1956) Die Abhängigkeit des Protonemawachstums und der Protonemapolarität bei Farnen vom Licht. *Planta* 47: 127-158.

Munõz V. and Butler W.L. (1975) Photoreceptor pigment for blue light in *Neurospora crassa. Plant Physiol.* 55: 421-426.

Oh-hama T. and Senger H. (1978) Spectral effectiveness in chlorophyll and 5-aminolevulinic acid formation during regreening of glucose-bleached cells of *Chlorella protathecoides. Plant Cell Physiol.* 19: 1295-1299.

Roldán J.M. and Bulter W.L. (1980) Photoactivation of nitrate reductase from *Neurospora crassa. Photochem. Photobiol.* 32: 375-381.

Ruyters G. (1984) Effects of blue light on enzymes. In: *Blue Light Effects in Biological Systems,* pp. 283-301, Senger H. (ed.) Springer-Verlag, Berlin.

Sachs J. (1864) Wirkungen des farbigen Lichts auf Pflanzen. *Bot. Z.* 353-358.

Saito N. and Werbin H. (1970) Purification of a blue-green algal deoxyribonucleic acid photoreactivation enzyme. An enzyme requiring light as physical cofactor to perform its catalytic function. *Biochemistry* 9: 2610-2620.

Schmid G.H. (1970) The effect of blue light on some flavin enzymes. *Hoppe-Seyler's Z. Physiol. Chem.* 351: 575-578.

Schmid R. (1984) Blue light effects on morphogenesis and metabolism in *Acetabularia*. In: *Blue Light Effects in Biological Systems,* pp. 419-432, Senger H. (ed.) Springer-Verlag, Berlin.

Schmidt W. (1984a) Blue light physiology. *Bioscience* 34: 698-704.

Schmidt W. (1984b) Blue light-induced, flavin-mediated transport of redox equivalents across artificial bilayer membranes. *J. Memb. Biol.* 82: 113-122.

Schmidt W., Galland P., Senger H. and Furuya M. (1990) Microspectrophotometry of *Euglena gracilis*. Pterin- and flavin-like fluorescence in paraflagellar body. *Planta* 182: 375-381.

Schuman Jorns M., Sancar G.B. and Sancar A. (1984) Identification of a neutral flavin radical and characterization of a second chromophore in *Escherichia coli* DNA-photolyase. *Biochemistry* 23: 2673-2679.

Senger H., ed. (1980) *The Blue Light Syndrome,* Springer-Verlag, Berlin.

Senger H., ed. (1984a). *Blue Light Effects in Biological Systems,* Springer-Verlag, Berlin.

Senger H. (1984b) Cryptochrome, some terminological thoughts. In: *Blue Light Effects in Biological Systems,* pp. 72, Senger H. (ed.) Springer-Verlag, Berlin.

Senger H., ed. (1987) *Blue Light Responses: Phenomena and Occurrence in Plants and Microorganisms,* Vol 1 and 2, CRC Press, Boca Raton, FA.

Senger H. and Briggs W.R. (1981) The blue light receptor(s): Primary reactions and subsequent metabolic changes. In: *Photochem. Photobiol. Rev. 6,* pp. 1-38, Smith K.C. (ed.) Plenum, New York.

Short T.W. and Briggs W.R. (1990) Characterization of a rapid, blue light-mediated change in detectable phosphorylation of a plasma membrane protein from etiolated pea (*Pisum sativum* L.) seedlings. *Plant Physiol.* 92: 179-185.

Short T.W., Porst M. and Briggs W.R. (1992) A photoreceptor system regulating *in vivo* and *in vitro* phosphorylation of a pea plasma membrane protein. *Photochem. Photobiol.* 55: 773-781.

Shropshire Jr. W. and Withrow R.B. (1958) Action spectrum of phototropic tip-curvature of *Avena. Plant Physiol.* 33: 360-365.

Smith H. (1975) *Phytochrome and Photomorphogenesis.* McGraw-Hill, London.

Sugai M., Tomizawa K., Watanabe M. and Furuya M. (1984) Action spectrum between 250 and 800 nanometres for the photoinduced inhibition of spore germination in *Pteris vittata*. *Plant Cell Physiol.* 25: (2), 205-212.

Warpeha M.F., Kaufmann L.S. and Briggs W.R. (1991a) A flavoprotein may mediate the blue light-activated binding of GTP to isolated plasma membranes of *Pisum sativum*. *Photochem. Photobiol.* 55: 595-603.

Warpeha M.F., Hamm H.E., Rasenick M.M. and Kaufmann L.S. (1991b) A blue-light-activated GTP-binding protein in the plasma membranes of etiolated peas. *Proc. Natl. Acad. Sci. USA* 88: 8925-8929.

Widell S., Caubergs R.J., and Larsson C. (1983) Spectral characterization of light-reducible cytochrome in a plasma membrane-enriched fraction and in other membranes from cauliflower inflorescences. *Photochem. Photobiol.* 38: 95-98.

Widell S. (1987) Membrane-bound blue light receptors - Possible connection to blue light photomorphogenesis. In: *Blue Light Responses: Phenomena and Occurrence in Plants and Microorganisms*, Vol. 2, pp. 89-98, Senger H. (ed.) CRC Press Inc., Boca Raton, Florida.

Widell S. and Larsson C. (1987) Plasma membrane purification. In: *Blue Light Responses: Phenomena and Occurrence in Plants and Microorganisms*, Vol. 2, pp. 99-107, Senger H. (ed.) CRC Press Inc., Boca Raton, Florida.

Yatsuhashi H., Hashimoto T. and Shimizu S. (1982) Ultraviolet action spectrum for anthocyanin formation in broom sorghum first internodes. *Plant Physiol.* 70: 735-741.

Zeiger E. (1984) Blue light and stomatal function. In: *Blue Light Effects in Biological Systems*, pp. 484-494, Senger H. (ed.) Springer-Verlag, Berlin.

Zurzycki J. (1967) Properties and localization of the photoreceptors active in displacement of chloroplasts in *Funaria hygrometrica*. I. Action Spectrum. *Acta Soc. Bot. Pol.* 36: 133-142.

5.2 Properties and transduction chains of the UV and blue light photoreceptors

Benjamin A. Horwitz

Department of Biology, Technion, Israel Institute of Technology, Haifa 32000, Israel

5.2.1 Introduction

Blue light (B) acts on a very wide range of processes and organisms (Chapter 5.1). Although action spectra do not always faithfully reflect the absorption spectra of the active pigments, it is still almost certain that a single chromophore cannot account for the varied array of action spectra that have been determined. Some organisms respond to most of the B and UV-A region, often following the spectral pattern that defines hypothetical B/UV-A photoreceptor(s), referred to as cryptochrome (Senger 1987), while the active wavelengths for others are confined to a single narrow peak (Chapter 5.1). Whereas many responses are apparent, the chemical identity of B perceiving chromophores is still only guessed at, despite years of study: flavins, pterins, haems, carotenoids or others. At first glance, there is considerable difficulty in discussing the transduction pathways of photoreceptors that have not yet been chemically characterized. However, detailed knowledge of the phytochrome chromophore and apoproteins has not yet led to knowledge of the steps in its transduction chain, so similar questions remain unanswered for photoresponses at both the red (R) and B ends of the visible spectrum. The problem of B transduction must be considered within the framework of the more general problem of signal transduction in plant cells. Stimuli other than B include: R, nutrition, gravity, stress, pathogens, *etc.* These stimuli start transduction chains leading to a variety of responses, some of which may be shared by B. Some of the B-perceiving organisms and responses, including plant and fungal phototropism and growth rate changes, induction of fungal sporulation and modified biosynthesis, are particularly amenable to genetic analysis.

Mutants can be used to ask the important question of whether a transduction step is upstream or downstream from the defect, or maybe not part of the pathway at all. Mutants of *Arabidopsis* with defects in phototropism and other responses to B are providing insight in this way (Section 5.2.5; Chapter 8.2). Fungi, traditionally

R.E. Kendrick & G.H.M. Kronenberg (eds.), Photomorphogenesis in Plants - 2nd Edition

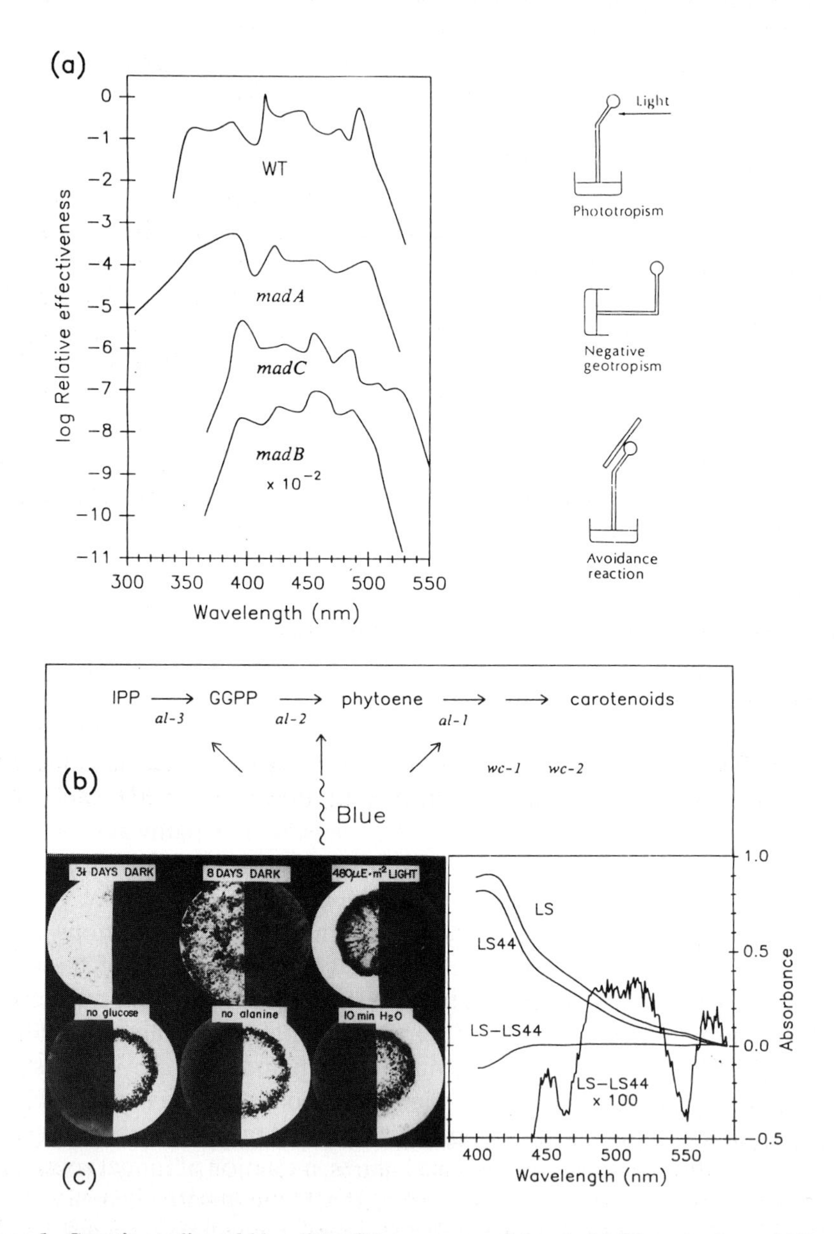

Figure 1. Genetic studies of blue light (B) responses of fungi. (a) Phototropism of *Phycomyces* sporangiophores. *Left* panel: modified action spectra show that *madB, C* are likely photoreceptor mutants [Galland P. and Lipson E.D. (1985) *Photochem. Photobiol.* 41: 331-335]. The *madB* curve is shifted 100-fold (2 log units) downward, for convenient comparison with *madC*. *Right* panel: the sporangiophore responds to light, gravity, and avoids barriers. The *madA, B, C* gene products are required for phototropism, but not the other tropisms. Mutants in *madD, E, G* loci are altered in all

studied along with plants, are evolutionarily distant and may have correspondingly different, perhaps unique, transduction pathways. Yet, the action spectra of some fungal responses are surprisingly similar to those in plants.

Genetic work on B responses of fungi began over 20 years ago when M. Delbrück's group focused their attention on the Zygomycete *Phycomyces blakesleeanus,* isolating *mad* mutants with defective phototropism (Fig. 1a; Cerdá-Olmedo and Lipson 1987). Mutants of the Ascomycete *Neurospora crassa* that produce carotenoids in the conidia (asexually produced spores), but not in the mycelia, were described even earlier, and the photobiological significance of these 'white collar' *(wc)* loci is now clear (Fig. 1b). The *dim* mutants of *Trichoderma harzianum* (conidial form of an Ascomycete) sporulate when stressed, but are less sensitive to B (Fig. 1c). These three B-sensitive fungi share an important property: there is an alternate pathway to stimulate a response without light (Fig. 1). Such pathways exist for higher plants as well (Chapter 8.2). Molecular genetics, combined with early biochemical changes such as protein phosphorylation, are providing the assays, so far lacking, in the search for B/UV-A photoreceptors and their transducers. This chapter will follow the B signal, beginning with the primary photochemistry of the likely candidates for the chromophores.

5.2.2 Excited state chemistry of the chromophores

5.2.2.1 Flavins

Flavins and flavoproteins are potentially active photochemically, transferring electrons from electron donors. Flavins have long-lived singlet and triplet excited

---

tropisms and growth responses. (b) Photoregulatory mutants of *Neurospora.* Carotene biosynthesis is constitutive in conidia, but requires light in the mycelium. The white collar *wc-1* and *wc-2* gene products are required for carotene synthesis in the mycelium, and also for other blue B effects. The *al* gene products are biosynthetic enzymes: *al-1* encodes phytoene dehydrogenase, *al-2* corresponds to phytoene synthetase, and *al-3* encodes geranylgeranylpyrophosphate (GGPP) synthetase [Perkins D.D., Glassey M. and Bloom B.A. (1962) *Can. J. Genet. Cytol.* 4: 187-205; Schmidhauser T.J., Lauter F.R., Russo V.E.A. and Yanofsky C. (1990) *Mol. Cell. Biol.* 10: 5064-5070.; Nelson M.A., Morelli G., Carattoli A., Romano N. and Macino G. (1989) *Mol. Cell. Biol.* 9: 1271-1276; Harding R.W. and Melles S. (1983) *Plant Physiol.* 72: 996-1000]. (c) Blue-light induction of conidiation in *Trichoderma. Left* panel: colonies conidiate in response to light, osmotic stress (10 min in H_2O) or nutritional stress (8 days growth in the dark; transfer to medium lacking carbon or nitrogen source). The *dim* mutants have a lowered sensitivity to light, and respond normally to stresses. *Right* panel: *dimY* mutants have altered absorption spectra and are likely photoreceptor mutants [previously unpublished *in vivo* spectra; see: Horwitz B.A., Gressel J., Malkin S. and Epel B.L. (1985) *Proc. Natl. Acad. Sci. USA* 82: 2736-2740]. LS44 is a *dimY* strain; LS is the isogenic control, with wild-type (WT) photoresponse. The normalized difference spectrum shows overproduction of a substance absorbing near 400 nm, and altered fine structure at 450, 480, 520 and 550 nm.

states, from which several reactions are possible (Fig. 2): photosensitized oxidations, electron transfer without the participation of molecular oxygen, and proton exchange. The pK of the proton at position N5 in the molecule is increased several units upon excitation from the ground state to the first triplet state. The singlet oxygen pathway might be damaging to the cell, and less specific than some of the other reactions. The reaction mechanism of DNA photolyase, the only B receptor that is chemically well-characterized, is electron transfer from the excited singlet of (reduced) flavin to the pyrimidine dimer (Jorns *et al.* 1990). An electron transfer of this kind could also work for B/UV-A photoreceptors (with flavin chromophores) although other possibilities remain open.

Association with membranes may significantly alter the photochemical properties of flavins, as is the case for other pigments. Model systems have been designed in which B can generate redox or pH gradients across a membrane containing lipid-anchored flavins. Flavins were incorporated into artificial membrane vesicles by covalently attaching them to hydrocarbon chains. The

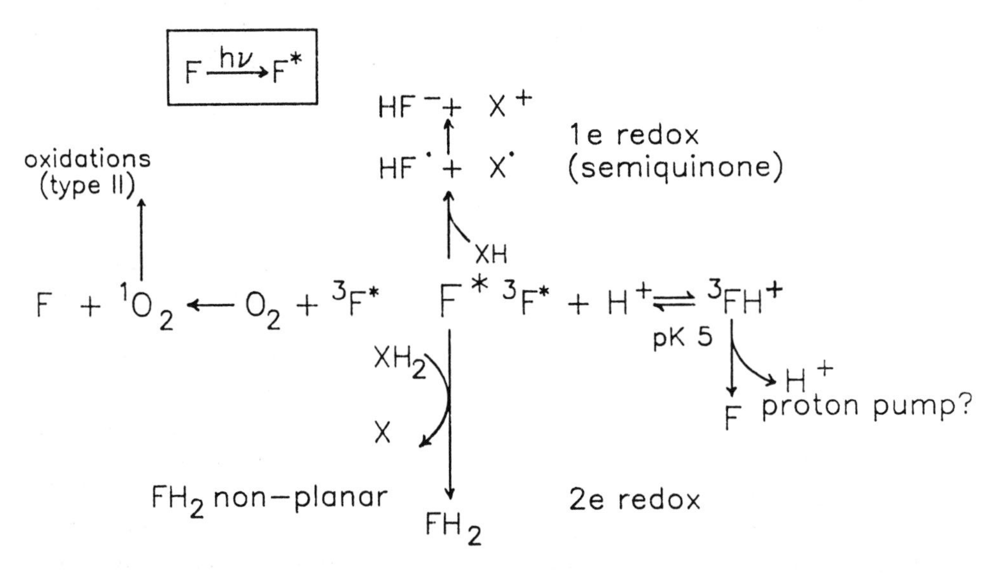

Figure 2. Possible routes for flavin excited state chemistry. F, oxidized flavin; FH_2, reduced flavin (dihydroflavin); X, hypothetical redox substrate; F^*, excited state, superscripts 1 and 3 indicate singlet or triplet. Flavin redox reactions proceeding *via* the semiquinone are one- electron reactions, in contrast to two-electron reactions in which there is no semiquinone intermediate. The two-electron reduction is the one normally used in (dark) enzymatic electron transfer. The semiquinone absorbs green and red, in addition to blue light (B) and UV-A; participation of a flavo-enzyme semiquinone in photoreception could explain some of the complex action spectra that are observed [Hertel R. (1980) In: *Photoreception and Sensory Transduction in Aneural Organisms,* pp. 89-105, Lenci F., Colombetti G. (eds.) Plenum, New York]. Excitation to the singlet or triplet makes the pK for protonation at position N5 more basic (pK = 5) than in the ground state [Song P.-S. (1968) *Photochem. Photobiol.* 7: 311-313]. This could allow transfer of protons across a membrane. Reduced flavin has a 'bent' configuration, in contrast to oxidized flavin which is planar. Photoreduction could therefore cause a conformational change in the (hypothetical) apoprotein.

vesicles were loaded with an electron acceptor (cytochrome c), and an electron donor was supplied outside. Upon illumination with B, redox equivalents traversed the membrane (Schmidt 1984).

The flavin chromophore has often been thought to be tightly bound to an apoprotein. However, free soluble flavins can be very active photosensitizers. *Neurospora* requires more flavin to respond to light than it needs for growth. Several lines of evidence, including flavin analogues, estimates of flavin concentrations, and the behaviour of flavin-requiring mutants, point to free cellular riboflavin, acting *via* the triplet state, as the photoreceptor (Fritz *et al.* 1990). In *Trichoderma,* the situation seems to be reversed: even a low flavin concentration (micromolar riboflavin in the growth medium) is sufficient for induction of sporulation, while the sporulation process itself needs more flavin (Horwitz and Gressel 1983). Phototropism mutants of *Phycomyces,* particularly *madA,* have altered flavin compositions, genetic evidence for flavin chromophores (Chapter 5.1). If free cellular riboflavin is the photoreceptor, it must interact with sites (binding proteins?) that confer the specificity to an otherwise general photosensitization reaction. Evidence for riboflavin-binding sites has been found in plant and fungal membranes, but the binding proteins have not yet been isolated (Nebenfuhr *et al.* 1991).

5.2.2.2 Carotenoids

Carotenoids have fewer known photochemical reactions than do flavins. Their excited states have very short lifetimes, and the excitation is rapidly lost by internal conversion: transitions between closely spaced vibrational and rotational levels dissipate the energy as heat. Therefore carotenoids appear non-fluorescent, compared to flavins. In contrast to the retinal chromophores of visual rhodopsin and bacteriorhodopsin, ß carotene does not photo-isomerize readily under physiological conditions. According to the most recent assignment of the lowest excited state of carotenes, the S_1 state is actually reached *via* an optically forbidden transition, while the major absorption band in the B would be attributed to the (allowed) S_0 to S_2 transition (Koyama 1991). If so, a primary photoreaction could proceed from the S_1 state, which would have a relatively long lifetime. Theoretical predictions based on this assignment of states suggest that carotene should be active at longer wavelengths, corresponding to direct excitation to S_1, probably in the R (Song in: Senger 1987). Carotenoids, if not primary acceptors, could still act as antenna pigments, transferring excitation energy to a photochemically active reaction centre. This is an important process in the photosynthetic membranes of bacteria and plants, and the combined action of carotenoids and chlorophyll might explain some combined effects of R and B that do not fit the phytochrome high irradiance response (HIR) pattern.

The genetic evidence is against participation of carotenoids in *Phycomyces* phototropism, because the carotene-free *carB,carR* strains respond normally. Yet, the *car* mutations have a dramatic effect on the threshold for sporangiophore development (Corrochano and Cerdá-Olmedo 1991).

5.2.2.3 Pterins

The spectroscopic properties and excited state chemistry of 2-amino-3H-pteridinone (known simply as pterin) have been studied (Chahidi *et al.* 1981). The pterin triplet energy is high enough to allow production of singlet oxygen, and also to react directly with hydrogen donors such as amino acids. Pterins are thus able to carry out at least two of the important photoreactions known for flavins (Fig. 2). The likely similarities between pterin and flavin photochemistry are underscored by the observation that phenylacetic acid and KI, long used as quenchers of flavin excited states, also quench pterin fluorescence (Warpeha *et al.* 1992). Pterins could transfer excitation energy to a flavin reaction centre. This is energetically feasible, and has been shown to occur in DNA photolyase where the 6-substituted dihydropterin chromophore transfers energy to the (reduced) flavin chromophore. Pterin mutants, as such, await isolation; several *Phycomyces mad* mutants have altered pterin profiles (Fig. 5 in Chapter 5.1). Defying a simple linear interpretation, pterins are not drastically reduced in *madB* and *madC,* where the UV-A peak is missing from the action spectra (Fig. 1a). Promising organisms, in addition to those already used (Chapter 5.1) for pterin studies would be fungi in which B and UV-A have opposite effects (Kumagai 1989).

5.2.3 Kinetic properties of the blue light photoreceptors

Photoresponses that obey first-order kinetics, can be straightforwardly described mathematically, as shown in Fig. 3. This model can simulate on the computer screen what the photoproduct concentration in the cell might look like during and after a pulse of B. The basic assumption is that the photoreceptor is converted to a physiologically active product that is slowly regenerated back to the photoreceptive form in the dark. How the photoproduct, 'p', is sensed during its presence in the cell still must be left to the imagination; possible transduction reactions are described later in this chapter. Kinetics of dark-decay can provide hints to the nature of the photoproduct, and its similarity in different organisms. The predicted values of 'k' for the photoreaction controlling stomatal opening in *Commelina communis,* and for maize coleoptile phototropism, are very similar (of the order of 0.001 s^{-1}) while for photocontrol of cell division in gametophytes of the fern *Adiantum,* the predicted dark decay of 'p' was about 3-fold slower (Iino *et al.* 1988). Thus, the photoproduct could be sensed during about 15 min, or much

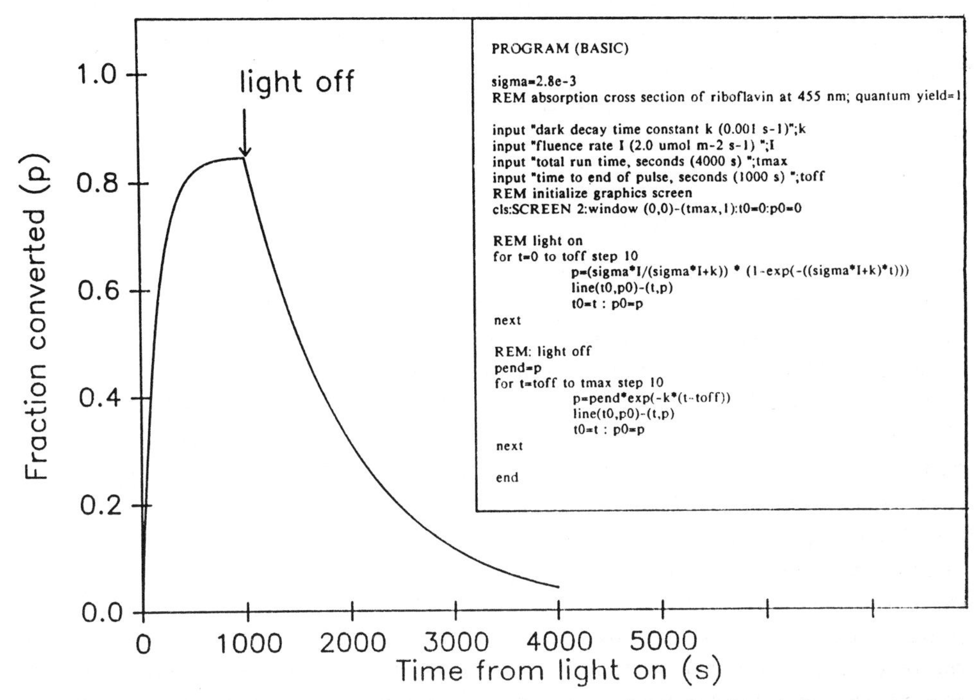

Figure 3. Kinetic modelling of blue light (B) reception. Data simulating the relative photoproduct level, 'p', were generated by a first-order model *(inset),* with the parameter values given in parentheses in the program: sigma is the effective cross section for photoconversion, 'I' is the fluence rate and 'k' is the first-order rate constant for dark regeneration. First-order models can explain B-induced absorbance changes [Lipson E.D. and Presti D. (1980) *Photochem. Photobiol.* 32: 383-391], the response of stomata [Iino M., Ogana T. and Zeiger E. (1985) *Proc. Natl. Acad. Sci. USA* 82: 8019-8024], and sporulation of *Trichoderma* [Horwitz B.A., Perlman A. and Gressel J. (1990) *Photochem. Photobiol.* 51: 99- 104]. The few lines of BASIC shown can be generalized to include fluence-rate gradients, multiple photoreceptors, second-order kinetics of a single photoreceptor, or integration of the photoproduct level over time.

less if the transduction reactions are very rapid. Can the first-order model explain all B responses? The answer is probably no: it does not take into account multiple photoreceptors, changes in sensitivity during illumination, second-order kinetics, and photochromic conversions. One could attempt, for example, to relate the dark adaptation level in phototropism directly to the value of 'p', the relative photoproduct concentration in the cell. Such a hypothesis can be excluded: the dark-adaptation kinetics of phototropism in *Phycomyces* and the coleoptiles of maize and oats show decay of more than one component (Galland 1990). Phototropism requires the storage of spatial information (Galland 1990; Nick and Schäfer 1991), which cannot be explained by photoreceptor kinetics and photoproduct gradients alone. Blue light might initiate a reaction-diffusion system. Such systems have the remarkable theoretical property that either a pulse or continuous stimulus can

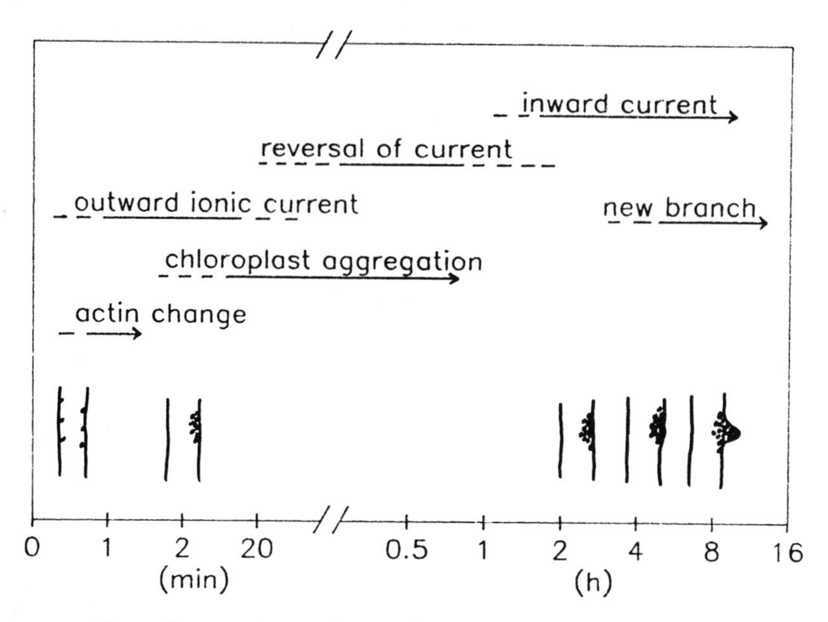

Figure 4. Schematic time course for blue-light induced morphogenesis in *Vaucheria.* Events preceding initiation of a new branch [Blatt M.R., Weisenseel M.H. and Haupt W. (1981) *Planta* 152: 513-526] are shown. These events are not necessarily part of the transduction chain to morphogenesis, though their timing suggests that they may be involved. The *left* half of the figure indicates 'early' changes, while those on the *right* are 'late.'

generate a stable pattern. The light-growth response of *Phycomyces,* because it varies continuously with the stimulus, is amenable to a technique known to engineers as systems analysis. This technique can predict the time course of a response to a stimulus in a general manner, based on a simple kinetic model for the relation between stimulus and response (Cerdá-Olmedo and Lipson 1987).

5.2.4 Rapid effects of blue light and their relevance to transduction

The earliest effects of B might reflect biochemical consequences of the photoreactions. The number of photons needed to trigger some responses to B is so small that amplification must take place. The amplification steps can be classified as 'early' and 'late' based on the time scale of development, as illustrated in Fig. 4 for B induction of branching in *Vaucheria.* This alga (belonging to the Chrysophyta) grows as a coenocytic filament, in which chloroplasts can move and aggregate in response to B. Rapid, or early, changes include: extracellular currents, cytoskeleton modifications, and chloroplast aggregation, all of which precede branching. Late amplifier steps, including gene expression, may well be shared by

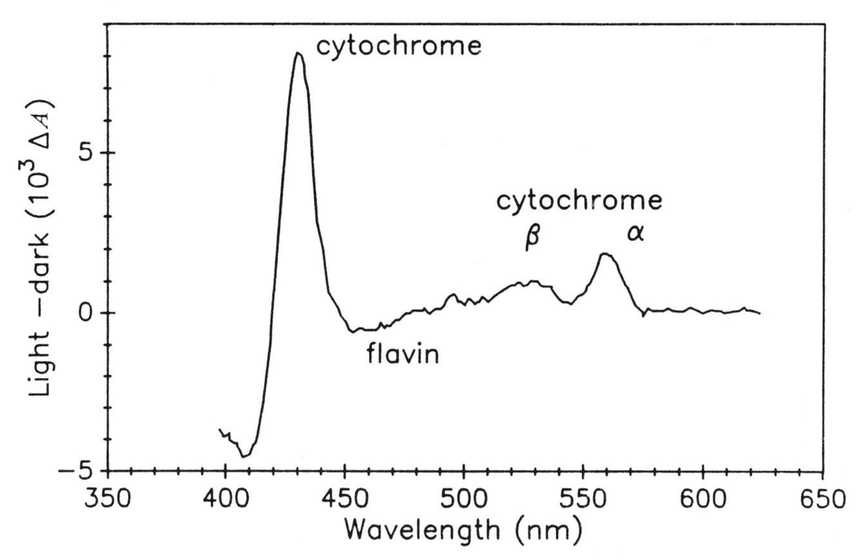

Figure 5. Light-induced absorbance change in purified cauliflower plasma membranes. The difference spectrum shows reduction of a b-type cytochrome. A dip in the spectrum near 450 nm corresponds to the absorbance change expected for flavin reduction, which is more or less evident in difference spectra recorded under different conditions. Redrawn from: [Asard H. and Caubergs R. (1992) In: *Biophysics of Photoreceptors and Photomovements in Microorganisms.* NATO ASI Series, Lenci F., Colombetti G. and Häder D.-P. (eds.) Plenum, New York].

developmental processes induced by B and R as well as those induced by stimuli other than light. An unresolved question is whether the rapid responses are early steps in photomorphogenesis, or merely unrelated effects of light. The rapid inhibition of growth in dicots (Fig. 7a; Chapter 9.1) and the light-induced growth responses of *Phycomyces* (Cerdá- Olmedo and Lipson 1987) are among the fastest known actions of B. The lag times are so short that they place strict time limits (seconds to minutes) on the transduction reactions.

5.2.4.1 Light-induced absorbance changes

Much effort has been invested to find rapid absorbance changes corresponding to the photoreactions, analogous to those of phytochrome. The spectra of the photoreceptors are likely to change as a result of the primary photoreactions. Indeed, an exposure to intense B leads to reduction of a b-type cytochrome (Fig. 5). Flavin reduction is often visible in, or least consistent with, the light-*minus*-dark difference spectra. Because of the way these photoreactions are detected, they are referred to as light-induced absorbance changes (LIACs). The action spectra sometimes, but not always, coincide with those for B responses in the same organisms. Furthermore, LIACs are not saturated by B fluences several orders of magnitude higher than those that saturate photomorphogenesis.

Observations of LIACs are quite universal, in extracts, tissues and cells with or without known photophysiology, and in *in vitro* model photochemical systems. These LIACs have been studied since their discovery by W.L. Butler and co-workers more than 20 years ago. Still, it is difficult to propose that flavin/ cytochrome photoreduction is the unique primary B reaction. Study of LIAC measurements will probably never provide as powerful an assay as phytochrome photoreversibility, but they continue to provide a biochemical direction in the search for B/UV-A photoreceptors. Any flavin/cytochrome containing enzyme could, in principle, be responsible for LIACs. Nitrate reductase may be the relevant enzyme for photoconidiation of *Neurospora* (Ninnemann 1991). With the help of cloned genes encoding this and other haem proteins participating in LIACs, it will be possible to use molecular genetics to separate LIACs from, or implicate them in, B responses.

If LIACs can be shown to be physiologically relevant, it will be important to identify the next step in transduction. A clue to this step might be a protein with homology to the Ras family of small GTP-binding proteins, associated with cytochrome b in human neutrophils (Quinn *et al.* 1989). Photoreduction of a b-type cytochrome might trigger a signalling protein in plants. Such a signal transducer might share homology with small GTP-binding proteins, or the heterotrimeric G-protein class. How a cytochrome could transmit a conformational signal is a matter for speculation. Alternatively, a short redox chain might be set up in or across the membrane, eventually leading to charge separation and a change in membrane potential. Transduction might proceed from the oxidized substrate, rather than from the photoreduced (flavo)cytochrome. The endogenous electron donors for the LIAC have not been identified. Recent electron spin resonance experiments with nitroxide probes indicate that these compounds can compete with the electron donors (Walczak *et al.* 1991).

5.2.4.2 Electrical consequences of blue light reception

Blue light has often been used as a stimulus eliciting changes in electrical potentials and currents. Blue light/UV-A photoreceptors could modulate electrical properties in a number of ways: the photoreceptor could itself be an ion channel or pump, it could be closely associated with a channel, act on a channel *via* G-proteins, or indirectly through a second messenger such as Ca^{2+} and a calcium-gated channel. Several ways to measure electrical parameters are illustrated for an idealized plant or fungal cell in Fig. 6. Electrical potential differences and currents have been measured across the plasma membrane of single cells, at the surface of aerial organs, or in a growth medium with low but measurable conductivity. The patch-clamp technique provides a unique opportunity to study channels in a membrane patch isolated from the rest of the cell; the composition of the solution bathing one or both sides of the membrane can

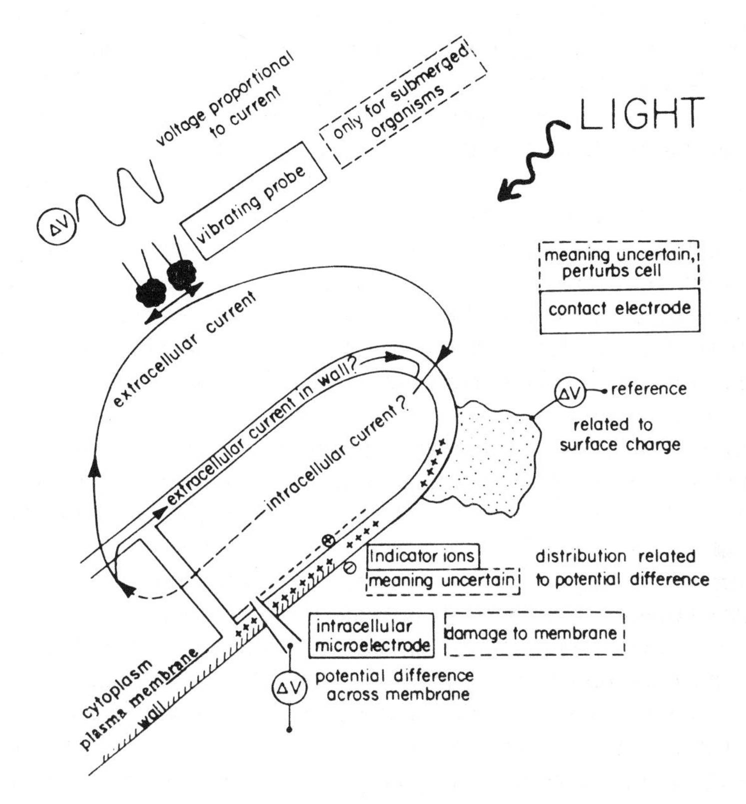

Figure 6. Techniques that have been used to study changes in electrical parameters following exposure to blue light (B): advantages and drawbacks. An idealized cell is illustrated, with internal compartments omitted for simplicity. Some of the problems with each technique are indicated (remarks enclosed by dashed lines). For review, see Blatt M.R. (1987) *Photochem. Photobiol.* 45: 933-938. Different electrical parameters are measured by each method. Examples of modulations by B are: surface potential [Schrank A.R. (1946) *Plant Physiol.* 21: 362-365]; membrane potential [Fig. 7; Cooke T.J., Racusen R.H. and Briggs W.R. (1983) *Planta* 159: 300- 307]; extracellular currents (Fig. 4). The patch-clamp technique, applied to protoplasts, allows the study of individual channels and the application of effectors that would not normally permeate the membrane [Assmann S.M., Simoncini L. and Schroeder J.I. (1985) *Nature* 318: 285-287].

be altered. The extracellular vibrating probe is the least destructive of these methods, but the molecular basis for the non-uniform distributions of ionic currents is not yet known: it probably reflects the distribution of ion channels and transporters, including those modulated by light.

The simplest (and first-used) method employed surface electrodes (Fig. 6). Phototropic stimulation of *Avena* coleoptiles sets up a large surface potential difference between the two sides preceding the redistribution of growth. It must be shown how the potential difference is related to distribution of plant growth regulators, and how they relate to bending, to make causal inferences. The electrical asymmetry in the coleoptile may be the expression of auxin redistribution, leading us back to ask how light started the chain, the eternal circle. The B inhibition of growth of cucumber hypocotyls is preceded by a large depolarization of the membranes of the parenchyma cells (Fig. 7a; Chapter 9.1). The electrical response peaks near where the first derivative of the growth rate is maximal, as if the deceleration of growth were proportional to how far the membrane potential is from the resting state.

A suggestive temporal correlation between B, electrical changes and morphogenesis, as opposed to changes in growth rate, was found in *Vaucheria* (Fig. 4; Chapter 9.4). A B-induced outward ionic current preceded, and continued during, aggregation of chloroplasts. The initiation of a new side branch followed chloroplast aggregation in response to point irradiations with B. Apparently, the chloroplasts aggregate because they become trapped in a fibrillar network formed in response to B. The biochemical changes between 20 min and 4 h (Fig. 4) are not yet known. Rapid effects of B on the cytoskeleton are known in higher plants as well: it has long been known that B can change cytoplasmic streaming and viscosity.

In fungi, B also alters electrical properties. Outward currents specific to the photo-induced state in *Trichoderma* appeared in the region of the hyphae formed 90-120 min after the short photo-inductive pulse (Horwitz *et al.* 1984). This is late enough to be after the first amplification steps and may be related to a stressed state of the photo-induced mycelia. Localized current loops on a scale of 30 μm or less can be inferred from the measurements. The membranes of the induced hyphae probably became leaky to ions in small discrete patches, just preceding branching. Unlike the outward currents of *Vaucheria,* the sites of outward current in *Trichoderma* could not be related to the sites of new branches. Large changes in membrane potentials were recorded upon irradiation with B in two fungi (Fig. 7b). *Phycomyces madA, madB* and *madC* mutants responded normally to a saturating B irradiation, while double mutants *madA,madB* and *madB,madC,* and a mutant defective in all three loci were blind. Strains with the same double and triple mutant genotypes also had increased thresholds for photo-induced sporangiophore development (Corrochano and Cerdá-Olmedo 1991). The B fluences in Fig. 7b, though, would have induced sporangiophore development even in the triple *madA,madB,madC* mutant, so the comparison is, as often occurs, a very non-linear one.

The patch-clamp method (Fig. 6) has been applied to isolated stomatal guard-cell protoplasts, where B causes a transient current flow. The current is attributed to a light-modulated electrogenic proton pump in the plasma membrane. The involvement of membranes is particularly evident for guard cells, because movement of these cells, whether light-induced or not, is an osmotic response. Ion flow changes are probably sufficient to complete the transduction chain in guard cells (Chapter 9.3).

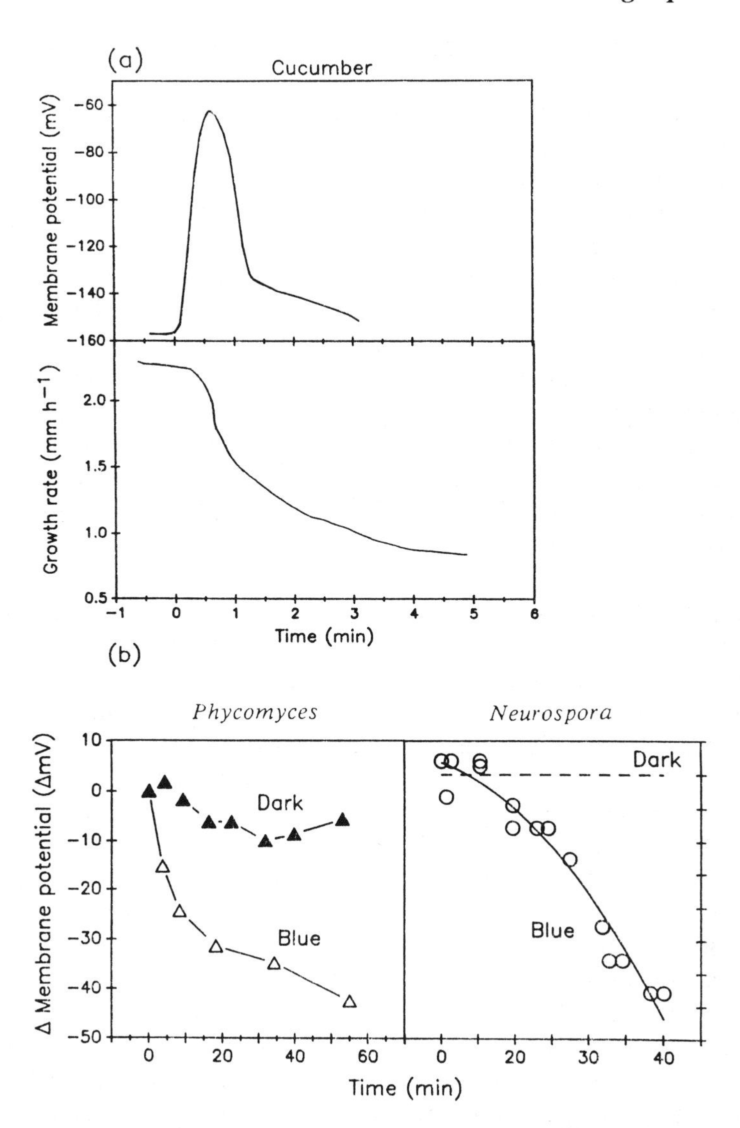

Figure 7. Blue light (B) effects on membrane potential. Examples of intracellular recordings from fungal and plant cells are shown. (a) *Top* panel, transient depolarization induced by B in cells of the elongating zone of cucumber hypocotyls [redrawn from Spalding E.P. and Cosgrove D.J. (1988) *Planta* 178: 407-410]. Continuous illumination started at time zero, with 10 μmol m^{-2} s^{-1} B. For comparison, the time course of rapid growth suppression is shown *(bottom* panel). A similar response in bean hypocotyls is noteworthy in that the UV-A portion of it requires the presence of carotenoids [Hartmann E. and Schmidt K. (1980) In: *The Blue Light Syndrome,* 221-227, Senger H. (ed.) Springer-Verlag, Berlin]. (b) Hyperpolarizations induced by B in hyphae of *Phycomyces,* 15 W m^{-2}; double and triple *mad* mutants are blind [Weiss J. and Weisenseel M.H. (1990) *J. Plant Physiol.* 136: 78-85], and *Neurospora,* 20 W m^{-2}; *wc-1* is blind [Potapova T.V., Levina N.N., Belozerskaya T.A., Kritsky M.S. and Chailakhian L.M. (1984) *Arch. Microbiol.* 137: 262-265; Levina N.N., Belozerskaya T.A., Kritsky M.S. and Potapova T.V. (1988) *Exp. Mycol.* 12: 77-79]. Continuous illumination began at time zero.

5.2.5 The biochemistry of transduction: intracellular signalling

5.2.5.1 Transmembrane signalling and transduction by G-proteins

Plant cells could, in principle, employ signalling pathways that are distinct from those used by animal cells. A simpler hypothesis, which has guided recent work, is that some of the elements are shared by all eukaryotic cells. The view that B responses are ancient and essential in evolution, yet span through to mammalian systems (Gressel and Rau in: Shropshire and Mohr 1983) favours a search for homology between signalling pathways.

It has long been thought that the photoreceptors might be embedded in, or weakly bound to, membranes. The principal support for this view is: (i) experiments indicating that some B/UV-A photoreceptors are dichroically orientated; (ii) analogy with chlorophyll and rhodopsin and (iii) electrical effects of B. All these arguments have counter-arguments. Dichroism is not observed in some organisms (*e.g. Trichoderma).* Conversely, a photoreceptor could be dichroic, and be attached to a subcellular structure other than a membrane (cytoskeleton, organelles). Analogies like (ii) are merely suggestive. Finally, soluble messengers (*e.g.* acetylcholine in nerves) certainly can change membrane potentials, and the photoreceptor need not itself be a gated ion channel.

The transmission of many hormonal signals (with notable exceptions such as steroids) into animal cells is initiated by the binding of a hormone to a cell surface receptor. The signal is then transmitted to a transducer system such as a second messenger-producing enzyme. Certain receptors, such as tyrosine kinases, may directly transmit a signal. With others, there is a separate transducer molecule. Such transduction processes are often mediated by G-proteins, a family of closely related, but distinct, proteins that bind guanine nucleotides (Simon *et al.* 1991). G-proteins are heterotrimers of α (39-52 kD), β (35-36 kD) and γ (8-10 kD) subunits. The α subunit contains the GTP-binding site, as well as sites for interaction with receptor and effector molecules, while the $\beta\gamma$ subunits are apparently involved in the mechanism by which receptors activate the G-protein. Members of the G-protein family transduce signals by a common mechanism. In the basal state, the α subunit contains bound GDP and is tightly associated with the $\beta\gamma$ subunits, whereas binding of GTP induces a decrease in the affinity of α for $\beta\gamma$ and its association with an effector molecule. Activated hormone receptors induce the exchange of bound GDP for GTP on the α subunit, and the GTP-bound α modulates the activity of the effector. G-proteins have an intrinsic GTPase activity which reverts the GTP-bound state to the inactive, GDP-bound state.

Light enters plant cells without a cell surface receptor, but the excited photoreceptor might well transfer its signal to a G-protein. Rhodopsins act in animal eyes in this manner, through transducin, a G-protein. G-protein activity was detected in plants and fungi by the binding and physiological activity of non-hydrolysable GTP analogues, and α subunit genes have been cloned from plants (Ma *et al.* 1990). Cholera and pertussis toxins can catalyse ribosylation of most G-proteins using ADP. These covalent modifications can result in either stimulation or inhibition of the activity of the corresponding G-protein. Such ADP-ribosylation, as well as the binding of a non-hydrolysable GTP analogue, point to the existence of G-proteins in membranes from the B-sensitive Basidiomycete *Coprinus congregatus* (Kozak and Ross 1991). There is preliminary evidence in the same report that B increases binding of the GTP analogue. Analogues and toxins can be delivered to the membranes while measuring the K^+ current in whole-cell patch-clamp studies of stomatal guard cells. Such studies on guard-cell protoplasts from fava bean leaves showed that treatments that should activate G-proteins decrease inward current, thus acting opposite to B (Fairley-Grenot and Assmann 1991).

More direct, enzymatic evidence for G-protein activation by B was found in membranes from the buds of etiolated pea seedlings. Gene expression in the apical buds of pea seedlings is modulated by B. This response is distinct from phototropism, which results from differential elongation in the growing zone of the epicotyl in response to asymmetric B. Purified plasma membranes from pea buds have an intrinsic, B-activated GTPase activity (Fig. 8a). The membranes contain a 40-kD polypeptide which can be ADP-ribosylated by cholera and pertussis toxins in a B-sensitive manner. Antibodies to the α subunit of mammalian transducin also recognized a 40-kD polypeptide which might be the same one that was ADP-ribosylated. Thus several lines of evidence suggest that there is G-protein transduction in responses of the buds to B. Curiously, R had little effect on the membrane GTPase activity, even though buds of dark-grown peas have very sensitive far-red light-absorbing form of phytochrome (Pfr) reactions.

5.2.5.2 Effectors and second messengers

The first step in the response chain is by definition photochemistry. A reasonable hypothesis for the second step, as outlined in the previous section, is transduction by a G-protein. For the third step, there are more possibilities. It is convenient to discuss intracellular messengers for B as effectors for G-protein activation, even though several of these pathways can be activated in other ways. In animal cells, a variety of effectors transfer signals that are initially transduced by G-proteins. These include modulation of ion channels, adenylate cyclase, and phospholipase C. Downstream from these steps, a protein kinase can often be found, modulated by Ca^{2+} or cyclic AMP (cAMP). Phosphorylation of the target proteins by B activated kinases would lead to the response, by changing enzyme activities and gene expression.

The evidence that the animal cell second messenger cAMP is involved in the photoresponses of plants and fungi is controversial at present (Trewavas and Gilroy 1991). Both cAMP and cGMP are serious second messenger candidates in

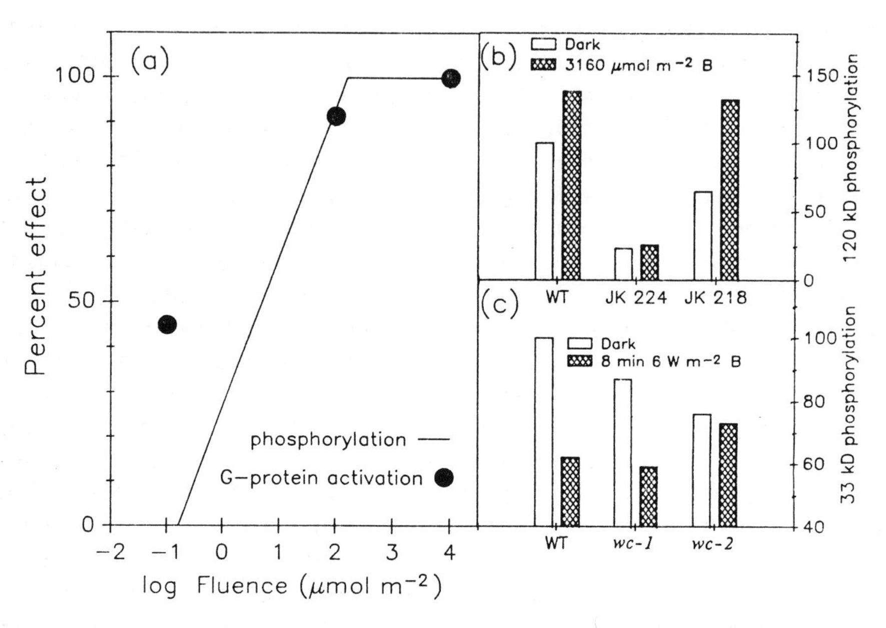

Figure 8. Possible blue light (B) transduction reactions. (a) Fluence-response data are compared for: G-protein activation in membrane fractions from apical buds [total GTPase activity, estimated by integration from: Warpeha K.M.F., Hamm H.E., Rasenick M.M. and Kaufman L.S. (1991) *Proc. Natl. Acad. Sci. USA* 88: 8925-8929]; inhibition of *in vitro* phosphorylation of a 120-kD membrane polypeptide by *in vivo* B irradiation of subapical hypocotyl segments [redrawn from: Short T.W. and Briggs W.R. (1990) *Plant Physiol.* 92: 179-185]. (b) Altered protein phosphorylation in photoresponse mutants of *Arabidopsis* [Reymond P., Short T.W., Briggs W.R. and Poff K.L. (1992) *Proc. Natl. Acad. Sci. USA* 89: 4718-4721.] and (c) *Neurospora* [Lauter F.-R., Russo V.E.A. (1990) *J. Photochem. Photobiol.* B 5: 95-103].

the photoresponses of filamentous fungi (Hasunuma *et al.* 1987). The first target of renewed interested in cAMP signalling in photomorphogenesis is thus the B region of the spectrum.

Several studies in plants have addressed the phosphoinositide (PI) cycle, in which hydrolysis of phosphatidylinositol 4,5-diphosphate (PIP_2) by phospholipase C in the plasma membrane releases the messengers inositol 1,4,5-trisphosphate (IP_3) and diacylglycerol (DAG) (Berridge and Irvine 1984). The IP_3 in turn causes the release of Ca^{2+} from intracellular stores, and DAG activates a protein kinase C. Evidence for phytochrome modulation of this pathway is discussed in Chapter 4.6. The stimulation of phosphoinositide turnover by white light in *Samanea pulvini* could well include both phytochrome and B/UV-A activation (Morse *et al.* 1987). Irradiation of sunflower hypocotyls with white light (of which the B may be the active part) decreased PIP_2 labelling upon *in vitro* phosphorylation of lipids in isolated plasma membranes (Memon and Boss 1990). There was no parallel increase in IP_3, and the effect on PIP_2 is explained by photo-inhibition of the phosphoinositide kinase. The control of stomatal aperture is the only specific B response where there is, at present, clear evidence for the phosphoinositide pathway. Even here, a direct correlation exists between IP_3 and the response, rather than between IP_3 and light. The release, inside guard cells, of chemically caged IP_3 or Ca^{2+} (Blatt *et al.* 1990; Gilroy *et al.* 1990) leads to closure. A fortunate correlate of the opposite actions of light and IP_3 is that the light pulse used to release the caged IP_3 would not be expected, on its own, to close the pore but rather to open it. Could the action of B be to lower intracellular IP_3 levels, promoting opening? Such a model does not fit the picture of IP_3 release from the membrane phosphoinositides as a transient, activating second messenger for the light signal.

In animal cells, protein kinases are modulated during signal transduction *via* G-proteins, and *via* pathways that do not depend on G-proteins. In plant cells too, the phosphorylation state of proteins is a likely link between early transduction steps and enzyme activities (Trewavas and Gilroy 1991). Studies of the pattern of phosphorylation in a membrane fraction from epicotyl segments from dark-grown peas led to the discovery of a phosphoprotein that may be closely related to B perception (Fig. 8a). Irradiation with B (but not R) modulates the phosphorylation of a membrane protein with a molecular mass of 120 kD. The 120-kD polypeptide was first detected when pea membrane fractions were incubated *in vitro* with γ-^{32}P ATP. A prominent band at 120 kD was not phosphorylated when the intact stem segments had received B prior to extraction. Irradiation of intact epicotyl sections with B thus prevented phosphorylation in the isolated membranes. The 120-kD protein was phosphorylated when the intact tissue was labelled with phosphate before B irradiation. In the first experimental design (light *in vivo,* labelled ATP *in vitro*), B apparently caused the phosphorylation sites to become occupied by unlabelled phosphate, preventing *in vitro* labelling (Short *et al.* 1992). Serine (and a small number of threonine) residues are phosphorylated, so that the tyrosine kinase pathway does not seem to be involved. Purified plasma-membrane fractions, even when partially solubilized by the detergent Triton X-100, respond to B, leading to the very promising conclusion that this isolated fraction contains the photoreceptor, the kinase and its substrate (Short *et al.* 1992). If membranes from, for example, maize are given B and then mixed with an equivalent amount of unirradiated membrane proteins from pea, the pea substrate is phosphorylated by the activated maize kinase. The basis for this experiment is that the apparent molecular mass of the substrate varies enough between plant species for them to be clearly separated on gels. Activated dicot kinase can also phosphorylate monocot substrate, but less efficiently than in the reverse experiment (Reymond *et al.* 1992). The stage has been set for purification of the kinase, because solubilized fractions can be separated on columns and assayed for their ability to phosphorylate the substrate. Formally, the kinase and substrate could still reside in the same polypeptide or protein.

Several lines of evidence suggest that the phosphorylation of the 120-kD membrane protein is part of the transduction chain for phototropism, rather than a ubiquitous, LIAC-type photoreaction. The fluence and time dependence are compatible with first-positive phototropism in pea. The tissue localization of the phosphorylated protein is in good agreement with that predicted for the photoreceptor for first-positive phototropism. Phosphorylation of the 120-kD species is much decreased in an *Arabidopsis* mutant having a 20-30 fold decrease in photosensitivity for first-positive phototropism (Fig. 8b). Thus, the mutant could be deficient in the 120-kD substrate, the kinase, or the ability of the kinase to be activated.

The B signal for phosphorylation might reach the kinase through a G-protein. Could the G-protein found in pea buds be responsible for the phosphorylation? The answer, in theory, could be yes, but the systems perceiving light would seem different. In Fig. 8a, the fluence dependence of the inhibition of *in vitro* phosphorylation of pea membranes by an *in vivo* B pulse is compared with that for G-protein activation. Although the *(in vitro)* G-protein activation requires a few-fold less light than the *in vivo* phosphorylation changes, they do occur in the same, low-fluence B range. If G-protein activation (which is assayed in isolated membranes) is compared with the phosphorylation changes induced by *in vitro* irradiation, the discrepancy is greater, because a 30-fold greater fluence is needed to phosphorylate the 120-kD polypeptide in plasma membranes than in intact tissue (Short *et al.* 1992). The phosphorylation of the 120-kD polypeptide is evident in epicotyls but not in buds. In contrast, G-protein activation was studied in the apical buds. Furthermore, quenching the flavin excited state by KI blocks the G-protein activation in the micromolar range where KI is specific for the triplet state (Warpeha *et al.* 1992). Conversely, phosphorylation of the 120-kD protein is inhibited by KI only in the millimolar range where it quenches singlet, as well as, triplet excited flavin. These data, taken together, make it unlikely that the phosphorylation of the 120-kD protein is mediated by the same sensor system found in the buds. There is much evidence for multiple B receptors, so it is possible that the same G-protein be activated by different B/UV-A photoreceptors in the buds and epicotyl segments.

Protein phosphorylation events also occur in other B sensitive organisms. When *Trichoderma* extracts were incubated with γ-^{32}P ATP, two polypeptides, 114 kD and 18 kD, were labelled in the light but not in the dark. Application of cAMP, albeit at concentrations higher than any found in the cell, could substitute for light (Gresik *et al.* 1989). Light decreased the phosphorylation of a 33-kD polypeptide in *Neurospora* mycelia, labelled with orthophosphate *in vivo* (Fig. 8b). Photo-mutants *wc-1* and *wc-2* had altered phosphorylation patterns. The 33-kD polypeptide was already less phosphorylated in the dark in *wc-2,* while in *wc-1,* the 33-kD protein lost its phosphate label following illumination. These findings are genetic evidence for involvement of protein phosphorylation in B transduction in fungi, as in phototropism of the molecular genetic 'model' plant *Arabidopsis* (Fig. 8b).

Calcium has been implicated as a second messenger in plants (Trewavas and Gilroy 1991), but the details are, to say the least, still obscure. Most of the evidence for (and against) a transducer role for Ca^{2+} and calmodulin in photomorphogenesis has come from studies of phytochrome responses, as discussed in Chapter 4.6. Blue light was used in some experiments, as illustrated by the following few examples. High levels of extracellular calcium reduce stomatal opening in the light (Schwartz 1985). Thus B could, theoretically, promote stomatal opening by lowering Ca^{2+} levels in the wall or inside the guard cells. Roblin *et al.* (1989) propose that the Ca^{2+} required for R effects on the pulvinar movements of *Cassia* leaves must enter the cells from outside, while for B effects the Ca^{2+} is mobilized from internal stores. These conclusions are based on studies with calcium-channel blockers.

Blue light will not inhibit the elongation of cucumber hypocotyl segments unless Ca^{2+} is present (Fig. 9). Since the seedlings were grown under continuous R the response must be specific to B. The Ca^{2+} probably acts in the cell wall, or perhaps at the outer surface of the plasma membrane, because the chelators do not enter the cell. Though the added calcium probably cannot alter the tightly regulated intracellular Ca^{2+} pool, the effective concentrations are low (Fig. 9). Sineshchekov and Lipson (1992) also applied calcium from outside the cell in experiments on

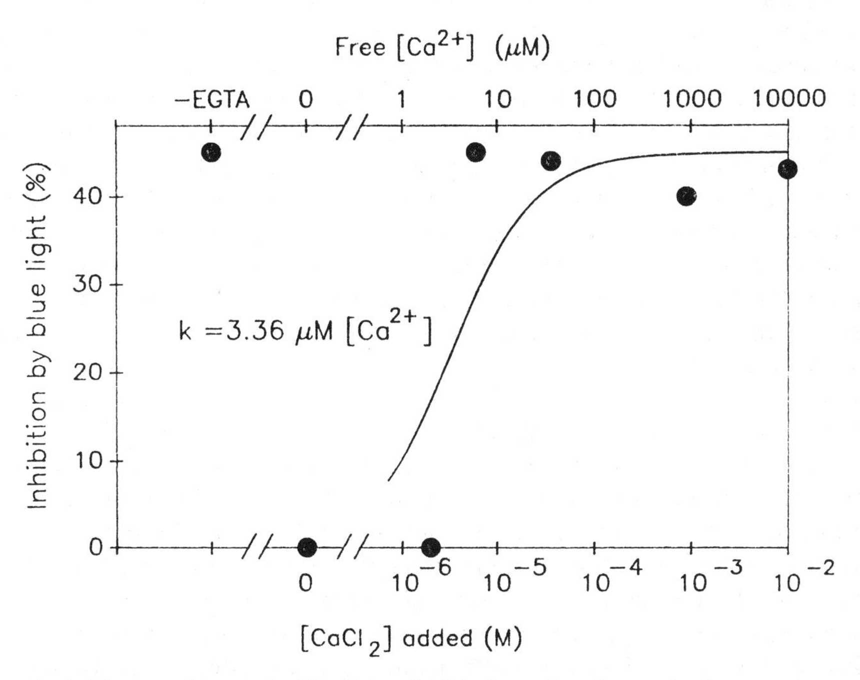

Figure 9. Calcium requirement for blue light to inhibit elongation of cucumber hypocotyl segments. Subapical segments were cut from hypocotyls of cucumber seedlings grown under weak continuous red light (R) and incubated in the presence of 2 *μM* indolyl acetic acid (IAA), 0.1 m*M* EGTA, and $CaCl_2$ at the indicated concentrations. The *x* axis at the top of the figure indicates the free calcium concentration, as calculated from the dissociation constants for EGTA [Shinkle J.R. and Jones R.L. (1988) *Plant Physiol.* 86: 960-966]. The effective $[Ca^{2+}]$ for half-maximal effect, k, is estimated by a least-squares fit to a hyperbolic saturation curve (solid line).

Phycomyces sporangiophores. They found that the phototropic bending rate of the sporangiophore is only weakly dependent on [Ca^{2+}]. In contrast, adaptation, measured by the phototropic latency, is Ca^{2+} dependent. The sensory adaptation process may reside in the cell membrane (Galland 1989), explaining how extracellular [Ca^{2+}] can affect latency. Calcium in the cell wall space of plant cells may have a profound influence on growth and morphogenesis. This unique aspect of calcium signalling in plant cells (Trewavas and Gilroy 1991) could apply to fungi as well. However, a study of calcium distributions in *Phycomyces* sporangiophores (Morales and Ruiz-Herrera 1989) did not detect any in the cell wall. Such questions can be investigated further with new imaging technology. Measurements of free cytoplasmic Ca^{2+} in plant cells have become possible in recent years, with the help of calcium-sensitive fluorescent dyes or luminescent systems. The excitation light for the fluorescent measurement system is in the UV-A, corresponding to peaks in various B/UV action spectra. Despite this built-in interference between measurement and physiology, Russ *et al.* (1991) were able to show that UV-A increases intracellular [Ca^{2+}] in *Mougeotia.* A faster chloroplast rotation, explained by lower cytoplasmic viscosity, was correlated with this increased [Ca^{2+}].

Light and plant hormones control many of the same processes, and the suggestion that light could act by altering the levels of plant growth substances is logical. A much studied example is first-positive phototropism in grass coleoptiles, where B acts by redistribution of auxin. Just how generally this Cholodny-Went model can be applied is discussed in Chapter 9.2. Could hormones act as intracellular messengers as well? The question is an open one. The exogenous cytokinin benzyladenine acts like B to inhibit hypocotyl elongation in dark-grown cucumber seedlings (Cohen *et al.* 1991), and there is a Ca^{2+} requirement analogous to that found for B. Another inhibitor of elongation, ethylene, acted independently of B in R-grown pea epicotyls (Laskowski *et al.* 1992).

5.2.5.3 Can blue light biochemically stress?

Many photo-induced processes, particularly in fungi, can be stimulated or strongly modified by other factors, especially those that apply a stress. Is light just another manifestation of stress and if so, how primary is the effect of the stress imposed by irradiation? From an evolutionary point of view, light clearly acts as an early warning system for many stresses as a fungus emerges from an enclosed substrate to lower humidity and to light containing UV. Blue light often induces UV-protective pigmentation. B may signal the changes in season or differential fluence rates in higher plants. The 'B as stress signal' hypothesis (Gressel and Rau in: Shropshire and Mohr 1983) could suggest mechanisms that might not be easily apparent from the analogies with animal cell signalling made in the previous sections. *Trichoderma* can be considered a good case in point. Stress, as an alternate pathway for morphogenesis, is useful for genetic studies, but can also suggest biochemical mechanisms. Irradiation with B leads to a rapid increase in oxygen uptake in *Trichoderma* and other fungi. Although oxygen is not necessary for the primary photo-act, at least in *Trichoderma*, the mitochondrial respiratory chain might be involved in transduction. An enhancement of respiration is one of the classical B effects in algae and plants (Senger 1987).

Nitrate, when serving as sole nitrogen source induces *de novo* synthesis of nitrate reductase, and favours conidiation of *Neurospora*. Growth is optimal on ammonium nitrate. Photo-induction of conidiation might be related to a nitrogen stress or change in pathway of nitrogen metabolism. *Neurospora* becomes more sensitive to photo-induction of conidiation when starved (no substrates). Under these conditions, nitrate reductase is expected to be suboptimal, but its subunits are present and may be photo-activated. Nitrate reductase is required for conidiation, but not phase-shifting of the circadian rhythm or carotene synthesis, as shown using mutants (Ninnemann 1991). The chromophores of nitrate reductase certainly qualify it to be a B/UV-A photoreceptor. Phototropism in plants could be thought of as a slowing down of growth on the (stressed?) highest irradiated side.

The stress hypothesis suggests many biochemical models for B control. Many flavo-enzymes are sensitive to B, including nitrate reductase and lactate dehydrogenase, though very high fluences are required to directly affect enzyme activities *in vitro*. The entire mitochondrial electron transport chain includes chromophores that might be photoperceptive. Free flavin could act as a photosensitizer (Fig. 2). The mechanism could be a type II photosensitization, in which singlet oxygen is formed, subjecting the cell to oxidative stress. Ways must be found to follow, *in vivo*, the fine tuning by light of metabolic flows at physiological fluences. The pathway of carbohydrate metabolism in several fungi changes after irradiation and/or during the change from vegetative to reproductive morphogenetic pathways (Gressel and Rau in: Shropshire and Mohr 1983).

5.2.6 Concluding remarks

Fragments of the transduction pathways in different organisms have been elucidated, but their integration into a whole presents major problems: are the G-proteins that have been found involved in phototransduction or in other processes? Is the intracellular or extracellular [Ca^{2+}], or both, important in plant cells? Do B and R share the same second messengers? The next few years will certainly see a generalization of the experiments described in this chapter to many more B-sensitive organisms, perhaps leading to a unifying hypothesis rather than an ever-growing diversity of responses and their phototransducers. It is almost unnecessary to recall here the element of speculation introduced by our lack of knowledge of the photoreceptors' identities. There may be homologies between signalling pathways in plant, fungal and animal cells, as well as between the enzyme sites that bind the B-absorbing chromophores (haem, flavin and pterin). Such sequence homology could be put to good use in order to shorten the route to discovery of B/UV-A photoreceptors and their transducers.

Acknowledgements. Thanks to Jonathan Gressel, co-author of the first edition of this chapter, for helpful discussions and for inspiring many of the ideas that continue to be expressed in this revision. I am also grateful to Winslow R. Briggs for helpful suggestions and access to data prior to publication. Some of the material on genetic approaches in fungi is adapted, with thanks to Maarten Koornneef and to the editors, from the first edition of this textbook. Preparation of this chapter was supported in part by a grant from the Israel Academy of Sciences.

5.2.7 Further reading

Cerdá-Olmedo E. and Lipson E.D. (1987) *Phycomyces,* Cold Spring Harbor Laboratory.

Lipson E.D. and Horwitz B.A. (1991) Photosensory receptors and their transduction, In: *Sensory Receptors and Signal Transduction,* Vol. 7, *Modern Cell Biology*, pp. 1-64, Satir B. (ed.) Alan R. Liss, New York.

Senger H. ed. (1987) *Blue Light Responses: Phenomena and Occurrence in Plants and Microorganisms,* CRC Press, Boca Raton, Florida.

Shropshire W. and Mohr H. eds. (1983) *Photomorphogenesis,* In: *Encyclopedia of Plant Physiology* New Series, Vols. 16A and 16B (Chapters 5, 22, 23, 26) Springer-Verlag, Berlin.

Trewavas A.J. and Gilroy S. (1991) Signal transduction in plant cells. *Trends in Genetics* 7: 356-361

5.2.8 References

Berridge N.J. and Irvine R.F. (1984) Inositol trisphosphate, a novel second messenger in cellular signal transduction. *Nature* 312: 315-321.

Blatt M.R., Thiel G. and Trentham D.R. (1990) Reversible inactivation of K^+ channels of *Vicia* stomatal guard cells following the photolysis of caged inositol 1,4,5-triphosphate. *Nature* 346: 766-769.

Chahidi C., Aubailly A., Momzikoff A., Bazin M. and Santus R. (1981) Photophysical and photosensitizing properties of 2-amino-4-pteridinone: a natural pigment. *Photochem. Photobiol.* 33: 641-649.

Cohen L., Gepstein S. and Horwitz B.A. (1991) Similarity between cytokinin and blue light inhibition of cucumber hypocotyl elongation. *Plant Physiol.* 95: 77-81.

Corrochano L.M. and Cerdá-Olmedo E. (1991) Photomorphogenesis in *Phycomyces* and in other fungi. *Photochem. Photobiol.* 54: 319-327.

Fairley-Grenot K. and Assmann S.M. (1991) Evidence for G-protein regulation of inward K+ channel current in guard cells of fava bean. *Plant Cell* 3: 1037-1044.

Fritz B.J., Kasai S. and Matsui K. (1990) Blue light photoreception in *Neurospora* circadian rhythm: Evidence for involvement of the flavin triplet state. *Photochem. Photobiol.* 51: 607-610.

Galland P. (1989) Photosensory adaptation in plants. *Bot. Acta* 102: 11-20.

Galland P. (1990) Phototropism of the *Phycomyces* sporangiophore: A comparison with higher plants. *Photochem. Photobiol.* 52: 233-248.

Gilroy S., Read N.D. and Trewavas A.J. (1990) Elevation of cytoplasmic calcium by caged calcium or caged inositol trisphosphate initiates stomatal closure. *Nature* 346: 769-771.

Gresik M., Kolarova N. and Farkas V. (1989) Light-stimulated phosphorylation of proteins in cell-free extracts from *Trichoderma viride. FEBS Lett.* 248: 185-187.

Hasunuma K., Funadera K., Shinohara Y., Furukawa K. and Watanabe M. (1987) Circadian oscillation and light-induced changes in the concentration of cyclic nucleotides in *Neurospora. Curr. Genet.* 12: 127-133.

Horwitz B.A. and Gressel J. (1983) Elevated riboflavin requirement for postphotoinductive events in sporulation of a *Trichoderma* auxotroph. *Plant Physiol.* 71: 200-204.

Horwitz B.A., Weisenseel M.H., Dorn A. and Gressel J. (1984) Electric currents around growing *Trichoderma* hyphae, before and after photoinduction of conidiation. *Plant Physiol.* 74: 912-916.

Iino M., Nakagawa Y. and Wada M. (1988) Blue light-regulation of cell division in *Adiantum* protonemata: Kinetic properties of the photosystem. *Plant Cell Environ.* 11: 547-561.

Jorns M.S., Wang B., Jordan S.P. and Chanderkar L.P. (1990) Chromophore function and interaction in *Escherichia coli* DNA photolyase: Reconstitution of the apoenzyme with pterin and/or flavin derivatives. *Biochemistry* 29: 552-561.

Koyama Y. (1991) Structures and functions of carotenoids in photosynthetic systems. *J. Photochem. Photobiol.* B 9: 265-280.

Kozak K.R. and Ross I.K. (1991) Signal transduction in *Coprinus congregatus:* Evidence for the involvement of G proteins in blue light photomorphogenesis. *Biochem. Biophys. Res. Comm.* 179: 1225-1231.

Kumagai T. (1989) Temperature and mycochrome system in near-UV light inducible and blue light reversible photoinduction of conidiation in *Alternaria tomato. Photochem. Photobiol.* 50: 793-798.

Laskowski M.J., Seradge E., Shinkle J.R. and Briggs W.R. (1992) Ethylene is not involved in the blue light-induced growth inhibition or red light-grown peas. *Plant Physiol.* 100: 95-99.

Ma H., Yanofsky M.F. and Meyerowitz E.M. (1990) Molecular cloning and characterization of GPA1, a G protein α subunit gene from *Arabidopsis thaliana. Proc. Natl. Acad. Sci. USA.* 87: 3821-3825.

Memon A.R. and Boss W.F. (1990) Rapid light-induced changes in phosphoinositide kinases and H^+-ATPase in plasma membrane of sunflower hypocotyls. *J. Biol. Chem.* 265: 14817-14821.

Morales M. and Ruiz-Herrera J. (1989) Subcellular localization of calcium in sporangiophores of *Phycomyces blakesleeanus. Arch. Microbiol.* 152: 468-472.

Morse M.J., Crain R.C. and Satter R.L. (1987) Light-stimulated phosphatidylinositol turnover in *Samanea saman* leaf pulvini. *Proc. Natl. Acad. Sci. USA* 84: 7075-7078.

Nebenfuhr A., Schäfer A., Galland P., Senger H. and Hertel R. (1991) Riboflavin binding sites associated with flagella of *Euglena:* A candidate for blue-light photoreceptor? *Planta* 185: 65-71.

Nick P. and Schäfer E. (1991) Induction of transverse polarity by blue light: An all-or-none response. *Planta* 185: 415-424.

Ninnemann H. (1991) Photostimulation of conidiation in mutants of *Neurospora crassa. J. Photochem. Photobiol.* B 9: 189-199.

Quinn M.T., Parkos C.A., Walker L., Orkin S.H., Dinauer M.C. and Jesaitis A.J. (1989) Association of a Ras-related protein with cytochrome b of human neutrophils. *Nature* 342: 198-200.

Reymond P., Short T.W. and Briggs W.R. (1992) Blue light activates a specific protein kinase in higher plants. *Plant Physiol.* 100: 655-661.

Roblin G., Fleurat-Lessard P. and Bonmort J. (1989) Effects of compounds affecting calcium channels on phytochrome- and blue pigment-mediated pulvinar movements of *Cassia fasciculata. Plant Physiol.* 90: 697-701.

Russ U., Grolig F. and Wagner G. (1991) Changes of cytoplasmic free Ca^{2+} in the green alga *Mougeotia scalaris* as monitored with indo-1, and their effect on the velocity of chloroplast movements. *Planta* 184: 105-112.

Schmidt W. (1984) Blue light-induced, flavin-mediated transport of redox equivalents across artificial bilayer membranes. *J. Membrane Biol.* 82: 113-122.

Schwartz A. (1985) Role of Ca^{2+} and EGTA on stomatal movements in *Commelina communis* L. *Plant Physiol.* 79: 1003-1005.

Short T.W., Porst M. and Briggs W.R. (1992) A photoreceptor system regulating *in vivo* and *in vitro* phosphorylation of a pea plasma membrane protein. *Photochem. Photobiol.* 55: 773-781.

Simon M.I., Strathmann M.P. and Gautam N. (1991) Diversity of G proteins in signal transduction. *Science* 252: 802-808.

Sineshchekov A.V. and Lipson E.D. (1992) Effect of calcium on dark adaptation in *Phycomyces* phototropism. *Photochem. Photobiol.* 56: 667-675.

Walczak T., Gabrys H. and Swartz H.M. (1991) Blue light photoreception in higher plants studied with ESR spectrophotometry. *J. Plant Physiol.* 137: 662-668.

Warpeha K.M.F., Kaufman L.S. and Briggs W.R. (1992) A flavoprotein may mediate the blue light-activated binding of guanosine 5'-triphosphate to isolated plasma membranes of *Pisum sativum* L. *Photochem. Photobiol.* 55: 595-603.

Part 6 Coaction between pigment systems

6. Coaction between pigment systems

Hans Mohr

Biological Institute II, Albert-Ludwig University
Schänzlestr. 1, 79104, Freiburg, Germany

6.1 Sensor pigments in higher plants*

All life on earth is fuelled by sunlight. In order to efficiently harvest the light quanta by the process of photosynthesis, plants must adapt to the light conditions of their particular habitat. In fact, development of photoautotrophic higher plants is 'opportunistic' in the sense that the developmental process is in part controlled by light. It is only the basic developmental patterns of plant construction which are strictly determined by the genes; within these limits fine tuning of developmental events is controlled by the actual light climate at the site where the plant has to grow (Mohr 1982).

In order to respond properly a plant has to continuously and accurately sense the light conditions in its environment. This implies that a plant must be capable of sensing the quality and quantity of light throughout the sun's spectrum as far as sunlight leads to electronic excitations (290-800 nm). For reasons of molecular physics it is improbable that a single photoreceptor can fulfil this task. Rather, we might expect that higher plants will use several *sensor pigments* to check the whole spectral range with the sensitivity and accuracy required.

As far as we know today, three different sensor pigments occur in higher plants (Mohr 1984). These are *phytochromes* (operating predominantly in the red (R) far-red (FR) spectral range), a blue (B)/UV-A photoreceptor, sometimes referred to as cryptochrome (Chapter 5.1), and a *UV-B photoreceptor*. The action spectrum related to the latter photoreceptor shows a single intense peak at 290 nm and no action at wavelengths longer than 350 nm (Yatsuhashi *et al.* 1982). Until recently no clear picture had emerged as to the mode of coaction and the relative importance of different sensor pigments in different species and in the different organs of a particular species. For example, in the case of

*In this chapter only intact higher plants will be considered. Results obtained with plant cell cultures will be dealt with in Chapter 9.5.

R.E. Kendrick & G.H.M. Kronenberg (eds.), Photomorphogenesis in Plants - 2nd Edition

light-mediated *anthocyanin synthesis* the mustard seedling cotyledons respond exclusively to phytochrome (Table 1 and Fig. 1a), while in the hypocotyl a specific stimulatory effect of B can be detected at higher fluence rates (Fig. 1b). Other plants (*e.g.* Schirokko wheat) seem to use the UV-B photoreceptor and phytochrome, or are capable (such as the milo seedling) of sensing the light conditions throughout the sun's spectrum from UV-B to the near infra-red (Mohr 1984).

6.2 A unifying model of coaction

Naturally, one must be prepared to find different evolutionary solutions of the problem of coaction in different species, with different photoresponses of the same plant and even with the same photoresponse in different parts of the same plant (Fig. 1). Moreover, the study of coaction is complicated by the fact that B/UV always activates phytochrome, since B/UV inevitably converts the R-absorbing form of phytochrome (Pr) to the FR-absorbing form (Pfr).

There are two hypothetical possibilities for the mode of coaction. Firstly, the different sensor pigments may operate independently of each other, eventually causing the same terminal photoresponse (*e.g.* anthocyanin synthesis in tomato hypocotyls). While an independent action would not necessarily lead to an *additive coaction* (one might expect a non-linear hyperbolic relationship between *total* stimulus and response), any single photoreceptor must be expected to elicit a significant response. Secondly, the sensor pigments depend on each other to bring about the final photoresponse. As an example, phytochrome in its Pfr form may be the only effector to operate on gene expression in photomorphogenesis; however, to establish or to increase and maintain respon-

Table 1. Induction and its reversion in the case of anthocyanin synthesis in seedling cotyledons of mustard (*Sinapis alba* L.). Red (660 nm, Pfr/P = 0.8), medium far-red (720 nm, Pfr/P = 0.03) and long-wavelength far-red (756 nm, Pfr/P < 0.01) light pulses were given to dark-grown seedlings 36 h after sowing. Anthocyanin was assayed 24 h after the light pulse treatment. Data from R. Schmidt. A 5 min light pulse suffices to establish the phytochrome photoequilibrium, Pfr/P, where P = total phytochrome (Pr + Pfr).

Light treatment	Relative amount of anthocyanin (%)
5 min 660 nm	100
5 min 720 nm	49
5 min 660 nm + 5 min 720 nm	51
5 min 756 nm	32
5 min 660 nm + 5 min 756 nm	32
5 min 720 nm + 5 min 756 nm	35
Dark control	9

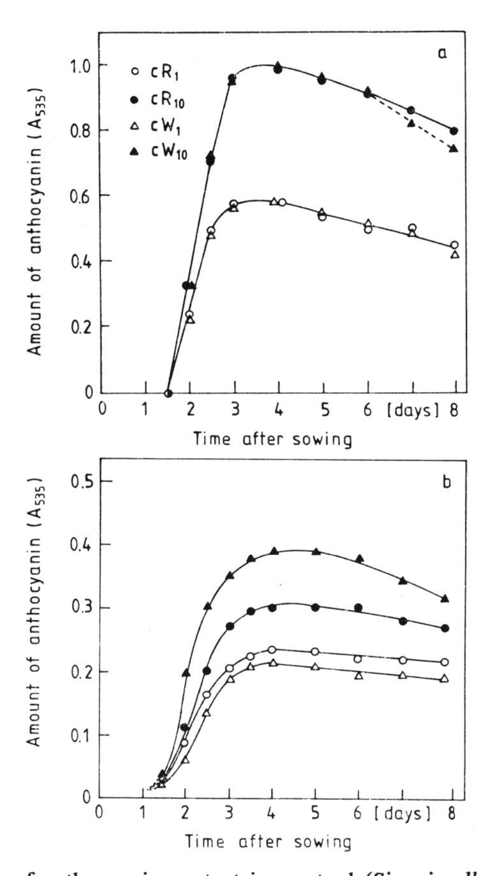

Figure 1. Time course of anthocyanin content in mustard (*Sinapis alba* L.) cotyledons (a) and hypocotyl (b) in continuous red (cR) and white light (cW) of different fluence rates (cR_1, 0.68 W m^{-2}; cR_{10}, 6.8 W m^{-2}; cW_1, 0.86 W m^{-2}; cW_{10}, 8.6 W m^{-2}). The W was applied with approximately the same photon fluence rate of R as present in the corresponding R fields. This requires an approximately 30% higher fluence rate in the W fields compared to the corresponding R fields. The phytochrome photoequilibrium Pfr/P in W is determined essentially by the R part of the spectrum and is thus almost the same as in pure R (Pfr/P > 0.7), where P = total phytochrome, Pr + Pfr. Data from H. Drumm-Herrel.

siveness in a plant cell towards Pfr, concomitant light absorption in the B/UV-A photoreceptor or the UV-B photoreceptor might be required. It seems that a coaction of this kind between phytochrome and the B/UV sensor pigments is characteristic for most photomorphogenesis (Fig. 2). This coaction must be considered as highly economic since a single 'effector' (namely Pfr) suffices to bring about the molecular events leading to photomorphogenesis, and yet information about the whole solar spectrum, as far as it is relevant to the plant, can contribute to the rate and extent of the photomorphogenic responses.

The following representative case studies show that this kind of 'interdependent coaction' in fact operates in nature.

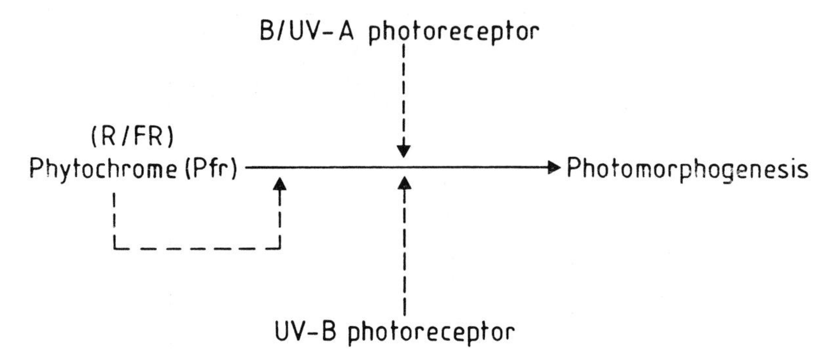

Figure 2. Suggested mode of coaction between phytochrome and the photoreceptors absorbing blue (B)/UV-A light. After Mohr (1986). This model generalizes the more specific model derived previously from the experiments on anthocyanin formation in the milo mesocotyl (see Fig. 5).

6.3 Photomorphogenesis of the milo seedling (*Sorghum vulgare Pers.*, cv. Weider-hybrid)

6.3.1 Accumulation of plastid GPD (glyceraldehyde-3-phosphate dehydrogenase, EC 1.2.1.13) in the shoot (mainly primary leaf)

Regulation by light of accumulation of this major enzyme of the chloroplast matrix documents the interdependent coaction of B/UV and phytochrome in light-mediated plastidogenesis.

It was found that responsiveness towards Pfr, established by single light pulses, is extremely weak in dark-grown milo shoots, while prolonged light treatments lead to a dramatic increase of responsiveness (degree of response per unit Pfr) (Oelmüller and Mohr 1984).

Figure 3 indicates that long-term light causes a rapid and strong increase of responsiveness (*responsiveness amplification*) which tends to saturate after approximately 6 h. Both B and UV are equally effective, and far more effective than R. Since light-mediated changes of *total phytochrome* (P) levels are the same in R, B, and UV, it is clear that UV and B cause a several times higher responsiveness than R. However, it is equally obvious that even R alone, operating exclusively through phytochrome, exerts a considerable effect on responsiveness to Pfr.

It has been concluded that responsiveness of GPD synthesis to Pfr strongly depends on the quality and quantity of the ambient light. The plant measures light throughout the spectrum and this information, obtained *via* phytochrome and the B/UV photoreceptors, determines the efficiency of Pfr action or, in other words, the actual responsiveness towards Pfr.

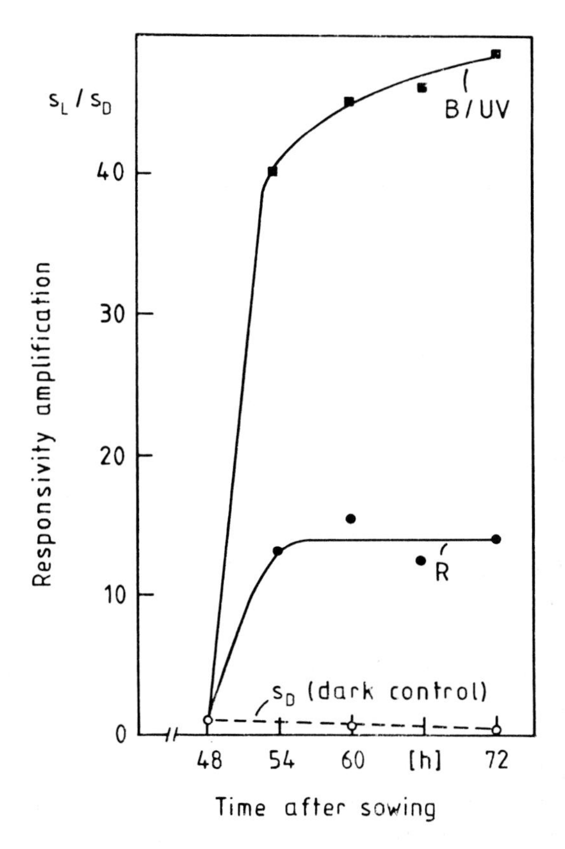

Figure 3. Time course of increase of responsiveness in long term red (R) and blue (B)/UV light (B/UV) for phytochrome-mediated glyceraldehyde-3-phosphate dehydrogenase (GPD) induction in the milo (*Sorghum vulgare* Pers.) shoot. The light treatment commenced 48 h after sowing. S_L, responsiveness to Pfr in light treated material; S_D, responsiveness to Pfr in dark-grown material. S_D expressed relative to S_D at 48 h. After Oelmüller and Mohr (1984).

It was shown further (Oelmüller and Mohr 1985a) that the specific effect of B/UV cannot be explained by an effect of light on gross protein synthesis. Rather, the pertinent data indicate that amplification of responsiveness to Pfr by B/UV is a specific process directly related to the mechanism of modulation of gene expression by phytochrome.

6.3.2 Synthesis of anthocyanin

Control by light of anthocyanin appearance takes place in the same way in all organs of the seedling which turn red (taproot, mesocotyl, coleoptile). The following detailed description is for the mesocotyl.

The *mesocotyl* of the milo seedling does not produce anthocyanin in complete darkness. As described originally by Downs and Siegelman (1963) even long-term R or FR do not lead to any anthocyanin synthesis while white light (W) or B/UV cause strong and rapid pigmentation (Table 2). Experiments with R and FR pulses given after an inductive W period of 3 h have shown that phytochrome can act once a B/UV effect has occurred. However, the expression of the B/UV effect is controlled by phytochrome.

In experiments with dichromatic irradiation (simultaneous irradiation with two kinds of light to strongly modulate the level of Pfr on a constant background of B/UV) it was found that the B/UV photoreaction as such is not affected by the presence or virtual absence of Pfr during the B/UV treatment (Drumm and Mohr 1978).

The tentative interpretation of the data obtained with milo was that Pfr is the effector which causes anthocyanin synthesis through activation of gene expression, while the B/UV effect was considered as establishing responsiveness towards Pfr. In the case of anthocyanin synthesis in the milo seedling there is no responsiveness without the operation of a B/UV photoreceptor.

The action of B/UV in anthocyanin synthesis of the milo mesocotyl shows the characteristics of an induction response. As shown in Fig. 4 even 5 min of UV-A suffice to induce responsiveness towards R, *i.e.* towards Pfr. However, UV-A is ineffective (as far as anthocyanin formation is concerned) without Pfr. This is shown by the ineffectiveness (with regard to appearance of anthocyanin) of the UV-A treatment up to 10 min, provided that virtually all Pfr is returned to Pr at the end of the UV-A treatment by a long-wavelength FR (RG9 filtered light) pulse [Pfr/P (where P = total phytochrome, Pr + Pfr) in RG9 light < 0.01]. Clearly, 5 min of UV-A achieves something, namely modified responsiveness to Pfr, but 5 min of UV-A *alone* does not suffice to cause anthocyanin synthesis. However, UV-A induces responsiveness to Pfr so rapidly that after 15 min light induction is no longer fully reversible. This means that in the presence of UV-A, Pfr can perform its *initial action** within 15 min even though it requires 3 h before anthocyanin appears.

In a series of further experiments (Drumm-Herrel and Mohr 1981; Oelmüller and Mohr 1985b) it was shown that besides the B/UV-A photoreceptor and phytochrome a UV-B photoreceptor is also involved. A comprehensive model was elaborated which describes the subtle coaction of the different sensor pigments in bringing about anthocyanin synthesis (Fig. 5). The major features of the model are the following: B/UV cannot mediate induction of anthocyanin synthesis in the absence of Pfr. Rather, the relatively fast action of B/UV (order of minutes) establishes responsiveness of the anthocyanin producing mechan-

*The term initial action designates the action of Pfr on some cell function which is no longer reversible by the removal of Pfr. The onset of the initial action is defined by the escape from full reversibility.

Table 2. Induction (or lack of induction) of anthocyanin in the mesocotyl of milo seedlings (*Sorghum vulgare* Pers.) by light of different qualities (W: Xenon arc light, similar to sunlight, 250 W m^{-2}). In the case of a 3 h light treatment the seedlings were kept in the dark for 24 h before extraction of anthocyanin. Red light (R, Pfr/P = 0.8), medium far-red light (FR, Pfr/P = 0.03), long-wavelength FR (756 nm, Pfr/P < 0.01). A 5 min light pulse suffices to establish the phytochrome photoequilibrium. After Drumm and Mohr (1978).

Treatment (onset 60 h after sowing)	Amount of anthocyanin (measurement: 87 h after sowing) (*A* at 510 nm)
27 h dark	0
27 h W	1.85
27 h R	0
27 h FR	0
3 h W	0.19
3 h B/UV	0.19
3 h W + 5 min R	0.19
3 h W + 5 min 756 nm	0.06
3 h W + 5 min 756 nm + 5 min R	0.20
3 h B/UV + 5 min R	0.19
3 h B/UV + 5 min 756 nm	0.05
3 h B/UV + 5 min 756 nm + 5 min R	0.19

ism to Pfr. In this system Pfr operates *via* two different channels. As the effector of the terminal response it sets in motion the signal-response chain which eventually leads to the appearance of anthocyanin. This is a slow process with a lag-phase of the order of 3.5 h. The second function of phytochrome is to determine the effectiveness of the effector Pfr in mediating anthocyanin synthesis. This is a very fast and highly sensitive mode of phytochrome action which can readily be detected within 1 min. However, as long as the plant has not received B/UV the strong effect of R on the effectiveness of Pfr remains cryptic. The effect of a R pretreatment and the effect of a B/UV pretreatment on responsiveness towards Pfr (or expressed differently, effectiveness of Pfr) were found to be totally independent of each other, even though it is the B/UV which permits operation of Pfr.

While it seems that the scheme in Fig. 5 represents the usual interdependence of B/UV and light absorbed by phytochrome, it must be emphasized that in most cases studied so far a B/UV treatment is not obligatory for a Pfr action to occur (see Fig. 3). In fact, B/UV causes an intensification of Pfr-mediated processes which occur even in R alone, albeit at a low rate. The absolute requirement for a B/UV treatment in this case allows a complete experimental separation of the effect of B/UV *per se* from that of phytochrome *per se*. So far we have no knowledge about the 'mechanism' of the fast reactions which determine the effectiveness of the effector Pfr.

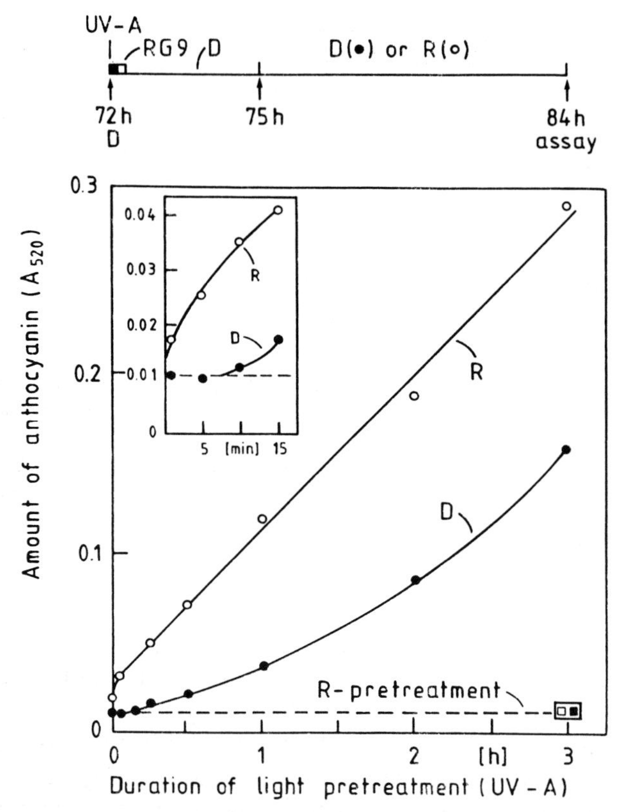

Figure 4. The effect of UV-A pretreatment on the accumulation of anthocyanin in subsequently given red light (R) (○) or darkness (D) (●). Dark-grown milo (*Sorghum vulgare* Pers.) seedlings 72-h-old were irradiated with UV-A of differing duration (abscissa). The UV-A treatment was terminated by a long-wavelength far-red (RG9-light) pulse which returns almost all Pfr to Pr. From 75-84 h after sowing seedlings were either kept in darkness (●) or in R (○). For UV-A treatment of less than 3 h, an appropriate dark interval separated the end of the (UV-A + RG9-light)-treatment and the onset of R. A typical experimental protocol is indicated above. The inset enlarges the results obtained during the first 15 min of light treatment (different set of experiments). R-pretreatment: R given instead of UV-A (□). This value is identical with the control in complete darkness (■). After Oelmüller and Mohr (1985b).

6.4 Photomorphogenesis of the Scots pine seedling (*Pinus sylvestris* L.)

6.4.1 Axis (hypocotyl) straight growth

As in all higher plants, photomorphogenesis is also a conspicuous feature of development in coniferous trees; in particular it is obvious that elongation of the monopodium (stem) is strongly affected by light (compare a solitary pine tree with a pine tree growing in a dense stand!). We have found that in Scots pine control of stem growth by light is clearly expressed at the seedling stage and can readily be studied under controlled conditions (Fernbach and Mohr 1990).

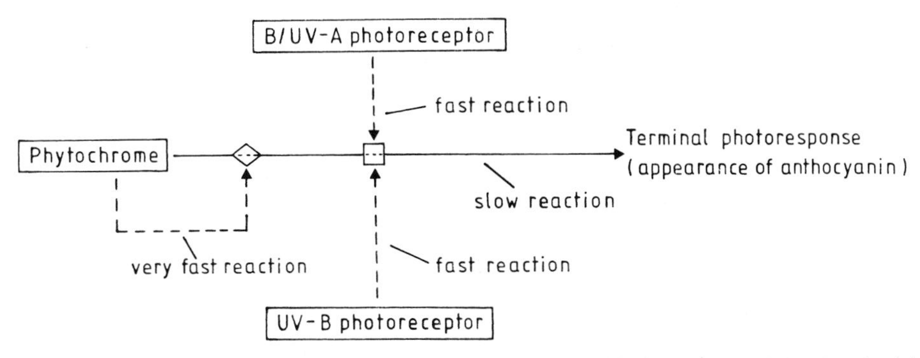

Figure 5. Suggested mode of coaction between blue/UV light (B/UV) and light absorbed by phytochrome in light-mediated anthocyanin formation in the milo (*Sorghum vulgare* Pers.) seedling. ——› : Temporal sequence of events set in motion by the effector Pfr and leading to the terminal response. – – –› : Light-dependent reactions which determine the effectiveness of the effector Pfr (in other words, the responsiveness of the anthocyanin producing mechanism towards Pfr). The point of action of UV-B, relative to the action of B/UV-A remains undecided at present. After Oelmüller and Mohr (1985b).

Pine trees are genetically adapted to strong light. This is reflected in the high light requirement to reduce axis growth, *i.e.* to prevent etiolation of seedlings. On the other hand, maximum control of axis elongation is achieved even under short-day conditions. A daily main light period of 8 h was chosen, and the axis (hypocotyl) growth was measured over a period of 8 days after sowing. The light treatment was given from days 4-8 after sowing. Measurements (end point determinations) were taken at the beginning of day 9. Up to 9 days, growth in darkness, as well as under short-day conditions is linear.

It was found that daily repeated R pulses had no significant effect on axis elongation in the pine seedling, although phytochrome can readily be measured in pine embryos and etiolating seedlings, and even whole short days with R reduced axis elongation only slightly compared to W of comparable irradiance (Table 3). On the other hand, B had a considerable effect although less than W (Fig. 6). It appeared that the photomorphogenic light operates mainly *via* the B/UV-A photoreceptor, a situation well known from the pteridophytes (Mohr 1972). However, experiments with dichromatic light led to a different picture. In these experiments strong R or FR (RG9-light) beams were applied simultaneously with B to establish a high or low Pfr:P ratio, and thus a high or low level of Pfr, during the light period in the presence of B/UV.

The results (Fig. 6) show that B exerted its effect on axis elongation only if Pfr was readily available. When the level of Pfr was kept very low, with simultaneous RG9-light, no effect of B was detectable. When a high level of Pfr was established, with simultaneous R, B reduced axis elongation to the same extent as W.

Table 3. Effect of light pulse treatments at the end of a short day (8 h light/day) on the length of hypocotyls of Scots pine (*Pinus sylvestris* L.) seedlings at the beginning of day 9 after sowing. The daily light pulses (5 min each) to establish the phytochrome photoequilibrium were given with red light (R, 40 W m^{-2}, $\varphi_{RG9} < 0.01$). The energy flux during the short day (8 h light/day) was as follows: white light (W, 80 W m^{-2}); red light (R, 40 W m^{-2}); blue light (B, 20 W m^{-2}); simultaneous B plus RG9-light ([B + RG9], 20 W m^{-2} and 40 W m^{-2}). ΔR = Response one − response two. Response two is the response if the seedlings receive a saturating 5-min R pulse (φ_R = 0.8) prior to the beginning of the dark period. Response 1 is the response if the saturating 5-min light pulse is with RG9-light ($\varphi_{RG9} < 0.01$). Δ R is a precise gauge for the responsiveness of the system towards Pfr. After Fernbach and Mohr (1990).

Light treatment	Hypocotyl length (mm) ± SE	ΔR
Dark control	47.0	
R pulse only	45.9	0
RG9-light pulse only	46.1	
8 h R	45.4	
8 h R + RG9-light pulse	47.1	2
8 h R + RG9-light pulse + R pulse	44.9	
8 h (B + R) + R pulse	31.2	
8 h (B + R) + RG9-light pulse	39.8	9
8 h (B + RG9) + R pulse	38.6	
8 h (B + RG90 + RG9-light pulse	46.6	8
8 h W	29.5	
8 h W + R pulse	28.8	
8 h W + RG9-light pulse	35.7	
8 h W + RG9-light pulse + R pulse	29.4	7

From these data it was concluded that B does not affect axis growth directly. Rather, B amplifies the responsiveness towards Pfr in accordance with the previously suggested model of coaction (Fig. 2). This conclusion was supported by the results obtained with end-of-day pulse treatments (Table 3).

In this type of experiment the value ΔR may be used as a gauge of responsiveness towards Pfr during a dark period (Oelmüller and Mohr 1984). The definition of ΔR is response one minus response two. Response two is the response (length of axis in the present case) if a seedling receives a 5 min R pulse, sufficient to establish the phytochrome photoequilibrium (φ_R = 0.8), at the end of the daily light period, *i.e.* immediately prior to the beginning of the dark period. Response 1 is the response if the saturating 5 min light pulse is RG9-light ($\varphi_{RG9} < 0.01$). As shown repeatedly, ΔR is a precise gauge for the responsiveness of the system towards Pfr.

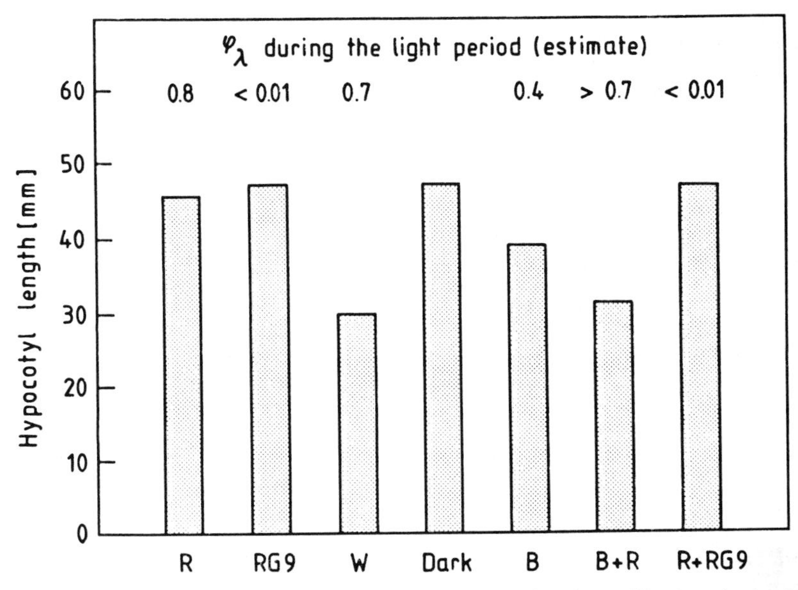

Figure 6. Length of hypocotyls of Scots pine *(Pinus sylvestris* L.) seedlings at the beginning of day 9 after sowing. The daily light period (8 h light/day) was given with different light qualities (see legend to Table 3). The φ-values can only be considered as estimates. After Fernbach and Mohr (1990).

6.4.2 Synthesis of plastid Fd-GOGAT (ferredoxin-dependent glutamate synthase, EC 1.4.7.1) in the cotyledonary whorl

The enzyme Fd-GOGAT plays a major role in nitrate/ammonium assimilation (Miflin and Lea 1976). Elmlinger and Mohr (1991) showed that in pine cotyledons the enzyme was strongly induced by light, whereas nitrate had only a minor inductive effect and ammonium no effect at all (data not shown). Up to 10 days after sowing the effect of W could be attributed to the action of phytochrome. Thereafter, there was an absolute requirement for B for a further increase in the enzyme level. At first sight it appeared that by 10 days the action of phytochrome was replaced by that of the B/UV-A photoreceptor. However, dichromatic experiments (simultaneous treatment of the seedlings with two light beams to vary the level of Pfr during B irradiation showed that B does not affect enzyme appearance if the Pfr level is kept low by simultaneously applied RG9-light (Table 4). It is concluded that B has no direct effect on Fd-GOGAT appearance. Rather the action of B seems to be restricted to maintaining responsiveness of Fd-GOGAT synthesis to Pfr more than 10 days after sowing.

This case study is particularly intriguing since it documents the changing light requirement of a plant during development, an important aspect for the plant physiologist, as well as for the designer of growth chambers.

Table 4. Action of dichromatic irradiation on the level of ferredoxin-dependent glutamate synthase, EC 1.4.7.1 (Fd-GOGAT) in cotyledonary whorls of water-grown pine seedlings after a 6-day blue-light (B) pretreatment after the onset of the experimental period (4 day after sowing, see Fig. 7). The B and red light (R) [(B + R), 10 W m^{-2} and 20 W m^{-2}, $\varphi = 0.7$] or B and RG9-light [(B + RG9), 10 W m^{-2} and 20 W m^{-2}, $\varphi < 0.05$] were given simultaneously for 2 days. Control experiments were carried out as indicated. Whorls were assayed 8 days after the onset of the experimental period (12 days after sowing). The values in the table are means ± SE from five to eight independent experiments. After Elmlinger and Mohr (1991).

Light treatment (after onset of experimental period)	Fd-GOGAT (pkat/whorl)
6 day B (pretreatment)	156 ± 6.2
6 day B + 2 day D	130 ± 2.0
6 day B + 2 day B	251 ± 2.6
6 day B + 2 day R	259 ± 4.7
6 day B + 2 day (B + R)	254 ± 3.5
6 day B + 2 day (B + RG9)	155 ± 4.0
6 day B + 2 day RG9	147 ± 1.5
6 day R	152 ± 5.2
6 day R + 2 day R	144 ± 5.9
8 day D	32 ± 1.8

In a further study on glutamine synthetase (GS), it was confirmed that appearance of both enzymes, plastid GS (GS_2) as well as Fd-GOGAT, are regulated co-ordinately in precisely the same way. Thus, the demands of optimum ammonium assimilation are met (Elmlinger and Mohr 1992). In the case of GS_2, the transcript level was also measured. It was found that the transcript level is controlled by phytochrome (Pfr) whereby B leads to a strong responsiveness amplification. If the Pfr level was kept low by simultaneously applied RG9-light [similar to (B + RG9) in Table 4] the transcript level as well as the rate of protein synthesis remained very low. However, transcript level and rate of protein synthesis were not correlated. Rather, besides its action at the transcript level, B exerted a strong and specific stimulatory effect on GS protein synthesis (again in analogy to Table 4) (Elmlinger *et al.* 1993).

6.5 Photosensors involved in light control of stem elongation in seedlings of angiosperm plants

6.5.1 Hypocotyl elongation in the mustard (Sinapis alba *L.*) *seedling*

In this species hypocotyl growth is exclusively controlled by phytochrome. Under our experimental conditions, the growth rate of the mustard hypocotyl in the dark can be considered to be constant between 48 and 78 h after sowing

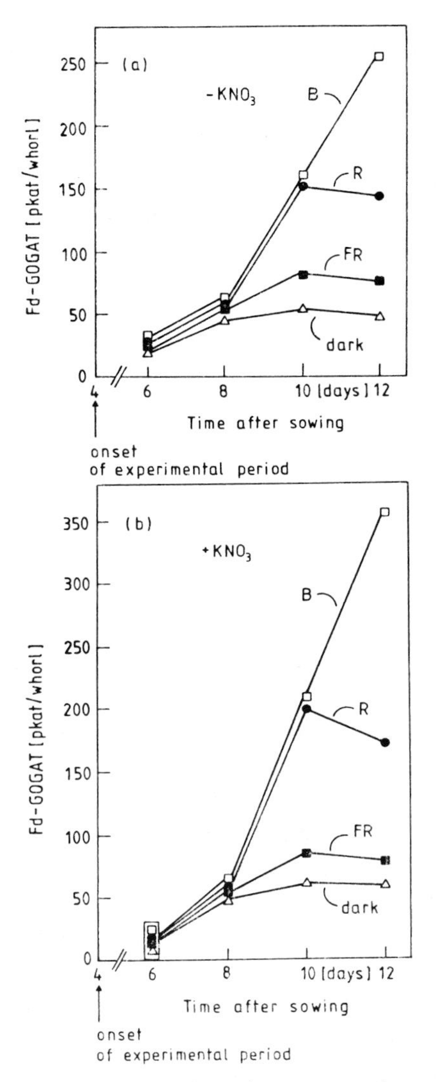

Figure 7. Time courses of ferredoxin-dependent glutamate synthase (Fd-GOGAT, EC 1.4.7.1) levels in cotyledons of Scots pine (*Pinus sylvestris* L.) seedlings either grown (a) on water ($-KNO_3$) or (b) on 15 m*M* KNO_3 ($+ KNO_3$) in continuous far-red light (FR, 3.5 W m^{-2}; ■–■), red light (R, 6.8 W m^{-2}; ●–●) or blue light (B, 10 W m^{-2}; □–□) or darkness (Δ–Δ). After Elmlinger and Mohr (1991).

(Fig. 8). Continuous FR (cFR, 3.5 W m^{-2}, operating through phytochrome, see Mohr 1984) results in a marked reduction in the growth rate. Growth in FR can be considered as a kind of 'base line' since this growth rate cannot be decreased further by any light treatment. The same reduction in growth rate as seen in cFR is *transiently* observed when light pulses are applied 48 h after sowing, and the seedling is then returned to darkness. The different light pulses (R, FR, RG9-

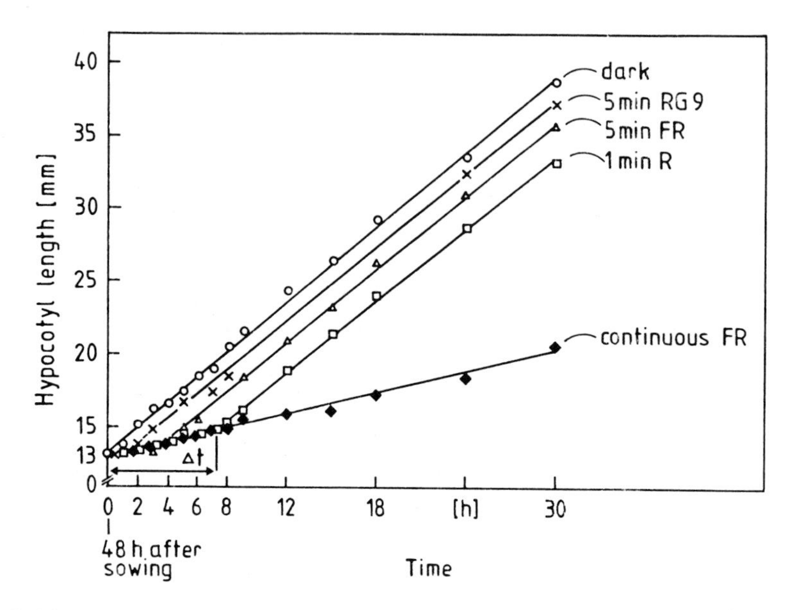

Figure 8. Time courses of hypocotyl elongation in mustard (*Sinapis alba* L.) seedlings in darkness (○), continuous far-red light (♦) and after saturating light pulses, which establish different Pfr levels: (×) $\varphi_{RG9} = 0.0014$; (Δ) $\varphi_{FR} = 0.023$; (□) $\varphi_R = 0.8$; onset of light treatment at 48 h after sowing. Data for phytochrome photoequilibria φ_λ *in vivo* as a function of wavelength are from Schäfer *et al.* (1975, Fig. 2). Δt, extrapolated duration of the time between the light pulse and the point of resumption of growth. After Oelze-Karow and Mohr (1989).

light) establish different phytochrome photoequilibria (see legend to Fig. 8). Apparently, the amount of Pfr established by a light pulse determines the length of time (Δt) before the growth rate characteristic for darkness is restored.

The data in Fig. 8 indicate that Pfr control of hypocotyl elongation in mustard acts as a threshold response. This implies that as long as the level of Pfr remains above the threshold, elongation is inhibited (*i.e.* elongation occurs at the 'base line' rate). As soon as the level of Pfr decreases below the threshold level, the dark rate of elongation is immediately restored. Phytochrome degradation (destruction) kinetics, measured in the hook part of the hypocotyl, are in *quantitative* agreement with this concept (Oelze-Karow and Mohr 1989).

6.5.2 Hypocotyl elongation in the cucumber (Cucumis sativus *L.*) *seedling*

Cucumber seedlings have often been used to study rapid inhibition of stem elongation by B (see Shinkle and Jones 1988, and references therein). The response is considered *not* to be mediated by phytochrome, and similar reactions are assumed to occur in other species. In fact stem growth in cucumber is strongly affected by B (Fig. 9). It is agreed that the action of B cannot be

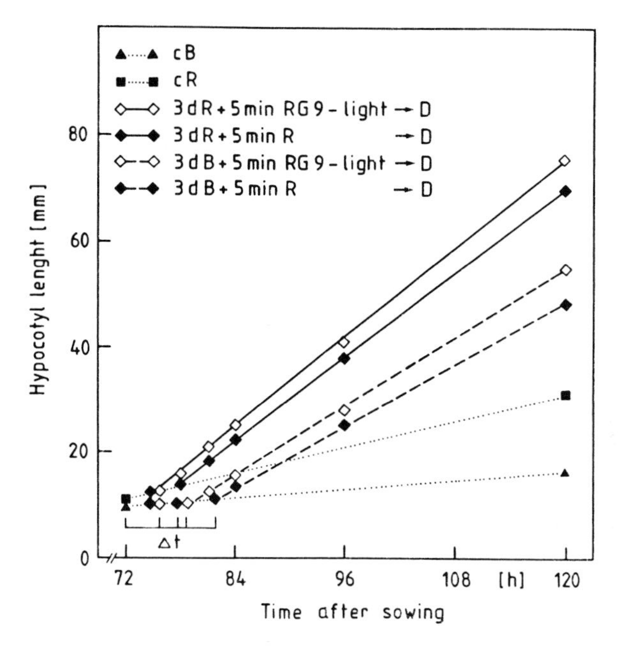

Figure 9. Time course of hypocotyl elongation in cucumber (*Cucumis sativus* L.). Seedlings were grown in continuous red (cR) or continuous blue (cB) from the time of sowing (········), or were transferred from light to darkness (D) 72 h (3 d) after sowing. Immediately prior to transfer they received 5 min light pulses to establish a high ($\varphi_R = 0.8$) or low ($\varphi_{RG9} < 0.01$) Pfr level. The time span between the light pulse (*e.g.* R) and the onset of the high growth rate is called Δt. Fluence rates: R, 6.8 W m^{-2}; B, 10 W m^{-2}; far-red light (FR), 3.5 W m^{-2}; RG9-light, 10 W m^{-2} (see Fig. 8). After Drumm-Herrel and Mohr (1991).

explained by B affecting the phytochrome system (Chapter 9.1; Cosgrove 1986; Mohr 1986). Moreover, there is agreement that B action on stem elongation is distinct from B mediation of phototropism (Chapter 9.2). The pending problem was to determine the kind of coaction between B and R, absorbed by phytochrome, in controlling stem elongation. As always, the study of coaction was complicated by the fact that B/UV inevitably operates on phytochrome photoconversion.

In light pulse experiments it was found that hypocotyl growth in the etiolated cucumber seedling responds weakly to 5 min R and B pulses whereas FR and RG9 light pulses had no effect. Blue and RG9-light pulses were given in sequence or 5 min B was applied simultaneously with high fluence rate R (phytochrome photoequilibrium, Pfr/P = $\varphi_{B+R} \approx 0.8$) or RG9-light ($\varphi_{B+RG9} < 0.03$) to maintain a high or low φ during application of the B pulse. The results showed that there was no expression of a B effect if the level of Pfr was kept low. A specific effect of B probably occurs, but cannot be expressed in the almost total absence of Pfr. Thus, the above model (Fig. 2) was perfectly verified in the case of light pulses.

In long term experiments hypocotyl elongation was strongly affected by continuous R, FR and B (Fig. 9). Light to dark transfer experiments suggest that growth in darkness of seedlings de-etiolated by R or B is controlled by phytochrome through a threshold mechanism, similar to mustard (Fig. 8).

The results of dichromatic experiments (long-wavelength FR, 756 nm, applied simultaneously with B or R) showed that the action of B on axis elongation is related to the level of Pfr, even during long-term irradiation. When $\varphi_{B+756} < 0.03$, the action of B was largely abolished. The small residual B effect cannot be attributed to a phytochrome-independent action of B on axis elongation, because a small residual effect was also observed with dichromatic R + 756 nm-light.

The strong effect of B on axis elongation in cucumber during long-term irradiation (Fig. 9) can readily be explained in terms of Fig. 2: B specifically stimulates responsiveness of the growing axis cells towards Pfr.

So far the crucifer *Arabidopsis thaliana* (L.) Heynh. is the only higher plant for which B-response mutants have been identified. Koornneef *et al.* (1980) described a *hy4* mutant which showed reduced hypocotyl inhibition in B while maintaining normal phytochrome levels and R/FR responses. Recently, Liscum and Hangarter (1991) have isolated four mutant lines that fail to show B-dependent inhibition of hypocotyl elongation, while their phytochrome-mediated inhibition of hypocotyl growth by FR appears to be normal. These findings confirm that two photosensory systems function in the hypocotyl growth response, and the observations reported so far appear to be fully compatible with the above coaction model (Fig. 2).

6.6 Gene expression in the tomato phytochrome-deficient *aurea* mutant

The *aurea* (*au*) mutant of *Lycopersicon esculentum* Mill., cv. Moneymaker, has been intensively investigated and identified as a phytochrome photoreceptor mutant (Adamse *et al.* 1988a, b). In dark-grown *au* seedlings the phytochrome level detected spectrophotometrically and immunologically is less than 5% of that in the isogenic wild type (see Oelmüller and Kendrick 1991 for review). The reduction in the phytochrome photoreceptor concentration causes a defective photoregulation of gene expression and reduced chlorophyll accumulation. Oelmüller and Kendrick (1991) have shown that the almost normal photomorphogenesis of the *au* mutant in W is to be attributed primarily to light absorption in the B/UV-A region of the spectrum. According to these authors, a B/UV-A photoreceptor that operates in concert with the residual amount of phytochrome allows survival of the mutant, while light absorption by phytochrome alone is not sufficient. This interpretation is obviously in accordance with the coaction model in Fig. 2.

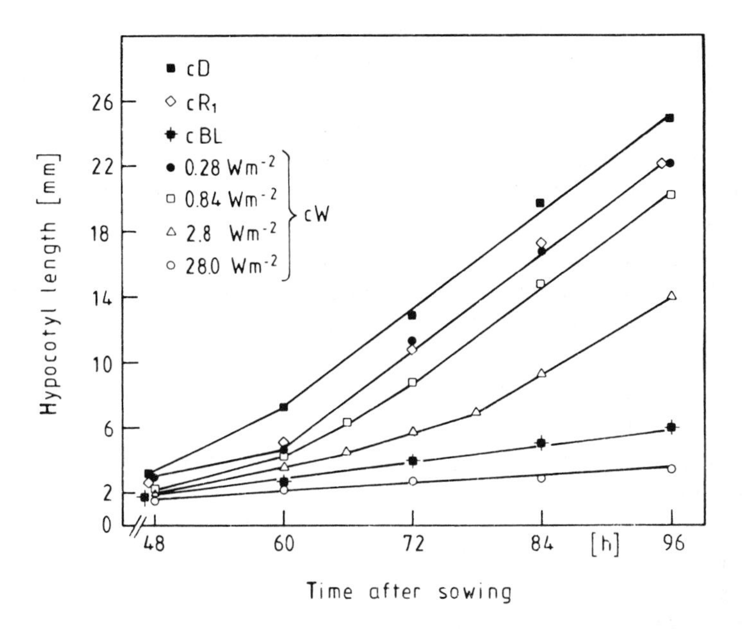

Figure 10. Time courses of hypocotyl elongation in the sesame (*Sesamum indicum* L.) seedling. cD, continuous darkness, cR, continuous low fluence rate red light (0.68 W m^{-2}); cB, continuous blue light (7 W m^{-2}); cW, continuous white light of different energy fluence rates (W m^{-2}). Light given from time of sowing. After Drumm-Herrel and Mohr (1984).

6.7 Coaction between photoreceptors in phototropism: sesame seedling, *Sesamum indicum* L.

6.7.1 Hypocotyl straight growth

In sesame, the rate of *hypocotyl elongation* is controlled by W and B, while R (continuous R, as well as repeated R pulses) is totally ineffective beyond 60 h after sowing (Fig. 10). Between 36 and 60 h after sowing the growth rate responds strongly to R pulses, the effect of which is fully reversible by FR. Thus, an action of phytochrome is indicated, but only up to 60 h (Drumm-Herrel and Mohr 1984).

Importantly no B effect on hypocotyl growth was detectable in weak W. Rather, longitudinal growth of the hypocotyl in weak W was found to be precisely the same as in R of a corresponding fluence rate (Fig. 10). A *threshold fluence rate* for the action of B was also reported in control of hypocotyl growth in *Cucumis sativus* L. (Gaba *et al.* 1984). This lack of a B effect at low fluence rate is amazing insofar as in dicot seedlings (including sesame) a very low fluence rate (*e.g.* 1 mW m^{-2}) of continuous B suffices to induce a strong phototropic growth response if applied *unilaterally* (Steinitz and Poff 1984).

In general dicot seedlings respond very sensitively, even towards low fluence rates of unilateral B or W. However, with regard to straight growth (in omnilateral light) the seedlings are 'B-blind' at low fluence rates. Obviously, a strong phototropic response can be elicited by unilateral B which does not have any lasting effect on straight growth if applied omnilaterally.

Direct photographic analysis supports this conclusion. Rich and Smith (1985) analyzed hypocotyl growth rates during phototropism in green *Sinapis alba* L. seedlings under experimental conditions where changes in phytochrome were minimized. Varying amounts of continuous B were added to background light from a low pressure sodium lamp, and changes in growth rates on the irradiated and shaded sides of the hypocotyl were followed photographically. Unilateral B which led to phototropic curvature simultaneously caused an inhibition of growth on the irradiated side and an acceleration on the shaded side but with no net change in growth (except at high fluence rates). In agreement with Fig. 10 plants showed no change in net growth when omnilateral B was applied at moderate fluence rates.

6.7.2 Phytochrome and phototropism

It was repeatedly shown, Shropshire and Mohr 1970, that in dicot seedlings the phototropic response can under no circumstances be elicited by unilateral R although the seedling responds sensitively towards unilateral B. On the other hand, the rate and extent of the phototropic response was found to be affected strongly by a pretreatment with R operating *via* phytochrome.

Phototropism in sesame is particularly interesting since it demonstrates unambiguously that the phytochrome-mediated effect of a R pretreatment on phototropism is unrelated to the control of straight growth by phytochrome. As documented above (Fig. 10), hypocotyl straight growth in sesame is not affected by R beyond 60 h. However, a R pretreatment (operating through phytochrome) can exert a strong effect on the rate of the phototropic curvature at a later time (Fig. 11). Thus, phytochrome strongly affects differential growth of the hypocotyl flanks (elicited by unidirectional B) even though it does not affect straight growth of the hypocotyl. Apparently, the effect of R on the growth potential of the hypocotyl cells remains cryptic until unilateral B is given.

In a series of technically sophisticated experiments it was found (Woitzik and Mohr 1988a) that during bending a high level of Pfr, such as established by omnilateral R, *inhibits* the rate of bending (for this reason the seedlings in Fig. 11 were placed in unilateral B with a minimum amount of Pfr, *i.e.* they received a saturating FR pulse before the onset of B). However, if R was applied unilaterally from the same direction as B, R *increased* the rate of curvature. When simultaneous R was given from above, or from the opposite direction to B, it

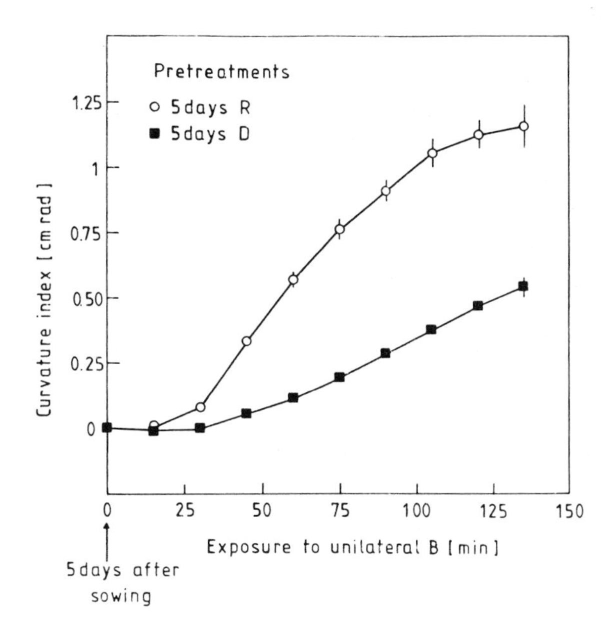

Figure 11. Phototropic curvature of the hypocotyl of the sesame (*Sesamum indicum* L.) seedling in continuous unilateral blue light (B, 8 mW m^{-2}). The 'curvature index' is a sensitive gauge of the phototropic response which considers not only the angle of curvature but also the length of the responding zone of the axis. The seedlings were either kept in darkness (D) for 5 days or for 5 days in weak red light (R, 0.68 W m^{-2}) until exposure to unilateral B. After Woitzik and Mohr (1988a).

reduced the rate of curvature compared to B alone. Obviously, the seedling is capable of detecting the direction of R relative to the direction of B. While a mechanistic explanation of these effects cannot be advanced at present, it is clear that the seedling is capable of superimposing information about the actual light conditions during bending on a 'memory' of the light conditions prior to the onset of bending. Thus, the previous as well as the actual light conditions determine its phototropic responsiveness.

The behaviour of the sesame seedling can be readily explained in *teleonomic* terms: the seedling is genetically programmed in such a way as to move its shoot by means of phototropism (and gravitropism, Woitzik and Mohr 1988b) to that segment of space where optimal light conditions are to be expected.

6.8 Conclusion

It appears that phytochrome as Pfr is the effector proper in bringing about photomorphogenesis (de-etiolation) in higher plants, whereas the B/UV-A photoreceptor and the UV-B photoreceptor (together with phytochrome) determine the plant's responsiveness towards Pfr. Phototropism, on the other hand, can only be elicited in dicot seedlings by B/UV. In this case, it is phytochrome which modulates the rate of the response. It seems that the same B/UV-A photoreceptor mediates phototropism in both, dicot and monocot plants (Baskin and Iino 1987).

6.9 Further reading

Mohr H. (1980). Interaction between blue light and phytochrome in photomorphogenesis. In: *The Blue Light Syndrome,* pp. 97-118, Senger H. (ed.) Springer-Verlag, Berlin.

Mohr H. (1984) Criteria for photoreceptor involvement. In: *Techniques in Photomorphogenesis*, pp. 13-42, Smith H. and Holmes M.G. (eds.) Academic Press, London.

Mohr H. and Drumm-Herrel H. (1983) Coaction between phytochrome and blue/UV light in anthocyanin synthesis in seedlings. *Physiol. Plant.* 58: 408-414.

Mohr H., Drumm-Herrel H. and Oelmüller R. (1984) Coaction of phytochrome and blue/UV light photoreceptors. In: *Blue Light Effects in Biological Systems,* pp. 6-19, Senger H. (ed.) Springer-Verlag, Berlin.

Schäfer E. and Haupt W. (1983) Blue-light effects in phytochrome-mediated responses. In: *Encyclopedia of Plant Physiology,* New Series, Vol. 16B, *Photomorphogenesis*, pp. 723-744, Shropshire Jr. W. and Mohr H. (eds.) Springer-Verlag, Berlin.

Woitzik F. and Mohr H. (1988a) Control of hypocotyl phototropism by phytochrome in a dicotyledonous seedling (*Sesamum indicum* L.). *Plant Cell Environ.* 11: 653-661.

6.10 References

Adamse P., Kendrick R.E. and Koornneef M. (1988a) Photomorphogenetic mutants of higher plants. *Photochem. Photobiol.* 48: 833-841.

Adamse P., Jaspers P.A.P.M., Bakker J.A., Wesselius J.C., Heeringa G.H., Kendrick R.E., Koornneef M. (1988b) Photophysiology of a tomato mutant deficient in labile phytochrome. *J. Plant Physiol.* 133: 436-440.

Baskin T.I. and Iino M. (1987) An action spectrum in the blue and ultraviolet for phototropism in alfalfa. *Photochem. Photobiol.* 46: 127-136.

Cosgrove D.J. (1986) Photomodulation of growth. In: *Photomorphogenesis in Plants,* pp. 341-366, Kendrick R.E. and Kronenberg G.H.M. (eds.) Martinus Nijhoff, Dordrecht.

Downs R.J. and Siegelman H.W. (1963) Photocontrol of anthocyanin synthesis in milo seedlings. *Plant Physiol.* 38: 25-30.

Drumm H. and Mohr H. (1978) The mode of interaction between blue (UV) light photoreceptor and phytochrome in anthocyanin formation of the *Sorghum* seedling. *Photochem. Photobiol.* 27: 241-248.

Drumm-Herrel H. and Mohr H. (1981) A novel effect of UV-B in a higher plant (*Sorghum vulgare*). *Photochem. Photobiol.* 33: 391-398.

Drumm-Herrel H. and Mohr H. (1984) Mode of coaction of phytochrome and blue light photoreceptor in control of hypocotyl elongation. *Photochem. Photobiol.* 40: 261-266.

Drumm-Herrel H. and Mohr H. (1991) Involvement of phytochrome in light control of stem elongation in cucumber (*Cucumis sativus* L.) seedlings. *Photochem. Photobiol.* 53: 539-544.

Elmlinger M.W. and Mohr H. (1991) Coaction of blue/ultraviolet-A light and light absorbed by phytochrome in controlling the appearance of ferredoxin-dependent glutamate synthase in the Scots pine (*Pinus sylvestris* L.) seedling. *Planta* 189: 374-380.

Elmlinger M.W. and Mohr H. (1992) Glutamine synthetase in Scots pine seedlings and its control by blue light and light absorbed by phytochrome. *Planta* 188: 396-402.

Elmlinger M.W., Batschauer A., Oelmüller R. and Mohr H. (1993) Coaction of blue light and light absorbed by phytochrome in control of glusamine synthetase gene expression in Scots pine (*Pinus sylvestris* L.) seedlings. *Planta* in press.

Fernbach E. and Mohr H. (1990) Coaction of blue/ultraviolet-A light and light absorbed by phytochrome in controlling growth of pine (*Pinus sylvestris* L.) seedlings. *Planta* 180: 212-216.

Gaba V., Black M. and Attridge T.H. (1984) Photocontrol of hypocotyl elongation in de-etiolated *Cucumis sativus* L. *Plant Physiol.* 74: 897-900.

Koornneef, M., Rolff E. and Spruit C.J.P. (1980) Genetic control of light-inhibited hypocotyl elongation in *Arabidopsis thaliana* (L.) Heynh. *Z. Pflanzenphysiol.* 100: 147-160.

Liscum E. and Hangarter R.P. (1991) *Arabidopsis* mutants lacking blue light-dependent inhibition of hypocotyl elongation. *Plant Cell* 3: 685-694.

Mohr H. (1972) *Lectures on Photomorphogenesis*, Chapter 22, Springer, Heidelberg, New York.

Mohr H. (1982) Principles in plant morphogenesis. In: *Axioms and Principles of Plant Construction*, pp. 93-111, Sattler R. (ed.) Martinus Nijhoff, The Hague.

Mohr H. (1986) Mode of coaction between blue/UV light and light absorbed by phytochrome in higher plants. In: *Blue Light Responses - Phenomena and Occurrence in Plants and Microorganisms, Vol. I*, pp. 133-144, Senger H. (ed.) CRC Press, Boca Raton.

Miflin B.J. and Lea P.J. (1976) The pathway of nitrogen metabolism in plants. *Phytochemistry* 15: 873-885.

Oelmüller R. and Mohr H. (1984) Responsivity amplification by light in phytochrome-mediated induction of chloroplast glyceraldehyde-3-phosphate dehydrogenase (NADP-dependent, EC 1.2.1.13) in the shoot of milo (*Sorghum vulgare* Pers). *Plant Cell Environ.* 7: 29-37.

Oelmüller R. and Mohr H. (1985a) Specific action of blue light on phytochrome-mediated enzyme syntheses in the shoot of milo (*Sorghum vulgare* Pers). *Plant Cell Environ.* 8: 27-31.

Oelmüller R. and Mohr H. (1985b) Mode of coaction between blue/UV light and light absorbed by phytochrome in light-mediated anthocyanin formation in the milo (*Sorghum vulgare* Pers.) seedling. *Proc. Natl. Acad. Sci. USA* 82: 6124-6128.

Oelmüller R. and Kendrick R.E. (1991) Blue light is required for survival of the tomato phytochrome-deficient *aurea* mutant and the expression of four nuclear genes coding for plastidic proteins. *Plant Mol. Biol.* 16: 293-299.

Oelze-Karow H. and Mohr H. (1989) An analysis of phytochrome-mediated threshold control of hypocotyl growth in mustard (*Sinapis alba* L.) seedlings. *Photochem. Photobiol.* 50: 133-141.

Rich T. and Smith H. (1985) Phytochrome and phototropism in light-grown plants. In: Book of Abstracts, *European Symposium on Photomorphogenesis in Plants*, p.108, Wageningen.

Shinkle J.R. and Jones R.J. (1988) Inhibition of stem elongation in *Cucumis* seedlings by blue light requires calcium. *Plant Physiol.* 86: 960-966.

Shropshire W. and Mohr H. (1970) Gradient formation of anthocyanin in seedlings of *Fagopyrum* and *Sinapis* unilaterally exposed to red and far-red light. *Photochem. Photobiol.* 12: 145-149.

Steinitz B. and Poff K.L. (1984) Phototropism in *Arabidopsis* seedlings. *Supplement to Plant Physiol.* 75: no. 1, 73.

Woitzik F. and Mohr H. (1988b) Control of hypocotyl gravitropism by phytochrome in a dicotyledonous seedling (*Sesamum indicum* L.). *Plant Cell Environ.* 11: 663-668.

Yatsuhashi H., Hashimoto T. and Shimizu S. (1982) Ultraviolet action spectrum for anthocyanin formation in broom *Sorghum* first internode. *Plant Physiol.* 70: 735-741.

Part 7 The light environment

7.1 Sensing the light environment: the functions of the phytochrome family

Harry Smith

Department of Botany, University of Leicester, Leicester, LE1 7RH, UK

7.1.1 Introduction

Why do plants have *photoreceptors*? What selective advantage does the possession of signal-transducing photoreceptors confer? Does the operation of the photoreceptors in plants growing in the natural environment reflect what we have learnt about them from laboratory experiments?

The above three questions were posed in the first edition of this book, and an attempt was made to arrive at reasonable answers based on the information currently available. Since that time, we have been fortunate to experience a massive increase in knowledge and understanding concerning the phytochromes, the most intensively investigated and arguably the most important signal-transducing photoreceptors in higher plants. The new level of awareness has meant that, although the above questions are no less relevant, the answers we can provide are in some senses more reliable and satisfying. Even so, the new knowledge has posed as many questions as it has answered. In particular, it has raised fundamental questions that relate directly to the functions of the different members of the phytochrome family: do the different phytochromes have different informational and physiological functions which provide adaptive value? In this chapter, these fundamental questions are explored in relation to the characteristics of the natural radiation environment, and to current evidence derived from physiological experiments using mutant and transgenic plants. Since photoperiodism is covered elsewhere in this book (Chapter 7.3), it is excluded from consideration in this chapter.

R.E. Kendrick & G.H.M. Kronenberg (eds.), Photomorphogenesis in Plants - 2nd Edition

7.1.2 The function of informational photoreceptors

The concept of 'function' is of crucial importance to this chapter. In photosynthesis, where photoreceptors serve to capture radiant energy, the function is obvious; however, in the case of *signal-transducing* photoreceptors, there are so many subtle variations of signal and response that they often conspire to make functional analyses extremely complex. For such photoreceptors it is logical to identify two types of function; an *informational* function, defined as the capacity of a photoreceptor to acquire information about the radiation environment (Smith 1983), and one or more *physiological* functions, defined as the role of the photoreceptor in evoking specific physiological responses to perceived environmental signals. Since informational photoreceptors have presumably evolved as a result of specific selection pressures, the informational function of each photoreceptor, or family of photoreceptors, should be separate and distinct, and related in some definable way to specific light signals in the natural environment. Put in more concrete terms, we should logically expect all the phytochromes, which have presumably evolved from a common ancient form, to have an informational function traceable to some common fundamental signal-transducing property. Even though there are multiple types of the phytochromes, each possibly exerting different physiological functions and thereby regulating different aspects of development, on *a priori* grounds we might expect their informational characteristics to remain more or less unchanged. Obviously, this becomes a question of photoreceptor evolution. Assuming the progenitor phytochrome had a photoperceptive function that provided adaptive value in the then current radiation environment, it is quite feasible that subsequent modifications of photoperceptive function could arise because of selection pressure from changes in the environment. Unfortunately, the term 'function' is used very loosely, and often with scant regard to the importance of the perception of information as outlined above. With knowledge of the physiological roles of different members of the phytochrome family growing at an incredible rate, we can now begin to consider whether the concept that the phytochromes have a common underlying photoperceptive function has any real validity.

The obvious approach to understanding the informational function of a specific photoreceptor is to match its characteristic properties with those of the natural radiation environment, asking the question: does the environment provide any signals which would be both ecologically relevant *and* detectable by the photoreceptor? A positive answer to this question would not, of itself, be sufficient; it is necessary then to carry out simulation experiments in which the environmental signals are disentangled from other interfering, but irrelevant, signals in order to show that plants respond as would be predicted. There are thus three steps in forming a view of the informational function of a photoreceptor: (i) analysis of the radiation environment; (ii) comparison with the properties of the putative photoreceptor; (iii) simulation of the natural signal: response system under conditions unconfounded by other environmental variables.

In this chapter, current evidence on the functions of phytochromes in the perception of the natural light environment is outlined in accordance with these three separate steps. This treatment is rendered difficult because the photophysical properties of all the members of the phytochrome family are not yet known. Consequently, we are reduced to the dangerous expedient of basing our conclusions on the properties of the one phytochrome that has been adequately investigated; *i.e.* phytochrome A, and in particular phytochrome A from oats. If, in the fullness of time, it is discovered that the other phytochromes have quite different photophysical characteristics, then our ideas on the perceptive functions of the phytochromes may have to change.

7.1.3 Information in the light environment

Before dealing specifically with the phytochromes, it will be useful to consider the nature of the information that can be derived from the natural light environment. In this context we restrict ourselves to the photobiologically-active portion of the radiant spectrum incident upon the earth's surface; this waveband (*ca.* 400-800 nm) essentially comprises those wavelengths in which the energy per photon is sufficient to initiate photochemistry, but not so large that it causes ionization. This region comprises upwards of 55% of all the radiation reaching the earth's surface and, *via* photosynthesis, powers almost all life on the earth; it is subject to variation in a number of parameters. The amount of light, its distribution across the spectrum, its timing and its direction are obviously ways in which daylight can vary. There are also more subtle ways in which it can vary, including the degree to which it is polarized, and the extent and nature of light scattering. From the light environment, therefore, it should theoretically be possible to derive information on: light quantity, light quality, light direction, light periodicity, and the degree of light polarization.

In theory, the acquisition of these differing types of information requires only a relatively limited range of perception mechanisms. The perception of light quantity must involve photon-counting. Similarly, the perception of the spectral distribution of radiation (*i.e.* light quality) must involve the estimation of ratios of photons in two or more wavelength bands; this of course depends upon photon-counting in these bands, plus some device for comparing the quantities thus measured. Perception of the direction of the actinic beam must depend upon the detection of photon gradients in space, again a matter of photon-counting, and comparison, at spatially-separated points in the organism. The perception of the duration of exposure to light depends on the timing of light-dark transitions; this may involve simply photon-counting, but also could involve other events that occur at the natural light-dark transitions, such as changes in the spectral distribution of radiation. Finally, the perception of the plane of polarization of natural radiation must clearly depend upon dichroic arrangements of photoreceptors. These concepts are brought together in Table 1.

7.1.4 Light-quantity perception: theoretical aspects

Since 'photon counting' lies at the basis of the perception of all information from the light environment, it is important to have some understanding of what the term means. The laws of photochemistry, specifically the Stark-Einstein Law of Photochemical Equivalence, state that the absorption of one quantum causes a photochemical change in one molecule (or atom, or electron). Thus, in principle, any photochemical reaction has the potential to act as a photon counter, as long as a metabolic mechanism exists for transduction of the *quantity* of photochemical product into a biological change. Photosynthesis operates as a photon counter, with a quantum yield (*i.e.* Φ = number of molecules affected divided by the number of photons absorbed) of 0.125; *i.e.* eight photons are required for every molecule of CO_2 fixed. Most photomorphogenic responses appear to have much higher quantum yields, apparently often more than 1, implying at first sight a departure from the law of photochemical equivalence. Such cases can only be understood if some mechanism of metabolic amplification of the initial photochemical reaction occurs, but in order for precise photon counting to be obtained, the degree of amplification must bear a quantitative relationship to the quantity of initial photoproduct. Thus, for photomorphogenic responses in which it is claimed that the *quantity* of light is perceived, the transduction chain must include amplification mechanisms that result in graded output directly related to the magnitude of the initial photoevent. On this basis, responses that display threshold, or all-or-none, relationships with fluence cannot operate as efficient light quantity detectors. In practice, most photomorphogenic responses bear relationships in which the biological response is linearly related to the logarithm of the fluence; these relationships in themselves are difficult to account for on the laws of photochemistry, as they imply amplification mechanisms that become progressively less amplificatory with increased fluence.

An important distinction should be made between mechanisms that simply count photons (*i.e.* respond to total fluence), and those that count the rate at which photons arrive (*i.e.* respond to fluence rate, or photon irradiance). The former should obey another of the laws of photochemistry, namely the Bunsen-Roscoe Reciprocity Law, in which the response is proportional to the quantity of photoproduct, irrespective of whether that quantity is produced by brief pulses of high photon irradiance, or longer periods of low photon irradiance.

Table 1. Information that can be derived from the light environment and the mechanism necessary for its perception.

Information	Perception mechanism
Light quantity	Photon counting
Light quality	Photon ratios
Direction of light	Photon gradients
Duration of light	Timing of light-dark transitions
Polarization	Dichroic photoreceptor arrangement

Photoresponses that are proportional to fluence rate will not show reciprocity, since the response will be greater the greater the rate at which photons are absorbed. In the natural environment, as opposed to the photobiology laboratory, actinic radiation is usually present for long periods of time, and consequently we might expect responses that are dependent on fluence rate to be of greater ecological significance than those that operate as a function of total fluence.

7.1.5 Light-quality perception: theoretical aspects

A moment's thought will confirm the view, stated above without supporting evidence, that light quality perception can only occur *via* a mechanism in which the amount of light in separate, discrete wavebands is compared. A sensor comprising only one element, capable only of estimating the number of photons falling within one waveband (broad or narrow) would be incapable of discriminating between overall photon irradiance changes and changes in the spectral distribution of radiation. The minimum requirement is for two comparative elements that interact in a manner so as to allow the relative photon irradiances in two separate wavebands to be estimated. In theory, a comparative bi-chromatic sensor would be automatically compensated for fluctuations in irradiance that did not affect spectral distribution. This is so because changing the overall irradiance would not alter the ratio of irradiances in any two wavebands, as long as the spectral distribution remained constant. Such a sensor would, of course, provide limited information, and multi-element sensors would be necessary for the resolution of detailed variations in spectral distribution; for example, most theories of colour vision in humans are based upon interactions between three comparative elements. One of the objectives of this chapter, however, is to demonstrate the sophistication which is possible using a simple bi-chromatic sensor of light quality.

In theory, a viable-light quality sensor should: (i) consist of at least two elements absorbing photons in well-separated wave-bands; (ii) have two elements capable of interacting in some way such that light absorption by both contributes appropriately to the signal transduction chain; (iii) have a mechanism which is compensated for fluctuations in light quantity. Since each of the two minimum elements must act as a photon counter, then the comparison between the two elements will represent a comparison between the photoproducts of two spectrally distinct photon counters. Perhaps the simplest possible comparative mechanism would be the case in which the photoproduct of one photon counter is removed by the action of another; the steady-state concentration of the first photoproduct would then effectively be a measure of the relative activities of the two photon counters.

Such theoretical considerations are useful and important, but it has to be said that they are essentially a rationalization in hindsight of the empirical findings resulting from the study of phytochrome over the last 40 years. Phytochrome acts as an ideal light-quality sensor, in which the relative rates of absorption of photons in the red (R) and far-red (FR) are compared by the photoconversion of the two forms of phytochrome, the R-absorbing form (Pr) and the FR-absorbing form (Pfr), such that the steady-state concentration of Pfr is a function of the relative quantities of R and FR photons. The significance of this for understanding the roles of the phytochromes in acquiring information from the natural light environment cannot be overstated.

7.1.6 The complexity of spectral information

Nowadays, the spectral distribution of radiation may be easily measured by the use of a spectroradiometer, an instrument that splits incoming light into a continuous spectrum, which is then scanned by a sensitive detector. Because of the variable geometry of plant parts in the natural environment, it is clearly impossible to specify a single photodetector arrangement which could be used to measure the light intercepted by vegetation. A number of different devices have been used, including long cylindrical detectors, spherical detectors, and flat-surface detectors. A degree of standardization has been achieved, by convention and convenience, rather than by agreement on the most appropriate arrangement. Plant photomorphogenesists tend to use flat, horizontal detectors which are cosine-corrected, *i.e.* corrected for the effect of angle of incidence. Although such a detector cannot be said to simulate any normal plant surface, the conventional standardization at least allows the direct comparison of data from different laboratories.

Figure 1 shows the typical spectral distribution of daylight; the data are expressed in photon, rather than energy terms, since photoreceptor action is dependent on the number of photons absorbed, not the amount of energy. This spectrum is quite complex, and the various fluctuations in daylight quality which occur each add further complications, making it difficult to handle the

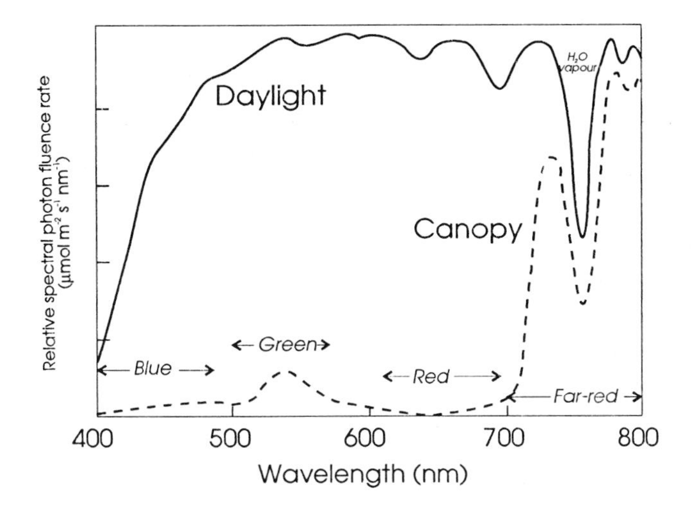

Figure 1. Typical spectral photon distributions of daylight above and within a vegetation canopy. The main spectral regions (*i.e.* blue, green, red, and far-red) are indicated by arrows. Note the almost complete lack of red and blue light within the canopy, and the relative enrichment of the far-red. Note also the deep trough in the daylight spectrum within the far-red caused by absorption of radiation by water vapour.

data so easily obtained by spectroradiometry. It is clear, therefore, that considerable simplification is necessary if we wish to search for variations in daylight quality that might be of informative value to plants. Firstly, we can entirely ignore the various peaks and hollows in the broad curve of Fig. 1; these are due to absorption by components of the atmosphere (mainly H_2O and O_2) and do not fluctuate significantly. The greatest simplification, however, comes from reducing the spectral data to a single parameter. Bearing in mind the theoretical considerations in Section 7.1.5, we should choose specific pairs of wavebands and calculate ratios between them, being careful to choose wavebands that are absorbed by the known putative photoreceptors. This argument is potentially circular, but nevertheless has been very rewarding, as will be seen.

7.1.6.1 R:FR, φ_e and φ_c; phytochrome related parameters

The absorption spectra of phytochrome purified from etiolated oats, both in the Pr form and after saturating R irradiation, is shown in Fig. 1 (Chapter 4.1). This phytochrome is phytochrome A, and as yet comparable spectra of the other species of phytochrome are not available. Consequently, we can only assume at present that the absorption spectra of all the phytochromes are similar. The absorption maxima are broad peaks centred near 660 nm and 730 nm respectively. Because of the photochromicity of phytochrome and the overlap of the absorption spectra of Pr and Pfr, continuous broad-band irradiation establishes an equilibrium between Pr and Pfr which is not affected by the irradiance (within wide limits), but is markedly affected by the relative amounts of R and FR. These properties of phytochrome fulfil the three requirements listed in Section 7.1.5 for the simplest form of light quality-sensor. Thus, the ratio of the photon irradiance in the R, to that in the FR, is a parameter of the light environment which is directly related to the spectral properties of phytochrome. Virtually any portion of the R between 600 and 700 nm, and any portion of the FR between 710 and *ca.* 780 nm, could be used to construct the R:FR photon ratio. In practice, however, although there is some difference of approach evident in the literature, there has been a tendency to use a standardized formula as follows:

$$\text{R:FR} = \frac{\text{photon irradiance between 655 and 665 nm}}{\text{photon irradiance between 725 and 735 nm}}$$

The Greek symbol ζ (*zeta*) was originally used for this particular R:FR ratio, but is rarely applied today. Some groups, particularly that of M. J. Kasperbauer, use the inverse ratio (*i.e.* FR:R) but of course the relationships arrived at are identical in principle.

The benefit of using R:FR as a simplified parameter of the spectral distribution of natural radiation lies in the fact that R:FR can be readily transformed into a measure of the relative proportions of Pr and Pfr present at photoequilibrium. Figure 2 shows the hyperbolic relationship which has been shown to exist between R:FR (as defined above) and Pfr/P (or φ) where P = total phytochrome, Pr + Pfr. This curve, constructed empirically, fits the equation for a rectangular hyperbola (Hayward 1985). The equation, and graphical relationship, are mathematical approximations which arise when only two wavelengths (R and FR) are involved. Using this relationship, any measured value of R:FR can be transformed to a value (φ_e) which represents the Pfr/P to be expected in the outer epidermis of the irradiated tissue (ignoring any light reflected or scattered back from within that tissue); the transformation is made simply by reading φ_e from the curve for any measured value of R:FR. The actual direct measurement of Pfr/P in light-grown plants yet eludes the advance of analytical technology, mainly because there is very little phytochrome present and its absorption is overwhelmed by that of chlorophyll. The parameter φ_e is merely a physiologically-relevant way of expressing the relative amounts of R and FR in the incident light. Actual Pfr/P within tissue is, of course, determined by the incident light, but it is quantitatively variable depending on the optical properties of the tissue.

The use of R:FR and φ_e as indicators of light quality has been very valuable, as will be seen below. An important caveat should be made, however, concerned with the blue (B) region of the spectrum. Fig. 1 (Chapter 4.1) shows that

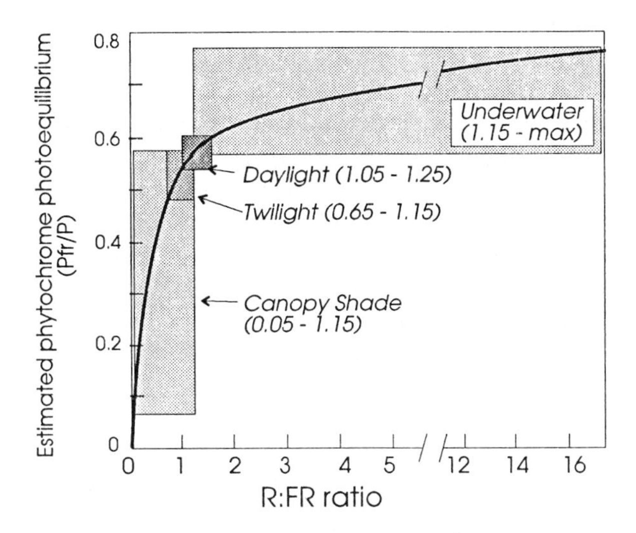

Figure 2. The relationship between red (R):far-red (FR) and calculated Pfr/P. From Smith (1982). The relationship is a rectangular hyperbola in which the steepest part of the curve lies within the R:FR range found within vegetation canopies. In contrast, even very large increases in R:FR underwater cause only small changes in Pfr/P, as these values lie near the asymptote.

Pr and Pfr also absorb B, but not to the same extent as R and FR. In theory, therefore, light having large amounts of B relative to R or FR would affect Pfr/P such that the function for φ_e would no longer hold. In practice, this turns out not to be a problem in most natural conditions, since such an imbalance between B and R and FR is rarely encountered, although it can obviously be very important with artificial lighting. There is, in any case, a way of dealing with this problem. Assuming complete spectroradiometric data for the 400-800 nm waveband are available, the spectral photon irradiances can be integrated with the extinction coefficients of Pr and Pfr and the quantum efficiencies of the Pr → Pfr and Pfr → Pr phototransformations over the whole waveband. In other words, the influence of light over the whole wavelength region absorbed by phytochrome on the photoconversion of Pr to Pfr, and *vice versa*, can be calculated, resulting in a more meaningful value for Pfr/P than can be derived from R:FR (Chapter 4.7). This value is known as φ_c, denoting *calculated* Pfr/P. The procedure sounds complicated, but in practice is simply performed with the aid of a computer program into which spectroradiometric data are automatically input. The spectroradiometric data are split into a series of discrete wavelength intervals and the average photon irradiance in each interval is multiplied with the extinction coefficients of Pr and Pfr, and the quantum efficiencies of the Pr → Pfr and Pfr → Pr photoconversions, specific for that wavelength interval (Chapter 4.7). The computer thus integrates the spectral photon irradiance data with the absorption cross section data [*i.e.* the product of the extinction coefficient (ε) and the quantum yield (Φ) at each wavelength] for both forms of phytochrome.

Although this method is quick and convenient, the parameter that results (φ_c) is subject to a number of caveats. In particular, the absorption cross section data that are available for oat phytochrome A differ, depending upon the source data, resulting in different values for φ_c. A more substantial point is that, up to now, absorption cross section data are only available for phytochrome A from cereals. Since (as will be shown below) phytochrome A is normally not involved in regulating processes that occur in light-grown plants, the use of φ_c to relate physiological responses quantitatively to phytochrome is fraught with danger. However, in spite of its flaws the method is the best we have, and the consistency of relationships that have been derived is, at least, encouraging. Until the photochemical and photophysical characteristics of the other phytochromes have been determined, we must use the only method available.

7.1.7 The natural radiation environment

It has been possible to identify at least four ecological conditions in which the characteristics of the light environment, particularly the spectral distribution, vary in ways that are both informative, and potentially detectable by a phytochrome, and thus the rest of this chapter concentrates on the putative informational and physiological functions of the phytochrome family of photoreceptors. First, general properties of the natural radiation environment are outlined that indicate very strongly the potential roles of phytochromes in acquiring environmental information. Following this broad outline, specific information relating to the possible ecological functions of individual phytochromes is presented.

7.1.7.1 The daylight spectrum

The spectrum of daylight in Fig. 1 was obtained using a cosine-corrected horizontal detector which allows light from the whole hemisphere above the sensor to be collected in the correct geometry. This, so-called, ‘global’ radiation comprises two main components; direct radiation from the solar disk and scattered, or diffuse, radiation which has interacted with molecules or particles in the atmosphere before reaching the sensor. The overall shape of the daylight spectrum is determined mainly by the relative contributions of direct and scattered radiation. The direct beam from the sun is essentially a broad band peaking in spectral photon irradiance at *ca.* 620 nm, falling off rather steeply towards the shorter wavelengths and more shallowly towards the infra-red (IR). Scattering of light by molecules and small particles in the atmosphere is inversely proportional to the fourth power of the wavelength (*Rayleigh scattering*). Thus, as wavelength is decreased, scattering increases greatly. This type of scattering is responsible for the blue colour of the sky. Since scattering occurs in all directions, some of the incident sunlight is reflected out from the atmosphere, giving the now well-known bluish halo around the earth when viewed from space; the moon, which has no atmosphere, has no halo. Atmospheric particles with diameters greater than the wavelength of the impinging light scatter more neutrally across the wavelength range. Consequently, the diffuse, or scattered, component of global radiation is very high in B but does contain some R and FR, whilst the direct beam is relatively low in B but high in R and FR.

Many hundreds of determinations of the spectral distribution of daylight have led to the generalization that the R:FR ratio is remarkably constant (Smith 1982). As long as the solar angle (*i.e.* the angle between the tangent to the earth's surface at the position of the sensor, and a line connecting the sensor to the solar disk) is greater than *ca.* 10°, R:FR averages 1.15, with a standard error of the mean of 0.02. Weather and cloud conditions have virtually no effect on R:FR, even though a heavily overcast sky can reduce total 400-800 nm irradiance by more than ten fold. The remarkable constancy of R:FR in daylight provides a standard value against which natural radiation, modified by ecologically-relevant factors, may be compared. As will be seen, there are very few terrestrial situations in which R:FR goes above the 1.15 daylight value.

7.1.7.2 Diurnal fluctuations in daylight quality

The daily march of the sun across the sky causes predictable diurnal fluctuations in spectral distribution. As the solar angle diminishes towards dusk, the contribution of direct, relative to diffuse, radiation declines, commonly leading to a pronounced relative peak in the B. Simultaneously, the direct beam traverses an increasingly long path through the atmosphere, enhancing atmospheric absorption and scattering, thus depleting the shorter wavelengths, leading to a small, but measurable, drop in R:FR; similar effects are observed at sunrise. With a large relative increase in B, it would be dangerous to use R:FR and Fig. 2 to derive the value φ_c. This is, therefore, a situation in which integration of all the spectral data to compute φ_c will give a more accurate indication of what might be expected to happen to phytochrome.

7.1.7.3 Light quality within vegetation canopies

The photosynthetic pigments, the chlorophylls and carotenoids, absorb light over almost the whole of the visible spectrum (*i.e.* 400-700 nm). Some of the light in the green is either transmitted or reflected, which is why leaves are green to our eyes. What is not so immediately obvious is that vegetation hardly absorbs any radiation between 700 and 800 nm. Thus, virtually all the incoming FR is either transmitted or reflected; *i.e.* the FR is scattered either through the leaf, or from the surface of the leaf. Since our visual systems are very insensitive to radiation beyond *ca.* 700 nm, we fail to recognise that leaves should look FR, rather than green! If we could see FR, then our perceptions of the natural world would be very different, 'this green and pleasant land' would be 'far-red and sinister', no doubt! More to the point, however, the fact that vegetation absorbs R yet is virtually transparent to FR, obviously has major effects on the R:FR within a canopy. Figure 1 shows both a typical daylight spectrum and a spectrum of daylight filtered through a vegetation canopy, demonstrating the very marked change in R:FR compared to unfiltered daylight. Values for global R:FR within various canopies range as follows: (i) oak woodland, 0.1-0.75; (ii) wheat crop 0.2-0.5; (iii) sugar beet crop, 0.11-0.45. A comprehensive list can be found in Morgan and Smith (1981).

The extent to which R:FR is depressed by vegetational shade varies, of course, with the density of the canopy and the depth of the sensor within that canopy. Thus, within regular canopies, such as a wheat crop, R:FR is more or less logarithmically related to leaf area index (LAI = the total leaf area intercepted within a projected unit ground area), with a sharp fall in R:FR with a relatively small increase in LAI. This means that R:FR would be a sensitive indicator of shading.

A more subtle effect of vegetation on R:FR depends on the direction of propagation of the radiation being measured, or perceived. Unfiltered solar radiation is propagated essentially vertically downwards, and is highly directional; in other words, the radiation is only slightly scattered. After interaction with the leaves of a vegetation canopy, multiple scattering occurs, causing the radiation to be propagated more randomly, although even at this stage it is not propagated equally in all directions. This means that if directional sensors are used, the radiation that is propagated more-or-less horizontally within a canopy will already have interacted with vegetation and will consequently be depleted in R and relatively enriched in FR; radiation within a canopy that is propagated more-or-less vertically downwards, however, will have a relatively high component of unfiltered daylight and will have a high R:FR (Smith *et al.* 1990). This point has a far-reaching significance, as will become evident later.

7.1.7.4 Light quality underwater

More than half of plant life on earth is underwater. The underwater light climate presents a number of interesting differences from that immediately above the surface. Refraction at the air-water discontinuity leads to the incident light from above being concentrated into a cone of half-angle 48.6°; consequently, a sensor facing upwards, below, but near to the surface inevitably receives a proportion of upwelling radiation reflected back down from the surface. More important phenomena, as far as light quality is concerned, are scattering and absorption by water itself, and by dissolved molecules or suspended particles. Rayleigh scattering results in the selective attenuation of the blue region of the spectrum of downwelling radiation. Water has strong absorption bands in the FR and in the near IR, and therefore the FR is also selectively attenuated. Thus, in clear water, downwelling radiation is effectively 'compressed' with increasing depth into a decreasingly narrow band of wavelengths, usually peaking at or around 500 nm. In turbid waters, major complications are evident. Most natural waters have varying amounts of organically-derived material, much of it soluble, which absorbs both B and R radiation; being yellow in appearance, this material is usually known by its original German term *Gelbstoff*. Very large amounts of chlorophyllous micro-organisms can also be present, which absorb strongly in the B and R. In consequence, underwater spectra can be extremely variable. The increase in R:FR with depth underwater can be very large, but is offset substantially if there is any shading vegetation within the water column above the sensor (Spence 1981; Morgan and Smith 1981). The R:FR ratio thus is a reliable indicator of depth only when the interfering effects of chlorophyllous shading materials are minimal.

7.1.7.5 The light environment under the soil

A crucial phase in the establishment of seedlings is the transition from growth entirely beneath the soil surface, when all life processes are dependent upon the utilization of stored reserves, and growth at and above the surface, when the provision of energy *via* photosynthesis becomes possible. Since this transition is triggered by the emergence of the seedling from the below-soil environment to the aerial environment, it is probable that the large change in the light conditions provides signals important for the developing seedling. Because soils are complex and differ widely, it is difficult to generalize on the light environment within the soil; however, it is not correct to assume that there is no light at all beneath the soil surface. Little research has been done on this question but, at least for the lighter sandy or loamy soils, amounts of light sufficient to induce photomorphogenic reactions may be present at surprising depths. The R:FR ratio is depressed below the soil, at least for some soils, but the effects on spectral distribution are small compared to those on light quantity. Thus, on *a priori* grounds, it might be expected that a useful signal for the approach to the soil surface would be a rapid increase in light quantity, rather than a change in R:FR; this question is taken up below.

7.1.8 The phytochromes as sensors of environmental R:FR

7.1.8.1 Sensitivity considerations

The analysis of predictive R:FR fluctuations in nature set out in Section 7.1.7 indicates that phytochrome photoreceptors, if capable of operating to monitor R:FR, may be able to provide the plant with information on: (i) the timing of the daily photoperiod; (ii) shading by other vegetation; (iii) depth of immersion in water. Other important signals exist within the natural light environment, particularly changes in the total amount of light below and above the soil surface; it is conceivable that phytochromes may also be able to detect changes in light quantity. Concentrating for the moment on R:FR perception, however, an important question is whether or not the environmental fluctuations are large enough to be detected by phytochrome. Put another way: are the phytochrome photoconversions sensitive enough to detect the changes that occur in nature?

One aspect of this question relates to the curve in Fig. 2. The asymptotic nature of the relationship between R:FR and Pfr/P means that, for R:FR values between 0 and 2, small changes in R:FR elicit relatively large changes in Pfr/P; in contrast, for values much above R:FR = 2, even large changes in R:FR can cause only small changes in Pfr/P. The implications for environmental R:FR perception are summarized in Fig. 2. The R:FR of shade environments lie on the steepest part of the curve, whereas those of underwater environments lie at or close to the asymptote. Thus, the phytochromes could, in principle, act as sensitive detectors of the R:FR changes associated with vegetation shade, but would be of little value as detectors of the depth of immersion in water. These correlations are suggestive, but do not constitute proof that phytochromes do act as R:FR detectors in plants growing in the natural environment, since in all conditions in which R:FR is reduced, there are simultaneous reductions in the total amount of light. The way to test the hypothesis is to investigate the growth of plants under controlled conditions, in which R:FR ratio is varied whilst the amount of photosynthetically-active radiation (*i.e.* PAR, 400-700 nm) is held constant. This is most conveniently achieved by growing plants in uniform PAR with varying amounts of additional FR. Before outlining the results and implications of such experiments, an understanding of the strategies of plants in response to the presence of competing vegetation is needed.

7.1.8.2 Plant strategies in response to shade

The acclimative responses of herbaceous plants to shade by other vegetation can be viewed in terms of two extreme strategies (Grime 1979). One strategy, that of *shade tolerance,* involves relatively slow growth rates, the conservation of energy and resources, perennation usually by vegetative processes, and the

Table 2. The shade avoidance syndrome.

Physiological process	Response to shade
Extension growth	Accelerated
Internode extension	Rapidly increased
Specific stem weight	Reduced
Petiole extension	Rapidly increased
Leaf/tendril extension	Marginally increased
Leaf development	Retarded
Leaf area growth	Reduced
Leaf thickness	Reduced
Chloroplast development	Retarded
Chlorophyll synthesis	Retarded
Chlorophyll a/b ratio	Balance changed
Apical dominance	Strengthened
Branching	Inhibited
Tillering (in grasses)	Inhibited
Flowering	Accelerated
Rate of flowering	Markedly increased
Seed set	Severe reduction
Fruit development	Truncated
Seed germinability	Severely reduced
Assimilate distribution	Marked change
Storage organ deposition	Severe reduction
Sink monopolisation	Markedly increased

development of photosynthetic structures that are especially efficient at low light levels; the latter phenomenon includes changes in the stoichiometry of thylakoid components, an enhanced ratio of light harvesting to CO_2-fixation, and the reduction of respiratory rates. The opposite extreme is *shade avoidance*, a syndrome of growth and developmental changes in which internode and petiole extension growth is favoured at the expense of leaf development. The major features of the shade-avoidance syndrome are summarized in Table 2. As the name suggests, if successful, shade avoidance has the overall effect of projecting the photosynthetic structures (usually leaves) into those parts of the environmental mosaic in which the resource of light is plentiful. This is a less anthropomorphic and more scientific way of saying what is common knowledge: that 'plants grow towards the light'. However, as will be seen, they do not, in fact, grow *towards* high light, they grow *away* from low R:FR.

Shade avoiders tend to be photosynthetically inefficient at low light levels, but have the capacity rapidly to direct growth potential from leaf development to shoot extension upon the first detection of incipient shading. Shade avoidance is an effective strategy for life in an herbaceous community, but has

Figure 3. Effect of R:FR on plant growth. These *Chenopodium album* (fat hen) seedlings were grown in four growth chambers which provided uniform photosynthetically-active radiation (400-700 nm), but different amounts of additional FR so as to vary R:FR, thereby simulating the light quality within a canopy, whilst keeping the amount of light for photosynthesis constant. The numbers attached to the plant pots indicate the R:FR, symbolized by ζ. Photograph taken in 1976 by Dr D. C. Morgan.

limitations for herbs growing on the floor of a dense forest. The two strategies, avoidance and tolerance, are not necessarily mutually exclusive, since some plants display intermediate strategies and appear to be able to adapt to life either in open or shaded habitats, whilst other plants can exhibit shade avoidance and shade tolerance at different points in their life cycle. In evolutionary terms, shade avoidance appears to be a relatively recent invention, since it is predominantly found in the angiosperms, although some gymnosperms show some shade-avoidance characteristics. Ferns, mosses and liverworts show little if any capacity to react to vegetational shading by characteristic avoidance reactions, and generally cope with shade by tolerance responses.

7.1.8.3 R:FR perception and the induction of shade-avoidance reactions

It is now well-established that growing shade-avoiding species in relatively high irradiance white light to which various amounts of FR have been added results in developmental responses essentially similar to those seen in natural canopy shade (Smith 1982; Casal and Smith 1989a). This experimental procedure allows R:FR to be varied over the range normally found in vegetation canopies whilst maintaining each set of test plants at a uniform level of PAR. Figure 3 shows seedlings of the shade-avoiding weed *Chenopodium album* grown in cabinets in which the PAR was held uniform but the R:FR was decreased by supplementation with varying fluence rates of FR. The most striking effect is the enhanced elongation growth at low R:FR, but all the other components of the shade-avoidance syndrome listed in Table 2 can be induced by such decreases in R:FR. Thus, this experiment indicates definitively that the induction of the shade-avoidance syndrome requires the perception of the spectral changes associated with shade, rather than the changes in total light quantity. This conclusion is rendered more forcible by plotting rates of extension against φ_c, the calculated phytochrome photoequilibrium (Fig. 4a). This linear relationship, which has been obtained for a wide range of species, provides convincing evidence that the perception of shade and the induction of shade-avoidance responses is phytochrome-mediated.

The ecological significance of phytochrome-mediated R:FR perception has been studied by determining the quantitative relationships between extension growth and φ_c for a number of species adapted either as shade avoiders or shade tolerators. The resultant relationships, summarized in Fig. 4b, show that all species tested detected the depression in R:FR, but only those genetically adapted as shade avoiders reacted by marked increases in extension growth rate.

Two further important characteristics of R:FR perception are its rapidity and its compensation for changes in irradiance. Using position-sensitive transducers to enable the continuous monitoring of stem extension rate, changes in extension rate caused by changes in R:FR ratio can be detected within minutes. If

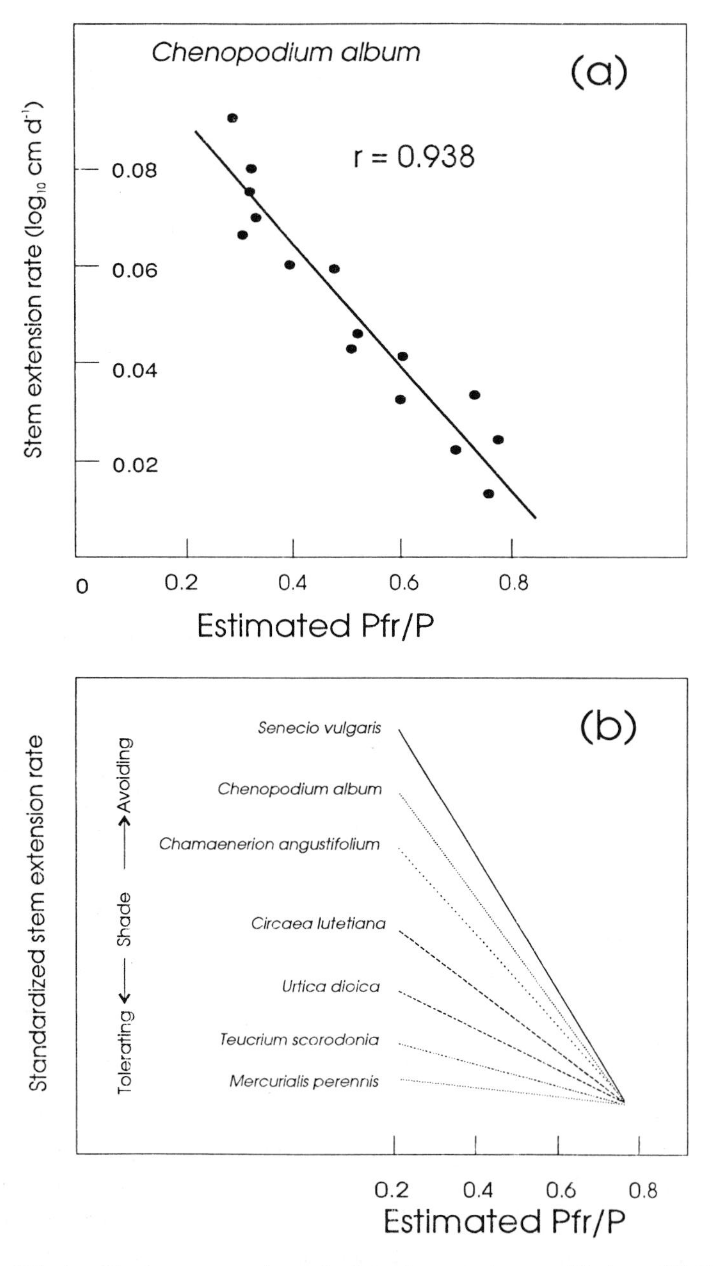

Figure 4. Relationship between estimated phytochrome photoequilibrium (Pfr/P, or φ_e) and extension growth. (a) Shows the detailed data for the *C. album* seedlings in Fig. 3. From Morgan and Smith (1976). (b) Similar data for a range of herbaceous seedlings, normalized at $\varphi_e = 0.75$. Note that all the species tested showed a linear relationship between φ_e and extension growth, but that the slope of the line was much greater for shade-avoiding, than for shade-tolerating, species. From Smith (1982).

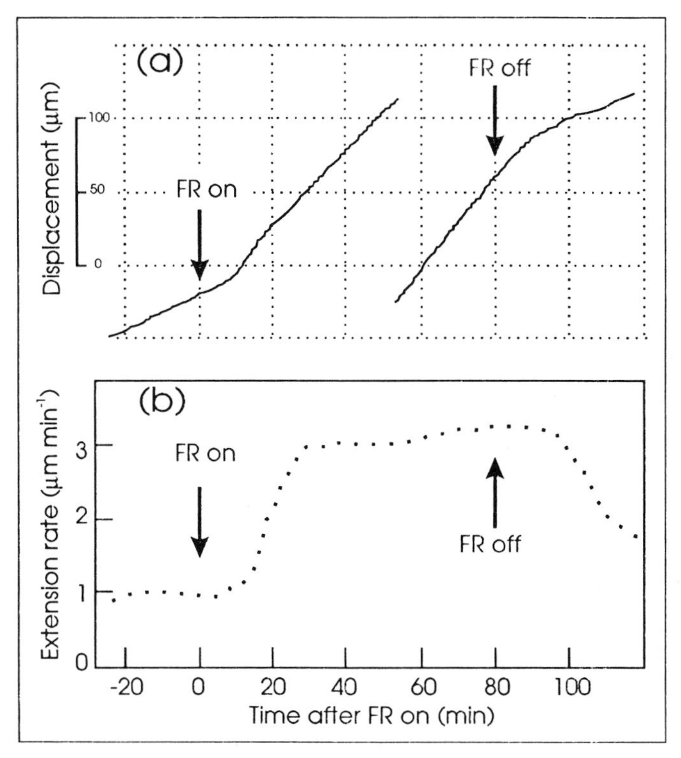

Figure 5. Rapidity of extension rate changes in response to alterations of R:FR. The extension rate of a *Sinapis alba* (white mustard) plant was monitored using a linear displacement transducer as the plant was grown in background white fluorescent light of high R:FR. At the first arrow, high irradiance FR was supplied *via* fibre-optic probes directed at the most apical growing internode; at the second arrow the FR was switched off. (a) Shows a tracing of the transducer chart recorder; in (b) the computed extension rates are shown. Note that the initial extension rate is very slow (1 μm min^{-1} is just over 1 mm day^{-1}), but within 20-30 minutes after the FR is switched on the rate has almost trebled.

plants are grown in fluorescent white light (which virtually lacks FR), growth rate is relatively slow; *i.e.* 2-week old mustard seedlings elongate at a rate of about 1-2 μm min^{-1} (about 2 mm per day). Applying high irradiance FR from fibre-optic probes directly to the terminal, growing internode, results in a large increase in extension rate (often up to three-fold) that begins within 10-15 min of the onset of FR (Fig. 5). Removal of the FR leads to a corresponding reduction in extension rate. If bifurcated fibre-optic probes are used so that R and FR can be mixed in known quantities before presentation to the growing internode, the extension growth rate at steady-state can be shown to be a direct linear function of φ_c calculated from the proportions of R and FR; this is true whether the reduction in R:FR is brought about by increasing the proportion of FR with a constant R, or decreasing the proportion of R with a constant FR. Furthermore, within wide limits, the extension rate is determined by the R:FR at the internode and is independent of the fluence rate of white light presented from above (Child and Smith 1987). These results indicate that the perception of R:FR is precisely quantitative and is compensated for variations in total irradiance.This means that, in principle, phytochrome-mediated R:FR perception should not only be able to operate at the light levels that exist within dense canopies, but should also function at the high irradiances present in sparse stands of plants that are not sufficiently close to cast actual shade.

One theoretical problem associated with R:FR perception at high irradiances results from the nature of the photoconversions, which proceed through a number of relatively long-lived intermediates in the two directions (Chapters 4.1 and 4.3). At low irradiances the initial photochemical reactions occur much more quickly than the thermochemical interconversions of the intermediate states. However, as irradiance is increased the decay of intermediates gradually becomes rate-limiting, so that at daylight levels upwards of 50% of the total phytochrome may be present as photoconversion intermediates (Kendrick and Spruit 1972; Smith and Fork 1992). The problem here is that, if the growth responses are a function of the steady-state concentration of Pfr, as the central theory of phytochrome action proposes, then at daylight irradiances the concentration of Pfr would be substantially reduced compared to shade irradiances because of the accumulation of intermediates. One test of this concern was to expose mustard seedlings attached to a transducer to very large and sudden fluctuations in fluence rate of mixed R and FR presented to the internodes *via* fibre-optic probes. The results suggested that transitions between very high and low fluence rates produced transitory fluctuations in extension growth even at a constant R:FR, but that once the transitory changes were finished, steady-state growth rate was a function of R:FR, and thus of φ_c (Smith 1990). This appears to indicate that extension rate is responsive to sudden changes in [Pfr] brought about by the accumulation of intermediates at high irradiance, but eventually adapts to the new state by a mechanism not yet understood. These data sit uncomfortably with the central concept that phytochrome action is mediated entirely and only through the concentration of Pfr.

7.1.8.4 Proximity perception, or the detection of neighbours

A strategy for the detection and avoidance of shade would obviously be most effective if it were able to operate *before* shading actually occurred. That this is likely was presaged by the experiments in which FR was presented to growing internodes *via* fibre-optic probes (Morgan, *et al.* 1980; Child and Smith 1987), which showed that plants could react to low R:FR from the side even when the rest of the plant was exposed to white light of high R:FR. That this phenomenon occurs in nature was suggested by studies of soybeans grown in either north-south or east-west rows (Kasperbauer *et al.* 1984). In these observations, it was shown that R:FR near the top of the north-south rows of plants was lower on the west side in the morning, and on the east in the evening; Kasperbauer *et al.* (1984) explained this by proposing that the adjacent rows act as FR reflectors when the sun is low in the sky, thus redirecting the FR back towards adjacent rows. Direct effects of reflected FR on plant growth in the field have been demonstrated by a series of experiments by the group of R. A. Sânchez, based in Buenos Aires (Ballaré *et al.* 1987, 1990). In the first experiments, seedlings of *Datura ferox*, a strongly shade-avoiding herb, were grown in the field close to, but on the northern side of, grass hedges that were either green, or bleached by being sprayed with a herbicide. The weed seedlings were not shaded by the grass hedges, because (being in the Southern Hemisphere) the direction of the solar beam was from the north. Those plants adjacent to unbleached, green hedges grew significantly faster than those near to the bleached hedges. Confirmation that the increased extension rate was due to low R:FR radiation reflected from the green hedges came from conceptually similar experiments in which mirrors that selectively reflected either R or FR were used in place of the hedges; mirrors reflecting FR caused increased growth whilst those reflecting R did not. *Datura ferox* seedlings, when inserted into a sparse canopy of similar seedlings not dense enough to cast actual shade, grew faster than in the open. If their growing internodes were surrounded by transparent collars containing dilute copper sulphate solution (which absorbs FR), there was no increase in growth. From these experiments it was concluded that phytochrome-mediated R:FR perception was sufficiently sensitive to allow the detection of reflected light from neighbouring vegetation; in other words, plants can detect the presence of neighbours by perceiving the reduced R:FR of light reflected from them.

The quantitative aspects of neighbour detection were elaborated by Smith *et al.* (1990) who measured the reflection signals from stands of tobacco, and also measured actual Pfr/P in samples of purified oat phytochrome exposed to the radiation reflected from the tobacco stands. It was clear from these data that, in principle, neighbour detection could operate over significant distances, and that the signals progressively increased as proximity to the neighbours increased. On this basis, phytochrome-mediated R:FR perception provides plants with the capacity for proximity perception; in other words, plants not only can detect their neighbours, they can effectively perceive how far away they are, and therefore are able to gauge the competitive threat posed. The ecological significance of these results is that the shade avoidance syndrome can be induced *before* actual shading occurs. The implication is obviously that, as discussed above, phytochrome-mediated R:FR perception must be capable of operating under conditions in which plants are exposed to full daylight from above, and to low irradiances of reflected radiation from the side.

7.1.8.5 End-of-day effects

Some of the earliest experiments on phytochrome action in light-grown plants were those of Downs *et al.* (1957) at Beltsville, in which it was shown that plants given short exposures to FR at the end of the daily light period grew faster than those not so treated. These so-called 'end-of-day' FR effects were fully reversible by subsequent R, and displayed normal fluence response relationships and reciprocity; consequently, end-of-day (EOD) responses appear qualitatively similar to the classic R/FR reversible, low fluence responses (LFR) extensively studied in dark-imbibed seeds and etiolated seedlings. The EOD responses are characterized by slow escape from reversibility; this has been taken as evidence that they are mediated by a form of phytochrome in which Pfr is stable and long-lived (see Smith and Whitelam 1990 for review). Superficially, EOD responses to FR are very similar to responses to long-term FR given as a supplement to daily white light; this raises the question as to whether proximity perception and shade avoidance in nature are merely manifestations of low R:FR experienced at the end of the day. The potential importance of twilight reductions in R:FR were highlighted by the experiments of Kasperbauer *et al.* (1984) mentioned above, in which the lowest R:FR in row-grown soybeans occurred when the sun was low in the sky. There is a logical problem here, however. The R:FR is generally low at twilight, irrespective of any interaction of solar radiation with vegetation. Indeed, examining the responses of shade-avoiding species to reduced R:FR (as seen in Fig. 4) would suggest that, even in the absence of neighbouring plants, shade avoiders should grow more rapidly in the early mornings and late evenings than they do during the day. Thus, responses to low R:FR at twilight and dawn might be expected to interfere with proximity perception and the induction of the shade-avoidance syndrome.

Apparently, plants in the natural environment have a subtle mechanism for 'ignoring' low R:FR signals at the end of the day. Experiments by Casal and Smith (1989b) showed that FR directed at the internodes of mustard seedlings attached to a growth transducer only caused enhanced extension growth if the rest of the plant was simultaneously exposed to relatively high irradiance white light. In the classic laboratory EOD experiments, FR pulses are given to plants immediately after the end of a period of light treatment at relatively high irradiances. In nature, the low R:FR of twilight is accompanied by very low irradiances. Casal and Smith (1989b) demonstrated that FR treatment of the internodes required some factor generated in leaves exposed to white light in order for the FR-mediated stimulation of extension growth to occur. The nature of the co-factor is unknown, and an obvious first candidate was some metabolite produced by photosynthesis. However, the component of white light that elicits the capacity to respond to EODFR turned out to be blue light. In other words, blue-light treatment of the leaves seems to be necessary for the production of the factor required for FR-mediated enhancement of elongation rate. An implication of these results is that, in nature, effective shade-avoidance responses to low R:FR reflected from neighbouring plants requires the leaves of the responding plant to be exposed to relatively high irradiances. This would mean that plants in open or sparse canopies would be able to respond sensitively and rapidly to the first signals of impending shading, whereas those growing in deeply-shaded environments would not waste growth potential in useless response to shade.

7.1.8.6 Seed germination and seedling establishment in nature

A crucial phase in every higher plant's life history is germination of the seed and establishment of the resultant seedling as an independent, photoautotrophic individual. Although large seeds with plentiful reserves often germinate well in total darkness, a remarkably large proportion of higher plant seeds require light for germination, particularly those herbs and pioneer species that produce large quantities of small seeds, each with correspondingly limited reserves of mobilizable resources. For most small seeds, imbibition in darkness does not lead to germination, but often only very small amounts of light are required to initiate the developmental processes that result in radicle protrusion. Subsequently, if light is withheld, the pattern of growth and development is characterized by etiolation, in which stem elongation is maximized at the expense of leaf and root development. The complete transition to the photoautotrophic mode (de-etiolation) requires exposure of etiolated seedlings to long periods of relatively high irradiance white light, although brief exposures to narrow wavebands (*e.g.* R) can initiate some of the partial processes involved in de-etiolation. That etiolation and de-etiolation occur in the natural environment, as well as in photomorphogenesis laboratories, is a matter of simple observation, although to find fully-etiolated seedlings one has to resort to turning over stones and looking underneath. The nature of the natural light signals that initiate germination and de-etiolation, however, is not clear, and the role of the phytochrome photoreceptors is arguable, at least.

The scenario is usually portrayed as follows. Mature seed released from the mother plant is dehydrated and dormant, preventing premature germination. The majority of the population of seed becomes incorporated into the upper layers of the soil together with detritus derived from vegetation, and the seeds then become fully hydrated and potentially capable of germination. Germination is from then on prevented by the lack of light penetrating into the soil (but see above regarding the below-soil light environment). It has long been known that many soils contain large populations of seeds of a wide range of species which can remain dormant for years (Wesson and Wareing 1969). The requirement for germination is soil disturbance and the consequent exposure of the seed to light, although other environmental variables such as temperature or nitrate availability can under certain circumstances override the light-stimulation of germination. The ecological significance is that survival of the species is enhanced by mechanisms that limit resource depletion whilst at the same time spread the probability of germination over wide time periods. Germination is an all-or-nothing, stochastic process, so conditions that partially favour germination will result in a proportion of the population germinating whilst the rest remains dormant; the ecological significance of this is also obvious, in that it spreads the chances of survival across the whole population.

The sensitivity of seeds to light is often very high; for example, the classic Beltsville experiment that first demonstrated R/FR photoreversibility showed that only a few minutes exposure to dim R was sufficient to induce imbibed lettuce seed to germinate (Borthwick *et al.* 1952). This high sensitivity can, in the case of many weed species, be increased by several orders of magnitude after a period of burial in soil (Taylorson 1972; Scopel *et al.* 1991). Thus, seeds in soil are poised to react to the very slightest exposure to light. On the other hand, the R:FR of the radiation incident upon the seed also regulates whether or not the seed will germinate; for the majority of small seeds, germination is inhibited by low R:FR. This inhibition by low R:FR may be a manifestation of the same R:FR perception processes responsible for proximity perception in growing plants, or it may be related to the high irradiance response in which germination is reduced by exposure to FR (Frankland 1981). Ecologically, the prevention of germination under low R:FR makes sense, since such seeds exposed to light beneath dense vegetation canopies under which normal photoautotrophic growth would be impossible would be prevented from germination. Thus, although buried seed may be stimulated to germinate by very small exposures to light, whether or not the seed will proceed to full germination is determined by the R:FR ratio of the light to which it is subsequently exposed. The overall control of germination may therefore involve two separate phytochrome-mediated responses.

In the natural environment, exposure of plants to light is normally for long periods, but soil disturbance represents a possible exception to this generalization. In some cases, as soil is disturbed by an animal or by cultivation, seed may be briefly exposed to light and then covered up again. It is in these circumstances that etiolation becomes of significance in the natural environment. It is probable, although perhaps never actually measured, that most small seeds germinate at or close to the soil surface, and the period of etiolation growth is thus very brief. Small seeds with low reserves that germinate after soil disturbance, but which are then deeply buried may never make it to the surface. Consequently, except for large seeds with ample reserves, etiolation growth is, in effect, a strategy of last resort. The processes that comprise de-etiolation are controlled by several photoreceptors, including blue light-absorbing photoreceptors and protochlorophyll; thus, although phytochrome-mediated processes are involved, it is not possible to account for de-etiolation entirely through the action of the phytochromes. Etiolation appears to be a developmental condition in which the normal (de-etiolation) pattern is repressed, as seems evident from the recent discovery of mutants that show the de-etiolated growth pattern (except for chlorophyll synthesis) in darkness. The *det* and *cop* mutants of *Arabidopsis* behave in darkness as if they were in the light (see Fig. 10 in Chapter 4.6; Chory *et al.* 1989; Deng and Quail 1992); they are recessive mutations, which implies that light acts to inactivate repressors of photomorphogenesis. Consistent with this view is that etiolation is only found in the evolutionary more advanced orders of terrestrial plants; bryophytes, ferns and many gymnosperms grow in the dark as they do in the light, even to the extent of making chlorophyll. It seems, therefore, that the angiosperms, in particular, have evolved mechanisms, presumably of substantial adaptive value, that operate to repress the normal growth pattern in the absence of light.

7.1.9 Eco-physiological functions of the members of the phytochrome family

The concept that the different members of the phytochrome family may have different eco-physiological functions (Smith and Whitelam 1990) is now firmly established. In some quarters, however, the idea is taken too literally, with the result that more sophisticated scenarios in which overlapping, shared or cooperative functions may exist tend to be overlooked. On the other hand, the concept of independent physiological functions for different phytochromes has at last allowed a rationalization of the different physiological response modes that have been ascribed to phytochrome. During the period when it was assumed there was only one phytochrome, the existence of several physiologically distinct modes of response was, to say the least, puzzling.

7.1.9.1 The physiological response modes

For dark-imbibed seeds and etiolated plants, at least three different phytochrome-mediated response modes have been identified and defined by their photobiological characteristics. These are the very low fluence responses (VLFR), the low fluence responses (LFR), and the high irradiance responses (HIR). The HIR may be further subdivided operationally into a R-mediated HIR and a FR-mediated HIR. The LFR represent the classical R/FR reversible responses as characterized by the Beltsville group in the 1950's; they are saturated at relatively low fluences and exhibit full reciprocity. The VLFR are only detectable in seeds or seedlings that have been imbibed or grown in total darkness, since even the low fluences of laboratory 'safelights' are sufficient to saturate response. The VLFR saturate at very low fluences and show reciprocity, but are not FR-reversible because even FR is absorbed by Pr to a large enough extent to establish sufficient Pfr to saturate the response. The HIR do not exhibit reciprocity but usually show a linear relationship with the logarithm of fluence rate. The FR-HIR is restricted to etiolated plants but as it disappears from de-etiolated plants, an underlying, persistent R-HIR can sometimes be detected.

In light-grown plants, three further modes of photoperception have been recognized, although there may be overlap between them; these are: EOD, R:FR effects, and photoperiodic effects. There is still confusion regarding the discreteness of these proposed phytochrome-mediated response modes. For example, EODFR treatment characteristically induces developmental changes that are similar to, but usually of lower magnitude, than those that result from exposing plants to a low R:FR during the whole of the day; thus EOD and R:FR responses may be manifestations of the same basic response mechanism. Furthermore, EOD responses are R/FR reversible, saturate at relatively low fluences, show full reciprocity (usually) and only differ from the 'classical' LFR of etiolated plants by having an unusually slow escape from reversibility. Thus, it may be logical to consider the possibility that the LFR of etiolated plants, and the EOD and R:FR responses of light-grown plants, are essentially the same.

7.1.9.2 The members of the phytochrome family

The germinal work of Sharrock and Quail (1989), who identified five genes for phytochromes in *Arabidopsis*, has laid the basis for a complete re-appraisal of ideas regarding the physiological functions of the phytochromes. The *Arabidopsis phy* genes (*phyA-phyE*) represent a small multigene family with 50-70% sequence homology. At least three of these genes (*phyA, phyB,* and *phyC*) are expressed in *Arabidopsis.* As far as is known, the chromophore for all the phytochromes is identical. The molecular characteristics of the *phy* genes and their products is detailed elsewhere in this book (Chapter 4.2). In the angiosperms (as exemplified by *Arabidopsis*) there are at least five phytochromes, up to seven distinct physiological response modes, and perhaps three or four recognizable ecological situations in which phytochrome-mediated perception of information from the natural radiation environment may be important. The challenge is to identify the processes that connect the photoreceptors, the response modes and the ecological functions.

7.1.9.3 Approaches for identifying the physiological functions of the phytochromes

In recent years, two principal approaches towards elucidating the physiological functions of the different phytochromes have been developed: (i) the use of mutant plants with reduced levels of individual phytochromes, and (ii) the use of transgenic plants transformed with individual *phy* genes driven by constitutive promoters and thereby expressed at high levels. Currently, this research has allowed the allocation of specific functions to both phytochrome A and to phytochrome B, but it has also revealed overlap between the functions of individual phytochromes, and that both the shade-avoidance and the de-etiolation syndromes may each be a composite of different processes regulated by different phytochromes. So, matters become more complicated as they become more simple!

A number of so-called photomorphogenic mutants exist, although the genetic lesion has been properly characterized in only a few (see Chapter 8.2). Photomorphogenic mutants may be caused by lesions in one of the *phy* genes, in genes coding for enzymes responsible for chromophore biosynthesis and assembly, or in genes coding for downstream transduction-chain components. Only mutants known to be deficient in specific photoreceptors can be of use in attempting to allocate physiological function. Several photomorphogenic mutants of *Arabidopsis,* selected for long hypocotyls when grown in white light, are known (the *hy* mutants). Of these, *hy3* is deficient in phytochrome B (Somers *et al.* 1991), and several alleles of *hy3* have been shown to have lesions in the *phyB* gene (Reed *et al.* 1993). Other mutants deficient in phytochrome B are known, although the genetic nature of the lesions is not yet established; these include the *lh* mutant of cucumber, and the *ein* mutant of *Brassica rapa.* Recently a new class of *Arabidopsis* photomorphogenic mutants which appear normal in white light, but are elongated under continuous FR have been isolated. Parks and Quail (1993) called these *hy8,* Nagatani *et al.* (1993) called them *fre* (for far-red elongated) and Whitelam *et al.* (1993) named them *fhy* (for far-red elongated hypocotyl); although these proposed names indicate the lack of cooperation between groups investigating the same phenomena, the mutants appear to be very similar selections. Whitelam *et al.* (1993) have been able to determine the genetic nature of a *phyA*-null; one of their selections has an inversion in the *phyA* gene. Transgenic plants that express to high levels introduced *phyA* genes (Keller *et al.* 1989; Boylan and Quail 1989) or *phyB* genes (Wagner *et al.* 1991) have become available in recent years.

7.1.9.4 The physiological function of phytochrome A

Phytochrome A is light-labile; in other words, phytochrome A Pfr is degraded rapidly, whereas the corresponding Pr is relatively stable, so that under conditions in which Pfr is continuously generated from Pr by light, the concentration of phytochrome A drops. In many cases examined, phytochrome A is lost under continuous white or R with a half life of less than 1 h (see Smith and Whitelam 1990 for review). In addition, phytochrome A synthesis is repressed by Pfr, so that although phytochrome A can accumulate to high levels in etiolated plants, its synthesis is switched off in the light. These facts have contributed to a widely-accepted generalization that phytochrome A is responsible for those parts of the de-etiolation syndrome that are phytochrome-mediated. This is almost certainly incorrect, and at present the only photomorphogenic response that can be allocated reliably to phytochrome A is the FR-mediated HIR.

Transgenic tobacco (McCormac *et al.* 1991, 1992a, b) and *Arabidopsis* (Whitelam *et al.* 1992) that express an introduced oat *phyA* gene under a constitutive viral promoter exhibit a persistent FR-mediated inhibition of extension growth when grown in white light. Being driven by a constitutive promoter, the introduced *phyA* gene is not repressed by Pfr and, additionally, the oat phytochrome A Pfr in tobacco and *Arabidopsis* is not degraded as rapidly as the endogenous Pfr. Thus, phytochrome A is present at quite high levels after de-etiolation in these transgenic plants. When such de-etiolated plants are subjected to a period of continuous FR their elongation growth is strongly inhibited, whereas the corresponding wild type is unaffected. One consequence of this is that *phyA* overexpressers, when exposed to white light of low R:FR exhibit *inhibition* of extension growth, in contrast to the classical strong stimulation seen in the wild types (Fig. 6; McCormac *et al.* 1991). These effects are interpreted as being due to the persistence in de-etiolated transgenic seedlings of a FR-mediated HIR inhibition of extension that, in wild-type plants, is restricted to etiolated seedlings.

Although the original tobacco *phyA* overexpressers (*ca.* 9-fold increase in phytochrome A content) are dwarfed when grown in white light, they exhibit a normal response to EODFR treatment; indeed, the magnitude of the response is greater than that of the wild type because the extension rate of the transgenics in white light alone is greatly reduced. If, however, the EODFR is extended from a brief pulse (15 min) to an all-night (*i.e.* 12 h) treatment, then the difference in response between the wild type and the transgenics is very striking. 'All-night' FR treatment of the wild types is not significantly different from EOD pulse FR, but in the *phyA* overexpressers, extension growth rate is massively reduced by 'all-night' FR (McCormac *et al.* 1992). Again, these data seem best interpreted in terms of the persistence in de-etiolated seedlings of a FR-HIR.

The overexpression of the introduced *phyA* gene also has effects on the photomorphogenesis of etiolated seedlings. The *Arabidopsis phyA* over-

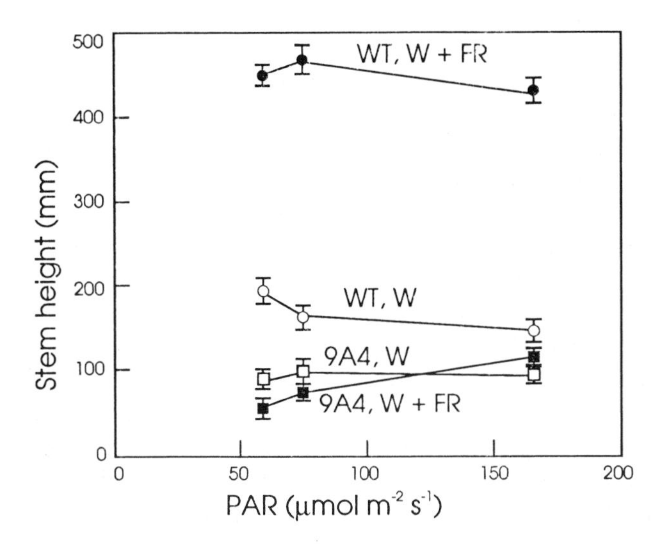

Figure 6. Effect of expression of an introduced oat phyA gene in transgenic tobacco on the extension rate responses to R:FR. Wild-type tobacco seedlings (WT) and *phyA* overexpressers (9A4) were grown at different PAR irradiances at either high R:FR (W), or low R:FR (W + FR) and plant height measured. Note that in the WT, low R:FR markedly increases plant height over the treatment period, whereas in the transgenic overexpressers, low R:FR either has no effect, or actually inhibits growth compared to high R:FR. These data can be interpreted if it assumed that FR-mediated inhibition of extension (analogous to the FR-HIR) persists in light-grown transgenic *phyA* overexpressers. From McCormac *et al.* (1991).

expressers exhibit an enhanced sensitivity of elongation growth to inhibition by continuous R, and they are also more sensitive to pulses of R as assayed by the acceleration of chlorophyll production in subsequent white light. Recent results have shown that cell cultures of transgenic tobacco overexpressing an introduced *phyA* gene exhibit phytochrome-mediated regulation of the chlorophyll *a*/*b*-binding protein (*Cab*) gene expression, a phenomenon that occurs in wild-type leaves but not in wild-type cell cultures (A.C. McCormac unpublished data). Thus, the presence of elevated levels of phytochrome A in transgenic plants enhances the sensitivity of the plant to R given continuously or as pulses, and this may be consistent with phytochrome A having some participatory role in the de-etiolation process. However, the persistent FR-mediated growth inhibition seen in light-grown plants is the most obvious effect, and is responsible for the severely dwarfed phenotype usually obtained with *phyA* overexpressers (Chapter 4.8).

Further evidence that phytochrome A mediates the FR-HIR comes from the selection of putative *phyA* nulls in *Arabidopsis* by screening for long hypocotyls under continuous FR (Parks and Quail 1993; Nagatani *et al.* 1993; Whitelam *et al.* 1993). These seedlings de-etiolate normally and preliminary observations indicate that the only photomorphogenic response that is aberrant is the FR-mediated inhibition of extension growth. As these seedlings are *phyA* null, then their apparently normal de-etiolation when transferred to white light indicates that phytochrome A is not absolutely required for the processes that make up the de-etiolation syndrome.

The conclusion that the physiological function of phytochrome A may be restricted to the FR-HIR poses difficult questions regarding the ecological significance of the phytochrome molecule about which we know most. Although this particular high irradiance reaction is recognised in the laboratory from its action maximum in the FR, in most plants the spectral dependency is found over a wide range, extending into the R. In the laboratory, the wavelength of maximum action is dependent upon the period of irradiation, with R giving greater inhibition than FR for short periods, and *vice versa* for long periods of light treatment. This is normally explained on the basis of the light-lability of phytochrome A Pfr, such that over long periods of irradiation, a higher steady-state [Pfr] occurs with FR than with R. In the laboratory, although the response is described as a 'high irradiance response', the actual fluence rate range over which irradiance dependence is observed is not great. In the classic experiments of Hartmann (1966), the fluence rate range for the FR-HIR in lettuce hypocotyl growth was 0.1-10 μmol m^{-2} s^{-1}. Later observations by Beggs *et al.* (1980) extended the range to *ca.* 30 μmol m^{-2} s^{-1}, for mustard seedlings. This is not a *high* irradiance, in comparison with daylight, but may represent a significant proportion of the FR radiation present in some natural radiation environments, particularly within vegetation canopies. Thus, to search for an ecological role for the FR-HIR, situations must be identified in which plants may be exposed to relatively long periods of relatively high fluence rate FR; such situations are rare.

The most likely scenario is that the FR-HIR may be important in seed germination and in early seedling establishment. As mentioned earlier, seed exposed briefly to light will be stimulated to germinate; however, if subsequently that seed is exposed to continuous FR, then germination is inhibited. In the soil, disturbed seed subsequently recovered with vegetation detritus will receive radiation that is predominantly FR, and such seed may be prevented from germinating. Similarly, seed exposed above the surface but under a vegetation canopy would be prevented from germinating; on a population basis, such canopy-induced inhibition of germination would favour the survival of the species. It is therefore conceivable that a FR-HIR may have ecological relevance in such circumstances; certainly, experiments to investigate this idea could be devised.

Another possibility relates to the point at which seedlings, emerging from the soil, redirect their developmental processes from the etiolated to the photoautotrophic growth pattern. Perception of the *quantity* of radiation *via* an HIR could be advantageous in reducing extension growth at this transition. The argument against a role for phytochrome A in this developmental switch-point is that the *phyA*-null mutants of *Arabidopsis* isolated by Parks and Quail (1993), Nagatani *et al.* (1993) and Whitelam *et al.* (1993) de-etiolate quite normally under white light in the laboratory, only showing long hypocotyls, indicative of incomplete de-etiolation, under continuous FR. Thus, a function for phytochrome A specific for FR radiation must be sought. A possibility is that seedlings emerging from the soil under a vegetation canopy may be advantaged if their extension growth is reduced at that point, although such an advantage would be limited in the case of seeds with small resources. There is no doubt that, with our present state of knowledge, it is difficult to ascribe a convincing ecological role for phytochrome A, if phytochrome A is only responsible for the FR-HIR. The obvious way to investigate this question further is to carry out competition/fitness experiments with wild type and *phyA* null plants; unfortunately, because of its rosette growth habit and rapid bolting phenology, *Arabidopsis* is particularly unsuited to such an approach.

7.1.9.5 The physiological function of phytochrome B

Studies of the cucumber *lh*, the *Arabidopsis hy3* and the *Brassica rapa ein* mutants have been used to obtain evidence on the putative functions of phytochrome B. In addition, two different *phyB* overexpressers have been investigated in detail (McCormac *et al.* 1993); the RBO strains are *Arabidopsis* plants transformed with a rice *phyB* gene, and the ABO are *Arabidopsis* transformed with an homologous *Arabidopsis phyB* gene. The RBO seedlings have about twice the normal content of phytochrome B, whilst the ABO seedlings have about 15 times the wild-type level (Wagner *et al.* 1991).

The first studies on the cucumber *lh* mutant indicated that it did not respond to EODFR treatment (see Kendrick and Nagatani 1991 for review). The *lh* mutant has massively elongated hypocotyls and when grown in white light of high R:FR takes on the phenotype of wild types treated with white light with very low R:FR; in other words, *lh* appears to behave as if it is in a constitutive shade-avoiding mode. Even so, when treated with white light of moderate irradiance supplemented with high fluence rate FR so as to provide a very low R:FR, it does exhibit a significant stimulation of extension growth (Whitelam and Smith 1991). The *lh* mutant also responds as rapidly to FR supplied to the hypocotyl *via* fibre-optic probes as does the wild type; although the relative response in the mutant was much smaller than in the wild type, the absolute increase in extension growth rate was virtually identical (Smith *et al.* 1992). In other words, the capacity to perceive R:FR and to respond by increased extension growth is not entirely lost in the cucumber *lh* mutant. Indeed, given appropriate fluence rates of FR in the laboratory, *lh* also shows a small but significant response to EODFR. These data seem to indicate that the absence of phytochrome B disables the capacity for R:FR perception, but not entirely.

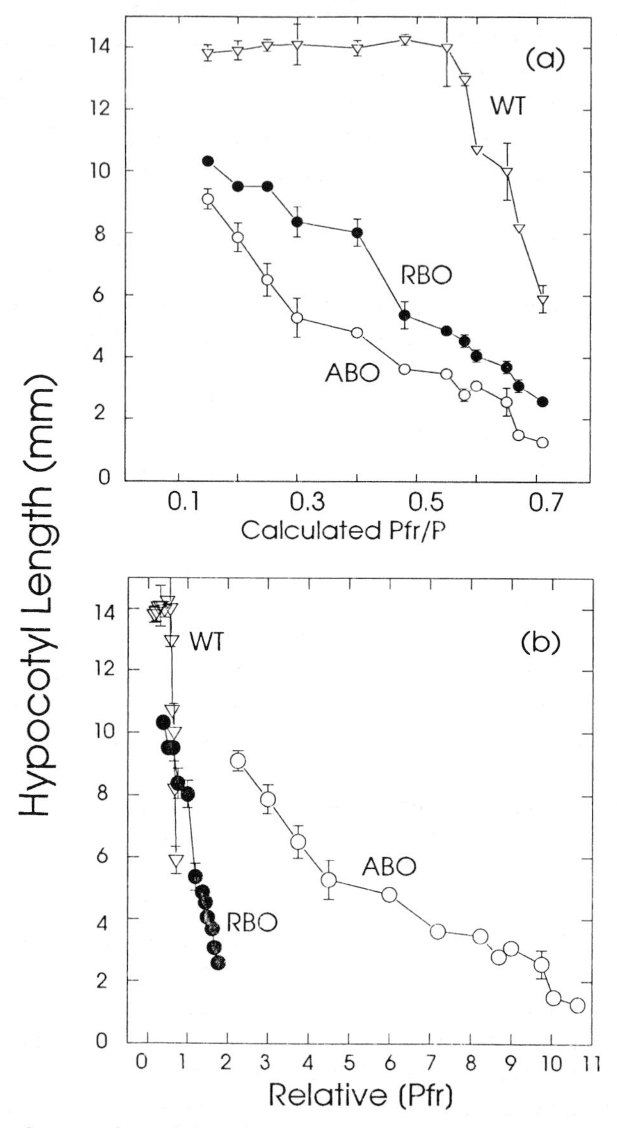

Figure 7. Effect of expression of introduced phyB genes in transgenic *Arabidopsis* on the extension rate responses to R:FR. (a) Wild-type *Arabidopsis* seedlings and two transgenic *phyB* overexpressers (RBO and ABO) were grown in treatment cabinets that provided uniform PAR but a range of R:FR; the length of the hypocotyls are recorded against the Pfr/P calculated from the spectral scans. Note that the wild type shows a very sensitive response to reduced R:FR, but the *phyB* overexpressers show reduced, though still graded, responses. (b) The data from (a) are here plotted against calculated [Pfr], based on assumed relative values for phytochrome B amounts of WT = 1. RBO = 2, and ABO = 15. From McCormac *et al.* (1992b).

The *Arabidopsis hy3* mutant is almost completely blind to low R:FR (Whitelam and Smith 1991). One thing is certain; the loss of phytochrome B in *hy3* has no effect at all on the capacity of etiolated seedlings to respond to continuous FR by extension growth inhibition, and consequently phytochrome B does not seem to be required for the FR-HIR (McCormac *et al.* 1993). The *Brassica rapa ein* mutant lacks phytochrome B, but although it has long hypocotyls and internodes, it nevertheless still shows a small extension growth stimulation when de-etiolated seedlings are grown in low R:FR (Devlin *et al.* 1992).

Arabidopsis phyB overexpressers have a dwarf phenotype (Wagner *et al.* 1992), but surprisingly they still show a graded response to supplementary FR given with white light (Fig. 7; McCormac *et al.* 1993). If phytochrome B were solely responsible for R:FR mediated effects on the extension growth of light-grown plants, it would be expected that a two-fold (RBO), or even more so, a 15-fold (ABO) excess of phytochrome B Pfr would lead to complete disablement of the response. Put another way, if phytochrome B Pfr inhibits extension growth rate in light-grown plants, then in transgenic plants with 15 times as much phytochrome B, it should not be possible to lower [Pfr] sufficiently with supplementary FR to relieve the inhibition. There are several possible explanations for these data, including unreliable estimations of phytochrome B levels and possible problems caused by ectopic expression of the introduced *phyB* genes. Taken at face value, however, these data do not sit easily with the central concept that Pfr is the sole active form of phytochrome B, and may perhaps indicate that the biological signal is related to the ratio of the concentrations of Pr and Pfr, rather than simply the concentration of Pfr itself. This view is unfashionable, but it has been proposed several times to account for the lack of correlation between response and measured or calculated Pfr levels.

An EODFR treatment stimulates hypocotyl extension in both wild-type and *phyB*-overexpressing lines. The relative stimulation is much greater in the overexpressers, because the extension rate in white light alone is relatively reduced, but there is no qualitative difference in response. It is instructive that overexpression of *phyB* does not lead to a situation in which continuous FR throughout the 'night' causes extension rate inhibition, as happens with *phyA* overexpressers, in which a similar treatment results in serious growth inhibition (McCormac *et al.* 1991, 1992a; Whitelam *et al.* 1992). Thus, it seems that phytochrome B cannot substitute for phytochrome A in mediating a FR-HIR in light-grown seedlings. It is known that EODFR pulses are without effect on the *hy3* mutant. The tentative conclusion may be drawn therefore, that phytochrome B is responsible for the EODFR effects on extension growth in *Arabidopsis.*

There is also evidence for specific roles of phytochrome B during de-etiolation. The *hy3* mutant does not exhibit extension growth inhibition in response to continuous R treatment, whilst the wild type does. Overexpression of either *phyA* or *phyB* in *Arabidopsis* increases sensitivity to continuous R (McCormac *et al.* 1993). The *hy3* mutant is also effectively 'blind' to pulses of R, but the *phyB* overexpressers show an enhanced response to R pulses that is totally reversed by pulses of FR. These data indicate perhaps that phytochrome B is required both for the LFR which is R/FR reversible and for the R-mediated HIR, but the possibility of overlapping and possibly shared functions for different phytochromes cannot be ruled out.

The current state of knowledge on the physiological functions of the different phytochromes is in considerable flux. As yet, no mutants deficient in other phytochromes (*e.g.* C to E) have been identified, nor have transgenic over- or underexpressers of these phytochromes been reported. An important caveat should in any case be made regarding the interpretation of transgenic experiments. The expression of introduced genes is known in many cases to be unusual or aberrant, which must engender caution regarding the meaning of physiological results. A similar, but often ignored, caveat should be placed on the interpretation of experiments using mutants. The simple demonstration that a physiological response is lacking in a mutant, even a mutant whose genetic basis is clear, should not be taken as definitive evidence that the gene product lacking in that mutant is necessarily directly responsible for the observed response. Like biochemical inhibitors, mutants are only really definitive when a particular response is still *present;* in such a case one can be confident that the missing gene product is *not* responsible for the observed physiological process. Taking this cautious approach, at present one can conclude no more than the following: (i) phytochrome A mediates the FR-HIR; (ii) phytochrome B is partially responsible for R:FR regulation of extension growth, and *may* mediate the LFR which is R/FR-reversible, the EOD response and the R-HIR.

7.1.9.6 Which phytochrome mediates the shade-avoidance syndrome?

The shade-avoidance syndrome is a composite of many different physiological responses all regulated by phytochrome through the perception of the R:FR ratio. One of the surprising findings of recent research with mutants deficient in phytochrome B is that not all of these phytochrome-mediated responses to the R:FR ratio are mediated by the same phytochrome. In particular, the marked acceleration of flowering induced by low R:FR in wild-type *Arabidopsis* is equally evident in the phytochrome B-deficient *hy3* mutant, if the *hy3* lesion is placed into a late-flowering background (G.C. Whitelam pers. comm.) Thus, although the extension growth responses to R:FR are largely disabled in *hy3,* the acceleration of flowering is not; this must mean that phytochrome B is not involved in the R:FR ratio regulation of flowering time. More subtly, even some components of internode and leaf growth are unaffected by the absence of phytochrome B in *hy3.* Although low R:FR has little effect on stem length in *hy3,* if specific stem weight (*i.e.* weight per unit length, or SSW) is measured, then the effect of reduced R:FR is not diminished at all in *hy3.* Indeed, the

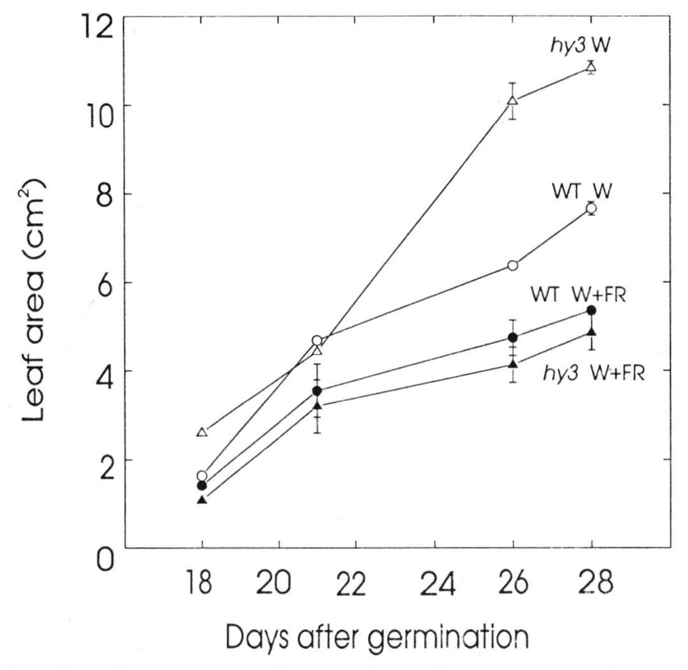

Figure 8. Retention of components of the shade avoidance syndrome in the *hy3* mutant of *Arabidopsis*. In this experiment, total leaf area was measured in *hy3* seedlings grown for 7 days in white light (W) then transferred to high or low R:FR conditions. Note that even in these seedlings, which are devoid of phytochrome B, low R:FR has a very marked effect on leaf area, as in the controls, demonstrating that phytochrome B cannot be solely responsible for all components of shade avoidance. From Robson *et al.* (1993).

reduction of SSW by supplementary FR is *greater* in *hy3* than in the wild type; furthermore, the effect of the *hy3* lesion (*i.e.* the loss of phytochrome-B) affects SSW in an opposite direction to treatment with supplementary FR (Robson *et al.* 1993). Leaf area expansion in wild-type *Arabidopsis* is reduced under low R:FR treatment; this effect is undiminished in *hy3* (Fig. 8) indicating again that, even in the absence of phytochrome B, certain components of the shade-avoidance syndrome remain unaffected. Similar results have been obtained with the phytochrome-B-deficient *Brassica rapa ein* mutant. These observations seem to indicate the intriguing possibility that different phytochromes may be responsible for regulating the extension growth and the radial expansion of plant cells. At the very least, the data show that not all the components of the shade-avoidance syndrome are mediated by phytochrome-B. We must therefore conclude that two or more phytochromes co-operate to perceive R:FR ratio and to induce the many integrated responses that together constitute proximity perception and shade avoidance. Present knowledge on the informational and physiological functions of the phytochrome family is summarized in Fig. 9; it is clear from this figure that a great deal remains to be elucidated.

7.1.10 Concluding remarks

The phytochrome family has arisen and diversified through evolution presumably as a result of environmental selection pressures. Knowledge of the phytochromes of lower plants is growing apace, and computer-aided sequence comparisons are already indicating evolutionary relationships. Sharrock and Quail (1989), when comparing the sequences of the then known phytochromes, concluded that phytochrome A was the most recent member of the family to evolve, in the angiosperms but somewhere before the split between the monocots and the dicots. The ecological functions of the phytochromes in lower plants have not yet been investigated, although there are many phytochrome-mediated physiological phenomena recorded for algae, mosses, liverworts, ferns and gymnosperms. A consideration of the phylogenetic distribution of phytochrome-mediated phenomena indicates that the phenomena about which we know most (de-etiolation and shade avoidance) are restricted to the most evolutionary advanced members of the plant kingdom. It is of considerable intellectual interest to speculate on the selection pressure that led to the original evolution of the unique photochromicity of the phytochromes. This R/FR photochromicity seems perfectly adapted to detecting the change in the relative proportions of R and FR in radiation after interaction with other chlorophyllous material, and yet it stretches the imagination to conceive of a scenario in which such proximity perception may be of adaptive value to free-living unicellular algae, in which the phytochromes must presumably have been evolved. Furthermore, since shade avoidance seems to have evolved with the higher green plants and to have become most effective in the angiosperms, we need to consider whether proximity perception is no more than a modern manifestation of an ancient function whose significance has not yet been realized. It is certainly true, and distinctly ironic, that the phytochrome whose properties upon which all our perceptions are based, *i.e.* phytochrome A from monocots, is the most advanced, and therefore likely to be the least like the ancient phytochromes. It is even more ironic that the function of phytochrome A, which alone of the phytochromes accumulates to levels sufficient for it to be extracted, purified and characterized, seems on present knowledge to be restricted to the arcane phenomenon of the FR-mediated HIR.

For at least thirty years after the classic observations of Sterling Hendricks and Harry Borthwick that led to the discovery of phytochrome, it was considered by most that only one phytochrome existed. The problems and paradoxes that this concentration on singularity posed have been well-documented, and it was with a sense of excitement and relief that photomorphogenesists embraced the realization that there is a family of phytochromes. The simplification that this discovery provided was that we could, at last, reconcile all the different modes of phytochrome-mediated responses by assuming that the different members of the phytochrome family had discrete physiological

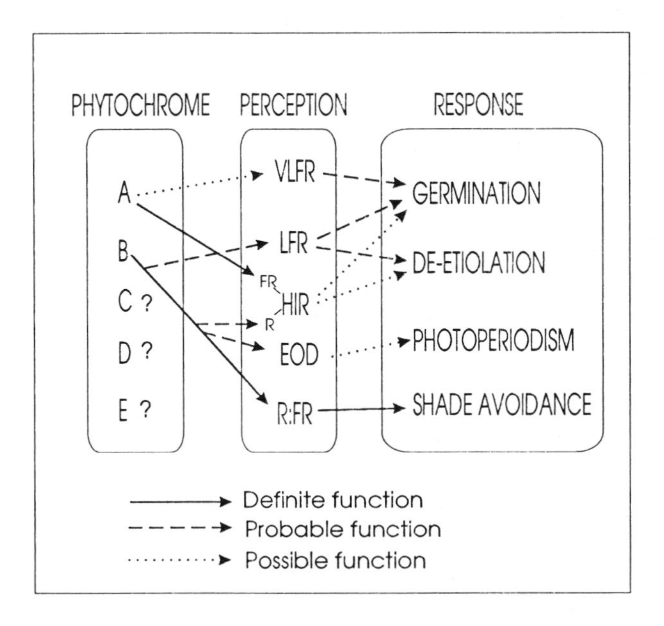

Figure 9. A summary diagram that attempts to show the present state of knowledge regarding the informational and eco-physiological functions of the members of the phytochrome family. The solid arrows indicate definite (but not necessarily sole) functions, the dashed lines indicate proposed functions for which some evidence exists, and the dotted lines indicate speculations based on limited data. The paucity of solid lines on the right-hand side of the diagram reflects the fact that very few photomorphogenecists have yet attempted to use proper ecological methods to justify claims for ecological functions.

functions and separate mechanisms of action. Attractive though this view is, it now seems likely to be an oversimplification. We should now be contemplating a further level of complexity in which different phytochromes perceive the same environmental signals, but act through the regulation of different partial processes in a cooperative manner to evoke the overall physiological responses that collectively confer ecological and adaptive value.

7.1.11 Further Reading

Casal J.J. and Smith H. (1989a) The function, action and adaptive significance of phytochrome in light-grown plants. *Plant Cell Environ.* 12: 855-862.

Grime J.P. (1979) *Plant Strategies and Vegetation Processes.* John Wiley, London.

Smith H. (ed.) (1981) *Plants and the Daylight Spectrum.* Academic Press, London.

Smith H. (1982) Light quality, photoperception and plant strategy. *Ann. Rev. Plant Physiol.* 33: 481-518.

Smith H. and Morgan D.C. (1983) The function of phytochrome in nature. In: *Encyclopedia of Plant Physiology*, New series, 16B, *Photomorphogenesis*, pp.401-517, Shropshire Jr. W. and Mohr H. (eds.) Springer-Verlag, Berlin.

Smith H. and Whitelam G.C. (1990) Phytochrome, a family of photoreceptors with multiple physiological roles. *Plant Cell Environ.* 13: 695-707.

Smith H. (1992) Ecology of photomorphogenesis: clues to a transgenic programme of crop plant improvement. *Photochem. Photobiol.* 56: 815-822.

Spence D.H.N. (1981) Light quality and plant responses underwater. In: *Plants and the Daylight Spectrum*, pp. 245-275, Smith H. (ed.) Academic Press, London.

7.1.12 References

Ballaré C.L., Sânchez R.A., Scopel A.L. and Ghersa C.M. (1987) Early detection of neighbour plants by phytochrome perception of spectral changes in reflected sunlight. *Plant Cell Environ.* 10: 551-557.

Ballaré C.L., Scopel A.L. and Sânchez R.A. (1990) Far-red radiation reflected from adjacent leaves: An early signal of competition in plant canopies. *Science* 247: 329-332.

Beggs C.J., Holmes M.G., Jabben M. and Schäfer E. (1980) Action spectra for the inhibition of hypocotyl growth by continuous irradiation in light- and dark-grown *Sinapis alba* L. *Plant Physiol.* 66: 615-618

Borthwick H.A., Hendricks S.B., Parker M.W., Toole E.H. and Toole V.K. (1952) A reversible photoreaction controlling seed germination. *Proc. Natl. Acad. Sci. USA*. 33: 662-666.

Boylan M.T. and Quail P.H. (1989) Oat phytochrome is biologically active in transgenic tomatoes. *Plant Cell* 1: 765-773.

Casal J.J. and Smith H. (1989b) The "end-of-day" phytochrome control of internode elongation in mustard. Kinetics, interaction with the previous fluence rate, and ecological implications. *Plant Cell Environ.* 12: 511-520.

Child R. and Smith H. (1987) Phytochrome action in light-grown mustard: Kinetics, fluence-rate compensation and ecological significance. *Planta* 172: 219-229.

Chory J., Peto C.A., Feinbaum R., Pratt L.H. and Ausubel F. (1989) *Arabidopsis thaliana* mutant that develops as a light-grown plant in the absence of light. *Cell* 58: 991-999.

Deng X.W. and Quail P.H. (1992) Genetic and phenotypic characterization of *cop-1* mutants of *Arabidopsis thaliana. Plant J.* 2: 83-95.

Devlin P.F., Rood S.B., Somers D.E., Quail P.H. and Whitelam G.C. (1992) Photophysiology of the elongated internode *(ein)* mutant of *Brassica-rapa ein* mutant lacks a detectable phytochrome B-like polypeptide. *Plant Physiol.* 100: 1442-1447.

Downs R.J., Hendricks S.B. and Borthwick H.A. (1957) Photoreversible control of elongation in Pinto beans and other plants under normal conditions of growth. *Bot. Gaz.* 118: 199-208

Frankland B. (1986) Perception of light quantity. In: *Photomorphogenesis in plants*, pp 219-236, Kendrick R. E. and Kronenberg G.H.M. (eds.) Martinus Nijhoff Publishers, Dordrecht.

Hartmann K.M. (1966) A general hypothesis to interpret 'high energy phenomena' of photomorphogenesis on the basis of phytochrome. *Photochem. Photobiol.* 5: 349-366

Holmes M.G. and Smith H. (1975) The function of phytochrome in plants growing in the natural environment. *Nature* 254: 512-514.

Holmes M.G. and Smith H. (1977) The function of phytochrome in the natural environment. II. The influence of vegetation canopies on the spectral energy distribution of natural daylight. *Photochem. Photobiol* 25: 539-545.

Kasperbauer M.J., Hunt P.G. and Sojka R.E. (1984). Photosynthate partitioning and nodule formation in soybean plants that received red or far-red light at the end of the photosynthatic period. *Physiol. Plant.* 74: 415-417.

Keller J.M., Shanklin J., Vierstra R.D. and Hershey H.P. (1989) expression of a functional monocotyledonous phytochrome in transgenic tobacco. *EMBO J.* 8: 1005-1012.

Kendrick R.E. and Spruit C.J.P. (1972) Light maintains high levels of phytochrome intermediates. *Nature (New Biol.)* 237: 281-282.

Kendrick R.E. and Nagatani A. (1991) Phytochrome mutants. *Plant J.* 1: 133-139.

McCormac A.C., Cherry J.R., Hershey H.P., Vierstra R.D. and Smith H. (1991) Photoresponses of transgenic tobacco plants expressing an oat phytochrome gene. *Planta* 185: 162-170

McCormac A.C, Whitelam G.C. and Smith H. (1992a) Light-grown plants of transgenic tobacco expressing an introduced oat phytochrome A gene under the control of a constitutive viral promoter exhibit persistent growth inhibition by far-red light. *Planta* 188: 173-181.

McCormac A.C., Whitelam G.C., Boylan M.T., Quail P.H.and Smith H. (1992b) Contrasting responses of etiolated and light-adapted seedlings to red-far-red ratio - a comparison of wild-type, mutant and transgenic plants has revealed differential functions of members of the phytochrome family. *J. Plant Physiol.* 140: 707-714.

McCormac A.C., Wagner D., Boylan M.T., Quail P.H., Smith H. and Whitelam G.C. (1993) Photoresponses of transgenic *Arabidopsis* seedlings expressing introduced phytochrome-B encoding cDNAs: evidence that phytochrome A and phytochrome B have distinct photoregulatory functions. *Plant J.* 4: 19-27.

Morgan D.C., O'Brien T. and Smith H. (1980) Rapid photomodulation of stem extension in light-grown *Sinapis alba* L. Studies on kinetics, site of perception and photoreceptor. *Planta* 150: 95-101.

Morgan D. C. and Smith H. (1976) Linear relationship between phytochrome photoequilibrium and growth in plants under simulated natural radiation. *Nature* 262: 210-212.

Morgan D.C. and Smith H. (1981) Non-photosynthetic responses to light quality. In: *Encyclopedia of Plant Physiology New Series* Vol 12A, pp. 109-134, Lange O.L., Nobel P.S., Osmond C.B. and Ziegler H. (eds.) Springer-Verlag Berlin.

Nagatani A., Reed J.W. and Chory J. (1993) Isolation and initial characterization of *Arabidopsis* mutants that are deficient in phytochrome A. *Plant Physiol.* 102: 269-277

Parks B.M. and Quail P.H. (1993) *hy8* a new class of *Arabidopsis* long hypocotyl mutants deficient in functional phytochrome A. *Plant Cell* 5: 39-48.

Reed J.W., Nagpal P., Poole D.S., Furuya M. and Chory J. (1993) Mutations in the gene for the red/far-red light receptor phytochrome B alter cell elongation and physiological responses throughout *Arabidopsis* development. *Plant Cell* 5: 147-157.

Robson P.R.H., Whitelam G.C. and Smith H. (1993) Evidence from phytochrome-B-deficient mutants that multiple phytochromes are required for the perception of R:FR ratio in the shade avoidance syndrome. *Plant Physiol.* in press

Scopel A.L., Ballaré C.L. and Sânchez R.A. (1991) Induction of extreme light sensitivity in buried weed seeds and its role in the perception of soil cultivations. *Plant Cell Environ.* 14: 501-508.

Sharrock R.A. and Quail P.H. (1989) Novel phytochrome sequences in *Arabidopsis thaliana*: structure, evolution, and differential expression of a plant regulatory photoreceptor family. *Genes Devel.* 3: 534-544.

Smith H. (1983) The natural radiation environment: Limitations to the biology of photoreceptors. Phytochrome as a case study. In *The Biology of Photoreceptors*, SEB Symposium 36, pp 1-18, Cosens D.J. and Vince-Prue D. (eds.) Cambridge University Press, Cambridge, UK.

Smith H. (1990) Phytochrome action at high photon fluence rates: rapid extension rate responses of light-grown mustard to variations in fluence rate and red:far-red ratio. *Photochem. Photobiol.* 52: 131-142.

Smith H and Fork D.C. (1992) Direct measurement' of phytochrome photoconversion intermediates at high photon fluence rates. *Photochem. Photobiol.* 56: 599-606.

Smith H. and Holmes M.G. (1977) The function of phytochrome in the natural environment. III. Measurement and calculation of phytochrome equilibrium. *Photochem. Photobiol* 25: 547-550.

Smith H., Casal J.J. and Jackson G.M. (1990) Reflection signals and the perception by phytochrome of the proximity of neighbouring vegetation. *Plant Cell Environ.* 13: 73-78.

Smith H., Turnbull M. and Kendrick R.E. (1992) Light-grown plants of the cucumber long hypocotyl mutant exhibit both long-term and rapid elongation responses to irradiation with supplementary far-red light. *Photochem. Photobiol.* 56: 607-610.

Somers D.E., Sharrock R.A., Tepperman J.M. and Quail P.H. (1991) The *hy3* long hypocotyl mutant of *Arabidopsis* is deficient in phytochrome-B. *Plant Cell* 3: 1263-1274.

Taylorson R.B. (1972) Phytochrome controlled changes in dormancy and germination of buried weed seeds. *Weed Sci.* 20: 417-422.

Wagner D., Tepperman J.M. and Quail P.H. (1991) Overexpression of phytochrome-B induces a short hypocotyl phenotype in transgenic *Arabidopsis. Plant Cell* 3: 1275-1288.

Wesson G. and Wareing P.F. (1969) The induction of light sensitivity in weed seeds by burial. *J. Exp. Bot.* 20: 414-425.

Whitelam G.C. and Smith H. (1991) Retention of phytochrome-mediated shade avoidance responses in phytochrome-deficient mutants of *Arabidopsis,* cucumber and tomato. *J. Plant Physiol.* 139: 119-125.

Whitelam G.C., McCormac A.C., Boylan M.T. and Quail P.H. (1992) Photoresponses of *Arabidopsis* seedlings expressing an introduced *phyA* gene: persistence of etiolated plant type responses in light-grown plants. *Photochem. Photobiol.* 56: 617-621

Whitelam G.C., Johnson E., Peng J., Carol P., Anderson M.L., Cowl J.S. and Harberd N.P. (1993) Phytochrome A null mutants of *Arabidopsis* display a wild-type phenotype in white light. *Plant Cell* in press.

7.2 Light direction and polarization

Manfred Kraml

Institut für Botanik, der Universität Erlangen-Nürnberg, Staudtstr. 5, 91058 Erlangen, Germany

7.2.1 Introduction

The perception of *light direction* yields important information enabling organisms to optimize their position in the natural environment by appropriate orientation movements. Well-known examples are *phototaxis* and *phototropism.* Phototaxis is the orientation movement of motile organisms with respect to light direction. The phenomenon can be demonstrated easily using phytoflagellates, desmids or cyanobacteria (Section 9.4.2). If unidirectional low irradiance light is applied, these photosynthetic micro-organisms usually move towards the light source (= positive phototaxis), whereas in high irradiance light they move away from it (= negative phototaxis). J. Buder (1919) showed in a basic experiment that the direction in which the light beam propagates is the important factor for orientation movements of the phototactic phytoflagellate *Euglena,* and not a gradient of irradiance: in a converging light beam organisms move against the irradiance gradient and towards the light source (Fig. 1).

Phototropism describes the phenomenon of growth movements of plants in relation to the light direction (Chapter 9.2). J.Buder (1920) investigated how the *unilateral light* stimulus acts and showed that in phototropism the unilateral stimulus is transformed into an internal light gradient, which is sensed by the plant. The light direction itself is of secondary importance for the bending reaction. Instead of irradiating the *Avena* coleoptile from the side, J. Buder irradiated only one half of it from above and produced a strong internal light gradient (Fig. 2). The coleoptile clearly bent towards the bright side despite the change in direction of propagation of the light stimulus. In an even more ingenious experiment, J.Buder irradiated the *Avena* coleoptile with a light guide unidirectionally from inside. The coleoptile bent as if this side had been irradiated from the outside. The major question which arises from Buder's experiment is: how does the organism transform light direction into differential light absorption in order to elicit a differential response (*e.g.* a greater growth rate of cells on one side of the coleoptile than on the other)? Two basic mechan-

R.E. Kendrick & G.H.M. Kronenberg (eds.), Photomorphogenesis in Plants - 2nd Edition

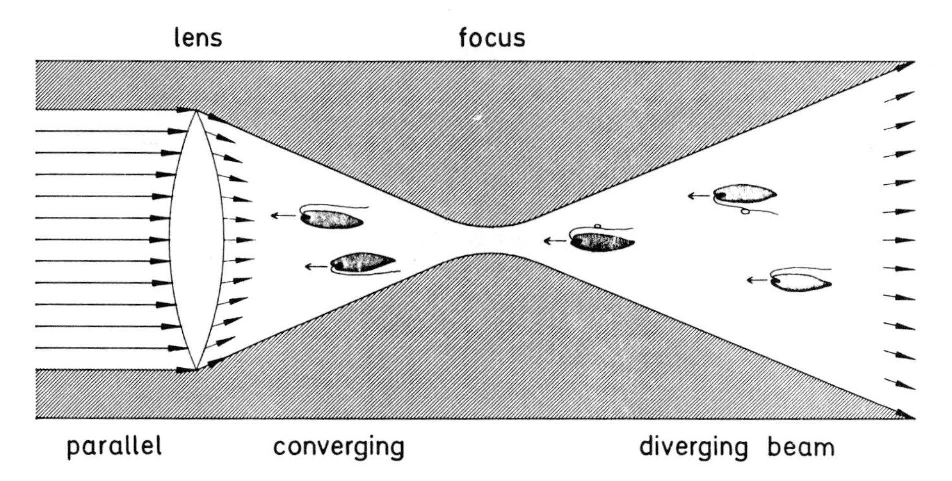

Figure 1. The behaviour of positive phototactic organisms in a focused light beam. The flagellates move through the focus and toward the light source even in the converging beam. Modified after Buder J. (1919) and Halldal P. (1959) *Physiol. Plant.* 12: 742-752.

isms for the perception of unilateral light will be considered in this chapter: (i) *light attenuation* (= *shading mechanism*) and (ii) *refraction* (= *lens effect*). Quite different from these mechanisms is the involvement of pigment dichroism, which is, strictly speaking, not a mechanism for the perception of light direction, only enabling identification of the plane in which the light propagates. The perception of *polarized light* is closely related to pigment dichroism and they are therefore discussed together. Biological examples will be limited to unicellular or filamentous organs which are very suitable for polarized light experiments. The aspect of light propagation in multilayered plants is discussed elsewhere (Chapter 7.4). First we will consider a few basic physical aspects of light direction and polarization.

7.2.2 Physical aspects of light direction and polarization

7.2.2.1 The rectilinear propagation of light

In an optic isotropic medium, light is transmitted with a high degree of directionality. Rays of light may therefore be considered as straight lines along which light energy travels from a source to a receptor. The fact that objects which interrupt light beams give fairly sharp shadows is a good demonstration of this principle. The theory of J.C. Maxwell, presented at the Royal Society in 1864 made it evident that light is composed of transverse *electromagnetic waves.* Figure 3 shows the distribution of the electric and magnetic vectors in a monochromatic wave. Both vectors describe sine waves around a common line.

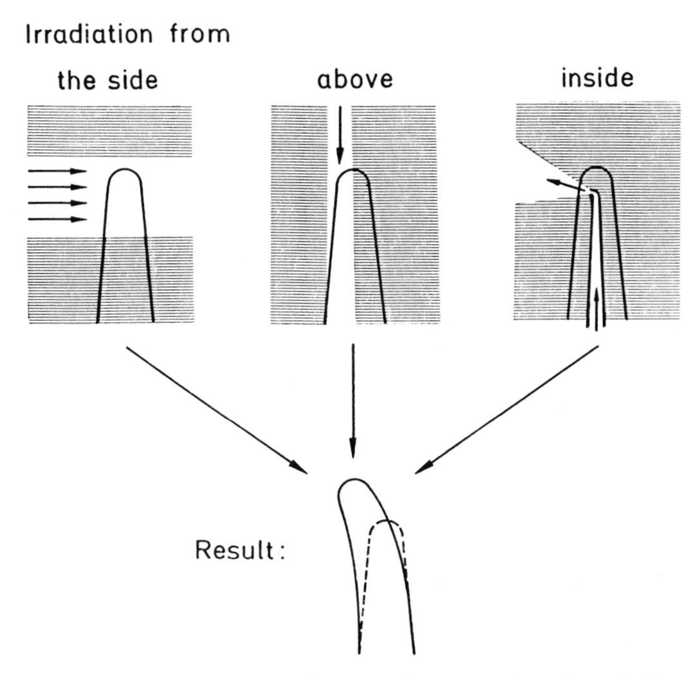

Figure 2. The diagrams illustrate some classical experiments used to show the importance of an internal light gradient for light-induced bending of the *Avena* coleoptile. The small arrows denote the propagation directions of the light stimuli.

The *magnetic vector* is always perpendicular to and in phase with the wave of the electric vector. The whole wave moves away from the light source rectilinearly and with the velocity of light ($c = 3 \times 10^8$ m s^{-1} in a vacuum). *Electromagnetic radiation* is quantized, *i.e.* it is composed of elementary particles known as photons, which are localized quanta of energy. The photons have zero rest mass and therefore very large numbers of low-energy photons can be present in a light beam. When electromagnetic waves of light pass through matter, the electric component (the E-vector) is important as it exerts a force on charged particles such as electrons. The electrons of molecules in the light path are set into oscillation at a frequency equal to that of the light. A photon of light may be absorbed if this frequency is in resonance with the natural frequency of oscillation of the electrons. Extremely monochromatic radiations, such as emitted by lasers, are composed of only one type of photons. The *photon energy* (E) corresponds to the wavelength (λ), *i.e.* the colour of the monochromatic radiation according to Planck's equation:

$$E = h \times \nu \text{ or } E = h \times c/\lambda$$

where h is the Planck constant, ν is the frequency and c is the velocity of light in a vacuum.

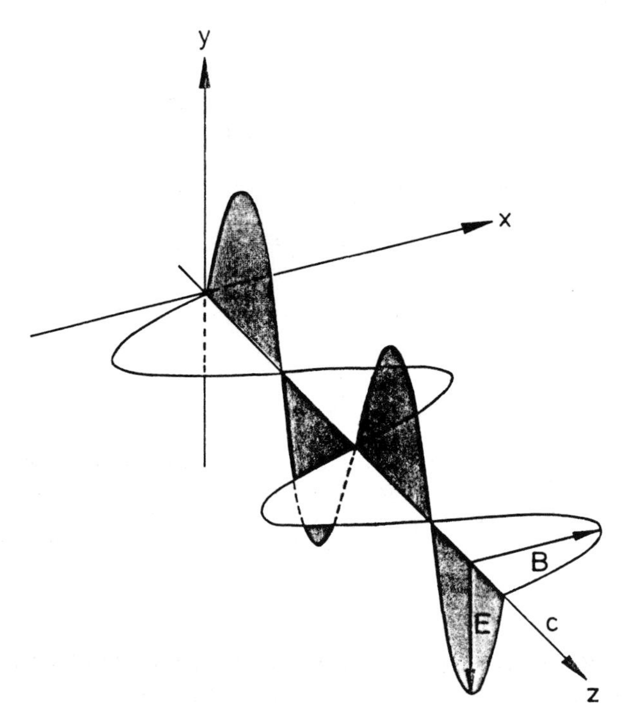

Figure 3. Schematic representation of orthogonal electric fields (E) and magnetic fields (B) in an electromagnetic wave, which moves away from the light source with the velocity of light (c). After Hecht E. and Zajak A. (1974) *Optics,* Addison-Wesley, London.

7.2.2.2 Polarization

Light can be *plane* polarized, *circularly* polarized, and *elliptically* polarized. In plane polarized light the orientation of the E-vector remains constant and only the amplitude varies periodically at any point. Therefore the wave motion is confined to a spatially fixed plane of vibration. The light is circularly polarized if the maximal amplitude remains constant and the orientation of the vector varies regularly. In the case of elliptically polarized light both maximal amplitude and orientation vary in such a way that the end of the vector moves around an ellipse.

Most of the well-known natural or artificial light sources emit nearly *non-polarized* radiation. This light is composed of a rapidly varying sequence of different polarization states, *i.e.* in a beam of light travelling toward the observer the orientation of the E-vector changes in time intervals of the order of 10^{-8} s. The distribution of photons in non-polarized light can be considered as statistically and rotationally symmetric. Another representation of non-polarized light has proved to be very useful. When the orientation of the E-vector is assumed to change randomly and rapidly, the net result is mathematically as

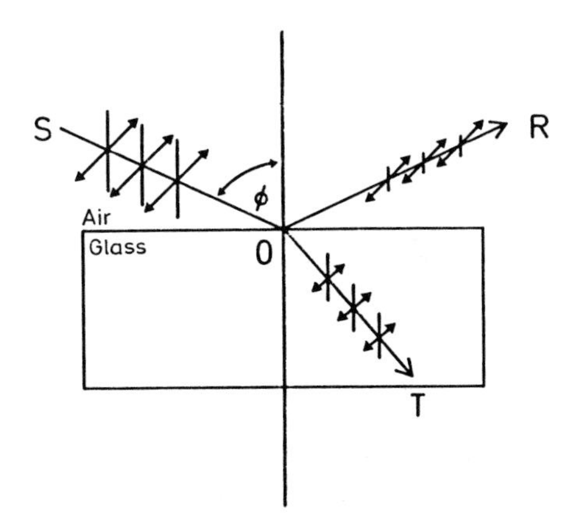

Figure 4. Polarization of light by reflection and refraction at the dielectric medium glass. SO = non-polarized light, incident at an angle ϕ to a glass surface; OR = reflected ray, partially plane-polarized (quantitatively plane-polarized in this case of ϕ = 57°); OT = refracted ray, partially plane-polarized. Modified after Jenkins F.A. and White H.E. (1976) *Fundamentals of Optics,* McGraw-Hill, Tokyo.

though there are two linearly polarized waves at right angles (orthogonal) to each other, incoherent and of equal amplitude. Linear *polarizers* are materials which separate these two rectangular components of non-polarized light, absorbing one and transmitting the other. They consist of strongly dichroic materials, *i.e.* minerals or organic compounds which have the property of absorbing selectively one of the two components of non-polarized light. Polarization filters appear grey because they absorb at least 50% of the incident light. A pair of crossed polarizers discard both rectangular components of the radiation and completely block a light beam.

Light may also become plane polarized by *reflection* at the surface of a dielectric medium such as glass (Fig. 4). From an incoming non-polarized wave, the component vibrating parallel to the surface (= normal to the incident plane) will be predominantly reflected, the other component predominantly refracted. This has to be borne in mind, *e.g.* when using unsilvered or backside-silvered mirrors for biological experiments. Moreover the polarization filter should always be positioned close to the biological object and not in between the light source and mirrors, which might bring about partial depolarization.

Polarized light is ubiquitous in the natural environment. It arises by reflection from surfaces such as water and by scattering in air. The zenith blue sky especially at dawn or dusk may be 90% polarized, and in clear water light near the surface may reach 60% polarization.

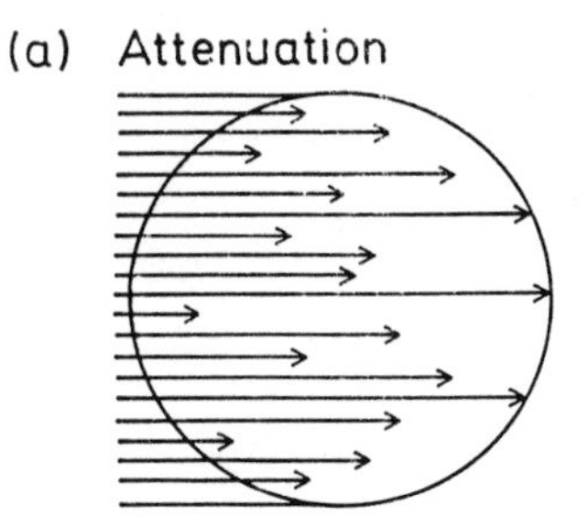

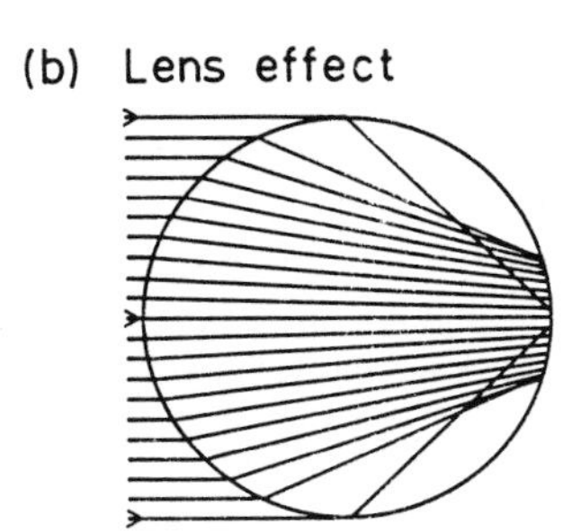

Figure 5. Schematic diagram of the light paths within cross sections of a cell when irradiated by a unilateral beam of normally incident parallel light. (a) Attenuation: the cell content is highly absorbing and scattering. In water (refractive index equal to that of the cytoplasm), refraction can be ignored. (b) Lens effect: the cytoplasm is transparent (*e.g. Phycomyces* sporangiophore); irradiation in air. After Castle E.S. (1965) *J. Gen. Physiol.* 48: 409-423.

7.2.3 Mechanisms for the perception of unilateral light

7.2.3.1 Perception of light direction by attenuation

When a light beam passes through living plant cells, the internal photon fluence rate decreases from the front side, where the light enters the cell, to the rear (Fig. 5a). The reasons are absorption and scattering. Light absorption requires the existence of pigments in the cell. Internal gradients of light absorption can be established by the photoreceptor pigment itself, which is present anyway and is able to perceive the wavelength of the actinic light. Alternatively, one or more screening pigments may be involved in addition to the photoreceptor (Chapter 7.4).

7.2.3.2 Perception of light direction by refraction (= lens effect)

Cells of eukaryotes are composed of material with a *refractive index (n)* higher than air. In addition they are large enough to act as spherical lenses, *i.e.* to focus a unilateral beam across the cell to the distal side. There are at least two necessary conditions for a good lens effect. (i) The difference in refractive index between the cells and the surrounding medium has to be great enough to cause appreciable focusing. This is always the case in air, but rarely so in water. Hence the perception of light direction *via* a lens effect is not a good mechanism for organisms which are living exclusively in water (*e.g.* flagellates and aquatic algae). (ii) The organisms should be transparent with a fairly homogeneous and colourless cytoplasm and vacuole, so that attenuation by absorption is minimal. Ideal candidates for a lens effect are found in fungi (*e.g.* *Phycomyces*).

Figure 5b illustrates the lens effect in a cross section of a *Phycomyces* sporangiophore is illustrated. Note that the internal fluence rate should be higher at the distal compared to the proximal side. A dependency of the physiological effect on the applied photon fluence rate can be derived by the following consideration: at fluence rates smaller than the saturation level at the distal side, the lens effect will cause greater stimulation of the photoreceptors per unit area at the distal side than at the proximal side. At a still higher fluence rate the number of excited pigment molecules at the proximal side will increase and the physiological response should decrease.

7.2.3.3 Spatial and temporal sensing of an internal light gradient

Shading mechanisms and lens effects establish a gradient of internal fluence rate within the organism, which brings about a spatial absorption difference in the photoreceptor pigment system of non-motile cells such as spores or slowly moving algae like pennate diatoms and desmids, which are gliding on a substrate without rotation during locomotion. The photoreceptors at different regions of the cell, for example at the proximal and the distal side, detect an internal fluence rate gradient by comparing light absorption at the same time (spatial sensing).

In contrast, *Euglena* and many other phytoflagellates move forward while rotating around the longitudinal axis of the cell. During the rotation, photoreceptive regions in the cell change their position relative to the laterally impinging light beam (for details see 9.4.2.2.). The photoreceptors therefore sense different light signals at different times (temporal sensing).

7.2.4 Biological examples for perception of light direction by attenuation and lens effect

Organisms which use the shading (attenuation) mechanism for the perception of the light direction should be preferentially photo-excited by a unilateral light stimulus at the proximal side of the cell. In the case of the lens effect more photoreceptors will be excited at the distal side. Thus it seems possible to discriminate between both strategies by irradiating only one half of the cell and shading the other, as done by J. Buder with *Avena* coleoptiles (Fig. 2b). If a shading mechanism is involved, the response to a unilateral light stimulus should be similar to shading of the distal side (Fig. 2a, b). The opposite should hold for a lens effect.

7.2.4.1 Induction of polarity by unilateral light

Many plant species release apolar spores or gametes, which are very useful objects for the study of *polarity* (Wettstein 1965; Weisenseel 1979). An excellent signal for the induction of polarity is a unilateral light stimulus. It builds up an *absorption gradient* and thus orientates the morphological axis for germination. The newly induced axis of polarity always has the same orientation as the largest absorption gradient in the cell.

In spores of the fern *Osmunda* (and some other ferns) the unilateral light stimulus elicits an outgrowth at the distal ('shaded') pole. The same is true for conidia of the fungus *Botrytis*. Despite this conformity, the basic mechanism for the perception of light direction is different in both systems. This is shown by partial irradiation from below (Fig. 6): whereas *Osmunda* spores germinate from the darker part of the spore, conidia of *Botrytis* germinate from the irradiated part of the spore. This indicates that the absorption gradient in

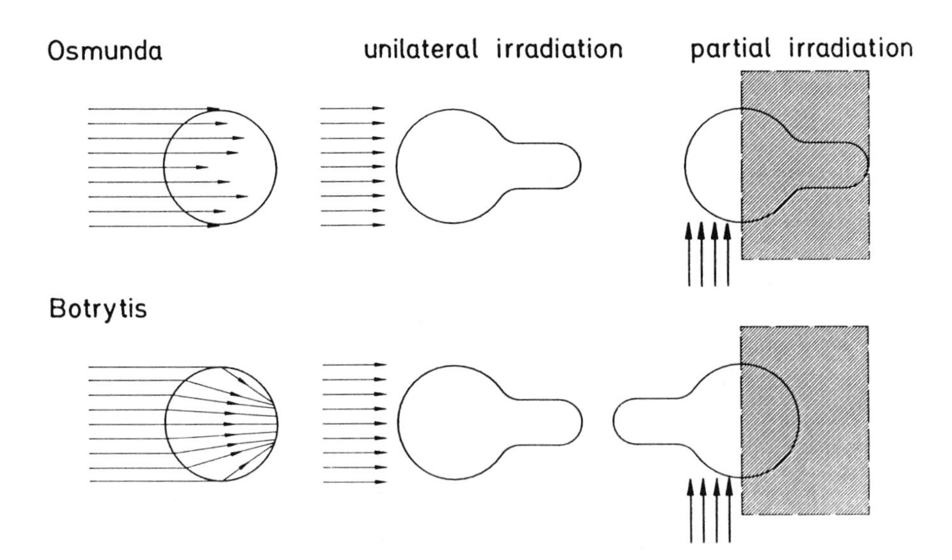

Figure 6. Outgrowth of spores of *Osmunda* and *Botrytis* due to unilateral irradiation (on the left the predicted light paths within cross sections of the spores are illustrated) or partial irradiation from below.

Osmunda is established by a shading mechanism, and in *Botrytis* a lens effect brightens the distal pole. An examination of the spores under the microscope supports this hypothesis: fern spores often contain carotenoids and many chloroplasts in the cytoplasm, whereas the conidia of *Botrytis* are colourless and hyaline with a smooth cell surface.

7.2.4.2 Phototropism of Phycomyces

Phycomyces is a coenocytic mould with typical sporangiophores: single-celled, straight, negative gravitropic aerial hyphae, which grow upwards rapidly in distinct developmental stages. Photobiological experiments are mostly performed with stage IVb sporangiophores: They are 2.5-3.5 cm long, have a mature black sporangium on the top and grow relatively constantly at a rate of 3 mm h^{-1}, until they reach a final length of 15-20 cm. During this growth a clockwise rotation of the sporangium with angular velocity of about 10° min^{-1} is typical. Elongation is confined to a small *growing zone* 0.1-2.5 mm below the sporangium. Microbeam experiments proved that the photosensitive zone parallels the growing zone in the region from 0.5-2.5 mm. Longitudinal and transverse cross sections of the sporangiophore can be seen in Fig. 7. The sporangiophore has a diameter of 0.1 mm, and a central vacuole is present along the entire structure, which is kept erect by a turgor pressure of *ca.* 2 bar (1 bar = 0.1 MPa). In unilateral blue (B) or UV-A light sporangiophores of *Phycomyces* bend towards the light source (Fig. 8a). Partial irradiation (Fig. 8b) indicates

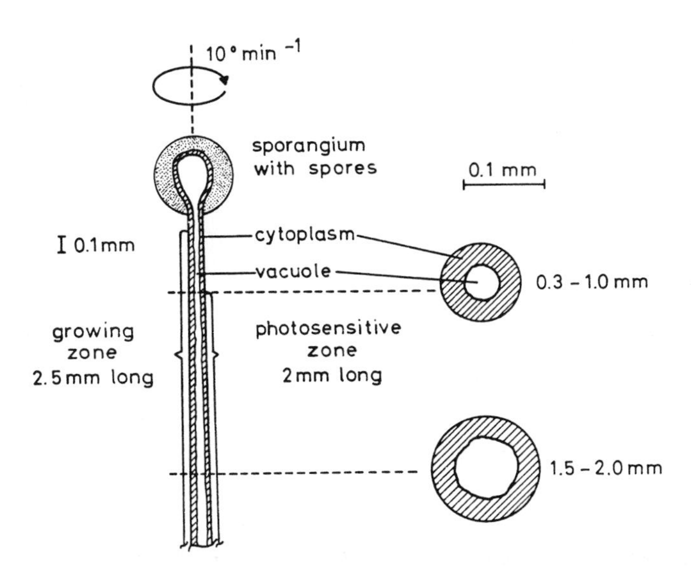

Figure 7. Longitudinal section and cross-sections of the sporangiophore of *Phycomyces.* Modified after Foster K.W. (1977) *Annu. Rev. Biophys. Bioeng.* 6: 419-443.

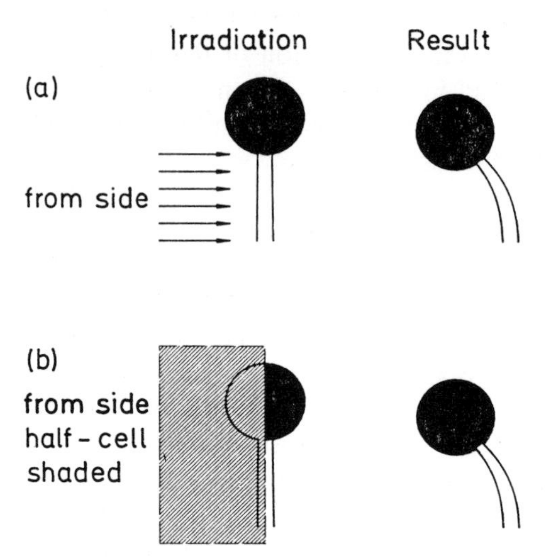

Figure 8. Schematic illustration of bending reactions of *Phycomyces* sporangiophores. (a) In unilateral blue light or UV-A. (b) After partial irradiation from the side (irradiation from above is not possible because of the black sporangium on top of the sporangiophore).

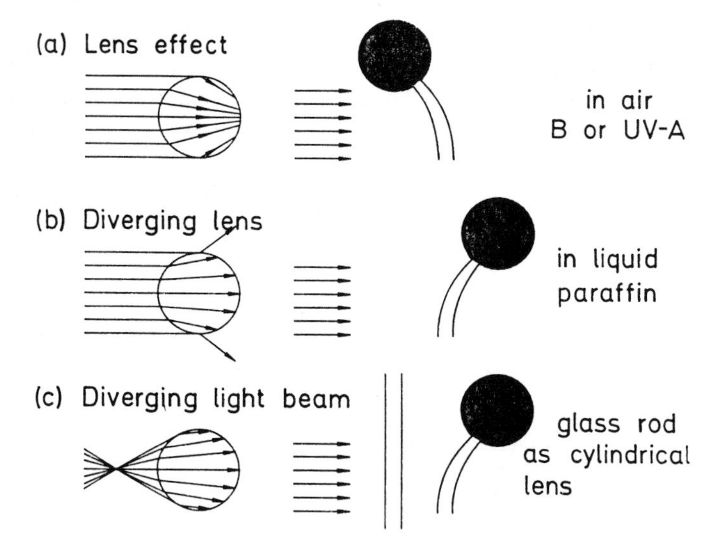

Figure 9. Experimental procedures for establishing irradiance and absorption gradients in *Phycomyces* sporangiophores. The diagrams on the *left* side illustrate light paths within cross-sections through the growing zone. On the *right* the bending reactions are shown. B = blue light.

that a lens effect is responsible for the absorption gradient, as the sporangiophore bends away from the irradiated half-cell.

That the lens effect operates in *Phycomyces* (Fig. 9a) was elegantly proved by the following experiments. (i) When the sporangiophore is immersed in a medium of higher refractive index, *e.g.* liquid paraffin (Fig. 9b) (refractive index 1.47), the cell becomes a diverging lens and phototropic bending is inversed (Buder 1918; Banbury 1952). (ii) Irradiation of a sporangiophore with a highly divergent light beam (Fig. 9c) also reverses the phototropic response. Shropshire Jr. (1962) performed a very elegant experiment: he placed a cylindrical glass rod (diameter, 0.16 mm) parallel to the sporangiophore at a distance of 0.14 mm. Parallel light was focused by the glass (focal point between glass rod and sporangiophore) so that the light path within the sporangiophore became divergent. The result was negative phototropism, as predicted.

Does attenuation also play a role in phototropism of *Phycomyces*? Of course, there is some scattering and absorption of unilateral B in the sporangiophore, which reduces the lens effect slightly. Therefore the *phototropic neutrality point* (*i.e.* the refractive index of a medium, in which sporangiophores are phototropically neutral) with a value of n = 1.295 is appreciably below the refractive index of either the cytoplasm (n = 1.36) or the vacuole (n = 1.34). This also agrees with the fact that sporangiophores immersed in water respond by negative phototropism in unilateral light (*cf.* Bergman *et al.* 1969).

Recently quantitative descriptions, as well as measurements of the light propagation in a sporangiophore were published. Steinhardt and Fukshansky (1987) elaborated accurate mathematical models of the light profile, derived from realistic assumptions of refractive-index values for the cell wall, cytoplasm and the vacuole. Dennison and Vogelmann (1989) measured the irradiance profile in unilateral light at the distal cell surface of the sporangiophore using a fibre-optic microprobe. The resulting profile consists of two steeply rising sides enclosing a central plateau which ranged in fluence rate from 1.6 to 2.2 times that of the incident light beam. In the theoretical modelling much greater contrast between the sides and the central portion of the lens profile was predicted. The differences have been discussed by Fukshansky and Richter (1990). Both methods are valuable approaches to get a better insight into phototropism of *Phycomyces*.

7.2.5 Action dichroism and polarized light

7.2.5.1 Characterization of dichroic pigment orientation by polarized light

Action dichroism is the dependence of a photoresponse on the vibration plane of linearly polarized light and indicates dichroic photoreceptor orientation. Dichroic pigments such as flavins, phytochrome and carotenoids have preferred electrical vector orientations for the absorption of light. If the E-vector of the exciting light beam is oriented parallel to the transition dipole moment of the

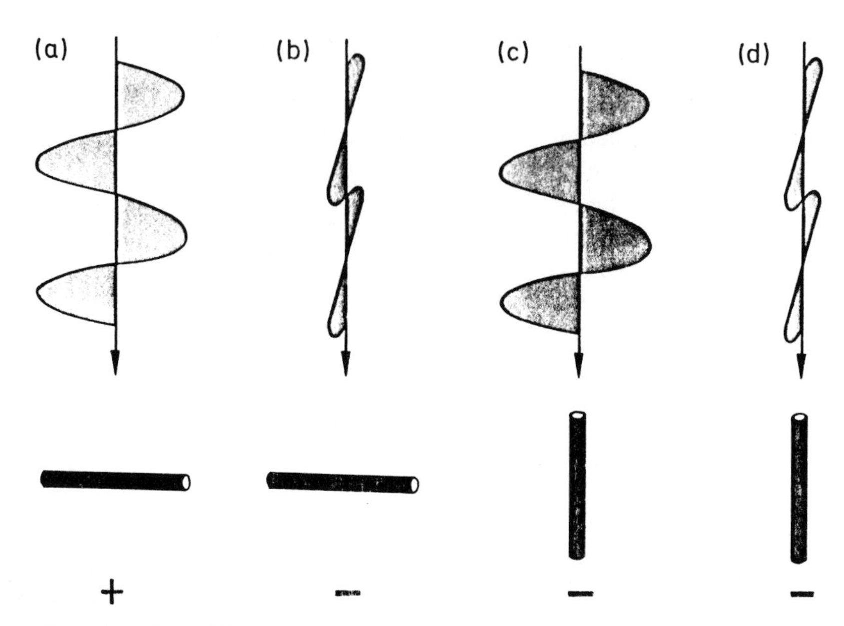

Figure 10. Absorption of linearly polarized light by a dichroic photoreceptor. The bars indicate the transition dipole moment of the pigment molecule (see text for details).

photoreceptor (Fig. 10a), the probability for absorption of an incident photon is very high (according to $A = \cos^2\alpha$; where A is the absorption of a molecule and α the angle between the electronic transition moment of the molecule and the electric vector of the incident light). It decreases to a minimum for photoreceptors orientated transversely to the E-vector (Fig. 10b). Pigment absorption is also low if the electronic transition moment of the molecule is parallel to the propagation direction of polarized light (Fig. 10c, d). The dichroic character of the individual photoreceptor molecules, however, do not necessarily bring about an absorption dichroism within a plant cell. If the photoreceptors are distributed randomly, the transition moments of all molecules cancel each other and the anisotropy of the individual molecules cannot be distinguished. Only if the dichroic photoreceptors are predominantly orientated with respect to each other in the cell, will the molecular anisotropy be transformed in a structural anisotropy, resulting in an *absorption dichroism* for linearly polarized light. A distinct orientation of photoreceptors in plant cells that exhibit protoplasmic streaming requires a structural base for the localization of dichroic pigments. This is very often the plasma membrane or the adjacent cortical cytoplasm.

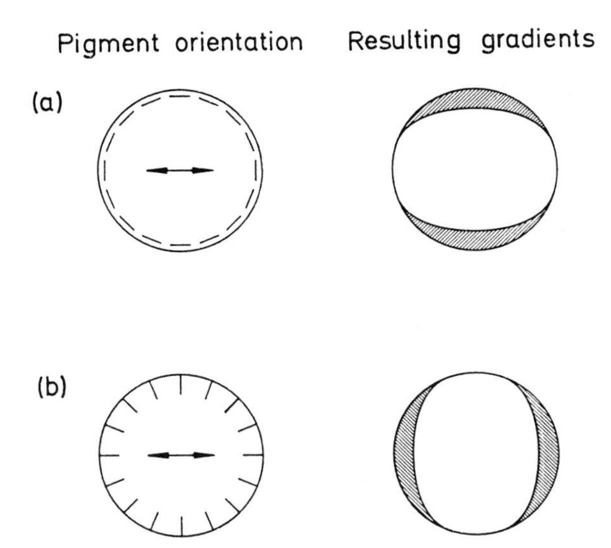

Figure 11. Absorption of linearly polarized light (E-vector parallel to the double-headed arrows) by orientated photoreceptors in spherical cells (cross-sections). Pigment orientation (a) parallel or (b) normal to the cell surface, respectively. The resulting absorption gradients are indicated by the shaded areas. The photoreceptors are represented by the dashes.

7.2.5.2 Perception of 'light direction' by dichroic orientated photoreceptors

The dichroic orientation of photoreceptor pigments does not enable plant cells *per se* to measure the propagation direction of a unilateral light stimulus. Hence dichroism is not directly comparable to attenuation and lens effect. Whereas both these mechanisms produce a bipolar absorption gradient by influencing the amount of light reaching the photoreceptor, dichroism produces *tetrapolar absorption gradients* (*e.g.* Fig. 11) due to the absorption of the photoreceptors with particular orientations. The fluence rate of unilateral light in dichroic systems remains unchanged all over the cell (if there is no additional attenuation or lens effect). For the formation of a tetrapolar gradient in a cylindrical cell it makes no difference whether the light beam is given from above or below. Only the orientation of the light path through the cell is important and not the propagation direction. This is clearly in contrast to the cases of attenuation and lens effect.

7.2.5.3 The formation of tetrapolar gradients by pigment dichroism

Examples of spatial different pigment absorption in dichroism are seen in spherical cells (*e.g.* spores) and cylindrical cells (*e.g.* filaments).

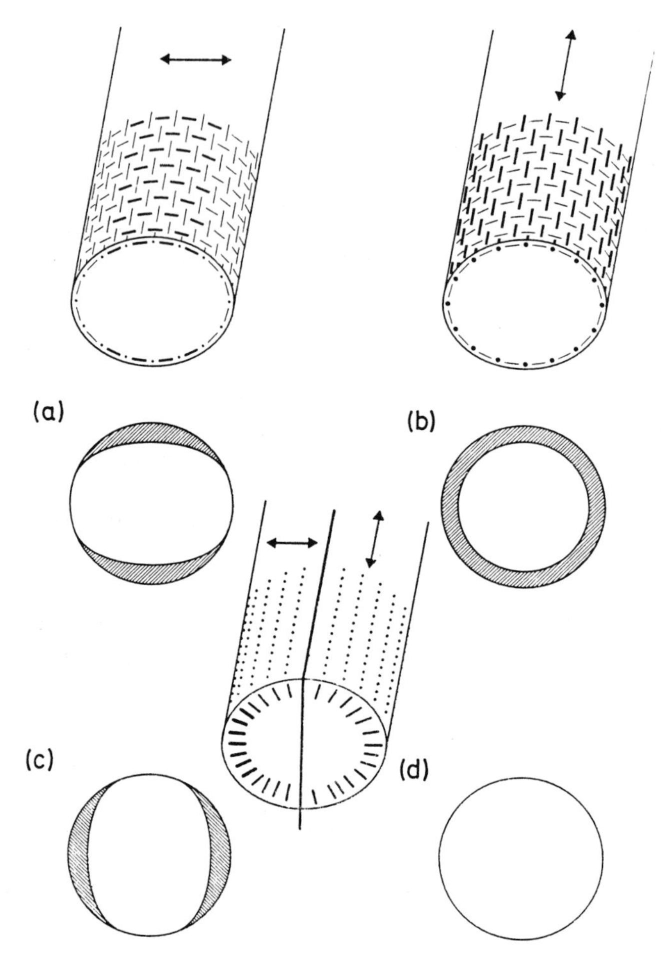

Figure 12. Absorption of linearly polarized light (E-vector parallel to the double-headed arrows) by orientated photoreceptors in cylindrical cells. The dashes represent the idealized vectors of the transition moments. The bold dashes symbolize pigment orientations more likely to absorb the impinging linearly polarized light (E-vector parallel to the double-headed arrows). The resultant absorption gradients are shown in cross-sections. (a) Photoreceptor molecules surface-parallel, E-vector perpendicular; (b) Photoreceptor surface-parallel, E-vector longitudinal; (c) photoreceptor surface-normal, E-vector perpendicular; (d) photoreceptor surface-normal, E-vector longitudinal.

7.2.5.3.1 Spherical cells. Assuming that the surface parallel photoreceptors are otherwise randomly orientated and that the light is applied normal to the plane of the page with the E-vector parallel to the double-headed arrow, the following tetrapolar absorption gradient (in the cross section of the cell) can be predicted (Fig. 11a): best excitation of the chromophores will occur at all regions parallel to the E-vector of the light. If pigment orientation is normal to the cell surface (Fig. 11b), maximum excitation can be predicted for chromophores at cell regions normal to the E-vector of the light. With non-polarized radiation no absorption gradient can be produced in either case.

7.2.5.3.2 Cylindrical cells. Surface parallel photoreceptors with random distribution in this plane have the following spatial absorption characteristics for plane polarized light, applied from above: an absorption gradient exists in polarized light with the E-vector vibrating perpendicular to the long axis of the cell (Fig. 12a). There is excellent absorption at the front and rear, but only weak absorption at the flanks. If the E-vector vibrates longitudinal to the cell axis (Fig. 12b), the photoreceptors are in an ideal position for absorption at the front and the rear as well as at the flanks, and hence there is no absorption gradient. Non-polarized light as a theoretical combination of transverse and longitudinal vibrating light is also able to establish an absorption gradient, due to the inefficient absorption of the transverse component at the flanks (Table 1). Orientation of the transition moments normal to the cell surface changes the situation. If the E-vector vibrates perpendicular (Fig. 12c), absorption is good at the flanks, but bad at the front and the rear. Longitudinal vibrating radiation (Fig. 12d) is not good at any position of the cell. Absorption gradients are again possible with non-polarized or perpendicularly polarized light, the tetrapolar gradient being contrary to the situation for surface parallel orientation (see Table 1).

7.2.6 Biological examples for action dichroism

The action dichroism of B photoreceptors is a broad field of discussion and cannot be presented here in detail (*cf.* Zurzycki 1980). However, a few examples for action dichroism in B and red light (R) will be referred to here.

Table 1. Formation of absorption gradients by linearly polarized or non-polarized light in cylindrical cells exhibiting dichroic pigment orientation. ⊥ and ‖: E-vector vibrates perpendicular or longitudinal to the cells; * = non-polarized light.

	E-vector	Front + Rear	Flanks	Gradient
Surface parallel	⊥	+	−	+
	‖	+	+	−
	*	++	+−	−
Surface normal	⊥	−	+	+
	‖	−	−	−
	*	−−	+	+

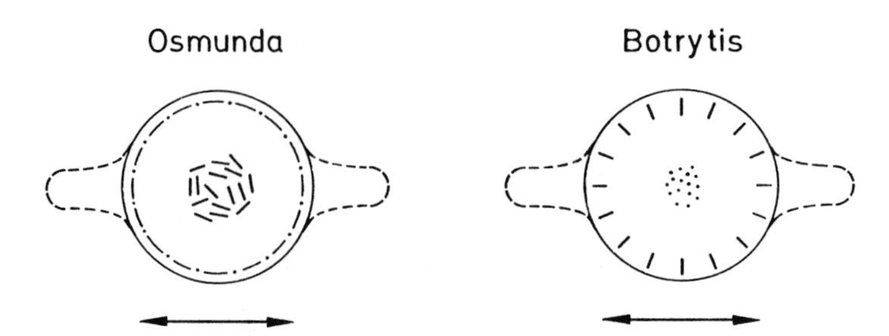

Figure 13. Diagrams of photoreceptor orientation in spores of *Osmunda* and conidia of *Botrytis*. The double-headed arrows represent the vibration plane of polarized light. The dashed humps indicate directions of most frequent germination (see text for details). After Jaffe L.F. and Etzold H. (1962) *J. Cell Biol.* 13: 13-31.

7.2.6.1 Dichroism and induction of polarity

It was indicated by partial irradiation that spores of *Osmunda* germinate from the darker half-cell, whereas conidia of *Botrytis* prefer the irradiated part (Fig. 6). Irradiation with linearly polarized light results in an outgrowth of both *Osmunda* and *Botrytis* parallel to the E-vector of the incident radiation. According to Figs. 6 and 11 this means that the chromophores are localized parallel to the surface in *Osmunda* and normal to it in *Botrytis* (Fig. 13).

7.2.6.2 Effects of polarized light in Phycomyces

Light growth response and phototropism can be stimulated more effectively by transversely polarized B, compared to polarized B with the E-vector parallel to the long axis of the sporangiophore (*e.g.* Jesaitis 1974). This indicates a dichroic orientation of the photoreceptors at the plasma membrane (Steinhardt *et al.* 1989). The precise kind of orientation is still under discussion.

The photoreceptor pigments in *Phycomyces* are also still unknown. During the past years research has indicated that the classical one-receptor model (*flavin*) for phototropism is too simplistic. Dichromatic pulse experiments (Löser and Schäfer 1986) and the use of mutants (*cf.* Galland and Lipson 1987) may be good tools to characterize the photoreceptors in more detail (Chapter 9.2).

7.2.6.3 Flip-flop dichroism of phytochrome

The dichroic orientation of phytochrome has been studied thoroughly during the past 30 years. Filamentous organisms are particularly favourable for the

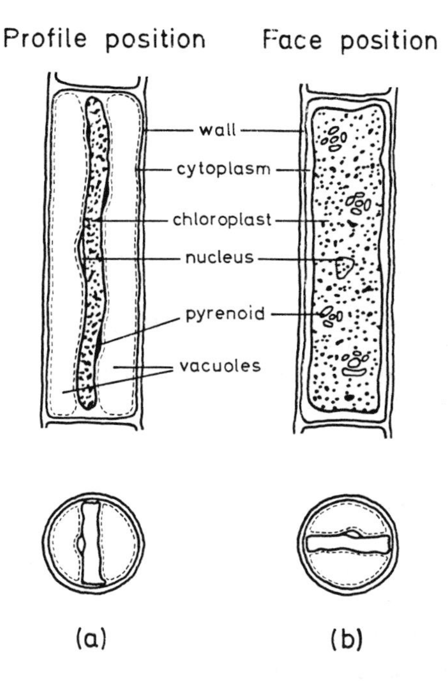

Figure 14. Mougeotia cell in surface view and in cross-section, showing different positions of the single flat chloroplast.

experiments with linearly polarized light, namely, protonemata of ferns and the green alga *Mougeotia.* In ferns and *Mougeotia,* the R-absorbing form of phytochrome (Pr) is orientated parallel to the cell surface and the far-red light (FR)-absorbing form of phytochrome (Pfr) is normal to it. Photoconversion therefore results not only in a spectral absorption change, but also in the *dichroic orientation change* of the photoreceptors. This flip-flop mechanism complicates the understanding of action dichroism in phytochrome mediated responses. Another difficulty for interpretations arises from the overlap of the absorption spectra of Pr and Pfr especially in the R spectral region (Chapter 4.1): not only Pr but also Pfr may be excited by R.

7.2.6.4 Action dichroism of phytochrome in Mougeotia

The cylindrical cells of *Mougeotia* contain a single axial chloroplast, which divides the cell in two halves, each with a long vacuole surrounded by cytoplasm. The cytology can be seen especially clearly in the profile (edge on) position of the chloroplast (Fig. 14a), whereas in the face position the flat side of the chloroplast is exposed to the observer (Fig. 14b).

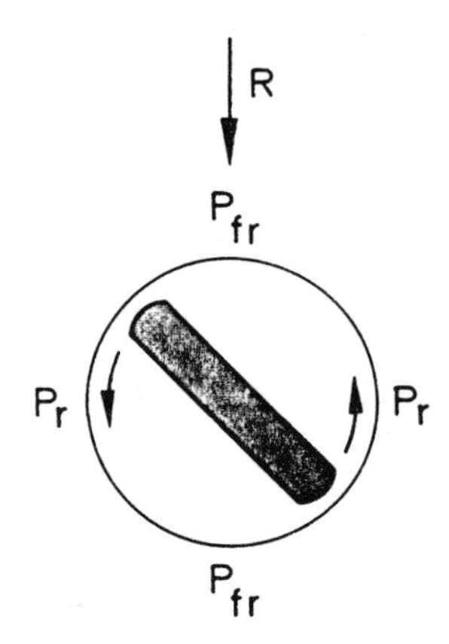

Figure 15. Schematic cross-section through a cell of *Mougeotia,* showing the movement of the chloroplast relative to a tetrapolar Pfr-gradient. R = red light. After Haupt W. (1972) *Acta Protozool.* 11: 179-188.

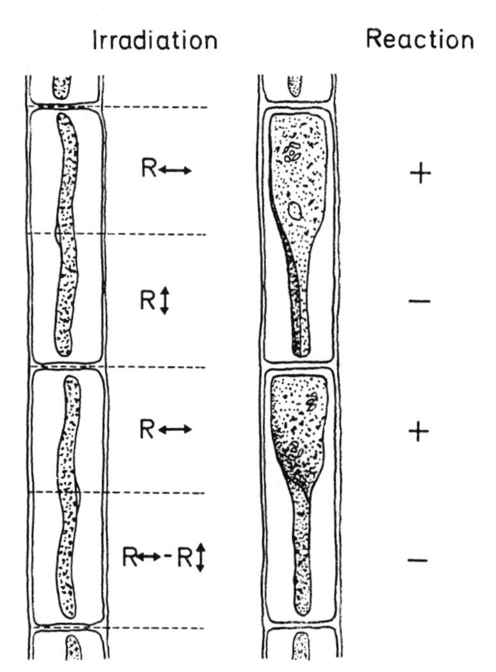

Figure 16. Induction of chloroplast movement in *Mougeotia* by linearly polarized red light (R) (E-vector parallel to the double-headed arrows). Both halves of the cell received distinct light stimuli, using a microscope for partial irradiation. (Kraml M. and Lauber P. unpublished data).

The R-induced *chloroplast movement* is mediated by phytochrome. The plastid re-orientates in a tetrapolar gradient of Pfr within 15 min. The long edges of the chloroplast, which are touching the cytoplasm at opposite parts of the cell (ideal for a tetrapolar gradient!), glide away from the regions of high Pfr-content (Fig. 15). Gradient formation and phytochrome orientation were analyzed by W. Haupt and co-workers (*cf.* Haupt 1982), using linearly polarized light.

In cells of *Mougeotia* with the chloroplast in the profile position, irradiation with linearly polarized R, with the E-vector transverse to the cell axis, induces chloroplast movement to the face position much more effectively than parallel vibrating light (Fig. 16). Moreover it is important that the latter is able to cancel Pfr-gradients (Fig. 16), indicating that it is absorbed as well by the dichroic photoreceptors. According to Table 1 this dichroic behaviour indicates that the Pr-molecules are orientated parallel to the cell surface, having transition moments in longitudinal as well as in perpendicular direction.

More detailed experiments indicated an orientation of the *Mougeotia* chromophores in the outer cytoplasm or at the plasma membrane, and along helical lines (Fig. 17). First data indicating the flip-flop dichroism of phytochrome in *Mougeotia* were published by Haupt (Fig. 18; Haupt 1960): in Pfr-containing *Mougeotia,* FR of low fluence rate and vibrating transversely to the cell axis is able to induce chloroplast movement, whereas longitudinally vibrating FR is ineffective. This result is only consistent with Pfr-transition moments normal to the cell surface (Fig. 12c, d and Table 1).

The flip-flop dichroism was confirmed by *microbeam* experiments of Haupt (1970) and recently by the author (Fig. 19), using a more modern apparatus. In Pr-containing cells R applied as a microbeam at the flanks is most effective with the E-vector vibrating parallel to the long axis of the cell. In contrast, FR microbeams were most effective vibrating transversely at the flanks, if the cells were pre-irradiated with R-microbeams.

A change in the orientation of phytochrome during photoconversion has also been found *in vitro,* although this amounts only to 32° or 148° rather than the postulated 90° (Sundqvist and Björn 1983).

An interesting point is the establishment of Pfr-gradients under saturating R conditions. This story is complicated by the fact that photoreversion of Pfr by R has to be included even in pulse treatments of less than 1 s. Even during this short R pulse, full photoreversibility in *Mougeotia* is possible (Kraml and Schäfer 1983) since the dichroic change from surface parallel (Pr) to surface normal (Pfr) has occurred (Kraml *et al.* 1984). Moreover the photoequilibrium state of phytochrome in dichroic systems depends not only on the relative spectral absorption of Pr and Pfr, but also on the dichroic orientation. It is well known (Part 4) that in phytochrome samples with randomly orientated photoreceptors the photoconversion by R (660 nm) is saturated at about 0.80 Pfr/P (where P = total phytochrome, Pr + Pfr). At the front and rear of *Mougeotia,* Pr is doubly favoured for the absorption of R vibrating perpendicular to the cell axis: spectral absorption as well as the dichroic orientation are favourable for photoconversion; in contrast both are unfavourable for Pfr. The photo-

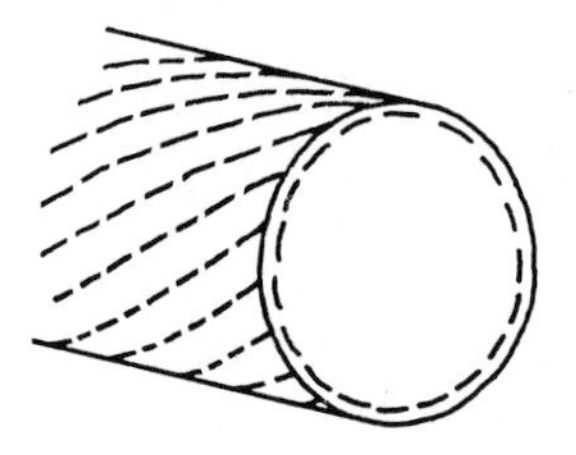

Figure 17. Scheme to illustrate the preferential orientation of dichroic Pr-chromophores in *Mougeotia* along a left-handed screw thread. The dissection of the transition moment (dashes) into vectors for the absorption of linearly polarized light vibrating longitudinally or perpendicular, respectively, results in schemes identical to Figs. 12a and 12b. After Haupt W. (1970) *Physiol. Veg.* 8: 551-563.

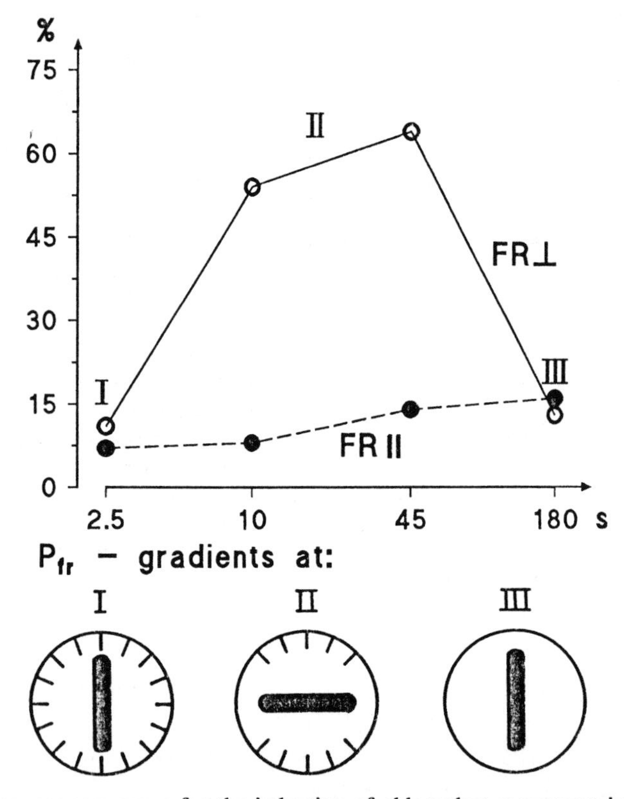

Figure 18. Fluence response curve for the induction of chloroplast movement in *Mougeotia* by linearly polarized far-red light (FR) vibrating perpendicular to the cell axis. The abscissa denotes irradiation time in FR, the ordinate is resulting proportion of chloroplasts in face position (%). The schematic cross-sections illustrate Pfr-gradients at different stages of the FR fluence-response curve. The Pr-molecules are not drawn. Modified after Haupt (1960).

equilibrium state (Pfr/P) therefore exceeds 0.80 and may attain a value of 0.95. At the flanks the transition moment of Pfr is ideally situated for the absorption

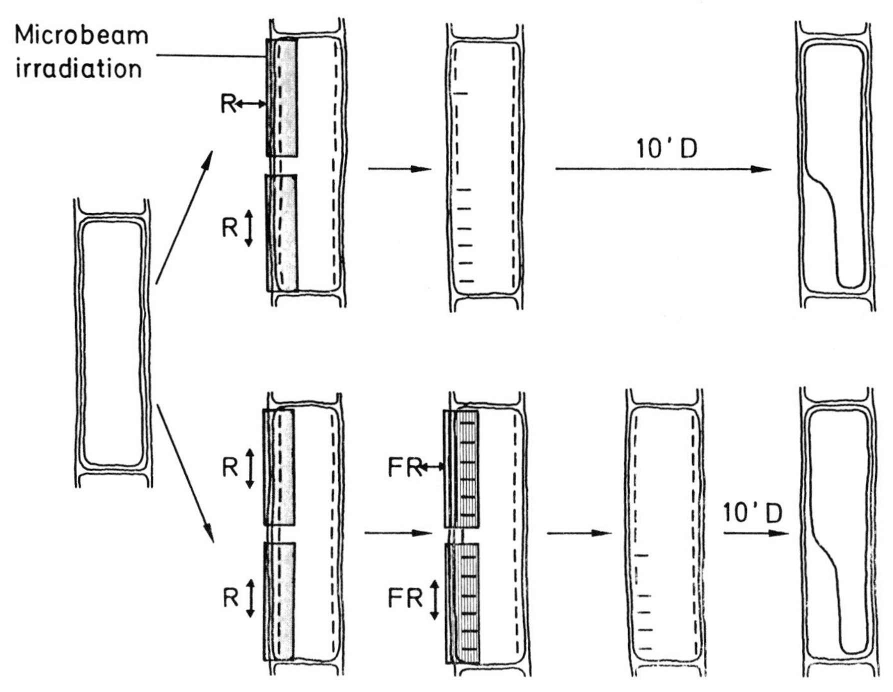

Figure 19. Demonstration of the dichroic orientation change during photoconversion of phytochrome in *Mougeotia* by microbeam irradiations. The experiments start with chloroplasts in face position (*left*) and phytochrome in the Pr-state after saturating far-red light (FR). Rectangular microbeams (for equipment see Fischer C. and Kraml M. (1990) *Photochem. Photobiol.* 52: 211-216) are applied to both halves of a cell using different E-vectors (double-headed arrows) of polarized red light (R) or FR. The dashes represent Pr and Pfr-molecules orientated parallel or normal to the cell surface, respectively. Local chloroplast movement is induced by R vibrating parallel rather than perpendicular. The induction by R is reversed by an immediate subsequent irradiation with FR vibrating perpendicularly rather than parallel. D = darkness. (Kraml M. unpublished data).

of light vibrating perpendicular to the cell axis, whereas the orientation of Pr is unsuitable for light absorption. This may result in a Pfr/P of about 0.5 despite the spectral advantage of Pr. In summary a strong Pfr-gradient can be maintained in *Mougeotia* even under continuous R conditions.

7.2.6.5 Phytochrome dichroism in fern and moss protonemata

The flip-flop dichroism of phytochrome was first postulated by Etzold (1965) in order to interpret polarotropism in linearly polarized light. Protonemata of *Dryopteris,* grown on agar and covered by a cover glass, so that growth is restricted to one plane, grow perpendicular to the E-vector of the incident light.

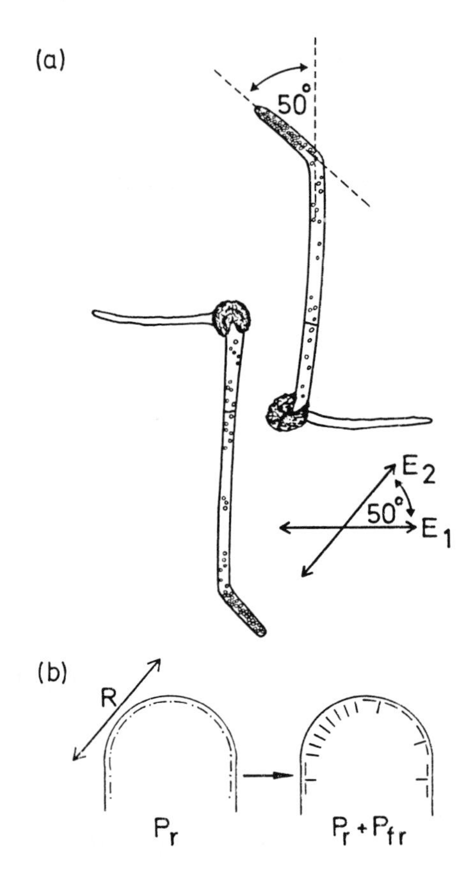

Figure 20. Polarotropism of fern protonemata. (a) Protonemata of *Dryopteris* were irradiated with linearly polarized red light (R) vibrating perpendicularly to the filaments (E1) for 3 days. Afterwards the plane of vibration was changed by 50° (E2). (b) Scheme of phytochrome orientation in tips of protonemata in darkness *(left)* and after irradiation with linearly polarized R (the double-headed arrow denotes the E-vector). The axis of maximum absorption of phytochrome turns by 90° during photoconversion from Pr to Pfr. All the Pr-molecules are converted to Pfr, whose transition moments (dashes) are oriented parallel to the E-vector. Pfr-molecules parallel to the E-vector are re-converted to Pr. Thus a Pfr-gradient is established even in continuous R. (a) After Steiner A.M. (1969) *Photochem. Photobiol.* 9: 493-506. (b) After Etzold (1965).

Changes of the E-vector result in corresponding changes of the growth direction (Fig. 20). As the site of maximal Pfr-content corresponds to the growth centre, transversely oriented R should be most effectively absorbed at the tip; *i.e.* the transition moments of Pr are orientated parallel to the surface. Additional experiments with FR indicated a change of the transition moments of 90°.

The flip-flop dichroism in fern protonemata was confirmed by microbeam experiments with *Adiantum* (Chapter 9.7; Kadota *et al.* 1982; Wada *et al.* 1983). Here, regions 5-15 μm behind the apex were irradiated, not the tip of the protonemata. A similar 90° re-orientation of the phytochrome chromophore was demonstrated using protonemata of the moss *Ceratodon* (Hartmann *et al.* 1983).

7.2.6.6 Wavelength-dependent action dichroism of flavin-mediated photo-responses

Action spectra of flavin-mediated photoresponses usually exhibit maxima in the B near 450 nm as well as in UV-A at about 370 nm (*c.f.* Galland and Senger 1988), corresponding to the absorption characteristics of flavoproteins. Interestingly action dichroism in the two spectral regions may give antagonistic effects for the same physiological response. An explanation for this phenomenon is possible without assuming that two different photoreceptors are involved. In the alloxazine system of flavins there exist two different electronic transition moments for B and UV-A. The angle between them is about 20-30°. Haupt (1984) has shown in a theoretical paper that inversion of action dichroism in the B and near UV can be predicted even for angles as small as 20°.

7.2.6.7 Wavelength-dependent action dichroism for B/UV-A in low-irradiance movement of Mougeotia

We have recently been able to demonstrate the phenomenon mentioned above for the low-irradiance movement in *Mougeotia.* In order to abolish the phytochrome-mediated response of B, strong FR irradiation was applied simultaneously to the short-wavelength light (Gabrys *et al.* 1984). Fluence-response curves for the induction of chloroplast movement were determined for linearly polarized B (Fig. 21a) as well as UV-A (Fig. 21b). In both cases an action dichroism is evident. Polarized B vibrating parallel to the cell axis is much more effective than vibrating perpendicular. The opposite is true for UV-A. All fluence-response curves shown in Fig. 21 are nearly parallel. Moreover the curves for B vibrating parallel and UV-A vibrating perpendicular are almost identical. This indicates the involvement of a single photoreceptor in both spectral regions with different electronic transition moments for B and UV-A. A good candidate for that photoreceptor would be a flavin [see Section 7.2.6.6; *cf.* also Schönbohm (1980) for the tonic B effect in high-irradiance movement of *Mougeotia*].

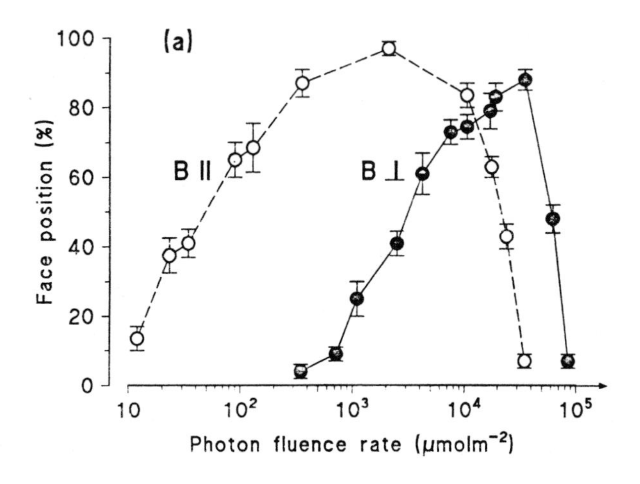

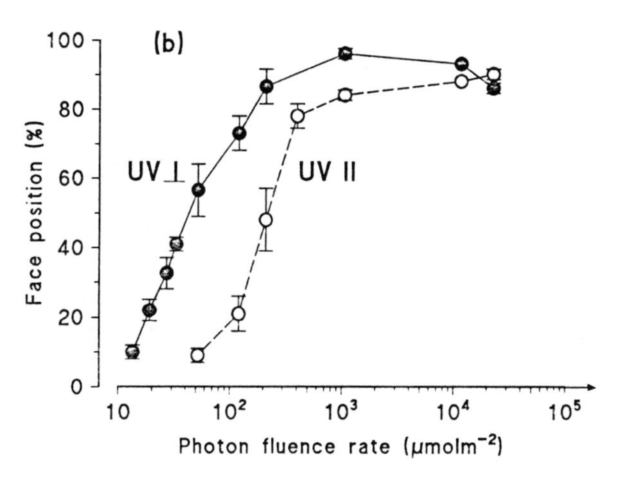

Figure 21. Fluence response curves for the induction of low-irradiance chloroplast movement in *Mougeotia* by: (a) linearly polarized blue light (B, maximum 450 nm); (b) linearly polarized UV-A (max 359 nm) and simultaneously applied strong far-red (RGN-9, 40 W m^{-2}) light. After Kraml M. and Schuller J. (1993) *Photochem. Photobiol.* in press.

7.2.6.8 Orientation of the B/UV-A photoreceptor in Mougeotia *as analyzed by microbeam irradiations*

Microbeam irradiation should provide more insight into the orientation of the electronic transition moments of the dichroic photoreceptor in the cell. Rectangular microbeams of polarized light were applied successively at both flanks of the algal cell to induce chloroplast orientation from edge to face position (Fig. 22). The whole *Mougeotia* cells were irradiated simultaneously by strong non-polarized FR. The B microbeams were most effective if the electrical

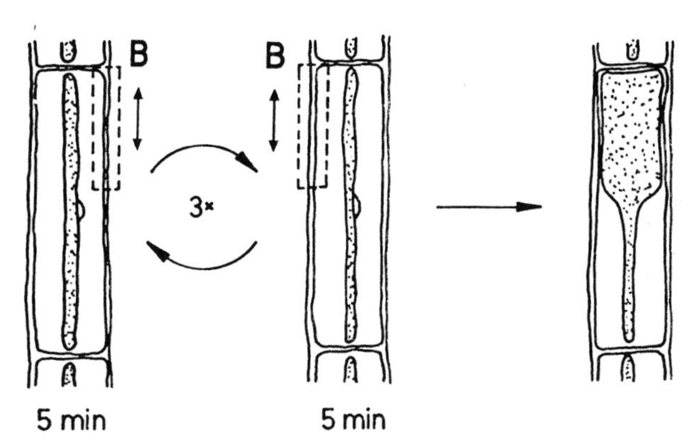

Figure 22. Schematic presentation of microbeam irradiation of *Mougeotia* with linearly polarized blue (B) or UV-A light applied alternately at both flanks of the cell and changing every 5 min. The whole preparation was simultaneously irradiated by strong far-red.

vector was oriented parallel to the cell axis (Fig. 23a). However, in the UV-A chloroplasts responded highly to microbeams vibrating perpendicular (Fig. 23b). These data indicate that the transition moments for the B absorption band of the photoreceptor involved in low-irradiance movement is orientated more or less parallel to the plasma membrane, whereas the transition moments for the UV-A absorption is orientated predominantly perpendicular. This is the first time that the difference in transition moments for B and UV-A have been demonstrated by microbeam irradiation. Recently, the transition moments for B in *Adiantum* protonemata were also shown to be parallel to the cell surface by microbeam experiments (Hayami *et al.* 1992).

7.2.7 Concluding remarks

Different strategies have been presented which are used by organisms to sense the direction of the incident light. The ecological significance of these mechanisms is evident in many cases. For example, phototaxis is a useful response to guide photosynthetic micro-organisms towards a favourable or away from a harmful light environment. Of course, the objection is correct that in direct sunlight it makes no sense to move phototactically a few meters towards the 'light source' or away from it. However, motile green micro-organisms usually live in an aquatic environment in which irradiance decreases drastically from the surface to deeper regions of the sea or lake. Many dinoflagellates are known to migrate vertically, down at night and up in the early morning. In several species, phototaxis in response to B is most effective during early morning, and is synchronized by a circadian rhythm. Orientation movements relative to the light direction are also important in partially shaded habitats. Phototropism as

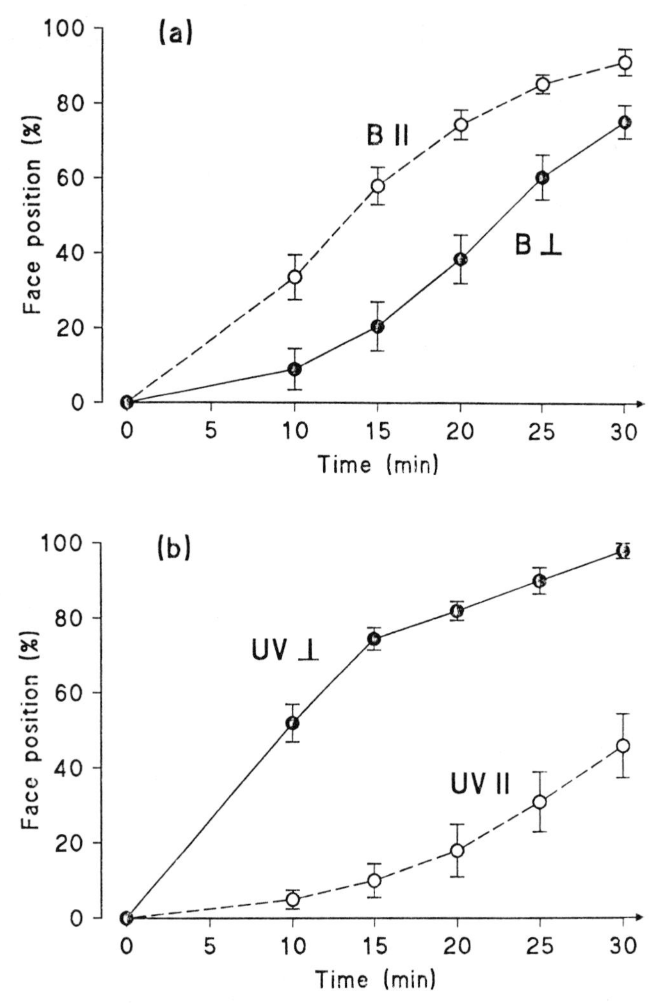

Figure 23. Induction of chloroplast movement in *Mougeotia* by microbeams of: (a) linearly polarized blue light (B maximum 450 nm), (b) linearly polarized UV-A (maximum 359 nm) light and simultaneously applied strong far-red. See also Fig. 22. E-vector in the direction of the double-headed arrows, vibrating parallel or perpendicular to the cell axis. After Kraml M. and Koch B. (1993) *Photochem. Photobiol.* in press.

well as phototaxis are effective mechanisms of photomovement as can be shown by young seedlings growing out of darker areas into the light. Moulds, which are non-photosynthetic organisms, use phototropic bending to maximize the dispersal of spores by orientating the growth of their fruiting bodies towards the light. A good example is *Pilobolus,* whose sporangia are aimed with remarkable accuracy into the direction of the impinging light and then discharged forcibly (Chapter 9.6).

Is polarized light important for the orientation in the natural environment? Examples from animals are well known: many insects can utilize polarization information from a patch of blue sky. For example, in the bee's eye there is an analyzer for polarized light. A variety of vertebrates are known to have an intra-ocular E-vector analyzer. For plants polarized light seems to be less important than for animals. The light environment of algae is, at best, only partially polarized. Nevertheless, dichroism is very important, since it enables light direction to be measured even in non-polarized light, as shown for chloroplast movement in *Mougeotia.*

Acknowledgements. Work in the author's laboratory was gratefully supported by the Deutsche Forschungsgemeinschaft.

7.2.8 Further reading

Cerdá-Olmedo E. and Lipson E.D. (eds.) (1987) *Phycomyces.* Cold Spring Harbor Lab., Cold Spring Harbor, New York, USA.

Dennison D.S. (1979) Phototropism. In: *Encyclopedia of Plant Physiology,* New Series, Vol. 7, *Physiology of Movements,* pp. 506-566, Haupt W. and Feinleib M.E. (eds.) Springer-Verlag, Berlin.

Haupt W. and Scheuerlein R. (1990) Chloroplast movement. *Plant Cell and Environ.* 13: 595-614.

Jenkins F.A. and White H.E. (1976) *Fundamentals of Optics.* McGraw-Hill, Tokyo.

Senger H. (ed.) (1987) *Blue Light Responses: Phenomena and Occurrence in Plants and Microorganisms.* CRC Press, Boca Raton, Florida.

Wagner G. and Grolig F. (1992) Algal chloroplast movements. In: *Algal Cell Motility,* pp. 39-72, Melkonian M. (ed.) Chapman and Hall, New York.

7.2.9 References

Banbury G.H. (1952) Physiological studies in the *Mucorales.* Part I. The phototropism of sporangiophores of *Phycomyces blakesleeanus. J. Exp. Bot.* 3: 77-85.

Bergman K., Burke P.V., Cerdá-Olmedo E., David C.N., Delbrück M., Foster K.W., Goodell E.W., Heisenberg N., Meissner G., Zalokar M., Dennison D.S. and Shropshire Jr. W. (1969) *Phycomyces. Bacteriol. Rev.* 33: 99-157.

Buder J. (1918) Die Inversion des Phototropismus bei *Phycomyces. Ber. Dtsch. Bot. Ges.* 36: 104-105.

Buder J. (1919) Zur Kenntnis der phototaktischen Richtungsbewegungen. *Jb. Wiss. Bot.* 58: 105-220.

Buder J. (1920) Neue phototropische Fundamentalversuche. *Ber. Dtsch. Bot. Ges.* 38: 10-19.

Dennison D.S. and Vogelmann T.C. (1989) The *Phycomyces* lens: measurement of the sporangiophore intensity profile using a fibre-optic microprobe. *Planta* 179: 1-10.

Etzold H. (1965) Der Polarotropismus und Phototropismus der Chloronemen von *Dryopteris filix mas* (L.) Schott. *Planta* 64: 254-280.

Fukshansky L. and Richter T. (1990) What do we know about the non-uniform perception of a phototropic stimulus in *Phycomyces*? *Planta* 182: 107-112.

Gabrys H., Walczak T. and Haupt W. (1984) Blue-light-induced chloroplast orientation in *Mougeotia*: evidence for a separate sensor pigment besides phytochrome. *Planta* 160: 21-24.

Galland P. and Lipson E.D. (1987) Light physiology of *Phycomyces* sporangiophores. In: *Phycomyces,* pp. 49-92, Cerdá-Olmedo E., Lipson E.D. (eds.) Cold Spring Harbor Laboratory, Cold Spring Harbor, New York, USA.

Galland P. and Senger H. (1988) The role of flavins as photoreceptors. *J. Photochem. Photobiol.* 1: 227-294.

Hartmann E., Klingenberg B. and Bauer L. (1983) Phytochrome-mediated phototropism in protonemata of the moss *Ceratodon purpureus Brid. Photochem. Photobiol.* 38: 599-603.

Haupt W. (1960) Die Chloroplastenbewegung bei *Mougeotia* II. Die Induktion der Schwachlichtbewegung durch linear polarisiertes Licht. *Planta* 55: 465-479.

Haupt W. (1970) Über den Dichroismus von Phytochrom-660 und Phytochrom-730 bei *Mougeotia. Z. Pflanzenphysiol.* 62: 287-298.

Haupt W. (1982) Light-mediated movement of chloroplasts. *Annu. Rev. Plant Physiol.* 33: 205-233.

Haupt W. (1984) Wavelength-dependent action dichroism: A theoretical consideration. *Photochem. Photobiol.* 39: 107-110.

Hayami J., Kadota A. and Wada M. (1992) Intracellular dichroic orientation of the blue light-absorbing pigment and the blue-absorption band of red-absorbing form of phytochrome responsible for phototropism of the fern *Adiantum* protonemata. *Photochem. Photobiol.* 56: 661-666.

Jesaitis A.J. (1974) Linear dichroism and orientation of the *Phycomyces* photopigment. *J. Gen. Physiol.* 63: 1-21.

Kadota A., Wada M. and Furuya M. (1982) Phytochrome-mediated phototropism and different dichroic orientation of Pr and Pfr in protonemata of the fern *Adiantum capillus-veneris* L. *Photochem. Photobiol.* 35: 533-536.

Kraml M. and Schäfer E. (1983) Photoconversion of phytochrome *in vivo* studied by double flash irradiation in *Mougeotia* and *Avena. Photochem. Photobiol.* 38: 461-467.

Kraml M., Enders M. and Bürkel N. (1984) Kinetics of the dichroic reorientation of phytochrome during photoconversion in *Mougeotia. Planta* 161: 216-222.

Löser G. and Schäfer E. (1986) Are there several photoreceptors involved in phototropism of *Phycomyces blakesleeanus*? Kinetic studies of dichromatic irradiation. *Photochem. Photobiol.* 43: 195-204.

Schönbohm E. (1980) Phytochrome and non-phytochrome dependent blue light effects on intracellular movements in freshwater algae. In: *The Blue Light Syndrome,* pp. 69-96, Senger H. (ed.) Springer, Berlin.

Shropshire Jr. W. (1962) The lens effect and phototropism of *Phycomyces. J. Gen. Physiol.* 45: 949-958.

Steinhardt A.R. and Fukshansky L. (1987) Spatial factors in *Phycomyces* phototropism: analysis of balanced responses. *J. Theor. Biol.* 129: 301-323.

Steinhardt A.R., Popescu T. and Fukshansky L. (1989) Is the dichroic photoreceptor for *Phycomyces* phototropism located at the plasma membrane or at the tonoplast? *Photochem. Photobiol.* 49: 79-87.

Sundqvist C. and Björn L.O. (1983) Light induced linear dichroism in photoreversibly photochromic sensor pigments. II. Chromophore rotation in immobilized phytochrome. *Photochem. Photobiol.* 37: 69-75.

Wada M., Kadota A. and Furuya M. (1983) Intracellular localization and dichroic orientation of phytochrome in plasma membrane and/or ectoplasm of a centrifuged protonema of fern *Adiantum capillus veneris* L. *Plant Cell Physiol.* 24: 1441-1447.

Weisenseel M.H. (1979) Induction of polarity. In: *Encyclopedia of Plant Physiology,* New Series, Vol. 7, *Physiology of Movements,* pp. 485-505, Haupt W. and Feinleib M.E. (eds.) Springer-Verlag, Berlin.

Wettstein D. (1965) Die Induktion und experimentelle Beeinflussung der Polarität bei Pflanzen. In: *Encyclopedia of Plant Physiology, Differentiation and Development.* 15/I, pp. 275-330, Lang A. (ed.) Springer-Verlag, Berlin.
Zurzycki J. (1980) Blue light-induced intracellular movement. In: *The Blue Light Syndrome,* pp. 50-68, Senger H. (ed.) Springer-Verlag, Berlin.

7.3 The duration of light and photoperiodic responses

Daphne Vince-Prue

Department of Botany, University of Reading, Whiteknights, Reading RG6 2AS, UK

7.3.1 Introduction

Exposure to alternating periods of light and darkness is a feature of the environment of most plants and animals. It is hardly surprising, therefore, to find that responses to the timing of these light/dark cycles are a nearly ubiquitous characteristic of life and that many of these responses represent useful adaptations to a fluctuating environment. A large number of biochemical events and several aspects of behaviour (such as 'activity periods' in certain animals and 'sleep movements' in leaves) have been found to occur at a particular time of the day or night. For example, in the unicellular alga, *Gonyaulax polyedra*, there are two different manifestations of bioluminescence: low-intensity glowing which peaks towards the end of the night and high-intensity flashing which peaks in the middle of the night (Fig. 1c). In order to be able to locate an event at a particular time of day, the organism must possess some kind of timekeeping mechanism and must also be able to discriminate between light and darkness. Thus, daily timekeeping requires both a *clock* and a *photoreceptor*. When examined in detail it appears that in most, if not all cases the underlying clock is an endogenous oscillator which has certain characteristic properties in relation to light/dark cycles (Fig. 1). The photoreceptor, in contrast, varies widely between organisms.

7.3.2 Circadian rhythms

As an example, let us take the daily 'sleep' movements of leaves. In many plants, leaves are held more or less horizontal during the day and assume either

R.E. Kendrick & G.H.M. Kronenberg (eds.), Photomorphogenesis in Plants - 2nd Edition

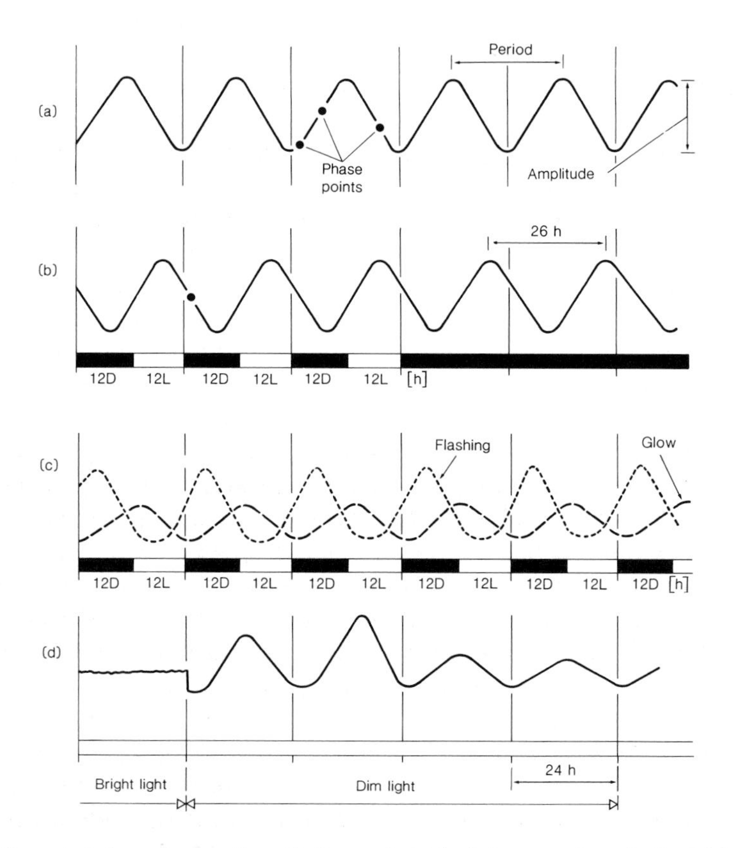

Figure 1. Characteristics of circadian rhythms. A typical free-running rhythm (a) is shown entrained (b) to 24 h light (L) /dark (D) cycles followed by reversion to the free-running period (26 h in this example) following transfer to continuous darkness. (c) Shows the phase relationships of two different rhythms in *Gonyaulax polyedra* entrained to 12 h L/12 h D cycles. (d) Shows suspension of the *Gonyaulax* 'flashing luminescence' rhythm in continuous bright light and release or re-starting of the rhythm following transfer to dim light.

an upward or a downward position at night. However, when plants are transferred from light/dark cycles to continuous darkness, the daily leaf movements persist for a time even though the changes between day and night no longer occur (Fig. 1b). It has been concluded, therefore, that the observed rhythm in leaf movements is based on an unseen, endogenous pacemaker which is self-sustaining and continues to oscillate without reference to the periodic light/dark changes. Thus, like the hands of a clock, the observed rhythm is coupled to an underlying oscillatory process, the *circadian oscillator* (Fig. 2). The operation of the oscillator is deduced from studies of the behaviour of the observed, coupled rhythm. In Fig. 2, two rhythms are shown coupled to the same oscillator (*e.g.* the *Gonyaulax* bioluminescence rhythms) but, in some cases (*e.g.* the leaf movement and photoperiodic rhythms in *Pharbitis nil)*, studies of the rhythms indicate that they are coupled to different circadian oscillators, *i.e.* there appears to be more than one clock driving circadian rhythms. In darkness, or continuous dim light, the rhythm is said to be *free-running* and its periodicity (Fig. 1b) is close to, but not exactly 24 h, hence the term *circadian* (from the Latin for 'about one day') is used. The period of the free-running rhythm has a Q_{10} close to 1; the clock mechanism is, therefore, temperature compensated so that timekeeping continues accurately under different temperature conditions.

Because the period of the free-running rhythm is not exactly 24 h, the rhythm drifts in relation to solar time. In our example (Fig. 1b) the free-running period is longer than 24 h, a characteristic of many diurnal organisms, and the peak time is later each day. Other rhythms have a period shorter than 24 h and so would gain time. The rhythm, therefore, has to be *entrained* (synchronized) to a 24 h-cycle by a periodic environmental signal, usually light (Fig. 1b), called a *Zeitgeber* (from the German for 'time-giver'). This action of light is considered to be on the circadian oscillator to which the observed rhythm is coupled (Fig. 2) and, under natural conditions, the main environmental Zeitgebers are the daily light/dark transitions at dawn and dusk. Although the rhythms are innate, they normally require an environmental signal, such as exposure to light or a change in temperature, to start them. Moreover, many rhythms damp out when the organism is in continuous darkness or in bright light and then require an environmental Zeitgeber to re-start them (Fig. 1d).

How does a circadian oscillator enable a coupled response to occur at a particular time of day? Figure 1c shows that different rhythms (in this case bioluminescence in *Gonyaulax)* have different *phase relationships* with the entraining light cycles, resulting in maximum 'flashing' luminescence near midnight, and maximum 'glowing' at the end of the night. Thus, daily timekeeping is a consequence of the phase relationships between the rhythm

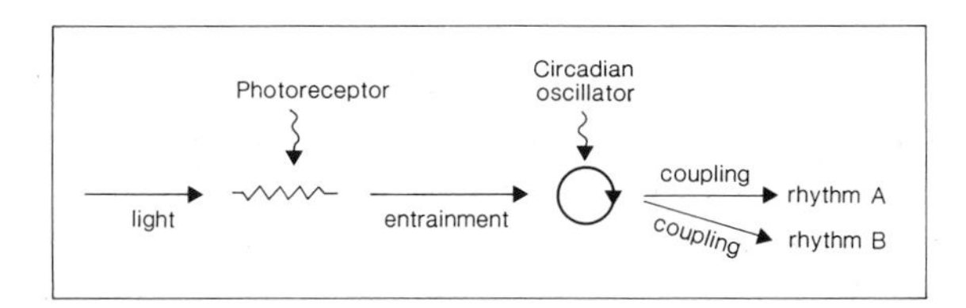

Figure 2. Action of light on a circadian oscillator and coupling to circadian rhythms. In this example, two different rhythms are shown coupled to the same circadian oscillator.

and the entraining environmental Zeitgebers. However, the daily durations of light and darkness are continually changing with the season and so the entraining light signals do not always occur at the same times of day. How, then, does the response remain 'on time'? As an example, we can take a particular phase point of the leaf movement rhythm already considered (the term *phase* is used for any point in the cycle (Fig. 1a) or, sometimes, for a portion of the cycle). In a 12-h light/12-h dark cycle, this phase point occurs 1 h after the onset of darkness (Fig. 1b). However, if the daylength is extended by 1 h, it will appear to have drifted forward in relation to the dusk (light-off) Zeitgeber and the appropriate response is a correcting backwards shift (or phase delay) such that the phase point still occurs 1 h after the end of the light period. Similarly, the appropriate answer to light perceived at a phase-point which would normally occur 1 h before dawn would be a phase advance. All circadian rhythms in animals and plants appear to operate in this way and exhibit characteristic *phase-response curves* to light. This phasing action of light is considered to be on the circadian oscillator (Fig. 2) and the usual experimental protocol employed to elucidate the oscillator's phase responses to dawn and dusk signals is to place the organism in continuous darkness, give a relatively short exposure to light at different phases of the free-running rhythm, and observe the direction and magnitude of the resulting phase shift. All agree in having delay responses when light is given at the beginning of the night-phase of the rhythm (the *subjective night*) with a sudden change to advance responses near the middle of the subjective night, and a prolonged dead zone of near insensitivity to light during the subjective day (Fig. 3). Experimentally determined phase-response curves of this kind allow a fairly accurate prediction of the relationships between the plant or animal's circadian timekeeping and local time, where the main environmental Zeitgebers are the light/dark transitions at dawn and dusk. However, the continuous period of daylight must also have some effect. Rhythms can be entrained by skeleton photoperiods which consist of two light pulses every 24 h, the first pulse being read as dawn (light-on) and the second as dusk (light-off). However, continuous light periods of more than 12 h cannot be simululated by skeletons as the organism always reads the shorter interval between the pulses as 'day'. For more details of the phase-response characteristics of circadian rhythms see Daan (1982).

Although the mechanism of entrainment to particular light/dark cycles is not understood, it is obvious that a photoreceptor is required for the perception of the entraining signals. Circadian rhythms occur widely in both plants and animals and a variety of different photoreceptor pigments appear to have been utilized in different organisms. For at least some rhythms in plants (*e.g.* CO_2 evolution in *Bryophyllum* and *Lemna* and leaf movements in *Phaseolus vulgaris* and *Samanea saman*) phytochrome has been shown to be the photoreceptor, but it is clearly not involved in photocontrol of the circadian system in many other cases, *e.g.* in *Gonyaulax*. Many organisms respond to blue light.

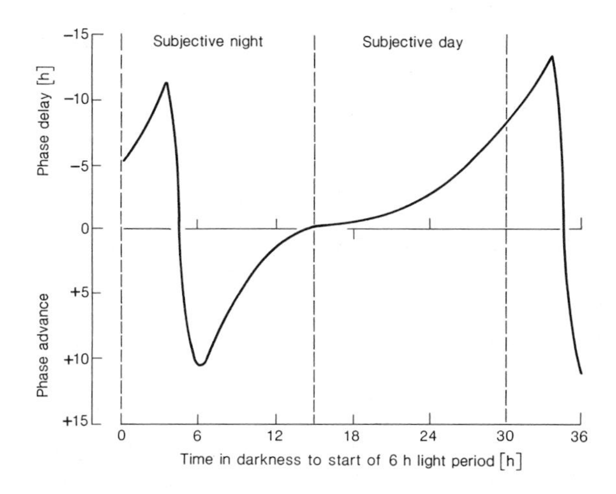

Figure 3. Phase-response curve for photoperiodic timekeeping in the flowering response of *Chenopodium rubrum*. Plants received a 6-h light interruption at different times in a 30-h free-running rhythm of flowering in darkness. The response curve shows the phase delay or advance compared with the uninterrupted control rhythm. From: King R.W. and Cumming B. (1972) *Planta* 103: 281-301.

7.3.3 Seasonal responses

In addition to establishing the relationship between the timing of an event and the time of day, the perception of the duration of light (or darkness) can locate an event in relation to the time of year. In this type of response, termed *photoperiodism*, an event occurs or fails to occur depending on the daily durations of light and/or darkness experienced by the organism. Photoperiodic responses are widespread throughout the plant and animal kingdom and the range of physiological responses is enormous (Vince-Prue 1975). In plants, the most widespread responses involve sexual reproduction (the initiation and development of flowers), asexual reproduction (the formation of bulbs, tubers, runners, *etc.*) and dormancy phenomena (development of winter resting buds in woody plants, bud dormancy in tubers, *etc.*) The adaptive value of seasonal photoperiodic responses is self evident. Daylength is the only environmental factor which from year to year gives completely reliable information about the passage of the seasons and there are many possible advantages to the organism in being able to time events in relation to the time of year. For example, daylength signals may synchronize the energy demands of reproduction with the season of greatest light receipt in summer or, as with autumnal short days for trees of high latitudes, induce bud dormancy and low-temperature hardiness to enable the plant to survive the following winter. In other species, the induction of resting structures may be induced by long-day conditions which accompany or precede a period of water stress. Even in tropical latitudes, many plants are responsive to daylength and may utilize this signal to synchronize flowering or other activities with seasonal events such as dry or wet periods. Clearly, in the tropics, timekeeping needs to be more precise than at higher latitudes because the seasonal changes in daylength are much smaller. One of the most valuable aspects of photoperiodism is the ability to be predictive (*e.g.* autumn short days precede winter cold) so that the necessary acclimation changes can take place before the environmental condition is experienced.

Despite the superficial differences between overt circadian rhythms and photoperiodic responses, there is now much evidence to suggest that photoperiodism is a special case of a circadian rhythm. It is thought that photoperiodic timekeeping depends on the fact that a circadian rhythm is coupled to the daily light Zeitgebers in such a way that the direction of the response is determined by the duration of light received in each 24-h cycle. The main part of this chapter will, therefore, consider this general hypothesis for the perception of light duration in photoperiodism with particular reference to the photoperiodic time-measuring process, the actions of light, and the identity and functioning of the photoreceptor(s). The discussion is restricted to the control of floral initiation as this has been the most intensively studied process in plant photoperiodism. However, other processes, such as tuber formation and dormancy, seem to be regulated by mechanisms that operate in a manner similar to that which controls floral induction, even though the ultimate biochemical pathways leading to the observed responses are unlikely to be the same. Most of the evidence discussed comes from experiments carried out with the small number of species where a single photoperiodic cycle of the correct length can effect the transition to reproduction (*e.g. Xanthium strumarium, Pharbitis nil, Chenopodium rubrum, Lolium temulentum* and *Hordeum vulgare*, cv. Wintex). Because it is easier to impose complex experimental protocols during a single cycle, such plants have been studied in considerable detail and many current ideas about the photoperiodic mechanism are based on studies of floral induction in them.

7.3.4 General aspects of photoperiodism

Following earlier experiments of Hans Klebs and Julien Tournois who independently recognized that it was the daily duration of light rather than its quantity that was important in the regulation of flowering time, it was in the early 1920's that Wightman Garner and Harry Allard saw clearly that flowering and many other responses in plants could be accelerated by either long days or short days depending on the species. In particular, observations on a variety of tobacco,

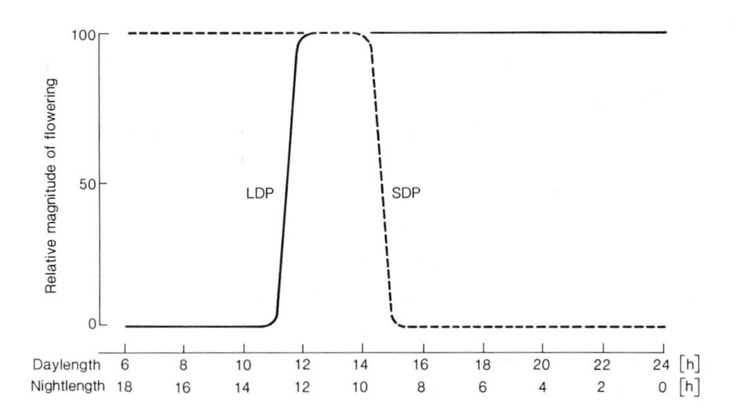

Figure 4. Characteristics of the photoperiodic response. Long-day plants (LDP) flower when the daylengths exceeds (or nightlength is less than) a critical duration. Short-day plants flower when the daylength is less than (or nightlength exceeds) a critical duration. The critical value varies between species and, in this example, both LDP and SDP would flower in photoperiods between 12 and 14 h duration.

Maryland Mammoth, which failed to flower under natural conditions in Washington, D.C., but could be induced to flower by artificially shortening the daylength, led to the discovery that flowering in this plant required exposure to daylengths less than a certain critical duration. Examination of a wide range of species allowed classification according to their flowering response to daylength. Two major categories were identified: *short-day plants* (SDP), where flowering occurs or is accelerated in short days (SD) and *long-day plants* (LDP) where flowering occurs or is accelerated in long days (LD). A few plants require exposure to both LD and SD but in a particular sequence; predictably these are called LD-SD plants or SD-LD plants according to the sequence in which the photoperiods must be presented. There are some other minor groups (Vince-Prue 1975) and there are also many plants where daylength does not play a regulatory role in flowering (*day-neutral plants*).

It is important to recognize that the factor which governs classification as a SDP or LDP is whether flowering occurs (or is accelerated) only when the daylength exceeds (LDP) or only when it is less than (SDP) a certain critical duration in each 24-h cycle (Fig. 4). This *critical daylength* varies between species. In some cases, LDP and a SDP may both flower when grown in a particular daylength (12-14 h for the examples shown in Fig. 4); only when the daylength is decreased or increased is the classification into LDP or SDP possible. Under natural conditions, this mechanism in LDP effectively delays flowering until the critical daylength of the plant is exceeded. It can, therefore, identify the lengthening days of spring or early summer and is associated with the summer-flowering habit. Many SDP, on the other hand, flower in autumn when the days shorten sufficiently. It is obvious, however, that daylength alone is an ambiguous seasonal signal near the equinoxes and various strategies are employed by plants to avoid this ambiguity. For example, there may be a dual daylength requirement for LD followed by SD (autumn), or a coupled temperature response as in the SDP, strawberry, where initiation of flowers in spring is prevented by the previous exposure to winter cold. There is little evidence, however, to support the suggestion that plants might avoid the ambiguity by measuring 'lengthening' *versus* 'shortening' days, although it is known that some animals do so. On the other hand, successive developmental stages sometimes have different critical daylengths. The florist's chrysanthemum, for example, requires a shorter daylength for the development of an inflorescence than for its initiation, thus ensuring that normal inflorescences and open flowers occur only in autumn.

Under natural conditions, daylength and nightlength are absolutely related to form a 24-h cycle of light and dark. Thus, the critical daylength could be detected by measuring either the duration of light, or darkness, or even their relative durations (Fig. 4). To establish which of these features is the controlling factor, it is necessary to vary the light and dark periods independently. Using this approach, Karl Hamner and James Bonner were able to show that photoperiodic timekeeping in the single-cycle SDP, *Xanthium strumarium*, is essentially a question of measuring the duration of darkness. Flowering occurred only when the dark period exceeded 8.5 h, irrespective of the relative durations of light and darkness in the experimental cycle. The importance of the duration of darkness is also well illustrated in the SDP, *Pharbitis nil* cv. Violet. In this plant, seedlings otherwise grown throughout in continuous white light (W) will flower in response to a single dark period, providing this exceeds about 9 h, and the response is essentially independent of the duration of the preceding period of continuous light. Thus, in SDP at least, it appears that the duration of darkness is the primary determinant of the photoperiodic response, while the effects of light can usually be interpreted as interactions with this essential component. Under natural cycles of light and darkness, the result is an apparent critical daylength which, if exceeded, prevents flowering in SDP, although it is in fact a *critical nightlength* (CNL) which is important (Fig. 4). The more difficult question of time-measurement and duration perception in LDP is deferred to a later section.

A feature which underlines the importance of the dark period is that it can be rendered ineffective by an interruption with a short light treatment or *night-break* and exposures of only a few minutes given in the middle of an inductive dark period of 14-16 h will prevent flowering in many SDP, including *Xanthium* and *Pharbitis*. Originally such night-breaks were thought to act by splitting the long night into two shorter, and hence ineffective dark periods.

This explanation is inadequate, however, since the time at which a night-break is most effective is not altered by prolonging the dark period. In both *Xanthium* (Fig. 10) and *Pharbitis* (Fig. 7), for example, a night-break given at the 8-9th h of a 48-h dark period effectively prevents flowering even though the subsequent 40-h dark period is much longer than the critical nightlength for these plants (8-9 h). Such experiments have established that the night-break response is a transient period of sensitivity to light which is related in time to the beginning of the dark period.

Despite the overriding importance of the duration of uninterrupted darkness, the light period is not without effect. Perhaps the most important attribute of the photoperiod is that a preceding exposure to light is necessary for a subsequent dark period to be effective. For example, in the single-cycle SDP *Xanthium*, an inductive (*i.e.* longer than critical) dark period is only effective if it is preceded by a minimum of 3-5 h of light. Similarly, in *Pharbitis*, dark-grown seedlings will flower if exposed to 5 min red light (R) and simultaneously sprayed with a cytokinin, followed by an inductive dark period. Controls without R do not flower, even though they have been exposed to darkness for 2-3 days (much longer than the critical nightlength) before transfer to continuous W. This light requirement is not confined to single-cycle plants, for *Kalanchoë* will flower with a few seconds of light each day over several cycles, whereas darkness for the same duration is without effect. It is evident from these and other similar experiments that both light and darkness are necessary components of the photoperiodic mechanism in SDP.

In single-cycle SDP, there seems to be no upper limit to the duration of the photoperiod, provided that the critical dark period is exceeded but this is not so when several cycles are required for induction. When *Glycine max* was given seven cycles composed of different light and dark durations (Fig. 5), there was a sharply defined critical nightlength of 10 h, which was independent of the photoperiod duration. However, with an inductive 16-h dark period, the longest photoperiod which allowed flowering was between 18 and 20 h. It is not clear if 18-20 h represents an absolute upper limit or whether the cycle lengths examined between 36 and 46 h (20-30 h light plus 16 h dark) were unfavourable. Later experiments (Fig. 6) tend to support the latter conclusion so that the apparent upper limit to daylength in multi-cycle plants is probably a property of the rhythmic nature of the time-measuring system.

The irradiance of light during the photoperiod is important in that there is a threshold before the photoperiod functions as such. This may be very low, as in *Xanthium*, where only 0.02 W m^{-2} produced a strong promoting effect on flowering, while increasing the irradiance above this threshold had relatively little further effect. It must be emphasized, however, that *Xanthium* is a single-cycle plant and that the photoperiod conditions were imposed for only 1 day. There are many examples where the magnitude of flowering is strongly influenced by light quantity and it has been suggested on several occasions that this

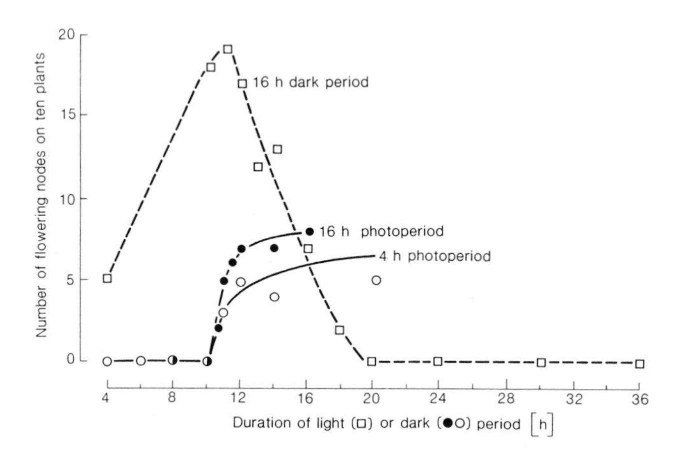

Figure 5. **The effect of varying the duration of the light and dark periods on flowering in soybean (*Glycine max*). Plants received 7 cycles in which the duration of the light period accompanying a 16-h dark period (□), or the dark period accompanying a 4-h (○) or 16-h (●) light period was varied. From Hamner K.C. (1940) *Bot. Gaz.* 101: 658-687.**

indicates a requirement for photosynthesis or photosynthetic products. This might explain why many SDP do not flower if the photoperiod is very short or the irradiance is low. It would indeed be surprising if there were no interaction between photosynthesis and flowering because of the fundamental role of photosynthesis in the plant. However, it seems unlikely that photosynthesis has any role in the photoperiodic mechanism itself for there are several SDP where the photoperiodic induction mechanism has been shown to operate with amounts of light which are photosynthetically insignificant. Some examples *(Pharbitis, Kalanchoë, Xanthium)* have already been given. The most likely conclusion is that the light requirement for photoperiodic induction in SDP is not photosynthetic.

From the results of these and many other early experiments, it appears that there are two essential components of the photoperiodic process in SDP. Time is measured in darkness and the critical factor for floral induction is a sufficiently long dark period or, in the majority of SDP, a succession of such dark periods. Exposure to a nightlength longer than a critical value does not, however, result in flowering unless it is preceded by a photoperiod whose duration does not appear to be critical. In the following section, we will consider the nature of the time-measuring system and how it is coupled in SDP to effect floral induction in longer than critical dark periods.

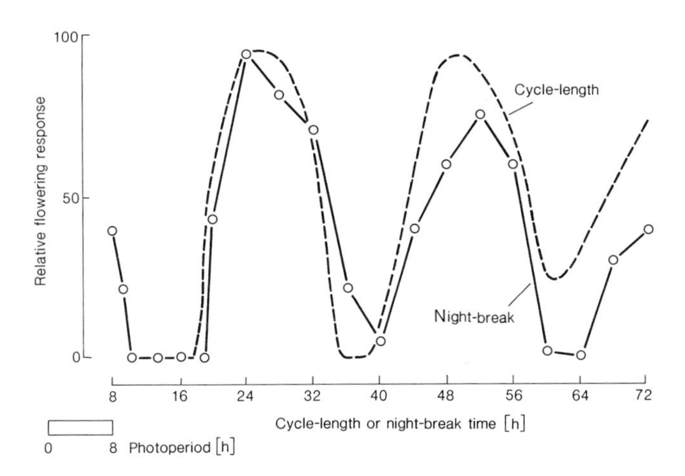

Figure 6. Rhythmic flowering response of soybean (*Glycine max*) to variations in cycle-length or night-break time. Plants received 7 cycles consisting of an 8-h photoperiod followed by various durations of darkness to generate different cycle lengths (*e.g.* 24-h cycle length = 8 h light + 16 h dark). In the night-break experiment, plants received 7 cycles of an 8-h photoperiod followed by 72 h darkness interrupted by a single 4-h night-break. The data are plotted for coincidence of the light-on signal (dawn); thus the time of beginning the night-break is shown. Night-break data from Coulter M.W. and Hamner K.C. (1964) *Plant Physiol.* 39: 848-856. Cycle length data from Hamner K.C. and Takimoto A. (1964) *Amer. Nat.* 98: 295-322.

7.3.5 Photoperiodic timekeeping

For many years, it was rather generally assumed that the duration of the dark period was measured by a kind of hourglass consisting of one or more light-sensitive reactions which must go to completion in order for floral induction to take place in SDP, or which would prevent floral induction in LDP. The time taken for this reaction(s) to occur would then represent the critical dark period. The most specific proposal for the nature of the hourglass was made in 1960 by Sterling Hendricks and co-workers. Observations in dark-grown seedlings had shown that the far-red light (FR)-absorbing form of phytochrome (Pfr) was lost by thermal reactions following transfer to darkness and so it was suggested that the time-measuring hourglass might be the time taken for Pfr to fall below a critical threshold below which it no longer inhibited flowering in SDP, nor promoted it in LDP. This attractive idea that, in plant photoperiodism, phytochrome is directly involved in both time-keeping and photoperception was rapidly shown to be incorrect, although it still appears in some text books! One of the main arguments against Pfr loss being the main photoperiodic timer is

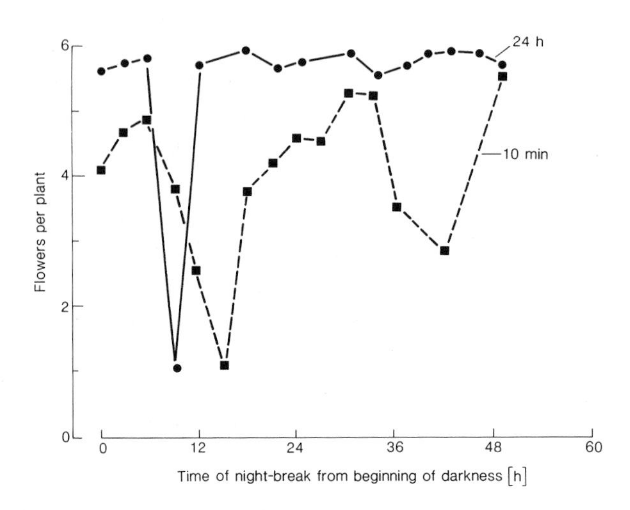

Figure 7. The effect of photoperiodic duration on the pattern of flowering response to a night-break in *Pharbitis nil.* Following a single photoperiod of 10 min (■) or 24 h (●), a dark period of 48 or 72 h was interrupted with a single 10-min night-break with red light at the time shown on the abscissa. Data of Lumsden P.J. (1984) D. Phil. Thesis, University of Sussex.

that lowering Pfr photochemically by exposing plants to FR before transfer to darkness has little effect on the critical nightlength (Vince-Prue 1983; Vince-Prue and Takimoto 1987). Other suggestions for an 'hourglass' timer have been much less precise.

Although the possibility of an hourglass component has not been entirely excluded, most attention is now focused on the hypothesis that photoperiodic time-measurement depends on a circadian oscillator of the same kind as that which underlies the endogenous biological rhythms discussed at the beginning of the chapter. Considering for the moment only SDP, the rest of this section will look at some of the evidence for the involvement of a circadian clock in photoperiodic timekeeping and consider how this clock may operate under natural day/night cycles to measure the critical nightlength and control the induction of flowering.

Many kinds of experiment have established that photoperiodism involves a rhythmic change of sensitivity to light. For example, if plants are given a relatively short light exposure (night-break) at different times during a very long inductive dark period, there is often a 24-h periodicity in the flowering response. Rhythms of this kind have been observed in several SDP, including *Glycine max* (Fig. 6), *Chenopodium rubrum* (Fig. 12) and *Kalanchoë blossfeldiana*. In contrast, *Xanthium* and *Pharbitis* are exceptions and, in these plants, a night-break is only inhibitory 8-9 h after transfer to darkness, with no further periods of inhibition (Fig. 10). Although, at first sight, this type of response seems to be an hourglass, there is evidence that a circadian rhythm is the underlying timer. When dark-grown seedlings of *Pharbitis* are exposed to 24 h W followed by 48 h darkness at 25°C, there is a single maximum in the inhibitory response to a night-break. However, if the 24 h photoperiod is replaced by 10 min R, the response to a night-break becomes rhythmic with minimum flowering occurring at circadian intervals (Fig. 7). As the timing mechanism is almost certainly the same in these treatments, the apparent hourglass must be caused by a damping of the rhythmic flowering response following the long photoperiod. In *Pharbitis*, this damping may be due to the fact that, when the photoperiod is sufficiently long, induction is complete in the first photoperiodic cycle and so no second inhibition point is possible. The same considerations probably also apply to *Xanthium* which, although apparently non-rhythmic with respect to the night-break response, exhibits a number of features such as phase-shifting, which are characteristic of circadian rhythms (Papenfuss and Salisbury 1967).

The rhythmic nature of the flowering response can also be demonstrated by the so-called 'resonance' experiments. In these, the flowering response changes rhythmically with increasing duration of the dark period which follows a short photoperiod (Fig. 6). It is evident that, if flowering is to occur in these plants, the light/dark pattern must be synchronized in some way with an internal oscillation. Other plants, including *Xanthium* and *Kalanchoë*, do not show clear-cut rhythms under these conditions although this need not imply differences in the underlying mechanism; *Kalanchoë* shows a rhythmic response to a night-break and, as already discussed, *Xanthium* exhibits other features associated with circadian rhythms.

A characteristic feature of overt circadian rhythms is that the phase of the rhythm can be advanced or delayed by exposure to light, the precise response depending on when the light is presented. Similar phase-shifts have been demonstrated in photoperiodism. Examples include the rhythm of flowering response to a 6-h night-break in *Chenopodium rubrum*, for which a complete phase-response curve has been constructed showing the classical features of phase delay by light given early in the subjective night and phase advance by light given late in the subjective night (Fig. 3). For other SDP, the information is less complete but similar phase shifts have been reported in *Xanthium* (Papenfuss and Salisbury 1967) and *Pharbitis* (Fig. 8).

There is now a considerable amount of evidence of the kind briefly outlined above, although admittedly for only a few species, which suggests that photoperiodic timekeeping in higher plants involves a circadian oscillator. However, despite vast and often confusing literature, the precise way in which this oscillator achieves the measurement of a critical nightlength in SDP is still unknown. It should also be recognized that the same mechanism may not

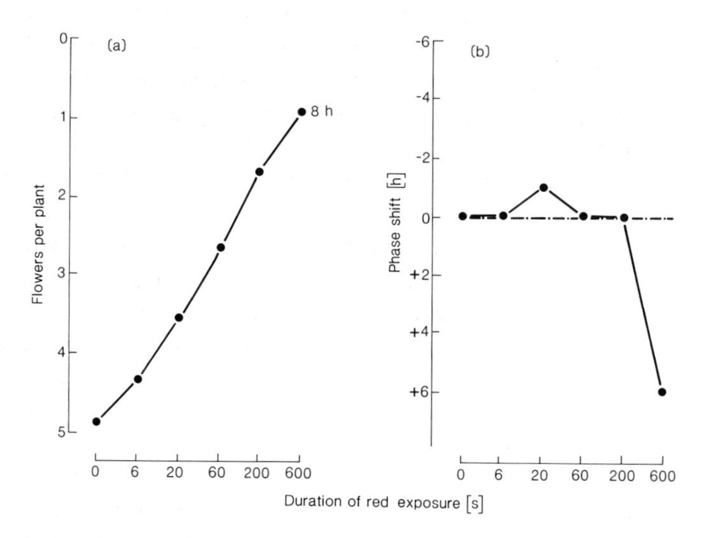

Figure 8. Fluence-response curves for (a) the inhibition of flowering and (b) phase-shifting of the photoperiodic rhythm by light given 8 h after the beginning of a 72-h dark period following a 24-h photoperiod. From Lumsden P.J. and Furuya M. (1986) *Plant Cell Physiol.* 27: 1541-1551.

operate in all plants. The original proposal made by Erwin Bünning in 1936 was that photoperiodic time measurement involves a regular oscillation of phases (*i.e.* portions of the rhythm) with different sensitivities to light. Bünning suggested that a 12-h light-requiring, or *photophile* phase alternated with a 12-h dark-requiring, or *skotophile* phase. Light was required during the photophile phase but flowering was inhibited by light given during the skotophile phase, thus giving rise to the inhibitory night-break effect in SDP and the need for both a critical duration of darkness and for light during the photoperiod. The LDP were accommodated in the model by suggesting that the photophile and skotophile phases were displaced compared with SDP. This rather general model has become known as *Bünning's hypothesis*. Since then, other more explicit models based on known properties of the circadian system have been developed.

7.3.5.1 External coincidence models

Much of the work has followed the lead set by Bünning in proposing that there is a light-sensitive phase in the photoperiodic rhythm (Bünning's skotophile phase). This type of scheme is known as *external coincidence*. It assumes that there is a single photoperiodic rhythm and that light has a direct effect to prevent the induction of flowering in SDP (or to induce flowering in LDP) when it is coincident with a particular light-sensitive phase of the rhythm; this is sometimes called the *inducible phase*. An alternative scheme (called *internal coincidence*) ascribes photoperiodic responses to the interaction of two rhythms with induction occurring only when critical phase points in the two rhythms coincide. This would take place only in particular light/dark cycles to give rise to LD or SD responses. Much of the evidence obtained with single-cycle SDP, particularly with *Pharbitis* and *Xanthium* is consistent with an external coincidence model and is discussed in more detail below.

One of the requirements for an external coincidence mechanism is that the inhibition of flowering and control of the phase of the rhythm are two distinct actions of light. This question has been explored in *Pharbitis* by examining in detail the fluence-response characteristics of the phase-shifting response and the night-break inhibition of flowering. For example, at the 8th h of darkness (when light was strongly inhibitory), a marked reduction of flowering was obtained with an exposure to light which had no effect on the phase of the rhythm (Fig. 8). At other times, *e.g.* at the 6th h, the rhythm could be phase-shifted with no effect on flowering. These results argue strongly that the inhibition of flowering by light is not a consequence of a shift in the phase of the rhythm(s) (as required by an internal coincidence mechanism) and affords strong support for the external coincidence theory of photoperiodic control in this plant. Unfortunately, this point has not yet been examined in other species. The control of the phase of the rhythm is considered to result from the action of light on the circadian oscillator, whereas the night-break inhibition of flowering results from a direct action of light on the coupled photoperiodic rhythm.

One of the most explicit external coincidence models for photoperiodic timekeeping in SDP has been developed from studies with *Pharbitis* (Vince-Prue and Lumsden 1987). This is a particularly useful subject for studying the way in which the photoperiodic rhythm is entrained because dark-grown seedlings will respond to a single light/dark cycle. This means that the relationship between photoperiod duration and dark timekeeping can be studied without the complicating problem of entrainment to non-inductive cycles during growth of the plant to reach experimental size. This problem can be partly overcome, as in experiments with *Xanthium*, by exposing plants to a 'neutral' non-inductive dark-period before transferring plants to the experimental cycle (Papenfuss and Salisbury 1967).

One approach to understanding how the circadian rhythm operates to control photoperiodic timekeeping has been to examine the time of maximum sensitivity to a night-break (NB_{max}) following a single photoperiod of varying duration. In *Pharbitis* (Fig. 9a), it was found that maximum sensitivity occurred at a constant time (15 h) from the *beginning* of the photoperiod, when this was less than about 6 h. When the photoperiod was longer than this, the time of night-

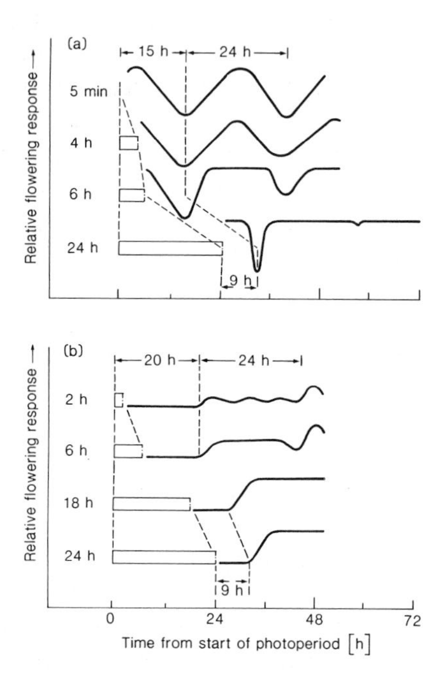

Figure 9. The effect of photoperiod duration on (a) the time of night-break sensitivity and (b) the critical nightlength in *Pharbitis nil.* In (a), plants received a single photoperiod of different durations as indicated on the figure, followed by an inductive dark-period interrupted by a 10-min night-break. Plants were returned to continuous white light (W) 72 h after the beginning of the photoperiod. In (b), plants received a photoperiod of varying duration followed by a dark period also of varying duration before transfer to continuous W. Both critical nightlength (CNL) and the time of maximum sensitivity to a night-break (NB_{max}) were constantly related in time to light-on with photoperiods of < 6 h duration. NB_{max} was constant (9 h) from light-off with photoperiods of > 6 h duration, but CNL was constant (9 h) from light-off only after photoperiods of 18 h or longer. From (a) Lumsden P.J., Thomas B. and Vince-Prue D. (1982) *Plant Physiol.* 70: 277-282, and (b) Lumsden P.J. (1984) D. Phil. Thesis, University of Sussex.

break sensitivity was delayed and NB_{max} always occurred at a constant time (9 h) after the *end* of the photoperiod. When plotted from the beginning of darkness, the shift in the time of NB_{max} following photoperiods of 10 min or 24 h is clearly evident (Fig. 7). These results have been interpreted in the following way. A single photoperiodic rhythm of sensitivity to light is initiated (or rephased) by transfer to light at dawn and the light-sensitive (inducible) phase of this rhythm occurs about 15 h after the light-on signal [*i.e.* at circadian time (CT) = 15]. This rhythm initially continues to run in continuous light so that, in real time, the NB_{max} always occurs 15 h after light on. A light-on signal is thus sufficient to induce flowering under these conditions unless a second pulse of light is given at the specific, light-sensitive phase of the rhythm. After about 6 h in continuous light (CT6), the rhythm appears to become 'suspended' and remains at CT6 for as long as the plant remains in continuous light. It is then released by the light-dark transition. Since the rhythm is suspended at CT6, the time of NB_{max} (in real time) is always about 9 h after transfer to darkness (*i.e.* at CT15), as would be predicted from the original light-on rhythm. At temperatures sufficiently high to maintain growth, the natural daylength would be longer than 6 h so that the flowering rhythm would always go into suspension during the photoperiod and be released by the light to dark transition at the end of the day; this would result in NB_{max} always occurring at a constant time from dusk. The *Pharbitis* experiments were carried out with dark-grown seedlings but very similar results have been obtained for normal light-grown plants of *Xanthium*, where the rhythm goes into 'suspension' after 5 h in the light and the time of NB_{max} occurs 8.5 h after transfer to darkness (Papenfuss and Salisbury 1967).

The apparent suspension of the rhythm at a particular phase point is not peculiar to photoperiodism. In many overt rhythms, rhythmicity is abolished after approximately 12 h in continuous light and is resumed at CT12 when the organism is returned to darkness (Lumsden 1991). It has been suggested that the apparent suspension in continuous light may result from a change in the dynamics of the circadian oscillator which migrates to a so-called *light-limit cycle*, effectively oscillating about one phase position. Following transfer to darkness, the circadian oscillator moves to a *dark-limit cycle* taking essentially the same time to reach it regardless of the starting position in the light-limit cycle. The rhythm would then appear to resume from a fixed phase point irrespective of the duration of the preceding continuous light. Alternatively, the oscillator may really stop in continuous light as, for example, if essential ion channels were not open.

How do these results relate to the measurement of the critical nightlength which, as we have seen, is the primary determinant of flowering in SDP? This question has been addressed in dark-grown *Pharbitis* seedlings by varying the durations of both the photoperiod and the (uninterrupted) dark period (Fig. 9b). A rhythmic response to the duration of darkness was observed after photoperiods of < 12 h duration establishing that nightlength timing is also associated with a circadian rhythm. The results from other experiments with *Pharbitis* confirm this important conclusion (Saji *et al.* 1984). After longer photoperiods, the rhythmic response to dark duration largely disappeared as was observed with the NB_{max} rhythm. When the photoperiod was up to 6 h, the sum of the light period and the CNL was constant at about 20 h suggesting that, as with the night-break sensitivity response, the rhythm began with the light-on signal and continued to run in the light. After photoperiods of 18 h, the CNL was constant (about 9 h) from the end of the light period. With photoperiods between 6 and

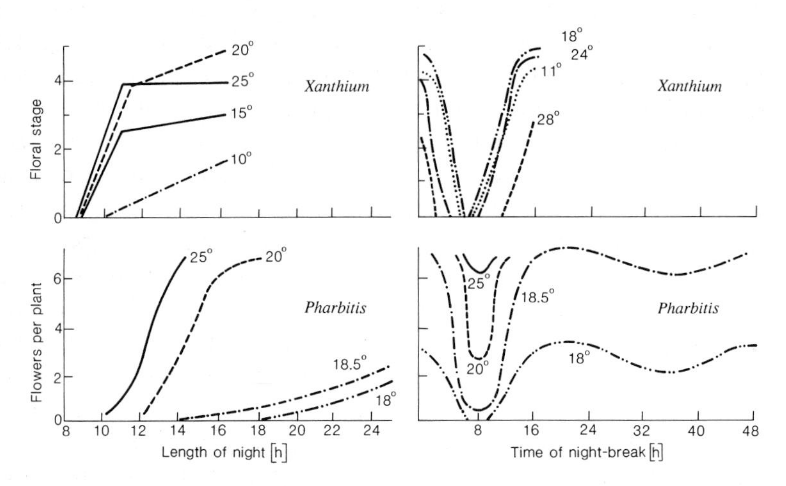

Figure 10. Effects of temperature on critical nightlength and time of night-break sensitivity in *Pharbitis nil* and *Xanthium strumarium*. *Left:* flowering responses as a function of nightlength. *Right:* flowering response of plants given a short night-break at different times in a 48-h dark period. From Salisbury F.B. and Ross C.W. (1969) *Plant Physiology* (1st edition) Wadsworth Publishing Company, Belmont.

18 h, the CNL showed no constant relationship with either light-on or light-off. With short photoperiods, the CNL was longer than the time to NB_{max}, indicating that, for floral induction, additional dark reactions must take place after the light-sensitive, inducible phase of the rhythm (NB_{max}) has been reached. With long photoperiods, the CNL and the time to NB_{max} essentially coincide. From these results, it seems a reasonable conclusion that the CNL is controlled by the same flowering rhythm that gives rise to the night-break response, although the relationships between CNL and the dawn and dusk signals are less clearly defined.

Experiments on temperature sensitivity suggest that the CNL may involve reactions additional to those of the rhythm which times NB_{max}. The *time* of NB_{max} is not affected by temperature in *Xanthium* and this also applies to CNL (Fig. 10). However, there appears to be a two-phase response; the first phase is relatively temperature insensitive leading to a common value for CNL, whereas the second phase is more temperature dependent. These results indicate that the response to increasing nightlength in *Xanthium* has two components. The first, temperature-insensitive component is concerned with timing and the second, temperature-sensitive component (reminiscent of an hourglass) appears to be concerned with the magnitude of the flowering response. In *Pharbitis*, the time of NB_{max} is similarly unaffected by temperature, but in contrast, the CNL is markedly temperature-dependent (Fig. 10). From the *Xanthium* results, it is possible that the temperature-dependent component of the CNL in *Pharbitis* merely represents the time required to achieve a measurable flowering response after a common temperature-insensitive timing point has been reached. The correspondence between CNL and NB_{max} under other conditions, together with the rhythmic nature of both processes and their apparently similar relationships with photoperiod duration make it likely that the timing component of both is the same circadian rhythm which, under natural conditions, is released at the transition to darkness. However, the nature of the additional dark reactions which appear to be involved in the induction of flowering await elucidation.

To summarize photoperiodic timekeeping in *Pharbitis nil*, it appears that the underlying time-measuring process is the circadian interval between a light-on signal (CT0) and an inducible, light-sensitive phase of the rhythm at CT15. This interval can vary in real time, as already discussed. A similar basis for photoperiodic time-measurement has also been proposed for the SDP, *Lemna paucicostata* (Oota 1983). Following photoperiods of several hours duration, the rhythm is set to a constant phase-point at the transition to darkness (at dusk) and the photoperiodic response of the plant (flowering or non-flowering) depends on whether the light-sensitive, inducible phase of the rhythm has been reached before the next exposure to light (dawn) occurs. Operationally, this mechanism measures a CNL which, if exceeded, results in floral induction. The light-on signal at dawn then re-phases the rhythm for the next cycle. While this specific external-coincidence hypothesis is largely based on results obtained with the two single-cycle plants *Xanthium* and especially *Pharbitis*, photoperiodic timekeeping in several other SDP has been shown to run from the end of the photoperiod and the model may, therefore, have more general validity.

7.3.6 Photoperception

It is evident that the measurement of the duration of light and darkness in photoperiodism is a property of the clock rather than of the photoreceptor. However, since the control of floral induction is a consequence of the timing of the light and dark periods, the interaction of the photoreceptor with the clock must be a crucial feature of the overall mechanism. Two actions of light have already been described; *i.e.* to control the phase of the photoperiodic rhythm and to inhibit flowering in the night-break response. Additionally, it is well documented that removal of the Pfr form of phytochrome early in the night can also inhibit flowering. The photoreceptor(s) thus has multiple actions in the photoperiodic control of flowering (Fig. 11).

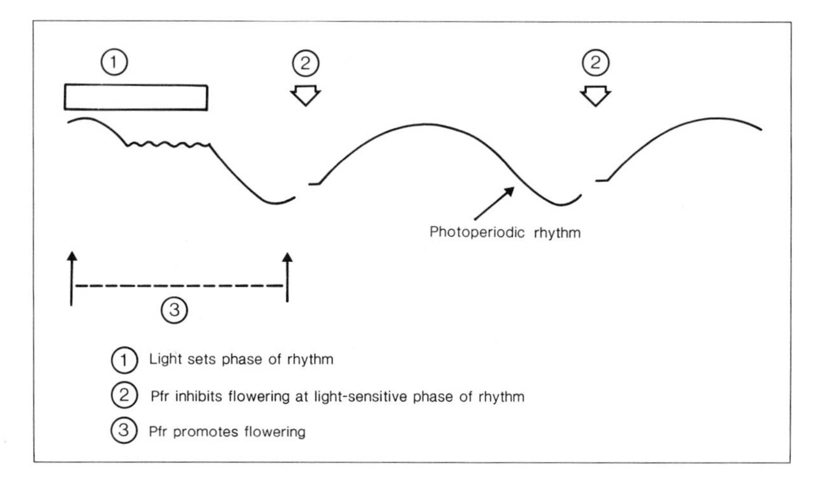

Figure 11. **Scheme showing multiple actions of light in the photoperiodic control of flowering in *Pharbitis nil*.**

7.3.6.1 The night-break reaction

The only phytochrome-mediated reaction unequivocally associated with the photoperiodic mechanism in SDP is the night-break effect to inhibit flowering. This R-FR reversible response was one of the first physiological processes shown to be under the control of phytochrome. In many SDP, a night-break is effective with only a relatively short exposure to light and reciprocity holds. Consequently action spectra can be constructed. The first action spectrum was determined in the 1940's at the Department of Agriculture Laboratory in Beltsville and showed maximum quantum effectiveness between 620 nm and 660 nm for the inhibition of flowering in *Xanthium;* subsequently, similar action spectra were obtained for other SDP (Vince-Prue 1983). Despite the shift in wavelength maximum from 660 nm (probably due to chlorophyll screening), the demonstration of reversibility by FR confirmed that phytochrome is the photoreceptor for the night-break effect. Thus, at certain times in the inductive dark period (determined by the circadian oscillator) the formation of Pfr prevents the flowering response. Although this action of Pfr can only be observed when floral initiation occurs, usually several weeks later, the available evidence shows that it is actually accomplished rapidly. For example, FR reversibility is lost within a few minutes in *Pharbitis* and *Kalanchoë* and within about 40 min in *Xanthium* (Vince-Prue 1983).

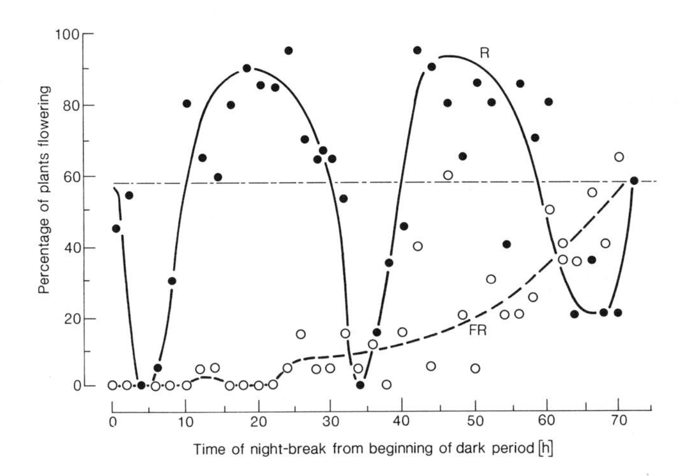

Figure 12. The flowering response of *Chenopodium rubrum* to periodic interruptions with red (R) or far-red light (FR). Plants were exposed to a 72-h dark period interrupted by 4 min R (●) or 10 s FR (○) at the time indicated on the abscissa. Curves were fitted by eye to the data of Cumming B.G., Hendricks S.B. and Borthwick H.A. (1965) *Can. J. Bot.* 43: 825-853.

Many experiments have been carried out on the night-break reaction in SDP, particularly on its timing. As already discussed, when a long dark period is scanned with a R night-break, a circadian rhythmicity is often observed in the response (Fig. 12), although sometimes this rhythmicity is seen only under certain conditions (Fig. 7). The first NB_{max} usually occurs at about the 7-9th h of darkness (depending on species) and, thereafter, inhibition occurs at circadian intervals. With shorter nights, such as would occur naturally, the NB_{max} also occurs at about the 7-9th h of darkness. Since (on the basis of the FR reversibility of the response) the inhibitory effect depends on the formation of Pfr, it is evident that there is insufficient Pfr present in the leaf to inhibit flowering after 7-9 h have elapsed in darkness. At the end of the photoperiod a considerable fraction of the total phytochrome would be in the Pfr form ($\approx$ 80% in R or fluorescent W as used in many experiments, and some 60% in natural daylight). The night-break response, therefore, seems to depend on phytochrome in which Pfr is relatively unstable. This point is returned to later in relation to the more stable Pfr involved in the Pfr-requiring reaction.

7.3.6.2 Phase-setting

The phase-setting action of light has been less clearly identified with phytochrome. One approach has been to determine the requirements for phase-shifting and, for such experiments, it is necessary to have some means of examining the timing response to the light given. In *Chenopodium*, the CNL has been used as an assay for the re-phasing action of light and it was found that the phase of the rhythm was advanced by a 6-h exposure to R, given after 9 h of darkness. However, under the conditions of this experiment, FR reversibility could not be demonstrated, although the effect of continuous light was partially satisfied by giving a 5-min pulse of R every 1.5 h suggesting the involvement of phytochrome (King and Cumming 1972). In dark-grown *Pharbitis* seedlings, however, a single 5-min pulse of R is sufficient to phase-shift the rhythm and reversibility by FR has been reported (Lumsden 1991). Thus, there is some direct evidence that the photoperiodic circadian oscillator is accessible to the action of phytochrome.

We have seen that, with relatively short light periods timing seems to begin from a light-on signal but, with longer photoperiods such as would occur naturally, timing begins from a light-off signal. Therefore, the function of light must not only be to initiate (or re-phase) the photoperiodic rhythm at dawn, but also subsequently to suspend it at a particular phase-point. In dark-grown *Pharbitis* seedlings, a saturating pulse of R was sufficient to start the flowering rhythm (Fig. 7); however, attempts to obtain FR reversibility were inconclusive because of the Pfr requirement for flowering under these conditions. Despite the failure to show FR reversibility, the sensitivity to very short exposures of R strongly suggests that phytochrome is the photoreceptor for the light-on signal which starts or re-phases the rhythm at dawn.

The identity and functioning of the photoreceptor involved in the apparent suspension of the rhythm in continuous light has hardly been investigated, although a few experimenters have approached the question by examining the light requirement for maintaining the rhythm in suspension. The usual protocol has been to give a photoperiod in W, which is long enough to suspend the rhythm, and to follow this with an experimental light period. When the photoperiod was extended with light of different wavelengths, the onset of dark timing was prevented most effectively (*i.e.* at lower irradiances) with R compared with other wavelengths (Takimoto 1967). Red was also more effective than R + FR, indicating the involvement of Pfr (Salisbury 1981). Another approach has been to extend the photoperiod with pulses of R. It was found that, over a period of 6 h, a brief pulse of R given every hour was sufficient to maintain the rhythm in suspension and prevent the onset of dark timing in *Pharbitis* (Vince-Prue and Lumsden 1987). However, the precise interpretation of this experimental result has proved difficult. A single pulse of R is sufficient to phase-shift the rhythm in *Pharbitis* and an analysis of the phase-shifting response indicated that the apparent effect of R pulses to maintain the rhythm in suspension could be explained by the effect of each pulse to phase-shift the already released rhythm. Thus, despite several indications that phytochrome is the photoreceptor involved in rhythm suspension, this still remains to be conclusively demonstrated.

Overall, the experimental results indicate that phytochrome is probably the photoreceptor for all actions of light to control the phase of the flowering rhythm, at least in *Pharbitis*. A single pulse of R can initiate a new rhythm, and advance or delay the phase of an existing rhythm (although this requires several hours in *Chenopodium* (King and Cumming 1972). Repeated pulses of light can apparently maintain the rhythm in suspension and generate a new light-off signal, although it is not certain that the effect of pulses is precisely the same as continuous light.

7.3.6.3 The dusk signal

A specific function of the photoreceptor is to sense the light-to-dark transition at dusk. Under natural conditions, the transition between day and night occurs during a period of twilight when the irradiance gradually decreases. There is also a change in the R:FR photon ratio, which usually decreases from the daylight value of about 1.1 to a value of about 0.7-0.8. Thus, either or both of these environmental changes might signal the end of the photoperiod and allow the flowering rhythm to be released. However, when compared with plants transferred to darkness from fluorescent light, which would establish a Pfr/P ratio of 0.8 [where P (the total phytochrome) is the sum of Pfr and the red-absorbing form, Pr], neither the CNL nor the time of NB_{max} was much affected when plants were exposed briefly to FR before transfer to darkness (Fig. 13, *Pharbitis*). In other words, the rapid photochemical reduction of Pfr level at the end of the photoperiod (as for example by the change in R:FR photon ratio during twilight) has little effect in advancing dark timekeeping compared with plants which have a high Pfr/P ratio when transferred to darkness. It has been shown that dark time-measurement occurs at the same rate as in darkness when plants are transferred to a sufficiently low irradiance of light, irrespective of its R:FR photon ratio (Fig. 13, *Xanthium;* Takimoto 1967; Lumsden and Vince-Prue 1984) and there is some evidence that a decrease in irradiance to a threshold value is the signal for the initiation of dark timing under natural radiation conditions (Salisbury 1981).

Although it is almost universally accepted that phytochrome is the photoreceptor for the perception of the light to dark transition at dusk, the supporting evidence is not entirely conclusive. Circumstantial evidence for the involvement of phytochrome is that the control system appears to function perfectly normally in R, with non-chlorophyll containing seedlings of *Pharbitis*. Perhaps

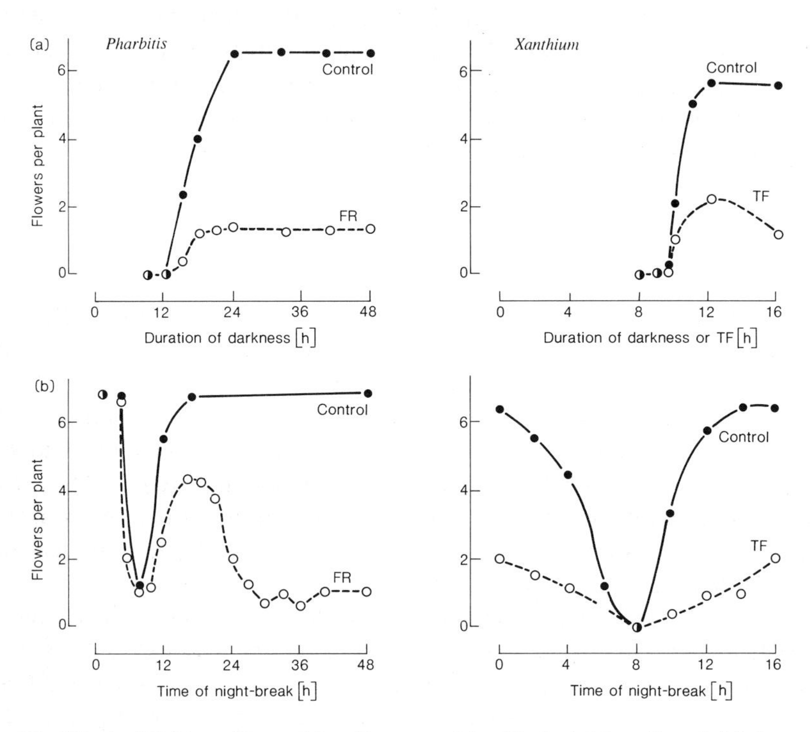

Figure 13. Effect of light quality and irradiance on (a) critical nightlength and (b) time of night-break sensitivity in ***Pharbitis nil*** (*left*) and ***Xanthium strumarium*** (*right*). ***Top left***: the effect of exposing plants to far-red light (FR) for 5 min at the end of the photoperiod. Control plants were transferred directly from white fluorescent light to darkness. ***Top right***: the effect of transferring plants to continuous tungsten-filament light at 0.1 W m^{-2} (TF). Control plants were transferred to darkness. In the night-break experiments (b), plants received an inductive period of 48 h dark (***Pharbitis***) or 16 h dark or TF (***Xanthium***): the 48 h and 16 h points respectively show the flowering values for plants receiving no night-break. Data for ***Pharbitis*** from Takimoto A and Hamner K.C. (1965) ***Plant Physiol.*** 40: 859-864, and for ***Xanthium*** from Salisbury F.B. and Ross C.W. (1969) ***Plant Physiology*** (1st edition), Wadsworth Publishing Company, Belmont.

the strongest evidence that a reduction in Pfr is indeed the 'light-off' signal comes from experiments with end-of-day FR. Although it appears to have no effect on timing in Fig. 13, the results of other more detailed experiments indicate that end-of-day FR does advance dark time-measurement by a small amount (Vince-Prue and Takimoto 1987). In *Pharbitis*, for example, the time of NB_{max} was advanced by 30-40 min by end-of-day FR compared with plants transferred to darkness from fluorescent light (Lumsden and Vince-Prue 1984).

Although few such detailed studies have been carried out, the results support the conclusion that the beginning of dark timekeeping is initiated by a reduction in Pfr; it may be that a particular threshold level of Pfr must be reached, but this has yet to be established.

If we are to account for the perception of the light-off signal entirely in terms of Pfr, it is necessary to assume that Pfr is lost rapidly through thermal reactions on transfer to darkness, since photochemical lowering of Pfr only advances timekeeping by some 30-40 minutes. The loss of Pfr must also become significant at low irradiances, since timekeeping appears to be initiated and proceed normally under these conditions (Fig. 13, *Xanthium*).

7.3.6.4 The Pfr-requiring reaction

One problem that is repeatedly encountered in studies of photoreceptor action in SDP and often complicates their interpretation is that there appears to be an additional component of the overall flowering response for which Pfr is required (Fig. 11). The need for Pfr can be demonstrated by giving an end-of-day exposure to FR, which has been shown to prevent flowering in many SDP when given under appropriate conditions. (Vince-Prue 1983). This effect is reversible by R showing that the Pfr form of phytochrome is required for flowering. Action spectra for the inhibition of flowering by FR and its re-induction by R confirm this conclusion.

There are two important questions regarding the Pfr-requiring reaction. Where is the reaction located and when does Pfr act? The location of the Pfr-requiring reaction for flowering in SDP has been determined only in *Pharbitis* and was shown to be in the cotyledons, *i.e.* in the organ where photoperiodic induction occurs (Knapp *et al.* 1986). With respect to the time of Pfr action, it has sometimes been suggested that the presence of Pfr is required during the inductive long night. This is clearly not so, as many experiments have shown that the Pfr-requirement for flowering can be satisfied during the photoperiod (Vince-Prue 1983). For example, when *Pharbitis* was given three 24-h cycles, flowering was not inhibited by end-of-day FR after photoperiods of 5 h or longer. Even with a single inductive cycle and a very long dark period (conditions which seem to favour the need for Pfr to be present during the dark period), flowering was not inhibited by end-of-day FR, provided that the photoperiod was sufficiently long (Vince-Prue and Lumsden 1987). Thus, whatever the events of the inductive night are, they do not seem to require the presence of Pfr, provided that some Pfr-requiring process has been completed beforehand. The duration of the Pfr requirement varies markedly and is influenced by experimental conditions. Results from a number of early experiments (see Vince-Prue 1983) indicate that Pfr is required for a shorter period in light than in darkness and, in the light, the requirement for Pfr continued longer when the photoperiod was given at a lower irradiance. These results suggest that the Pfr-requiring reaction may require substrates that are generated more rapidly in the light and are irradiance dependent.

Under natural conditions with longer photoperiods at high irradiances and repeated cycles, the Pfr requirement is satisfied in the light and does not continue into the dark period. However, if it has not been completed during the preceding photoperiod, the Pfr-dependent process can continue during the inductive night and in some cases may take a very long time for completion. One of the most important early experiments was with *Chenopodium rubrum,* where FR given even after 50 or 60 h of darkness substantially decreased the flowering response (Fig. 12). In *Pharbitis,* FR completely prevented flowering when given up to 12 h into the dark period following an 11-h photoperiod, and up to 18 h following a 10-min photoperiod (Vince-Prue 1983).

The Pfr-requiring reaction does not appear to be associated with a circadian rhythm. In general, the inhibitory effect of FR is greatest at the beginning of the night and gradually decreases with time. This contrasts markedly with the inhibitory effect of a night-break which shows a circadian rhythm. This was elegantly demonstrated in the *Chenopodium* experiment already cited (Fig. 12) and subsequently reported for other SDP, including *Pharbitis* (King *et al.* 1982). Moreover, although flowering is strongly depressed, removing Pfr at the beginning of an inductive dark period does not change the time of NB_{max} nor the CNL (Fig. 13). Thus the Pfr-requiring process does not seem to be part of the circadian time-measuring process of photoperiodism in SDP, although it is clearly necessary for floral induction.

These experiments on the inhibition of flowering by FR have shown that the Pfr associated with this reaction is extremely stable in darkness. For example, in *Chenopodium,* it was apparently still present and active after 60 h in darkness at physiological temperatures (Fig. 12). This contrasts with the more unstable Pfr involved in the night-break response and dusk signal. A consideration of Fig. 12 shows that, at certain times, flowering can be inhibited either by removing Pfr (active Pfr is present) or by the formation of Pfr (active Pfr is absent). This paradox has been recognized for many years and, because of it, several workers have advanced the theory that two different pools of phytochrome are involved in the control of flowering and that these have different kinetics for the loss of Pfr by thermal reactions (Vince-Prue 1983; Takimoto and Saji 1984). The simpler hypothesis that a single phytochrome pool is involved with different thresholds for the Pfr-promotion and Pfr-inhibition responses seems to be eliminated by the results of experiments with dark-grown seedlings of *Pharbitis*, where all wavelengths between 660 nm and 750 nm were inhibitory to flowering at certain times (Vince-Prue 1983). Moreover, the FR inhibition was readily reversed by a subsequent exposure to R at any time throughout the dark period *except* at the time of sensitivity to a night-break (Saji *et al.* 1983). A study with *Lemna* (Lumsden *et al.* 1987) has shown that, under conditions where the Pfr-requiring reaction continued during the dark period, an action spectrum for the inhibition of flowering at the beginning of the night corresponded to the absorption spectrum of Pfr and was similar to others at that time. However, at the time of NB_{max}, the action spectrum extended throughout R and FR, with the FR tail beyond 720 nm corresponding almost exactly with the action spectrum for the inhibition of flowering at the beginning of the night. It was concluded that both removal of Pfr (the Pfr-requiring reaction) and the formation of Pfr (the night-break reaction) inhibited flowering, supporting the concept that two pools of phytochrome with differently stable Pfr are involved in the control of flowering in *Lemna*. Using similar techniques, the same conclusion was reached for *Pharbitis* (Saji *et al.* 1983). However, here, the action spectrum at the time of NB_{max} corresponded to the absorption spectrum of Pr and showed no FR tail.

7.3.6.5 Do different phytochromes control flowering in SDP?

A brief summary at this stage may be helpful. The night-break inhibition of flowering in SDP is clearly dependent on Pfr; thus it seems that the Pfr present at the end of the photoperiod declines to a level which is not effective in the night-break reaction before the light-sensitive phase of the photoperiodic rhythm is reached (some 8-9 h after transfer to darkness in many plants). The Pfr-requiring reaction, which is essential for flowering but may not be specific to the photoperiodic process, is also clearly dependent on Pfr. In this case, however, the Pfr is highly stable in darkness. The dusk signal, which initiates dark time-measurement, is usually attributed to a thermochemical reduction in Pfr, although this has not unequivocally been proved. If Pfr is involved in the dusk signal, it must be highly unstable in darkness since, at physiological temperatures, dark time-measurement appears to begin within less than an hour after transfer to darkness. Thus the phytochrome that is associated directly with the photoperiodic mechanism (dusk signal perception and night-break inhibition) and that controlling the Pfr-requiring reaction must be different at least in the stability of Pfr.

7.3.6.5.1 Phytochrome destruction. Based on immunological studies, it is now well established that the phytochrome system of higher plants consists of at least two populations of photoreversible chromoproteins with the same chromophore but different protein moieties. One population (type I), which is relatively labile as Pfr, is synthesized in darkness and decreases rapidly in the light. The second population (type II) in which Pfr is much more stable, remains at a relatively constant level in light and darkness. Recent studies of the phytochrome proteins have shown that two stable type II phytochromes can be distinguished immunochemically (Pratt *et al.* 1991).

Phytochromes I and II differ in the stability of Pfr (Jordan *et al.* 1986). Following exposure to light, phytochrome I undergoes rapid destruction from Pfr [half-life ($t_{1/2}$) = 1-2 h] with proteolytic degradation of the molecule. In contrast, phytochrome II is only slowly degraded from Pfr ($t_{1/2}$ = 7-8 h). Although the two types of phytochrome have differentially stable pools of Pfr, there is no direct evidence concerning the involvement of either type in the photoperiodic responses of SDP. Nevertheless, the stability of Pfr in the Pfr-requiring reaction (especially in *Chenopodium,* where Pfr apparently remains active for 60 h or more in darkness (Fig. 12), strongly suggests that a type II phytochrome is involved. For the night-break and dusk signal the situation is ambiguous. The destruction of type I Pfr is sufficiently rapid to account for loss of activity by the time of night-beak sensitivity. The dusk signal appears to occur within 30 min or so after transfer to darkness and certainly some Pfr decay of type I phytochrome would have occurred within this time; since we do not know at what threshold time-measurement is coupled (there is even the possibility that the signal might require only a *decrease* in Pfr), the involvement of type I phytochrome is not excluded simply on the basis of the kinetics of Pfr destruction. Probably the strongest argument against the involvement of type I phytochrome is the fact that the photoperiodic mechanism continues to operate normally under conditions when little or no unstable type I phytochrome is present (*e.g.* after several days in continuous fluorescent W in *Pharbitis*, where down regulation is very strong and leads to very low levels of phytochrome I in the light). Operation of the photoperiodic mechanism through Pfr-destruction would require the re-synthesis of sufficient Pr to maintain sensitivity to the night-break and dawn signals.

7.3.6.5.2 Phytochrome reversion. The relative stabilities of type I and type II phytochromes concern the extent to which their Pfr is subject to destruction, with consequent loss of the protein moiety. Another way in which the level of Pfr may be reduced in darkness is through the process known as reversion, in which Pfr is thermochemically converted to Pr without any loss of total phytochrome; the need for re-synthesis to maintain sensitivity is thus obviated. Partly for this reason, it has often been suggested that reversion of Pfr might account for the dusk signal and it could also account for the need for light in the night-break reaction. One problem has been that reversion has not been observed *in vivo* either in the Gramineae or the Centrospermae. However, most of the measurements were carried out on dark-grown seedlings, in which type I phytochrome predominates.

Compared with Pfr destruction, dark reversion may be rapid at physiological temperatures with, for example, a $t_{1/2}$ of approximately 8 min being recorded for etiolated seedlings of *Sinapis alba* (Jordan *et al.* 1986). Rapid dark reversion has also been reported for light-grown seedlings of *Pharbitis.* After 24 h in W at 18°C, the total phytochrome pool appeared to be stable in darkness over a period of at least 5 h at 27°C. However, all of the Pfr reverted to Pr with $t_{1/2}$ of 15 min (Vince-Prue *et al.* 1978). This is consistent with measurements made on herbicide-treated seedlings where, after 36 h W (3 × 12 h exposures), $t_{1/2}$ for reversion was 19 min at 22°C (Rombach 1986). Dark-destruction kinetics indicated the presence of a stable phytochrome in both herbicide-treated plants given 24 h light and in dark-grown plants following a single R pulse.

Rombach (1986) has shown that, in cotyledons of *Pharbitis nil,* a stable Pfr and a pool of Pfr which undergoes rapid reversion are present simultaneously and has suggested that these may be equated with the two postulated Pfr pools in photoperiodism. The identity of the phytochrome species in these two pools is unknown. It is impossible to say whether the same phytochrome is present in both the stable and unstable pools, perhaps with differences in the cellular environment, or if they are different molecules. At least two different type II phytochromes have been recognized and one of these is characterized by an extended N-terminal domain. It is known that, for type I phytochrome, the N-terminal domain is crucial for the thermal stability of Pfr and its deletion in genes used for transformation leads to phytochrome that undergoes rapid dark reversion in the resultant transformed plants. It is possible, therefore, that the modification to the N-terminus in one of the type II phytochromes could alter its dark reversion characteristics compared with the other. Thus, different type II phytochromes are possible candidates for the processes where fast reversion (photoperiodic timekeeping) and slow reversion (Pfr-requiring reaction) appear to be important *in vivo* factors (Thomas 1991). The availability of appropriate mutants might help to resolve the problem.

The possibility that the two postulated pools might consist of the same type of phytochrome, but with different reversion characteristics is by no means excluded. For example, the rate of Pfr reversion of type I phytochrome *in vitro* can be differentially altered by monoclonal antibodies which react with different epitopes on the protein (Lumsden *et al.* 1985). Although the relationship between *in vitro* reversion and events *in vivo* is unknown, the fact that reversion may be affected in this way suggests that differently conjugated pools *in vivo* might show different reversion characteristics. Another pertinent observation concerns the behaviour of phytochrome dimers. A re-examination of the dark reversion data in *Sinapis* has shown a correspondence between the pool size undergoing dark reversion and the calculated pool size of the heterodimer PfrPr (Brockmann *et al.* 1987). The homodimer PfrPfr, in contrast, appears to be relatively stable. These differences in their dark reversion rates led to the suggestion (Thomas 1991) that different dimers of type II phytochrome might account for the two physiological pools in the control of flowering, with the night-break and dusk signals being mediated by the fast-reverting heterodimer and the Pfr-requiring reaction being mediated by the stable homodimer.

Whatever the identity of the phytochrome molecules undergoing slow and fast reversion as assayed spectrophotometrically in *Pharbitis* (Rombach 1986), there is as yet no direct evidence which links these two 'spectrophotometric' pools with the two postulated physiological pools in photoperiodism. The best correlation appears to be with the dusk signal. Physiological experiments of different kinds indicate that dark timing appears to be coupled within 30-60 min of transfer to darkness (Vince-Prue 1983); by this time, dark reversion would be essentially complete (Vince-Prue *et al.* 1978; Rombach 1986). As has been pointed out, however, the dusk and night-break signals have not been unequivocally established as being associated with a reduction in the amount of Pfr. Some time ago, the concept of loss of sensitivity to a stable Pfr was proposed to explain the observation that repeated exposures to R were needed to maintain chloroplast movement, even though Pfr was still present in the tissue. It has been pointed out that the dusk signal and night-break reaction could be explained in the same way if, for example, the action of Pfr in the perception of the light signals requires association with a receptor and this Pfr-receptor complex is highly unstable (Thomas and Vince-Prue 1987). The slight acceleration of dark timekeeping by end-of-day FR, which seems to indicate that a reduction in Pfr itself is the controlling event, would be explained by the immediate loss of effectiveness of the phytochrome-receptor complex when the Pfr component is removed photochemically. There is, however, no direct evidence for the existence of such complexes.

It is evident that much is yet uncertain regarding the identity of the phytochrome(s) involved in photoperiodism and that there is even uncertainty about the precise nature of some of the photoperiodic signals. Is the dusk signal due to a reduction in Pfr and, if so, does this occur through reversion? Why does rhythm 'suspension' require continuous light? Although physiological experiments have clearly established that there are differences in the apparent stability of the Pfr involved in the perception of photoperiodic signals and in the Pfr-requiring reaction, we are still a long way from understanding how these differences arise. The most likely suggestion, at the present time, is that a sub-population of type II phytochrome undergoes dark reversion. In the future, the availability of new mutants or transgenic plants may help to determine whether or not this sub-population is a distinct molecular species. Few useful phytochrome mutants are yet available for SDP but, in *Sorghum*, the mutant ma_3^R has been found recently to lack a type II phytochrome; this mutant is also insensitive to photoperiod (Deitzer 1993).

7.3.7 Photoperiodic induction under long photoperiods

Long-day plants flower when the daylength *exceeds* a certain critical value in a 24-h cycle (Fig. 4). The scheme outlined for SDP would account for the responses of LDP if the response groups were mirror images of a common mechanism. This would mean that, in LDP, the photoperiod would set the phase of a circadian rhythm and flowering would only occur if light were given at the light-sensitive phase of the rhythm, resulting in induction either with long-day/short-night cycles or with a night-break given at the appropriate time (the inducible phase) in a long dark period. However, a problem for understanding the controlling mechanism for LDP is that many of them show a negligible response to a brief night-break which would be expected to saturate the photoconversion of Pr to Pfr. In contrast, they show a strong flowering response to a long daily exposure to light (Vince-Prue 1983). A distinction between plants in which flowering appears to be controlled primarily by dark processes which can be prevented by a short night-break at the appropriate time and those in which a long light period appears to be of major importance has been made by referring to them as *dark-dominant* and *light-dominant* response types, respectively. They correspond largely, but not exactly to the classification of species as SDP and LDP. However, there are a few LDP (*e.g. Fuchsia hybrida* cv. Lord Byron) where flowering is responsive to a brief night-break as in SDP. There are also a few SDP (*e.g.* strawberry) which respond poorly or not at all to a brief night-break and where the *inhibition* of flowering requires exposures to long photoperiods.

7.3.7.1 Responses to light quantity

Many light-dominant plants show a more or less quantitative relationship between the duration and/or the irradiance of exposure to light and the magnitude of the flowering response. However, there are considerable differences between species in the way in which they respond to light (Vince-Prue 1983). In some cases, such as carnation, the flowering response is a function of the light integral, irrespective of whether this is given intermittently or continuously during a 16-h day-extension following an 8-h main light period in daylight. In carnation, it is necessary to irradiate plants over the entire 16-h period for maximum effect whereas, in *Brassica campestris,* a reduction in the duration of continuous light can be partly compensated for by an increase in irradiance. A third type of response is seen in *Lolium temulentum,* where only the duration of light seems important and, for a day-extension with incandescent light, the response was saturated at 1.0 W m^{-2}.

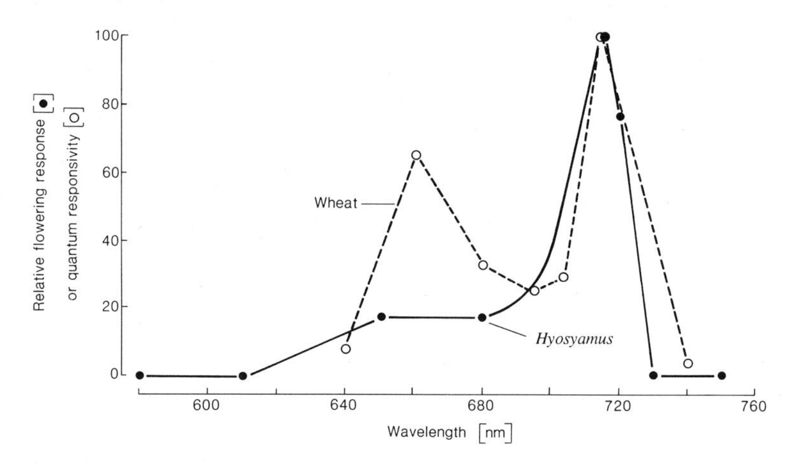

Figure 14. Action spectra for the promotion of flowering in long-day plants with long daily exposures to light. *Hyoscyamus niger* given light (66 J m^{-2} s^{-1}) during the middle 8 h of a 16-h dark period. Wheat (*Triticum aestivum*) was given light at different fluence rates throughout the 16-h period. Data for *Hyoscyamus* from Schneider M.J., Borthwick H.A. and Hendricks S.B. (1967) *Amer. J. Bot.* 54: 1241-1249. Data for wheat from Carr-Smith H.D, Johnson C.B and Thomas B. (1989) *Planta* 179: 428-432.

7.3.7.2 Responses to light quality

With respect to light quality, the responses of light-dominant plants show three features which appear to be characteristic. These are: (i) an action spectrum with maximum effect near 720 nm; (ii) increased flowering when FR is added to a day-extension with R and (iii) a change in response to FR and R during the course of each daily cycle.

In several early studies with LDP, the action spectrum for a night-break which promoted flowering was found to be the same as the action spectrum for an inhibitory night-break in SDP, indicating phytochrome as the photoreceptor in both response groups. Reversibility by FR was also demonstrated. However, brief night-breaks are often rather ineffective in LDP and may control flowering only under certain conditions. For example, the first action spectra for barley and *Hyoscyamus* were obtained using 11.5- or 12-h dark periods; with 16-h dark periods, such brief night-breaks were ineffective. With longer night-breaks, the flowering response increases and the action spectra commonly show a maximum near 720 nm (Fig. 14). A maximum near 720 nm is characteristic of the *high irradiance response* (HIR) observed in dark-grown seedlings under conditions of prolonged irradiation. As with the flowering response in many

LDP, the magnitude of the HIR in dark-grown seedlings depends on both the duration and the irradiance of the exposure to light. It has been proposed that the HIR in etiolated tissues deviates from the Pr-absorption spectrum as a consequence of Pfr instability. If this interpretation is correct it would implicate type I phytochrome in the HIR response. However, from many recent experiments we would expect that the bulk of the phytochrome in light-grown plants responding to long exposures to light would be stable, type II phytochrome, for which the action maximum should be closer to 660 nm. So far, only one experiment has attempted to address this problem directly. Using monoclonal antibodies specific for type I phytochrome, it has been found that there are measurable amounts of type I phytochrome in light-grown seedlings of wheat and that the amount increases in FR, which is favourable for flowering in this LDP (Carr-Smith *et al.* 1993). Photoprotection at higher irradiances might increase the pool of type I phytochrome. Consequently, it was suggested that this phytochrome may operate in the photoperiodic control of flowering in wheat, accounting for both the anomalous action spectrum and the irradiance dependency (Thomas 1991).

Light-dominant plants frequently show a much greater flowering response when FR is added to R or W during a long photoperiod. Unlike SDP, where flowering is inhibited by any mixture of R and FR which would establish a sufficiently high level of Pfr, the maximum flowering response in light-dominant plants occurs at intermediate R:FR ratios (close to those in natural daylight) and decreases sharply as the content of R is increased (Fig. 15). However, the promotion of flowering by FR varies with the time of day. For example, in a 16-h photoperiod, FR strongly promoted flowering during the second half of the day and had little, or sometimes no effect during the early part of the day (Vince-Prue 1983). When 8 h R + FR was followed by 8 h R, the resulting 16-h photoperiod was often no more effective than an 8-h day, even though the 8-h dark period was substantially shorter than the CNL for the species concerned, when grown under natural conditions. The results confirm the importance of the long photoperiod and demonstrate that the duration of darkness is not the only controlling factor in light-dominant plants.

7.3.7.3 Timekeeping

There is a good deal of general evidence of the kind already outlined for SDP that a circadian clock is involved in the perception of light and/or dark duration in LDP, although fewer species have been examined. Cycle-length experiments in *Hyoscyamus* and night-break experiments in *Lolium* (Fig. 16a) and *Hyoscyamus* have both revealed a circadian oscillation in the flowering response to R or W light and the underlying rhythm seems to be out of phase with that in SDP (as would be predicted from Bünning's original hypothesis).

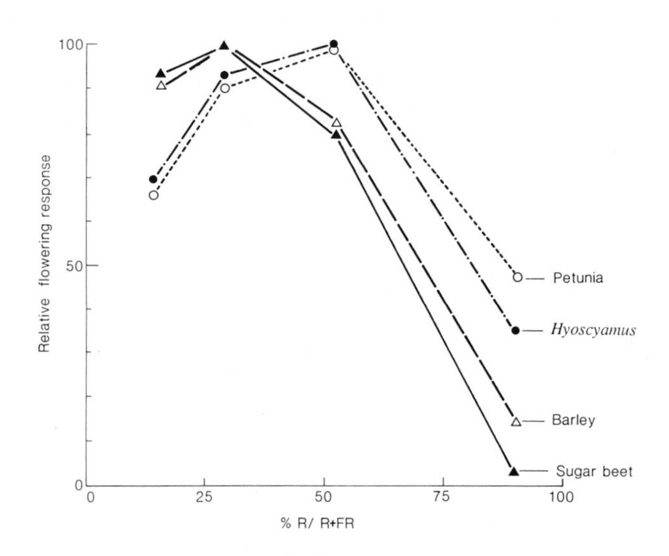

Figure 15. Effect of the red (R):far-red (FR) ratio of light given as an 8-h day extension on the flowering response in long-day plants. The experimental treatments were given daily following an 8-h period in daylight. From Lane H.C., Cathey H.M. and Evans L.T. (1965) *Amer. J. Bot.* 52: 1006-1014.

For example, a light exposure promoted flowering in the LDP *Lolium* (Fig. 16) at times when it inhibited flowering in the SDP *Glycine* (Fig. 6), while cycle lengths of 24 h and 48 h were most inhibitory to flowering in the LDP, *Hyoscyamus* but were optimal for flowering in *Glycine* (Fig. 6).

A rhythm in the promotion of flowering by FR added to R or fluorescent W has also been observed in two species of LDP and, in both, the rhythm clearly continues in the light for at least 48 h (Fig. 16b). In *Hordeum*, it has been shown that the phase of the rhythmic flowering response to FR can be shifted by an exposure to FR (Deitzer *et al.* 1982). In this, there seems to be a parallel between the behaviour of the rhythms in SDP and LDP. In the former, a rhythm of flowering response (inhibition) to R during darkness can be phase-shifted by R while, in LDP, a rhythm of response (promotion) to FR during the light can be phase-shifted by FR. In both cases, the rhythmic response is accompanied by large changes in the sensitivity to light. However, the significance, if any, of these observations for the circadian clock control of flowering in LDP is unknown.

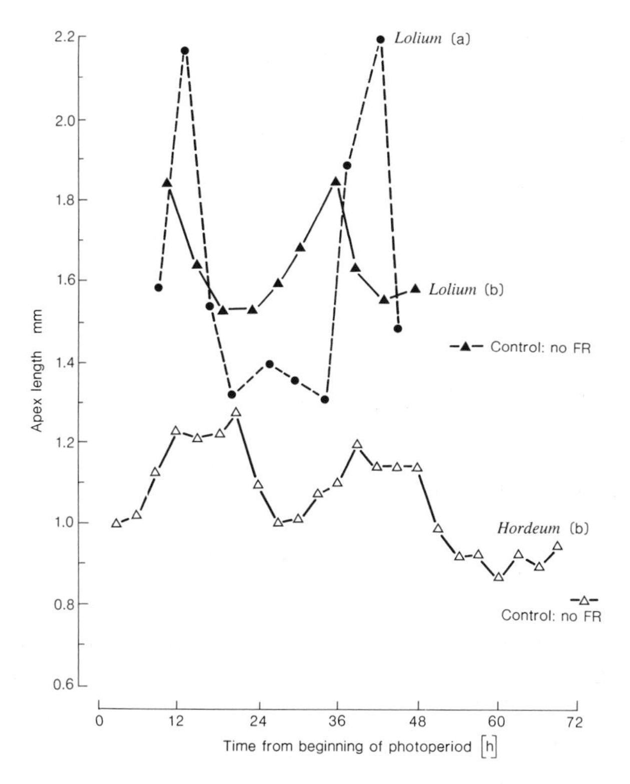

Figure 16. Rhythmic responses to red (R) and far-red light (FR) in long-day plants. (a) Effect on flowering in *Lolium temulentum* of a 4-h R night-break given at different times during a 42-h dark period. (b) Effect on flowering of 4 or 6 h FR added at different times to 40 h R (*Lolium*) or 72 h white fluorescent light (*Hordeum*). Data for *Lolium* from Vince-Prue D. (1975); for *Hordeum* from Deitzer G.F., Hayes R. and Jabben M. (1979) *Plant Physiol.* 64: 1015-1021.

7.3.7.4 Possible mechanisms

It is evident that the responses of LDP are much less well understood than those of SDP, both in terms of the identity and action of the photoreceptor and in the way in which the observed circadian rhythms may operate to control the flowering response under natural conditions. Early experiments with non-24-h cycles and the responsiveness to brief R night-breaks under certain conditions have demonstrated that LDP measure a critical nightlength which, if exceeded, prevents flowering. The night-break rhythm experiments indicate that the mechanism for this time-measuring process may be similar to that in SDP, with flowering dependent on whether or not light is given at an inducible phase of the rhythm. Results with *Lemna*, in which a light-on rhythm results in flowering only when a second pulse of light is given at a particular circadian time (Oota and Nakashima 1978) support this conclusion. However, the way in which this rhythm might measure a critical duration of darkness under natural conditions remains to be determined. It is possible that this occurs in the same way as in SDP, with the rhythm being released at dusk to run from a phase point that is established during the preceding photoperiod in continuous light: attempts by the author (unpublished data) to investigate whether this pattern of entrainment operates in *Lolium* were, however, unsuccessful and yielded ambiguous results.

The requirement for long exposures to be effective as a night-break could be explained if Pfr is highly unstable in darkness and is required over an extended portion of the rhythm. Although the effectiveness of intermittent lighting schedules supports this suggestion, other results indicate that this is not the explanation. For example, in *Lolium temulentum,* intermittent lighting was found to be much more effective with pulses of R and FR than with pulses of R alone. Other characteristic features of LDP such as the anomalous action spectra and the changing responsiveness to FR also remain to be explained. The effectiveness of added FR on flowering might arise from an increase in total type I phytochrome, as suggested by the immunoassay results obtained with wheat (Carr-Smith *et al.* 1993). However, this explanation seems less likely for barley where, even after 48 h in continuous fluorescent W when little type I phytochrome would be expected, the addition of FR for only a few hours strongly promoted flowering (Fig. 16). Another possible explanation is that the addition of FR to fluorescent W, while decreasing Pfr/P would also lead to an increase in the PfrPr heterodimer. A changing sensitivity to added FR would then be observed if there were a circadian rhythm of response to the heterodimer, possibly because of the circadian regulation of a specific receptor (Thomas 1991).

Although *Lolium temulentum* resembles barley in showing a rhythmic response to FR (Fig. 16b) it seems unlikely that this can be a consequence of heterodimer formation. A detailed series of experiments with *Lolium* showed that, at the time when FR promoted flowering most strongly, a period of darkness was also highly effective (Holland and Vince 1971). Moreover, the promotion of flowering obtained with several hours of darkness was decreased by a brief pulse of R given near the middle and increased by a brief exposure to FR given at the beginning: reversibility was obtained in both cases, demonstrating that the promotion of flowering by FR was largely, if not entirely due to a reduction in Pfr and not to heterodimer formation. When taken together with the rhythmic promotion of flowering by R (Fig. 16a), these results suggest that, in *Lolium,* Pfr inhibits flowering at some phases of a circadian rhythm and promotes it at others (a concept very close to Bünning's skotophile and photophile phases). The wavelength maximum at 720 nm and the optimal flowering response at a particular R:FR ratio might then result from establishing a Pfr/P value below the inhibitory threshold but sufficiently high to allow the promoting reaction to take place.

Despite several suggestions, it is still not known why FR is required for optimal flowering in LDP. It is also not understood how the observed rhythms in sensitivity to R and FR relate to photoperiodic timekeeping in these plants. Whatever the mechanism, however (and it may not be the same in all species), it is clear that LDP are well adapted to flowering under long days at the R:FR ratio present in natural daylight (Fig. 15).

The increasing number of phytochrome mutants for the LDP *Arabidopsis thaliana* may help to identify which type(s) of phytochrome is associated with the photoperiodic control of flowering in this response group. As discussed earlier, direct measurements in wheat have indicated that type I phytochrome may be important, accounting for the increased flowering response to low R:FR (Carr-Smith *et al.* 1993). In contrast, studies with mutants implicate type II phytochrome in the enhancement of flowering by added FR since, in *Arabidopsis hy3* (which lacks type II phytochrome) flowering in continuous light is essentially independent of the R:FR ratio. However, from experiments with *hy3* and late-flowering/*hy3* double mutants, it was concluded that both phytochrome II and another phytochrome species are responsible for the flowering response to changes in R:FR ratio (Whitelam 1993). The identity of this phytochrome is not known and the possible involvement of type I phytochrome (as indicated for wheat) cannot be excluded. *Arabidopsis* mutants lacking type I phytochrome which are now available may resolve the situation (Chapters 4.2 and 7.1).

Mutants have already provided some insight into which phytochrome (s) may be associated with the flowering response in LDP and this approach is likely to prove increasingly valuable in the future as further well-defined phytochrome mutants become available and the physiology of the photoperiodic response in LDP is better characterized. A detailed review of the flowering behaviour of mutants of several photoperiodically-sensitive species (mostly LDP) has recently been made (Deitzer 1993).

7.3.8 The action of phytochrome in photoperiodism

The photoperiodic process takes place in the leaves, even though the response is expressed elsewhere: for example, at the shoot apices for the initiation of flowering (Vince-Prue 1983). This was originally established by presenting the inductive photoperiod to the leaves while the apex remained in non-inductive daylengths and it has been confirmed for the major photoperiodic response groups for flowering, as well as for the induction of dormancy, bulbing and tuber formation. Exposure of only a single leaf is effective in some plants and, where several repeated cycles are required to effect flowering, it appears that these must be given to the same leaf. Photoperiodic events can also be shown to occur in leaves which have been separated from the rest of the plant. A leaf of the SDP, *Perilla,* for example, can be detached from the vegetative parent plant, exposed to short photoperiods and grafted to a vegetative receptor plant, which subsequently flowers in response to the signal from the grafted leaf. Thus, the process of photoperiodism appears to be concerned only with events taking place in the leaf, although processes occurring in other parts of the plant may modify the final response. The events which occur in the leaf do not involve morphological changes and are referred to as *induction;* when these are under the control of daylength we speak about photoperiodic induction (of flowering, *etc.*). Since daylength perception and floral induction are closely linked within the leaf and are largely independent of the final response at the apex, the processes involved with perception and induction can be considered separately from those involved with the translation of induction in the leaf into the final observed response. Only the former will be considered here.

Much evidence points to the conclusion that the timekeeping component of photoperiodism involves a circadian rhythm of sensitivity to light. For SDP, a major action of light is to entrain the rhythm in such a way that it is released at dusk to form the basis of measuring the CNL. It is assumed that this flowering rhythm is coupled to an underlying circadian oscillator and that it is this pacemaker which responds to the phase-controlling light signals (Fig. 2). Although there are formal descriptions of phase-control in terms of light- and dark-limit cycles, neither the mechanism of the circadian oscillator nor the molecular basis for the phase controlling action of light are known (Lumsden 1991).

The most immediately obvious action of light in photoperiodism is the night-break response. Because a similar pulse of light can also shift the phase of the photoperiodic rhythm, it is important that these two effects of light are clearly discriminated. This is not particularly difficult since the night-break effect occurs only at certain times, while a phase-shift can be obtained by light given at most times throughout the inductive dark period. A biochemical response obtained *only* at NB_{max} is, therefore, likely to be peculiar to the night-break effect. The two effects can also be differentiated in some cases by their fluence-response characteristics (Fig. 8).

The nature of the circadian rhythm that underlies photoperiodic timekeeping has not yet been identified. One possibility is that the system responds differently to light at different times because of a rhythmic change in some cellular component or function. An alternative to the proposal that the system responds differently to phytochrome at different phases of the rhythm is the suggestion that the circadian oscillator acts to change the properties of phytochrome and, in this way, affects its ability to function as a transducer. Some evidence for a change in phytochrome properties has been obtained in dark-grown seedlings of

Pharbitis, where changes in the flowering response to a R pulse were accompanied by apparent variations in the quantum yield for photoconversion between Pr and Pfr (King *et al.* 1982). However, the changes in quantum yield were quite small whereas the rhythmic flowering response was accompanied by changes of 20-40 times in the photon irradiance required for half maximum response to a R pulse; these are probably much too great to be accounted for solely in terms of photoreceptor properties. However, it is evident that changes in sensitivity do occur, *i.e.* considerably more light is required to saturate the response at some times than others, and these need to be explained (Thomas 1991).

7.3.8.1 Changes in gene expression

There have been relatively few attempts to study the molecular changes that are involved in the switch to the induced state in the leaves, or to study the action of Pfr in preventing (SDP) or promoting (LDP) this at the time of night-break sensitivity. In some SDP (*e.g. Perilla*) the induced state is extremely long-lived. Such a persistent developmental change would be expected to involve the expression of new genetic information and much recent work on floral induction has concentrated on this possibility. The night-break would then act to down-regulate (SDP) or up-regulate (LDP) the appropriate floral genes in the leaf.

Claims have been made for the formation of different proteins in leaves exposed to different photoperiods in both LDP and SDP. Following induction, changes in the products of *in vitro* translation of RNA from the leaves were observed in *Hyoscyamus* (Warm 1984). Unfortunately, the induction treatment consisted of 58 h of continuous W and, without other controls (for example, can the same changes be detected in non-vernalized plants which would not flower?), it is not possible to ascribe these changes with certainty to floral induction rather than to a non-specific effect of the long light treatment. One species in which effects of the immediate daylength environment can be separated from induction is the SDP, *Perilla,* where the induced state in the leaf persists even when plants are returned to non-inductive LD. In this case, no consistent qualitative or quantitative differences were found between translation products from induced and non-induced leaves when they were in the same daylength environment (Kimpel and Doss 1989). In *Pharbitis,* a single mRNA encoding a 28 kD polypeptide was reported to be quantitatively increased in SD, relative to plants receiving a night-break. However, such changes were not observed by other workers using similar treatments. Screening of cDNA libraries from *Pharbitis* seedlings exposed to a single LD or SD cycle has revealed only subtle differences; although there were a few differentially expressed genes, none were regulated qualitatively and the changes were not dramatic when compared with those observed in other developmental processes (O'Neill 1989, 1992). Thus, at present, it is not possible to say whether or not photoperiodic floral induction in the leaf involves changes in gene expression under phytochrome control.

7.3.8.2 Changes at the membrane level

Although it is well established that phytochrome regulates the expression of many genes, an alternative proposal for its action in some morphogenetic processes is that it modifies the properties of cellular membranes, perhaps leading to the release of a second messenger. Studies on turgor changes associated with the rhythmic sleep movements of leaves (Morse *et al.* 1987), or with the (non-rhythmic) light-induced swelling of wheat protoplasts (Bossen *et al.* 1990) have indicated that these involve alterations in inositolphospholipid turnover and intracellular calcium levels in a manner analogous with certain sensory transduction pathways in animals. Since Pfr action in a night-break can be extremely rapid (*e.g.* in *Pharbitis,* where the action of a R night-break cannot be reversed after < 2 min), it has been proposed that the control of flowering by phytochrome may also be dependent on a modification of membrane properties. It is known that some circadian rhythms, such as the sleep movements of leaves, involve periodic changes in the properties of membranes in the specialized cells of the pulvinus. Thus a possible working hypothesis is that circadian changes in the properties of a particular cellular membrane in the leaf allow induction to occur at the inducible phase and that the action of Pfr prevents this from taking place in SDP, but is required in LDP.

In an early experiment, it was found that daily transfer to distilled water inhibited flowering in the SDP, *Lemna paucicostata.* The distilled water transfer was most inhibitory 1-2 h after the time of NB_{max} and was partially overcome by adding Ca^{2+} ions (Halaban and Hillman 1970). Certain membranes thus apparently leak ions at certain times and this is somehow associated with the flowering response.

The possible involvement of calcium has also been investigated in *Pharbitis* (Friedman *et al.* 1989). The application of the calcium chelator, EGTA, to the cotyledons depressed flowering and this was reversed when calcium was applied within 30 min. In contrast, the calcium ionophore A23187 promoted flowering. In both cases, the greatest response was obtained at the beginning of, or before the inductive dark period and there was essentially no effect at the time of NB_{max}. If the action of phytochrome in the night-break reaction is to activate Ca^{2+} channels, as proposed for protoplast swelling (Bossen *et al.* 1988), the application of calcium ionophores would be expected to mimic R and *inhibit* rather than promote flowering when applied at, or before, the time of NB_{max} (assuming that uptake and transport to the active site occurs fairly rapidly). However, using the ionophore Bay K-8644 (which is expected to penetrate more easily than A23187) no significant inhibition of flowering in *Pharbitis* was obtained at any time before or during an inductive dark period (Tretyn *et al.* 1990). Neither did calcium-channel antagonists promote flowering when applied to *Pharbitis* seedlings at the time of NB_{max}, with or without a R night-break (Tretyn *et al.* 1990; D. Vince-Prue unpublished data). Thus, there seems to be little evidence from these experiments that phytochrome acts to inhibit flowering in SDP by activating calcium channels. An alternative possibility is that the process of floral induction itself is associated with the (rhythmic?) activation of calcium channels. In this case, it would be expected that flowering would be promoted by calcium ionophores and inhibited by channel blockers. However, there was essentially no effect on flowering in *Pharbitis* seedlings from application of the ionophore Bay K-8644 and only a slight inhibition from channel blockers; there was no clear evidence that the timing of the chemical application had any effect in either case (Tretyn *et al.* 1990). Although it can be argued that uptake and transport of externally applied chemical agents may obscure any time-based effects, it is evident from a number of experiments that calcium ionophores, channel blockers and calmodulin antagonists do not modify flowering in the SDP, *Pharbitis,* when applied at a time when they would be expected to interact directly with the phytochrome-dependent night-break reaction or with the process of photoperiodic induction.

The inhibitory effect of EGTA on flowering (Friedman *et al.* 1989) was confirmed (although to a lesser degree) by Tretyn *et al.* (1990). In both cases, maximum inhibition was obtained towards the end of the photoperiod suggesting a possible interaction with the Pfr-requiring reaction in the leaf, although an effect on timekeeping is not excluded since the CNL was increased when EGTA was applied immediately before the dark period; unfortunately the effect on the time of NB_{max} (which is a more precise indicator of timekeeping) was not examined. However, EGTA also strongly inhibited flowering when applied only to the shoot tips or after the end of the CNL when induction had already taken place.

The involvement of inositolphospholipid turnover has also been investigated in a more direct manner. The components of the inositolphospholipid pathway were found to be present in the cotyledons of *Pharbitis* and a night-break resulted in an increase in the second messenger, inositol 1,4,5-trisphosphate (P.J. Lumsden pers. comm.). As there was no effect at other times (*i.e.* when the action of light is to phase-shift the photoperiodic rhythm), it seems that the observed changes in inositolphospholipid turnover were a consequence of the night-break action of light at the inducible phase of the rhythm and were not associated with the effect of light on the underlying circadian oscillator. If inositolphospholipid turnover is involved in the night-break response, the subsequent transduction pathway from Pfr could lead to changes in calcium levels and/or to modifications in gene expression. However, this approach to the control of flowering is in its early stages and the evidence for changes in inositolphospholipids is still very limited. As discussed above, the results from the application of a variety of chemicals have not indicated that the effect of the night-break is to alter the level of endogenous calcium.

7.3.8.3 The nature of the floral stimulus

Once photoperiodic induction has taken place in the leaf or cotyledon, this must lead to the production or release of a stimulus which moves to the apex. Studies concerned with the movement of the stimulus from leaf to apex have shown that a relatively slow-moving chemical rather than an electrical message passes between the two. However, the chemical identity of the stimulus remains unknown, even though grafting experiments have demonstrated that the floral stimulus is the same, or at least is physiologically equivalent in plants of all photoperiodic response groups. Similarly it has been shown that the stimulus appears to be the same in several different plant species and genera. The effectiveness of the putative stimulus can, however, only be assessed by grafting a florally-induced donor leaf to a vegetative receptor plant and, since only closely related plants can be cross-grafted, it cannot be demonstrated that the same stimulus is effective in all plants. Grafting experiments have also shown that inhibitors of flowering are produced in some, but not all daylength-sensitive plants (Lang *et al.* 1977) and flowering could, therefore, depend on a balance between inhibiting and promoting substances. However, the chemical identification of such a flowering inhibitor has met with no more success than that of the postulated floral stimulus.

The failure to isolate any substance that is unambiguously and specifically associated with floral initiation has led to the proposal that a variety of known growth regulators (including hormones and sugars) may be involved in the regulation of flowering, with probably considerable differences between species (Bernier 1988). This could explain the wide range of substances that can promote flowering in different species and under different conditions. However, it is clear from many physiological experiments that the stimulus exported from a photoperiodically induced leaf differs in some as yet unidentified way from that exported from a non-induced leaf. It remains a major challenge to determine where that difference lies, as well as how it originates. Similarly, in other photoperiodic responses such as dormancy and tuber formation, there is, as yet, no complete identification of the substances that are under photoperiodic control in the leaves.

7.3.9 Further Reading

Lumsden P.J. (1991) Circadian rhythms and phytochrome. *Annu. Rev. Plant Physiol. Plant Mol. Biol.* 42: 351-371.

Vince-Prue D. (1975) *Photoperiodism in Plants*. McGraw Hill, Maidenhead.

Vince-Prue D. (1983) Photomorphogenesis and flowering. In: *Encyclopedia of Plant Physiology*, New Series, Vol. 16B; *Photomorphogenesis*, pp. 457-490, Shropshire W. and Mohr H. (eds.) Springer-Verlag, Berlin.

Vince-Prue D. and Takimoto A. (1987) Roles of phytochrome in photoperiodic floral induction. In: *Phytochrome and Photoregulation in Plants*, pp. 259-275, Furuya M. (ed.) Academic Press, Tokyo.

7.3.10 References

Bernier G. (1988) The control of floral evocation and morphogenesis. *Annu. Rev. Plant Physiol. Plant Mol. Biol.* 39: 175-219.

Bossen M.E., Kendrick R.E. and Vredenberg W.J. (1990) The involvement of a G-protein in phytochrome-regulated Ca^{2+}-dependent swelling of etiolated wheat protoplasts. *Physiol. Plant.* 80: 55-62.

Brockmann J., Rieble S., Kazarinova-Fukshansky N., Seyfried M. and Schäfer E. (1987) Phytochrome behaves as a dimer *in vivo*. *Plant Cell Environ.* 10: 105-111.

Carr-Smith H., Johnson C.B., Plumpton C. and Butcher W.G. (1993) The kinetics of Type I phytochrome in green, light-grown wheat *(Triticum aestivum* L). *Planta* (in press)

Daan S. (1982) Circadian rhythms in animals and plants. In: *Biological Timekeeping*, SEB Seminar Series, pp.11-32, Brady J. (ed.) Cambridge University Press, Cambridge.

Deitzer G.F. (1993) Molecular genetics of photoperiodic induction. In: *Plant Photoreceptors and Photoperception*, Proc. Int. Sympos. British Photobiology Society (in press).

Deitzer G.F., Hayes R.G. and Jabben M. (1982) Phase-shift in circadian rhythm of floral promotion by far-red energy in *Hordeum vulgare* L. *Plant Physiol.* 69: 597-601.

Friedman H, Goldschmidt E.E. and Halevy A.H. (1989) Involvement of calcium in the photoperiodic flower induction process of *Pharbitis nil*. *Plant Physiol.* 89: 530-534.

Halaban R and Hillman W. (1970) Response of *Lemna perpusilla* to periodic transfer to distilled water. *Plant Physiol.* 46: 641-644.

Holland R.W.K. and Vince D. (1971) Floral initiation in *Lolium temulentum* L: the role of phytochrome in the responses to red and far-red light. *Planta* 98: 232-243.

Jordan B.R., Partis M.D. and Thomas B. (1986) The biology and molecular biology of phytochrome. In: *Oxford Surveys of Plant Molecular and Cell Biology* Vol. 3, pp. 315-362, Miflin B.J. (ed.) Oxford University Press.

Kimpel J.A. and Doss R.P. (1989) Gene expression during floral induction by leaves of *Perilla crispa*. *Flowering Newsletter* 7: 20-25.

King R.W. and Cumming B.G. (1972) The role of phytochrome in photoperiodic time-measurement and its relation to rhythmic time-keeping in the control of flowering in *Chenopodium rubrum* L. *Planta* 108: 39-57.

King R.W., Schäfer E., Thomas B. and Vince-Prue D. (1982) Photoperiodism and rhythmic response to light. *Plant Cell Environ.* 5: 395-404.

Knapp P.H., Sawhney S., Grimmett M.M. and Vince-Prue D. (1986) Site of perception of the far-red inhibition of flowering in *Pharbitis nil* Choisy. *Plant Cell Physiol.* 27: 1147-1152.

Lang A., Chailakhyan M. Kh. and Frolova I.A. (1977) Promotion and inhibition of flower formation in a day-neutral plant in grafts with a short-day plant and a long-day plant. *Proc. Natl. Acad. Sci. USA* 74: 2412-2416.

Lumsden P.J. and Vince-Prue D. (1984) The perception of dusk signals in photoperiod time-measurement. *Physiol. Plant.* 60: 427-432.

Lumsden P.J, Saji H. and Furuya M. (1987) Action spectra confirm two separate actions of phytochrome in the induction of flowering in *Lemna paucicostata* 441. *Plant Cell Physiol.* 28: 1237-1242.

Lumsden P.J., Yamamoto K.T., Nagatani A and Furuya M. (1985) Effect of monoclonal anti-bodies on the *in vitro* Pfr dark reversion of pea phytochrome. *Plant Cell Physiol.* 26: 1313-1322.

Morse M.J., Crain R.C. and Satter R.L. (1987) Light-stimulated inositolphospholipid turnover in *Samanea saman* leaf pulvini. *Proc. Natl. Acad. Sci. USA* 84: 7075-7078.

O'Neill S.D. (1989) Molecular analysis of floral induction in *Pharbitis nil.* In: *Plant Reproduction: from Floral Induction to Pollination*, pp.19-28, Lord E. and Bernier G. (eds.) Am. Soc. Plant Physiol. Symposium Series, Vol. 1.

O'Neill S.D. (1992) The photoperiodic control of flowering: progress toward understanding the mechanism of induction. *Photochem. Photobiol.* 56: 789-801.

Oota Y. (1983) Physiological structure of the critical photoperiod of *Lemna paucicostata* 6746. *Plant Cell Physiol.* 24: 1503-1510.

Oota Y. and Nakashima H. (1978) Photoperiodic flowering in *Lemna gibba* G3; time measurement. *Bot. Mag. Tokyo,* Special Issue 1: 177-198.

Papenfuss H.D. and Salisbury F.B. (1967) Properties of clock re-setting in flowering of *Xanthium. Plant Physiol.* 42: 1562-1568.

Pratt L.H., Cordonnier M-M., Wang Y-C., Stewart S.J. and Moyer M. (1991) Evidence for three phytochromes in *Avena.* In: *Phytochrome Properties and Biological Action*, pp.39-55, Thomas B. and Johnson C.B. (eds.) NATO ASI Series H: Cell Biology, Vol 50.

Rombach J. (1986) Phytochrome in Norflurazon-treated seedlings of *Pharbis nil. Physiol. Plant.* 68: 231-237.

Saji H., Furuya M. and Takimoto A. (1984) Role of the photoperiod preceding a flower-inductive dark period in dark-grown seedlings of *Pharbitis nil* Choisy. *Plant Cell Physiol.* 25: 715-720.

Saji H., Vince-Prue D. and Furuya M. (1983) Studies on the photoreceptors for the promotion and inhibition of flowering in dark-grown seedlings of *Pharbitis nil* Choisy. *Plant Cell Physiol.* 25: 1183-1189.

Salisbury F.B. (1981) Twilight effect: initiating dark time-measurement in photoperiodism of *Xanthium. Plant Physiol.* 67: 1230-1238.

Takimoto A. (1967) Studies on the light affecting the initiation of endogenous rhythms concerned with photoperiodic responses in *Pharbitis nil. Bot. Mag. Tokyo* 80: 241-247.

Takimoto A. and Saji H. (1984) A role of phytochrome in photoperiodic induction: two phytochrome-pool theory. *Physiol. Plant.* 61: 675-682.

Thomas B. (1991) Phytochrome and photoperiodic induction. *Physiol. Plant.* 81: 571-577.

Thomas B. and Vince-Prue D. (1987) Phytochrome and photoperiodic induction in short-day and long-day plants. In: *Models in Plant Physiology and Biochemistry* Vol.II, pp. 121-125, Newman D.W. and Wilson K.G. (eds.) CRC Press, Boca Raton.

Tretyn A., Cymerski M., Czaplewska J., Lukasiewicz H., Pawlak A. and Kopcewicz J. (1990) Calcium and photoperiodic flower induction in *Pharbitis nil. Physiol. Plant.* 80: 388-392.

Vince-Prue D. and Lumsden P.J. (1987) Inductive events in the leaves: time-measurement and photoperception in the short-day plant, *Pharbitis nil.* In: *Manipulation of Flowering,* pp. 255-268, Atherton J.G. (ed.) Butterworths, London.

Vince-Prue D., King R.W. and Quail P.H. (1978) Light requirement, phytochrome and photoperiodic induction of flowering of *Pharbitis nil* Chois. II. A critical examination of spectrophotometric assays of phytochrome transformations. *Planta* 141:9-14.

Warm E. (1984) Changes in the composition of *in vitro* translated leaf mRNA caused by photoperiodic flower induction of *Hyoscyamus niger. Physiol. Plant.* 61: 344-350.

Whitelam G.C. (1993) The phytochrome molecules. In: *Plant Photoreceptors and Photoperception*, Proc. Int. Sympos. British Photobiology Society (in press).

7.4 Light within the plant

Thomas C. Vogelmann

Botany Department, University of Wyoming,
Laramie, WY 82071 USA

7.4.1 Introduction

A detailed knowledge of the optical properties of plants is necessary to understand how plants detect their light environment and how they may perceive light direction, light quantity, and spectral quality. However, a detailed description of what happens to light after it enters plant tissues is complicated by a number of optical phenomena such as lens effects, light scattering, and the sieve effect. Light scattering, in combination with internal reflection, creates a light trap so that fluence rates within plants can exceed by 3-4 times that of incident light. Internal fluence rates change with increasing depth within the tissues as the light is attenuated by absorption and scattering. This creates a light gradient which is necessary for the perception of light direction in phototropism. Other mechanisms for perception of light may depend upon other optical effects.

This chapter will review some of what is known about the optical properties of plants and how light is propagated across tissues and organs. Light gradients and their possible role in photomorphogenesis are discussed.

7.4.2 Physical aspects of light propagation in plants

7.4.2.1 Light as a particle versus *wave; limitations when considering plant optics*

A rigorous mathematical description of what happens to light after it enters a plant is limited in part by the complexity of the optics and also by the conceptual view of light that we choose to use. It is well-known that light has both particle and wave properties. It is convenient from the standpoint of mathematics to separate these properties and to treat light either as a wave or a particle (photon). For example, a wave function is used to describe the behaviour of light when it passes through a narrow slit and generates a series of interference

R.E. Kendrick & G.H.M. Kronenberg (eds.), Photomorphogenesis in Plants - 2nd Edition

fringes. However, because of difficulties in scaling up the mathematical equations, it is not yet possible to use the wave description to provide information on how much light is present within plant tissues or how it is distributed. For this type of problem, light is usually treated as a stream of particles or rays. For example, the possible paths that photons take when they pass through different layers of a leaf can be constructed from ray-tracing diagrams (*cf.* Fig. 16) which give a two-dimensional view of light propagation in a three-dimensional object. This method can provide some information of the paths that photons take when they are refracted and reflected between cellular layers of different refractive indices.

Unfortunately, the particle concept has limitations, especially when dealing with events at the microscopic scale where wave-phenomena begin to predominate. For example, when examining the leaf surface under high magnification to observe the patterns of light that are reflected from individual epidermal cells, it is possible to see a spot of light reflected from each cell and each spot is surrounded by a series of interference fringes which is clearly a wave-phenomenon (McClendon 1984). Following this further, most epidermal cells will focus light (Section 7.4.2.2) and it is possible to use ray-tracing diagrams to estimate focal lengths and intensifications in a manner similar to that used by engineers who design lenses for cameras and microscopes. But, ray-tracing is of questionable value for estimating the focal properties of epidermal cells because wave phenomena, observable at the leaf surface, will also be present inside the leaf. In short, it is difficult to know how much the wave properties influence results predicted by ray-tracing especially at the microscopic level where the wave properties of light can supersede the particle properties.

The particle concept may also lead to significant errors when considering how light is absorbed on a molecular scale. When theoretically describing how a chlorophyll molecule with a known cross-section absorbs light, the ray model leads to a 10% underestimate (Latimer 1984). The error results from ignoring the wave nature of light; otherwise the chlorophyll molecule appears to 'reach out and grasp' the photon. Thus, there are restrictions on how to treat the behaviour of light when considering events that occur at the microscopic scale.

7.4.2.2 Cells as lenses

Many plant and fungal cells can act as lenses that focus light. These cells have a curved outer surface similar to a planoconvex lens or, in some cases, a cylindrical lens. When surrounded by air on one side, there is a significant difference between the refractive index of the air (1.0) and that of the cell wall (1.45) and cytoplasm (1.33). Consequently, when light passes through these cells, its direction of travel is bent by refraction and the light is focused on microscopic areas within the cell.

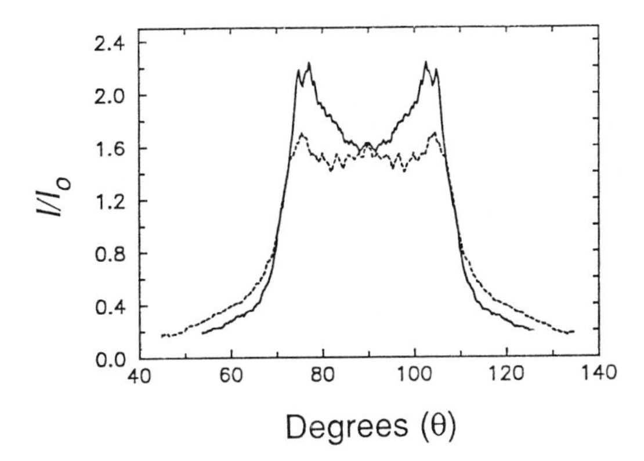

Figure 1. The *Phycomyces* sporangiophore as a cellular lens. A sporangiophore was irradiated from one side with collimated light and the amount of light measured on the distal cell wall with a fibre optic microprobe. I/I_o values show the amount of light (I) relative to the fluence rate of the incident light (I_o). Values for (θ) correspond to the azimuthal position on the cell where 90° is directly opposite the irradiated surface. Note that the *Phycomyces* sporangiophore concentrates light 1.5-2.2 fold. *Solid line*: intensity profiles within the growing zone located 1 mm beneath the base of the sporangium. *Broken line*: intensity profiles in the nongrowing zone, 10 mm beneath the sporangium. After Dennison and Vogelmann (1989).

One of the best documented examples of a cellular lens is the *Phycomyces* sporangiophore (Chapter 7.2). The sporangiophore of this organism consists of a spherical sporangium which develops on a single-celled stalk. The stalk is an exceptionally large cell about 200 μm in diameter and up to several centimetres in length. When parallel light strikes one side (proximal) of this cylindrical cell, the direction of travel is altered by refraction so that the light becomes focused on the opposite side (distal). Fluence rates on the distal side of the cell are around 1.6-2.2 times that of the incident light (Fig. 1; Dennison and Vogelmann 1989; Steinhart 1991). This depends upon a number of variables such as the angle of collimation, the angle of incidence of the actinic light, and the amount of light scattering within the sporangiophore.

Focusing of light within the sporangiophore plays a key role in phototropism. When the sporangiophore, which grows upright in air, is irradiated from one side, it bends towards the light. Altering the lens properties of this cell changes its phototropic behaviour. This can be done by eliminating the ability to focus light by immersing the sporangiophore in mineral oil which has a refractive index that is similar to that of the cell (*cf.* Dennison 1979). Irradiating the sporangiophore from one side under these conditions causes the sporangiophore to bend away from the light source. These results make sense if one proposes that the direction of phototropic curvature is determined by the side of the cell that receives the most light. When the sporangiophore is submerged in oil, the proximal side of the cell receives the highest fluence rate, whereas the opposite

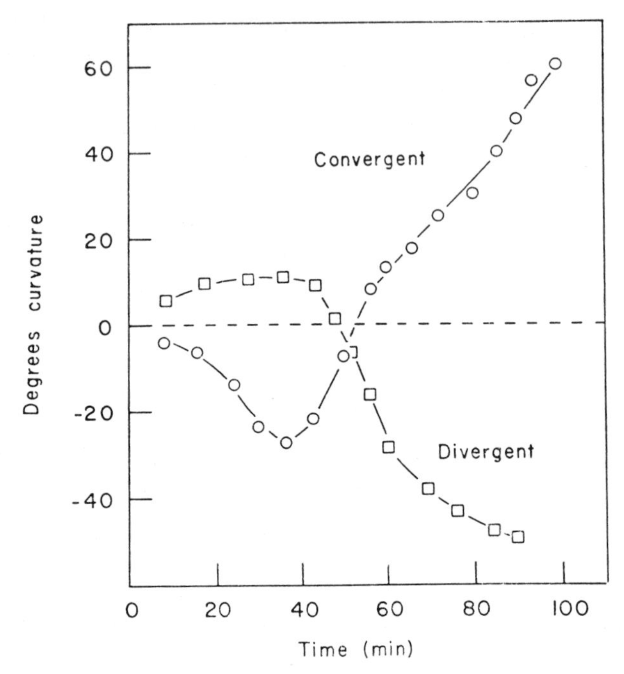

Figure 2. Reversal of first positive and negative curvature in *Avena* coleoptiles by submersion in paraffin oil. Normally, when coleoptiles are exposed to a small fluence of unilateral light, they bend towards the light (first positive curvature); exposure to increased fluences can cause negative curvature. However, when they are immersed in paraffin oil and irradiated, the phototropic responses are reversed. Small fluences now cause negative curvature and larger amounts of light cause positive curvature. After Meyer (1969).

occurs in air when cellular focusing creates a higher fluence rate on the distal cell wall.

Although largely overlooked, lens effects are widespread among land plants. They occur in structures consisting of one or several cells such as rhizoids of the liverworts *Marchantia* and *Lunularia.* Rhizoids are similar to small roots in that they normally grow downwards and absorb nutrients from the soil. When irradiated from one side, they focus light in a manner similar to *Phycomyces,* but the direction of their phototropic response is different. Rhizoids bend away from the light when they are in air and towards the light when submerged in mineral oil (Humphry 1966).

Lens effects may also play a role in the phototropic responses of higher plants but the supporting evidence is problematic. Similar to *Phycomyces,* experiments have been done by submerging *Avena* coleoptiles in mineral oil and irradiating them from one side. Under these conditions, low fluences that normally give rise to first positive curvature now caused negative curvature; higher fluence rates that normally result in first negative curvature elicited positive curvature (Fig. 2; Humphry 1966; Meyer 1969). The phototropic behaviour is consistent with a focusing mechanism, but it is difficult to know whether these responses were affected by the penetration of oil into the coleoptile tissues or the period of anoxia that occurred during submersion.

Another way to test for the presence of lens effects is to control the direction of light given during the photo-inductive period. Normally, experiments with phototropism have been done with light that is fairly well collimated. This light will be focused by a planoconvex lens, similar to the shape of many epidermal cells (see below); by placing such a lens in the light path, it is possible to redirect the light so that it converges towards a focal point. The light can also be redirected so that it becomes divergent instead of convergent, which can be done by inserting a cylindrical lens, such as a glass rod, in the light path. Here, the light first converges towards a focal point near the distal surface of the rod and then rapidly diverges away from the focal point. Irradiating *Avena* coleoptiles with convergent *versus* divergent light gave very different phototropic responses. Irradiation with convergent light resulted in normal positive phototropic curvature whereas irradiation with divergent light caused negative curvature (Fig. 3; Shropshire 1974). Although these results are suggestive of a lens mechanism within the coleoptile, it is difficult to control the fluence rate which becomes nonuniform as soon as a lens is placed in the light path. For example, when light exits the distal surface of a cylindrical lens, it is distributed unequally with azimuth (*cf.* Fig. 1) so that it is difficult to know whether the phototropic responses observed in *Avena* were a result of light direction or fluence rate (Iino 1990).

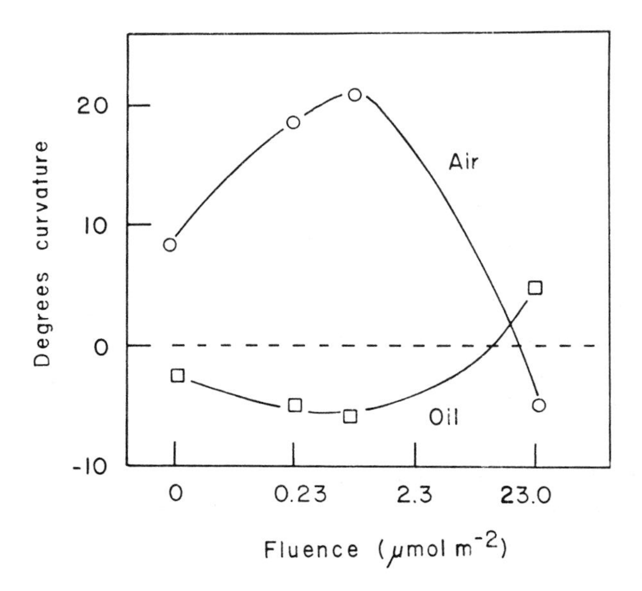

Figure 3. Reversal of phototropic curvature in *Avena* by irradiation with divergent light. The coleoptiles bend towards the light when irradiated with convergent light and away with divergent light. After Shropshire (1974).

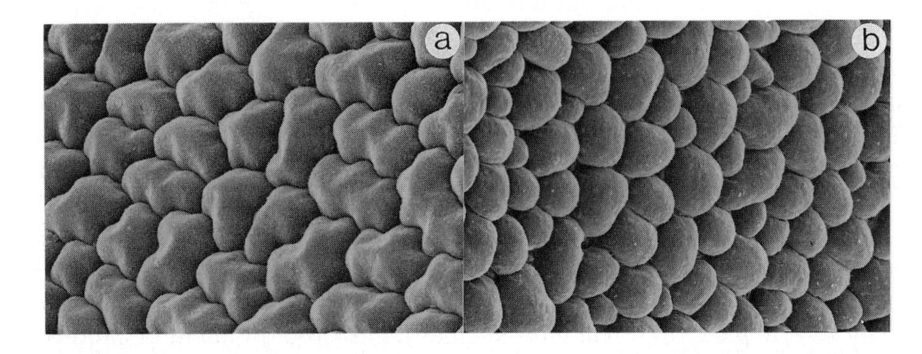

Figure 4. Epidermal cells that act as lenses in leaves of *Oxalis europea.* (a) Adaxial (*upper*) epidermis; (b) Abaxial (*lower*) epidermis. Scale bar: 50 μm. From Poulson and Vogelmann (1990).

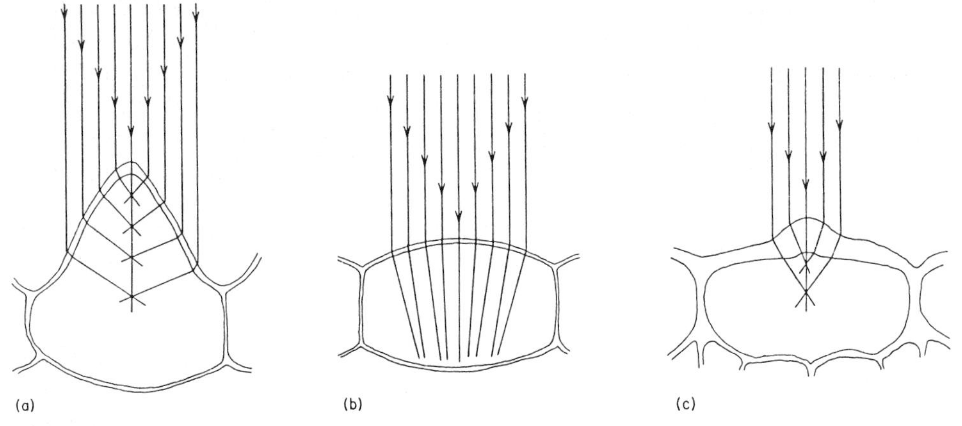

Figure 5. Ray-tracing diagrams of epidermal focusing in leaves of different plants. Each epidermal cell type can focus light in areas either inside or outside the cell: (a) *Anthurium leuconeurum*; (b) *Colocasia antiquorum*; (c) *Aquilegia vulgaris*. After Haberlandt (1914).

In addition, it is difficult to imagine how focusing could occur in multicellular organs such as a coleoptile where a large amount of light scattering would seem to eliminate the kind of lens action found in *Phycomyces.* However, mathematical modelling of light propagation through a maize mesocotyl indicated that despite light scattering, light can be focused on the distal surface of the mesocotyl (Steinhart 1991). Depending upon the shape of the individual epidermal cells, the lens effects could occur near the surface of phototropic organs. Immersion in mineral oil or irradiation with divergent light would alter the pattern of light distribution both across the coleoptile and near the surface. Although the bulk of the experimental work to date suggests that light gradients are the primary means of determining light direction, the possible role of lens effects in phototropism needs further scrutiny.

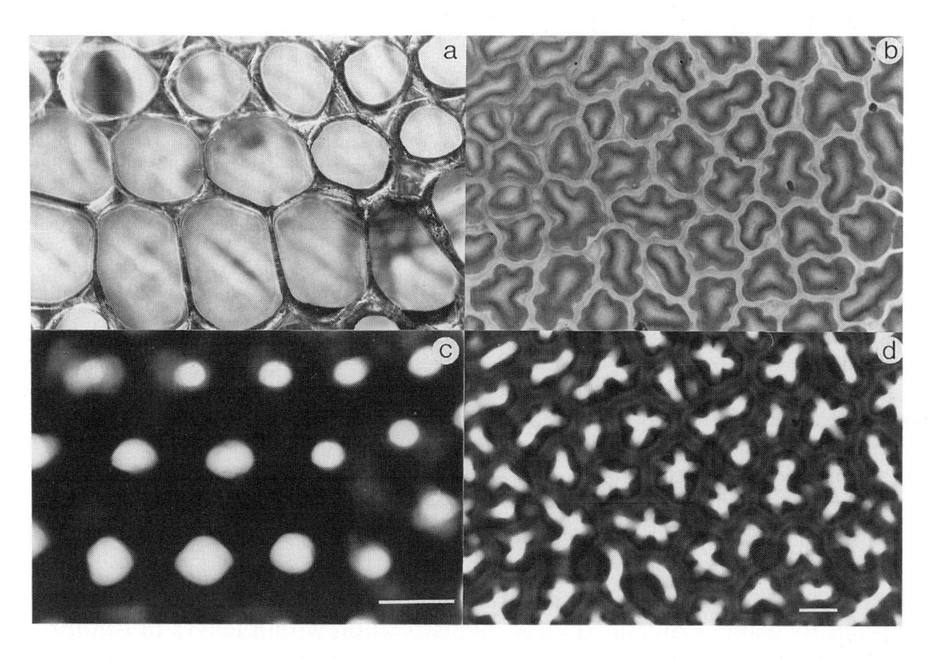

Figure 6. Focusing of light by epidermal cells from the upper leaf surface. Epidermal cell outlines from (a) *Zebrina pendula*; (b) *Oxalis europea* leaves. (c-d) Focusing of light in regions that correspond to the palisade layer of each leaf. Scale bar = 50 μm.

Cells that act as lenses are also found on the surfaces of many leaves (Figs. 4-6). The phenomenon was originally described in the early part of this century by Haberlandt (1914) when he showed that the epidermis of a number of plants could focus light. He thought that epidermal focusing could provide the basis for the perception of light direction so that leaves could optimize the position of their lamina to intercept light for photosynthesis. This hypothesis was later disproved and interest in this subject was lost until recently where the focusing ability of the leaf epidermis has gained attention from the standpoint of light capture for photosynthesis.

Plants that have some of the most striking lens effects are native to the tropical understorey such as *Anthurium, Begonia* and *Selaginella* (Bone *et al.* 1985). These plants grow under very low levels of ambient light, *ca.* 5 μmol $m^{-2}s^{-1}$ PAR (photosynthetically active radiation between 400-700 nm) which is about 0.3% of full sunlight at noon and very close to the compensation point for photosynthesis of C3 plants. It has been proposed that the highly convex epidermal cells may serve to concentrate light on some of the chloroplasts within the leaf (Bone *et al.* 1985; Lee 1986). Epidermal focusing would occur when the incident light is collimated, as it is when direct sunlight passes through gaps in the forest canopy. Otherwise, the light would be diffused by

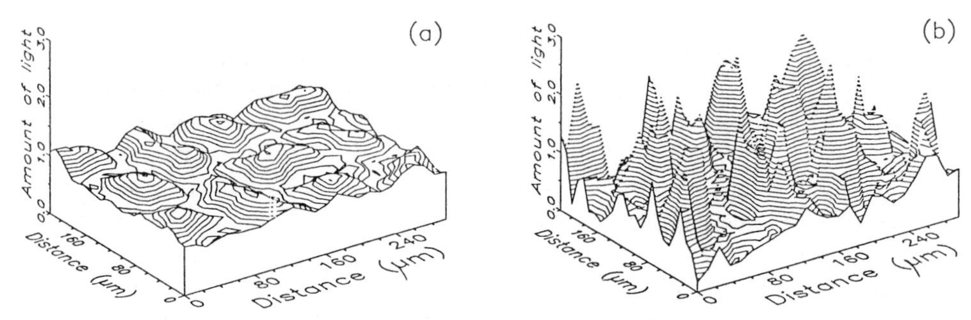

Figure 7. Uniformity of the radiation field after passing through the epidermis of *Oxalis europea* leaves. The adaxial epidermis was removed by peeling, placed on a microscope slide and irradiated with diffuse (a) or collimated (b) light. Using image analysis, uniformity of the radiation field was measured immediately beneath the epidermis in a region that corresponds to the location of the palisade. When irradiated with diffuse light, the radiation field is relatively uniform (a), whereas irradiation of the same sample with collimated light (b) results in concentration of light up to 2 times that of incident via focusing.

light scattering within the canopy and diffuse light is concentrated much less by epidermal focusing. Calculated focal intensification within leaves of *Anthurium* suggest that chloroplasts may be exposed to fluence rates that are as low as 2 and possibly as high as 10 times that of incident light (Bone *et al.* 1985).

Epidermal lens effects also occur in the leaves of temperate plants such as *Oxalis*, *Trifolium* and *Medicago sativa* (Martin *et al.* 1989, 1991; Poulson and Vogelmann 1990). In these plants, the epidermis severely distorts the uniformity of the radiation field and light is concentrated up to 4 times as it enters the palisade layer (Figs. 6 and 7). In the absence of absorption and light scattering, the light can be focused much more, up to 10 times, but this level of intensification is probably seldom reached inside a leaf because the light is attenuated by absorption and scattering before it converges at the focal point.

Whether or not epidermal focusing serves any physiological purpose remains to be seen. By focusing light upon some chloroplasts, others will receive less light. On one hand, this could be of some advantage when light is severely limiting for photosynthesis as it is within the tropical understorey. It could be that epidermal focusing allows at least some of the chloroplasts within the leaf to function despite conditions of extremely low light. On the other hand, in temperate plants such as *Medicago* and *Trifolium*, which usually grow in open fields and meadows, epidermal focusing can concentrate full sunlight to 2 times or more within the leaf. Such intense light may cause photoinhibition and damage the photosynthetic system. The amount of light that is absorbed by individual chloroplasts could be regulated to some extent by the ability to position the chloroplasts within the leaf. Although usually studied in algae, chloroplast movements (Chapter 9.4) have been described within the leaves of higher plants (Gabry's and Walczak 1980; Zurzycki 1961) and can be easily observed in understorey plants such as *Oxalis*.

Although it can be proposed that epidermal focusing imparts some sort of physiological advantage to the plant, it can also be argued that focusing is a secondary consequence of other selective processes. For example, it may be that convexly shaped cells help reduce specular reflection from the leaf surface. This could be advantageous in the understorey where much of the light is diffuse and strikes the leaf at low angles. Such light would normally be reflected from the surface of a completely flat leaf. It could also be that, given the turgor pressure of plant cells, it is difficult to make an epidermis that is topologically flat. Whatever proves to be the case, lens effects are common within plants and should not be overlooked when investigating processes mediated by light.

7.4.2.3 Absorption and the sieve effect

After passing the plant surface, the amount of light and its spectral quality may be altered by absorption. When pigments are uniformly dissolved within a spectrophotometer cuvette, absorbance is directly proportional to pigment concentration (Beer's law). Unfortunately, this relationship does not hold for pigments within plant tissues. The reason is that most pigments are packaged within organelles such as chloroplasts or vacuoles so that the distribution of pigment is not uniform within the tissues. This causes a geometrical problem of light absorption known as the *sieve effect.* This can be illustrated by a hypothetical experiment in which light is passed through a cuvette that contains a solution of dye that transmits 50% of incident light (Fig. 8a). The dye molecules can then be swept to one side of the cuvette by a semipermeable membrane, so that one half of the cuvette transmits 100%, and the pigmented half transmits only 25% (Fig. 8b). Apparent transmission through the whole cuvette is now 62.5% instead of the original 50%. Transmission continues to increase as the dye is concentrated in volumes of ever decreasing size (Fig. 8c). Thus, the sieve effect can lead to an underestimation of the amount of pigment present within tissues.

The matter is further complicated by the fact that both light scattering and the sieve effect depend upon the positioning of the organelle that contains the pigments. Such is the case with chloroplasts which, in many algae, bryophytes and higher plants, change their position within the cell in response to the amount of ambient light (Chapter 9.4) or a circadian rhythm. For example, the green alga *Ulva* has a circadian rhythm in which the chloroplasts move between the face and side walls of the thallus at different times of the day (Fig. 9; Britz 1979). This alters the amount of transmission through the thallus (Fig. 10) and

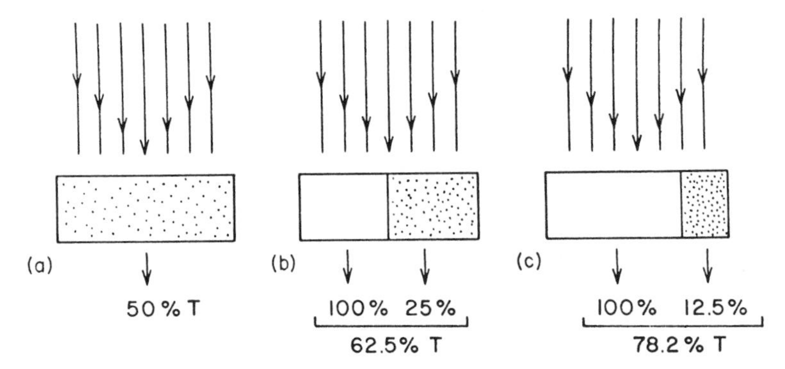

Figure 8. Nonhomogeneous pigment localization and the sieve effect. (a) 50% transmission (*T*) when pigment distribution is homogeneous; (b) sweeping the pigment to one half of the cuvette allows 100% *T* on one half of the cuvette and 25% *T* on the other resulting in an combined *T* of 62.5%; (c) greater concentration of pigment leads to higher transmission overall. The corresponding effect upon absorbance (*A*) can be calculated from the relationship: $A = -\log T$.

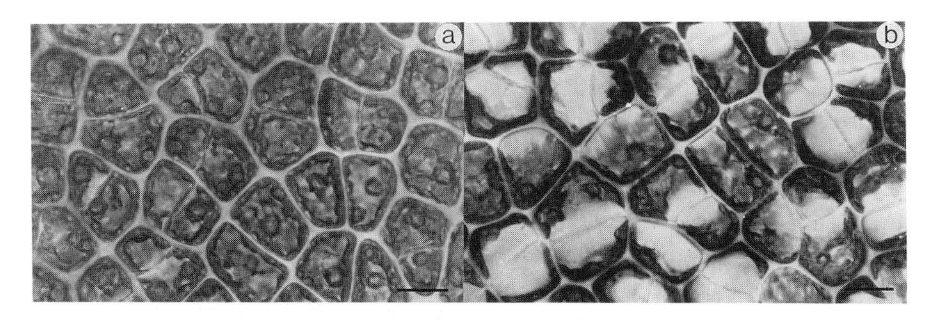

Figure 9. Chloroplast positioning in the green alga *Ulva*. Chloroplast location is regulated by a circadian rhythm in which the chloroplasts migrate periodically between the face (a) and profile (b) walls of the cells. This migration causes a corresponding circadian change in the transmission of light through the thallus (Fig. 10) *via* the sieve effect. Magnification bars = 10 μm. Photographs courtesy of S.J. Britz.

coincides with the circadian rhythm in photosynthetic capacity (Britz *et al.* 1976).

The sieve effect is greatest when pigment content is high and at wavelengths near the absorption maxima. Such is the case in leaves and macrophytic algae where the sieve effect is greatest at wavelengths that correspond to the largest absorption by the photosynthetic pigments (Fig. 11). The sieve effect is counter-balanced to some extent by light scattering. Whereas the sieve effect causes an underestimate of the amount of pigment present, light scattering increases the pathlength that photons travel and thus increases absorption per unit of pigment. The balance between these two phenomena depends upon the optical properties of the tissue and is not easily quantified.

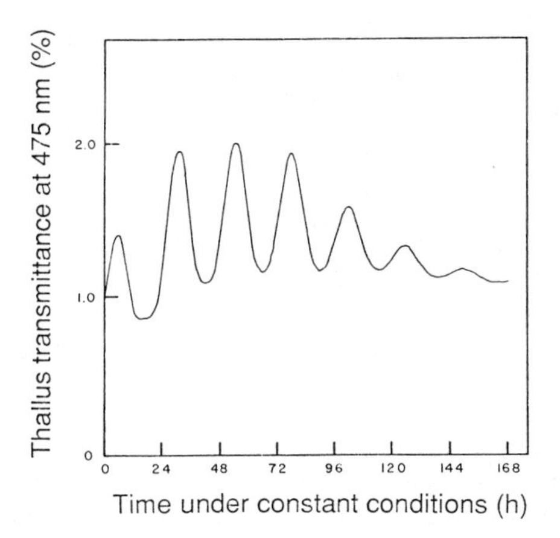

Figure 10. Circadian changes in *Ulva* thallus transmission caused by the sieve effect and migration of chloroplasts between the face and profile cell walls. After Britz *et al.* (1976).

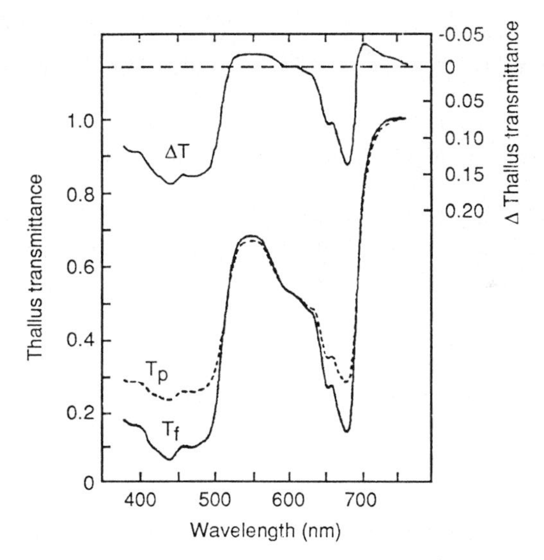

Figure 11. The sieve effect at different wavelengths within *Ulva.* Transmission spectra were measured when chloroplasts were in the face or profile position (Fig. 9.). Note that the largest changes in transmission occurred at wavelengths where the photosynthetic pigments have highest absorbance. After Britz S.J. and Briggs W.R. (1987) *Acta Physiol. Plant.* 9: 149-162.

7.4.2.4 Fluorescence effects

Light that is absorbed by compounds within plant tissue can be re-emitted as fluorescence and the spectral quality of ambient light can be altered as light of

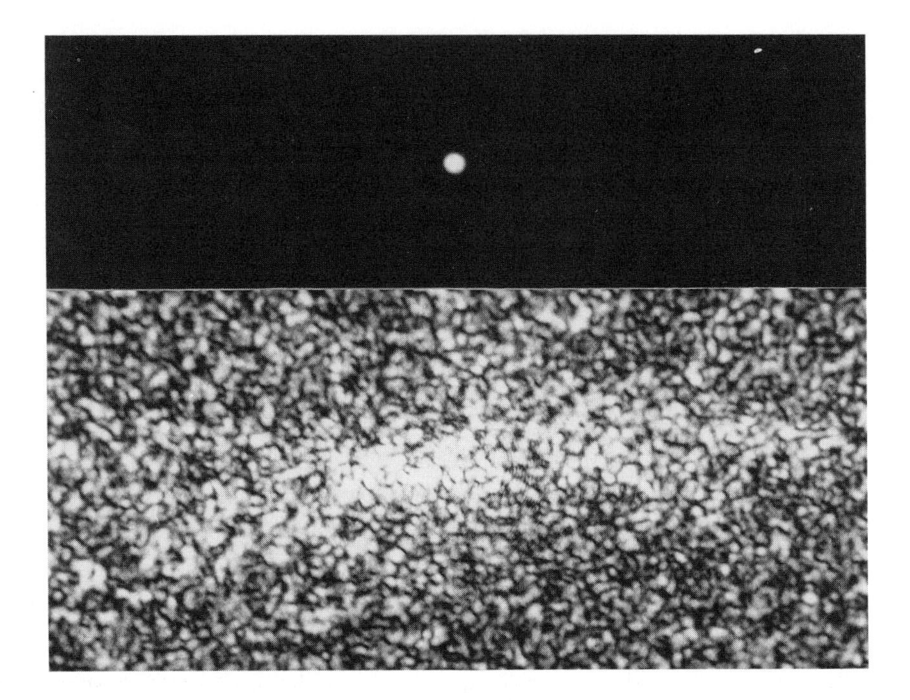

Figure 12. Scattering of a coherent beam of laser light by an onion epidermis. (a, *top*) Image of the laser, (b, *bottom*) scattering after passage through the epidermis.

shorter wavelengths is converted or pumped into spectral regions of longer wavelengths. Since chlorophyll *a* has an *in vivo* fluorescence maximum near 688 nm, fluorescence could affect phytochrome photoequilibrium (Chapter 4.7) within green tissues and lead to artifacts when attempting to predict the value of the photoequilibrium. However, the fluorescence effects are thought to be small and, as shown by mathematical modelling of the spectral environment within a green leaf (Lork and Fukshansky 1985), the amount of the deviation of Pfr/P [where P = total phytochrome, the red-light (R)-absorbing form of phytochrome (Pr) plus the far-red light (FR)-absorbing form of phytochrome (Pfr)], maintained at equilibrium under conditions with and without chlorophyll fluorescence, was maximally 15%.

7.4.2.5 Light scattering

When describing the interaction of light with single particles it is possible to distinguish between refraction, diffraction, and reflection. However, when describing optical dispersion in complex media such as suspensions of cells or in tissues, it is more convenient to collect these phenomena under the designa-

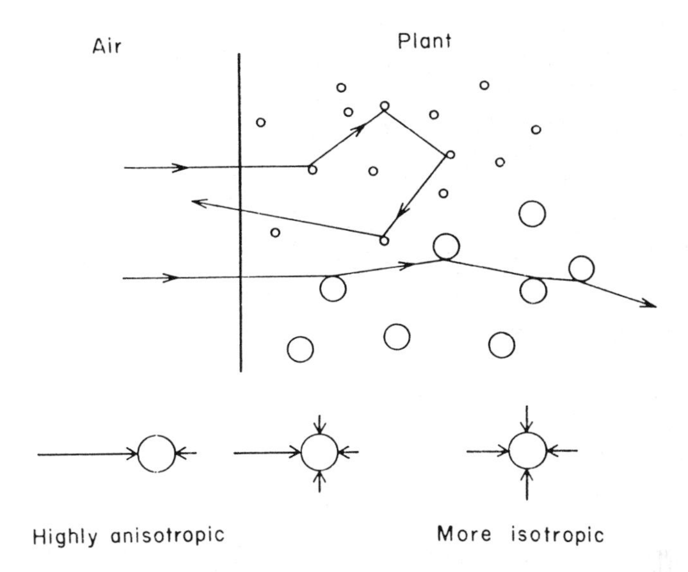

Figure 13. Particle size and light scattering. As collimated light enters a plant, particles that are smaller than the wavelength of light will scatter the light isotropically. Particles that are much larger than the wavelength of light will scatter at low angles so that there are only minor deflections in the path of the light beam. Reflecting boundaries at the plant surface (not shown) will trap photons inside the plant so that they undergo multiple scattering and internal reflection. Outside the plant the light has a highly anisotropic distribution. Scattering leads to greater isotropy with increasing depth but may not lead to a completely isotropic condition.

tion light scattering. This more general definition is used throughout this chapter.

As a beam of collimated light enters a plant it can be deflected numerous times by inhomogeneities within and between cells (Figs. 12 and 13). When light is scattered in this manner, its direction of propagation is translated from one to three dimensions; the more thorough this translation, the more diffuse it becomes. Near the surface of a plant, the distribution of light is highly anisotropic, but it becomes more isotropic at increasing depths. For reasons discussed below, it is doubtful that light ever becomes completely isotropic within most plant organs.

The size of particles that cause scattering within plants can be roughly grouped into two classes: those that are smaller than the wavelength of light and those that are larger. It happens that chloroplasts and mitochondria lie between these two classes. The manner in which light is scattered depends upon particle size. Relatively small particles tend to scatter light isotropically in all directions and the scattering is inversely proportional to the fourth power of the wavelength. Thus, blue light is scattered more than R. This type of scattering is called molecular or Rayleigh scattering and is the reason why our sky is blue rather than white or black. However, it is doubtful whether molecular scattering contributes much to the internal light environment within plant tissues. Instead, most of the scattering appears to be caused by the numerous intercellular air spaces. They comprise 30-40% of the volume of most plant tissues and fall into the class of large particles. Scattering of light by large particles occurs independently of wavelength and primarily at small angles close to the original direction of propagation of the incident beam of light. Deflection of light at small angles can be observed when collimated light passes through fog or a suspension of cells. In these situations, light is not scattered isotropically but rather is propagated mostly at low angles. In fact, a collimated beam of light fails to assume an isotropic distribution even after it passes through a dense yeast cell suspension where most light rays strike an average of 35 cells (Latimer 1982).

There is a large difference between the refractive indices of air (1.0) and plant cell walls (1.45). Consequently, the plant-air surface and the numerous cell-intercellular air spaces form reflecting boundaries. A significant proportion of rays that strike these boundaries are reflected in the same manner as when they strike the surface of a pool of water. The air-tissue surface can also reflect a significant proportion of light into the interior of the tissue (internal reflection), thereby increasing the pathlength and the probability of absorption. The light scattering capacity of intercellular air spaces is well known to microscopists and can be eliminated in plant material by infiltration with a medium such as oil that closely matches the refractive index of cell walls. This treatment can make an originally opaque tissue transparent. Water, which has a lower refractive index, has a similar effect, but is less effective (Fig. 14).

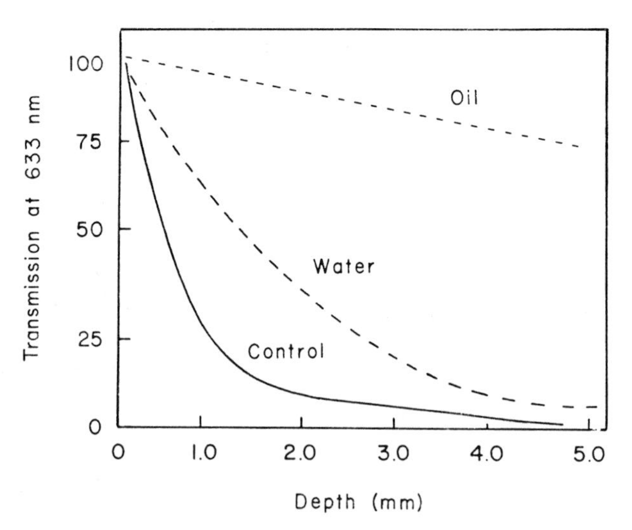

Figure 14. The role of intercellular air spaces in the attenuation of light. Sunflower hypocotyls were infiltrated with water or oil. Transmission of red light of 633 nm was measured at different depths within the hypocotyl. Matching the refractive index of the cell walls by replacing the intercellular air spaces with oil eliminates almost completely the attenuation of light by scattering. After Parsons A., Macleod K., Firn R.D. and Digby J. (1984) *Plant Cell Environ.* 7: 325-332.

In summary, a working view of what happens to a photon once it enters a plant should take into account the possibility of multiple internal reflection, multiple scattering and the statistical probability of absorption by pigments which are distributed nonuniformly within the tissues. Thus, a photon can be absorbed as soon as it passes into the first cell of a tissue, or it may be reflected back and forth so that it's pathlength is increased 4-5 times over what it would have been if it passed straight through the sample.

From the above discussion, it can be seen that absorbance is greatly increased by multiple scattering and internal reflection. These phenomena are sometimes exploited by plants to maximize absorbance. For example, multiple scattering and internal reflection enhance the appearance of colour within flowers (Kay *et al.* 1981) which is important for attraction of pollinators. Most of the anthocyanins, betalains and flavonoids that give petals their characteristic colour are located in the vacuoles of the epidermis. The epidermal cells of petals are usually papillose and are backed by a multicellular tapetal layer that lacks pigments, but is especially rich in intercellular air spaces. The tapetum acts as a reflector so that pathlength and absorption within the epidermis are greatly increased as light is reflected between the atmospheric boundary on one side and the tapetal layer on the other. Ray-tracing diagrams indicate that raised or papillose epidermal cells (Fig. 15) greatly increase the frequency of internal reflections, which in turn enhances absorption (Bernhard *et al.* 1968). Thus, the spectrum of re-emitted light from flowers is greatly altered from that which would occur in the absence of light scattering.

A similar mechanism is present within leaves to optimize light absorption for photosynthesis. In most leaves the palisade is subtended by an air-rich spongy mesophyll layer. After passing through the epidermis, the path of a single ray of

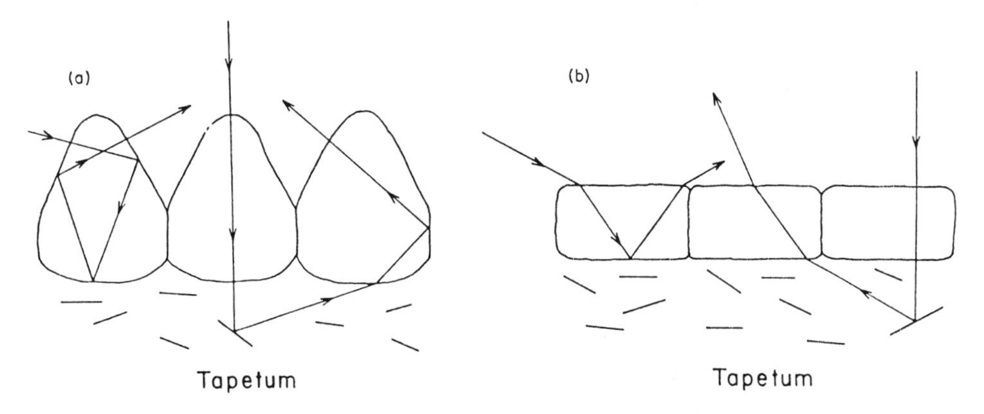

Figure 15. Schematic drawings of a petal surface with (a) and without (b) papillae. The presence of papillae allows greater absorption per unit amount of pigment *via* multiple reflection of light and increased pathlength within the cells. The tapetum is rich in intercellular air spaces and serves as a reflector that redirects down-welling light towards the petal surface. Pathlength is reduced in the absence of papillae, resulting in lower absorption.

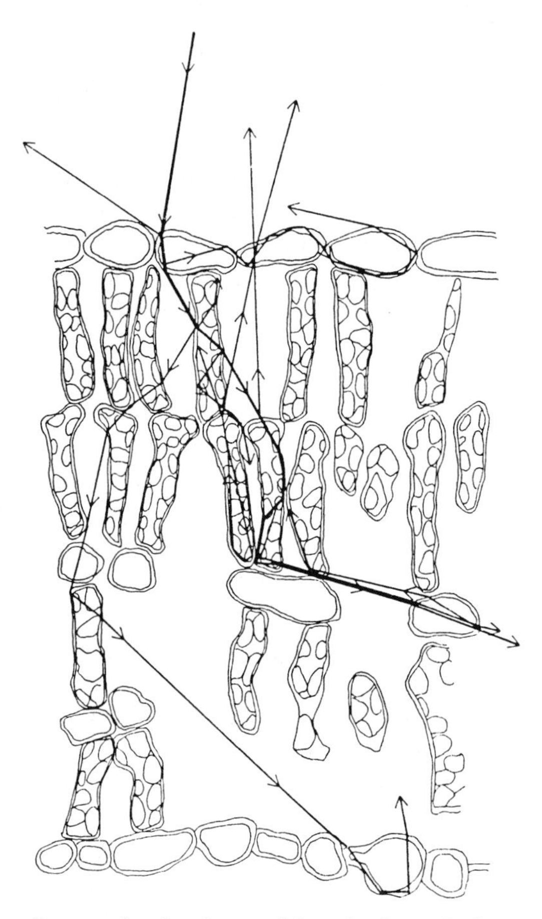

Figure 16. Ray-tracing diagram showing the possible paths that a single ray takes after it enters a soybean leaf. The path was calculated according to Snell's law and using the refractive indices of the cell wall, chloroplasts, cell sap and air. The ray was discontinued when its calculated value dropped below 1.8% of the energy contained in the incident ray. After Kumar R. and Silva L. (1973) *Appl. Opt.* 12: 2950-2954.

light can be tortuous as it is scattered between air, cell wall, chloroplast and cytoplasmic interfaces as shown by a ray-tracing diagram of a soybean leaf (Fig. 16). Although ray-tracing diagrams give some idea of the number of possible routes a photon can take through a leaf, they are a two-dimensional representation of a three-dimensional phenomenon and they do not provide information about the angular distribution of scattered light, the probability of absorption, or the amount of light at different depths.

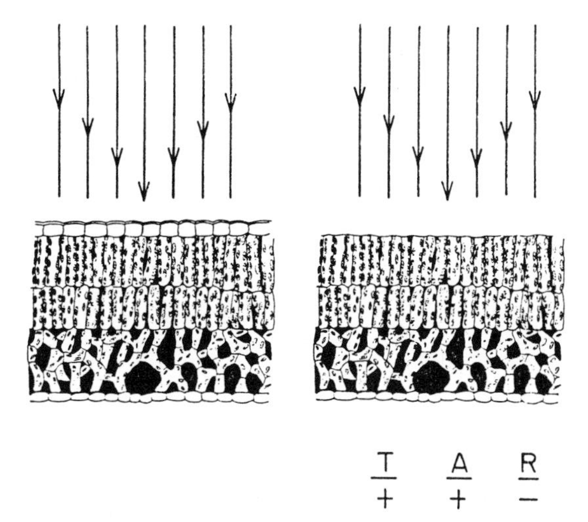

Figure 17. The epidermis as a reflecting layer on a leaf. Removal of the lower epidermis changes the amount of transmittance (*T*), absorbance (*A*) and reflectance (*R*) of visible light. The observed changes indicate that the lower epidermis acts as a reflector that returns photons into the leaf interior.

Several kinds of experiments have shown that each cell layer of a leaf can contribute to the absorption of light. For example, by peeling away the upper and lower epidermis, it is possible to show that these layers act as reflecting boundaries that return potentially escaping photons to the leaf interior. Removal of either upper or lower epidermis also results in increased transmittance and decreased absorbance within the visible region of the spectrum (Fig. 17; Lin and Ehleringer 1983). The spongy mesophyll appears to function as both a diffuser and reflector. This can be demonstrated by measuring transmission as thin paradermal slices are removed from the leaf (Terashima and Saeki 1983). In *Camellia* leaves there is less chlorophyll per unit thickness in the spongy mesophyll than in the palisade layer, but the spongy mesophyll attenuates more light at 550 and 680 nm than does the palisade (Fig. 18). Attenuation within the spongy mesophyll is largely mediated by reflection which indicates that, within a leaf, light that escapes absorption in the palisade has a good chance of being returned by the spongy mesophyll for a second chance. Changes in leaf anatomy can control the penetration of light into the photosynthetic layers. This will be discussed further in Section 7.4.3.5.

7.4.2.6 Plants as optical waveguides

When there are jumps in the refractive index between surfaces, light will be reflected away from either surface. When surfaces are parallel to one another,

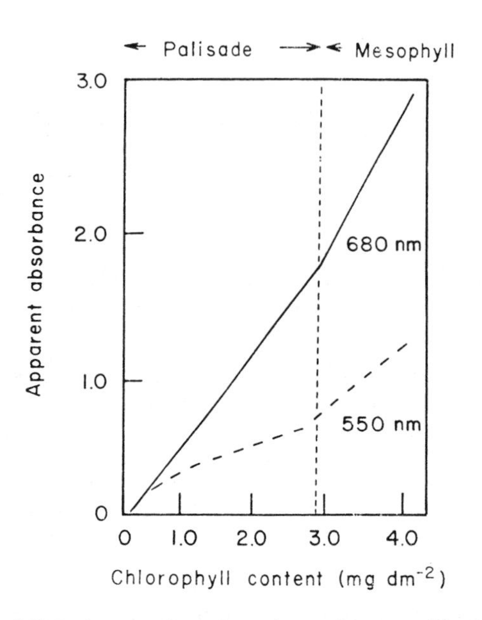

Figure 18. Attenuation of light by paradermal sections of a *Camellia* leaf as a function of its chlorophyll content. The upper surface of a leaf was irradiated with collimated light and transmission measured as thin slices were removed from the bottom of the leaf. Chlorophyll content of the slices was measured and apparent absorbance plotted against chlorophyll content (mg dm^{-2}) at the different depths. The change in the slopes of the lines show that, on the basis of chlorophyll content, there was greater attenuation of 680 and 550 nm light by the spongy mesophyll than the palisade layer. This indicates higher absorbance due to scattering within the spongy layer. After Terashima and Saeki (1983).

light can be reflected back and forth between them so that it travels in a direction parallel to the surfaces. This is the principle of light guiding and how light is propagated through optical fibres. Etiolated plant organs such as hypocotyls, roots and probably most single cells can guide light to some degree. However, there are some inherent differences between the light guiding capacity of optical fibres and plant cells so that a brief discussion of the principles of wave guiding is appropriate.

There are several different kinds of optical fibre but one of the most simple is a step index fibre. This is composed of a solid core of material with a relatively high refractive index and an outer layer of cladding with a lower refractive index (Fig. 19). For efficient light guiding, the difference between the refractive indices between these layers need not be large and is usually about 0.05. Optical fibres are made from materials with extremely high purity and their fabrication is subject to strict quality control since any optical discontinuities in the core or cladding lead to light loss. In a step index fibre, light that enters one end is reflected between the core and cladding and can be propagated through the fibre with an efficiency that approaches 100%. Light can also be guided through a strand of fibre that has only one layer. In this case the cladding is whatever

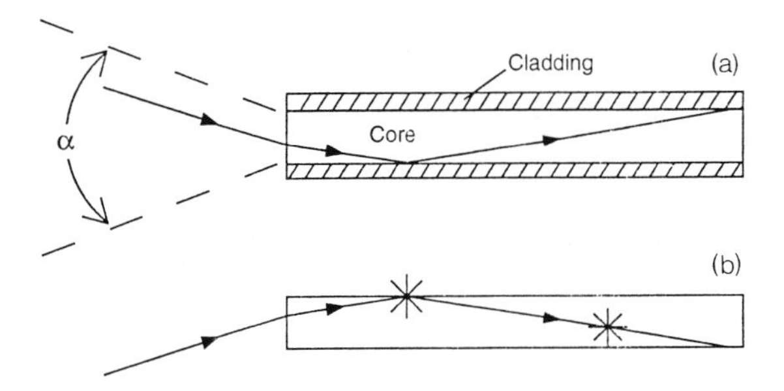

Figure 19. Propagation of light through optical fibres. (a) A step index fibre; (b) a fibre without cladding. Light enters one end of the fibre only within a limited acceptance angle (α). Light travels through the fibre by multiple reflection between two layers of different refractive indices. In (a) light is reflected between the core and cladding whereas in (b) light is contained within the fibre only when there is a difference in refractive index between the fibre and its surrounding. In (b) light escapes from the fibre when it is immersed in a liquid with a similar refractive index or when the light encounters a particle on the surface or interior of the fibre.

medium surrounds the fibre (*e.g.* air). Here, light guiding is very sensitive to inhomogeneities on the surface and a single speck of dust is sufficient to cause appreciable loss of light from the fibre.

Within the fibre, there is a critical angle of internal reflection: rays of light that intersect the boundary layers obliquely are reflected, whereas those that approach the boundary more perpendicularly can pass through and escape the optical fibre. This means that there is a relationship between the angle of light incident upon the receiving end of the fibre and the amount of light that enters and is carried to the opposite end. The maximum angle at which light can enter and then be propagated through the fibre is called the acceptance angle (Fig. 19). Light can also be propagated through bundles of optical fibres and if the individual fibres are parallel to one another, an image can be carried coherently from one end to the other. In summary, some important characteristics of fibre optic wave guides are: (i) two or more layers of homogeneous materials that have different refractive indices; (ii) an acceptance angle.

How well do plant cells meet the prerequisites for light guiding? Most etiolated organs such as hypocotyls consist of files of cells. The length of the longitudinal axis of these cells is usually 3-4 times their diameter and each cell is separated top and bottom from its neighbours by two cross walls. Between the walls are intercellular spaces that are usually filled with air; within each cell there is a large central vacuole and surrounding cytoplasm that contains particles such as plastids, mitochondria, a nucleus and other particles that will scatter light. Whereas optical fibres are made from very homogeneous materials, cells and plant organs are highly heterogeneous.

For light to be guided efficiently between the cell wall and cytoplasm or between the cytoplasm and vacuole, a difference of only 0.05 between refractive indices is needed and this is within the range observed between the different compartments of plant cells. Assuming that light scattering is not too severe, plant cells and tissues should be able to guide light. But how effectively do they do so? Measurements with segments of etiolated coleoptiles and mung bean hypocotyls have shown that they do in fact guide light, but with low efficiency (Mandoli and Briggs 1982a). Light guiding was observed over distances up to 4.5 cm within these samples, but the efficiency of transmission was only about 1% that of optical fibres. Despite this low efficiency, as will be discussed below, light guiding within etiolated tissues has important photomorphogenic consequences.

Exactly how light is propagated longitudinally through plant tissues is not clear, but it probably passes through the vacuole and cytoplasm and is reflected between cell walls. This was shown by experiments in which plant segments were infiltrated with liquids of different refractive index. Infiltration with water increased light-guiding whereas liquids that more closely matched the refractive index of the cell wall decreased it (Mandoli and Briggs 1984). If light is scattered in forward directions by intercellular air spaces it may also play a role in the longitudinal transmission of light through plant segments.

The evidence that plants have fibre optic properties is as follows: (i) light guiding can occur from one end of a plant segment to the other even when the segment is bent; (ii) light enters plant parts within given acceptance angles: 47° for mung bean hypocotyls, 59° for oat mesocotyl and 53° for maize roots; (iii) light is transmitted with some coherency. In other words, an image can be transmitted through a segment of mung bean hypocotyl (Fig. 20) similar to the way in which an image is transmitted through a bundle of optical fibres in which the strands are carefully aligned.

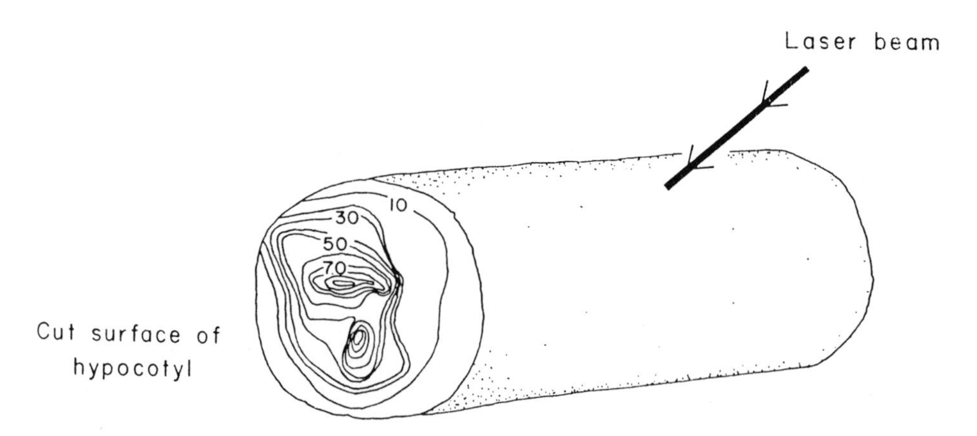

Figure 20. Coherent transfer of light through a segment of mung bean hypocotyl. Etiolated plant tissues are able to transmit an image over short distances. In this example the hypocotyl was irradiated obliquely with a laser beam of small diameter. The contour lines show the pattern of light as it emerges from the cut end of the hypocotyl. After Mandoli D.F. and Briggs W.R. (1984) *Sci. Amer.* 251: 90-98.

The ability to transmit light over relatively long distances (centimetres) can have important developmental consequences for the plant. The growth of oat seedlings provides a good example. The shoot of an oat seedling consists of a coleoptile, mesocotyl and node. In complete darkness, most of the shoot growth is accomplished by elongation of the mesocotyl. Exposure to minute amounts of R inhibits mesocotyl elongation and simultaneously stimulates elongation of the coleoptile and the enclosed primary leaves. Red light also stimulates greening of the shoot. The amount of light needed to change the rates of coleoptile and mesocotyl elongation is very small, the latter being a classic example of a very low fluence response (Chapter 4.7). The purpose of these responses is to position the node near the soil surface while launching the upper regions of the oat shoot on a development path that will give rise to photosynthetic competence.

There are two photoperceptive sites within the shoot that control mesocotyl and coleoptile elongation. The photoperceptive site for inhibition of mesocotyl elongation is near the top of the mesocotyl immediately beneath the node. Both this site and one immediately above the node control the stimulation of coleoptile elongation (Mandoli and Briggs 1982b). Thus, as the shoot of an oat seedling nears the soil surface, light travels longitudinally down the shoot to the photoreceptive sites which modulate growth rate and prepare the shoot for emergence. Axial transmission of light may mediate other physiological responses as well, such as root formation and positioning at specific depths within the soil.

7.4.3 Plants as light traps

7.4.3.1 Internal fluence rates within plants

Fluence rates can be easily measured outside plants, but it is important to choose a sensor with the appropriate geometry. When the light is collimated, fluence rate can be measured with a flat cosine-corrected sensor positioned so that it is perpendicular to the light. In this case, the numerical values for irradiance and fluence rate are the same (Chapter 2). This situation changes when the light is diffuse. A flat cosine-corrected sensor can still be used to measure irradiance, but under conditions where the light is completely diffuse, the irradiance will be only one quarter of times the fluence rate. It is more appropriate to measure fluence rate with a spherical sensor which is not sensitive to qualitative differences in light direction.

Because of light scattering, much of the light within the plant is diffuse and therefore it is appropriate to think in terms of the light that would be measured by a spherical sensor, placed at different locations within the plant tissues. If it were possible to do such an experiment, one would find that in some situations

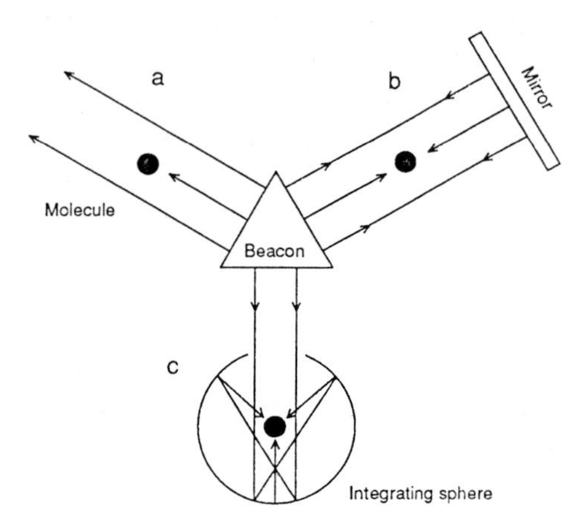

Figure 21. How to increase fluence rate without violating the law of conservation of energy. Three beams of light, each with a fluence rate of 1.0, are emitted in three directions (a, b, c) from a beacon. Inserting a mirror into beam 'b' doubles the fluence rate incident upon a molecule passing within the beam compared to 'a'. If both the molecule and the beam of light pass into the interior of an integrating sphere then the fluence rate incident upon the molecule can be increased 10 or more times. After Vogelmann T.C. and Björn L.O. (1986) *Physiol. Plant.* 68: 704-708.

the internal fluence rate would exceed the external fluence rate by 3-4 fold! At first glance, these values seem to be exceptionally high. Even more importantly, they might also seem to violate the law of conservation of energy. As will be discussed below, such values for internal fluence rates are not unreasonable for plant tissues, nor do they violate the laws that describe the behaviour of the universe.

Internal fluence rates can be increased in a number of ways without increasing the total number of photons emitted by a source. Consider the following example: A particle can be irradiated in space with a beam of light with a fluence rate of 1 W m^{-2} (Fig. 21). By inserting a mirror behind the particle, the fluence rate is increased to 2 W m^{-2}. In this case, we have doubled the fluence rate in the space that surrounds the particle without doubling the number of photons emitted by the light source. Now we can increase the complexity of the problem by asking what fluence rate we might expect to find within an integrating sphere in which there is multiple reflection of light between the curved surfaces. The exact numerical answer to this problem depends upon the reflectivity of the coating on the inside of the sphere, where the particle is positioned, and a number of other parameters such as the size of the hole in which the light beam enters. Assuming a reflectivity of 98% for the coating of the inner walls, internal fluence rates within an integrating sphere have been calculated to be 10 times or more that of incident light (Wendlandt and Hecht 1966).

Plants are not integrating spheres, nor are they mirrors, but they have two properties essential for the creation of a light trap. First, there is a substantial jump in refractive index between the air-cuticle interface (1.00 → 1.45). This creates a reflecting surface that returns potentially escaping light back into the interior of the plant. Second, the numerous intercellular air spaces and organelles scatter light intensely. This increases pathlength and extends the interval of time that photons dwell within the tissue. Rather than passing straight through the tissue, the photons bounce back and forth several times until they are absorbed or escape from the reflecting surface. Thus, as one photon enters the tissue and is trapped by surface reflection and internal scattering, so are subsequent photons. As a result, internal fluence rates within plants can exceed by several times the fluence rate of incident light (Table 1). For most plant tissues, the maximum theoretical internal fluence rate is between 3-4 times that of incident light. This number depends upon a number of variables such as the amount of light scattering, absorption and the refractive index of the cuticle.

It should be mentioned that plants are not unique in their light-trapping capacity. In fact, any object that has a refractive index substantially greater than air and scatters light internally will also trap light. This includes, for example, white plexiglass, in which internal fluence rates as high as 4.57 times incident light have been estimated (Kaufmann and Hartmann 1988). Enhanced internal fluence rates similar to those found within plants have been found in benthic mats (Jørgensen and Des Marais 1988), a number of animal tissues (Marijnissen *et al.* 1985; Star *et al.* 1987) and should be present in household items as commonplace as cheese.

What does light trapping by plants and light gradients mean in terms of photomorphogenesis? One obvious consequence of light trapping is that light would be much more effective in causing photoexcitation of photoreceptors within plants than with the same photoreceptors in a test tube. Internal fluence rates within etiolated plant tissues are commonly 3-4 times that of incident light, and consequently, the sensitivity to light may be increased 3-4 fold. At very low fluence rates this may increase the threshold at which light can be detected. At the other end of the scale, it may render full sunlight all the more harmful by increasing the probability of absorption and resulting photochemical damage.

Beneath the irradiated surface of the tissue, light is attenuated by absorption and scattering so the internal fluence rate decreases. This is referred to as the light gradient (Fig. 22) which plays an important role in detection of light direction (Section 7.4.4) and other physiological processes. Despite the fundamental importance of light gradients, they are difficult to measure and calculate, and analytical procedures have been developed only recently to examine light gradients in plant tissues. This area has made impressive advancements within the last 10 years.

Table 1. Relative internal fluence rates (I/I_o) near the irradiated surface of various plant organs.

Species	Organ	Wavelength (nm)	I/I_o	Method	Reference
Cucurbita pepo	cotyledon (e)	410	1.22	KM	3
		660	2.38		
		730	3.09		
	cotyledon (g)	410	1.21	KM	4
		660	1.21		
		730	2.86		
Zea mays	coleoptile (e)	450	2.2	FOP	5
Medicago sativa	leaf (g)	450	1.6	FOP	7
		550	2.0		
		680	1.6		
Crassula falcata	leaf (g)	550	1.2	FOP	6
		750	2.9		
Kalanchoe marmorata	leaf (g)	548	2.01	SFOS	2
		672	1.69		
		730	3.26		
		756	3.46		
	leaf (g)	548	1.26	SFOS	2
Philodendron scandens		672	1.12		
		730	2.50		
		756	2.87		
	leaf (g)	450	2.25	FOP	1
		550	2.40		
Spinaca oleracea		680	2.20		

FOP = fibre-optic probe, KM = Kubelka Munk theory, SFOS = spherical fibre-optic sensor, e = etiolated, g = green. 1 = Cui *et al.* 1991, 2 = Kaufmann and Hartmann 1988, 3 = Kazarinova-Fukshansky *et al.* 1985, 4 = Seyfried and Fukshansky 1983, 5 = Vogelmann and Haupt 1985, 6 = Vogelmann and Björn 1984, 7 = Vogelmann *et al.* 1989.

7.4.3.2 Calculation of light gradients within tissues

It is possible to calculate the distribution of light that exists within plant tissues. The complexity of the calculation depends upon the optical properties of the plant tissues and the scale of the research question. A common *misconception* is that it is possible to use Beer's law to model the distribution of light within tissues. This assumption is wrong for the following reasons. Beer's law relates absorbance with concentration by the following relationship:

$$A = \varepsilon c l$$

where: ε = extinction coefficient (L mol^{-1} cm^{-1}); c = pigment concentration (mol L^{-1}); l = pathlength (cm).

For this relationship to be valid: (i) the pigment must be distributed uniformly within the sample; (ii) the pathlength should be known. If both of these conditions are fulfilled, as they would for a cuvette that contains a solution of chlorophyll dissolved in 80% acetone, then it is possible to calculate the amount of light at any position within the cuvette by the Lambert formula:

$$I_x = I_o e^{-\varepsilon x}$$

where: I_x = fluence rate at depth x (μmol m^{-2} s^{-1}); I_o = incident fluence rate (μmol m^{-2} s^{-1}); ε = extinction coefficient (L mol^{-1} cm^{-1}); x = depth (cm).

However, the conditions that make these formulae valid are not met within plant tissues where: (i) pigments are sequestered nonhomogeneously in organelles (causing the sieve effect); (ii) light scattering increases pathlength by an unknown amount. Moreover, there is no single value for pathlength at a

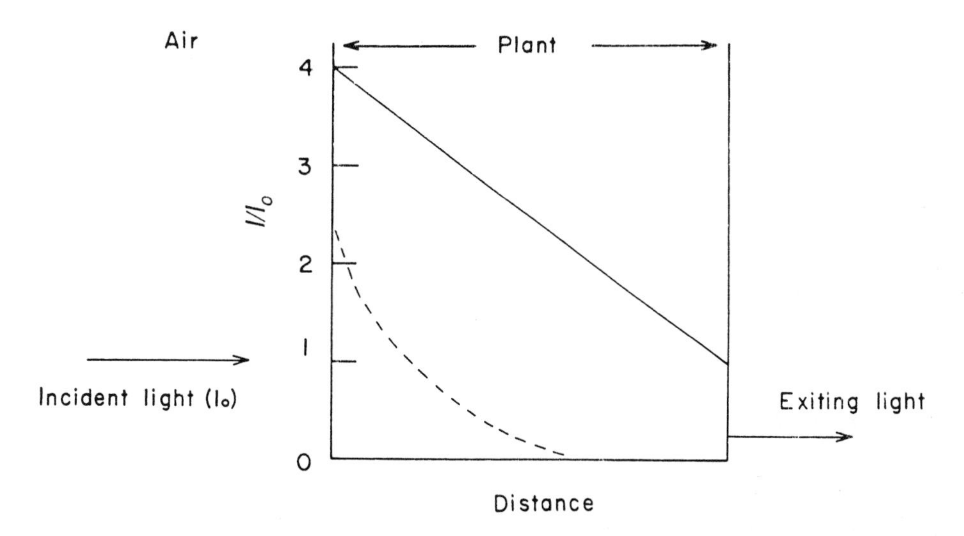

Figure 22. Hypothetical light gradient across a plant organ. The shape of the light gradient depends upon the balance between absorption and scattering. Most plant tissues scatter light intensely and, in the absence of strong absorption, the light gradient will be linear (solid line). Increasing absorption creates a light gradient that is more exponential in shape (broken line). Light scattering and internal reflection can increase the internal fluence rate near the irradiated surface 3-4 times that of incident light (I_o).

given wavelength within a particular plant tissue. Instead, pathlength is a matter of statistics: some photons will have a relatively short pathlength within the tissue, whereas others undergo multiple scattering and have a much longer pathlength. The mathematical relationship that relates the proportion of photons with a particular pathlength resembles a Poisson distribution (Fukshansky *et al.* 1992).

Light distribution within plant tissues can be calculated using a set of equations that were originally developed by Kubelka and Munk (KM; Kubelka 1948, 1954) to describe light propagation in light scattering media. A detailed description of the application of the KM theory to plant tissues has been described by Seyfried (1989). In order to apply the KM equations, it is necessary to measure the amount of light at a particular wavelength that is transmitted (T) and reflected (R) by the sample. This is usually done with the aid of an integrating sphere, which eliminates geometrical problems associated with distributing the light to a detector such as a photomultiplier tube. The relationships between reflectance (R), transmittance (T), sample thickness and several constants are given by:

$$R = 1/\{\mathrm{a} + \mathrm{b}\,[\coth(\mathrm{b}Sd)]\}$$

$$T = \mathrm{b}/[\mathrm{a}\sinh(\mathrm{b}Sd) + \mathrm{b}\cosh(\mathrm{b}Sd)]$$

$$\mathrm{a} = (1 + R^2 - T^2)/2R$$

$$\mathrm{b} = \sqrt{\mathrm{a}^2 - 1}$$

where: a,b = intermediate coefficients; d = thickness of the sample (mm); cosh, coth and sinh = hyperbolic cosine, cotangent and sine functions, respectively; K = coefficient for absorbance (mm^{-1}); S = coefficient for scattering (mm^{-1}).

From these equations, it is possible to derive two coefficients, one for scattering (S), and one for absorbance (K):

$$S = [\coth^{-1}(1 - \mathrm{a}R)/\mathrm{b}R]/\mathrm{b}d$$

$$K = S(\mathrm{a} - 1)$$

Using these coefficients it is then possible to calculate the amount of light at any depth (x) within a sample by:

$$I_x = I_o\,[(1 + R)\cosh(S\mathrm{b}x) - [(\mathrm{a} + 1)/\,\mathrm{b}](1 - R)\sinh(S\mathrm{b}x)]$$

where: I_x = internal fluence rate at depth x
I_o = external fluence rate

In practice, application of the KM equations is more complicated than described here. Firstly, the measurements of R and T need to be corrected for specular reflection from the surface of the sample. Otherwise, R will be overestimated and T underestimated. A procedure for making this correction has been described (Seyfried *et al.* 1983). Secondly, the KM calculations need to be modified to account for reflecting boundaries at the sample surface. Otherwise, internal fluence rates will be underestimated. Thirdly, since most plant samples consist of tissue layers with different optical properties, values of S and K should be known for each layer. This necessitates making optical measurements of paradermal slices of the sample, which may be easy for relatively thick samples (millimetres thick) and virtually impossible for thin ones (micrometers thick). Once R and T are known for each layer, it is possible to use an extended version of the KM theory to calculate, from the component layers, the light gradient that exists within the intact plant organ (Seyfried and Fukshansky 1983). In some cases, it may be appropriate to apply simplified versions of the KM equations (Seyfried 1989).

It is important to note that the KM approach is valid only under a special set of circumstances. First of all, the sample must be irradiated with diffuse light. Thus, the KM approach is not valid for samples irradiated with collimated light as in the natural environment where plants are exposed to direct sunlight. Secondly, light scattering within the sample must occur isotropically. Given the fact that most of the light scattering particles are rather large (intercellular air spaces), there are questions about the randomness of the light scattering. Nonetheless, the KM approach has provided both interesting and useful information about light gradients within plant tissues.

Light gradients have been calculated for zucchini cotyledons, *Cucurbita pepo* (Seyfried and Fukshansky 1983; Seyfried and Schäfer 1983) which are several millimetres thick, have a relatively homogeneous tissue composition and scatter light intensely. At 730 nm, a wavelength where there was little absorption, calculated internal fluence rates (I/I_o) were 2.8 times the incident light near the irradiated surface. Thereafter, the amount of light declined linearly with depth in the cotyledon (Fig. 23). At 660 nm where there was more absorption by chlorophyll, the internal fluence rate near the irradiated surface was 1.4; thereafter the amount of light decreased exponentially. Thus, both the values for maximum internal fluence rate and the shape of the light gradient are strongly influenced by the amount of absorption within the tissues.

The estimates of internal fluence rates near the irradiated surface of *Cucurbita pepo* agree well with those in other plant leaves measured with a microscopic spherical sensor that was placed close to the leaf surface. For example, in leaves of *Philodendron scandens* and *Kalanchoë marmorata,* I/I_o values ranged

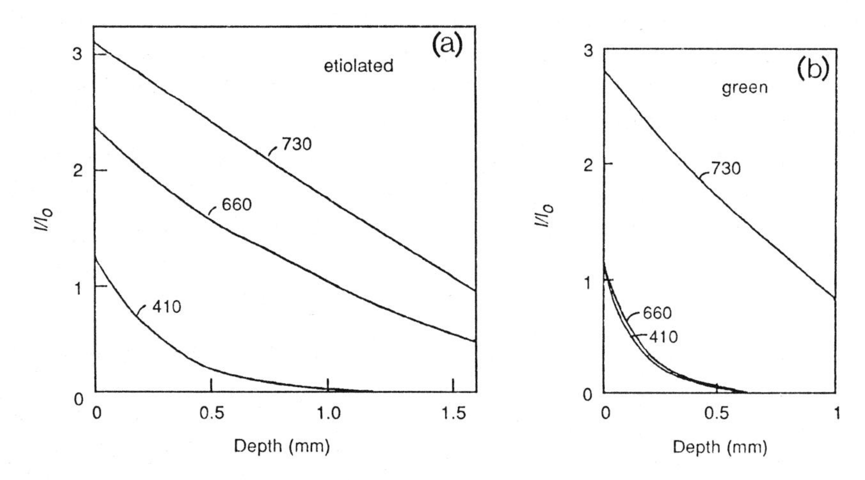

Figure 23. Calculated light gradients in etiolated and green cotyledons of *Cucurbita pepo*. Light gradients (I/I_o) were calculated from transmission and reflection measurements using the Kubelka-Munk theory. Light at 730 nm is absorbed weakly in etiolated and green cotyledons and the light gradient is near linear at this wavelength. Carotenoids absorb light in the blue at 410 nm which causes exponential light gradients in both sets of cotyledons. The synthesis of chlorophyll in green cotyledons causes a change in the 660 nm light gradient, from linear (a) to exponential (b). I = internal fluence rate within the cotyledon, I_o = incident fluence rate. After Kazarinova-Fukshansky *et al.* (1985).

from 1.12 at 616 nm to 3.46 at 756 nm (Table 1; Kaufmann and Hartmann 1988). In addition, the shape of the light gradient in *Cucurbita pepo* estimated by the KM theory agreed well with measurements of conversion of phytochrome from Pr to Pfr within paradermal slices taken from the cotyledon (Seyfried and Schäfer 1985), and with light measurements made with a fibre-optic probe (Section 7.4.3.3) that was inserted into the cotyledon (Knapp *et al.* 1988). Thus, from several different experimental approaches, a consensus has developed that I/I_o values of 3 and higher are not uncommon within plant tissues. Depending upon optical properties, light gradients decline either linearly or exponentially within the tissue.

Calculation of the parameters of light propagation within plant tissues has been developed further using a radiative transfer approach so that it is now possible to calculate light gradients in samples that are irradiated with collimated light. This necessitates knowledge of how rapidly a collimated beam of light is scattered by the sample and in which directions the light is deflected (phase function). It also requires more detailed knowledge of the anatomical characteristics of the tissues, for example, the number and distribution of chloroplasts and their pigment content. This allows light propagation to be treated statistically and to correct for the sieve effect. The latest refinements require more intensive calculations, but the radiative transfer approach is much more powerful than that of KM. In addition to light gradients, it is now possible to estimate photon transit times within the tissue (and thus pathlength), and to correct *in vivo* absorption spectra that are distorted by the sieve effect. A complete description of this approach is beyond the scope of this chapter and interested readers should refer to the original literature (Fukshansky 1991; Fukshansky *et al.* 1992). In addition, much research on radiation transport in animal tissues has been summarized in a recent review (Cheong *et al.* 1990), and some of these approaches may be applied to problems of photon transport in plants.

7.4.3.3 Experimental measurement of light gradients with a fibre-optic probe

It is possible to measure light gradients with a fibre-optic microprobe that is inserted into plant tissues (Vogelmann and Björn 1984; Vogelmann *et al.* 1991). The principle behind this technique is relatively simple. Single stranded optical fibre can be heated and stretched so that it forms a tapered point. The tapered end is optically sealed to prevent light leaks along the sides and then ground and polished so that light can now enter the extreme tip (Fig. 24). The light-sensing tip is extremely small and can be made so that it is only a few

Figure 24. Microscopic fibre-optic sensors for measuring light within tissues. Single stranded optical fibre can be tapered to a fine point by heating and stretching. The tip can then be truncated to allow light entry into the fibre. The resulting sensor has an acceptance angle and the probe can be used to measure partial light fluxes within plant tissues (Fig. 25). Then, light gradients can be reconstructed from the measurements. Spherical sensors that directly measure the fluence rate can be made by attaching a light scattering sphere on one end of an optical fibre. Such sensors have been used in studies with animal tissues and microbial mats but thus far it has not been possible to make a sensor small enough for use in plant tissues.

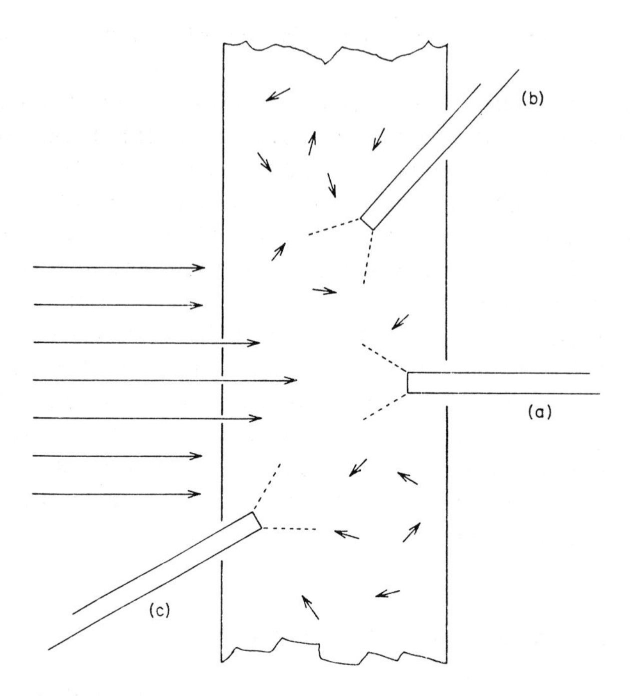

Figure 25. Use of a fibre-optic microsensor for measuring distribution of light within plant tissues. Optical fibres are directional sensors that capture light from directions that lie within their acceptance angle. Upon irradiating a sample with collimated light, the fibre-optic sensor can be inserted from the shaded side and advanced towards the irradiated surface (a). This measures the distribution of collimated and scattered light within the sample. By changing the orientation of the fibre it is possible to measure the distribution of scattered light. In (b), light that is scattered in the forward direction is measured. Note that the collimated light falls outside of the acceptance angle of the fibre and thus, is not measured. In (c), light that is backscattered within the sample is measured. By knowing the amount of collimated and diffuse light that is propagated in different directions it is possible to calculate the internal fluence rate at different depths within the sample (Fig. 26).

micrometers in diameter and considerably smaller than the diameter of an average plant cell (*ca.* 50 μm). The probe can then be inserted to various depths within plant tissue and light measurements made.

The advantage of this technique is that it is possible to make light measurements within tissue that is irradiated either with collimated or diffuse light. It is also possible to make measurements of a diverse range of samples without the necessity of having prior knowledge about their optical properties. In addition, the probes have high spatial resolution so that it is possible to measure light gradients within thin leaves and in some cases single cells. The fibres also have a limited field of view or acceptance angle (Figs. 19 and 24) so that it is possible to measure the directions in which light is scattered once it enters the tissue. This allows determination of the phase function which can be used to calculate how collimated light is scattered within tissues (Martinez v. Remisowsky *et al.* 1992).

On the other hand, the probes are very sensitive to their surrounding light micro-environment within the tissue. Shading by individual chloroplasts or local lens effects, created by individual epidermal cells, affect the light readings. This leads to high variability between replicate measurements, but this heterogeneity probably represents the real situation within plant tissues. Another point is that no matter how small the probes are, they are intrusive and it is difficult to know how much this affects the light measurements. However, results obtained with this technique agree reasonably well with those obtained by other approaches so that this particular limitation does not seem to be too severe. Finally, the probes have an angular sensitivity to light. Although this can be used to some advantage, it does not permit direct measurement of internal fluence rates. Instead, representative measurements of partial light fluxes must be made in several directions (Fig. 25) and I/I_o values calculated from these measurements (Fig. 26).

7.4.4 Light gradients and photomorphogenesis

Knowledge about light gradients is crucial for understanding mechanisms of light perception. Tissues that comprise plant organs often have different optical properties and it follows that light gradients are related to anatomy. Since

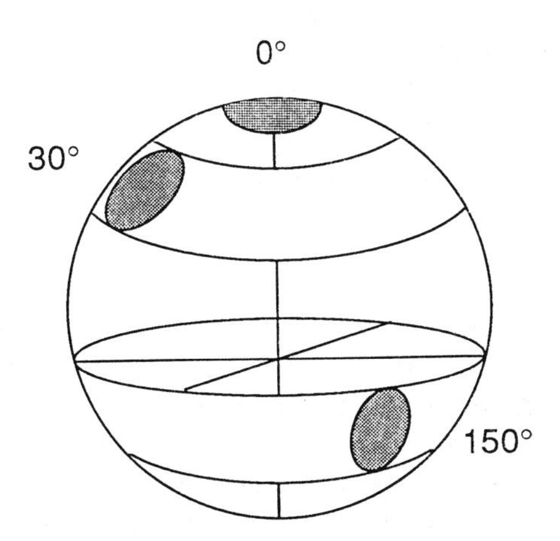

Figure 26. Method for calculating light gradients from measurements made with a fibre-optic probe. The probe measures only the light that falls within its acceptance angle which is represented by the shaded area on the surface of an imaginary sphere (fluence rate). It is then possible to calculate the remaining light that the fibre did not measure (unshaded parts). Light gradients can be reconstructed from measurements that are made in representative directions to sample adequately the distribution of collimated and scattered light within the sample. After Vogelmann and Björn (1984).

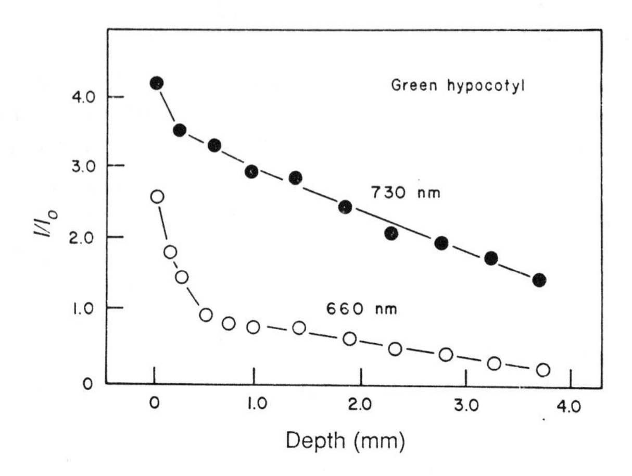

Figure 27. Light gradients in the red (660nm) and far-red (730) light in a green *Phaseolus vulgaris* hypocotyl. The hypocotyl was irradiated unilaterally with collimated light and light gradients measured with a fibre optic probe. Most of the chlorophyll was present within the 0.5 mm near the surface so that the light gradient at 660 nm declined exponentially within these tissues and linearly thereafter. In an etiolated hypocotyl, the gradient at 660 nm was similar to that for 730 nm (not shown).

anatomy varies widely, so do the characteristics of the light gradients. Sometimes plants control the penetration of light into their interiors. This is done by altering the amount of absorption by specific cell layers or by changing the amount of scattering by modifying cell shape and arrangement (Section 7.4.4.5). Thus, plant optics becomes all the more interesting. Each of the following sections is based around one or more example light gradient(s) and is followed with a brief discussion of some of the implications for photomorphogenesis and related areas.

7.4.4.1 Light gradients and phytochrome

Plant organs that are composed of relatively homogeneous tissues have relatively uniform light gradients. An example is *Phaseolus vulgaris* hypocotyls where light gradients do not have any remarkable features beyond the fact that they appear to follow the predictions of the KM theory. In etiolated hypocotyls the gradients at 660 nm and 730 nm were linear and similar to one another (Fig. 27). Internal fluence rates near the irradiated surface approached 4 which is probably very close to the theoretical maximum for plants. In green hypocotyls, the gradient at 730 nm remained relatively unchanged whereas the presence of chlorophyll created an exponential gradient at 660 nm (Fig. 27).

These and similar light gradients have several implications. First of all, the different shapes of the light gradients can be used to explain distortions in

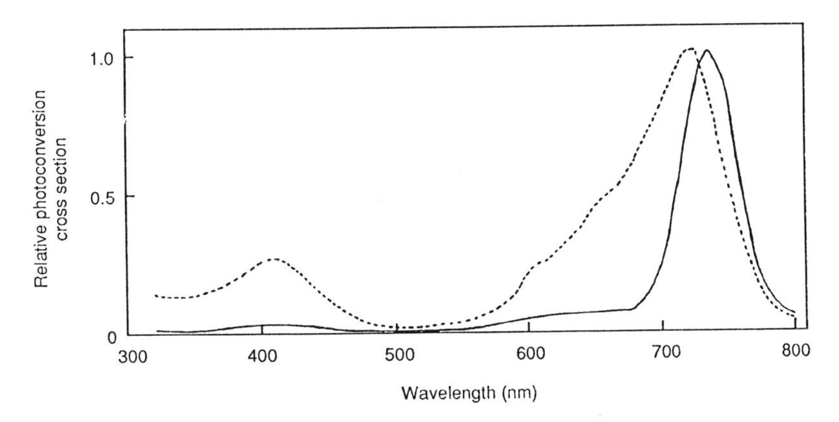

Figure 28. Possible distortions in action spectra for phytochrome-mediated responses in green and etiolated tissues. In the absence of screening pigments, calculated action spectra resemble the absorption spectrum of phytochrome (broken line). The presence of chlorophyll in green tissue reduces the penetration of blue and red light so that light in the far red becomes more effective. Consequently, the action maximum for a hypothetical response is shifted towards the far-red region and there is no action within the blue (solid line). The shapes of action spectra depend not only on the slope of the light gradients but also the location of phytochrome within the tissues. After Kazarinova-Fukshansky *et al.* (1985).

action spectra, which in pigmented tissues can deviate from the absorption spectrum of the photoreceptor. Hypocotyl elongation in *Phaseolus* and other plants is mediated by phytochrome and the action spectrum for elongation of etiolated hypocotyls has an action maximum near the absorption maximum of phytochrome at 660 nm. However, action spectra for elongation of green hypocotyls has an action maximum at 620 nm instead of 660 nm (Jose and Schäfer 1978). The 40 nm shift towards the blue is caused by the fact that 620 nm light penetrates further into the hypocotyl than 660 nm light and is therefore proportionally more effective in converting phytochrome to Pfr.

The relationships between action spectra, light gradients and the location of the photoreceptor within the tissues have been described in detail (Kazarinova-Fukshansky *et al.* 1985). In tissues where there is appreciable screening by chlorophyll and other pigments, action spectra can be greatly distorted so that in some cases they have little resemblance to the absorption spectrum of the photoreceptor *in vivo* (Fig. 28). In fact, understanding the origins of the distortions of action spectra can help identify the tissue layers that contribute to a photomorphogenic response. For example, for responses mediated by phytochrome in green plants, the action spectra should become more distorted the deeper the phytochrome is located within the tissues.

A corollary to these observations is that there may be boundaries on those tissues that are capable of detecting qualitative changes in spectral quality of light within the environment. Most visible light is strongly attenuated within

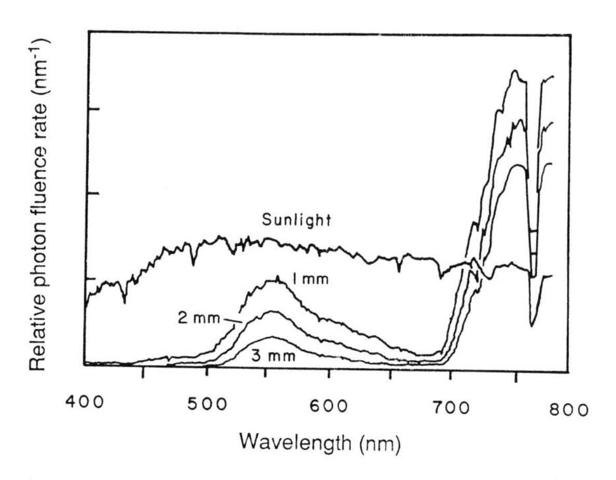

Figure 29. Internal photon fluence rates within green leaves of the succulent *Crassula falcata* when irradiated with sunlight. Light was measured with a fibre-optic probe inserted 1, 2 and 3 mm within the leaf.

green tissues by absorption, but internal fluence rates within the FR are commonly 3 times that of incident light (Fig. 29). It could be that the Pfr/P ratio is close to zero within the interior of green tissues so that the photoperceptive tissues may be near the surface.

Elevated internal fluence rates are also important when considering cycling rates between Pr and Pfr. The final photostationary Pfr/P ratio is determined by light quality, but the rate of cycling is determined by light quantity. Increased internal fluence rates should increase the cycling rates of phytochrome (Chapter 4.7) and the amount of cycling depends upon the location of phytochrome relative to the irradiated surface (Fig. 30; Kazarinova-Fukshansky *et al.* 1985). Given that there are more than one species of phytochrome within a given species of plant (Chapter 4.2), this problem assumes a new dimension.

7.4.4.2 Light gradients and phototropism

Historically, phototropism and light gradients have been closely linked together (Chapter 9.2; Iino 1990) but little is known about light gradients within most phototropic organs. Light gradients have been measured within maize coleoptiles and are influenced by anatomy which is more complex than for most plant organs. The coleoptile is a hollow cylinder composed of 8-15 cells on each side. Near the node, it surrounds the folded primary leaves that are rich in carotenoids. There are air-tissue boundaries within the primary leaves, and between the leaves and coleoptile. The presence of several epidermal-air boundaries, cells of different size and absorption properties creates a blue light (B) gradient

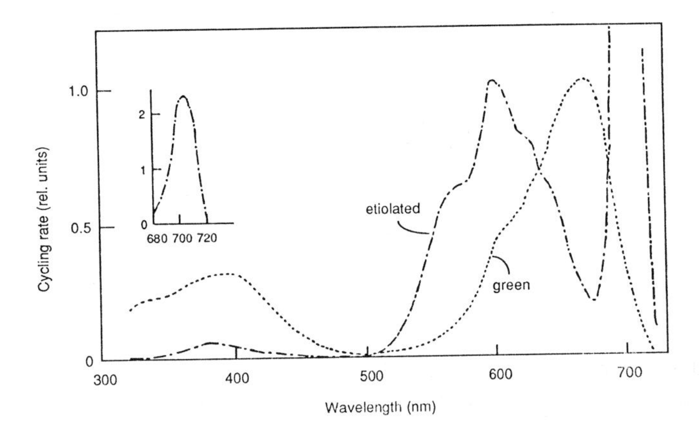

Figure 30. Calculated action spectra for a response mediated by cycling of phytochrome in green and etiolated tissues. The presence of chlorophyll shifts the action maximum near 600 nm, towards the far-red region. Calculated action spectra depend both upon light gradients and the location of phytochrome within the tissues. After Kazarinova-Fukshansky *et al.* (1985).

that differs between the different regions of the coleoptile at the tip, mid-region and base.

The B (450 nm) gradient in the basal region of the coleoptile is affected by the carotenoid-rich primary leaves (Vogelmann and Haupt 1985). When the coleoptile is irradiated unilaterally, the primary leaves absorb the B as it passes through the interior of the shoot. However, the coleoptile has a much lower carotenoid content and scatters some of the light before it enters the primary leaves. This light is scattered equatorially around the coleoptile so that there is more light present on the shaded side behind the primary leaves than within them (Fig. 31a, d).

The characteristics of the B gradient in this region differ from that in the mid-region above the primary leaves. As collimated B enters the coleoptile sheath, passes through the hollow centre and then into the shaded side of the coleoptile, it encounters several air-tissue boundaries. At these locations some light is reflected so that step-wise transitions are created in the light gradient (Fig. 31b, c).

The tip of the coleoptile is more homogeneous and is 12-15 cells in diameter. This region is the most sensitive to light and therefore the light gradient across the tip is of special interest. Measurements have shown that light is scattered as it passes across the tip creating a steep gradient (Fig. 31c, f).

Reconstructing the B gradients from these measurements shows that they are relatively steep throughout the different regions of the coleoptile (Fig. 32). The

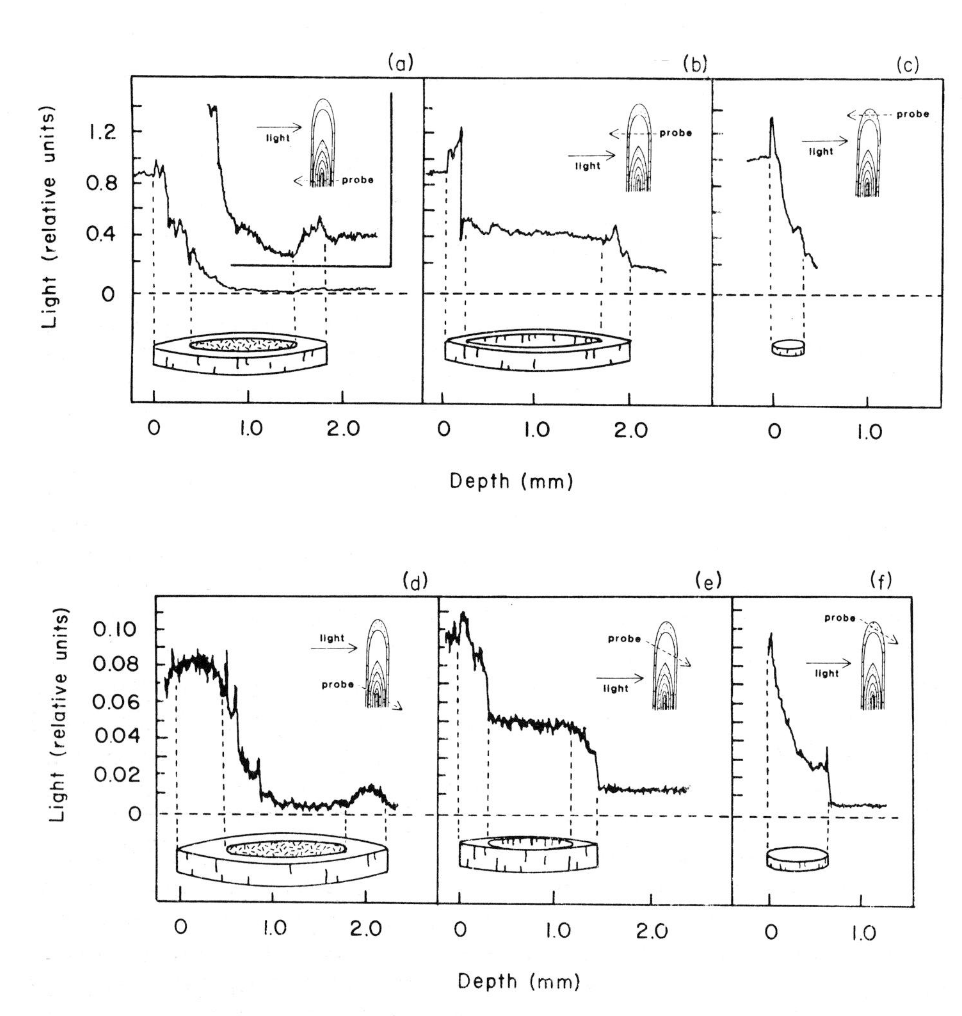

Figure 31. Distribution of transmitted (a-c) and scattered (d-f) blue light of 450 nm in different regions of an etiolated maize coleoptile. The coleoptile was irradiated unilaterally with collimated light. (a, d) The basal portion of the shoot that contains the coleoptile and primary leaves. In (a) the y-axis of the insert is expanded 10-fold. The peak of light on the shaded side in (a) and (d) is the result of equatorial scattering of light around the coleoptile periphery. (b, e) The region of the coleoptile above the primary leaves. The stepwise transitions are due to reflection between the epidermis-air interfaces. (c, f) The tip of a maize coleoptile. After Vogelmann and Haupt (1985).

ratio of the internal fluence rates near the irradiated and shaded surfaces was 4:1 in the tip and mid-region and 8:1 in the lower region that contains the primary leaves. This information is useful because there is good experimental evidence that the ability of coleoptiles to detect light direction is related to the steepness of the light gradient. The steeper the gradient, the more they bend towards the

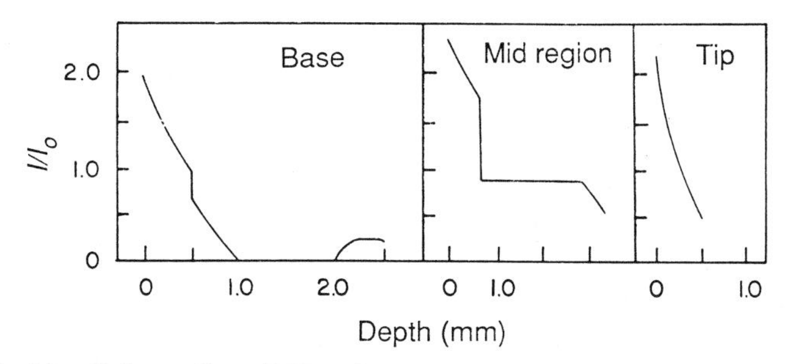

Figure 32. The blue-light gradient (450 nm) across representative regions of a maize coleoptile. The gradients were reconstructed from data presented in Fig. 31. The ratio of the internal fluence rates between the irradiated and shaded surfaces are 4:1 for the tip and mid-region of the coleoptile and 8:1 at the base. After Vogelmann and Haupt (1985).

light (*cf.* Dennison 1979; Piening and Poff 1988; Vierstra and Poff 1981). Thus far, there is no direct evidence that the tip of the coleoptile can focus light. However, this may depend upon scale. It seems unlikely that focusing could occur within a light-scattering tissue, 15 cells thick. But it could occur at the extreme tip of the coleoptile which is only a few cells in diameter.

Hopefully, as soon as more information becomes available about the photochemical and molecular properties of the B photoreceptor it will be possible to make the important link between light gradients and signal transduction in phototropism. Recent evidence suggests that one of the primary events in transduction is linked to massive phosphorylation of a membrane bound protein complex (Short and Briggs 1990).

It may seem obvious that the surface of plants would be optimal for detecting light quality, quantity and direction. However, even under the best of circumstances this may not be optimal from the standpoint of optics. When a plant is irradiated with collimated light, perhaps one half of the light signal crossing an epidermal cell is comprised of scattered light that originates from the interior of the tissues. This obscures the quality of the directional information that a single cell could perceive. There are several ways that light direction could be perceived in tissues where there is significant light scattering. One is *via* a lens effect where light is focused in a manner similar to that observed in single cells. This could create highly unequal fluence rates on opposing cell walls which leads to unequal absorption and the initiation of the transduction chain. A second way would be to detect light direction *via* the light gradient across the tissue. In this situation, single cells may not detect the light gradient that is across them, but they could respond individually to the fluence they receive. Thus, cells near the irradiated surface would respond more than cells near the shaded surface so that the phototropic response of the organ would be the integrated response of all the component cells, or perhaps just those cells within the epidermis.

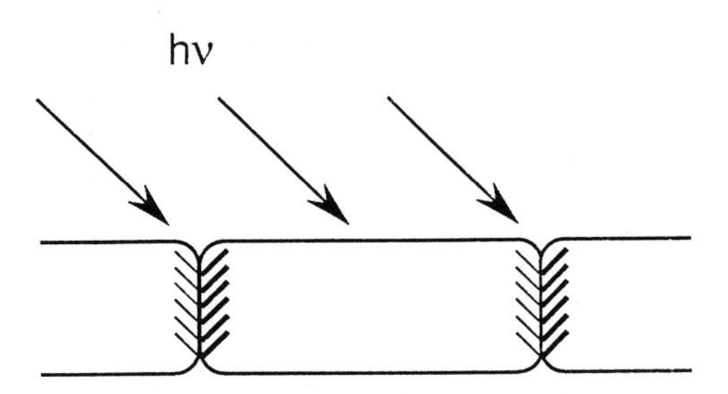

Figure 33. Detection of light direction by anisotropic orientation of photoreceptor molecules. Photoreceptors can be aligned near the end walls of cells so that only the photoreceptors on one side absorb light that comes from one direction (dark lines). Changing the direction of light by 90° would allow the photoreceptors on the other side of the cell to absorb the light. Either case could stimulate the polar transmission of a signal through the file of cells and give rise to a phototropic response. Thus, it is possible to detect light direction independent of a light gradient and this appears to be the way that leaves of suntracking plants determine the location of the sun. After Koller D., Ritter S., Briggs W.R. and Schäfer E. (1990) *Planta* 181: 184-190.

These two examples rely upon the use of light gradients for the perception of light direction (a lens effect creates an intracellular light gradient, albeit one where direction is reversed from the one in tissues). In both cases the light gradient results in a gradient in absorption which is the starting point for signal transduction. Yet, there is a third way to create a gradient in absorption without utilizing a light gradient. This can be accomplished by aligning the photoreceptors anisotropically within the cells. Such an arrangement would allow more absorption of light when it comes from certain directions over others (Fig. 33). This appears to be the mechanism that leaves of some plants use to track the sun. Members of the mallow family (Malvaceae) such as *Malva* and *Lavatera,* certain legumes (Fabaceae) such as *Lupinus* and many other plants in several other families position their leaves so that they are close to perpendicular to the rays of the sun throughout the day (Koller 1990). Thus, on clear days they follow the sun from sunrise to sunset. This phenomenon is known as suntracking and, in addition to leaves, is found in some flowers (Kevan 1975). The sites for photoperception in leaves appears to reside in the veins for *Malva* and *Lavatera* and near the base of the leaflets of *Lupinus* (Koller 1990). The site of photoperception in suntracking flowers is not yet known. The anisotropic orientation of photoreceptors provides an effective means of sensing light direction vectorially. It is not known how widespread the mechanism is, but in theory it could be used to sense light direction in phototropic responses.

7.4.4.3 Light gradients in leaves and photosynthesis

The distribution of visible light in leaves is important for providing the driving force for photosynthesis. In contrast to cotyledons and other plant organs which

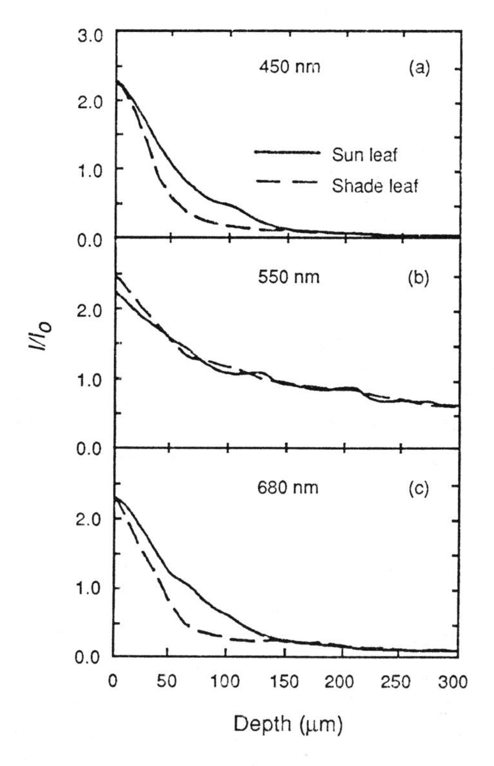

Figure 34. Light gradients within spinach leaves at wavelengths where there was strong and weak absorption. Light gradients were measured with a fibre-optic probe. After Cui *et al.* (1991).

are several millimetres thick, leaves of most plants are rather thin and usually range between 100-400 μm. Because they are only a few cell layers thick, they present a special challenge for the measurement of light gradients. In spite of the technical difficulties, light gradients have been measured within the leaves of several plants including spinach (Cui *et al.* 1991), alfalfa (Vogelmann *et al.* 1989) and *Smilacina* (Liliaceae) and *Thermopsis* (Fabaceae) (Vogelmann and Martin 1993).

A remarkable point about light gradients in leaves is that they are exceptionally steep at wavelengths corresponding to or near the absorption maxima of the photosynthetic pigments (Fig. 34). For example, in a spinach leaf, 90% of the light within the B (450 nm) and R (680 nm) is absorbed by the initial 140 μm of the leaf mesophyll. This distance is equivalent to about two palisade cells. Green light and other less strongly absorbed wavelengths penetrate much further and could be important for photosynthesis in cell layers located deeper within the leaf. It remains to be seen how closely the photosynthetic capacity of leaves corresponds to the light gradient. It is interesting to note that as the amount of light decreased within the leaf, chlorophyll content increased (Cui *et*

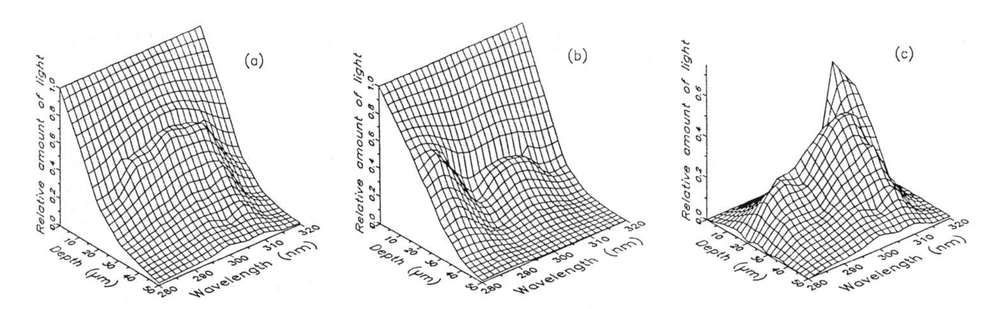

Figure 35. Light gradients within the UV-B in *Brassica campestris* leaves. Leaves were grown in the absence (a) or presence of UV-B (b). The steep light gradients in leaves grown under both conditions is caused by flavonoids in the epidermis which act as a UV screen. The difference between the light gradients (c) shows enhanced absorption of UV in UV-B-treated plants. The increased UV screening was caused by higher absorption within the epidermis. Data courtesy of C. Ålenius and J.F. Bornman.

al. 1991). This suggests that there may be a compensatory mechanism to equalize light absorption, and hence photosynthetic capacity, throughout the palisade. These relationships may also depend upon other factors such as the direction of light incident upon the leaf, whether it is collimated or diffuse, leaf thickness, chloroplast movement, and the PAR. Thus, the relationship between the photosynthetic capacity of the cell layers and the light gradients may not be as straight-forward as is commonly assumed.

7.4.4.4 How plants control the penetration of light

There are a number of situations in which it is advantageous to control how much light penetrates into plant tissues. This can be done by controlling the amount of absorption or light scattering. A good example of control of light penetration by changing absorption involves defense responses against UV. Radiation within the UV-B (280-320 nm) is especially damaging to plant tissues because the high energy per photon can result in the breakage or re-arrangement of carbon bonds within key molecules such as nucleic acids and proteins. Consequently, plants have evolved several defense mechanisms against UV-B. The simplest way to avoid damage within the tissues is to place screening pigments near the surface. In fact, most plants synthesize flavonoids which are sequestered within the vacuole of the epidermis and these pigments preferentially absorb UV-B while allowing penetration of longer wavelengths of light. Consequently, gradients of UV-B within leaves are very steep (Fig. 35) and in some cases do not extend beyond the epidermis (Day *et al.* 1993). Flavonoid synthesis is usually induced by UV; whereas etiolated tissues may be virtually defenceless against penetration of UV-B, plants exposed to elevated

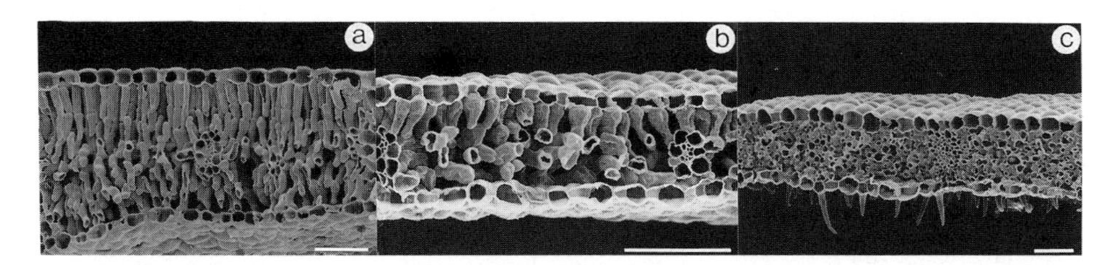

Figure 36. Anatomy of sun and shade leaves of *Thermopsis montana* (a, b) and *Smilacina stellata* (c). Leaves of *T. montana* grown in the sun (a) have a well-developed palisade layer in comparison to leaves grown in the shade (b). Leaves of the shade plant *S. stellata* consist of spongy mesophyll only (c). Columnar palisade cells in sun leaves facilitate the penetration of directional light (Fig. 37).

levels of this radiation synthesize more pigments and create steeper UV radiation gradients within the tissues (Fig. 35c).

A second way to influence the penetration of light is to alter the amount of light scattering within the tissues. This can be accomplished by changing the positioning of the organelles (Section 7.4.2.3) or by changing cell size and shape, a response commonly observed when leaves are grown under different fluence rates. Leaves grown under low light (shade leaves) often have a poorly differentiated layer of palisade and the cells are relatively amorphous in shape (Fig. 36a, b). Shade leaves are also thinner than their counterparts grown in the sun. Leaves grown under high light (sun leaves) are much thicker and they often have a well developed palisade that consists of one or more layers of tubular cells (Fig. 36c).

Since the palisade is an anatomical feature found in most leaves, to what advantage is it to possess tubular cells? The answer may depend upon the directional quality of light and the penetration of light into the leaf. Within the natural environment, high light is synonymous with direct sunlight which is highly collimated. However, the light that occurs underneath plant canopies has a low fluence rate and is predominately diffuse, although it can be punctuated by sunflecks, which are small beams of collimated light that leak through gaps in the canopy. Collimated light penetrates further into a leaf that has tubular palisade cells than it does in leaves with similar pigment content but containing spongy tissue only (Fig. 37a). Because sun leaves are substantially thicker than their shade counterparts, the facilitated penetration would distribute the light energy more evenly to cells throughout the leaf thus allowing for more uniform rates of photosynthesis and perhaps greater photosynthetic capacity overall. However, palisade cells do not facilitate the penetration of diffuse light (Fig. 37b). This may explain the poor development or absence of this layer in many shade leaves which develop under predominantly diffuse light. Depending on the amount of light during leaf development, mesophyll anatomy can be shade-like, sun-like or anything in between. Thus, leaves fine-tune the steep-

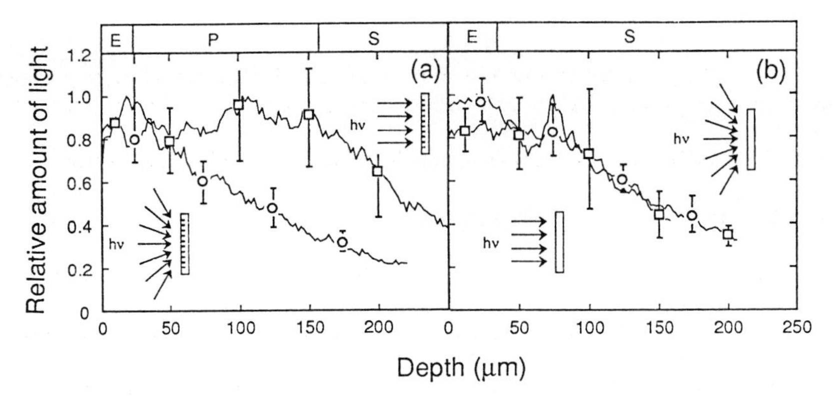

Figure 37. The influence of palisade tissue on the penetration of directional light. In (a) leaves with palisade were irradiated with directional (○) or diffuse light (□). The same experiment was repeated for leaves that lack palisade (b). Measurement of light gradients at 550 nm showed that the palisade facilitated the penetration of directional but not diffuse light. E = epidermis, P = palisade, S = spongy mesophyll. From Vogelmann and Martin (1993).

ness of the light gradient and in doing so, presumably optimize the photosynthetic capacity at the level of the whole leaf.

7.4.5 Summary

From the standpoint of optics, plants are complicated but great progress has been made in recent years towards solving some key problems. Whereas many experimental questions were unapproachable a few years ago, the development of new experimental techniques and capabilities for mathematical modelling of photon migration in plant tissues have made it possible to describe with ever increasing precision, optical phenomena that reside within plant cells and tissues. This should make it possible to examine, more rigorously, questions that pertain to mechanisms of light perception. There is reason for optimism that exciting and new discoveries will be forthcoming. A detailed knowledge of the optical properties of plants will contribute to the triad that is necessary for a complete understanding of photomorphogenesis: (i) the light environment within cells and tissues; (ii) the nature and location of photoreceptors; (iii) the mechanisms and end products of signal transduction. This triad comprises the responses that allow plants to interface their growth and development with their light environment.

7.4.6 Further reading

Fukshansky L. (1991) Photon transport in leaf tissue: Applications in plant physiology. In: *Photon-Vegetation Interactions*, pp. 253-302, Myneni R.B. and Ross J. (eds.) Springer-Verlag, Berlin.

Vogelmann T.C. (1993) Plant tissue optics. *Annu. Rev Plant Physiol. Mol. Biol.* in press.

7.4.7 References

Bernhard C.G., Gemne G. and Moller A.R. (1968) Modification of specular reflexion and light transmission by biological surface structures. *Quart. Rev. Biophys.* 1: 89-105.

Bone R.A., Lee D.W. and Norman J.M. (1985) Epidermal cells functioning as lenses in leaves of rain-forest shade plants. *Appl. Opt.* 24: 1408-1412.

Britz S.J. (1979) Chloroplast and nuclear migration. In: *Encyclopedia of Plant Physiology*, New Series, Vol. 7, *Physiology of Movements,* pp. 170-198, Haupt W. and Feinleib M.E. (eds.) Springer-Verlag, Berlin.

Britz S.J., Pfau J., Nultsch W. and Briggs W.R. (1976) Automatic monitoring of a circadian rhythm of change in light transmittance in *Ulva. Plant Physiol.* 58: 17-21.

Cheong W., Prahl, S.A. and Welch A.J. (1990) A review of the optical properties of biological tissues. *IEEE J. Quant. Elec.* 26: 2166-2185.

Cui M., Vogelmann T.C. and Smith W.K. (1991) Chlorophyll and light gradients in sun and shade leaves of *Spinacia oleracea. Plant Cell Environ.* 14: 493-500.

Day T.A., Vogelmann T.C. and DeLucia E.H. (1993) Differential penetration of ultraviolet-B radiation in leaves measured with a fiber-optic microprobe. *Oecologia,* in press.

Dennison D.S. (1979) *Phototropism.* In: *Encyclopedia of Plant Physiology,* New Series, Vol. 7, *Physiology of Movements,* pp. 506-560, Haupt W. and Feinleib M.E. (eds.) Springer-Verlag, Berlin.

Dennison D.S. and Vogelmann T.C. (1989) Intensity profiles in *Phycomyces* sporangiophores: Measurement with a fiber optic probe. *Planta* 179: 1-10.

Fukshansky L., Martinez v. Remisowsky A., McClendon J., Ritterbusch A., Richter T. and Mohr H. (1992) Absorption spectra of leaves corrected for scattering and distributional error: A radiative transfer and absorption statistics treatment. *Photochem. Photobiol.* 55:857-869.

Gabry's H. and Walczak T. (1980) Photometric study of chloroplast phototranslocations in leaves of land plants. *Acta Physiol. Plant.* 2: 281-290.

Haberlandt G.F. (1914) *Physiological Plant Anatomy* fourth ed., pp. 613-630, Macmillan, London.

Humphry V.R. (1966) The effects of paraffin oil on phototropic and geotropic responses in *Avena* coleoptiles. *Ann. Bot.* 30: 39-45.

Iino M. (1990) Phototropism: Mechanisms and ecological implications. *Plant Cell Environ.* 13: 633-650.

Jørgensen B.B. and Des Marais D.J. (1988) Optical properties of benthic photosynthetic communities: Fiber-optic studies of cyanobacterial mats. *Limnol. Oceanogr.* 33: 99-113.

Jose A.M. and Schäfer E. (1978) Distorted phytochrome action spectra in plants. *Planta* 138: 25-28.

Kay Q.O.N., Daoud H.S. and Stirton C.H. (1981) Pigment distribution, light reflection and cell structure in petals. *Bot. J. Linn. Soc.* 83: 57-84.

Kazarinova-Fukshansky N., Seyfried M. and Schäfer E. (1985) Distortion of action spectra in photomorphogenesis by light gradients within the plant tissue. *Photochem. Photobiol.* 41: 689-702.

Kaufmann W.F. and Hartmann K.M. (1988) Internal brightness of dish-shaped samples. *J. Photochem. Photobiol.* 1: 337-360.

Kevan P.G. (1975) Sun-tracking solar furnaces in high arctic flowers: significance for pollination and insects. *Science* 189: 723-726.

Knapp A.K., Vogelmann T.C., McClean T.M. and Smith W.K. (1988) Light and chlorophyll gradients within *Cucurbita* cotyledons. *Plant Cell Environ.* 11: 257-263.

Koller D. (1990) Light-driven leaf movements. *Plant Cell Environ.* 13: 615-632.

Kubelka P. (1948) New contributions to the optics of intensely light-scattering materials. Part I. *J. Opt. Soc. Amer.* 38: 448-457.

Kubelka P. (1954) New contributions to the optics of intensely light-scattering materials. Part II. *J. Opt. Soc. Amer.* 44: 330-335.

Latimer P. (1982) Light scattering and absorption as methods of studying cell population parameters. *Ann. Rev. Biophys.* 11: 129-150.

Latimer P. (1984) A wave optics effect which enhances light absorption by chlorophyll in vivo. *Photochem. Photobiol.* 40: 193-199.

Lee, D.W. (1986) Unusual strategies of light absorption in rain-forest herbs. In: *On the Economy of Plant Form and Function*. pp. 105-126, T.J. Givinish (ed.), Cambridge University Press, Cambridge, UK.

Lin Z.F. and Ehleringer J. (1983) Epidermis effects on spectral properties of leaves of four herbaceous species. *Physiol. Plant.* 59: 91-94.

Lork W. and Fukshansky L. (1985) The influence of chlorophyll fluorescence on the light gradients and the phytochrome state in a green model leaf under natural conditions. *Plant Cell Environ.* 8: 33-39.

Mandoli D.F. and Briggs W.R. (1982a) Optical properties of etiolated plant tissues. *Proc. Natl. Acad. Sci. USA* 79: 2902-2906.

Mandoli D.F. and Briggs W.R. (1982b) The photoperceptive sites and the function of tissue light-piping in photomorphogenesis of etiolated oat seedlings. *Plant Cell Environ.* 5: 137-145.

Mandoli D.F. and Briggs W.R. (1984) Fiber-optic plant tissues: Spectral dependence in dark-grown and green tissues. *Photochem. Photobiol.* 39: 419-424.

Marijnissen H.P, Star W.M., van Delft J.L. and Franken N.A. (1985) Light intensity measurements in optical phantoms and *in vivo* during HPD-photoradiation treatment using a miniature light detector with isotropic response. In: *Photodynamic Therapy of Tumors and other Diseases,* pp. 387-390, Jori G. and Perria C. (eds.) Libreria Progetto, Padua.

Martin G., Josserand S.A., Bornman J.F. and Vogelmann T.C. (1989) Epidermal focusing and the light microenvironment within leaves of *Medicago sativa. Physiol. Plant.* 76: 485-492.

Martin G., Myers D.A. and Vogelmann T.C. (1991) Characterization of plant epidermal lens effects by a surface replica technique. *J. Exp. Bot.* 42: 581-587.

Martinez v. Remisowsky A.J., McClendon J. and Fukshansky L. (1992) Estimation of the optical parameters and light gradients in leaves: Multi-flux versus two-flux treatment. *Photochem. Photobiol.* 55:857-865.

McClendon J.H. (1984) The micro-optics of leaves. I. Patterns of reflection from the epidermis. *Amer. J. Bot.* 71: 1391-1397.

Meyer A.M. (1969) Versuche zur 1. positiven und zur negativen phototropishen Krummung der *Avena* koleoptile: II. Die Inversion durch Paraffinol. *Z. Pflanzenphysiol.* 61: 129-134.

Piening C.J. and Poff K.L. (1988) Mechanism of detecting light direction in first positive phototropism in *Zea mays* L. *Plant Cell Environ.* 11:143-146.

Poulson M.E. and Vogelmann T.C. (1990) Epidermal focusing and photosynthetic light-harvesting in leaves of *Oxalis. Plant Cell Environ.* 13: 803-811.

Seyfried M. (1989) Optical radiation interactions with living tissue. In: *Radiation Measurement in Photobiology*. pp. 191-223, Academic Press, London.

Seyfried M. and Fukshansky L. (1983) Light gradients in plant tissue. *Appl. Opt.* 22: 1402-1408.

Seyfried M., Fukshansky L. and Schäfer, E. (1983) Correcting remission and transmission spectra of plant tissue measured in glass cuvettes: a technique. *Appl. Opt.* 22: 492-496.

Seyfried M. and Schäfer E. (1983) Changes in the optical properties of cotyledons of *Cucurbita pepo* during the first seven days of development. *Plant Cell Environ.* 6: 633-640.

Seyfried M. and Schäfer E. (1985) Phytochrome macrodistribution, local photoconversion and internal photon fluence rate for *Cucurbita pepo* L. cotyledons. *Photochem. Photobiol.* 42: 309-318.

Short T.W. and Briggs W.R. (1990) Characterization of a rapid, blue light-mediated change in detectable phosphorylation of a plasma-membrane protein from etiolated pea (*Pisum sativum* L.) seedlings. *Plant Physiol.* 92:179-185.

Shropshire Jr. W. (1974) Phototropism. In: *Proceedings of the VI International Congress on Photobiology*, pp. 1-6, Schenk G.O. (ed.) Mulheim, Inst. Strahlenchemie im Max-Planck-Inst. Kohlenforsch.

Star W.M., Marijnissen H.P, Jansen H., Keijzer M. and van Gemert M.J. (1987) Light dosimetry for photodynamic therapy by whole bladder wall irradiation. *Photochem. Photobiol.* 46: 619-624.

Terashima I. and Saeki T. (1983) Light environment within a leaf. I. Optical properties of paradermal sections of *Camellia* leaves with special reference to differences in the optical properties of palisade and spongy tissues. *Plant Cell Physiol.* 24: 1493-1501.

Vierstra R.D. and Poff K.L. (1981) Role of carotenoids in the phototropic response of corn seedlings. *Plant Physiol.* 68: 798-801.

Vogelmann T.C. and Björn L.O. (1984) Measurement of light gradients and spectral regime in plant tissue with a fibre optic probe. *Physiol. Plant.* 60: 361-368.

Vogelmann T.C. and Haupt W. (1985) The blue light gradient in unilaterally irradiated maize coleoptiles: Measurement with a fibre optic probe. *Photochem. Photobiol.* 41: 569-576.

Vogelmann T.C. and Martin G. (1993) The functional significance of palisade tissue: penetration of directional vs diffuse light. *Plant Cell Environ.* 16:65-72.

Vogelmann T.C., Bornman J.F. and Josserand S. (1989) Photosynthetic light gradients and spectral regime within leaves of *Medicago sativa. Proc. Phil. Trans. R. Soc. Lon. (B)* 323: 411-421.

Vogelmann T.C., Martin G., Chen G. and Buttry D. (1991) Fiber optic microprobes and measurement of the light microenvironment within plant tissues. *Adv. Bot. Res.* 18:256-296.

Wendlandt W.W. and Hecht H.G. (1966) *Reflectance Spectroscopy,* pp. 253-263, Wiley Interscience Pub., New York.

Zurzycki J. (1961) The influence of chloroplast displacements on the optical properties of leaves. *Acta Soc. Bot. Pol.* 30: 503-527.

7.5 Modelling the light environment

Lars Olof Björn

Section of Plant Physiology, Lund University,
Box 7007, S-220 07 Lund, Sweden

7.5.1 Introduction to natural light

Natural light at the surface of the earth is almost synonymous with light from the sun. Light from other stars has, as far as is known, photobiological importance only for the navigation by night-migrating birds. Moonlight, which originates from the sun, is important for the setting of some biological rhythms. It has been claimed that a full moon may perturb the photoperiodism of some short-day plants, and also synchronize rhythms in some marine animals. However, the majority of photobiological phenomena are ruled by daylight, and the remainder of this chapter will be devoted to this topic.

7.5.2 Modification of sunlight by the earth's atmosphere

Radiation from the sun is spectrally very similar to 'black-body' radiation of 6000 K. However, there are some deviations in the basic shape of the spectrum. Another characteristic of sunlight is the almost complete lack of some wavelength components due to reabsorption (Fraunhofer lines) of light by gases in the higher, cooler, layers of the sun. The Fraunhofer lines have no direct photobiological importance, but can be exploited in various ways with different measurements. Modelling of daylight begins with the sun's radiation. The next step is to consider the effect of the earth-sun distance, which varies throughout the year. The third step is to consider the effect of the earth's atmosphere and the elevation of the sun above the horizon, and finally effects of ground reflection, vegetation, and, in the case of aqueous environments, the effect of penetration into water.

The earth's atmosphere reflects, refracts, scatters, and partially absorbs the radiation from the sun, and thereby changes its spectral composition considerably. Part of the absorption and Rayleigh scattering is due to the main gases in the atmosphere, the concentration of which can be regarded as constant.

R.E. Kendrick & G.H.M. Kronenberg (eds.), Photomorphogenesis in Plants - 2nd Edition

Another part is due to ozone and water vapour, which occur in highly variable amounts. A third part is due to aerosols (man-made and natural), which is also highly variable. The absorption causes loss of light, while scattering causes some light to be lost to space, and some to appear as diffuse light (skylight). Light is also reflected by clouds, and thereby mostly lost to space. Light reflected from the ground is partly scattered downwards or reflected from clouds, and reappears at the surface as diffuse light, meaning that the ground reflectivity has some effect on skylight.

Daylight is also strongly dependent on the elevation of the sun above the horizon (90° minus the solar elevation is called the zenith angle of the sun), because the lower the sun, the more air the rays must pass before they reach the ground.

There are two basic ways of modelling daylight using computer programs. In the first, basic physical relationships are used to model radiative transfer by processes of absorption and scattering. A rigorous theory was worked out long ago and first applied to stellar atmospheres (Chandrasekhar 1950). For practical use of the general theory certain special cases are considered. For instance, the atmosphere can be treated as a flat layer above a flat ground (however, a spherical solution can also be found). Because the properties of the atmosphere change with elevation, the atmosphere has to be divided into a number of horizontal layers having different absorption and scattering coefficients. The scattering phase function, *i.e.* the probability distribution of direction changes in single scattering events, must also be known. It varies with the type of scattering particle. Using a Monte Carlo approach a statistical distribution of photon fates can be found. These approaches have the advantage of being founded on sound physical principles and, properly applied, are capable of producing accurate results. The main disadvantage is that they are very computer intensive, and not suitable for simple desk-top computers.

The other approach is a semi-empirical one. The whole depth of the atmosphere is treated as one unit, with different mathematical functions modelling the effects of Rayleigh scattering, aerosol scattering, and absorption. These approaches are not capable of such accurate results. Usually the light is divided into direct (coming from the direction of the sun) and diffuse (scattered light or skylight), and these components are modelled separately before being added up. Although several of the programs have a high degree of sophistication, they are easy to handle and produce results quickly. For most biological purposes the accuracy is sufficient, particularly in the visible and near infra-red regions. The next section will deal in some detail with one popular model. We shall not treat the UV-B region here, but refer to my earlier treatise on the subject (Björn 1989) and the much more sophisticated model by Smith *et al.* (1992).

As an example of the kinds of simplifications that are performed when going from a more rigorous procedure requiring large computers to one suitable for desk-top computers, let us look at the relations used for the attenuation due to Rayleigh scattering (*cf.* Teillet 1990). The basic equation is:

$$d_R = [8\pi^3(n^2 - 1)^2N_c/(3L^4N_s^2)](6 + 3g)/(6 - 7g)(p/p_0)(T_0/T)$$

where d_R is the Rayleigh optical depth, n the refractive index of air, g depolarization factor, N_c the columnar number density (air molecules per unit ground area), N_s the molecular number density (air molecules per unit volume), L wavelength (in cm), p pressure (p_0 = 101.325 kPa or 1013.25 mb for standard conditions), and T temperature (T_0 = 273.15 K for standard conditions). N_s, n, p, and T all vary with altitude, which necessitates separate treatment for the different air layers. In addition n varies with wavelength, so there is a 'hidden' wavelength dependence of d_R in addition to the factor L^4 in the nominator. In the 'desk top model' SPCTRAL2 (see below) all this is lumped together as:

$$d_R = L^{-4}(115.6406 - 1.3366L^{-2})^{-1}$$

for the whole thickness of the atmosphere. A slightly better and only marginally more complicated expression would have been:

$$d_R = 0.008569L^{-4}(1 + 0.0113L^{-2} + 0.00013L^{-4})$$

(Hansen and Travis 1974; Teillet 1990).

7.5.3 The SPCTRAL2 model of Bird and Riordan

I will present here a method based largely on a paper by Bird and Riordan (1986), describing a model called SPCTRAL2, which is applicable to the visible and adjoining spectral regions [an alternative approach for this part of the spectrum is that of Green and Chai (1988)]. The mode of model building and development of formulae are described by Bird and Riordan (1986) and Björn (1989), and I shall only very briefly outline the principles here. After that I shall concentrate on the practical use of the model and some developments of it. A slightly modified version of SPCTRAL2, which we call DAYLIGHT VIS-IR written in QuickBasic is listed in the Appendix. It is intended for use in the 300 to 800 nm region. As it is written, the amount of ozone has to be entered. However, ozone has little or no effect over most of the spectral region and a way of approximate automatic ozone estimation is described by Björn (1989).

Figure 1 is an example of input to and output from Daylight VIS-IR. The output panel shows three spectral irradiance spectra, representing direct sunlight, skylight (diffuse radiation), and their sum, the so-called 'global

```
YEAR, MONTH, DATE, TIME (h, min)? 1993,8,8,12,0
LATITUDE (N+, S-) AND LONGITUDE (E+, W-)? 55.8,13.4
OZONE, atm cm? .3
PRECIPITABLE WATER, cm? 5
SOLAR ELEVATION IS  49.63934 DEGREES
INCIDENCE ANGLE, degrees? 40.36066
BAROMETRIC PRESSURE, mbar? 1000
AMOUNT OF AEROSOL? 2
TILT ANGLE: ZERO FOR HORIZONTAL, 90 FOR VERTICAL, degrees? 0
GROUND ALBEDO? .2
LOWER WAVELENGTH LIMIT, MICROMETERS? .3
UPPER WAVELENGTH LIMIT, MICROMETERS? .8
SCALE? 6
```

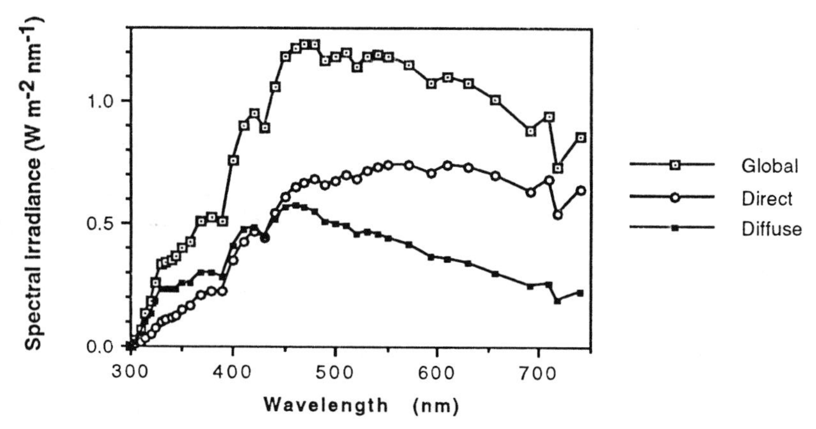

Figure 1. Daylight spectral irradiance at noon on a horizontal plane computed with the program DAYLIGHT VIS/IR. On top the input screen is shown. The parameters after the question marks are entered by the operator, the rest printed by the program.

radiation'. The vertical scale at the top is indicated by a short horizontal line on the vertical axis and a value of spectral irradiance in W m^{-2} nm^{-1}.

Note that the diffuse light has its maximum moved towards shorter wavelengths compared to the direct sunlight. This corresponds to the fact that the sky appears blue in colour, and also to the fact that Rayleigh scattering is inversely proportional to the fourth power of the wavelength.

In Fig. 2 the same values have been chosen again, except that the time has been chosen to be just before sunset instead of noon, and the tilt of the reference plane is such that the light from the sun impinges perpendicularly. The scale factor has been chosen such that the graph for global radiation fills the panel. Note how deep the absorption bands for water vapour and oxygen have become, because the light must pass so much air when the sun is so low in the horizon.

```
YEAR, MONTH, DATE, TIME (h, min)? 1993,8,8,19,0
LATITUDE (N+, S-) AND LONGITUDE (E+, W-)? 55.8,13.4
OZONE, atm cm? .3
PRECIPITABLE WATER, cm? 5
SOLAR ELEVATION IS  4.585979 DEGREES
INCIDENCE ANGLE, degrees? 0
BAROMETRIC PRESSURE, mbar? 1000
AMOUNT OF AEROSOL? 2
TILT ANGLE: ZERO FOR HORIZONTAL, 90 FOR VERTICAL, degrees? 85.414021
GROUND ALBEDO? .2
LOWER WAVELENGTH LIMIT, MICROMETERS? .3
UPPER WAVELENGTH LIMIT, MICROMETERS? .8
SCALE? 6
```

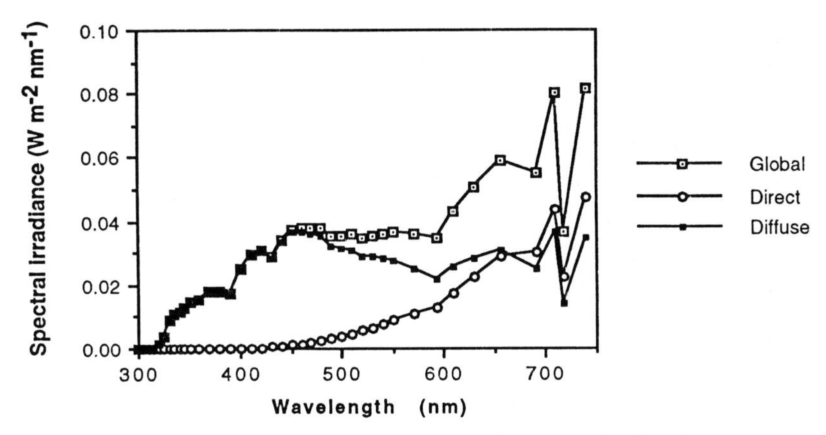

Figure 2. Daylight spectral irradiance on a plane perpendicular to the direction to the sun in the evening. Otherwise as Fig. 1.

7.5.4 Computation of photosynthetically active radiation

DAYLIGHT VIS-IR can be modified (DAYLIGHT PAR) to provide photon irradiance integrated over a certain interval, usually 400 to 700 nm, so-called 'photosynthetically active radiation' (PAR). As an example of what can be done with this program, I have computed PAR falling on leaves with different orientations under a clear sky at different times of the day (Fig. 3). For comparison I have also plotted the plant-active UV-B component of daylight computed using another program (*cf.* Björn 1989). The only radiation falling on a vertical leaf with the flat surfaces in the east-west direction at noon is the skylight component. The dip at noon is much less pronounced for UV radiation because a greater proportion of it comes from the sky than is the case for visible light.

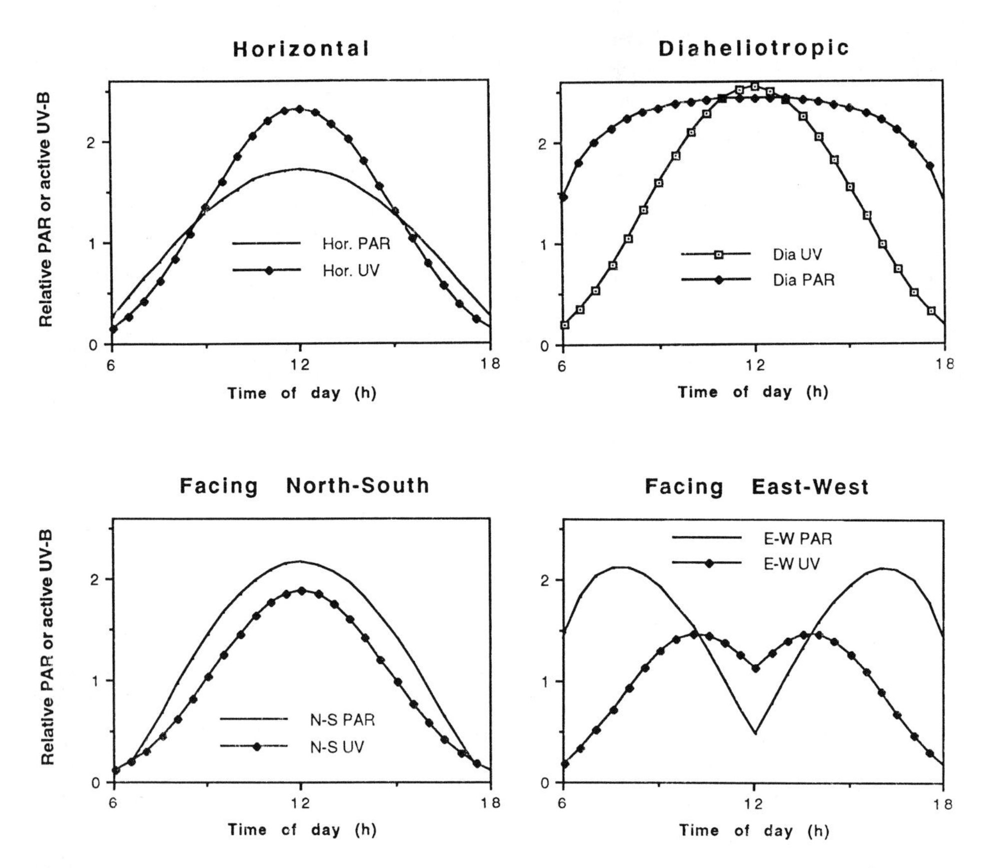

Figure 3. Daily course of photosynthetically active radiation (PAR, in mol photon m^{-2} h^{-1}, solid lines) and plant-active UV-B (UV-B, in Caldwell weighted J m^{-2} h^{-1}, lines with symbols, height adjusted for convenient plotting, but same scale in all frames) incident on leaves with different orientations. For the meaning of Caldwell weighted UV-B, see Chapter 2. North-South means that the surfaces of the leaves are in these directions (*i.e.* edges pointing in east and west). Values are computed for cloudless conditions in Lund, Sweden, on June 15. From Björn 1993, reprinted with permission of Elsevier Science Publishers BV.

7.5.5 Underwater daylight

Pure water transmits best in the UV-A to B region of the spectrum. The estimation of underwater light by computation is difficult and uncertain for several reasons. Natural waters do not consist of chemically pure water. Fresh water bodies and coastal sea water contain dissolved substances and particles that attenuate UV and B light more strongly than green, and therefore the light penetrating such waters contains less UV and B than that penetrating into clear ocean water. In clear ocean water it is water itself that is the main attenuator. Jerlov (1970) described several types of natural waters with respect to the penetration of visible light. For the penetration of UV-B into different types of natural water, the reader is referred to Smith and Baker (1979) as well as Baker and Smith (1982). An alternative for the visible region to the models described here is that of Morel (1988).

To compute the underwater light or UV radiation according to the model of Baker and Smith (1982) and Smith and Baker (1978), we should first know separately the direct beam (from the sun) and the scattered (diffuse) radiation above the water, as well as the solar elevation or zenith angle. In the case of a clear sky, these can be computed with 'DAYLIGHT VIS-IR'. The direct irradiance on a horizontal surface is cos(Z) times the direct beam *IDL* and the diffuse light is *ISL* , Z being the zenith angle of the sun.

We next compute the transmission through the water surface of the two irradiance components separately, add them up, and finally compute the transmission of light down through the water to the depth in which we are interested.

7.5.5.1 Transmission through the water surface

For computing the transmission of light through the water surface, several cases may be distinguished. A smooth horizontal water surface is theoretically simpler than one with ripples or waves, but not so common in real life. For a smooth water surface we may apply Fresnel's equation for unpolarized light. For the diffuse light we have to integrate over all directions above the water. The result of this integration, however, can be put equal to 93.4% penetration for all wavelengths. Strictly speaking, the refractive indices, and thus the reflectivity and transmittivity of the water surface, depends on wavelength, temperature, and salinity of the water, and on the air pressure. However, for all practical purposes these variations can be neglected, since the approximation that the water surface is flat is a much cruder one.

When the sky is overcast and the water surface smooth, and assuming the spectral irradiance above the water surface is known, the transmission factor (transmittivity) 0.948, *i.e.* 94.8% may be used. The reason that this is slightly different from the factor for diffuse radiation from a clear sky is that the overcast sky does not have the same brightness in all directions (a so-called 'cardioidal brightness distribution' has been assumed).

Several investigators, *e.g.* Preisendorfer and Mobley (1986), have tried to model transmission through a wind-roughened sea surface. Results can be summarized as follows: transmittance of skylight is only slightly affected (increase by 1.2% for clear sky and 0.5% for overcast sky). The transmission of direct radiation is virtually unaffected by wind to a zenith angle of 70°, after which it remains near 84% (*i.e.* at higher values than for a smooth water surface).

The change in irradiance as we proceed down through the water below the surface is governed by absorption and scattering of the light by the water itself and dissolved and suspended matter in it, and by the re-reflection down from the water surface of light that has been scattered up towards it. To handle the combined effect of absorption and scattering the so-called 'diffuse spectral attenuation coefficient'(K_T) is used. This has been experimentally determined by light measurements in natural water bodies (mostly the ocean). It differs from the absorption coefficient in that it includes the effect of scattering and also the fact that the rays of the downwelling light are not all vertical. It can be computed as a sum of three components, one being due to the water itself, one dependent on the amount of chlorophyll (including chlorophyll-like pigments and substances co-varying with chlorophyll-like pigments), and one dependent on other impurities.

7.5.5.2 Water types

It has been found convenient to separate the natural waters into two main categories, those containing less than 1 μg/dm^3 of chlorophyll-like pigments, and those containing more than this concentration. In those waters containing little chlorophyll, the organisms are mostly alive, and therefore all absorbing and scattering 'impurities' in the water are well correlated to the amount of chlorophyll. In waters with high amounts of chlorophyll, many organisms have died due to nutrient depletion and other causes, and it is appropriate to add a special term for the effect of matter which is not proportional to the amount of chlorophyll. Smith and Baker (1978) therefore obtain:

$$K_T(\lambda) = K_W(\lambda) + k_1(\lambda)\cdot C_K \qquad \text{for } C_K < 1$$

$$K_T(\lambda) = K_W(\lambda) + K_{x2}(\lambda) + k_2(\lambda)\cdot C_K \qquad \text{for } C_K > 1$$

Here C_K denotes the concentration of chlorophyll-like pigments in μg/dm^3 and postscript (λ) wavelength dependence. $K_W(\lambda)$ is the diffuse attenuation coefficient for the water itself and $K_{x2}(\lambda)$ that of the 'impurities' that are not proportional to the amount of chlorophyll. For the chlorophyll-associated attenuators, different proportionality factors, k_1 (λ) and $k_2(\lambda)$, must be chosen depending on the chlorophyll concentration. The wavelength-dependent constants $K_{x2}(\lambda)$, k_1 (λ) and $k_2(\lambda)$ are chosen in such a way that continuity of $K_T(\lambda)$ is obtained for $C_K = 1$.

7.5.5.3 Modelling water properties

Smith and Baker (1978) have listed values for $K_W(\lambda)$, $K_{x2}(\lambda)$, $k_1(\lambda)$ and $k_2(\lambda)$. However, for most applications it should be sufficient and more convenient to use analytical (polynomial) approximations rather than tables. I have found the following polynomials to give reasonable approximations (units are nm for λ, m^{-1} for $K_W(\lambda)$ and K_{x2}, and m^{-1} ($\mu g/dm^3)^{-1}$ for k_1 and k_2):

$K_W = 1.0153 - 0.0095\lambda + 3.782 \cdot 10^{-5}\lambda^2 - 7.009 \cdot 10^{-8}\lambda^3 + 4.916 \cdot 10^{-11}\lambda^4$ (300 nm $< \lambda <$ 700 nm).

$K_{x2} = -19.7971 + 0.1876\lambda - 6.816 \cdot 10^{-4}\lambda^2 + 1.198 \cdot 10^{-6}\lambda^3 - 1.018 \cdot 10^{-9}\lambda^4 + 3.335 \cdot 10^{-13}\lambda^5$ ($350 < \lambda < 575$ nm).

$K_{x2} = 1.819 \cdot 10^4 - 137.6747\lambda + 0.4146\lambda^2 - 6.207 \cdot 10^{-4}\lambda^3 + 4.616 \cdot 10^{-7}\lambda^4 - 1.363 \cdot 10^{-10}\lambda^5$ (575 nm $< \lambda <$ 670 nm).

$K_{x2} = 0$ (670 nm $< \lambda <$ 700 nm).

$k_1 = -1.7331 + 0.0152\lambda - 3.681 \cdot 10^{-5}\lambda^2 + 2.760 \cdot 10^{-8}\lambda^3$ ($350 < \lambda < 575$ nm).

$k_1 = -9037.2943 + 74.8275\lambda - 0.247\lambda^2 + 4.065 \cdot 10^{-4}\lambda^3 - 3.334 \cdot 10^{-7}\lambda^4 + 1.090 \cdot 10^{-10}\lambda^5$ (575 nm $< \lambda <$ 700 nm).

$k_2 = 0.3139 - 0.0012\lambda + 1.645 \cdot 10^{-6}\lambda^2 - 9.529 \cdot 10^{-10}\lambda^3$ (350 nm $< \lambda <$ 575 nm)

$k_2 = 694.3722 - 5.7408\lambda + 0.019\lambda^2 - 3.125 \cdot 10^{-5}\lambda^3 + 2.572 \cdot 10^{-8}\lambda^4 - 8.450 \cdot 10^{-12}\lambda^5$ (575 nm $< \lambda <$ 700 nm).

Based on the above, I have computed horizontal spectral irradiance for one very clear water, and one containing phytoplankton at a concentration of 0.5 μg chlorophyll per litre (Figs. 4 and 5).

An alternative way of modelling the underwater light attenuation, which is valid for the wavelength range 440-670 nm, is described by Voss (1992).

The UV-B region has to be treated separately. The reader is referred to Smith and Baker (1979). Their total diffuse attenuation coefficient for ultraviolet radiation for the clearest ocean water can be approximated by:

$27.8734 - 0.2457\lambda + 7.294 \cdot 10^{-4}\lambda^2 - 7.271 \cdot 10^{-7}\lambda^3$ m^{-1} (280 nm $< \lambda <$ 340 nm).

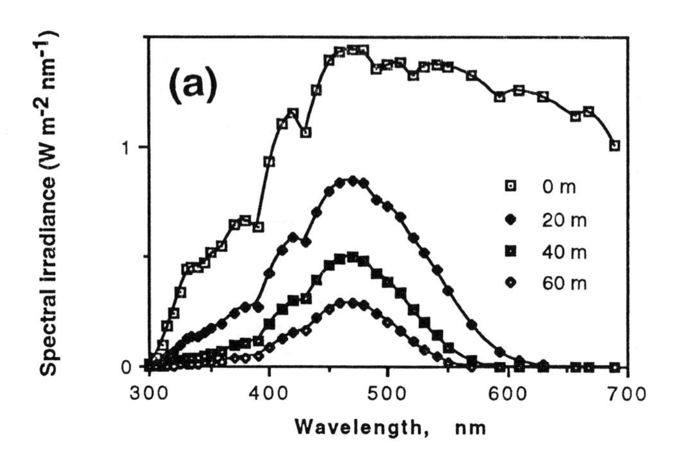

Figure 4(a). Spectral irradiance in clear water (no phytoplankton) computed for noon on June 15, at a point in Skagerrak (58°N, 10°E), clear sky.

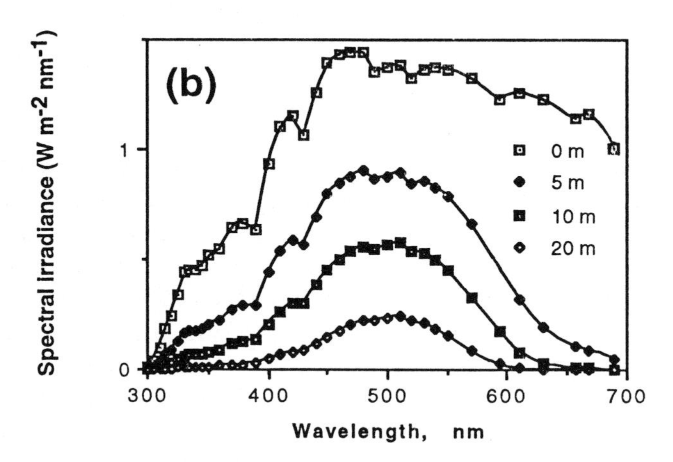

Figure 4(b). Same as in Fig. 4(a), except that the water contained phytoplankton corresponding to 0.5 μg chlorophyll per litre.

7.5.5.4 Underwater light direction

As everyone who has been swimming underwater knows, the underwater sky, due to refraction in the water surface, is confined to a solid angle much smaller than the sky seen from above the water (about a third of the sky solid angle above water). In calm weather virtually only light reflected from the bottom reaches the eye from outside this solid angle of 2.13 steradians (corresponding to a refraction angle of 48.6°).

Within this solid angle (the 'Snell window') the light direction close to the water surface depends on the position of the sun and the cloud conditions (*cf.*

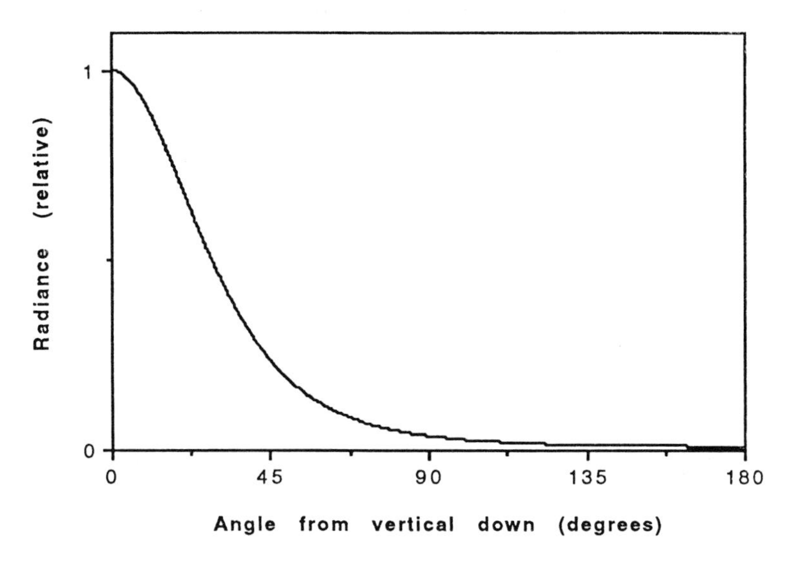

Figure 5. The angular distribution of radiance under typical sea conditions at a depth where the effect of surface angular distribution of radiance has vanished.

Jerlow and Fukuda 1960). Deeper down, however, the distribution of light directions, because of absorption and scattering effects, becomes independent of these factors and also on water surface structure. If water were only to absorb and not scatter, the light direction would, with increasing depth, approach the vertical, since components with all other directions would have longer path lengths through the water and be more strongly absorbed. If water were only to scatter light and not absorb, the light would become more and more isotropic with depth, *i.e.* an equal amount of light would strike an underwater sphere from all sides. In reality water both absorbs and scatters light, and the relation between the two processes varies with wavelength and with the type of natural water. With increasing depth, in real water, the angular light distribution approaches a non-isotropic distribution. Preisendorfer (1959) derived a theoretical expression for the limiting angular distribution, but Tyler (1960) found a slightly different expression to agree better with measurements, *i.e.* $I_\phi = I_o/(1 - \alpha \cdot \cos\phi)^\beta$, where I_o is the amount of light propagating in the vertical direction and I_ϕ the light propagating in a direction forming the angle ϕ with the vertical. Both α and β are constants depending on wavelength and water type (Preisendorfer's expression is a special case of this, with $\beta = 1$). Denton (1970) and Denton *et al.* (1972), for instance, used $\alpha = 0.71$ and $\beta = 2.68$ for the blue-green band (475-480 nm) penetrating deep into the sea in which they studied the matching of camouflaging bioluminescence and fish reflectance patterns to underwater daylight. This distribution, symmetrical around the vertical direction, which should be typical of clear sea water, is depicted in Fig. 5.

In the Gullmaren Fjord in Sweden, Jerlov and Fukuda (1960) found the symmetrical distribution to be almost (but not completely) reached at a depth of 30 m. In clear seawater a depth ten times greater may be required.

Knowledge of the angular distribution of light is necessary for converting measurements or modelling of under-water irradiance to the biologically more relevant fluence rate. For the distribution shown in Fig. 5 the fluence rate is 1.44 times the horizontal downward irradiance (*i.e.* with the sensor pointing up), and the horizontal upward irradiance is 4.3% of the horizontal downward irradiance.

7.5.6 Effects of ground and vegetation

Reflection from the ground is particularly important in the UV, since UV reflected upward by the ground is partially scattered downward again by the atmosphere, and the ground cover thus affects also downwelling radiation. This effect has been incorporated into the programs described above, although the spectral variation of the reflectivity is not considered. The effect of reflection from the ground is greatest when it is covered by snow.

The spectral characteristics of reflection from the ground can be quite important for plant growth, as shown by Hunt *et al.* (1985) and Kasperbauer and Hunt (1987) (*cf.* Chapter 7.1)

Penetration of light into the ground is important for the germination of seeds. Soil transmission generally increases with increasing wavelength, thus giving buried seeds a far-red light (FR) biased environment (Kasperbauer and Hunt 1988).

Plant canopies absorb visible light and UV radiation, but reflect and transmit FR and near infra-red radiation. Light in or under green vegetation is therefore strongly biased towards the longer wavelengths, a fact which is of paramount importance to the plants subjected to this regime (Chapter 7.1).

It is now also possible to measure light *inside* plants and animals (Chapter 7.4). For modelling of light inside leaves, the reader is referred to Björn (1992).

Many models have been developed for how plant canopies affect daylight. Most of them are not concerned with photomorphogenesis, but deal either with PAR, or remote sensing of vegetation (see Björn 1992 for a literature survey). It is beyond the scope of this chapter to describe the more advanced models here, and I shall concentrate instead on one of the first and simplest, the Monteith (1965) model.

Monteith (1965) divides the canopy into a number of horizontal layers, each of unit leaf area index (LAI). The LAI is the combined one sided area of the leaves divided by the corresponding ground area (in case of needles and other non-flat leaves usually the projected leaf area is used). Thus, each layer in the

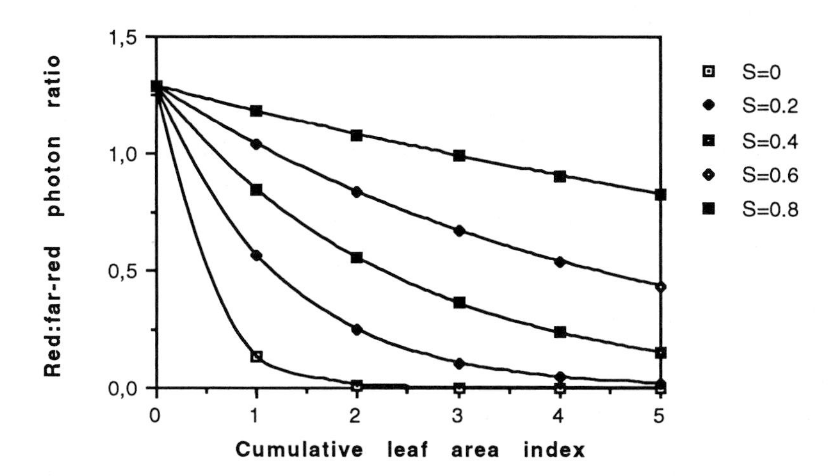

Figure 6. Red/far-red photon ratios in a canopy as computed with a combination of the DAYLIGHT VIS-IR model and the Monteiths (1965) canopy model. The parameter S is the fraction of the light that passes a canopy layer of unit leaf area index without encountering any leaf. The lowest curve corresponds to the rather unrealistic case that the leaves within each layer of unit leaf area index form a coherent sheet without any gaps. The highest curve is for a canopy where the leaves are to a large extent parallel to the main light direction, or highly clustered and overlapping.

Monteith model is selected such that the one-sided leaf area within the layer is equal to the ground area. The symbol S stands for the fraction of light passing through each layer without encountering any leaf and T the transmittance of a leaf. At a level where the cumulative leaf area index from the top is L, irradiance is $[S + (1 - S)T]^L$ times the irradiance at the top of the canopy. Two shortcomings of this model are that it does not consider light reflected from leaves, and that it does not take into account the fact that some leaves overlap within the same LAI layer, and that light is incident on the canopy from several directions. Nevertheless it provides an idea of how irradiance within the canopy varies with S and L. As an example I have computed the dependence of the R/FR ratio on S and L (Fig. 6; *cf.* Holmes and Smith 1977). The wavelengths 656 nm (R) and 724.4 nm (FR) were chosen for simplicity, since they are standard in the Bird and Riordan data set and thus need not be interpolated. The T values of 0.042 (656 nm) and 0.415 (724.4 nm) were used.

In more sophisticated canopy models, the distribution of leaf angles (angle to the horizontal and azimuth angle of tilt) must be known or assumed. Several models allow for such different distributions within different canopy layers. For an interesting example of what can be achieved with a detailed model in a real case, the reader is referred to Ryel *et al.* (1990), Beyschlag *et al.* (1990) and Barnes *et al.* (1990). This group studied how the modification of canopy architecture by UV radiation in two grass species affected the mutual shading, and thereby the competition between the species.

7.5.7 Effects of clouds

Clouds have a great influence on daylight, but there is no way of modelling the effect on the instantaneous daylight without knowing the type of cloud cover and the position of the clouds on the sky. The cloud effect on monthly averages is easier to model. A few suggestions are referenced by Björn (1989).

7.5.8 Further reading

Björn L.O. (1989) Computer programs for estimating ultraviolet radiation in daylight. In: *Radiation measurements in photobiology*, pp. 162-189, Diffey B.L. (ed.) Academic Press, London.

Björn L.O. (1992) Interception of light by plant leaves. In: *Crop Photosynthesis: Spatial and Temporal Determinants*, pp. 253-276, Baker N.R. and Thomas H. (eds.) Elsevier Science Publishers, Amsterdam.

Fouquart Y, Irvine W.M. and Lenoble J. (eds.) (1980) *Standard Procedures to Compute Atmospheric Radiative Transfer in a Scattering Atmosphere.* Vol. II. Radiation Commission (IAMAP), Boulder, Colorado.

Iqbal M. (1983) *An Introduction to Solar Radiation.* Academic Press, New York.

Lenoble J. (1977) *Standard Procedures to Compute Atmospheric Radiative Transfer in a Scattering Atmosphere.* Vol. I. Radiation Commission (IAMAP), Boulder, Colorado.

Jerlov N.G. (1968) *Optical Oceanography.* Elsevier Publishing Company, Amsterdam.

Jursa A.S. (ed.) (1985) *Handbook of Geophysics and the Space Environment.* Air Force Geophysics Laboratory, Air Force Systems Command, United States Air Force. (Can be obtained from National Technical Information Service, 5285 Port Royal Road, Springfield, VA 22161, USA. The Document Accession Number ADA 167000 should be quoted).

Myneni R.B. and Ross J. (eds.) (1991) *Photon-Vegetation Interactions.* Springer, Berlin.

7.5.9 References

Baker K.S. and Smith R.C. (1982) Biooptical classification and model of natural waters II. *Limnol. Oceanogr.* 27: 500-509.

Barnes P.W., Beyschlag W., Ryel R., Flint S.D. and Caldwell M.M. (1990) Plant competition for light analyzed with a multispecies canopy model. 3. Influence of canopy structure in mixtures and monocultures of wheat and wild oat. *Oecologia* 82: 560-566.

Beyschlag W., Barnes P.W., Ryel R., Caldwell M.M. and Flint S.D. (1990) Plant competition for light analyzed with a multispecies canopy model. 2. Influence of photosynthetic characteristics on mixtures of wheat and wild oat. *Oecologia* 82: 374-380.

Bird R.E. and Riordan C. (1986) Simple solar spectral model for direct and diffuse irradiance on horizontal and tilted planes at the earth's surface for cloudless atmospheres. *J. Climate and Appl. Meteorology* 25: 87-97.

Björn L.O. (1989) Computer programs for estimating ultraviolet radiation in daylight.In: *Radiation measurements in photobiology,* pp. 162-189, Diffey B.L. (ed.) Academic Press, New York.

Björn L.O. and Murphy T.M. (1985) Computer calculation of solar ultraviolet radiation at ground level. *Physiol. Vég.* 23: 555-561.

Chandrasekhar S. (1950) *Radiative transfer.* Oxford University Press.

Denton E.J. (1970) On the organization of reflecting surfaces in some marine animals. *Phil. Trans. Roy. Soc. Lond.* Series B. 258: 285-313.

Denton E.J., Gilpin-Brown J.B., Widder E.A., Latz M.F. and Case J.F. (1972) The angular distribution of the light produced by some mesopelagic fish in relation to their camouflage. *Proc. Roy. Soc. Lond.* B 225: 63-97.

Green A.E.S. and Chai S.-T. (1988) Solar spectral irradiance in the visible and infra-red regions. *Photochem. Photobiol.* 48: 477-486.

Hansen J.E. and Travis L.D. (1974) Light scattering in planetary atmospheres. *Space Sci. R.* 16: 527-610.

Holmes M.G. and Smith H. (1977) Spectral distribution of light within plant canopies. In: *Plants and the Daylight Spectrum,* pp. 147-158, Smith H. (ed.) Academic Press, New York.

Hunt P.G., Kasperbauer M.J. and Matheny T.A. (1985) Effect of soil surface color and *Rhizobium japonicum* strain on soybeen seedling growth and nodulation. *Agronomy Abstr.* 85: 157.

Jerlov N.G. (1970) Light: General introduction. In: *Marine Ecology,* Vol. 1, Part 1, pp. 95-102, Kinne O. (ed.) Wiley-Interscience.

Jerlov N.G. and Fukuda M. (1960) Radiance distribution in the upper layers of the sea. *Tellus* 12: 348-353.

Kasperbauer M.J. and Hunt P.G. (1987) Soil color and surface residue effects on seedling light environment. *Plant Soil* 97: 295-298.

Kasperbauer M.J. and Hunt P.G. (1988) Biological and photometric measurement of light transmission through soils of various colors. *Bot. Gaz.* 149: 361-364.

Monteith J.L. (1965) Light distribution and photosynthesis in field crops. *Ann. Bot. N.S.* 29: 19-37.

Morel A. (1988) Optical modelling of upper ocean in relation to its biogenous matter content (Case 1 waters). *J. Geophys. Res.* 93: 10749-10768.

Preisendorfer R.W. (1959) Theoretical proof of the existence of characteristic diffuse light in natural waters. *Contr. Scripps Instn. Oceanogr.* no. 1094: 1-9.

Preisendorfer R.W. and Mobley C.D. (1986) Albedos and glitter patterns of a wind-roughened sea surface. *Proc. SPIE (Internat. Soc. Optical Eng.)* 1637: 1293-1316.

Ryel R.J., Barnes P.W., Beyschlag W., Caldwell M.M. and Flint S.D. (1990) Plant competition for light analyzed with a multispecies canopy model.1. Model development and influence of enhanced UV-B conditions on photosynthesis in mixed wheat and wild oat canopies. *Oecologia* 82: 304-310.

Smith R.C. and Baker K.S. (1978) Optical classification of natural waters. *Limnol. Oceanogr.* 23: 260-267.

Smith R.C. and Baker K.S. (1979) Penetration of UV-B and biological dose rates effective in natural waters. *Photochem. Photobiol.* 29: 311-323.

Smith R.C., Wan Z. and Baker K.S. (1992) Ozone depletion in Antarctica: modelling its effect on solar UV irradiance under clear-sky conditions. *J. Geophys. Res.* 97: 7383-7397.

Teillet P.M. (1990) Rayleigh optical depth comparisons from various sources. *Appl. Optics* 29: 1897-1900.

Tyler J.E. (1960) Radiance distribution as a function of depth in an underwater environment. *Bull. Scripps Instn. Biol. Res.* 7: 363-412.

Voss K.J. (1992) A spectral beam model of the beam attenuation coefficient in the ocean and coastal areas. *Limnol. Oceanogr.* 37: 501-509.

7.5.10 Appendix

Listing of the program Daylight Vis-IR for computing daylight spectra under various conditions. It is based on the SPCTRAL2 model of Bird and Riordan (1986). The program is written in Microsoft QuickBasic, and is a basic form giving plots of the sunlight, skylight and global (total) spectral irradiance at ground surface. To yield neat plots, an appropriate scale factor must be entered (try, *e.g.* 10 to start with). Other forms of the program, which give spectral fluence rate, or PAR photon irradiance, or underwater spectral irradiance in tabulated form, or in a form for transfer to other programs, have also been constructed. In exchange for an empty high density 3.5" disk the author can supply a variety of such programs, either for interpreter or in compiled form. The standard will be for Macintosh Operating System 7. They do not function properly on operating systems with lower numbers; if such are to be used the system number must be specified.

```
REM THIS PROGRAM IS BASED ON THE ATMOSPHERIC MODEL
DESCRIBED
REM BY BIRD & RIORDAN, J OF CLIMATE AND APPL METEOR
REM VOL 25, PP87-97, 1986
REM LONGITUDE INPUT HAS NO FUNCTION HERE, BUT CAN BE
COUPLED
REM TO COMPUTERIZED OZONE ESTIMATION (SEE BJÖRN & MURPHY
1985)
1 : INPUT "YEAR, MONTH, DATE, TIME (h, min)";YE,MO,DATE,H,MIN
DELTA=YE-1980
PI=3.14159
IF MO<2 THEN DA=DATE:GOTO 200
IF MO<3 THEN DA=DATE+31:GOTO 200
IF MO<4 THEN DA=DATE+59:GOTO 200
IF MO<5 THEN DA=DATE+90:GOTO 200
IF MO<6 THEN DA=DATE+120:GOTO 200
IF MO<7 THEN DA=DATE+151:GOTO 200
IF MO<8 THEN DA=DATE+181:GOTO 200
IF MO<9 THEN DA=DATE+212:GOTO 200
IF MO<10 THEN DA=DATE+243:GOTO 200
IF MO<11 THEN DA=DATE+273:GOTO 200
IF MO<12 THEN DA=DATE+304:GOTO 200
IF MO<13 THEN DA=DATE+334:GOTO 200
IF DELTA MOD 4<>0 GOTO 200
IF MO>2 GOTO 200
DA=DA+1
200
:PSI=2*PI*(DA-1)/365:ES=1.00011+.034221*COS(PSI)+.00128*SIN(PSI)+.0
00719*COS(2*PSI)+.000077*SIN(2*PSI)
```

```
ED=.398*SIN((DA-80)*2*PI/365+.0335*(SIN(DA*PI/365)-SIN(1.3771)))
DI=ATN(ED/SQR(1-ED*ED))
INPUT "LATITUDE (N+,S-) AND LONGITUDE (E+, W-)" ; LA,LONG
INPUT "OZONE, atm cm";O
INPUT "PRECIPITABLE WATER, cm ";w
CZ=ED*SIN(LA*PI/180)+COS(DI)*COS(LA*PI/180)*COS((H+MIN/60-12)*PI/
12)
SH=ATN(CZ/SQR(1-CZ*CZ))
PRINT "SOLAR ELEVATION IS "SH*180/PI " DEGREES"
INPUT "INCIDENCE ANGLE, degrees ";TH
INPUT "BAROMETRIC PRESSURE, mbar ";P
TH=TH*PI/180
Z=(PI/2)-SH
ES=1.00011+.034221*COS(PSI)+.00128*SIN(PSI)+.000719*COS(2*PSI)+.00
0077*SIN(2*PSI)
M=1/(CZ+.15*(93.885-Z*180/PI)^(-1.253))
M1=M*P/1013
M11=1.8*P/1013
ALFA=1.0274
INPUT "AMOUNT OF AEROSOL";aerosol
BE=aerosol*.1
INPUT "TILT ANGLE: ZERO FOR HORIZONTAL, 90 FOR VERTICAL,
degrees ";TT
TT=TT*PI/180
INPUT "GROUND ALBEDO ";RGL
DIM HO(130),AW(130),AO(130),AU(130)
INPUT "LOWER WAVELENGTH LIMIT, MICROMETERS ";LBMIN
INPUT "UPPER WAVELENGTH LIMIT, MICROMETERS "; LBMAX
INPUT "SCALE";scale
CLS:CALL MOVETO(0,250):CALL LINETO(500,250)
CALL MOVETO(15,20):CALL LINETO(25,20):PRINT "  ".24*scale"
W/m2/nm"
CALL MOVETO(15,20):CALL LINETO(15,250)
FOR K1=0 TO 5:PSET(25+4*K1,40):CALL MOVETO(55,40):PRINT
"GLOBAL":NEXT K1
FOR K2=0 TO 4:CIRCLE(21+6*K2,55),1:CALL MOVETO(55,55):PRINT
"DIFFUSE":NEXT K2:CALL MOVETO(20,55): CALL LINETO(52,55)
CALL MOVETO(22,70):CALL LINETO(52,70):CALL
MOVETO(55,70):PRINT"DIRECT"
FOR I=0 TO 9
CALL MOVETO(15+50*I,250):CALL LINETO(15+50*I,240)
NEXT I:xx=2:yy1=250:yy2=250
300 :READ LB,HO(LB),AW(LB),AO(LB),AU(LB)
IF LB<LBMIN GOTO 300
TRL=EXP(-M1/((LB^4)*(115.6406-1.335/(LB*LB))))
TRL1=EXP(-M11/((LB^4)*(115.6406-1.335/(LB*LB))))
IF LB >= .5 THEN ALFA=1.206:IF LB >= .5 THEN BE=.088356*aerosol
```

```
TAL=EXP(-BE*LB^(-ALFA) *M)
TWL=EXP(-.2385*AW(LB)*w*M/(1+20.07*AW(LB)*w*M)^.45)
TWL1=EXP(-.2385*AW(LB)*w*1.8/(1+20.07*AW(LB)*w*1.8)^.45)
M0=(1+22/6370)/(CZ*CZ+2*22/6370)^.5
TOL=EXP(-AO(LB)*O*M0)
TUL=EXP(-1.41*AU(LB)*M1/(1+118.93*AU(LB)*M1)^.45)
TUL1=EXP(-1.41*AU(LB)*M11/(1+118.93*AU(LB)*M11)^.45)
IDL=HO(LB)*ES*TRL*TAL*TWL*TOL*TUL
WL=.945*EXP(-.095*(LOG(LB/.4))*(LOG(LB/.4)))
TL=BE*LB^(-ALFA)
TAAL=EXP(-(1-WL)*TL*M)
TAAL1=EXP(-(1-WL)*TL*1.8)
IRL= HO(LB)*ES*CZ*TOL*TUL*TWL*TAAL*(1-TRL^.95)*.5
TASL=EXP(-WL*TL*M)
TASL1=EXP(-WL*TL*1.8)
CTHETA=.65
ALG=LOG(1-(CTHETA))
AFS=ALG*(1.459+ALG*(.1595+ALG*.4129))
BFS=ALG*(.0783+ALG*(-.3824-ALG*.5874))
FS=1-.5*EXP((AFS+BFS*CZ)*CZ)
IAL=HO(LB)*ES*CZ*TOL*TUL*TWL*TAAL*(TRL^1.5)*(1-TASL)*FS
FS1=1-.5*EXP((AFS+BFS/1.8)/1.8)
REM The following formula is modified
RSL=TOL*TWL1*TAAL1*(.5*(1-TRL1)+(1-FS1)*TRL1*(1-TASL1))
IGL=(IDL*CZ+IRL+IAL)*RSL*RGL/(1-RSL*RGL)
ISL=IRL+IAL+IGL
CS=(LB+.55)^1.8
IF LB >.45 THEN CS=1
ISL=ISL*CS
ITL1=IDL*COS(TH)+ISL*((IDL*COS(TH)/(HO(LB)*ES*CZ))+.5*(1+COS(TT))*(
1-IDL/(HO(LB)*ES)))+.5*(IDL*CZ+ISL)*RGL*(1-COS(TT))
CALL MOVETO(xx,yy1):CALL
LINETO(1000*LB-285,250-IDL*COS(TH)/scale)
CALL MOVETO(xx,yy2):CALL
LINETO(1000*LB-285,250-(ITL1-IDL*COS(TH))/scale):CIRCLE(1000*LB-28
5,250-(ITL1-IDL*COS(TH))/scale),1
PSET(1000*LB-285,250-ITL1/scale)
IF LB>LBMAX-.03 THEN 1000
xx=1000*LB-285:yy1=250-IDL*COS(TH)/scale:yy2=250-(ITL1-IDL*COS(TH)
)/scale
GOTO 300
DATA .3,535.9,0,10,0
DATA .305,558.3,0,4.8,0
DATA .310,622,0,2.7,0
DATA .315,692.7,0,1.35,0
DATA .320,715.1,0,.8,0
DATA .325,832.9,0,.38,0
```

```
DATA .330,961.9,0,.16,0
DATA .335,931.9,0,.075,0
DATA .340,900.6,0,.04,0
DATA .345,911.3,0,0.019,0
DATA .350,975.5,0,.007,0
DATA .360,975.9,0,0,0
DATA .370,1119.9,0,0,0
DATA .380,1103.8,0,0,0
DATA .390,1033.8,0,0,0
DATA .400,1479.1,0,0,0
DATA .410,1701.3,0,0,0
DATA .420,1740.4,0,0,0
DATA .430,1587.2,0,0,0
DATA .440,1837.0,0,0,0
DATA .450,2005.0,0,0.003,0
DATA .460,2043.0,0,0.006,0
DATA .47,2043,0,0.009,0
DATA .48,2027,0,0.014,0
DATA .49,1896,0,0.021,0
DATA .5,1909,0,0.03,0
DATA .51,1927,0,0.04,0
DATA .52,1831,0,0.048,0
DATA .53,1891,0,0.063,0
DATA .54,1898,0,0.075,0
DATA .55,1892,0,0.085,0
DATA .57,1840,0,0.120,0
DATA .593,1768,.075,.119,0
DATA .61,1728,0,.12,0
DATA .63,1658,0,.09,0
DATA .656,1524,0,.065,0
DATA .6676,1531,0,.051,0
DATA .69,1420,.016,.028,.15
DATA .71,1399,.0125,.018,0
DATA .718,1374,1.8,.015,0
DATA .7244,1373,2.5,.012,0
DATA .74,1298,.061,.01,0
DATA .7525,1269,.0008,.008,0
DATA .7575,1245,.0001,.007,0
DATA .7625,1223,.00001,.006,4
DATA .7675,1205,.00001,.005,.35
DATA  .78,1183,.0006,0,0
DATA  .8,1148,.036,0,0
1000 : CALL MOVETO(0,265):PRINT "300 nm
                 700 nm"
INPUT ww:END
```

Part 8 A molecular and genetic approach to photomorphogenesis

8.1 The molecular biology of photoregulated genes

Alfred Batschauer[1], Philip M. Gilmartin[2], Ferenc Nagy[3] and Eberhard Schäfer[1]

[1]Institut für Biologie II, Albert-Ludwigs-Universität, Schänzlestr. 1, 79104 Freiburg, Germany,
[2]Centre for Plant Biochemistry and Biotechnology, University of Leeds, Leeds LS2 9JT, UK,
[3]Friedrich-Mischer-Institute, P.O. Box 2543, CH-4002 Basel, Switzerland

8.1.1 Introduction

In the last two decades a great deal of data has accumulated which shows that *gene expression* is very often under phytochrome control during de-etiolation and re-etiolation of young seedlings (for recent reviews see chapters in: Thomas and Johnson 1991). It is also well documented that some *photomovements* (polarotropism of some ferns and mosses and chloroplast orientation in the alga *Mougeotia*) are under phytochrome control (Chapters 7.2, 9.4 and 9.8). In these cases an action dichroism can be demonstrated which has been taken as an indication that phytochrome may act as a membrane effector. Wagner and his colleagues (1984) and Serlin and Roux (1984) have demonstrated the involvement of Ca^{2+}, calmodulin and actin filaments in the *signal transduction* process between the far-red light (FR)-absorbing form of phytochrome (Pfr) and chloroplast turning. Phytochrome localization (Chapter 4.5), photomovement (Chapter 9.4), and phytochrome and membranes (Chapter 4.6), will be discussed elsewhere in this book. Here we will concentrate on the question as to what is known about the mode of phytochrome action on gene expression. Before treating this problem, some general kinetic properties of phytochrome, signal transduction and underlying developmental programs will be discussed.

R.E. Kendrick & G.H.M. Kronenberg (eds.), Photomorphogenesis in Plants - 2nd Edition

8.1.2 Properties of phytochrome action

Generally, in photomorphogenesis one deals with analysis of an input-output relationship, whereby light acts on the input, and the output is the measurable response. To overcome the difficulties of kinetic analysis of a complex input-output system various research groups are searching for very fast responses in order to come as close as possible to the input photoreceptor (Quail 1983). As long as the first reaction after Pfr formation is not measurable we are working with a black-box system. The best we can do is to get as much information as possible about: (i) the kinetics of the phytochrome system itself; (ii) the kinetics of signal transduction; (iii) the kinetics of the response. Only then the 'full problem' of phytochrome action can be resolved (Fukshansky and Schäfer 1983).

There is not enough space in this chapter to discuss all the problems concerning the kinetics of phytochrome and signal transduction, but some basic information is necessary.

8.1.2.1 Kinetics of the phytochrome system

Only relatively simple *phytochrome models* based on spectrophotometric measurements have been published. These models take into account *de novo* synthesis of the red light (R)-absorbing form of phytochrome (Pr), the two light reactions between Pr and Pfr, and Pfr *dark reversion* and *destruction*. As discussed by Fukshansky and Schäfer (1983) these models explain neither the kinetics of dark reversion nor the influence of dark reversion on the Pfr level. In addition, a Pfr induced destruction of Pr has been described for some monocot (Dooskin and Mancinelli 1968; Stone and Pratt 1979) and dicot seedlings (Schäfer 1981; Heim and Schäfer 1984). Furthermore, *autoregulation* of phytochrome *synthesis* has been shown for grasses (Colbert *et al.* 1983; Gottmann and Schäfer 1982; Otto *et al.* 1983), and evidence is accumulating that two pools of phytochrome can be distinguished with respect to their destruction kinetics (Heim *et al.* 1981; Brockmann and Schäfer 1982) and immunological properties (Chapter 4.2; Pratt *et al.* 1991; Wang *et al.* 1991). Last but not least many different genes for phytochromes are expressed in plants (Sharrock and Quail 1989; Dehesh *et al.* 1991). Based on these findings, we have to conclude, that at the moment no phytochrome model exists, which describes precisely the kinetic properties of phytochrome after light pulses and during continuous irradiation. Additional experimental and theoretical work is necessary to combine all of this information in a reliable and testable phytochrome model.

8.1.2.2 Kinetics of signal transduction

Although no reliable phytochrome model exists so far, experiments have been performed to obtain information about kinetics of phytochrome signal transduction. For example one can ask, how long after a R pulse the photoresponse can be reverted to a defined degree. In this type of experiment the plants receive an inductive R pulse, are transferred to darkness for various periods of time, receive a reverting FR pulse, and are subsequently transferred back to darkness to allow complete gene expression. The kinetics of loss of *photoreversibility* describe aspects of the kinetics of signal transduction. Two types of plots were chosen for these experiments: (i) response (R) – response (Δt) divided by response (R) – response (FR) is plotted as a function of the delay time, Δt, between the R and FR pulse (Fig. 1); (ii) the response (Δt) can also be plotted as a function of Δt. In addition, this kind of experiment can answer the question, whether or not the amount of Pfr formed by a single light pulse correlates with the rate of the observed photoresponse. A re-evaluation of reversibility kinetics for phytochrome-controlled inhibition of hypocotyl growth (Schäfer *et al.* 1983) and accumulation of anthocyanin in cotyledons of mustard seedlings (Schmidt and Mohr 1983) indicates, that the photoresponse does indeed depend strongly on the amount of Pfr.

It was often thought that Pfr-response relationships, measured after inductive light pulses, are not affected by Pfr kinetics and reflect differences in the *primary reactions*. Unfortunately this is not true. The response which can be observed after a light pulse reflects the result of several interacting Pfr responses, *i.e.* Pfr-induced sensitization or adaptation, Pfr destruction, dark reversion, *etc.* and signal transduction. This problem will be discussed in more detail in Chapter 4.7, as well as in Chapter 6.

A search for fast phytochrome responses was chosen as a general approach to attack the problem of the primary reactions. The most rapid phytochrome mediated reactions described so far are the Pfr-induced pelletability and sequestering (McCurdy and Pratt 1986 a, b; Pratt and Marmé 1976) and Pfr-induced protein phosphorylation (Chapter 4.6; Otto and Schäfer 1988). To address the question whether sequestering is an integral part of signal transduction or only part of the degradation machinery, purification of the sequestered areas of phytochrome was performed (Hofmann *et al.* 1991). Recently a method was developed to isolate cytosol from plant suspension cell cultures. This cytosol still showed *in vitro* Pfr-induced phytochrome pelletability, as well as rapid light mediated protein phosphorylation. Although, there is no proof that sequestering and protein phosphorylation is part of the signal transduction chain, this method seems to be a powerful tool for the future to address the problem of primary reactions of phytochromes (K. Harter pers. comm.).

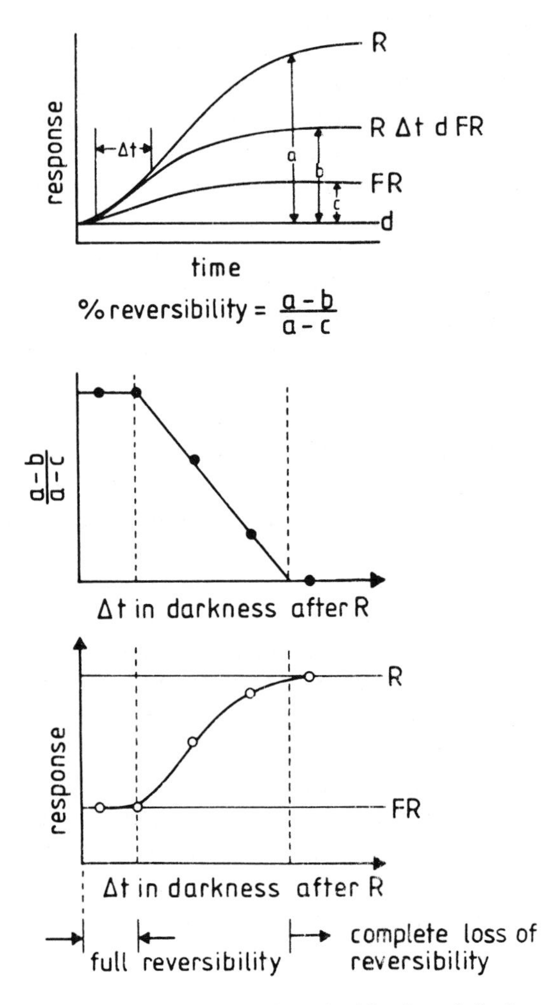

Figure 1. Scheme illustrating the method to analyse the kinetics of the loss of reversibility for phytochrome mediated responses. *Top*: studies of the time course of signal expression after an inductive red light (R) (a); far-red light (FR) pulse (c); or after a FR pulse given Δt hours after an initial R pulse (b). Response in darkness, d. *Middle*: plot of % reversibility as a function of the delay time between the R and FR pulse. *Bottom*: plot of the expression of the response as a function of the delay time between the R and FR pulse.

Studying signal transduction in photomorphogenesis it must be taken into account that photoregulation of gene expression also depends on the developmental state of the cell. Light can only trigger the pattern realization based on the endogenous controlled pattern of competence (Mohr 1983). These temporal and spacial patterns of competence have been described at the enzyme and mRNA level (Schopfer 1984; Wenng *et al.* 1990). The description of the underlying mechanisms which determine the temporal and spacial pattern of competence have not yet reached the molecular level.

It is obvious, that the full problem of photomorphogenesis can only be approached if additional information about phytochrome kinetics and the kinetics of signal transduction are available. Since, so far, the latter can only be measured indirectly by analysing the kinetics for loss of reversibility, one should try to use responses as closely related to the primary reaction of Pfr as possible. Therefore, although extremely cumbersome, measurements should not only be made for 'late' responses such as anthocyanin or chlorophyll (Chl) accumulation, but also for mRNA *content* or, even better, the rate of *transcription* as an initial response. Recent studies demonstrate that the expression of about 100-125 genes in higher plants is regulated by light and that this light-dependent expression is mediated in concert by different photoreceptors. Molecular biology and plant transformation techniques are powerful tools not only to measure transcription rates, but also to identify light regulatory elements (LREs) involved in light induced gene expression. Historically, the first plant genes cloned were those encoding for the most abundant protein components of the photosynthetic apparatus [for the small subunit of the ribulose-1,5-bisphosphate carboxylase/oxygenase (*RbcS*) and for the Chl *a*/*b*-binding proteins (*Cab*)]. In addition, it has also been shown that the expression of some of these genes is further modulated by an endogenous biological (circadian) oscillator. Our knowledge of the molecular mechanisms, by which light modulates gene expression *via* photoreceptor initiated signal transduction pathways, is largely, but not exclusively based on the expression properties of these two photosynthetic genes, as well as on the chalcone synthase (*Chs*) genes, encoding the first enzyme of the flavonoid pathway. Therefore, in this chapter we will describe the molecular approaches in analysing the underlying mechanisms for photoregulated gene expression in three case studies: The *Chs*, the *Cab*, and the *RbcS* genes.

8.1.3 Regulation of chalcone synthase expression in mustard and parsley

Flavonoids play an important role in the interaction of plants with their environment in protection against UV (Hahlbrock and Scheel 1989) and pathogens (phytoalexins) (Dixon 1986), the attraction of insects (flower pigmentation) (Mol *et al.* 1989) and the interaction with microbes (nodulation of legumes) (Peters *et al.* 1986). Chalcone synthase (CHS) is the key enzyme of the flavonoid pathway (Heller and Hahlbrock 1980). Therefore, regulation and accumulation of CHS is expected to occur in all cells which produce flavonoids. In mustard seedlings, flavonoid formation is visible as the accumulation of anthocyanin in the subepidermal layer of the hypocotyl and the lower epidermis of the cotyledons.

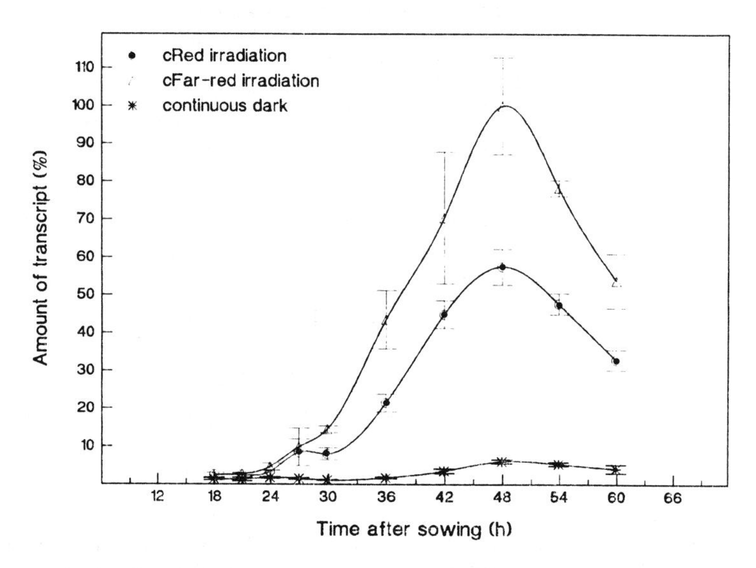

Figure 2. Kinetics of *Chs*-mRNA accumulation in cotyledons of mustard seedlings. Plants were irradiated from sowing with continuous (c) R or FR, and harvested at time points indicated for RNA isolation and Dot-blot analysis (from Batschauer *et al.* 1991).

Photoregulation of anthocyanin accumulation in mustard has been studied extensively by H. Mohr and his colleagues. For inductive light pulses a linear relationship between the amount of Pfr and the amount of anthocyanin accumulated in the cotyledons was observed (Drumm and Mohr 1978). The accumulation kinetics followed a probit curve (Lange *et al.* 1971). The question, whether CHS is the rate limiting enzyme for anthocyanin formation (Brödenfeldt and Mohr 1988) is complicated by the fact, that in addition to anthocyanin in the lower epidermis, quercitin is formed in the upper epidermis (Wellmann 1971; Beggs *et al.* 1987) and that CHS is encoded by a small gene family consisting of about four members in mustard (Batschauer *et al.* 1991). To obtain tools to solve these problems, cDNAs for different *Chs* genes were cloned from mustard (Ehmann and Schäfer 1988). In addition, one of the cDNAs was used to obtain a fusion protein in *E. coli* (T. Kretsch unpublished data), which enabled antibodies to be raised. Genomic clones for *Chs* were isolated (Batschauer *et al.* 1991), which were used for detailed studies of promoter elements.

Using gene-specific probes from the 3'non-coding sequences of the *Chs* cDNAs, Ehmann *et al.* (1991) demonstrated, that the analysed transcripts show very similar expression characteristics in mustard seedlings. Under continuous R and FR a starting point at 24-27 h after sowing was observed (Fig. 2). Etiolated seedlings showed an increase in dark expression 36-42 h after sowing. As reported for anthocyanin accumulation (Oelmüller and Mohr 1984) this increase in darkness can be partially inhibited by a long wavelength FR pulse

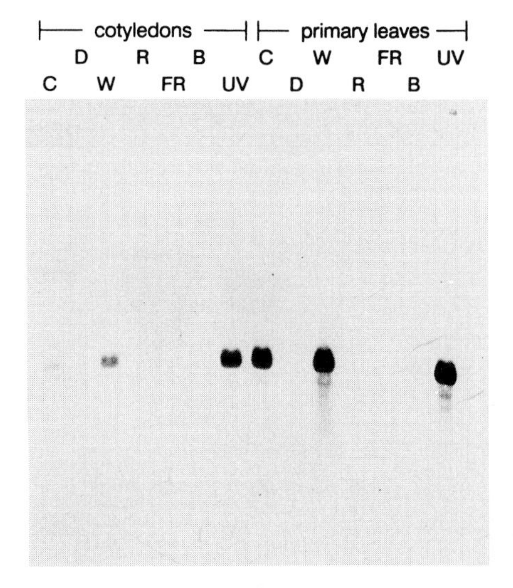

Figure 3. Light regulation of *Chs* expression in mature mustard plants. Plants were grown for 3 weeks in a light/dark cycle (16 h white light/8 h darkness) and then transferred to darkness for 32 h. Afterwards, plants were treated with different light for 12 h. 15 μg per lane of total RNA was separated on agarose-formaldehyde gels transferred to nitrocellulose filters and hybridized with a 747 bp fragment of the coding region of *Chs1* gene from mustard (Batschauer *et al.* 1991). Abbreviations: C: white light/dark cycle for 3 weeks, sample taken at the end of the light phase; D: plants kept in darkness for 32 h after the last light phase; W: white light; R: red light; FR: far-red light; B: blue light; UV: UV-light (Batschauer A. unpublished data).

given 6-21 h after sowing. This indicates that the *Chs* mRNA accumulation is under the control of both, type I and type II phytochromes and that they have in addition different competence points.

Whereas phytochrome seems to be the dominant if not the only photoreceptor controlling *Chs* mRNA accumulation in etiolated mustard seedings, blue-light (B)/UV photoreceptors control the expression in later stages of development (Fig. 3). However, beside phytochrome, B/UV photoreceptors also control *Chs* expression in seedlings, if they are kept in a light/dark cycles for 3 days followed by an extended dark period prior to irradiation with monochromatic light for 12 h (Frohnmeyer *et al.* 1992). This indicates, that the two *Chs* genes tested in mustard are under the control of type I and type II phytochromes; the high irradiance response; B and UV-B photoreceptors.

To test the organ specific gene expression, RNA can be prepared from different tissues and analysed in Northern blots using gene-specific probes (Fig. 4). Both tested *Chs* genes are expressed in cotyledons, epicotyls, primary leaves and flower buds. Cell specific gene expression was analysed at the RNA level by *in situ* hybridization and at the protein level by immunocytological techniques. Figure 5 shows that both *Chs* genes tested are expressed in mustard

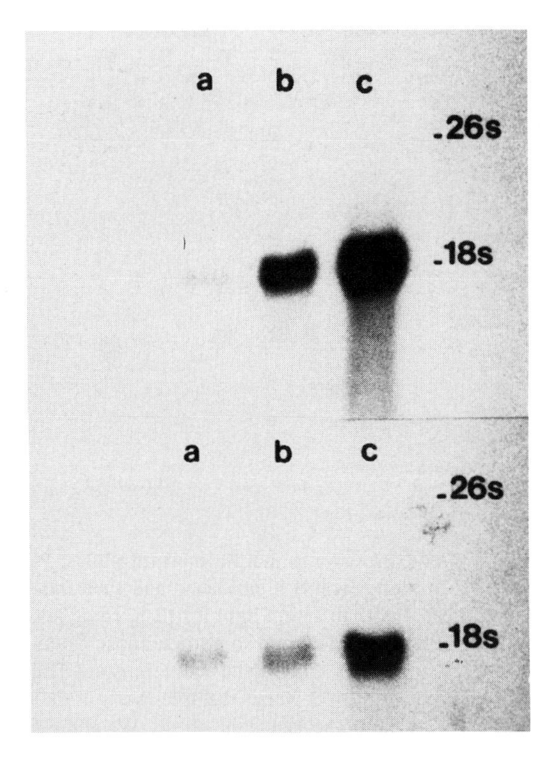

Figure 4. Expression of two different *Chs* genes from mustard in different organs, detected by Northern-blotting. Primary leaves were harvested 11 days after sowing, epicotyls and young flower buds 19 days after sowing. Plants were grown in a 16/8 h light/dark cycle at 25°C on vermiculite. *Upper* panel: specific probe for *Chs1* gene; *lower* panel: specific probe for *Chs2* gene. Lane a: epicotyl; lane b: primary leaves; lane c: flower buds. Positions of 18S and 26S rRNAs are indicated (from Ehmann *et al.* 1991).

in the lower and upper epidermis of the cotyledons, which is consistent with the localization of flavonoids.

Another technique to analyse light regulation and cell specific gene expression of defined genes is a molecular approach using transgenic plants. For this purpose we have cloned and characterized a *Chs* gene from mustard (Batschauer *et al.* 1991). A 1 kbp fragment 5'of the transcription start site of the *Chs1* gene and subfragments thereof were cloned in front of the *uidA* gene (Jefferson *et al.* 1986), encoding bacterial β-glucuronidase (GUS). Different substrates can be used to measure the GUS activity, including ones for histochemical detection. These constructs were used in transient expression assays (see below) and to transform tobacco and *Arabidopsis via* the *Agrobacterium*-mediated gene transfer. The methods used for tobacco and *Arabidopsis* transformation differ in some aspects, but the principal shown in Fig. 6 is the same.

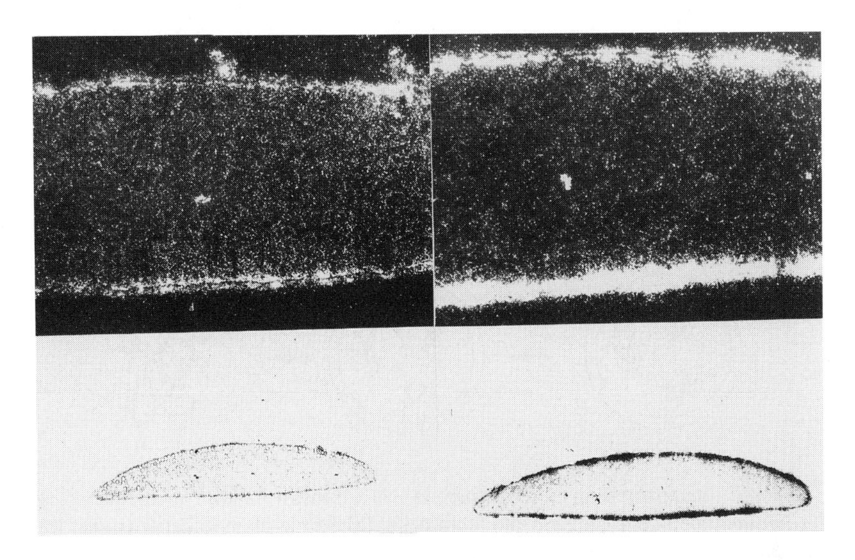

Figure 5. ***In situ*** hybridizations using specific probes for two *Chs* genes from mustard (*left* part, probe for *Chs1*; right part, probe for *Chs2*) with sections of a cotyledon grown under far-red light (FR) for 42 h. Sections were taken proximal (*left* part) or distal (*right* part) to the central vascular vein. Details of the sections are shown in the lower part of the picture (epifluorescence technique) (from Ehmann *et al.* 1991).

The 1 kbp fragment of the *Chs* promoter from mustard mediates both, light-induced gene expression and tissue specificity in transgenic *Arabidopsis* seedlings. Although, experiments are still in progress to identify *cis*-acting elements within the mustard *Chs* promoter, we conclude from the results obtained so far that most if not all elements mediating the *Chs* expression characteristics reside within the 1 kbp fragment located upstream of the transcription start site.

As well as mustard, parsley was analysed most intensively in respect to the regulation of flavonoid accumulation and *Chs* expression. These analyses were facilitated by the fact that suspension cultures from parsley show light-regulated formation of flavonoids and *Chs* expression (Hahlbrock and Scheel 1989) and even protoplasts derived from these cell cultures are able to react to light (Dangl *et al.* 1987), allowing transient expression studies of constructs bearing promoters of light-regulated genes (Lipphardt *et al.* 1988). In addition, only a single *Chs* gene is present in these cells, omitting all the problems in the interpretation of data obtained by measuring the expression of multigene families, at least with probes which are not specific for single members of the gene family. In parsley-cell cultures *Chs* expression was analysed on the mRNA level by Northern and Dot-blot analyses, and on the transcriptional level by run-off transcription measurements. The accumulation of flavonoids after the UV treatment is preceded by an increase in *Chs* mRNA and transcription in a

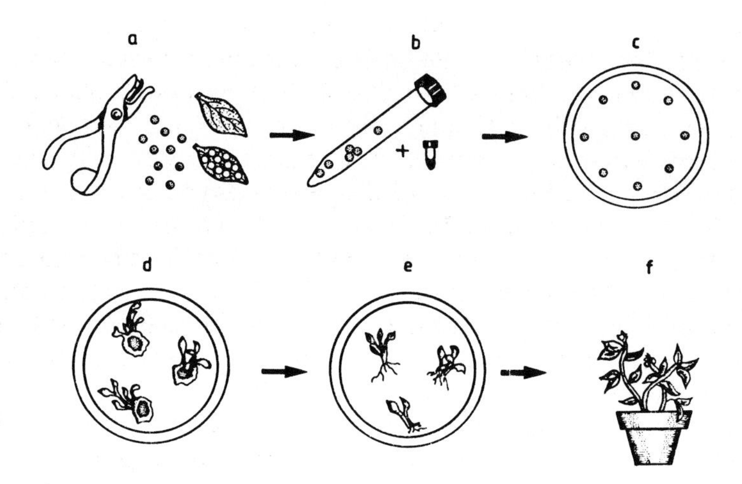

Figure 6. Scheme demonstrating the principle of transformation techniques used for *Agrobacterium*-mediated gene transfer into plant cells. (a) Sterile plant material (roots, leaf discs) is used for the transformation; (b) incubation of explants with *Agrobacterium*, carrying the construct to be transferred to the plant cells; (c) explants transferred to petri dishes on solid medium, containing antibiotics for selection against growth of *Agrobacteria* and for plant cells carrying the selectable marker gene. Hormones are included for shoot formation; (d) formation of green calli and shoots on selective medium; (e) transfer of shoots on fresh medium for root formation; (f) plantlets on soil to set flowers and seeds. (redrawn from Mohr H. and Schopfer P. (1992) *Lehrbuch der Pflanzenphysiologie.* Springer, Heidelberg).

temporal pattern, which would be expected if CHS is the rate-limiting step for flavonoid accumulation. It was demonstrated that UV is necessary for *Chs* expression in these cell cultures and that other photoreceptors have modulating effects (Bruns *et al.* 1986; Ohl *et al.* 1989). A pretreatment with B, for example, abolishes the lag phase, which is observed for *Chs* expression after an inductive UV pulse. The B-induced signal alone leads only to a marginal increase in *Chs* expression but persists for at least 20 h (Ohl *et al.* 1989).

The molecular events involved in *Chs* expression in parsley-cell cultures were reviewed recently by Weisshaar *et al.* (1991). Therefore, we will discuss these findings very briefly. Using the *in vivo* footprinting technique, Schulze-Lefert *et al.* (1989) detected four regions in the *Chs* promoter from parsley (named box I-IV in the order of their distance to the arbitrarily defined transcription start site) which were protected against DMS-induced methylation, most likely by binding of protein factors. All the protected regions reside within 275 bp upstream of position +1. The protection is induced by UV with a similar kinetic to the one for *Chs* transcription. Therefore, binding of these factors might be the last event in the signal transduction leading from the photoreceptors to the *cis*-acting elements of the *Chs* gene.

As mentioned already, protoplasts from parsley-cell cultures show light-induced gene expression. This allowed functional characterization of *Chs* promoters from parsley and mustard using the transient expression system. For transient expression analysis, parsley-cell cultures were treated with cellulase and pectinase in order to remove the cell wall. The protoplasts obtained by this procedure were used to introduce plasmid DNAs, carrying the *Chs* promoter-GUS reporter fusions and light-induced expression was analysed by the fluorimetric GUS assay. In the case of the parsley promoter the boxes defined by *in vivo* footprinting are indeed important for function. For example, box I and II together form a functional unit (unit 1) which is sufficient for light-induced expression (Weisshaar *et al.* 1991). Single-point mutations and small deletions within this unit 1 defined not only the necessity of defined basepairs in box I and box II precisely, but also demonstrated, that even slight changes in their distance can abolish function (Block *et al.* 1990).

The more upstream boxes III and IV also form a unit (unit 2) which can mediate light-induced expression if unit 1 is deleted, although unit 1 acts much more strongly (Schulze-Lefert *et al.* 1989). The redundancy of light-regulatory elements is also found in other genes, such as *RbcS* genes (Section 8.1.4). The mustard *Chs* promoter also mediates UV-induced expression in the transient parsley system (Fig. 7). A sequence with high homology to unit 1 is present in the mustard gene. Deletions in this region result in a loss of light-induced transient expression, indicating that similar, if not identical, elements in the signal-transduction chain are used for transcriptional control of *Chs* genes from parsley and mustard. Whereas the box I was so far identified only in promoters of *Chs* and chalcone isomerase (*Chi*) genes (both encoding enzymes for the flavonoid pathway), box II (identical to the G-box) was found in most *RbcS* promoters (Section 8.1.4) which are light-induced, but also in promoters of genes which are not light-regulated.

Using Southwestern screening (Singh *et al.* 1989), cDNAs were recently cloned encoding polypeptides, which bind specifically to these G-box motifs (Guiltinan *et al.* 1990; Weisshaar *et al. 1991*; Oeda *et al.* 1991; Schindler *et al.* 1992). All of them encode putative transcription factors with structural similarity. Common to all is a basic region (b), most likely the DNA-binding motif in close proximity to a leucine-zipper motif (ZIP) which functions as a dimerization domain. It was demonstrated that these (bZIP) factors form homo and hetero dimers which bind to defined G-box motifs with different affinities (see below; Weisshaar *et al.* 1991; Schindler *et al.* 1992). Since G-box elements are found in several light-regulated genes which differ in their patterns of expression and in the mode of light regulation and G-box elements are also found in genes which are not light-regulated (Williams *et al.* 1992), it is evident that G-box binding factors (GBFs) alone can not account for the complex regulation patterns of all the G-box-containing genes, although the formation of GBF (homo and hetero) dimers probably allows a high variability for binding to the target sequences. Although not directly proven, we assume that the light-induced *Chs* expression requires the concerted action of more than one type of

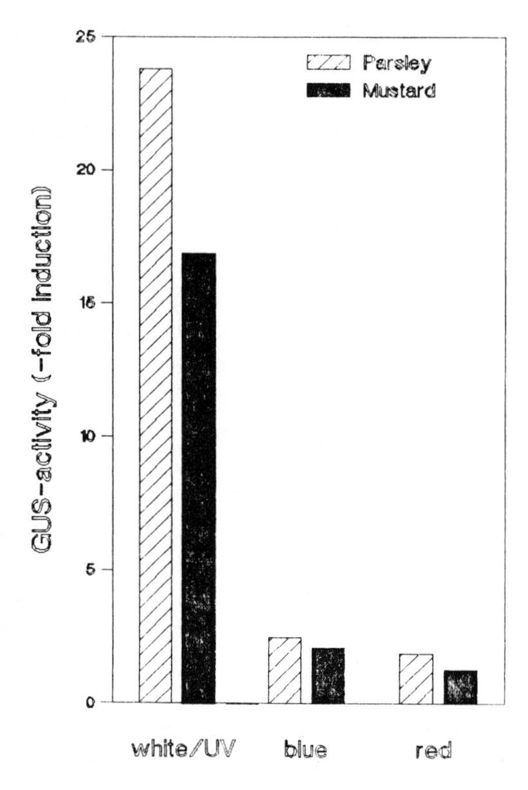

Figure 7. Transient expression of two chimeric constructs bearing promoter fragments of *Chs* genes from mustard and parsley. The protoplasts were transformed with constructs bearing 'full-length' promoters from mustard (solid bars) or parsley (open bars) fused to the β-glucuronidase (GUS) reporter gene. The transformed protoplasts were irradiated for 9 h and protein extracted after the end of the light treatment for measurements of GUS activity. The fluence rates were 4.2 and 3.7 W m^{-2} for UV-containing white light and blue light, respectively (from Frohnmeyer *et al.* 1992).

transcription factor. The light-regulatory elements of *Chs* genes (unit 1) are in our opinion suitable candidates for analysing such concerted actions of *trans*-factors.

8.1.3.1 Summary

Although molecular approaches could make an important impact and lead to progress in the analysis of mechanisms for transcriptional control of light-regulated genes, we are still a long way from drawing a conclusive picture. It is clear that: (i) more than one binding site is necessary for light-induced *Chs* transcription; (ii) different factors, which bind to these LREs are involved; (iii) bZIP transcription factors are involved in *Chs* transcription and that these factors are encoded by multi-gene families whose gene products can form homo- and hetero-dimers.

One can imagine that the complexity of factors involved in transcriptional control creates on the one hand, a high degree of flexibility for gene regulation and on the other makes it unlikely that the regulation of a single transcription factor alone, either on the transcriptional and/or post-transcriptional level can control the whole transcriptional event.

8.1.4 Regulation of ribulose-1,5-bisphosphate carboxylase/oxygenase

8.1.4.1 Ribulose-1,5-bisphosphate carboxylase/oxygenase (Rubisco) gene (RbcS) *organization and expression*

Genes encoding *RbcS* have attracted much interest due to the role of this enzyme in photosynthetic carbon assimilation, its abundance, the correspondingly high levels of *RbcS* mRNA in leaves and the light-dependent accumulation of these transcripts. The large amounts of Rubisco in leaves facilitated its biochemical characterization, including the demonstration that the two subunit types, large and small, are encoded within the chloroplast and nuclear genomes respectively (Ellis 1981). The relatively high abundance of *RbcS* transcripts in light-grown leaves as opposed to dark-grown leaves facilitated the isolation of *RbcS* cDNA sequences (Bedbrook *et al.* 1980; Broglie *et al.* 1981). These cDNA sequences have formed the basis for studies into the organization and regulation of the *RbcS* genes (for review see Tobin and Silverthorne 1985; Kuhlemeier *et al.* 1987; Manzara and Gruissem 1988; Dean *et al.* 1989a). The *RbcS* genes from all species examined to date, with the exception of cucumber which appears to contain only a single *RbcS* gene (Greenland *et al.* 1987) are organized in multi-gene families. The size of these families ranges from at least three in *Phaseolus vulgaris* (Knight and Jenkins 1992) and four in *Arabidopsis*, to approximately 13 in *Lemna* (for reviews see Manzara and Gruissem 1988; Dean *et al.* 1989a).

Much emphases has been placed on the analysis of *RbcS* gene expression patterns. The isolation of *RbcS* cDNA sequences has enabled two lines of research to advance an understanding of *RbcS* gene regulation. The first of these has involved the use of *RbcS* cDNA clones as molecular probes to measure the abundance of *RbcS* mRNA transcripts at various stages of development and in different plant tissues following exposure of the plant to specific wavelengths of light. The second line of research facilitated by the availability of these *RbcS* cDNA clones was the isolation and characterization of the nuclear genes that encode these sequences.

Initial studies of *RbcS* gene regulation relied on the measurement of *RbcS* steady state transcript levels by hybridization techniques such as Northern and

RNA filter binding assays. These studies, in several species, demonstrated that *RbcS* gene expression is organ specific, with the leaves containing the highest levels, and the abundance of the transcripts modulated by light. These observations were not entirely unexpected considering the organ specific and light mediated accumulation of Rubisco (for reviews see Manzara and Gruissem 1988; Dean *et al.* 1989a). The *RbcS* transcript levels in most species examined are dramatically reduced both in seedlings germinated in complete darkness, and also the leaves of light-grown plants that have been transferred to the dark for 2-3 days. Following illumination of both these etiolated and dark-adapted plants, the *RbcS* expression levels increase dramatically, indicating photocontrol of *RbcS* transcript accumulation (for reviews see Tobin and Silverthorne 1985; Kuhlemeier *et al.* 1987; Manzara and Gruissem 1988; Dean *et al.* 1989a). Two possible explanations could account for these observations. The first is that the *RbcS* genes are expressed constitutively and the altered levels of steady state transcript levels arise through the differential stability of the *RbcS* transcripts in response to a light signal. The second possibility is that the *RbcS* genes are transcriptionally regulated in response to light; such that in both etiolated seedlings and in light-grown plants transferred to the dark, the *RbcS* genes are transcriptionally silent in the absence of a light stimulus. It has been demonstrated in some species that post-translational regulation of *RbcS* mRNA abundance is involved in the light-responsive changes in *RbcS* transcript levels (Wanner and Gruissem 1991; Thompson *et al.* 1992; Kuhlemeier 1992), but it is transcriptional regulation of the *RbcS* genes that play a predominant role in the modulation of *RbcS* transcript abundance in response to light (for reviews see Tobin and Silverthorne 1985; Kuhlemeier *et al.* 1987; Manzara and Gruissem 1988 and Dean *et al.* 1989a).

From measurements of steady-state transcript levels, it is not possible to distinguish between changes in transcription rate and altered mRNA stability. The demonstration that the changes in *RbcS* mRNA levels were caused by modulation of *RbcS* transcription rates has been reported in several species including pea, *Lemna*, soybean, petunia and tomato by measuring *in vitro* transcription in isolated nuclei (Kuhlemeier *et al.* 1987; Manzara and Gruissem 1988; Dean *et al.* 1989a and references therein). These nuclear *run-on* experiments involved the isolation of nuclei from etiolated seedlings that had been treated with a pulse of light or from mature light-grown plants. Isolation of the nuclei prevents further initiation of transcription; transcripts initiated prior to disruption of the cell can, however, be transcribed to completion under appropriate assay conditions. Incubation of these isolated nuclei in the presence of buffer and ribonucleotide triphosphates including ^{3}H- or ^{32}P-UTP results in the incorporation of ^{3}H or ^{32}P into these initiated transcripts. The abundance of specific mRNA molecules can be measured by hybridization of the radiolabelled mRNA to *RbcS* cDNA copies immobilized onto nitrocellulose filters (see Kuhlemeier *et al.* 1987; Manzara and Gruissem 1988; Dean *et al.* 1989a and references therein). These experiments demonstrated that the light-induced increase in *RbcS* transcript abundance was due primarily to an increase in transcription. In addition, it has been reported in several species, including *Lemna*, soybean and tomato that post-transcriptional processing can play a role in *RbcS* gene regulation (Thompson *et al.* 1992; Wanner and Gruissem 1991). One further level of light regulation has been reported in *Amaranthus*. In this plant light mediated translational initiation of the small subunit of Rubisco has been demonstrated (Berry *et al.* 1990). There are, therefore, multiple tiers of regulation in the control of events linking *RbcS* gene activation and small subunit polypeptide accumulation.

The regulation of *RbcS* genes in response to light has been further dissected to identify the wavelength components of white light that mediate the response. Again there are clearly different levels of control since both R and B can mediate effects (Tobin and Silverthorne 1985; Kuhlemeier *et al.* 1987; Jenkins 1988). The induction of *RbcS* expression in immature etiolated tissues of several plant species is induced by R; the response is abrogated by a subsequent flash of FR indicating a typical phytochrome mediated response. Through the use of run-on transcription studies in *Lemna* and soybean, it has been demonstrated that the phytochrome-mediated response is modulated at the level of transcription (for reviews see Tobin and Silverthorne 1985; Kuhlemeier *et al.* 1987; Jenkins 1988; Nagy *et al.* 1988b). In contrast to the observations with etiolated tissue that phytochrome is the primary photoreceptor involved in mediating *RbcS* expression, increased *RbcS* levels observed in mature leaves are in response to B in association with phytochrome (Kuhlemeier *et al.* 1987). This complex pattern of regulation could suggest that distinct regulatory circuits mediate responses to different stimuli in these different tissues. It is possible that the observed expression patterns represent overlapping as opposed to distinct regulatory pathways; regulation of *RbcS* expression in mature leaves by B in concert with phytochrome would argue for this latter possibility.

Early studies on the expression of *RbcS* genes used whole *RbcS* cDNA sequences as probes in Northern hybridization and run-on transcription assays. The highly conserved nature of the protein coding sequence of *RbcS* genes therefore provided an expression profile for the entire gene family. The contribution of individual gene family members to the overall expression levels can only be determined by using gene-specific probes. Characterization of additional cDNA clones derived from different *RbcS* genes of several species demonstrated that there is sequence variability in the 5' and 3' non-coding regions of the different nuclear genes. These sequence differences have enabled the design of gene-specific probes which have been used in several ways to reveal the contribution of individual family members to the overall expression pattern seen for the gene family. Studies at the level of Northern analysis and RNA filter hybridization studies in petunia, maize and tomato (Dean *et al.* 1989a) and *Lemna* (Silverthorne *et al.* 1990) have revealed quantitative differences in the expression patterns of individual family members in these species. In addition, qualitative differences in the expression patterns of individual *RbcS* genes have been demonstrated in tomato (Manzara and Gruissem 1988; Wanner and Gruissem 1991).

The technique of 5' and 3' S1 nuclease analysis has been applied to the expression studies in pea (Kuhlemeier *et al.* 1987). Primer extension with gene-specific oligonucleotide primers has enabled the determination of individual *RbcS* expression levels in petunia, as has estimation of expression levels by analyses of specific cDNA abundance in cDNA libraries (Dean *et al.* 1989a). These studies have demonstrated that an individual *RbcS* family member may account for approximately 45-50% of total *RbcS* transcript levels in the leaves of some species, for example pea, petunia, and maize (Kuhlemeier *et al.* 1987; Dean *et al.* 1989a; Sheen and Bogorad 1986).

Dissection of the five member *RbcS* family from tomato has identified differences between the expression dynamics of the tomato *RbcS* gene family and other *RbcS* gene families. In contrast to the observations that mRNA abundance of all *RbcS* gene family members in pea and petunia are modulated by light, transcripts derived from two of the five tomato *RbcS* genes are maintained at high levels in plants transferred from the light to the dark (Manzara and Gruissem 1988). These mRNA steady state measurements suggest that the corresponding *RbcS* genes are not regulated in a strictly light-responsive manner. Further studies in etiolated seedlings using both steady state measurements of mRNA abundance and by nuclear run-on experiments demonstrate that three of the five tomato *RbcS* genes are transcriptionally active in etiolated tissue (Wanner and Gruissem 1991). This observation indicates that a light stimulus is not required to trigger the activation of these specific gene family members. Analyses of the organ-specific expression patterns of the tomato *RbcS* genes have also demonstrated differences in the organ-specificity of the five genes (Wanner and Gruissem 1991).

Similar expression dynamic studies of *RbcS* transcripts by analyses of RNA isolated from several organs of petunia have shown that the relative transcript ratios of individual *RbcS* genes in petunia does not vary between different organs (Dean *et al.* 1989a). However, similar studies in pea have demonstrated some differences between the relative *RbcS* transcript abundance in different tissues (Kuhlemeier *et al.* 1987). Such observations with gene-specific probes clearly distinguishe the individual contribution of specific family members to the total *RbcS* mRNA pools. Observation of the expression of three maize *RbcS* genes in bundle sheath cells at various stages of greening and between etiolated mesophyll cells demonstrated differential expression of these three genes (Sheen and Bogorad 1986). The developmental differences in expression observed between some individual *RbcS* genes suggests that in addition to regulatory DNA elements that mediate common patterns of expression, they contain regulatory elements that modulate the distinct expression patterns.

There is evidence from some plant species that the expression of *RbcS* genes is tightly coupled to the presence of chloroplast (Dean *et al.* 1989a). This observation suggests an additional level of control on the regulation of *RbcS* and other genes encoding proteins destined for the chloroplast. However, observations from tomato suggest that whilst light can modulate the quantitative expression levels of all five *RbcS* genes in cotyledons and leaves, the expression of some of these genes can be uncoupled from chloroplast development (Wanner and Gruissem 1991). Similar conclusions can be drawn from observations of *Arabidopsis* plants carrying the mutation *det2* (Chory *et al.* 1991). If these plants are grown from seed in complete darkness, the seedlings are de-etiolated and develop with a light-grown phenotype; however, chloroplast development does not occur. Associated with this phenotype is the expression of normally photoresponsive genes, including *RbcS* (Chory *et al.* 1991). The expression of these *RbcS* genes, therefore, occurs without a light stimulus and in the absence of differentiated chloroplast. An additional component of *RbcS* gene regulation is that of metabolite repression (Sheen 1990). Suppression of *RbcS* and other photosynthetic genes is brought about by carbohydrate end products of photosynthetic anabolism. This anabolic repression apparently overrides the organ-specific and light-responsive regulation of these genes (Sheen 1990). These combined observations, therefore, highlight the complexity of overlapping regulatory mechanisms that modulate *RbcS* gene expression.

Differences exist in the expression dynamics of specific *RbcS* gene family members in different species between etiolated seedlings and mature light-grown leaves transferred to the dark. Not only are there dramatic developmental differences between these two tissues, but the photoresponses of *RbcS* genes in these two tissues differ. For example, the influence of R alone on *RbcS* gene expression in etiolated tissues, and the interactions between R and B in *RbcS* photocontrol in mature leaves (Kuhlemeier *et al.* 1987). These observations are supported by the *Arabidopsis* photoresponse mutants *det1* (Chory *et al.* 1989), *det2* (Chory *et al.* 1991) and *cop1* (Deng *et al.* 1991). Plants carrying these mutations demonstrate a light-grown phenotype following germination in complete darkness. Additionally, the regulation of *RbcS* gene expression in seedlings grown in the absence of light is affected such that *RbcS* genes are activated without the requirement for a light stimulus. In light-grown plants transferred to darkness, the *RbcS* control during light/dark transitions is not affected in *det1* (Chory *et al.* 1989), but is affected in *det2* and *cop1* (Chory *et al.* 1991; Deng *et al.* 1991). These *Arabidopsis* mutants in combination with observations in other species therefore demonstrate a clear distinction between the control of *RbcS* gene expression in etiolated seedlings and dark adapted leaf tissue. A further level of developmental regulation is, therefore, superimposed on the control mediated by light, organ-specificity, metabolites and chloroplast differentiation. In conclusion, there are many complex factors that interact to modulate expression of *RbcS* genes, although it is light responsiveness that has received the most attention to date.

8.1.4.2 Light-responsive element localization and dissection

The availability of characterized *RbcS* cDNA sequences has facilitated the isolation of numerous *RbcS* nuclear genes from many species (Manzara and Gruissem 1988; Dean *et al.* 1989a). In an attempt to unravel the complex patterns of *RbcS* gene regulation, substantial effort has been directed towards the identification and characterization of *cis*-regulatory DNA elements that mediate the diverse responses; specifically attention has focused on the light-responsive DNA elements (Kuhlemeier *et al.* 1987; Manzara and Gruissem 1988; Dean *et al.* 1989a; Gilmartin *et al.* 1990). The expression characteristics of the various *RbcS* genes described above likely arise from combinations of discrete regulatory DNA elements. Indeed, DNA sequence comparisons between different *RbcS* genes reveals regions of upstream sequence that are conserved not only between individual genes within a species, but also between species (Manzara and Gruissem 1988; Dean *et al.* 1989a). It is likely that these conserved regions play a role in the modulation of responses common to all *RbcS* genes. It is presumed that additional distinct regulatory elements modulate the patterns of expression specific to individual genes.

Studies to define the *cis*-regulatory elements that mediate transcriptional responses to light have relied heavily on the construction of *RbcS* gene promoter fusions and the assay of these chimeric constructs in transgenic plants (for reviews see Kuhlemeier *et al.* 1987; Gilmartin *et al.* 1990). Initial observations that the transfer of intact *RbcS* genes from various species into heterologous transgenic hosts, such as tobacco and petunia, demonstrated not only the viability of such assays to identify regulatory DNA elements, but also that the regulatory mechanisms that control light-responsive expression of *RbcS* genes are conserved between species (Kuhlemeier *et al.* 1987; Benfey and Chua 1989). For example, *RbcS* genes isolated from pea remain light-responsive and organ-specific when transferred into transgenic tobacco and petunia plants (see Kuhlemeier *et al.* 1987). One reported difference in the observed conservation of regulatory mechanisms between species comes from the expression of wheat *RbcS* in transgenic tobacco. Wheat *RbcS* transcripts did not accumulate when the gene was transferred into transgenic tobacco (see Kuhlemeier *et al.* 1987). An explanation for this observation came from the demonstration that the splicing of the monocot intron occurs inefficiently in dicots (Kuhlemeier *et al.* 1987). The observation that a wheat *Cab* gene, that contains no introns, is correctly regulated in transgenic tobacco demonstrates that there is some conservation of regulatory information between monocots and dicots (Kuhlemeier *et al.* 1987).

The regulatory elements that modulate *RbcS* gene expression have, in several cases, been localized to the DNA sequences located 5', or upstream, of the coding region of the gene. This localization of the regulatory elements was achieved by construction of chimeric constructs using varying amounts of 5'-flanking sequence from different *RbcS* genes and the subsequent transfer of these gene fusions to transgenic plants (Kuhlemeier *et al.* 1987; Benfey *et al.* 1989). Such studies in transgenic plants and in transient assays not only confirmed the transcriptional control of *RbcS* genes, but have enabled the localization of DNA elements responsible for light responsiveness to upstream regulatory elements of *RbcS* genes from several species including soybean, tobacco, petunia (Dean *et al.* 1989a, b), *Arabidopsis* (Donald and Cashmore 1990), *Lemna* (Rolfe and Tobin 1991) and pea (Kuhlemeier *et al.* 1987; Gilmartin *et al.* 1991). However, in petunia it has also been demonstrated that regulatory sequences are also located downstream of the translation start site (Dean *et al.* 1989c). Comparisons between *RbcS* sequences and *in vivo* analyses have highlighted similarities, as well as differences, in the organization of the regulatory DNA sequences between *RbcS* genes from different species.

In pea *RbcS-3A*, multiple regulatory elements have been identified that mediate a transcriptional response to light (reviewed by Kuhlemeier *et al.* 1987; Gilmartin *et al.* 1991). A promoter containing 410 bp upstream of the transcriptional start site contains sufficient regulatory information for faithful *RbcS* gene expression in transgenic tobacco plants. Dissection of this sequence further revealed a redundancy of regulatory information such that at least three regions of this 410 bp element could independently activate transcription in a light-responsive manner. The shortest promoter fragment capable of such a response is only 166 bp in length. In contrast to the activity of this relatively short pea *RbcS* promoter, studies with the tomato *RbcS-3A* promoter demonstrate that 1100 bp of sequence are necessary for transcriptional activity in transgenic tobacco. Deletion of 100 bp from the 1100 bp tomato promoter resulted in a dramatic reduction in transcriptional activity (Ueda *et al.* 1989).

Most of the work to define the *RbcS* regulatory elements has focused on sequences located upstream of the transcription start site. However, work on petunia *RbcS* genes has demonstrated that sequences downstream of the transcription start site are involved in gene regulation. By the use of run-on transcription assays it has been shown that this effect is on the level of transcription and not at the level of mRNA stability (Dean *et al.* 1989c). However, studies in pea have not identified a role for downstream sequences in the differential regulation of *RbcS* genes (Kuhlemeier *et al.* 1988a). Differing expression patterns of *RbcS* genes from various species, and the differences observed with specific promoter fragments emphasizes the organizational complexity of the *cis*-acting regulatory sequences that mediate the intricate patters of *RbcS* expression.

The use of transgenic plants to assay expression patterns of *RbcS* gene promoter fragments has provided much information regarding the location and organization of these *cis*-acting regulatory DNA elements. Such transgenic plant experiments are, however, very costly in terms of both time and resources. A suitable transient assay for light-responsive *RbcS* gene expression studies would greatly facilitate the definition and characterization of the regulatory components of light-responsive elements. Unfortunately there has not, until recently, been a viable transient alternative to augment transgenic plant assays. However, recent advances towards a transient protoplast assay have demonstrated that *RbcS* gene activity is repressed by the end products of photosynthetic carbon assimilation (Sheen 1990). The presence of such sugars in the protoplast isolation medium has been shown to suppress photosynthetic gene expression. This observation opens the door to the use of a transient protoplast assay for *RbcS* light-responsive gene expression studies. An additional transient approach to studying *RbcS* promoter function is based upon the use of the particle gun (see Rolfe and Tobin 1991). The ballistic introduction of *RbcS* promoter fusions into plant cells that are recalcitrant to conventional transformation procedures, such as *Lemna*, has enabled homologous functional assays of *RbcS* promoter fragments in this plant (Rolfe and Tobin 1991).

The multiple levels of regulation that influence *RbcS* transcription involve distinct regulatory elements that confer specific aspects of the observed patterns of expression. It is possible that some regulatory elements contribute to just one defined aspect of the total promoter activity and yet others may mediate more complex patterns of expression in response to the interactions of different signals. For example, it is not clear whether specific light-responsive elements respond to only one photoreceptor, such that specific elements act as the termini for distinct signal-transduction pathways or whether more than one signal-transduction chain converges to act on a single *cis*-acting element. Additionally, the possibility exists that a specific regulatory element may influence more than one regulatory parameter. For example, *RbcS* expression in chloroplast containing cells and responsiveness to a particular photoreceptor may be immutably coupled. It is, therefore, necessary to define not only the expression characteristics of a specific *RbcS* promoter, but also those aspects of the expression profile conferred by smaller defined promoter fragments. The identification of the regulatory *trans*-acting proteins that interact with the *cis*-regulatory elements may help to unravel the regulatory complexity of *RbcS* gene expression.

8.1.4.3 Identification and characterization of trans-*acting factors*

From sequence comparisons between *RbcS* upstream sequences it has become apparent that specific sequence motifs are common to many *RbcS* genes (Manzara and Gruissem 1988). Several of these conserved DNA motifs have

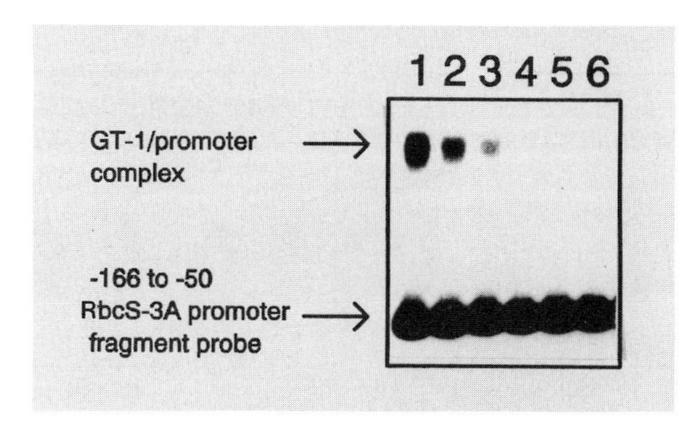

Figure 8. Gel shift competition assays for GT-1 binding. The *RbcS-3A* –166 to –50 promoter fragment was radio-labelled and incubated with tobacco nuclear extract in the absence or presence of unlabelled –166 to –50 promoter fragment as competitor DNA. Lane 1, no competitor; lane 2, 6-fold excess; lane 3, 12-fold excess; lane 4, 25-fold excess; lane 5, 50-fold excess; lane 6, 100-fold excess. The promoter fragment probe and the GT-1/promoter complex are shown.

subsequently been identified and characterized as the target sequences for specific nuclear DNA-binding proteins (Gilmartin *et al.* 1990). The techniques employed in the identification of proteins that interact with the *RbcS* upstream sequences are typically gel-retardation, DNAse I footprinting and methylation interference studies. Gel retardation assays involve the incubation of a radio-labelled promoter fragment with an extract of nuclear proteins and subsequent fractionation of the reaction by gel electrophoresis. A specific interaction between a nuclear protein and the radio-labelled DNA fragment can be detected as a result of reduced electrophoretic mobility of the DNA-protein complex in comparison to the mobility of the uncomplexed DNA fragment. Specificity of interactions can be assayed by competition with and excess of unlabelled factor binding sites (Fig. 8). The techniques of DNAse I footprinting and methylation interference provide information on the specific nucleotide sequences involved in a DNA-protein complex. In the case of DNAse I footprinting, this arises from the protection afforded those nucleotides bound by the protein from DNAse I digestion (Fig. 9). Methylation interference operates through the impediment of DNA-protein complex formation by the addition of a methyl group to specific G residues. Through a combination of these techniques, the precise location of a nuclear protein factor binding site can be identified within a promoter element.

The *RbcS* genes that have been studied in the greatest detail with regard to interaction with nuclear protein factors are those of *Arabidopsis* (Giuliano *et al.* 1988a; Schindler and Cashmore 1990), *Lemna* (Buzby *et al.* 1990), tomato (Manzara *et al.* 1991) and pea (for example, Green *et al.* 1987; Datta and Cashmore 1989; Lam *et al.* 1990; Sarokin and Chua 1992; Gilmartin *et al.* 1990) using nuclear extracts prepared from these species and tobacco. Some of the nuclear proteins identified that interact with specific DNA elements from

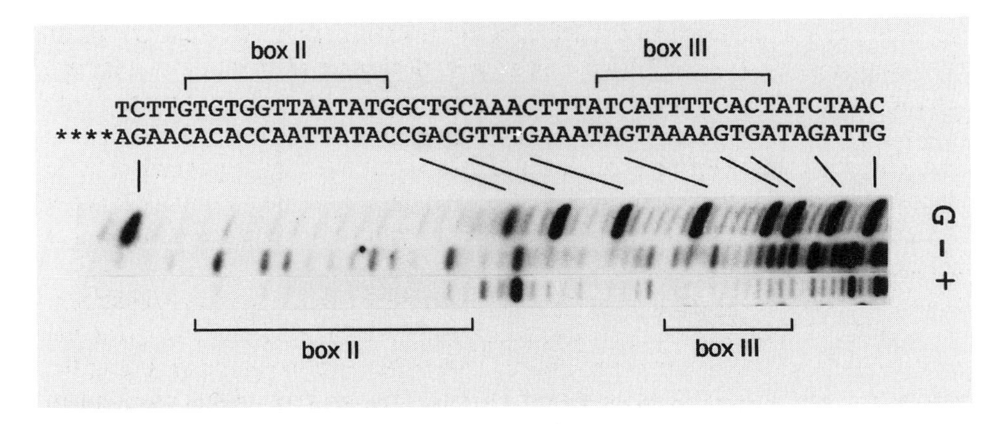

Figure 9. DNase I footprinting of GT promoter complexes with the pea *RbcS-3A* promoter. The *RbcS-3A* –166 to –50 promoter fragment was end labelled and used as a probe. The chemical sequencing G reaction lane is indicated (G). DNase digestion was performed in the absence (–) or presence (+) of tobacco nuclear extract. The regions of GT1 protection over boxes II and III are indicated as is the sequence of the promoter fragment. The **** indicates the labelled DNA strand in the reaction.

RbcS genes have been identified in more than one species. This observation likely reflects the conservation of regulatory mechanisms observed during the analyses of chimeric *RbcS* genes in heterologous transgenic hosts. However in some cases, nuclear protein factors defined in extracts prepared from one species appear to be absent from extracts prepared from other species. This could be due to differences in the procedures used to isolate nuclear proteins (Gilmartin *et al.* 1990). Alternatively, the specific binding site for a protein defined in one species may be absent from the promoter under study from another species; for example if a DNA-protein interaction is involved in an aspect of *RbcS* gene regulation specific to a particular family member or species.

Many of the nuclear protein binding motifs identified in *RbcS* gene promoters have been identified in other light responsive promoters, for example GT and GATA binding sites. However, not all conserved DNA elements in *RbcS* genes are unique to photoregulated genes, for example the G-box. These observations can be reconciled in several ways. The DNA elements may appear similar, yet subtle differences in their sequence may determine that they interact with distinct regulatory proteins: the discrimination by a DNA binding protein between two highly similar DNA motifs may occur as a consequence of differences in the DNA sequences flanking the conserved element. It is also possible that similar DNA elements may mediate diverse responses in different genes as a consequence of the differential abundance or activity of distinct regulatory proteins within specific cell types or at different developmental stages. Alternatively a conserved DNA element present within a light-responsive promoter may mediate a specific response as a consequence of its associations with other regulatory elements: the same element in a different sequence environment could lead to a completely different regulatory role (Schindler and Cashmore 1990). One final explanation for conservation of DNA elements between light-responsive and light-insensitive genes could be that these sequences may be involved in an aspect of gene regulation that is not affected by light, for example, organ specificity or as a general enhancer activity.

The characterization of nuclear proteins with specificity for *RbcS* upstream elements has been extended in several cases towards defining a role for the specific DNA-protein interactions. These studies have relied on determining the contribution of the specific sequence element to *RbcS* promoter function in two ways. First, by deletion or site directed mutation of the element within the *RbcS* promoter and subsequent assay of the activity of the resulting promoter in transgenic plants. Second, by multimerization of the specific DNA element and assay of their transcriptional activity in transgenic plants. We will focus here on four groups of DNA-binding proteins which have been characterized in the greatest detail.

8.1.4.3.1 GT-binding proteins. Studies with pea *RbcS-3A* led to the identification of conserved DNA elements that act as binding sites for the nuclear protein GT-1 (Green *et al.* 1987). This protein interacts with six GT rich motifs upstream of pea *RbcS-3A* (reviewed by Gilmartin *et al.* 1990). These sequences are conserved within many *RbcS* genes and are also present in other light-responsive genes (Gilmartin *et al.* 1990). The GT-1 binding sites of pea *RbcS-3A* show considerable sequence variation, but all appear to bind the same protein *in vitro*. Mutational analysis of GT-1 binding sites within the full 410 bp pea *RbcS-3A* promoter demonstrates a functional redundancy for these elements (Kuhlemeier *et al.* 1988b; Gilmartin *et al.* 1991). *Arabidopsis RbcS-1A* also contains binding sites for GT-1. Mutational analysis of these elements in the context of the full *RbcS-1A* promoter did not dramatically affect promoter activity (Donald and Cashmore 1990), as with pea this may be as a consequence of functional redundancy through the presence of multiple GT-boxes. A role for these conserved elements in pea *RbcS-3A* was uncovered when the mutational analyses were undertaken in the context of a truncated pea *RbcS-3A* promoter of 175 bp that contains only two GT-1 binding sites (Kuhlemeier *et al.* 1988a, b). Mutations in the GT boxes of this promoter that abolished binding of the nuclear protein *in vitro*, led to the loss of transcriptional activity conferred by this element *in vivo* (Kuhlemeier *et al.* 1988b; for reviews see Gilmartin *et al.* 1990, 1991). However, the complexity of the regulatory elements within this minimal promoter is emphasized by the recent observations that factor binding sites for the nuclear proteins 3AF3 and 3AF5 flank one of the GT-1 binding sites and appear to play a critical role in *RbcS* expression *in vivo* (Sarokin and Chua 1992). Definition of the critical role for one of the GT-1 binding sites in light responsive gene expression came from gain of function experiments in which one of these GT-1 binding sites (GTGTGGTTAATATG) was fused in four copies to a light-insensitive truncated cauliflower mosaic virus promoter. This chimeric promoter was assayed in transgenic plants and shown to be light-responsive (Lam and Chua 1990). These studies, therefore, demonstrated that the GT-1 binding site is a regulatory component, or molecular light switch, of the *RbcS* promoter.

Similar motifs are present in *RbcS* genes from other species as well as several other light inducible genes including tobacco *Cab-E* (Schindler and Cashmore 1990). The GT motifs are also present in the light-responsive phytochrome (*phyA*) promoter (Kay *et al.* 1989; Dehesh *et al.* 1990). This promoter confers transcriptional activity in the dark, but not in the light; it therefore exhibits the opposite pattern of expression to *RbcS* promoters.

The isolation of genes encoding DNA binding proteins with specificity for the GT motif present within the *RbcS* (Gilmartin *et al.* 1992; Perisic and Lam 1992) and *phyA* promoters (Dehesh *et al.* 1990) has provided a possible explanation for the presence of such closely related regulatory elements within promoters that confer opposite patterns of expression. Screening of cDNA expression libraries with labelled GT binding-site probes derived from pea *RbcS-3A* and rice *phyA* has led to the isolation of cDNA sequences that encode DNA binding proteins with specificity for each of these two elements. The sequence isolated from a tobacco cDNA expression library by virtue of the ability of its gene product to bind the pea GT element box II, GTGTGGTTAATATG, shows similar sequence specificity to nuclear GT-1 and has been named tobacco *GT-1a* (Gilmartin *et al.* 1992) and *B2F* (Perisic and Lam 1992). The rice protein which interacts specifically with the rice *phyA* GT element GCGGTAATT is termed GT-2 (Dehesh *et al.* 1990). These two proteins share a region of amino-acid homology only in the presumed DNA binding domain of the proteins (Gilmartin *et al.* 1992; Dehesh *et al.* 1992). These two proteins interact with similar DNA sequences, yet each shows specificity for its specific target site. It is, therefore, likely that GT binding proteins are present as a small family of related proteins that interact specifically with related sequences to mediate diverse patterns of gene expression. This observation could provide an explanation for the role of similar sequence elements within genes of unrelated expression characteristics; the sequence could be bound by distinct rather than the same nuclear protein.

8.1.4.3.2 G-box binding proteins. The G-box binding factor GBF was originally identified as an activity that interacts with a conserved motif containing the core CACGTG, present within the *RbcS* promoters from *Arabidopsis*, tomato and pea (Giuliano *et al.* 1988a). This motif has subsequently been identified within many genes, both light responsive and light insensitive (Schindler *et al.* 1992; Williams *et al.* 1992). A role of the G-box within the *Arabidopsis RbcS-1A* promoter has been demonstrated. Site specific mutation of the G-box element within the full 1.7 kbp promoter resulted in a dramatic reduction of transcriptional activity demonstrating a critical role for this element in *RbcS-1A* transcriptional activity (Donald and Cashmore 1990). It has recently become apparent through studies of DNA protein interactions within *RbcS* and other genes, as well as through the isolation of genes encoding G-box factors, that several related GBF activities exist. The binding specificity of different G-box factors is likely mediated through the sequences flanking the core element (Schindler *et al.* 1992; Williams *et al.* 1992). These observations would suggest that the different protein-protein and DNA-protein interactions between distinct G-box factors and similar target DNA sequences in different genes are responsible for the distinct regulatory patterns observed for the range of genes containing G-box elements. As with the GT-binding proteins, the complexity of *RbcS* promoter architecture is compounded by the existence of discrete, related nuclear proteins with specificity for closely related yet distinct *cis*-regulatory sequences.

8.1.4.3.3 GATA-binding proteins. Several nuclear proteins from various species have been identified that interact with a number of motifs containing the sequence GATA (Gilmartin *et al.* 1990; Buzby *et al.* 1990; Schindler and Cashmore 1990; Sarokin and Chua 1992). Numerous GATA containing elements have been identified by sequence comparisons among *RbcS* and other light-responsive genes. One of these elements, with a core of GATAAGG is present in the majority of light-responsive genes. This element, the I box, within the *Arabidopsis RbcS-1A* promoter has been shown to be the target of a protein GA-1 present within *Nicotiana plumbaginifolia* nuclear extracts (Schindler and Cashmore 1990). A similar sequence within an *RbcS* gene of *Lemna* has been identified as the binding site of a *Lemna* protein, LRF-1, the abundance of which is light regulated (Buzby *et al.* 1990). The corresponding sequence within pea *RbcS-3A* is bound by a factor GAF-1, present within *Nicotiana tabacum* nuclear extracts prepared from light-grown plants (Gilmartin *et al.* 1990). Differences in the binding specificities and abundance of these three proteins, which all interact with similar DNA sequences, could be due to the fact that they are distinct but related proteins, or perhaps as a consequence of the different assay conditions used by different researchers. This latter point highlights the possibility that different experimental conditions may lead to the identification of novel DNA-binding proteins, but it should also be considered that the same protein may behave differently under different assay conditions.

Another conserved GATA motif present in several light-responsive genes including some *RbcS* genes, contains two tandem GATA elements separated by two base pairs. This arrangement is typically present within many *Cab* promoters and a variant of it is found within the constitutive cauliflower mosaic virus promoter (Lam and Chua 1989). It is also found within the *Arabidopsis RbcS-1A* (Schindler and Cashmore 1990) and variants of it are located in three of the tomato *RbcS* genes (Manzara and Gruissem 1988). Different proteins demonstrating distinct binding characteristics have been identified from various species that can interact with these GATA elements. One of these, ASF-2, was identified as a tobacco nuclear protein with specificity for the GATA motif of the cauliflower mosaic virus 35S promoter. It was also shown to bind to the tandem GATA elements of a petunia *Cab* promoter (Lam and Chua 1989). The protein GA-1, identified as a *Nicotiana plumbaginifolia* DNA binding activity not only binds to the GATAAGG motif of tomato *RbcS-1A* and the *Nicotiana plumbaginifolia Cab-E* gene, but can interact with the tandem GATA motif (Schindler and Cashmore 1990). These binding specificities distinguish it from the proteins GAF-1; GAF-1 cannot interact with the tandem GATA element (Gilmartin *et al.* 1990). These observations would suggest that there are different yet related proteins that can interact with the GATAAGG motif (for example GAF-1 and LRF-1), the tandem GATA element (ASF-2) or both (GA-1). Investigations into the role of the GATA elements in *RbcS* transcription have demonstrated a quantitative role for some such GATA elements (for review see Gilmartin *et al.* 1990). Mutation of two GATAAGG motifs within the *Arabidopsis RbcS-1A* promoter led to a severe reduction in activity of the promoter (Donald and Cashmore 1990). These observations suggest that the GATA binding factors play a positive role in the light-responsive expression of *RbcS* genes. Since these are loss of function experiments, it is not clear whether the GATAAGG motifs are regulatory elements or play a purely quantitative role. Studies with multimerized copies of the tandem GATA element present within the cauliflower mosaic virus promoter and *Cab* promoters indicate that this element is involved in promoting leaf-specific expression, but does not play a role in light-regulated gene expression (Lam and Chua 1989).

8.1.4.3.4 AT-rich binding proteins. Nuclear proteins have been identified that interact with AT rich elements in pea and tomato *RbcS* genes. The detailed characterization of two of these proteins, AT-1 (Datta and Cashmore 1989) and 3AF-1 (Lam *et al.* 1990) have been reported. The protein AT-1 is interesting in that its ability to interact with its target sequence is modulated by phosphorylation; phosphorylation of AT-1 cancels its ability to interact with DNA. A specific role for this modification in *RbcS* gene expression has not been determined, but the element, which is present in both pea and tomato *RbcS* genes, appears to play a role in tomato *RbcS* gene expression. As discussed earlier, a deletion of the tomato *RbcS-3A* promoter from 1100 bp to 1000 bp results in a dramatic reduction in expression derived from this gene. The removal of this 100 bp disrupts the AT-1 binding site. Furthermore, if the 100 bp fragment is fused back to the remaining 1000 bp but in the opposite direction, the AT-1 site remains disrupted and promoter activity is not recovered (Ueda *et al.* 1989). This observation argues strongly for a positive role for the AT-1 binding site in *RbcS* gene expression.

The DNA-binding protein 3AF-1 interacts with a DNA element present within the pea *RbcS-3A* promoter. This element does not appear to be involved in light-responsive aspects of *RbcS* gene expression since fusion of four copies of this element to a minimal promoter does not result in a light-responsive element, but produces one that confers transcriptional activity in several cell types, independently of light (Lam *et al.* 1990). It is, therefore, likely that the 3AF-1 binding is involved in positive expression, but not involved in light-mediated responses. A gene encoding a DNA-binding protein with specificity for the 3AF-1 binding site has been isolated from tobacco. Northern analyses of RNA from different tissues with this cDNA as a probe indicate that, as with the GT and G-box binding proteins, 3AF-1 also exists as a member of a family of related DNA-binding proteins (Lam *et al.* 1990).

8.1.4.4 Summary

The organization of *RbcS* genes in multigene families gives rise not only to quantitative differences in the expression of individual family members, but in some species qualitative differences have been uncovered. Several levels of regulation control *RbcS* transcript accumulation, including different wavelengths of light, developmental and cell specific cues, an influence of chloroplast development and metabolic regulation. The complexity of this regulation is reflected in the complex organization of *RbcS* upstream sequences. The diverse patterns of expression are mediated through an array of *cis*-acting regulatory elements and the *trans*-acting regulatory proteins that interact with these regulatory DNA elements are themselves typically encoded by small gene families. Continuing attempts to unravel the process of *RbcS* gene expression will have to consider these various levels of complexity in order to provide insight into the specific components that link photoperception to a transcriptional response.

8.1.5 Circadian-clock, tissue-specific and light-regulated expression of *Cab* genes in higher plants

8.1.5.1 Function, structure and organization

8.1.5.1.1 Function. Higher plants have developed an elaborate apparatus for the efficient conversion of light into chemical energy *via* photosynthesis. All photosynthetic organisms that produce oxygen have two photosystems, PSI and PSII, respectively. Each photosystem is associated with 100-300 Chl molecules. Most of these, except a special pair of Chl *a* molecules, are not involved in photosynthesis *per se*, but are organized in light-harvesting antennae complexes. These surround the reaction centres of PSI and PSII and transfer light energy to them. Within these antennae the Chls are non covalently bound to specific, intrinsic membrane proteins. These Chl-protein complexes can contain only Chl *a* or both Chl *a* and *b* molecules; however both types contain several carotenoid molecules, which are thought to protect these complexes from the damaging effect of excessive light. Polypeptides of the Chl *a* antenna complexes are chloroplast encoded. Polypeptides of the Chl *a*/*b*-antenna complexes (Cab polypeptides) are encoded by nuclear genes and synthesized on free cytoplasmic ribosomes as a larger precursor (Grossman *et al.* 1980). After synthesis, these precursors are imported into the chloroplast by an energy-dependent process and processed correctly to the mature form. The mature polypeptides are then integrated into the thylakoids where they bind Chls *a* and *b* to form the above described protein-Chl *a*/*b*-antenna complexes (Abad *et al.* 1989). These protein-chlorophyll *a* and *b* complexes of PSI and PSII are frequently called light-harvesting complexes, LHCI and LHCII, respectively. Both LHCs contain different types of Cab polypeptides. The LHCI contains LHCI type I-IV polypeptides while LHCII contains LHCII type I-II and CP24, CP29(I) polypeptides. These eight types of Cab polypeptides, except LHCII type I and II, are substantially divergent, therefore it is believed that the gene duplication that gave rise to them occurred very early in the evolution of Cab-polypeptide containing organisms. In contrast, LHCII type I and II polypeptides are only 15% divergent from each other indicating a much later event during evolution (for review see Green *et al.* 1991). An almost complete set of *Cab* genes has been isolated and characterized from tomato, although the sequence of more than 60 *Cab* genes (mostly *Cab* genes encoding for LHCII type I Cab protein) from various species has been reported. The LHCII type I Cab proteins are encoded by multigene copies in most species. Within species these genes are very similar encoding identical or nearly identical proteins. In contrast, the genes encoding LHCII type II, CP29 and CP24 Cab proteins are present as single or duplicate copies in tomato (for a complete list of characterised *Cab* genes see Jansson *et al.* 1992).

8.1.5.1.2 Structure and organization. With the exception of the LHCII type I *Cab* genes all members of the family contain introns. The number of introns vary among the different types of the *Cab* genes and insufficient sequence data precludes tracing the evolution of introns within these genes. The protein sequences deduced from DNA sequences reveal that these proteins are made up of modules. Some of the modules appear to be almost identical in all the Cab polypeptides while others seem to have diverged probably more rapidly. However, all Cab proteins, irrespective of the similarity of modules, have three hydrophobic regions (transmembrane helixes) to span the thylakoid membrane. In all Cab proteins the two regions preceding the first and the third transmembrane helixes have a very distinctive pattern (zig-zag) and these regions are the most conserved parts of these proteins. The Cab proteins exhibit substantially more diversity in the region of the second transmembrane helix and even more diversity in the C-terminal regions, which are quite dissimilar (for a detailed description see Green *et al.* 1991).

8.1.5.2 Light-regulated expression of Cab *genes*

It has recently been reported that expression of different *Cab* genes from various monocot and dicot species was induced by light. *In vitro* nuclear run-off experiments provided evidence that light regulates the expression of *Cab* genes mainly, if not exclusively, at the level of transcription (Tobin and Silverthorne 1985). In addition, it has also been established that this light-induced transcription is mediated by different photoreceptors. Here we attempt to summarize results obtained about the expression characteristics of different *Cab* genes by emphasizing: (i) the involvement of different photoreceptors; (ii) the differences among the expression patterns of various *Cab* genes; (iii) the characteristics of *cis*- and *trans*-regulatory factors which may mediate this light-induced gene expression.

8.1.5.2.1 Phytochrome-mediated Cab *gene expression.* It was shown in the late 1970's that the amounts of Cab polypeptide and *Cab* mRNA increase in response to light treatments. It was also relatively well documented that the increase in steady-state mRNA level is transcriptionally regulated and at least partially mediated by phytochrome (Silverthorne and Tobin 1987). Previous studies have also established that the collective fluence response of *Cab* genes to R is biphasic, *i.e.* the levels of *Cab* mRNA are affected by light in the very-low fluence (VLF) range but there is also a response in the range, characteristic of ordinary low fluence (LF) phytochrome responses (Kaufman *et al.* 1984). While these studies clearly established the basic characteristics of *Cab* mRNA accumulation in response to R treatments, they provided little or no information about the expression of individual *Cab* genes. Very recently, by employing gene-specific probes, the expression characteristics of various types of *Cab* genes from *Arabidopsis* and pea were determined. The most detailed studies were performed by White *et al.* (1992). These authors showed that two type I genes (*Cab-8* and *AB96*), a type II (*Cab-215*) and a type III gene (*Cab-315*) (all encoding LHCII type Cab proteins) all respond similarly to brief R treatments. They found that a short R treatment led to a strong increase of steady-state mRNA levels of these *Cab* genes. By contrast, three other type I genes (*Cab-9*, *AB80* and *AB66*) (also encoding for LHCII type Cab proteins) showed little or no response to R. Similar studies were carried out by Sun and Tobin (1990) analysing the expression of three LHCII typeI Cab genes (*Cab1*, *Cab2* and *Cab3*) of *Arabidopsis thaliana*. In this latter case, all three *Cab* genes appear to have a relatively strong response to R, but at least in the *Arabidopsis* ecotype Columbia the specific transcripts accumulate to different levels. It should be noted that the qualitatively and quantitatively different expression patterns of LHCII type *Cab* genes are consistent with the hypothesis that individual gene family members may encode proteins for specialized functions (Smith 1990). Beside the above mentioned cases, fluence-response measurements for LHCI and LHCII type *Cab* mRNA accumulation were also performed in etiolated tomato and tobacco seedlings. Wehmeyer *et al.* (1990) demonstrated that both types of *Cab* genes show biphasic fluence responses to R. In addition, these authors also showed that the expression of the LHCII type *Cab* genes can be regulated by the so called 'high irradiance response' (HIR = fluence rates above 10 μmol m^{-2} s^{-1}). These latest results are of special importance, since they can serve as a basis for a detailed photobiological and molecular study of *Cab* promoters in transgenic tobacco and tomato seedlings.

8.1.5.2.2 Blue-light induced Cab *gene expression.* Photomorphogenesis occurs in higher plants largely through the concerted action of two types of photoreceptors; phytochrome responding primarily to R and the family of yet unidentified photoreceptors responding to B and near UV. Excitation of these B receptors is known to control a wide range of developmental processes such as inhibition of stem elongation (Chapter 9.1), chloroplast differentiation and phototropic curvature (Chapter 9.2). Many of these processes have biphasic fluence response curves. The biphasic nature of the collective fluence response to B indicates the involvement of B-low fluence (BLF, fluence rates below 10^{-1} μmol m^{-2} s^{-1}) and B-high fluence rate (BHF, fluence rates above 10^{1-3} μmol m^{-2} s^{-1}) responses, respectively. The same B system was also shown to regulate the expression of several pea nuclear genes, including *Cab* genes encoding for LHCII type proteins. It has recently been reported that excitation of the photomorphogenetic system responsible for the BLF response led to an increased *Cab* mRNA transcription (Marrs and Kaufman 1989). Excitation of the photomorphogenic system responsible for the BHF response resulted in an increased turnover rate of *Cab* mRNA (Marrs and Kaufman 1991). In addition to these experiments B-induced accumulation of LHCI and LHCII type *Cab* mRNAs, with similar kinetics, was also demonstrated in tomato and tobacco seedlings (Wehmeyer *et al.* 1990). Interestingly, the BLF response regulated *Cab* gene transcription may occur without the expression of other genes or the translation of pre-existing transcripts. Cycloheximide, an inhibitor of cytoplasmic protein synthesis, has no effect on the altered rates of *Cab* gene transcription in pea (Marrs and Kaufman 1991). This latter finding further underlines the difference between the R and B initiated signal-transduction pathways, since cycloheximide was shown to inhibit R-induced transcription of *Cab* genes in green transgenic tobacco or wheat seedlings (Lam *et al.* 1989b). The observations described above strongly emphasize the importance of B in regulating *Cab* gene expression. It should be noted, however, that a detailed analysis of B-induced expression of an individual *Cab* gene has not yet been published. This clearly indicates that despite recent advances there is still a lot of work to be done in this area.

8.1.5.2.3 Circadian clock-regulated Cab *gene expression.* The energy source for plants, sunlight, varies in intensity and in quality over a diurnal cycle. If plants are removed from light/dark cycles and placed under constant environmental conditions, processes at many levels continue to function rhythmically with periodicities close to 24 h. This 24-h rhythm is the hallmark of regulation by the circadian clock. This endogenous oscillation controls many plant functions ranging from stem elongation to the rhythmic expression of several nuclear genes. The circadian regulation in plants shares many properties with the circadian rhythms of other organisms, including temperature compensation of the period in constant (free-running) conditions and phase-shifting by environmental stimuli. The principal resetting stimulus for most circadian systems is light, which ensures synchronization of the oscillator to the natural dark/light cycles.

About 10 years ago it was first reported that steady state levels of *Cab* transcripts fluctuate diurnally in pea plants grown under light/dark cycles (Kloppstech 1985). Moreover this oscillatory pattern was maintained under constant environmental conditions indicating the regulation by a circadian clock. Thus the expression of *Cab* genes provides a relatively simple model, in contrast to multigene-regulated phenomenons such as flowering, to study circadian clock-regulated gene expression in higher plants. Recent data obtained by analysing *Cab* mRNA levels in different species further supported this hypothesis. Steady-state mRNA levels of *Cab* gene transcripts showed the characteristic oscillatory pattern in tomato, tobacco and in several other species (Giuliano *et al.* 1988b). In addition, results of *in vitro* nuclear run-off experiments indicated that the circadian clock controls the *Cab* gene expression at the level of transcription (Fejes *et al.* 1990; Millar and Kay 1991). Furthermore, it was established that phytochrome exerts a dual regulation on *Cab* gene expression since a short R pulse was sufficient to induce high level and rhythmic accumulation of *Cab* mRNA in dark-grown pea and tobacco seedlings (Tavladoraki *et al.* 1988; Wehmeyer *et al.* 1990). These experiments employed probes that detected mRNAs for all the genes of the LHCI and/or LHCII type families. Consequently, this approach did not distinguish any differences in regulation between family members, which at least in some cases turned out to be considerable (Millar and Kay 1991). Recently, by employing gene-specific probes and transgenic plants the expression patterns of three *Arabidopsis Cab* genes and that of the wheat *Cab1* gene were studied in detail. Figure 10 shows the circadian clock- and phytochrome-regulated expression of the wheat *Cab1* gene in etiolated wheat seedlings. These latest studies should facilitate the identification of *cis*-acting elements from these promoters (see below) which in turn should permit the molecular analysis of the mechanism mediating circadian-responsive transcription in higher plants.

8.1.5.3 The regulated expression of Cab *genes by* cis *and* trans-*acting elements*

It is apparent that the steady-state level of different *Cab* gene transcripts is regulated in harmony by the: (i) LF phytochrome system; (ii) VLF phytochrome system; (iii) HIR; (iv) BLF; (v) BHF and finally (vi) the circadian clock. In addition to the photoregulation, expression of the *Cab* genes is tissue/cell specific and developmentally programmed (Brusslan and Tobin 1992). At least four of these systems can regulate transcription (LF, VLF, BLF and the circadian clock) and at least one was shown to affect mRNA turnover (BHF). Transcriptional activation is primarily mediated through transcription factors that interact with specific DNA sequences. The identification of *cis*-regulatory elements and the isolation of the interacting transcription factors constitute crucial steps toward understanding light-induced *Cab* gene expression. Recently, by employing the transgenic technology, functional studies of *Cab* promoters became feasible. The combination of *in vitro* methods (DNA footprinting, gel-retardation assays, mutagenesis) and *in vivo* analysis of expression of *Cab*-promoter regulated chimeric genes in transgenic plants has yielded a considerable amount of data. Among the studied *Cab* promoters by far the best characterized are the *Arabidopsis Cab1*, *Cab2*, *Cab3* and the wheat *Cab1* promoters. It was shown that a 1.34 kb promoter region of the *Cab1* gene from *Arabidopsis* is sufficient to maintain phytochrome and developmentally regulated expression in transgenic plants (Brusslan and Tobin 1992). In addition, Millar *et al.* (1992) showed that the *Cab1* promoter-mediated transcription of a reporter gene is regulated by the circadian clock in transgenic plants. More

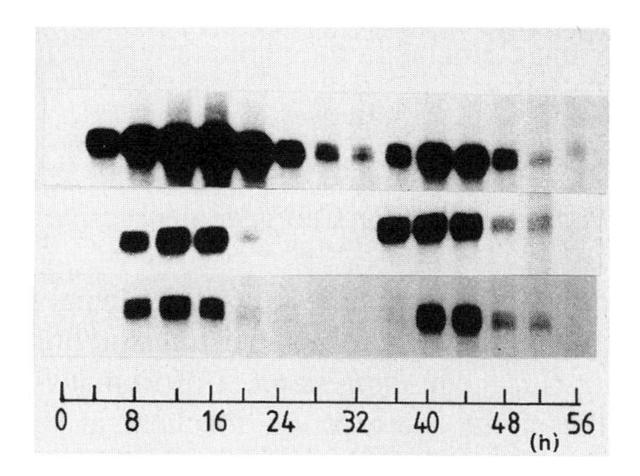

Figure 10. Northern blot analysis of light-induced, rhythmic *Cab-1* mRNA accumulation. The 3.5-day-old etiolated wheat seedlings were irradiated with either 5 min R (*upper* line), FR (*middle* line) or RG9-filtered light (*lower* line) and then returned to darkness. Samples were collected every 4 h. Each lane contains 20 μg of total RNA.

interestingly, however, they found that steady-state levels of *Cab1* mRNA does not fluctuate with a 24-h periodicity (Millar and Kay 1991). This finding indicates that the turnover rate of *Cab* mRNA may also be regulated by this endogenous oscillator. The *Cab2* and *Cab3* promoters from *Arabidopsis* were also characterized. Mitra *et al.* (1989) showed that a short 90 bp fragment of the *Cab3* promoter mediates shoot-specific and white light-inducible transcription in transgenic plants. Millar and Kay (1991) demonstrated that short fragments of the *Cab2* and *Cab3* promoters (319 bp and 420 bp, respectively) are sufficient for phytochrome, leaf specific and circadian clock-regulated expression in transgenic plants. In addition, they showed that transcription mediated by these promoters is regulated by the LF and VLF phytochrome system. Promoter sequences of the wheat *Cab1* gene involved in the regulated expression were also analysed. Nagy *et al.* (1988a) demonstrated that a 357 bp promoter region of this gene maintains maximal level and regulated expression in transgenic plants. Analysis of 5' deletion mutants and different chimeric genes demonstrated that this short promoter region contains multiple *cis*-acting regulatory elements. In addition it was also established that an enhancer-like element located within this region contains most of the regulatory elements for phytochrome, leaf specific and circadian clock-regulated expression (Fejes *et al.* 1990). The complex structure of *Cab* promoters is not unexpected. Multiple *cis*-acting regulatory elements were found in the promoter region of several *RbcS* genes. *In vitro* foot-printing experiments combined with the analysis of gene expression in transgenic plants firmly established the importance of several conserved sequences for light-induced expression of the *RbcS* genes (for details see Section 8.1.4.3). These conserved sequences, namely the G, GT-1 and GATA-boxes, are also present in the promoters of different *Cab* genes

(Castresana *et al.* 1988; Schindler and Cashmore 1990). The detailed characterization of these *cis*-acting elements in the regulation of different *Cab* promoters has yet to be accomplished.

8.1.5.4 Signal-transduction chains for Cab *gene expression*

The diversity and subtlety of *Cab* gene responses to changes in light quality and quantity indicates that a network of signal-transduction chains are likely involved in mediating *Cab* gene expression. Unfortunately, rather little is yet known concerning the signal-transduction mechanism that alters *Cab* gene expression. Consequently, this short summary is therefore limited to emphasizing what is known about this mechanism rather than giving a conclusive review of this field.

8.1.5.4.1 Photoreceptors. A large variety of photomorphogenic mutants which show deviation in the level, turnover-rate and the assembly of phytochrome and B photoreceptors have been described (Chapter 8.2). The altered characteristics of *Cab* gene expression in most of these mutants underlines the involvement of these photoreceptors in the regulated expression.

8.1.5.4.2 Second messengers. Among the earliest steps in the light-induced signal transduction chain can be the alteration of Ca^{2+} flux. It was shown that Ca^{2+} ionophores can substitute for Pfr in some phytochrome-mediated growth responses. Consistent with this finding Lam *et al.* (1989a) showed that calmodulin antagonists could lock light-induced expression of *Cab* genes in dark-adapted soybean suspension cultures. Heterotrimeric GTP-binding proteins are ubiquitous in animal cells. These molecules play an important role by modulating the levels of second messengers (cAMP, cGMP) *via* regulation of the activity of different enzymes (cAMP-cyclase, cGMP-phosphodiesterase *etc.*). Recent evidence indicates that such a GTP-binding protein may participate in the B signal transduction in plants.

8.1.5.4.3 Transcription factors. Binding of transcription factors to *cis*-regulatory sequences of promoters is considered to be the terminal step of the signal-transduction chains. Unfortunately, to date, no transcription factor has been isolated which shows specific interaction with any of the known *Cab* promoters. The isolation and characterization of such transcription factors could be the further step to elucidate the molecular mechanism by which the circadian clock and different photoreceptors control *Cab* gene expression in higher plants.

8.1.5.5 Summary

Studies aimed at unravelling the molecular mechanism by which light controls the expression of different *Cab* genes have already yielded valuable data. Analysis of transgenic plants *in vivo* combined with biochemical characterization of putative components of the signal-transduction cascade *in vitro* will undoubtedly give further information. The isolation of novel mutants affecting the circadian clock controlled gene expression of *Cab* genes could be especially rewarding. Such mutants could be essential in elucidating the molecular architecture of the biological clock and also in identifying some, yet unknown members of the pathway by which light regulates plant gene expression and development.

8.1.6 Further reading

Chory J., Peto C.A., Feinbaum R., Pratt L. and Ausubel F. (1989) *Arabidopsis thaliana* mutant that develops as light grown plant in the absence of light. *Cell* 58: 991-999.

Dean C., Pichersky E. and Dunsmuir P. (1989a) Structure, evolution and regulation of *RbcS* genes in higher plants. *Annu. Rev. Plant Physiol.* 40: 415-439.

Gilmartin P.M., Sarokin L., Memelink J. and Chua N.-H. (1990) Molecular light switches for plant genes. *Plant Cell* 2: 369-378.

Kay S.A. and Millar A.J. (1992) Circadian regulated *Cab* gene transcription in higher plants. In: *The Molecular Biology of Circadian Rhythms*, pp. 73-89, Young M. (ed) Marcel Dekker, New York.

Kuhlemeier C., Green P.J. and Chua N.-H. (1987) Regulation of gene expression in higher plants. *Annu. Rev. Plant Physiol.* 38: 221-257.

Manzara T. and Gruissem W. (1988) Organization and expression of the genes encoding ribulose 1,5-bisphosphate carboxylase in higher plants. *Photosynth. Res.* 16: 117-139.

Quail P.H. (1991) Phytochrome: A light-activated molecular switch that regulates plant gene expression. *Annu. Rev. Genet.* 25: 389-409.

Thompson W.F. and White M.J. (1991) Physiological and molecular studies of light-regulated nuclear genes in higher plants. *Annu. Rev. Plant Physiol. Plant Mol. Biol.* 42: 423-466.

Weisshaar B., Block A., Armstrong G.A., Herrmann A., Schulze-Lefert P. and Hahlbrock K. (1991) Regulatory elements required for light-mediated expression of the *Petroselinum crispum* chalcone synthase gene. In: *Symposia of the Society for Experimental Biology*. vol 45, pp. 191-210, Jenkins G.I. and Schuch W. (eds.) SEB Symposia Series.

8.1.7 References

Abad M.S., Clark S.E. and Lamppa G.K. (1989) Properties of a chloroplast enzyme that cleaves the chlorophyll *a/b* binding protein precursor. *Plant Physiol.* 90: 117-124.

Batschauer A., Ehmann B. and Schäfer E. (1991) Cloning and characterization of a chalcone synthase gene from mustard and its light-dependent expression. *Plant Mol. Biol.* 16: 175-185.

Bedbrook J.R., Smith S.M. and Ellis R.J. (1980) Molecular cloning and sequencing of cDNA encoding the precursor to the small subunit of chloroplast ribulose-1,5-bisphosphate carboxylase. *Nature* 287: 692-697.

Beggs C.J., Kuhn K., Böcker R. and Wellmann E. (1987) Phytochrome-induced flavonoid biosynthesis in mustard (*Sinapis alba* L.) cotyledons. Enzymic control and differential regulation of anthocyanin and quercitin formation. *Planta* 172: 121-126.

Benfey P.N. and Chua N.-H. (1989) Regulated genes in transgenic plants. *Science* 244: 174-181.

Berry J.O., Breiding D.E. and Klessig D.F. (1990) Light-mediated control of translational initiation of ribulose-1,5-bisphosphate carboxylase in amaranth cotyledons. *Plant Cell* 2: 795-803.

Block A., Dangl J.L., Hahlbrock K. and Schulze-Lefert P. (1990) Functional borders, genetic fine structure, and distance requirements of *cis* elements mediating light responsiveness of the parsley chalcone synthase promoter. *Proc. Natl. Acad. Sci. USA* 87: 5387-5391.

Brockmann J. and Schäfer E. (1982) Analysis of Pfr destruction in *Amaranthus caudatus* L. Evidence for two pools of phytochrome. *Photochem. Photobiol.* 35: 555-558.

Bröedenfeldt R. and Mohr H. (1988) Time courses of phytochrome-induced enzyme levels in phenylpropanoid metabolism (phenylalanine ammonia-lyase, naringenine-chalcone synthase) compared with time courses for phytochrome-mediated end-product accumulation (anthocyanin, quercitin). *Planta* 176: 383-390.

Broglie R., Bellemare G., Bartlett S.G., Chua N.-H. and Cashmore A.R. (1981) Cloned DNA sequences complementary to mRNAs encoding precursors to the small subunit of ribulose-1,5-bisphosphate carboxylase and a chlorophyll *a/b*-binding polypeptide. *Proc. Natl. Acad. Sci. USA* 78: 7304-7308.

Bruns B., Hahlbrock K. and Schäfer E. (1986) Fluence dependence of the ultraviolet-light-induced accumulation of chalcone synthase mRNA and effects of blue and far-red light in cultured parsley cells. *Planta* 169: 393-398.

Brusslan J.A. and Tobin E.M. (1992) Light-independent developmental regulation of *cab* gene expression in *Arabidopsis thaliana* seedlings. *Proc. Natl. Acad. Sci. USA* 89: 7791-7795.

Buzby J.S., Yamada T. and Tobin E.M. (1990) A light-regulated DNA binding activity interacts with a conserved region of a *Lemna gibba rbcS* promoter. *Plant Cell* 2: 805-814.

Castresana C., Garcia-Luque I., Alonso E., Malik, V.L. and Cashmore A.R. (1988) Both positive and negative regulatory elements mediate expression of a photoregulated *CAB* gene from *Nicotiana plumbaginifolia*. *EMBO J.* 7: 1929-1936.

Chory J., Nagpal P. and Peto C.A. (1991) Phenotypic and genetic analysis of *det2*, a new mutant that affects light-regulated seedling development in *Arabidopsis*. *Plant Cell* 3: 445-459.

Colbert J.T., Hershey H.P. and Quail P.H. (1983) Autoregulatory control of translatable phytochrome mRNA levels. *Proc. Natl. Acad. Sci. USA* 80: 2248-2252.

Dangl J.L., Hauffe K.D., Lipphardt S., Hahlbrock K. and Scheel D. (1987) Parsley protoplasts retain differential responsiveness to UV light and fungal elicitor. *EMBO J.* 6: 2551-2556.

Datta N. and Cashmore A.R. (1989) Binding of a pea nuclear protein to promoters of certain photoregulated genes is modulated by phosphorylation. *Plant Cell* 1: 1069-1077.

Dean C., Favreau M., Bedbrook J. and Dunsmuir P. (1989b) Sequences 5' to translation start regulate expression of petunia *rbcS* genes. *Plant Cell* 1: 209-215.

Dean C., Favreau M., Bond-Nutter D., Bedbrook J. and Dunsmuir P. (1989c) Sequences downstream of translation start regulate quantitative expression of two petunia *rbcS* genes. *Plant Cell* 1: 201-208.

Dehesh K., Bruce W.B. and Quail P.H. (1990) A *trans*-acting factor that binds to a GT-motif in the phytochrome gene promoter. *Science* 250: 1397-1399.

Dehesh K., Tepperman J., Christensen A.H. and Quail P.H. (1991) PhyB is evolutionarily conserved and constitutively expressed in rice-seedling shoots. *Mol. Gen. Genet.* 225: 305-313.

Dehesh K., Hung H., Tepperman J.M. and Quail P.H. (1992) GT-2: A transcription factor with twin autonomous DNA-binding domains of closely related but different target sequence specificity. *EMBO J.* 11: 4131-4144.

Deng C.-W., Caspar T. and Quail P.H. (1991) *cop1*: a regulatory locus involved in light-controlled development and gene expression in *Arabidopsis. Genes Develop.* 5: 1172-1182.

Dixon R.A. (1986) The phytoalexin response: elicitation, signalling and control of host gene expression. *Biol. Rev.* 61: 239-291.

Donald R.G.K. and Cashmore A.R. (1990) Mutation of either G box or I box sequences profoundly affects expression from the *Arabidopsis rbcS-1A* promoter. *EMBO J.* 9: 1717-1726.

Dooskin R.H. and Mancinelli A.L. (1968) Phytochrome decay and coleoptile elongation in *Avena* following various light treatments. *Bull. Torrey Bot.Club* 95: 474-487.

Drumm H. and Mohr H. (1978) The mode of interaction between blue (UV) light photoreceptor and phytochrome in anthocyanin formation of the *Sorghum* seedling. *Photochem. Photobiol.* 27: 241-248.

Ehmann B. and Schäfer E. (1988) Nucleotide sequences encoding two different chalcone synthases expressed in cotyledons of SAN 9789 treated mustard (*Sinapis alba* L.). *Plant Mol. Biol.* 11: 869-870.

Ehmann B., Ocker B. and Schäfer E. (1991) Developmental- and light-dependent regulation of the expression of two different chalcone synthase transcripts in mustard cotyledons. *Planta* 183: 416-422.

Ellis R.J. (1981) Chloroplast proteins - synthesis, transport and assembly. *Annu. Rev. Plant Physiol.* 32: 111-137.

Fejes E., Pay A., Kanevsky I., Szell M., Adam E., Kay S. and Nagy F. (1990) A 268-bp upstream sequence mediates the circadian clock regulated transcription of the wheat cab-1 gene in transgenic plants. *Plant Mol. Biol.* 15: 921-932.

Frohnmeyer H., Ehmann B., Kretsch T., Rocholl M., Harter K., Nagatani A., Furuya M., Batschauer A., Hahlbrock K. and Schäfer E. (1992) Differential usage of photoreceptors during plant development for chalcone synthase expression. *Plant J.* 2: 899-906.

Fukshansky L. and Schäfer E. (1983) Models in photomorphogenesis. In: *Encyclopedia of Plant Physiology*, New Series 16A, pp. 69-95, Shropshire Jr.W. and Mohr H. (eds.) Springer, Berlin.

Gilmartin P.M., Memelink J. and Chua N.-H. (1991) Dissection of the light-responsive elements of pea rbcS-3A. In: *Phytochrome properties and biological action.* Thomas B. and Johnson C.B. (eds). NATO ASI series, 50: 141-155.

Gilmartin P.M., Memelink J., Hiratsuka K., Kay S.A. and Chua N.-H. (1992) Characterization of a gene encoding a DNA binding protein with specificity for a light-responsive element. *Plant Cell* 4: 839-849.

Giuliano G., Pichersky E., Malik V.S., Timko M.P., Scolnik P.A. and Cashmore A.R. (1988a) An evolutionarily conserved protein binding sequence upstream of a plant light-regulated gene. *Proc. Natl. Acad. Sci. USA* 85: 7089-7093.

Giuliano G., Hoffman N.E., Ko K., Scolnik P.A. and Cashmore A.R. (1988b) A light entrained circadian clock controls transcription of several plant genes. *EMBO J.* 7: 3635-3642.

Gottmann K. and Schäfer E. (1982) In vitro synthesis of phytochrome apoprotein directed by mRNA from light and dark-grown *Avena* seedlings. *Photochem. Photobiol.* 35: 521-525.

Green P.J., Kay S.A. and Chua N.-H. (1987) Sequence-specific interactions of a pea nuclear factor with light-responsive elements upstream of the *rbcS-3A* gene. *EMBO J.* 6: 2543-2549.

Green B.R., Pichersky E. and Kloppstech K. (1991) Chlorophyll *a/b*-binding proteins: An extended family. *Trends Biochem. Sci.* 16: 181-186.

Greenland A.J., Thomas M.V. and Walden R.M. (1987) Expression of two nuclear genes encoding chloroplast proteins during early development of cucumber seedling. *Planta* 170: 99-110.

Grossman A., Bartlett S. and Chua N.-H. (1980) Energy-dependent uptake of cytoplasmically-synthesized polypeptides by chloroplasts. *Nature* 285: 625-628.

Guiltinan M.J., Marcotte W.R. and Quatrano R.S. (1990) A plant leucine zipper protein that recognizes an abscisic acid responsive element. *Science* 250: 267-271.

Hahlbrock K. and Scheel D. (1989) Physiology and molecular biology of phenypropanoid metabolism. *Annu. Rev. Plant Physiol. Plant Mol. Biol.* 40: 347-369.

Heim B. and Schäfer E. (1984) The effect of red and far-red light in the high irradiance reaction of phytochrome (hypocotyl growth in dark-grown *Sinapis alba* L.). *Pant Cell Environ.* 7: 39-43.

Heim B., Jabben M. and Schäfer E. (1981) Phytochrome destruction in dark- and light-grown *Amaranthus caudatus* seedlings. *Photochem. Photobiol.* 34: 89-93.

Heller W. and Hahlbrock K. (1980) Highly purified 'flavanone synthase' from parsley catalyses the formation of naringenine chalcone. *Arch. Biochem. Biophys.* 200: 617-619.

Hofmann E., Grimm R., Harter K., Speth V. and Schäfer E. (1991) Partial purification of sequestered particles of phytochrome from oat (*Avena sativa* L.) seedlings. *Planta* 183: 265-273.

Jansson S., Pichersky E., Bassi R., Green R.E., Ikeuchi M., Melis A., Simpson J.D., Spandfort M., Staehelin A.L. and Thornber P.J. (1992) A nomenclature for genes encoding the chlorophyll *a/b*-binding proteins of higher plants. *Plant Mol. Biol. Reporter* 10: 242-253.

Jefferson R.A., Burgess S.M. and Hirsh D. (1986) β-Glucuronidase from Escherichia coli as a gene-fusion marker. *Proc. Natl. Acad. Sci. USA* 83: 8447-8451.

Jenkins G. (1988) Photoregulation of gene expression in plants. *Photochem. Photobiol.* 48: 821-832.

Kay S.A., Keith B., Shinozaki K., Chye M.-L. and Chua N.-H. (1989) The rice phytochrome gene: structure, autoregulated expression, and binding of GT-1 to a conserved site in the 5' upstream region. *Plant Cell* 1: 351-360.

Kaufman L.S., Thompson W.F. and Briggs W.R. (1984) Different red light requirements for phytochrome induced accumulation of *Cab* RNA and *rbcS* RNA. *Science* 226: 1447-1449.

Kloppstech K. (1985) Diurnal and circadian rhythmicity in the expression of light induced plant nuclear messengers. *Planta* 165: 502-506.

Knight M.R. and Jenkins G.I. (1992) Genes encoding the small subunit of ribulose-1,5-bisphosphate carboxylase/oxygenase in *Phaseolus vulgaris* L.:nucleotide sequence of cDNA clones and initial studies of expression. *Plant Mol. Biol.* 18: 567-579.

Kuhlemeier, C. (1992) Transcriptional and post-transcriptional regulation of gene expression in plants. *Plant Mol. Biol.* 19: 1-14.

Kuhlemeier C., Fluhr R. and Chua N.-H. (1988a) Upstream sequences determine the difference in transcript abundance of pea *rbcS* genes. *Mol. Gen. Genet.* 212: 405-411.

Kuhlemeier C., Cuozzo M., Green P., Goyvaerts E., Ward K. and Chua N.-H. (1988b) Localization and conditional redundancy of regulatory elements in *rbcS-3A,* a pea gene encoding the small subunit of ribulose-bisphosphate carboxylase. *Proc. Natl. Acad. Sci. USA* 85: 4662-4666.

Lam E. and Chua N.-H. (1989) ASF-2: A factor that binds to the cauliflower mosaic virus 35S promoter and a conserved GATA motif in *Cab* promoters. *Plant Cell* 1: 1147-1156.

Lam E. and Chua N.-H. (1990) GT-1 binding site confers light responsive expression in transgenic tobacco. *Science* 248: 471-474.

Lam E., Benedyk M. and Chua N.-H. (1989a) Characterization of phytochrome regulated gene expression in a photoautotrophic cell suspension: possible role for calmodulin. *Mol. Cell. Biol.* 9: 4819-4823.

Lam E., Green P.J., Wong M. and Chua N.-H. (1989b) Phytochrome activation of two nuclear genes requires cytoplasmic protein synthesis. *EMBO J.* 8: 2777-2783.

Lam E., Kano-Murakami Y., Gilmartin P.M., Niner B. and Chua N.-H. (1990) A metal-dependent DNA-binding protein interacts with a constitutive element of a light-responsive promoter. *Plant Cell* 2: 857-866.

Lange H., Shropshire Jr.W. and Mohr H. (1971) An analysis of phytochrome-mediated anthocyanin synthesis. *Plant Physiol.* 47: 649-655.

Lipphardt S., Brettschneider R., Kreuzaler F., Schell J. and Dangl J.L. (1988) UV-inducible transient expression in parsley protoplasts identifies regulatory *cis*-elements of a chimeric *Antirrhinum majus* chalcone synthase gene. *EMBO J.* 7: 4027-4033.

Manzara T., Carrasco P. and Gruissem W. (1991) Developmental and organ-specific changes in promoter DNA-protein interactions in the tomato *rbcS* gene family. *Plant Cell* 3: 1305-1316.

Marrs K.A. and Kaufman L.S. (1989) Blue light regulation of transcription for nuclear genes in pea. *Proc. Natl. Acad. Sci. USA* 86: 4492-4495.

Marrs K.A. and Kaufman L.S. (1991) Rapid transcriptional regulation of the *Cab* and *pEA207* gene families in peas by blue light in the absence of cytoplasmic protein synthesis. *Planta* 183: 327-333.

McCurdy D.W. and Pratt L.H. (1986a) Kinetics of intracellular redistribution of phytochrome in *Avena* coleoptiles after its photoconversion to the active, far-red-absorbing form. *Planta* 167: 330-336.

McCurdy D.W. and Pratt L.H. (1986b) Immunogold electron microscopy of phytochrome in *Avena*: Identification of intracellular sites responsible for phytochrome sequestering and enhanced pelletability. *J. Cell Biol.* 103: 2541-2550.

Millar A.J. and Kay S.A. (1991) Circadian control of cab gene transcription and mRNA accumulation in *Arabidopsis*. *Plant Cell* 3: 541-550.

Millar A.J., Short R.S., Chua N.-H. and Kay A.S. (1992) A novel circadian phenotype based on firefly luciferase expression in transgenic plants. *Plant Cell* 4: 1075-1087.

Mitra A., Choi H.K. and An G. (1989) Structural and functional analyses of *Arabidopsis thaliana* chlorophyll *a/b*-binding protein *(cab)* promoters. *Plant Mol. Biol.* 12: 169-179.

Mol J.N.M., Stuitje A.R. and van der Krol A. (1989) Genetic manipulation of floral pigmentation genes. *Plant Mol. Biol.* 13: 287-294.

Mohr H. (1983) Pattern specification and realization in photomorphogenesis, pp 336-357 In: *Photomorphogenesis*, Shropshire W. Jr and Mohr H. (eds) Springer, Berlin New York.

Nagy F., Kay S.A. and Chua N.-H. (1988a) A circadian clock regulates transcription of a wheat *Cab-1* gene. *Genes Develop.* 2: 376-382.

Nagy F., Kay S.A. and Chua N.-H. (1988b) Gene regulation by phytochrome. *Trends in Genet.* 4: 37-42.

Oeda K., Salinas J. and Chua N.-H. (1991) A tobacco bZIP transcription activator (TAF-1) binds to a G-box-like motif conserved in plant genes. *EMBO J.* 10: 1793-1802.

Oelmüller R. and Mohr H. (1984) Responsivity amplification by light in phytochrome-mediated induction of chloroplast glyceraldehyde-3-phosphate dehydrogenase (NADP-dependent, EC 1.2.1.13) in the shoot of milo (*Sorghum vulgare* Pers.). *Plant Cell Environ.* 7: 29-37.

Ohl S., Hahlbrock K. and Schäfer E. (1989) A stable blue-light-derived signal modulates ultraviolet-light induced activation of the chalcone synthase gene in cultured parsley cells. *Planta* 177: 228-236.

Otto V. and Schäfer E. (1988) Rapid phytochrome-controlled protein phosphorylation and dephosphorylation in *Avena sativa*. *Plant Cell Physiol.* 29: 1115-1121.

Otto V., Mösinger E., Sauter M. and Schäfer E. (1983) Phytochrome control of its own synthesis in *Sorghum vulgare* and *Avena sativa*. *Photochem. Photobiol.* 38: 693-700.

Perisic O. and Lam E. (1992) A tobacco DNA binding protein that interacts with a light-responsive box II element. *Plant Cell* 4: 831-838.

Peters N.K., Frost J.W. and Long S.R. (1986) A plant flavone, luteolin, induces expression of *Rhizobium meliloti* nodulation genes. *Science* 233: 977-980

Pratt L.H. and Marmé D. (1976) Red-light enhanced phytochrome pelletability. Reexamination and further characterization. *Plant Physiol.* 58: 686-692.

Pratt L.H., Stewart S.J., Shimazaki Y., Wang Y.-C. and Cordonnier M.-M. (1991) Monoclonal antibodies directed to phytochrome from green leaves of *Avena sativa* L. cross-react weakly or not at all with the phytochrome that is most abundant in etiolated shoots of the same species. *Planta* 184: 87-95.

Quail P.H. (1983) Rapid action of phytochrome in photomorphogenesis. In: *Encyclopedia of Plant Physiology*, New Series 16A, Photomorphogenesis, pp.178-212, Shropshire, Jr.W. and Mohr, H. (eds.) Springer, Berlin.

Rolfe S.A. and Tobin E.M. (1991) Deletion analysis of a phytochrome-regulated monocot rbcS promoter in a transient assay system. *Proc. Natl. Acad. Sci. USA* 88: 2683-2686.

Sarokin L. P. and Chua N.-H. (1992) Binding sites for two novel phosphoproteins, 3AF5 and 3AF3, are required for *rbcS-3A* expression. *Plant Cell* 4: 473-483.

Schäfer E. (1981) Phytochrome and daylight. In: *Plants and the Daylight Spectrum*, pp. 461-480, Smith H. (ed.) Academic Press, London.

Schäfer E., Löser G. and Heim B. (1983) Formalphysiologische Analysen der Signaltransduktion in der Photomorphogenese. *Ber. Dtsch. Bot. Ges.* 96: 497-509.

Schindler U. and Cashmore A.R. (1990) Photoregulated gene expression may involve ubiquitous DNA binding proteins. *EMBO J.* 9: 3415-3420.

Schindler U., Terzaghi W., Beckmann H., Kadesch T. and Cashmore A.R. (1992) DNA binding sites preferences and transcriptional activation properties of the *Arabidopsis* transcription factor GBF1. *EMBO J.* 11: 1275-1289.

Schmidt R. and Mohr H. (1983) Time course of signal transduction in phytochrome mediated anthocyanin synthesis in mustard cotyledons. *Plant Cell Environ.* 6: 235-238.

Schopfer P. (1984) Photomorphogenesis. In: *Advanced Plant Physiol.*, pp. 380-407, Wilkins M.B. (ed.) Pittman, London.

Schulze-Lefert P., Dangl J.L., Becker-Andre M., Hahlbrock K. and Schulz W. (1989) Inducible *in vivo* DNA footprints define sequences necessary for UV light activation of the parsley chalcone synthase gene. *EMBO J.* 8: 651-656.

Serlin B.S. and Roux S.J. (1984) Modulation of chloroplast movement in the green alga *Mougeotia* by the Ca^{2+} ionophore A23187 and by calmodulin antagonists. *Proc. Natl. Acad. Sci. USA* 81: 6368-6372.

Sharrock R.A. and Quail P.H. (1989) Novel phytochrome sequences in *Arabidopsis thaliana*: Structure, evolution and different expression of a plant regulatory photoreceptor family. *Genes Develop.* 3: 1745-1757.

Sheen J. (1990) Metabolic repression of transcription in higher plants. *Plant Cell* 2: 1027-1038.

Sheen J. and Bogorad L. (1986) Expression of the ribulose-1,5-bisphosphate carboxylase large subunit and three small subunit genes in two cell types of maize leaves. *EMBO J.* 5:3417-3422.

Silverthorne J. and Tobin E. (1987) Phytochrome regulation of nuclear gene expression. *Bioessays* 7: 18-23.

Silverthorne J., Wimpee C.F., Yamada T., Rolfe S.A. and Tobin E.M. (1990) Differential expression of individual genes encoding the small subunit of ribulose-1,5-bisphosphate carboxylase in *Lemna gibba*. *Plant Mol. Biol.* 15: 49-58.

Singh H., Clerc R.G. and LeBowitz J.H. (1989) Molecular cloning of sequence-specific DNA-binding proteins using recognition site probes. *Bio Techniques* 7: 252-261.

Smith H. (1990) Signal perception, differential expression within multigene families and the molecular basis of phenotypic plasticity. *Plant Cell Environ.* 13: 585-594.

Stone H.J. and Pratt L.H. (1979) Characterization of the destruction of phytochrome in the red absorbing form. *Plant. Physiol.* 63: 680-682.

Sun L. and Tobin E.M. (1990) Phytochrome-regulated expression of genes encoding light-harvesting chlorophyll *a/b*-protein in two long hypocotyl mutants and wild type plants of *Arabidopsis thaliana*. *Photochem. Photobiol.* 52: 51-56.

Tavladoraki P., Kloppstech K. and Argyroudi-Akoyunoglou J. (1989) Circadian rhythm in the expression of the mRNA coding for the apoprotein of the light-harvesting complex of photosystem II. *Plant Physiol.* 90: 665-672.

Thomas, B. and Johnson C.B. (eds.) (1991) *Phytochrome properties and biological action.* NATO ASI Series, Vol. 50; Springer, Berlin.

Thompson D.M., Tanzer M.M. and Meagher R.B. (1992) Degradation products of the mRNA encoding the small subunit of ribulose-1,5-bisphosphate carboxylase in soybean and transgenic petunia. *Plant Cell* 4: 47-58.

Tobin E.M. and Silverthorne J. (1985) Light regulation of gene expression in higher plants. *Annu. Rev. Plant Physiol.* 36: 569-5693.

Ueda T., Pichersky E., Malik V.S. and Cashmore A.R. (1989) Level of expression of the tomato *rbcS-3A* gene is modulated by a far upstream promoter element in a developmentally regulated manner. *Plant Cell* 1: 217-227.

Wagner G., Valentin P., Dieter P. and Marmé D. (1984) Identification of calmodulin in the green alga *Mougeotia* and its possible function in chloroplast reorientation movement. *Planta* 162: 62-67.

Wang Y.-C., Stewart S.J., Cordonnier M.-M. and Pratt L.H. (1991) *Avena sativa* L. contains three phytochromes, only one of which is abundant in etiolated tissue. *Planta* 184: 96-104.

Wanner L.A. and Gruissem W. (1991) Expression dynamics of the tomato *rbcS* gene family during development. *Plant Cell* 3: 1289-1303.

Wehmeyer B., Cashmore A.R. and Schäfer E. (1990) Photocontrol of the expression of genes encoding chlorophyll *a/b* binding proteins and small subunit of ribulose-1,5-bisphosphate carboxylase in etiolated seedlings of *Lycopersicum esculentum* (L) and *Nicotiana tabacum* (L). *Plant Physiol.* 93: 990-997.

Wellmann E. (1971) Phytochrome-mediated flavone glycoside synthesis in cell suspension cultures of *Petroselinum hortense* after preirradiation with ultraviolet light. *Planta* 110: 283-286.

Wenng A., Batschauer A., Ehmann B. and Schäfer E. (1990) Temporal patterns of gene expression in cotyledons of mustard (*Sinapis alba* L.) seedlings. *Bot. Acta.* 103: 240-243.

White J.M., Fristensky B.W., Falconet D., Childs L.C., Watson J.C., Alexander L., Roe B.A. and Thompson W. (1992) Expression of the chlorophyll *a/b*-protein multigene family in pea (*Pisum sativum* L.). *Planta* 188: 190-198.

Williams M., Foster R. and Chua N.-H. (1992) Sequences flanking the hexameric G-box core CACGTG affect the specificity of protein binding. *Plant Cell* 4: 485-496.

8.2 Photomorphogenic mutants of higher plants

Maarten Koornneef[1] and Richard E. Kendrick[2,3]

Departments of [1]Genetics and [2]Plant Physiology,
Wageningen Agricultural University,
Dreijenlaan 2, NL-6703 HA Wageningen, The Netherlands
[3]Laboratory for Photoperception and Signal Transduction,
Frontier Research Program,
Institute for Physical and Chemical Research (RIKEN),
Hirosawa 2-1, Wako City, Saitama 351-01, Japan

8.2.1 Introduction

The regulation of plant development by light is mediated by several different photoreceptors. After photoperception, a transduction chain relays the signal to the terminal response(s). The complexity of photomorphogenesis is due to the different photoreceptors and the multiple steps of the transduction chain(s), which may involve such aspects as changes in gene expression, interaction with plant hormones, membrane changes *etc.*, which remain largely unknown. The components of these transduction chains can either be different or similar. A further complication arises because of the co-action of several photoreceptive systems regulating the same process (Chapter 6) or because of the multiple effects induced by a single photoreceptor.

The availability of genotypes (often as induced mutants) in which certain parts of the morphogenetic pathway are eliminated provides useful tools for the study of photomorphogenesis. To be able to draw the 'right' conclusion from mutant studies it is necessary that the primary defect caused by the mutation is known, so that the various effects observed in the mutant can all be traced back to this primary cause. Since light is the inducer of a chain of events which ultimately result in a physiological effect, the response itself can be derived indirectly from the light signal. For example the effect on growth may be a consequence of changes in hormone levels which are under the control of the photoreceptor.

The complexity of many physiological and developmental responses means that the conclusions from work with mutants are not always straightforward and analysis of mutants is not only required at the genetic and molecular level to

R.E. Kendrick & G.H.M. Kronenberg (eds.), Photomorphogenesis in Plant - 2nd Edition

identify the primary defect, but must also be accompanied by a thorough physiological analysis.

The use of mutants in photomorphogenesis research is relatively new, but is expanding rapidly. Many new mutants are now available, more is known about their primary defects and in a few cases the molecular nature of the lesions are known. Reviews describing the various mutants in higher plants are those of Kendrick and Nagatani (1991) and Chory (1991, 1993).

In addition to mutants defective in specific morphogenic steps, the cloning of genes involved in these processes and the possibility of introducing them into plants has provided genotypes that overexpress these genes. Such genotypes can be considered as another class of mutants which are expected to be 'mirror-images' of deficiency mutants.

8.2.2 General aspects of the genetic and molecular analysis of mutants

8.2.2.1 The cloning and transfer of genes

The primary effect of a mutation is the defectiveness or altered expression of a gene. For the physiological interpretation of mutants it is important to know what the function of the gene is at the DNA and protein level. When ideas about the defect in the mutant are present, these mutants can be analyzed for the respective proteins or DNA sequences. However, mutants themselves have become important tools to isolate (clone) the respective genes, even without knowledge about their biochemical nature. These mutant-based cloning strategies include:

(i) *Map-based* or *positional cloning.* This procedure is based on an accurate location of the mutant on a genetic map close to a DNA marker. Assuming that genetic experiments have shown that the gene resides on a particular limited stretch of DNA, this DNA can be used to transform the mutant. When the wild-type (WT) gene is present on the DNA used for transformation, the mutation is complemented, that is the WT phenotype is restored. Positional cloning is only feasible in species with a relatively small genome and with a detailed genetic map such as *Arabidopsis,* tomato and rice.

(ii) *Genomic subtraction.* The genomic subtraction procedure allows the cloning of genes where the mutation is due to a small deletion. The technique allows the isolation of DNA present in the WT, but absent in the mutant.

(iii) *Tagging.* The insertion of a piece of DNA *e.g.* by transformation into a gene of the host may interrupt the reading frame of that gene and lead to a mutation. Since the transformed DNA is usually well characterized this may serve as a 'tag' to identify the DNA flanking this insertion and thereby the cloning of sequences of the disrupted gene. In *Arabidopsis* there are already many cases of successful tagging of genes by randomly inserted (*Agro-*

bacterium tumefaciens derived) T-DNA (Feldmann 1991). Tagging can also be achieved by mobile DNA elements, called *transposons.* Transposons have been identified and used to clone genes in maize and *Antirrhinum,* but have also been transferred to other species, where they are 'active' and can induce mutations.

(iv) *Differential screening* of mRNAs specific for the WT. Mutations especially those due to deletions and insertions result in the absence of the mRNA that is normally produced by the WT allele. The identification of this particular mRNA is a way to clone the gene involved. The success of this procedure depends on the abundance of that mRNA and the absence of such a mRNA in the mutant.

(v) *Shotgun complementation.* By transforming a mutant with the WT gene, the mutant phenotype is restored to WT. This complementation by transformation is an important proof (gain-of-function) that the right gene has been cloned, but can also be used to identify the gene in question when this complementation experiment is performed with random WT DNA. This procedure requires both a very efficient transformation procedure and a rapid identification of the WT phenotype.

Genes cloned on the basis of their product, which can either be the isolated protein (*e.g.* phytochrome) or an abundant mRNA can be introduced into plants such as *Arabidopsis,* tobacco and tomato, by well established transformation procedures either in the sense or antisense orientation. This results in overexpression or suppression of the introduced gene and endogenous gene, respectively.

8.2.2.2 General aspects of mutant isolation

Since mutants are an important tool both for the cloning of genes and in their functional analysis, the basic principles of mutant isolation for a particular organism have to be known.

Most induced mutations (by radiation or chemical treatment) are recessive: a gene loses its function by the mutation, but one WT allele in a diploid organism provides sufficient gene product to mask such a mutation. When one starts with diploid WT homozygotes, which is the case in most higher plants, one needs two generations to detect the mutation (Fig. 1). In organisms which are haploid during part of their life cycle (*e.g.* mosses) or during their whole life cycle except the meiotic cells (*e.g.* fungi), recessive mutations are not masked by dominant alleles and therefore a meiotic generation is not necessary for the detection of a recessive mutant (Fig. 1). In allopolyploid (*e.g.* tobacco, wheat) and tetraploid (*e.g.* potato) species, at least two copies of each gene are present in the gametes, which means that mutation of one of the two copies of the gene will not result in a mutant phenotype, because the second gene performs the

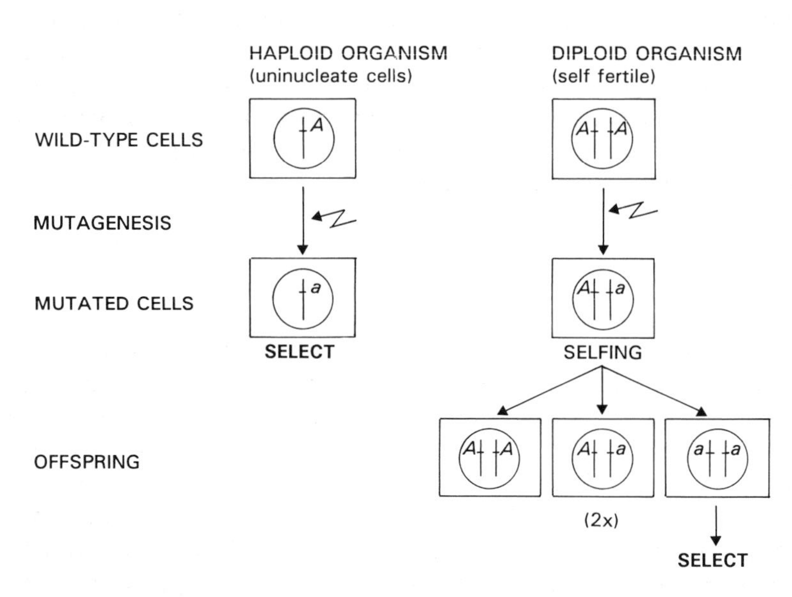

Figure 1. The isolation of recessive mutants in haploid and diploid organisms. In higher plants growing out of mutagen-treated seed or the treated plant itself is called the M_1 generation and the next generation in which the recessive mutants occur the M_2 generation.

same function as the mutated gene. Genetic redundancy also makes it difficult to identify mutants for multi-gene families.

In addition to mutants, genetic variation can also be found among the 'natural' genetic variation within a species, within a collection of ecotypes or cultivars. This type of genetic variation is limited because it will not include extreme variants with a strongly reduced chance of survival (in nature) or with reduced yield (for cultivated species).

After the mutant has been isolated or an interesting 'natural' variant has been identified, a genetic characterization has to be performed. Such an analysis involves the following steps: (i) the determination of the number of genes involved and allelism tests with previously described mutants with a similar phenotype; (ii) the determination of the dominance relationships between the various alleles; (iii) the location of genes on the linkage map and (iv) the analysis of epistatic relationships between similar mutants at different loci. Epistatic relationships are derived from the analysis of double mutants. When a mutant at a specific locus (*aa*) obscures the phenotype of the second gene *B* (*i.e.* no difference between *aaBB* and *aabb*), gene *A* is epistatic to gene *B*. The biochemical and physiological interpretation of epistasis is that the gene product of *A* is required for the functioning of gene *B*. In a case where the effect of the two mutations is additive it is assumed that the two genes affect pathways that proceed independently from each other. However, in the case where the two mutations are 'leaky', *i.e.* they are only partial defective, an additive phenotype is also expected in the double mutant, even when the genes act in the same pathway. For the analysis of the function of a gene it is important to be sure which characters are controlled by that gene. A gene mutation may affect different characters (pleiotropism). However, pleiotropism may be mimicked by a second mutated gene in the mutant.

8.2.3 Photomorphogenic mutants

Photomorphogenic mutants would be expected to have a strong pleiotropic phenotype, since many responses are regulated by light and are expected to be modified when the mutation affects the photoreceptor itself or steps immediately following the perception of light. The difference between photoreceptor mutants and so-called 'transduction-chain mutants' is that the first class has a defect in the photoreceptor. In case of phytochrome this can be analyzed by spectrophotometry, immunology or RNA and DNA analysis. Transduction-chain mutants for a photoreceptor should have the same pleiotropic phenotype as the photoreceptor mutant, but should possess the photoreceptor itself. Mutations down stream of the transduction chain are classified as response mutants (Adamse *et al.* 1988c).

The similarity of the phenotype in white light (W) of specific mutants with aspects of etiolated plants such as an elongated hypocotyl, led to the identification of the first photoreceptor mutants in *Arabidopsis* (Koornneef *et al.* 1980). Mutant screens have also been performed in broad-band red (R) and far-red light (FR) and have yielded additional mutants. Specific screens in broad-band B and screens of specific responses (*e.g.* phototropism) led to mutants such as the B-insensitive mutants in *Arabidopsis* (Khurana and Poff 1989; Liscum and Hangarter 1991).

Recently, mutants with an altered flowering behaviour have also been correlated to defects in phytochrome (Childs *et al.* 1991). Apart from mutants with a reduced light response, mutants have been isolated with an enhanced response to light. The high-pigment mutants of tomato, which can effectively be selected in conditions that have a non-saturating response in WT are an example of this latter group. Extreme cases of this group are those mutants that need no light at all. Chory *et al.* (1989b) were the first to identify such mutants, which are characterized by a partial de-etiolated phenotype in darkness.

The isolation and characterization of various types of mutants affected in their light response are discussed in the following sections.

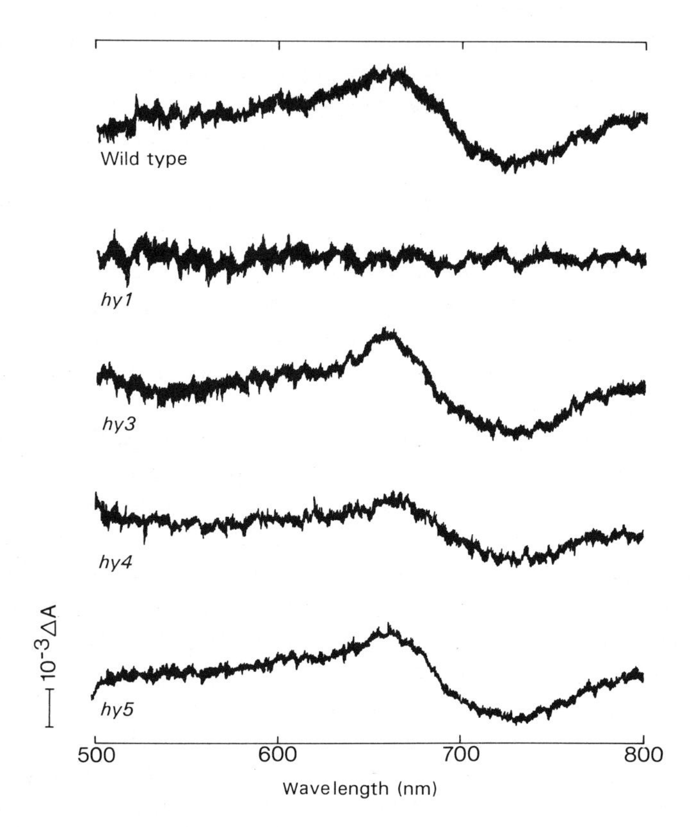

Figure 2. Difference spectra for the phytochrome phototransformation in dark-grown seedlings of wild type and hypocotyl mutants of *Arabidopsis*. The seedling were first irradiated with red light and the recordings show the spectral change after a saturating exposure to far-red light. Note no detectable phytochrome in *hy1*, yet an apparently wild type level in *hy3*, *hy4* and *hy5* (after Parks *et al.* 1989).

8.2.3.1 Phytochrome-deficient mutants

Since it has been established that various phytochrome types exist and that different genes code for components of these phytochrome variants (Sharrock and Quail 1989) a number of predictions about mutations in the photoreceptor can be made. Mutations in specific phytochrome genes allow the analysis of the physiological functions of the different phytochromes. Mutations in any individual phytochrome gene will not result in plants that are completely deficient in phytochrome. In cases where more than one phytochrome acts on the same processes, at most, mild phytochrome-deficiency symptoms can be expected. Since the various phytochromes most likely carry the same chromophore, mutations affecting chromophore biosynthesis would be expected to lead to reduced levels of all phytochrome type molecules.

A number of mutants have now been recognized as being directly affected in phytochrome itself (Fig. 2). These mutants have as a common phenotype: an elongated hypocotyl and also often elongated internodes in W and in specific broad-band light sources. In addition, many mutants are characterized by their pale-green colour. Mutants that are deficient in spectrophotometrically-detectable phytochrome in etiolated seedlings and where the mutant phenotype can be complemented by feeding precursors of the chromophore, such as biliverdin, are classified as chromophore mutants (Parks *et al.* 1992).

Cucumber (López-Juez *et al.* 1992), *Arabidopsis* (Nagatani *et al.* 1991a; Somers *et al.* 1991) and *Brassica* (Devlin *et al.* 1992) mutants with a rather similar elongated phenotype, but with phytochrome detectable in etiolated seedlings (Fig. 2) were shown to lack a phytochrome protein that cross reacts with phytochrome B-specific antibodies (Chapter 4.2). In the case of the *Arabidopsis hy3* mutant, mutations have been detected within the *phyB* gene by sequence analysis of a number of independently isolated *hy3* mutants (Reed *et al.* 1993).

The various phytochrome-related mutants known in different plant species are listed in Table 1. This table indicates that many mutants affecting the chromophore are known and several affect a light-stable phytochrome. Recently phytochrome-A mutants have been identified in *Arabidopsis,* by selection for long hypocotyls in continuous FR (Parks and Quail 1993; Nagatani *et al.* 1993).

The tomato *aurea (au)* mutant has now been tentatively listed as a chromophore mutant (Table 1 and Fig. 3), although for a long time it seemed the only clear example of a phytochrome A-type mutant because it lacked both spectrophotometrically and immunological detectable phytochrome in etiolated seedlings (Parks *et al.* 1987), but contained functional phytochrome in light-grown plants (López-Juez *et al.* 1990b). However, it has not been possible to rescue this mutant with biliverdin, which is the case for the long-hypocotyl mutants [*hy1, hy2* and *hy6* (Koornneef *et al.* 1990; Chory *et al.* 1989b; Parks *et al.* 1989)] of *Arabidopsis.* However, *phyA* mRNA is present and yields apparently normal phytochrome apoprotein after transcription *in vitro.* It appears that the *au* and the similar yellow-green 2 (*yg-2)* mutations of tomato are epistatic to phytochrome overexpression of the phytochrome A protein, which can only be explained when there is a shortage of available chromophore (A. van Tuinen unpublished data). A possible explanation that chromophore mutants behave as phytochrome-A mutants is that the chromophore biosynthesis is the rate-limiting step in these presumably leaky mutants at the time when apophyto-

Table 1. Summary of phytochrome mutants identified in various plant species. See text for references.

Species	Mutant classification		
	Chromophore	Phytochrome A type	Phytochrome B type
Arabidopsis	hypocotyl *(hy1)* hypocotyl *(hy2)* hypocotyl *(hy6)*	FR elongated *(fre)* hypocotyl *(hy8)*	hypocotyl *(hy3)*
Brassica rapa			elongated internode *(ein)*
Cucumber			long hypocotyl *(lh)*
Nicotiana plumbaginifolia	etiolated *(eti1)* etiolated *(eti2)*		
Sorghum bicolor			maturity *(ma3)*†
Tomato	aurea *(au)** yellow green *(yg-2)**		

*Provisional.
†Is deficient in a light-stable phytochrome.

chrome A is present in largest quantities, *i.e.* in etiolated seedlings, whereas in de-etiolated plants chromophore availability may be less limiting.

Recently several tomato mutants at the locus *tri* (temporarily R insensitive) have been isolated, which have a reduced inhibition of hypocotyl growth specifically in R (A. van Tuinen pers. comm.). In these mutants, this lack of inhibition is only temporary and lasts for 2 days on transfer of etiolated seedlings to R. A study of anthocyanin biosynthesis in the *tri* mutant revealed that at very low R fluence rates a similar level to WT was found, whereas at high R fluence rates the level was severely reduced (L.H.J. Kerckhoffs and A. van Tuinen pers. comm.). While the level of phytochrome A has been shown to be normal in this mutant, recently it has been shown that it lacks a light-stable phytochrome (L.H.J. Kerckhoffs pers. comm.).

Mutants deficient in phytochrome can also, in principle, be obtained by transforming a plant with an antisense construct of the phytochrome gene, whereafter the expression of the antisense gene can suppress the expression of the endogenous gene. Although this approach has been employed successfully for several plant genes it has not been reported for phytochrome. However, 'transgenic' phytochrome-deficient mutants were obtained by expression of a gene encoding a synthetic antibody derivative that binds to phytochrome A in tobacco (Owen *et al.* 1992). A slightly reduced time to germinate and reduced inhibition by R and FR in this transgenic plant are the predicted phenotype for a genotype with reduced phytochrome A content. It is not clear how phyto-

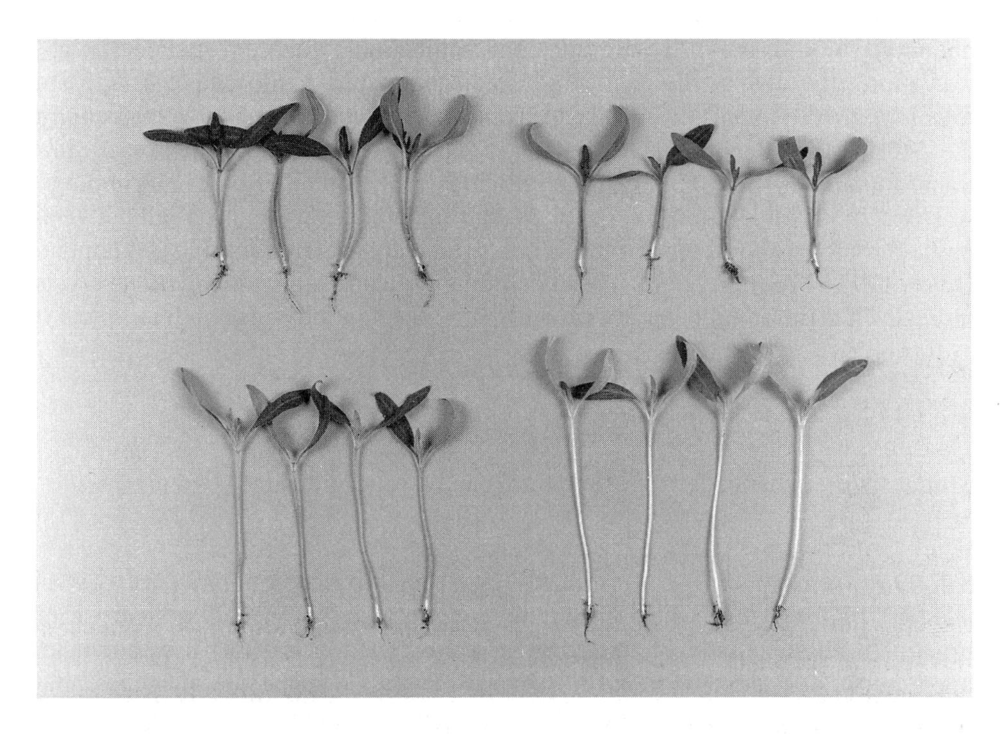

Figure 3. The phenotype of wild type and photomorphogenic mutants of tomato. Two-week-old white light-grown seedlings. *Top left*, wild type; *top right*, *hp* mutant; *bottom left*, *au,hp* double mutant; *bottom right*, *au* mutant. The *au* mutant has a long hypocotyl and the cotyledons are yellow. The *hp* mutant has a short hypocotyl and has high anthocyanin levels. The *au,hp* double mutant is more like *au* than *hp*. The *au* and the *au,hp* mutants lack the bulk labile pool of phytochrome in dark-grown seedlings, whereas in the *hp* mutant this pool is quantitatively similar to that in the wild type.

chrome-type specific this approach can be and how far phytochrome levels can be reduced.

8.2.3.2 Phytochrome overexpressors

By transforming plants with *phy* genes it is possible to obtain plants with enhanced levels of phytochrome. This could be achieved, both for *phyA* (Boylan and Quail 1989,1991) and *phyB* (Wagner *et al.* 1991) genes, by overexpressing monocot and dicot *phy* genes under the control of constitutive promotors (Chapter 4.8), in dicot plants such as tobacco (Keller *et al.* 1989; Nagatani *et al.* 1991), tomato (Boylan and Quail 1991) and *Arabidopsis*

(Boylan and Quail 1991; Wagner *et al.* 1991). Not only enhanced levels of the phytochrome protein, but also high levels of spectrophotometric activity could be detected. Depending on the level of phytochrome overexpression a dwarf phenotype was associated with this trait in all cases, although reduced height was more obvious at the hypocotyl stage compared to internode length. In *Arabidopsis* phytochrome overexpressors and those obtained by transforming the SR1 tobacco genotype (Nagatani *et al.* 1991) (in contrast to cv. *Xanthi* transformants, Keller *et al.* 1989) no obvious effect on adult plant morphology was observed. In addition, no major overall differences were observed between phytochrome A and phytochrome B overproducers in *Arabidopsis* (Boylan and Quail 1991, Wagner *et al.* 1991), although detailed physiological analysis indicated a differential behaviour for a number of treatments (see later).

8.2.3.3 Light-response mutants

Mutants with enhanced response to light can be subdivided into two groups.

8.2.3.3.1 Mutants with a constitutive light phenotype. Chory *et al.* (1989) isolated mutants that when grown in complete darkness had a number of properties characteristic of light-grown plants, such as a short hypocotyl, an open hook, the development of primary leaves, expression of genes that normally are light induced *etc.* Mutants with this phenotype in *Arabidopsis* have been called de-etiolated (*det*) (Chory *et al.* 1989a, 1991) constitutively photomorphogenic (*cop*) (Deng *et al.* 1991; Deng and Quail 1992; Wei and Deng 1992) and light-responsive dwarfs (*lrd*) (K. Feldmann pers. comm.). Approximately 7-12 different loci have been identified by the various groups. Some of these mutants were initially isolated in different mutant screens on the basis of their aberrant phenotype in the light. For instance, *cop1* is allelic with the old *fus1* = *emb168* mutants, selected on the basis of a red-coloured embryo (D. Meinke pers. comm.). The phenotypic characteristics of the mutants differ depending on locus and allele and are summarized for the four best described mutants (*det1, det2, cop1 and cop9)* in Table 2. A similar mutant to *cop1* and *det1* has recently been reported in pea, which has light-independent photomorphogenesis (*lip*) (Shannon *et al.* 1992). An additional feature of the phenotype different to that reported for the *Arabidopsis* mutants is a reduced level of phytochrome in etiolated seedlings, although this might be anticipated since the *phyA* gene in pea is negatively controlled by light, *via* phytochrome (Furuya *et al.* 1991).

The interpretation of these recessive (assumed loss-of-function) mutants is that the WT-gene products prevent de-etiolation and that light leads to their depletion or inhibits their action. Recently, the *cop1* gene has been cloned and

Table 2. Summary of the phenotypic effects of the *det* and *cop* mutants of *Arabidopsis* compared to wild type (WT). See text for references.

Genotype	WT	*det1*	*det2*	*cop1*	*cop9*
Phenotype in darkness					
Phytochrome control of seed germination	+	–	?	+	+
Hypocotyl length	long	short	short	short	short
Cotyledon expansion	–	+	+	+	+
Leaf development	–	+	–	–	–
Chloroplast differentiation	–	±	–	±	±
Anthocyanin accumulation	–	+	+	+	+
Presence of light-induced mRNAs	–	++	++	++	++
Phenotype in light					
Flowering	normal	normal	delayed	earlier	lethal
Plant height	normal	dwarf	dwarf	dwarf	max. 4 leaves
Tissue specific gene expression affected	–	+	–	?	?
Dark adaptation of light-induced mRNAs	+	+	–	–	–

appears to code for a protein containing a zinc-finger motif and a domain homologous to the WD-40 repeat motif of G_b proteins and may function as a negative transcriptional regulator capable of direct interaction with components of a G-protein signalling pathway (Deng *et al.* 1992).

As *det1*, *det2* and *cop1* are epistatic to both phytochrome deficient *(hy1-3 and 6)* and *hy4* (tested for *det* mutants only; Chory, 1992) mutants it is suggested that both phytochrome A and B can decrease DET activity or its synthesis.

8.2.3.3.2 Light hyper-responsive mutants. A number of mutants have been shown to be more responsive to light than the corresponding WT. Since these mutants have a normal etiolated phenotype in darkness they represent a class different from the constitutive group described above. Examples of this class are the tomato high-pigment mutants represented by two loci (*hp-1*) (Peters *et al.* 1989) and (*hp-2*)(Peters *et al.* 1991) and the pea *lw* mutant (Weller and Reid 1993). This hyper-responsiveness is especially obvious in tomato in such processes as anthocyanin synthesis and hypocotyl growth inhibition under green and yellow light which is relatively inefficient in inducing phytochrome responses in WT. Detailed fluence-rate response analysis indicated a 6-fold responsiveness amplification to R (Peters *et al.* 1992). It has been suggested that the *HP* gene product works in the same way as B and reduces or counteracts an inhibitor of phytochrome action. In addition to the enhanced responsivity to R, light-grown plants, especially of the extreme *hp-1* alleles are characterized by very dark-green immature fruits, dark leaves and reduced plant height. This phenotype resembles the light exaggerated phenotype of the phytochrome A overexpressors (Chapter 4.8) and is the opposite of that of phytochrome-deficient mutants. However, phytochrome A levels are normal in the *hp-1* mutants, examined so far (Peters *et al.* 1989). The observation that the *au* mutant (Adamse *et al.* 1989; Peters *et al.* 1992) is epistatic (Fig. 3), although not completely, to *hp-1* indicates that phytochrome is required for the expression of the *hp-1* mutation and differs from the *det* mutants where the epistatic relationship is reversed, indicating that DET bypasses the phytochrome requiring step.

The *lw* mutant of pea is a dwarf, with a reduced response to gibberellins (GA). Flowering is strongly delayed in short days whereas in continuous light the difference compared to the WT is much less extreme. Dwarfing of the *lw* mutant is more extreme in continuous R compared to FR and B and furthermore exhibits an exaggerated response to end-of-day FR (EODFR). These characteristics indicate that the *lw* mutant is hyper-responsive to phytochrome, similar to the tomato *hp* mutants (Weller and Reid 1993).

8.2.3.4 Blue-light mutants

The first mutant described with a specific B defect was the *hy4* mutant of *Arabidopsis.* Whereas this mutant has only a slightly longer hypocotyl than WT in W, in continuous B the hypocotyl is much longer than WT (Koornneef *et al.* 1980). Further studies (Jenkins *et al.* 1993; Chory 1992) showed that the cotyledon area of the mutant is reduced under all wavelengths tested and that petiole length is longer in B compared to WT. Anthocyanin production and chalcone synthase are also somewhat reduced. However, other B-induced physiological responses such as stomatal opening and phototropism appear normal and no obvious whole-plant phenotype is observed. R. Hangarter and co-workers (pers. comm.) have isolated a more extreme *hy4* allele, which in contrast to the original *hy4* alleles is completely insensitive to broad-band B. Recently A.R. Cashmore's group have cloned the gene corresponding to *HY4*, by T-DNA tagging. This shows significant homology to known flavoproteins, suggesting it encodes a B photoreceptor (Ahmad *et al.* 1993).

Liscum and Hangarter (1991) screened *Arabidopsis* M_2 populations specifically for mutants insensitive to broad-band B and identified three additional loci (*blu1, blu2, blu3*), which in contrast to *hy4* have no obvious phenotype in W, although at lower fluence rate and in short days these mutants have longer hypocotyls than WT. However, in B hardly any inhibition is observed. This lack of inhibition is specific for high irradiance B in the wavelength range 410-500 nm (Young *et al.* 1992). It appears that *hy4* mutants are less inhibited, not only in the B, but also in the green part of the spectrum, which may explain their more elongated hypocotyl in W (R. Hangarter pers. comm.)

Khurana and Poff (1989) isolated *Arabidopsis* mutants modified specifically in B mediated phototropism. Although no detailed genetic analysis has been reported, the various mutant phenotypes suggest that mutations at a number of different loci affect phototropism. In some of these mutants the gravitropic response is also reduced (Khurana *et al.* 1989), indicating that these genes affect the growth-curvature response itself as is the case for the 'stiff' mutants of *Phycomyces*. Mutants that are specific for phototropism and that change the sensitivity to B are potential candidates for photoreceptor mutants. The *Arabidopsis* JK224 mutant, in which the light threshold of the first-positive phototropism is shifted to higher fluence has been suggested to be a photoreceptor mutant (Fig. 4). However, in this mutant the fluence response of the second-positive curvature is normal (Konjevic *et al.* 1992). Other specific phototropism mutations (*e.g.* JK218) abolish phototropism completely. Since *blu* and *hy4* mutants behave identically to WT with respect to phototropism; the phototropism mutants show a normal hypocotyl inhibition by B, and the JK218,*blu1* double mutant has both defects, these mutants convincingly show that phototropism and hypocotyl elongation are controlled by different B photoreceptors (Liscum *et al.* 1992).

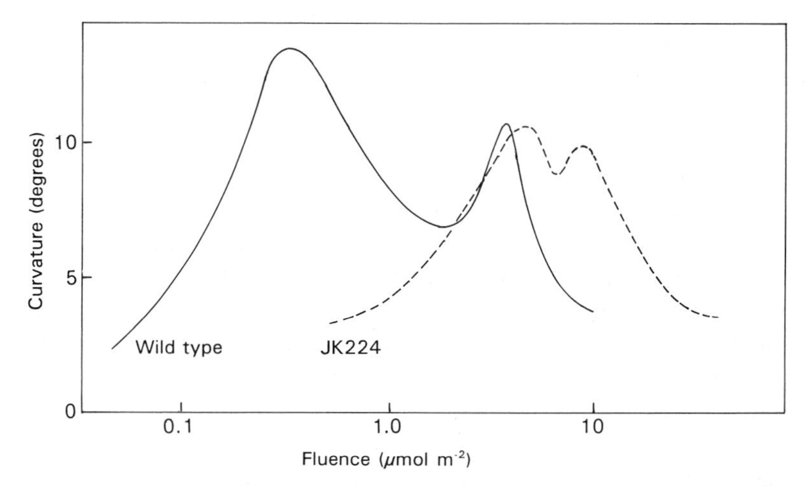

Figure 4. Fluence-response relationships for blue light-induced hypocotyl curvature of the JK224 mutant and wild type of *Arabidopsis* (after Konjević *et al.* 1992).

8.2.3.5 Putative transduction-chain mutants with reduced light responsiveness

In cases where mutants are defective in a photoresponse, but show an active photoreceptor they might be affected in the transduction chain. A problem with this concept is that the photoreceptor molecule can only be analysed for phytochrome, so this classification is not yet possible for other photoreceptors. Despite the fact that many phytochrome defective mutants have been found, only one mutant in *Arabidopsis*, one in pea and one in tomato can be classified tentatively as transduction chain mutants. The *Arabidopsis hy5* mutant has a reduced photo-inhibition of hypocotyl growth by R and FR, but phytochrome levels are comparable to those in WT.

The fact that the *hy5,det1* double mutant is almost indistinguishable from WT, *i.e.* the mutant phenotypes are additive, suggests that the genes have independent effects on the same process, which act in opposite ways. The *HY5* gene product may be the primary target of the critical molecules repressed by the *DET* products (Chory 1992), which are suggested to be factors related to cytokinin synthesis or action, since the addition of cytokinins to etiolated plants phenocopies the *det* mutation.

The pea *lv* mutant, which initially was characterized as a GA hyper-responsive mutant was subsequently shown to be a photomorphogenic mutant (Nagatani *et al.* 1990) because it lacked typical stable phytochrome (phytochrome B) responses such as R inhibition of elongation and an EODFR elongation response. However, since active phytochrome B could be detected, *lv* was suggested to block the transduction of this light-stable phytochrome. Recently, Weller and Reid (1993) pointed out that it cannot be excluded that other phytochrome proteins which have yet to be studied corresponding to the phytochrome C, D, or E of *Arabidopsis* could be mutated.

8.2.4 The use of mutants in understanding photomorphogenesis

Well characterized mutants in which certain parts of the photomorphogenic pathway are eliminated provide an efficient tool to identify the various components of such pathways. In fact such genotypes will exhibit a simpler photomorphogenesis than their corresponding isogenic WT. The relevance of the deleted part in the mutant is directly indicated by its difference from the WT.

Physiological experiments do not allow a clear distinction between the various phytochrome types. Thus far, specific B-high irradiance response (HIR) effects have been hard to distinguish from phytochrome HIR effects since the latter pigment also absorbs in the B region of the spectrum. The interaction between different B- and R-absorbing photoreceptors provided another complication in studying the contribution of the various photoreceptors in specific physiological processes (Chapter 6).

The available mutants affecting one phytochrome species are the most direct way to determine functional differences between the various phytochrome types. These phytochrome mutants together with specific B and UV mutants will allow a further insight in the concerted action of photoreceptors in photomorphogenesis.

A number of physiological responses, where mutants have increased our understanding of photomorphogenesis are discussed below.

8.2.4.1 Seed germination

Many plant species require light for seed germination, a process which has been shown to be stimulated by the FR-absorbing form of phytochrome (Pfr). Mutants in which phytochrome is non-functional should therefore fail to germinate. However, the Pfr requirement for germination can be extremely low or can be supplemented by another germination stimulatory factor, such as GA. Further complications are that phytochrome also affects the light requirement (sensitivity to Pfr), 'set' during seed development. The effect of the spectral composition of the light source during growth of *Arabidopsis* plants was demonstrated by McCullough and Shropshire (1970) and Hayes and Klein (1974), who found that these conditions influenced the sensitivity of the seeds to R and the level of dark germination. Seeds obtained from plants grown in fluorescent light (relatively high R:FR photon ratio) had a higher dark germination, but a lower sensitivity of R, to induce germination than seeds from plants grown in conditions with additional incandescent lamps (lower R:FR ratio). High dark germination correlates to the amount of Pfr at equilibrium 'set' during seed development, whereas the sensitivity for light to promote germination is correlated to the degree of dormancy.

The *hy1, hy2* and *hy3* mutants (Koornneef *et al.* 1980) show an altered responsiveness to light and have a reduced phytochrome content in their seeds (Spruit *et al.* 1980). The fluence-response curves for the induction of germination by R are more shallow for the mutants than those of WT, but dark germination is often relatively high (Cone and Kendrick 1985). These fluence-response curves are compatible with a reduced phytochrome content combined with a so-called germination-promoting, light-independent 'overriding factor' (Cone and Kendrick 1985). This can be interpreted as a reduced light requirement (reduced dormancy) due to phytochrome deficiency during seed development. Recent experiments by Whitelam (1992) indicated that phytochrome B overexpressors in *Arabidopsis* have a high dark germination when the mother plants are grown in low R:FR in contrast to phytochrome A overexpressors which behave similarly to WT, whereas the *hy3* mutant has a low dark germination both in high and low R:FR conditions, suggesting that phytochrome B is the type of phytochrome predominantly responsible for these effects during seed development. The cucumber *lh* mutant shows no differences in its germination behaviour compared to WT. In tomato, where germination takes place in darkness, the predominantly phytochrome A deficient *au* mutant shows a reduced dark germination, probably because a relatively high level of Pfr normally present in the mature WT seed is absent. Dark germination can be prevented by continuous FR in the WT, but not in *au*. However, a stimulation by R was observed in some seed batches (Georghiou and Kendrick 1991). The observation that phytochrome A overexpression in tobacco results in a decrease in the inhibitory effect of FR is an indication that phytochrome A functions in seed germination (McCormac *et al.* 1991). These data suggest the involvement of both phytochrome A and phytochrome B type phytochromes in the regulation of seed germination.

8.2.4.2 The inhibition of hypocotyl and internode elongation

The regulation of elongation growth by light is probably one of the most complicated light-induced physiological responses. This is not only because multiple photoreceptors are involved, but because the photomorphogenic effects depend on the light pretreatment and physiological competence of the tissue. Especially important is if a plant is etiolated or de-etiolated in this respect. Differences also exist between plant species. Some of the photomorphogenic effects can be classified as low fluence responses (LFRs) others as either R- or FR-HIRs (Chapter 4.7). The ecological significance of the various effects is described in Chapter 7.1. Since differences in elongation growth represent the most obvious difference between photomorphogenic mutants and their WTs, the dissection of this pathway has benefitted greatly from the genetic approach to understanding photomorphogenesis.

8.2.4.2.1 The role of the light-labile and light-stable phytochrome. An example of the complicated photomorphogenic effects on elongation growth is the effect of FR on hypocotyl elongation. Etiolated plants are not only inhibited by continuous FR, but also by continuous R, although to a lesser extend in many (but not all) species. However, de-etiolated plants respond by an enhanced elongation growth to a decrease in R:FR ratio or the application of EODFR, the latter being reversible by a subsequent R pulse. A summary of R and FR effects in *Arabidopsis* WT and a number of genotypes with aberrant phytochrome levels is given in Table 3.

When comparing the FR effects on the phytochrome B deficient *hy3* mutants with WT it is clear that FR is inhibitory in etiolated plants, however in the de-etiolated state the plants behave as low R:FR ratio-grown plants which lack a further promotive effect of either EODFR or low R:FR ratio. The phytochrome

B mutants of cucumber *(lh)* and *Brassica (ein)* show a very similar behaviour. In the pea *lv* mutant increasing the R:FR ratio even leads to stronger elongation, which is the opposite effect seen in WT (Weller and Reid 1993). These observations and the fact that phytochrome B overexpressors behave similarly to WT, contrast with the enhanced inhibition observed at higher R:FR ratio for phytochrome A overexpressors. This, together with the absence of the FR-HIR in phytochrome A deficient mutants clearly suggests that this FR-HIR is exclusively under the control of phytochrome A, whereas FR acting on phytochrome B promotes elongation because Pfr levels are reduced (LFR and R-HIR). How FR promotes inhibition *via* phytochrome A is explained by the light-labile character of phytochrome A-Pfr, which means that as soon as Pfr is formed it is broken down resulting in a very low absolute Pfr concentration. Since FR leads to less Pfr, a significant pool remains present for a longer time period and enables a low, but significant amount of Pfr to be maintained for Pfr action (Hartmann 1966). For the light-stable phytochrome B, only the R:FR ratio appears to be important, resulting in R being most effective where stable phytochrome predominates (in de-etiolated plants). The observation that the phytochrome A deficient *Arabidopsis hy8* and *fre* mutants lack the FR-HIR, but have a normal response to R confirm the exclusive role of phytochrome A in this response and that of phytochrome B for the EODFR effect (Fig. 5) and the R-HIR (Table 3).

Table 3. Summary of the effects of different light treatments on hypocotyl elongation in *Arabidopsis* genotypes with deviating phytochrome content compared to wild type (WT). See text for references.

	Genotype					
	WT	*hy1*	*hy8*	AAO*	*hy3*	ABO†
Content‡						
Phytochrome A	+	–	–	+++	+	+
Phytochrome B	+	nd	+	+	–	+++
Treatment‖						
R continuous	–	o	–	––	o	––
FR continuous	–	o	o	––	–	nd
W + supplementary FR	+	nd	nd	–	o	nd
EODFR	+	+	+	+	o	+
W + 'night-time' FR	+	nd	nd	–	nd	+

*AAO = *Arabidopsis* phytochrome A overproducer.
†ABO = *Arabidopsis* phytochrome B overproducer.
‡Phytochrome present (+) or absent (–).
‖Treatment promotes (+), inhibits (–), inhibits strongly (––) or has no (o) effect on hypocotyl or internode elongation compared to the dark or untreated control.
nd = not determined.

Since phytochrome B is a major light-stable phytochrome type it is predicted to be active in light-grown plants, a prediction supported by the phenotype of phytochrome B-deficient mutants. The small response to low R:FR ratio observed in these mutants has been explained by the fact that phytochromes C, D and E, which also appear to be light stable, may play a similar, but much smaller role in this process.

8.2.4.2.2 The effect of B and UV-A photoreceptors. The tomato *au* mutant, which is primarily phytochrome A deficient, also exhibits a reduced response in the B spectral region. This is explained by the fact that in tomato, an important

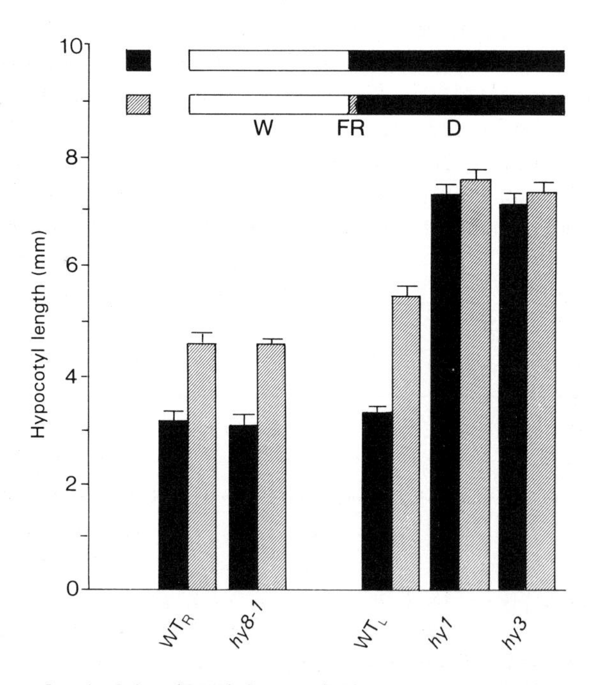

Figure 5. The effect of end-of-day (EOD) far-red (FR) given at the end of the daily white light (W) photoperiod before darkness (D) on hypocotyl elongation of hypocotyl mutants of *Arabidopsis*. The mutants are compared to their isogenic wild type: *hy8-1* to ecotype RLD (WT_R) and *hy8-1* and *hy3* to ecotype Landsberg (WT_L). Note that the phytochrome A-deficient *hy8-1* exhibits a normal response, whereas the chromophore-deficient *hy1* and phytochrome B-deficient *hy3* have long hypocotyls under the control conditions and exhibit no further elongation with EODFR (after Parks *et al.* 1993).

role played by B is to make the hypocotyl more responsive to phytochrome, and therefore phytochrome deficiencies would obscure this co-action. However, as shown in fluence-rate response curves for hypocotyl elongation in *Arabidopsis* (Fig. 6), B is clearly effective in phytochrome mutants of this species. In addition the B insensitive *blu1* mutants respond normally to R. These fluence rate-response curves indicate that both pigment systems can act independently and additively, which is also suggested by the additive effect of the two mutations in W. Surprisingly the *blu1* and *hy6* mutants and the *blu1,hy6* double mutant were as sensitive to UV-A as the WT (Fig. 6), which provides the first convincing indication that the classical B/UV-A absorbing photoreceptor may in fact represent two independent pigment systems (Young *et al.* 1992).

8.2.4.3 Phototropism

An important conclusion from the existence of specific B mutants for defects in inhibition of hypocotyl elongation and phototropism is that there must be two independent transduction chains leading to these responses (Liscum *et al.* 1992). The genetic separation of the first- and second-positive curvature in phototropism is another indication of the complexity of B perception in higher plants (Konjevic *et al.* 1992). Since the nature of the lesions involved are not established it cannot be excluded that these separations result from a branching of the transduction chain from a single photoreceptor. However, in such a case, mutants defective in both responses would be expected and the shift of the threshold and saturation response of the first positive curvature suggest that the photoreceptor itself is involved in the mutation of JK224 (Konjevic *et al.* 1992). Recently, Reymond et al. (1992) have shown that JK224 showed a much lower degree of phosphorylation by B of a specific 120-kD protein compared to WT, suggesting that this phosphorylation might be an early step in the phototropism transduction pathway.

Simulated phototropism, which is observed when one cotyledon of a de-etiolated dicot seedling is kept in darkness and the seedling is irradiated from above with W is not observed in the phytochrome B-deficient cucumber *lh* mutant (Adamse *et al.* 1987) indicating a role of phytochrome B in this process. In addition to the dominant role played by B-absorbing receptors, a role is also played by phytochrome B, detecting the R:FR gradient across the hypocotyl leading to phototropic curvature away from nearby vegetation, as has been shown by the analysis of the *lh* mutant (Ballaré *et al.* 1992).

8.2.4.4 Chlorophyll synthesis and chloroplast development

Chloroplast development and the induction of many of the genes coding for proteins involved in photosynthesis (both nuclear and plastidic) are under the control of phytochrome and a B/UV-A photoreceptor. Many of these processes are uncoupled from light in the *det* and *cop* mutants, although no 'greening' is observed. This is because chlorophyll synthesis in higher plants involves a photoreduction step.

Mutants affecting the phytochrome chromophore (*hy1, hy2, hy6* of *Arabidopsis* and presumably *au* and *yg-2* of tomato) when grown in W are pale green, have a reduced chlorophyll content, an increased chlorophyll *a/b* ratio and reduced thylakoids and grana formation (Koornneef *et al.* 1985). However, phytochrome B-deficient mutants are only slightly paler green compared to WT, suggesting that phytochrome B is not, or only to a minor extent, involved in chloroplast development. It is interesting to note that W-grown seedlings of the recently isolated *Arabidopsis* phytochrome A mutants (Parks and Quail 1993; Nagatani *et al.* 1993) also have chlorophyll levels practically indistinguishable from WT. These results suggest that either both phytochrome A and B can independently regulate greening or another phytochrome gene family member is involved.

The levels of light-induced mRNAs have been studied extensively in the *au* mutant. During de-etiolation seedlings of this mutant under R have severely reduced mRNAs [*e.g.* chlorophyll *a/b*-binding proteins (CAB) and several other light-induced genes encoding plastidic proteins] (Sharrock *et al.* 1988, Oelmüller *et al.* 1989). However, in B no difference with WT was observed (Oelmüller and Kendrick 1991) which has been explained by the fact that B makes the plant so responsive to the reduced level of phytochrome in the *au* mutant, that it is no longer limiting (Fig. 7). In transgenic tobacco over-expressing phytochrome A, evidence has been presented for a modification of the circadian rhythm in *Cab* mRNA abundance (Kay *et al.* 1989).

8.2.4.5 The induction of flowering

Photomorphogenic mutants have been described for species that are either long-day (LD) plants (*Arabidopsis*, pea), daylength-insensitive (tomato) or short-day (SD) plants (*Sorghum bicolor*).

In all daylength-sensitive species, phytochrome deficiency results in early flowering, especially in those conditions where the WT exhibits a delay in flowering. These conditions are SD and high R:FR in *Arabidopsis* (Goto *et al.* 1991, Whitelam and Smith 1991) and pea (Weller and Reid 1993) and LD in *Sorghum.* This effect renders these mutants more or less insensitive to daylength. The general conclusion that can be drawn from these observations is that

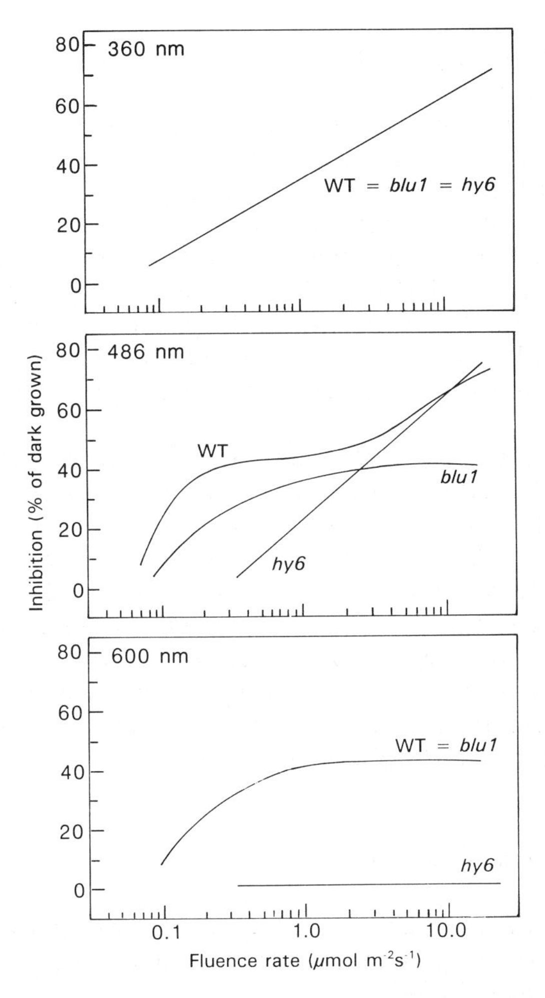

Figure 6. Fluence rate-response relationships for the inhibition of hypocotyl growth of wild type and the *hy6* and *blu1* mutants of *Arabidopsis* in UV-A (360 nm), blue (486 nm) and red (600 nm) light. Indicates the existence of a least three photosystems involved in the inhibition of hypocotyl growth (after Young *et al.* 1992).

Pfr exerts an inhibitory effect on flower initiation, especially when the conditions for flowering are less than optimal. Although in the *Sorghum ma3* mutant

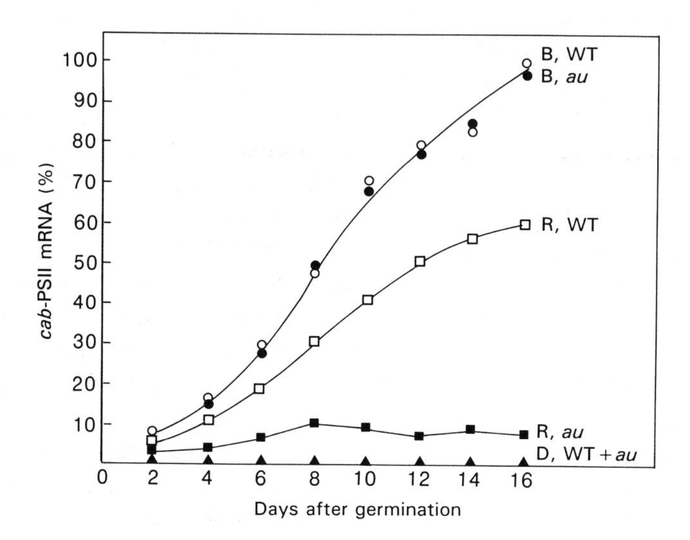

Figure 7. The relative amounts of chlorophyll *a*/*b*-binding protein of photosystem II (*cab*-PSII mRNA) in cotyledons of 2-16 day-old wild-type (WT, open symbols) and *au*-mutant (closed symbols) tomato seedlings. The seedlings were grown in either in red (R) or blue light (B) (after Oelmüller and Kendrick 1991).

(Childs *et al.* 1992) and the *Arabidopsis hy3* mutant this is correlated with a light-stable phytochrome the effect of other phytochrome types cannot be excluded since *hy2,hy3* double mutants are earlier than the monogenic mutants (M. Koornneef unpublished data). Phytochrome hyper-responsivity results in delayed flowering in the pea *lw* mutant (Weller and Reid 1993). However, no effects on flowering were observed for phytochrome A and phytochrome B overexpressors in *Arabidopsis* (Whitelam 1992), which seems to be an additional adult plant characteristic where in *Arabidopsis* no dramatic phenotype due to overexpression of these genes is observed. It should also be emphasized that mutations in other genes, not directly associated with phytochrome, can lead to daylength insensitivity in *Arabidopsis* (Koornneef *et al.* 1991) and pea (Murfet 1989) indicating that the primary receptor of daylength is not phytochrome.

8.2.4.6 Photomorphogenic mutants and plant hormones

Many of the mutants now shown to be defective in inhibition by light were initially identified as GA overproducers (*ein, ma3, yg-6 = au*) or GA hypersensitives (*lv*) because the phenotype of these mutants resembled plants sprayed with GAs. A possible mechanism for phytochrome action with respect to elongation growth might be a repression of GA biosynthesis. In a number of cases elevated levels of GAs have been found (Rood *et al.* 1990, Beall *et al.* 1990). In tomato the effect of GA application has an effect on chlorophyll content and leaf shape which is not completely the same as the phenotype of *au* mutants.

No increased levels of GAs were found in the pea *lv* mutants, but a strong increase in the sensitivity to GAs was observed (Reid and Ross 1988). These data seem somewhat conflicting and although a role for GA in mediating phytochrome-regulated growth changes is clear it could result from an affect on GA levels and/or GA sensitivity.

The observation in the *lv* mutant that auxin levels in epidermal peels are increased, as they are in the WT after FR irradiation suggests that IAA may also be involved in phytochrome-mediated growth responses (Behringer *et al.* 1992). The long-hypocotyl phenotype in IAA overproducers (H. Klee pers. comm.) leads to a similar conclusion. The similarity between the *det* phenotype in *Arabidopsis* and the application of cytokinins to dark-grown plants was reported by Chory *et al.* (1991). No biochemical data are available that describe the cytokinin level or cytokinin sensitivity in these mutants.

8.2.5 Conclusions

Considerable progress has been made during the past 5 years in the identification and characterization of photomorphogenic mutants. Those related to phytochrome have been especially well studied. The discovery that phytochrome is not a single molecule, but is a mixture of related proteins encoded by a small gene family has made mutants modified with respect to specific phytochrome types extremely useful tools to determine their functions. This has led to suggested functions of phytochrome A and B. The function of phytochromes C, D and E still remains unclear. If these proteins have a similar function to phytochrome B, but are expressed at lower levels, they will explain the leakiness that has sometimes been reported for phytochrome B mutants. However, it will be difficult to find mutants of these genes, unless these phytochromes have specific functions.

The lack of knowledge about the B/UV-A photoreceptors makes the classification of B insensitive mutants more complex than is the case for phytochrome mutants. However, the large number of B mutants identified already in *Arabidopsis* and the high specificity, both for response type and wavelength dependence, already suggest that perception of B is complex and diverse photoreceptors play a role. It is predicted that the B mutants in *Arabidopsis* will lead to the cloning of the respective genes and lead to the chemical identification of these photoreceptors.

The large number of genes involved in photomorphogenesis, at least 20 in *Arabidopsis,* when daylength-insensitive flower-initiation genes are included, indicates a degree of complexity even more than expected on the basis of physiological research alone. This complex network means that the simplification of photomorphogenesis expected from mutant analysis is not always so obvious.

8.2.6 Further Reading

Chory J. (1993) Out of darkness: mutants reveal pathways controlling light-regulated development in plants. *Trends in Genet.* Vol. 9, 5: 167-172.

Kendrick R.E. and Nagatani A. (1991) Phytochrome mutants. *Plant J.* 1: 133-139.

8.2.7 References

Adamse P., Jaspers P.A.P.M., Kendrick R.E. and Koornneef M. (1987) Photomorphogenetic responses of a long hypocotyl mutant of *Cucumis sativus* L. *J. Plant. Physiol.* 127: 481-491.

Adamse P., Jaspers P.A.P.M., Bakker J.A., Wesselius J.C., Heeringa G.H., Kendrick R.E. and Koornneef M. (1988a) Photophysiology of a tomato mutant deficient in labile phytochrome. J. *Plant Physiol.* 133: 436-440.

Adamse P., Kendrick R.E. and Koornneef M. (1988b) Photomorphogenetic mutants of higher plants. *Photochem. Photobiol.* 48: 833-841.

Adamse P., Peters J.L., Jaspers P.A.P.M., van Tuinen A. Koornneef M. and Kendrick R.E. (1989) Photocontrol of anthocyanin synthesis in tomato seedlings: a genetic approach. *Photochem. Photobiol.* 50: 107-111.

Ahmad M., Lin C., Chan J.W.Y. and Cashmore A.R. (1993) A mutant of *Arabidopsis thaliana* defective in blue light responses: the sequence of the *HY4* gene is indicative of a blue light photoreceptor. *Book of Abstracts - European Symposium Photomorphogenesis in Plants,* p. 7, Pisa.

Ballaré C.L., Scopel A.L., Radosevich S.R. and Kendrick R.E.(1992) Phytochrome-mediated phototropism in de-etiolated seedlings. *Plant Physiol.* 100: 170-177.

Beall F.D., Morgan P.W., Mander L.N., Miller F.R. and Babb K.H. (1991) Genetic regulation of development in sorghum. V. The ma_{3R} allele results in gibberellin enrichment. *Plant Physiol.* 94: 116-125.

Behringer F.J., Davies P.J. and Reid J.B. (1992) Phytochrome regulation of stem growth and indole-3-acetic levels in the *lv* and *Lv* genotypes of *Pisum. Photochem. Photobiol.* 56: 677-684.

Boylan M.T. and Quail P.H. (1989) Oat phytochrome is biologically active in transgenic tomatoes. *Plant Cell* 1: 765-773.

Boylan M.T. and Quail P.H. (1991) Phytochrome A overexpression inhibits hypocotyl elongation in transgenic *Arabidopsis. Proc. Natl. Acad. Sci. USA* 88: 10806-10810

Childs K.L., Pratt L.H. and Morgan P.W. (1991) Genetic regulation of development in *Sorghum bicolor*: VI. The ma_3^R allele results in abnormal phytochrome physiology. *Plant Physiol.* 97: 714-719.

Chory J. (1991) Light signals in leaf and chloroplast development: photoreceptors and downstream responses in search of a transduction pathway. *New Biologist* 3: 538-548.

Chory J. (1992) A genetic model for light-regulated seedling development in *Arabidopsis. Development* 115: 337-354.

Chory J., Peto C.A., Ashbaugh M., Saganich R., Pratt L. and Ausubel F. (1989a) Different roles for phytochrome in etiolated and green plants deduced from characterization of *Arabidopsis thaliana* mutants. *Plant Cell* 1: 867-880.

Chory J., Peto C., Feinbaum R., Pratt L. and Ausubel F. (1989b) *Arabidopsis thaliana* mutant that develops as a light-grown plant in the absence of light. *Cell* 58: 991-999.

Chory J., Nagpal P. and Peto C.A. (1991) Phenotypic and genetic analysis of *det2*, a new mutant that affects light-regulated seedling development in *Arabidopsis. Plant Cell* 3: 445-459.

Chory J., Aguilar N. and Peto C. (1991) The phenotype of *Arabidopsis det1* mutants suggests a role for cytokinins in greening. In: *Molecular Biology of Plant Development,* Symp. Soc. Exp. Biol. XLV, pp. 21-29, Jenkins G.I. and Schuch W. (eds.) The Company of Biologists Ltd, Cambridge.

Cone J.W. and Kendrick R.E. (1985) Fluence-response curves and action spectra for promotion and inhibition of seed germination in wildtype and long-hypocotyl mutants of *Arabidopsis thaliana* L. *Planta* 163: 43-54.

Deng X.-W. and Quail P.H. (1992) Genetic and phenotypic characterization of *cop1* mutants of *Arabidopsis thaliana. Plant J.* 2: 83-95.

Deng X.-W., Caspar T and Quail P.H. (1991) *Cop*1: a regulatory locus involved in light-controlled development and gene expression in *Arabidopsis. Genes Develop.* 5: 1172-1182.

Deng X.-W., Matsui M., Wei N., Wagner D., Chu A.M., Feldmann K.A. and Quail P.H. (1992) *COP1*, an *Arabidopsis* regulatory gene, encodes a protein with both a zinc-binding motif and a G_b homologous domain. *Cell* 71: 1-20.

Devlin P.F., Rood S.B., Somers D.E., Quail P.H. and Whitelam G.C. (1992) Photophysiology of the elongated internode *(ein)* mutant of *Brassica rapa. Plant Physiol.* 100: 1442-1447.

Feldmann K.A. (1991) T-DNA insertion mutagenesis in *Arabidopsis*: mutational spectrum (1). *Plant J.* 1: 71-82.

Frances S, White M.J., Edgerton M.D., Jones A.M., Elliot R.C. and Thompson W.F. (1992) Initial characterization of a pea mutant with light-independent photomorphogenesis. *Plant Cell* 4: 1519-1530.

Furuya M., Ito N., Tomizawa K. and Schäfer E. (1991) A stable phytochrome pool regulates the expression of the phytochrome I gene in pea seedlings. *Planta* 183: 218-221.

Georghiou K. and Kendrick R.E. (1991) The germination characteristics of phytochrome-deficient *aurea* mutant tomato seeds. *Physiol. Plant.* 82: 127-133.

Goto N., Kumagai T.K and Koornneef M. (1991) Flowering responses to light-breaks in photomorphogenic mutants of *Arabidopsis thaliana*, a long-day plant. *Physiol. Plant.* 83: 209-215.

Hartmann K.M. (1966) A general hypothesis to interpret 'high energy phenomena' of photomorphogenesis on the basis of phytochrome. *Photochem. Photobiol.* 5: 349-366.

Hayes G.R. and Klein W.H. (1974) Spectral quality influence of ligth during development of *Arabidopsis thaliana* plants in regulating seed germination. *Plant and Cell Physiol.* 15: 643-653.

Jenkins G.I., Jackson J.A., Shaw M.J. and Urwin N.A.R. (1993) A genetic approach to understanding responses to UV-A/Blue light. In: *Plant Photoreceptors and Photoperception,* Holmes M.G. (ed.), British Photobiology Society, in press.

Kay S.A., Nagatani A., Keith B., Deak M., Furuya M. and Chua N.-H. (1989) Rice phytochrome is biologically active in transgenic tobacco. *Plant Cell* 1: 775-782.

Keller J.M., Shanklin J., Vierstra R.D. and Hershey H.P. (1989) Expression of a functional monocotyledonous phytochrome in transgenic tobacco. *EMBO J.* 8: 1005-1012.

Khurana J.P. and Poff K.L. (1989) Mutants of *Arabidopsis thaliana* with altered phototropism. *Planta* 178: 400-406.

Konjevic R., Khurana J.P. and Poff K.L. (1992) Analysis of multiple photoreceptor pigments for phototropism in a mutant of *Arabidopsis thaliana. Photochem. Photobiol.* 55: 789-792.

Koornneef M., Rolff E. and Spruit C.J.P. (1980) Genetic control of light-inhibited hypocotyl elongation in *Arabidopsis thaliana* (L.) HEYNH. *Z. Pflanzenphysiol.* 100: 147-160.

Koornneef M., Cone J.W., Dekens R.G., O'Herne-Robers E.G., Spruit C.J.P. and Kendrick R.E. (1985) Photomorphogenic responses of long hypocotyl mutants of tomato. *J. Plant Physiol.* 120: 153-165.

Koornneef M., Hanhart C.J. and van der Veen J.H. (1991) A genetic and physiological analysis of late flowering mutants in *Arabidopsis thaliana. Mol. Gen. Genet.* 229: 57-66.

Kraepiel Y., Jullien M., Caboche M. and Miginiac E. (1992) Etiolated mutants of *Nicotiana plumbaginifolia. J. Exp. Bot.* 43: 23.

Lipucci di Paola M., Collina Grenci F., Caltavuturo L., Tognoni F. and Lercari B. (1988) A phytochrome mutant from tissue culture of tomato. *Adv. Hort. Sci.* 2: 30-32.

Liscum E. and Hangarter R.P. (1991) *Arabidopsis* mutants lacking blue light-dependent inhibition of hypocotyl elongation. *Plant Cell* 3: 685-694.

Liscum E., Young J.C., Poff K.L. and Hangarter R.P. (1992) Genetic separation of phototropism and blue light inhibition of stem elongation. *Plant Physiol.* 100: 267-271.

López-Juez, E., Nagatani A., Buurmeijer W.F., Peters J.L., Kendrick R.E. and Wesselius J.C. (1990b) Response of light-grown wild-type and *aurea*-mutant tomato plants to end-of-day far-red light. *J. Photochem. Photobiol.* B: Biology 4: 391-405

López-Juez E., Nagatani A., Tomizawa K.-I., Deak M., Kern R., Kendrick R.E. and Furuya M. (1992) The cucumber long hypocotyl mutant lacks a light-stable PHYB-like phytochrome. *Plant Cell* 4: 241-251.

McCormac A.C., Cherry J.R., Hershey H.P., Vierstra R.D. and Smith H. (1991) Photoresponses of transgenic tobacco plants expressing an oat phytochrome gene. *Planta* 185: 162-170.

McCormac A, Whitelam G and Smith H. (1992) Light-grown plants of transgenic tobacco expressing an introduced oat phytochrome A gene under the control of a constitutive viral promoter exhibit persistent growth inhibition by far-red light. *Planta* 188: 173-181.

McCullough, J.M. and Shropshire W. (1970) Physiological predetermination of germination responses in *Arabidopsis thaliana* (L) Heynh. *Plant Cell Physiol.* 11: 139-148.

Murfet IC (1985) *Pisum sativum* L. In: *Handbook of Flowering* IV, pp. 97-126, Halevy A.H. (ed) CRC Press, Boca Raton, Florida.

Nagatani A., Reid J.B., Ross J.J., Dunnewijk A. and Furuya M. (1990) Internode length in *Pisum.* The response to light quality, and phytochrome type I and II levels in *lv* plants. *J. Plant Physiol.* 135: 667-674.

Nagatani A., Chory J. and Furuya M. (1991a) Phytochrome B is not detectable in the *hy3* mutant of *Arabidopsis*, which is deficient in responding to end-of-day far-red light treatments. *Plant Cell Physiol.* 32: 1119-1122.

Nagatani, A., Kay S.A., Deak M., Chua N.-H. and Furuya M. (1991b) Rice type I phytochrome regulates hypocotyl elongation in transgenic tobacco seedlings. *Proc. Natl. Acad. Sci. USA* 88: 5207-5211.

Nagatani A, Reed J.W. and Chory J. (1993) Isolation and initial characterization of *Arabidopsis* mutants that are deficient in phytochrome A. *Plant Physiol.* in press

Oelmüller R. and Kendrick R.E. (1991) Blue light is required for survival of the tomato phytochrome-deficient *aurea* mutant and the expression of four nuclear genes coding for plastidic proteins. *Plant Mol. Biol.* 16: 293-299.

Oelmüller, R., Kendrick R.E. and Briggs W.R. (1989) Blue-light mediated accumulation of nuclear-encoded transcripts coding for proteins of the thylakoid membrane is absent in the phytochrome-deficient *aurea*-mutant of tomato. *Plant Mol. Biol.* 13: 223-232.

Owen M., Gancecha A., Cockburn W. and Whitelam G.C. (1992) Synthesis of a functional anti-phytochrome single-chain Fv protein in transgenic tobacco. *Bio/Technology* 10: 790-794.

Parks B.M. and Quail P.H. (1991) Phytochrome-deficient *hy*1 and *hy*2 long hypocotyl mutants of *Arabidopsis* are defective in phytochrome chromophore biosynthesis. *Plant Cell* 3: 1177-1186.

Parks B.M. and Quail P.H. (1993) *hy8* a new class of *Arabidopsis* long hypocotyl mutants deficient in functional phytochrome A. *Plant Cell* 5: 39-48.

Parks B.M., Jones A.M., Adamse P., Koornneef M., Kendrick R.E. and Quail P.H. (1987) The *aurea* mutant of tomato is deficient in spectrophotometrically and immunochemically detectable phytochrome. *Plant Mol. Biol.* 9: 97-107.

Parks B.M., Shanklin J., Koornneef M., Kendrick R.E. and Quail P.H. (1989) Immunochemically detectable phytochrome is present at normal levels but is photochemically nonfunctional in the *hy 1* and *hy 2* long hypocotyl mutants of *Arabidopsis. Plant Mol. Biol.* 12: 425-437.

Peters J.L., van Tuinen A., Adamse P., Kendrick R.E. and Koornneef M. (1989) High pigment mutants of tomato exhibit high sensitivity for phytochrome action. *J. Plant Physiol.* 134: 661-666.

Peters J.L., Wesselius J.C., Georghiou K.C., Kendrick R.E., van Tuinen A. and Koornneef M. (1991) The physiology of photomorphogenetic tomato mutants. In: *Phytochrome Properties and Biological Action.* NATO ASI series H: Cell Biology, Vol. 50, pp. 237-247, Thomas B.and Johnson C.B. (eds.) Springer-Verlag, Berlin.

Peters J.L., Schreuder M.E.L., Verduin S.J.W. and Kendrick R.E. (1992) Physiological characterization of a high pigment mutant of tomato. *Photochem. Photobiol.* 56: 75-82.

Quail P.H. (1991) Phytochrome: a light-activated molecular switch that regulates plant gene expression. *Annu. Rev. Genet.* 25: 389-409.

Reed J.W., Nagpal P., Poole D.S., Furuya M. and Chory J. (1993) Mutations in the gene for red/far-red light receptor phytochrome B alter cell elongation and physiological responses throughout *Arabidopsis* development. *Plant Cell* 5: 147-157.

Reid J.B. and Ross J.J. (1988) Internode length in *Pisum.* A new gene, *lv*, conferring an enhanced response to gibberellin A_1. *Physiol. Plant.* 72: 595-604.

Reymond P., Short T.W., Briggs W.R. and Poff K.L. (1992) Light-induced phosphorylation of a membrane protein plays an early role in signal transduction for phototropism in *Arabidopsis thaliana. Proc. Natl. Acad. Sci. USA* 89: 4718-4721.

Rood S.B., Zonewich K.P. and Bray D. (1990) Growth and development of Brassica genotypes differing in endogenous gibberellin content. II Gibberellin content, growth analysis and cell size. *Physiol. Plant.* 79: 679-685.

Shannon F., White M.J., Edgerton M.D., Jones A.M., Elliot R.C. and Thompson W.F. (1992) Initial characterization of a pea mutant with light-independent photomorphogenesis. *Plant Cell* 4: 1519-1530.

Sharrock R.A. and Quail P.H. (1989) Novel phytochrome sequences in *Arabidopsis thaliana*: structure, evolution, and differential expression of a plant regulatory photoreceptor family. *Genes Development* 3: 1745-1757.

Sharrock R.A., Parks B.M., Koornneef M. and Quail P.H. (1988) Molecular analysis of the phytochrome deficiency in an *aurea* mutant of tomato. *Mol. Gen. Genet.* 213: 9-14.

Somers D.E., Sharrock R.A., Tepperman J.M., and Quail P.H. (1991) The *hy3* long hypocotyl mutant of *Arabidopsis* is deficient in phytochrome B. *Plant Cell* 3: 1263-1274.

Spruit C.J.P., van der Boom A. and Koornneef M. (1980) Light induced germination and phytochrome content of seeds of some mutants of *Arabidopsis. Arabid. Inf. Serv.* 17: 137-141.

Wagner D., Tepperman J.M. and Quail P.H. (1991) Overexpression of phytochrome B induces a short hypocotyl phenotype in transgenic *Arabidopsis. Plant Cell* 3: 1275-1288.

Wei N. and Deng X.W. (1992) *Cop9*: a new genetic locus involved in light-regulated development and gene expression in *Arabidopsis. Plant Cell* 4: 1507-1518.

Weller J.L. and Reid J.B. (1993) Photoperiodism and photocontrol of stem elongation in two photomorphogenic mutants of *Pisum sativum* L. *Planta* 189: 15-23

Whitelam G.C. (1993) The phytochrome molecules. In: *Plant Photoreceptors and Photoperception.* Holmes M.G. (ed.), British Photobiology Society, in press.

Whitelam G.C. and Smith H. (1991) Retention of phytochrome-mediated shade avoidance response in phytochrome-deficient mutants of *Arabidopsis*, cucumber and tomato. *J. Plant Physiol.* 139: 119-125.

Young J.C., Liscum E. and Hangarter R.P. (1992) Spectral-dependence of light-inhibited hypocotyl elongation in photomorphogenic mutants of *Arabidopsis*: evidence for a UV-A photosensor. *Planta* 188: 106-114.

Part 9 Selected topics

9.1 Photomodulation of growth

Daniel J. Cosgrove

Department of Biology, Pennsylvania State University,
University Park, PA 16802, USA

9.1.1 Introduction

9.1.1.1 Light gives the plant cues about the environment

Light, and its absence, has a remarkable influence on the growth and form of plants. In the dark, the new growth of plants develops a peculiar appearance: stems become long and spindly, leaves remain folded and small, and the apical part of the stem often forms a hook (Fig. 1). Such an appearance is referred to as etiolated. Other characteristics of etiolation include suppressed chloroplast development, reduced pigmentation (both photosynthetic and non-photosynthetic pigments), and reduced levels of many enzymes and other substances. In this chapter the modulation of plant size and shape by light is examined.

There are two general means by which light modifies plant growth. The first is *via* photosynthesis. Long-term growth requires a sufficient supply of energy, which in green plants is provided by the capture of light energy in photosynthesis. Photosynthesis also provides assimilated carbon, used in the synthesis of cell wall polysaccharides, membrane lipids, proteins, and other cellular constituents. Photosynthesis provides these basic resources (energy and carbon building blocks) needed for sustained growth of green plants.

Photocontrol of growth also occurs *via* the photomorphogenic pigments. These include phytochrome and the still unidentified blue light (B) and UV absorbing receptors (Parts 4 and 5). These pigments provide neither the energy nor the carbon used in plant growth. Rather, they detect one or more conditions in the environment (such as the quality, quantity, direction and duration of light) and modulate growth in response to these conditions. To bring about the marked effects of light on plant size and shape (Fig. 1), photomorphogenic pigments regulate the rate and directionality of growth of the various plant organs. There may be complex interactions between photosynthesis and photomorphogenic processes, but this chapter will focus on growth control by the photomorphogenic pigments, which offer ample complexity without the

R.E. Kendrick & G.H.M. Kronenberg (eds.), Photomorphogenesis in Plants - 2nd Edition

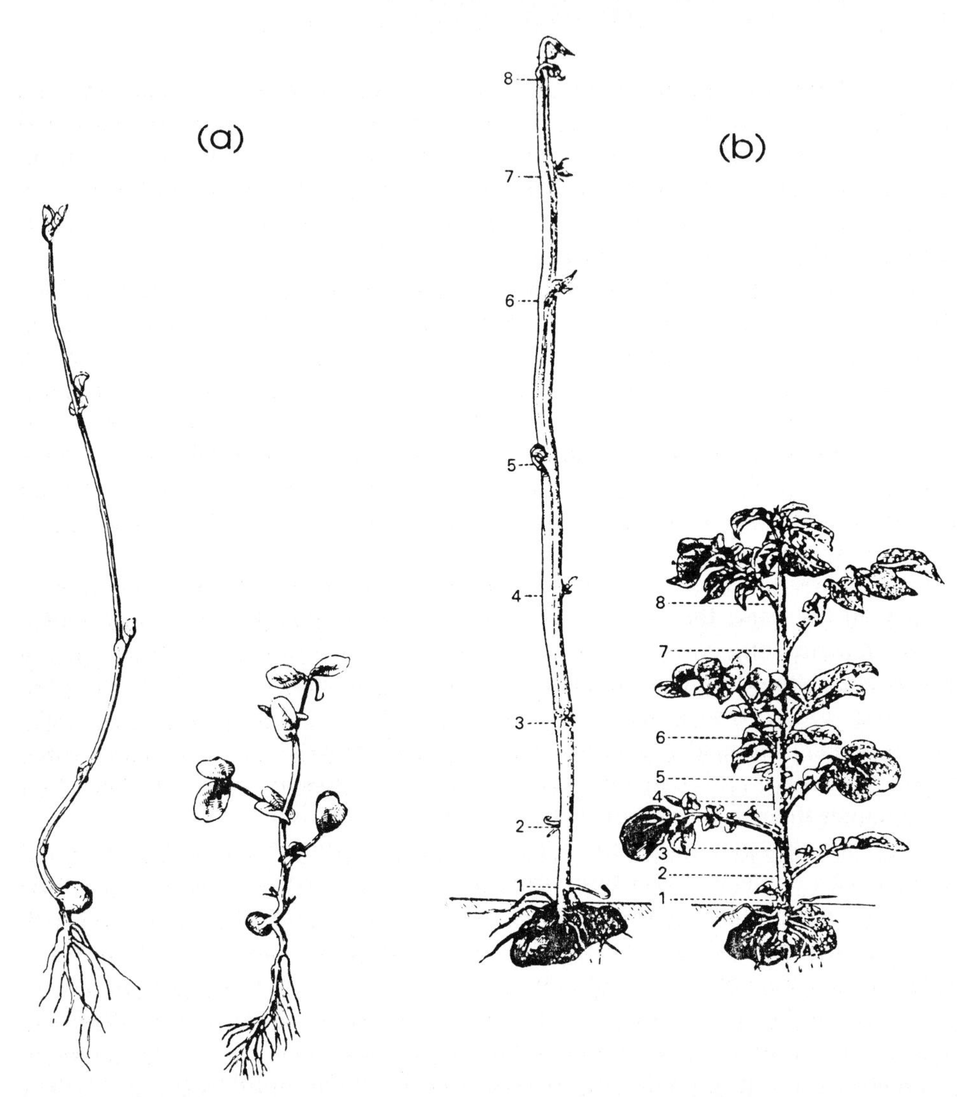

Figure 1. Comparison of the growth and form of pea seedlings (a) and potato plants (b) grown in the dark *(left)* and in the light *(right)*. Note the etiolation of the dark-grown plants: internodes are long and lanky, leaves are unexpanded, and the stem forms an apical hook. From Pfeffer W. (1904) *The Physiology of Plants,* 2nd Edition, transl. by A.J. Ewart, Clarendon Press, Oxford

addition of photosynthesis as a variable. In short, photosynthesis provides the essential raw materials needed for growth but the photomorphogenic pigments modulate and control the growth process itself.

9.1.1.2 Light effects on growth depend on the type of organ

Leaves, stems, flowers and roots respond to light in diverse ways. Bear in mind that light is not a simple stimulus. It can vary in several ways, including spectral distribution, quantity, direction, duration and periodicity. The response of plant organs to such a complex stimulus is not always easy to describe without oversimplification. Nevertheless, 'typical' light-growth responses of the various plant organs can be briefly outlined.

Stems growing in darkness tend to develop very long, thin internodes. Such dark-grown stems typically elongate at a relatively fast rate, drawing upon stored reserves (*e.g.* from the cotyledons, endosperm, or root tissues) for the energy and carbon requirements needed to support such growth. When dark-grown plants are moved into the light, stems often slow elongation growth and become thicker in girth, increasing in their mechanical strength. Apical hook opening is a special type of light response which involves a complicated pattern of differential growth on the two sides of the stem, such that the curved region of the stem straightens.

Leaves behave quite differently. In the dark, leaf primordia may be initiated on the flanks of the apical meristem and develop to an embryonic, unexpanded state. Without light, further development is suspended, or at least greatly suppressed; leaves remain small in the dark. There are exceptions, such as the leaves of grasses which may elongate but remain rolled up, and the leaf stalks of some dicotyledons (*e.g.* petioles of rhubarb). When irradiated with visible light, however, leaves are triggered to unfold, expand, and complete their development.

A specialized leaf found in grass seedlings is the coleoptile; it is a closed hollow cylinder that forms a protective sheath around the young leaves and shoot apex, enabling those parts of the plant to be pushed through the soil with little physical damage. Charles Darwin, in his influential monograph *The Power of Movement in Plants*, chose the coleoptile for his classical studies of phototropism, which eventually led to the discovery of auxin by Fritz Went and others (Chapter 9.2). Although the coleoptile is sometimes referred to as a stem, morphologically it is a leaf and its response to light is more like that of a leaf than a stem. In the dark, *Avena* and *Zea* coleoptiles stay relatively small, but when exposed to light, their growth rates accelerate transiently.

Flower organs also exhibit light-growth responses. These responses have not been well characterized, but are probably responsible for at least some of the daily opening and closing action of petals and petal-like organs.

Roots are not particularly noted for their light-growth responses. In general light inhibits root elongation, but there are two other types of light responses that deserve mention. Firstly, the roots of some plants grown in complete darkness are not responsive to gravity, but will grow in a random direction. Only after exposure to light do they begin to respond gravitropically (geotropically), growing downwards. Light is also known to modify the gravitropic responses of stems, but not to such a remarkable extent. Secondly, the formation of adventitious roots in certain species is affected by light.

Gametophyte development of some ferns requires light for the transition from the protonemal stage, where the plant grows as a single filament, to the prothallial stage, where it is typically heart-shaped (Fig. 3 in Part 1 and Chapter 9.7).

From this description of light effects, it should be evident that the nature of the growth response to light depends upon the tissue. Plant physiologists have preferentially studied some types of light responses, to the relative neglect of others. In particular, elongation growth of grass coleoptiles and of stems of young seedlings has been studied in greatest detail. The light-growth responses of leaves and other organs have been studied much less. In the sections below we will examine what is known about the modulation of growth by light: the photoreceptors involved in these responses, the cellular and physical mechanisms by which light alters growth in some organs, and the possibility that light may act through the mediation of one or more growth hormones. It must be kept in mind, however, that the diversity of the growth responses to light makes it likely that a diverse set of mechanisms may be involved in such responses. Phototropism, which is a special case of growth modification in response to a light gradient, is covered in Chapter 9.2.

9.1.1.3 Light-growth responses are adaptive and controlled developmentally

When seedlings and other types of plants lie buried beneath the soil surface, they cannot photosynthesize or disperse their pollen and seeds to the winds until their shoots reach the soil surface. The etiolated habit, with its rapid stem elongation, apical hook, and unexpanded leaves (Fig. 1), is an effective means for seedlings and other propagules buried in the soil to bring their leaves to the surface in the shortest time while damaging them as little as possible. Once the shoot reaches the surface of the soil and is in full sunlight, the developmental pattern changes: the leaves expand and become photosynthetically competent, and the stem acts as a physical support to hold the leaves in a favourable position to intercept light. Thus light-growth responses of stems and leaves serve as a means to change the form of the plant from one suited to growth through the subterranean environment to one suited to growth in the open air in sunlight.

Plants may also use light as a means to detect shading by other plants. In response to such shading, the stem elongates at a faster rate, the petioles elongate, and the leaf angle may become steeper (McCormac *et al.* 1991). This may be beneficial when the competition between plants for full sunlight is severe, as in dense crop canopies (Chapter 7.1).

In roots, the potentiation of gravitropism by light provides a means whereby roots can grow extensively in the richest soil horizons (the upper regions), yet still avoid the soil surface, which is prone to drying out and physical disturbance.

These adaptive responses of stems, leaves and roots are themselves developmentally controlled. For example, exposure to light often alters the sensitivity of the plant to further light exposures. This is most marked in the transition from the etiolated state to the light-grown state (Fig. 1), where light elicits many developmental changes, including alteration of the transcription of the phytochrome genes (Chapter 4.2) and modification of other photosensory responses. Adaptation of the light-sensing machinery probably occurs throughout the life of the plant, in response to environmental signals and developmental changes within the plant. We are still at the early stages of understanding the mechanisms of such light adaptation, which may involve changes in photoreceptor gene expression or in the transduction machinery that connects photoreceptors with developmental and physiological processes (Galland 1991). Genetic mutants and transgenic plants with defined alterations in the photosensory machinery (*e.g.* McCormac *et al.* 1991; Reed *et al.* 1992) may prove valuable in understanding the mechanisms of such light adaptation.

9.1.2 Photobiology of plant growth

In their natural environment, plants react to broad spectrum light using more than one pigment (Chapters 6. and 7.1). The literature on light-growth responses of plants documents a bewildering array of plant responses. Response characteristics depend on the plant species, age, organ type, and the history of light treatments. In this section, the characteristics of the most studied light-growth responses are reviewed.

For convenience photoresponses are sometimes divided into categories based on their light requirements; they may be classified as a low fluence response (LFR), a very low fluence response (VLFR), or a high irradiance response (HIR). Growth responses to continuous, bright light (*e.g.* sunlight) typically fall into the HIR category and do not obey the Bunsen-Roscoe Reciprocity Law. Plants in natural environments are usually exposed to long periods of sunlight at relatively high fluence rates, conditions which elicit HIRs. These light responses have the greatest effect on plant growth and account for most of the differences in the growth pattern between plants in the dark and in the light. Growth responses to brief light pulses (few seconds to minutes) generally fall into the LFR category and often follow the reciprocity law. Such responses are called inductive because brief irradiations induce a response lasting many hours or longer. Sometimes these responses are also called 'end-of-day' (EOD) responses (Chapter 7.3), because, within limits, it is only the last irradiation

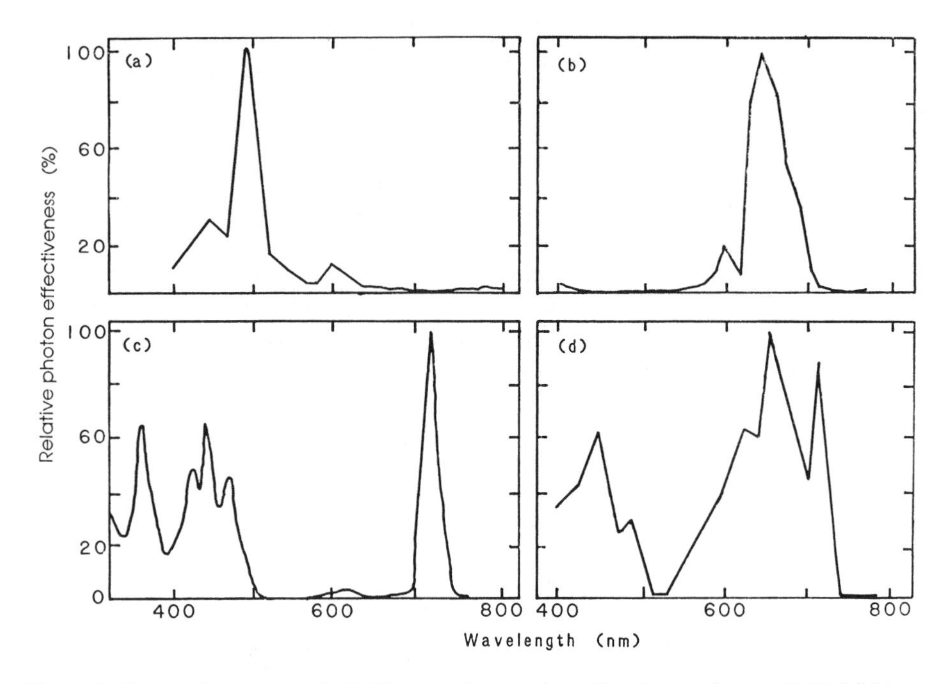

Figure 2. Four action spectra for inhibition of stem elongation by continuous light. Light-grown (a) *Chenopodium rubrum* and (b) *Sinapis alba;* dark-grown (c) *Lactuca sativa* and *Sinapis alba.* Data redrawn from (a) Holmes M.G. and Wagner E. (1982) *Plant Cell Physiol.* 23: 745-750; (b) and (d) from Beggs C.J., Holmes M.G., Jabben M. and Schäfer E. (1980) *Plant Physiol.* 66: 615-618; and (c) from Hartmann (1967).

before a dark period which determines the subsequent growth rate in the dark for many hours. Phytochrome LFRs generally exhibit the classical red light (R)/far-red light (FR) photoreversibility that is one hallmark of a phytochrome response, whereas VLFRs generally lack such reversibility (as do HIRs). In natural environments, LFRs and VLFRs are probably important for modulating subterranean growth. The LFRs are also important for plants that compete for sunlight by attempting to overtop neighbouring plants (Chapter 7.1). The stems of such plants elongate faster when their light environment is enriched in FR or when they are exposed to a FR pulse at the end of a photoperiod.

9.1.2.1 Action spectra reveal the diversity of continuous light-growth responses

Many action spectra have been constructed for inhibition of stem growth by continuous light. Generally such spectra show the greatest activity in the B and R regions; spectra from dark-grown seedlings often have a peak in the FR region. Figure 2 illustrates four such action spectra.

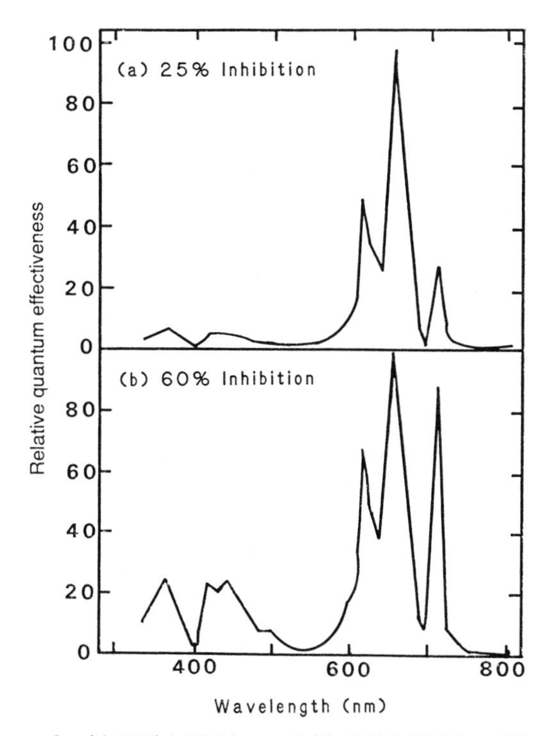

Figure 3. Action spectra for (a) 25% inhibition and (b) 60% inhibition of hypocotyl elongation of dark-grown *Sinapis alba.* Data redrawn from Holmes M.G. and Schäfer E. (1981) *Planta* 153: 267-272.

Several aspects of these action spectra deserve mention. Firstly, note the considerable variation between them. Light-grown *Sinapis* exhibits very little activity in the B region, whereas light-grown *Chenopodium* shows the greatest activity with B, and only minor activity in the R. Dark-grown *Lactuca* exhibits prominent activity in a narrow FR band and in the B region. Secondly, note that the action spectrum of dark-grown *Sinapis* shows a pronounced peak in the B region, which disappears upon treatment with continuous light. This is an example of how the responsiveness of plants to light may change during development. As a consequence, the particular form of an action spectrum may depend on the age of the plant, the light pretreatment and the duration of the light treatment used to obtain the action spectrum. Thirdly, note the very sharp peaks in the action spectrum for *Lactuca.* Such sharp peaks suggest a photochromic pigment (a photoreversible pigment, such as phytochrome) or the interaction of two or more pigments.

In addition to the variations that arise from differences among species, action spectra can also depend somewhat on the level of the growth response chosen to construct the spectrum. This point is illustrated in Fig. 3, where action spectra are plotted for 25% and 60% inhibition of hypocotyl elongation in dark-grown

Sinapis alba. Comparing these two action spectra, we see that for small growth inhibitions (25%), R is by far the most effective, whereas for larger inhibitions B and FR grow in importance. To understand this, it is helpful to examine the fluence rate-response curves from which Fig. 3 was constructed. Two such curves are plotted in Fig. 4, where we can see that R (653 nm) is most effective for responses less than about 65% inhibition, but it saturates (or at least greatly flattens out) at a fluence rate of about 3 μmol m^{-2} s^{-1}. In contrast, B (446 nm) is not saturated even at 150 μmol m^{-2} s^{-1}. If we constructed an action spectrum for 80% inhibition, we would expect the relative quantum effectiveness of the B region to exceed that of the R region. The difference between these two fluence rate-response curves suggests that different photoreceptors are involved in the responses to R and B.

9.1.2.1.1 Phytochrome. Phytochrome is believed to mediate the HIRs in both the R and FR regions of the spectrum, but the exact mode of phytochrome action is not yet resolved. The level of the FR-absorbing form of phytochrome (Pfr) appears to be important for HIRs and could give rise to the R peak. In addition, the phytochrome cycling rate may be important and could give rise to the FR peak. Cycling refers to the interconversion of the R-absorbing form of phytochrome (Pr) and Pfr by light. It has been hypothesized that a short-lived phytochrome intermediate might participate in a reaction affecting growth; at higher rates of photoconversion, this intermediate would build up to greater levels and thus confer a fluence-rate dependence on the HIR.

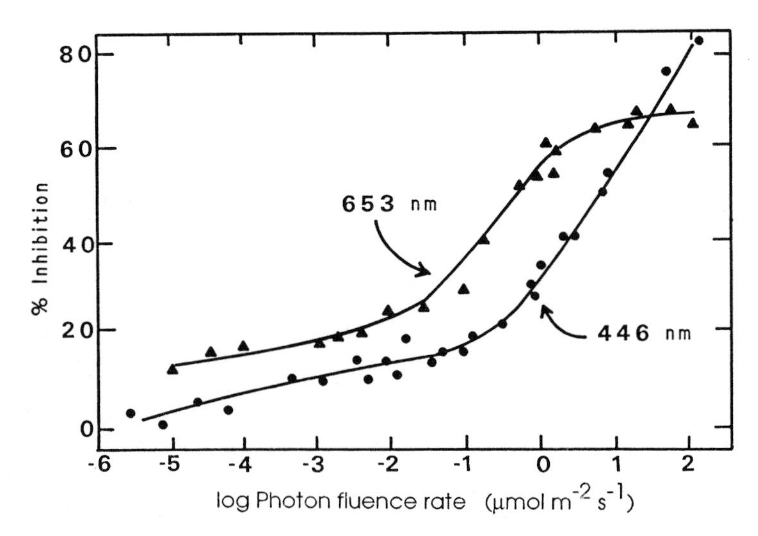

Figure 4. Photon fluence-rate response curves for the inhibition of hypocotyl elongation in dark-grown *Sinapis alba* hypocotyls. (a) 446 nm light; (b) 653 nm light. Data redrawn from Holmes M.G. and Schäfer E. (1981) *Planta* 153: 267-272.

Hartmann (1967) proposed that the FR peak in the action spectrum for *Lactuca* (Fig. 2) is due entirely to phytochrome. According to his thesis, the peak at 716 nm reflects a compromise between two competing reactions: one reaction in which Pfr inhibits growth, and a second in which Pfr is destroyed (Chapter 4.7). If seedlings are irradiated with R to establish a high Pfr/P (where P = total phytochrome, Pr + Pfr) ratio (φ), then growth will initially be strongly inhibited. However, the dark destruction of Pfr over time reduces the amount of Pfr, and thus reduces the effectiveness of photo-inhibition of growth. As a result, the most effective light should be that which establishes a (low) level of Pfr to minimize the rate of Pfr destruction, yet gives sufficient Pfr to inhibit growth. Such light would, according to K.M. Hartmann, maintain the greatest amount of Pfr during the prolonged light treatment, and thus cause the greatest inhibition. It should be recalled that dark-grown seedlings contain a large amount of phytochrome which is greatly reduced by Pfr destruction (Chapter 4.4) when the seedlings are exposed to light. Light-grown plants, in contrast, contain much less phytochrome, but it is relatively stable and arises from transcription of (a) different member(s) of the phytochrome gene family (Chapter 4.2). This alteration in phytochrome gene transcription has probably complicated many of the early studies of light-growth in which plants were irradiated for many hours or days.

To test the idea that phytochrome can account for the 716 nm peak in dark-grown *Lactuca,* K.M. Hartmann developed an ingenious experimental approach in which seedlings were irradiated with two wavelengths of light. Two examples will serve to illustrate these dichromatic irradiation experiments. When monochromatic light of 658 nm or 766 nm is applied separately to the seedlings, little inhibition of growth rate occurs, even at high fluence rates, but when the two wavelengths are applied simultaneously, a very large effect is found. Separately, the monochromatic irradiations convert phytochrome predominantly into the Pfr or Pr forms, respectively. Together they establish the intermediate amount of Pfr most effective for prolonged growth inhibition. In a second test, K.M. Hartmann showed that the inhibition caused by 716 nm light could be greatly reduced by simultaneous irradiation with high-fluence rate 658 nm light. The 658 nm irradiation pushes φ away from the most effective value established by 716 nm light. The results from each of these dichromatic irradiations indicate that phytochrome mediates the sharp FR peak in the action spectrum for *Lactuca.*

In light-grown plants, the phytochrome pool is much smaller and more stable and Pfr destruction is less significant. It seems likely, therefore, that the sharp FR peak noted by K.M. Hartmann is transitory and occurs only as the plant undergoes the transition from the etiolated to the light-grown state. In light-grown plants, phytochrome appears to mediate the response to R. Photosynthesis has been ruled out by the use of chlorophyll-free seedlings, which still show prominent activity in R. The best way to account for the activity of R seems to be *via* its influence on the amount of Pfr in the plant (Chapter 7.1).

Let us examine the role of phytochrome in the growth responses of light-grown *Chenopodium* hypocotyls. In one study, *Chenopodium* seedlings were grown under three different fluence rates of white fluorescent light supplemented with different amounts of FR to establish different values of φ. The effect of such a change in light quality on the stem growth of this weedy species (Fig. 3 in Chapter 7.1) indicates that stem elongation rates are a function of φ (Fig. 5). Clearly there is a good correlation between the value of φ and the growth rate under all three backgrounds of white light (W). Such correlation is consistent with the notion that Pfr inhibits stem elongation.

Figure 5 also brings out an important point about the fluence-rate dependence of growth responses in continuous light. Fluence rates were selected to induce different rates of interconversion of Pr and Pfr, that is, different cycling rates. There is a striking downward displacement of the growth rate as the background fluence rate increases. Is this effect of fluence rate due to phytochrome cycling? Evidently not, because when the irradiation consists exclusively of monochromatic R, there is little effect of fluence rate on growth (Fig. 5). Evidently the B component of the spectrum is responsible for the fluence-rate

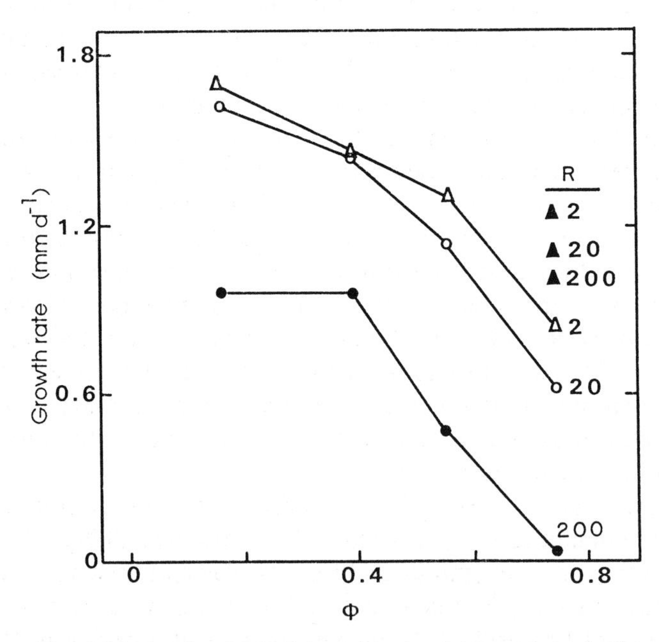

Figure 5. The effect of increasing Pfr/P (φ) on hypocotyl elongation of *Chenopodium* at three different rates of phytochrome cycling. The value of φ was controlled by adding FR to background white light. The numbers represent the cycling rate (in min^{-1}). Solid triangles show the growth rate as a function of three fluence rates of R. Data from Ritter A., Wagner E. and Schäfer E. (1981) *Planta* 153: 556-560, as adapted by Holmes (1983).

dependence of the response to W in Fig. 5, whereas R inhibits stem growth in light-grown *Chenopodium* primarily by establishing a high level of Pfr, not by phytochrome cycling. This mechanism may be thought of as a continuous or repeated phytochrome induction response, or LFR. Adding support to this idea, studies with dark-grown *Sinapis* (Heim and Schäfer 1982) found that more than 90% of the effect of continuous R could be substituted by hourly 5-min pulses if equal total fluences were given in the two treatments. The effect of the 5-min pulses was at least partially reversible by subsequent FR pulses.

The situation is different in light-grown *Sinapis,* which exhibits a marked fluence-rate dependence in R. How to account for the fluence-rate dependence of this phytochrome response is not yet resolved. One suggestion is that there are one or more dark (thermal) reactions which modify Pfr. If these dark reactions are sufficiently rapid, then photoequilibrium is never reached because Pfr is removed by the dark reactions nearly as fast as it is formed by photoconversion from Pr. Thus the amount of phytochrome in the Pfr form will depend on the rate of photoconversion of phytochrome, *i.e.* on the fluence rate. The nature of these hypothetical rapid dark reactions remains to be worked out.

9.1.2.1.2 Specific blue light responses. Since phytochrome absorbs B as well as R and FR, B will stimulate some phytochrome activity. However, several lines of evidence lead to the conclusion that a B photoreceptor, distinct from phytochrome, controls growth. The identity of the B photoreceptor in these light-growth responses has remained elusive (Chapter 5.1).

As described above, the response of *Chenopodium* to different fluence rates of W at any given φ was attributed primarily to the B photoreceptor. Another, and more direct, way to see the influence of B on growth is to irradiate plants with low fluence rate B in a background of high fluence rate yellow light. The yellow light overrides any influence of B on phytochrome. In Fig. 6 we see the results of two such experiments. Yellow light was obtained from low pressure sodium lamps. Its energy is concentrated in a narrow peak at 589 nm, establishing φ at about 0.74. The value of φ was not affected by the small amounts of added B, yet stem elongation was strongly reduced by the added B. Note that inhibitions in excess of 50% were possible with B amounting to less than 0.5% of the background fluence rate.

Other experimental work using *Sinapis* indicates that the contribution of B to phytochrome-mediated responses is small (Jabben *et al.* 1982). This lack of B effect might seem surprising considering the B peak in the absorption spectrum of phytochrome. *In vivo,* B has a low effectiveness because it is attenuated within the tissue (Chapter 7.4). Moreover, at room temperature (25°C) rapid dark reactions reduce the amount of Pfr and thereby prevent the attainment of photoequilibrium under low fluence rate light. Indeed, at fluence rates of B below 8 μmol m^{-2} s^{-1}, φ in *Sinapis* cotyledons depends strongly on fluence rate.

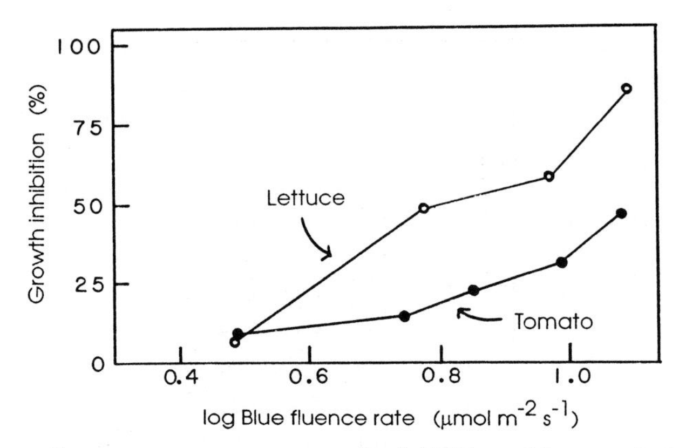

Figure 6. **Photon fluence-rate response curves for inhibition of hypocotyl elongation in lettuce and tomato seedlings by blue light added to a background of high irradiance yellow light (150 μmol m^{-2} s^{-1}) which maintained φ nearly constant. Redrawn from Thomas B. and Dickinson H. (1979) *Planta* 146: 545-550.**

This is not the case at 0°C, where the dark reactions are suppressed. Jabben *et al.* (1982) concluded that B contributes less than 3% to phytochrome-mediated responses in the natural environment.

Growth inhibitions due to the B photoreceptor may be distinguished from those mediated by phytochrome by several additional criteria. (i) In *Cucumis* and *Raphanus* seedlings, R is detected by phytochrome in the cotyledons, whereas B is perceived by the growing region of the stem (Cosgrove 1981; Jose 1977). (ii) The B responsiveness is selectively lost during development of some species, such as *Sinapis* and *Lactuca.* (iii) The time course of growth responses to B and R differs in many plants. For example, when stem elongation is measured using high sensitivity growth transducers, B is found to inhibit growth within 60 s, whereas the R response begins 15-90 min after start of irradiation, depending on species (Cosgrove 1981). Growth rate recovers from B inhibition quickly after the plant is returned to darkness. In contrast, phytochrome induction responses persists for many hours in the dark. (iv) Specific B responses appear to require higher fluence rates than R responses mediated by phytochrome. For example, inhibition of *Cucumis* hypocotyls begins at about 2 μmol m^{-2} s^{-1} and increases in a log-linear fashion with increasing fluence rates (Gaba *et al.* 1984). So far, light saturation of B growth responses has not been detected because the high fluence rates required for saturation are greater than conventional lamps can provide. (v) Finally specific B responses have been separated genetically (Chapter 8.2).

From the above discussion, we see that the growth responses of plants to continuous light are not simple. They depend on both light quality and quantity, they involve at least two photoreceptors, and they change during plant development. In addition, there are likely to be interactions between the various

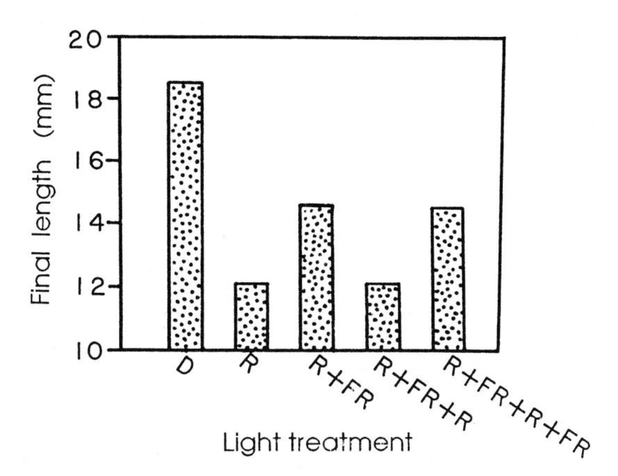

Figure 7. Red(R)/far-red light (FR) reversibility of rice coleoptile growth. Plants were kept in the dark (D) or were irradiated with various schedules of R and FR (3 min each). Note that the rice coleoptiles respond to light quite differently from oat and maize coleoptiles which typically elongate more quickly after R. Data from Pjon C. and Furuya M. (1967) *Plant Cell Physiol.* 8: 709-718.

photoresponses induced when plants are irradiated with W. We are still far from a complete understanding of the changes in light sensitivity exhibited by plants. Possible reasons for some of the sensitivity changes include reduced phytochrome content due to Pfr destruction, altered phytochrome gene expression, and screening due to synthesis of photosynthetic pigments, as well as alteration of the coupling between photoreceptor and growth processes.

9.1.2.2 Plants also exhibit growth responses to brief light pulses

Brief irradiation with R inhibits stem elongation in many plants, and this effect is at least partially reversed when followed immediately by FR. Such responses fall into the LFR category and Fig. 7 illustrates a particularly strong LFR for elongation of intact rice coleoptiles. When R and FR irradiations are given sequentially, only the last irradiation is important in controlling the subsequent growth rate. Such R/FR reversibility is the hallmark of an inductive type of phytochrome response. The FR treatment does not completely restore the growth rate to that of the dark-control plants because a small proportion of the phytochrome remains in the Pfr form after FR irradiation (Chapters 4.1 and 4.7).

In most plants the effect of a single inductive irradiation on growth is rather small, particularly in comparison with the pronounced effects reported for induction of flowering and germination in some especially sensitive species (Chapters 4.7 and 7.3). Commonly the stems of etiolated dicots show only weak inductive growth responses, but they develop stronger responses after a period of continuous or intermittent irradiation. In contrast, etiolated grass seedlings are exceptionally sensitive to light, which is why they have been used so extensively to study light-growth responses.

Careful studies of the light-growth responses of coleoptiles have found that not all the cells in the organ respond in the same way. The basal cells of the coleoptile grow more slowly after irradiation, whereas the apical cells elongate faster. The total response of the coleoptile, then, is a sum of both inhibitory and stimulatory light responses, and therefore may vary with age. Similar observations have been made for some hypocotyls.

Careful studies with seedlings grown in complete darkness have found that plant growth is sensitive to such incredibly small amounts of light that the response saturates at fluences producing no Pfr detectable by spectrophotometry. Indeed, the light emitted by a single flash of a firefly is sufficient to saturate these sensitive growth responses. Such VLFRs are also thought to be mediated by phytochrome, but they do not show R/FR reversibility, evidently because the amount of Pfr produced with FR is sufficient to elicit the growth response. Some plants are so sensitive to light that exposure to dim green 'safelights' can elicit and even saturate a VLFR. It is likely that many VLFRs have been overlooked because of the use of green safelights during plant handling.

9.1.2.3 Other photoreceptors

There are numerous reports of light-growth responses with peak activities in spectral regions which are difficult to ascribe either to phytochrome or to the B photoreceptor. In some cases such displaced peaks might be due to screening from other pigments, especially the photosynthetic pigments, or due to interaction between pigments. Further work is needed to characterize these light responses.

9.1.3 Mechanisms of action

In the last section we described the *photo* side of light-growth responses. We now turn our attention to the *growth* side of these responses. Growth in this context is defined as an irreversible increase in plant size. We first look at the issue of growth by cell expansion versus cell division, then examine the biophysical basis for light's action on stem and leaf growth. Finally, we consider some of the intermediates which may connect light perception with altered growth biophysics.

9.1.3.1 Light can affect cell number and cell size

Growing organs are made up of cells, so it is natural to consider that an increase in organ size might be accomplished by an increase in cell size without an effect on cell number. Bean-leaf expansion, when induced by W, can occur in this manner (review in Van Volkenburgh *et al.* 1987). On the other hand, light could affect the number of cells in an organ, without influencing the average cell size. This might occur, for example, if light affected the cell division rate, and if expansion processes were co-ordinately affected so that average cell size was unchanged.

It is often stated, incorrectly, that growth may occur by cell division or by cell expansion. Such statements confuse the processes of cell division and expansion with their end results, cell number and cell size. Cell division, *per se,* does not increase the size of an organ. It just divides the volume of the organ into smaller compartments (cells). Cell division without expansion does occur during some developmental processes, but it does not result in growth. If an organ increases in size, it does so ultimately because cells are expanding, with or without concurrent cell division. Without cell division, average cell size simply increases in parallel with organ size. If cells are dividing at the same time that they are expanding, then the average cell size will depend on the history of the relative rates of cell division and expansion. When the rates of division and expansion are exactly matched (a rarity), *average* cell size stays constant, despite active cell expansion. In this special case, light could affect cell number in a organ without affecting average cell size. Nevertheless, expansion processes would still be affected by light.

Although some studies show that light may affect growth without effects on cell division (cell number), more commonly both cell division and expansion are affected, usually unevenly, so that average cell size changes. Typically, the stems of light-grown plants are made up of smaller cells than are found in the stems of dark-grown plants. This observation indicates that light inhibits cell expansion more than cell division. However, it is unusual for cell number to be independent of light. This indicates that light also affects cell division rates, but the mechanism of such action could be quite indirect. For example, if cell expansion stimulates cell division, then an inhibition of cell expansion might secondarily inhibit division. We do not know enough about the controls of cell division and expansion processes to make strong statements on this point.

In growing axes such as stems, which show indeterminant growth and which exhibit a large developmental gradient along the axis, a thorough analysis of light-growth effects requires a kinematic analysis, in which local rates of cell division activity and cell expansion activity are measured as a function of position along the stem (Silk 1984). Such information is needed because the cells along the stem axis vary in age, developmental stage and, at least in some cases, growth response to light. For example, Jose (1977) found that B stimu-

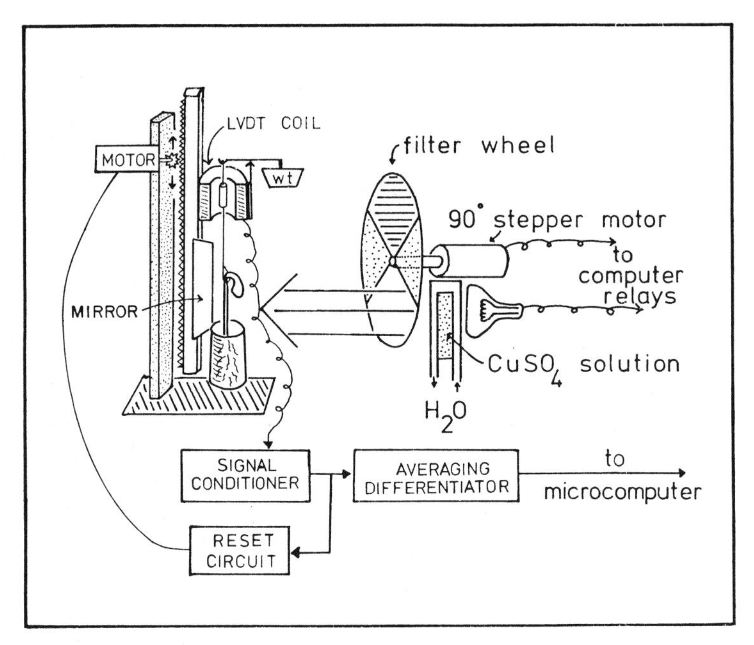

Figure 8. Apparatus to measure rapid light-growth responses. Stem elongation is measured with a linear variable differential transformer (LVDT) which consists of a set of specially-tapered primary and secondary coils and a moveable metal cylinder or core. The core is attached to the top of the plant and a small upward tension is applied *via* a counterweight assembly. As the stem lengthens, the core moves upwards and the output of the LVDT changes. This signal is differentiated to produce a voltage proportional to the elongation rate of the plant. A motor is used to keep the LVDT coil within the linear range of the transducer. Light duration and irradiance may be programmed and controlled by a computer, which also digitizes and stores the growth signal. Redrawn from Cosgrove D.J. and Green P.B. (1982) *Plant Physiol.* 68: 1447-1453.

lated stem elongation in some parts of the radish hypocotyl, but inhibited elongation in other parts. Likewise, Shinkle *et al.* (1992) reported that R changed the distribution of growth along the hypocotyl of cucumber seedlings, in this case without affecting the total elongation rate of the hypocotyl.

9.1.3.2 Light may affect growth quickly (in seconds) or more slowly (hours)

Clues to the mechanisms by which light modulates growth can be gained from the kinetics of the response, *i.e.* the lag between onset of illumination and onset of the growth response, and the speed of the response after the initial lag period. Fast growth responses are best measured with electronic position transducers (see example in Fig. 8). The world's speed record for light-growth responses is probably held by etiolated cucumber seedlings (Cosgrove 1981; Spalding and Cosgrove 1989). About 20-30 s after the onset of B, the cucumber hypocotyl

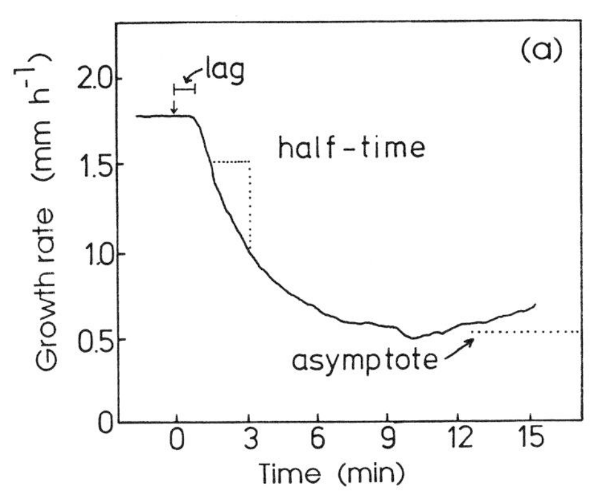

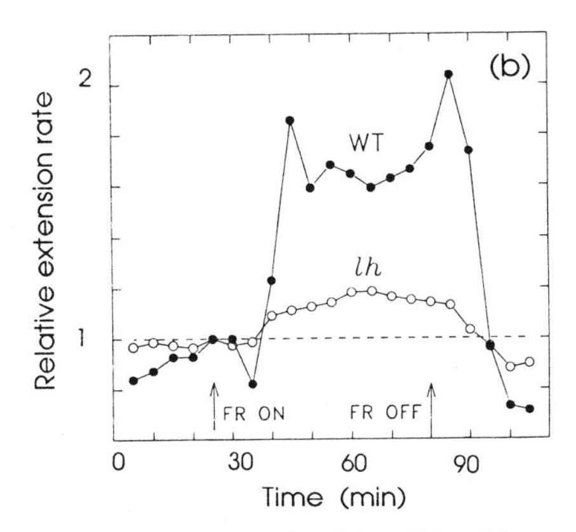

Figure 9. Rapid inhibition of growth in etiolated sunflower hypocotyls by blue light (a) or stimulation of hypocotyl growth in light-grown cucumber hypocotyls by far-red light (FR) (b). In both cases stem extension rates were measured with position transducers of the type shown in Fig. 8. In (b), a long-hypocotyl mutant of cucumber *(lh)* is measured, in addition to a wild-type (WT) cucumber line. Adapted from Cosgrove D.J. and Green P.B. (1981) *Plant Physiol.* 68: 1447-1453 and from Smith *et al.* (1992).

starts to reduce its elongation rate; the elongation rate then decays quickly. Figure 9a shows a similar response in sunflower seedlings, where the lag is about 60 s.

In comparison, phytochrome-mediated growth responses seem to have a longer lag; a 10-15 min lag is commonly observed in light-grown seedlings (Fig. 9b). Harry Smith *et al.* (1992) used fibre optics to supplement background W with local FR irradiations to the stems of light-grown cucumber seedlings, while simultaneously measuring the stem growth rate. They found that growth in W was reversibly stimulated by FR supplementation, with a lag of 10-15 min. Similar results were also observed with other species. The difference in lag times between R and B responses constitutes one piece of evidence, among many, that growth responses to W are complex and involve light perception by both phytochrome and the B photoreceptor.

Most studies indicate that the light stimulus is sensed by the growing stems. This conclusion is based on light-shielding experiments or local irradiations with fibre optics. There are also cases in which the leaves detect a light stimulus and transmit some unidentified signal to modulate stem growth. Thus, light-growth responses may involve signals from at least two photoreceptors and from two or more regions of the plant.

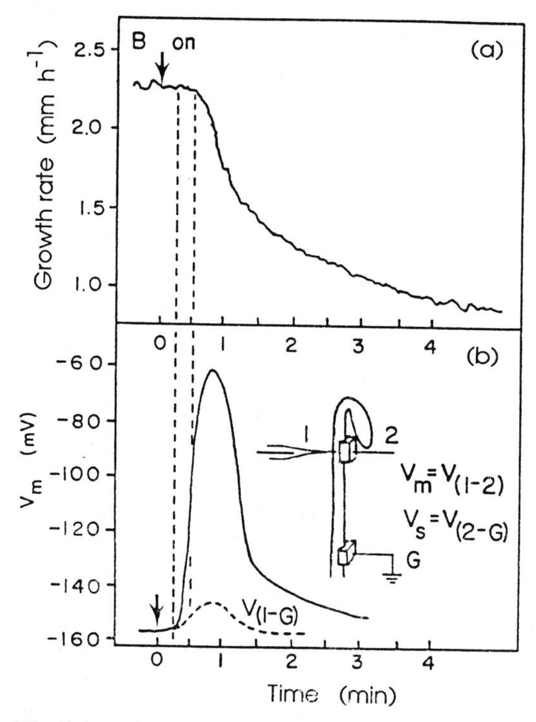

Figure 10. Blue light (B) elicits a large growth inhibition (a) and an electrical depolarization of the plasma membrane (b) in etiolated cucumber seedlings. The growth rate in (a) was measured with position transducers, as in Fig. 8, with simultaneous measurements of electrical potentials from intracellular electrodes and surface electrodes. The membrane potential is shown as V_m; the surface potential as V_s. Note that the membrane starts to depolarize before the growth starts to decline. Adapted from Spalding and Cosgrove (1989).

9.1.3.3 Electrical and biochemical changes in membranes often precede or accompany light-growth responses

Changes in membrane properties have long been suspected of playing a role in light responses of plants. These membrane changes can be classified into two broad categories: electrical changes, as a result of altered ion transport, and biochemical changes, typically as a result of enzymatic modification of one or more of the membrane components.

In many plant tissues, light elicits small changes in surface potential or cell-membrane potential. These changes are indicative of altered transport of one or more ionic species across the plasma membrane. For example, if light activated the H^+ pump, the plasma membrane would hyperpolarize (electrical potential inside the cell becomes more negative); in contrast, activation of K^+ channels or Cl^- channels would typically depolarize the membrane by bringing the potential closer to the Nernst reversal potential for K^+ or Cl^-.

A remarkable membrane depolarization precedes the rapid growth inhibition of cucumber seedlings by B (Spalding and Cosgrove 1989). Fifteen seconds after B is turned on, the plasma membrane begins to depolarize (becomes less negative). This depolarization is transient and can be detected with intracellular micro-electrodes or with surface electrodes (Fig. 10). As the membrane reaches its maximum depolarization, the growth rate begins to fall. This electrical depolarization and the ensuing growth inhibition show similar fluence-response characteristics; moreover, both electrical and growth responses are speeded up at higher fluences. These observations suggest that the electrical depolarization is part of the transduction pathway leading to the B growth inhibition. The mechanism of depolarization appears to involve both inactivation of the plasma membrane H^+ pump and activation of a depolarizing current (perhaps by opening anion channels; see Spalding and Cosgrove 1992).

Light also causes rapid changes in the phosphorylation state of membrane phospholipids and proteins. For example, a 10-s pulse of W causes rapid changes in phosphoinositide kinases and H^+-ATPases in sunflower hypocotyl membranes (Memon and Boss 1990). In maize leaves, brief R treatment elicits fast changes in the levels of phospholipids and phosphoinositides (Guron *et al.* 1992). Furthermore B also induces changes in the phosphorylation state of one or more membrane proteins (Reymond *et al.* 1992). These and related observations suggest that kinases are involved in the early stages of light-growth responses, but more work is needed to determine how such phosphorylation changes are elicited by light and how they might lead to modulation of growth processes.

9.1.3.4 Light affects cell wall yielding properties

When light modulates plant growth, it does so *via* intermediates which affect the physical and chemical processes that comprise cell expansion. Altered kinase activity may be part of this chain of intermediates, as suggested above. In the following sections we first review the basic mechanisms underlying plant cell expansion, and then consider how light modulates those processes. As we will see, most evidence indicates that light acts by altering wall loosening and yielding.

Plant cells may be thought of as membrane-bound sacks filled with a watery solution and compressed by a rigid cell wall. Viewed from this admittedly over-simplified perspective, growth requires two things: the cells must absorb water from their environment in order to increase in size, and the cell wall surrounding the cells must irreversibly expand to accommodate the water influx (Fig. 11). Cells may also expand or contract reversibly, as occurs during the opening and closing of guard cells, but such reversible changes in cell size are distinct from growth. During sustained and steady growth, water uptake and

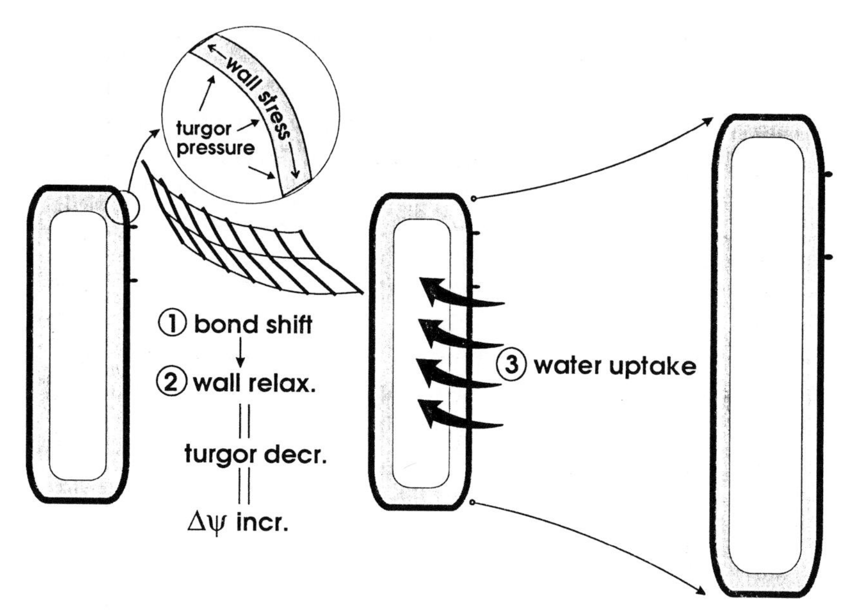

Figure 11. Simplified model of the biophysical processes that underlie plant cell expansion. Cell turgor pressure results in a substantial tensile stress in the cell wall (*inset*). Growth begins with a biochemical modification of the load-bearing bonds in the wall (1). This results in a reduction, or relaxation (relax.), of the wall tension and cell turgor pressure (2). The decrease (decr.) in cell turgor creates the increase in water potential difference ($\Delta\psi$ incr.) needed for water absorption by the cell (3).

wall expansion occur simultaneously and at equal rates, but they are nevertheless distinct processes.

The wall expands irreversibly during growth when the wall is biochemically modified (loosened) to allow it to yield to the internal hydrostatic pressure (turgor pressure) of the cell. The growth of plant cells and tissues is often expressed with the relation:

$$\text{growth rate} = m(T - Y)$$

where T is turgor pressure, Y is the yield threshold and m is the wall extensibility. The yield threshold is defined as the minimum turgor pressure required to sustain wall growth. Wall extensibility is a measure of how readily the wall irreversibly expands in response to the wall stress, which is expressed in terms of turgor pressure. Actually, wall extensibility depends on how rapidly the wall is being loosened by biochemical processes, and so it is not a simple physical constant.

Work in the 1960's and 1970's gave indirect evidence that R inhibited stem growth by causing a stiffening of the cell wall. More recent studies (*e.g.* Kigel and Cosgrove 1990) used direct measurements of cell turgor pressure and wall

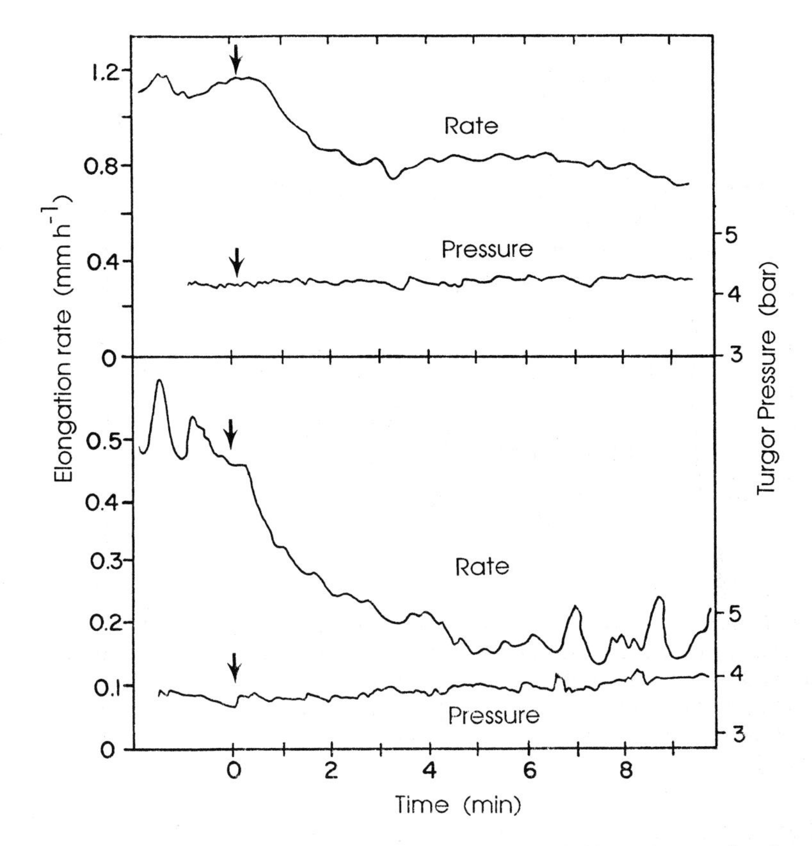

Figure 12. Cell turgor pressure remains constant during the inhibition of cucumber hypocotyl elongation by blue light (*arrow* indicates start of irradiation). Two examples are shown here, in which growth rate was measured with a position transducer at the same time that cell turgor pressure was measured with a cell pressure probe. The results show that growth is not reduced *via* a reduction in cell turgor. From Cosgrove (1988).

relaxation characteristics to show that R and B slowed stem growth in etiolated pea seedlings by slowing wall loosening (relaxation). Similar methods were used to examine the rapid inhibition of growth by B in cucumber seedlings. By direct measurements of turgor pressure with the cell pressure probe, Cosgrove (1988) found that turgor pressure remained steady during the B growth inhibition (Fig. 12). In contrast, B slowed the rate of wall relaxation. These and related results showed that B altered the wall yielding properties, principally by a change in wall extensibility (m), with little effect on the yield threshold (Y).

Van Volkenburgh and co-workers have investigated the light-induced expansion of *Phaseolus* leaves by W. They concluded that light did not directly influence Y or T, but appeared to stimulate expansion, at least in part, *via* an increase in m (Van Volkenburgh *et al.* 1987).

From these and related studies, light appears to modulate stem and leaf growth by modulating the wall-loosening properties of the growing cells. However, We do not yet know how this is accomplished at the biochemical level. Evidently light does not directly modify the wall. The fastest light-growth responses, those to B, have lag times of 15-30 s. For R and FR responses, lag times may be as short as 10-15 min and sometimes longer than 90 min. By comparison, photochemical and enzymatic reactions occur much more quickly (about 10^{-9} and 10^{-3} s, respectively). Hence it is unlikely that photoreceptors directly interact with the cell wall, but rather modify cellular processes which control the yielding characteristics of the wall.

At present we have only an incomplete picture of how light modifies the yielding properties of the wall. Several possibilities have been put forward, usually as separate mechanisms. For example, some researchers have hypothesized that light acts on growth by altering growth hormone synthesis, destruction, transport or conjugation. Others have suggested that light alters the ionic environment of the cell wall, the rates of synthesis of cell wall components, or the activity of hypothetical wall-loosening enzymes. Still others have advocated that light acts directly on transcription and translation of gene products necessary for growth. While many of these activities listed above do change in response to light, it is not clear which are primary events that stimulate growth, and which are secondary events consequent to the altered growth rates. Since light operates through several photosensory mechanisms, it seems likely that growth may be influenced *via* more than one mechanism.

9.1.3.5 Hormones may be involved in some light-growth responses

Since the earliest studies on auxin, researchers have looked for a causal link between hormones and light-growth responses. The discovery of auxin can be traced back to Charles Darwin's observations about phototropism of grass coleoptiles. He found that irradiation of the tip of the coleoptile resulted in subsequent bending of the lower (non-irradiated) portions of the coleoptile. This observation suggested that a stimulus was transmitted from the tip to the rest of the coleoptile, and eventually led to the Cholodny-Went hypothesis, which explained phototropism as due to lateral asymmetry in auxin distribution (Chapter 9.2).

Evidence of a transmissible stimulus has similarly been found for some, but not all, light-growth responses of stems. For example, when the cotyledons of *Cucumis, Raphanus* and other species are irradiated with R, the elongation of the stem below the cotyledons is subsequently inhibited. Although other types of transmissible stimuli are possible, growth hormones have attracted considerable attention and the evidence for hormone-mediation of light responses is worth examining.

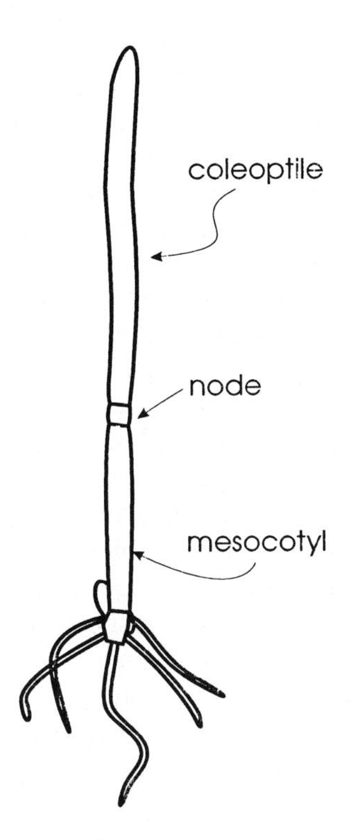

Figure 13. Diagram of an oat seedling, showing the coleoptile and the mesocotyl.

9.1.3.5.1 Auxin. When grass seedlings are grown in the dark, the shoot consists of two principal organs: the coleoptile, which encloses the unexpanded leaves, and the mesocotyl (Fig. 13). The growth of both these organs is sensitive to light. In the 1930's I. Van Overbeek found that light reduced the yield of diffusible auxin from excised coleoptile tips, and he proposed that the growth of the subtending mesocotyl was dependent on the supply of auxin from the coleoptile. Time course studies (Iino 1982) showed that after a brief R treatment the growth rate of the mesocotyl is reduced within 4 h to about 20% of the dark controls. Simultaneously the auxin concentration in the mesocotyl falls to 50% of the dark controls. When the growth begins to recover after about 10 h, the auxin concentration also rises. These changes in auxin concentration in the mesocotyl seemed to be due to changes in auxin supply from the coleoptile and they correlated with the growth rate of the mesocotyl.

Mesocotyl growth in *Avena* seedlings is controlled by two phytochrome-mediated responses (Iino 1982). One is a VLFR in which growth is proportional to the logarithm of the fluence, starting around 10^{-4}, saturating at 3×10^{-2} μmol m^{-2} of R for *Avena.* The VLFR results in a 50% inhibition of *Avena* mesocotyl growth at saturation. The second is an LFR which operates between

1-100 μmol m^{-2} of R and results in 80% suppression of *Avena* mesocotyl growth. Iino (1982) found that the fluence-response curves for mesocotyl growth correlated fairly well with the fluence-response curves for yield of diffusible auxin from coleoptile tips.

For the observed reductions in auxin to be causally related to the inhibition of mesocotyl growth, it must be shown that the endogenous auxin is limiting growth. The work of M. Iino suggests that the endogenous auxin in the mesocotyl is in the linear stimulatory range, but near saturation for growth. Thus a reduction in auxin supply should reduce the growth rate. However, not all of the growth effect of R can be attributed to reduction in auxin supply. Some effect of R could still be seen when auxin was supplied to the ends of mesocotyls by agar blocks containing 1-10 μM auxin. Moreover, the 80% inhibition of growth 4 h after a 10 min R treatment is greater than expected from the 50% inhibition of endogenous auxin concentration. Thus a large part (perhaps half) of the inhibition of mesocotyl growth may be attributed to reduction in the supply of auxin from the coleoptile, but other mechanisms also seem to be operating. The nature of these other effects of R is still unclear.

Another point about the light-growth responses of grass seedlings is noteworthy. At the same time that R inhibits mesocotyl growth, it also stimulates coleoptile growth. This stimulation of coleoptile growth occurs despite the reduction in auxin concentration by R. Endogenous auxin is not supraoptimal (*e.g.* a reduction in auxin concentration does not increase growth, as is sometimes the case with roots). Therefore the coleoptile response must be mediated by some mechanism other than by a change in auxin level.

In dicots, there is much less evidence for photomodulation of stem growth *via* auxin. Perhaps the best evidence comes from a study by Behringer *et al.* (1992), who found that auxin (indole-3-acetic acid) levels in the epidermis of pea stems increased 40% after EODFR treatments. Parallel experiments indicated that this EODFR treatment doubled stem growth rate. These results suggest that part of the EODFR response might be the result of increased auxin levels in the stem epidermis. A photomorphogenic mutant was also studied, one in which the phytochrome signal may be blocked. They found that the mutant lacked the EODFR growth response and similarly lacked the increase in auxin levels. These results are promising, but more work remains to assess the significance of these local auxin changes for the growth of the stem.

9.1.3.5.2 Gibberellins. Gibberellin (GA) stimulates stem elongation in many plants, and the biosynthesis of active GAs is a prime determinant of stem length. Many genetic dwarfs have turned out to be deficient in GA synthesis. Studies in the 1950's and 1960's suggested that light could modify growth *via* this growth hormone in three potential ways: (i) a reduction GA synthesis; (ii) an increase in GA destruction, or (iii) a reduction in the plant's responsiveness to GA.

There is some evidence for regulation of GA synthesis by light, but results are complicated by the fact that there are many different forms of GA, of which only a subset are active for control of stem growth. In spinach, long days (LD) induce bolting (rapid elongation of the rosette stem) in parallel with increases in the active forms of GAs (Talon *et al.* 1991). Likewise, petiole elongation increases in LD. The effect of LD can be suppressed with GA biosynthesis inhibitors and can be mimicked, at least in part, by applications of GAs. Light also affects the metabolism of applied GAs, indicating that one or more steps in GA metabolism are influenced by photoperiod. Thus there is some reason to believe that the bolting induced by LD is mediated by altered GA levels. However, it is not clear that the increase in GAs is sufficient to account for all of the growth stimulation; changes in responsiveness of the plant to GA likely also play a role (Zeevaart *et al.* 1990).

In nonbolting plants such as peas, there is indirect evidence that light likewise affects the metabolism of GAs. For example, pea seedlings of a dwarf variety (Progress) responded strongly to GA_1 and GA_{20} when they were kept in the dark, but selectively lost their response to GA_{20} after R treatment (Campbell and Bonner 1986). This result was interpreted to mean that R interfered with the conversion of GA_{20} (inactive) to GA_1 (active). However, direct measurements of GA in plants grown under various light conditions or in dark do not support the hypothesis that light affects stem elongation by altering the levels of active GAs (Reid *et al.* 1990; Ross *et al.* 1992). These latter authors attributed the effect of light on growth to changes in responsiveness to GAs. Some caution is needed in interpreting 'hormone responsiveness'. This could be anything from a change in reception and transduction of a hormonal signal to a change very remote from hormone action. For example, if light enhanced a biochemical crosslinking of the wall, the plant would be 'less responsive' to GA, yet the action of light would be unrelated to GA signalling.

9.1.3.6 Other possible mediators of light-growth responses

Light might also take more direct action on the biochemical processes that give rise to plant growth. For example, it is known that the pH of the cell wall free space is important in controlling wall expansion. If light directly altered H^+ extrusion, it might alter growth. The stimulation of leaf expansion by W is accompanied by increased H^+ extrusion by the cells of the leaf (Van Volkenburgh *et al.* 1987). Light stimulation of coleoptile elongation is likewise accompanied by increased H^+ extrusion. Light inhibition of stem growth may follow a reduction in the rate of H^+ extrusion; this evidently has not been measured directly, but is implied from the studies of membrane depolarizations (Spalding and Cosgrove 1992) and membrane H^+-ATPase activity (Memon and Boss 1990). In other light responses, changes in wall peroxidase activity have been detected (Casal *et al.* 1990), but the significance of these effects for growth need further evaluation.

9.1.4 Summary

Plants use light as an environmental cue to alter their development and modify their growth. The R and FR responses are mediated by phytochrome, whereas responses to the B part of the spectrum largely operate *via* a specific B photoreceptor. Phytochrome involvement is complex, involving a VLFR which responds to 'negligible' quantities of light, a LFR which shows the classical R/FR reversibility, and a HIR which probably operates by two or more mechanisms.

The inhibition of stem growth and stimulation of leaf growth by light occurs by an alteration of the yielding characteristics of the cell wall. Rapid light-growth responses are sometimes associated with rapid changes in the membrane potentials and membrane phosphorylation. There is reason to suspect that these biochemical changes may serve as parts of the transduction of the light responses, but more work is needed to establish the exact role of such membrane changes. Alterations in hormone levels and hormone sensitivity have also been implicated in some light-growth responses. Perhaps the use of genetic mutants and transgenic plants (Chapters 4.8 and 8.2) will help to unravel the detailed mechanisms of light modulation of plant growth.

Much work remains to be done before the control of growth by light will be fully understood. The most attractive responses for mechanistic studies of how light alters growth are those in which light rapidly alters growth, *e.g.* the rapid responses of stems to B and to the R:FR photon ratio. These light responses act so quickly that secondary effects of light are minimized and there is some hope of tracking the mechanism of the growth response all the way back to the photoreceptor. Indeed, the control of growth by light is such an important aspect of plant development, and moreover is relatively tractable to experimental manipulation and analysis, that more research directed along these lines ought to provide considerable insight into the control of plant growth and development.

9.1.5 Further reading

Cosgrove D.J. (1993) Wall extensibility: its nature, measurement, and relationship to plant cell growth. *New Phytol.* 124: 1-23.

Galland P. (1991) Photosensory adaptation in aneural organisms. *Photochem. Photobiol.* 54: 1119-1134.

Holmes M.G. (1983) Perception of shade. *Phil. Trans. R. Soc. Lond.* B 303: 503-521.

Reed J.S., Nagpal P. and Chory J. (1992) Searching for phytochrome mutants. *Photochem. Photobiol.* 56: 833-838.
Silk W.K. (1984) Quantitative descriptions of development. *Annu. Rev. Plant Physiol.* 35: 479-518.

9.1.6 References

Behringer F.J., Davies P.J. and Reid J.B. (1992) Phytochrome regulation of stem growth and indole-3-acetic acid levels in the *lv* and *Lv* genotypes of *Pisum. Photochem. Photobiol.* 56: 677-684.
Campbell B.R. and Bonner B.A. (1986) Evidence for phytochrome regulation of gibberellin A20 3ß-hydroxylation in shoots of dwarf *(le le) Pisum sativum* L. *Plant Physiol.* 82: 909-915.
Casal J.J., Whitelam G.C. and Smith H. (1990) Phytochrome control of extracellular peroxidase activity in mustard internodes: correlation with growth and comparison with the effect of wounding. *Photochem. Photobiol.* 52: 165-172.
Cosgrove D.J. (1981) Rapid suppression of growth by blue light: Occurrence, time course, and general characteristics. *Plant Physiol.* 67: 584-590.
Cosgrove D.J. (1988) Mechanism of rapid suppression of cell expansion in cucumber hypocotyls after blue-light irradiation. *Planta* 176: 109-116.
Gaba V., Black M. and Attridge T.H. (1984) Photocontrol of hypocotyl elongation in de-etiolated *Cucumis sativus* L.: Long term fluence-rate-dependent responses to blue light. *Plant Physiol.* 74: 897-900.
Guron K., Chandok M.R. and Sopory S.K. (1992) Phytochrome-mediated rapid changes in the level of phosphoinositides in etiolated leaves of *Zea mays. Photochem. Photobiol.* 56: 691-695.
Hartmann K.M. (1967) Ein Wirkungsspectrum der Photomorphogenese unter Hochenergiebedingungen und seine Interpretation auf der Basis des Phytochrome (Hypokotylwachstumshemmung bei *Lactuca sativa* L.). *Z. Naturforsch.* 22b: 1172-1175.
Heim B. and Schäfer E. (1982) Light-controlled inhibition of hypocotyl growth in *Sinapis alba* L. seedlings. *Planta* 154: 150-155.
Iino M. (1982) Inhibitory action of red light in the growth of the maize mesocotyl: evaluation of the auxin hypothesis. *Planta* 156: 388-395.
Jabben M., Beggs C.J. and Schäfer E. (1982) Dependence of Pfr/P_{tot} - ratios on light quality and light quantity. *Photochem. Photobiol.* 35: 709-712.
Jose A. (1977) Photoreception and photoresponses in the radish hypocotyl. *Planta* 136: 125-129.
Kigel J. and Cosgrove D.J. (1990) Photoinhibition of stem elongation by blue and red light: effects on hydraulic and cell wall properties. *Plant Physiol.* 95: 1049-1056.
McCormac A.C., Cherry J.R., Hershey H.P., Vierstra R.D. and Smith H. (1991) Photoresponses of transgenic tobacco plants expressing an oat phytochrome gene. *Planta* 185: 162-170.
Memon A.R. and Boss W.F. (1990) Rapid light-induced changes in phosphoinositide kinases and H^+-ATPase in plasma membrane of sunflower hypocotyls. *J. Biol. Chem.* 265: 14817-14821.
Reid J.B., Hasan O. and Ross J.J. (1990) Internode length in *Pisum.* Gibberellins and the response to far-red-rich light. *J. Plant Physiol.* 137: 46-52.
Reymond P., Short T.W. and Briggs W.R. (1992) Blue light activates a specific protein kinase in higher plants. *Plant Physiol.* 100: 655-661.
Ross J.J., Willis C.L., Gaskin P. and Reid J.B. (1992) Shoot elongation in *Lathyrus odoratus* L.: Gibberellin levels in light and dark-grown tall and dwarf seedlings. *Planta* 187: 10-13.
Shinkle J.R., Sooudi S.K. and Jones R.L. (1992) Adaptation to dim-red light leads to a non-gradient pattern of stem elongation in *Cucumis* seedlings. *Plant Physiol.* 99: 808-811.
Smith H., Turnbull M. and Kendrick R.E. (1992) Light-grown plants of the cucumber long hypocotyl mutant exhibit both long-term and rapid elongation growth responses to irradiation with supplementary light. *Photochem. Photobiol.* 56: 607-610.

Spalding E.P. and Cosgrove D.J. (1989) Large membrane depolarization precedes blue light inhibition of growth in cucumber hypocotyls. *Planta* 178: 407-410.

Spalding, E.P., Cosgrove, D.J. (1992) Mechanism of blue-light induced plasma-membrane depolarization in etiolated cucumber hypocotyls. *Planta* 188: 199-205.

Talon M., Zeevaart J.A.D. and Gage D.A. (1991) Identification of gibberellins in spinach and effects of light and darkness on their levels. *Plant Physiol.* 97: 1521-1526.

Van Volkenburgh E., Cleland R.E. and Schmidt M.G. (1987) The mechanism of light-stimulated leaf cell expansion. *Soc. Exp. Biol. Seminar Series* 27: 223-238.

Zeevaart J.A.D., Talon M. and Wilson T.M. (1990) Stem growth and gibberellin metabolism in spinach in relation to photoperiod. In: *Gibberellins*, pp. 273-279, Takahashi N., Phinney B.O. and MacMillan J. (eds.) Springer, Berlin.

9.2 Phototropism

Richard D. Firn

Department of Biology, University of York, Heslington, York, Y01 5DD, UK

9.2.1 What is phototropism?

Most people are familiar with the sight of a young seedling bending towards a window or the brightest source of light to which it is exposed (Fig. 1). This directional growth response, induced by unequal irradiation with wavelengths detected by the blue light (B) photoreceptor, is known as phototropism* and has fascinated plant physiologists for more than 150 years.

Phototropism, unlike so many other well known physiological responses, is a non-essential response evoked only in some young seedlings growing under certain conditions. You only have to walk round a garden or through a farmer's field to observe that very few mature plants show any evidence of a current or previous phototropic response. A plant can grow and reproduce successfully without utilizing the phototropic response. So why do plants have a capacity to develop a phototropic response? In young seedlings, there may be some advantage in the growing axis responding phototropically in order to orientate the young cotyledons or leaves so that they are perpendicular to the strongest source of light to optimize light interception and photosynthesis. However, as the plant grows and develops, it produces leaves on petioles and the power to orientate the photosynthetic structures is devolved to the petioles or to the leaf pulvini and the main axis looses its phototropic sensitivity (Fig. 2).

*The term phototropism is also used to describe growth of single cells of lower plants towards light (such as the sporangiophore of the fungus *Phycomyces* or the protomena of the moss *Funaria* or *Physcomitrella*). However, there are good reasons to believe that the phototropism of these optically simple cells differs physiologically from that of higher plants. Such responses will not be considered further in this chapter. Higher plants grown under certain conditions can also respond to a gradient of the far-red absorbing form of phytochrome (Pfr) across them and a phototropic response is generated by the gradient of Pfr-induced growth inhibition. There are no good reasons to assume that such a phototropic response is physiologically similar to B-induced phototropic response hence such phytochrome-induced responses will not be discussed in this chapter.

R.E. Kendrick & G.H.M. Kronenberg (eds.), Photomorphogenesis in Plants - 2nd Edition

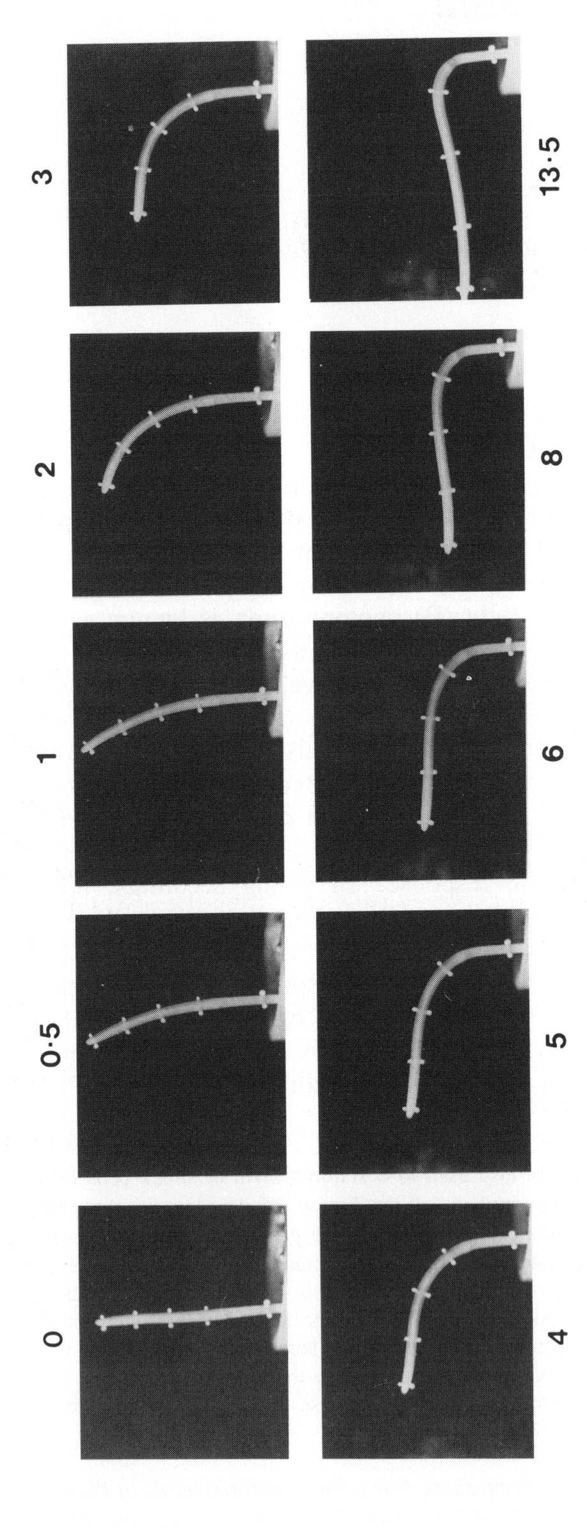

Figure 1. The various phases of phototropism shown by an etiolated *Avena* coleoptile subjected to continuous unilateral blue light. Surface markers define zones and it is evident that curvature under these conditions is evident in all elongating zones even after only 30 min of stimulation. Note also that the second zone beneath the tip is curved during the first few hours but begins to straighten after 4 h due to autotropism. Photographs taken by J.M. Franssen.

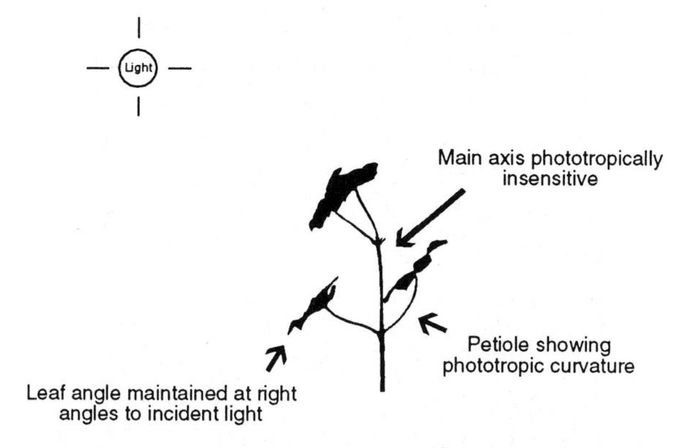

Figure 2. Light-induced organ movements: phototropism is common in young seedlings, but rare in older plants. Petiole curvature and leaf pulvinule-directed movement of leaves serve to orient leaves to the optimum position in older plants.

Given how physiologically unimportant phototropism seems to be, one might wonder why so much attention has been paid to this response and question whether it deserves such close scrutiny in future. There are, in fact, two good reasons why phototropism has been studied so extensively and these reasons remain valid:

(i) Phototropism is caused by differential growth and the dramatic changes in cell elongation evoked by a phototropic stimulus are a very attractive system in which to study the control of cell elongation. The phototropic stimulus is easily applied, it can be given to defined regions of the shoot and the magnitude of the stimulus can be varied (these are distinct advantages compared for example to gravitropism).

(ii) Phototropism is a response controlled by the B photoreceptor and this tropic response has many attractive features if one is interested in studying events controlled by this photoreceptor (*i.e.* phototropism is rapid, dramatic and occurs in easy-to-grow young seedlings).

Phototropism is therefore studied not so much because of the importance of the process itself, but more because the component parts of the process, B perception and differential cell extension, are of fundamental importance. It must also be added that phototropism has an intricacy, subtlety and beauty which gives it a strong intellectual appeal for study *per se*. Furthermore, useful phototropic experiments can still be conducted using simple methods and minimal facilities hence the subject is accessible to all.

9.2.2 Scope of this chapter

Those coming afresh to the subject of phototropism face two considerable problems: a vast literature stretching back more than 100 years and the lack of a widely accepted, unifying model to explain any aspect of the phenomena. These are also formidable problems to those attempting to introduce others to the subject. Hence in this chapter, instead of reviewing this extensive literature I will try to provide some simple basic information and will introduce the reader to some current debates.

The nature of the B photoreceptor is discussed elsewhere in Chapter 5.2 hence the current chapter will concentrate on the growth co-ordination mechanisms which are used to bring about differential growth. The way in which light gradients are generated in organs, an important but until recently a neglected area of phototropic research, is discussed more fully in Chapters 7.2 and 7.4 .

9.2.3 An historical summary of some key concepts of phototropism

Phototropism, or *heliotropism* as it was then called, was studied by a number of workers in the last century; A.P de Candolle, J. Wiesner, J. Sachs, W. Rothert and F. Czapek. These studies were just part of the exciting new subject of plant physiology, a subject being studied with particular vigour and skill in some German and Central European universities. By the start of the 20th century much useful information about phototropic responses had been gathered. For instance, it had already been shown that the wavelengths in the blue region of the spectrum were most effective at inducing the response and it was clear that differential cell expansion across the responding organ caused curvature. The phenomena was very appropriate for study at that time because there was much to be learned by simple experiments. Such was the simplicity of the experiments, that normal laboratory facilities were not even needed and one fundamental piece of work was done in a private house. This was work conducted by one of the great scientific figures of the day, Charles Darwin (aided by his son Francis). Darwin's interest in plant physiology in his old age gave the subject an added importance, especially in the English speaking world where botanical studies until that time had been dominated by anatomy, morphology and taxonomy. Darwin was fascinated by all types of organ movements. As well as gravitropism and phototropism, he studied *tendril coiling*, the sleep movement of leaves and flowers (*nyctonasty*) and the swaying oscillations of elongating organs (*circumnutation*). Only a small section in Darwin's book *The Power of Movement in Plants* reported studies on phototropism but one finding was to catch the imagination of generations of plant physiologists that followed. Darwin noted that in *Phalaris* coleoptiles the sites of phototropic perception and phototropic response were separate and this led him to conclude: *"that when the seedlings are freely exposed to a lateral light, some influence is transmitted from the upper to the lower part, causing the latter to bend"*. It is often overlooked, however, that Darwin was aware that this conclusion did not hold for *Avena* coleoptiles. He stated quite clearly that Avena coleoptiles: *"offer a strong exception to the rule that illumination of the upper part determines the curvature of the lower part"*. Despite this latter observation, confirmed and extended by Rothert, a number of workers in the first two decades of the 20th century often chose to work on *Avena* coleoptiles when investigating how the tip controlled coleoptile elongation (H. Fitting, P. Boysen-Jensen, A. Paál, P. Stark and H. Söding).

In parallel with this work on the importance of the tip* in controlling elongation in coleoptiles, studies on the role of the root tip in controlling root growth had also led to the view that the tip was able to influence cell elongation in the extension zone (T. Ciesielski, C. Darwin and N. Cholodny). In an attempt to formulate a unifying view of the control of plant organ extension, ideas from work on root and coleoptiles were independently drawn together by N. Cholodny and F.W. Went. This resulted in the Cholodny-Went (C-W) model of plant tropisms, which proposed that the *redistribution of endogenous growth substances* in the tip was transmitted to the elongation zone where *differential growth* was induced. The search for the growth substance involved resulted in Went, using his *Avena* curvature bio-assay, discovering *auxin*. Most of the work on plant tropisms that followed in the 1930's (especially the elegant, extensive studies by H.E. Dolk) centred around the C-W theory and the evidence supporting this model was marshalled very effectively by F.W. Went and K.V. Thimann in 1937 in their influential book *Phytohormones*.

When plant physiology entered a new era of growth after the lull caused by the 1939-45 war, the C-W model was firmly established and for the next 30 years most studies of phototropism accepted the theory as the best theoretical framework for experimental design and interpretation of results.

However, by no means all workers studying phototropism during the first half of this century accepted the concept of the paramount importance of the tip. As noted by C. Darwin and subsequently investigated more thoroughly by W. Rothert, some organs had extensive regions of phototropic sensitivity. Many workers recognized that at low fluence rates of unilateral light, only the most sensitive cells near the tip could perceive the phototropic stimulus but there was evidence that at higher fluence rates cells in the elongation zone also perceived the stimulus. The most precise alternative to the C-W model of phototropism

*The term tip is used rather than the word apex in order to emphasize that the structures which were originally ascribed a special role in the *Cholodny-Went (C-W) model* were actually very different anatomically, morphologically and physiologically. The only feature that a root apex and a coleoptile tip share in common is that they are positioned at the end of an elongating organ.

arose from studies by A.H. Blaauw in the first two decades of this century when he studied the effects of light on hypocotyl elongation. He argued that phototropism was nothing more than a direct inhibition of cell growth by light. To support this view, he showed a good correlation between the amount of light striking cells in unilaterally irradiated organs and the growth rate at various positions across the organ. About 10 years later similar studies were conducted on *Avena* coleoptiles by C. van Dillewijn who felt that *Blaauw's model* could explain certain aspects of coleoptile phototropism. Some support for this view was offered by H.G. Du Buy and E. Nuernbergk who produced the first accurate analysis of the differential growth causing phototropism. They found a marked growth inhibition at the irradiated side, a finding consistent with Blaauw's model. Indeed when F.W. Went and K.V. Thimann summarized the literature on phototropism in 1937, they expressed the view that the Blaauw model explained sunflower hypocotyl phototropism better than the C-W model and it was felt that the two models were possibly complementary.

During the 1950's and 1960's, when plant hormones became the most fashionable area of plant physiology, doubts about the universal validity of the C-W model were largely forgotten, especially in elementary teaching of the subject. The C-W model was seen as the foundation stone of hormonal physiology, and the model became a dogma. However, the C-W model began to stagnate and during the 1970's no significant advances in understanding were made as a result of following the model. By the 1980's, doubts began to be raised about the validity of the C-W model (R.D. Firn, J. Digby, J. Bruinsma, A.J. Trewavas) but the model was strongly defended (W.R. Briggs, M. Iino, B. Pickard).

9.2.4 Ways of inducing a phototropic response

Experimentally it is convenient to study phototropism in seedlings exposed to unilateral light, but it must always be remembered that plants in a natural environment rarely if ever receive such an extreme treatment. Plant organs which are phototropically sensitive have evolved to respond phototropically if the fluence rates received by the flanks are not uniform. A radial asymmetrical light gradient must be established but there is no requirement for unilateral light nor need the direction of the response be towards the stronger source (Fig. 3). If seedlings are exposed to unequal bilateral illumination, a phototropic response will be induced if the ratio of the two sources exceeds 1:1.2. Hence the commonly used unilateral light treatment is a very extreme, unnatural form of phototropic stimulus. Furthermore, by only thinking of phototropism as a response to unilateral light, as so often seems to be the case, restraining conceptual and experimental limitations are imposed.

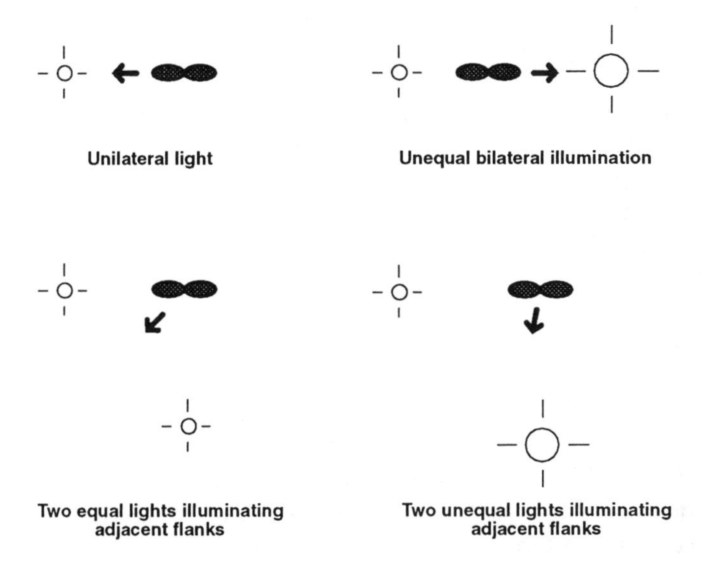

Figure 3. Phototropism can be induced by a single unilateral light or by unequally illuminating a seedling with two light sources and placing the seedling equidistant from the sources. The relative intensity of the lamps is indicated by the diameter of the circle and the arrow shows the direction of phototropic curvature.

The physiological state of the plant before a phototropic stimulation is given can have a profound effect on the response. Of the two popular experimental systems, 4-7-day-old grass coleoptiles and dicot seedlings, quite different growth conditions are usually used. Coleoptiles are normally completely etiolated or grown under very dim red light (R). However, young dicot seedlings are not very responsive when etiolated and are usually used after de-etiolation (transferred to light at least 24 h before being given a phototropic stimulus then replaced in darkness for some hours before use in a phototropic experiment). All these various treatments can obviously influence, not only the growth rate of the seedlings (hence any subsequent phototropically induced patterns of growth-rate changes), but also the gravitropic sensitivity of the seedlings hence have a secondary effect on the phototropic response (see Section 9.2.9)

9.2.5 Measuring the phototropic response

The phototropic response seems so easy to measure; a simple measurement of how much the organ has bent seems sufficient. This simplicity is very misleading. A phototropically curving organ is a heterogeneous population of cells and

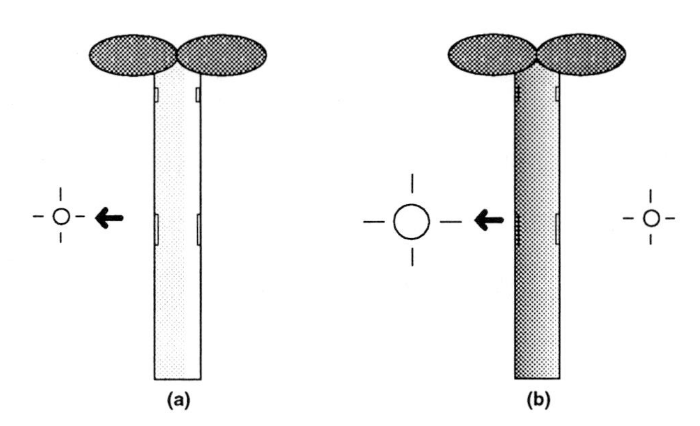

Figure 4. Phototropism is caused by the unequal photostimulation of cells which may be at different stages of their development. Cells entering the elongation zone are smaller than cells leaving the elongation zone. The density of hatching represents the approximate light gradient in (a) a unilaterally illuminated hypocotyl; (b) a hypocotyl exposed to the unequal bilateral exposure capable of causing phototropism.

the organ only responds to a phototropic stimulus if the cells at different positions in the organ have responded unequally (Fig. 4). A single measurement such as degree of organ curvature can record the response of the organ but no such single figure could possibly describe the different responses that must be occurring at the cell level at different positions across the organ. By thinking of phototropism in terms of the simple response of the organ, instead of thinking about the varied response of the individual cells, numerous problems and ambiguities have arisen.

9.2.5.1 The inadequacy of angle of curvature measurements

The traditional way of measuring organ curvature is to assess the *angle of curvature,* the angle through which the axis of the apical region has passed through relative to the fixed axis of the base following phototropic stimulation (Fig. 6). However, there are four reasons why this measurement cannot fully describe the complex temporal and spatial events giving rise to organ curvature; hence are unsuitable for all but preliminary studies.

9.2.5.1.1 The same degree of curvature can be produced by many different types of response. Organ curvature is caused by differential growth, *i.e.* different rates of cell elongation at the convex and concave sides of the curving organ. In the simplest case where one is changing a straight organ into a curved

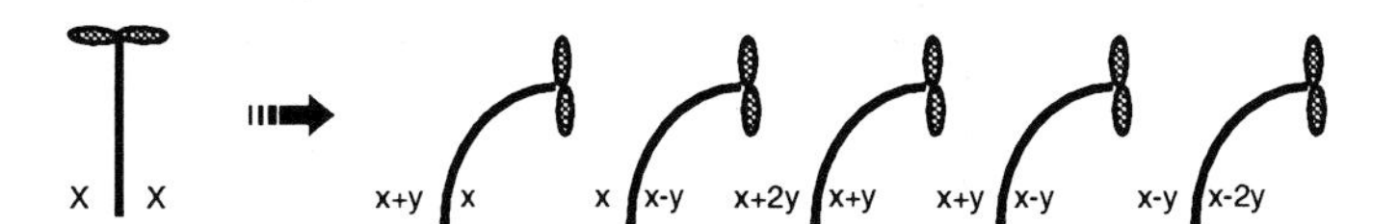

Figure 5. Very different patterns of organ curvature can produce curvature. A hypocotyl elongating at $x\%$ h^{-1} in dim light is exposed to unilateral light and the maximum change in growth rate is $y\%$ h^{-1}. Five different growth-rate patterns are theoretically possible.

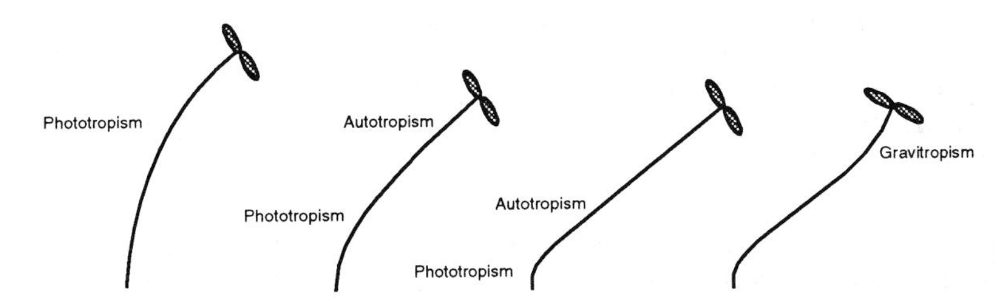

Figure 6. After a phototropic stimulation is given, the organ shape is initially determined by the phototropic response, but autotropism eventually straightens the upper regions. If the phototropic stimulation is withdrawn, gravitropism becomes evident. Note that the angle of curvature fails totally to give a meaningful description of organ curvature.

organ shaped like a perfect arc, the angle of curvature is related directly to the difference in elongation rates between the two sides. Curvature can therefore be brought about by altering the growth rates of the two sides in five ways (Fig. 5). Quite different patterns of growth-rate changes could give rise to identical angle of curvature changes. Hence measurements of the elongation rates at the convex and concave flanks of the organ after giving it a phototropic stimulus are a much better way of characterizing the overall response because such measurements allow one to determine exactly what the cells are doing. The simplest way of making these measurements is to measure the displacement of surface markers, placed at various positions along the elongation zone. Such measurements were first carried out on enlarged images of time-lapse photographs of a plant before and during the development of a phototropic response (Fig. 1). These measurements were exacting and labour intensive (which perhaps explains why angle of curvature measurements have been so popular), but computer analysis of photographs or video-images is now possible to ease the labour of this task and more automated video-image analysis is under development.

9.2.5.1.2 The location of organ curvature. Differential elongation causing phototropic curvature is confined to the elongation zone (once again a difference from gravitropic curvature which can involve a re-initiation of elongation in cells having recently left the elongation zone). There is evidence that cells in different parts of the elongation zone may not respond homogeneously or synchronously when given a phototropic stimulus. Consequently, organs with very similar angles of curvature could in fact have very different shapes, indicative of different responses (Fig. 6).

9.2.5.1.3 Temporal complexity. The extent of curvature reached at any time after the start of the phototropic stimulation period will depend on the lag time before the initiation of differential growth, the magnitude of differential growth induced and the duration of the period of differential growth. Consequently two temporal variables (lag and duration) contribute to the curvature measured and it is obvious that a single measurement of curvature cannot define two variables.

9.2.5.1.4 The autotropic straightening response. It has often been observed that regions of an organ showing an obvious tropistically induced curvature subsequently straighten (note how the most apical regions of the coleoptile shown in Fig.1 are curved after 2 h but are straight again 6 h after unilateral irradiation). This straightening response is called *autotropism* and it commences at the free end (usually the apical end) of the curving organ and occurs subsequently in more basal regions. Hence, in a phototropically stimulated organ, once curvature has developed to some extent, the autotropic response beginning at the tip can be reducing the 'angle of curvature' whilst the phototropic response in the more basal zones can be increasing this angle. Thus, because of autotropism an organ can maintain a fixed angle of curvature whilst still showing a phototropic response (note the similar angle of curvature of the tip of the coleoptile shown in Fig. 1 between 3 h and 6 h)!

The four reasons outlined for the ambiguities inherent in simple angle of curvature measurements should make it clear that more precise measurements of the response should be used where possible.

9.2.5.2 The phototropic responses of individual cells within the organ

Although growth-rate measurements of cell patches approaches the ideal of measuring the response of cells rather than the organ, it should be remembered that the growth of cells within an intact organ must be constrained by simple physical links resulting from all cells being part of an organ. This concept is easily appreciated by considering what would happen if only the cells at one flank of an organ start to grow more rapidly. The organ would start to curve and the concave flank would soon be under compression and cells on that flank might change their apparent elongation rate simply as a result of this physically transmitted effect.

One interesting possibility just on the horizon is that phototropically-induced intracellular changes will soon be measurable and it may then be possible to characterize the initial phototropic response at a cell level, irrespective of the constraints imposed by the fact that cells within an organ cannot behave with complete independence due to mechanical factors. For instance, B-induced changes in cytoplasmic [Ca^{2+}] (Gehring *et al.* 1990) or specific mRNA populations [*e.g.* the mRNAs that T. Guilfoyle's group has shown to be rapidly regulated in auxin-treated or tropistically stimulated organs (Li *et al.* 1991)] may soon be quantifiable in cells or groups of cells.

9.2.6 The basic elements of the phototropic response

Four distinct phases of the overall response seem to exist*:

(i) The *perception phase* is the period when the photoreceptor is being activated by photons and it is a period of 'physical' asymmetry within the organ.

(ii) The *latent period* or lag phase is a period when a cellular or chemical asymmetry is established within the organ but growth-rate changes are not yet evident.

(iii) The *differential growth* or curvature phase is when the organ curvature develops.

(iv) The *autotropic phase* when the organ begins to straighten again (Fig. 6).

However, this orderly distinction of phases is rarely quite so clear. For instance, it will only be evident if the duration of the perception phase is very short, otherwise perception will also be occurring during the latent period or even during the later periods. Likewise, the response of the organ may not be homogeneous hence the last two phases can, for instance, be co-incident but spatially separated. However, it is conceptually useful to consider each of these phases separately, even if this is rarely experimentally the case.

---

* Readers will notice that the author has not included the so-called *transduction* phase. The transduction phase is a vague concept because the same term can be used in quite different ways. The term could describe events that occur at a molecular, cell, tissue or organ level and in theory more than one transduction phase could thus occur during phototropism (Firn 1991).

9.2.6.1 The perception phase

A number of questions need to be asked about this phase. Where does the photoperception take place, at an organ, tissue, cell or subcellular level? The question is unanswered at present except that we know that B can be perceived in at least some cells throughout the elongation zone of coleoptiles and hypocotyls. Until the B photoreceptor involved in phototropism is identified and spatially located by immunocytochemical means it will be impossible to locate the responsive cells and the site of perception in the cell.

When it is known where the B causing phototropism is perceived it will be possible to consider how the signal is processed to provide a phototropic response. An electronics engineer building a system to detect and respond to the relative levels of incident light falling on a vertical, somewhat opaque structure would simply place light detectors around the periphery and would use a simple comparator circuit to monitor the signal strengths in order to detect regions of maximum and minimum response. In other words, a comparative system of some means would be needed that would operate at the whole structure level. However, plants seem not to possess such a simple comparator system. For instance, phototropically sensitive organs do not detect and respond to the strongest signal in their environment (Fig. 3). It seems more likely that a plant organ responds to a phototropic stimulus simply by means of the collective individual response of many cells within the organ. Indeed a simple model based on the amount of light attenuation across an organ and a second order photochemical reaction at positions across the organ has been constructed (Iino 1987). This model, which fits experimental observations of maize coleoptile phototropism, supports the concept of localized perception at all positions across the organ.

Proponents of some of the variants of the C-W model of phototropism (see Section 9.2.8) need to argue that light gradients are perceived across cells because such models require that individual cells become directionally polarized and move auxin in response to that polarization. The detection of the very small light gradient across a single cell at the shaded side of an optically complex organ such as a coleoptile poses a serious challenge to such models and the spatially very complex gradients recorded across organs (see Chapter 7.4) would give rise to further complications which have not been fully addressed.

9.2.6.2 The latent period

Measurements of organ displacement using highly sensitive opto-electronic methods are easily made but the complex nutational movements of the organ before and after stimulation make it very difficult to decide exactly when the

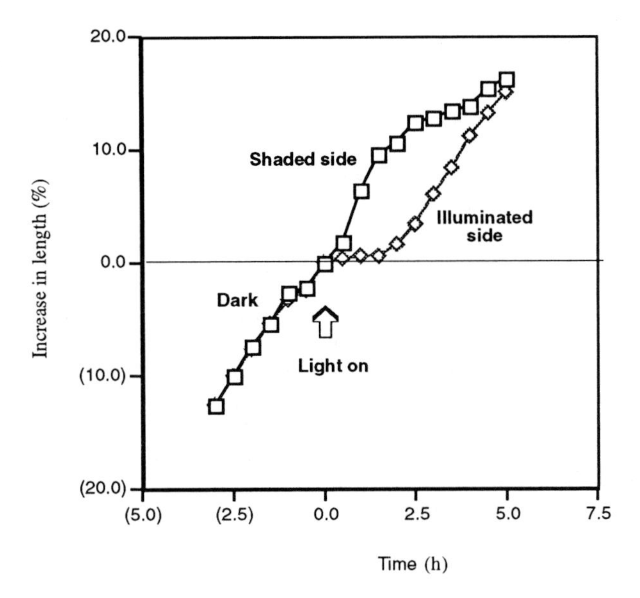

Figure 7. Growth-rate patterns causing phototropism.

phototropic response actually begins. At high fluence rates a latent period of about 4 min has been reported in coleoptiles, but the latent period at low fluence rates seems longer (up to 30 min). Whether these different values are the extremes of a continuum or two different responses is unknown. Interestingly in the case of gravitropism, the rather similar latent period is sufficiently long to allow stimulus-induced differential gene expression. Full studies of gene expression during the latent period of phototropism are awaited.

9.2.6.3 Patterns of differential growth

The growth-rate changes at the convex and concave sides of phototropically stimulated organs have been determined by several groups and some representative data is shown in Fig. 7. A number of conclusions can be drawn:

(i) A severe reduction or cessation of elongation at the irradiated side is a consistent feature of all forms of phototropism (Franssen *et al.* 1981; Hart *et al.* 1982; Rich *et al.* 1985).

(ii) The growth-rate changes at the shaded side are less consistent. The elongation may be unchanged, may increase or may decrease then increase.

(iii) At low fluence rates in coleoptiles, the growth-rate changes are evident first at the tip and the responses occur, later in more basal zones (Iino and

Briggs 1984). However, in coleoptiles and hypocotyls subjected to continuous unilateral stimulation the growth-rate changes apparently occur simultaneously throughout the growing zone (Franssen *et al.* 1982; Hart *et al.* 1982).

9.2.6.4 The autotropic phase

Some hours after the phototropic response has begun, the bending organ enters a new phase as the apical zone changes its pattern of differential growth. The irradiated side starts to elongate again, its newly restored elongation rate exceeding that of the shaded side, hence curvature in the apical zone declines (Fig. 6). This straightening reaction is subsequently measurable in the adjacent zones and the region of organ curvature becomes confined to the basal part of the organ. This autotropic reaction is common to phototropism and gravitropism. No satisfactory explanation for the initiation of autotropism exists, yet the growth-rate changes which occur are as dramatic as those causing organ curvature. Autotropism occurs after phototropism, even if the plant is on a clinostat hence autotropism is not simply negative gravitropism as once argued by some workers.

9.2.7 Fluence-response curves for phototropism

Many studies have been made of this relationship during the last 60 years in the belief that such information will aid our understanding of the mechanisms involved. Even the simplest studies yield a complex relationship (Dennison 1979) and the fullest studies have revealed a daunting degree of complexity (Blaauw and Blaauw-Jansen 1970). The precise relationship found depends on the type of plant used, the history of the plant before stimulation and whether the fluence rate is varied or the irradiation time is changed.

When studying coleoptiles, it is often found that the *fluence-response curves* show two distinct phases*. The 'first-positive' response is induced by short duration, low fluences and the 'second-positive' response results from longer duration, higher fluence rates. The first-positive response obeys the *Bunsen-Roscoe Reciprocity Law* (*i.e.* the response depends only on the number of photons causing the response and is independent of the fluence rate used) while the second positive does not. A similar two-phase fluence-response curve has been measured in dicot seedlings.

---

*More complicated fluence-response curves have been reported with *third positive* and *first negative* phases being identified. However, such phases have only been reported in certain types of plants grown under special conditions and seem not to be an important general feature of phototropism.

Before considering the meaning of such fluence-response curves, some limitations of the approach must be noted, limitations that have their basis in factors already considered:

(i) Fluence-response relationships are only likely to be simple if all the responding cells have identical sensitivities and show a synchronous response. In coleoptiles and hypocotyls these requirements are unlikely to be met. It has been shown, for instance, that cells at the apex are more sensitive than those in the basal region of the coleoptile. Consequently as the fluence rate to an organ is increased one would expect more cells, both longitudinally and radially, to reach the threshold for a response.

(ii) Angle of curvature measurements are ambiguous and it is possible that measurements of growth-rate changes would yield simpler fluence-response curves. Some evidence in support of this possibility has been published (Iino and Briggs 1984). They provide fluence-response curves using angles of curvature measurements and growth-rate changes at the irradiated and shaded sides of coleoptiles subjected to first-positive stimulation. Comparing these curves, it is apparent that they are not identical.

One must ponder exactly what a fluence-response relationship really means in the case of phototropism and what one can hope to learn from it. If one seeks a fluence-response relationship that is valid at the cell level, with the aim that it will tell one about the mechanisms involving light perception at a molecular and cell level, then measuring the response of the organ may be imperfect. However, despite this considerable reservation about the actual measurements being made, recent models to explain the fluence-organ responses have been advanced by Iino (1987, 1990) for coleoptiles and by Steinitz and Poff (1986) and Janoudi and Poff (1990) for dicots. It has been shown that for second-positive curvature two thresholds operate, both of which must be exceeded in order for a response to occur. There is a minimum number of quanta (the same minimum as for the first-positive response) and a minimum duration of exposure. Furthermore, *adaptation* has been shown to occur. Blue light absorption (whether as part of the phototropic response or independently) desensitizes the plant to subsequent B exposure and after a pulse of B the plant only slowly (10-20 min.) recovers its sensitivity. In *Arabidopsis* the desensitization is an effect on the amplitude of the subsequent response (Janoudi and Poff 1991) and it is not a shift in the sensitivity threshold but it is reported by Iino (1988) that in corn the adaptation is partly a shift in threshold sensitivity. The minimum duration of exposure and the long periods of the B-induced desensitization explain why prolonged durations of exposure are needed to maximize the second-positive response.

9.2.8 Models of phototropism

9.2.8.1 The Cholodny-Went models

As explained in Section 9.2.3 the quest for a hormonal explanation of phototropism was a landmark in plant physiology. However, one of the problems that has beset the evaluation of involvement of hormones in phototropism has been that several very different versions of the Cholodny-Went model have co-existed. Giving these very different models the same name has confused attempts to refute the original model and, contrary to the usual progression of science, the C-W model has become less precise with time (Firn 1992). It is instructive to briefly consider the various versions of the model in order to illustrate the very different concepts that the model has been thought to encompass over the years.

9.2.8.1.1 Auxin as a longitudinal messenger: model 1. The original C-W model was based on the view that the tip played a special role in controlling organ extension and that the sites of phototropic perception and phototropic response were separate. It has now been shown beyond doubt that the tip is *not* the site of phototropic perception in dicots and in coleoptiles responding to a second positive stimulation (Fig. 8). Thus the basic principle on which the original C-W model was based (the role of the tip) is now thought to apply only to phototropic curvature of coleoptiles towards very low fluence rates of short duration. Unfortunately, this grave deficiency of the original model was not always recognized by subsequent workers, many of whom carried out work assuming that the tip was the site of events of paramount importance. Sadly such approaches are at best illogical and at worst misleading if the tip has no special role in the type of phototropism being studied.

Recently it has been shown (Hasegawa *et al.* 1989) that the 'classic' Went phototropism experiments which provided the main support for the C-W model were misleading due to the fact that the bio-assay used to measure 'auxin' was apparently non-specific. When indole-3-acetic acid (IAA) was measured by physicochemical means rather than by bioassay, no evidence was found for an unequal flow of IAA from the illuminated and shaded sides of unilaterally illuminated *Avena* coleoptile tips.

9.2.8.1.2 Auxin as a lateral co-ordinator: model 2. Given that it was already known at the time of the inception of the original C-W model that a phototropic response could arise without the tip acting as the sole light sensing region, one might wonder why the model was so highly regarded. It seems as if the unifying nature of the model, which attempted to explain phototropism and both root and

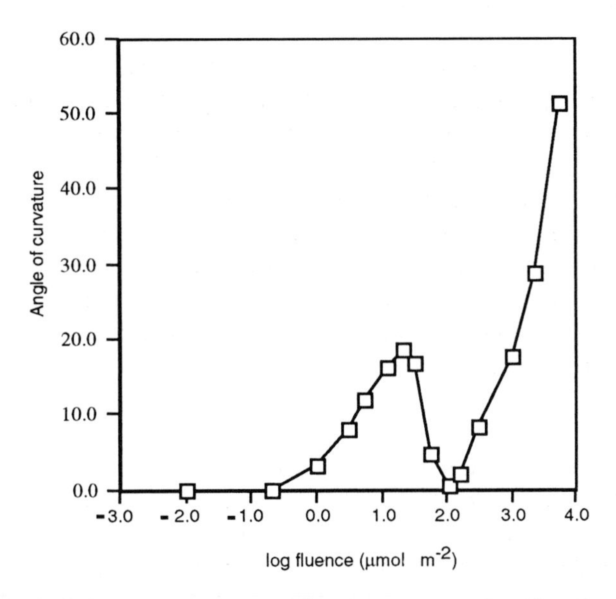

Figure 8. A typical fluence-response curve for *Zea mays* coleoptile phototropic curvature subjected to unilateral blue light (B). The duration of exposure to 0.4 μmol m^2 s^{-1} B was varied to alter the fluence. Macleod K., Firn R.D. and Digby J. (unpublished data).

shoot gravitropism, was just too tempting. Rather than reject the model simply because some organs could show a good phototropic response even when the tip was not illuminated, the first of many accommodations were made without realizing that the model was not simply being modified slightly but was being fundamentally changed. It seemed a minor change to suggest that a phototropic stimulus 'polarized' cells in the growing zone, as well as in the tip. However, the modified model now ascribes a quite different role to auxin. Auxin is no longer carrying essential information from cells involved in perception to cells in the elongation zone, instead auxin is being moved in a way dependent on the light gradient in the organ. The cells moving auxin laterally must have perceived and responded to the light gradient and such cells are not gaining information from auxin. The only role for auxin would be as a co-ordinator of growth across the organ. However, in the case of gravitropism there is evidence against the need for such co-ordination because longitudinally bisected pieces of hypocotyl or coleoptile can show independent gravitropic responses, consequently the need for this in phototropism remains to be demonstrated. In other words, this modified C-W model has not been derived to explain a known phenomena, indeed it seeks a phenomena to explain. This version of the model is challenged experimentally by the failure of modern analytical techniques to detect a transorgan gradient of IAA in phototropically stimulated coleoptiles (Feyerabend and Weiler 1988), although one recent study reported extremely small gradients (Iino 1991).

9.2.8.1.3 Differential rates of longitudinal auxin movement giving rise to auxin gradients: model 3. This version of the model accepted that the tip has no role, but retained the view that the supply of auxin to the elongation zone is the controlling factor in organ extension. The model was simply another possible way of altering the flow of auxin to the growing zone, but quite how the phototropic stimulation regulated the basipetal auxin transport system was never fully explained. If for instance it was argued that the rate of auxin transport was simply dependent on the B fluence at any position and that light gradients simply set up auxin flow gradients, it would be predicted that B given equilaterally should modulate organ extension; a prediction that is false (see later). This model also failed to gain much experimental support when rates of longitudinal auxin movement were measured.

9.2.8.1.4 Differential rates of auxin destruction giving rise to auxin gradients: model 4. The previous three models although very different, at least shared one idea: light somehow co-ordinated organ growth by means of auxin supply. However, model 4 completely abandons the concept of auxin movement being a part of the phototropic response. Instead B-induced IAA degradation was postulated to regulate the [IAA], hence control the rate of organ elongation. This model was not derived to explain some aspect of phototropism physiology which the previous models could not explain, rather it just seemed like a nice alternative. However, the [IAA] content of phototropically stimulated organs does not fall as the model predicts hence this model can be rejected.

9.2.8.1.5 Auxin as a local messenger: model 5. If, as some researchers believe, the peripheral cell layers play a crucial role in controlling organ extension, the failure to find transorgan movement of auxin in some phototropically stimulated organs (model 1 and 2) need not disprove an involvement of auxin in the phototropic response. It could be argued that IAA simply moves locally from the cortex into the peripheral cell layers and the regulator is a local messenger. In a sense this model is simply a refinement of model 2, operating on a more local scale. However, very careful measurements of the [IAA] in the peripheral cell layers and in the cortex of phototropically stimulated organs failed to find any support for this model (Feyerabend and Weiler 1988).

9.2.8.1.6 Auxin sensitivity arguments: model 6. Within years of the discovery of auxin it was noted that the correlation between [IAA] and rate of cell elongation was poor in some organs and it was suggested that cells might vary in their sensitivity to auxin. If one is convinced that auxin is involved in phototropism but all measurements fail to find the changes in [IAA] predicted by the above models, the temptation to evoke 'sensitivity' arguments is obviously overwhelming. However, any model based on a change in sensitivity of cells to a constant supply of IAA is a model very different indeed to models 1-5. What such sensitivity models usually ignore is that there would not really be any obvious reason to choose IAA from any other cell regulator in such a model: indeed auxin is reduced in its role from extracellular messenger to a intracellular regulator. A mechanism would also be needed to explain how the cells could develop the differential sensitivity needed to give rise to differential growth. So far no such model has been advanced to explain phototropism in these terms.

9.2.8.1.7 Multi-regulator control: model 7. Following the track leading from model 1, through the various models 2-6 but not necessary in any particular order, one eventually arrives at the final, debased concluding version. It is admitted that there is always some evidence inconsistent with every explanation based solely on auxin but rather than abandon the idea of any auxin involvement, it is suggested that auxin is part of a consortium of regulators: including H^+, Ca^{2+}, auxin, gibberellins. Whenever, there is no evidence against auxin being the controller, auxin is regarded as the key regulator. However, in other cases, other members of the consortium substitute for auxin. The problems with such a model are many and various. Firstly, being so vague and many faceted, it fails the basic requirement of any scientific model, in that it is not experimentally falsifiable: you might disprove the involvement of any one compound by a particular experiment but the model allows such a result hence can never be disproved. The second major problem is that it is somewhat unlikely on evolutionary grounds that such a vague model could arise. To have two regulators linked into the same perception-response chain seems unlikely but to have more seems implausible in evolutionary terms.

9.2.8.2 The Blaauw model of phototropism

The essence of the Blaauw model is that the cells that respond to light also perceive the light. The model equates the B inhibition of cell elongation with the B induction of phototropism. Unilateral light treatment would create a light gradient across an organ and a gradient of photo-inhibition would be created. Blaauw measured the light gradient, underestimating it slightly (Chapter 7.4) and obtained a fluence-response curve for the effect of light on straight growth. Combining this information he concluded that his simple model could account for phototropism of sunflower hypocotyls. Measurements of the growth-rate changes causing phototropism towards unilateral light provided evidence which was reasonably consistent with the model in that the growth rate at the irradiated side slowed down. While the C-W model 1 seemed strongest when applied to the phototropic responses activated by the tip, the Blaauw model seemed most suited to phototropic responses in which the areas of photoperception(s) and photoresponse(s) were identical, as is the case in hypocotyl phototropism. Although the C-W Model 2 was advanced to explain types of phototropism which Blaauw also sought to explain, the Blaauw and C-W models were sometimes thought of as being complementary (Went and Thimann 1937, p 174).

The mechanism by which light directly controls the growth of cells could of course involve endogenous cell regulators and specific models involving such regulators have been proposed. Indeed, certain variants of the C-W model (*e.g.* 5, 6 and 7 above) could actually be regarded as being derived more from the central idea of the Blaauw model than from the original C-W model 1. Because B usually inhibits cell elongation, the possibility that B-induction of growth inhibitors underlie the Blaauw model has received recent attention. The most developed version of this concept comes from the work of Hasegawa and co-workers (K. Hasegawa *et al.* 1987) who have provided evidence for an asymmetrical distribution of growth inhibitors (raphanusanins) in photostimulated radish.

However, the Blaauw model suffers from some inconsistencies. For instance Macleod *et al.* (1985) showed that coleoptiles subjected to simultaneous unequal bilateral illumination did not show the growth-rate changes predicted (differential inhibition of elongation was predicted but not found, although the mechanical independence of the two sides of the coleoptile was possibly overestimated). Cosgrove (1985) showed that in cucumber hypocotyls, the B-induced inhibition of elongation was much faster than any detectable B-induced phototropic curvature. The fact that two types of B-insensitive *Arabidopsis* mutants have been reported, phototropism defective mutants (Khurana and Poff 1989) and long hypocotyl mutants whose elongation is insensitive to B (Liscum and Hangarter 1991), neither of which seem to be defective in both phototropism and B-induced inhibition of elongation, supports the view that the simple version of the Blaauw model cannot be correct (Chapter 8.2). However, if the B-induced inhibition of elongation is present in an organ one would expect it to contribute to phototropism and if it does not do so then a further mystery awaits an explanation.

9.2.9 Phototropism and gravitropism

Phototropism is often discussed in articles alongside *gravitropism* (the directional growth response in relation to the gravity vector) and it is important to appreciate the interrelationship between the two. The relationship between the two tropisms is often set within the framework of the C-W model of tropisms because that model was based on a unifying concept. However, by viewing the relationship from that prospective only, many interesting differences are usually overlooked (Firn and Myers 1989). Some of the interesting differences and relationships are:

(i) The non-essential nature of seedling phototropism is in marked contrast to very obvious importance of gravitropism, which is seen in most plants and at most stages of their development.

(ii) The phototropic response is initiated by the application of *unequal* stimuli at various positions around the growing zone. The gravitropic response results in the detection of an *equal* signal at different locations across the organ: in the case of gravitropism either the direction of the stimulus is detected or the uniform signal is transduced into another signal with a magnitude dependent on the cell's orientation relative to the *g* vector.

(iii) As far as is known, all phototropically responsive organs are gravitropically sensitive but the converse is not true. It seems likely that gravitropism evolved before phototropism.

(iv) As an organ begins to develop a phototropic curvature it will also begin to experience gravitropic stimulation, which if expressed as a gravitropic response would counteract the phototropic curvature. Hence a mechanism must exist to enable the phototropic response to either over-ride the gravitropic response or to dominate it. This is an area that requires exploration because the final curvature developed after a period of sustained phototropic stimulation is greater if the organ is on a clinostat (hence is responding to a uniform gravity vector). This suggests that *all* studies of phototropic curvature which are not conducted on a clinostat are actually studies of an interaction between phototropism and gravitropism. This fact has considerable implications for any studies that attempt to demonstrate factors (other light treatments, chemical treatments or genetic changes) which influence the magnitude of the phototropic response. Obviously any factor which influences gravitropism might indirectly influence phototropism *via* this interaction.

9.2.10 Conclusions

Two themes run through this chapter:

(i) Phototropism should be thought of in terms of the responses of cells. We must recognise that studies of the phototropic response of an organ, although more easily made , should only be undertaken if the study can clearly be related to a cellular model.

(ii) Too much of our knowledge comes from studies where the response studied has been measured in an ambiguous manner.

This chapter started with a picture of a seedling prettily bending towards the light. The response looked so simple and so easy to measure. By now the reader should realise that the response is not simple and that unambiguous measurements are far from easy. One hundred years ago it was difficult to study cells and that was a good reason to study organs. Phototropism seems to have become bogged down with concepts and beliefs that have their roots in that era. Now so much more is known about cells that it should be possible to think about phototropism at the cell level. It is time to move on.

9.2.11 Further reading

Curry G.M. (1969) Phototropism. In: *Physiology of Plant Growth and Development*, pp. 245-273,Wilkins M.B. (ed.) McGraw-Hill, London.

Dennison D.S. (1979) Phototropism. In: *Encyclopedia of Plant Physiology*, New Series, 7, *Physiology of Movements*, pp. 506-566, Haupt W. and Feinleib M. E. (eds.) Springer-Verlag, Berlin.

Firn R.D. and Digby J. (1980) The establishment of tropic curvatures in plants. *Annu. Rev. Plant Physiol.* 31: 131-148.

Hart J.W. (1990) *Plant Tropism,* pp. 1-208, Unwin Hyman, London.

Iino M. (1990) Phototropism: mechanisms and ecological implications. *Plant Cell Environ.* 13: 633-650.

Jost L. (1907) *Lectures in Plant Physiology,* Transl. R.J.H. Gibson, Clarendon Press, Oxford.

Pohl U. and Russo V.E.A. (1984) Phototropism. In *Membranes and Sensory Transduction*, pp. 231-329, Colombetti G. and Lenci F. (eds.) Plenum Press, New York..

Went F.W. and Thimann K.V. (1937) *Phytohormones,* Macmillan, New York.

9.2.12 References

Blaauw O.H. and Blaauw-Jansen G. (1970) The phototropic responses of *Avena* coleoptiles. *Acta Bot. Neerl.* 19: 755-763.

Cosgrove D.J. (1985) Kinetic separation of phototropism from blue-light inhibition of stem elongation. *Photochem. Photobiol.* 42: 745-751.

Feyerabend M. and Weiler E.W. (1988) Immunological estimation of growth regulator distribution in phototropically reacting sunflower seedlings. *Physiol. Plant.* 74: 185-193.

Firn R.D. and Myers A.B. (1989) Plant movements caused by differential growth - unity or diversity of mechanisms? *Environ. Exp. Bot.* 29: 47-55.

Firn R.D. (1991) Phototropism - the need for a sense of direction. *Photochem. Photobiol.* 52: 255-260.

Firn R.D. (1992) What remains of the Cholodny-Went models? Which models? *Plant Cell Environ.* 15: 769-770

Franssen J.M., Cook S.A., Digby J. and Firn R.D. (1981) Measurements of differential growth causing phototropic curvature of coleoptiles and hypocotyls. *Z. Pflanzenphysiol.* 103: 207-216.

Franssen J.M., Firn R.D. and Digby J. (1982) The role of the tip in the phototropic curvature of *Avena* coleoptiles: positive curvature under conditions of continuous illumination. *Planta* 155: 281-286.

Gehring A.A., Williams D.A., Cody S.H. and Parish R.W. (1990) Phototropism and geotropism in maize coleoptiles are spatially correlated with increases in cytoplasmic free calcium. *Nature* 345: 528-530.

Hart J., Gordon D.C. and Macdonald I.R. (1982) Analysis of growth during phototropic curvature of cress hypocotyls. *Plant Cell Environ.* 5: 361-366.

Hasegawa K., Sakoda M. and Bruinsma J. (1989) Revision of the theory of phototropism in plants: a new interpretation of a classical experiment. *Planta* 178: 540-544.

Iino M. (1987) Kinetic modelling of phototropism maize coleoptiles *Planta* 171: 110-126.

Iino M. (1988) Desensitization by red and blue light of phototropism in maize coleoptiles. *Planta* 176: 183-188.

Iino M. and Briggs W.R. (1984) Growth distribution during first positive phototropic curvature of maize coleoptiles. *Plant Cell Environ.* 7: 97-104.

Janoudi A.K. and Poff K.L. (1990) A common fluence threshold for first positive and second positive phototropism in *Arabidopsis thaliana*. *Plant Physiol.* 94: 1605-1608.

Janoudi A.K. and Poff K.L. (1991) Characterisation of adaptation in phototropism of *Arabidopsis thaliana*. *Plant Physiol.* 95: 517-521.

Khurana J.P. and Poff K.L. (1989) Mutants of *Arabidopsis thaliana* with altered phototropism. *Planta* 178: 400-406.

Konjevic R., Steinitz B. and Poff K.L. (1989) Dependence of the phototropic response of *Arabidopsis thaliana* on fluence rate and wavelength. *Proc. Natl. Acad. Sci. USA* 86: 9876-9880.

Li Y., Hagen G. and Guilfoyle (1991) An auxin-responsive promoter is differentially induced by auxin gradients during tropisms. *Plant Cell* 3: 1167-1175.

Liscum E. and Hangarter R.P. (1991) *Arabidopsis* mutants lacking blue light-dependent inhibition of hypocotyl elongation. *Plant Cell* 3: 685-694.

Macleod K., Brewer F., Firn R.D. and Digby J. (1984) The phototropic response of *Avena* coleoptiles following localised continuous unilateral illumination. *J. Exp. Bot.* 35: 1380-1389.

Macleod K., Digby J. and Firn R.D. (1988) Evidence inconsistent with the Blaauw model of phototropism. *J. Exp. Bot.* 36: 312-319.

Rich T.C.G., Whitelam G.C. and Smith H. (1985) Phototropism and axis extension in light-grown mustard (*Sinapis alba* L.) seedlings. *Photochem. Photobiol.* 42: 789-792.

Steinitz B. and Poff K.L. (1985) A single positive phototropic response induced with pulse light in hypocotyls of *Arabidopsis thaliana* seedlings. *Planta* 168: 305-315.

9.3 The photobiology of stomatal movements

Eduardo Zeiger

Department of Biology, University of California at Los Angeles, Los Angeles, California 90024-1606, USA

9.3.1 Introduction

Stomata, from the Greek for mouth, are pores in the epidermis of aerial organs of plants. In the broadest sense, stomata are to leaves what membranes are to cells; that is, they help maintain a closed organization of the organ while allowing a selective interchange between the plant and its environment. Stomata are a crucial adaptation to terrestrial habitats. Plants living in dry land have developed an impermeable cuticle that helps prevent excessive water loss. This cuticle, however, is also impermeable to carbon dioxide (CO_2), which is required for *photosynthesis.* In the absence of differential permeability to water and CO_2, stomata provide temporal control over the diffusion of these essential gases. The stomata open at times of high photosynthetic activity and abundant water and close when water is limiting or photosynthesis is not occurring, such as at night.

Typical leaves have a highly impermeable epidermis with stomata interspersed in it. Water reaches the leaf *via* the vascular tissue and saturates the intercellular air spaces. The rate of water evaporation from the leaf depends on its thermal load, the humidity gradient across the leaf, and the actual dimensions of the stomatal pores. Carbon dioxide shares its diffusion path with water, entering the leaf through the stomatal pore and diffusing to the site of photosynthesis in the *mesophyll.* The dimensions of the stomatal pores are controlled by the *guard cells,* a pair of specialized structures that surround the pore.

Functionally, guard cells are best described as multisensory turgor valves. Stomata respond to many internal and external stimuli, including light, relative humidity, intercellular CO_2 concentrations, temperature, phytohormones, and air pollutants. Specific responses to these stimuli can be demonstrated in both the intact leaf and isolated stomata, indicating that guard cells have the capacity to perceive all these stimuli and transduce them into tightly modulated stomatal apertures. This capacity of guard cells to integrate different stimuli into optimal stomatal apertures leads to their description as multisensory devices.

R.E. Kendrick & G.H.M. Kronenberg (eds.), Photomorphogenesis in Plants - 2nd Edition

The dimension of the stomatal pore is controlled by the degree of *turgor* in the guard cells. Upon sensing an appropriate stimulus, the guard cells increase their uptake of ions, particularly K^+. Electroneutrality is maintained by simultaneous uptake of Cl^- and endogenous synthesis of malate^{2-} from carbon skeletons originating from starch breakdown. In addition, guard cells accumulate osmotically active sugars, arising from photosynthetic carbon fixation and starch hydrolysis. The ensuing increase in the *osmotic potential* of the guard cells results in water uptake and increased turgor and, because of the mechanical properties of the guard cell walls, a widening or opening of the stomatal pore. The reverse process results in stomatal closing. It is thus apparent that the modulation of stomatal movements in response to prevailing stimuli is the result of metabolic processes in the guard cells transducing perceived signals into osmoregulation and turgor changes.

9.3.2 Light as an environmental signal for stomatal movements

A time-course of changes in *net photosynthesis, stomatal conductance, intercellular CO_2 concentrations* and *photon irradiance* in a leaf of *Gerea* growing in Death Valley, California, during an early spring morning is shown in Fig. 1. Both stomatal conductance and net photosynthesis increase sharply with irradiance, illustrating the major role of light in the stimulation of these two processes. In fact, several lines of evidence indicate that, in natural environments, under non-limiting water supply, the stomatal response to light is a central mechanism regulating stomatal movements.

The data in Fig. 1 also illustrate the marked coupling between net photosynthesis and stomatal conductance in the leaf. This coupling is evident from the stabilization of intercellular CO_2 concentrations after sunrise. Under constant ambient CO_2 concentrations, the intercellular concentration of CO_2 depends on its rate of diffusion into the leaf, which is a function of stomatal aperture, and the rate of CO_2 fixation in the mesophyll. The maintenance of stable intercellular CO_2 concentrations while both net photosynthesis and stomatal conductance increase can only result from synchronous changes in both CO_2 diffusion and fixation. In fact, a tight coupling between rates of mesophyll photosynthesis and stomatal conductance has been demonstrated under several experimental conditions and constitutes a basic physiological property of leaves.

Beyond its intrinsic interest for leaf physiology, this coupling poses an important question for stomatal biologists. If a change in photosynthetic rate in response to an environmental signal such as light is always accompanied by a change in stomatal conductance, can it be shown that the observed stomatal response is a specific reaction to the environmental stimulus rather than an obligatory, direct tracking of photosynthesis? The co-ordination of

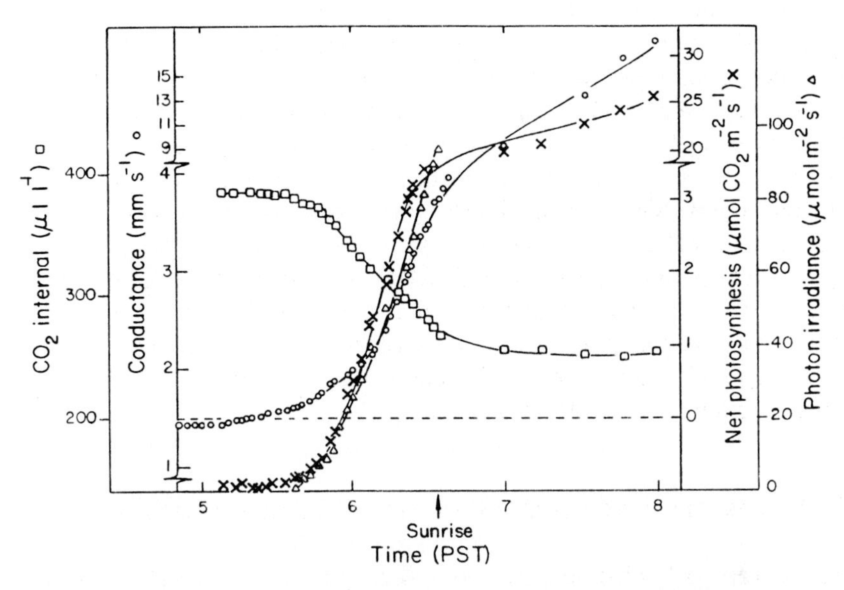

Figure 1. Time-courses of stomatal conductance (○), net photosynthesis (×), calculated intercellular CO_2 concentrations (□) and incident photon irradiances (Δ) in an attached leaf of *Gerea canescens* growing in Death Valley, California in March, 1980. PST: Pacific Standard Time. The leaf was enclosed in a gas exchange cuvette the night before, with the glass window of the chamber facing southeast. Measurements were made with a portable gas exchange system; photon irradiances were measured with a Li-Cor quantum sensor. After Zeiger E., Field C. and Mooney, H.A. (1981) In: *Plants and the Daylight Spectrum,* pp. 391-407, Smith H. (ed.) Academic press, New York, with permission.

photosynthetic rates with appropriate stomatal conductances by a hypothetical 'mesophyll messenger' has been suggested, but several lines of evidence, indicate that this mechanism is unlikely to play a role in stomatal responses.

9.3.3 Direct response of stomata to light

One piece of evidence that stomata respond directly to environmental stimuli rather than to a signal from the mesophyll is the finding that stomata, both in the intact leaf and in isolation, can respond directly to light. Evidence for a specific light response of isolated stomata arises from work with isolated guard-cell *protoplasts.* When guard cells are treated with cellulolytic enzymes such as cellulase, which digest the cell wall, the protoplast is released (Fig. 2). Several research groups have now confirmed initial work by P.K. Hepler and myself (Zeiger and Hepler 1977) showing that these guard-cell protoplasts behave as osmometers, swelling under conditions leading to stomatal opening and shrinking when exposed to closing signals such as the water stress hormone,

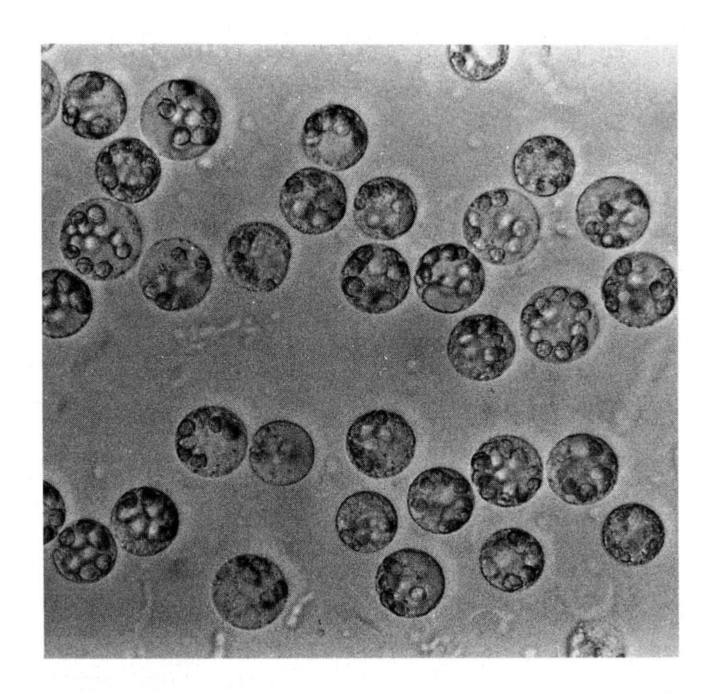

Figure 2. Guard-cell protoplasts isolated from *Vicia faba*. Protoplasts are about 16 μm in diameter. After Shimazaki and Zeiger (1985), with permission.

abscisic acid. Guard-cell protoplasts respond to blue light (B) by swelling. This swelling is specific to guard cells, neighbouring epidermal cells do not swell in response to B, and requires the presence of K^+ in the incubation medium. The capacity of guard-cell protoplasts to respond to B provides evidence for a specific light response of stomata, independent of any other component of the leaf.

A demonstration of a direct light response of stomata in the intact leaf is more difficult, because it is necessary to discriminate the specific response of stomata from that of the mesophyll. Sharkey and Raschke (1981) inhibited the photosynthetic response of the mesophyll with the herbicide cyanazine and, using gas exchange techniques, showed that stomata responded to light in the absence of CO_2 fixation by the mesophyll. Zeiger and Field (1982) used a different approach and irradiated leaves with monochromatic red light (R) from a laser, of sufficient fluence rate to saturate the photosynthetic response. After rates of both photosynthesis and conductance were at steady-state, a low irradiance B laser beam was used to stimulate the leaf. A clear-cut increase in stomatal conductance was observed without any immediate photosynthetic response, indicating that the stomata were responding directly to B. We can therefore conclude that, both in the intact leaf and in isolation, stomata respond directly to light, and this response is transduced into changes in stomatal aperture. The demonstration of intrinsic stomatal responses to light further indicates that the observed coupling between mesophyll photosynthesis and stomatal responses in response to changing light conditions is the result of coordinated responses of the stomata and the mesophyll, rather than the expression of a 'mesophyll messenger'.

9.3.4 The photobiological components of the light response of stomata

A characterization of the photobiological properties of the stomatal response to light is an important prerequisite for the understanding of how photoreception in guard cells is transduced into changes in stomatal aperture. Unlike most, but not all, epidermal cells, guard cells almost always have chloroplasts (the achlorophyllous stomata of *Paphiopedilum* are an exception, see below). This nearly universal presence of chloroplasts in guard cells argues for a functional role of these organelles in stomatal movements and leads to the prediction that, with chlorophyll as the primary photoreceptor, action spectra for stomatal opening should closely match the absorption spectrum of chlorophyll, with strong peaks in the B and R. However, we have already discussed results showing that under certain conditions, guard-cells responded to B but not R, a finding inconsistent with chlorophyll being the absorbing pigment. In fact, light-dependent stomatal opening can be shown to have two distinct components, one reflecting the activity of the guard-cell chloroplasts and the other a specific B response.

These features of the light response of stomata become apparent when stomata in epidermal peels of *Commelina communis* are irradiated with broadband B and R. Figure 3 shows the fluence rate-dependence of stomatal opening under the two light regimes, with the responses to B and R contrasting in two important ways. Firstly, there is a strong response to B at low fluence rates, whereas the response to R at the same fluence rate is indistinguishable from the opening observed in the dark in the absence of CO_2. Secondly, the magnitude of the B response is much greater than that of the R response. Both features can be explained by postulating that in R, only the photosynthetic response of guard-cell chloroplasts is activated, this response having a high fluence rate activation threshold. In B, both the photosynthetic response and the B photosystem of guard cells are stimulated, with the photosynthetic component presumably being identical to the R response and the additional opening being the expression of the B photosystem. The dotted line in Fig. 3 shows the difference between the B and R responses, giving an estimate of the response of the B photosystem. This partitioning of the light response of stomata into a B and a photosynthetic component is supported by the experimental data shown in Fig. 4. The stomata were irradiated with saturating fluence rates of R and, after the completion of the response, exposed to a simultaneous beam of low fluence

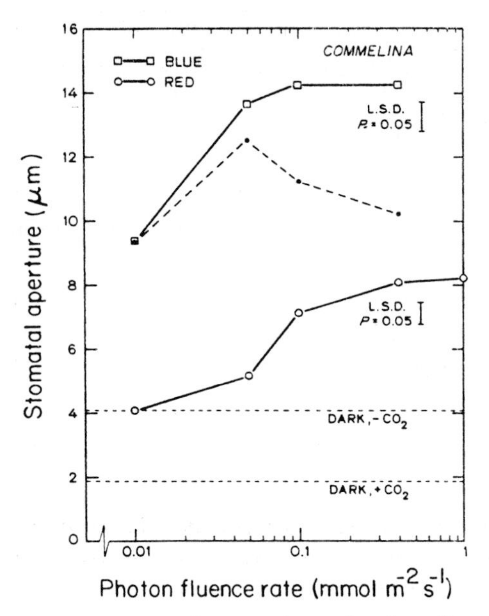

Figure 3. Fluence-rate response curves of stomatal opening in epidermal peels of *Commelina communis* irradiated with broad-band blue (□) or red (○) light. Peels were incubated in 30 m*M* KCl, 0.1 *M* $CaCl_2$ and 10 m*M* 2-[N-morpholino]ethane sulphonic acid (MES), pH 6.1 for 3 h at 26° C. The solution was bubbled with CO_2-free air. Subtraction between aperture values in blue and red light (●); broken straight lines, aperture values at the end of a 3 h incubation in darkness under normal or CO_2-free air; L.S.D., least significant difference. After Schwartz A. and Zeiger E. (1984) *Planta* 161: 129-136, with permission.

rate B. The sharp increase in stomatal opening in response to the added B provides evidence for the operation of two different photoreceptors in guard cells since, with photosynthesis saturated with R, no further stimulation of photosynthesis by B would be expected. Identical conclusions can be drawn when intact leaves are irradiated with monochromatic B and R from lasers as described above.

Further evidence from the dual-photoreceptor hypothesis comes from work with stomata from the orchid *Paphiopedilum.* This genus of the Orchidaceae is unique in the lack of chlorophyll in its guard cell plastids. Although a lack of guard cell chloroplasts has been reported in other instances, such as chimeras in *Pelargonium, Paphiopedilum* stands as the only reported case in which this unusual condition has been observed at a generic level. *Achlorophyllous* stomata have been reported in eight species of *Paphiopedilum,* while a survey of 26 other orchid species showed chloroplasts in all their guard cells (D'Amelio and Zeiger 1988).

In their original report, Nelson and Mayo (1975) measuring *transpiration* in intact leaves of *Paphiopedilum,* found responses to B and R, and concluded that

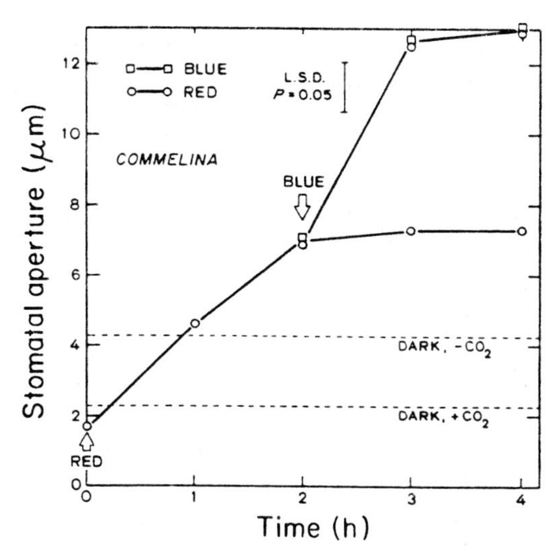

Figure 4. Stomatal response in epidermal peels of *Commelina communis* to the addition of 0.01 mmol m^{-2} s^{-1} broad-band blue light (arrow) to saturating (0.2 mmol m^{-2} s^{-1}) red light. Incubation conditions, broken lines and credit as in Fig. 3.

guard-cell chloroplasts did not play a role in the light response of stomata. However, transpiration measurements in the intact leaf are difficult to interpret, because of the complex interplay between the different stimuli affecting stomatal opening. Zeiger *et al.* (1983) used epidermal peels to measure stomatal opening in response to B and R and found that, in isolation, *Paphiopedilum* stomata opened in response to B but not R. These findings are consistent with the dual photoreceptor hypothesis; in the absence of guard-cell chloroplasts, stomata would be expected to lack a R response but be capable of responding to B because of the activity of the B photoreceptor. Gas exchange experiments with intact leaves of *Paphiopedilum* have confirmed that the limited opening observed in response to R was an indirect response to the reduced intercellular levels of CO_2 resulting from the photosynthetic activity of the mesophyll. In contrast, the gas exchange results under B were similar to those obtained with white light.

9.3.5 Action spectroscopy

The spectral properties of the stomatal response to light provide further information on the involved sensory transduction processes. Hsiao and his co-workers (1973) studied the wavelength-dependence of stomatal opening and K^+ uptake in *Vicia faba* and showed that, at low fluence rates (*ca.* 10 μmol m^{-2} s^{-1}) only a single peak was observed at 455 nm. At higher fluence rates, the B peak was unchanged but a second, smaller peak at 650 nm was also resolved. These spectral features are remarkably similar to that of the light-stimulated increases of stomatal conductance in the intact leaf of ***Xanthium strumarium,*** which shows a major peak around 450 nm and a small shoulder at 660 nm (Sharkey and Ogawa 1987). It is generally accepted that these spectral features reflect the absorption spectrum of chlorophyll, with the absorption spectrum of a hitherto unidentified B photoreceptor superimposed on the chlorophyll absorption in the blue waveband.

The spectral properties of the B response can be more clearly resolved under saturating fluence rates of R. Karlsson (1986) measured an action spectrum for enhancement of stomatal conductances of wheat seedlings by B, in a background of R irradiation (Fig. 5). The obtained spectrum, showing peaks at 450 and 470 nm and a shoulder at 420 nm, is remarkably similar to the absorption spectrum of guard-cell carotenoids.

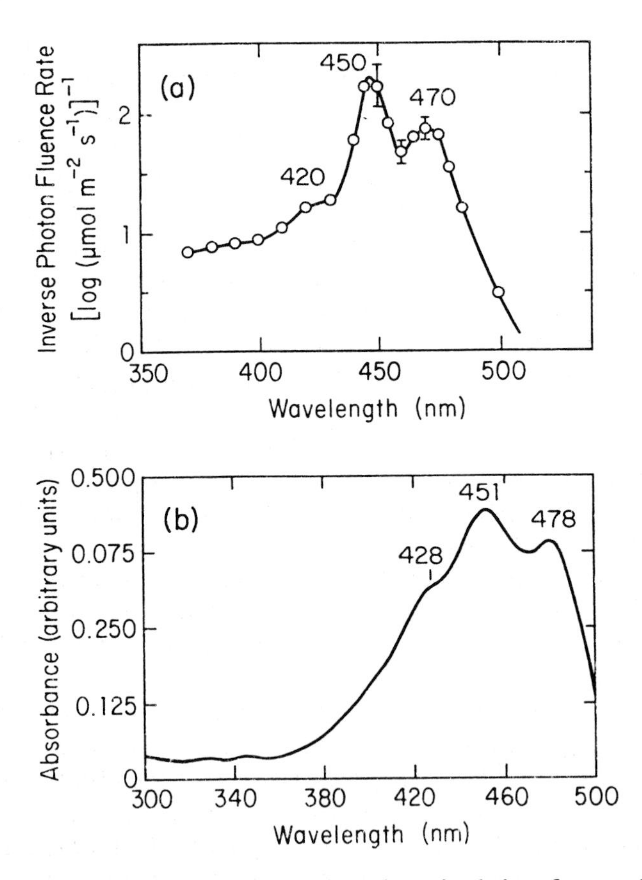

Figure 5. (a) Action spectrum for the blue light-dependent stimulation of stomatal opening under saturating fluence rates of red light in wheat seedlings (After Karlsson (1986) with permission. (b) Absorption spectrum from β-carotene isolated from guard-cell protoplasts of *Vicia faba*. (After Karlsson *et al.* (1992), with permission.

9.3.6 Properties of the guard-cell chloroplast

The involvement of both the guard-cell chloroplast and a B photosystem in the response of stomata to light makes it necessary to characterize the specific properties of each photosystem and their role in signal transduction during stomatal movements. It has long been established that chloroplasts are a highly conserved feature of guard cells, and some of the classical theories on stomatal function, such as the starch-sugar theory, have postulated a central role of the chloroplast in the modulation of guard-cell turgor. Guard-cell chloroplasts are smaller than their mesophyll counterparts and have about 1/80 of their chlorophyll content. In addition, guard-cell chloroplasts have distinctly large starch deposits. In mesophyll chloroplasts, starch is synthesized in the light and hydrolysed in the dark; in the guard-cell chloroplast, starch is hydrolysed in the light and synthesized in the dark. Light-dependent starch hydrolysis in the guard-cell chloroplast is regulated by B.

Guard-cell chloroplasts show typical photosynthetic properties, including photosystem I (PSI) and II (PSII) activity, cyclic and non-cyclic photophosphorylation, oxygen evolution, and photosynthetic carbon fixation (Fig. 6).

Several lines of evidence indicate that guard-cell chloroplasts play an important role in stomatal function. Stomatal opening in epidermal peels stimulated by R is inhibited by the photosynthetic inhibitor 3-(3, 4-dichlorophenyl)-1, 1-dimethylurea (DCMU), but it is insensitive to the respiratory poison KCN. A role of guard-cell chloroplasts in stomatal movements is also evident from the observation that the achlorophyllous stomata from the orchid *Paphiopedilum* fail to open in response to R. Some of the enzymes of the Calvin cycle, NADP-glyceraldehyde 3-phosphate dehydrogenase, 3-phosphoglycerate kinase and triose phosphate isomerase are very active in the guard-cell chloroplast, and these high activities favour the transport of triose phosphates from the chloroplast into the cytosol. This transport of triose phosphates results in an increase in the ATP supply in the cytosol, as evident from the observation that R causes a 2-3 fold increase in the ATP/ADP ratio of guard-cell protoplasts (Shimazaki *et al.* 1989). Proton pumping at the guard-cell plasma membrane requires ATP to fuel the H^+ ATPase, and photophosphorylation in the guard-cell chloroplast appears to be a primary energy source for light-stimulated stomatal opening.

A patch clamp study has provided direct evidence for a stimulation of proton pumping by the guard-cell chloroplast. Pump activity was measured as an outward electrical current at the guard-cell plasma membrane, upon R irradiation. The electrical currents were DCMU, vanadate and CCCP sensitive, indicating that photosynthetic activity at the guard cell chloroplast stimulated a H^+ ATPase at the plasma membrane. These patch-clamp experiments have also shown that, besides ATP, a second hitherto unidentified photosynthetic product is required for pump stimulation (Serrano *et al.* 1988).

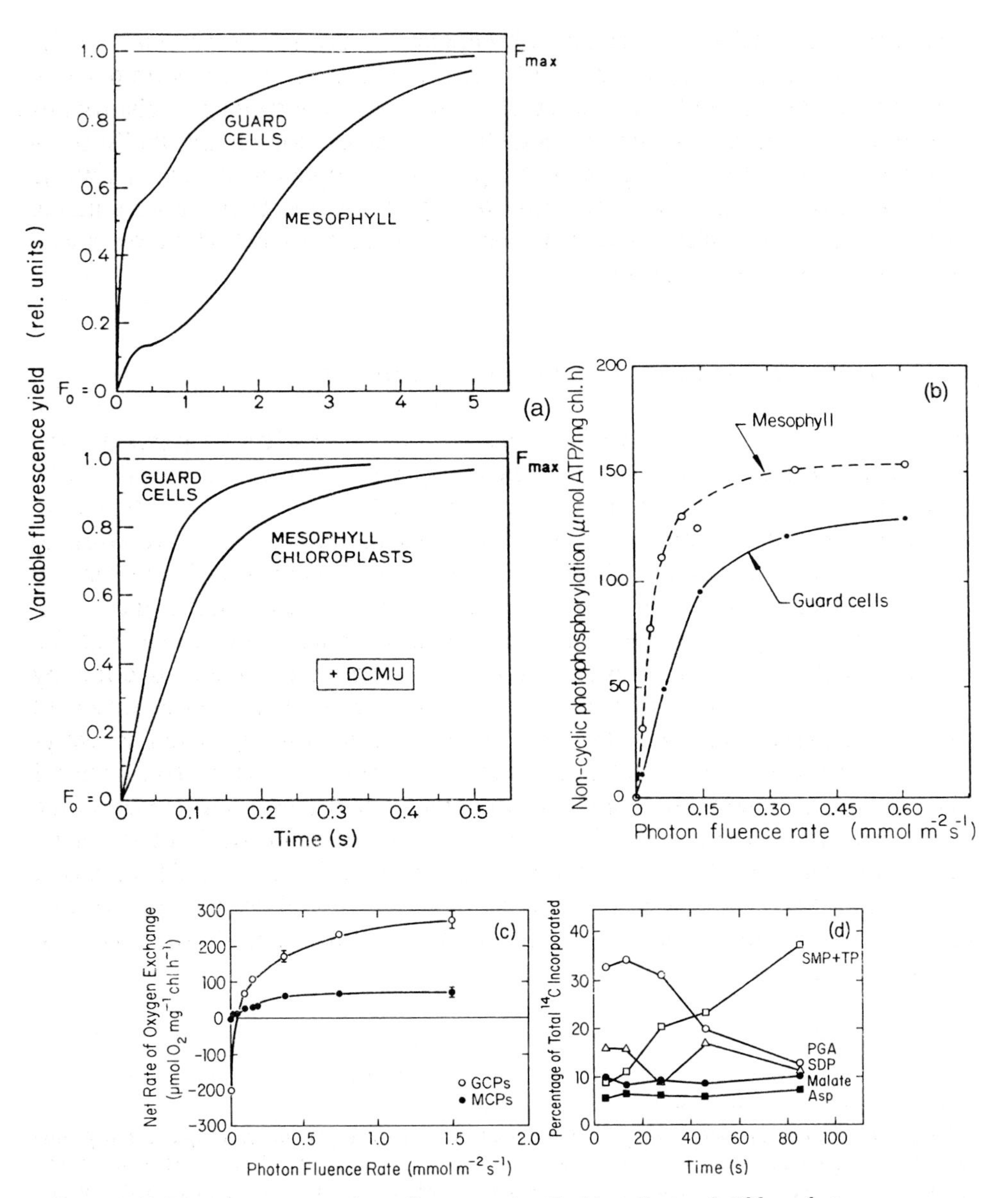

Figure 6. (a) Fluorescence transients from guard-cell chloroplasts of ***Chlorophytum comosum.*** The albino portions of the variegated leaves of *C. comosum* are completely devoid of mesophyll chloroplasts but have green guard-cell chloroplasts and are thus a good source of uncontaminated guard-cell chloroplasts. The traces show room temperature, chlorophyll *a* variable fluorescence of mesophyll and guard-cell chloroplasts. Both preparations exhibited kinetics typically associated with electron transport (*top*), including the faster transients observed in the presence of the photosynthetic inhibitor, 2-(3,4-dichlorophenyl)-1, 1-dimethylurea (DCMU) (*bottom*). The data were normalized to the values of the initial fluorescence yield (F_0 = 0) and the maximum fluorescence yield F_{max} (F_{max} = 1), Zeiger E., Armand P. and Melis A. (1981) *Plant Physiol.* 67:

Biochemical analysis of guard cells opened under R has shown that the concentration of glucose, fructose and sucrose increase in parallel with opening (Talbott and Zeiger 1993). These results indicate that the guard-cell chloroplast can modulate guard-cell osmotic potentials *via* three distinct metabolic pathways: K^+ uptake driven by proton pumping and coupled to starch hydrolysis and malate biosynthesis, accumulation of sugars ensuing from photosynthetic carbon fixation mediated by the Calvin cycle, and accumulation of sugars ensuing from starch hydrolysis.

9.3.7 Properties of the stomatal response to blue light

Irradiation of stomata with B stimulates both the photosynthetic pigments and the B-dependent photosystem. Thus, the responses to B represent the combined output of the two photoresponses of guard cells. As a result of this interaction, stomatal opening in response to B is consistently higher than that in response to R, on an equal quantum basis (Fig. 3). Specific responses to B are usually assayed under saturating fluence rates of R, to eliminate interference from photosynthetic pigments (Fig. 4). Increases in stomatal conductance (Fig. 7), enhancement of malate biosynthesis (Fig. 8) and medium acidification by guard-cell protoplasts (Fig. 9) are typical examples of B responses observed under a dual-beam protocol. Other stomatal responses to B, including early-morning increases in stomatal conductance in intact leaves (Fig. 10), stomatal opening in epidermal peels (Fig. 3), and starch hydrolysis, can be observed under low fluence rates of B given in the absence of R. Photosynthetic responses of guard cells have a relatively high stimulation threshold (Shimazaki and Zeiger 1985) and low fluence rates of R fail to stimulate stomatal opening (Fig. 3). Starch hydrolysis, on the other hand, is insensitive to R, irrespective of fluence rates.

17-20, with permission. (b) Non-cyclic photophosphorylation in isolated mesophyll and guard-cell chloroplasts from *Vicia faba,* as a function of photon fluence rate of orange (Cinemoid 5A) light. After Shimazaki and Zeiger (1985), with permission. (c) Rates of photosynthetic oxygen evolution as a function of fluence rates of red light from mesophyll (MCPs) and guard cell protoplasts (GCPs) from *Vicia faba.* From Mawson B.T. (1993) *Plant Cell Environ.* in press. (d). Percentage of total radioactivity in different fractions extracted from guard-cell protoplasts from *Vicia faba.* The illuminated protoplasts were exposed to $^{14}CO_2$ at time zero. Radioactivity was incorporated into the organic acids malate and aspartate (Asp), 3-phosphoglyceric acid (PGA), sugar monophosphates and triose phosphates (SMP + TP) and sugar diphosphates (SDP). The percentage of radioactivity in PGA decreased with time, whereas that of sugar monophosphates increased. This indicates that PGA is a primary carboxylation product, as expected from the operation of the Calvin cycle. From Gotow K., Taylor S. and Zeiger E. (1988) *Plant Physiol.* 86: 700-705.

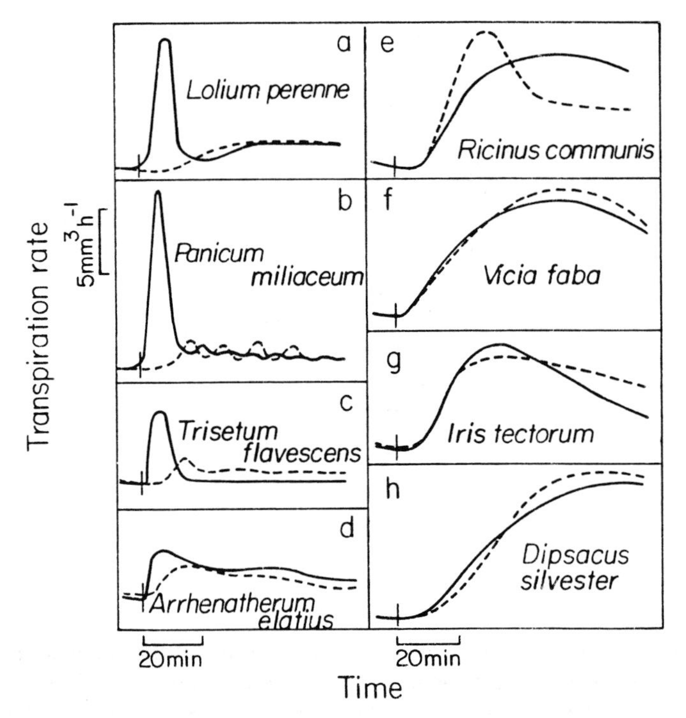

Figure 7. (a-d) Increase in transpiration in four grasses in response to a blue light (B) step (—) given as indicated by the small vertical lines. Responses to a red light (R) step (----). (e-h) Transpiration response to B (—) and R (----) steps in four non-graminaceous species. Note the absence of the B response. From Johnsson M., Issaias S., Brogard T., and Johnsson A. (1976) *Physiol. Plant.* 36: 229-232, with permission.

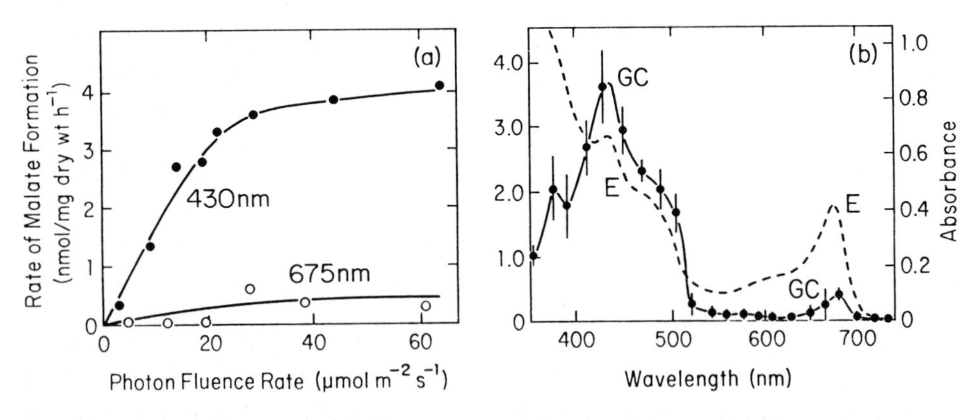

Figure 8. (a) Rate of malate formation in guard cells from epidermal peels of *Vicia faba* irradiated with monochromatic blue (430nm) or red (675 nm) light, as function of photon fluence rate. (b) Action spectrum for malate formation in *Vicia* guard cells (GC) and the absorption spectrum of isolated epidermal peels (E). From Ogawa T., Ishikawa H., Shimada K. and Shibata K. (1978) *Planta* 161: 129-136, with permission.

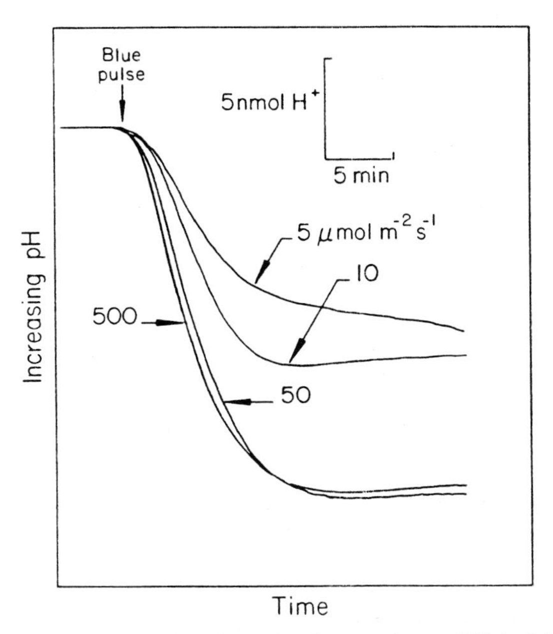

Figure 9. Acidification of a suspension of guard-cell protoplasts of *Vicia faba* in response to a 30 s blue light (B) pulse given in a background of saturating (1000 μmol m^{-2} s^{-1}) red light. Fluence rates of B are indicated in the figure (5-500 μmol m^{-2} s^{-1}). From Shimazaki K., Iino M. and Zeiger E. (1986) *Nature* 319: 324-326, with permission.

Application of B pulses in a background of saturating R has resolved some kinetic properties of the stomatal response to B (Zeiger *et al.* 1985; Iino *et al.* 1985). In these gas exchange experiments with attached leaves of *Commelina communis,* R saturated the photosynthetic responses of the guard-cell chloroplasts and the mesophyll and, under these conditions, a pulse (1-100 s) of B (250 μmol m^{-2} s^{-1}) elicited a transient increase in stomatal conductance without any direct effect on photosynthesis. The stomatal response started shortly after the pulse, peaked within 15 min, and returned to baseline levels after 50-60 min. Upon completion of the response to the first pulse, a second pulse caused a response indistinguishable from the first. Pulses of R on a background of continuous R had no effect, confirming that the response was specific to B and independent of photosynthetic activity.

Measurements of increases in conductance as a function of pulse duration showed that the response increased with pulse length up to 30 s (Fig. 11) where it showed saturation (Zeiger *et al.* 1985). The responses to pulses of varying durations were compared to the response to a saturating pulse and showed apparent exponential kinetics with a half-time of about 9 s. Reciprocity was tested by choosing a given fluence and changing either the duration of the pulse or the fluence rate. The response was found to be proportional to total fluence

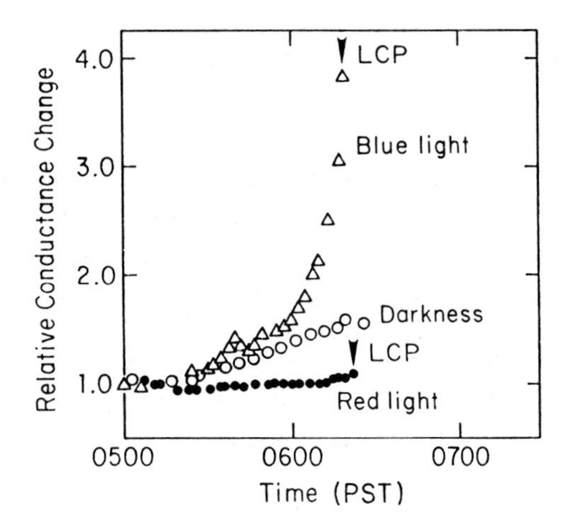

Figure 10. Relative changes in stomatal conductance as a function of light quality in an attached leaf of *Malva parviflora* enclosed in a gas exchange cuvette. Relative changes were calculated from data obtained with the same leaf, on three consecutive days between 4 am and 5 am Pacific Standard Time (PST). The glass window of the cuvette was covered with a blue (◇) or red (●) filter. (□) Stomatal conductance in darkness; LCP, light compensation point of photosynthesis in the mesophyll. From Zeiger E., Field, C. and Mooney H.A.(1981) In: *Plants and the Daylight Spectrum*, pp. 391-407, Smith H. (ed.) Academic Press, New York, with permission.

and independent of duration or fluence rate. Hence, as many photobiological reactions, the B-dependent photosystem in stomata responds to the number of incident photons (Iino *et al.* 1985; Karlsson 1986).

The observation that two saturating pulses, 50 min apart, generated two complete responses, whereas a single pulse twice as long as the saturating pulse did not cause any further increase in stomatal conductance, indicate that the capacity to respond to B was restored slowly with time. The kinetic properties of this recovery were characterized by giving two saturating pulses at increasing intervals (Fig. 12). The response to two pulses increased with the interval between them and a plot of the area under the conductance curves relative to that in response to two saturating pulses 50 min apart also showed apparent exponential kinetics with a half-time of about 9 min.

These data led to the development of a model (Iino *et al.* 1985) postulating that two interconvertible components of the transduction process, *A* and *B*, mediate the photoreception of B. An inactive form of the photoreceptor (*A*) is assumed to be converted to *B* by B, with the increases in stomatal conductance proportional to the available amount of *B*. The photoproduct *B* is converted back to *A* in a thermal reaction. The reaction can be written as:

$$A \underset{k_d}{\overset{k_l}{\rightleftharpoons}} B$$

where k_l and k_d are rate-constants for the light and thermal reaction, respectively. In this model, a pulse of B converts some amount of *A* to *B*, and the reaction saturates when all the available *A* has been photoconverted to *B*. In the two-pulse experiments, all available *A* is converted to *B* by the first saturating pulse and, with time, reverts by a thermal reaction back to *A*. The longer the time between pulses, the higher is the amount of regenerated *A* available at the time of the second pulse and the larger is the integrated response to the two pulses.

These kinetic properties of the B response and the ensuing predictions from the model should be useful for the identification of the B photoreceptor and of intervening sensory transduction steps. The model predicts that under the continuous B irradiation typical of natural environments, both the light and

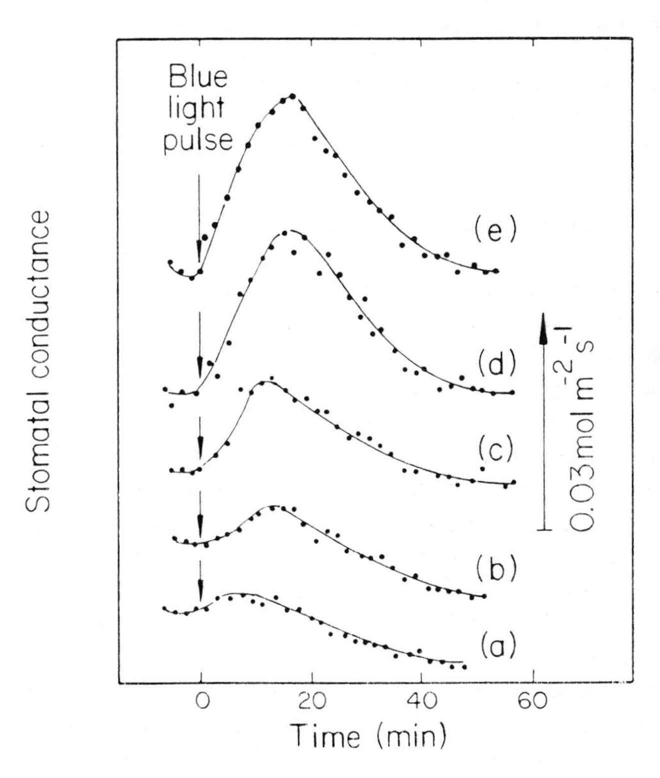

Figure 11. Changes in stomatal conductance in response to a blue light pulse (250 μmol m^{-2} s^{-1}) in a background of saturating red light. Numbers at the ends of the traces indicate pulse duration. Attached leaves of *Commelina communis* were enclosed in a gas exchange cuvette, and stomatal conductance and net photosynthesis were simultaneously measured with a data acquisition system at 2 min intervals. From Zeiger *et al.* (1985), with permission.

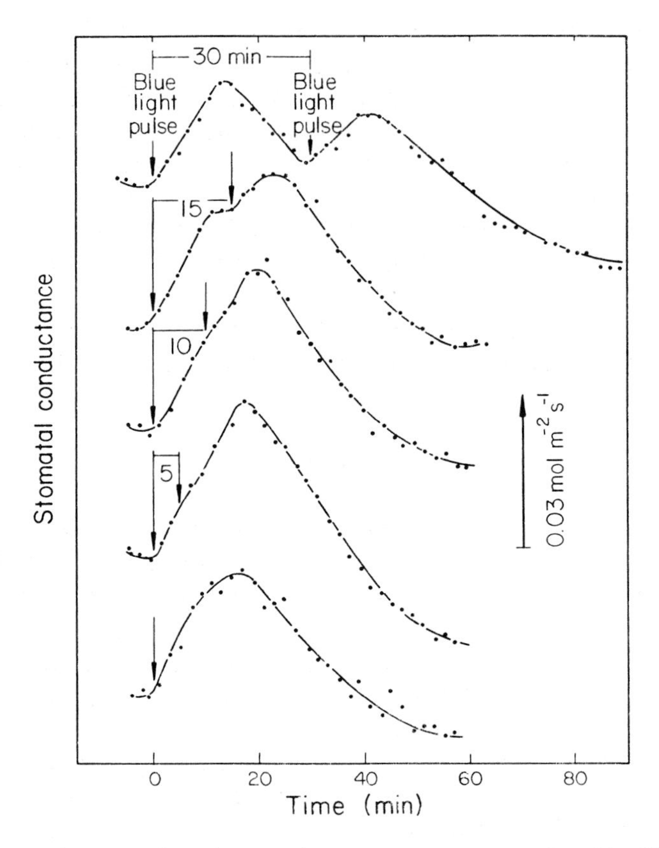

Figure 12. **Changes in stomatal conductance in response to two saturating, 30 s blue light pulses (250 μmol m^{-2} s^{-1}) in a background of saturating red light. The two pulses were given at increasingly longer time intervals, as indicated. Other experimental conditions as in Fig. 11. From Zeiger *et al.* (1985), with permission.**

thermal reactions will occur, resulting in photostationary steady-state concentrations of *B* which will determine the extent of stomatal opening depending on the *B* component. The actual amount of *B* would depend on the concentration of *A*, the fluence rates of B, and the specific rate-constants for the light and thermal reactions. To date, these parameters have only been measured in *Commelina,* but qualitative studies on wheat, *Vicia,* soybean, sugarcane, *Phaseolus vulgaris,* and *Paphiopedilum harrissianum* have revealed a capacity to respond to B pulses in all cases. It will be interesting to learn whether the kinetic properties of the B response are species-specific or common to all leaves and whether, within a given species, they change with developmental stage and environmental conditions.

Pulses of B given in a background of saturating R also stimulate proton pumping at the guard-cell plasma membrane. In suspensions of guard-cell protoplasts, stimulation of proton pumping by B results in a fluence-rate dependent acidification of the suspension medium (Fig. 9). In patch-clamp experiments with single protoplasts, B elicits outward electrical currents, typical of an electrogenic ion pump (Assmann *et al.* 1985). The photobiological characteristics of these responses are homologous to the stomatal responses to B in the intact leaf, indicating that proton pumping is a component of the sensory transduction process transducing B perception into modulated stomatal apertures. Stimulation of proton pumping by B and R are both sensitive to vanadate, a specific inhibitor of plasma membrane H^+ ATPases (Serrano *et al.* 1988; Amodeo *et al.* 1992). Thus, the sensory transduction processes modulating the two light responses of guard cells appear to converge into the stimulation of a H^+ ATPase that controls proton fluxes at the guard-cell membrane.

9.3.8 Localization of the blue light photoreceptor

The predominance of the hypothesis of flavins as putative B photoreceptors (Part 5) has influenced the analysis of the stomatal response to B (Zeiger *et al.* 1987). On the other hand, the recent isolation of carotenoids from guard cells and their chloroplasts (Karlsson *et al.* 1992; Srivastava and Zeiger 1992) has underscored the remarkable similarity between the action spectrum for the stimulation of stomatal conductances by B (Karlsson 1986), and the absorption spectrum of the guard-cell carotenoids.

Carotenoids in the guard-cell chloroplast could play a role in the sensory transduction of stomatal responses to B. Carotenoids are known to respond to the electrical potential and pH gradient at the thylakoid membrane and they could act as molecular switches. Depending on the energy status of the thylakoid, carotenoids could funnel photons into the antenna pigments at the light-harvesting complexes of the guard-cell chloroplast, and function as accessory pigments in photosynthetic reactions. In an alternative mode, changes in the energy state of the thylakoid could result in a separation of the carotenoids from the chlorophyll molecules, with the larger intermolecular distances precluding further energy transfer to the chlorophylls. In this mode, B absorbed by the carotenoids would be transduced into reactions resulting in B responses. In the case of the B-stimulated starch hydrolysis, the sensory transduction process would require the activation of hydrolytic enzymes at the stroma. In the case of the stimulation of the H^+ATPase at the guard-cell plasma membrane, the process would require the involvement of a second messenger, analogous to the one suggested by the patch-clamp experiments studying H^+ATPase stimulation by R (Serrano *et al.* 1988).

One of the implications of a possible role of carotenoids in the stomatal responses to B is that guard-cell chloroplasts should exhibit an intrinsic B response. Indirect evidence for such a response has been obtained in studies of chlorophyll *a* fluorescence transients from chloroplasts in intact guard cells.

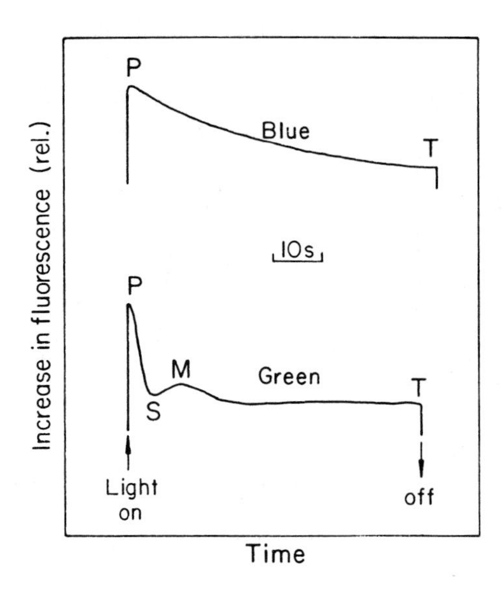

Figure 13. Chlorophyll *a* fluorescence transients from a single guard-cell pair in an epidermal peel from *Vicia faba*, obtained with a microfluorospectrophotometer. Actinic blue light induces transients devoid of fine structure whereas actinic green light induces a faster quenching rate and an M peak associated with Calvin cycle activity. From Mawson B. and Zeiger E. (1991) *Plant Physiol.* 96: 753-760.

Typical transients from dark-adapted mesophyll chloroplasts induced by actinic B show an initial rise in fluorescence yield and a subsequent quenching associated with photophosphorylation. A second rise in fluorescence, the so-called 'M peak', is associated with CO_2 fixation. In contrast, transients from guard-cell chloroplasts, show a very rudimentary quenching and no M peak. However, a recent study on the sensitivity of the fluorescence transients from guard-cell chloroplasts to light quality has shown that these rudimentary transients do not reflect intrinsic properties of the chloroplasts, as originally thought, but rather a response to the actinic B (Mawson and Zeiger 1991). This is evident from results showing that fluorescence transients induced by actinic green light have the fine structure typical of mesophyll chloroplasts (Fig. 13). Direct evidence for a B response of the guard-cell chloroplast has been recently obtained in a study with isolated guard-cell chloroplasts (Srivastava and Zeiger 1992). In this case, the B-induced transients show a more pronounced quenching than that seen in response to R. A B-induced enhancement of the R-stimulated quenching from chloroplasts of cotton guard cells has an action spectrum consistent with the involvement of carotenoids in B photoreception (Fig. 14).

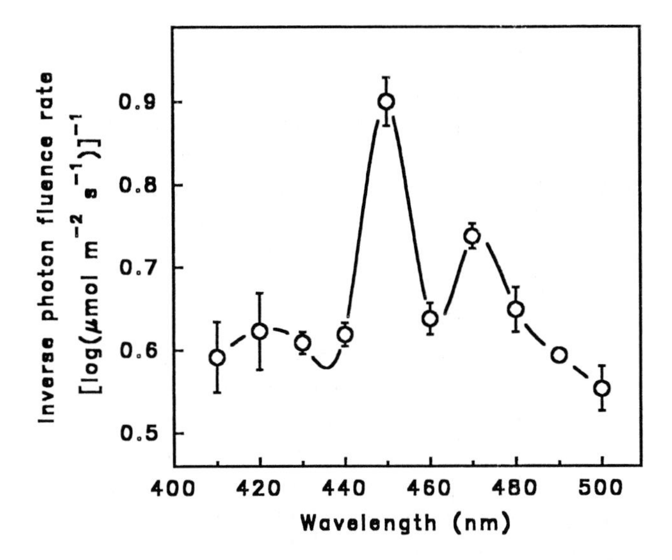

Figure 14. Action spectrum for the enhancement of the red light (R) induced chlorophyll fluorescence quenching by blue (B), in a suspension of abaxial guard cells from cotton (*Gossypium barbadense*). Four different fluence rates of monochromatic B, between 3 and 25 μmol m^{-2} s^{-1}, were used at each wavelength, in a background of 500 μmol m^{-2} s^{-1} of broad-band R. From Quiñones M.A., Lu Z. and Zeiger E. (1993) *Plant Cell Environ.* in press.

9.3.9 Sensory transduction of the stomatal response to light

Our understanding of the sensory transduction of the stomatal response to light is increasing rapidly. Light absorbed by the antenna pigments of the guard-cell chloroplast and the B-photoreceptor is transduced into a stimulation of a H^+ATPase at the guard-cell plasma membrane. The electrochemical gradient resulting from proton pumping drives active K^+ uptake *via* ion channels (Serrano and Zeiger 1989). Potassium ions are balanced by Cl^- uptake and the biosynthesis of malate. Starch hydrolysis, which provides carbon skeletons for malate biosynthesis, and the biosynthesis of malate proper are stimulated by B. Osmotically active sugars produced from starch hydrolysis and photosynthetic carbon fixation also play a role in guard-cell osmoregulation.

Emphasis of current research centres around intervening steps in sensory transduction, including G-proteins (Fairley-Grenot and Assmann 1991) and the precise relationship between light perception and osmoregulation.

9.3.10 Regulatory aspects of the light response of stomata in the intact leaf

As discussed in Section 9.3.1, leaves growing in natural conditions under non-limiting water supply reach nearly saturating levels of photosynthesis and stomatal conductance early in the morning. The insensitivity of photosynthetic reactions in guard cells to low fluence rates (Fig. 6) and the high sensitivity of the stomatal response to B (Figs. 3 and 11) point to a dominant role of the B response in the early phases of opening, as demonstrated in the B dependence of stomatal opening at dawn (Fig. 10). With increasing fluence rates, photophosphorylation in guard-cell chloroplasts is activated and the B response saturates, so stomatal opening later in the day primarily depends on photosynthetic activity in guard-cell chloroplasts. At these stages, the CO_2 response *via* modulation of photophosphorylation and photosynthetic carbon fixation in guard cells are likely to become important.

Two sets of data support this integrated view of the light response of stomata. Firstly, calculations of the apparent *quantum yield* of stomatal conductances as a function of fluence rates show that at low irradiances the B response is very effective. On the other hand, at moderate to high irradiances, B and R are equally effective (Fig. 15), indicating a dependence on photosynthetic activity at the guard-cell chloroplast. Secondly, maximal stomatal conductances in the orchid *Paphiopedilum* are very low, indicating that a stomatal system relying solely on the B response could be functionally stable only in ecophysiological conditions requiring low conductance levels.

It should be noted however that the B response is functionally important not only at low fluence rates. Stomatal conductances are always suboptimal under R, even at saturation, implying a functional role of the B response at all fluence rates. Although insufficiently characterized, this role appears to include both metabolic and environmental aspects.

Recent studies also implicate a role played by *phytochrome* in stomatal function. Immunocytochemical localization has revealed the presence of phytochrome in guard cells, and studies with both epidermal peels and intact leaves have provided some evidence for R-far-red light (FR) reversible effects on stomatal movements. More information is needed to fully characterize the functional role of this potentially important, third photoreceptor system in stomata.

9.3.11 Ecophysiological and agricultural implications of the light response of stomata

The primary role of stomata in the function of leaves and the impact of light on stomatal responses have both basic and applied implications. Ecophysiological studies usually include measurements of stomatal conductances in the analysis of plant-environment interactions. Emerging information on the cellular mechanisms regulating stomatal responses should enhance our understanding of the role of stomata in plant acclimations and adaptations. In the case of the stomatal response to light, a better understanding of the process of light percep-

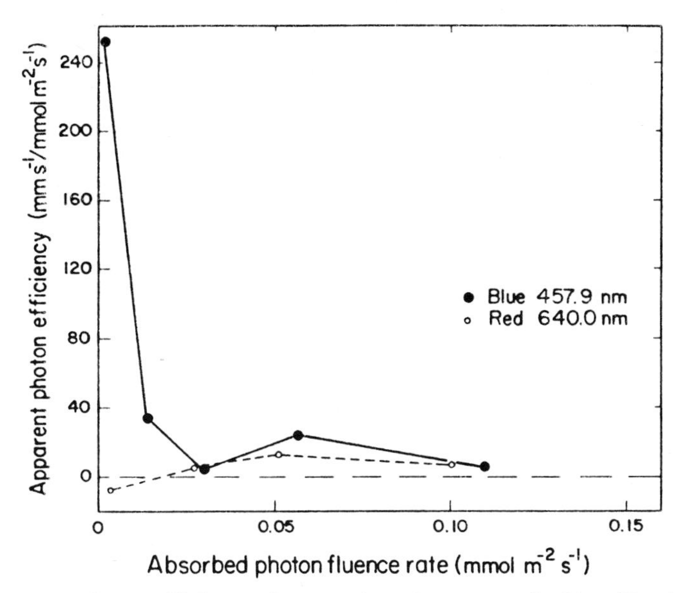

Figure 15. Apparent photon efficiency of stomatal conductance under blue (B, ●) or red (R, ○) light. The data were calculated as point to point slopes from plots of stomatal conductance *versus* photon fluence rates in leaves of *Malva parviflora.* The higher the quantum efficiency, the higher the increase in conductance sustained by a given number of quanta. Note the large quantum efficiency of B at low fluence rates and the similar efficiency of B and R at moderate to high fluence rates. From Zeiger and Field (1982), with permission.

tion and transduction should facilitate studies on leaf adaptations to light quality and quantity, and its impact on leaf gas exchange. Recent work in my laboratory has shown that abaxial stomata from cotton leaves respond more to R than to B, whereas adaxial stomata show the opposite response (Lu *et al.* 1993). This differential sensitivity to light quality resulted in contrasting adaxial/abaxial ratios of stomatal conductances in leaves growing in a greenhouse and a growth chamber. These observations probably underlie modes of leaf acclimation to different light environments which deserve further investigation.

Our increasing understanding of stomatal function could also have an impact on agricultural practices, by allowing specific manipulations of stomatal properties aimed at improved crop adaptations and higher yields. Recent work with Pima cotton (*Gossypium barbadense*) has shown that advanced lines selected for higher yields and heat resistance have higher levels of stomatal conductances. The enhanced leaf transpiration ensuing from the higher conductances increases evaporative cooling and lowers leaf temperatures. Thus, breeding for higher yields and heat resistance in Pima cotton has imposed a selection pressure for higher stomatal conductances, probably associated with an adaptative advantage ensuing from lower leaf temperatures. Because cotton is grown under intensive irrigation, the higher transpiration rates do not entail a

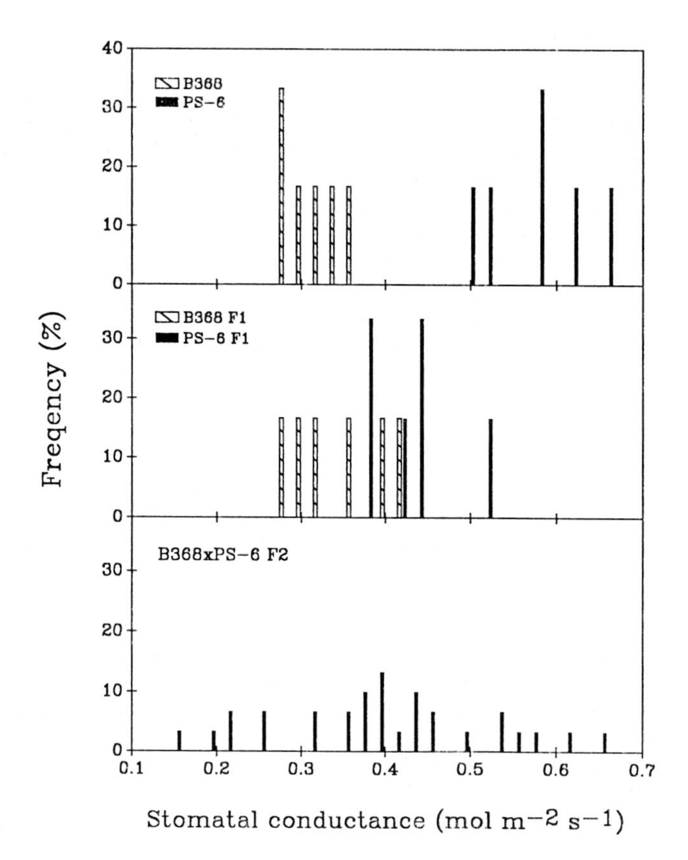

Figure 16. Genetic variation of stomatal conductances in a primitive cultivar of Pima cotton (*Gossypium barbadense*), B368, and the advanced Pima line PS-6, selected for higher yields and heat resistance. Stomatal conductances were measured in greenhouse plants with a Li-Cor 1600 steady-state porometer. *Top panel:* distribution of stomatal conductances in the two parental populations. *Middle panel:* distribution of stomatal conductances in the two reciprocal F_1 crosses (the maternal line in each cross in indicated). *Bottom panel:* distribution of stomatal conductances in the F_2 population. From Lu Z. and Zeiger E. (1993) *Plant Physiol.* in press.

risk of water stress. Crosses between primitive cultivars and advanced lines show that the high conductance trait is genetically determined (Fig. 16). These findings indicate that selection for specific physiological traits, including stomatal properties, could increase the efficiency of breeding programmes. Pima cotton provides us with a model system for the use of stomatal properties, such as the light response, in the design of breeding strategies seeking higher yields and improved crop adaptations to different environments.

9.3.12 Further reading

Sharkey T.D. and Ogawa T. (1987) Stomatal responses to light. In: *Stomatal Function*, pp. 195-208, Zeiger E., Farquhar G. and Cowan I. (eds.) Stanford University Press, Stanford.
Tenhunen J.D., Pearcy R.W. and Lange O.L. (1987) Diurnal variations in leaf conductance and gas exchange in natural environments. In: *Stomatal Function*, pp. 323-352, Zeiger E., Farquhar G. and Cowan I. (eds.) Stanford University Press, Stanford.
Zeiger E. (1983) The biology of stomatal guard cells. *Annu. Rev. Plant. Physiol.* 34: 441-475.
Zeiger E., Iino M., Shimazaki K. and Ogawa T. (1987) The blue light response of stomata: mechanism and function. In: *Stomatal Function*, pp. 209-228, Zeiger E., Farquhar G. and Cowan I. (eds.) Stanford University Press, Stanford.
Zeiger E. (1990) Light perception in guard cells. *Plant Cell Environ.* 13: 739-747.

9.3.13 References

Amodeo G., Srivastava A. and Zeiger E. (1992) Vanadate inhibits blue light-stimulated swelling of *Vicia* guard cell protoplasts. *Plant Physiol.* 100:1567-1570.
Assmann S.M., Simoncini L. and Schroeder J. (1985) Blue light activates electrogenic ion pumping in guard cell protoplasts of *Vicia faba*. *Nature* 318: 285-287.
D'Amelio E. and Zeiger E. (1988) Diversity in guard cell plastids of the Orchidaceae. *Can. J. Bot.* 66: 257-271.
Fairley-Grenot K. and Assmann S.M. (1991) Evidence for G-protein regulation of inward K^+ channel current in guard cells of faba bean. *Plant Cell* 3: 1037-1044.
Hsiao T.C., Allaway W.G. and Evans L.T. (1973) Action spectra for guard cell Rb^+ uptake and stomatal opening in *Vicia faba*. *Plant Physiol.* 51: 82-88.
Iino M., Ogawa T. and Zeiger E. (1985) Kinetic properties of the blue light response of stomata. *Proc. Nat. Acad. Sci. USA* 82: 8019-8023.
Karlsson P.E. (1986) Blue light regulation of stomata in wheat seedlings. II. Action spectrum and search for action dichroism. *Physiol. Plant.* 66: 207-210.
Karlsson P.E., Bogomolni R.A. and Zeiger E. (1992). High performance liquid chromatography of pigments from guard cell protoplasts and mesophyll tissue of *Vicia faba* L. *Photochem. Photobiol.* 55: 605-610.
Lu Z., Quiñones M.A. and Zeiger E. (1993) Abaxial and adaxial stomata from Pima cotton differ in their pigment content and sensitivity to light quality. *Plant Cell Environ.* in press.
Mawson B.T. and Zeiger E. (1991) Blue light-modulation of chlorophyll *a* fluorescence transients in guard cell chloroplasts. *Plant Physiol.* 96: 753-760.
Nelson S.D. and Mayo J.M. (1975) The occurrence of functional non-chlorophyllous guard cells in *Paphiopedilum* spp. *Can. J. Bot.* 53: 1-7.
Serrano E.E., Zeiger E. and Hagiwara S. (1988) Red light stimulates an electrogenic proton pump in *Vicia* guard cell protoplasts. *Proc. Natl. Aca. Sci. USA* 85: 436-440.
Serrano E.E. and Zeiger E. (1989) Sensory transduction and electrical signalling in guard cells. *Plant Physiol.* 91: 295-299.
Sharkey T.D. and Raschke K. (1981) Separation and measurement of direct and indirect effects of light on stomata. *Plant Physiol.* 68: 33-40.
Shimazaki K. and Zeiger E. (1985) Cyclic and non-cyclic photophosphorylation in isolated guard cell chloroplasts from *Vicia faba* L. *Plant Physiol.* 78: 211-214.
Shimazaki K., Terada J., Tanaka K. and Kondo N. (1989) Calvin-Benson cycle enzymes in guard cell protoplasts from *Vicia faba*. *Plant Physiol.* 90: 1057-1054.
Srivastava A. and Zeiger E. (1992) Fast fluorescence quenching from isolated guard cell chloroplasts of *Vicia faba* is induced by blue light and not by red light. *Plant Physiol.* 100: 1562-1566.

Talbott L. and Zeiger E. (1993) Sugar and organic acid accumulation in guard cells of *Vicia faba* during opening: red and blue light separate photosynthetic and non-photosynthetic pathways of osmoregulation. *Plant Physiol.* in press.
Zeiger E. and Hepler P.K. (1977) Light and stomatal function: blue light stimulates swelling of guard cell protoplasts. *Science* 196: 887-889.
Zeiger E. and Field C. (1982) Photocontrol of the functional coupling between photosynthesis and stomatal conductance in the intact leaf. *Plant Physiol.* 70: 370-375.
Zeiger E., Assmann S.M. and Meidner H. (1983) The photobiology of *Paphiopedilum* stomata: opening under blue but not red light. *Photochem. Photobiol.* 37: 627-630.
Zeiger E., Iino M. and Ogawa T. (1985) The blue-light response to stomata: pulse kinetics and some mechanistic implications. *Photochem. Photobiol.* 42: 759-763.

9.4 Photomovement

Wolfgang Haupt and Donat-P. Häder

Institut für Botanik und Pharmazeutische Biologie, der Universität, Erlangen-Nürnberg, Staudtstr. 5, 91058 Erlangen, Germany

9.4.1 Introduction

Plant movement can be induced or controlled by light. In a broad sense such responses are called photomovement (photomovement *sensu lato*). However, the present chapter excludes the growth curvature- and turgor-based cell-shape changes (which are indeed plant movements), discussed in Chapters 9.2 and 9.3. Thus, here, photomovement *sensu stricto* comprises locomotion of whole organisms (mostly unicells or cell colonies) and intracellular movement (displacement of cell organelles) under the regulatory influence of light. Both phenomena will be discussed separately, and afterwards similarities and differences between them will be pointed out.

9.4.2 Photomovement in motile micro-organisms

In many motile micro-organisms a number of external factors affect motility, including gravity, chemical and thermal gradients, electric and magnetic fields, as well as mechanical stimuli. Light certainly plays an important role for orientation not only in photosynthetic micro-organisms. These external cues are used to optimize the position of the organisms in their environment, to control the amount of light they are exposed to, and to improve their chances for reproduction.

Photomovement responses of motile micro-organisms can be formally classified as three reaction types (Häder 1979):

(i) *Phototaxis* describes a movement with respect to the light direction. Positive phototaxis is a movement toward the light source, negative phototaxis away from it. Some organisms move perpendicular to the light direction, which is called diaphototaxis or transversal phototaxis, or at another angle.

(ii) *Photokinesis* is defined as a steady state-dependence of the linear velocity of an organism on the ambient fluence rate. The reference value is the velocity

R.E. Kendrick & G.H.M. Kronenberg (eds.), Photomorphogenesis in Plants - 2nd Edition

measured in darkness: a velocity higher than the dark value is described as positive photokinesis. Some organisms stop in darkness (*Dunkelstarre*) or at high fluence rates (*Lichtstarre*).

(iii) A sudden step-up or step-down in the fluence rate induces a transient *photophobic response*, which may be a stop, a reversal or a turn in the movement direction, depending on the organism. In contrast to phototaxis, both phobic responses and photokinesis are independent of the light direction.

9.4.2.1 Motile bacteria

Many bacteria show some kind of motility, either gliding or by means of bacterial flagella. Most photomotile behaviour is observed in photosynthetic bacteria and archaebacteria which also utilize light as a source of energy (Armitage 1991). In addition, some enteric bacteria respond to extremely bright flashes of light. While only cyanobacteria have been found to be capable of phototaxis most photomotile bacteria show exclusively photophobic responses (both step-up and/or step-down) and some photokinesis. However, recently the nomenclature has been incorrectly used in several cases. For example, photophobic responses of halobacteria are incorrectly described as phototaxis, since although repetitive phobic responses may eventually lead to an accumulation pattern in nonuniformly illuminated fields, the responses are definitely independent of the light direction.

Halobacterium halobium possesses a rod-shaped cell flagellated with a tuft at one or both ends. In contrast to enteric bacteria, which show smooth swimming interrupted by short tumbling sequences, the cells change direction of movement at regular intervals of about 10 s moving equally well in either direction. When *Halobacterium* encounters an increase in the fluence rate of green or orange light, the cells respond by suppressing the next reversal for some time. The same behaviour is induced by a step-down in blue (B) or UV light. In contrast, a premature reversal is elicited by a step-up in B or UV radiation or a step-down in green or orange light. The cells respond similarly in light gradients, a behaviour which causes them to accumulate in regions of optimal light conditions with sufficient intensities for light energy conversion and to avoid areas of damaging UV radiation. This is comparable with responses of enteric bacteria in chemical gradients; therefore the response to a step-up in green/orange light or a step-down in B/UV has been called an attractant response, while the opposite is termed a repellent response.

Early investigations of the action spectrum had led to the assumption that bacteriorhodopsin (bR), which is a light-driven proton pump, doubles as a photoreceptor for the photophobic response. This hypothesis was proved wrong since mutants lacking bR showed a perfect phobic reaction. Likewise, halorhodopsin (hR), a light-driven chloride pump, could be ruled out. Finally, a

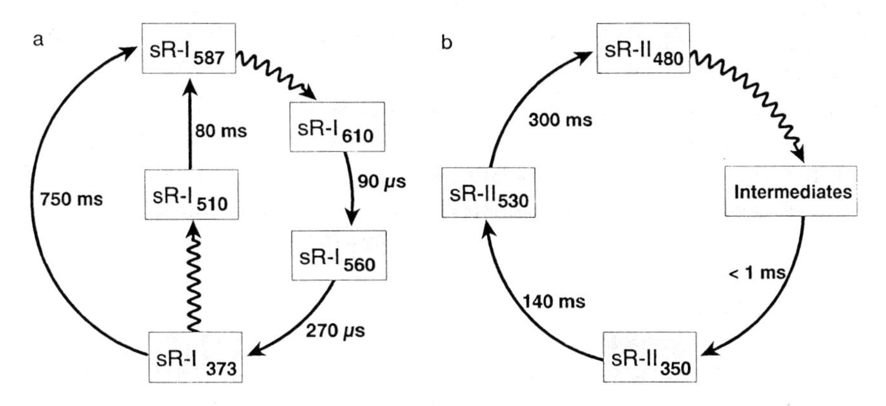

Figure 1. Photocycle of sensory rhodopsin II (sR-II) (a) and I (sR-I) (b) from *Halobacterium halobium.* After Oesterhelt and Marwan (1990).

mutant with a third chromophore was isolated, which was called sensory rhodopsin I (sR-I). Like phytochrome, sR-I is a photochromic pigment which undergoes a photocycle when stimulated by light (Fig. 1a). In its unexcited form the pigment absorbs at 587 nm and operates as an attractant receptor. After absorption of a suitable quantum of light the molecule forms a long-lived species absorbing maximally at 373 nm, which after absorption of a UV quantum is thought to mediate the repellent response. However, a number of observations conflict with this simple hypothesis of a dual function of sR-I, *e.g.*, during development of the cells the attractant response was found to be established before the B/UV response was present. This conflict was eventually resolved when Takahashi (1992) found a forth rhodopsin, sensory rhodopsin II (sR-II), also called phoborhodopsin, which is an exclusive B/UV receptor (Fig. 1b).

The sensory transduction chain of photophobic responses converges with that of chemosensory responses also present in halobacteria. The sR-I is not an ion pump and does not affect the membrane potential, as do bR and hR. It is associated with the 97 kDa membrane-bound, methyl-accepting phototaxis protein (MPP). Methyl-accepting proteins have been identified in chemotactically responding eubacteria, where they are involved in the adaptation process. The sR-II has been found to activate a methylation system in the membrane, and it is speculated that both sensory rhodopsins relay their signals to methyl-accepting proteins by protein-protein interaction. The crucial step in photosensory transduction is the coupling of retinal isomerization to the generation of a conformational change in the protein thought to be based on a steric interaction between the chromophore and the 13-methyl group of the protein. Fumarate seems to be involved in the final step of the transduction chain and reverses the direction of flagellar rotation (Marwan and Oesterhelt

1991). Thus, signals from the chemoreceptors, as well as signals from both photoreceptors are processed in a complex manner.

Purple photosynthetic bacteria form another group of photomotile prokaryotes, observed more than a century ago. Photokinesis is observed in several genera including *Rhodospirillum* and *Chromatium.* The action spectra for these responses resemble those of photosynthesis, and photokinesis is effectively impaired by inhibitors of the photosynthetic electron transport chain and by uncouplers, indicating that the increased velocity is the result of enhanced metabolic energy made available to the motor apparatus of the cell. Similar interpretations seem to hold for cyanobacteria and some gliding algae such as desmids.

When the helically shaped bipolarly flagellated *Rhodospirillum* swims from an illuminated area into a dark zone it undergoes a phobic response and reverses direction of movement as if 'frightened' (*phobos* = greek for fright). In an irradiated spot fringed by a darker field the organisms are trapped since they back up each time they encounter a decrease in irradiance. Photophobic responses in purple bacteria are also linked to the photosynthetic machinery, and they are affected by inhibitors of the photosynthetic electron transport chain. Mutants which lack the photosynthetic reaction centres, but otherwise possess the full complement of photosynthetic pigments are impaired in photophobic responses, indicating that the light-mediated responses are coupled to the photosynthetic electron transport. One hypothesis assumes that the photosynthetically generated proton motive force is measured by a 'protometer', which could be part of the flagellar motor, or alternatively by a membrane-bound cytochrome.

Filamentous, gliding cyanobacteria show photophobic responses, which have been investigated to some extent in the genus *Phormidium*. Here the photosynthetic pigments also double as photoreceptor molecules. In light, protons are pumped from the cytoplasm into the thylakoid vesicles. When the filament moves into a dark area or the whole filament is darkened, the proton gradient breaks down, which results in a small electric potential change (depolarization of the cytoplasm). This in turn is thought to operate voltage-dependent, calcium-specific ion channels in the cytoplasmic membrane, which allow a transient passive Ca^{2+} influx into the cytoplasmic compartment. This further depolarizes the membrane potential and thus is an amplification mechanism. The direction of movement has been found to be controlled by an electric potential gradient between the two ends of the filament. During a phobic response the former front end is depolarized by the mechanism described above, so that the electrical gradient along the length of the filament reverses polarity, thus causing a reversal of movement (Häder 1987a).

Cyanobacteria in the Oscillatoriaceae family show a primitive phototaxis based on a trial-and-error mechanism: the filaments reverse the direction of movement on an irregular basis. When moving toward the light source, reversal is suppressed and when moving away from it, a premature reversal is initiated during positive phototaxis. Negative phototaxis is brought about by the opposite behaviour. However, how the direction of light is detected, is not known. Members of the Nostocaceae family, in contrast, show phototaxis with true steering. The filaments either move with one leading end or in a U-shaped fashion. In lateral light they actively turn toward or away from the light direction. Positive and negative phototaxis in *Anabaena* have different action spectra. The switch from a positive to a negative response is thought to be mediated by a high concentration of singlet oxygen produced by the transfer of excess energy from the photoreceptor pigments.

9.4.2.2 Photosynthetic flagellates

In addition to photokinesis and phobic responses, phototaxis is an ecologically significant mechanism of orientation in photosynthetic flagellates. Many organisms show a positive phototaxis at lower and negative phototaxis at higher fluence rates. In *Chlamydomonas* the direction of the response depends (in addition to the fluence rate) on the ionic composition of the medium, especially the [Ca^{2+}], and in a *Cryptomonas* species the degree of pigmentation controls the crossover point between positive and negative phototaxis.

In *Euglena*, the photoreceptor for phototaxis is thought to be located in the paraflagellar body (PFB), a swelling at the basis of the emerging flagellum inside the reservoir (Fig. 2a). Action spectra of phototaxis have been interpreted to indicate the involvement of a flavin as a photoreceptor molecule. The involvement of flavins was also suggested by microspectrofluorometric analysis of the PFB, as well as quencher studies. In fluorescence microscopy, the paraflagellar body appears as a blue spot when excited by UV radiation (Fig. 2b) and as a green spot when a blue excitation radiation is used. In electron microscopic photographs the PFB shows a quasi-crystalline structure, suggesting a dichroic orientation of the photoreceptor pigments. This finding is further supported by the fact that *Euglena* shows a pronounced polarotactic orientation, *i.e.* the cells move with respect to the plane of polarized light (Häder 1987b).

Recently, a technique was developed to isolate the flagella of *Euglena* with the PFBs still attached. Biochemical analysis showed that the PFB contains at least four major proteins with molecular masses in the range of 27 to 33 kDa. All proteins carry a pterin and one of them an additional flavin as indicated by excitation and emission fluorescence spectroscopic measurements. Thus, it can be speculated that the pterins operate as antenna pigments which funnel their energy to the flavin which is the photoreceptor proper.

In the past, several researchers have proposed a periodic shading mechanism to be the basis for light direction detection in *Euglena*. Like many other flagel-

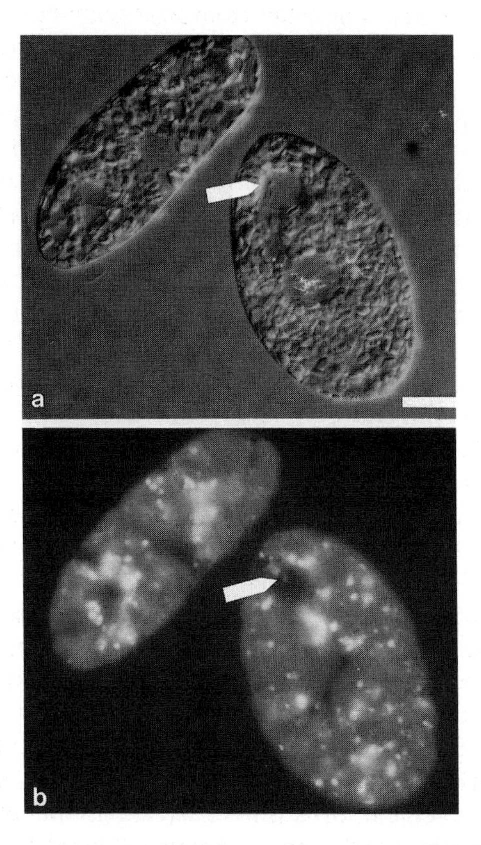

Figure 2. The paraflagellar body of *Euglena gracilis* photographed using interference contrast (a) and in fluorescence microscopy when excited by UV radiation. After Brodhun and Häder (1990).

lates, the cells rotate helically around their long axis during forward locomotion. In a lateral light beam, the stigma will cast a shadow onto the PFB once during the rotation because of the geometrical configuration of the organelles in the front end of the cell. The photoreceptor is thought to elicit a signal to swing out the flagellum when the irradiance decreases, which turns the front end toward the stigma and thus toward the light source by some angle. This process is repeated until the long axis of the cone around which the cell rotates is aligned with the light direction. This shading hypothesis, however, was found to be incompatible with a number of recent results including the observation that cells which have no stigma can orient in a lateral light beam. An alternative explanation is based on the dichroic orientation of the photoreceptor pigments deduced from experiments using polarized light (see above). Because the cell rotates during forward locomotion, a lateral light beam causes a modulation of the signal received by the photoreceptor which eventually causes the cell to align its path with the light direction. Additional shading from the stigma and the rear end allow it to distinguish between light coming from left or right and front or rear, respectively. Strains with no stigma lack this information and therefore often orient in two directions perpendicular to the light beam.

In contrast to *Euglena*, the action spectrum for phototaxis in the green flagellate *Chlamydomonas* shows two maxima at 440 and 503 nm. However, when the action spectrum was obtained on the basis of the threshold values, the 440-nm peak was missing (Hegemann 1991). A mutant, FN68, which shows phototaxis only at very high fluence rates after growth in darkness, can be reconstituted by supplementation of the cells with retinal or retinal analogues. Reconstitution with ^3H labelled retinal identified only one protein in a membrane preparation with a molecular mass of about 32 kDa, assumed to be a rhodopsin. A further proof for the presence of rhodopsin in *Chlamydomonas* is the detection of its gene in the cell; however, it still needs to be shown experimentally that the rhodopsin is indeed located in the photoreceptor organelle which is thought to be associated with the stigma.

K.W. Foster and R.D. Smyth have proposed a hypothesis according to which the stigma operates as a spatially sensitive antenna using interference reflection. When a light beam hits the stigma, which may be composed of several layers of carotene-stained lipid droplets, it is partially reflected at each interface between layers with different refraction indices. Since the layers are spaced at about a quarter of the wavelength of the action spectrum maximum, the reflected wave interferes constructively with the incoming wave, forming a maximum near the membrane overlaying the reflecting stigma. Indeed, the amplified reflected light can be visualized in an epimicroscope, provided the stigma faces the observer and thus the light source. The interference reflector hypothesis is backed by electrical measurements which show a maximal amplitude when the stigma faces the light source. When the cell is rotated while being held by a suction pipette, a minimum in the electrical potential occurs when the chloroplast shades the stigma; this observation is compatible with a shading hypothesis, which can still be discussed for *Chlamydomonas* in contrast to *Euglena.*

Not only vegetative cells, but also gametes of *Chlamydomonas eugametos* are phototactic depending on the strain: one mt^- strain was found to be non-phototactic while another mt^- and an mt^+ strain were found to be strongly positive phototactic. After sexual fusion, the mt^+ strain powers the pair, but the direction of phototactic orientation is reversed, showing negative phototaxis at all effective fluence rates irrespective whether the mt^- partner was formerly phototactic or not. This behaviour can be interpreted on an ecological basis: positive phototaxis brings the gametes to the surface which increases their chances for finding a mating partner, while negative phototaxis guides the pairs to the bottom where the zygotes develop.

Calcium ions seem to be essential for signal transduction of phototaxis in *Chlamydomonas.* They also strongly influence the beating pattern of the flagella: experiments with detergent-extracted cell models indicate that positive phototaxis is elicited by a transient increase and negative phototaxis by a transient decrease of intracellular calcium with respect to the unstimulated level; furthermore, calcium-channel blockers adversely affect phototaxis. In addition, Ca^{2+} may have a role in adaptation, in continuous light as well as after flash stimulation. Other experiments also showed the involvement of potassium in transmembrane signalling. Potassium is the major ion responsible for the generation of the membrane potential. This fact suggests an electrical event in the sensory transduction chain for phototaxis in *Chlamydomonas,* which has not been found in *Euglena,* despite intensive investigations.

Electric measurements using intracellular micro-electrodes have been successful only in a few organisms, which may be due to the mechanical damage caused by the inserted micro-electrode. In an alternative approach the cells are sucked onto a pipette, so that the part inside the pipette is electrically insulated from the remainder of the cell. After a B pulse a transient electrical potential change can be measured, which is positive (depolarization) when the pipette covers a membrane patch overlaying the stigma, or negative when it covers another part of the cell. The light pulse induces a cascade of electrical events in *Chlamydomonas* which can be measured either by micropipettes or by macro-electrodes in a population. More detailed electrical measurements were performed in the green flagellate *Haematococcus.* Within a few milliseconds after an intense flash a primary potential difference occurs, which is followed by a permanent late potential difference (Fig. 3). The latter has a lower amplitude and dissipates after some time with a time constant of 15-40 ms. Both potentials depend on the fluence rate in a graded fashion and are followed by an all-or-none regenerative response if the flash intensity exceeds a certain threshold level. Removal of Ca^{2+} from the medium strongly reduces the amplitudes of the electric responses, although other ions also seem to be involved in the current generation. When using a microbeam, photo-induced electric responses are generated only when the stigma region of the cell is irradiated. The action spectrum of the electric events is similar to that of phototaxis even in its fine structure.

In *Chlamydomonas,* the morphologically similar two flagella respond differently to light stimuli: near the threshold fluence rate the *cis*-stigma flagellum responds when the light is switched on while the *trans*-stigma flagellum does not. This is also shown by high speed cinematographic analysis of the beating pattern after light stimulation. *Haematococcus* shows an even more asymmetric beating pattern between the two flagella which is thought to be responsible for the change of direction underlying phototactic reorientation.

Judging from the action spectrum, the freshwater alga, *Peridinium gatunense* employs a still different set of photoreceptors. Maximal activity is found with several peaks in the red region. Lack of effect of photosynthetic electron transport inhibitors, as well as a total lack of activity in the range of the Soret bands of the chlorophylls argue against the involvement of the photosynthetic

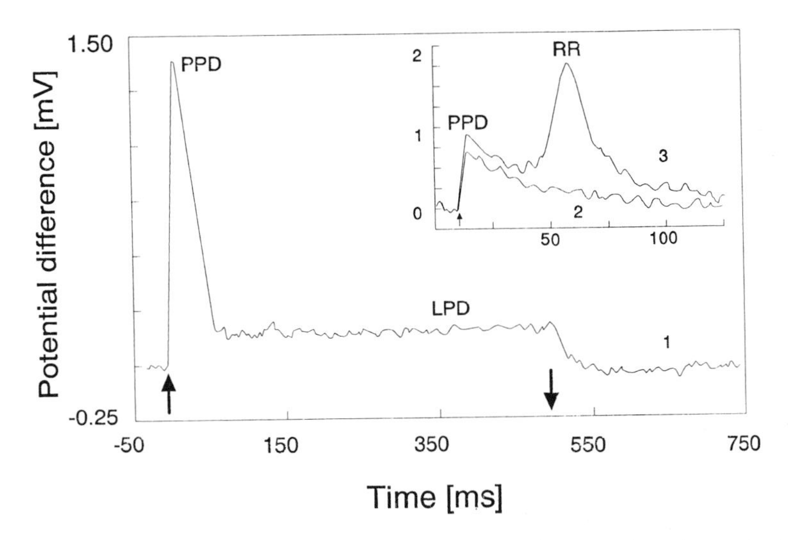

Figure 3. Electrical membrane potential changes measured with a suction pipette in *Haematococcus pluvialis.* After a blue light source is switched on (*up arrow,* time = 0) a primary potential difference (PPD) occurs followed by a late potential difference (LPD) which dissipates when the light is switched off (*down arrow*) (curve 1). Insert: When the flash intensity exceeds a certain threshold, a regenerative response (RR) occurs on top of the LPD (curve 3) which is not seen when the flash intensity is insufficient (curve 2). After Sineshchekov (1991).

apparatus. Recent biochemical analysis has revealed the presence of at least four different chromoproteins separated from a membrane fraction absorbing at 580, 638, 667 and 710 nm which correspond with the maxima in the action spectrum for phototaxis. Light energy absorbed by shorter wavelength pigments is emitted as fluorescence at wavelengths which are absorbed by pigments with maxima at longer wavelengths. These results are indicative of a cascade of light harvesting pigments which eventually funnel their absorbed energy into a pigment absorbing near 710 nm. The nature of the pigments involved has not yet been revealed.

9.4.2.3 Slime moulds

The cellular slime mould, *Dictyostelium discoideum,* has a life cycle which alternates between unicellular amoebae and multicellular pseudoplasmodia, called slugs. In addition to chemotaxis and thermotaxis both life stages show phototactic orientation, which, however, is based on different photoreceptor pigments and on a different mechanism of light direction detection in amoebae and slugs. The amoebae show a pronounced phototaxis which is positive at low fluence rates and negative at higher fluence rates. The action spectra of both responses are similar to each other and to that for photo-inhibition of aggrega-

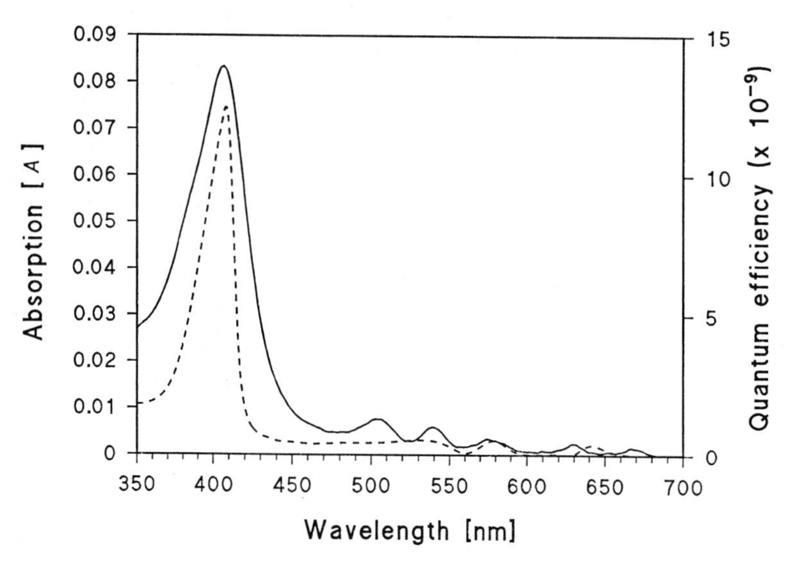

Figure 4. Absorption spectrum of a mitochondrial membrane fraction enriched in a pigment which contains protoporphyrin IX as a chromophoric group from AX2 amoebae (*solid line*) and action spectrum for amoebal positive phototaxis (*dashed line*) in *Dictyostelium discoideum* (after Vornlocher and Häder 1992; Häder and Poff 1979).

tion with a peak at 405 nm and further maxima throughout most of the visible spectrum up to 640 nm (Fig. 4).

Low fluence rate microbeams cause the development of pseudopodia at the irradiated site, while microbeams of high fluence rates cause pseudopodia to be withdrawn when irradiated and new ones to be produced in a different part of the amoeba. These observations indicate that in *Dictyostelium* amoebae the perception of the light direction is based on a comparison of the readings of multiple photoreceptors located on the periphery of the cell.

Biochemical analysis of amoebae has identified a 45.5 kDa protein with a protoporphyrin IX bound as a chromophoric group. While most of the pigment is bound to the mitochondria where it might operate as a shading pigment (Fig. 4), a smaller fraction was detected in the cytoplasmic membrane where it might function as a photoreceptor for phototaxis. If this hypothesis is substantiated *Dictyostelium* amoebae represent a rather rare case where shading and receptor pigment are identical.

When the amoebae have depleted their food supply the cells start to aggregate guided by chemotaxis toward excreted cAMP. In light, slugs orientate exclusively positively phototactically with an action spectrum which differs significantly from that of the amoebae; it has two peaks at 420 and 440 nm and smaller additional maxima up to 610 nm. The difference between slug and amoebal phototaxis is further stressed by the fact that a mutant lacking slug phototaxis shows clear amoebal phototaxis.

The mechanism of perception of the light direction in slugs is based on a lens effect. Laterally impinging light is focused to the distal side due to the refractive index of the cytoplasm which is sufficiently higher than that of the surrounding air. At wavelength < 300 nm the lens effect is nullified by high internal absorption leading to negative phototaxis; this behaviour can also be induced experimentally in visible light by introducing neutral red into the cells which is readily taken up by amoebae before aggregation. P. Fisher and K. Williams (1981) assumed that the focused light induces the production of a low molecular weight metabolite, termed slug turning factor (STF), which is supposed to operate as a chemical repellent. According to this hypothesis, positive phototaxis is the result of negative chemotaxis. However, the molecule has not yet been isolated.

Careful analysis has shown that slug phototaxis is bimodal, *i.e.*, the slugs deviate right and left from the light direction. In wild-type cells this angle can be rather small, while some mutant strains have large deviation angles. The experimental conditions also affect the deviation angle. An optical model has been developed to explain this phenomenon: a beam of light hitting the slug parallel to its long axis is focused to the rear. Since light perception is restricted to the front end of the slug it has to turn from the light direction by a certain angle until the focused beam is perceived.

9.4.2.4 Ecological consequences of photomovement

Micro-organisms employ photomovement for a number of ecologically important purposes (Häder 1988). Photosynthetic micro-organisms utilize photoorientation to solve an inherent dilemma of their environment: on the one hand they rely on solar radiation to satisfy their energetic needs. On the other hand they do not possess the protective capabilities of higher plants to avoid the radiation damage of visible and UV spectral bands of unfiltered solar radiation at the surface of the body of water they occupy. Exposure to direct sunlight has been found to impair photo- and gravi-orientation, negatively affect motility and to bleach photosynthetic pigments, as well as damage other cellular components.

Photosynthetic flagellates orientate in the water column by using antagonistic stimuli which allows them to move to levels of appropriate irradiation and to adjust their position to the constantly changing environmental conditions. In darkness the cells move upward (*e.g.* guided by negative gravitaxis) to warrant that they are near the surface of the water body at sunrise. This is also of an ecological advantage for photosynthetic organisms when the orientation to light fails, as in turbid waters. In *Euglena* it could be shown that the earth's gravitational field is responsible for the upward movement, since in a space experiment under microgravity conditions the cells moved in random directions. Movement upwards is supported by positive phototaxis in weak light (below 1.4 W m^{-2}). In strong light the upward movement is counterbalanced by a precise negative phototaxis. These antagonistic responses warrant that the organisms move close enough to the surface to receive sufficient light for photosynthesis and simultaneously avoid excessive visible and UV radiation. Other flagellates utilize different strategies to orientate within the water column: several *Peridinium* species (both freshwater and marine) have been found to show positive phototaxis as a means of upward movement up to a certain irradiance, which in *P. gatunense* is below 180 W m^{-2} (Fig. 5a). Above that threshold the cells swim diaphototactically (Fig. 5b), *i.e.* perpendicular to the light direction; this behaviour is observed at irradiances up to > 1200 W m^{-2} which is higher than tropical solar radiation at the surface at noon.

Many flagellates undergo daily vertical migrations of several meters to adapt to the changing conditions in the water column. Furthermore, light modulates the precision of gravitactic orientation: in darkness the cells show a high degree of upward orientation while in light the precision of gravitaxis decreases. The nature of the photoreceptor for this regulatory behaviour has still not been determined.

Gliding and swimming prokaryotes utilize photophobic responses to move to and stay in areas of suitable light conditions for growth and survival. The simple reversal at boundaries to dark or too bright areas is an effective strategy to adapt to the constantly changing light conditions in the habitat. The motile slime moulds also employ photomovement responses to improve their chances for finding food and augment their reproductive capabilities. The amoebae optimize their position within the top soil using positive and negative phototaxis to gain access to the bacteria which occupy the decaying leaf zone. After depleting the food supply and aggregation to multicellular slugs the organisms are guided by exclusive positive phototaxis to the surface, where the spores, eventually developed in the mature sorocarps, have a better chance for distribution to other habitats. While the individual responses to light and other stimuli (thermotaxis, chemotaxis and gravitaxis) have been studied in detail in the laboratory, the interplay between the various responses under ecological conditions in the natural habitat is still not fully resolved.

9.4.3 Photoregulation of intracellular movement

Movement of cell organelles is controlled by light in either of two ways. One concerns cytoplasmic streaming, which does not change the overall distribution of organelles; rotational streaming (*cyclosis*) being the type with the highest degree of regularity. In this case light may stimulate streaming of a previously

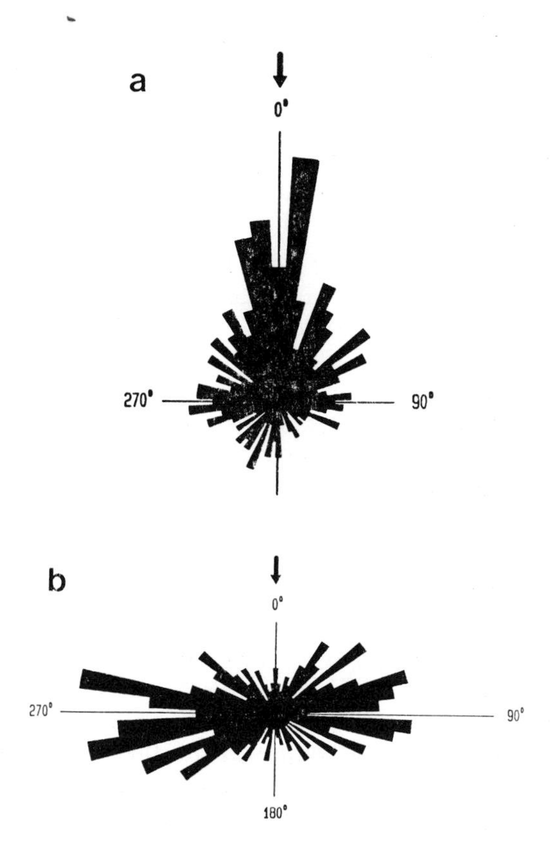

Figure 5. Circular histograms showing the number of *Peridinium gatunense* cells moving in each of 64 sectors indicating positive phototaxis at 10 W m^{-2} (a) and diaphototaxis at 1100 W m^{-2} (b). After Liu *et al.* (1990).

resting cytoplasm, or it may accelerate or intensify existing streaming. This so-called *photodinesis* is well investigated in the fresh-water plants *Vallisneria* and *Elodea*. The second type of light-controlled movement is the redistribution of chloroplasts in unidirectional light, which results in well-defined patterns of arrangement depending on irradiance and direction. In some examples, common features of both types of movement have been found.

9.4.3.1 Photodinesis

In the mesophyll cells of *Vallisneria spiralis* some rotational streaming can be observed irrespective of the presence of light, but this primarily concerns cytoplasm with organelles below the size of chloroplasts. With increasing irradiance this streaming is accelerated, and above a certain threshold the

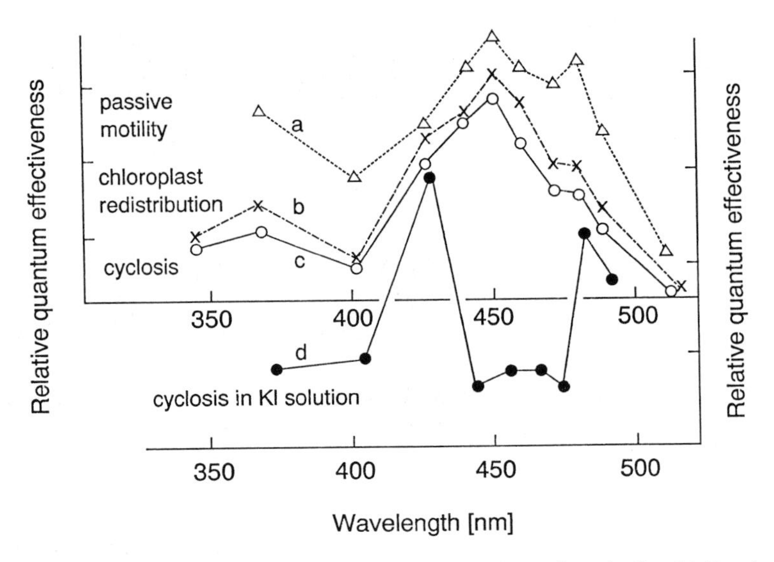

Figure 6. Action spectra of different cell responses in *Vallisneria spiralis.* (a) Passive motility (effect of centrifugation); (b) chloroplast redistribution (high-irradiance movement); (c) photodinesis (light-induced cyclosis of chloroplasts); (d) photodinesis in 0.2 *M* potassium iodide solution (KI). Modified after Seitz (1967a, b).

chloroplasts start to participate in the movement. Instead of continuous light, a short pulse (seconds or minutes) can be used to induce a photodinetic effect. In either case, the effect is only transient and depends on the irradiance of the continuous light or on the fluence of the pulse. This transiency is due to an adaptation to the new condition.

To a first approximation, the action spectrum of photodinesis (Fig. 6, curve c) is consistent with a flavin as the photoreceptor pigment. Moreover, the response is inhibited by potassium iodide (KI) solutions (Fig. 6, curve d). Although inhibitions by KI can be interpreted not only by its quenching the triplet-excited state of flavins and thus are no proof of flavin as the photoreceptor pigment, in this case the KI inhibition is important because it appears to be wavelength-specific, *i.e.* the main peaks of the putative flavin are more strongly depressed than the responses in neighbouring spectral regions. Whatever the mechanism of the KI effect, the latter observation clearly points to a second photoreceptor pigment.

This view is supported by comparative irradiance-response curves, which are steeper in those spectral regions where the additional pigment is assumed to contribute to the effect. For this second photoreceptor pigment, the most likely candidate is the B-absorbing band of chlorophyll *a* and accordingly a marginal effect of red light (R) can be observed.

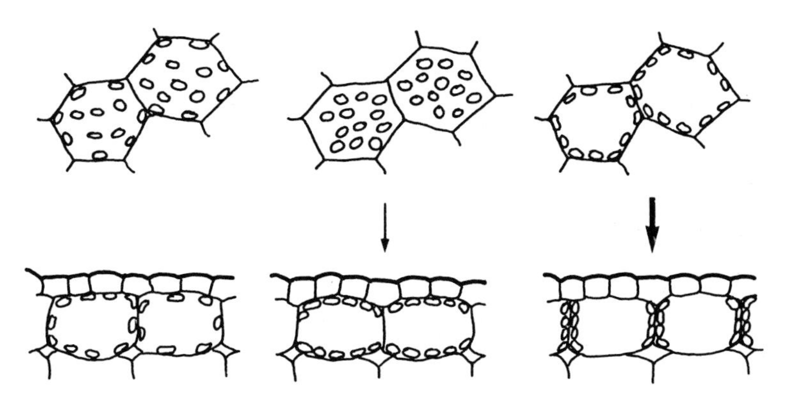

Figure 7. Chloroplast distribution in 'mesophyll' cells of *Lemna* in darkness (*left*), low irradiance (*middle*) and high irradiance (*right*); surface view (*above*) and cross section (*below*). After Haupt and Wagner (1984).

This conclusion has a very interesting consequence. Two photoreceptor systems can independently start the reaction chain leading to photodinesis, although they are localized in two different cell compartments: flavin in the cytoplasm, close to or even bound to the cell membrane (see below), and chlorophyll in the plastids.

To help elucidate the mechanism of light regulation of streaming, a second light effect in *Vallisneria* is important: The sedimentation of chloroplasts by centrifugal force is facilitated by light; action spectrum (Fig. 6, curve a) and fluence-response curves of this effect are very similar to those of photodinesis. Thus, light affects the anchoring of the chloroplasts to structures in the cortical cytoplasm, and it is tempting to relate this anchoring to actin microfilaments. However, for the motive force to be generated, actin has to interact with myosin as has been found for cytoplasmic streaming in Characean algae. As to the regulation of this actin-myosin interaction by light, no comprehensive model has yet been worked out, but bio-electric phenomena (see below) and effects *via* ATP may be involved.

9.4.3.2 Light-regulated chloroplast redistribution

Leaves of the moss *Funaria* and fronds of the duckweed *Lemna* are among the most thoroughly investigated higher plants with respect to chloroplast redistribution. As shown in Fig. 7, at low and medium irradiances chloroplasts always gather at those cell walls which are parallel to the surface (periclinal walls), *i.e.* perpendicular to the incident light; at high irradiances (*e.g.* direct sunlight), instead, they prefer the anticlinal walls, *i.e.* parallel to the incident light. These are the regions, where internal fluence rates are highest and lowest,

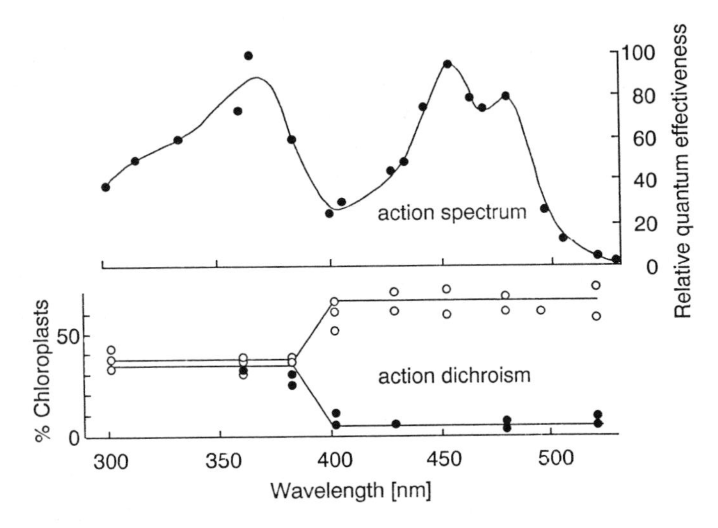

Figure 8. Action spectrum and wavelength-dependent action dichroism of chloroplast redistribution in *Funaria hygrometrica.* The action spectrum refers to the 'low-irradiance movement', the spectral action dichroism to the 'high-irradiance movement'. For the latter, percentage of chloroplasts per unit length of the anticlinal cell wall is given, with the electrical vector perpendicular (○) and parallel (●) to the wall, respectively. After Zurzycki in Haupt and Wagner (1984).

respectively, as can be shown in model calculations (Section 7.2.3.2). Thus, a gradient of light absorption in the cytoplasm is the orientating signal for the chloroplasts to move to their 'low-irradiance arrangement' or 'high-irradiance arrangement'. Accordingly, the main photoreceptor pigment resides in the cytoplasm rather than in the chloroplasts.

The action spectra for low-irradiance and high-irradiance movement are identical in *Lemna,* and the same is true in several other species. Moreover, there is close similarity between action spectra of different species (*cf.* Fig. 6, curve (b) with Fig. 8, top), indicating a common photoreceptor pigment. The maxima near 370 and 450 nm suggest that a flavin is involved This fits nicely the conclusion for photodinesis in *V. spiralis* (see above), taken together with the observation that in this species the action spectra for chloroplast redistribution, for photodinesis, and for light-facilitated centrifugation effects have almost identical maxima and minima (*cf.* Fig. 6, curves a, b and c).

Additional information is obtained by an action dichroism (Chapter 7.2), found in nearly all species, whenever the effect of polarized light has been tested. From the observed patterns of chloroplast distribution it can be concluded that absorption of linearly polarized light is highest along those anticlinal cell walls parallel to the electrical vector. This points to a dichroic orientation of the pigment molecules, which means that they are associated with stable cell structures, close to the cell membrane. Interestingly, in *Funaria* this action dichroism disappears at wavelengths shorter than 400 nm (Fig. 8, *bottom*), fitting the fact that in flavin molecules the transition moment for near UV absorption has an orientation different from that for the B absorption.

The relationship between chloroplast redistribution and light control of cytoplasmic streaming has been investigated in more detail in the coenocytic alga *Vaucheria.* In this alga, chloroplast distribution is very similar to that mentioned above, with low-irradiance and high-irradiance arrangement under respective conditions and random distribution in darkness. In addition, since a coenocyte is characterized by lack of cross walls, chloroplasts can also move over long distances in a longitudinal direction and thus migrate to irradiated regions of the tube: they accumulate at the irradiated part until, eventually, the whole vacuole in this region may be replaced by chloroplast containing cytoplasm.

To understand the response, we start with homogeneously irradiated *Vaucheria* tubes. Here the cytoplasmic layer exhibits streaming, visualized by movement of chloroplasts and other organelles, along near-longitudinal filamentous structures, which have been shown to be bundles of actin microfilaments. Apart from rotational streaming, cytoplasmic streaming in *Vaucheria* is not continuous, but reverses its direction every few minutes. If only a small part of the tube is irradiated (microbeam irradiation), cytoplasmic movement in this region stops, after a lag period of seconds or minutes, but continues in the adjacent dark regions. Thus, cytoplasm with chloroplasts, which enters the field, is trapped, resulting in chloroplast accumulation. Only the short-wavelength region is effective, and the photoreceptor pigment is probably localized in the fixed layer of cytoplasm close to the surface.

The first measurable effect of microbeam irradiation is an outward current of protons from the light field, thus causing a local hyperpolarization (Fig. 9). This effect has a fluence dependence and an action spectrum comparable to the chloroplast accumulation. Following hyperpolarization a structural change can be observed in the cytoplasm. The filamentous structures disperse and form a network instead (Fig. 10). It is proposed that this reticulation indicates inactivation of actin microfilaments, causing the cessation of chloroplast movement in the light field. After switching off the light, the reticulation disappears and the chloroplast aggregation is lost. Thus, light-induced pattern formation of chloroplasts is causally connected with a photodinetic effect, but interestingly, here we are dealing with a negative photodinesis, *i.e.* inhibition of streaming by light.

Thus, in *Vaucheria* we have at least an outline of a transduction chain, linking the light signal to the observed response. How light absorption by a yellow pigment (presumably a flavin) results in proton efflux and inactivation of actin is not yet known. Moreover, it is not advisable to generalize from this highly specific example to other systems showing chloroplast redistribution.

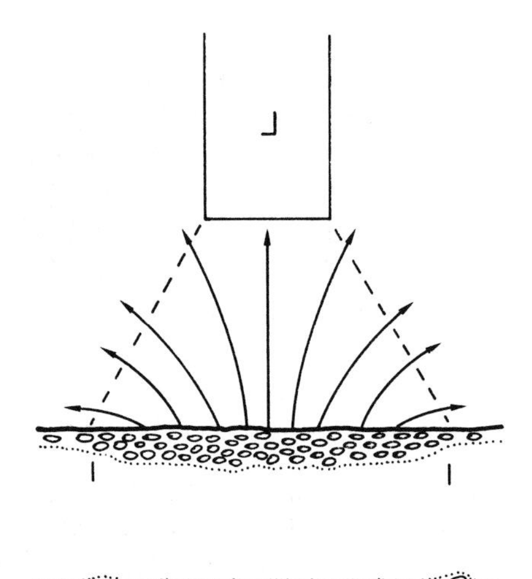

Figure 9. Electric field outside a 'cell' of *Vaucheria sessilis,* which is locally irradiated by an optical fiber (L), the light cone of which is indicated by the dashed lines. The local accumulation of chloroplasts is shown. After Blatt *et al.* (1981).

9.4.3.3 Mougeotia, *a special case of chloroplast movement*

The green fresh-water alga *Mougeotia scalaris* is particularly interesting for several reasons. In contrast to most green plants, there is no redistribution of chloroplasts in the cell, but the single, large, ribbon-shaped chloroplast rotates in the cell so as to expose its face to light at low and medium irradiance, or its edge to light at high irradiances. Moreover, phytochrome is the main photoreceptor pigment (Fig. 15 in Chapter 7.2), and accordingly, short pulses or flashes of light, even as short as 1 ms or less, can induce a full low-irradiance orientation during a subsequent dark period of, *e.g.* 15 min (Section 7.2.6.4). Thus extrapolation from other systems to *Mougeotia* and *vice versa* is hardly possible, and the transduction chain has to be analyzed separately. In the following paragraphs knowledge of the low-irradiance response is summarized.

As previously explained, phytochrome in *Mougeotia* is located in the cortical cytoplasm, *i.e.* in that layer which does not participate in the movement. Its dichroic orientation enables the cell to transform the light direction into a bisymmetric gradient of the far-red (FR)-absorbing active form of phytochrome (Pfr), and the chloroplast orientates in that gradient: its edge slides away from those regions with the highest Pfr level (see Fig. 15 in Chapter 7.2). As a result, the chloroplast ceases to rotate when its edges have reached the regions with the lowest Pfr level. Since Pfr is relatively stable for more than 1 h, and since it obviously does not diffuse in the cell, its gradient can guide the chloroplast in complete darkness to a precise orientation.

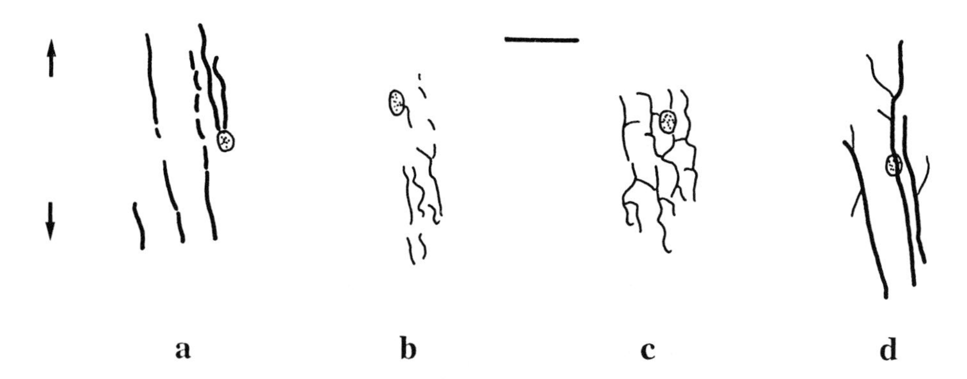

Figure 10. Pattern of actin microfilament bundles in *Vaucheria sessilis* under the influence of light. Surface view (grazing optical section); the longitudinal axis of the *Vaucheria* filament is indicated by two arrows. (a) At the onset of actinic irradiation, the dark configuration is seen. (b), (c) 30 s and 180 s, respectively, after light-on, disintegration of the fibres and reticulation proceeds. (d) 10 min after light-off, the reticulation disappears, and cable-like fibres reappear. One chloroplast is shown in each figure. Bar = 10 μm. Redrawn from microphotographs of Blatt and Briggs (1980).

In *Mougeotia* there is evidence that actin microfilaments play a role in the movement. These microfilaments can be identified and demonstrated by several means. They are mainly found in the cytoplasm close to the chloroplast edges, and anti-actin drugs efficiently inhibit the orientation movement. In contrast, failure of inhibition by colchicine makes it very unlikely that microtubules are involved in the generation of motive force.

A most interesting problem, then, is the transduction chain, which links the primary effect of light with the final response, *i.e.* the Pfr gradient with the activity of actin microfilaments (Haupt 1991). In other words: how is actin activity controlled by Pfr?

It is well known that calcium is a main controlling factor of actin activity, thus calcium may also be a good candidate for an internal signal in *Mougeotia.* Indeed, under certain conditions, depletion of calcium inhibits the response, and this inhibition parallels the loss of that fraction of calcium which is not easily extracted by water (bound calcium). Responsivity is restored by providing the cells with calcium. Moreover, calcium is concentrated in cytosomes, specific membrane-coated vesicles, which are also known as 'tannin vesicles'. The highest density in the cytoplasm is found near the chloroplast's edge (Fig. 11), *i.e.* near to the actin microfilaments. Finally, there is good evidence that the calcium-binding protein, calmodulin, is involved in the response. Yet, all these observations taken together do not prove that calcium is a link in the transduction chain in the sense of a directional internal signal. Instead, to-date more complicated models are under discussion (*e.g.* Fig. 12). It is obvious that generation of motive force by the actin-myosin system requires anchoring of

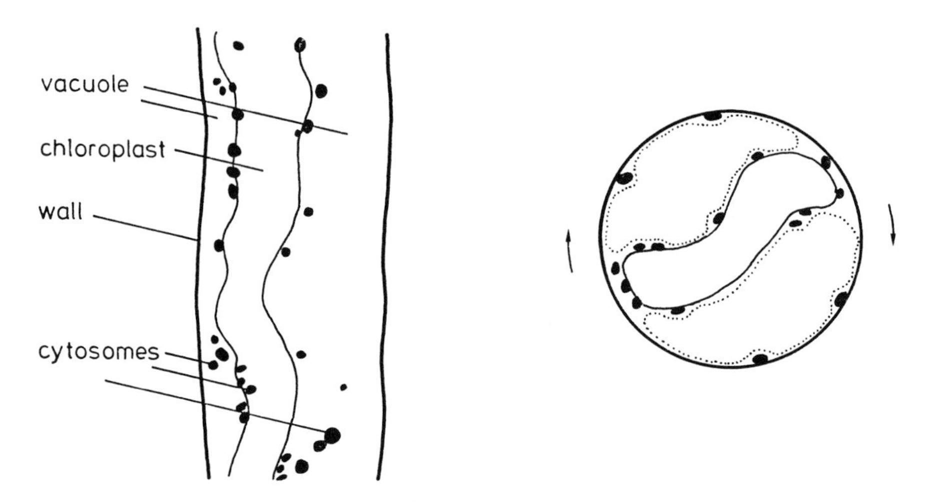

Figure 11. Calcium distribution in *Mougeotia scalaris* *Left:* combined surface view and longitudinal section, showing calcium (black) in cytosomes close to the chloroplast's edge. *Right:* cross section, showing the Ca^{2+}-containing cytosomes most abundant at the chloroplast's edge. Section was taken from a cell while the chloroplast performed its orientation movement (arrows). After Wagner and Rossbacher; Wagner and Klein in Haupt (1983).

actin microfilaments to the respective structures in addition to actin-myosin interaction. Both factors may be under the control of light in different ways. In *Mougeotia,* Pfr may control actin-myosin interaction *via* calcium-calmodulin in a scalar way, *i.e.* irrespective of a gradient; but this interaction can result in orientated movement, only if Pfr additionally controls, as a gradient, anchoring of actin to cell structures. No model can yet explain how this vectorial Pfr signal is transduced to the anchoring sites in the cell.

Two observations with B effects make the story even more complicated. (i) The low-irradiance response can be induced by B, as well as by R, although with a much lower efficiency. Interestingly, relatively low irradiances of B are effective even with a high-irradiance background of FR (Fig. 13, *bottom*), which maintains nearly all phytochrome in the red-absorbing form (Pr) (thus completely abolishing the effect of a R irradiation, Fig. 13, *top*). This means that B certainly acts *via* a separate photoreceptor pigment, which may well be comparable to the B photoreceptor in the other systems (*e.g. Vaucheria, Lemna, Funaria, Vallisneria,* judging from the action spectrum of the B effect). Not much is yet known about possible coaction or interaction of the two photoreceptor systems. (ii) Blue light is a prerequisite for the high-irradiance response. Although perception of light direction for this response also relies on a Pfr gradient, an additional absorption of strong B is required in order to reverse an edge-to-face movement to a face-to-edge movement in the same Pfr gradient, *i.e.* the chloroplast's edge is now attracted by Pfr rather than repelled.

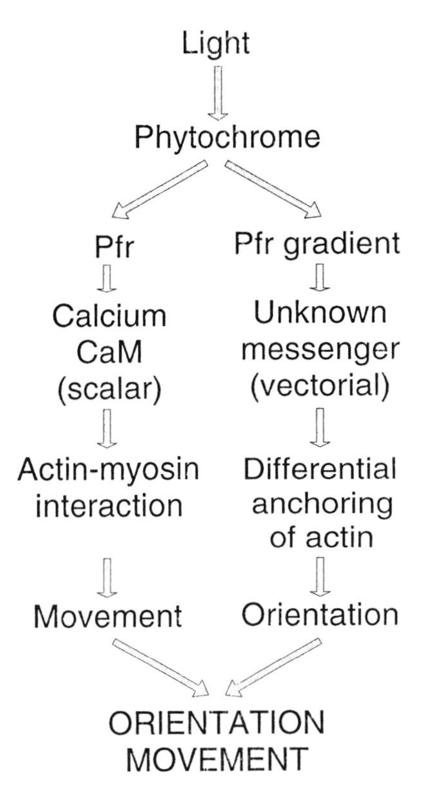

Figure 12. A model for a possible transduction chain of light-orientated movement of the *Mougeotia* chloroplast. The signal in general (scalar) and its direction (vectorial) are transduced by two different series of steps, with their integration resulting in the orientated movement. CaM = calmodulin. Partly after Grolig and Wagner in Haupt (1991).

Interestingly, the direction of this additional B has no bearing on the orientation of the chloroplast (Fig. 14a). The action spectrum of this B effect also has some similarity with that of the other responses mentioned before (*cf.* Fig. 14b with Figs. 6 and 8). It is not yet known whether a relationship exists between the two B systems acting in *Mougeotia.*

9.4.4 Synopsis

Four types of light-regulated intracellular movement have been discussed in this chapter: photodinesis as a non-orientated response; chloroplast redistribution to unidirectional light; chloroplast accumulation in *Vaucheria* upon microbeam irradiation; and orientated chloroplast rotation in *Mougeotia.* The first three types are typical B responses (more precisely: B/UV-A responses), and there is

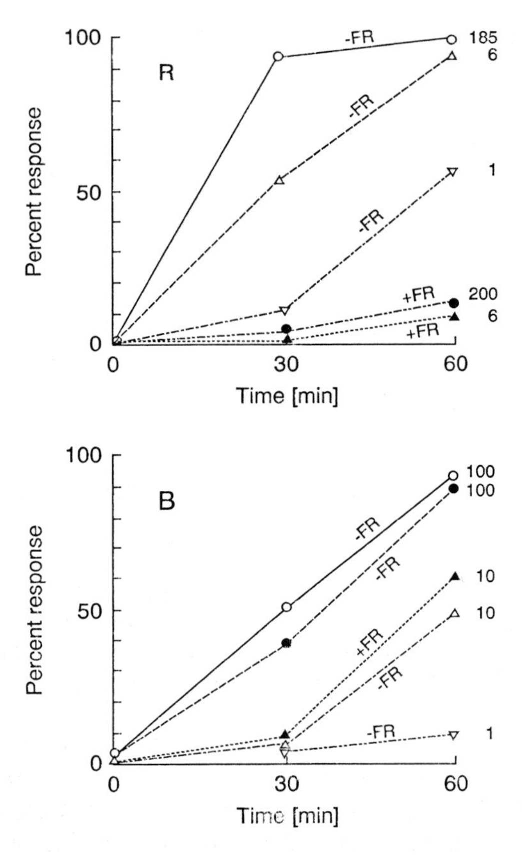

Figure 13. Chloroplast orientation in *Mougeotia* to red (R, *above*) and blue (B, *below*) light, respectively, with and without a strong far-red (FR) background irradiation. The fluence rates of R and B are indicated at the curves (W m^{-2}). The abscissa denotes the time after onset of continuous R or B, on the ordinate the % chloroplasts in face position is shown. After Gabryś *et al.* (1984)

no argument against a common photoreceptor pigment for all of them. Moreover, the B effects in *Mougeotia* may be attributed to the same system, but that must await confirmation. In addition, a few systems use phytochrome as a photosensory pigment for chloroplast orientation. In *Mougeotia*, phytochrome is much more effective than the B system, but in fern gametophytes (*Adiantum capillus veneris, Dryopteris sparsa*) both photoreceptor systems are about equally effective.

Remarkably, phytochrome shares with the B systems the dichroic orientation of the photoreceptor molecules and their likely location at or close to the cytoplasmic membrane. It may well be that association of a pigment with membraneous structures is required for its function in transforming light signals to internal signals as part of a transduction chain. Another interesting feature

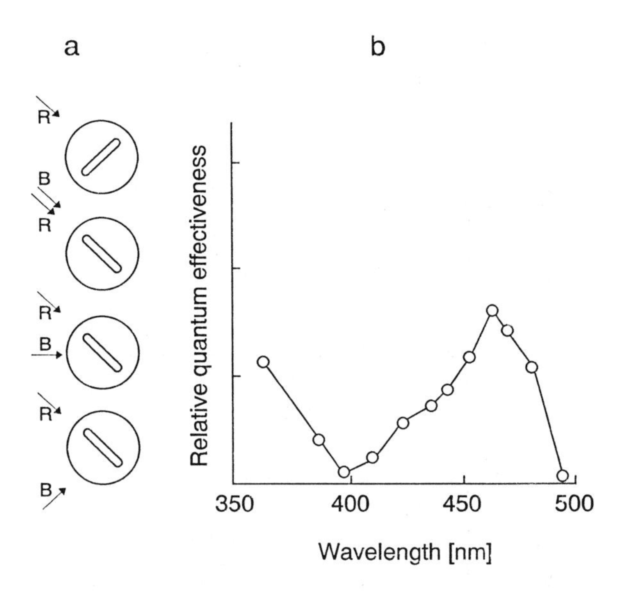

Figure 14. High-irradiance orientation in *Mougeotia.* (a) Cross-sections, showing the chloroplast orientation in red (R) and blue (B) light from different directions: face orientation to R alone, but profile to R, if B is given simultaneously from any direction. (b) Action spectrum of the 'switch function' of B. After Schönbohm in Haupt (1983) and Haupt and Wagner (1984).

which has been demonstrated is the duplicity of photoreceptor systems: photodinesis in *Vallisneria* independently makes use of two photoreceptor systems, which are located in different cell compartments, and in *Mougeotia*, a B system has been demonstrated to act independently of the most important phytochrome system. These observations may suggest that the photoreceptor systems in question were not evolved primarily to control intracellular movement, but that during evolution the latter has coupled to the already existing photoreceptor (and transduction?) systems.

In all examples so far, the moving force is generated by actin microfilaments, most probably by their interaction with myosin. However, much detail is still unknown, especially concerning the directed movements. Involvement of microtubules is occasionally proposed. This can most easily be understood as a mechanical inhibition of movement by these cytoskeleton elements.

As to the orientation of chloroplasts, two fundamentally different principles have been reported in this chapter. The response in *Vaucheria* is close to a trial-and-error mechanism: those chloroplasts that move by chance into the appropriate region are 'trapped' there, which eventually results in an accumulation pattern. In contrast, in *Mougeotia* there appears to be a strictly orientated movement, which brings the chloroplast into the new position. The final result, *viz.* optimization of absorption for photosynthesis, can obviously be achieved equally well by either principle.

9.4.5 Comparative conclusions

Superficially there is similarity between photomovement of organisms and that of cell organelles. Light-induced redistribution of chloroplasts especially resembles phototaxis of motile organisms, and indeed, it is frequently called 'chloroplast phototaxis'; yet, there is a fundamental difference. In true phototaxis, the moving organism itself has to perceive the light direction and to orientate to it, even if perception of direction shows a broad diversity both in principles and in complexity (see below). Contrastingly, in orientated chloroplast redistribution light direction is perceived by the 'environment' of the moving organelles. As a consequence, the organelles do not really 'orientate' to light, but to a so-called 'cytoplasmic gradient', which is the result of light absorption in this cytoplasmic environment. Thus, for chloroplast movements the term phototaxis should be avoided.

This difference in perception has consequences for the 'memory' of the responding system. A cytoplasmic gradient, most probably residing at the cell membrane, is not disturbed by the organelle movement; it can therefore control the movement for some time after a short light pulse. This after-effect is only limited by the life time of the localized products of light absorption, and hence of different duration: membrane hyperpolarization in *Vaucheria* lasts for a few minutes, but the Pfr gradient in *Mougeotia* is still effective after some hours. In contrast, motile organisms steadily change their spatial relation to the light source, they depend on the need to continually make new measurements, thus they have to clear their memory very fast. In consequence, short pulses have after-effects only insofar as without further stimuli, movement of the organism continues in the previous direction.

Thus, photomovement of organisms and photomovement of organelles are two completely different responses. Especially in the former a great diversity has been found concerning the photoreceptor pigments, whereas in chloroplast orientation two main pigment systems appear to serve light perception. Moreover, in the photomovement of organisms the same final response (*e.g.* phototaxis) can be achieved by various mechanisms, if different species are considered; this diversity comprises perception, transduction, and response proper. For example perception of light direction can be restricted to a primitive 'front-rear distinction' (*Phormidium*), or it can be a true recognition of the direction, be it by an absorption gradient at a given moment (*Dictyostelium, Anabaena*), or by periodic changes of light absorption at a single photoreceptor site, due to rotation of the cell (*Euglena*).

9.4.6 Further reading

Colombetti G. and Lenci F. eds. (1984) *Membranes and Sensory Transduction*, Plenum Press, New York.

Colombetti G., Lenci F., and Song P.-S. (eds.) (1985) *Sensory Perception and Transduction in Aneural Organisms,* Plenum Press, New York.
Douglas R. H., Moan J., and Dall'Acqua F. (eds.) (1988) *Light in Biology and Medicine,* Plenum Press, New York, London.
Häder D.-P. (1979) Photomovement. In: *Encyclopedia of Plant Physiology,* New Series, vol. 7, *Physiology of Movements,* pp. 268-309, Haupt W. and Feinleib M.E. (eds.) Springer-Verlag, Berlin.
Häder D.-P. and Tevini M. (1987) *General Photobiology,* Pergamon Press, Oxford.
Haupt W. and Feinleib M.E. (eds.) (1979) *Encyclopedia of Plant Physiology,* New Series, Vol. 7, *Physiology of Movements,* Springer-Verlag, Berlin.
Haupt W. and Scheuerlein R. (1990) Chloroplast movement. *Plant Cell Environ.* 13: 595-614.
Lenci F., Ghetti F., Colombetti G., Häder D.-P. and Song P.-S. (eds.) (1991) *Biophysics of Photoreceptors and Photomovements in Micro-organisms,* NATO-ASI Series Plenum, New York.
Nultsch W. and Häder D.-P. (1988) Photomovement in motile micro-organisms II. *Photochem. Photobiol.* 47: 837-869.
Wada M., Grolig F. and Haupt W. (1993) Light-oriented chloroplast positioning. Contribution to progress in photobiology. *J. Photochem. Photobiol. B.* 17: 3-25.

9.4.7 References

Armitage J.P. (1991) Photoresponses in eubacteria. In: *Biophysics of Photoreceptors and Photomovements in Micro-organisms,* pp. 43-52, Lenci F., Colombetti G., Ghetti F., Häder D.-P. and Song P.-S. (eds.) NATO-ASI Series Plenum, New York.
Blatt M.R. and Briggs W.R. (1980) Blue light induced cortical fiber reticulation concomitant with chloroplast aggregation in the alga *Vaucheria sessilis. Planta* 147: 355-362.
Blatt M.R., Weisenseel M.H. and Haupt W. (1981) A light-dependent current associated with chloroplast aggregation in the alga *Vaucheria sessilis. Planta* 152: 513-526.
Brodhun B. and Häder D.-P. (1990) Photoreceptor proteins and pigments in the paraflagellar body of the flagellate, *Euglena gracilis. Photochem. Photobiol.* 52: 865 - 871.
Fisher P.R. and Williams K.L. (1981) Bidirectional phototaxis by *Dictyostelium discoideum* slugs. *FEMS Microbiol. Lett.* 12: 87-89.
Gabryś H., Walczak T. and Haupt W. (1984) Blue-light-induced chloroplast orientation in *Mougeotia.* Evidence for a separate sensor pigment besides phytochrome. *Planta* 160: 21-24.
Häder D.-P. (1987a) Photosensory behavior in procaryotes. *Microbiol. Rev.* 51: 1-21.
Häder D.-P. (1987b) Polarotaxis, gravitaxis and vertical phototaxis in the green flagellate, *Euglena gracilis. Arch. Microbiol.* 147: 179-183.
Häder D.-P. (1988) Ecological consequences of photomovement in micro-organisms. *J. Photochem. Photobiol. B.* 1: 385-414.
Häder D.-P., Poff K.L. (1979) Light-induced accumulations of *Dictyostelium discoideum* amoebae. *Photochem. Photobiol.* 29: 1157-1162.
Haupt W. (1983) Movement of chloroplasts under the control of light. *Prog. Phycol. Res.* 2: 227-281.
Haupt W. (1991) Introduction to photosensory transduction chains. In: *Biophysics of Photoreceptors and Photomovements in Micro-organisms,* pp. 7-19, Lenci F., Colombetti G., Ghetti F., Häder D.-P. and Song P.-S. (eds.) NATO-ASI Series Plenum, New York.
Haupt W. and Wagner G. (1984) Chloroplast movement. In: *Membranes and Sensory Transduction,* pp. 331-375, Colombetti G. and Lenci F. (eds.) Plenum Press, New York.
Hegemann P. (1991) Photoreception in *Chlamydomonas.* In: *Biophysics of Photoreceptors and Photomovements in Micro-organisms,* pp. 223-229, Lenci F., Colombetti G., Ghetti F., Häder D.-P. and Song P.-S. (eds.) NATO-ASI Series Plenum, New York.

Liu S.-M., Häder D.-P. and Ullrich W. (1990) Photoorientation in the freshwater dinoflagellate, *Peridinium gatunense* Nygaard. *FEMS Microbiol. Ecol.* 73: 91-102.

Marwan W. and Oesterhelt D. (1991) Light-induced release of the switch factor during photophobic responses of *Halobacterium halobium*. *Naturwiss.* 78: 127-129.

Oesterhelt D. and Marwan W. (1990) Signal transduction in *Halobacterium halobium*. In: *Biology of the Chemotactic Response.* Vol. 46, pp. 219-239, Armitage J.P. and Lackie J.M. (eds.) Cambridge University Press.

Seitz K. (1967a) Wirkungsspektren für die Starklichtbewegung der Chloroplasten, die Photodinese und die lichtabhängige Viskositätsänderung bei *Vallisneria spiralis* ssp. *torta*. *Z. Pflanzenphysiol.* 56: 246-261.

Seitz K. (1967b) Eine Analyse der für die lichtabhängigen Bewegungen der Chloroplasten verantwortlichen Photorezeptorsysteme bei *Vallisneria spiralis* ssp. *torta*. *Z. Pflanzenphysiol.* 57: 96-104.

Sineshchekov O.A. (1991) Electrophysiology of photomovements in flagellated algae. In: *Biophysics of Photoreceptors and Photomovements in Micro-organisms,* pp. 191-202, Lenci F., Colombetti G., Ghetti F., Häder D.-P. and Song P.-S. (eds.) NATO-ASI Series Plenum, New York.

Takahashi T. (1992) Automated measurements of movement responses in halobacteria. In: *Image Analysis in Biology,* pp. 315-328, Häder D.-P. (ed.) CRC Press, Boca Raton.

Vornlocher H.-P. and Häder D.-P. (1992) Isolation and characterization of the putative photoreceptor for phototaxis in amoebae of the cellular slime mold, *Dictyostelium discoideum*. *Bot. Acta.* 105: 47-54.

9.5 Photocontrol of flavonoid biosynthesis

Christopher J. Beggs[1] and Eckard Wellmann

Institut für Biologie II/Botanik, Albert-Ludwigs-Universität, Schänzlestr. 1, 79104 Freiburg, Germany
[1]Shell Forschung GmbH, Zur Propstei, Postfach 100, D-6501 Schwabenheim, Germany

9.5.1 Introduction

The *flavonoids* are derived from the flavan or isoflavan skeleton (Fig. 1) and comprise a large group of secondary metabolites from higher plants (Harborne *et al.* 1975). In the literature the term flavonoid is often used to mean all flavonoids except *anthocyanins* and for the sake of convenience we have followed this usage. Anthocyanins are 3- or 3,5-glycosides of anthocyanidins (Fig. 2) and include the principal red (R), violet and blue (B) plant pigments (λ_{max} 520-545 nm).

The first modern observations on photocontrol of flavonoid synthesis were recorded in the late 18th century by J. Senebier, who noticed that, in certain plants, R pigments (now known to be anthocyanins) were only produced under the influence of light, whereas in other plants these pigments were always present, regardless of the irradiation conditions. In the 19th century these results were confirmed and enlarged upon by J. Sachs and H.C. Sorby. In the present century, interest began to centre more on which spectral wavebands were responsible for these light effects. Early work (see Arthur 1936 for a summary) considered that UV-B and B/UV-A were the most effective wavebands in stimulating anthocyanin synthesis. However, it was later noted that light of longer wavelengths, in particular R, was also effective (Withrow *et al.* 1953). Similar observations were made for flavonoids other than anthocyanins for both UV and R (Piringer and Heinze 1954). As these observations were being made, interest was growing in the general problem of control of plant morphogenesis by light and the mechanisms and photoreceptors involved (Borthwick 1972). It was not long before anthocyanin synthesis was also being used as a useful model system for photomorphogenic studies. This led again to more rapid progress in understanding the mechanisms involved in light-stimulated anthocyanin and flavonoid synthesis.

R.E. Kendrick & G.H.M. Kronenberg (eds.), Photomorphogenesis in Plants - 2nd Edition

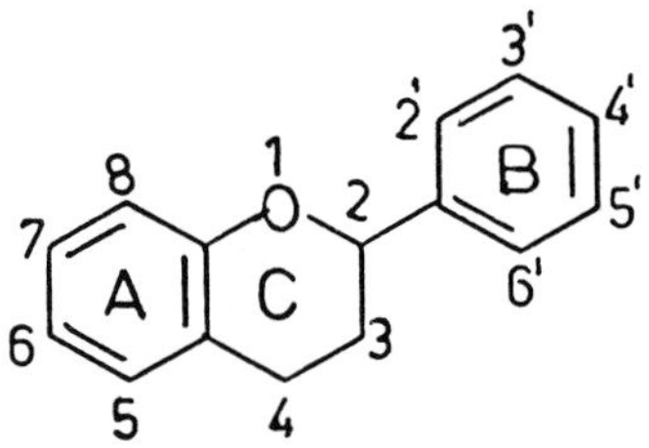

Figure 1. Flavan (1) and isoflavan (2) skeleton. Numbering in isoflavan is the same as in flavan.

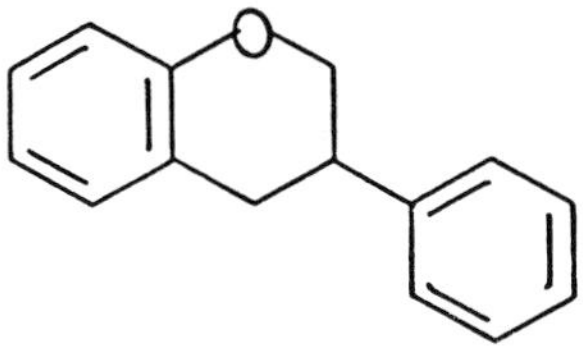

9.5.2 Flavonoid biosynthesis

The biosynthesis of flavonoids has been covered in recent reviews (Hahlbrock and Scheel 1989; Harborne 1993) in which details concerning individual reactions can be found. In this chapter pathways are only described which are known to be influenced by light. As shown in Fig. 3 the carbon atoms in the flavan skeleton originate from two pathways: ring B with carbon atoms 2, 3 and 4 is derived from a phenylpropane unit and the carbon atoms of ring A from a head to tail condensation of acetate units. Early chemogenic studies of flower colour had indicated that, in the formation of the various flavonoid classes, there is competition for a common precursor. Tracer experiments and enzymatic studies proved that this common precursor is a chalcone (Fig. 2) originating in the plant from 4-coumaroyl-CoA and three molecules of malonyl-CoA by action of the enzyme *chalcone synthase* (CHS)(Fig. 4). Further transformations of the chalcone to various flavonoids are outlined in Fig. 2. Whereas aurone is probably derived directly from chalcone, all other flavonoids, including the isoflavones are synthesized *via* the isometric (2S) flavanone. Dihydroflavonols are biosynthetic intermediates for anthocyanidins (*via* flavan 3,4-diols), flavonols, and catechins. The known enzymes catalysing these transformations are listed in Fig. 2.

Coumaroyl-CoA, which is one of the substrates of chalcone synthase, is derived from the so called 'general' phenylpropanoid metabolism. This is

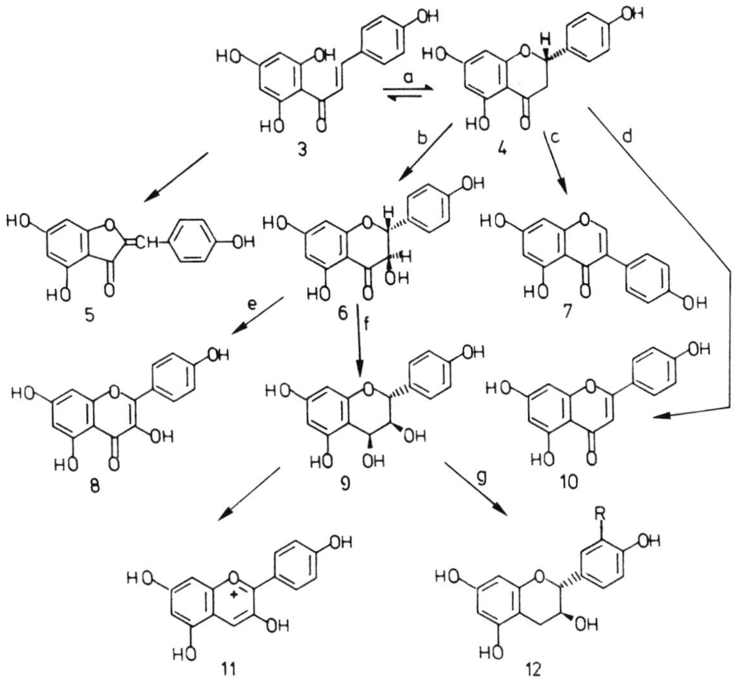

Figure 2. Transformation of tetrahydroxychalcone (3) to various flavonoids. 4, flavanone (2S naringenin); 5, aurone; 6, dihydroflavonol (dihydrokaempferol); 7, isoflavone (genistein); 8, flavonol (kaempferol); 9, flavan-3,4-diol (leucopelargonidin); 10, flavone (apigenin); 11, anthocyanidin (pelargonidin); 12, catechin (R = OH);. Known enzymes: a, chalcone isomerase; b, flavanone 3-hydroxylase; c, 'isoflavone synthase' (probably two enzymes); d, flavone synthase; e, flavonol synthase; f, dihydroflavonol 4-reductase; g, flavan-3-ol synthase.

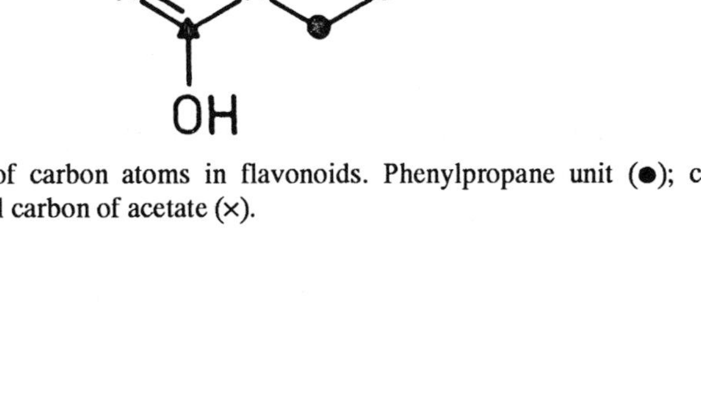

Figure 3. Origin of carbon atoms in flavonoids. Phenylpropane unit (●); carboxyl carbon of acetate (▲); methyl carbon of acetate (×).

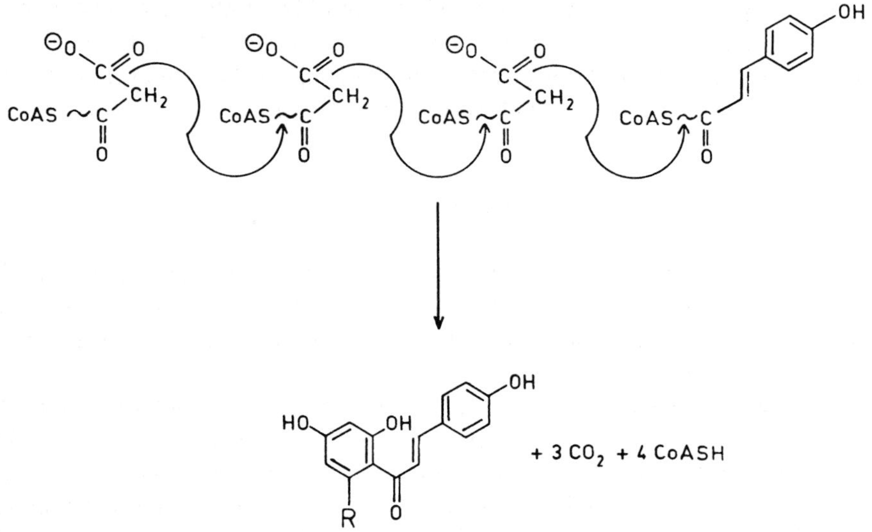

⊖O C O CH2 CoAS ~ C O
⊖O C O CH2 CoAS ~ C O
⊖O C O CH2 CoAS ~ C O
CoAS ~ C O OH

HO OH OH R O + 3 CO_2 + 4 CoASH

Figure 4. Chalcone synthase reaction (R = OH). The deoxychalcone synthase leading to deoxychalcone (R = H) is not known.

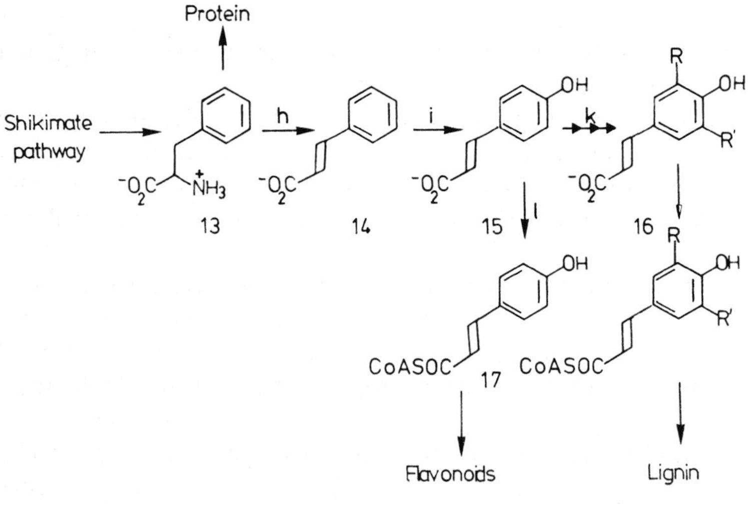

Protein
Shikimate pathway
h i k l
^-O_2C $\overset{+}{N}H_3$ 13
^-O_2C 14
^-O_2C OH 15
^-O_2C R OH R' 16
CoASOC OH 17
CoASOC R OH R'
Flavonoids
Lignin

Figure 5. General phenylpronanoid metabolism. 13, L-phenylalanine; 14, *trans* cinnamic acid; 15, 4-coumaric acid; 16, ferulic (R = OCH_3) acid; 16, sinapic (R = R' = OCH_3) acid; 17, 4-coumaroyl-CoA. Enzymes: h, Phenylalanine ammonia-lyase; i, cinnamate 4-hydroxylase; k, hydroxylases and methylases; l, 4-coumarate: CoA ligase.

defined as the sequence of reactions involved in the conversion of L-phenylalanine to substituted cinnamoyl-CoA esters (Fig. 5).

Chalcone synthase, the key enzyme of flavonoid biosynthesis and some other enzymes of the flavonoid pathway have been studied at the level of gene activities leading to a profound understanding of light action on primary regulative processes of flavonoid biosynthesis (Chapter 8.1).

9.5.3 The photoreceptors and effective wavebands

This section concerns the wavebands which are most effective in stimulating flavonoid synthesis and, where known, the photoreceptors involved. Properties of the photoreceptors are not discussed and the reader is referred to the relevant chapters of this book.

Although the B/UV responses were the first discussed by Arthur (1936), the first light effects to be rigorously studied were those to R and far-red light (FR) involving the R-FR photoreversible pigment, *phytochrome,* as photoreceptor. The property whereby a pulse of R led to a photoresponse and a subsequent pulse of FR could negate the response had led to the discovery of this photoreceptor by the H.A. Borthwick and S.B. Hendricks' group at Beltsville USA in the 1940's (Borthwick 1972). Curiously, the first pigment system for which phytochrome control was shown was not for the red anthocyanins, but for a flavonoid in tomato *(Lycopersicon esculentum)*-fruit skin (Piringer and Heinze 1954). Several years later, phytochrome control of anthocyanin synthesis in mustard *(Sinapis alba)* (Mohr 1957) and red cabbage (Siegelman and Hendricks 1957) seedlings was demonstrated.

Studies on photocontrol of anthocyanin synthesis at this time became much involved with discussions on the nature of the photoreceptor for the so-called *high irradiance responses* (HIRs). It had been observed by Mohr (1957) and Siegelman and Hendricks (1957) that, although when pulse irradiation was applied, R was most effective for inducing anthocyanin synthesis and this effect could be reversed by a FR pulse, this was only the case for short term pulse irradiation. *Action spectra* (Fig. 6) for light-induced anthocyanin synthesis in various plant systems, where long-term irradiation was applied, showed maximal stimulation in the FR with considerably less effectiveness in the R region of the spectrum. Later observations showed that similar action spectra for long-term irradiation also occurred for other photoresponses: *e.g.* hypocotyl elongation inhibition. The experiments of K.L. Hartmann showed conclusively that phytochrome and not a separate FR-absorbing pigment was the photoreceptor for these responses to FR. A further problem is the relevance of the FR reaction under natural conditions. Continuous monochromatic FR is not effective in most photomorphogenic responses of white light (W)-grown tissue. For further discussion of HIRs the reader is referred to Chapter 4.7.

As mentioned above, many action spectra for continuous light photomorphogenic responses (including those for anthocyanin and flavonoid synthesis) show

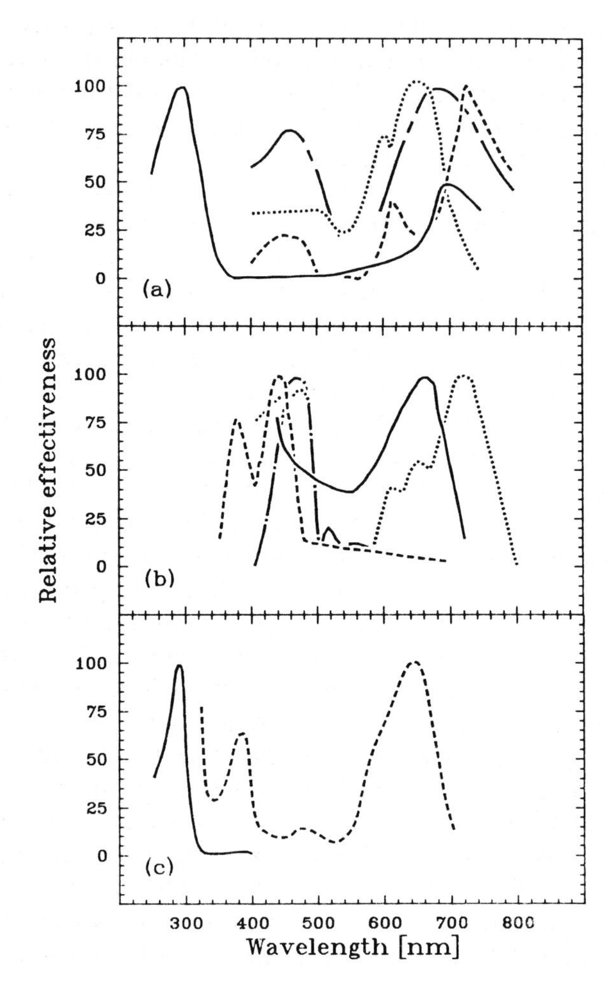

Figure 6. Some examples of action and response spectra for anthocyanin synthesis under continuous irradiation (a) ——— = *Spirodela oligorrhiza*; — — — = red cabbage (*Brassica oleracea*); – – – = turnip (*Brassica rapa*); ········· = apple skin (*Malus* sp.). (b) ——— = *Haplopappus gracilis* cotyledons; —·— = *Sorghum vulgare*; – – – = *Haplopappus gracilis* seedlings; ········· = mustard (*Sinapis alba*). (c) *Sorghum bicolor*, ——— = UV-B effectiveness, – – – = UV-A to red effectiveness. UV-B is drawn to a separate logarithmic scale.

considerable effectiveness in the B/UV waveband (Fig. 6). It has long been a problem in photomorphogenesis research, whether the photoreceptor for such responses is phytochrome, which also absorbs in the B or if they are due to a separate B receptor (Chapter 5.1). Phytochrome, although absorbing in the B, shows low effectiveness with B pulses which may, in part, be due to optical screening effects. For continuous light responses, the situation remains unclear, but recent studies on HIR-mediated growth responses make it very likely that a separate B receptor is at least partially involved. In some plants, *e.g. Sorghum vulgare,* B is effective in inducing anthocyanin formation where R is ineffective (Fig. 6; Downs and Siegelman 1963). Here it is even more certain that a separate photoreceptor is involved. A further group of photoreceptors are those absorbing UV-B radiation. Although the effect of UV-B on anthocyanin has long been known (Arthur 1936) and later experiments on effects of sunlight on flavonoid and anthocyanin formation showed that sunlight filtered through window glass was often less effective than unfiltered sunlight, a definite characterisation of the effect was first published by E. Wellmann. He showed that in parsley cell suspension cultures and seedlings, flavonoids could be induced by UV-B radiation. The UV-B was an essential prerequisite for flavonoid synthesis and other photoreceptors (phytochrome and a B photoreceptor) were only effective if UV-B had also been applied. An action spectrum (Fig. 7), showed that maximal effectiveness was at about 300 nm with little effectiveness above 320 nm. Furthermore, the effect showed a linear fluence-response relationship. It was later possible to show that such responses were widespread in plant systems. Table 1 shows examples of plant systems where UV-B has been shown to be effective in inducing flavonoid or anthocyanin synthesis and Fig. 7 shows several typical action spectra for UV-B induced responses. A problem remains as to the identity of the photoreceptor. Phytochrome also absorbs in the UV-B, but this would not explain the frequent absence of a response to R or B given alone. A detailed investigation of light-controlled anthocyanin synthesis in *Sorghum bicolor* was made by Yatsuhashi *et al.* (1982). These authors published an action spectrum (Fig. 6) for the response from 250 to 700 nm and found peaks of effectiveness at 290, 385, 480, and 650 nm. They were able to nullify all responses by subsequent FR with the exception of that to 290 nm radiation. They thus argued that the 290 nm peak was not due to phytochrome, although the others probably were. If one accepts this viewpoint one is left with the problem of what the UV-B receptor might otherwise be.

In the Leguminosae, DNA itself can probably act as a photoreceptor for stimulation of flavonoid synthesis. Members of this family can synthesize *isoflavonoids* either constitutively or in response to stress or disease. One such stress is UV. In those legumes tested, irradiation with short-wavelength UV-B or UV-C leads to formation of isoflavonoids. The action spectrum (Beggs *et al.* 1985 and Fig. 7) is shifted somewhat to shorter wavelengths when compared with that for UV-B induction of flavonoids or anthocyanins (Wellmann 1983; Yatsuhashi *et al.* 1982; Beggs and Wellmann 1985) and the effects can be photoreversed (photorepaired) by simultaneous or subsequent irradiation with UV-A or B (Beggs *et al.* 1985). This property is similar to the phenomenon of *photoreactivation* (Chapter 5.1) and implies that the primary photoreceptor is DNA itself in which cyclobutane-type pyrimidine dimers are formed by the action of the UV-B or -C and which can subsequently be repaired (monomer-

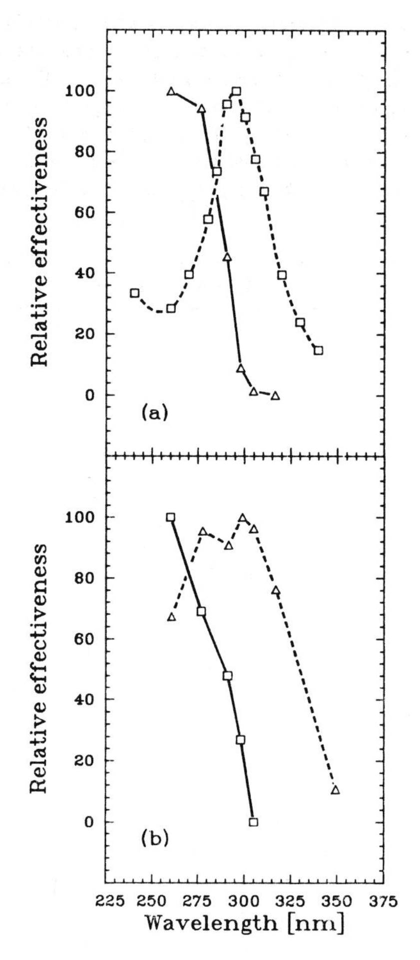

Figure 7. Some examples of action and response spectra for UV-B effects on anthocyanin and flavonoid synthesis. (a) Δ——Δ = UV-B inhibition of phytochrome-induced anthocyanin synthesis in mustard (*Sinapis alba*), an example of a UV-B/UV-C damage action spectrum (Wellmann *et al.* 1984), □– – –□ = flavonoid synthesis in parsley (*Petroselinum hortense*) cell cultures (E. Wellmann unpublished data). (b) □——□ = Isoflavonoid (coumestrol) synthesis in bean (*Phaseolus vulgaris*) primary leaves (Beggs *et al.* 1985), Δ– – –Δ = anthocyanin synthesis in maize (*Zea mays*) coleoptiles (Beggs and Wellmann 1985).

ized) by the B/UV-A dependent photoreactivating enzyme (photolyase). Although probably mediated *via* damage to DNA, the response should still be seen as a specific one, where a particular biosynthetic pathway is activated by this damage. It is not the same as the effect of UV (of the same wavelengths) in inhibiting phytochrome-induced anthocyanin synthesis in mustard (Fig. 7; Wellmann *et al.* 1984) or UV induced anthocyanin synthesis in broom *Sorghum*

Table 1. Some examples of UV-B induced anthocyanin and flavonoid formation in various plant species: Column three shows the approximate irradiation time required to induce significant pigment formation when irradiated with monochromatic UV (298 nm) at a fluence rate of 0.25 $\mu mol \cdot m^{-2} s^{-1}$. Pigments were extracted 24 h after the beginning of the irradiation. Between irradiation and extraction the material was kept in the dark at 25°C. A = anthocyanin, F = flavonoid. (From Beggs C.J., Schneider-Ziebert U. and Wellmann E. (1986) In: *Stratospheric Ozone Reduction. Solar Radiation and Plant Life,* pp. 235-250, Worrest R.C. and Caldwell M.M. (eds.) Springer, Berlin.)

Plant system	Pigment induced	Irradiation time
Petroselinum crispum		
hypocotyl	F	min
seedling root	F	min
cell cultures	F	s
Anethum graveolens		
hypocotyl	F	h
seedling root	F	min
cell cultures	F	min
Daucus carota		
hypocotyl	F	h
seedling root	F	min
cell cultures	F	min
Rumex patientia		
hypocotyl	A	h
Sinapis alba		
hypocotyl	F, A	h
seedling root	F	min
cell cultures	F	min
Lepidium sativum		
seedling root	F	min
Zea mays		
coleoptile	A	s
Triticum aestivum		
coleoptile	F, A	min, h
Secale cereale		
coleoptile	F, A	min
Anthirrhinum majus		
hypocotyl	F	h
seedling root	F	h
Nicotiana tabacum		h
hypocotyl	F	h
seedling root	F	

(Hashimoto *et al.* 1991) which is also photoreactivable, but is probably a simple damage effect.

Similar responses to UV have been described for stilbene formation in vine and peanuts (*Arachis hypogaea*) (a chemotaxonomically aberrant member of the Leguminosae). Stilbenes have a biosynthetic pathway initially in common with that of flavonoids and isoflavonoids.

Many isoflavonoids show marked antifungal properties and are induced in leguminous plants by fungal attack and a number of so-called 'elicitors'. The relevance of UV in this response, is, however, not clear as the wavelengths responsible do not appear in sunlight at the earth's surface. It is more likely that the response is part of the plant's general response to stress, whereby phytoalexins are produced when the plant is weakened and therefore more susceptible to fungal attack.

9.5.4 Coactions between the photoreceptors

The situation in *S. vulgare* and parsley, where a B or UV-B treatment respectively is required before the other photoreceptors can act has been briefly referred to in the previous section. Such coactions appear to be fairly frequent in plant photomorphogenesis. Coaction refers to the situation where one photoreceptor appears to modify the response to another, so that the response to both is greater (or, in principle, also less) than the sum of the responses to each photoreceptor alone (Chapter 6.).

In *S. vulgare* and parsley, it could be shown (Downs and Siegelmann 1963; Wellmann 1983) that there was no effect of R or FR on anthocyanin or flavonoid synthesis when these wavelengths were given alone. On the other hand, irradiation with B or UV-B gave a response. However, this response was then under the control of phytochrome. Blue light or UV-B followed by FR was considerably less effective than B or UV-B alone or than these wavebands followed by R. In certain varieties of maize, the situation appears to be the reverse of that in *S. vulgare* (Beggs and Wellmann 1985). Here, UV-B is ineffective in inducing anthocyanin synthesis unless the plants have previously been irradiated with continuous R (which itself has no effect when given alone, although continuous FR alone is very effective). In other varieties, UV-B is effective alone, but a R pretreatment increases the response to UV-B. Recently, Lercari *et al.* (1989) reported considerable differences between cabbage varieties in the responsiveness to UV and R with respect to anthocyanin formation.

As an example of the possible complexity of interactions between photoreceptors, Duell-Pfaff and Wellmann (1982), using a parsley cell culture, investigated, in more detail, the situation in flavonoid synthesis. Red light, B and FR given alone for 6 h have no effect, whereas 10 min UV is very effective. If given for 6 h after the UV, all three wavebands increased the response to UV in a fluence rate-dependent response. Blue light was by far the most effective waveband. A short FR pulse given after the UV decreases the response to UV, as might be expected, but continuous FR increases the response. If given before the UV pulse, the three wavebands also increased the response, in the case of B and FR more so than when given after the UV treatment. The effects of these continuous B or R pre- or post-treatments could be substituted for by several hourly-spaced pulses. This was not the case for FR. The effect of UV appeared to require a specific level of the FR-absorbing form of phytochrome (Pfr), because when saturating FR pulses establishing different Pfr levels were given after the UV, the response was dependent on the Pfr level. Furthermore, it appears that the effect of B is, at least in part, due to a separate B photoreceptor (Duell-Pfaff and Wellmann 1982).

Little is known concerning the mechanisms involved in the interactions between the UV-B and B/UV-A absorbing photoreceptors and phytochrome. One of the problems receiving the most attention has been whether phytochrome in the Pfr form is necessary for a response to UV or B. As described above, some of the responses to B or UV can be reversed to some extent by a subsequent FR pulse. This implies a requirement for Pfr. The problem is that frequently the response cannot be fully reversed by FR and in some cases no reversion can be demonstrated. Thus flavonoid production in parsley-cell cultures can be reduced to about 45% of the UV or UV + R level by a subsequent FR pulse and even to about 10% in the case of low UV fluence (Duell-Pfaff and Wellmann 1982). In the other well studied system, anthocyanin synthesis in *Sorghum,* the situation is also confusing. Yatsuhashi *et al.* (1982) found no reversion of UV-B induced anthocyanin synthesis in *S. bicolor* whereas in *S. vulgare* reversion was possible (Drumm and Mohr 1978). Oelmüller and Mohr (1985) were able to show that, in *S. vulgare,* complete reversion was obtained if the period of UV irradiation was short (less than 10 min). If the UV treatment was longer than about 10 min complete reversion could not be attained. This result is in harmony with the situation in parsley-cell cultures where greater reversion was attainable if the UV fluence applied was low. Yatsuhashi *et al.* (1982), however, found no reversion for both 2 and 10 min of UV irradiation in *S. bicolor.* By means of simultaneous irradiation with B and FR, Drumm and Mohr (1978) were able to show that, in *S. vulgare,* Pfr was apparently not necessary during a 3 h period of B irradiation. A R pulse given after the B treatment was sufficient to satisfy the apparent requirement for Pfr. The often incomplete reversion reported has often been explained by either a saturation of the response to Pfr at very low Pfr levels or a very fast loss of photoreversibility whereby the Pfr effect already occurs during the UV or B irradiation period. The second possibility is neither supported by the simultaneous irradiation experiments of Drumm and Mohr (1978) nor by the fact that even very long UV or B treatments can often be significantly reversed by FR. The first possibility is almost impossible to exclude fully as UV and B always produce some Pfr and 100% reversion to the R-absorbing form of phytochrome (Pr) by FR cannot be achieved. The existence of systems where 100% reversion can be achieved (Oelmüller and Mohr 1985) suggests that it is at least not a universally applicable explanation. In summary, the answer to the problem is unclear. In most systems, however, it would seem that Pfr is at least necessary for full expression of the response. Where activation of more than one photoreceptor is obligatory, no information is available as to how they coact. Oelmüller and Mohr (1985) have suggested that B/UV somehow modify the plant system so as to allow Pfr to act, *i.e.* phytochrome is the photoreceptor which actually 'switches on' the flavonoid synthesis machinery and the B/UV photoreceptors confine their action to establishing a situation where Pfr can perform this task. Nevertheless, the evidence, at present available, equally permits the reverse situation, that Pfr establishes the necessary circumstances for the UV-B and B photoreceptors to be able to 'switch on' flavonoid synthesis.

9.5.5 Mode of action of light-induced flavonoid synthesis

The greatest progress in elucidating the mode of action of light-induced flavonoid synthesis has been made by K. Hahlbrock and co-workers using the parsley-cell suspension culture system. These cultures produce flavonoids only upon irradiation. The work on parsley was carried out using a broadband W source with a high B and UV content. Thus it is not possible to make specific statements with respect to the role of particular photoreceptors in stimulating the effects described above. It is known from the work of E. Wellmann that phytochrome, a B receptor and a UV-B photoreceptor are involved in the parsley response and all three photoreceptors would be stimulated by the light source used in these experiments. Nevertheless, the work described below delivers an almost complete picture of the events leading from gene activation to flavonoid accumulation in a single system.

Earlier experiments (Hahlbrock and Scheel 1989) showed that irradiation of the cells led to a coordinated increase in the activity of the enzymes involved in the biosynthetic pathway leading from phenylalanine to the flavone glycosides malonylapiin and graveobioside B. Enzymes of the general phenylpropanoid pathway (called group I by Hahlbrock and leading from phenylalanine to p-coumaroyl CoA (Fig. 5); reached maximal activity at about 15 h after irradiation started. The so-called group II enzymes catalysing the formation of a chalcone from p-coumaroyl CoA and 3 malonyl CoAs and later steps in the specific flavonoid pathway (see Fig. 2) reached their maxima several hours later. It was also possible to show that several enzymes not related to flavonoid metabolism showed no response to the light treatment. Later experiments showed that, for some of the enzymes (the others were not tested) and including representatives from both groups, the changes in activity were due to changes in the activities of the mRNAs coding for these enzymes and that these activity changes were again due to increases in mRNA amounts. To complete the story it was then possible to demonstrate that for two enzymes *phenylalanine ammonia lyase* (PAL, group I) and CHS (group II) these mRNA increases could

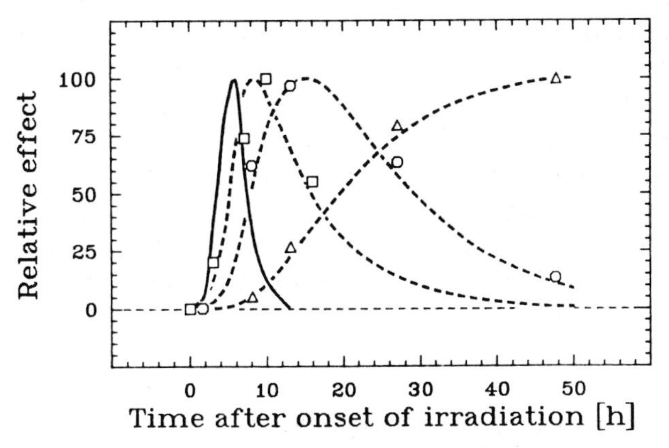

Figure 8. Time courses for light-induced chalcone synthase (*Chs*) gene transcription (——), *Chs* mRNA (□), CHS activity (○) and flavonoid accumulation (Δ). Dotted lines represent expected changes in mRNA, enzyme activity and flavonoid amounts calculated from observed changes in *Chs* gene transcription rates (from Chappell and Hahlbrock 1984).

be attributed to a transient increase in the transcription rates of the relevant genes (Chappell and Hahlbrock 1984). Figure 8 shows the time courses for light-induced effects on transcription of the gene for CHS, as well as changes in mRNA for *Chs*, activity of the enzyme and accumulation of flavonoids. Thus for one system it has been possible to describe much of the chain from the light stimulus to flavonoid synthesis. Two problems arise in extrapolating these results from this system to others: (i) cell cultures often differ somewhat from intact plants; (ii) the functions of the three photoreceptors involved cannot be separated in this type of experiment. It should be noted, however, that the interaction described above between UV-B and phytochrome for flavonoid synthesis in parsley can also be followed at the level of enzyme activities (Wellmann 1983) and transcription of the relevant genes (Chapter 8.1).

9.5.6 The problem of correlation between enzyme activities and flavonoid accumulation

K. Hahlbrock and co-workers were able to show for parsley-cell suspension cultures, that the enzymes of the flavonoid biosynthetic pathway were induced by light in a co-ordinated fashion. They suggested that, for cells at the beginning of the stationary phase of growth, PAL could be the rate-limiting enzyme. Not only did PAL reach its maximum activity at the same time as the maximum rate of synthesis of flavonoids, but also integration of the curve for PAL activity values gave a curve corresponding to that for flavonoid accumulation. Other workers, however, could find no such correlations in other systems.

A possible explanation of such contradictions was suggested by the work of E. Wellmann (Beggs *et al.* 1987): in the case of phytochrome-induced flavonoid synthesis in mustard cotyledons flavonoid and anthocyanin synthesis were spatially separated. Flavonoid synthesis took place almost exclusively in the upper epidermis whereas anthocyanin synthesis was almost completely confined to the lower epidermis. Furthermore distribution of PAL activity (after irradiation) was such that 65 % of the activity for the whole cotyledon was found in the upper epidermis. Thus, on taking the whole cotyledon, by far the greatest proportion of the induced PAL activity was associated with flavonoid synthesis in the upper epidermis and only a relatively small proportion with anthocyanin synthesis in the lower epidermis. In the mustard cotyledon the accumulation kinetics for anthocyanins and flavonoids are very different, flavonoid accumulation reaches its peak several hours after anthocyanin. Thus it would not be expected that a correlation between total PAL and anthocyanin accumulation should occur even if PAL were to be the rate limiting enzyme. The importance of the spatial separation is also strengthened by the existence of different modes of phytochrome induction. Anthocyanin synthesis (and increased activity of the lower epidermis enzyme) is favoured by R pulses, whereas flavonoid synthesis (and activity of the upper epidermis enzyme) is favoured by continuous FR. A good temporal correlation between PAL activity and products is obtained if one compares enzyme activity from the lower or upper epidermis alone with anthocyanin or flavonoid accumulation respectively. The results therefore suggest that information concerning the spatial distribution of the induced pigments must first be obtained for the tissue being investigated before any conclusions concerning possible correlations or rate limiting steps can be drawn. In the case of the cell-suspension culture system, one is almost certainly dealing with a homogeneous system and thus comparison of total enzyme activity with total product is probably justified. For mustard cotyledons one must compare the formation of specific products with the related enzyme pools. In plants other than mustard, little is yet known concerning the spatial separation of pigment synthesis. It has, of course, been known for a long time that flavonoid and anthocyanin synthesis are often concentrated in the epidermal and sub-epidermal layers but to what extent flavonoids are produced in the one epidermis and anthocyanins in the other is not known. What is known is that the enzymes involved may give rise to many products, often simultaneously (Graham 1991). Thus enzymes of the flavonoid pathway may be involved in the synthesis of flavonoids and/or anthocyanins or (in the Leguminosae) of isoflavonoids. The variety of products originating from the general phenylpropanoid pathway is, as might be expected, even larger. These enzymes may also be involved in the synthesis of lignins, simple phenylpropanoids and their esters or various phytoalexin-like compounds (*e.g.* furanocoumarins and stilbenes). Furthermore, some products may be constitutive, some light induced and some induced by other factors. The enzymes involved often belong to separate pools and are induced by separate mechanisms, or, in the case of constitutive products, are themselves constitutive enzymes. Examples of such separate systems are the case of mustard described above or the situation in parsley-cell cultures when treated with light or a fungal elicitor (Hahlbrock and Scheel 1989). Upon light treatment PAL, 4-cinnamoyl CoA ligase and CHS are all induced and lead to flavonoid synthesis. On treatment with a fungal elicitor, PAL and the ligase, but not CHS, are induced and the end products are furanocoumarins and not flavonoids. Furthermore the activity kinetics of the light-induced enzymes are completely different from those of the elicitor-induced enzymes and naturally the elicitor-induced enzymes show no correlation with flavonoid accumulation. Lignins and simple phenylpropanoids and their esters are often constitutive in the plant (or at least occur at particular times in the plants development without an external signal) and thus the enzymes involved in their synthesis will show a similar behaviour. There is thus always the danger that relatively small rises in induced pools may be masked by large constitutive pools. Similar circumstances may explain the confliction between results where enzymes were extracted from whole plant organs. In summary, a prerequisite for attempts to correlate enzyme activities with product accumulation is a detailed study of both spatial distribution of enzyme and product and of the possible existence of other products than the one being studied. Unless it can be clearly shown that the enzyme activity measured relates only to the product measured it is not possible to draw any firm conclusions.

9.5.7 Significance of light induction of flavonoids and anthocyanins

The functions of flavonoids and anthocyanins in plants has always been a matter for much speculation. Although many possible physiological roles have been postulated, in most cases these have been based on observations of the effects of isolated flavonoids on various plant processes. Speculation on such possible functions has often been reviewed (*e.g.* Harborne *et al.* 1975). The functions of these compounds which can be considered to have been more convincingly demonstrated are their roles in flower and fruit colouration, as screening pigments, as protection against herbivores and as phytoalexins.

The extent to which light induction is involved in these functions is, however, with the exception of screening, less clear. It is obvious that it is of no advantage to the plant to produce pigments for flowers or fruits until these organs have reached the developmental stage at which attraction of animals is of importance. Although light may be involved, internal factors and temperature are probably of more importance in determining the stage of development at which pigmentation occurs.

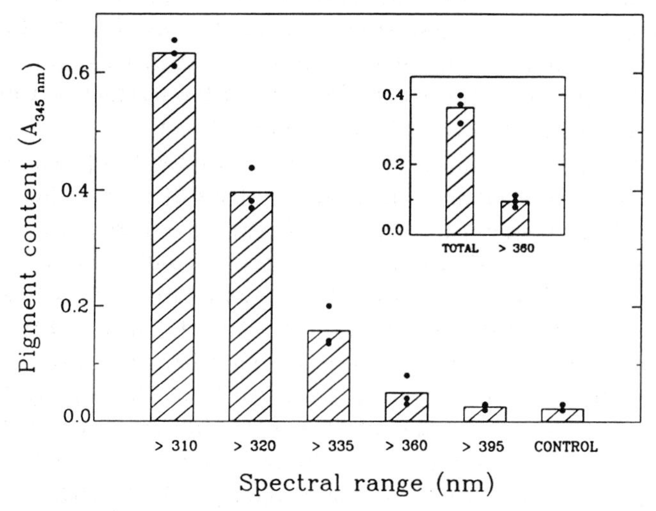

Figure 9. Accumulation of flavonoid pigments in the epidermal layers of white cabbage leaves. Plants were grown under UV free white light and transferred for pigment induction to a UV/white light source for 2 days. The leaves were covered with UV transmission cut-off filters which gradually absorb increasing parts of the UV special range. Insert: Pigment content after 2 days of reduced (1/3) solar irradiation under quartz (total) and 360 nm transmission cut-off filter.

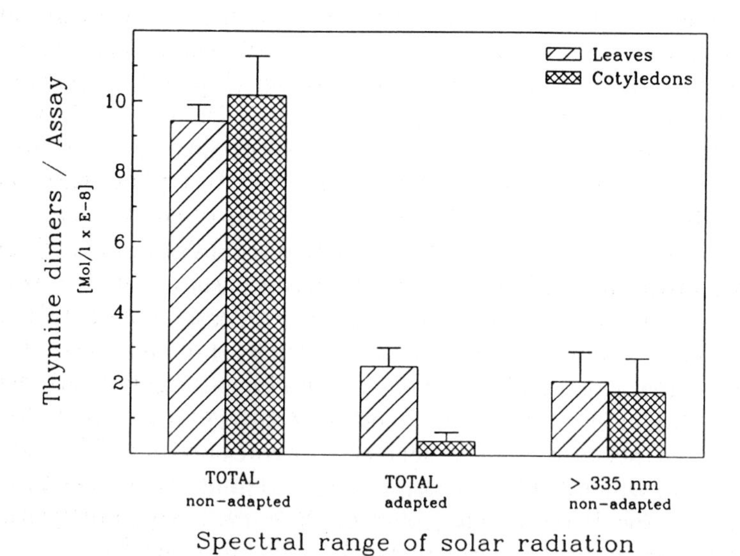

Figure 10. Thymine dimer formation in the lower epidermal layer of white cabbage leaves after 4 h of solar irradiation at 2°C. Irradiations were carried out under quartz (total) or, excluding the major part of DNA damaging UV, under 335 nm transmission cut-off filters. The plants were pretreated with UV/white light as shown in Fig. 9 under 310 nm cut-off (adapted, *i.e.* high pigment content) or 395 nm cut-off (non-adapted, *i.e.* without pigments).

The suggestion that flavonoids and anthocyanins may act as screening pigments which protect the plant from damage by solar UV is based originally on observations that the pigments are often restricted to the epidermal and sub-epidermal layers and that mountain plants often contain considerably more flavonoids than those growing at low altitudes. The idea was supported by the findings that UV-B is frequently an important stimulant of pigment production. Correlations exist between flavonoid concentrations in epidermal cells and the inhibition of photosynthesis by high UV-B loads (Tevini *et al.* 1991). Further aspects of the theory of UV protection of flavonoid/anthocyanin screening have been reviewed by Caldwell (1981) and Wellmann (1983). The absorption spectra of most common flavonoids are optimal for absorbing potentially damaging solar UV radiation, and, together with the fast response to UV with its linear fluence-rate characteristics, this supports the theory. In the case of the anthocyanins, the situation is less clear, because these pigments do not absorb efficiently in the UV waveband. It is more probable that their role is in the photoprotection of B-absorbing pigments such as the flavoenzymes or in the protection of the photosynthetic apparatus during early stages of development.

Figures 9 and 10 show recent results (L. Riegger and E. Wellmann unpublished data) from experiments comparing the role of solar UV with respect to the potential risk of DNA damage and the protective reaction of flavonoid formation in white cabbage leaves. The short-wavelength UV part of solar radiation is essential for pigment formation (flavonol glycosides) both in the lower and upper epidermal layer. This UV adaptation correlates with an increase in UV-B tolerance against DNA damage (Fig. 10). Therefore thymine dimer formation in response to high solar UV-B irradiations is significantly reduced as compared to flavonoid free controls. This dimer formation only occurs if the photoreactivating repair enzyme, DNA photolyase, is experimentally inhibited by cooling. Under normal growth conditions, a protective effect of flavonoids would not become obvious, because of the high efficiency of photoreactivation, leaving no trace of dimers. With regard to solar UV-B increases which are to be expected in consequence of stratospheric ozone depletion, the protective function of flavonoid pigments should not be overestimated. Such spectral changes (limited to the UV-B range) would increase the potential of DNA damage to a much higher extent than contribute to a further stimulation of flavonoid biosynthesis. This can be concluded from the different action spectra for both responses (Fig. 7) in comparison with solar energy distribution in the UV-B range. The capacity of DNA photolyase, a light-induced enzyme (Langer and Wellmann 1990), rather than flavonoid induction, which can be sensitive to UV-B damage itself (Fig. 7), should play a greater role in a plant's protection against UV-B damage. There is no doubt, however, that the UV screening function of these pigments protects a great variety of cell compounds from photodestruction.

9.5.8 Further reading

Hahlbrock K. and Scheel D. (1989) Physiology and molecular biology of phenylpropanoid metabolism. *Annu. Rev. Plant Physiol. Plant Mol. Biol.* 40: 347-369.

Harborne J.B., Mabry T.J. and Mabry H. (1975) *The Flavonoids*. Chapman and Hall, London.

Harbone J.B. (1993) *The Flavonoids. Advances in Research (1988-1991)*. Chapman and Hall, London.

Mancinelli A.L. (1985) Light dependent anthocyanin synthesis: A model system for the study of plant photomorphogenesis. *Bot. Rev.* 51: 107-157.

9.5.9 References

Arthur J.M. (1936) Radiation and anthocyanin pigments. In: *Biological Effects of Radiation* 2, pp. 109-1118, Duggan B. M. (ed.) McGraw-Hill, New York.

Beggs C.J. and Wellmann E. (1985) Analysis of light-controlled anthocyanin formation in coleoptiles of *Zea mays* L.: The role of UV-B, blue, red and far-red light. *Photochem. Photobiol.* 41: 481-486.

Beggs C.J., Stolzer-Jehle A. and Wellmann E. (1985) Isoflavonoid formation as an indicator of UV-stress in bean (*Phaseolus vulgaris* L.) leaves. The significance of photorepair in assessing potential damage by increased solar UV-B radiation. *Plant Physiol.* 79: 630-634.

Beggs C.J., Schneider-Ziebert U. and Wellmann E. (1986) UV-B and adaptive mechanisms in plants. In: *Stratospheric Ozone Reduction. Solar Ultraviolet Radiation and Plant Life*, pp. 235-250, Worrest R.C. and Caldwell M.M. (eds.) Springer-Verlag, Berlin.

Beggs C.J., Kuhn K., Böcker R. and Wellmann E. (1987). Phytochrome-induced flavonoid biosynthesis in mustard (*Sinapis alba* L.) cotyledons. Enzymic control and differential regulation of anthocyanin and quercetin formation. *Planta* 172: 121-126.

Borthwick H.A. (1972) History of Phytochrome. In: *Phytochrome*, pp. 3-22, Mitrakos K. and Shropshire Jr. W. (eds.) Academic Press, London.

Caldwell M.M. (1981) Plant response to solar ultraviolet radiation. In: *Encyclopedia of Plant Physiology*, pp. 169-197, New Series. 12A, *Physiological Plant Ecology I*, Lange. O.L., Nobel P.S. Osmond C.B. and Ziegler H. (eds.) Springer-Verlag, Berlin.

Chappell J. and Hahlbrock K. (1984) Transcription of plant defence genes in response to UV-light or fungal elicitor. *Nature* 311: 76-78.

Downs R.J. and Siegelman H.W. (1963) Photocontrol of anthocyanin synthesis in milo seedlings. *Plant Physiol.* 38: 25-30.

Drumm H. and Mohr H. (1978) The mode of interaction between blue (UV) light photoreceptor and phytochrome in anthocyanin formation of the *Sorghum* seedling. *Photochem. Photobiol.* 27: 241-248.

Duell-Pfaff N. and Wellmann E. (1982) Involvement of phytochrome and a blue light photoreceptor in UV-B induced flavonoid synthesis in parsley (*Petroselinum hortense* Hoffm.) cell suspension cultures. *Planta* 136: 213-217.

Graham T.L. (1991) Flavonoid and isoflavonoid distribution in developing soybean seedling tissues and in seed and root exudates. *Plant Physiol.* 95: 594-603.

Harbone J.B. (1976) Functions of flavonoids in plants. In: *Chemistry and Biochemistry of Plant Pigments*, 2nd Ed. Vol. 1, pp. 736-779, Goodwin T.W. (ed.) Academic Press, London.

Hashimoto T., Shichijo C. and Yatsuhashi H. (1991) Ultraviolet action spectra for the induction and inhibition of anthocyanin synthesis in broom *Sorghum* seedlings. *J. Photochem. Photobiol. B Biol.* 11: 353-364.

Langer B. and Wellmann E. (1990) Phytochrome induction of photoreactivating enzyme in *Phaseolus vulgaris* L. seedlings. *Photochem. Photobiol.* 52: 861-863.

Lercari B. Sodi F. and Sbrana C. (1989) Comparison of photomorphogenetic responses to UV light in red and white cabbage (*Brassica oleracea* L.). *Plant Physiol.* 90: 345-350.

Mohr H. (1957) Der Einfluss monochromatischer Strahlung auf das Längenwachstum des Hypocotyls und auf die Anthocyanbildung bei Keimlingen von *Sinapis alba* L. (= *Brassica alba* Boiss.). *Planta* 49: 389-405.

Oelmüller R. and Mohr H. (1985) Mode of coaction between blue/UV light and light absorbed by phytochrome in light-mediated anthocyanin formation in the Milo (*Sorghum vulgare* Pers.) seedling. *Proc. Natl. Acad. Sci.* USA 82: 6124-6128.

Piringer A.A. and Heinze P.H. (1954) Effect of light on the formation of a pigment in the tomato fruit cuticle. *Plant Physiol.* 29: 467-472.

Siegelman H.W. and Hendricks S.B. (1957) Photocontrol of anthocyanin formation in turnip and red cabbage seedlings. *Plant Physiol.* 32: 393-398.

Siegelman H.W. and Hendricks S.B. (1958) Photocontrol of anthocyanin synthesis in apple skin. *Plant Physiol.* 33: 185-190.

Tevini M., Braun J. and Fieser G. (1991) The protective function of the epidermal layer of rye seedlings against UV-B radiation. *Photochem. Photobiol.* 53: 329-334.

Wellmann E. (1983) UV radiation in Photomorphogenesis In: *Encyclopedia of Plant Physiology*, pp. 745-756, New Series. 16B. *Photomorphogenesis.* Shropshire Jr. W. and Mohr H. (eds.) Springer-Verlag, Berlin.

Wellmann E., Schneider-Ziebert U. and Beggs C.J. (1984) UV-B inhibition of phytochrome-mediated anthocyanin formation in *Sinapis alba* L. cotyledons. Action spectrum and the role of photoreactivation. *Plant Physiol.* 75: 997-1000.

Withrow R.B., Klein W.H., Price L. and Elstad V. (1953) Influence of visible and near infra-red radiant energy on organ development and pigment synthesis in bean and corn. *Plant Physiol.* 28: 1-14.

Yatsuhashi H., Hashimoto T. and Shimizu S. (1982) Ultraviolet action spectrum for anthocyanin formation in broom *Sorghum* first internodes. *Plant Physiol.* 70: 735.

9.6 Photomorphogenesis in fungi

Gérard Manachère

Laboratory of Fungal Differentiation,
Claude Bernard University of Lyon I,
Bât. 405-43, Bd. du 11 Novembre 1918,
F-69622 Villeurbanne Cedex, France

9.6.1 Introduction

Many aspects of general growth and development of fungi are affected by light. This chapter will cover the morphogenic effects of light, particularly in regard to the developmental features of sporophores. Formation of such structures represents sophisticated examples of integration of successive differentiations in fungi and offers numerous models of photomorphogenic effects. Non-morphogenic effects of light such as the eventual influence of light on the synthesis of some compounds, such as pigments, *e.g.* carotenoids are not considered.

The initiation and subsequent development of sporophores from a vegetative mycelium depend firstly on the genetic ability of such a thallus to fruit, regardless of whether it is a lower or higher fungus, or if it has asexual or sexual sporogenesis. The ability of a fungal thallus to change to a reproductive state results from a genetic competence responding to specific physico-chemical environmental changes. In all experimental cases, it is first necessary to define the conditions for adequate preliminary vegetative growth, and then to determine conditions required to reach the state of maturity. Schematically, when such maturity is attained by mycelia, two successive phases can be theoretically distinguished in the formation of sporophores: (i) an initiation phase of fruiting release; (ii) a morphogenic phase, including the differentiation of sporophores and achieved by a fundamental subphase of sporogenesis, itself ended by sporulation. The accomplishment of such phases is dependent on various external factors, (chemical and physical) and more or less defined endogenous factors (genetic and physiological) (see for review, Manachère 1980, 1985, 1988).

The initiation phase and subsequent morphogenetic phase of sporophores often appear to be photocontrolled. However, light is never necessary for

R.E. Kendrick & G.H.M. Kronenberg (eds.), Photomorphogenesis in Plants - 2nd Edition

normal growth of vegetative mycelia. Furthermore, in most cases light appears to reduce growth. Such a relative inhibitory effect of light on vegetative growth is not surprising, from an ecophysiological point of view. It can be seen by the naked eye that mycelia developed in darkness, in controlled cultures, often produce more aerial hyphae than those of the same species developed under uninterrupted light and characterized by a 'fluffy' appearance. Thus, mycelial colonies growing in Petri dishes under a photoperiodic regime show alternate rings corresponding to the successive phases of light and darkness.

Fungi can be classified according to their photosensitivity in relationship to fruiting (Table 1). When there is no other limiting factor, light appears necessary for the fruiting of numerous species. This was known and described more than a century ago by O. Brefeld and confirmed by numerous authors (see for review, Manachère 1988). The mycelia of such species remain sterile in darkness: *e.g. Coprinus congregatus* and various other fungi such as other basidiomycetes of different groups, ascomycetes, or species producing sporangiophores, pycnidia or various conidial forms.

However, light is not indispensable for sporophore initiation in numerous species. The vegetative mycelia of such fungi produce either normal or subnormal sporophores in darkness (*e.g.* conidiophores of *Monilia fructicola* or fruit-bodies of some basidiomycetes such as the cultivated mushroom *Agaricus bisporus*) or more frequently, particularly when macromycetes are considered, abnormal primordia with thin stipes and reduced pilei (*e.g. Coprinus 'cinereus'**, *Asterophora parasitica, Asterophora lycoperdoides*). Sometimes, in the absence of light, such etiolated sporophores carry abnormal lateral branches which appear just like any other etiolated sporophore bud (*e.g. Lentinus tigrinus, Asterophora parasitica*). In all these latter cases, only an adequate light regime of suitable duration and irradiance at the appropriate stages of development allows normal morphogenesis.

Firstly, it must be emphasized that, if in numerous cases, light is indispensable to a normal evolution of successive developmental phases of sporophores (initiation, morphogenesis, sporogenesis, sporulation) a relative excess of light (irradiance or/and duration) may inhibit the initiation or progress of fruiting (Section 9.6.3).

9.6.2 Photo-induction of fruiting

9.6.2.1 General data

As mentioned above, an absolute requirement for light is seen for initiation of fruiting in various species of lower and higher fungi. Fungi of this type occur in

**Coprinus cinereus* (Schaeff. ex Fr. Gray sensu Konr.): modern usual denomination of a complex group of *Coprinus,* also called: *C. lagopus, C. macrorhizus.* (Table 1).

Table 1. Light and fruiting: some classical examples. Adapted from Manachère (1970) *Ann. Sci. Nat. Biol. Vég.* Série 12, 11: 1-95.

(a) No fruiting in darkness; fruiting under adequate light conditions		
Basidiomycetes (basidiocarps)		
Cyathus stercoreus	Lu (1965)	*Amer. J. Bot.* 52: 432-437.
Clavicorna pyxidata	James and Mc Laughlin (1988)	*Mycologia* 80: 89-98.
Phellinus contiguus	Butler and Wood (1988)	*Trans. Br. mycol. Soc.* 90: 75-83.
Favolus arcularius	Kitamoto *et al.* (1968)	*Plant Cell Physiol.* 9: 797-805.
Merulius lacrymans	Zur Lippe and Nesemann (1959)	*Arch. Mikrobiol.* 34: 132-148.
Poria ambigua	Robbins and Hervey (1960)	*Mycologia* 52: 231-247.
Sphaerobolus stellatus	Friederichsen and Engel (1960)	*Planta* 55: 313-326.
	Alasoadura (1963)	*Ann. Bot.* (G.B.) 27: 123-145.
Coprinus congregatus	Manachère (1961)	*C.R. Acad. Sci.* Série D (Paris) 252: 2912-2913.
Coprini spp.	reviewed in Manachère (1988)	*Cryptogamie Mycol.* 9: 291-323.
Ascomycetes (ascocarps or conidia)		
Scutellinia umbrarum	Schrantz (1980)	*Cryptogamie Mycol.* 1: 241-250.
Phomopsis phaseoli (conidial form of *Diaporthe phaseolorum*)	Sar *et al.* (1979)	*Phytopathol. med.* (Fr.) 18: 10-20
Other groups (sporangiophores)		
Pilobolus spp.	Jacob (1961)	*Flora* 151: 329-344.
	Uebelmesser (1954)	*Arch. Mikrobiol.* 20: 1-33.
(b) Fruiting in darkness and under adequate light conditions		
(i) Normal or sub-normal sporophores in darkness		
Basidiomycetes (basidiocarps)		
Agaricus bisporus (cultivated mushroom)	Koch (1958)	*Arch. Mikrobiol.* 30: 409-432.
Nia vibrissa (marine gasteromycete)	Doguet (1968)	*Bull. Soc. Mycol. Fr.* 84: 343-351.
Pterula sp.	Mc Laughlin and Mc Laughlin (1972)	*Mycologia* 64: 599-608.
Other groups (conidiophores)		
Monilia fructicola	Jerebzoff (1961)	*Thèse Sciences,* Univ. Toulouse.
(ii) Abnormal primordia (basidiocarps)		
Coprinus 'cinereus'	Borriss (1934a, b)	*Planta* 22: 28-69; 644-684.
	Madelin (1956)	*Ann. Bot.* (G.B.) 20: 467-480.
Lentinus tigrinus	Schwantes (1968)	*Mushroom Sci.* 7: 257-272.
Flammulina velutipes	Plunkett (1956)	*Ann. Bot.* (G.B.) 20: 563-586.
Pleurotus ostreatus	Zadrazil (1974)	*Mushroom Sci.* 9: 621-652.
Asterophora parisitica	Viot (1970)	*C.R. Acad. Sci.* Série D (Paris) 270: 790-792.
Asterophora lycoperdoides	Mc Meekin (1991)	*Mycologia* 83: 220-223.

most of the major groups, *e.g. Pilobolus* sp. and *Choanephora curcubitarum* among the zygomycetes, *Pyronema omphalodes* and various other discomycetes among the ascomycetes, some species of *Fusarium* and some members of the sphaeropsidales among the fungi imperfecti (*cf.* Hawker 1966, Tan 1978) but also various species of different groups among the basidiomycetes (*cf.* Manachère 1988). In some cases, photo-induction is effective only if accompanied by complementary inductive factors. Photo-initiation of sexual sporophores (apothecia) and of asexual spores (chlamydo-spores) of *Scutellinia umbrarum* is only possible if stimulatory bacteria (*Aeromonas* and *Arthrobacter* sp.) are present (Table 1). Moreover, vegetative mycelia of *S. umbrarum* remain sterile under uninterrupted light. Inhibitory effects of various aspects of fruiting (initiation, but also morphogenesis, sporogenesis) by relatively excessive light (duration or irradiance) are not unusual, and this example is only one of many.

In some species, levels of cyclic AMP (cAMP) appear to be correlated with fruiting, photo-induced or otherwise (Gressel and Rau 1983; Manachère 1985). In *Coprinus 'cinereus'* it was observed that fruit-body initiation was photodependent. Simultaneously, the specific activities of adenylate cyclase and cAMP phosphodiesterases increased rapidly after the onset of irradiation, but this was not observed in dark-grown cultures (Uno and Ishikawa 1982). Increase in the level of cAMP in relation to the induction of primordia by light has been observed in some other species (*e.g. Schizophyllum commune,* Yli-Mattila 1987). However, beyond the particular question of photocontrol of fruiting, the inducing power of cAMP and of other presumed fruiting-inducing substances (FIS) remain hypothetical. Moreover, it seems that the concept of FIS covers a large spectrum of molecules having a more or less defined role in general metabolism rather than a specific power in reproductive processes in fungi (Manachère 1985, 1988).

9.6.2.2 Photo-induction in basidiomycetes

In relation to initiation of sporophores of typical higher fungi (*e.g.* agaricales) various *Coprini* exhibit a clear photodependance. For instance, vegetative mycelia of *Coprinus congregatus* remain sterile when cultivated in darkness at 25°C. Irradiated cultures of the same species produce a ring of primordia just behind the front of the growing hyphae at the time of irradiation when developed on a minimal liquid medium, no subsequent part of the thallus being induced (Durand 1983a). Moreover, microscopical observation of primordia is possible 10 h after the beginning of the light (300 mW m^{-2} at the level of the cultures). Localization of photo-inducibility at the level of hyphae growing in the dark has been also reported by Ross (1982) in a sporeless mutant (pale phenotype) of the same species developed on solid media and illuminated for 12 h. However, under such cultural conditions, no primordia were initiated until the mycelia had reached the edge of the culture plate. Similar results have been reported for the basidiomycete *Schizophyllum commune* (Raudaskoski and Yli-Mattila 1985). In the Ross experiments, photo-induction appears to be memorized until vegetative growth is achieved. It has also been established that photo-induction can also be memorized by fully developed mycelia (Manachère and Bastouill-Descollonges 1985).

Fruiting of dark-grown cultures of our strain of *C. congregatus** is induced by a single 12 h white light (W) break with irradiation at the level of the culture of about 300 mW m^{-2}. After a few days in darkness, cultures present typical anomalous primordia with thin stipes and reduced pilei (etiolated form). However, it is noticeable that such primordia can achieve normal development, but with a correlative delay if submitted to a suitable light regime [*e.g.* light (L), dark (D) = 12 h, 12 h] before abortion in prolonged darkness. Moreover, in this strain of *C. congregatus,* it has been established that photo-induction is also frequently memorized without macroscopically recognizable effects for a period of up to 9.5 days in darkness, against only up to 7.5 days for the visibly etiolated photo-induced primordia described above.

'Memory' of photo-induction of fruiting appears not only possible but indispensable in *Lentinula edodes* (the edible mushroom *Shiitake*): cultures are receptive to light during the first vegetative growth phase, and show memory of the exposure by fruiting later in darkness (Leatham and Stahmann 1987). It is noticeable that irradiation of short duration (few minutes to 1 h) and/or of low irradiance (*e.g.* 10 mW m^{-2}) are often efficient in inducing fruiting of most species, *e.g. C. congregatus.* Effects of wavelengths on the initiation of fruiting will be briefly discussed later (Section 9.6.5) in relation to the general problem of photoreceptors involved in successive photodependent fruiting phases.

9.6.2.3 Variability of light requirements: interaction of factors

The light requirement for initiation of fruiting is not always absolute. Acknowledgement of such relative effects appears useful for a better understanding of apparently contradictory results in various fungi of all systematic groups. Although *Pyronema domesticum* normally requires light to produce apothecia,

*The original strain of *C. congregatus* studied in our laboratory produces normal sporulating sporophores of a 'dark' phenotype according to the Ross nomenclature (Ross 1982). We have also isolated a natural mutant strain (Bastouill-Descollonges and Manachère 1990a) producing, at 25°C, sporeless sporophores of a 'pale' phenotype according to Ross nomenclature. Our strains of *C. congregatus* are generally cultivated on malt agar media. The best synchronization of fruiting is obtained when cultures are transferred to an alternating light, dark: 12 h, 12 h regime after a preliminary vegetative growth in darkness.

apothecial formation occurred in darkness with supplementary aeration (Moore-Landecker and Shropshire Jr. 1982). Such results can be compared to observations related to *Schizophyllum commune:* the shortening of the cells observed in fruiting mycelia indicated that the primary factor controlling cell length was aeration and subsequently light (Raudaskoski and Virtanen 1982).

Some results indicate that the light requirement can also vary in relation to the presence or absence of more or less defined nutrients; age of the irradiated mycelia and to temperature. Vegetative mycelia of *Merulius lacrymans* remain generally sterile in darkness, but fruiting is possible in the absence of light, if a suitable nitrogen source, DL-α-alanine, is used (Table 1). In *Coprinus 'cinereus',* light is not indispensable but can accelerate initiation of fruiting (Table 1). One hour exposure of growing mycelia of *Poria ambigua* to W is necessary and sufficient to determine not only initiation of fruiting but also full morphogenesis of fruit bodies, sporogenesis and sporulation. However, such induction is effective only if inocula of such cultures have been submitted to a preliminary irradiation of about 15 min (Table 1). It seems probable that temporary elevations of temperature sometimes act as a 'shock' and induce fruiting in darkness for species usually light dependant from the point of view of the initiation of sporophores (*Coprinus congregatus,* G. Manachère, unpublished data).

The results discussed here emphasize the necessity of clearly considering the concept of interactions of factors, when the roles of external, chemical and physical factors are studied. This is not only true for initiation of fruiting, but also for consecutive development of sporophores, including sporogenesis and sporulation.

9.6.3 Photomorphogenesis of sporophores and photosporogenesis

9.6.3.1 General data

It appears that light, effective at initiation, is also sufficient for a full accomplishment of sporophore morphogenesis and correlative sporogenesis for numerous species, more particularly lower fungi and some macromycetes such as the basidiomycetes *Schizophyllum commune* (Yli-Mattila 1990) or *Poria ambigua* (Table 1),

However, it also appears that in some cases if light is not absolutely necessary for a basically normal evolution of sporophores, complementary light may nevertheless be beneficial; *e.g.* allowing a more rapid achievement of sporophore development or determining a differential morphogenesis. For instance, conidiophores of *Monilia fructicola* show a different morphogenesis in full darkness and in continuous light: cultures of this fungus submitted to alternative periods of light and darkness exhibit zonations of sporophores which are short (darkness effect) and tall (light effect)(Table 1). This is one example of a typical exogenous circadian rhythm (*cf.* Section 9.6.4.1).

Variability of the influence of light on fruiting has been described by Hawker (1966), who gave examples of fungi producing spores equally well in darkness, continuous light or alternating darkness and light. However, light may influence the size and shape of spores and spore-bearing structures of fungi which are able to sporulate freely in the dark, as with *Sordaria fimicola,* which produce equivalent numbers of perithecia in darkness or in light, although those subjected to irradiation are significantly larger. The macroconidia of some species of *Fusarium* are also longer in irradiated cultures. The length of the sporangiophores of *Phycomyces* and some other members of the mucorales are also influenced by light, even in species which produce sporangia freely in continuous darkness. These, and many other spore-bearing structures are strongly phototropic: such a response is of great importance in spore discharge (*cf.* Section 9.6.3.2.2). The type of spores produced may also be controlled by light. Variability in the influence of light on fruiting has also been described in some macromycetes. Apothecial formation of *Pyronema domesticum* normally requires light, but occurs in darkness with adequate aeration: fruit bodies produce immature or fertile ascospores depending on the nutrients present in the substrate (Moore-Landecker and Shropshire Jr. 1982).

9.6.3.2 Photomorphogenesis and photosporogenesis in basidiomycetes

In numerous basidiomycetes, irrespective of whether fruit initiation is photodependent or not, light is often necessary for a normal morphogenesis and sporogenesis. In darkness, the vegetative mycelia of such species produce abnormal primordia of various types (*cf.* Section 9.6.1). In such cases, only a light regime of suitable duration and irradiance at appropriate stages of development allows a normal morphogenesis.

9.6.3.2.1 Coprinus congregatus: *a model species.* Among the photosensitive basidiomycetes, *C. congregatus* appears as one of the more complex (Fig. 1). This species exhibits all the typical photoresponses observed independently in other fungi. Therefore it can be considered as a model and used for general reference. Mycelia of this species cultivated in darkness remain sterile. At 25°C, light determines the sequence of morphogenesis during the course of successive defined phases. From morphological and physiological points of view, successive stages and phases of development and the nuclear behaviour of hymenial cells up to sporogenesis, are well defined and determined by the daily light and dark periods. Each stage is defined, at 25°C in most observations, by the number of hours before sporophore maturity, which is the 0-h

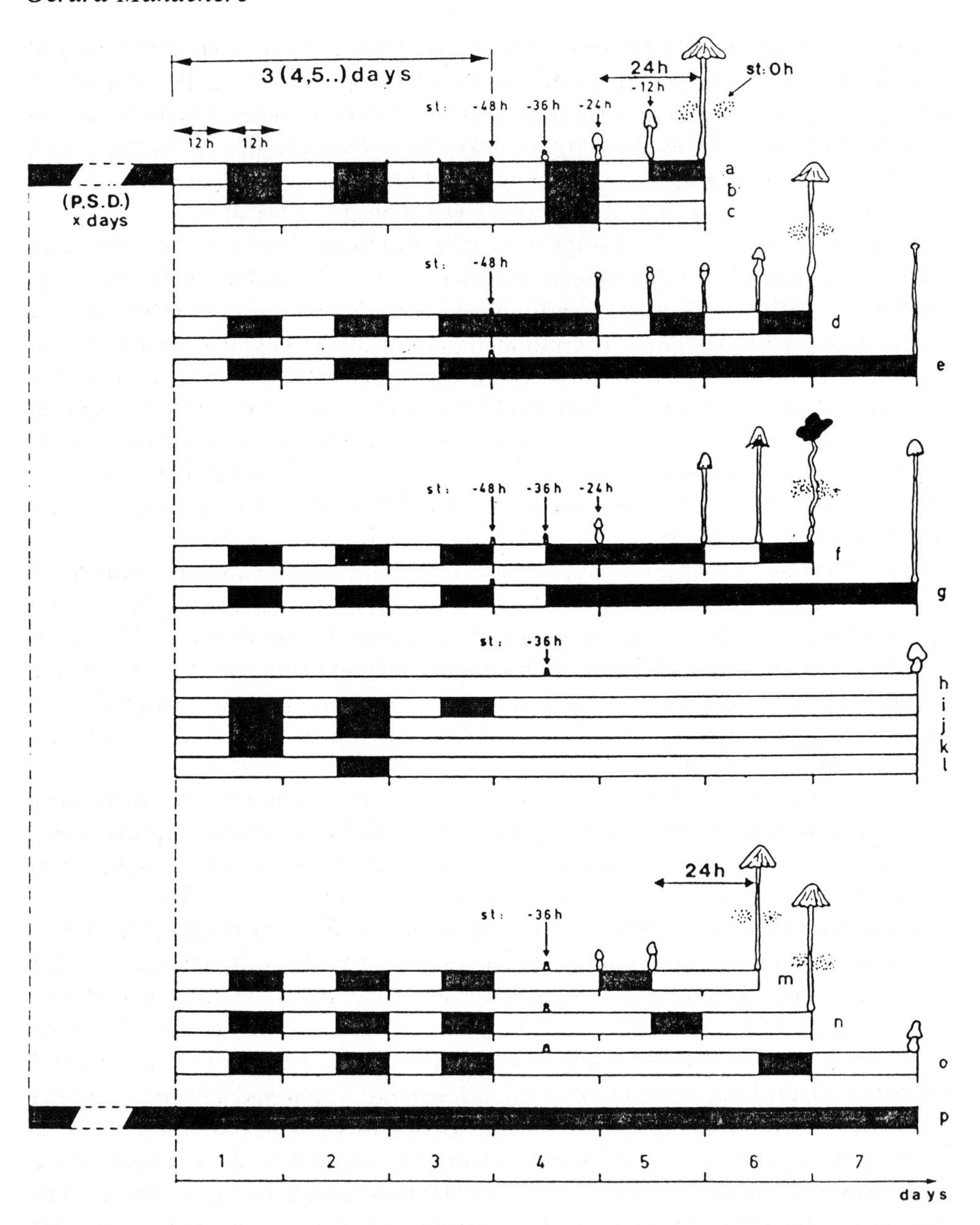

Figure 1. Normal and anomalous morphogenesis of *Coprinus congregatus* fruit bodies in different light regimes. After Manachère (1970) *Ann. Sci. Nat. Bot. Biol.* Vég. Série 12, 11: 1-95 (see text for details). P.S.D., preliminary stay in darkness; st: –48 h, –36 h *etc.* are stages observed 48 h, 36 h before autolysis of caps (st: 0 h).

stage, at the beginning of the photoperiod where it is observed (*e.g.* '–36-h stage' represents 36 h before the 0-h stage). With respect to the cytological state, and particularly to the development of meiosis, there is no karyogamy at the –36 h stage: young basidia still possess the typical two nuclei state of the dikaryotic phase. At the –24-h stage, 50% of basidia have a typical diploid nucleus, and all basidia usually reach karyogamy between the –20-h and –16-h stages. The final meiotic divisions begin at the –16 h-stage and are completed at the –12 h-stage (Manachère and Bastouill-Descollonges 1982).

However, at 25°C, there is a photo-inhibition phase between the –36-h and –24-h stages, light being necessary before and after this phase. More precisely, in continuous light after the –36-h stage (stage of sensitivity to darkness) one can essentially observe an inhibition of stipe elongation and an arrest of meiosis at the diploid stage; correlatively, one can also notice a thickening of stipes and no opening of pilei (Fig. 2). A similar inhibitory effect of continuous light on stipe elongation and meiosis, *i.e.* arrest at the diploid stage was also observed on primordia of various other *Coprini*: *C. 'cinereus'* at 35°C (Lu 1972), dikaryotic sporophores at 28°C (Kamada *et al.* 1978) and monokaryotic sporophores of a mutant strain of the same 'species' at 25°C (Miyake *et al.* 1980).

The photo-inhibition phase of primordia development of *C. congregatus* is of particular interest not only in the control of achievement of morphogenesis and sporogenesis of fruit bodies, but also from a temporal point of view. The end of the dark period indispensable to the accomplishment of this phase, *i.e.* the beginning of the indispensable consecutive photoperiod, acts as a signal. This 'light-on signal' determines the final events leading to the defined 0 h-stage, approximately 24 h later.

More generally, various fungi belonging to different groups reach their development and project their spores at fixed hours under regular daily light/dark regimes: it would seem that the determination of these programmed phenomena is similar to the case of *C. congregatus* (*cf.* Section 9.6.4).

It was also established with *C. congregatus* that a W break (Manachère 1968) or a blue light (B) break (Durand 1982a, 1983b) provided at a suitable moment during the dark period, necessary and sufficient, as defined above, can inhibit final morphogenesis and sporogenesis. Various experiments were conducted to more fully understand the inhibitory effect of light during the dark-induced process of fruit-body maturation in *C. congregatus* (Durand 1983b). Initiated primordia were subjected to different dark periods interrupted by a short B break at different times. The fruiting response depended on the duration of the dark period following the light break. For any inductive dark period longer than 3.5 h, a period of darkness lasting half as long as the inductive night completely inhibited fruit-body maturation when given after the light break (*dark inhibitory process*). Longer dark periods after the light break caused recovery of the maximal fruiting response (*dark recovery process*). The effects of the dark inhibitory and dark recovery processes were alternatively reversible, the fruiting response depending on the duration of darkness after the last light break. Study of the time course for terminal fruiting showed that a new dark-

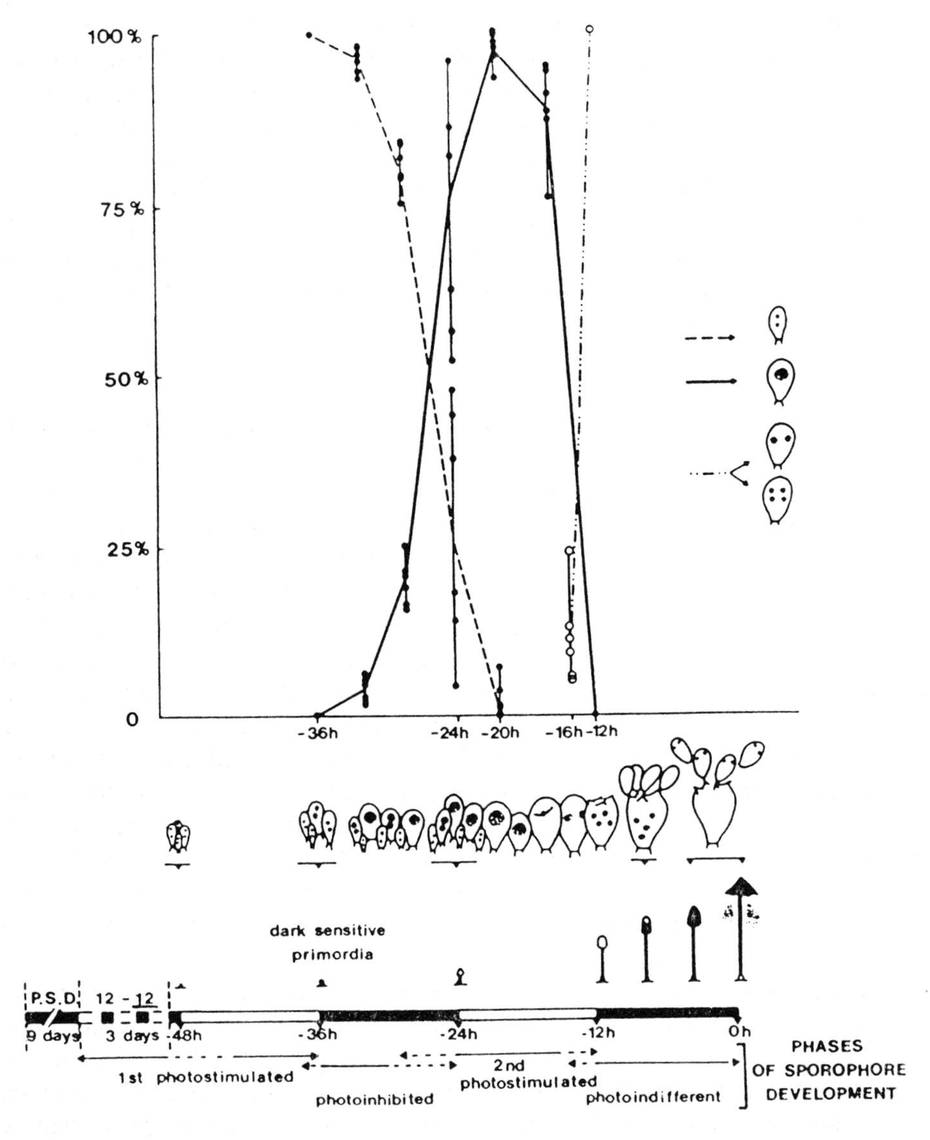

Figure 2. Meiosis and sporogenesis evolution in *Coprinus congregatus* under photoperiodic regime (light) L, (dark) D = 12 h, 12 h. Culture temperature: 25°C. Statistical representation of the evolution of meiosis from the dark sensitive stage of primordia (*cf.* Fig. 1) until the formation of the 4 meiotic nuclei; comparative study of cytological and morphological evolution of sporophores. After Manachère and Bastouill-Descollonges (1982). P.S.D., preliminary stay in darkness.

induced process of fruit-body maturation was initiated by the beginning of the dark period after the light break (light-off signal).

It is also noticeable that the principal characteristics of fruit photo-induction and photomorphogenesis of sporophores described and precisely defined for *C. congregatus* (Fig. 1) were also described for *C. 'cinereus'* (Ballou and Holton 1985). This is confirmation of the general value of the *C. congregatus* model from the point of view of fruiting control of most photosensitive macromycetes. It should be borne in mind that lower fungi and also particularly basidiomycetes such as *Schizophyllum commune,* characterized by practically stipeless sporophores, exhibit the simplest light requirement: in such species, reduced light induces not only initiation, but also a whole development of sporophores, including sporogenesis (all-or-none response).

9.6.3.2.2 Control of fruit-body growth by hymenial cells: a possible relationship to phototropic curvature. Growing sporophores of numerous lower and higher fungi are often characterized by a positive phototropism, at least during the juvenile stages (Gooday 1985). However, it is noticeable that, at least in the case of some macromycetes *e.g. Coprinus congregatus,* fruit bodies become negatively gravitropic at the end of their development [between the –24-h and –12-h stages (G. Manachère, unpublished data)]. Therefore, the tip of the growing stipes tends towards the vertical in such a manner that, at the time of sporulation, the sporogenous lamellae of agaricales or the hymenial pores of polyporales will be in a vertical position, so that the basidiospores have the best chance to be projected and dispersed in the atmosphere. More generally, if a negative gravitropism is a fundamental characteristic of growing sporophores of most species, a positive phototropism can often intervene as a complementary process, in order to direct sporulation out of the substrates. A positive phototropism can orient asci or the apical zone of mature ascocarps (*Sordaria* spp., *Dasyobolus* spp.), sporangiophores of *Pilobolus* spp. and other micromycetes (Fig. 3; Ingold 1965).

In basidiomycetes, particularly agaricales, studies on phototropism and gravitropism of growing sporophores could give useful indications about internal correlations implicated in the morphogenesis of such structures (Gooday 1985; Manachère 1988). It has been demonstrated that in *C. congregatus,* under standard light conditions inducing no positive phototropic curvature, *i.e.* vertical irradiance, the removal of half the cap resulted in a curvature of the stipe when the operation was performed more than 12 h before the end (sporulation, cap autolysis) of the fruit-body development (Fig. 4). Such a curvature is similar to the phototropic response of a young sporophore to directed light. No phototropic curvature was observed on older growing sporophores, which were characterized only by basic negative gravitropism. Complementary experiments consisting of joining separated caps and stipes of different physiological stages have shown the pattern of growth substance production in *C. congregatus* (Table 2, Fig. 5). In most cases, stimulation of stipe elongation was maximal

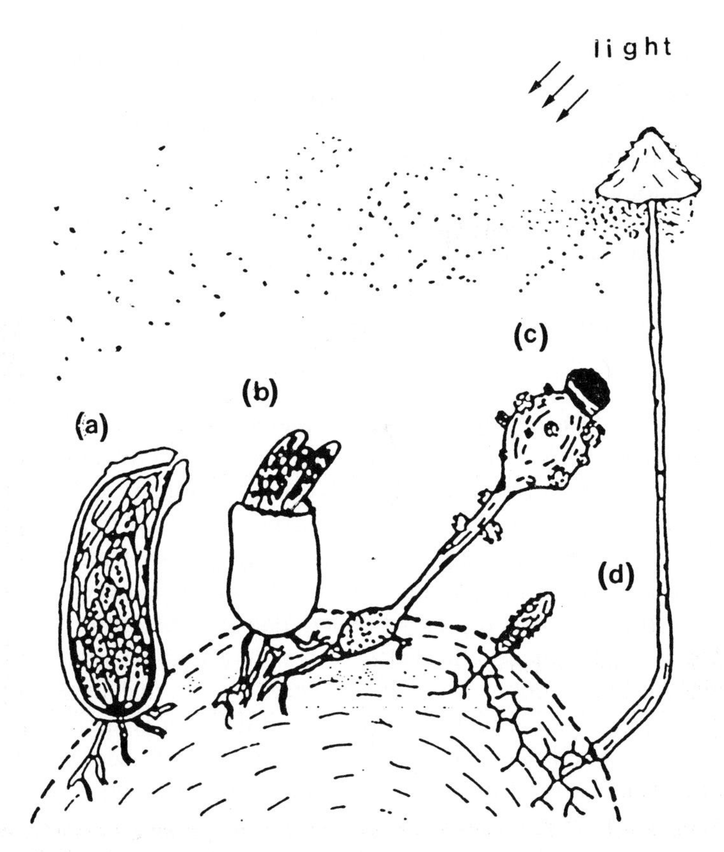

Figure 3. Phototropism of sporophores in relationship to spore discharge. (a) *Sordaria fimicola;* (b) *Dasyobolus immersus*; (c) *Pilobolus* sp.; (d) *Coprinus* sp. By courtesy of Professor C.T. Ingold.

when the deposited caps came from fruit bodies that were still 16 h away from the end of their development. It is noticeable that the so-called –16-h stage under our standard conditions is characterized by the beginning of meiotic divisions following karyogamy (*cf.* Section 9.6.3.2.1). Recently, it has been established that general photomorphogenic characteristics of sporeless sporophores of a mutant form of *C. congregatus* [pale phenotype (Bastouill-Descollonges and Manachère 1990b)] were practically identical to those of normal sporulating sporophores (dark phenotype). From a cytological point of view, evolution of fully developed pale sporophores does not go beyond the diploid fusion nucleus and sporogenesis is, of course, impossible. Therefore in normal sporulating sporophores, it can be hypothesized that mechanisms involved in their final photocontrolled morphogenesis would be practically independent from evolution of basidia beyond the diploid nucleus stage. In other words, the mechanisms involved in final morphogenesis of sporophores

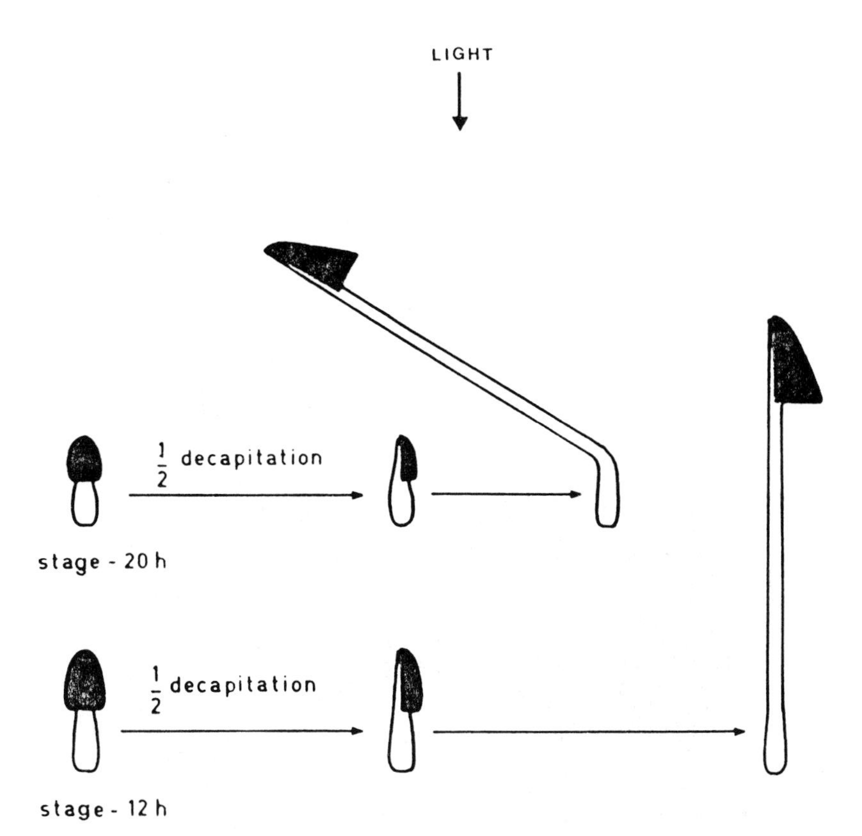

Figure 4. Consequence of the removal of half the cap of growing primordia of *Coprinus congregatus* at different physiological stages. After Bret in Manachère (1978) *Rev. Mycol.* 42: 191-252.

Table 2. Role of the cap on stipe elongation of *Coprinus congregatus.* After Robert and Bret (1987) *Can. J. Bot.* 65: 505-508. Inhibitory effect of caps from photo-inhibited primordia on elongation of decapitated stipes from primordia whose maturation was induced. Development stage: time before autolysis of caps (0-h stage). Control: decapitated stipes only. Experiment: caps from primordia developed for 5 days in continuous light were then applied, following decapitation; n: number of treated stipes. *NB*: for more details see Fig 5.

Developmental stage	Control			Experiment			Inhibition
(h:min)	n	Li	ΔL	n	Li	ΔL	%
−19:35	36	5.6 ± 0.2	4.8 ± 0.5	57	6.1 ± 0.2	3.8 ± 0.3	19.9
−17:45	34	6.3 ± 0.2	20.8 ± 1.9	47	6.5 ± 0.2	8.6 ± 0.8	58.3
−15:40	42	7.1 ± 0.2	47.7 ± 1.3	65	7.2 ± 0.2	21.9 ± 1.3	53.5
−13:50	51	8.2 ± 0.2	48.9 ± 0.9	57	8.7 ± 0.2	31.9 ± 1.2	34.8

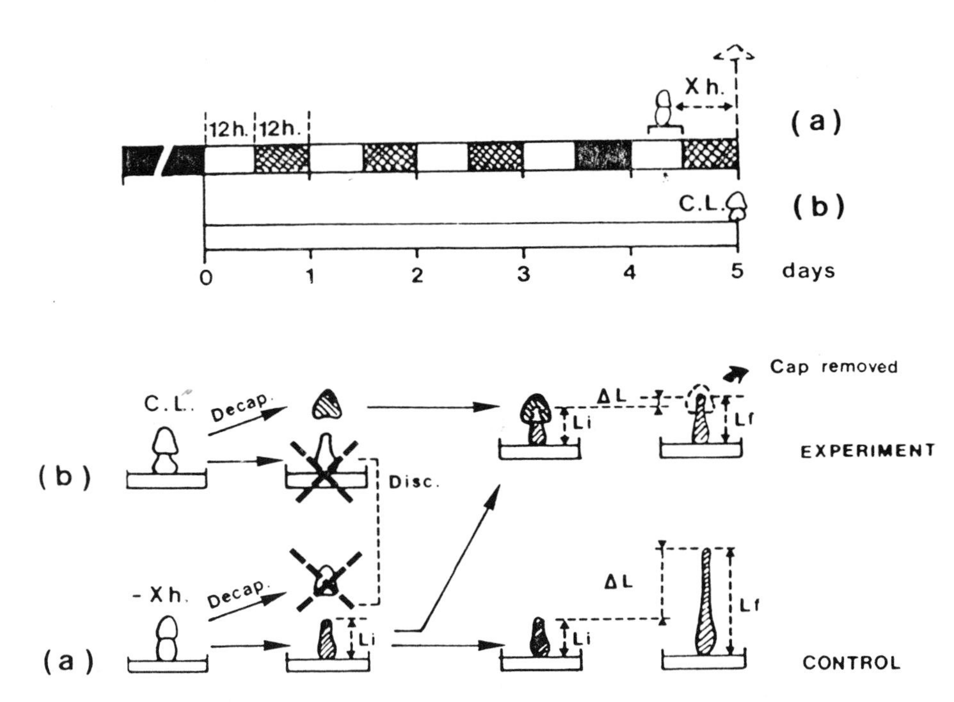

Figure 5. Light conditions and method used to show the presence of an inhibitor of stipe elongation in the cap of primordia of *Coprinus congregatus* developed under continuous light. Decap., decapitation: separation of pileus from stipe at the beginning of the experiment. After Robert and Bret (1987) *Can. J. Bot.* 65: 505-508. (a) X h: stages of development of operated primordia (*cf.* Table 2): X h before autolysis. (b) C.L., abortive primordia produced under continuous light. Li, initial length, measured just after decapitation. Lf, final length, measured just after a minimum of 24 h in the light and, then, cap removal when necessary. ΔL, residual growth of only decapitated stipes (control) or receiving caps from C.L. primordia (experiment). disc. : discarded parts. *NB*: hatched periods are where darkness is not necessary for normal development. The black periods correspond to an absolute dark requirement at 25°C (culture temperature). For more details: see Table 2.

of *C. congregatus* would be independent of those involved in final photosporogenesis (Fig. 6).

Several experimental results imply the production of growth-promoting factors in young lamellae of *C. congregatus,* as well as other agaricales, practically up until the meiotic divisions. By analogy with the classical Cholodny-Went model (*cf.* Chapter 9.2) one can hypothesize that phototropic and gravitropic curvatures eventually observed during growth of young sporophores would be determined by asymmetrical distribution of such growth-promoting factors produced by lamellae and controlling elongation of the tip of the stipes. The identity of such substances remains unknown.

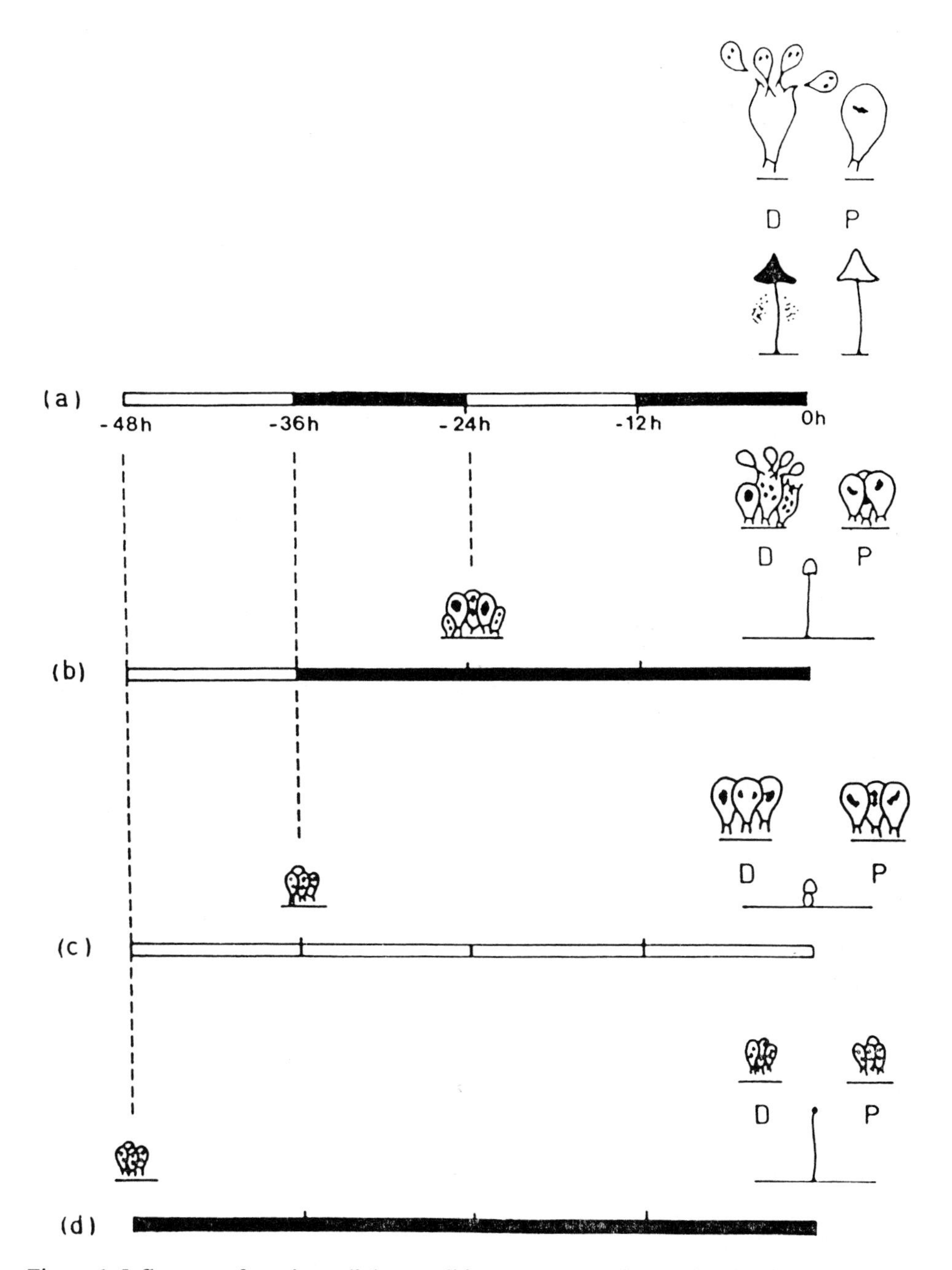

Figure 6. Influence of various light conditions on general morphogenesis, meiosis and sporogenesis of normal sporulating (D, dark form) and asporogenous (P, pale form) sporophores of *Coprinus congregatus*. Definitions of physiological stages: *cf.* Fig. 1 and text. After Bastouill-Descollonges and Manachère (1990b).

9.6.3.2.3 Interactions of light and temperature: consequences for photoperiodic responses. It can be readily demonstrated in the laboratory that the effect of a particular external factor on reproduction may be modified by changes in other factors. For instance, sufficient aeration and suitable light conditions are frequently necessary for minimal fruiting. Modification of one factor can generally give a better fruiting if no other factor is limiting. In *Flammulina velutipes,* a minimal aeration permitted primordia initiation and stipe elongation, but cap expansion was prevented despite the presence of light (Table 1).

As previously described, primordia of *C. congregatus* developed at relatively elevated temperatures (20-25°C) are characterized by a fundamental dark requirement when a defined stage of sensitivity to darkness is reached (–36-h stage). Furthermore, the dark requirement varies with temperature (Fig. 7). This is not found at temperatures below 17.5°C, and normal development of primordia was obtained in continuous light (300 mW m^{-2}) by lowering the basal temperature of culture from 25 to 10°C for 6 h (Fig. 8). More precisely, at low levels of irradiance (1.5 mW m^{-2}) a slight decrease in temperature, 25 to 22.5°C, was sufficient to release primordia from photo-inhibition (Durand 1982b). Meiosis and consecutive sporogenesis are also strictly dark dependent at 25°C, but can proceed in continuous light at 10°C. It appears that the dark requirement triggering meiosis and basidiocarp maturation in several *Coprini,* such as *C. 'cinereus'* (Lu 1972) or *C. congregatus* (Manachère and Bastouill-Descollonges 1982) is not absolute, and is temperature dependent.

Furthermore, in certain lower and higher fungi, an indispensable dark period determining the maturation of sporophores, including sporogenesis, at a relatively elevated temperature, can be replaced by lowering the temperature. For instance, *Alternaria solani* and *A. tomato* failed to develop conidia under continuous irradiation at 25°C, but the inhibitory effect of light was no longer present at 15°C (Aragaki 1961; Lukens 1966). However, incubation temperature and irradiance affected conidial size in *A. cichorii* (Vakalounakis and Christias 1985b), but irradiance affected conidial morphology only at low temperatures (15°C): at higher irradiances *ca.* 1240 lx* against 160 lx) length and width of conidia were increased; temperature changes affected conidial size at all irradiances tested.

Such photoperiodic phenomena are observed in other groups of plants and animals. Some of the developmental features cited in *C. congregatus* have their analogy in the photoperiodic responses of some extremely sensitive short-day plants (*cf.* Chapter 7.3), particularly those which can be induced to flower by a single dark period, such as *Xanthium strumarium, Pharbitis nil, Lemna*

---

*The illumination of a surface by W is often measured in lux (lumens m^{-2}). Such units of illuminance cannot be converted directly into irradiance units (fluence rate) and can be used only for comparing light sources of similar spectral quality. 1 lx at 555 nm is equivalent to 1.6 mW m^{-2} or 7 nmol m^{-2} s^{-1}.

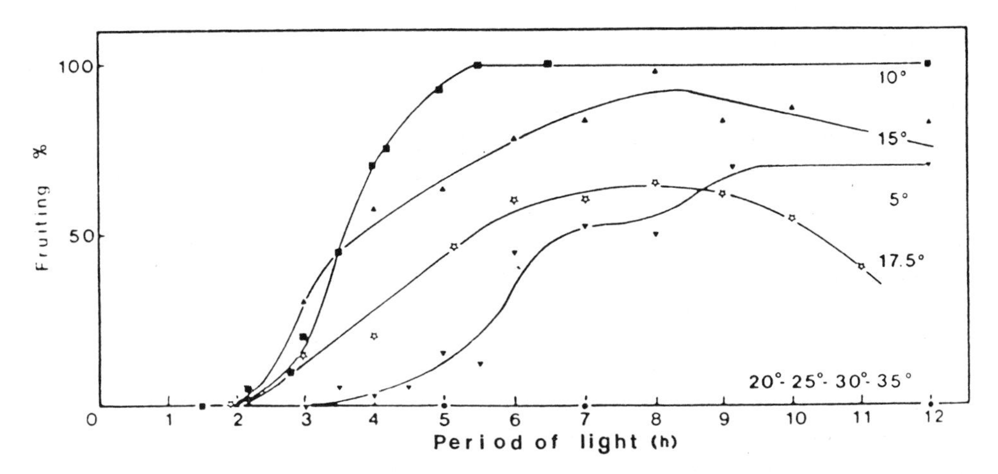

Figure 7. Light and temperature interactions during fruiting in *Coprinus congregatus*. Effect of different temperature treatments in continuous light, in place of the usual inductive dark period on dark sensitive primordia (*cf.* Fig.1). After Robert (1971) C.R. Acad. Sci. Sér. D. (Paris) 273: 154-157 and Robert and Durand (1979) *Physiol. Plant.* 46: 174-178.

perpusilla, Perilla ocymoides (see for review Manachère 1985, 1988). Therefore, floral initiation, conidia formation, or fruit-body maturation, including sporogenesis, can be induced by a number of alternative pathways and are not strictly dependent upon a specific photoperiodic process. In other words, the problem of interactions of various factors, temperature and others, with light, and eventual additive effects, should also be considered.

9.6.4 Photocontrol of fruiting rhythms

Sporophores of various fungi of different groups reach their development and project their spores at fixed hours under regular daily light regimes. The best known examples correspond to circadian rhythms and involve a photoperiodic regulation. However, some non-circadian rhythmical production of sporophores, and correlative sporulations, can be also controlled by daily light-dark cycles. Fungi therefore can be useful model systems for the general study of rhythms (Millet and Manachère 1979).

9.6.4.1 Circadian rhythms

Daily spore liberation is often regulated by rainfall and daily variations of temperature and photoperiodic control can be involved. Influence of daily light-dark cycles on various aspects of fungal reproduction has been established

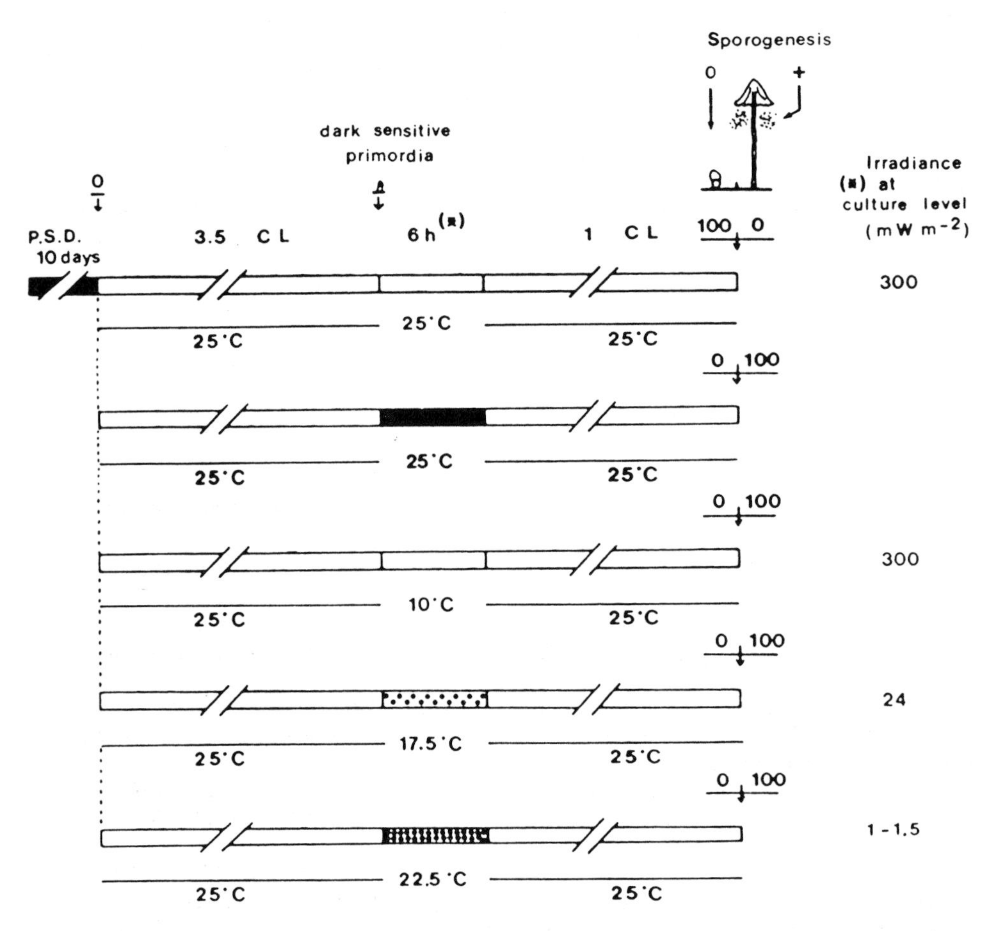

Figure 8. Effects of light and temperature interactions on maturation of sporophores of *Coprinus congregatus.* Cultures grown in darkness for 10 days (P.S.D., preliminary stay in darkness) were then exposed to white light (300 mW m^{-2}) without interruption for 3.5 days (C.L., continuous light). At the end of this period, they have produced dark sensitive primordia (*cf.* Fig. 1). They were then submitted to various thermal and light regimes for 6 h. Percentage of cultures with aborted primordia only (no basidiospores produced) and of cultures with normal sporulating sporophores was evaluated 24 h after the end of the eventual maturating regime: for simplicity, only conditions allowing fruiting are reported here. After Robert (1971) *C.R. Acad. Sci.* Sér. D (Paris) 273: 154-157 and Durand (1983) Thèse Sciences, Lyon.

experimentally. Schematically, regular succession of daily light and dark periods often leads to sporulation (strictly exogenous circadian rhythms of sporulation, *e.g. Sphaerobolus stellatus, Pilobolus crystallinus*) or to a better synchronization, so-called entrainment (non-autonomous typical endogenous circadian rhythms of sporulation, *e.g. Pellicularia filamentosa, Daldinia concentrica, Pilobolus sphaerosporus*) of periodical sporophore maturation and correlative discharge of sexual or asexual spores (Table 3). However, daily

Table 3. Fruiting rhythms in fungi: some classical examples. After Manachère (1968) *Bull. Soc. Mycol. Fr.* 84: 603-619 and Manachère (1978) *Bull. Soc. Bot. Fr.* 125: 243-262. *Only under daily light; **under daily light; in continuous light and/or in continuous darkness.

(a) Circadian rhythms		
(i) Exogenous circadian rhythms*		
Production of sporophores		
Aspergillus ochraceus	Jerebzoff (1961)	Thèse Sciences, Univ. Toulouse
Aspergillus niger	Jerebzoff (1963)	*C.R. Acad. Sci.* Série D (Paris) 256: 759-761.
Differential growth of sporophores		
Monilia fructicola	Jerebzoff (1961)	*Thèse Sciences,* Univ. Toulouse
Discharge of asexual spores		
Pilobolus crystallinus	Uebelmesser (1954)	*Arch. Mikrobiol.* 20: 1-33.
Pilobolus kleinii	Page (1956)	*Mycologia* 48: 206-224.
	Jacob (1961)	*Flora* 151: 329-344.
Pilobolus umbonatus	Jacob (1961)	*Flora* 151: 329-344.
Discharge of ascospores		
Sordaria fimicola	Ingold and Dring (1957)	*Ann. Bot.* (G.B.) 21: 465-477.
Podospora tetraspora	Walkey and Hervey(1967)	*Trans. Br. mycol. Soc.* 50: 229-240.
Hypoxylon investiens	Kramer and Pady (1970)	*Mycologia* 62: 1170-1186.
Discharge of basidiospores		
Sphaerobolus stellatus	Friederichsen and Engel (1960)	*Planta* 55: 313-326.
	Alasoadura (1963)	*Ann. Bot.* (G.B.) 27: 123-145.
(ii) Endogenous circadian rhythms**		
Differential growth of sporophores (in continuous darkness + addition of yeast extract)		
Monilia fructicola	Jerebzoff (1961)	*Thèse Sciences,* Univ. Toulouse
Discharge of asexual spores		
Pilobolus sphaerosporus	Schmidle (1951)	*Arch. Mikrobiol.* 16: 80-100.
	Uebelmesser (1954)	*Arch. Mikrobiol.* 20: 1-33.
Discharge of ascospores		
Daldinia concentrica	Ingold and Cox(1955)	*Ann. Bot.* (G.B.) 19: 201-209.
Lopadostoma turgidum	Walkey and Harvey (1967)	*Trans. Br. mycol. Soc.* 50: 229-240.
Bombardia fasciculata	Pady and Kramer (1969)	*Trans. Br. mycol. Soc.* 53: 449-454.
Hypoxylon rubiginosum	Kramer and Pady (1970)	*Mycologia* 62: 1170-1186.
Hypoxylon truncatum		
Discharge of basidiospores		
Pellicularia filamentosa	Carpenter (1949)	*Phytopathol.* 39: 980-985.
Sphaerobolus stellatus (in continuous darkness, rarely)		
	Engel and Friederichsen (1964)	*Planta* 61: 361-370.
(b) Low frequency rhythms: periods (days)		
(i) Photodependent species		
Coprinus congregatus	4 to 6 days (20 to 25°C) 12 to 24 days (10°C)	
	Manachère and Robert (1972)	*J. Interdiscip. Cycle Res.* 3: 135-143.
Sphaerobolus stellatus	10 to 12 days (20 to 22°C)	
	Friederichsen and Engel (1960)	*Planta* 55: 313-326
	Alasoadura (1963)	*Ann. Bot.* (G.B.) 27: 123-145.
(ii) Aphotic species		
Agaricus bisporus	4 to 8 days (primordia picked) 10 to 12 days (primordia unpicked)	
(cultivated mushroom)	Cook and Flegg (1962)	*J. Hort. Sci.* 37: 167-174.

light-dark periods can also regulate maturation of reproductive structures and consecutive sporulations, without correlative circadian rhythms (Section 9.6.4.2).

Attention has been given to daily regulation of various species, most of them associated with plant diseases and recognized by aerobiologists, *e.g. Cladosporium, Phytophthora, Ustilago, Deightoniella, Peronospora* spp. (Ingold 1965). In *Sordaria fimicola,* a circadian sporulation rhythm occurs under a L, D = 12 h, 12 h regime, with a maximum in the middle of the light period. An analogous rhythm has been reported by Ingold in various other ascomycetes (*Podospora curvula)* and by other authors for species from other groups (mucorales such as *Pilobolus sphaerosporus* and *P. crystallinus;* small basidiomycetes such as *Sphaerobolus stellatus,* Table 3). Such species can be defined as diurnal sporulators, while other species can be defined as nocturnal sporulators. Under a L, D = 12 h, 12 h regime, more spores of *Sordaria verruculosa, Hypoxylon fuscum* or *Pellicularia filamentosa,* a basidiomycete causing leaf spot of *Hevea,* are usually discharged in the dark periods (Ingold 1965). In most cases, experiments demonstrated that this is not the result of an inhibitory effect of light. Sporulation of such species generally remains possible under uninterrupted light, some showing a circadian rhythmicity for several days (*endogenous circadian sporulation rhythms*) others becoming arythmical (*exogenous circadian sporulation rhythms*).

Adequate experiments, with controlled changes in photoperiodic conditions, demonstrate that peaks of sporulation are normally determined by changes from dark to light (light-on signals). More precisely, regulation of rhythmicity of both diurnal and nocturnal sporulators appears similar and related to the existence of a defined temporal interval between the reception of the stimulus ((light-on signal) and the maximum corresponding response (peak of sporulation). Such a temporal relationship, delay between signal and induced effect, has been noticed for several species: 28-30 h in *Pilobolus sphaerosporus* and *P. crystallinus;* 28 h in *Sphaerobolus stellatus;* 8 h in *Daldinia concentrica;* 8-12 h in *Sordaria verruculosa* (Ingold 1965). Such delays generally correspond to culture temperatures of 20-25°C, but are probably thermo-dependant. From a physiological point of view, it appears that differences between diurnal and nocturnal sporulators are related to a different lapse of time between the inductive light-on signal and the correlative sporulation peak: the underlying mechanism clearly being similar. From an adaptative point of view, Ingold (1965) postulated that "*nocturnal sporulation might be an advantage for stem or leaf pathogens with spores which quickly lose their viability and depend on infection drops for germination. Conversely, in other species, coprophilous for instance* (Pilobolus *spp,* Sphaerobolus stellatus), *shooting of spores or sporangia during the day time might ensure that it takes place at the time when the ejected spores can be aimed by the quickly adjusting phototropic response of sporophores which, in temperate latitudes, helps to scatter such spores or sporangia over the grass surrounding the dung. However, generally speaking, it is by no means easy to suggest the possible biological value of rhythms of spore discharge".*

9.6.4.2 Low frequency rhythms

Annual fruiting rhythms are a well known characteristic of higher fungi, with a peak of carpophore production in early autumn and a subsidiary one in spring in temperate climates. Rainfall and changes in temperature can generally be considered the most essential factors involved. Some results, however, suggest that photoperiodic determination of fruiting is possible, at least at the level of maturation of sporophores and control of sporogenesis and sporulation (*cf.* Section 9.6.3.2.3). Alternatively, fruiting of some higher fungi is characterized by an endogenous rhythm of production of fruit bodies, with a periodicity of some days (Manachère 1980). Periods of such low frequency rhythms are unrelated to fluctuation in external factors and are temperature dependent (Table 3). In a few photodependent species (*Coprinus congregatus, Sphaerobolus stellatus*) fruiting appears as a consequence of an adjustment of the underlying endogenous rhythm, at least at the level of maturation of sporophores caused by daily light regimes.

In *C. congregatus* it has been established that the underlying fruiting rhythm usually observed in daily irradiated cultures is also present under continuous light (Fig. 9). At 25°C, under uninterrupted light, primordia abort. However, one single dark period, for instance of 12 h, can be sufficient to determine a normal development of sporophores, provided that it intervenes at an appropriate moment. Under such conditions, each dark period reveals cultures with primordia sensitive to darkness at the moment of interruption of the light and the cultures produce mature sporophores about 24 h after the end of the inductive dark period. It appears that the percentage of fruiting cultures varies periodically according to the length of the light period preceding the dark interruption. The periodicity noticed (about 4 days) is practically identical to that observed for most of the cultures submitted to a classical L, D =12 h, 12 h regime. It therefore appears that in usual photoperiodical conditions, daily dark periods progressively 'reveal' the primordia which have become sensitive to darkness and permit their development about 24 h after the end of the dark period.

Other data confirm such regulation of the endogenous rhythm of production of sporophores of *C. congregatus* by daily photoperiods. When cultures were maintained at 25°C under a L, D = 1 h, 23 h regime, the first flush was delayed on average by 2 days in comparison with cultures under a L, D = 12 h, 12 h regime. Furthermore, this flush generally showed only one fruit body/culture, against an average of three fruit bodies/culture under a L, D = 12 h, 12 h

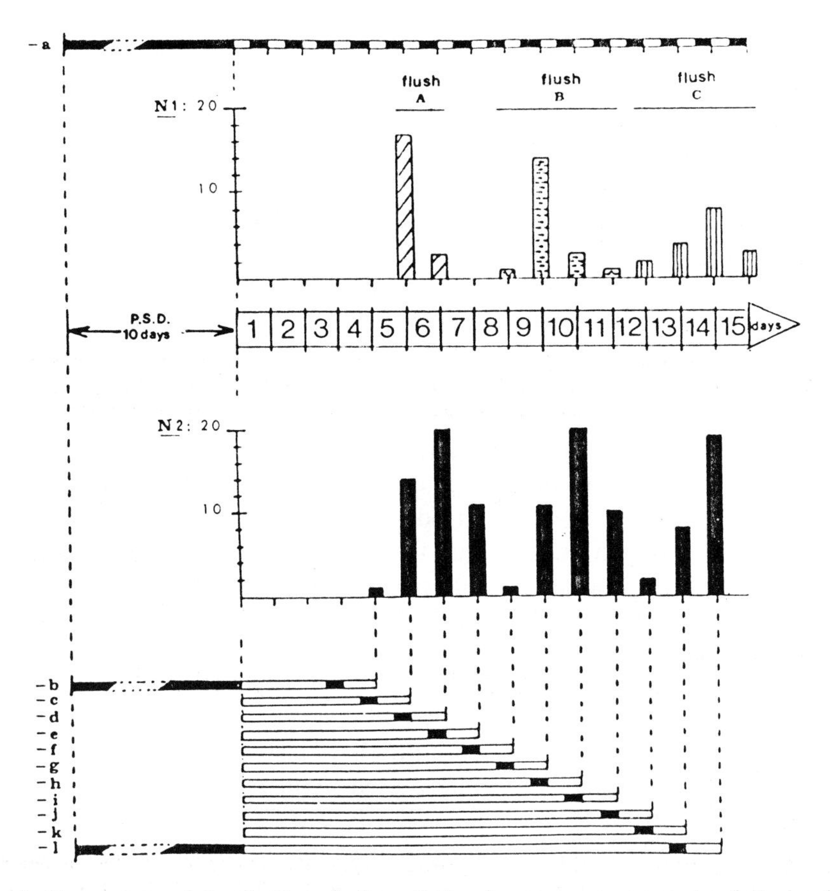

Figure 9. Comparison of the fruiting rhythm of *Coprinus congregatus* under daily irradiation light (L), dark (D) = 12 h, 12 h regime, series (a) and under continuous light (series b-l) interrupted by only a single dark period (12 h) provided at different times. Number of cultures/series: 20. N1, N2: number of cultures/series with mature sporophores (0-h stage) at the indicated moment. Days 1-15 : number of days of irradiation, after a preliminary stay in darkness (P.S.D.) of 10 days. After Manachère (1967) *C.R. Acad. Sci.* Sér. D (Paris) 265: 1485-1488.

regime. Nevertheless, if dry weight was considered, rather than number of fruit bodies, productivity appeared identical in the two series of cultures: reduction in daily photoperiod resulted in fewer fruit bodies which were taller and heavier than those formed under longer photoperiods (Fig. 10). Similar results were obtained when the irradiation was lowered from 300 to 10 mW m^{-2}, the cultures remaining under a L, D = 12 h, 12 h regime (Robert 1982). The reduction of daily light is compensated by regulations at the level of the whole fungal thallus. From an ecophysiological point of view, one can consider that, together with negative gravitropism and positive phototropism (*cf.* Section 9.6.3.2.2), this give an additional opportunity to produce primordia for sporulation.

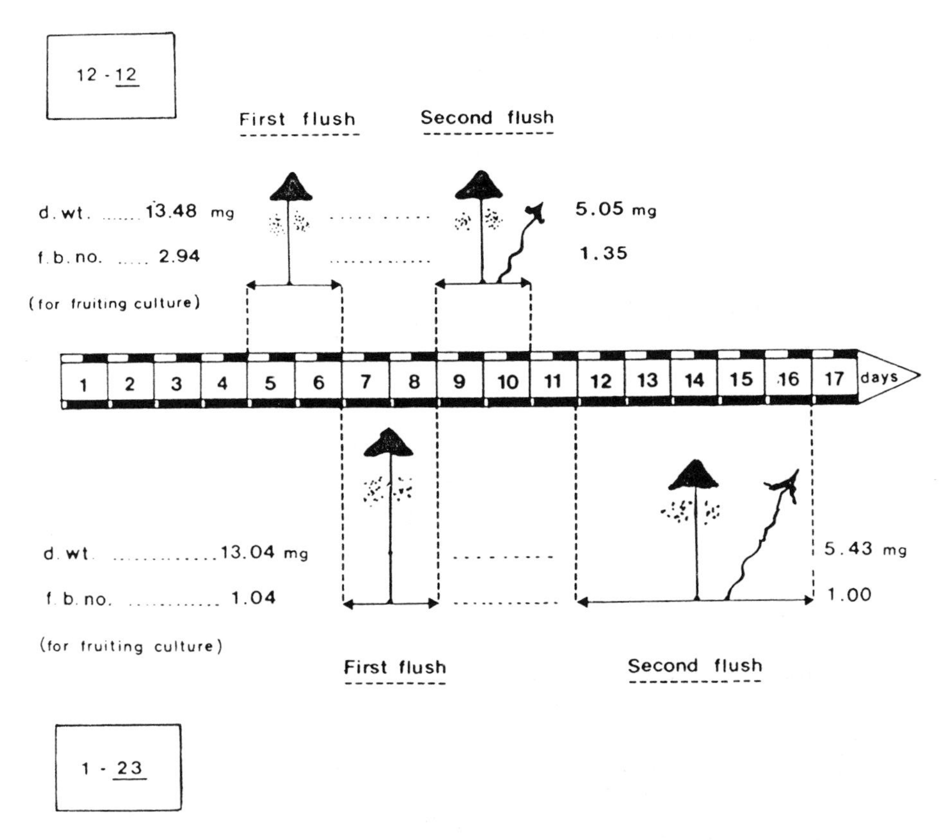

Figure 10. Comparison of fruiting of *Coprinus congregatus* under two distinct photoperiodic treatments, (light) L, (dark) D = 12 h, 12 h regime (control series) and L, D = 1 h, 23 h regime. Number of cultures/series: 25. Mean dry weight (d.wt.) and mean fruit body number (f.b.no.) for fruiting cultures and by flushes are indicated. Culture temperature: 25°C. Irradiance: 300 mW m^{-2} (white light). After Manachère and Bastouill-Descollonges (1983) *Trans. Br. mycol. Soc.* 81: 630-633.

9.6.5 Action spectra and photoreceptors

9.6.5.1 Survey

Fungi have for a long time been convenient and useful organisms as experimental material in photobiology. It seems evident that several photoreceptors exist in fungi. Fungal responses are produced by wavelengths from the UV to the red (R) region of the spectrum (Tan 1978; Durand 1985; Manachère 1985). Schematically, the effects of monochromatic light on fruiting may be divided into three categories:

(i) Fungi which exhibit responses only elicited by UV-B. In such species, only wavelengths below 330 nm are generally effective (Gressel and Rau

1983). It is noticeable that such non-lethal responses to UV generally concern conidiation of phytopathogenic fungi [*Pyricularia oryzae, Botrytis cinerea, Ascochyta pisi,* various species of *Alternaria* and *Helminthosporium,* reviewed by Tan (1978)] or morphogenesis of perithecia of some ascomycetes [*Pleospora herbarum,* Leach and Trione (1966); *Nectria galligena,* Dehorter (1976); Dehorter *et al.* (1980)]. The chemical nature of the photoreceptors remain obscure. A compound, named P 310, was suggested as the photoreceptor by Leach (1965). The chemical and physical characteristics of P 310, (*mycosporine*), were then elucidated (Arpin *et al.* 1979). However, although the presence of mycosporine(s) is, in various cases, associated with sporogenesis, there are no definitive experimental data to suggest that such molecule(s) are photoreceptor(s).

(ii) Fungi which exhibit responses elicited by UV-A and B (300-520 nm) (Kumagai 1988; Durand 1987). The so-called B/UV-A responses are common among all groups of fungi as well as among other micro-organisms, plants and animals (Part 5). The B action spectra of numerous species generally exhibit a primary sensitivity in the B region (about 450 nm) often with a second smaller peak in the UV-A region of the spectrum (370 nm) and no action beyond 520 nm. The pigment system responsible for various B/UV-A processes has sometimes been named cryptochrome (Chapter 5.1). Action spectra have been used to support the 'flavin-hypothesis' to explain photoreception in various fungi (*cf.* Part 5). Moreover, in some species [*Helminthosporium oryzae, Alternaria tomato, Botrytis cinerea, Alternaria cichorii: cf.* Tan (1978); Kumagai (1988); Gressel and Rau (1983); Vakalounakis and Christias (1985a)] sporulation is controlled by a reversible reaction in the B and UV-A. A photoreversible pigment named mycochrome, has been assumed to be involved in the control of sporogenesis of such fungi (*cf.* Part 5).

(iii) A few fungi which exhibit responses to the longer wavelengths of the visible spectrum, above 520 nm. In *Verticillium agaricinum,* phytochrome has been postulated to be a photoreceptor in photo-induction of carotenoid synthesis (Valadon *et al.* 1982). However, other investigations on the same species, have shown that R did not affect carotenogenesis (Björn 1986). Such isolated reports about the hypothetical involvement of phytochrome in fungal development require further investigation.

9.6.5.2 Photoreception in basidiomycetes

Light not only induces fruiting and promotes achievement of morphogenesis, including sporogenesis and sporulation of sporophores of *Coprinus congregatus,* but can also inhibit development of primordia and progression of meiosis, when provided at an excessive level at the so-called 'stage of sensitivity to darkness' (Section 9.6.3.2.1). The action spectra for the initiation phase of

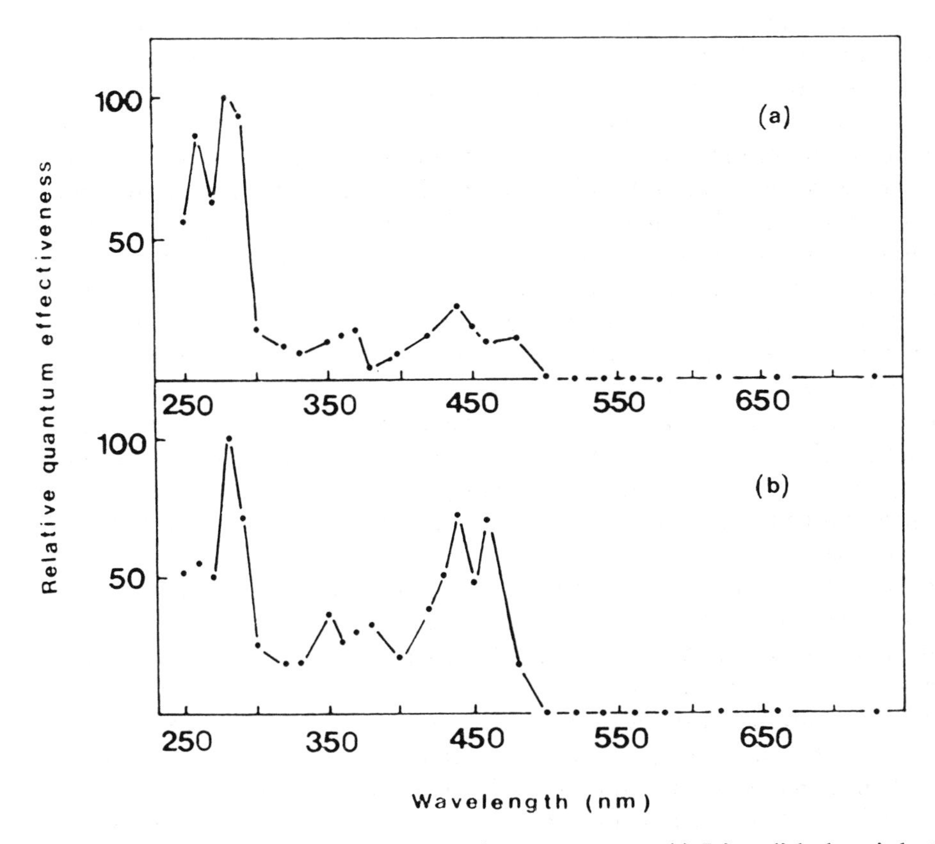

Figure 11. Action spectra in fruiting of *Coprinus congregatus.* (a) Primordial photo-induction (b) Photo-inhibition of development of dark sensitive primordia (*cf.* Fig. 1). The ordinate value for the most effective wavelength was set at 100. After Durand and Furuya (1985).

this species showed peaks of effectiveness at 260, 280, 370 and 440 nm. However, the spectral sensitivity for inhibition of primordial development was quite similar and suggested a common photoreceptor during each successive morphogenetical phase (Fig. 11). A similar action spectrum was established for the fruit-body formation of another basidiomycete *Schizophyllum commune* (Yli-Mattila 1985), with a specific 340-360 nm peak higher than the peaks in the B region. Over the last 20 years, basically analogous action spectra have been described for several other basidiomycetes, with peaks in both UV-A, 320-400 nm, and B, 400-520 nm [*e.g. Favolus arcularius,* Kitamoto *et al.* (1974); *Psilocybe cubensis,* Badham (1980); *Pleurotus ostreatus,* Richartz and Mac Lellan (1987)]. Furthermore, fairly precise data indicate that carpophore initiation and development in basidiomycetes can also be eventually inhibited by 'excessive' B/UV-A (see for review Durand 1985).

The nature of the hypothetical B/UV-A photoreceptor(s) of *C. congregatus* remains unclear. From the action spectra and some results related to the effect of various inhibitors, they seem to be flavin photoreceptors rather than carotenoids (Durand and Furuya 1985; Durand 1985, 1987). Experiments using inhibitors known to react with illuminated flavins (potassium iodide, phenylacetic acid) demonstrated that these inhibitors, while having other effects, do not act at the photoreceptor level and can no longer be regarded as specific inhibitors of the B/UV-A response.

No recent data consider the effectiveness of yellow or red radiation on fruiting of macromycetes. However, more than 20 years ago, C.T. Ingold and collaborators described the influence of yellow-red light on fruiting of *Sphaerobolus stellatus.* In this species, irradiations below 550 nm were effective in allowing the full accomplishment of a first morphogenetic phase of sporophores (more than half of the approximately 2-week developmental period). However, for the terminal phase (maturation phase ended by basidiospores discharge) yellow-red radiation was effective. Blue light retarded this last phase. Furthermore, the stimulatory effect of yellow light could be reversed by B and repeated reversibility was possible by alternating short exposures of B and yellow light (Alasoadura 1963; Ingold and Nawaz 1967; Ingold and Peach 1970). As emphasized by Tan (1978) *"it is a pity that studies were not extended to the UV-A region; otherwise we would know whether the reversibility demonstrated for* S. stellatus *is similar to the mycochrome system or not".* Characteristics of *S. stellatus* seem exceptional among basidiomycetes and, indeed, among other fungi. They await confirmation by precise action spectra.

9.6.5.3 Variability in light-quality requirements

Some results suggest that light requirements from a qualitative point of view may vary in relation to other environmental factors and, eventually, in relation to successive morphogenetical phases.

In the ascomycete *Gelasinospora reticulispora,* three different stages of photosensitivity for perithecia formation were recognized during growth of vegetative mycelia (Inoue and Furuya 1974; Furuya 1986). In the first stage, hyphae require an inductive dark period of 30 h at 25°C which can be interrupted by a short exposure to B or UV-A. In the second stage, brief irradiation with B or UV-A triggers the perithecial initiation in the dark-induced hyphae, but irrespective of timing of this irradiation, perithecial primordia are not visually formed until 54 h after the inoculation, this being the time required for the second stage hyphae to reach maturity. In the third stage, hyphae that are matured in the dark can produce their primordia immediately after a brief irradiation with B or UV-A. The photocontrol of perithecial differentiation in *Gelasinospora* can only take place when the hyphae are grown on corn meal agar. This suggests that the photocontrol system does not always regulate the processes of reproductive differentiation, but operates under the influence of other cellular factors.

It has also been established that incubation temperature and irradiance interact and affect conidial morphogenesis in *Alternaria cichorii* (Section 9.6.3.2.3). Two distinct phases can be distinguished in sporulation of this species: an inductive phase leading to formation of conidiophores, and a 'terminal phase' leading to formation of conidia (Vakalounakis and Christias 1986). The inductive phase was induced by UV irradiation below 340 nm and proceeded most efficiently at high temperatures, while the terminal phase was inhibited by B/UV-A/UV-B (310-420 nm) and B/UV-A (360-420 nm) and was operative at lower temperatures. Inhibition of conidiation by B/UV-A/UV-B was caused by the B wavelengths and it increased with increase in temperature. The inhibition was temporary under continuous B/UV-A/UV-B irradiation but permanent under continuous B alone. Extended exposure (3 days or more) to B/UV-A/UV-B promoted sporulation in a temperature-dependent way.

Finally, light and aeration are essential for the fruiting of the basidiomycete *Lentinula edodes, alias* Shiitake (Leatham and Stahmann 1987). An inadequate aeration, probably resulting in CO_2 accumulation, near the time of fruiting, resulted in failure of the primordia to expand. Furthermore, the stimulatory wavelengths were dependent on the composition of the growth medium. Red wavelengths (620-680 nm) stimulated and B wavelengths (400-500 nm) inhibited fruiting on low calcium media. Conversely, B wavelengths stimulated fruiting on high calcium media and R was not sufficient. However, stimulation of fruiting of *L. edodes* by R on low calcium media only allowed development of subnormally pigmented button-stage fruit bodies. These results remain exceptional and do not modify the views on the inefficiency of yellow-red radiations in inducing fruiting of fungi, particularly basidiomycetes, in most cases.

9.6.6 Concluding remarks

Fundamental studies on different aspects of photomorphogenesis in fungi, from primordia initiation to actual sporogenesis, are still required. *Coprini* remain one of several useful model systems for such work, for instance on the nature of the photoreceptors involved. More than ever, the study of lower and higher fungi should contribute to a better understanding of various physiological problems in plants, such as: photoperiodic control of reproduction; the nature of the B and UV photo-induced or photo-inhibited responses; interactions of light with other external factors, particularly temperature, controlling morphogenesis; internal correlations controlling phototropic curvatures in growing structures; rhythmical processes of differentiation of reproductive structures.

Acknowledgement. The author wishes to thank Prof. G.W. Gooday, University of Aberdeen, for valuable comments and critical review of this chapter.

9.6.7 Further reading

Durand R. (1987) A photosensitive system for blue/UV light effects in the fungus *Coprinus*. In: *Blue Light Responses: Phenomena and Occurrence in Plants and Microorganisms,* Vol.1, pp. 31-41, Senger H. (ed.) C.R.C. Press Inc., Boca Raton.

Gressel J. and Rau W. (1983) Photocontrol of fungal development. In: *Encyclopedia of Plant Physiology,* New series 16B, *Photomorphogenesis,* pp. 603-639, Shropshire Jr. W. and Mohr H. (eds.) Springer-Verlag, Berlin.

Horwitz B.A. and Gressel J. (1987) First measurable effects following photoinduction of morphogenesis. In: *Blue Light Responses: Phenomena and Occurrence in Plants and Microorganisms,* Vol. 2, pp. 53-70, Senger H. (ed.) C.R.C. Press Inc., Boca Raton.

Kumagai T. (1988) Photocontrol of fungal development. *Photochem. Photobiol.* 47: 199-203.

Millet B. and Manachère G. (1983) Introduction à l'étude des rythmes biologiques. Vuibert (ed.) Paris.

9.6.8 References

Alasoadura S.O. (1963) Fruiting in *Sphaerobolus* with special reference to light. *Ann. Bot.* 27: 123-145.

Aragaki M. (1961) Radiation and temperature interaction on the sporulation of *Alternaria tomato. Phytopathol.* 51: 803-805.

Arpin N., Curt R. and Favre-Bonvin J. (1979) Mycosporines : mise au point et données nouvelles concernant leurs structures, leur distribution, leur localisation et leur biogenèse. *Rev. Mycol.* 43: 247-257.

Badham E.R. (1980) The effect of light upon basidiocarp initiation in *Psilocybe cubensis. Mycologia* 72: 136-142.

Ballou L.R. and Holton R.W. (1985) Synchronous initiation and sporulation of fruit bodies by *Coprinus cinereus* on a defined medium. *Mycologia* 77: 103-108.

Bastouill-Descollonges Y. and Manachère G. (1990a) Control of basidiospore production by a nuclear gene in the fungus *Coprinus congregatus. Sex. Plant Reprod.* 3: 103-108.

Bastouill-Descollonges Y. and Manachère G. (1990b) Experimental evidence for photomorphogenesis and photosporogenesis dissociation in *Coprinus congregatus. Crypt. Bot.* 1: 399-408.

Björn L.O. (1986) Reinvestigation of the proposed red light effect on carotenogenesis in the fungus *Verticillium agaricinum. Physiol. Plant.* 68: 648-657.

Dehorter B. (1976) Induction des périthèces de *Nectria galligena* Bres. par un photocomposé mycélien absorbant à 310 nm. *Can. J. Bot.* 54: 600-604.

Dehorter B., Jacques R. and Lacoste L. (1980) Photoinduction des périthèces du *Nectria galligena* II. Influence de la qualité de la lumière. *Can. J. Bot.* 58: 2212-2217.

Durand R. (1982a) Photoperiodic response of *Coprinus congregatus:* effects of light breaks on fruiting. *Physiol. Plant.* 55: 226-230.

Durand R. (1982b) Fruiting of *Coprinus congregatus.* Interacting effect of radiant flux density and temperature. *Experientia* 38: 341-342.

Durand R. (1983a) Effects of inhibitors of nucleic acid and protein synthesis on light-induced primordia initiation in *Coprinus congregatus. Trans. Br. mycol. Soc.* 81: 553-558.

Durand R. (1983b) Light breaks and fruit-body maturation in *Coprinus congregatus:* dark inhibitory and dark recovery process. *Plant Cell Physiol.* 24: 899-905.

Durand R. (1985) Blue U.V. - light photoreception in fungi. Review. *Physiol. Vég.* 23: 935-943.

Durand R. and Furuya M. (1985) Action spectra for stimulatory and inhibitory effects of U.V. and blue light on fruit-body formation in *Coprinus congregatus. Plant Cell Physiol.* 26: 1175-1183.

Furuya M. (1986) Photobiology of fungi In: *Photomorphogenesis in Plants,* Ist ed. pp. 503-520, Kendrick R.E. and Kronenberg G.H.M. (eds.) Martinus Nijhoff Publishers, Dordrecht.

Gooday G.W. (1985) Elongation of the stipe of *Coprinus cinereus.* In: *Developmental Biology of Higher Fungi,* pp. 311-332, Moore D., Casselton L.A., Wood D.A. and Frankland J. (eds.) Cambridge University Press, Cambridge.

Hawker L.E. (1966) Environmental influences on reproduction. In:*The Fungi,* Vol. 2, *The Fungal Organism,* pp. 435-469, Ainsworth G.C.and Sussman A.S. (eds.) Academic Press, New York.

Ingold C.T. (1965) *Spore Liberation,* Clarendon Press, Oxford.

Ingold C.T. and Nawaz M. (1967) Sporophore development in *Sphaerobolus:* effect of blue and red light. *Ann. Bot.* 31: 469-477.

Ingold C.T. and Peach J. (1970) Further observations on fruiting in *Sphaerobolus* in relation to light. *Trans. Brit. mycol. Soc.* 54: 211-220.

Inoue Y. and Furuya M. (1974) Perithecial formation in *Gelasinospora reticulospora* II. Promotive effects of near-ultraviolet and blue light after dark incubation. *Plant Cell Physiol.* 15: 195-204.

Kamada T., Kurita R. and Takemaru T. (1978) Effects of light on basidiocarp maturation in *Coprinus macrorhizus. Plant Cell Physiol.* 19: 263-275.

Kitamoto Y., Horikoshi T. and Suzuki A. (1974) An action spectrum for photoinduction of pileus formation in a basidiomycete *Favolus arcularius. Planta* 119: 81-84.

Leach C.M. (1965) Ultraviolet-absorbing substances associated with light-induced sporulation in fungi. *Can. J. Bot.* 43: 185-200.

Leach C.M. and Trione E.J. (1966) Action spectra for light-induced sporulation of the fungi *Pleospora herbarum* and *Alternaria dauci. Photochem. Photobiol.* 5: 621-630.

Leatham G.F. and Stahmann M.A. (1987) Effect of light and aeration on fruiting of *Lentinula edodes. Trans. Br. mycol. Soc.* 88: 9-20.

Lu B.C. (1972) Dark dependence of meiosis at elevated temperatures in the Basidiomycete *Coprinus lagopus. J. Bacteriol.* 111: 833-834.

Lukens R.J. (1966) Interference of low temperatures with the control of tomato early blight through use of nocturnal illumination. *Phytopathol.* 56: 1430-1431.

Manachère G. (1968) Influence d'illuminations au cours de la nyctipériode unique nécessaire à la maturation des carpophores de *Coprinus congregatus* Bull. ex Fr. développés sous un éclairage quotidien de 12 h. *C.R. Acad. Sci.* Série D (Paris) 267: 1454-1457.

Manachère G. (1980) Conditions essential for controlled fruiting of macromycetes. A review. *Trans. Br. mycol. Soc.* 75: 255-270.

Manachère G. (1985) Sporophore differentiation of higher fungi: a survey of some actual problems. *Physiol. Vég.* 23: 221-230.

Manachère G. (1988) Regulation of sporophore differentiation in some macromycetes, particularly in Coprini : on overview of some experimental studies, from fruiting initiation to sporogenesis. *Cryptogamie Mycol.* 9: 291-323.

Manachère G. and Bastouill-Descollonges Y. (1982) Recherches cytophysiologiques sur la sporogenèse de *Coprinus congregatus* Bull. ex Fr. : introduction à l'étude du déroulement de la méiose en rapport avec les conditions lumineuses et thermiques. *Cryptogamie Mycol.* 3: 391-408.

Manachère G. and Bastouill-Descollonges Y. (1985) An analysis of photo-induced fruiting in *Coprinus congregatus*. *Photochem. Photobiol.* 42: 725-729.

Millet B. and Manachère G. (1979) Morphogenèse rythmée chez les végétaux. *Bull. Soc. Bot. Fr.* 126, *Actual. Bot.*: 51-74.

Miyake H., Tanaka K. and Ishikawa T. (1980) Basidiospore formation in monokaryotic fruiting bodies of a mutant strain of *Coprinus macrorhizus*. *Arch. Mikrobiol.* 126: 207-212.

Moore-Landecker E. and Shropshire W. Jr. (1982) Effect of aeration and light on apothecia, sclerotia and mycelial growth in the discomycete *Pyronema domesticum*. *Mycologia* 74: 1000-1013.

Raudaskoski M. and Virtanen H. (1982) Effect of aeration and light on fruit-body induction in *Schizophyllum commune*. *Trans. Br. mycol. Soc.* 78: 89-96.

Raudaskoski M. and Yli-Mattila T. (1985) Capacity for photoinduced fruiting in a dikaryon of *Schizophyllum commune*. *Trans. Br. mycol. Soc.* 85: 145-151.

Richartz G. and Mac Lellan A.J. (1987) Action spectra for hyphal aggregation, the first stage of fruiting in the basidiomycete *Pleurotus ostreatus*. *Photochem. Photobiol.* 45: 815-820.

Robert J.C. (1982) Corrélations physiologiques régularisant la morphogenèse des carpophores de *Coprinus congregatus* Bull. ex Fr.: influence des facteurs externes sur divers aspects métaboliques; introduction à la connaissance des mécanismes du rythme de fructification. *Thèse Sciences*, Université Lyon I.

Ross I.K. (1982) Localization of carpophore initiation in *Coprinus congregatus*. *J. Gen. Microbiol.* 128: 2755-2762.

Tan K.K. (1978) Light-induced fungal development. In: *The Filamentous Fungi*, Vol 3, *Developmental Mycology*, pp. 334-357, Smith J.E. and Berry D.R. (eds.) Arnold, London.

Uno I. and Ishikawa T. (1982) Biochemical and genetic studies on the initial events of fruit-body formation. In: *Basidium and Basidiocarp: Evolution, Cytology, Function and Development.*, pp. 113-123, Wells K. and Wells E.K. (eds.) Springer-Verlag, New York.

Vakalounakis D.J. and Christias C. (1985a) Blue-light inhibition of conidiation in *Alternaria cichorii*. *Trans. Br. mycol. Soc.* 85: 285-289.

Vakalounakis D.J. and Christias C. (1985b) Light intensity, temperature and conidial morphology in *Alternaria cichorii*. *Trans. Br. mycol. Soc.* 85: 425-430.

Vakalounakis D.J. and Christias C. (1986) Light quality, temperature and sporogenesis in *Alternaria cichorii*. *Trans. Br. mycol. Soc.* 86: 247-254.

Valadon L.R.G., Osman M., Mummery R.S., Jerebzoff-Quintin S. and Jerebzoff S. (1982) The effect of monochromatic radiation in the range 350 to 750 nm on the carotenogenesis in *Verticillium agaricinum*. *Physiol. Plant.* 56: 199-203.

Yli-Mattila T. (1985) Action spectrum for fruiting in the basidiomycete *Schizophyllum commune*. *Physiol. Plant.* 65: 287-293.

Yli-Mattila T. (1987) The effect of UV-A light on cAMP level in the basidiomycete *Schizophyllum commune*. *Physiol. Plant.* 69: 451-455.

Yli-Mattila T. (1990) Photobiology of fruit-body formation in the basidiomycete *Schizophyllum commune*, *Thesis*, Turku University, Reports Dept. of Biology.

9.7 Photobiology of ferns

Masamitsu Wada[1] and Michizo Sugai[2]

[1]*Biology Department, Tokyo Metropolitan University,*
Minami Ohsawa 1-1, Hachioji-shi, Tokyo 192-03, Japan
[2]*Department of Biology, Toyama University,*
Gofuku 319, Toyama 930, Japan

9.7.1 Introduction

Fern gametophytes are the haploid phase during the alternation of generations in the life cycle of ferns. They are simple independent organisms, starting from single cells (spores), which develop into heart shaped prothalli after an initial filamentous stage of a few linearly aligned cells. Antheridia and archegonia develop on a three dimensional tissue (called the mat) at the central part of gametophytes and fertilization results in the sporophytic generation. The gametophytic stage, although short and showing simple organization, is indispensable for the ferns' sexual reproduction.

Gametophytes possess, although in a simplified form, almost all the basic physiological phenomena found in higher plants. The developmental processes of fern gametophytes usually depend on light. A new light signal is always needed for each process in development, from spore germination to cell elongation, phototropic curvature, cell division, two-dimensional differentiation, antheridium formation. Without these photosignals, further development is severely depressed, so that the developmental processes are easily experimentally (or artificially) manipulated by light (Fig. 1). Physiological phenomena such as chloroplast movement are also easily induced by light (Fig. 2). This means that fern gametophytes provide a good model system to study development and physiology of plants at the cellular or even subcellular level, because photoregulation of the development, as well as their simple organization are quite unique in the plant kingdom. Furthermore, gametophytes can grow autotrophically in an inorganic medium without surrounding tissues, giving us advantages of easy cultivation and direct observation.

In contrast, photoresponses of fern sporophytes have only been studied in a few cases, not only because synchronous culture is very difficult, but also

R.E. Kendrick & G.H.M. Kronenberg (eds.), Photomorphogenesis in Plants - 2nd Edition

Figure 1. Letters created with protonemata by phototropic and polarotropic responses under red light.

because it is hard to overcome the temporal and spacial problems of getting sufficient sporophytic material.

In this chapter, the photocontrol of each step of the gametophyte development is described and results obtained from recent analytical studies are reviewed.

9.7.2 Spore germination

9.7.2.1 Photoregulation

Spores of most fern species require light for germination (Miller 1968). In *Pteris vittata,* spore germination was induced by brief irradiation with red light (R) and inhibited by subsequent far-red light (FR). This effect is repeatedly R-FR reversible showing the involvement of the phytochrome system. Phytochrome has been detected spectrophotometrically in imbibed spores of the fern, *Lygodium japonicum.* The amount of phytochrome in *Lygodium* spores increases during imbibition and this increase is closely correlated with an increased sensitivity to R (Fig. 3). This increase was completely inhibited by gabaculine (Manabe *et al.* 1987), an inhibitor of phytochrome biosynthesis, but the inhibition could not be found in *Anemia* (Schraudolf 1987). These contradictory results make it difficult to say whether the increase of phytochrome results from biosynthesis or simply by rehydration.

The sensitivity of spores to R is affected by nitrate in the culture medium at concentrations higher than 1 μM, the optimum being 1 m*M*. Nitrate works when it is added in the presence of the FR-absorbing form of phytochrome (Pfr) or just after the reversion of Pfr to the R-absorbing form (Pr) by FR. This suggests that the nitrate acts during early steps in the Pfr-induced signal-transduction

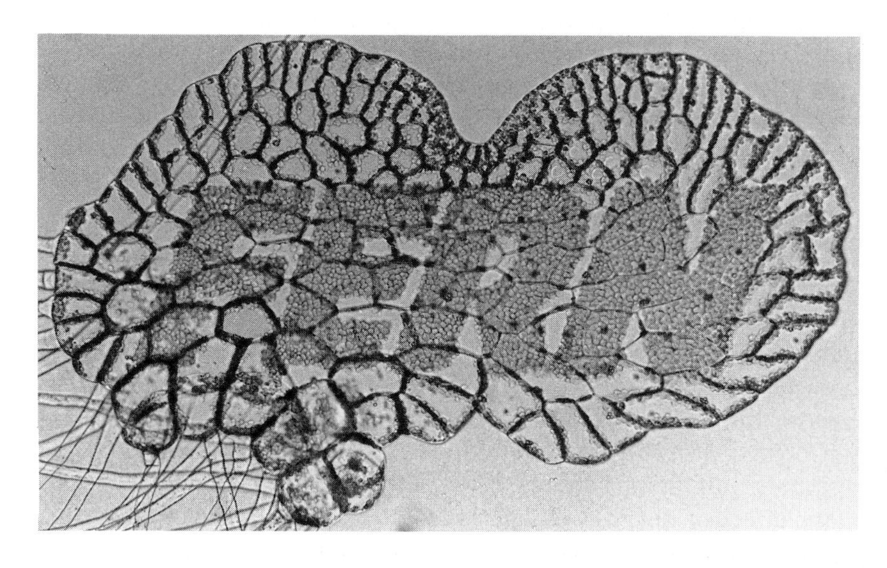

Figure 2. Chloroplast photo-orientation under high fluence of blue light (outside of the letters). Chloroplasts moved to anticlinal wall of prothallial cells.

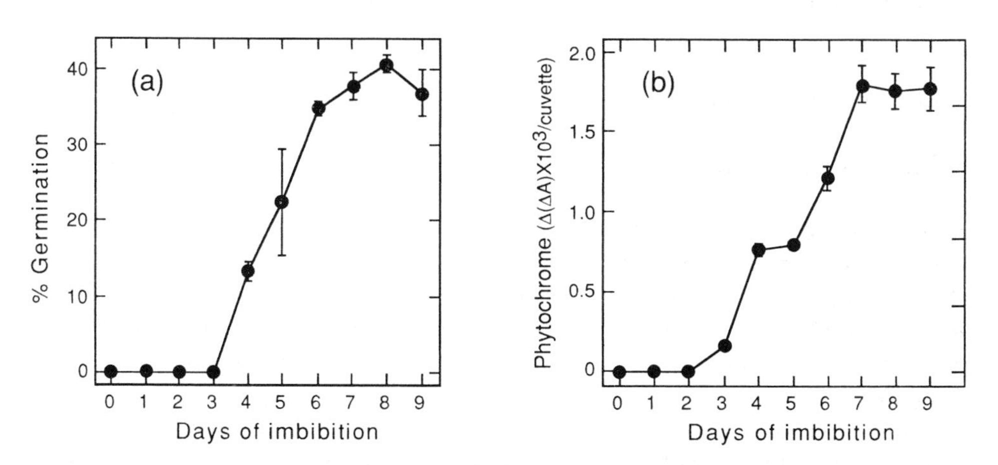

Figure 3. Changes of germination rate (a) and phytochrome content (b) during imbibition of spores in *Lygodium japonicum.* After: (a) Tomizawa K. Sugai M. and Manabe K. (1983) *Plant Cell Physiol.* 24: 1043-1048; (b) Tomizawa K. Manabe K. and Sugai M. (1982) *Plant Cell Physiol.* 23: 1305-1308.

chain. However, the nitrate effect is only clear under non-saturating R conditions, being fully substituted by continuous irradiation with R in the absence of nitrate (Haas and Scheuerlein 1990), suggesting that the nitrate is not involved as an essential step of the phytochrome-transduction chain.

Red light-induced fern-spore germination was inhibited by brief irradiation of blue light (B) and this inhibition was not cancelled by subsequent R. This evidence suggested that the so-called 'B-induced photoreaction' was involved in the germination process of many ferns such as *Pteris, Lygodium, Adiantum* and *Ceratopteris.* The action spectra for the inhibition in *Pteris* and *Adiantum* have three peaks at around 280, 380, and 440 nm. The peak in the UV-C is higher than those of UV-A and B after correction for transmittance of the spore coat (Fig. 4). The shape of the action spectra of the inhibition is in accordance with the action spectra for phototropisms of *Avena* and other B effects (Chapter 5.1). These photoreactions have led to carotenoids, flavins and/or pterins being postulated as possible photoreceptor pigments, but the nature of these pigments is still unclear. It would be worthwhile studying spore germination in *Ceratopteris* whose life cycle is relatively short and where mutants can be selected easily (Cooke *et al.* 1987).

Respiratory inhibitors such as NaN_3 and KCN and ethanol cancelled the inhibitory effect of near UV-A and B on R-induced spore germination. However UV-C induced inhibition was not cancelled by these chemicals (Sugai and Furuya 1990). These results may suggest that different photoreceptor pigments absorb UV-C and UV-A and B.

9.7.2.2 Signal transduction

In several species, such as *Onoclea, Adiantum, Lygodium* and *Dryopteris,* spore germination was not induced by R when using Ca^{2+} free media, but could be recovered by Ca^{2+} supplementation. In *Onoclea* the calcium ionophore A23186 induced germination in total darkness, although it has not been confirmed in other ferns. The Ca^{2+} is exclusively required for phytochrome mediated spore germination. However, kinetic studies indicate that timing of the Ca^{2+} requirement is not the first step of signal transduction of phytochrome but is required at a late stage, 10 h after the escape from reversibility by FR in *Adiantum,* as well as in *Dryopteris* (Dürr and Scheuerlein 1990).

In some species belonging to Schizaeaceae such as *Lygodium* and *Anemia,* spore germination is induced in darkness by gibberellins (GAs) instead of R. Dark germination of many Polypodiaceous ferns is also induced by antheridiogen, the antheridium-inducing factor secreted from mature gametophytes of the bracken fern, *Pteridium aquillinum.* In *Lygodium* this GA-induced germination was not inhibited by AMO-1618, an inhibitor of GA biosynthesis, while R-induced germination was completely inhibited by AMO-1618 (Kagawa and Sugai 1991). This evidence suggests that, in *Lygodium,* R induces the biosynthesis of GA *via* the phytochrome system, and subsequently, GA induces spore germination.

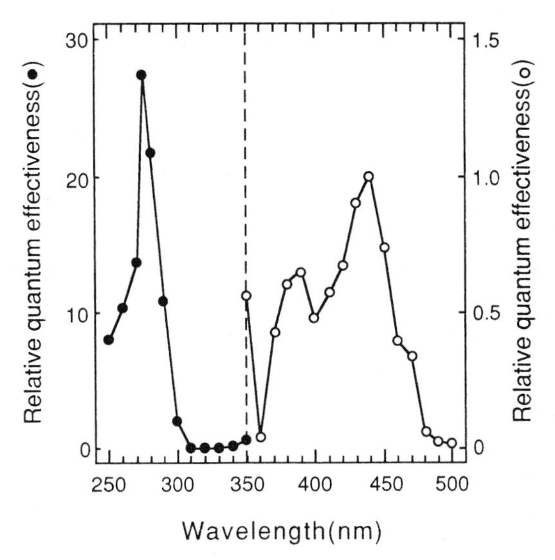

Figure 4. Action spectrum for photo-inhibition of red light-induced germination in *Adiantum capillus-veneris* corrected at each wavelength, taking into account the transmission spectrum of the spore coat. After Sugai M. and Furuya M. (1985) *Plant Cell Physiol.* 26: 953-956.

The time lag between germination (cell protrusion through the spore coat) and light treatment differs depending on the fern species from 24 h for *Osmunda* to about 6 days for *Pteris.* The events that occur during this period are largely unknown, except for phenomena such as chlorophyll synthesis, nuclear migration and an unequal cell division for rhizoid development.

9.7.3 Protonemal growth

9.7.3.1 Apical growth

In homosporous ferns, after spore germination, but before two-dimensional differentiation, a filamentous stage usually exists as the protonema(ta) (Fig. 5). Protonemata have a strong polarity and grow at the tip. The duration of this stage and critical cell number for the next stage (differentiation to a prothallus) are variable depending on species and environmental conditions. Under R or very low irradiance white light (W), protonemata grow longer and thinner with a lower rate of cell division than under B or high irradiance W.

Under appropriate R conditions fern protonemata usually grow toward a light source as long single cells. In *Adiantum* they grow at a constant rate (10 μm h^{-1}) and a constant width (*ca.* 15-20 μm) for several days, although the growth rate varies from species to species and is also influenced by R fluence rate. During

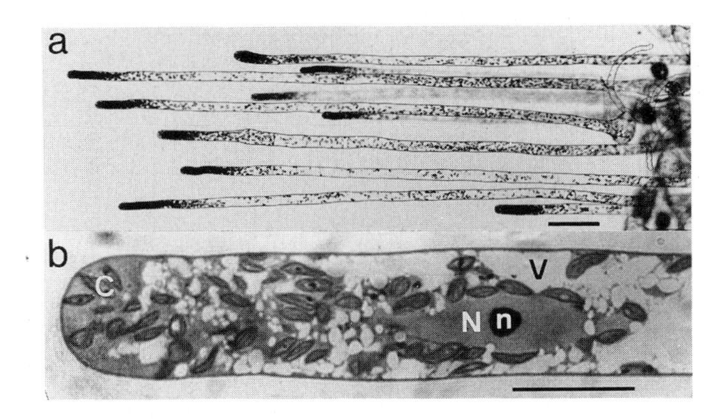

Figure 5. ***Adiantum*** protonemata grown under continuous red light. (a) Grown for 10 days. After Yatsuhashi H. Kadota A. and Wada M. (1985) *Planta* 165: 43-50. (b) Light micrograph of a longitudinal section. C = chloroplast, N = nucleus, n = nucleolus, V = vacuole. After Wada M. (1988) *Bot. Mag. Tokyo* 101: 519-528.

growth, a nucleus localizes some distance from the tip (about 60 μm in *Adiantum*). An exception is *Pteris* in which protonemata show linear growth under R but no tropic responses. *Lygodium* and *Ceratopteris* lack a protonemal stage and become two-dimensional even under R, although linear growth occurs in darkness in *Lygodium.*

The photoreceptor for the R effect on protonemal growth is thought to be phytochrome. However, the different responses to R in different species make it difficult to understand on the basis of a simple phytochrome mechanism. Further analysis is required to see whether fern phytochrome consists of several molecular species, has different binding partners, or has different transduction chains. It is important that the intracellular location of the physiologically active phytochrome is resolved.

9.7.3.2 Apical swelling

Growth retardation occurs when R-grown protonemata are transferred to darkness, B or W. Continuous B (or W) induces cell swelling at the apical region within 1-2 h (Fig. 6). The photoreceptor is the so-called 'B/UV-A photoreceptor', sometimes referred to as cryptochrome (Chapter 5.1). Dichroic effects of B on cell swelling and partial cell swelling induced by partial irradiation with a microbeam revealed that the photoreceptor was localized at the cell periphery close to the plasma membrane and that the photosignal remains localized.

During the first hour after the onset of B, the protonemal apex continues to grow at the same rate with the same width as in R. Thereafter, an increase of

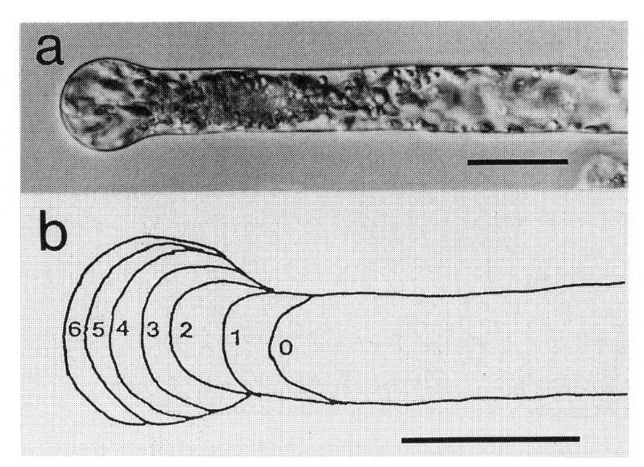

Figure 6. Apical swelling of *Adiantum* protonemata under continuous blue light (B). (a) Photomicrograph, irradiated with B for 4 h, bar = 50 μm. (b) Time-lapse sequence of 1 h interval after the onset of B, bar = 50 μm. After Murata and Wada (1989).

cell width becomes detectable. Cell swelling occurs at the newly grown portion under B, but not in the region already grown under preceding R (Murata and Wada 1989).

9.7.3.3 Cytoskeleton and microfibrils

During linear growth under R, a band (15-20 μm in width) of cortical microtubules and microfilaments runs perpendicular to the cell axis around the subapical part of a protonema (Fig. 6a, b). This band may play an important role in regulating cell width and may control the direction of cellulose microfibrils, perpendicular to the cell axis, in the innermost layer of the cell wall. The resulting cell wall may be strong enough to maintain the cell width at a constant diameter.

When a protonema was transferred from R to B, microfilaments and microtubules in the band were disrupted (Fig. 7) and the pattern of microfibril arrangement became random. These intracellular changes of cytoskeletons start within 15 min and are complete about 1 h after the onset of B, before a change of cell shape becomes detectable under a microscope. Consequently, the change of microfibril arrangement at the newly grown area may weaken the apical area, resulting in cell swelling. The cell wall at the cylindrical region including the subapex of the protonemata which have been grown under R may be too strong to swell. The lag period before apical swelling may be the preparatory stage for reconstruction of a new and weaker region at the apex. A similar relationship between cytoskeleton and microfibrils was observed in tropic responses as shown in Section 9.7.4.1.

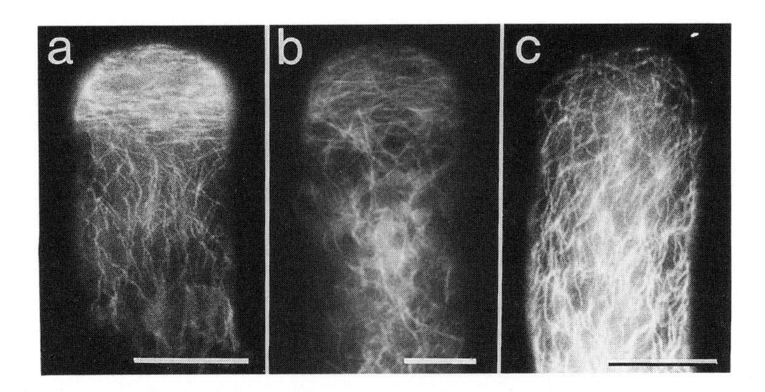

Figure 7. Cortical cytoskeletons of apical part of *Adiantum* protonemata. (a) A band of microtubules under red light (R). (b) A band of microfilaments under R. (c) Microtubules 3 h after blue light irradiation. Microtubules were stained by indirect fluorescence with antitubulin antibody, microfilaments were stained with rhodamine-labelled pharoidin, bar = 10 μm. (a,c) After Murata T. and Wada M. (1989) *Planta* 178: 334-341. (b) Kadota A. and Wada M. (1989) *Plant Cell Physiol.* 30: 1183-1186.

9.7.4 Phototropism and polarotropism

9.7.4.1 Tropic responses

Phototropism and polarotropism are induced by non-polarized and polarized light, respectively. In the former, protonemata show a tropic response towards the light source, but in the latter cells grow perpendicularly to the vibration plane and the incident ray of the polarized light (Fig. 8). In polarotropism, the tropic response becomes detectable about 1 h after turning the electrical vector, irrespective of the turning angle size (Fig. 9).

In many cases so far tested, photoreceptors of the tropic response of ferns are phytochrome and the B photoreceptor, but the extent of involvement of both pigment systems is different in each species. *Dryopteris* responds to both pigments equally, but *Pteris vittata* lacks the phytochrome-mediated tropic response (Kadota *et al.* 1989). In *Adiantum* the B response is rather weak and can only be detected when one side of a protonema is irradiated with a B microbeam. In this species the B response is composed of two components induced by B absorption of phytochrome and the B photoreceptor itself. The response mediated by the latter pigment is quite weak and saturated at a low fluence. However, the phytochrome-mediated B response is active at higher fluences (Fig. 10).

Interestingly, young leaves of *Adiantum* show a phytochrome-mediated tropic response which is unusual in higher plants, although the B photoreceptor-mediated response, similar to that found in higher plants, is also observed.

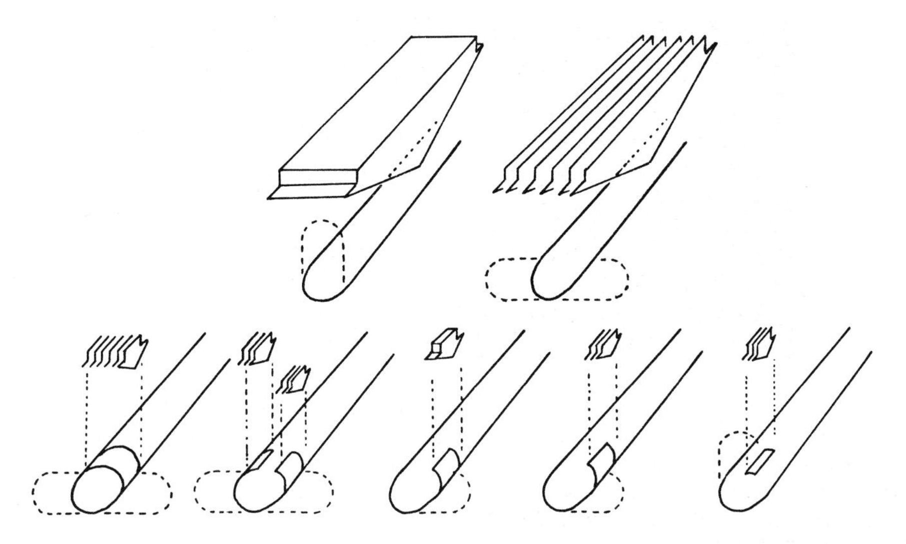

Figure 8. Scheme showing phototropic and polarotropic responses induced by whole cell *(top)* and local *(bottom)* irradiations. Single three-dimensional arrow head: polarized light, multiple parallel arrow heads: unpolarized light. After Wada M. and Kadota A. (1987) In: *Phytochrome and Photoregulation in Plants,* pp. 239-248, Furuya M. (ed.) Academic Press, Tokyo.

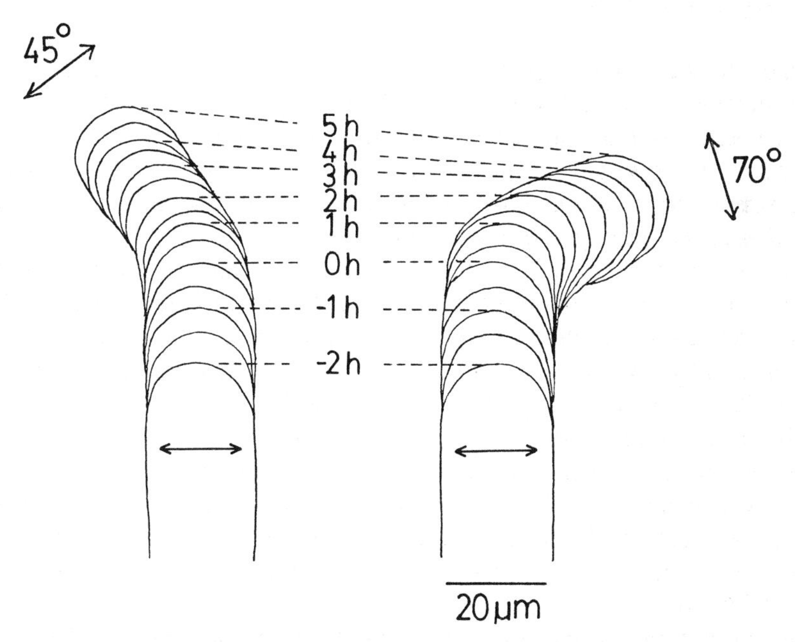

Figure 9. Time-lapse sequence of polarotropism induced by turning electrical vector of polarized red light 45° or 70° at hour 0. After Wada M. Murata T. and Shibata M. (1990) *Bot. Mag. Tokyo* 103: 391-401.

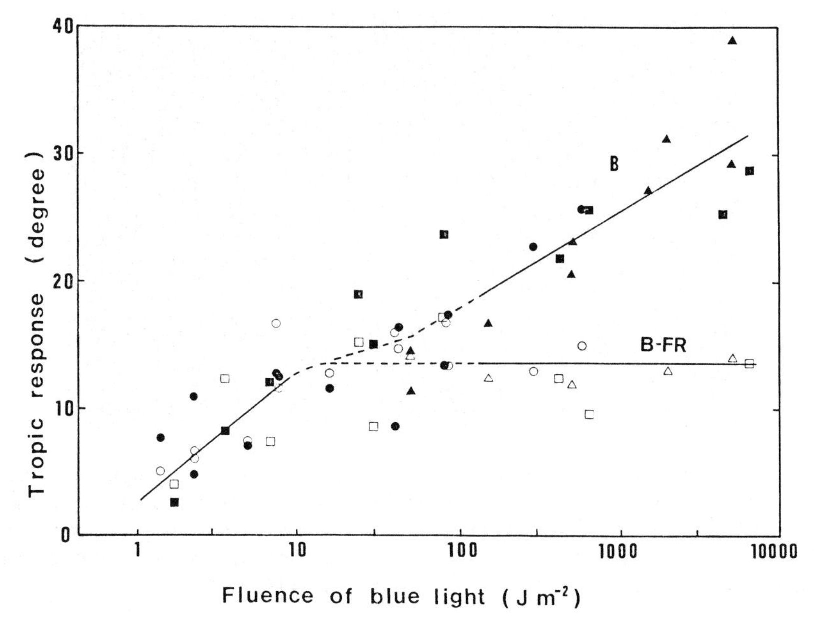

Figure 10. Phototropic response induced by half-side irradiation with blue light (B) in red light-grown *Adiantum* protonemata. One half of a protonema was irradiated with B of various fluences (closed symbols), or followed by far-red light (FR) (open symbols) to cancel the effect of B absorption of phytochrome. For details see Hayami J. Kadota A. and Wada M. (1986) *Plant Cell Physiol.* 27: 1571-1577.

9.7.4.2 Localization and orientation of photoreceptors

Local irradiation of a protonema with a microbeam of polarized R revealed that the photoreceptive site of R in polarotropism was the subapical part and not the tip of the protonema. Moreover, the light was perceived at both sides of the subapex and not the front surface facing the light. Actually, the phototropic response could be induced by a R microbeam (even non polarized light) on one side of the protonemal subapex. When the tip was irradiated with a R microbeam, the protonema continued to grow straight without a change in the direction of growth. Consequently, a protonema may grow at a site with the highest Pfr concentration. Thus, when the site of highest Pfr concentration changes from the tip to one-side, the cell shows a tropic response. Similar results were obtained in a centrifuged protonema (Fig. 11) in which endoplasm had been spun down (Fig. 12), indicating that phytochrome must be localized very close to the plasma membrane. The photoreceptive site of B for phototropism was also shown to be the subapex of the protonemata.

Microbeam irradiation on one side of the subapex with a polarized R vibrating by various planes showed that the light with the electrical vector parallel to the growing axis (or plasma membrane) is most effective. The results

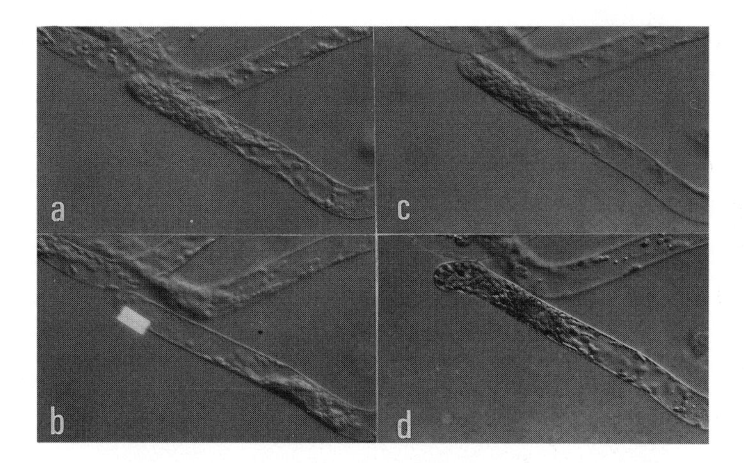

Figure 11. Phototropic response induced by local irradiation with a red light microbeam in a centrifuged *Adiantum* protonema. (a) Before centrifugation, (b) centrifuged cell with a microbeam (10 x 20 μm), (c) after irradiation, cytoplasm was returned by centrifugation, (d) a day after microbeam irradiation, tropic response is obvious towards the irradiated side. For detail see Wada M. Kadota A. and Furuya M. (1983) *Plant Cell Physiol.* 24: 1441-1447.

mean that the transition moment of the Pr is more or less parallel to the plasma membrane as shown in *Mougeotia* (Chapter 7.2). In addition, Pfr is predicted to be perpendicular to the cell surface. The B photoreceptor, as well as the B-absorbing band of phytochrome has also been shown to be parallel to the membrane (Hayami *et al.* 1992).

9.7.4.3 Recognition of light direction

In the natural environment, protonemata grow towards a light source. However, they do not recognize the light direction itself. When one side of the subapex is irradiated with a microbeam, the protonema grows towards the irradiated side, but not to the light source. What is recognized in this case is a difference in the concentration of activated photoreceptors between the front and rear or both sides, as is shown with simultaneous irradiation on both sides with two microbeams of different fluence rates (Iino *et al.* 1990). In other words, they recognize the point with the highest concentration of activated photoreceptors. The angle of tropic response is dependent on the gradient established.

Protonemata growing in the air show phototropism towards the light source, meaning that the lens effect (Chapter 7.2) is not strong enough to show a negative phototropism in this organism. This is probably because the tip of the cell is filled with many chloroplasts which absorb the light preventing any lens effect.

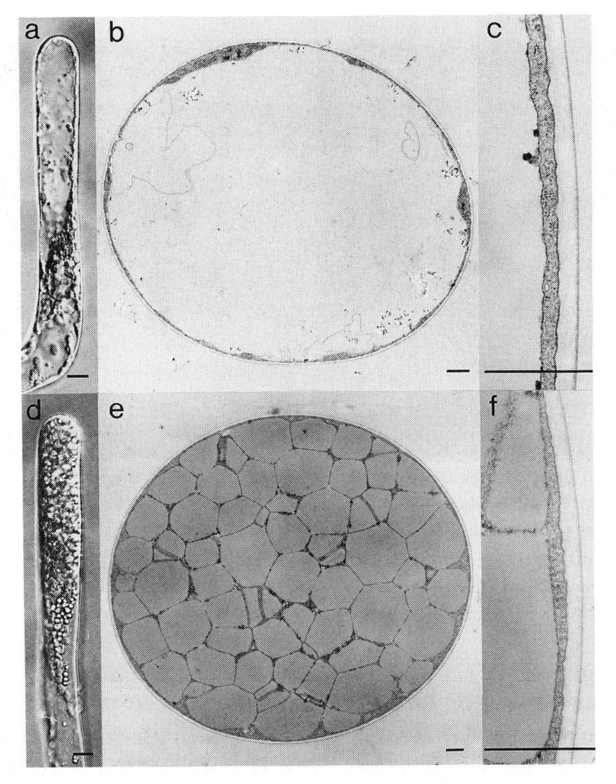

Figure 12. Light and electron micrographs of centrifuged cells with a large vacuole (a, b, and c) or filled with oil droplets (d, e and f) at the apical part of the protonemata. (b, e) Electron micrographs of cross sections, (c, f) high magnification of the cells. V-shaped protonemata were centrifuged three times each for chasing out or filling up oil droplets and for taking down cytoplasm, bar = 1 μm except those in (a and d) where bar = 10 μm.

9.7.4.4 Mechanism of tropic response

The precise mechanism of phototropism is not yet known. Time lapse recording of polarotropism under a microscope (Fig. 9) and structural changes in the process revealed fragmentary evidence for part of a transduction chain. Preceding a shape change of the protonemal apex, 1 h after the inductional treatment of phototropism, the bands of microtubules and microfilaments at the subapical part change their direction towards the new growing side (within 15 min of the onset of phototropism)(Fig. 13). Thereafter the microfibrils at the innermost layer of the cell wall also change their orientation to parallel with the cytoskeletal bands (Wada *et al.* 1990). These structural changes may weaken plasticity or elasticity of the cell wall in the newly growing zone, resulting in cell protrusion towards the new direction.

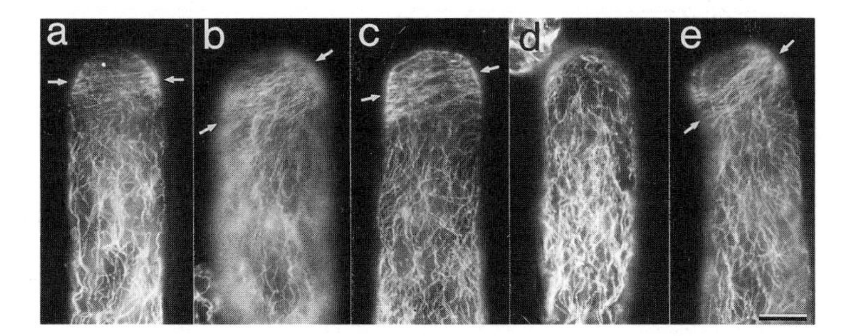

Figure 13. Immunofluorescence micrographs of cortical microtubules at the subapical part of *Adiantum* protonemata during polarotropism. (a) A band of microtubules before the induction of polarotropism. (b) 1 h after turning the electrical vector of polarized red light 45°. (c-e) 20 min, 1 h and 3 h after turning electrical vector 70°, respectively. Arrows indicate the orientation of the microtubule band, bar = 10 μm. After Wada M. Murata T. Shibata M. (1990) *Bot. Mag. Tokyo* 103: 391-401.

9.7.5 Cell division and its orientation

9.7.5.1 Cell cycle

The timing, direction and the site of cell division are very important for the appropriate development of gametophytes. In fern protonemata the cell cycle progression (timing of cell division) is controlled by phytochrome and the B photoreceptor (Fig. 14). Under R protonemata grow keeping their cell cycle at the beginning of the G1 phase. When protonemata were transferred from R to darkness or B, the G1 phase starts to progress, the speed being more rapid in B than in darkness. However, it is not known, whether the G1 in B and in darkness progress *via* the same route with different speeds or by different routes. The latter may be plausible because G1 in darkness can be reversed by R when a protonema is irradiated before entering the S phase, but G1 under B becomes irreversible at an early stage. No decisive answer can be given before precise experiments are carried out.

However, the G2 phase is prolonged by FR given before progression of the G1 phase. The photoreceptor is phytochrome. The S and M phases, at least in *Adiantum,* are not influenced by light even if it is given before or during the cell cycle.

9.7.5.2 Localization of photoreceptors

The B photoreceptor mediating cell division is localized on, or very close to, the nucleus. Only a microbeam of B focused at the nuclear region can induce cell

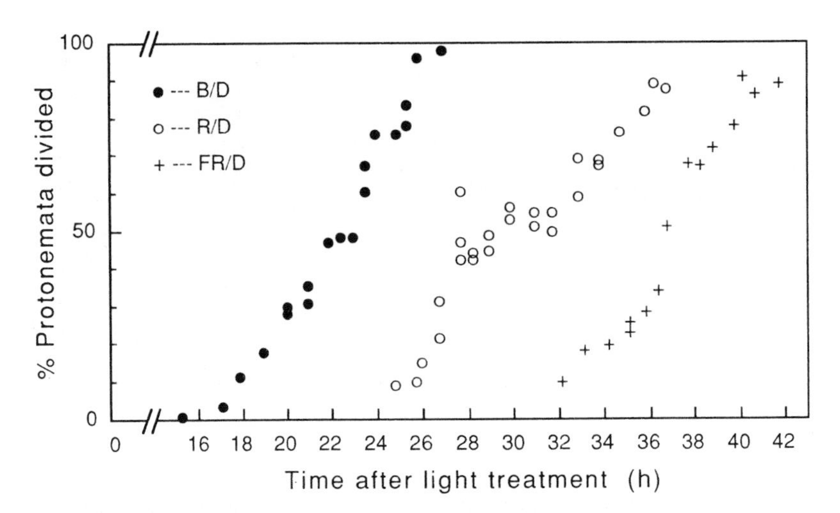

Figure 14. Time courses of cell division of red light-grown protonemata of *Adiantum* in the dark (D) after short irradiation with blue [(B), 0.85 W m^{-2} for 10 min], red [(R), 0.85 W m^{-2} for 10 min] and far-red light [(FR), 30 W m^{-2} for 10 min] at 0 h. After Wada M. and Furuya M. (1972) *Plant Physiol.* 49: 110-113.

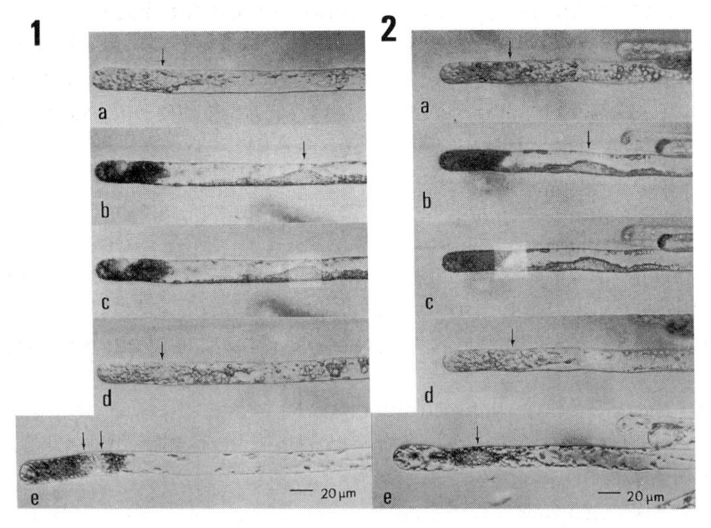

Figure 15. Induction of cell division with a blue microbeam (30 x 30 μm^2) at the nuclear region (1) or 45-75 μm from the tip (2) in centrifuged cells. (a) Before centrifugation. (b) After basipetal centrifugation. Oil drops accumulate in the apex. (c) Microbeam irradiation (17 W m^{-2} for 30 s). (d) Acropetal centrifugation after microbeam irradiation. (e) 30 h after microbeam irradiation. Arrows indicate the position of a nucleus. In 1e, cell division was induced and a new cell plate is observed. After Kadota A. Fushimi Y. and Wada M. (1986) *Plant Cell Physiol.* 27: 989-995.

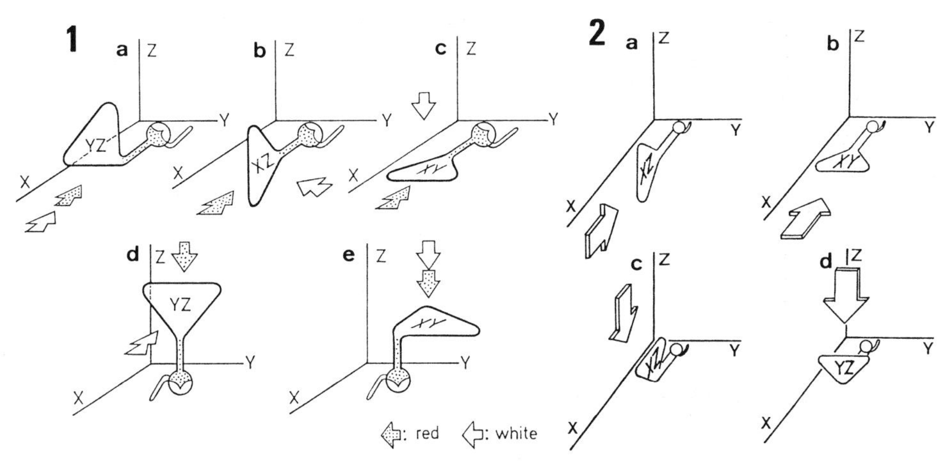

Figure 16. Diagrams showing prothallial expansion of *Adiantum* gametophytes under unpolarized (1) and polarized white light (2). Protonemata grown under red light were irradiated continuously with white light from various directions (shown by the arrows). (1) After Wada M. and Furuya M. (1971) *Planta* 98: 177-185. (2) Kadota A. and Wada M. (1986) *Plant Cell Physiol.* 27: 903-910.

division. Moreover, the photoreceptive site remains associated with the nucleus even when it is centrifuged down to the basal part (Fig. 15). The FR effect on the G2 phase is mediated by phytochrome, whose localization is close to the plasma membrane and spreads over the cell. The R effect on G1 maintenance is mediated by phytochrome, since the response shows typical R-FR reversibility. However, the intracellular localization of the phytochrome is not yet known.

9.7.5.3 Orientation of cell division

The direction of the two dimensional growth of gametophytes is controlled by the light direction or the vibration plane of polarized W (Fig 16). Prothalli spread facing a light source of non-polarized W, *i.e.* perpendicular to the incident rays, or parallel to the plane of the E-vector and the incident ray. The orientation response is observed from a very early stage of the two-dimensional prothallial development, suggesting the direction of cell division may be controlled by light.

When R-grown protonemata of *Adiantum* are transferred to continuous W, to induce successive cell divisions, the first division occurs transversely, but the second or the third division is longitudinal (parallel to the cell axis) and parallel to the incident ray. The transition from transverse to longitudinal division is the turning point from a one dimensional filament to a two dimensional prothallus. The number of cell divisions to reach the turning point is dependant on the species and the light conditions. The first sign of the two dimensional differenti-

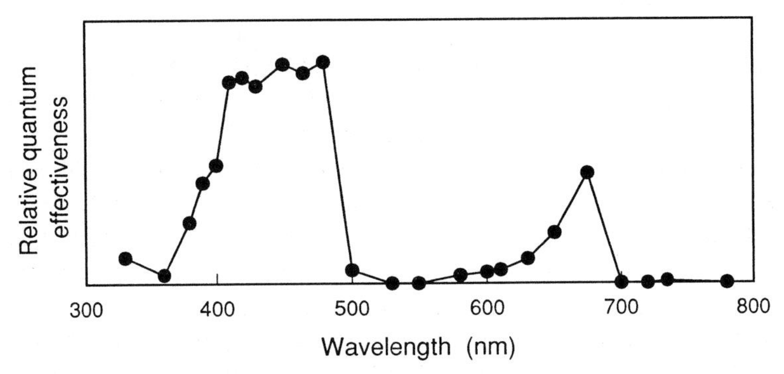

Figure 17. Action spectrum of chloroplast photo-orientation in *Adiantum*. After Yatsuhashi H. Kadota A. and Wada M. (1985) *Planta* 165: 43-50.

ation is apical cell swelling of the protonemata as described above. Before longitudinal cell division the apical cell becomes flat so that the side facing the light becomes wider than it is thick (Wada and Murata 1988), suggesting that cell division occurs according to the minimal area or pressure hypotheses (Miller 1980). However, pressing the apical cell with glass rods from both sides making the cell thicker and narrower, could not change the direction of the cell division (T. Murata and M. Wada unpublished data). The controlling factor of the orientation of cell division may not be so simple as cell shape or pressure. Changes in intracellular structure such as cytoskeletons should be studied precisely during the cell plate orientation.

9.7.6 Chloroplast photo-orientation

9.7.6.1 Survey

Chloroplasts move to the light-irradiated side under a low fluence rate (low fluence response) or to the anticlinal wall to avoid light when the fluence rate of the light is too high (high fluence response). The former response may occur to achieve a high efficiency of photosynthesis, and the latter to avoid photobleaching of photosynthetic pigments. The B photoreceptor(s) is(are) the main photoreceptor for both the high and low fluence responses in a wide variety of plants (Wada *et al.* 1993). However, phytochrome is the main photoreceptor in *Mougeotia* and *Mesotaenium* which have a single large, ribbon-shaped chloroplast in each cell (Haupt and Scheuerlein 1990) (Chapters 7.2 and 9.4). Chloroplast photo-orientation in ferns has not been studied extensively except in the protonemata of *Adiantum,* in which phytochrome, as well as the B photoreceptor mediates chloroplast photo-orientation.

9.7.6.2 Low fluence response

Both B and R are equally effective in the low fluence response in *Adiantum* when a cell is irradiated with polarized light vibrating perpendicularly to the cell axis or partially with a microbeam. Polarized light irradiated from the cell tip is very effective, irrespective of the direction of the electrical vector. The photo-orientation movement to polarized light can be explained if the transition moments of photoreceptors are assumed to be parallel to the cell surface, as in the case of the phototropic response. In *Pteris,* as in the case of polarotropism in which B, but not R is effective, the protonema does not show R-induced chloroplast movement, although B does induce the response, suggesting that the same phytochrome molecules are used both in chloroplast photo-orientation and polarotropism.

9.7.6.3 High fluence response

The high fluence response is also induced both by R and B in *Adiantum.* The same pigment systems as those in the low fluence response are thought to be involved (Yatsuhashi and Wada 1990). Blue light higher than 10 W m^{-2} and R of about 600 W m^{-2} lead to these responses. However, the mode of action is somewhat different in the high fluence responses to B and R. Under a B microbeam, chloroplasts moved out of the blue spot but gathered on both sides of the microbeam. Under high irradiance with R, chloroplast accumulation was not observed at the edges of the R beam. In other experiments, when B was given to a whole protonemal cell, chloroplasts gathered on both sides of the cell, but in R chloroplasts showed similar patterns to the low fluence response, *i.e.* they gathered at the front surface of the cell. The R response was first thought to be mediated by some sort of photobleaching or (destruction) of phytochrome, but it may not be so simple. Further analyses are required before decisive conclusions can be drawn. The mechanism switching between the high and low fluence responses is not yet known.

9.7.6.4 Mechanism of photomovement

The irradiation with two adjacent microbeams of different fluence rates induces chloroplast movement from one spot to another (Fig. 18). In the range of the low fluence response, if the ratio between the two R microbeams are more than 1.2 -1.5, the difference can be recognized and chloroplasts move to the spot of higher fluence rate.

Inhibitor studies revealed that cytochalasin-B and N-ethylmaleimide (NEM) inhibited the photomovement, but colchicine did not, indicating that the

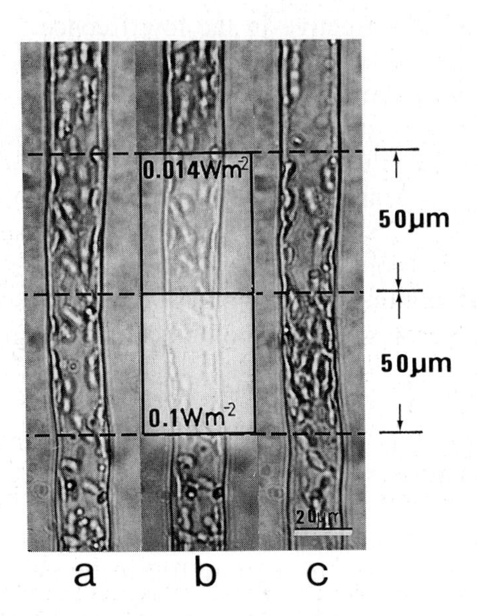

Figure 18. Chloroplast photo-orientation induced by two adjacent microbeams with different fluence rates (*upper* part 0.014 W m^{-2}: *lower* part 0.1 W m^{-2}) of red light. (a) Before irradiation. (b) Two microbeams. (c) 3 h after the onset of irradiation. Chloroplasts accumulated in the microbeam with a higher fluence rate, bar = 20 μm.

movement is mediated by the actomyosin system, but not by microtubules (Kadota and Wada 1992b). However, any change of the cortical meshwork of microfilaments could not be detected during photomovement. After photo-orientation a circular structure of microfilaments is detected between the chloroplast and the plasma membrane (Fig. 19). The structure may anchor chloroplasts to appropriate positions (Kadota and Wada 1992a).

9.7.7 Concluding remarks

Fern gametophytes are a good experimental system to study photomorphogenesis, in that cells can be observed and responses followed under the microscope. Other interesting phenomena such as the photocontrol of antheridium formation (Gemmrich 1986), have not been mentioned because of the space limitation in this chapter. The greatest problem of using ferns as experimental material is acquiring sufficient tissue. This is a problem for biochemistry, which requires a large amounts of tissue for extraction of proteins or other compounds. Further work on fern photomorphogenesis may not be simple, but experiments

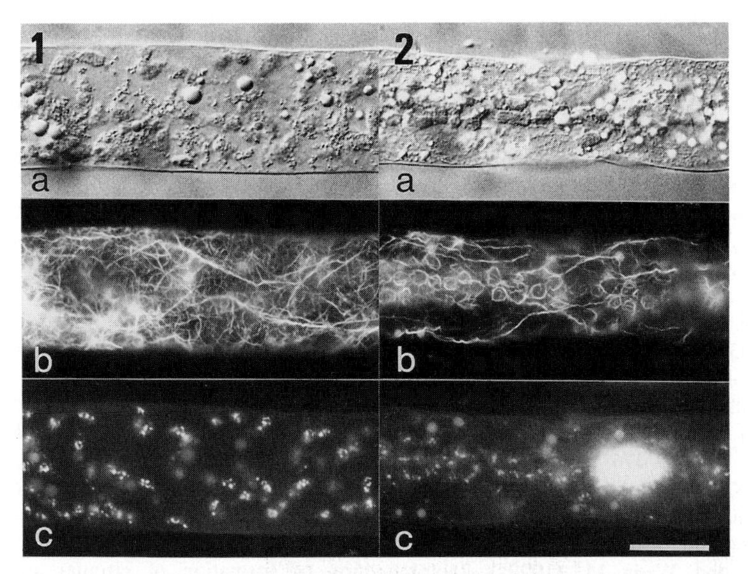

Figure 19. Circular structures of microfilament appeared at the periphery of chloroplasts after photo-orientation. 1. Dark control, 2. After photo-orientation. (a) Light micrographs taken under Nomarski optics. (b) Fluorescence microscopy of microfilament stained with rhodamin-pharoidin. (c) Dapi staining of chloroplast and nuclear DNA, bar = 10 μm. After Kadota A. and Wada M. (1989) *Protoplasma* 151: 171-174.

must be performed in conjunction with molecular biological techniques on the basis of mutant studies (Hickok *et al.* 1987). Fern gametophytes are potentially one of the best materials for studying the primary actions and transduction chains of phytochrome and the B photoreceptor(s).

9.7.8 Further reading

Dyer A.F. (1979) *The Experimental Biology of Ferns.* Academic Press, London. pp 657.

Furuya M. (1983) Photomorphogenesis in ferns. In: *Encyclopedia of Plant Physiology*, New Series, 16B *Photomorphogenesis*, pp.569-600, Shropshire Jr. W. and Mohr H. (eds.) Springer-Verlag, Berlin.

Miller J.H. (1968) Fern gametophytes as an experimental material. *Bot. Rev.* 34: 361-440.

Raghavan V. (1989) *Developmental Biology of Fern Gametophytes.* Cambridge University Press, Cambridge. pp 361.

Wada M. and Kadota A. (1989) Photomorphogenesis in lower green plants. *Annu. Rev. Plant Physiol. Plant Mol. Biol.* 40: 169-191.

9.7.9 References

Cooke T.J. Racusen R.H. Hickok L.G. and Warne T.R. (1987) The photocontrol of spore germination in the fern *Ceratopteris richardii. Plant Cell Physiol.* 28: 753-759.

Dürr S. and Scheuerlein R. (1990) Characterization of a calcium-requiring phase during phytochrome-mediated fern-spore germination of *Dryopteris paleacea* Sw. *Photochem. Photobiol.* 52: 73-82.

Gemmrich A.R. (1986) Antheridiogenesis in the fern *Pteris vittata*. I. Photocontrol of antheridium formation. *Plant Sci.* 43: 135-140.

Haas C.J. and Scheuerlein R. (1990) Phase-specific effect of nitrate on phytochrome-mediated germination in spores of *Dryopteris filix-mas* L. *Photochem. Photobiol.* 52: 67-72.

Haupt W. and Scheuerlein R. (1990) Chloroplast movement. *Plant Cell Environ.* 13: 595-614.

Hayami J. Kadota A. and Wada M. (1992) Intracellular dichroic orientation of the blue light-absorbing pigment and the blue-absorbtion band of the red-absorbing form of phytochrome responsible for phototropism of the fern *Adiantum* protonemata. *Photochem. Photobiol.* 56: 661-666.

Hickok L.G. Warne T.R. and Slocum M. (1987) *Ceratopteris richardii*: applications for experimental plant biology. *Amer. J. Bot.* 74: 1304-1316.

Iino M. Shitanishi K. Kadota A. and Wada M. (1990) Phytochrome-mediated phototropism in *Adiantum* protonemata. I. Phototropism as a function of the lateral Pfr gradient. *Photochem. Photobiol.* 51: 469-476.

Kadota A. Kohyama I. and Wada M. (1989) Polarotropism and photomovement of chloroplasts in the protonemata of the ferns *Pteris* and *Adiantum*: Evidence for the possible lack of dichroic phytochrome in *Pteris*. *Plant Cell Physiol.* 30: 523-531.

Kadota A. and Wada M. (1992a) Photoinduction of formation of circular structures by microfilament on chloroplasts during intracellular orientation in protonemal cells of the fern *Adiantum capillus-veneris*. *Protoplasma* 167: 97-107.

Kadota A. and Wada M. (1992b) Photoorientation of chloroplasts in protonemal cells of the fern *Adiantum* as analyzed by use of a video-tracking system. *Bot. Mag. Tokyo* 105: 265-279.

Kagawa T. and Sugai M. (1991) Involvement of gibberellic acid in phytochrome-mediated spore germination of the fern *Lygodium japonicum*. *J. Plant Physiol.* 138: 299-303.

Manabe K. Ibushi N. Nakayama A. Takaya S. and Sugai M. (1987) Spore germination and phytochrome biosynthesis in the fern *Lygodium japonicum* as affected by gabaculine and cycloheximide. *Physiol. Plant.* 70: 571-576.

Miller J.H. (1980) Orientation of the plane of cell division in fern gametophytes: The roles of cell shape and stress. *Amer. J. Bot.* 67: 534-542.

Murata T. and Wada M. (1989) Organization of cortical microtubules and mcirofibril deposition in response to blue-light-induced apical swelling in a tip-growing *Adiantum* protonemal cell. *Planta* 178: 334-341.

Schraudolf H. (1987) The effect of gabaculine on germination and gametophyte morphogenesis of *Anemia phyllitidis* L. Sw. *Plant Cell Physiol.* 28: 53-60.

Sugai M. and Furuya M. (1990) Photo-inhibition of red-light-induced spore germination in *Pteris vittata:* Cyanide, azide and ethanol counteracts restorable inhibitory action of near UV and blue-light but not that of far UV. *Plant Cell Physiol.* 31: 415-418.

Wada M. Grolig F. and Haupt W. (1993) Light-oriented chloroplast positioning. Contribution to progress in photobiology. *J. Photochem. Photobiol. B: Biol.* 17: 3-25.

Wada M. and Murata T. (1988) Photocontrol of the orientation of cell division in *Adiantum.* IV. Light-induced cell flattening preceding two-dimensional growth. *Bot. Mag. Tokyo* 101: 111-120.

Wada M. Murata T. and Shibata M. (1990) Changes in microtubule and microfibril arrangement during polarotropism in *Adiantum* protonamta. *Bot. Mag. Tokyo* 103: 391-401.

Yatsuhashi H. and Wada M. (1990) High-fluence rate responses in the light-oriented chloroplast movement in *Adiantum* protonemata. *Plant Science* 68: 87-94.

Index

abscisic acid 685, 686
absorbance spectra (see absorption spectra)
absorption
  characteristics 19, 279
  dichroism 428
  gradient 424-426, 429-431, 528, 722, 730
  spectra 40, 41, 43, 52, 54, 55, 56, 110, 111, 118, 125, 165, 166, 283, 312, 318-320, 327-329, 333, 335, 383, 384, 433, 473, 519, 523, 641, 687, 690, 694, 699, 716, 749
acceptance angle see optical fibre
Acetabularia
  hair whorl formation 304, 306
acetylcholine 203, 340
achlorophyllous stomata 687, 688, 691
acidification 693
actin filaments 559, 721, 723, 725, 726, 729
actin-myosin interaction 721, 725-727, 729, 800
actinometry 18, 19, 493
action dichroism 40, 41, 427, 431, 433-442, 559, 722
action spectra 8, 9, 11, 23, 24, 30, 31, 33, 34, 168, 236, 245-247, 255, 301-311, 314, 318-321, 327-329, 332, 335, 406, 439, 466, 469, 471, 473, 478, 479, 482, 523, 525, 636-639, 687, 689, 690, 694, 699-701, 708, 710, 711, 713-716, 720-723, 726, 727, 737-740, 749, 775-780, 786, 787, 798
action spectroscopy 33, 689, 711
adenylate cyclase 341, 756
ADP ribosylation 189, 341
Adiantum 35, 39, 132, 177, 178, 332, 438, 441, 784, 786-790, 793, 795-799
  capillus veneris 728
aequorin 4
Aeromonas 756
agaricales 756, 763, 766
Agaricus bisporus 754, 755, 771
Agrobacterium tumefaciens 566, 568, 602
akinetes 11
DL-α-alanine 758
alfalfa 529
algae 34, 52, 60, 73, 177, 306, 307, 334, 347, 423, 433, 443, 447, 498-500, 559, 710, 714, 723, 724
alkaline phosphatase 168
all or nothing responses (see also phytochrome) 763
alleles 74, 83, 403, 603, 604, 610-612
alloploid species 603
Alternaria 305, 776
  cichorii 312, 768, 776, 779
  solani 768
  tomato 305, 307, 310, 768, 776
Amaranthus caudatus 143-145, 230, 573
5-amino-1,3-cyclohexadienyl-carboxylic acid (see gabaculine)
4-amino-5-hexynoic acid 107
2-amino-3H-pteridinone (see pterins)
5-amino levulinic acid (ALA) 64, 65, 107, 304, 307, 310
cAMP 190, 341, 342, 344, 592, 716, 756
  cyclase 592
  phosphodiesterases 756
amyloplasts 179
Anabaena variabilis 711, 730
Anacystis nidulans 309
Anemia 784, 786
Anethum graveolens 741
angle of curvature (see phototropism)
8-anilinonaphthalene-1-sulphonate (ANS) 125, 126
bis-anilinonaphthalene 8-sulphonate (*bis*-ANS) 115, 116, 126
anisotropic (see light)
annexin (Ca^{2+}-binding protein) 202, 204
antagonism between photoreceptor systems 83, 439, 717
antenna pigments 11, 331, 586, 711
antheridium formation 800
antheridogen 786
anthocyanidins 733-735
anthocyanins 505, 733, 737, 740, 749
  function 747, 749
  spectral responsivity for synthesis 321,

361, 737-742
synthesis 36, 204, 245-257, 278, 285, 312, 321, 354-361, 561, 563, 564, 611-614, 733, 738-743, 746, 747
Anthurium leuconeurum 496-498
antibody 166-168, 172-174, 180-183, 277, 276, 341, 564, 790
anti-fungal properties 742
Antirrhinum majus 603, 741
apical hook opening 86, 87, 631-634
apolar spores 424
apoprotein 74, 81, 106, 108-115, 166, 168
apple (see *Malus* sp.)
Aquilegia vulgaris 496
Arabidopsis thaliana 72-88, 106, 108, 154-156, 243, 260, 272-275, 279, 281, 284, 287, 289, 291, 294, 368, 408-411, 483, 566, 567, 571, 577, 579-584, 590, 591, 602, 603, 606, 607, 610, 615-617, 620, 622, 673
mutants (see also mutants) 81, 82, 85, 106, 112, 342, 368, 402, 404, 405, 407-409, 575, 605-612, 678
Arachis hypogaea 741
archaebacteria 708
Arrhenatherum elatius 694
Arthrobacter sp. 756
artichoke 233
ascocarps 755, 763
Ascochyta pisi 776
ascomycetes 329, 754-756, 772, 776, 778
Aspergillus
niger 771
ochraceus 771
Asterophora
lycoperdoides 755
parasitica 754, 755
atomic absorption spectroscopy 198
ATP 148, 149, 153, 188, 193, 196, 343, 344, 691, 721
ATPase activities 178, 196, 203, 649, 656, 691, 699, 701
aurone 734, 735
autoradiographic analysis 193, 198
autotropism 667
auxin (see indole acetic acid)
Avena sativa 5, 6, 76-78, 87-89, 93, 94, 96-99, 144-157, 165, 166, 169, 170, 172, 174-175, 181, 182, 190, 194, 204, 212, 227, 231, 232, 260, 274, 276, 302, 304, 307, 337, 404, 405, 417, 419, 424, 494, 495, 510, 511, 633, 643, 653, 654, 659, 663, 664, 674, 786
electric potential induction in coleoptiles 312
phytochrome 52, 73, 107, 108, 110, 113-115, 120-122, 127-129, 170, 214, 215, 269, 273-287, 290-294, 379, 383, 397
seedlings tetrapyrrole-deficient 109

bacteria 105, 106, 708
bacteriorhodopsin 331, 708
barley (see *Hordeum vulgare*)
basidiocarps 755, 768
basidiomycetes 754-759, 763, 772, 776, 777, 779
bathochromic shift 60, 62, 118
bean (see *Phaseolus vulgaris*)
Beer's law (see Lambert-Beer relationship)
Begonia 497
betalains 505
bi-chromatic sensor 381
bile pigment 51-53, 55
cyclic-helical conformations 55
bilin 51, 110, 281
biliprotein 53, 54, 211
bilirubin oxidase 126
bilitrienes 109
biliverdin (IX) 64, 107, 109, 112, 127, 607
bioluminescence 447-449, 548
biphasic fluence response 235, 243, 587, 588
Bird and Riordan model 539, 549
Blaauw theory (see phototropism)
black body 51, 537
black box processes 262, 560
blue-green algae (see Cyanobacteria)
blue light 9-12, 30, 35, 169, 201, 219, 227, 244-247, 255-259, 301-313, 318, 320, 321, 327-348, 354, 356-359, 361-372, 383, 384, 387, 389, 398, 425-427, 432, 439-442, 450, 503, 523, 525, 526, 529, 543, 565, 568, 570, 573, 575, 588, 589, 592, 605, 611, 612, 621, 622, 636-638, 640-642, 645-649, 651, 652, 656, 659, 660, 662, 669, 670, 673, 675-678, 689-703, 708, 711, 714, 715, 720, 723, 726-729, 733, 738-740, 742-744, 761, 777-779, 785-790, 792, 795, 796, 798, 799
action spectra 301-311, 720, 722, 727, 729, 776, 777, 786, 792
electrical effects of 336, 338-340, 648,

649, 699-701, 715, 723
gradient 389, 524, 525, 527, 543, 670
-high irradiance response (B-HIR) 246, 254, 354, 588, 590, 729, 737, 785
photoreceptor(s) 5, 10, 11, 35, 220, 247, 255, 301-322, 327-348, 353-372, 400, 431, 432, 527, 588, 592, 612, 619, 620, 631, 641, 642, 656, 659, 661, 662, 670, 689, 690, 697, 700, 701, 726-728, 738, 739, 743, 744, 745, 790, 793, 795, 798, 801
(as) stress signal hypothesis 346
threshold fluence for action 369, 370
transcriptional control by 588-590
translational control by 589
/UV effects (responses) 301-322, 327-348, 353-372, 588, 709, 727, 737, 738, 740, 776, 779, 786
/UV effects (responses) directional 302, 659
/UV effects (responses) metabolic 302
/UV effects (responses) morphogenic 302
/UV effects (responses) non-directional 302
/UV-A /(UV-B)photoreceptor 5, 11, 35, 88, 247, 254-258, 301-322, 327-348, 353-372, 440, 565, 588, 619, 620, 623, 709, 727, 743, 744, 776-778
/yellow light effects 776, 778, 788
bolometer 18
Bombardia fasciculata 771
Botrytis 424, 425, 432
cinerea 776
Brassica 144, 192
campestris 477, 530
oleracea 737, 738
rapa 247, 606, 608, 738
broccoli 196
Bryopyllum 450
bryophytes 401, 499
bulb(ing) 170, 451, 483
Bünning's hypothesis 460, 479, 482
Bunsen-Roscoe reciprocity law (see Roscoe-Bunsen)

C-5 pathway of tetrapyrrole biosynthesis 64
cabbage 227, 247, 248, 250, 252, 254, 256, 257, 742, 748, 749
calcium 3, 4, 178, 188, 193, 195-203, 207, 259, 336, 341-347, 486-488, 559, 592, 669, 677, 710, 711, 713, 714, 725-727, 779, 786
-activated ATPase 193
cytoplasmic 188-204, 207, 346, 710
calmodulin 190, 193, 198, 202, 345, 559, 725-727
antagonists 487, 592
inhibitor 193, 201
Calothrix sp. 110
Calvin cycle 278, 571, 691, 693, 700
Camellia 507, 508
carbonyl cyanide 3-chlorophenyl hydrazone (CCCP) 691
ß-carotene 320, 331, 690
carotenoids 309, 318, 321, 331, 339, 387, 425, 427, 518, 524, 525, 586, 690, 699, 713
(as) photoreceptors 11, 314, 327, 427, 699, 778, 786
synthesis (carotenogenesis) 304, 309, 312, 318, 321, 329, 347, 776
casein-kinase 196
Cassia 345
catechin 735
cauliflower 145, 233, 275, 306, 335, 581, 583, 584
c-DNA (see DNA)
cell aggregation 716, 718
cell division 644, 783, 787, 795-797
control by phytochrome and B-photoreceptor 795
cell elongation 278, 411, 661, 663, 666, 669, 677, 678, 783
cell expansion 644, 649, 650, 655, 788, 789, 798
cell microprojectile-mediated introduction of promoter constructs 94
cell pressure probe 651
cell self fertile 604
cell suspension cultures 405, 739-747
cell uninucleate 604
cell wall 177, 181, 345, 346, 427, 433, 492, 504, 506, 509, 510, 569, 631, 649-652, 685, 721, 789, 794
extensibility 203, 650, 651
loosening 649-652
pH of 655
properties 318, 503, 504, 649, 650, 656
proton extrusion into (acidification) 655
yield threshold 649-652, 656
cellulase 569, 685
Centrospermae 233, 474

Ceratodon purpureus 73, 79-81, 100, 131, 274, 279, 439
Ceratopteris 786, 788
mutants 786
cereal crop plants 386
chalcone 734, 744
chalcone isomerase (CFI) 569, 735
chalcone synthase (CHS) 563, 568, 612, 734, 736, 737, 744, 745, 747
Chamaenerion angustifolium 394
channels 336, 337, 341, 345, 463, 486, 487, 648, 649, 701, 710, 714
characean algae 721
chemotaxis 709, 715-718, 741
Chenopodium 636, 637, 640, 641
album 393-394
rubrum 451, 452, 458, 459, 467-469, 472, 474
Chlamydomonas 13, 314, 711, 713, 714
interference reflector hypothesis 713 713
eugametos strains 713
Chlorella 304, 309, 312
protothecoides 307
chloronemata 305, 310
chlorophyll 142, 145, 166-167, 224, 277-279, 285, 308, 309, 319, 331, 340, 384, 387, 389, 412, 466, 492, 507, 508, 515, 517, 522, 523, 525, 529, 544-546, 620, 687-691, 699-701, 714, 721
synthesis 64, 107, 156, 170, 224, 233, 305, 307, 309, 310, 368, 391, 401, 405, 511, 518, 563, 620, 622, 787
chlorophyll *a* 320, 502, 586, 692, 699-701, 720
chlorophyll *a*/*b*-binding proteins (Cab) 310, 563, 586, 587, 620, 622
polypeptides of 586, 587
ratio 391, 620
chlorophyll *b* 320, 586
chlorophyll RCI 320
Chlorophytum comosum 692
chloroplast(s) 11, 179, 278, 302, 308, 312, 334, 338, 356, 391, 425, 433-437, 497-499, 501, 503, 506, 518, 521, 571, 575, 578, 585, 620, 631, 687, 691, 713, 725-727, 783, 785, 788, 793
(in) guard cells 687-689, 691-693, 695, 699-702
high irradiance arrangement 720-723, 785, 798-800
low irradiance arrangement 721-724, 798, 799
mesophyll 691-693, 695, 700
orientation 11, 177, 308, 440, 559, 724, 728, 729, 785, 798-801
photomovement 35, 60, 177, 197, 200-202, 236, 308, 346, 433-443, 476, 498, 499, 719, 720, 723, 724, 727, 729, 783, 785, 798-801
rearrangement (redistribution) 302, 303, 308, 498-501, 531, 719, 721-723, 727, 730, 783, 785, 799-801
sedimentation 721-724, 727
chlorpromazine 193
Choanephora curcubitarum 756
cholera toxin 189, 190, 194, 341
Cholodny-Went hypothesis/model (see phototropism)
Chou-Fasman prediction 114, 128
Chromatium 710
chromic acid-ammonia degradation (see tetrapyrrole)
chromopeptide 53, 54
chromophore 8, 40, 41, 51, 54, 55, 107, 281, 289, 321, 347, 348, 430, 432, 435, 473, 606, 607
attachment 281-284, 289, 294
conformational changes 40, 55, 60
chromosomes 72, 81
Chrysanthemum 454
cinnamate 4-hydroxylase 736
trans cinnamic acid 736
cinnamoyl-CoA 736
4-cinnamoyl CoA ligase 747
circadian rhythm 13, 31, 187, 242, 278, 302, 318, 347, 441, 447-449, 450, 452, 455, 458-466, 468, 472, 473, 477, 479, 481, 482, 484, 487, 499-501, 537, 563, 586, 589-593, 620, 759, 769-772, 774
circadian time 462, 463, 482
Circaea lutetiana 394
circumnutation 662
Cladosporium 772
clinostat 672, 679
coaction of photoreceptors (see photoreceptors, interaction)
codon(s) (see gene(s))
colchicine 725, 799
coleoptile(s) (elongation)8, 169-175, 181, 204, 260, 274, 337, 338, 346, 357, 417, 495, 496, 510, 511, 514, 524-526, 633,

644, 652-655, 662-665, 668, 670-676, 678, 741
collimated light (see light parallel)
colloidal gold 167, 181, 182
Colocasia antiquorum 496
Commelina communis 332, 687-689, 695, 697, 698
computerization 24, 31, 36-39, 41, 43, 155, 165, 333, 385, 412, 537-550, 646, 667
conidia 755, 759, 769
conidiation 302, 306, 321, 327, 754, 769, 775, 779
 in *Alternaria* 305, 310, 312, 768
 in *Neurospora* 306, 312, 315, 318, 329, 336, 347
 in *Trichoderma* 329, 331
conidiophores 424, 754, 755, 758
Convolvulus arvensis 169
Coprini spp. 755, 756, 764, 768
Coprinus
 cinereus 754, 756, 758, 761, 763, 768
 congregatus 33, 341, 754-779
 congregatus growth-promoting factors 766
 congregatus monokaryotic fruiting bodies 761
 congregatus mutant 764, 767
 congregatus role of cap on stipe elongation 763-766
 lagopus 754
 macrorhizus 754
corn (see *Zea mays*)
corn mayweed (see *Matricaria indora*)
cortical fibre reticulation (see *Vaucheria*)
cortical microfilaments 789, 790, 794, 800, 801
cortical microtubules 789, 790, 794, 795, 800
cosine-corrected sensor 226, 382, 386
cotyledonary whorl 363, 364
cotyledon(s) (expansion growth) 82, 253, 354, 355, 363, 365, 471, 475, 486-488, 514, 517, 518, 528, 561, 563-567, 575, 612, 641, 659, 738, 746, 748
coumarate CoA-ligase 736
4-coumaric acid 736
4-coumaroyl-CoA 734, 736, 744
Crassula falcata 514, 524
cress (see *Lepidium sativum*)
critical daylength 453, 454
critical night length 453-459, 462-464, 468-472, 479, 481, 482, 484, 487
 temperature sensitivity 457, 464, 465
cryptochrome 5, 321, 353, 776, 779, 788
Cryptomonas 711
cucumber (*Cucumis sativus*) 88, 227, 230, 253, 275, 338, 339, 345, 346, 366-369, 571, 642, 646-649, 651, 652, 678
 mutant 275, 607, 608
Cucurbita pepo 43, 73, 88, 141-145, 149, 155, 156, 215, 226, 232, 274, 276, 514, 517, 518
curvature (see phototropism)
cyanazine 686
Cyanidium caldarium 64
cyanobacteria 11, 52, 110, 309, 417, 708, 710
Cyathus pyxidata 755
cyclic photophosphorylation 691
cyclins 151
cyclobutane ring 308, 739
cycloheximide 589
cyclosis 718, 720
cysteine (residue) 53, 105, 108, 109, 281
cytochalasin-B 799
cytochrome 710
cytochrome b 305, 306, 315, 335, 336
cytochrome c 320, 331
cytokinin 346, 455, 614, 623
cytoplasmic gradient 34, 177, 722, 730
cytoplasmic region 109, 181-183, 716, 717, 721, 723-725, 728
cytoplasmic ribosomes 586
cytoplasmic streaming 338, 718, 719, 721, 723
cytoskeletal competence 202, 340, 729
cytoskeletons 789, 790, 794, 798
cytosol (isolation) 4, 561
cytosomes (tannin vesicles) 725, 726

Daldinia concentrica 770-772
dark-limit cycle 463
dark relaxation 60
Dasyobolus spp. 763, 764
Datura ferox 397
Daucus carota 741
day extension 261, 477
day-length sensitivity (see photoperiodic response types)
day-light 5, 220, 383, 396, 537, 538
 diurnal fluctuation in quality 387
 spectral distribution of 383, 537, 540, 541

day-neutral plants (see photoperiodic response types)
day-night transition (see light-off signal)
de-etiolation 75, 83, 159, 196, 204, 205, 213, 228, 236, 246, 253, 261, 262, 368, 371, 384, 386, 399-407, 409, 412, 413, 457, 459, 461, 463, 475, 479, 518, 522-525, 559, 571, 572, 574, 575, 583, 590, 610, 616-620, 636, 639, 641, 645, 647, 665
Deightoniella 772
depolarization (see membrane)
desmids 417, 423, 710
deuterium isotope effect of 60
deuterium lamp 18
diacyl glycerol (DAG) 192, 342
diaminovalerate 65
diaphototaxis (see phototaxis transversal)
Diaporthe phaseolorum 755
diatoms 423
3-(3,4-dichlorophenyl)-1,1-dimethyl urea (DCMU) 254, 691, 692
dichroic
  orientation change 131, 433, 435-437
  pigment orientation 29, 132, 177, 178, 340, 380, 418, 423, 427-433, 435, 436, 440, 528, 722, 724, 728, 788
dichroism (circular) 60, 114, 126, 128, 130, 132, 133, 421, 443
dichromatic irradiation 218, 221, 232, 248, 257, 258, 358, 361, 363, 364, 368, 432, 639, 788
Dictyostelium discoideum 715, 716, 730
dihydroflavin 317
dihydroflavonol 4-reductase 735
dihydroflavonols 734, 735
dill (see *Anethum graveolens*)
diploid 603, 604, 761, 764
Dipsacus silvester 694
dithionite 125
DNA 76, 93, 94, 99, 132, 195, 206, 308, 314, 315, 330, 332, 569, 576, 581, 584, 602, 603, 605, 739, 740, 748, 749, 801
  marker 602
  photoreactivation 304, 308, 309, 314, 315, 739-741, 749
  -protein complex(es) 579-582
  sequences 72, 402, 576-580, 583, 587, 590, 602
c-DNA 72, 88, 90, 107-109, 114, 133, 195, 274, 485, 564, 569, 571, 573, 574, 576, 582, 585
  recombinant 107, 109
T-DNA 603, 612
DNAse I footprinting 568, 579, 580, 590, 591
dormancy 451, 452, 483, 488, 615
dry seeds (see seeds)
Dryopteris 200, 437, 438, 786, 790
  morphogenetic index 305, 310
  filix-mas 12, 305
  paleacea 201
  sparsa 728
Dunkel-/Lichtstarre 708

ecotypes 275, 604
edge-of-dish effect 757
electric pulses 19
electromagnetic radiation 418, 419
electromagnetic wave 418
electron microscopy (EM) 115, 116, 146, 168, 181-183, 278, 711, 794
electron spin resonance (ESR) 336
electronic position-sensitive transducers 393, 395, 396, 398, 642, 646-648, 651
electronic transition moments 428, 430, 431, 435, 436, 438-441, 793, 799
ELISA (see phytochrome)
Elodea 719
elongation zone 174, 660
end-of-day FR 236, 237, 398, 402, 404, 407, 409, 470, 471, 476, 612, 614, 616-618, 636, 654
end-of-day response 235, 236, 241, 260, 362, 398, 402, 410, 413, 457, 471, 635
endoplasmic reticulum 34, 179, 181
energy fluence (see fluence)
entrainment of a rhythm 449, 450, 461, 482, 484, 770
epicotyl hook 175
epidermal peels 498, 507, 623, 687-689, 691, 693, 694, 700, 702
epidermis 175, 384, 492, 495-499, 502, 505, 507, 521, 524, 527, 530, 532, 563, 564, 566, 654, 683, 686, 687, 746, 748, 749
epi-fluorescence microscope 35, 567, 713
escape curve (escape time) 561
escape from far-red light reversibility (see far-red light reversibility
Escherichia coli 109, 133, 273, 314, 306
ethylene 346
N-ethylmaleimide (NEM) 799

etiochloroplasts 65, 178
etiolated plants (forms) 8, 9, 75, 82-88, 94, 141-159, 165, 166, 169-175, 181-183, 212, 213, 223, 224, 228, 230, 232, 233, 245, 250-254, 260-262, 272, 274, 277-278, 280, 282, 284, 286, 289, 293, 341, 361, 367, 383, 398-402, 404, 406, 409, 463, 468, 474, 475, 478, 479, 484, 508-511, 513, 518, 522, 523, 525, 526, 530, 564, 565, 572-575, 590, 591, 605, 607, 608, 610, 616, 631-639, 641, 644-648, 651, 665, 757
etioplasts 65, 178
Euglena 39, 417, 423, 711, 714, 717, 730
  gracilis 44, 45, 312, 314-316, 712, 713
eukaryotes 146, 147, 149, 422
eukaryotic algae 11
extension growth (see growth)
external coincidence 460, 461, 465
extinction coefficient 41, 42, 54, 62, 385, 515
eye, human spectral sensitivity of 23, 381
eyespot (stigma) 714
eyespot in *Chlamydomonas* 714
eyespot in *Euglena* 712, 713

far-red light 7-10, 30, 35, 37, 40, 74, 83, 85, 86, 90, 91, 141, 144, 150, 153, 157, 165, 166, 178, 182, 187, 189, 191-195, 199, 201, 211, 212, 218-222, 225-227, 230-252, 256, 257, 259, 261, 274, 311, 353, 358-370, 382-384, 387-390, 393, 392, 395-400, 402, 403, 435-442, 458, 467-472, 478-483, 522-524, 548, 561, 562, 564, 567, 591, 605, 612, 614, 617, 623, 636-641, 643, 644, 647, 652, 726, 728, 737, 739, 742, 743, 746, 784, 792, 795-797
  gradient 522-524
  -high irradiance response (FR-HIR) 82, 83, 246, 254, 260, 261, 401, 402, 404-407, 409, 410, 412, 413, 637, 639, 737
  reversibility 7, 9, 30, 42, 43, 88, 105, 108-111, 154, 165, 166, 190, 193, 197, 201, 211, 212, 226, 229, 235-242, 245, 247, 249, 253, 260, 262, 263, 279, 311, 368, 369, 398, 400-402, 410, 412, 466-468, 471, 472, 478, 482, 561-563, 573, 636, 641, 643, 644, 656, 702, 737, 743, 784, 786, 792, 797
fava bean 341
Favolus arcularius 755, 777
ferns 9-13, 41, 73, 132, 177, 200, 306, 310, 332, 393, 401, 412, 424, 437, 438, 559, 634, 728, 783-801
  antheridi(a)(um) formation 783
  gametophytes 12, 783, 786, 795, 797, 800, 801
  photobiology of 11, 12, 783-801
  photoreceptors 792, 793, 795, 799
  phototropism 783, 784, 788-794, 799
  polarotropism 790, 791, 794, 795, 799
  prothallical stage 433, 437, 438, 783, 785, 787, 797
  protonemal stage 41, 132, 177, 178, 634, 784, 787-799
  spore germination 12, 200, 306, 783-787
  spore imbibition of 784-787
  sporophytes 783, 784
  three-dimensional differentiation (mat) 783
  two-dimensional differentiation 787, 788, 797
ferredoxin-dependent glutamate synthase (Fd-GOGAT) 363-365
ferredoxin-$NADP^+$ reductase 64
fibre optic probe (see optical fibre)
firefly, flash 644
flagellates (see phytoflagellates)
flagellum 44, 45, 708, 710-714
Flammulina velutipes 755, 768
flash photolysis 40, 42, 61, 62, 119, 123, 132
flavan-3-ol synthase 735
flavan skeleton 733-735
(2S) flavanone 734, 735
flavanone 3-hydroxylase 735
flavin(s) 45, 258, 310, 315-318, 320, 329, 330-332, 335, 336, 344, 348, 427, 432, 439, 699, 711, 720, 721, 723
  electronic transition moments 723
  mediated transport of redox equivalents 312, 330
  as photoreceptor 11, 258, 310, 314, 327, 330, 347, 432, 439, 699, 711, 720, 776, 778, 786
  triplet state 329, 330-332, 720
flavin adenine dinucleotide (FAD) 306, 314, 315
flavin adenine mononucleotide (FMN) 44, 312, 316
flavoenzymes 330, 749

flavone 735, 744
flavone synthase 735
flavonoid(s) 505, 530, 566, 733-737, 739, 749
  biosynthesis (pathway) 530, 531, 563, 567-569, 733-749
  function of 530, 563, 747, 748
flavonol 735
flavonol synthase 735
flavoprotein 316, 439, 612
floral genes 485
floral stimulus 485, 488
flower induction and suppression 5, 6, 8, 190, 229, 235, 241, 242, 245, 261, 302, 391, 410, 451, 452, 454-473, 475, 477-488, 589, 612, 620, 622, 623, 643, 768, 769
flower petals 505, 633
fluence 17, 20, 311, 312, 380, 641, 653, 743, 792
fluence rate 10, 17, 20-22, 31, 33, 163, 311-314, 333, 346, 354, 367, 369, 370, 380, 381, 396, 402, 406, 407, 423, 427, 429, 478, 491, 493-495, 498, 511-513, 515, 519, 521, 527, 531, 548, 570, 587, 639-642, 664, 671, 672, 686-695, 698, 699, 701-703, 707, 711, 713-716, 721, 728, 741, 768, 787, 793, 799, 800
fluence rate-response curve(s) 235, 239, 240, 243, 311, 436, 439, 440, 587, 611, 613, 619, 621, 638, 642, 688, 742
fluence response curve (relationships) 204, 256, 342, 398, 460, 461, 484, 615, 649, 654, 672, 673, 675, 678, 720, 721, 723, 739
fluorescein 168
fluorescence 35, 43-45, 62, 126, 315-317, 319, 331, 332, 346, 501, 502, 569, 699, 700, 711, 712, 715, 790, 795, 801
fluorescence transients 692, 699-701
Förster energy transfer 126
Fourier transform infra-red spectra 45, 54, 60, 119
Fraunhofer lines 537
Fresnel's equation for unpolarised light 543
Fuchia hybrida cv. Lord Byron 477
fumarate 709
Funaria 659, 721, 722, 726
Funaria chloroplast rearrangement 304, 308, 721, 722
  hygrometrica 304, 308, 722
fungi(al) 11, 13, 306, 307, 315, 327, 332, 338-340, 342, 344, 346, 347, 423, 424, 492, 603, 659, 742, 753-780
  action spectra 34, 329, 775-780
  annual fruiting rhythms 773
  antagonistic effects of light 768
  caps 763-768
  circadian sporulation rhythms 769-773
  classification 754
  dark inhibitory process 761
  dark recovery process 761
  effect of aeration 768, 779
  elicitor 742, 747
  fluence-response curves
  fruiting inducing substances (FIS) 756
  hyphae 754, 756, 778
  mutants 756, 757, 761, 764
  photocontrol of development 754, 757, 763, 764, 767, 768
  photocontrol of reproduction 754, 756, 769, 779
  photo-inhibition of cap formation 768
  photo-induction of fruit body primordia 753, 754, 756-766, 768, 769, 777-779
  photo-induction memory of 757
  photo-induction and temperature 768-770
  photomorphogenesis 753-780
  photoperiodism 754, 761, 768-775, 780
  photoreceptors 775-780
  photoreceptors P310 (mycosporine) 776
  photosensitivity 754
  photosporogenesis 758, 759
  phototropism 327, 328, 332, 333, 346, 759, 763, 764, 766, 774, 780
  phytopathogenic 775
  reproductive differentiation 759-763
  sporophores/sporogenesis 753-761, 761-764, 767-773, 776
  stage of sensitivity to darkness 759-763, 777, 778
furanocoumarins 747
Fusarium 309, 756, 759

G-box binding factor(s) (GBFs) 196, 569, 580, 582, 583
gabaculine 64, 107, 784
gametes 12, 332, 424, 603, 634, 713, 728
gas exchange (techniques) 684-686, 696, 703
gel retardation assays 579
Gelasinospora reticulispora 778

perithecial formation in 33, 304, 307, 312, 778
Gelbstoff 389
gene(s) 7, 71-101, 105, 164, 171, 204, 262, 353, 402, 652, 713, 737, 744
activation 89, 744
antisense 608
autoregulation 89
boxes 76-78, 93, 94-99, 568, 569, 580, 591
Cab 190, 200, 405, 576, 582, 584, 586, 587, 589-593, 620
Cab-photosystem 211 200, 206, 278, 588, 589, 622
Chi 569
Chs 563-568, 570, 745
cloned 71, 84, 106, 156, 160, 206, 262, 271, 294, 336, 340, 563, 564, 566, 602, 603
copies 76, 603
coding for phytochrome (*phyA-E*) 5, 71-101, 106, 133, 141, 272, 274, 275-279, 291, 294, 402, 403, 475, 606-611
codon 77, 80, 108, 109
down regulation 85, 87, 106, 403, 485
epistatic 604, 611
exons 76-81
expression 71-101, 105, 197, 200, 204, 258, 259, 272, 275-278, 285, 334, 341, 368, 402, 403, 405, 408, 478, 488, 559, 561, 563, 565-582, 584-593, 601, 610, 611, 635, 643, 671
expression control by *cis*-acting elements 93-99, 567, 568, 576-578, 583, 585, 587, 590-592
expression control by transacting factors 93-99, 570, 578, 585, 587, 590
families 5, 71, 72, 564, 567, 571, 573-575, 580, 585, 587, 604, 623, 639
FedA ferredoxin 206
floral 485
ß-glucuronidase 86
(in)directly regulated 89
introns 76-80, 272, 274, 276, 576, 587
nomenclature of 73, 74
N-terminus in 108, 109, 291
overexpression 71-85, 106, 108, 109, 112, 261, 262, 272, 274, 276-280, 404, 405, 407-410, 602, 603, 607-611, 622
phyA 72-85, 90-99, 141-145, 156, 158, 169, 173, 174, 177, 180, 183, 190, 274-276, 279, 285, 402-405, 409, 582, 607, 609, 610
phyA expression 279, 402, 404
phyA-null 403, 405-407
phyB 72-86, 142, 144, 145, 160, 275, 279, 402, 403, 407-410, 607, 609
phyC 72, 75, 79, 85, 86, 142, 143, 160, 402, 403, 410
phyD, *phyE* 72, 75, 142, 160, 403, 410
product 410, 571, 582, 652
product COP 205
product HP 611
product HY5 614
product DET 614
promoter/coding region 71, 72, 86, 91-97, 274, 568, 577, 591
promoter elements 94-97, 274, 275, 280, 284, 403, 564, 567, 569, 570, 576-585, 590-592
RbcS 563, 569, 571, 577-585, 591
recombinant 108, 110
-specific probes 85, 86, 90, 564, 565, 567, 573, 574, 590
suppression 575, 603
tagging 602, 603, 612
transcription 109, 163, 164, 272, 275, 569-571, 573, 576-578, 581-585, 587-592, 635, 639, 652, 745
transfer of 566, 568, 602, 603
uidA 566
genetic analysis 327, 329, 603, 604, 605
genetic map 72, 81, 603
genetic variation 604
genetically programmed 371, 704
genome 72, 81, 258, 571, 602
genomic clone sequence (analysis) 274
genomic substraction 602
genotypes 338, 601, 610, 611, 616, 617
geotropism (see gravitropism)
Gerea canescens 684, 685
germination (see also seed germination) 3, 11, 201, 237, 238, 244, 304, 306, 312, 313, 413, 424, 432, 575, 615, 643
gibberellic acid (GA)/gibberellins (GAs) 178, 190, 612, 615, 622, 623, 654, 655, 677, 786
ß-glucuronidase (GUS) 86, 87, 566, 569, 570
glutamate 1-semi-aldehyde (GSA) 65
glutamine synthetase (GS) 364
glutamyl-tRNA 65

glyceraldehyde-3-phosphate dehydrogenase (GPD) (see NADP)
Glycine max 31, 396, 398, 455-458, 480, 506, 572, 573, 577, 592, 698
cGMP 341, 592
cGMP-phosphodiesterase 592
Gonyaulax polyedra 447-450, 456
Gossypium barbadense 700, 701, 703, 704
grafting experiments 484, 488
Gramineae 474
grass mesocotyl (seedlings) 644, 653, 654
gravitropism 8, 169, 305, 310, 312, 327, 328, 371, 425, 613, 633-635, 661, 662, 665, 667, 668, 670, 672, 675, 679, 717, 718, 763, 766, 774
green light 11, 383, 388, 700, 708
growth 35, 211, 302, 322, 334, 335, 338, 363, 635, 644, 654, 655, 788
  biochemical mechanism 652-656
  differential 341, 633, 661-664, 666, 668, 669, 671, 672, 677
  elongation (extension) 10, 156-159, 169, 170, 217, 219, 236, 237, 241, 278, 360, 361, 364, 367-372, 390-399, 404-411, 588, 633, 634, 636, 645, 660, 667, 668, 671, 672
  elongation zone 339, 371, 425, 426, 493, 642, 663, 666-668, 670, 675, 676
  expansion phase 633
  lag times of 335, 646, 647, 652, 668
  light adaptation 3, 262, 378, 390, 401, 635
  light-exaggeration (habit) 277, 280, 285, 286, 289, 293
  photoreceptive sites for 34, 169, 425-427, 511, 647, 792, 795-797
  photomodulation 4, 601, 631-656
  physical nature of 649-652
  -rate 9, 327, 338, 365, 366, 390, 393, 394-398, 511, 633, 636, 640, 643, 646-653, 664, 665, 667-669, 671-673, 678, 787, 788
  responses changing during plant development 634-642
  substances 203, 652-655, 663, 763, 766
  and wall expansion 650
  and water uptake 649, 650
GTP 188-190, 196, 315, 336, 340, 341, 592
GTPase activity 189, 190, 340-342
guard cell(s) 176, 338, 343, 345, 649, 683-694, 698-700
  carbon metabolism in 684, 688, 693
  protoplasts 338, 341, 685, 686, 690, 693, 695, 698

haem (IX) 64, 107, 319, 327, 348
haem oxygenase 64
Haematococcus pluvialis 714, 715
Haemophilis 309
halo 387
halobacteria 708, 709
Halobacterium halobium 708, 709
halorhodopsin 708
haploid 603, 604, 783
Haplopappus gracilis 738
Hartmann's hypothesis (see phytochrome)
Helianthus annuus 230, 233, 342, 504, 647, 649, 664, 678
heliotropism 662
Helminthosporium oryzae 776
heparin 196
high irradiance response 7, 10, 235, 236, 239, 241, 244-255, 260, 311-313, 318, 331, 396, 400-402, 410, 478, 479, 565, 590, 614-618, 635, 636, 638, 656, 737, 738, 799
  action spectra 247, 305, 402, 636-641
high performance liquid chromatography (HPLC) 317
homozygotes 286, 603
Hordeum vulgare 87, 88, 243, 452, 478-482
hormones 203, 287-289, 302, 340, 346, 601, 622, 634, 652, 656, 663, 664, 674, 683
  floral 488
  responsiveness to 655
horse daisy (see *Matricaria indora*)
hydropathy (index) 98, 114
Hyoscyamus niger 478-480, 485
hyperpolarization (see membranes)
hypocotyl(s) 43, 86, 170, 176, 192, 225, 339, 342, 354, 355, 361-363, 369, 405, 504, 508-510, 522, 563, 649, 664, 666, 670, 672, 673, 675, 678, 741
  elongation 82, 83, 219, 229, 246, 249, 255, 278, 289, 361-370, 407, 409, 523, 607, 613, 617-619, 642, 644, 646, 647, 652, 654, 655, 664, 666, 667
  growth inhibition of 82, 158, 250, 253, 256, 257, 260, 279, 285, 286, 290, 291, 294, 305, 338, 345, 346, 361-366, 406, 407, 561, 608, 611, 616, 621, 636-644, 647-652, 655, 656, 737

straight growth 360, 369, 370
Hypoxylon fuscum 772
investiens 771
rubiginosum 771
truncatum 771

illuminance 23, 768
imbibition (see seeds)
immunoaffinity chromatography 204
Impatiens parviflora 170
inclusion bodies 109, 146, 151
indole-3-acetic acid (IAA) 302, 304, 308, 338, 345, 346, 623, 633, 652, 654, 663, 669, 670, 674-677
inducible phase 460, 462, 464, 465, 477, 482
inductive dark period 455, 471, 472, 768, 773
inductive responses 262, 471, 472, 635
infra-red light 34, 35, 37, 39, 43, 354, 386, 389, 538
infra-red spectroscopy 128, 133
inositol phospholipid 188, 486-488
inositol 1,4,5-trisphosphate (IP_3) 192, 206, 207, 342, 343, 487
instrumentation 29-45
integrating sphere 512, 513, 516
intercellular air spaces 504, 505, 509, 510, 513, 517, 683
internal coincidence 461
internal fluence rate 417, 422, 423, 491, 511-518, 521, 522, 524, 526, 527
internode 169, 391, 607, 616
ion efflux 10, 197, 648, 684, 699
ionomycin 200
ionophore(s) 200, 486, 487, 592, 786
Iris tectorum 694
iron 310
iron chelatase 107
irradiance 20-22, 29, 235, 417, 456, 469, 470, 479, 511, 540, 685, 712, 757, 759, 768, 770
isoflavan skeleton 733, 734
isoflavone 735
isoflavone synthase 735
isoflavonoids 734, 739, 740, 742, 743
isomerism 119, 127
isomerism *cis-trans* 58
isomerism Z,E 57-59, 119, 122
isotropic (see light)

Kalanchoë
blossfeldiana 455, 456, 458, 459, 466
marmorata 514, 517
karyogamy 761, 764
kinases 131, 193, 343, 344, 649
Kubelka-Munk equations or theory 514, 516-519, 522

lactate dehydrogenase 312, 347
Lactuca sativa 7, 30, 219, 239-241, 246, 249, 255, 305, 311, 400, 406, 636, 637, 639, 642
lambda repressor 284
Lambert-Beer relationship 499, 514
Lavatera 528
law of conversion of energy 512
leaf 170-173, 388, 391, 398, 399, 405, 410, 483-488, 492, 496-500, 505, 507, 508, 514, 517, 520, 528-532, 541, 542, 571, 572, 574, 575, 584, 631, 633-635, 645, 647, 649, 651, 655, 656, 659, 661, 683-686, 688-690, 692, 695-699, 701-703, 740, 748, 790
leaf area index 388, 411, 549
leaf canopy 549
leaf sleep movements 236, 241, 242, 259, 302, 447-450, 486, 633, 662
leaflet movement 661
leaky mutants 67, 605, 607, 623
Leguminosae 203, 563, 739, 742, 746
Lemna 450, 571-573, 577-579, 583, 721, 722, 726
gibba 482
paucicostata 189, 190, 465, 472, 473, 486
perpusilla 768, 769
lens effect 418, 422-424, 426, 427, 429, 491, 494-499, 521, 527, 528, 717, 793
lenses 492, 493, 495, 496
Lentinula eodus see also *Shiitake* 757, 779
Lentinus tigrinus 754, 755
Lepidium sativum 171, 172, 741
lettuce (see *Lactuca sativa*)
light
absorbance 544, 547, 548
actinic 165, 422, 379, 381, 700, 725
anisotropic 428, 503, 528
attenuation (see blue light, gradient)
bilateral exposure 659, 664-666, 678, 679, 783
blue:red photon ratio (see blue light)
circularly polarized 17, 420

coherent 17, 502, 509, 510
collimated (see light, parallel)
cyclic irradiations 249-251
light/dark cycle 230, 249, 251, 261, 447-451, 454-459, 461, 463, 466, 467, 477, 479, 565, 566, 575, 589, 754, 757, 759-761, 769-774
light/dark transition (see light, -off signal)
diffraction 502
diffuse 19-22, 386, 497-499, 511, 517, 520, 530-532, 538, 540, 541, 543
diffuse spectral attenuation coefficient 544
direct 386, 388, 538, 540, 541, 543, 717, 721
direction 4, 11, 17, 19, 20, 29, 177, 319, 379, 381, 388, 397, 417-443, 491, 495, 496, 511, 513, 526-531, 547, 631, 633, 708, 711, 712, 715-717, 719, 730, 793, 797
direction perception by attenuation (see spatial sensing)
direction perception by refraction (see lens effect)
divergent 418, 427, 495, 496
electric vector of 418-420, 427-432, 434, 435, 437, 438, 440-442, 722, 790-792, 795, 797, 799
elliptically polarized 420
energy 17, 18, 22, 418
environment 10, 13, 81, 83, 322, 353, 386, 393, 417, 491, 502, 504, 532, 537-550, 684, 707-730
global (see radiation global)
gradients 417-419, 423, 491, 496, 513-532, 634, 641, 642, 662, 664, 666, 670, 675-677
-growth responses (adaptive significance of) 211, 377-402, 633-636
guiding 417, 507-510
-harvesting chlorophyll *a/b* 391, 586, 699
-harvesting pigments 302, 318
-induced absorbance changes (LIACs) 303-306, 311-315, 318, 335, 336, 343
intensity 17
internal reflection 491, 503-505, 509
isotropic 17, 21, 22, 418, 503, 504, 517, 547
laser 17, 40, 42, 123, 419, 502, 510, 686, 688
light-limit cycle 463
magnetic vector of 418-420
measuring probes 519
molecular scattering of 387, 389, 503, 537-540
nomenclature of (see light terminology of)
non-polarized 420, 421, 431, 440, 443, 543, 790-792, 797
-off signal (dusk) 379, 381, 387, 398, 449, 450, 462-465, 468-474, 476, 482, 484, 761, 762
omnilateral 370
-on signal (dawn) 262, 387, 398, 449, 450, 457, 462-465, 468, 474, 482, 702, 717, 761, 772
parallel 17, 19-21, 422, 427, 493, 495, 497, 498, 503, 504, 508, 511, 518, 520-522, 526, 527, 530-532
particle properties of 419, 491, 492
perception of 3-5, 17-25, 29, 51, 353, 377-413, 417, 422, 423, 465, 491, 511, 519, 521, 530-532
phase function 518, 520, 538
plane (linear) polarized 10, 17, 41, 42, 60, 418, 420, 421, 427-443, 722, 790, 791, 797
polarization by reflection 421
polarization of 17, 29, 35, 41, 42, 164, 379-381, 417-443, 712, 722, 790-792, 795, 797, 799
polarizer (linear) 421
power of 17
propagation 169, 176, 417, 418, 427, 429, 491, 492, 496, 503, 509-511, 516, 518, 520
quality of 4, 29, 159, 163, 217, 220, 234, 236, 251, 252, 286, 310, 353, 356, 363, 379, 381, 382, 387, 389, 392, 470, 478, 491, 499, 501, 523, 524, 527, 531, 592, 631, 640, 642, 696, 703, 778
quality of, indicators for 384
quantification of 17, 22, 419
quantity of 4, 17, 29, 159, 163, 286, 353, 356, 379-382, 389, 390, 393, 406, 455, 477, 491, 499, 522, 524, 527, 531, 592, 631, 633, 642, 664, 703, 707
quantity, instantaneous 17
quantity time-integrated 17
red:far-red photon ratio of (see red light)
reflection 17, 225, 226, 384, 388, 397, 399, 492, 499, 502, 505, 507, 508, 512,

513, 526, 537, 538, 548
reflectivity 538, 548
refraction 388, 418, 422, 492, 493, 502, 537
scattering of 30, 31, 115, 166, 225, 319, 379, 384, 386, 388, 389, 427, 491, 493, 496, 498-500, 502-506, 508, 510, 511, 513-518, 520, 522, 525-527, 531, 537, 538, 544, 547
scattering particles 43, 386, 389, 502, 503, 509, 517, 538
sensing (see light, perception of)
sieve effect 491, 499-501, 518, 519
small angle scattering of 504
SOX radiation 256-258
spectral quality of 4, 19, 379, 537, 615
stress 761
terminology of 17, 20, 768
time factor 17, 379, 390
transmission/transmittance 388, 543, 544, 548, 549, 786, 787
trap 491, 511, 513
unilateral 369-371, 417, 418, 422-429, 493, 494, 522, 525, 526, 659, 660, 663-668, 672, 674, 675, 677, 678, 719, 727
velocity of 18, 419, 420
wave properties 418, 491, 492
linear variable differential transformer (see electronic position-sensitive transducer)
lipid membrane 318
liposomes 130, 132
liverworts 393, 412, 494
Lolium
perenne 694
temulentum 452, 477, 479-482
long-day plants (see photoperiodic response types)
long-hypocotyl mutants (see mutants)
Lopadostoma turgidum 771
Lorentz-Mie scattering (see light, small angle scattering)
low fluence response (LFR) 7, 170, 177, 235, 236, 239, 241, 243, 247, 253, 286, 312-314, 318, 344, 369, 398, 401, 402, 410, 413, 587, 588, 590, 591, 616, 617, 635, 636, 641, 643, 653, 656, 663, 687, 724, 799
lumen 23
Lunularia 494
Lupinus 528
lux 23, 768
Lycopersicon esculentum 87, 106, 108, 156, 237-240, 256, 257, 273, 275, 277-279, 284, 285, 287, 290, 291, 294, 354, 368, 572-575, 579, 582, 584, 585, 586, 589, 602, 603, 609, 610, 620, 622, 642, 737
Lygodium japonicum 784-786, 788

macromycetes 754, 758, 759, 763, 778
maize (see *Zea mays*)
malate biosynthesis 684, 693, 694, 701
Malus sp. 738
Malva parviflora 528, 696, 703
Marchantia 494
mast cells 132
Matricaria indora 6
Medicago sativa 498, 514
meiosis 761, 762, 764, 766-768, 776
meiotic generation 603
membrane(s) 190, 330, 331, 338-343, 601, 649, 656, 710, 721, 722
depolarization 338, 339, 648, 649, 655, 710
hyperpolarization 339, 648, 723, 730
permeability 315, 337, 338, 648
potential 10, 259, 336-340, 648, 656, 709, 714, 715
site of phytochrome action 189, 192, 259, 486
Mercuriales perennis 394
Merulius lacrymans 755, 758
mesocotyl 172, 173, 274, 357-359, 496, 511, 653
growth 169, 170, 203, 511, 653, 654
mesophyll 191, 505, 507, 508, 529, 531, 532, 574, 683, 689, 696, 719, 721
chloroplasts 689, 691, 700, 721
messengers 685, 687
photosynthesis 683-685, 692
Mesotaenium 798
5,10-methenyl-tetrahydrofolyl-polyglutamate (see pterin(s))
N-methyl mesoporphyrin IX 107
microbeam (experiments) 29, 34-40, 164, 169, 170, 176, 177, 425, 435, 437, 438, 440-442, 714, 716, 723, 727, 788, 790, 792, 793, 795, 796, 799, 800
microfibrils 789, 794
microspectrophotometry 36-38, 43-45, 178, 700, 711
milo (see *Sorghum vulgare*)
Mimosa pudica 241, 242

mitochondria 34, 178-182, 347, 503, 716
Monilia fructicola 754, 755, 758, 771
monoclonal antibodies (see phytochrome)
moonlight 537
Monte Carlo approach 538
Monteith canopy model 548, 549
mosses 73, 128, 131, 274, 279, 308, 393, 412, 437, 559, 603, 659, 721
  protonemal stage 279, 437, 439, 659
Mougeotia 35, 60, 73, 79, 132, 177, 178, 197, 201, 202, 346, 433-437, 439-443, 559, 724-730, 793, 798
  model for chloroplast movement 727
  scalaris 724, 726
mRNA (see RNA)
mung bean 510
murexide 198
mustard (see *Sinapis alba*)
mutagen 604
mutagenesis 94, 100, 156, 271, 275, , 590, 604
mutagenesis site-specific 271, 275
mutants 13, 67, 74, 76, 81, 82, 147, 151, 204, 258-260, 272-275, 288-294, 307, 327, 342, 347, 377, 393, 401, 403, 407, 408, 410, 432, 601-603, 608, 635, 656, 710, 713, 717, 756, 757, 761, 764, 786, 801
  bacteriorhodopsin 708
  carotenoid 67
  chromophore-deficient 106, 607
  deficiency 13, 81, 275, 403, 410, 602
  deletion 108, 281-285, 290-293
  double 318, 338, 483, 604, 605, 609, 613, 614, 619, 621
  induction and isolation 84, 593, 601, 603-605
  light-response 610, 611
  loci of 604, 611
  M_1, M_2 generations 604, 612
mutants of *Arabidopsis*
  blu B-insensitive 327, 368, 605, 612, 613, 619, 621, 623, 678
  cop constitutive-photomorphogenetic 205, 401, 575, 610, 611, 620
  det de-etiolated 204, 205, 401, 575, 610, 611, 620, 623
  fhy FR elongated hypocotyl 403
  fre FR elongated 403, 608, 618
  hy long hypocotyl 71-85, 112, 206, 289, 290, 368, 403, 407-411, 483, 607-608, 611-613, 615-621, 678
  JK218 phototropism 327, 342, 344, 613
  JK224 phototropism 327, 342, 344, 613, 619
  lrd light-responsive dwarfs 279, 610
mutants of *Brassica*
  ein elongated internode 403, 407, 409, 411, 608, 617, 622
mutants of *Coprinus* 764, 767
mutants of cucumber
  lh long hypocotyl 81, 82, 403, 407, 608, 616-619, 647
mutants of *Neurospora*
  al albino 318, 329
  band 318
  wc white collar 329, 339, 342, 344
mutants of *Nicotiana plumbaginifolia*
  eti etiolated 608
mutants of pea
  lip light-independent 610
  lw dwarf 611, 612, 622
  lv long stemmed (internodes) 206, 614, 617, 622, 623
mutants of *Phycomyces*
  car carotene-free 332
  mad phototropic 329, 328, 331, 332, 338, 339
  'stiff' 613
mutants of *Physcomitrella*
  ptr phototropic and polarotropic
mutants of *Scenedesmus obliquus* 307, 309, 318
mutants of *Sorghum bicolor* 608
  ma3 phytochrome type II deficient 476
mutants of tomato
  au (aurea) 83, 112, 368, 607-609, 616, 619-623
  hp high pigment 605, 609-612
  tri temporarily R-insensitive 608
  yg yellow-green 607, 608, 620, 622
mutants of *Trichoderma*
  dim dim sighted 329
mutants
  photomorphogenetic 259-262, 403, 592, 601-624, 654
photoreceptor 13, 81, 206, 329, 328, 368, 605
  phytochrome 76, 84, 112, 160, 259, 282-284, 288, 289, 294, 403, 475, 476, 483, 592, 623, 654
  phytochrome-deficient 76, 81, 83, 100,

275, 285, 289, 368, 403, 410, 607-608, 618
point mutation 108
primary defect 601, 602
recessive (assumed loss-of-function) 401, 603, 604, 610
response 605
shotgun complementation 578, 603
transduction chain 259, 605, 614
myceli(a)(um) 306, 329, 338, 344, 753, 754, 756-759, 778
mycochrome system 310, 321, 776, 778
myoglobin 128

NAD kinase 202, 203
NADH dehydrogenase activity 178
NADP glyceraldehyde-3-phosphate dehydrogenase 356, 357, 691
NADP reduction 178
NADPH (-cofactor) 64, 65, 203
naphthalene sulphonamides 199
(the) natural radiation environment 81, 158, 184, 301, 353-355, 377-413, 417, 421, 441, 443, 447, 467, 469, 472, 479, 481-483, 497, 517, 524, 531, 537, 589, 635, 636, 642, 659, 664, 684, 685, 697, 698, 701, 718, 737, 793
Nectria galligena 776
Nernst potential for K^+ 197, 648
Neurospora 331, 344
crassa 304, 306, 309, 312, 315, 318, 329, 336, 339, 347
mutants (see mutants)
nitrate reductase 304, 306, 347
Nia vibrissa 755
Nicotiana
plumbaginifolia 4, 583, 584, 608
tabacum 106, 108, 149, 150, 156-159, 272-275, 280-286, 289, 312, 397, 404, 405, 452, 453, 566, 576, 577, 580, 581, 583, 585, 589, 590, 603, 608, 609, 620, 741
tabacum SR1 genotype 275-278, 609
tabacum cv. Xanthi 275-278, 286-293, 609
night-break 7, 168, 235, 236, 242, 245, 450, 454, 455, 457, 458-467, 469, 470, 472-474, 476-479, 481, 482, 484-488, 761
nitrate reductase 304, 306, 312, 315, 336, 347
nitroxide probes 336
Norflurazon 67, 145, 170
Northern blot analysis 45, 87, 90, 565-567, 571, 573, 585, 591
NTPase 203
nuclear magnetic resonance 54, 57, 133
nuclear protein factors 93, 94, 196, 579-585
nuclear run-off experiments 587, 589
nuclear run-on experiments 572-574, 577
nuclease (analysis) 574
nucle(i)(us) 34, 39, 90-96, 164, 178-182, 190, 193, 195, 196, 203, 509, 571, 572, 576, 761, 762, 764, 787, 788, 795-797
nuclei isolation of 178
nyctinasty (see leaf sleep movements)
oak 388

oat (see *Avena sativa*)
Occam's razor 122
Oenothera lamarckiana 312
one-instant mechanism (see spatial sensing)
onion 502
Onoclea 786
optical fibre(s) 427, 493, 508-510, 520, 724
acceptance angle 509, 510, 519-521
probe 395, 396, 407, 514, 518, 519, 520-522, 524, 529, 647
opto-electronic methods 670
oscillator (see pacemaker)
oscillator strength ratio 42, 117
osmotic potential (pressure) 684, 693, 701
Osmunda 424, 425, 432, 787
overriding factor (see seed germination)
Oxalis europa 496-499
oxygen singlet 711
oxygen tension/uptake 347

^{32}P incorporation 91, 193, 195, 343, 344, 572
pacemaker (endogenous) 448, 459, 484
palisade layer 497, 498, 505, 507, 508, 531, 532
Panicum miliaceum 694
Paphiopedilum harrissianum 687-689, 691, 698, 702
papillose epidermal cells 497, 505
paraflagellar body 44, 45, 314-316, 711, 712
parallel light (see light, parallel)
parsley (see *Petroselinum hortense*)
patch-clamp technique 197, 336-338, 341, 691, 699, 714, 715

pattern of temporal and spacial competence 562, 563
pea (see *Pisum sativum*)
peanut (see *Arachis hypogaea*)
Pectinophora 312
Pelargonium 688
Pellicularia filamentosa 770-772
Peridinium gatunense 714, 715, 718, 719
chromoproteins 715
Perilla 484, 485
ocymoides 769
perithecial formation 33, 35, 307, 318, 759, 776, 778
Peronospora spp. 772
peroxidase 168, 175, 178, 656
pertussis toxin 341
petioles 170, 391, 612, 633, 655, 659, 661
Petroselinum
crispum 741
hortense 563, 567-570, 739, 740, 742-746
Petunia 86, 87, 480, 572-574, 576, 577, 584
Phalaris 662
Pharbitis nil 144, 225, 242, 447, 452, 454-456, 458, 459, 461-476, 485-487, 768
phase response (relationship) 448-451, 459
Phaseolus 8, 236, 237, 339
vulgaris 450, 522, 523, 571, 645, 651, 698, 740
phase-shifting 302, 312, 347, 459-461, 468, 469, 480, 484, 487, 589
Phellinus contiguus 755
phenothiazines 199
phenotype(s) 112, 156, 159 204-206, 262, 271-280, 285-294, 405, 407, 409, 575, 602-607, 609-614, 618, 622, 623, 756, 757, 764
L-phenylalanine 736, 744
phenylalanine ammonia-lyase (PAL) 736, 744-747
N-phenylimine S-23142 107
phenylpropane 735
phenylproponoid metabolism pathway 734, 736, 744, 746, 747
Philodendron scandens 514, 517
phoborhodopsin (see sensory rhodopsin II)
Phomopsis phaseoli 755
Phormidium 710, 730
phosphatidylinositol 4,5-bisphosphate (PIP_2) 192, 193, 207, 342
phosphodiestrase(s) 756
3-phosphoglycerate kinase 691
phosphoinositide (PI) cycle 342, 343, 649
phospholipase C 190, 192, 207, 341, 342
phosphoprotein 189, 194, 291, 649
phosphoprotein antibodies 195
photo-accumulation 727-730
photo-autotrophic plants 399, 400, 406, 783
photochemical cyclization 55
photochromicity 51, 384, 412, 637, 709
photochromobilin 108
photoconversion 51, 213-217, 234, 330-333, 382, 390, 396
photodetector 17, 19
photodinesis 719-723, 727, 729
photoerythrobilin 105
photokinesis 707, 708, 710, 711
photokinesis positive 708
photolyase 308, 309, 314, 315, 330, 332, 740, 749
photomorphogenesis 3-11, 37, 211, 258, 335, 354-356, 360, 368, 371, 389, 401, 491, 513, 521, 522, 560, 562, 563, 588, 631, 632, 733
photomovement 3, 4, 39, 338, 442, 559, 707-730
photon 19, 382, 419, 420, 491, 492, 500, 505, 507, 512-513, 516, 519, 538, 669, 672, 699
counting 18, 19, 37-40, 379-382
energy 18, 379, 419
flux density (PFD) 20, 22
gradients in space 379, 381
irradiance 20, 22, 380, 381, 384-386, 485, 684
number of 18, 696
Poisson distribution of 516
photoperiodic(ism) 4-7, 13, 168, 262, 377, 413, 447-488, 754, 768, 769, 772
in fungi 759, 760, 768, 769
photoperception 262, 447-488
timekeeping 447, 450-452, 454, 457-471
response types 235, 237, 402, 451-453, 620, 622
response types dark dominant 477
response types day neutral 453
response types light dominant 477-479
types long day 261, 452-454, 457, 460, 461, 477-483, 485, 486
response types short day 5, 361-363, 452-461, 463, 465-467, 471-474, 476-482, 484-487, 768
photophile 460, 482, 537

photophobic response 708, 710, 711, 718
attractant/repellant 708, 709
photophosphorylation 649, 656, 691, 693, 700, 702
photoprotectants 330, 332, 747, 749
photoreactivation (see DNA)
photoreceptor(s) 5, 11, 29, 34, 40, 51, 71, 81, 163, 168, 192, 211, 255, 274, 276-280, 301-322, 327-348, 353, 377-379, 422, 447, 450, 452, 481, 485, 513, 523, 528, 563, 601, 605, 619, 634, 635, 638, 644, 652, 670, 687, 707-730, 733, 737, 757, 775-780, 786, 790, 792, 793, 795, 799
(see) bacteriorhodopsin
(see) blue light
coaction (see photoreceptor(s) interaction)
(see) cryptochrome
dichroic arrangement 380, 381, 427, 428, 435, 711-713, 728
(see) flavin(s)
function of 71, 100, 141, 378
green light 318
(see) high irradiance response
interaction 5, 9, 220, 254, 255, 258, 309-311, 321, 353-372, 588, 601, 615, 619, 636, 637, 640-643, 647, 720, 721, 726
orientation of (see dichroism)
(in) photosynthesis 377
(see) phytochrome
(in) signal transduction 34, 89, 315, 377, 378, 528, 723-726
sites (see growth)
(see) UV-A, UV-B absorbing
photoresponse modes 258, 759
photoreversibility (see far-red light reversibility)
photosynthesis 11, 29, 52, 170, 253, 254, 278, 303, 308, 309, 311, 318, 353, 378-380, 389, 391, 398, 417, 441, 456, 497, 498, 500, 505, 507, 511, 528-532, 563, 571, 575, 578, 586, 631, 632, 634, 639, 659, 683, 684, 686, 688, 689, 691, 695, 699, 701, 702, 707, 710, 714, 715, 717, 718, 729, 749, 798
inhibition of 24, 247, 254, 498, 710, 749
net 684, 685, 697
photosynthetic
apparatus adaptation 312, 322, 391
bacteria 708, 710
electron transport 692, 699, 710, 714
light compensation point 497, 696
photon flux density (PPFD) 20, 23
pigments 387, 500, 501, 529, 631, 643, 644, 693, 710, 717, 798
photosynthetically active radiation (PAR) 23, 24, 390, 392, 393, 405, 408, 497, 530, 541, 542, 548
photosystem I and II 586, 622, 691
phototaxis 11, 39, 302, 303, 315, 417, 441, 442, 707, 708, 710, 711, 713-717, 730
methyl-accepting protein (MPP) 709
negative 417, 707, 711, 713-715, 717, 718
positive 417, 418, 707, 711, 713-716, 718, 719
transversal 707, 719
photothermal device(s) 18, 19
phototropism 4, 10, 132, 169, 244, 302-304, 307, 308, 311, 312, 315-319, 322, 327, 328, 332, 333, 337, 341, 343, 346, 347, 367-372, 417, 425-427, 432, 441, 442, 491, 493-496, 524, 526-528, 588, 605, 612, 613, 619, 633, 634, 652, 659-680, 759, 763, 764, 766, 783, 786, 789-792, 799
adaptation in 673
angle of curvature 307, 308, 370, 371, 494, 660-662, 665-668, 672, 673, 675, 783, 793
autotropic phase 660, 667-669, 672
Blaauw theory of 308, 664, 677, 678
Cholodny-Went hypothesis of 346, 652, 663, 664, 670, 674, 678, 679, 766
curvature index 371
(in) dicotyledons 369-372, 665, 672-674
first positive response 311, 312, 344, 346, 494, 495, 612, 613, 619, 672, 673, 763, 774
growth curvature phase 669
lag (latent) period 668-670
measurement of 665-669
models of 662, 674
negative 427, 494, 495, 672
neutrality point 427
perception phase 495, 669, 670
perception sites 662, 674
second positive response 613, 619, 672-674
threshold light level for response 673
tip response 663, 668, 674, 676, 678
phycocyanin 52-57

conformational changes 55
crystalline 55
phycocyanobilin 52-55, 66, 109-112, 119, 281
phycoerythrobilin 64, 111
Phycomyces 309, 312, 315, 423, 425-427, 432, 494, 496
blakesleeanus 329, 339
light growth reaction 334, 335, 425-427, 659
mutants (see mutants)
phototropism 303, 304, 311, 312, 318, 328, 332, 333, 346, 422-427, 493
sporangiophore 303, 312, 314, 317, 318, 328, 332, 338, 346, 422, 423, 425-427, 432, 493, 659, 754, 759
phylogenetic relationship 72, 73, 301, 412
Physcomitrella patens 659
phytoalexins 563, 742, 747
phytochrome 5-10, 11, 35, 51, 163-184, 187, 198, 211-269, 271-294, 307, 311, 318, 327, 335, 342, 353, 354, 356, 358, 359, 361, 363-372, 377, 378, 382-387, 390, 398, 400-403, 427, 433-435, 450, 466, 522, 559-563, 565, 603, 605, 631, 636-639, 642, 647, 654, 656, 709, 724-729, 737-740, 742-746, 776, 784-786, 788, 790, 792, 793, 795, 797-799, 801
phytochrome A (see also phytochrome types) 40, 74-92, 100, 154-157, 174, 177, 272, 276-284, 290, 291, 294, 379, 383, 386, 403-407, 409, 410, 412, 413, 608-611, 616-620, 623
apoprotein (PHYA) 74, 106, 108-115, 156, 272, 279, 281, 327
mutants 83, 607, 619
overexpress(ing)(ors) 274, 275, 284, 286, 288, 291, 404, 405, 409, 609, 610, 612, 616, 617, 620, 622
phytochrome
absorbance dipole 177
absorption spectrum 51, 52, 54, 111, 116-118, 123, 165, 166, 213, 295, 383-385, 641
active form 51, 105, 158, 169, 176, 178, 180, 211, 226, 271-295
all or nothing responses 36, 380, 400
amino-acid homology 107, 402
amino-acid sequence 72-76, 78, 107, 108, 132, 155, 281-283, 285, 402
(as an) antigen 166, 167, 172
apoprotein (PHY) 64, 105-107, 110-114, 118, 125, 133, 166, 168, 272, 289
apoprotein recombinant 105, 109
assays 163-184, 271, 275
autoregulation of synthesis 404, 560
phytochrome B (see also phytochrome types) 75-76, 80-86, 100, 154, 155, 272, 275, 284, 403, 407-411, 413, 609, 610, 614, 617-620, 623
apoprotein (PHYB) 74, 81
mutants 81, 275, 410, 619, 620
overexpress(ing)(ors) 274, 407-410, 609, 615, 617, 622
specific antibodies 607
phytochrome
binding 130, 179, 788
bio-assays 164, 168, 176, 271, 275
biosynthesis 9, 106, 113, 159, 229, 230, 261, 262, 404, 784
(as) blue light receptor 321, 384, 385, 439
phytochrome C 75-76, 80-86, 155, 272, 410, 413, 614, 618, 623
phytochrome
C-terminus 75, 77-81, 100, 110, 112, 115, 116, 130, 131, 133, 282-284, 293, 294
chromophore oxidative degradation 53, 54
chromophore spatial re-orientation 41, 60, 125, 126, 131, 177, 438
conformation changes 55, 57, 58, 105, 117, 126, 127, 129, 130, 132, 271, 281, 282, 290
control of enzyme activity 178, 203, 307
control of gene expression/transcription 9, 71, 84, 85, 87, 89-96, 132, 144, 164, 178, 183, 195, 197, 205, 259, 354, 357, 358, 364, 368, 485, 486, 559, 561-563, 573, 587, 588, 590, 591, 620
control of gibberellin levels 178, 786
control of ion transport (fluxes) 178, 197, 259
control of membrane permeability 164, 183, 197, 206, 259
control of protein phosphorylation 194-197
crystallographic analysis 117, 119, 133
cycling (rate) 10, 63, 144, 146, 151-154, 159, 216-225, 233, 253, 254, 524, 525, 638, 640, 641
cytoplasmic distribution 34, 146
phytochrome D 75, 82, 83, 272, 410, 413,

614, 618, 623
phytochrome
dark reactions 223, 228, 231, 639, 641
dark reversion (see phytochrome Pfr)
degradation 53, 141-160, 366, 473
dehydration 61, 63
denaturation 55, 799
destruction (see phytochrome Pfr and Pr)
dichroism 117, 122, 126, 432-437, 724
(in) dicotyledons 142-145, 156, 158, 170, 274
differences between Pr and Pfr 5-9
difference spectrum 43, 110, 111, 150, 152, 157, 165, 166, 277
(as a) dimer 40, 105, 109, 115, 130, 131, 223, 224, 233, 243, 261, 271, 273, 277, 281, 283, 284, 290, 294, 475, 476, 482
discovery of 5-9, 30, 43, 142, 211, 212, 263, 400, 412, 737
domains (molecular) 130, 277 276, 281-284, 288, 290, 294
dose (-response curves) 279, 286-289, 293
phytochrome E 75, 82, 83, 272, 410, 413, 614, 618, 623
phytochrome
ecological relevance 386-413
electronic transitions 116, 117
electrostatic interactions 116, 125
(as an) enzyme 184, 286
enzyme-linked immunosorbent assay (ELISA) 166-168, 172-174, 287
epitopes (binding sites) 130, 149-155
family (functions of) 5, 377-413
flip-flop dichroism of 432-438
fluorescence 59, 118, 120-122, 129, 171, 172, 282
fluorescence lifetimes 121, 122
flux (see phytochrome cycling rate)
full-length 54, 56, 109, 110, 116, 124, 147, 276, 277, 281, 283, 285, 287, 291-294
G-domain 116
genes (*phy*) 5, 71-101, 105, 195, 271, 272, 274-279, 560, 607, 639, 643
(in) green tissue (light-grown plants) 82, 166, 168, 272, 276, 278, 522-524
(in) guard cells 176
Hartmann model of action 249, 260, 639
holoprotein (spectrally active phytochrome) 74, 105, 165, 168
hydropathy profile 80
hydrophillic regions of 80, 290
hydrophobic regions of 80, 290
immobilized 60, 131, 180
immunochemical assays (detection) 75, 76, 84, 86, 106, 109, 141-147, 150, 157, 164, 166, 168, 169, 176, 179, 180, 228-231, 276, 279, 281, 282, 286, 368, 473, 474, 482, 560, 605
immunocytochemistry 34, 45, 91, 151, 167, 168, 172, 174-181, 213, 274, 702
immunofluorescence assay 172
intercellular distribution of 163, 164, 168, 229
intermediates 10, 40-42, 60-63, 118-125, 132, 223, 234, 396, 638
intermediates absorption spectra 125
in vitro responses 178, 271
isoelectric point 114, 127
isolation and purification 93, 154, 212, 213, 263, 271, 274, 282, 397, 412, 561
Kendrick and Frankland model of action 254, 255
(as a) kinase 188
kinetics of action 559, 560
labile (PI, type I) 106, 74, 75, 141-155, 160, 173, 213, 228, 233, 235, 254, 260, 261, 268, 269, 272, 404, 406, 473-475, 479, 482, 483, 565, 609, 616, 617
large 63, 115, 124, 290
(in) light-grown plants 142, 146, 174, 213, 274, 457, 459, 475, 479, 639
light stable (PII, type II) 74, 75, 90, 106, 141-146, 152, 155, 160, 173, 174, 213, 228, 233, 260, 261, 272, 473-476, 479, 483, 565, 607, 608, 614-620, 639
localization 10, 34, 132, 163-184, 187, 229, 523-525, 559, 788, 792, 797
lumi-F 61, 62, 123
lumi-R 61-63, 120-125
mechanism of action 163
(and) membranes 10, 34, 35, 130, 132, 146, 178-182, 189, 259, 486, 559
meta-F 61, 62, 123
meta-Ra 61-63, 123-125
meta-Rb 61-63
meta-Rc 61-63, 123-125
(and) mitochondria 178
models 254, 255, 560, 561
molar conversion cross section 213-215, 225, 259, 268, 269, 385, 386

molecular biology of action 71, 163, 183
molecular mass 154, 157, 174, 276, 277, 279, 280
molecule 132
monoclonal antibodies (MAP) 76, 82, 84, 126, 128, 130, 131, 157, 166, 167, 172-174, 180, 182, 276, 287, 475, 479
(in) monocotyledons 142-145, 156, 170, 274, 275
monomer 105, 115, 147, 174
N-terminus 78-81, 100, 110, 112, 126-130, 133, 282, 283, 290, 291, 475
native (see phytochrome full-length)
nomenclature (see terminology)
oligomer 115
(and) organelles 178, 179
orientation 435, 438
'paradoxes' 226, 241, 243, 244, 412, 472
particulate 179
Pbl (see phytochrome, meta-Rb)
pelletability 146, 152, 179, 183, 229, 561
PEST sequence 117, 129, 147, 155, 284
Pfr absorption spectrum 8, 52, 57, 65, 116, 118, 126, 211-214, 433, 473
Pfr chromophore 51, 59, 105, 126, 271
Pfr dark reversion 106, 124, 125, 144, 145, 223, 228, 230, 233, 234, 277, 283, 457, 465-471, 473-476, 485, 560, 561
Pfr dark reversion inverse 234
Pfr destruction (decay) 9, 75, 85, 92, 106, 141, 143, 145-158, 183, 223, 228-234, 250, 251, 254, 260-262, 271, 277, 280, 281, 284, 285, 404, 474, 475, 485, 560, 561, 617, 639, 641-643
Pfr extinction coefficient 10, 213, 214, 385, 386
Pfr initial action 89, 163, 187, 188, 202, 207, 228, 254, 258, 259, 341, 354, 358, 559-561, 563, 725
Pfr reaction partner of 40, 228, 243, 245, 254, 255, 259, 260, 476
Pfr response relationships 89, 190, 193, 197, 211, 241, 371, 398, 402, 561, 564
Pfr responsiveness 275, 280, 354-363, 368, 372, 482, 562
Pfr under steady state conditions 382
photochemical properties 119, 213, 223, 386
photoequilibrium 10, 51, 63, 120, 143, 159, 215-220, 234, 236, 247, 268, 355, 359, 361, 362, 366, 367, 384, 385, 390, 393, 394, 396, 397, 408, 435, 436, 469, 482, 483, 502, 524, 639-642
(and) photoperiodism 450, 457, 458, 465-473, 475, 476, 478, 482-488
photophysics 119, 379, 386
photoprotection of 233, 479
photoreversibility 65, 105, 120, 165, 176, 178, 211, 212, 223, 231, 273, 277, 281, 282, 290, 292, 311, 336, 358, 435, 466, 482, 560-562
photostationary state 234
phototransformation 7-10, 40-42, 52, 59, 92, 105, 119, 120, 123, 129-132, 141-145, 151-155, 166, 178-182, 189, 211-221, 224, 225, 228-230, 234, 235, 249, 259, 271, 282-284, 290, 353, 360, 382, 396, 404, 433, 435, 437, 438, 458, 477, 485, 518, 523, 638, 641, 743, 784
phototransformation quantum yield 120, 485
physicochemical properties 8, 263, 271, 281, 294
physiology of action 75, 131, 158, 211-269, 271, 377-413
(and) plastids 34
polypeptide(s) map 72
pools 143, 144, 147-150, 166, 171, 176, 226, 472-476, 479, 560, 639
Pr absorption spectrum 8, 52, 54, 55, 116, 118, 211-214, 433, 473, 479, 523
Pr chromophore 51, 59, 105, 106, 271
Pr denaturation 54, 55
Pr destruction 9, 144, 230, 232, 277, 560
Pr extinction coefficient 10, 54, 213, 214, 385, 386
Pr flash photolysis 62
Pr protonation 54, 57, 127
Pr specific monoclonal antibodies 131
Pr synthesis 9, 159, 229, 230, 474, 560
prelumi-R 121
primary functions 163, 179, 561, 801
primary structure 114, 189
proteolysis *in vitro* 53, 54, 63, 115, 129, 159, 281, 283
purification (see phytochrome isolation)
quantum yield for photoconversion 59, 213-215, 385, 386
receptor(s) 132, 133
regulation (see phytochrome control)
related parameters 213, 215-227, 254, 266, 267, 383-398

relative cycling rate (see phytochrome cycling)
residue(s) 57, 115, 126, 128-130
responses 87, 89, 187, 188, 235, 262, 274, 400, 401, 523, 559-562, 635-641, 659
Schäfer model of action 254, 255, 560
secondary and tertiary structure 114, 115
(as a) sensor of R:FR 384-402
sequences (in databases) 80, 81, 281, 282
sequestering 132, 146, 147, 151-154, 159, 160, 183, 229, 271, 561
small 63, 124
spectrofluorometric assays (detection) 172
spectrophotometric assays (detection) 8, 9, 43, 75, 86, 106, 107, 141, 145, 164-172, 176, 179, 213, 223-230, 233, 234, 241, 278, 280, 281, 286, 287, 292, 368, 476, 560, 605, 607, 610, 644, 784
structural domains 115-117, 147, 271
structure 51-67, 105, 115, 116, 126-131, 271-295
structure tripartite 'Y' shaped 115
subcellular fractionation 177, 179, 180, 183, 184
synthesis (see phytochrome biosynthesis)
terminology 74, 104, 105-134, 141
tetrapyrrole (linear) chromophore 51, 274
threshold responses 279, 286, 288, 380, 457, 483
types 5, 10, 71-76, 100, 106, 112, 133, 141, 145, 164, 165, 168, 172-174, 187, 212, 225, 228-230, 258, 261, 262, 271, 378, 473, 474, 483, 524, 565, 606-608, 614, 615, 620, 788
types co-operation of 411
undegraded 118
unstable (dark, see labile phytochrome)
VanDerWoude model of action 131, 223, 255, 261
phytochromobilin 51, 52, 57, 64-66, 105, 107, 109-112, 290
iron pathway 64, 107
phytoene 329
phytoflagellates 11, 39, 311, 417, 418, 423, 711-714, 717, 719
phytopathogenic fungi 775
Phytophthora 772
phytoplankton 544-546
pigment 29, 51, 310, 353, 499, 500, 505, 515, 518, 523, 530, 531, 631, 733
orientation (see dichroic)
Pilobolus crystalinus 770-772
kleinii 771
sphaerosporus 770-772
sporangiophore 755, 763
spp. 312, 442, 755, 756, 763, 764
umbonatus 771
Pima cotton (see *Gossypium barbadense*)
Pinus 72
sylvestris L. 360-365
Pisum sativum 73, 78, 79, 86, 88, 97, 106, 107, 109, 114, 121, 127, 129, 133, 155, 171, 174, 193, 196, 203, 204, 226, 273, 274, 281, 315, 341, 343, 344, 346, 572, 574, 576, 577, 579-582, 584, 588-590, 632, 651, 654, 655
dwarf 654, 655
'paradox' 226, 241
Planck's constant 18, 419
plant gene expression regulation by light 559-593
plant-growth rooms 30, 393, 392
plant optics 417-442, 491-532
plasma membrane 177, 179, 181-183, 197, 306, 335, 336, 338, 341, 342, 345, 428, 432, 435, 441, 648, 649, 691, 698, 699, 701, 788, 792, 793, 797
plasmid 109, 569
plastidogenesis 356
pleiotropism 605
Pleospora herbarum 776
Pleurotus ostreatus 755, 777
Podospora
curvula 772
Podospora tetraspora 771
polarity induction 424, 432
polarized light (see light)
polarotropism 177, 132, 437, 438, 559, 711, 784, 787, 790, 794, 795, 799
polymerase(s) 151
polynomials 545
Poria ambigua 755, 758
Porphyridium cruentum 110
potato (see *Solanum tuberosum*)
probit(s) curve 564
protease inhibitor cocktail 194
proteases 148
protein kinases 78-81, 100, 188-192, 196-198, 291, 293, 341-343
protein phosphatases 194-197
protein phosphorylation 81, 188, 189, 192-

197, 291, 315, 329, 341-344, 527, 561, 584
proteolysis 53, 65, 143, 147-151, 155, 156, 281, 283, 474
proteolytic complex 148
prothallial stage (see ferns)
protochlorophyll(ide) 43, 166, 307, 309, 318, 400
proton extrusion 648, 649, 691, 693, 695, 698, 699, 701, 723
proton motive force 710
protonemal stage (see ferns and mosses)
protoplasmic streaming 428
protoplasts 190-192, 199, 337, 567-570, 578, 685, 686, 693, 699
(of) guard cells (see guard cells)
(as) osmometers 685
swelling 190-192, 199, 486, 686
protoporphyrin (IX) 64, 107, 716
pseudopodia 716
Psilocybe cubensis 777
Pteridium aquillinum 786
pteridophytes 361
pterin(s) 11, 44, 45, 314-317, 327, 332, 348, 711, 786
Pteris vittata 304, 306, 312, 784, 786-788, 790, 799
Pterula sp. 755
pulvinule 345, 486, 659, 661
pumps 336-338, 648, 649, 691, 693, 698, 699, 701, 708, 709
Pyricularia oryzae 776
pyrimidine dimers 308, 314, 330, 739
Pyronema
domesticum 757, 759
omphalodes 756

quanta(um) 380, 685
quantum yield (efficiency) 10, 19, 303, 306, 314, 380, 385, 386, 466, 638, 702
quercitin 564
Quin-2 198

radiance 547
radiation 20, 22, 518, 519
global 386, 540, 541
natural 33, 311, 313, 322, 353, 355, 378, 379, 497, 498
radiator hollow heat 18
Raphanus sativus 642, 646, 652, 678
raphanusanins 678
ray-tracing diagrams 492, 496, 505, 506
Rayleigh scattering (see light, molecular scattering of)
reciprocity law (see Roscoe-Bunsen)
red cabbage (see *Brassica oleracea*)
red clover (see *Trifolium pratense*)
red light 5-10, 35, 41, 43, 74, 76, 82-96, 141, 143, 144, 149-151, 156, 157, 165, 166, 169, 177, 178, 182, 187-203, 211, 212, 218-222, 227-252, 257-261, 274, 279, 306, 309, 311, 313, 318, 327, 330, 331, 335, 341, 345-347, 353, 355, 357-371, 382-384, 387-389, 395-400, 404-406, 409, 410, 412, 431, 433-435, 437, 438, 455, 458-460, 467-472, 475-487, 503, 504, 511, 522, 523, 529, 561, 562, 564, 570, 573, 575, 587, 588-591, 605, 611-614, 621, 622, 636-643, 646-656, 665, 686-703, 720, 726, 728, 729, 733, 737-739, 742, 743, 746, 775, 776, 778, 779, 784-800
:far-red light photon ratio 82, 220-222, 236, 379, 384-392, 395-400, 402, 404, 405, 407-413, 469, 479, 480, 483, 615-620, 656
:far-red light photon ratio indicator of shading 388
:far-red light photon ratio and neighbour detection 397, 398
/far-red light reversibility (see far-red light reversibility)
gradient 522-524
-high irradiance response (R-HIR) 246, 248, 254, 261, 401, 402, 410, 413, 639, 640
redox reactions 303, 305, 315
reflecting boundaries 397, 504, 506-508, 513, 517
refractive index 422, 423, 427, 492, 493, 504, 506-510, 513, 538, 539, 543, 713, 717
remote sensing 548
resonance experiments 54, 459
resonance-Raman spectra 54, 60, 119, 127, 133
respiration 304, 307, 309, 318, 322, 347, 391, 691, 786
responsiveness amplification 321, 356, 357, 362, 364, 612
reticulocyte lysate 109, 147
retinal 331, 709, 713

rhizoid development 787
rhodamine 168, 181, 182, 790, 801
rhodopsin 5, 8, 11, 123, 314, 318, 331, 340, 713
Rhodospirillum 710
rhubarb 633
rhythmicity study of 769
riboflavin 331
ribonucleoprotein particles 180
ribulose-1,5-bisphosphate carboxylase-/oxygenase (Rubisco) 278, 571, 572
ribulose-1,-5 bisphosphate carboxylase/-oxygenase (Rubisco) LSU 571
ribulose-1,5-bisphosphate carboxylase/-oxygenase (Rubisco) SSU 563, 571, 573
Riccinus communis 694
rice 72, 73, 78, 79, 80, 86, 93, 94, 97, 99, 108, 154-156, 273-279, 284, 291, 293, 407, 582, 643
mRNA 9, 77, 81, 82, 85-92, 96, 143, 148, 196, 200, 202, 272, 274, 275, 485, 560, 562-565, 567, 571, 572, 574, 577, 587-591, 603, 605, 607, 622, 669, 744, 745
  differential screening 603
rRNA 566, 620
Robertson-Berger meter 24
Roscoe-Bunsen reciprocity law 235, 236, 250, 251, 303, 311, 380, 381, 398, 401, 402, 466, 635, 672, 695
Rumex patientia 741
rye (see *Secale cereale*)

Saccharomyces 314
safelight (green) 31, 34, 35, 91, 166, 218, 243, 401, 644
Samanea
  pulvini 342
  saman 450
'sandwich' ELISA (see phytochrome)
scalar irradiance (see fluence rate)
Scenedesmus 311, 312, 319
  chlorophyll synthesis in 305, 309, 312, 319
  obliquus 307
  mutant (see mutants)
  respiration 304, 307
scentless mayweed (see *Matricaria indora*)
Schirokko wheat 354
Schizophyllum commune 756-758, 763, 777
screening effects 422, 466, 643, 644, 738, 747, 749
Scutellinia umbrarum 755, 756
Secale cereale 120, 149, 214, 269, 741
second messengers 89, 259, 336, 340-344, 347, 486, 487, 592, 699
secondary metabolites 733
seed(s)
  dehydration of (maturation) 212, 399
  dormancy 399
  dry 229, 234
  imbibition of 7, 239, 241, 398-400
  germination 5, 7, 30, 63, 212, 217, 236, 239-245, 260, 285, 302, 399, 406, 548, 615
  germination in dark 7, 234, 399, 615, 616
  germination overriding factor 400, 615
  germination photo-induction of 43, 238, 240, 399, 400, 406
  germination photo-inhibition of 239, 240, 400
Selaginella 73, 78, 81, 274, 497
semiquinones (flavoenzyme semiquinones) 330
Senecio vulgaris 394
sensor pigments (see photoreceptors)
sensory rhodopsin (I, II) 709
sesame (*Sesamum indicum*) 369-371
shade 82, 221, 236, 388, 393, 396, 441, 550, 634
  avoidance (intolerance) 236, 391, 393, 394, 397-399, 403, 407, 410-413
  plants 531
  responses 390, 399
  tolerance 390, 393, 394
Shemin-pathway 64
Shiitake 757, 779
shikimate pathway 736
short day plants (see photoperiodic response types)
sieve effect (see light)
signal transduction G-proteins 132, 188, 189-192, 206-207, 315, 336, 340-344, 347, 611, 701
signal transduction G-proteins effectors 341
signal transduction phenomena (chains) 3, 4, 13, 40, 89, 91, 93, 95, 100, 130, 132, 133, 163, 169, 170, 183, 187-207, 258, 259, 289, 294, 301-303, 313, 315, 316, 318, 333-346, 377, 380, 382, 486, 527, 528, 559-563, 568, 569, 578, 589, 592, 593, 601, 614, 619, 649, 656, 677, 689,

691, 696, 697, 699, 701, 709, 713, 714, 723-728, 745, 784-786, 788, 794, 801
Sinapis alba 36, 170, 224, 233, 246, 250, 255, 261, 321, 354, 355, 364-366, 370, 395, 396, 398, 406, 475, 561, 563-570, 636-638, 641, 642, 737, 738, 740, 741, 746, 747
skeleton photoperiod 450
skotophile 460, 482
skylight 537-539, 543, 544
sleep movements (see leaf sleep movements)
slime moulds 715-718
slug turning factor (STF) 717
Smilacina stellata 529, 531
snapdragon (see *Antirrhinum majus*)
sodium borohydride 126, 128
sodium dodecyl sulphate (SDS) 128, 147, 194, 276, 282
Solanum tuberosum 72, 73, 78, 274, 603, 632
solar angle 387, 538, 543, 544, 547
solar tracking 528
solid angle 546, 547
Sordaria
  fimicola 759, 764, 771, 772
  spp. 763
  verruculosa 772
Soret bands 117, 309, 319, 320, 714
Sorghum
  bicolor 608, 620, 738, 739, 743
  vulgare 312, 321, 354-361, 738-740, 742, 743
sorocarps 718
Southern blot analysis 72
soybean (see *Glycine max*)
space irradiance (see internal fluence rate)
spatial sensing 422-424, 427, 429
  one-instant mechanism 423
  shading mechanism 418, 422-424
spatial separated pigment synthesis 746, 747
spectral attenuation coefficient (diffuse) 544, 545
spectral photon distribution 322, 379-384, 491, 633, 768
  canopy shade light 383, 385, 387, 388, 390, 393, 392, 396, 400, 406, 548, 549, 634
  daylight 301, 379, 382, 383, 385-388, 539, 540, 543
  (in) leaves 548
  (in) soil 389, 548
  twilight 385, 398, 541
  underwater 385, 388, 390, 542, 544-546, 548
spectral responsiveness 578
  changes during development 576, 577, 636, 637
spectrographs
  Beltsville 30, 31
  Okazaki 30-33
spectrophotometer(s) 9, 165
  Beltsville 8, 43
spectroradiometer 24, 226, 382, 383
Sphaerobolus stellatus 755, 770-773, 778
spherical fibre-optic sensor 514, 517, 519
spherical irradiance 22
spinach (see *Spinacia oleracea*)
Spinacia oleracea 514, 529, 655
  bolting 655
Spirodela oligorrhiza 738
sporangiophore (see *Phycomyces* and *Pilobolus*)
sporulation (see also conidiation) 310, 312, 313, 327, 331, 333, 753, 754, 758, 763, 772
  temporal interval after stimulus 772
sporulators, diurnal and nocturnal 772
standard lamp 18
starch synthesis/breakdown 307, 309, 684, 691, 693, 699, 701
Stark-Einstein law of photochemical equivalence 380
staurosporine 200
stentorin 318
steric exclusion chromatography 15
stigma (see eye spot)
stilbene 741, 742, 746
Stokes' radius 115
stomata(l) 11, 649, 683-704
  abaxial 701, 703
  adaxial 703
  air pollutant effects on 683
  blue light effect on 322, 332, 333, 342, 345, 612, 686, 687, 689-691, 693-696, 699-701
  conductance 684-686, 690, 693, 695-699, 702-704
  dual photoreceptor hypothesis of 687-689, 693
  guard cells (see guard cells)

intercellular carbon dioxide effects on 683-685, 689
phytochrome effects on 702
phytohormone effects on 683
potassium cyanide effects on 691
pre-dawn opening 693, 702
red light effects on 685-689, 702
relative humidity effects on 683
three photoreceptor system in 702
strawberry 254, 454, 477
stress conditions 327, 338, 346, 347, 452, 704, 739, 742
stress conditions sporulation under 329
stress relaxation (cell wall) 650, 651
subtilisin 115
succinate uptake 178
succinyl-CoA 64, 107
sugar beet 388, 480
sugar cane 698
sugars 693, 701
sun plants 531
sunburn meters 23, 24
sunflower (see *Helianthus annuus*)
sunlight 537-540, 589, 634-636, 739, 742
sun-tracking (see solar tracking)
surface charge 337, 648
surface marker 667

Tanada effect (see surface charge)
tapetum 505
tautomerism 58, 59
temperature shock 758
temporal sensing 36-39, 423
two instant mechanism 711, 712
tendril coiling 662
tetrahydroxychalcone 735
tetranitromethane 126, 128
tetraploid species 603
tetrapolar absorption gradient 429-431, 434
tetrapyrrole 56
tetrapyrrole biosynthesis 64, 289, 307
tetrapyrrole chromic acid-ammonia degradation 53
tetrapyrrole open chain 8, 51, 105, 111, 274
tetrapyrrole phytochrome chromophore 7, 51, 113, 274, 281
Teucrium scorodonia 394
thermal vibrations 19
thermopile 18
Thermopsis montana 529, 531
thermotaxis 715, 718
threshold boxes 31, 33
threshold response 366, 368, 369, 457, 469, 471, 472, 693
thylakoid 318, 391, 586, 620, 699, 710
thymine dimers 748, 749
time-lapse recording 36, 37, 667, 789, 791, 794
time measurement (in photoperiodism) 7, 447, 449, 456-458, 461, 469, 470, 473-476, 479, 482-484, 487
endogenous oscillator 447, 455, 458, 459
hour-glass 457-459, 464
tobacco (see *Nicotiana tabacum*)
tomato (see *Lycopersicon esculentum*)
transaminase 64, 65, 107
transcription (see also phytochrome control of) 81
transcription factor(s) 98, 99, 569, 579
transcription *in vitro* 178
transducers (see electronic position sensitive)
transduction chain (see signal transduction chain)
transformation (genetic) 275
transgenic plants 81, 83-87, 100, 106, 108, 109, 112, 115, 156, 159, 160, 258, 260-262, 271-294, 377, 403-405, 407-410, 476, 566, 567, 576-578, 581, 582, 589-591, 593, 608, 620, 635, 656
translation 275
transmembrane helixes 587
transpiration 649, 688, 689, 694, 703
transposons 603
Trichoderma harzianum (*viride*) 329, 331, 333, 338, 340, 344, 346, 347
Trifolium pratense 498
triose phosphate isomerase 691
Trisetum flavescens 694
Triticum aestivum 190-192, 199, 321, 478, 479, 482, 483, 486, 576, 589-591, 603, 690, 698, 741
Triton X-100 343
turgor pressure 486, 499, 650, 651, 684, 691
turnip (see *Brassica rapa*)
two-instant mechanism (see spatial sensing)

ubiquitin(ation) 129, 147-157, 160, 277, 284
Ulva 499-501
uncouplers 710
Urtica dioica 394

Ustilago 772
UTP 572
UV 11, 33, 282, 301, 309, 310, 346, 356-359, 367, 372, 530, 543, 548, 549, 563, 565, 567-570, 708, 709, 711, 712, 717, 718, 733, 739-744, 748, 749, 775, 776, 779
UV-A 11, 246, 301, 302, 307, 320, 330, 339, 346, 354, 358, 360, 425, 426, 439-442, 542, 621, 723, 733, 738-740, 776-778, 786
UV-A photoreceptor 220, 321, 327, 353, 354, 356, 358, 619, 631, 743, 776
UV-B 11, 23, 246, 301, 305, 307, 310, 320, 327, 530, 538, 541, 542, 545, 733, 738-742, 749, 775
UV-B photoreceptor 5, 11, 301, 321, 353, 354, 356, 358, 620, 631, 739, 743-745
UV-C 301, 314, 739, 740, 786
UV damage 24, 308, 739, 740
UV-high irradiance response (UV-HIR) 31, 246, 254

Vallisneria spiralis 719-722, 726, 729
vanadate 691, 699
Vaucheria sessilis 302, 304, 308, 334, 723, 725-727, 729, 730
  cortical fibre reticulation 302, 304, 308, 312, 338, 723, 725
  cytoskeleton 334, 338
  electrical changes induced by blue light 312, 338, 723, 724, 730
vector irradiance (see irradiance)
vegetative/floral transition 236
vernalization 454
Verticillium agaricinum 776
very low fluence response (VLFR) 9, 34, 165, 170, 218, 235, 242-245, 313, 401, 402, 413, 511, 513, 587, 590, 591, 608, 635, 636, 644, 653, 656, 674
vibrating probe 337
Vicia faba 686, 689, 690, 693-695, 700
video tape recording 35, 37, 39, 40
vine (see *Vitis vinifera*)
visible spectrum 379, 387, 538
vision (human) 4, 23, 34, 51, 381
Vitis vinifera 741

water potential (difference) 650
wave guiding (see light guiding)
weeds 393
Western blot analysis (immunoblotting) 45, 166, 190
wheat (see *Triticum aestivum*)

X-ray analysis 55, 56, 115, 133, 148
Xanthium strumarium 242, 452, 454-456, 458, 459, 461, 463-466, 469-471, 690, 768
xylem development 203

yeast 105, 106, 109, 133, 149, 151, 273, 281, 504
yellow light 641, 642, 778, 779

Zea mays 73, 78, 79, 88, 93, 94, 96, 142, 145, 149, 155, 170, 172, 173, 192, 193, 230, 244, 274, 312, 343, 496, 573, 574, 603, 742
  coleoptile 169, 311, 312, 314, 332, 333, 514, 524, 526, 527, 633, 643, 649, 670, 673-675, 740, 741
  'paradox' 243, 244
  roots 305, 310, 312, 510
Zebrina pendula 497
Zeitgeber 449, 450, 452
zinc-finger motif 206, 611
Zinnia 203
zucchini (see *Cucurbita pepo*)
zygomycetes 329, 756